P. Angelov · K.T. Atanassov · L. Doukovska
M. Hadjiski · V. Jotsov · J. Kacprzyk
N. Kasabov · S. Sotirov · E. Szmidt
S. Zadrożny
Editors

Intelligent Systems'2014

Proceedings of the 7th IEEE International Conference Intelligent Systems IS'2014, September 24–26, 2014, Warsaw, Poland, Volume 1: Mathematical Foundations, Theory, Analyses

Editors
P. Angelov
School of Computing and Communications
Lancaster University
Lancaster, United Kingdom

K.T. Atanassov
Institute of Biophysics and Biomedical Engineering
Bulgarian Academy of Sciences
Sofia, Bulgaria

L. Doukovska
Intelligent Systems Department
Institute of Information and Communication Technologies
Bulgarian Academy of Sciences
Sofia, Bulgaria

M. Hadjiski
University of Chemical Technology and Metalurgy
Sofia, Bulgaria

V. Jotsov
University of Library Studies and IT (ULSIT)
Sofia, Bulgaria

J. Kacprzyk
Polish Academy of Sciences
Systems Research Institute
Warsaw, Poland

N. Kasabov
Knowledge Engineering and Discovery Research Institute
Auckland University of Technology
Auckland, New Zealand

S. Sotirov
Faculty of Technical Sciences
Intelligent Systems Laboratory
Prof. Assen Zlatarov University
Bourgas, Bulgaria

E. Szmidt
Polish Academy of Sciences
Systems Research Institute
Warsaw, Poland

S. Zadrożny
Polish Academy of Sciences
Systems Research Institute
Warsaw, Poland

ISSN 2194-5357 ISSN 2194-5365 (electronic)
ISBN 978-3-319-11312-8 ISBN 978-3-319-11313-5 (eBook)
DOI 10.1007/978-3-319-11313-5

Library of Congress Control Number: 2014949165

Springer Cham Heidelberg New York Dordrecht London

Printed on acid-free paper

Springer is part of Springer Science+Business Media (www.springer.com)

Foreword

This two volume set of books constitutes the proceedings of the 2014 IEEE 7th International Conference Intelligent Systems (IS), or IEEE IS'2014 for short, held on September 24–26, 2014 in Warsaw, Poland. Moreover, it contains some selected papers from the collocated IWIFSGN'2014 - Thirteenth International Workshop on Intuitionistic Fuzzy Sets and Generalized Nets held on September 25, 2014.

The conference was organized by the Systems Research Institute, Polish Academy of Sciences, Department IV of Engineering Sciences, Polish Academy of Sciences, and PIAP – Industrial Institute of Automation and Measurements. It was sponsored by the Poland Section of the IEEE (institute of Electrical and Electronics Engineers) CIS (Computational Intelligence Society), and the Polish Operational and Systems Research Society. The technical sponsors included the IEEE Systems, Man and Cybernetics Society, the IEEE Computational Intelligence Society, the Polish Neural Network Society, the Polish Association of Artificial Intelligence, and the Polish Operational and Systems Research Society.

The conference had a special session devoted to the presentation of results obtained within the "International PhD Projects in Intelligent Computing" and these two volumes contain the respective papers. The conference gathered both the Ph.D. students participating in the Project and their supervisors from Poland and abroad, as well as some foreign partners involved in the Project providing an excellent opportunity for the publication and dissemination of contributions by young researchers. This project was supported by the Foundation for Polish and financed from the European Union within the Innovative Economy Operational Programme 2007-2013 and European Regional Development Fund.

The papers included in the two proceedings volumes have been subject to a thorough review process by three highly qualified peer reviewers. Comments and suggestion from them have considerable helped improve the quality of the papers but also the division of the volumes into parts, and assignment of the papers to the best suited parts.

Thanks are due to many people and parties involved. First, in the early stage of the preparation of the conference general perspective, scope, topics and coverage, we have received an invaluable help from the members of the International Program Committee, notably the chairs responsible for various aspects of the Conference, as well as people from the IEEE, notably Mr. Kevin Uherek, who helped make formal arrangements that would be most proper for this particular Conference. In fact, an overwhelming majority of members of the International Program Committee had been the members of all our sponsors which had provided an extraordinary situation of a synergy between the International Program Committee and all the sponsoring parties. That help during the initial planning stage had resulted in a very attractive and up to date proposal of the scope and coverage that had clearly implied a considerable interest of the international research communities active in the areas covered who had submitted a large number of very interesting and high level publications. The role of many session organizers should also be greatly appreciated. Thanks to their vision and hard work, we had been able to collect many papers on focused topics which had then resulted, during the Conference, in very interesting presentations and stimulating discussions at the sessions.

The members of the International Program Committee, together with the session organizers, and a large group of other anonymous peer reviewers had undertaken a very difficult task of selecting the best papers, and they had done it excellently. They deserve many thanks for their excellent job for the entire community who is always concerned with the quality and integrity.

At the stage of the running of the Conference, many thanks are due to the members of the Organizing Committee, chaired by Ms. Krystyna Warzywoda and Ms. Agnieszka Jóźwiak, and supported by their numerous collaborators.

And last but not least, we wish to thank Dr. Tom Ditzinger, Dr. Leontina di Cecco and Mr. Holger Schaepe for their dedication and help to implement and finish this large publication project on time maintaining the highest publication standards.

July 2014

P. Angelov
K.T. Atanassov
L. Doukovska
M. Hadjiski
V. Jotsov
J. Kacprzyk
N. Kasabov
S. Sotirov
E. Szmidt
S. Zadrożny

Contents

Part III: Multicriteria Decision Making and Optimization

Part IV: Issues in Intuitionistic Fuzzy Sets

Part V: Fuzzy Cognitive Maps and Applications

Part VI: Issues in Logic and Artificial Intelligence

Part VII: Group Decisions, Consensus, Negotiations

Part VIII: Issues in Granular Computing

Part XI: Issues in Data Analysis and Data Mining

Part XII: Issues in Generalized Nets

Part XV: Perception, Judgment, Affect, and Sentiment Analyses

Part I
Counting and Aggregation

Intelligent Counting – Methods and Applications

Maciej Wygralak

Faculty of Mathematics and Computer Science, Adam Mickiewicz University
Umultowska 87, 61-614 Poznań, Poland
wygralak@amu.edu.pl

Abstract. This paper deals with intelligent counting, i.e. counting performed under imprecision, fuzziness of information about the objects of counting. Formally, this collapses to counting in fuzzy sets. We will show that the presented methods of intelligent counting are human-consistent, and reflect and formalize real, human counting procedures. Other applications of intelligent counting in intelligent systems and decision support, including questions of similarity measures and time series analysis, will also be outlined or mentioned.

Keywords: Intelligent counting, Fuzzy set cardinality, Sigma *f*-count, Fuzzy cardinality, Applications.

1 Introduction

It seems that counting belongs to the most basic and frequent mental activities of human beings. Its result, the cardinality of a collection of elements, is a basis for making a decision in a lot of situations and dimensions of our life. However, speaking about counting, one should distinguish between two very different cases.

- The objects of counting are *precisely specified*, e.g. "How many apples are there in the basket?". This is the trivial case collapsing to the usual counting in a set by means of the natural numbers ("do not think, just count the apples!").
- Those objects are *imprecisely, fuzzily specified*, e.g. "How many *big* apples are there in the basket?". What we deal with is now a more sophisticated and advanced situation as each apple in the basket is big to a degree. So, what and how to count? Also only to a degree? This counting under information imprecision, as one sees, thus requires thinking and intelligence and, consequently, can be termed *intelligent counting*. Formally, it collapses to counting in a fuzzy set.

We will discuss two approaches to intelligent counting: the scalar approach and the fuzzy one which lead to scalar and fuzzy cardinalities, respectively. They are reflections and formalizations of real, human counting procedures. We like to look at old, well-known constructions from cardinality theory of fuzzy sets in another way: just from the viewpoint of human counting. Moreover, main applications of intelligent counting in many areas of intelligent systems and decision support will be mentioned.

P. Angelov et al. (eds.), *Intelligent Systems'2014*,
Advances in Intelligent Systems and Computing 322, DOI: 10.1007/978-3-319-11313-5_1

2 Two Basic Human Procedures of Counting under Information Imprecision

Looking at human procedures of counting under information imprecision, two of them seem to be fundamental.

- **Counting by thresholding** (*CAC, cut-and-count*).

Assume we ask “How many *warm* days were there last (calendar) summer?”. The standard human way of doing is then the following:

- establish a threshold temperature, say, 22°C,
- count up all the summer days with temperature ≥ 22°C or > 22°C.

Speaking formally, this collapses to defining the cardinality $|A|$ of a fuzzy set A as

$$|A| = |A_t| \quad \text{or} \quad |A| = |A^t|,$$

where A_t and A^t, respectively, denote the usual t-cut set and sharp t–cut set of A, respectively; $A^t = \{x\colon A(x) > t\}$. CAC thus leads to *scalar cardinalities* of fuzzy sets (see Section 3). The cardinality of A is then a single nonnegative real number.

- **Counting by multiple thresholding** (*MCAC, multiple cut-and-count*).

The answer to queries like “How many *warm* days were there last (calendar) summer?” is now given in a more advanced form involving many threshold, e.g.

“52 days were at least *fairly warm* (≥ 22°C), including 35 *definitely warm* days (≥ 25°C) of which 16 were *totally warm* (≥ 28°C)”.

This cardinal information can be modeled as

$$\begin{aligned}|A| &= \mathit{totally}/16 + \mathit{definitely}/35 + \mathit{fairly}/52\\ &= 1/16 + 0.8/35 + 0.4/52\\ &= 1/|A_1| + 0.8/|A_{0.8}| + 0.4/|A_{0.4}|.\end{aligned}$$

The result of counting is now a fuzzy set in $\{0, 1, 2, \ldots\}$. MCAC thus leads to *fuzzy cardinalities*, to fuzzy sets of nonnegative integers.

3 Scalar and Fuzzy Cardinalities

As to scalar counting procedures, their suitable formalization seems to be the notion of the *sigma f–count* $\sigma_f(A)$ of a fuzzy set A in a finite universe $\boldsymbol{U}$ (see [30, 31]):

$$\sigma_f(A) = \sum_{x \in \mathrm{supp}(A)} f(A(x)).$$

f: [0, 1] → [0, 1] is here a *weighting function* understood as a non-decreasing function with $f(0) = 0$ and $f(1) = 1$. Each weight $f(a)$ forms a *degree of participation* in the counting process in A assigned to an element x whose membership degree in A is $a \in [0, 1]$. $\sigma_f(A)$ is viewed as the cardinality of A. Obviously, $\sigma_{id}(A)$ collapses to the classical sigma count of A (see e.g. [5, 34]). Let us give a few simplest examples of weighting functions and the resulting sigma f–counts.

- **Counting by thresholding:** $f(a) = (1 \text{ if } a \geq t, \text{ else } 0)$ with $t \in (0, 1]$. Then

$$\sigma_f(A) = |A_t|.$$

- **Counting by sharp thresholding:** $f(a) = (1 \text{ if } a > t, \text{ else } 0)$ with $t \in [0, 1)$. Now

$$\sigma_f(A) = |A^t|.$$

- **Counting by joining:** $f(a) = a^p$, $p > 0$. Then

$$\sigma_f(A) = \sum_{x \in \mathrm{supp}(A)} (A(x))^p$$

  and $p = 1$ generates the usual sigma count of A.

- **Counting by thresholding and joining:** $f(a) = (a^p \text{ if } a \geq t, \text{ else } 0)$, $t \in (0, 1]$ and $p > 0$. So, $p = 1$ gives

$$\sigma_f(A) = \sum_{x \in A_t} A(x).$$

  The reader is referred to [30, 31] for more examples.

Let us move on to the MCAC method. Going to extreme, the counting person can use all possible threshold values from (0, 1], and combine all the results of counting. Consequently, one defines

$$|A| = \sum_{t \in (0,1]} t / |A_t| = \sum_{k \geq 0} \sup\{t : |A_t| = k\} / k.$$

What we get is the *basic fuzzy count* BF(A) of A introduced in [32], the oldest and a bit forgotten type of fuzzy cardinality:

$$\mathrm{BF}(A) = \sum_{k \geq 0} \sup\{t : |A_t| = k\} / k.$$

Let us look at the following simple example with

$$A = 0.6/x_1 + 1/x_2 + 0.9/x_3 + 0.8/x_4 + 1/x_5 + 0.6/x_6 + 0.6/x_7 + 0.2/x_8.$$

Then

$$\mathrm{BF}(A) = 1/2 + 0.9/3 + 0.8/4 + 0.6/7 + 0.2/8.$$

As one sees, BF(A) is generally nonconvex and can be viewed as a dynamic and compact piece of information about all possible results of counting in A by thresholding. If BF(A)(k) = 0, k is impossible as a result of that counting. BF(A)(k) > 0 means that BF(A)(k) is a maximum threshold t giving k as a result of counting.

A slight modification of BF(A) leads to convex fuzzy cardinalities:

$$|A| = \sum_{k\geq 0} \sup\{t\colon |A_t| \geq k\}/k = \sum_{k\geq 0} [A]_k/k$$

with $[A]_k$ denoting the kth greatest membership degree in A; $[A]_0 = 1$. This gives

$$\mathrm{FG}(A) = \sum_{k\geq 0} [A]_k/k,$$

the *FGCount of* A, the most commonly known type of fuzzy cardinality (see [1, 33, 34]). For A from the previous example, we obtain

$$\mathrm{FG}(A) = 1/0 + 1/1 + 1/2 + 0.9/3 + 0.8/4 + 0.6/5 + 0.6/6 + 0.6/7 + 0.2/8.$$

Speaking generally and practically, FG(A) forms a ranking list of membership degrees in A. Let us notice that

$$|A_t| = |(\mathrm{FG}(A))_t| - 1,$$

i.e. FG(A) can be viewed as a compact piece of information about the results of counting by thresholding in A with all possible thresholds.

If more convenient, counting in a fuzzy set A can be replaced with *indirect* counting, i.e. counting in the complement A'. Its result is then subtracted from the cardinality of the whole universe. This simple trick was a motivation for introducing in [34] the concept of the *FLCount of* A, which is dual to that of the FGCount:

$$\mathrm{FL}(A) = \sum_{k\geq 0} (1 - [A]_{k+1})/k,$$

We get

$$|(A')_t| = |(\mathrm{FL}(A))_t| - 1.$$

FL(A) is thus a compact piece of information about all possible results of counting by thresholding in A'.

Finally, one has to mention the idea of the *FECount of* A which forms an aggregation of FG(A) and FL(A) (see [29, 34]). Then

$$\mathrm{FE}(A) = \mathrm{FG}(A) \cap \mathrm{FL}(A), \quad \text{i.e. } \mathrm{FE}(A) = \sum_{k\geq 0} \min([A]_k, 1 - [A]_{k+1})/k.$$

FE(A) can be viewed as a compact piece of information about all possible results of counting by thresholding in the intersection $A \cap A'$, i.e. in the fuzzy set of "embarassing" elements being simultaneously in A and A' and, in other words, satisfying to a degree both an imprecise property and its opposite:

$$|(A \cap A')_t| = \max(0, |(\mathrm{FE}(A))_t| - 1)$$

for each t from (0, 1]. This suggests a close connection between FECounts and classification issues (see [31] for details). A worth mentioning type of cardinal information about A seems to be the scalar information $\mathrm{COG}_{\mathrm{FE}(A)}$ derived from FE(A).

4 A Brief Overview of Applications

In the previous section, we have shown that scalar and fuzzy cardinalities of fuzzy sets reflect and formalize real, human procedures of counting under imprecision of information about the objects of counting. This section presents various examples of applications of intelligent counting in intelligent systems and decision support.

4.1 Cardinality-Based Similarity, Distance, and Fuzziness Measure

Let us focus our attention on the well-known *Jaccard coefficient* for sets **A** and **B**:

$$J(\boldsymbol{A}, \boldsymbol{B}) = |\boldsymbol{A} \cap \boldsymbol{B}| / |\boldsymbol{A} \cup \boldsymbol{B}|.$$

It is possible to extend that coefficient to scalar cardinalities of fuzzy sets A and B:

$$J(A, B) = \sigma_f(A \cap B)/\sigma_f(A \cup B).$$

And this extension, too, forms a similarity measure which can be used to define the *Jaccard metric*

$$D(A, B) = 1 - J(A, B),$$

and the *Jaccard fuzziness measure*

$$Fuzz(A) = 1 - D(A, A')$$

whenever f is strictly increasing (see [3, 4, 31] for details). Analogous extensions involving scalar cardinalities of fuzzy sets can be done for other classical cardinality-based coefficients, e.g. the inclusion, overlap, and matching coefficients.

4.2 Time Series Analysis

Assume we have two time series $Y = \{ y_i: i = 1, \ldots , k\}$ and $Z = \{ z_i: i = 1, \ldots , k\}$, i.e. two sequences of numerical observations at equidistant time moments, subjected to a prior normalization. So, $y_i, z_i \in [0, 1]$ and, thus, Y and Z can be treated as two fuzzy sets. Let us construct the corresponding time series $Diff(Y, Z)$ of differences:

$$Diff(Y, Z) = \{ d_i : i = 1, \dots, k\} \quad \text{with} \quad d_i = |y_i - z_i|.$$

The cardinality of $Diff(Y, Z)$, especially, its scalar cardinality, can be a useful basis for decision making. If, say, Y and Z, respectively, represent actual demand for a product and forecasted demand, respectively, then $|Diff(Y, Z)|$ can be used to select a suitable probabilistic model or to evaluate forecast accuracy (see [31]).

4.3 References to Other Applications

This final subsection presents selected bibliographical references to other applications of intelligent counting in selected areas of intelligent systems and decision support.

- Computational approach to linguistic quantifiers (see e.g. [6, 11, 15, 24, 25, 34]).
- Flexible querying in databases, information retrieval ([13, 20, 23, 35]).
- Discovery of association rules and sequential patterns ([7, 8, 12]).
- Linguistic summarization in databases ([2, 22, 27]).
- Linguistic summarization of time series ([17, 18]).
- Text categorization ([27, 36]).
- Control and decision making ([10, 15, 19]).
- Group decision making ([14, 16, 26]).
- Consensus reaching ([21]).
- Social networks and trust propagation ([9, 28]).

5 Conclusions

The subject of this paper was intelligent counting understood as counting under imprecision of information about the objects of counting. Speaking more formally, this collapses to counting in a fuzzy set. We showed that well-known scalar and fuzzy methods of that counting are human-consistent: are reflections and formalizations of real, human counting procedures. In particular, this is the case of FGCounts, FLCounts and FECounts of fuzzy sets. Using the conventional interpretation of these constructions in many-valued logic, each membership degree $FG(A)(k)$, $FL(A)(k)$, and $FE(A)(k)$, respectively, is a degree to which a fuzzy set A has at least, at most, and exactly k elements, respectively. This paper shows that those fuzzy cardinalities form a compact piece of information about all possible results of counting by thresholding. Finally, applications of intelligent counting in various areas of intelligent systems and decision support were mentioned.

References

1. Blanchard, N.: Cardinal and ordinal theories about fuzzy sets. In: Gupta, M.M., Sanchez, E. (eds.) Fuzzy Information and Decision Processes, pp. 149–157. North-Holland, Amsterdam (1982)
2. Bosc, P., Dubois, D., Pivert, O., Prade, H., De Calmès, M.: Fuzzy summarization of data using fuzzy cardinalities. In: Proc. 9th International Conference on Information Processing and Management of Uncertainty in Knowledge-Based Systems (IPMU2002), Annecy, France, pp. 1553–1559 (2002)
3. De Baets, B., De Meyer, H.: Transitivity-preserving fuzzification schemes for cardinality-based similarity measures. Europ. J. Oper. Research 160, 726–740 (2005)
4. De Baets, B., De Meyer, H., Naessens, H.: On rational cardinality-based inclusion measures. Fuzzy Sets and Systems 128, 169–183 (2002)
5. De Luca, A., Termini, S.: A definition of a non-probabilistic entropy in the setting of fuzzy sets theory. Inform. and Control 20, 301–312 (1972)
6. Delgado, M., Sánchez, D., Vila, M.A.: Fuzzy cardinality based evaluation of quantified sentences. Int. J. Approx. Reasoning 23, 23–66 (2000)
7. Dubois, D., Prade, H., Sudkamp, T.: On the representation, measurement and discovery of fuzzy associations. IEEE Trans. on Fuzzy Systems 13, 250–262 (2005)
8. Dyczkowski, K.: A less cumulative algorithm of mining linguistic browsing patterns in the World Wide Web. In: Proc. 5th EUSFLAT Conf., Ostrava, Czech Rep., pp. 129–135 (2007)
9. Dyczkowski, K., Stachowiak, A.: A recommender system with uncertainty on the example of political elections. In: Greco, S., Bouchon-Meunier, B., Coletti, G., Fedrizzi, M., Matarazzo, B., Yager, R.R. (eds.) IPMU 2012, Part II. CCIS, vol. 298, pp. 441–449. Springer, Heidelberg (2012)
10. Felix, R.: Decision making with interactions between goals. In: Proc. 13th Zittau Fuzzy Coloquium, Zittau, Germany, pp. 3–10 (2006)
11. Glőckner, I.: Fuzzy Quantifiers - A Computational Theory. Springer, Heidelberg (2006)
12. Hong, T.-P., Lin, K.-Y., Wang, S.-L.: Mining fuzzy sequential patterns from quantitative transactions. Soft Computing 10, 925–932 (2006)
13. Jaworska, T., Kacprzyk, J., Marín, N., Zadrożny, S.: On dealing with imprecise information in a content based image retrieval system. In: Hüllermeier, E., Kruse, R., Hoffmann, F. (eds.) IPMU 2010. LNCS, vol. 6178, pp. 149–158. Springer, Heidelberg (2010)
14. Kacprzyk, J.: Group decision making with a fuzzy linguistic majority. Fuzzy Sets and Systems 18, 106–118 (1986)
15. Kacprzyk, J.: Multistage Fuzzy Control. A Model-Based Approach to Fuzzy Control and Decision Making. John Wiley, Chichester (1997)
16. Kacprzyk, J., Nurmi, H., Fedrizzi, M.: Group decision making and a measure of consensus under fuzzy preferences and a fuzzy linguistic majority. In: Zadeh, L.A., Kacprzyk, J. (eds.) Computing with Words in Information/Intelligent Systems– Applications, vol. 2, pp. 243–269. Physica-Verlag, Heidelberg (1999)
17. Kacprzyk, J., Wilbik, A.: Temporal linguistic summaries of time series using fuzzy logic. In: Hüllermeier, E., Kruse, R., Hoffmann, F. (eds.) IPMU 2010. CCIS, vol. 80, pp. 436–445. Springer, Heidelberg (2010)
18. Kacprzyk, J., Wilbik, A., Zadrożny, S.: Linguistic summarization of time series using a fuzzy quantifier driven aggregation. Fuzzy Sets and Systems 159, 1485–1499 (2008)
19. Kacprzyk, J., Yager, R.R.: "Softer" optimization and control models via fuzzy linguistic quantifiers. Information Sciences 34, 157–178 (1984)

20. Kacprzyk, J., Zadrożny, S.: Computing with words in intelligent database querying: standalone and internet-based applications. Information Sciences 134, 71–109 (2001)
21. Kacprzyk, J., Zadrożny, S.: Soft computing and Web intelligence for supporting consensus reaching. Soft Computing 14, 833–846 (2010a)
22. Kacprzyk, J., Zadrożny, S.: Computing with words is an implementable paradigm: fuzzy queries, linguistic data summaries and natural language generation. IEEE Trans. on Fuzzy Systems 18, 461–472 (2010b)
23. Lietard, L., Rocacher, D.: Complex quantified statements evaluated using gradual numbers. In: Castillo, O., Melin, P., Ross, O., Cruz, R., Pedrycz, W., Kacprzyk, J. (eds.) Theoretical Advances and Applications of Fuzzy Logic and Soft Computing. ASC, vol. 42, pp. 46–53. Springer, Heidelberg (2007)
24. Liu, T., Kerre, E.E.: An overview of fuzzy quantifiers, Part I: Interpretations. Fuzzy Sets and Systems 95, 1–21 (1998a)
25. Liu, T., Kerre, E.E.: An overview of fuzzy quantifiers, Part II: Reasoning and applications. Fuzzy Sets and Systems 95, 135–146 (1998b)
26. Pankowska, A., Wygralak, M.: General IF-sets with triangular norms and their applications to group decision making. Information Sciences 176, 2713–2754 (2006)
27. Pilarski, D.: Linguistic summarization of databases with Quantirius: a reduction algorithm for generated summaries. Inter. Jour. of Uncertainty, Fuzziness and Knowledge-Based Systems 18, 305–331 (2010)
28. Stachowiak, A.: Trust propagation based on group opinion. In: Hüllermeier, E., Kruse, R., Hoffmann, F. (eds.) IPMU 2010. CCIS, vol. 80, pp. 601–610. Springer, Heidelberg (2010)
29. Wygralak, M.: Fuzzy cardinals based on the generalized equality of fuzzy sets. Fuzzy Sets and Systems 18, 143–158 (1986)
30. Wygralak, M.: Cardinalities of Fuzzy Sets. Springer, Heidelberg (2003)
31. Wygralak, M.: Intelligent Counting under Information Imprecision. Applications to Intelligent Systems and Decision Support. Springer, Heidelberg (2013)
32. Zadeh, L.A.: A theory of approximate reasoning. In: Hayes, J.E., Michie, D., Mikulich, L.I. (eds.) Machine Intelligence, vol. 9, pp. 149–194. John Wiley&Sons, New York (1979)
33. Zadeh, L.A.: Fuzzy probabilities and their role in decision analysis. In: Proc. 4th MIT/ONR Workshop on Command, Control and Communications, pp. 159–179. MIT Press, Cambridge (1981)
34. Zadeh, L.A.: A computational approach to fuzzy quantifiers in natural languages. Comput. and Math. with Appl. 9, 149–184 (1983)
35. Zadrożny, S., Kacprzyk, J.: FQUERY for Access: towards human consistent querying user interface. In: Proc. ACM Symp. on Applied Computing (SAC 1996), Philadelphia, USA, pp. 532–536 (1996)
36. Zadrożny, S., Kacprzyk, J.: Computing with words for text processing: An approach to the text categorization. Information Sciences 176, 415–437 (2006)

Recommender Systems and BOWA Operators

Przemysław Grzegorzewski[1] and Hanna Łącka[2]

[1] Systems Research Institute, Polish Academy of Sciences,
Newelska 6, 01-447 Warsaw, Poland
[2] Interdisciplinary PhD Studies at the Polish Academy of Sciences,
Jana Kazimierza 5, 01-248 Warsaw, Poland
pgrzeg@ibspan.waw.pl,
h.lacka@phd.ipipan.waw.pl

Abstract. When making recommendations based on the aggregated correlations both their absolute values and signs are meaningful and important. This is the reason why traditional aggregation operators, like OWA functions, may not be satisfactory. Therefore, a generalization of OWA operators which might be useful in aggregating a bipolar information is proposed and examined.

Keywords: Aggregation function, association measure, bipolarity, collaborative filtering, correlation coefficient, ordering, OWA operator, preferences, ranks, rating, recommender system.

1 Introduction

The main goal of a recommender system is to generate some meaningful recommendations to an active user, advising on what he might like. The most common recommender systems are applications predicting what movies a user would like to see, or what product a customer might be interested to buy.

Although different techniques are applied in recommender systems, the most common is the *collaborative filtering*, which indicates items basing on similarity measures between users and/or items. In other words, items recommended to an active user are those preferred by similar users (see, e.g., [6]).

The most popular setting in which a recommender system is considered is a matrix (sometimes called a *utility matrix*) with rows corresponding to users, columns corresponding to items and cells containing values that represents ratings given to items by the users.

In this paper we consider a much more complicated setting where the utility matrix contains attribute columns with possibly different domains and where the entries of the cells for each user-attribute pair contain rankings made by the user for items/objects belonging to the domain of a given attribute. Therefore, instead of similarity measures between users based on binary correlations between pairs of rankings produced by users on a set of items under study, we need a measure of association between two groups of rankings. Consequently, it seems that some aggregation-like operators might be helpful to generate a desired recommendation based on such type of data.

P. Angelov et al. (eds.), *Intelligent Systems'2014*,
Advances in Intelligent Systems and Computing 322, DOI: 10.1007/978-3-319-11313-5_2

Since in the case of correlation both its absolute value and its sign is meaningful and important, traditional aggregation operators, like OWA functions, may not be satisfactory. To be more specific, a desired measure might be treated as a member of a family of semi-aggregation operators, which behave as traditional aggregation functions when we restrict our consideration to the absolute values of their arguments, but simultaneously they do not loose information about the signs of those arguments. An example of such operator was proposed and examined by Łącka and Grzegorzewski in [5]. In this paper we introduce a new family of operators, called BOWA (bipolar OWA functions), and show that the function discussed in [5] is a member of that interesting family of operators.

The paper is organized as follows: In Sec. 2 a model for the data representation discussed is suggested. Next, in Sec. 3 we discuss the desired properties of the requested measure of association between two sets of rankings. Then a new family of operators is suggested in Sec. 4.1, while in Sec. 4.2 we discuss the properties of BOWA operators and in Sec. 4.3 we consider the operator proposed in [5] in the framework of BOWA operators.

2 Data Representation

Let $\mathcal{X}$ denote a set of consumers and let $\mathcal{Y}$ be a set of attributes. Without loss of generality we assume that $\mathcal{Y}$ is a finite set of size n. Moreover, we assume that $\mathcal{U}_j$ is a domain of the attribute $Y_j \in \mathcal{Y}$. Each domain consists of objects, with respect to which the consumers express their preferences. Therefore, it is assumed that the domain of each attribute is finite.

Hence, for any consumer $A \in \mathcal{X}$ we get n rankings corresponding to successive attributes, so the observation related to A might be perceived as a vector $[R_{A1}, R_{A2} \ldots, R_{An}]$, where R_{Aj} is a ranking of objects belonging to the domain of the j-th attribute. An example of so prepared data set is given in Table 1.

Table 1. Exemplary data set of clients' preferences

A	R_{A1}	R_{A2}	$\ldots$	R_{An}
B	R_{B1}	R_{B2}	$\ldots$	R_{Bn}
$\ldots$	$\ldots$	$\ldots$	$\ldots$	$\ldots$

Consider a ranking R_{Aj}. Since it reflects the user's preferences on variants belonging to the domain $\mathcal{U}_j$ of the attribute $Y_j \in \mathcal{Y}$, it is also a vector, i.e.

$$R_{Aj} = (r_{Aj}^{(1)}, r_{Aj}^{(2)}, \ldots, r_{Aj}^{(l_j)}), \tag{1}$$

where $r_{Aj}^{(k)}$, $k = 1, \ldots, l_j$ is a rank assigned to k-th object belonging to $\mathcal{U}_j$ and where l_j stands for the size of the domain $\mathcal{U}_j$.

Since our goal is to identify both similar and dissimilar consumers, we are interested in defining a measure of association between every pair of consumers.

3 Desired Measure of Association between Two Groups of Rankings

Spearman's r_S or Kendall's τ (see, e.g., [3]) are classical measures of association (or correlation) between two rankings. However, now our goal is to measure association not between two rankings but between two sets of rankings $[R_{A1}, R_{A2} \ldots, R_{An}]$ and $[R_{B1}, R_{B2} \ldots, R_{Bn}]$. Moreover, particular rankings may correspond to attributes with different domains.

Before we suggest a family of candidates for such measures, let us try to list the desired properties of the requested measure of association between two sets of rankings. Denote such a hypothetical measure (coefficient) by S. It seems that S should satisfy at least the following requirements (see [5]):

R1. $S : \mathcal{X} \times \mathcal{X} \to [-1, 1]$.
R2. S should assume its maximal value if and only if all rankings are pairwise perfectly concordant, i.e. $S = 1$ if and only if (R_{Aj}, R_{Bj}) are perfectly concordant for all $j = 1, \ldots, n$.
R3. S should assume its minimal value if and only if all rankings are pairwise perfectly discordant, i.e. $S = -1$ if and only if (R_{Aj}, R_{Bj}) are perfectly discordant for all $j = 1, \ldots, n$.
R4. S should be commutative, i.e. $S(A, B) = S(B, A)$.
R5. S should not depend on the permutation of attributes.
R6. S should depend on pairwise correlations between rankings calculated for the same attribute. Moreover, the increase in pairwise correlations for all attributes should result in the increase of S, and conversely, the decrease in pairwise correlations for all attributes should result in the decrease of S.
R7. Any pairwise correlation between rankings that is different from zero should be rewarded in S, i.e. the higher the absolute value of the pairwise correlation between rankings, the stronger influence of that correlation on the value of S.
R8. S should assume value equal to zero in case of lack of pairwise correlation for all attributes or when positive and negative correlations balance, i.e. for each positive correlation there exists a negative one with equal absolute value.

Let s denote any pairwise correlation measure between two rankings, taking values in $[-1, 1]$, like Kendall's τ or Spearman's r_S (see e.g. [3]). Moreover, let $(s^1_{AB}, s^2_{AB}, \ldots, s^n_{AB})$ be a vector of pairwise correlations obtained for all attributes under study for two consumers $A, B \in \mathcal{X}$, i.e. $s^j_{AB} = s(R_{Aj}, R_{Bj})$, $j = 1, \ldots, n$.

Taking into account postulates R1–R8 discussed above one may define the desired measure of association between two groups of rankings corresponding to consumers A and B, $A, B \in \mathcal{X}$, as

$$S(A, B) = F(s^1_{AB}, s^2_{AB}, \ldots, s^n_{AB}), \tag{2}$$

where $F : [-1, 1]^n \to [-1, 1]$ is a suitable function.

As we look on (2) we may expect that F should be an appropriate aggregation function, since its goal is to aggregate several correlations to a single value. Moreover, postulates R1–R8 suggest it might be an OWA operator, especially that we want to reward higher correlations (see R7). Unfortunately, F cannot be a typical aggregation function. Why? As it is well known, an aggregation function should have at least two fundamental properties: the preservation of bounds and the monotonicity condition (see, e.g., [1,2,4]). In our case there is no problem with the preservation of bounds, i.e. $F(-1,-1,\ldots,-1)=-1$ and $F(1,1,\ldots,1)=1$, which coincides with postulates R2 and R3, respectively. However, the monotonicity condition means that $\boldsymbol{x} \leq \boldsymbol{y}$ implies $F(\boldsymbol{x}) \leq F(\boldsymbol{y})$ for all $\boldsymbol{x}, \boldsymbol{y} \in [-1,1]^n$, where $\boldsymbol{x}=(x_1,\ldots,x_n)$, $\boldsymbol{y}=(y_1,\ldots,y_n)$ and where $\boldsymbol{x} \leq \boldsymbol{y}$ means that each component of $\boldsymbol{x}$ is not greater than the corresponding component of $\boldsymbol{y}$. And here is the problem, since, by R7, we want to promote higher correlations regardless of their signs, so F cannot be monotone on the whole interval $[-1,1]$. Hence we need another type of aggregation operator that would be monotone not on the whole domain but for absolute values of arguments while still keeping track of the signs. Such family of operators taking into account bipolarity will be suggested in the next section.

4 BOWA Operators

4.1 Definition

An ordered weighted averaging operator (OWA) is a function $F:[0,1]^n \to [0,1]$ having the following form

$$F(x_1,\ldots,x_n)=\sum_{j=1}^{n} w_j \cdot x_{(j)}, \tag{3}$$

where $\boldsymbol{w}=[w_1,\ldots,w_n]$ is a vector of weights such that $w_j \geq 0$ for $j=1,\ldots,n$ and $\sum_{j=1}^{n} w_j = 1$, and where $x_{(j)}$ denotes the j-th largest element of the collection of aggregated objects $x_1,\ldots,x_n$. This notion, proposed by Yager [8], is also known in statistics as the so-called L-statistic.

In some situations a slightly different notation is used. Namely, having any vector $\boldsymbol{x}=[x_1,\ldots,x_n]$ let $\boldsymbol{x}\searrow$ denote the vector obtained from $\boldsymbol{x}$ by arranging its components in the non-increasing order $x_{(1)} \geq \cdots \geq x_{(n)}$. Then we may denote (3) as

$$F(x_1,\ldots,x_n)=\langle \boldsymbol{w}, \boldsymbol{x}\searrow \rangle, \tag{4}$$

i.e. we obtain $F(\boldsymbol{x})$ as the scalar product of vectors $\boldsymbol{w}$ and $\boldsymbol{x}\searrow$.

Several generalizations of OWA operators are suggested in the literature, like WOWA (i.e. Weighted OWA, [7]), Neat OWA, GenOWA (i.e. Generalized OWA), OWG (i.e. Ordered Weighted Geometric functions), OWH (i.e. Ordered Weighted Harmonic functions), Power-based GenOWA, T-OWA, S-OWA, ST-OWA, OWMAx, etc. (see, e.g. [1,2,4]), but none of those function complies with bipolarity of the input data. Before we define a desired function let us introduce the following notation.

For given sample $x_1, \ldots, x_n$ let us consider the sequence of pairs $(x_1, 1), \ldots, (x_n, n)$. Then let us order x's in the non-increasing order keeping in mind second parameters of each pair, i.e. the original labels of observations. Hence we get $(x_{(1)}, i_1), \ldots, (x_{(n)}, i_n)$, where $x_{(j)} = x_{i_j}$. This way we define order statistics applied in various fields and, in particular, for defining OWA function (3).

Similarly, let us consider the following sequence of pairs: $(|x_1|, 1), \ldots, (|x_n|, n)$ and then let us arrange absolute values of x's in the non-increasing order keeping track of their original labels. We obtain $(|x|_{(1)}, i_1), \ldots, (|x|_{(n)}, i_n)$, where $|x|_{(1)} \geq \ldots \geq |x|_{(n)}$ and $|x|_{(j)} = |x_{i_j}|$. Hence

$$|\boldsymbol{x}| \searrow = [|x|_{(1)}, \ldots, |x|_{(n)}] \tag{5}$$

denotes the vector obtained from $|\boldsymbol{x}|$ by arranging its components in the non-increasing order. Moreover, assuming that $x^*_{(j)} = x_{i_j}$ we obtain

$$\boldsymbol{x}^* \searrow_B = [x^*_{(1)}, \ldots x^*_{(n)}], \tag{6}$$

i.e. the sequence of x's ordered identically as their absolute values. Here the symbol $\searrow_B$ indicates a non-increasing ordering of elements obtained for their absolute values and thus ignoring their signs. We will call such ordering *bipolar*. Keeping in mind this notation we are ready for defining the desired operator.

Because of the problems that may appear because of possible ties let us firstly restrict our attention to data which appear as outcomes of the experiments described by the continuous distributions. This way we may assume that there are no such two observations x_i and x_j that $|x_i| = |x_j|$. One may think that such assumption is quite artificial and makes a bad impression. However, it is a common practice in statistics (especially in nonparametric statistics where we deal with ranks) which allows for a more elegant expression of the definitions and theorems. This is also the reason why we start from this very assumption. But we also explain what to do in the presence of ties, i.e. if the above mentioned assumption does not hold.

Definition 1. *Let $\boldsymbol{w} = [w_1, \ldots, w_n]$ be a vector of weights such that $w_j \geq 0$ for $j = 1, \ldots, n$ and $\sum_{j=1}^n w_j = 1$. Suppose that $x_1, \ldots, x_n$ are realizations of the continuous random variable defined on the interval $[-1, 1]$. A function $F : [-1, 1]^n \to [-1, 1]$ defined as*

$$F(x_1, \ldots, x_n) = \sum_{j=1}^{n} w_j \cdot x^*_{(j)} \tag{7}$$

is called the Bipolar OWA function (BOWA), where $x^*_{(j)}$ denotes the j-th largest absolute value of element in the collection of aggregated objects $x_1, \ldots, x_n$ multiplied by the original sign of that element.

Please notice, that using a convenient notation applied in (4), we may express the BOWA operator as follows

$$F(x_1, \ldots, x_n) = \langle \boldsymbol{w}, \boldsymbol{x}^* \searrow_B \rangle, \tag{8}$$

where $\langle \cdot, \cdot \rangle$ is the scalar product of vectors.

Example 1. Consider the following vector $\boldsymbol{x} = [0.3, -0.1, 0.8, -0.7]$ and a vector of weights $\boldsymbol{w} = [w_1, \ldots, w_4]$. We get $|\boldsymbol{x}| \searrow= [0.8, 0.7, 0.3, 0.1]$ and $\boldsymbol{x}^* \searrow_B= [0.8, -0.7, 0.3, -0.1]$ so by (8) we calculate the value of the BOWA function as follows

$$F(0.3, -0.1, 0.8, -0.7) = w_1 \cdot 0.8 + w_2 \cdot (-0.7) + w_3 \cdot 0.3 + w_4 \cdot (-0.1).$$

□

Now, as we have promised, we explain how to define a BOWA operator in the presence of ties.

Suppose that after arranging the absolute values of observations $x_1, \ldots, x_n$ into the non-increasing order we get a sequence $|\boldsymbol{x}| \searrow= [|x|_{(1)}, \ldots, |x|_{(n)}]$ with ties, i.e. there exist a subsequence such that $|x|_{(j)} = |x|_{(j+1)} = \cdots = |x|_{(j+d-1)}$. Then, if $d > 1$ and if not all $w_j, \ldots, w_{j+q-1}$ are equal, it would be not clear which weights attribute to particular values of $|x|^*_{(j)}, \ldots, |x|^*_{(j+d-1)}$, especially, they they may differ in their signs.

Therefore, to obtain a general definition of a BOWA operator available also for ties, let us assume that let us assume that there are g groups in our data set $x_1, \ldots, x_n$, where $1 \leq g \leq n$, such that members of given group do not differ in their absolute values. In other words we have $|x|_{(1)} = \cdots = |x|_{(d_1)} < |x|_{(d_1+1)} = \cdots = |x|_{(d_1+d_2)} < \ldots < |x|_{(d_{g-1}+1)} = \cdots = |x|_{(d_1+d_2+\ldots+d_g)}$, where $d_1, d_2, \ldots, d_g$ denote frequencies of the successive groups and, of course, $d_1 + d_2 + \ldots + d_g = n$.

Now, according to the tradition well-established in statistics, which is also well justifiable in the presence of ties, we will attribute to each value in a the same group identical weight which are computed as the average of weights that would be gathered by elements of that group if they had not been tied. It means that values $|x|^*_{(1)} = \cdots = |x|^*_{(d_1)}$ are multiplied by $(w_1 + \ldots + w_{d_1})/d_1$, $|x|^*_{(d_1+1)} = \cdots = |x|^*_{(d_1+d_2)}$ are multiplied by $(w_{d_1+1} + \ldots + w_{d_1+d_2})/d_2$, and so on.

Therefore, assuming notation introduced above, the BOWA operator in the presence of ties might be defined more formally as follows

$$F(x_1, \ldots, x_n) = \sum_{j=1}^{n} \widetilde{w}_j \cdot x^*_{(j)}, \tag{9}$$

where

$$\widetilde{w}_j = \begin{cases} \frac{1}{d_1}(w_1 + \ldots + w_{d_1}) & \text{if} \quad 1 \leq j \leq d_1 \\ \frac{1}{d_2}(w_{d_1+1} + \ldots + w_{d_1+d_2}) & \text{if} \quad d_1 + 1 \leq j \leq d_1 + d_2 \\ \ldots & \ldots \\ \frac{1}{d_g}(w_{n-d_g+1} + \ldots + w_n) & \text{if} \quad n - d_g + 1 \leq j \leq n \end{cases} \tag{10}$$

One may easily notice that (7) is a particular case of (9) when there are no ties, i.e. $d_1 = \ldots d_g = 1$.

Example 2. Consider the following vector $\boldsymbol{x} = [0.3, -0.1, 0.8, -0.8, -0.3, -0.8]$ and a vector of weights $\boldsymbol{w} = [w_1, \ldots, w_6]$. We get $|\boldsymbol{x}| \searrow= [0.8, 0.8, 0.8, 0.3, 0.3, 0.1]$

so two observations are tied and we have $g = 3$ groups such that $d_1 = 3$, d_2 and $d_3 = 1$. By (10) we obtain $\widetilde{w}_1 = \widetilde{w}_2 = \widetilde{w}_3 = \frac{w_1+w_2+w_3}{3}$, $\widetilde{w}_4 = \widetilde{w}_5 = \frac{w_4+w_5}{2}$ and $\widetilde{w}_6 = w_6$ and thus by(9) we get

$$\begin{aligned} F(0.3,-0.1,0.8,-0.8,-0.3,-0.8) &= \frac{w_1+w_2+w_3}{3}\cdot 0.8 + \frac{w_1+w_2+w_3}{3}\cdot(-0.8) \\ &+\frac{w_1+w_2+w_3}{3}\cdot(-0.8) + \frac{w_4+w_5}{2}\cdot 0.3 \\ &+\frac{w_4+w_5}{2}\cdot(-0.3) + w_6\cdot(-0.1) \\ &= \frac{w_1+w_2+w_3}{3}\cdot(-0.8) + w_6\cdot(-0.1). \end{aligned}$$

□

4.2 Properties

In this section we examine basic properties of BOWA operators. After examining some general properties of those operators we consider whether they satisfy the postulates required by a suitable measure of association between two sets of rankings, discussed in Sec. 3.

Lemma 1. *A BOWA function is idempotent, i.e. $F(a,\ldots,a) = a$ for any $a \in [-1,1]$.*

Proof. When discussing $F(a,\ldots,a)$, for any $a \in [-1,1]$, we have to apply formula (9) because of ties. Actually, then we have a single group of observation so for any vector of weights $[w_1,\ldots,w_n]$ by (10) we get $\widetilde{w}_1 = \ldots = \widetilde{w}_n = \frac{1}{n}\sum_{j=1}^{n} w_j = \frac{1}{n}$. Therefore, $F(a,\ldots,a) = \sum_{j=1}^{n}\frac{1}{n}\cdot a = a$, which proves the lemma. ■

By Lemma 1 we conclude immediately that each BOWA operator preserves its bounds, i.e. $F(-1,\ldots,-1) = -1$ and $F(1,\ldots,1) = 1$.

The proof of the following lemma is straightforward.

Lemma 2. *A BOWA function is symmetric, i.e.*

$$F(x_1,\ldots,x_n) = F(x_{\sigma(1)},\ldots,x_{\sigma(n)}) \tag{11}$$

for any permutation $\{\sigma(1),\ldots,\sigma(n)\}$ of $\{1,\ldots,n\}$.

Let us consider for a moment some specific weighting vectors. If all weights are equal, i.e. $\boldsymbol{w} = [\frac{1}{n},\ldots,\frac{1}{n}]$, then the corresponding BOWA operator becomes the arithmetic mean. Indeed, we get

$$F(x_1,\ldots,x_n) = \sum_{j=1}^{n}\frac{1}{n}\cdot x^*_{(j)} = \frac{1}{n}\sum_{j=1}^{n} x^*_{(j)} = \frac{1}{n}\sum_{j=1}^{n} x_j = \overline{x}. \tag{12}$$

Naturally, it is the only weight vector for which the ordering of elements does not influence the function's value. In this very case BOWA operator returns the same results as OWA and therefore keeps all OWA's properties.

If $\boldsymbol{w} = [1, 0, \ldots, 0]$ then

$$F(x_1, \ldots, x_n) = \arg\max_x |\boldsymbol{x}|, \tag{13}$$

while if $\boldsymbol{w} = [0, \ldots, 0, 1]$ then

$$F(x_1, \ldots, x_n) = \arg\min_x |\boldsymbol{x}|, \tag{14}$$

Generally, if $w_k = 1$ and for all $j = 1, \ldots, n$ except the k-th $w_j = 0$, then

$$F(x_1, \ldots, x_n) = x^*_{(k)}, \tag{15}$$

i.e. we obtain the k-th order statistic of the *bipolar* order.

Lemma 3. *Each BOWA function is homogeneous, i.e.* $F(\lambda \cdot x_1, \ldots, \lambda \cdot x_n) = \lambda \cdot F(x_1, \ldots, x_n)$, *for all* $\lambda \in [-1, 1]$ *and for all* $(x_1, \ldots, x_n) \in [-1, 1]^n$.

Proof. This property implies immediately from the fact that for any $|x_i| \geq |x_j|$ we get $|\lambda \cdot x_i| \geq |\lambda \cdot x_j|$, which means that the order of absolute values for all arguments after multiplying those arguments by a constant, is preserved. ■

Although BOWA operators posses many properties typical to aggregation functions, they are not generally aggregation operators (according to popular definitions, like e.g. given in [1,4]), because of the lack of monotonicity. Moreover, BOWA operators are also not shift-invariant.

Example 3. Consider a vector of weights $\boldsymbol{w} = [0.1, 0.9]$ and two data vectors: $\boldsymbol{x} = [-0.8, 0.5]$ and $\boldsymbol{y} = [-0.6, 0.7]$. It is obvious that $\boldsymbol{x} < \boldsymbol{y}$. However, $F(x_1, x_2) = 0.37 > F(y_1, y_2) = -0.47$, which shows that F is not monotone.

Moreover, one may notice that $\boldsymbol{y} = \boldsymbol{x} + 0.2$. However, $F(x_1 + 0.2, x_2 + 0.2) = F(y_1, y_2) = -0.47$ does not equal to $F(x_1, x_2) + 0.2 = 0.37 + 0.2 = 0.57$ which means that F is not shift-invariant. □

The following fact (although its proof is straightforward) is worth noting:

Lemma 4. *Each BOWA function for absolute values of the arguments, i.e.* $F(|x_1|, \ldots, |x_n|)$, *is an OWA operator.*

Hence, consequently, $F(|x_1|, \ldots, |x_n|)$ possesses all the properties typical to OWA operators, so it is non-decreasing, shift invariant, etc.

It might be also proved, that BOWA operators, similarly as OWA functions, do not have neutral or absorbing elements, except for the special cases. Namely, zero is a neutral element of the max operator (13) and an absorbing element of the min operator (14).

Now, knowing basic properties of BOWA operators, let as consider whether they actually seem to be useful for constructing measures of association between two groups of rankings described in Sec. 3.

Just by Def. 1 and further generalization allowing ties, each BOWA function takes values in the interval $[-1, 1]$, which fulfills postulate R1.

As a conclusion of Lemma 1 we have obtained $F(1, \ldots, 1) = 1$. Hence if all rankings are pairwise perfectly concordant then a BOWA operator reaches its maximal value. Similarly, $F(-1, \ldots, -1) = -1$ means that if all rankings are pairwise perfectly discordant then a BOWA operator reaches its minimal value. These two remarks are in favor to postulates R2 and R3, respectively. To fulfill them perfectly, i.e. to get the inverse implication, we have to add a requirement that a vector of weights has no zero elements, i.e. $w_i > 0$ for $i = 1, \ldots, n$.

Requirement R4 holds just by the definition of BOWA operators, whole R5 is guaranteed by Lemma 2.

Postulate R6 is fulfilled due to the fact each BOWA operator behaves as OWA operator for absolute values of the arguments, i.e. when we consider $F(|x_1|, \ldots, |x_n|)$. It is important since the OWA operator is monotonic. Hence when we increase or decrease the absolute values of the arguments of F, the value of our measure also increases or decreases, respectively, what has been postulated in R6. This way, by an adequate choice of the weights, we may also reward higher correlations whatever are their signs, according to postulate R7 (an example of such vector of weights has been suggested in [5] and is also discussed below). Finally, R8 implies directly from (9) and (10).

4.3 Example

Consider two consumers A and B who specified their preferences by assigning ranks to several variants belonging to the domain of four attributes under study. Moreover, assume that the correlation between their preferences for each separate attribute was calculated and as a result we received the following four numbers: $s^1_{AB} = 0.3$, $s^2_{AB} = 0.8$, $s^3_{AB} = -0.3$ and $s^4_{AB} = 0.1$.

To aggregate these four coefficients the following operator $F_{LG} : [-1, 1]^n \to [-1, 1]$ was suggested in [5]

$$F_{LG}(x_i, \ldots, x_n) = \frac{2}{n(n+1)} \sum_{j=1}^{n} r(|x_j|) \cdot x_j, \tag{16}$$

where $r : [0, 1] \to \mathbb{R}^+$ is a function such that

$$r(z) = \frac{1}{2} + \sum_{i=1}^{n} c(z - |x_i|) \tag{17}$$

and where c is defined as

$$c(u) = \begin{cases} 0 & \text{if} \quad u < 0 \\ \frac{1}{2} & \text{if} \quad u = 0 \\ 1 & \text{if} \quad u > 0. \end{cases} \tag{18}$$

By (17) we get $r(|0.3|) = \frac{1}{2} + (\frac{1}{2} + 0 + \frac{1}{2} + 1) = 2.5$, $r(|0.8|) = \frac{1}{2} + (1 + \frac{1}{2} + 1 + 1) = 4$, $r(|-0.3|) = \frac{1}{2} + (\frac{1}{2} + 0 + \frac{1}{2} + 1) = 2.5$ and $r(|0.1|) = \frac{1}{2} + (0 +$

$0+0+\frac{1}{2})=1$. Then, keeping in mind (2) and substituting arguments of (16) by the correlation coefficients, we obtain the following measure of association between two groups of rankings delivered by two consumers A and B: $S(A,B) = F_{LG}(0.3, 0.8, -0.3, 0.1) = 0.33$.

On the other hand, let us consider given correlation coefficients as a vector $\boldsymbol{x} = [0.3, 0.8, -0.3, 0.1]$. Hence we get a vector of absolute values $|\boldsymbol{x}| \searrow = [0.8, 0.3, 0.3, 0.1]$ with ties. It is evident that, according to (9) and (10), tied observation should obtain identical weights. In our case, using ranks given by (17), we may compute a vector of weights $\widetilde{w} = [0.4, 0.25, 0.25, 0.1]$ and therefore, by (9) we get

$$F(0.3, 0.8, -0.3, 0.1) = 0.4 \cdot 0.8 + 0.25 \cdot 3 + 0.25 \cdot (-3) + 0.1 \cdot 0.1 = 0.33,$$

which, of course, coincides with the result stated above. It is not surprising since one can easily prove that the suggested operator (16) is a member of a family of BOWA operators. □

5 Conclusions

In this paper with have introduced a new family of BOWA operators that may be useful for aggregating a bipolar information when both the absolute values of arguments and their signs should be taken into account keeping in mind their specific roles, i.e. strength of the possible relation (association, correlation, etc.) and its direction. This way BOWA operators generalize well-known OWA operators. As we have shown, the suggested family of operators possess many properties indicating that they may be useful in recommender systems for constructing measures of association between groups of rankings.

Acknowledgment. Study was supported by research fellowship within "Information technologies: research and their interdisciplinary applications" project co-financed by European Social Fund (agreement no. POKL.04.01.01-00-051/10-00).

References

1. Beliakov, G., Pradera, A., Calvo, T.: Aggregation Functions: A Guide for Practitioners. Springer, Heidelberg (2007)
2. Calvo, T., Kolesarova, A., Komornikova, M., Mesiar, R.: Aggregation operators: Properties, classes and construction methods. In: Calvo, T., Mayor, G., Mesiar, R. (eds.) Aggregation Operators. New Trends and Applications. SCI, vol. 97, pp. 3–104. Springer, Heidelberg (2002)
3. Gibbons, J.D., Chakraborti, S.: Nonparametric Statistical Inference. Marcel Dekker Inc., New York (2003)
4. Grabisch, M., Pap, E., Marichal, J.L., Mesiar, R.: Aggregation Functions, Cambridge (2009)

5. Łącka, H., Grzegorzewski, P.: On measuring association between groups of rankings in recommender systems. In: Rutkowski, L., Korytkowski, M., Scherer, R., Tadeusiewicz, R., Zadeh, L.A., Zurada, J.M. (eds.) ICAISC 2014, Part II. LNCS, vol. 8468, pp. 423–432. Springer, Heidelberg (2014)
6. Melville, P., Sindhwani, V.: Recommender systems. In: Encyclopedia of Machine Learning, Springer, Heidelberg (2010)
7. Torra, V.: The wiegthted OWA operator. Int. J. Intelligent Systems 12, 153–166 (1997)
8. Yager, R.R.: On ordered weighted averaging aggregation operators in multicriteria decisionmaking. IEEE Transactions and Systems, Man and Cybernetics 18, 183–190 (1988)

Aggregation Process and Some Generalized Convexity and Concavity

Barbara Pękala

Faculty of Mathematics and Natural Sciences,
University of Rzeszów,
Pigonia 1, 35-310 Rzeszw, Poland

Abstract. The averaging aggregation operators are defined and some interesting properties are derived. Moreover, we have extended concave and convex property. The main results concerning aggregation of generalized quasiconcave and quasiconvex functions are presented and some properties of aggregation operators are derived and discussed. We study the class of concavity and convexity of two variable aggregation operators that preserve these properties.

Keywords: Aggregation, fuzzy set, interval-valued fuzzy set, convex, concave property.

1 Introduction

In this paper we are interested in aggregation of a finite number of real numbers into a single number and its use in designing new classes of generalized convex functions that may be useful in optimization theory and decision analysis. In decision making, aggregated values are typically preference or satisfaction degrees restricted to the unit interval [0, 1]. Here, we consider a decision problem in X, i.e., the problem to find a best decision in the set of feasible decisions X with respect to several criteria functions.

Convexity is one of the most important aspects connected with the study of geometric properties of not only crisp, but also fuzzy sets and interval-valued fuzzy sets. Some generalization of convexity in mult-expert decision problems by penalty function is used (see [1], [2]). Various definitions of convexity and its generalized versions are widely used, mostly in optimization problems (see [8]). An important property of convex sets is that convexity is preserved under intersections. We will study a similar question for the case of quasiconvex, quasiconcave and T-quasiconcave, S-quasiconvex interval-valued fuzzy sets. However, instead of the intersection we will consider an arbitrary aggregation operator on the space of images. As the intersection of fuzzy sets is defined by a triangular norm, which is a particular case of an aggregation operator, our result will be valid also for intersections. Together with the aggregation of lattice elements we will consider also an aggregation of interval-valued fuzzy sets.

The paper is structured as follows. In Sectction 2 triangular norms and conorms are introduced and suitable definitions and properties are mentioned.

P. Angelov et al. (eds.), *Intelligent Systems'2014*,
Advances in Intelligent Systems and Computing 322, DOI: 10.1007/978-3-319-11313-5_3

Then, aggregation functions and their basic properties are defined, averaging aggregation operators are defined and some interesting properties are derived. In Section 3 and 4 we consider some extend concave, convex and quasiconcave and quasiconvex functions T-quasiconcave and S-quasiconvex functions. Finally the results concerning aggregation of generalized concave and convex functions are presented and discussed.

2 Basic Definitions

The fuzzy set theory turned out to be a useful tool to describe situations in which the data are imprecise or vague. The idea of a fuzzy sets was defined in [9].

Definition 1 (cf. [9]). *A fuzzy set (relation) A on a universe X is a mapping*

$$A : X(X^2) \to [0,1].$$

Special classes of operations are t-norms and t-conorms which are useful in approximate reasoning, e.g. in medical diagnosis and information retrieval.

Definition 2 (cf. [6]). *A triangular norm T on $[0,1]$ is an increasing, commutative, associative operation $T : [0,1]^2 \to [0,1]$ with a neutral element* 1.
A triangular conorm S on $[0,1]$ is an increasing, commutative, associative operation $S : [0,1]^2 \to [0,1]$ with a neutral element 0.

Example 1. The following are most popular examples of t-norms:

$$T_M(x,y) = \min(x,y),$$
$$T_P(x,y) = xy,$$
$$T_L(x,y) = \max(0, x+y-1), S_M(x,y) = \max(x,y),$$
$$S_P(x,y) = x+y-xy,$$
$$S_L(x,y) = \min(1, x+y), x,y \in [0,1].$$

For every triangular norm T, the function $T^* : [0,1]^2 \to [0,1]$ defined by $T^*(x,y) = 1 - T(1-x, 1-y), x,y \in [0,1]$ is a t-conorm. The converse statement is also true.
Now, we put the crucial definition for investigations in this paper. Aggregation by itself is an important task in any discipline where the fusion of information is of vital interest. In case of aggregation operators it is assumed that a finite number of inputs from the same (numerical) scale, most often the unit interval, are being aggregated. The following is a specialization to our framework of the definition of aggregation functions from the recent book [3].

Definition 3. *An operation $\mathcal{A} : ([0,1])^n \to [0,1]$ is called aggregation if is increasing and*

$$\mathcal{A}(\underbrace{0,...,0}_{n\times}) = 0, \quad \mathcal{A}(\underbrace{1,...,1}_{n\times}) = 1.$$

Definition 4. *An aggregation A in $[0,1]^n$ is called*
- *conjunctive if*

$$A(x_1, ..., x_n) \leq T_M(x_1, ..., x_n),$$

- *disjunctive if*

$$A(x_1, ..., x_n) \geq S_M(x_1, ..., x_n),$$

- *averaging if*

$$T_M(x_1, ..., x_n) \leq A(x_1, ..., x_n) \leq S_M(x_1, ..., x_n).$$

Averaging aggregations are for example: arithmetic mean, geometric mean, harmonic mean or root-power mean. We recall some interesting properties of binary relations in a lattice L.

Definition 5. *Let $f, g : L^2 \to L$.*
- *f commute with g if*

$$f(g(a,b), g(c,d)) = g(f(a,c), f(b,d)),$$

- *f is min-transitive if*

$$\min(f(a,b), f(b,c)) \leq f(a,c),$$

- *f is dually max-transitive if*

$$\max(f(a,b), f(b,c)) \geq f(a,c),$$

$a, b, c, d \in L$.

3 Quasiconcavity, Quasiconvexity and T-quasiconcavity, S-quasiconvexity

In this section and the following sections we shall deal with our main problem, that is, the aggregation of generalized concave and convex functions. First, we will look for sufficient conditions that secure some properties of quasiconcavity, quasiconvexity and some of its generalizations. The concepts of quasiconcavity, quasiconvexity and T-quasiconcavity, S-quasiconvexity of a function can be introduced in several ways. The following definitions are suitable for our purpose. From now on, the set X is supposed to be a nonempty subset of R^n, we shall denote it $X \subseteq R^n$.

Definition 6. *Let $X \subseteq R^n$. A function $f : X \to L$ is called*
- *quasiconcave on X if $f(\lambda x + (1-\lambda)y) \geq f(x) \wedge f(y)$,*
- *quasiconvex on X if $f(\lambda x + (1-\lambda)y) \leq f(x) \vee f(y)$,*

for every $x, y \in X$ and every $\lambda \in [0,1]$ with $\lambda x + (1-\lambda)y \in X$.

We know that if a function is concave (convex), then it is quasiconcave (quasiconvex), but not vice-versa.
In contrast to the above part of the section, now we restrict our attention to functions on R^n whose range is included in the unit interval [0, 1] of real numbers. Such functions can be interpreted as membership functions of fuzzy subsets of R^n, important in fuzzy optimization and decision making. In this context they are called fuzzy criteria. We therefore use several terms and some notation of fuzzy set theory.

Definition 7 ([7]). *Let $X \subseteq R^n$, T (S) be a triangular norm (conorm). A function $f : X \to [0,1]$ is called*
- *T-quasiconcave on X if $f(\lambda x + (1-\lambda)y) \geq T(f(x), f(y))$,*
- *S-quasiconcave on X if $f(\lambda x + (1-\lambda)y) \leq S(f(x), f(y))$,*

for every $x, y \in X$, $x \neq y$ and every $\lambda \in [0,1]$ with $\lambda x + (1-\lambda)y \in X$.

Directly by above definition and by $T \leq \min$, $\max \leq S$ we obtain that if a function is quasiconcave (quasiconvex), then it is T-quasiconcave (S-quasiconvex), but not vice-versa.

Proposition 1 ([7]). *Let $X \subseteq R^n$, T be a triangular norm and $f : X \to [0,1]$ be T-quasiconcave on X. Then $f^* = 1 - f$ is T^*-quasiconvex on X, where T^* is t-conorm dual to T.*

4 Aggregation and Quasiconcave, Quasiconvex and T-quasiconcave, S-quasiconvex

In what follows we shall investigate the properties of aggregations of fuzzy criteria given by membership functions of fuzzy sets in R^n with the help of a t-norm (t-conorm) or aggregation function. Particularly, we deal with some sufficient conditions such that the aggregations of fuzzy criteria will become T-quasiconcave (S-quasiconvex) in some extension of the fuzzy sets theory. The generalized concavity properties are very important in optimization or decision making as they secure any local maximizer to be also a global one.
Some extension of the fuzzy sets theory is interval-valued fuzzy sets theory introduced, independently by Sambuc in 1975 [8] and Gorzałczany in 1987 [4]. In which to each element of the universe a closed subinterval of the unit interval is assigned which approximates the unknown membership degree.

Definition 8. *An interval-valued fuzzy set (relation) A on X is a mapping*

$$A : X(X^2) \to L^I,$$

where

$$L^I = \{[x_1, x_2] : x_1, x_2 \in [0,1] : x_1 \leq x_2\}$$

$$[x_1, x_2] \leq_{L^I} [y_1, y_2] \Leftrightarrow x_1 \leq y_1 \wedge x_2 \leq y_2, \tag{1}$$

with operations

$$[x_1, x_2] \wedge [y_1, y_2] = [\min(x_1, y_1), \min(x_2, y_2)],$$

$$[x_1, x_2] \vee [y_1, y_2] = [\max(x_1, y_1), \max(x_2, y_2)].$$

$(L^I, \leq_L)$ is a complete lattice with the units $1_{L^I} = [1, 1]$ and $0_{L^I} = [0, 0]$.

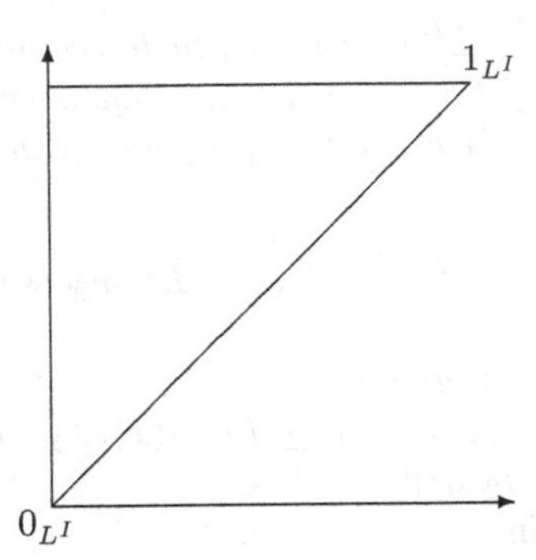

Fig. 1. Lattice L^I

Now we recall some special operation

Definition 9. *An operation $\mathcal{F} : (L^I)^2 \to L^I$ is called decomposable if there exist operations $F_1, F_2 : [0, 1]^2 \to [0, 1]$ such that for all $x = [x_1, x_2], y = [y_1, y_2] \in L^I$*

$$\mathcal{F}(x, y) = [F_1(x_1, y_1), F_2(x_2, y_2)]. \quad (2)$$

From the definition above we obtained the another method of construction of binary aggregation - representable aggregation build with $A_1 A_2$:

Theorem 1. *An operation $\mathcal{A} : (L^I)^2 \to L^I$ is aggregation, called representable aggregation, if and only if there exist aggregations $A_1, A_2 : [0, 1]^2 \to [0, 1]$ such that for all $x = [x_1, x_2], y = [y_1, y_2] \in L^I$ and $A_1 \leq A_2$*

$$\mathcal{A}(x, y) = [A_1(x_1, y_1), A_2(x_2, y_2)]. \quad (3)$$

Proof. Let A_1, $A_2 : [0, 1]^2 \to [0, 1]$ be aggregations and $x \leq z$, $y \leq t$. So, $x_1 \leq z_1$, $y_1 \leq t_1$, $x_2 \leq z_2$, $y_2 \leq t_2$. Then $A_1(x_1, y_1) \leq A_1(z_1, t_1)$, $A_2(x_2, y_2) \leq A_2(z_2, t_2)$. Thus $\mathcal{A}(x, y) \leq \mathcal{A}(z, t)$ which means that $\mathcal{A}$ is increasing.
$\mathcal{A}(\mathbf{0}, \mathbf{0}) = [A_1(0, 0), A_2(0, 0)] = \mathbf{0}$, $\mathcal{A}(\mathbf{1}, \mathbf{1}) = [A_1(1, 1), A_2(1, 1)] = \mathbf{1}$. Conversely, let $\mathcal{A}$ be increasing and $x_1 \leq z_1$, $y_1 \leq t_1$, $x_2 \leq z_2$, $y_2 \leq t_2$. Let $x = [x_1, x_2]$, $y = [y_1, y_2]$, $z = [z_1, z_2]$, $t = [t_1, t_2]$. Then $x \leq z$, $y \leq t$ and by the monotonicity of $\mathcal{A}$ we have $[A_1(x_1, y_1), A_2(x_2, y_2)] = \mathcal{A}(x, y) \leq \mathcal{A}(z, t) = [A_1(z_1, t_1), A_2(z_2, t_2)$. Directly from this we obtain $A_1(x_1, y_1) \leq A_1(z_1, t_1)$ and $A_2(x_2, y_2) \leq A_2(z_2, t_2)$. It means that A_1 and A_2 are increasing.
$\mathcal{A}(\mathbf{0}, \mathbf{0}) = \mathbf{0}$, so $[A_1(0, 0), A_2(0, 0)] = [0, 0]$ and $\mathcal{A}(\mathbf{1}, \mathbf{1}) = \mathbf{1}$, so $[A_1(1, 1), A_2(1, 1)] = [1, 1]$. Thus $A_1(0, 0) = 0$, $A_2(0, 0) = 0$ and $A_1(1, 1) = 1$, $A_2(1, 1) = 1$.

Especially we have representable T-norm (S-conorm), i.e.
$T(x,y) = [T_1(x_1,y_1), T_2(x_2,y_2)]$, where T_1, T_2 are fuzzy t-norms
$(S(x,y) = [S_1(x_1,y_1), S_2(x_2,y_2)]$, where S_1, S_2 are fuzzy t-conorms).

We can observe following condition for T-quasiconcavity (S-quasiconvexity) of interval-valued fuzzy sets.

Lemma 1. *Let $T : (L^I)^2 \to L^I$ be representable triangular norm. The interval-valued fuzzy set $f = [f_1, f_2] : X \to L^I$ is T-quasiconcave (S-quasiconvex) if and only if f_1 is T_1-quasiconcave and f_2 T_2-quasiconcave (S_1, S_2-quasiconvex, respectiviely).*

Proof. We assume that $f = [f_1, f_2] : X \to L^I$ interval-valued fuzzy sets is T-quasiconcave, i.e.
$f(\lambda x + (1-\lambda)y) \geq T(f(x), f(y)) \Leftrightarrow$
$[f_1(\lambda x + (1-\lambda)y), f_2(\lambda x + (1-\lambda)y)] \geq T([f_1(x), f_2(x)], [f_1(y), f_2(y)]) =$
$[T_1(f_1(x), f_1(y)), T_2(f_2(x), f_2(y))]$.
Directly with (1) we obtain
$f_1(\lambda x + (1-\lambda)y) \geq T_1(f_1(x), f_1(y))$ and $f_2(\lambda x + (1-\lambda)y) \geq T_2(f_2(x), f_2(y))$.
Which means that f_1 is T_1-quasiconcave and f_2 T_2-quasiconcave.
The converse statement is obvious. S-quasiconvexity we may prove in analogical way.

Moreover directly from definitions 5 and 9 we obtain connection between commutativity and generations in representable operation.

Lemma 2. *Let $T : (L^I)^2 \to L^I$ be representable triangular norm, $\mathcal{A} : (L^I)^2 \to L^I$ be representable aggregation. $\mathcal{A}$ commute with T if and only if A_1 commute with T_1 and A_2 commute with T_2, where $\mathcal{A} = [A_1, A_2]$*

Similar to [5] we obtain

Proposition 2. *Let $X \subseteq R^n$, $\mathcal{A} : (L^I)^2 \to L^I$ is representable aggregation, $f, g : X \to L^I$ are T-quasiconcave (S-quasiconvex) interval-valued fuzzy sets, where $T(S) : (L^I)^2 \to L^I$ is representable triangular norm (conorm).*
$\mathcal{A}$ preserves T-quasiconcavity (S-quasiconvexity) if and only if its representatives A_1, A_2 preserve T_1-quasiconcavity, T_2-quasiconcavity (S_1-quasiconvexity, S_2-quasiconvexity), respectively.

Now we consider problem preserving T-quasiconcavity (S-quasiconvexity) by aggregation. In the following proposition the necessary and sufficient condition for such an aggregation operator to preserve quasi-convexity is presented.

Proposition 3. *Let $X \subseteq R^n$, $\mathcal{A} : (L^I)^2 \to L^I$ is representable aggregation, $f, g : X \to L^I$ are T-quasiconcave interval-valued fuzzy sets, where $T : (L^I)^2 \to L^I$ is representable triangular norm and $\mathcal{A}$ commute with T, then $\mathcal{A}(f,g)$ is T-quasiconcave.*

Proof. We assume that $\mathcal{A}$ commute with T, i.e.
$\mathcal{A}(T(f(x), f(z)), T(g(x), g(z))) = T(\mathcal{A}(f(x), g(x)), \mathcal{A}(f(z), g(z)))$ and

$f, g : X \to L^I$ are T-quasiconcave interval-valued fuzzy sets
For $y = \lambda z + (1 - \lambda x)$ we examine
$\mathcal{A}(f,g)(y) = \mathcal{A}([f_1(y), f_2(y)], [g_1(y), g_2(y)]) = [A_1(f_1(y), g_1(y)), A_2(f_2(y), g_2(y))]$.
By Lemma 1 we have
$[A_1(f_1(y), g_1(y)), A_2(f_2(y), g_2(y))] \geq$
$[A_1(T_1(f_1(x), f_1(z)), T_1(g_1(x), g_1(z))), A_2(T_2(f_2(x), f_2(z)), T_2(g_2(x), g_2(z)))]$.
By assumption about commutativity $\mathcal{A}$ and T and Lemma 2 we obtain
$[A_1(T_1(f_1(x), f_1(z)), T_1(g_1(x), g_1(z))), A_2(T_2(f_2(x), f_2(z)), T_2(g_2(x), g_2(z)))] =$
$[T_1(A_1(f_1(x), g_1(x)), A_1(g_1(z), g_1(z))), T_2(A_2(f_2(x), g_2(x)), A_2(g_2(z), g_2(z)))] =$
$T([A_1(f_1(x), g_1(x)), A_2(f_2(x), g_2(x))], [A_1(g_1(z), g_1(z)), A_2(g_2(z), g_2(z))]) =$
$T(\mathcal{A}(f(x), g(x)), \mathcal{A}(f(z), g(z)))$. So $\mathcal{A}$ preserve T-quasiconcavity.

And dually we obtain

Proposition 4. *Let $X \subseteq R^n$, $\mathcal{A} : (L^I)^2 \to L^I$ is representable aggregation, $f, g : X \to L^I$ are S-quasiconvex interval-valued fuzzy sets, where $S : (L^I)^2 \to L^I$ is representable triangular norm and $\mathcal{A}$ commute with S, then $\mathcal{A}(f, g)$ is S-quasiconvex.*

Corollary 1. *Let $X \subseteq R^n$, $g, h : X \to L^I$ are T-quasiconcave interval-valued fuzzy sets, where $T : (L^I)^2 \to L^I$ is representable triangular norm. Then $\varphi : X \to L^I$ defined by $\varphi(x) = T(g(x), h(x))$ is T-quasiconcave.*

Corollary 2. *Let $X \subseteq R^n$, $g, h : X \to L^I$ are S-quasiconvexity interval-valued fuzzy sets, where $S : (L^I)^2 \to L^I$ is representable triangular norm. Then $\varphi : X \to L^I$ defined by $\varphi(x) = S(g(x), h(x))$ is S-quasiconvex.*

Now we consider quasiconcavity (quasiconvexity) in cartesian product, so for relations.

Definition 10. *Let $X \subseteq R^n$. A function $f : X^2 \to L^I$ is called*
- *quasiconcave on X^2 if $f(\lambda x + (1 - \lambda)y, \lambda z + (1 - \lambda)t) \geq \min(f(x, z), f(y, t))$,*
- *quasiconvex on X^2 if $f(\lambda x + (1 - \lambda)y, \lambda z + (1 - \lambda)t) \leq \max(f(x, z), f(y, t))$,*

for every $x, y, z, t \in X$ and every $\lambda \in [0, 1]$ with $\lambda x + (1 - \lambda)y$, $\lambda z + (1 - \lambda)t \in X$.

Interesting connection between quasiconcavity (quasiconvexity) and min-transitivity (dually max-transitivity) we observe in following theorems.

Proposition 5. *Let $X \subseteq R^n$, $\mathcal{A} : (L^I)^2 \to L^I$ is representable aggregation, $f, g : X^2 \to L^I$ are quasiconcave, min-transitive interval-valued fuzzy relations. If $\mathcal{A}$ preserve min-transitivity, then $\varphi : X^2 \to L^I$ defined by $\varphi(x, y) = \mathcal{A}(f(x, y), g(x, y))$ is quasiconcave.*

Proof. We assume that f, g are min-transitive and $\mathcal{A}$ preserve min-transitivity, i.e. for $z, k, m \in X$
$f(z, k) \wedge f(k, m) \leq f(z, m)$, $g(z, k) \wedge g(k, m) \leq g(z, m)$ and

$$\mathcal{A}(f(z, k), g(z, k)) \wedge \mathcal{A}(f(k, m), g(k, m)) \leq \mathcal{A}(f(z, m), g(z, m)). \tag{4}$$

But in especially: $f(z,k) \wedge f(k,m) = f(z,m)$, $g(z,k) \wedge g(k,m) = g(z,m)$, then

$$\mathcal{A}(f(z,m), g(z,m)) = \mathcal{A}(f(z,k) \wedge f(k,m), g(z,k) \wedge g(k,m)). \tag{5}$$

By properties of lattice operation $\wedge$ we obtain

$$\mathcal{A}(f(z,k) \wedge f(k,m), g(z,k) \wedge g(k,m)) \leq \mathcal{A}(f(z,k), g(z,k)) \wedge \mathcal{A}(f(k,m), g(k,m)). \tag{6}$$

So by (4)-(6) we have

$$\mathcal{A}(f(z,k) \wedge f(k,m), g(z,k) \wedge g(k,m)) = \mathcal{A}(f(z,k), g(z,k)) \wedge \mathcal{A}(f(k,m), g(k,m)). \tag{7}$$

Now by quasiconcavity of f, g and monotonicity of $\mathcal{A}$ and by (7) for $x = \lambda z + (1-\lambda)k$, $y = \lambda k + (1-\lambda)m$ we obtain

$$\varphi(\lambda z + (1-\lambda)k, \lambda k + (1-\lambda)m) =$$

$$\mathcal{A}(f(\lambda z + (1-\lambda)k, \lambda k + (1-\lambda)m), g(\lambda z + (1-\lambda)k, \lambda k + (1-\lambda)m)) \geq$$

$$\mathcal{A}(f(z,k) \wedge f(k,m), g(z,k) \wedge g(k,m)) =$$

$$\mathcal{A}(f(z,k), g(z,k)) \wedge \mathcal{A}(f(k,m), g(k,m)) =$$

$$\varphi(z,k) \wedge \varphi(k,m).$$

So φ is quasiconcave.

And dually we obtain

Proposition 6. *Let $X \subseteq R^n$, $\mathcal{A} : (L^I)^2 \to L^I$ is representable aggregation, $f, g : X^2 \to L^I$ are quasiconvex interval-valued fuzzy relations. If $\mathcal{A}$ preserve dually max-transitivity, then $\varphi : X^2 \to L^I$ defined by $\varphi(x,y) = \mathcal{A}(f(x,y), g(x,y))$ is quasiconvex.*

5 Conclusion

We considered some generalisations of convexity and concavity and preserving these properties by aggregations. In the future research we want examine another generalisations of convexity and concavity. For example, instead of $\lambda x + (1 - \lambda y)$ we can use averaging $A(x,y)$. Moreover we can consider connection of some properties of representable aggregations with some linear order defined in L^I.

Acknowledgements. This work was partially supported by the Center for Innovation and Transfer of Natural Sciences and Engineering Knowledge in Rzeszów.

References

1. Bustince, H., Jurio, A., Pradera, A., Mesiar, R., Beliakov, G.: Generalization of the weighted voting method using penalty functions constructed via faithful restricted dissimilarity functions. European J. Operational Research 225, 472–478 (2013)
2. Bustince, H., Barrenechea, E., Calvo, T., James, S., Beliakov, G.: Consensus in multi-expert decision making problems using penalty functions defined over a Cartesian product of lattices. Information Fusion 17, 56–64 (2014)
3. Grabisch, M., Marichal, J.-L., Mesiar, R., Pap, E.: Aggregation functions. Encyclopedia of mathematics and its applications, vol. 127. Cambridge University Press (2009) ISBN 978-0-521- 51926-7
4. Gorzałczany, M.B.: A method of inference in approximate reasoning based on interval-valued fuzzy sets. Fuzzy Sets Syst. 21(1), 1–17 (1987)
5. Janiš, V., Král, P., Renčová, M.: Aggregation operators preserving quasiconvexity. Inf. Sci. 228, 337–344 (2013)
6. Klement, E.P., Mesiar, R., Pap, E.: Triangular norms. Kluwer Acad. Publ., Dordrecht (2000)
7. Ramík, J., Vlach, M.: Aggregation function and generalized convexity in fuzzy optimization and decision making. Ann. Oper. Res. 195, 261–276 (2012)
8. Sambuc, R.: Functions ϕ-floues. Application á l'aide au diagnostic en pathologie thyroidienne. Ph. D. Thesis, Universitde Marseille, France (1975)
9. Zadeh, L.A.: Fuzzy sets. Information and Control 8, 338–353 (1965)

Some Class of Uninorms in Interval Valued Fuzzy Set Theory

Paweł Drygaś

Faculty of Mathematics and Natural Sciences, University of Rzeszów,
35-959 Rzeszów, ul. Pigonia 1, Poland
paweldrs@gmail.com, paweldrs@ur.edu.pl

Abstract. Uninorms are an important generalization of triangular norms and triangular conorms. Uninorms allow the neutral element to lie anywhere in the unit interval rather than at zero or one as in the case of a t-norm and a t-conorm.

Since interval valued fuzzy sets, Atanassov's intuitionistic fuzzy sets and L^I-fuzzy sets are equipollent, therefore in this paper we describe a generalization of uninorms on L^I. For example, we describe the structure of uninorms, discuss the possible values of the zero element for uninorms and of the neutral element, especially for decomposable uninorms.

Keywords: Interval-valued fuzzy set, lattice L^I, uninorm, decomposable operations.

1 Introduction

The triangular norms and conorms play an important role in the fuzzy set theory. They are used for the generalization of intersection and union of fuzzy sets, for extension the composition of fuzzy relations and for many other concepts.

Uninorms are important generalizations of triangular norms and triangular conorms. Uninorms allow a neutral element to lie anywhere in the unit interval rather than at zero or one as in the case of a t-norm and a t-conorm.

Interval valued sets are generalization of fuzzy sets in which to each element of the universe a closed subinterval of the unit interval is assigned which approximates the unknown membership degree. Another extension of fuzzy set is Atanassov's intuitionistic fuzzy set introduced by Atanassov [1]. Since it is shown that Atanassov's intuitionistic fuzzy set theory is equipollent to interval valued fuzzy set theory and that both are equivalent to L-fuzzy set theory in the sense of Goguen [12] w.r.t. a lattice L^I we consider a generalization of uninorm on the field of interval valued fuzzy set theory.

In Section 2, we put the definition of a fuzzy set, an Atanasov's intuitionistic fuzzy set, an interval valued fuzzy set and L-fuzzy set. Next, we recall the relationship between them. In Sections 3, we recall the properties of uninorms in $[0, 1]$. Next, the definition and properties of uninorms on L^I are given. Additionally, we put the description of decomposable uninorms and we discuss the possible values of the neutral element and zero element for these uninorms.

P. Angelov et al. (eds.), *Intelligent Systems'2014*,
Advances in Intelligent Systems and Computing 322, DOI: 10.1007/978-3-319-11313-5_4

Moreover, since L^I is a lattice we present different structure of uninorms than in [7] and some properties of such operations.

2 Interval Valued and Atanassov's Intuitionistic Fuzzy Sets

First we recall the notion of fuzzy set theory and some its extensions.

Definition 1 ([17]). *A fuzzy set A in a universe X is a mapping*

$$A : X \to [0, 1].$$

Fuzzy set describe the degree to which a certain point belongs to a set. A is also called a membership function and $A(x)$ is called membership degree of $x \in X$.

The natural way of extension the operations on sets to fuzzy sets is by membership functions

$$(A \cup B)(x) = \max(A(x), B(x)),$$

$$(A \cap B)(x) = \min(A(x), B(x)),$$

for $x \in X$.

There are many other generalizations of these operations. Some of them are based on triangular norms and triangular conorms (cf.[15]) which we may use instead of operations min, max.

Definition 2 ([15]). *A triangular norm T is an increasing, commutative, associative operation $T : [0, 1]^2 \to [0, 1]$ with neutral element 1.*
A triangular conorm S is an increasing, commutative, associative operation $S : [0, 1]^2 \to [0, 1]$ with neutral element 0.

Example 1 ([15]). Well-known t-norms and t-conorms are:
$T_M(x, y) = \min(x, y)$, $S_M(x, y) = \max(x, y)$,
$T_P(x, y) = x \cdot y$, $S_P(x, y) = x + y - xy$,
$T_L(x, y) = \max(x + y - 1, 0)$, $S_L(x, y) = \min(x + y, 1)$,

Intuitionistic fuzzy sets were introduced by Atanassov as an extension of the fuzzy sets in the following way.

Definition 3 (cf. [1], [2]). *An Atanassov's intuitionistic fuzzy set A on a universe X is a triple*

$$A = \{(x, \mu(x), \nu(x)) : x \in X\}, \tag{1}$$

where $\mu, \nu : X \to [0, 1]$ and $\mu(x) + \nu(x) \leq 1$, $x \in X$.
$\pi_A(x) = 1 - \mu_A(x) - \nu_A(x)$ is called the hesitation degree of x.

An Atanassov's intuitionistic fuzzy set assigns to each element of the universe not only a membership degree $\mu(x)$ but also a nonmembership degree $\nu(x)$, $x \in X$.

Another extension of fuzzy sets are interval valued fuzzy sets introduced by Zadeh [18]. In interval valued fuzzy sets to each element of the universe a closed subinterval of the unit interval is assigned and this is the way of describing the unknown membership degree.

Definition 4 ((cf. [13])). *An interval valued fuzzy set A in a universe X is a mapping $A : X \to \mathrm{Int}([0,1])$, where $\mathrm{Int}([0,1])$ denotes the set of all closed subintervals of $[0,1]$, i.e. a mapping which assigns to each element $x \in X$ the interval $[\underline{A}(x), \overline{A}(x)]$, where $\underline{A}(x), \overline{A}(x) \in [0,1]$ and $\underline{A}(x) \leq \overline{A}(x)$.*

An interval valued fuzzy set A on X can be represented by the L^I-fuzzy set A in the sense of Goguen. Namely

Definition 5 (cf. [12]). *An L-fuzzy set A on a universe X is a function $A : X \to L$ where L is a lattice.*

In this paper by $(L^I, \leq_{L^I})$ we mean the following complete lattice

$$L^I = \{[x_1, x_2] : x_1, x_2 \in [0,1] : x_1 \leq x_2\}, \tag{2}$$

with following order

$$[x_1, x_2] \leq_{L^I} [y_1, y_2] \Leftrightarrow x_1 \leq y_1 \wedge x_2 \leq y_2. \tag{3}$$

$(L^I, \leq_L)$ is a complete lattice with operations

$$[x_1, x_2] \wedge [y_1, y_2] = [\min(x_1, y_1), \min(x_2, y_2)],$$

$$[x_1, x_2] \vee [y_1, y_2] = [\max(x_1, y_1), \max(x_2, y_2)].$$

and the boundary elements $1_{L^I} = [1,1]$ and $0_{L^I} = [0,0]$.

Deschrijver and Kerre [5] (see also [4]) showed that Atanassov intuitionistic fuzzy sets are equipollent to interval valued fuzzy sets. The isomorphism assign the Atanassov intuitionistic fuzzy set the interval value fuzzy set as follows: $(x, \mu_A(x), \nu_A(x)) \mapsto [\mu_A(x), 1 - \nu_A(x)]$.

In this article we will develop our investigations for $(L^I, \leq_L)$, since in this case we have the product order and it will be easier to prove the main result.

Definition 6 ([7]). *A triangular norm $\mathcal{T}$ on L^I is an increasing, commutative, associative operation $\mathcal{T} : (L^I)^2 \to L^I$ with a neutral element 1_{L^I}.*
A triangular conorm $\mathcal{S}$ on L^I is an increasing, commutative, associative operation $\mathcal{S} : (L^I)^2 \to L^I$ with a neutral element 0_{L^I}.

Example 2. The following are examples of t-norms on L^I

$$\inf(x, y) = [\min(x_1, y_1), \min(x_2, y_2)],$$
$$\mathcal{T}(x, y) = [\max(0, x_1 + y_1 - 1), \min(x_2, y_2)],$$

and t-conorm on L^I

$$\sup(x, y) = [\max(x_1, y_1), \max(x_2, y_2)].$$

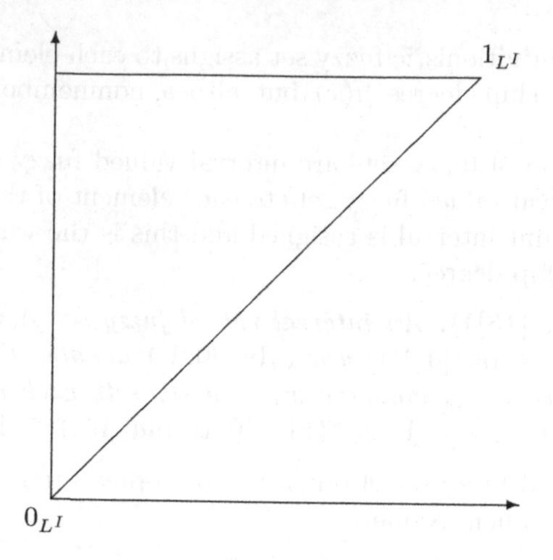

Fig. 1. Lattice L^I

Now, we recall one of the crucial definition for investigations in this paper.

Definition 7 ([9]). *An operation $\mathcal{F} : (L^I)^2 \to L^I$ is called decomposable if there exist operations $F_1, F_2 : [0,1]^2 \to [0,1]$ such that for all $x, y \in L^I$*

$$\mathcal{F}(x,y) = [F_1(x_1,y_1), F_2(x_2,y_2)], \tag{4}$$

where $x = [x_1, x_2]$, $y = [y_1, y_2]$.

The following lemma characterize certain family of decomposable operations

Lemma 1 (cf. [9]). *Increasing operations $F_1, F_2 : [0,1]^2 \to [0,1]$ in (4) gives a decomposable operation $\mathcal{F}$ if and only if $F_1 \leq F_2$.*

Remark 1. If we use the triangular norms in the construction of decomposable operation, then we obtain decomposable triangular norm. The same situation we have if we use triangular conorms, uninorms or nullnorms. Moreover decomposable triangular norms, triangular conorms, uninorms, nullnorms are also called t-representable triangular norms, triangular conorms, uninorms and nullnorms. The other properties of decomposable operations can be found in [9].

3 Uninorms

In this section we recall the definition and properties of a uninorms on $[0,1]$ and next we describe the uninorms on L^I. First we show the relationship with t-norms and t-conorms on L^I and next we describe some properties of representable uninorms on L^I, e.g. we discuss the possible values of the neutral element and zero element for these uninorms some properties of these operations.

Definition 8 ([16]). *Operation $U : [0,1]^2 \to [0,1]$ is called a uninorm if it is commutative, associative, increasing and has the neutral element $e \in [0,1]$.*

Theorem 1 ([10]). *If a uninorm U has the neutral element $e \in (0,1)$, then there exist a triangular norm T and a triangular conorm S such that*

$$U(x,y) = \begin{cases} T^*(x,y) & \text{if } x, y \leq e, \\ S^*(x,y) & \text{if } x, y \geq e, \end{cases} \tag{5}$$

where

$$\begin{cases} T^*(x,y) = \varphi^{-1}\left(T\left(\varphi(x), \varphi(y)\right)\right), \ \varphi(x) = x/e, & x, y \in [0,e] \\ S^*(x,y) = \psi^{-1}\left(S\left(\psi(x), \psi(y)\right)\right), \ \psi(x) = (x-e)/(1-e), & x, y \in [e,1] \end{cases}. \tag{6}$$

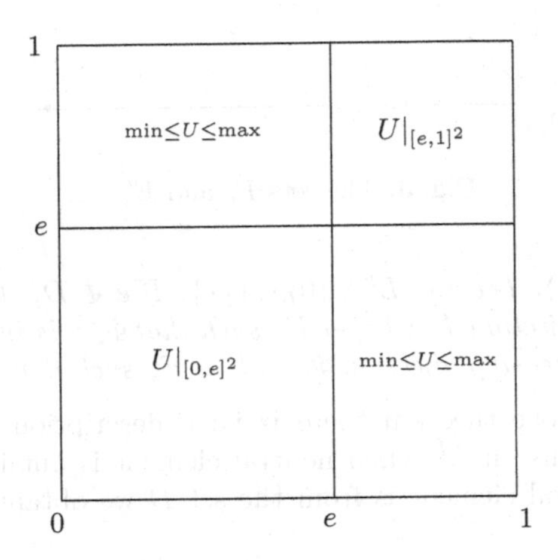

Fig. 2. The structure of uninorms on $[0,1]$

Lemma 2 (cf. [10]). *If U is a uninorm with the neutral element $e \in (0,1)$ then*

$$\min(x,y) \leq U(x,y) \leq \max(x,y) \text{ for } x \leq e \leq y \text{ or } y \leq e \leq x. \tag{7}$$

Lemma 3 (cf. [10]). *If U is an uninorm with the neutral element $e \in (0,1)$ then $U(0,1) \in \{0,1\}$ and $U(0,1)$ is the zero element of operation U.*

Definition 9 (cf. [6]). *Operation $\mathcal{U} : (L^I)^2 \to L^I$ is called a uninorm if it is commutative, associative, increasing and has the neutral element $e \in L^I$.*

In Theorem 1 there is given the structure of uninorms on $[0,1]$ which show the relationship with t-norms and t-conorms. To provide a similar description we define the following sets on L^I:

$$E_e = \{x \in L^I : x \leq_{L^I} e\},$$
$$E_e' = \{x \in L^I : x \geq_{L^I} e\},$$
$$D = \{[x,x] : x \in [0,1]\}.$$

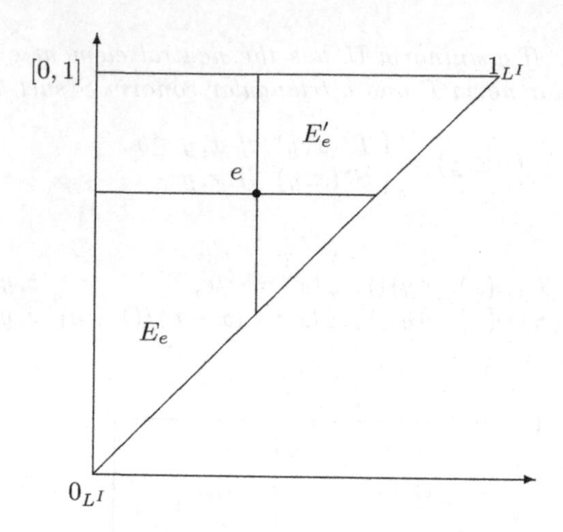

Fig. 3. The sets E_e and E'_e

Theorem 2 (cf. [6]). *Let $e \in L^I \setminus \{0_{L^I}, 1_{L^I}\}$. If $e \notin D$, then there does not exist an increasing bijection $\Phi_e : L^I \to E_e$ such that Φ_e^{-1} is increasing and there does not exist an increasing bijection $\Psi_e : L^I \to E'_e$ such that Ψ_e^{-1} is increasing.*

Because of the above theorem there is no a description of uninorms with t-norms and t-conorms on L^I when neutral element is outside the the set D. However, if the neutral element is from the set D we obtain the following description

Theorem 3 (cf. [6]). *If a uninorm $\mathcal{U}$ has the neutral element $e = [e_1, e_1] \in D \setminus \{0_{L^I}, 1_{L^I}\}$, then there exist a t-norm $\mathcal{T}$ and a t-conorm $\mathcal{S}$ on L^I such that*

$$\mathcal{U}(x, y) = \begin{cases} \mathcal{T}^*(x, y) \text{ if } x, y \leq e \\ \mathcal{S}^*(x, y) \text{ if } x, y \geq e \end{cases}, \tag{8}$$

where

$$\begin{cases} \mathcal{T}^*(x, y) = \Phi_e^{-1}\left(\mathcal{T}\left(\Phi_e(x), \Phi_e(y)\right)\right), \\ \quad \Phi_e(x) = (e_1 x_1, e_1 x_2), \ x, y \in E_e; \\ \mathcal{S}^*(x, y) = \Psi_e^{-1}\left(\mathcal{S}\left(\Psi_e(x), \Psi_e(y)\right)\right), \\ \quad \Psi_e(x) = (e_1 + x_1 - e_1 x_1, e_1 + (1 - e_1) x_2) \ x, y \in E'_e. \end{cases} \tag{9}$$

Similar to the Lemma 2 we have

Lemma 4 (cf.[8]). *If $\mathcal{U}$ is a uninorm with the neutral element $e \in L^I$ then for all $x, y \in L^I$ such that $x \leq e \leq y$ we have*

$$x \leq \mathcal{U}(x, y) \leq y.$$

Lemma 5 (cf.[8]). *If $\mathcal{U}$ is a uninorm with the neutral element $e \in L^I$ then for all $x, y \in L^I$ such that $x \leq e \leq y$ or $y \leq e \leq x$ we have*

$$\min(x, y) \leq \mathcal{U}(x, y) \leq \max(x, y).$$

But, for zero element we obtain more posibilities

Lemma 6 (cf. [6]). *If $\mathcal{U}$ is a uninorm with the neutral element $e \in L^I \setminus \{0_{L^I}, 1_{L^I}\}$ then for all $x \in L^I$ we have*

$$\mathcal{U}(0_{L^I}, 1_{L^I}) = \mathcal{U}(\mathcal{U}(0_{L^I}, 1_{L^I}), x),$$

i.e. $\mathcal{U}(0_{L^I}, 1_{L^I})$ is a zero element of uninorm $\mathcal{U}$.

Lemma 7 (cf. [6]). *If $\mathcal{U}$ is a uninorm with the neutral element $e \in L^I \setminus \{0_{L^I}, 1_{L^I}\}$ then $\mathcal{U}(0_{L^I}, 1_{L^I}) = 0_{L^I}$ or $\mathcal{U}(0_{L^I}, 1_{L^I}) = 1_{L^I}$ or $\mathcal{U}(0_{L^I}, 1_{L^I}) \| e$.*

Example 3. Let U_1, U_2 be uninorm given by

$$U_1(x,y) = \begin{cases} \max(x,y), & \text{if } x,y \in [0.1,1], \\ \min(x,y) & \text{elsewhere,} \end{cases}$$

$$U_2(x,y) = \begin{cases} \min(x,y), & \text{if } x,y \in [0,0.1], \\ \max(x,y) & \text{elsewhere,} \end{cases}$$

then for uninorm

$$\mathcal{U}(x,y) = [U_1(x_1,y_1), U_2(x_2,y_2)]$$

we have $\mathcal{U}(0_{L^I}, 1_{L^I}) = [U_1(0,1), U_2(0,1)] = [0,1]$ and $\mathcal{U}$ is neither conjunctive nor disjunctive and this is an example of decomposable uninorm.

Example 4. Let U be an arbitrary uninorm. Operation

$$\mathcal{U}(x,y) = [\min(U(x_1,y_2), U(y_1,x_2)), U(x_2,y_2)]$$

is not decomposable.

For arbitrary uninorm the zero element is equal 0_{L^I}, 1_{L^I} or it is incomparable with neutral element. If we consider decomposable uninorm then we have the following results

Lemma 8. *If $\mathcal{U}$ is a decomposable uninorm with the neutral element $e \in L^I$ then $\mathcal{U}(0_{L^I}, 1_{L^I}) = 0_{L^I}$ or $\mathcal{U}(0_{L^I}, 1_{L^I}) = 1_{L^I}$ or $\mathcal{U}(0_{L^I}, 1_{L^I}) = [0,1]$.*

Proof. Since $\mathcal{U}$ is decomposable, then there exist U_1 and U_2, such that

$$\mathcal{U}(x,y) = [U_1(x_1,y_1), U_2(x_2,y_2)].$$

We consider the four possible cases:

- $U_1(0,1) = 0$, $U_2(0,1) = 0$, then $\mathcal{U}(0_{L^I}, 1_{L^I}) = [U_1(0,1), U_2(0,1)] = [0,0] = 0_{L^I}$
- $U_1(0,1) = 0$, $U_2(0,1) = 1$, then $\mathcal{U}(0_{L^I}, 1_{L^I}) = [U_1(0,1), U_2(0,1)] = [0,1]$
- $U_1(0,1) = 1$, $U_2(0,1) = 1$, then $\mathcal{U}(0_{L^I}, 1_{L^I}) = [U_1(0,1), U_2(0,1)] = [1,1] = 1_{L^I}$
- $U_1(0,1) = 1$, $U_2(0,1) = 0$ not occur, according to the Lemma 1.

Remark 2. We cannot use the pair of disjunctive and conjunctive uninorms for construction of a decomposable uninorm, because this leads to the fourth case in the above lemma.

If we consider decomposable uninorms, then we obtain some dependencies between neutral elements of its components uninorms.

Theorem 4. *If a uninorm $\mathcal{U}$ is decomposable then $e_1 \geq e_2$, where e_1 and e_2 are the neutral element of uninorms U_1 and U_2.*

Proof. Let e_1 and e_2 be the neutral element of uninorms U_1 and U_2. Then

$$\mathcal{U}([e_1, e_1], [e_2, e_2]) = [U_1(e_1, e_2), U_2(e_1, e_2)] = [e_2, e_1] \in L^I.$$

So, $e_2 \leq e_1$.

Since $[e_1, e_2] \in L^I$, then directly from above we obtain

Theorem 5. *If $\mathcal{U}$ is a decomposable uninorm with a neutral element $e = [e_1, e_2]$, then $e \in D$.*

Corollary 1. *Every decomposable uninorm has a representation given by* (8)

Remark 3. Let us note, that due to [14] operation no need to fulfill the condition $e \in D$ to be a uninorm.

Example 5 (cf [14]). Operation $U : (L^I)^2 \to L^I$ defined as follows:

$$\mathcal{U}([x_1, x_2], [y_1, y_2]) = \begin{cases} [0, x_2 \wedge y_2], & \text{if } x_1 = y_1 = 0 \\ [x_1 \vee y_1, x_2 \vee y_2], & \text{if } (x_1 = 0, y_1 > 0 \text{ and } y_2 = 1) \\ & \text{or } (x_2 = 1, x_1 > 0 \text{ and } y_1 = 0) \\ [y_1, y_2], & \text{if } x_1 = 0 \text{ and } [y_1, y_2] \parallel [0, 1] \\ [x_1, x_2], & \text{if } y_1 = 0 \text{ and } [x_1, x_2] \parallel [0, 1] \\ [1, 1], & \text{otherwise} \end{cases}.$$

is a uninorm with the neutral element $[0, 1]$.

For a description of the uninorms as in above example we need the following facts

Lemma 9. *The sets E and E' with order given by* (3) *are bounded lattices.*

Lemma 10. *Let $\mathcal{U}$ be a uninorm with neutral $e \in L^I \setminus \{0_{L^I}, 1_{L^I}\}$. Then $\mathcal{U}|_E$ is a t-norm on E and $\mathcal{U}|_{E'}$ is a t-conorm on E'.*

Theorem 6 (cf [14]). *Let $e \in L^I \setminus \{0_{L^I}, 1_{L^I}\}$ and $\mathcal{U}$ be a uninorm on L^I with a neutral elment e. Then then there exist a t-norm $\mathcal{T}$ on the lattice E and a t-conorm $\mathcal{S}$ on the lattice E' such that*

$$\mathcal{U}(x, y) = \begin{cases} \mathcal{T}(x, y) \text{ if } x, y \in E \\ \mathcal{S}(x, y) \text{ if } x, y \in E' \end{cases}, \tag{10}$$

Of course lemmas 4-7 works for such uninorms. Moreover we have that (compare with [14])

Lemma 11. *Let $\mathcal{U}$ be a uninorm with neutral $e \in L^I \setminus \{0_{L^I}, 1_{L^I}\}$. Then it holds:*

(a) $\mathcal{U}(x, y) \leq x$ for $x \in L^I, y \in E$
(b) $\mathcal{U}(x, y) \leq y$ for $x \in E, y \in L^I$
(c) $\mathcal{U}(x, y) \geq x$ for $x \in L^I, y \in E$
(d) $\mathcal{U}(x, y) \geq y$ for $x \in E, y \in L^I$

We may also construct a uninorm with a given neutral element, e.g.

Theorem 7. *Let $e \in L^I \setminus \{0_{L^I}, 1_{L^I}\}$ and $\mathcal{U}$ be given by*

$$\mathcal{U}([x_1,x_2],[y_1,y_2]) = \begin{cases} [x_1 \wedge y_1, x_2 \wedge y_2], & \text{if } x_1, y_1 \le e_1,\ x_2, y_2 \le e_2 \\ [x_1 \vee y_1, x_2 \vee y_2], & \text{if } (x_1 \le e_1 \le y_1 \text{ and } x_2 \le e_2 \le y_2) \\ & \text{or } (y_1 \le e_1 \le x_1 \text{ and } y_2 \le e_2 \le x_2) \\ [y_1, y_2], & \text{if } x_1 \le e_1, x_2 \le e_2 \text{ and } [y_1,y_2] \parallel [e_1,e_2] \\ [x_1, x_2], & \text{if } y_1 \le e_1, y_2 \le e_2 \text{ and } [x_1,x_2] \parallel [e_1,e_2] \\ [1,1], & \text{otherwise} \end{cases},$$

is a uninorm with the neutral element e.

By Lemma 7 the zero element of uninorm is in the set $\{0_{L^I}, 1_{L^I}\}$ (the case of conjunctive or disjunctive uninorm) or it is incomparable with neutral element. We put the following question

Open problem Let $e \in L^I \setminus \{0_{L^I}, 1_{L^I}\}$ and $z \in L^I$, such that $z \| e$ be fixed. There exist a uninorm $\mathcal{U}$ with a neutral element e and zero element z?

If neutral element is from the set D, then in [7] there is given a construction of such uninorms .

Let φ and ψ be given as in (6)

Theorem 8 (cf [7] Theorem 25). *Let $e = [e_1, e_1] \in D \setminus \{0_{L^I}, 1_{L^I}\}$, $z = [z_1, z_2] \in L^I$; T_1 and T_2 be t-norms, S_1 and S_2 be t-conorms on $[0,1]$ such that*

(i) $z \| e$,
(ii) there exist t-norms T_{1a} and T_{1b} such that T_1 is an ordinal sum of T_{1a} and T_{1b} with intervals $[0, \varphi(z_1)]$, $[\varphi(z_1), 1]$,
(iii) there exist t-conorms S_{1a} and S_{1b} such that S_1 is an ordinal sum of S_{1a} and S_{1b} with intervals $[0, \psi(z_2)]$, $[\psi(z_2), 1]$,
(iv) $T_1(x_1, y_1) \le T_2(x_1, y_1)$ and $S_1(x_1, y_1) \le S_2(x_1, y_1)$, for all $x_1, y_1 \in [0,1]$.

Define the mapping $\mathcal{U} : (L^I)^2 \to L^I$ by, for all $x, y \in L^I$

$$(\mathcal{U}(x,y))_1 = \begin{cases} z_1, & \text{if } \quad (x_1 < z_1 \text{ and } y_1 \ge z_1 \text{ and } y_2 > e_1 \\ & \quad \text{or} (y_1 < z_1 \text{ and } x_1 \ge z_1 \text{ and } x_2 > e_1, \\ U_1(x_1, y_1), & \text{else}; \end{cases}$$

$$(\mathcal{U}(x,y))_2 = \begin{cases} z_2, & \text{if } \quad (x_2 > z_2 \text{ and } y_2 \le z_2 \text{ and } y_1 < e_1 \\ & \quad \text{or} (y_2 > z_2 \text{ and } x_2 \le z_2 \text{ and } x_1 < e_1, \\ U_2(x_2, y_2), & \text{else}; \end{cases}$$

where, for all x_1, y_1, x_2, y_2 in $[0,1]$

$$U_1(x_1, y_1) = \begin{cases} \varphi^{-1}(T_1(\varphi(x_1), \varphi(y_1))), & \text{if } \max(x_1, y_1) \le e_1, \\ \psi^{-1}(S_1(\psi(x_1), \psi(y_1))), & \text{if } \min(x_1, y_1) \ge e_1, \\ \min(x_1, y_1), & \text{else}, \end{cases}$$

$$U_2(x_2, y_2) = \begin{cases} \varphi^{-1}(T_2(\varphi(x_2), \varphi(y_2))), & \text{if } \max(x_2, y_2) \le e_1, \\ \psi^{-1}(S_2(\psi(x_2), \psi(y_2))), & \text{if } \min(x_2, y_2) \ge e_1, \\ \max(x_2, y_2), & \text{else}. \end{cases}$$

Then $\mathcal{U}$ is a uninorm on L^I with neutral element e for which $\mathcal{U}(0_{L^I}, 1_{L^I}) = z$.

4 Conclusion

Inspired by papers [7] and [14], we introduce the concept of uninorms on L^I whithout the assumption, that the neutral element is on the diagonal. Of course we can not describe such uninorms using t-norms and t-conorms on L^I, but we obtain similar description (using t-norms and t-conorms defined on the other lattices). In this paper we put some properties of uninorms with neutral element as arbitrary point on L^I and there is given the construction of such uninorms. Moreover, since there is more possibilities for the value of the zero element of uninorm on L^I than for ordinary unionorm we recall the construction of uninorm with given neutral element on the diagonal and given a zero element incomparable with neutral element, and next put some open problem connected with this.

Acknowledgements. This work was partially supported by the Centre for Innovation and Transfer of Natural Sciences and Engineering Knowledge.

References

1. Atanassov, K.T.: Intuitionistic fuzzy sets. Fuzzy Sets Syst. 20, 87–96 (1986)
2. Atanassov, K.T.: Intuitionistic Fuzzy Sets. Springer, Heidelberg (1999)
3. Atanassov, K.: On Intuitionistic Fuzzy Sets Theory. Springer, Heidelberg (2012)
4. Atanassov, K., Gargov, G.: Interval-valued intuitionistic fuzzy sets. Fuzzy Sets and Systems 31, 343–349 (1989)
5. Deschrijver, G., Kerre, E.E.: On the relationship beetwen some extensions of fuzzy set theory. Fuzzy Sets Syst. 133, 227–235 (2003)
6. Deschrijver, G., Kerre, E.E.: Uninorms in L^*-fuzzy set theory. Fuzzy Sets Syst. 148, 243–262 (2004)
7. Deschrijver, G.: Uninorms which are neither conjunctive nor disjunctive in interval-valued fuzzy set theory. Information Sciences 244, 48–59 (2013)
8. Drygaś, P.: On the structure of uninorms on L*. In: Magdalena, L., Ojeda-Aciego, M., Verdegay, J.L. (eds.) Proceedings of IPMU 2008, Torremolinos (Malaga), pp. 1795–1800 (2008)
9. Drygaś, P., Pękala, B.: Properties of decomposable operations on same extension of the fuzzy set theory. In: Atanassov, K.T., et al. (eds.) Advences in Fuzzy Sets, Intuitionistic Fuzzy Sets, Generalized Nets and Related Topics. Foundations, vol. I, pp. 105–118. EXIT, Warszawa (2008)
10. Fodor, J.C., Yager, R.R., Rybalov, A.: Structure of uninorms. Internat. J. Uncertain. Fuzziness Knowledge-Based Syst. 5, 411–427 (1997)
11. Fuchs, L.: Partially Ordered Algebraic Systems. Pergamon Press, Oxford (1963)
12. Goguen, A.: L-fuzzy sets. J. Math. Anal. Appl. 18, 145–174 (1967)
13. Gorzałczany, M.B.: A method of inference in approximate reasoning based on interval-valued fuzzy sets. Fuzzy Sets Syst. 21, 1–17 (1987)

14. Kracal, F., Mesiar, R.: Uninorms on bounded lattices. Fuzzy Sets Syst. (accepted, 2014)
15. Klement, E.P., Mesiar, R., Pap, E.: Triangular norms. Kluwer Acad. Publ., Dordrecht (2000)
16. Yager, R., Rybalov, A.: Uninorm aggregation operators. Fuzzy Sets Syst. 80, 111–120 (1996)
17. Zadeh, L.A.: Fuzzy sets. Inform. Control 8, 338–353 (1965)
18. Zadeh, L.A.: The concept of a linguistic variable and its application to approximate reasoning. Inform. Sci. 8, 199–249, 301–357 (1975); 9, 43–80

The Modularity Equation in the Class of 2-uninorms

Ewa Rak

Faculty of Mathematics and Natural Sciences, University of Rzeszów,
35-310 Rzeszów, ul. Prof. S. Pigonia 1, Poland
ewarak@ur.edu.pl

Abstract. This paper is mainly devoted to solving the functional equations of modularity of special class of aggregation operators with 2-neutral elements. Our investigations are motivated by modular logical connectives and their generalizations used in fuzzy set theory e.g. triangular norms, conorms, uninorms and nullnorms. In this work the modularity of two binary operations from the family of 2-uninorms ($\mathbf{U}_{k(e,f)}$) which generalizes both uninorms and nullnorms is considered. In particular, all of the possible solutions for one of the three classes of these operations depending on the position of its absorbing and neutral elements are characterized.

Keywords: Triangular norm, Triangular conorm, Nullnorm, Uninorm, Modularity equation, 2-uninorm.

1 Introduction

The functional equations involving aggregation operators (e.g.[2], [3], [4], [13], [14]) play an important role in theories of fuzzy sets and fuzzy logic. A new direction of investigations is concerned of modularity equation for uninorms and nullnorms ([7], [14]). Uninorms, introduced by Yager and Rybalov [17], and studied by Fodor et al. [8], are special aggregation operators that have proven to be useful in many fields like fuzzy logic, expert systems, neural networks, aggregation, and fuzzy system modeling (see [6], [9], [11], [16]). Uninorms are interesting because their structure is a special combination of t-norms and t-conorms [10] having a neutral element lying somewhere in the unit interval. The first notions of nullnorms and t-operators appeared respectively in [3], [12]. Both of these operations, whose also generalize concepts of t-norms and t-conorms, where an absorbing element is from the whole unit interval, have been studied further in (e.g. [5], [15]). Moreover, in [3] it is proved that nullnorms and t-operators have the same structures so they are equivalent.

Our consideration was motivated by logical connectives which generalize both uninorms and nullnorms. Such generalization was introduced by P. Akella in [1]. A 2-uninorm F belongs to the class of binary operations denoted by $U_{k(e,f)}$ in which for some absorbing element $k \in (0,1)$, where $e \leq k \leq f$: $F(e,x) = x$ for all $x \leq k$ and $F(f,x) = x$ for all $x \geq k$.

P. Angelov et al. (eds.), *Intelligent Systems'2014*,
Advances in Intelligent Systems and Computing 322, DOI: 10.1007/978-3-319-11313-5_5

This paper is organized as follows. In section 2, we mention the algebraic structures of uninorms, nullnorms and 2-uninorms. Then, the functional equation of modularity is recalled (section 3). In section 4, the partial solutions of modularity equations for described one of subclasses of 2-uninorms are characterized.

2 Some Classes of Binary Operators

We start with basic definitions and facts.

Definition 1 ([17]). *Let* $e \in [0,1]$. *Operation* $U : [0,1]^2 \to [0,1]$ *is called a uninorm if it is associative, commutative, increasing with respect to both variables and fulfilling*

$$\forall_{x\in[0,1]}\ U(x,e) = x. \qquad (\textit{neutral element } e \in [0,1])$$

By $\boldsymbol{U}_e$ *we denote the family of all uninorms,* $e \in [0,1]$.

Now we recall the general structure of a uninorm (for more details see [8]). We use the following notation $D_e = [0,e) \times (e,1] \cup (e,1] \times [0,e)$ for $e \in (0,1)$.

Theorem 1 (cf. [8]). *Let* $e \in [0,1]$. $U \in \boldsymbol{U}_e$ *if and only if*

$$U = \begin{cases} T_U \text{ in } [0,e]^2 \\ S_U \text{ in } [e,1]^2, \\ C \ \text{ in } D_e \end{cases} \tag{1}$$

where T_U *and* S_U *are operations respectively isomorphic with some triangular norm and triangular conorm and increasing* $C : D_e \to [0,1]$ *fulfils*

$$\min(x,y) \le C(x,y) \le \max(x,y) \ \ \textit{for} \ \ (x,y) \in D_e.$$

In cases $e = 1$ or $e = 0$ the uninorm U is reduced to a triangular norm and triangular conorm, respectively (see [10], p. 4, 11).
Moreover, for any uninorm we have $U(0,1) \in \{0,1\}$. With the additional assumption of continuity we obtain the following characterization of uninorms.

Theorem 2 (cf. [8]). *Let* $e \in (0,1)$ *and* $U : [0,1]^2 \to [0,1]$ *be a uninorm.*
If $U(0,1) = 0$ *and the function* $U(x,1)$ *is continuous except for the point* $x = e$ *then* $C = \min$ *in (1) and the class of such uninorms is denoted by* $\boldsymbol{U}_{\mathrm{e}}^{\min}$.
If $U(0,1) = 1$ *and the function* $U(x,0)$ *is continuous except for the point* $x = e$ *then* $C = \max$ *in (1) and the class of such uninorms is denoted by* $\boldsymbol{U}_{\mathrm{e}}^{\max}$.

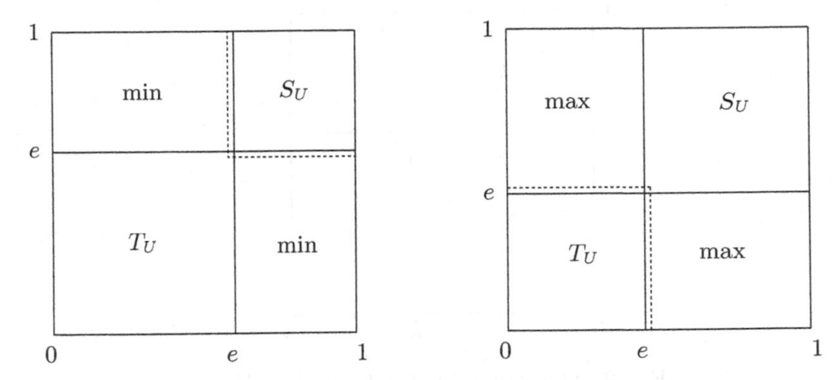

Fig. 1. The structure of uninorms from the class of $\mathbf{U}_e^{\min}$ and $\mathbf{U}_e^{\max}$

Definition 2 ([3]). *Let $k \in [0,1]$. Operation $V : [0,1]^2 \to [0,1]$ is called a nullnorm if it is associative, commutative, increasing with respect to both variables and fullfiling*

$$Z1)\ \forall_{x\in[0,1]}\ V(k,x) = V(x,k) = k, \qquad (\textit{zero element } k \in [0,1])$$

$$Z2)\ \forall_{x\in[0,k]}\ V(0,x) = V(x,0) = x, \qquad (\textit{neutral element } e = 0 \textit{ on } [0,k])$$

$$Z3)\ \forall_{x\in[k,1]}\ V(1,x) = V(x,1) = x. \qquad (\textit{neutral element } e = 1 \textit{ on } [k,1])$$

By $\mathbf{V}_k$ we denote the family of all nullnorms, $k \in [0,1]$.

More general families of operations with absorbing (or zero) element are examined in [15].

We have the following characterization of a nullnorm [3] (equivalently t-operator [12]).

Theorem 3 ([3]). *Let $k \in [0,1]$. Operation $V : [0,1]^2 \to [0,1]$. $V \in \mathbf{V}_k$ if and only if*

$$V = \begin{cases} S_V & in\ [0,k]^2 \\ T_V & in\ [k,1]^2 \\ k & in\ D_k \end{cases},$$

where S_V and T_V are operators isomorphic with some triangular conorm and triangular norm, respectively.

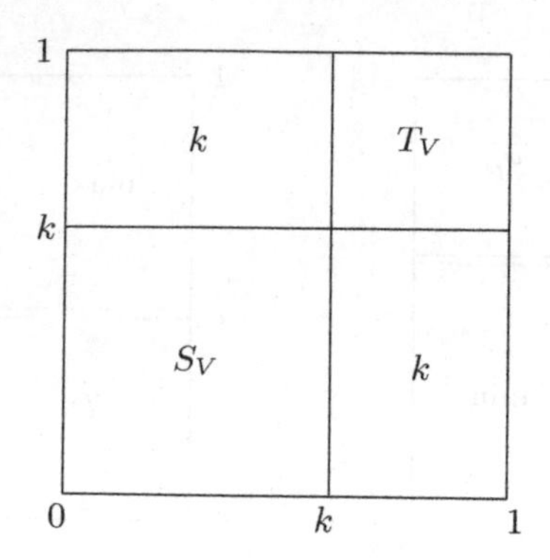

Fig. 2. The structure of nulnorm $V \in \mathbf{V}_k$

Now we recall the definition and some results about the class of 2-uninorms $\mathbf{U}_{k(e,f)}$.

Definition 3 (cf. [1]). *Let $k, e, f \in [0,1]$. By $\boldsymbol{U}_{k(e,f)}$ we denote the class of all 2-uninorms $F : [0,1]^2 \to [0,1]$ which are commutative, associative, increasing with respect to both variables, fulfilling $\forall_{x \le k}\ F(e,x) = x$ and $\forall_{x \ge k}\ F(f,x) = x$.*

Remark 1. • If $0 = k \le f = 1$ then 2-uninorm F is a triangular norm T with neutral element $f = 1$.
• If $0 = e \le k = 1$ then 2-uninorm F is a triangular conorm S with neutral element $e = 0$.
• If $0 = k \le f \le 1$ or $0 \le e \le k = 1$ then 2-uninorm F is an arbitrary uninorm U with neutral element f or e, respectively.
• If $e = 0$ and $f = 1$ then 2-uninorm F is a nullnorm V (t-operator) with zero element k.
• Moreover, such operators as: S-uninorms, T-uninorms, Bi-uninorms [15] are examples of 2-uninorms.

In order to exclude the subclasses of T, S, U, V from the class of 2-uninorms we assume that $0 \le e < k < f \le 1$.

Lemma 1 (cf. [1]). *Let $F \in \boldsymbol{U}_{k(e,f)}$ be a 2-uninorm, $0 \le e < k < f \le 1$. Then two mappings $U_1, U_2 \in \boldsymbol{U}_e$ defined by*

$$U_1(x,y) = \frac{F(kx, ky)}{k} \quad \textit{for} \quad x, y \in [0,1],$$

$$U_2(x,y) = \frac{F(k + (1-k)x, k + (1-k)y)}{1-k} \quad \textit{for} \quad x, y \in [0,1]$$

are uninorms with neutral elements $\frac{e}{k}$ and $\frac{f-k}{1-k}$, respectively.

Lemma 2 (see [1]). *Let $F \in \boldsymbol{U}_{k(e,f)}$ be a 2-uninorm, $0 \le e < k < f \le 1$. Then we have*

i) $F(x,0)$ is discontinuous at the point e if and only if $U_1(x,0)$ is discontinuous at $\frac{e}{k}$.
ii) $F(x,1)$ is discontinuous at the point f if and only if $U_2(x,1)$ is discontinuous at $\frac{f-k}{1-k}$.

Lemma 3 (see [1]). *Let $F \in \boldsymbol{U}_{k(e,f)}$ be a 2-uninorm, $0 \leq e < k < f \leq 1$. Then $F(0,1) \in \{0,k,1\}$.*

From the above lemmas we obtain three subclasses of operations in $\mathbf{U}_{k(e,f)}$ based on its zero element $F(0,1)$, denoted by $\mathbf{C}^0_{k(e,f)}$, $\mathbf{C}^k_{k(e,f)}$, $\mathbf{C}^1_{k(e,f)}$ (or simplifying them $\mathbf{C}^0, \mathbf{C}^k, \mathbf{C}^1$).

Representation of 2-uninorms $F \in \mathbf{C}^0, \mathbf{C}^k, \mathbf{C}^1$ under discontinuity at the points e and f, where $U^c \in \mathbf{U}_e^{\min}$ and $U^d \in \mathbf{U}_e^{\max}$ (cf. [1]).

Theorem 4. *Let $F \in \boldsymbol{U}_{k(e,f)}$ be a 2-uninorm, $0 \leq e < k < f \leq 1$, where $F(x,1)$ is discontinuous at the points e and f.*
$F \in C^0$ and $F(1,k) = k$ if and only if F has the following form

$$F = \begin{cases} U^c & \text{in } [0,k]^2 \\ U^c & \text{in } [k,1]^2 \\ \min & \text{in } (k,1] \times [0,e) \cup [0,e) \times (k,1] \\ k & \text{in } [k,1] \times [e,k] \cup [e,k] \times [k,1] \end{cases}.$$

Theorem 5. *Let $F \in \boldsymbol{U}_{k(e,f)}$ be a 2-uninorm, $0 \leq e < k < f \leq 1$, where $F(x,1)$ is discontinuous at e and $F(x,e)$ is discontinuous at f. $F \in C^0$ and $F(1,k) = 1$ if and only if*

$$F = \begin{cases} U^c & \text{in } [0,k]^2 \\ U^d & \text{in } [k,1]^2 \\ \min & \text{in } (k,1] \times [0,e) \cup [0,e) \times (k,1] \\ \max & \text{in } (f,1] \times (e,k) \cup (e,k) \times (f,1] \\ k & \text{in } [k,f] \times [e,k] \cup [e,k] \times [k,f] \end{cases}.$$

Theorem 6. *Let $F \in \boldsymbol{U}_{k(e,f)}$ be a 2-uninorm, $0 \leq e < k < f \leq 1$, where $F(x,0)$ is discontinuous at the points e and f.*
$F \in C^1$ and $F(0,k) = k$ if and only if F has the following form

$$F = \begin{cases} U^d & \text{in } [0,k]^2 \\ U^d & \text{in } [k,1]^2 \\ \max & \text{in } (f,1] \times [0,k) \cup [0,k) \times (f,1] \\ k & \text{in } [k,f] \times [0,k] \cup [0,k] \times [k,f] \end{cases}.$$

Theorem 7. *Let $F \in \boldsymbol{U}_{k(e,f)}$ be a 2-uninorm, $0 \leq e < k < f \leq 1$, where $F(x,f)$ is discontinuous at e and $F(x,0)$ is discontinuous at f. $F \in C^1$ and*

$F(0,k)=0$ *if and only if*

$$F=\begin{cases} U^c & in\ [0,k]^2 \\ U^d & in\ [k,1]^2 \\ \min & in\ (k,f]\times[0,e)\cup[0,e)\times(k,f] \\ \max & in\ (f,1]\times[0,k)\cup[0,k)\times(f,1] \\ k & in\ [k,f]\times[e,k]\cup[e,k]\times[k,f] \end{cases}.$$

Theorem 8. *Let $F\in\mathcal{U}_{k(e,f)}$ be a 2-uninorm, $0\le e<k<f\le 1$, where $F(x,0)$ and $F(x,1)$ are discontinuous at the points e and f, respectively. Then $F\in C^k$ if and only if F has the following form*

$$F=\begin{cases} U^d & in\ [0,k]^2 \\ U^c & in\ [k,1]^2 \\ k & in\ [k,1]\times[0,k]\cup[0,k]\times[k,1] \end{cases}. \tag{2}$$

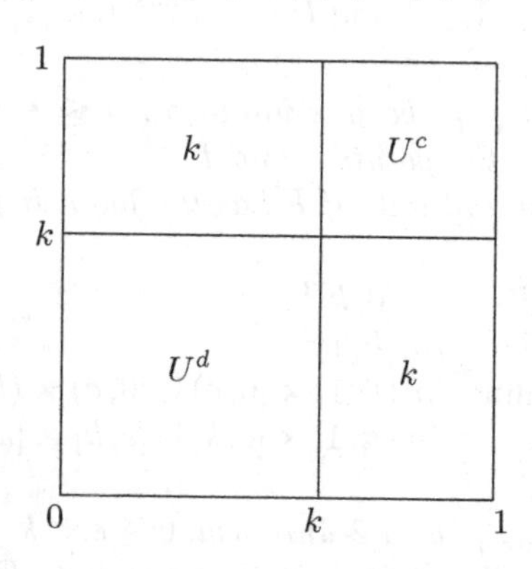

Fig. 3. Structure of 2-uninorm $F\in C^k$.

3 Functional Equation of Modularity

Now, we consider the modularity equation of two binary operators. Let us remind some of the most important facts relating to this topic.

Definition 4 (cf. [14]). *Let $F,G:[0,1]^2\to[0,1]$ be commutative operations. We say that F and G are modular, if*

$$\forall_{x,y,z\in[0,1]}\ \ z\le x\Rightarrow F(x,G(y,z))=G(F(x,y),z). \tag{3}$$

The modularity equation (3) involving triangular norms and conorms has the following trivial solutions:

Theorem 9 (cf. [14]).
i) Let $F=T_1$ and $G=T_2$ be triangular norms. Then they fulfil (3) if and only if $T_1=T_2$.
ii) Let $F=S_1$ and $G=S_2$ be triangular conorms. Then they fulfil (3) if and only if $S_1=S_2$.

iii) Let $F = T$ and $G = S$ be a triangular norm and triangular conorm, respectively. Then they fulfil (3) if and only if $T = \min$ and $S = \max$.
iv) Let $F = S$ and $G = T$ be a triangular conorm and triangular norm, respectively. Then there do not exist the solutions of equation (3).

In case of modular uninorms from the class of $\mathbf{U}_e$, $e \in (0, 1)$ we have

Lemma 4. *Let F and G be uninorms in $\boldsymbol{U}_e$. If $F = U_{e_1}$ and $G = U_{e_2}$ are modular then $e_2 \leq e_1$.*
Moreover, if $e_1 = e_2$ then $U_{e_1} = U_{e_2}$.

Theorem 10 (cf. [14]). *Uninorms $F = U_{e_1}$, $G = U_{e_2}$, where $e_2 < e_1$ satisfy (3) if and only if one of the following cases holds*
i) $e_2 = 0$ and $U_{e_1} \in \boldsymbol{U}_{e_1}^{\min}$ for $e_1 \in (0, 1]$ then U_{e_1} is a triangular norm and U_{e_2} is a triangular conorm;
ii) $e_2 = 0$ and $U_{e_1} \in \boldsymbol{U}_{e_1}^{\max}$ for $e_1 \in (0, 1)$ then

$$U_{e_1} = \begin{cases} \min & in\ [0, e_1]^2 \\ S & in\ [e_1, 1]^2 \\ \max & otherwise \end{cases},$$

$$U_{e_2} = \begin{cases} S & in\ [e_1, 1]^2 \\ \max & otherwise \end{cases};$$

iii) $e_1 = 1$ and $U_{e_2} \in \boldsymbol{U}_{e_2}^{\min}$ for $e_2 \in (0, 1)$ then

$$U_{e_1} = \begin{cases} T & in\ [0, e_2]^2 \\ \min & otherwise \end{cases},$$

$$U_{e_2} = \begin{cases} T & in\ [0, e_2]^2 \\ \max & in\ [e_2, 1]^2 \\ \min & otherwise \end{cases};$$

iv) $e_1 = 1$ and $U_{e_2} \in \boldsymbol{U}_{e_2}^{\max}$ for $e_2 \in [0, 1]$ then U_{e_1} is a triangular norm and U_{e_2} is a triangular conorm.

4 The Modularity Equation of 2-uninorms

The following table includes all of presented subclasses of 2-uninorms, for which we consider the modularity condition.
By $\star$ we mark already considered cases (25), known from the literature.
By $\bullet$ we mark the case which will be examined below.
The others (74) are proved in papers prepared to publication.

Table 1. All possible cases (100) to examine the modularity of operations from the class of 2-uninorms

$\mathbf{F} \setminus \mathbf{G}$	$\mathbf{C^k}$	$\mathbf{C^0_k}$	$\mathbf{C^0_1}$	$\mathbf{C^1_0}$	$\mathbf{C^1_k}$	$\mathbf{T}$	$\mathbf{S}$	$\mathbf{U^c}$	$\mathbf{U^d}$	$\mathbf{V}$
$\mathbf{T}$						⋆	⋆	⋆	⋆	⋆
$\mathbf{S}$						⋆	⋆	⋆	⋆	⋆
$\mathbf{U^c}$						⋆	⋆	⋆	⋆	⋆
$\mathbf{U^d}$						⋆	⋆	⋆	⋆	⋆
$\mathbf{V}$						⋆	⋆	⋆	⋆	⋆
$\mathbf{C^k}$	•									
$\mathbf{C^0_k}$										
$\mathbf{C^0_1}$										
$\mathbf{C^1_0}$										
$\mathbf{C^1_k}$										

Now we consider the modularity equation of $F, G \in C^k$.

Theorem 11. *Let $F \in C^{k_1}_{k_1(e_1,f_1)}$, $0 \le e_1 < k_1 < f_1 \le 1$ and $G \in C^{k_2}_{k_2(e_2,f_2)}$, $0 \le e_2 < k_2 < f_2 \le 1$ be 2-uninorms. Then F and G are modular if and only if $k_1 = k_2 = k$ and these 2-uninorms have the following structures (see Fig. 3)*

$$F = \begin{cases} S & in\ [e_1, k]^2 \\ T & in\ [k, f_2]^2 \\ \max & in\ [e_1, k] \times [0, e_1] \cup [0, e_1] \times [e_1, k], \\ k & in\ [k, 1] \times [0, k] \cup [0, k] \times [k, 1] \\ \min & otherwise \end{cases} \tag{4}$$

and

$$G = \begin{cases} S & in\ [e_1, k]^2 \\ T & in\ [k, f_2]^2 \\ \min & in\ [e_1, k] \times [0, e_1] \cup [0, e_1] \times [e_1, k]. \\ k & in\ [k, 1] \times [0, k] \cup [0, k] \times [k, 1] \\ \max & otherwise \end{cases} \tag{5}$$

Proof. Let $F \in \mathbf{C}^{k_1}_{k_1(e_1,f_1)}$ and $G \in \mathbf{C}^{k_2}_{k_2(e_2,f_2)}$ be modular 2-uninorms, where $k_2 \le k_1$. Taking $x = k_1$, $z = k_2$ and $y \in [0, 1]$ in (3) we get
$k_1 = F(k_1, k_2) = F(k_1, G(y, k_2)) = G(F(k_1, y), k_2) = G(k_1, k_2) = k_2$.
Thus $k_1 = k_2 = k$.
Now on account of the structures (2) of 2-uninorms F and G on $[0, k]^2$ and $[k, 1]^2$ we have
$F|_{[0,k]^2} = U^{(1)d}_{e_1}$, $G|_{[0,k]^2} = U^{(2)d}_{e_2}$ and $F|_{[k,1]^2} = U^{(1)c}_{f_1}$, $F|_{[k,1]^2} = U^{(2)c}_{f_2}$.
Using twice Lemma 4 and Theorem 10 ii) and iii) we obtain that
$e_2 = 0$ and for $e_1 \in (0, k)$

$$U^{d(1)}_{e_1} = \begin{cases} \min & in\ [0, e_1]^2 \\ S & in\ [e_1, k]^2 \\ \max & otherwise \end{cases}, \quad U^{d(2)}_{e_1} = \begin{cases} S & in\ [e_1, k]^2 \\ \max & otherwise \end{cases}$$

as well as $f_1 = 1$ and for $e_2 \in (0, k)$

$$U^{c(1)}_{f_2} = \begin{cases} T & in\ [0, e_2]^2 \\ \min & otherwise \end{cases}, \quad U^{c(2)}_{f_2} = \begin{cases} T & in\ [0, e_2]^2 \\ \max & in\ [e_2, 1]^2 \\ \min & otherwise \end{cases},$$

which together with the general structures of 2-uninorms proves (4) and (5).

Conversely, let 2-uninorm F be given by (4) and 2-uninorm G be given by (5), where $e_1 < k < f_2$.
To prove the equation (3) we have to consider 40 cases, however we can integrate some of them:
• if $0 \le x, y, z \le e_1$, then the restrictions F and G to set $[0, e_1]^2$ are modular because of associativity of S;
• if $e_1 \le x, y, z \le k$, then the restrictions F and G to set $[k, f_2]^2$ are modular directly from Theorem 9 iii);
• if $f_2 \le x, y, z \le 1$, then the restrictions F and G to set $[f_2, 1]^2$ are modular because of associativity of T.
In other cases let us denote $L = F(x, G(y, z))$, $R = G(F(x, y), z)$. Therefore
• if $e_1 \le z \le k \le x, y$, then L=$F(x, k) = k = G(\min(x, y), z) = R$;
• if $e_1 \le y \le k \le x, z$, then L=$F(x, k) = k = G(k, z) = R$;
• if $e_1 \le x, z \le k \le y$, then L=$F(x, k) = k = G(k, z) = R$;
• if $e_1 \le y, z \le k \le x$, then L=$F(x, S(y, z)) = k = G(k, z) = R$, where $S(y, z) \in [e_1, k]$;
• if $k \le \min(y, z) \le f_2 \le \max(y, z), x$, then L=$F(x, \min(y, z)) = \min(x, \min(y, z)) = \min(y, z) = G(\min(x, y), z) = R$;
• if $k \le \min(x, y), z \le f_2 \le \max(x, y)$, then L=$T(x, z) = G(\min(x, y), z) = R$, where $T(x, z) \in [k, f_2]$;
• if $\min(x, y), z \le e_1 \le \max(x, y) \le k$, then L=$F(x, \max(y, z)) = \max(x, \min(y, z)) = \min(y, z) = G(\min(x, y), z) = R$;
• if $y \le e_1 \le x, z \le k$, then L=$F(x, \max(y, z)) = F(x, z) = S(x, z) = G(x, z) = G(\max(x, y), z) = R$, where $S(x, z) \in [e_1, k]$;
• if $z \le e_1 \le x, y \le k$, then L=$F(x, \max(y, z)) = F(x, y) = S(x, y) = \max(S(x, y), z) = G(S(x, y), z) = R$, where $S(x, y) \in [e_1, k]$;
• if $\min(y, z) \le e_1, x, \max(y, z) \ge k$, then L=$F(x, k) = k = G(k, z) = R$;
• if $x, z \le e_1$, $y \ge k$, then L=$F(x, k) = k = G(k, z) = R$;
• if $y, z \le e_1$, $x \ge k$, then L=$F(x, \max(y, z)) = k = G(k, z) = R$;
• if $\min(y, z) \le e_1 \le \max(y, z)$, $x \ge f_2$, then L=$F(x, \max(y, z)) = k = G(k, z) = R$;
• if $z \le e_1 \le x \le f_2 \le y$, then L=$F(x, k) = k = G(x, z) = G(\min(x, z), k) = R$;
• if $z \le e_1 \le y \le k \le x \le f_2$, then L=$F(x, \max(y, z)) = F(x, y) = T(x, y) = \max(T(x, y), z) = G(T(x, y), z) = R$, where $T(x, y) \in [e_1, k]$;
• if $y \le e_1$, $k \le z \le f_2 \le x$, then L=$F(x, k) = k = G(k, z) = R$;
• if $z \le e_1$, $k \le y \le f_2 \le x$, then L=$F(x, k) = k = G(y, z) = G(\min(x, y), z) = R$;
• if $z \le e_1 \le x \le k \le y \le f_2$, then L=$F(x, k) = k = G(k, z) = R$;

• if $y \leq e_1 \leq z \leq k \leq x \leq f_2$, then L=$F(x, \max(y, z)) = F(x, z) = k = G(k, z) = R$.
In all considered cases we get $L = R$, which proves (3).

5 Conclusion

Inspired by uninorms and nullnorms, we examined the functional equation of modularity in the class of 2-uninorms. We gave the partial characterization of equation (3), namely the case when both operations F and G are from the subclass of $\mathbf{C}^k_{k(e,f)}$. This contribution and the other prepared papers will involve all possible cases which characterize the solutions of modularity equations for operations F and G from the class of 2-uninorms.

References

1. Akella, P.: Structure of n-uninorms. Fuzzy Sets Syst. 158, 1631–1651 (2007)
2. Calvo, T., De Baets, B.: On the generalization of the absorption equation. J. Fuzzy Math. 8(1), 141–149 (2000)
3. Calvo, T., De Baets, B., Fodor, J.C.: The functional equations of Frank and Alsina for uninorms and nullnorms. Fuzzy Sets Syst. 120, 385–394 (2001)
4. Drewniak, J., Drygaś, P., Rak, E.: Distributivity equations for uninorms and nullnorms. Fuzzy Sets Syst. 159, 1646–1657 (2008)
5. Drygaś, P.: A characterization of idempotent nullnorms. Fuzzy Sets Syst. 145, 455–461 (2004)
6. Dubois, D., Prade, H.: Fundamentals of Fuzzy Sets. Kluwer Academic Publishers, Boston (2000)
7. Feng, Q.: Uninorm solutions and (or) nullnorm solutions to the modilarity condition equations. Fuzzy Sets Syst. 148, 231–242 (2004)
8. Fodor, J.C., Yager, R.R., Rybalov, A.: Structure of uninorms. Internat. J. Uncertainty, Fuzzines Knowledge-Based Syst. 5, 411–427 (1997)
9. Gabbay, D., Metcalfe, G.: Fuzzy logics based on [0, 1)-continuous uninorms. Arch. Math. Logic 46, 425–449 (2007)
10. Klement, E.P., Mesiar, R., Pap, E.: Triangular norms. Kluwer Acad. Publ., Dordrecht (2000)
11. Klir, G.J., Yuan, B.: Fuzzy Sets and Fuzzy Logic, Theory and Application. Prentice Hall PTR, Upper Saddle River (1995)
12. Mas, M., Mayor, G., Torrens, J.: t-operators. Internat. J. Uncertainty, Fuzziness Knowledge-Based Syst. 7, 31–50 (1999)
13. Mas, M., Mayor, G., Torrens, J.: The distributivity condition for uninorms and t-operators. Fuzzy Sets Syst. 128, 209–225 (2002)
14. Mas, M., Mayor, G., Torrens, J.: The modularity condition for uninorms and t-operators. Fuzzy Sets Syst. 126, 207–218 (2002)
15. Mas, M., Mesiar, R., Monserat, M., Torrens, J.: Aggregation operations with annihilator. Internat. J. Gen. Syst. 34, 1–22 (2005)
16. Yager, R.R.: Uninorms in fuzzy system modeling. Fuzzy Sets Syst. 122, 167–175 (2001)
17. Yager, R.R., Rybalov, A.: Uninorm aggregation operators. Fuzzy Sets Syst. 80, 111–120 (1996)

Part II
Linguistic Summaries Counting and Aggregation

Dealing with Missing Information in Linguistic Summarization: A Bipolar Approach

Guy De Tré[1], Mateusz Dziedzic[1,2,3], Daan Van Britsom[1], and Sławomir Zadrożny[2]

[1] Dept. of Telecommunications and Information Processing, Ghent University, Sint-Pietersnieuwstraat 41, B-9000 Ghent, Belgium
[2] Systems Research Institute, Polish Academy of Science, ul. Newelska 6, 01-447 Warsaw, Poland
[3] Department of Automatic Control and Information Technology, Cracow University of Technology, ul. Warszawska 24, PL–31–155 Kraków, Poland
{Guy.DeTre,Daan.VanBritsom}@UGent.be, mdziedzic@pk.edu.pl, zadrozny@ibspan.waw.pl

Abstract. Linguistic summaries of databases provide quick insight in the stored data and are important facilities for understanding and grasping the meaning of large data collections. This is especially relevant in the context of big data. However, large data collections often suffer from incomplete data, which are in the case of relational databases modelled by so-called null-values. In this paper we propose a novel soft computing technique for measuring the quality of a linguistic summary in the case of missing information. More specifically we describe and illustrate how bipolar satisfaction degrees can be used to model both the validity of the summary and the hesitation about this validity that might be caused due to missing information. The extra information about the hesitation provides the users with a semantically richer description of the summarization results, which is important in view of a correct interpretation.

Keywords: Linguistic summaries, databases, fuzzy sets, bipolarity.

1 Introduction

A linguistic database summary provides users with a concise, representative description—in (quasi) natural language—of some (preselected) characteristic features of the stored data. As data collections getting larger and larger, linguistic summaries gain more importance because they help grasping the essence of the information which is encapsulated in the data in a fast, useful and comprehensible way. This is why it makes sense to equip database management systems with facilities for linguistic data abstraction and generalisation.

A fuzzy logic based technique using linguistically quantified propositions for linguistic data summarization has been first proposed by Yager [15,16] and further developed in [5,6]. An example of a summary obtained with this technique is '*most senior workers have a high salary*'. Later, the technique has been generalised using protoforms [7] and using interval-valued fuzzy sets [12]. Other works

P. Angelov et al. (eds.), *Intelligent Systems'2014*,
Advances in Intelligent Systems and Computing 322, DOI: 10.1007/978-3-319-11313-5_6

focus on the linguistic summarization of time series (e.g., [8,1]). A different kind of linguistic summaries are expressed by if-then rules like '*if seniority is long then salary is high*' [14]. Other related work includes [9,17,13]. Here, we study the approach based on linguistically quantified propositions as proposed in [15].

An important issue in summarization is the quality of the summary. There are different quality aspects including: validity, relevance, length, coverage, specificity, compatibility and unambiguity. All of them have been briefly discussed in, e.g., [3]. In this paper, we specifically deal with validity and coverage. Without a doubt, summarization should yield valid summaries. Hence, techniques based on linguistically quantified propositions compute the validity of different propositions and select only those with the largest validity to present them to the user. Coverage is also important. A set of summaries should cover as much of the data set as possible, because the non-covered part could be relevant and negatively influence the quality of the summarization.

A common cause of incomplete coverage of linguistic database summaries is missing information. The handling of missing information in databases has been widely addressed in research; cf., e.g., [4]. The most commonly used technique is to model missing data with a pseudo value, called '*null*' [2]. However, studying the impact of the presence of null-values on the overall quality of linguistic database summaries is still an open research challenge.

In this work we investigate how soft computing techniques can be used to handle missing information while generating linguistic database summaries. We assume that null-values cause hesitation about the validity of a summary. Bipolar data modelling techniques are applied to quantify this hesitation. More specifically, we describe how traditional validity measures can be extended to bipolar satisfaction degrees [10] which quantify both the validity of a summary and its associated hesitation. This provides users and applications with extra information about the overall quality of a summary. To measure overall quality, different aggregation methods for validity and hesitation are proposed.

The remainder of the paper is structured as follows. In Section 2 some preliminaries are given. Some general issues related to linguistic database summarization using linguistically quantified propositions are explained. Next, some basic concepts and definitions of bipolar satisfaction degrees are described. In Section 3, validity measures for linguistic summaries are discussed. Our new proposal for measuring the hesitation that is caused by missing information is presented in Section 4. Section 5 deals with some techniques to measure the overall quality of a summary based on its validity and hesitation. Next, some illustrative examples, demonstrating the presented approach are given in Section 6. Finally, some conclusions are reported in Section 7.

2 Preliminaries

2.1 Linguistic Database Summarization

Assume that $D = \{r_1, \ldots, r_n\}$ is a finite set of database records, which are characterized by a finite list $A = [A_1 : T_1, \ldots, A_m : T_m]$ of attributes. Each attribute $A_i : T_i$, $i = 1, \ldots, m$ consists of a name A_i and an associated data type T_i which determines the domain dom_{A_i} of allowed values for A_i. A domain can comprise numeric or non-numeric values and the null-value $\perp$ is used to denote that the actual value of A_i is missing [2]. An example of an attribute list characterising employee records is $[name, age, seniority, salary, department]$.

The actual value of an attribute A_i in a record r_j will be denoted by $A_i[r_j]$. Hence, the seniority of an employee represented by a record r with values $[John, 35, 5, \perp, Admin]$ is $seniority[r] = 5$ years and John's salary is missing, i.e., $salary[r] = \perp$.

With these considerations, following Yager's basic data summarization approach [15,16], a linguistic summary S of a record set (or database) D can be specified using linguistic quantifiers [18] as follows:

$$S = \text{Among all } r \text{ with } R,\ Q \text{ are}|\text{have } P \tag{1}$$

where

- Q is a linguistic quantifier, which expresses a quantity in agreement like, e.g., 'most'.
- P is a summariser specified over an attribute A_k which refers to a linguistic value (fuzzy predicate) that is defined on the domain dom_{A_k} and the attribute's name A_k, e.g., 'low salary'.
- R is an optional linguistic qualifier specified over a different attribute $A_l \in A$, $l \neq k$ which refers to a linguistic value that is defined on the domain dom_{A_l} and the attribute's name A_l, e.g., 'long seniority'.
- $\mathcal{J}$ is a validity measure, expressing the validity of the summary. Traditionally, for a given summary description S, $\mathcal{J}(S)$ is a number from the unit interval $[0, 1]$ with extreme values 0 ('not at all'), and 1 ('fully'). Usually, for considering the summary as being useful, $\mathcal{J}(S)$ is required to be large enough.

The summary obtained by Eq. (1) summarises the values of attribute A_k hereby considering all records of the record set D. If the optional qualifier R is used, only the fuzzy subset of records of D satisfying the constraint imposed by R is considered for the summarization. In case no qualifier R is used, Eq. (1) simplifies to

$$\mathcal{J}(\text{Among all } r,\ Q \text{ are}|\text{have } P). \tag{2}$$

An example of a linguistic summary without the optional qualifier is

$$\mathcal{J}(\text{Among all employees, most have low salary}) = 0.8$$

and with a qualifier for *seniority*

$$\mathcal{J}(\text{Among all employees with long seniority, most have moderate salary}) = 0.9.$$

In the current approaches, the validity $\mathcal{J}(S)$ of a summary S is expressed by a single number of the unit interval $[0, 1]$. The computation of the validity will be discussed in Section 3. Using a single number for expressing validity does not permit us to adequately cope with hesitation originating from incomplete data coverage.

2.2 Bipolar Satisfaction Degrees

To efficiently cope with hesitation, bipolar satisfaction degrees (BSDs) are used [10] for expressing validity. The term *bipolar* is hereby used to denote that a BSD provides both positive and negative information, respectively indicating what we know about the validity and invalidity of the summary. A BSD is a couple

$$(s, d) \in [0, 1]^2 \tag{3}$$

where s is the *satisfaction degree* and d is the *dissatisfaction degree.* Both s and d take their values in the unit interval $[0, 1]$ reflecting to what extent the BSD represents satisfaction, resp. dissatisfaction. The extreme values are 0 ('not at all'), and 1 ('fully'). The values s and d are independent of each other. A BSD can be used to express the validity of a summary in which case s (resp. d) denotes to which extent the summary is considered to be valid (resp. invalid).

Three cases are distinguished:

1. If $s + d = 1$, then the BSD is *fully specified.* This situation corresponds to traditional unipolar reasoning.
2. If $s + d < 1$, then the BSD is *underspecified.* In this case, the difference $h = 1 - s - d$ reflects the *hesitation* about the validity of the summary.
3. If $s + d > 1$, then the BSD is *overspecified.* In this case, the difference $c = s + d - 1$ reflects the *conflict* in the validity considerations.

With the understanding that i denotes a t-norm (e.g., min) and u denotes its associated t-conorm (e.g., max), the basic operations for BSDs (s_1, d_1) and (s_2, d_2) are [10]:

- *Conjunction.* $(s_1, d_1) \wedge (s_2, d_2) = (i(s_1, s_2), u(d_1, d_2))$
- *Disjunction.* $(s_1, d_1) \vee (s_2, d_2) = (u(s_1, s_2), i(d_1, d_2))$
- *Negation.* $\neg(s_1, d_1) = (d_1, s_1)$.

3 Validity Measures for Linguistic Summaries

Computing the validity $\mathcal{J}(S)$ of a summary S boils down to evaluating the truth of the proposition expressing this summary with respect to the record set under consideration.

In Yager's summarization approach [15,16], a non-decreasing proportional linguistic quantifier Q is modelled by a fuzzy set whose membership function μ_Q takes values in the unit interval as presented in [18]. For each $x \in [0, 1]$, $\mu_Q(x)$ expresses the grade to which a proportion of $100 \cdot x\%$ records is assumed to correspond to the semantics of the linguistic term modelled by the quantifier Q.

The summarizer P and the optional qualifier R are also modelled by fuzzy sets whose membership functions μ_P and μ_R are resp. defined on dom_{A_k} and dom_{A_l}. For each $x \in dom_{A_k}$ (resp. $x \in dom_{A_l}$), $\mu_P(x)$ (resp. $\mu_R(x)$) expresses to which extend x is considered to be compatible with the linguistic term modelled by P (resp. R).

With the above considerations, the validity $\mathcal{J}(S)$ of a linguistic summary S

$$S = \text{Among all } r \text{ with } R,\ Q \text{ are|have } P$$

can be computed by

$$\mathcal{J}(S) = \mu_Q \left(\frac{\sum_{j=1}^{n} i(\mu_R(A_l[r_j]), \mu_P(A_k[r_j]))}{\sum_{j=1}^{n} \mu_R(A_l[r_j])} \right) \tag{4}$$

where i denotes a t-norm operator (e.g., min). For simpler summaries of the form

$$S = \text{Among all } r,\ Q \text{ are|have } P$$

Eq. (4) reduces to

$$\mathcal{J}(S) = \mu_Q \left(\frac{1}{n} \sum_{j=1}^{n} \mu_P(A_k[r_j]) \right). \tag{5}$$

4 Hesitation Measurement in Linguistic Summaries

For the computations in Eq. (4) and (5) all records r_j, $j = 1, \ldots, n$ of the data set D are considered, including those for which the attributes A_k or A_l have a null-value $\perp$. The basic way to deal with null values is to neglect records containing them what may distort the meaning of linguistic summaries. In fact, usually $\mu_P(\perp) = \mu_R(\perp) = 0$ due to which null-values are considered being fully incompatible with the linguistic terms that are represented by P and R. Hence, records containing null-values influence the computations of the validity in such a way that they artificially lower the sums in the equations. This can lead to inaccurate summaries. For example, consider an employee data set where all salaries are represented by null-values. In such a database, the summary

$$\mathcal{J}(\text{Among all employees, none have low salary})$$

will have a high validity, although nothing is known about the salary of any employee.

To help solving these kind of problems caused by null-values, we propose to express the validity of a summary using a BSD (s, d). With a BSD, hesitation is modelled as $h = 1 - s - d$.

Considering an attribute A and a data set $D = \{r_1, \ldots, r_n\}$, we propose to express the hesitation h_A^D about the value of A over D by

$$h_A^D = \frac{card(\{r | r \in D \wedge A[r] = \perp\})}{n} \tag{6}$$

where *card* denotes the cardinality of the set. Hence, the hesitation is considered to be the proportion of records of D that have a null-value for A.

Based on these findings, we propose the following alternative for Eq. 5. With the understanding that $D_A^* \subseteq D$ is the subset of D consisting of all records that do not have a null-value for attribute A, i.e., $D^*_A = \{r | r \in D \wedge A[r] \neq \perp\}$, we state that

$$\mathcal{J}(\text{Among all } r,\ Q \text{ are} |\text{ have } P) = (s, d) \quad (7)$$

where

$$s = (1 - h^D_{A_k}) \cdot \left(\mu_Q \left(\frac{1}{card(D^*_{A_k})} \sum_{r \in D^*_{A_k}} \mu_P(A_k[r]) \right) \right)$$

and

$$d = (1 - h^D_{A_k}) \cdot \left(1 - \mu_Q \left(\frac{1}{card(D^*_{A_k})} \sum_{r \in D^*_{A_k}} \mu_P(A_k[r]) \right) \right).$$

With Eq. 7, we obtain that the hesitation h of the BSD (s, d) is $h = 1 - s - d = h^D_{A_k} \in [0, 1]$, such that $s + d = 1 - h \leq 1$. Hence, the hesitation about the validity of the summary is considered to be the proportion of records of D that have a null-value for the attribute A_k on which the summariser P is defined. For the computation of the satisfaction degree s, only records having non-null values for A_k are considered. The dissatisfaction degree is obtained from $d = 1 - h - s$. Therefore, s is assumed to be the fraction of $1 - h$ that corresponds with the validity obtained from Eq. 5 using the reduced data set $D^*_{A_k}$. In the case of no null-values, no hesitation is considered and computing s is the same as computing the validity with Eq. 5.

Using a similar approach, the following alternative for the validation of the more general summaries, cf. Eq. 4, is obtained. First, we let $h^D_{A_k \cup B_l}$ be the hesitation about the values of two attributes A_k and A_l over a data set $D = \{r_1, \ldots, r_n\}$. This hesitation is considered to be the proportion of records of D that have a null-value for either A_k or A_l, i.e.,

$$h^D_{A_k + A_l} = \frac{card(\{r | r \in D \wedge (A_k[r] = \perp \vee A_l[r] = \perp)\})}{n}. \quad (8)$$

Second, we consider $D^*_{A_k + A_l} \subseteq D$ to be the subset of D consisting of all records that do not have a null-value for A_k, nor A_l, i.e., $D^*_{A_k + A_l} = \{r | r \in D \wedge A_k[r] \neq \perp \wedge A_l[r] \neq \perp\}$. Then, assuming that the summarizer P and qualifier R refer to the attributes A_k and A_l, respectively, we have:

$$\mathcal{J}(\text{Among all } r \text{ with } R,\ Q \text{ are}|\text{have } P) = (s, d) \quad (9)$$

where

$$s = (1 - h^D_{A_k + A_l}) \cdot \left(\mu_Q \left(\frac{\sum_{r \in D^*_{A_k + A_l}} i(\mu_R(A_l[r_j]), \mu_P(A_k[r_j]))}{\sum_{r \in D^*_{A_k + A_l}} \mu_R(A_l[r_j])} \right) \right)$$

and

$$d = (1 - h^D_{A_k + A_l}) \cdot \left(1 - \mu_Q \left(\frac{\sum_{r \in D^*_{A_k + A_l}} i(\mu_R(A_l[r_j]), \mu_P(A_k[r_j]))}{\sum_{r \in D^*_{A_k + A_l}} \mu_R(A_l[r_j])} \right) \right)$$

where i again denotes a t-norm operator (e.g., min).

Now, the hesitation h of the BSD (s, d) is $h^D_{A_k+A_l} \in [0, 1]$. So, the hesitation about the validity of the summary is considered to be the proportion of records of D that have a null-value for the attribute A_k on which the summarizer P is defined or have a null-value for the attribute A_l on which the optional qualifier R is defined. The computation of s is then done using only those records with non-null values for A_k and A_l. Similar to the previous case, s is considered to be the fraction of $1 - h$ that corresponds to the validity obtained from Eq. 4 using the reduced data set $D^*_{A_k+A_l}$.

5 Overall Quality Measurement in Linguistic Summaries

Using BSDs (s, d) provides the user with two quality measures s and d, which resp. reflect to what extent the summary is considered to be valid and invalid. If $s + d < 1$, the hesitation about the validity $h = 1 - s - d$ caused by missing information is larger than zero.

In practice, users or applications might require a single, overall quality measure for summaries. For that purpose, BSDs have to be ranked using a ranking function. Reconsider that $h = 1 - s - d$. If one looks for summaries with a validity that is as high as possible and an invalidity that is as low as possible, the ranking function

$$\begin{aligned} \rho : [0, 1]^2 &\rightarrow [0, 1] \\ (s, d) &\mapsto \frac{s + (1 - d)}{2} \end{aligned} \tag{10}$$

can be used. Other ranking functions are discussed in [11]. If required a threshold value $\delta_\rho \in [0, 1]$ can be used. If this is the case, only summaries S with $\rho(\mathcal{J}(S)) \geq \delta_\rho$ will be considered as being acceptable. Another option is to work with two threshold values $\delta_s, \delta_d \in [0, 1]$. In such a case, only summaries with a BSD (s, d) satisfying $s \geq \delta_s$ and $d \leq \delta_d$ are considered being acceptable.

6 Illustrative Examples

To illustrate our approach, consider the relation *Employee* presented in Table 6. The table shows the employee ID, Age, seniority, salary and department of ten employees of a company. Null-values occur and are allowed for all attributes except the employee ID which is the primary key field. Next, consider the linguistic summary

S = Among all employees, most have low salary.

Assume that the linguistic quantifier 'most' and the linguistic term 'low' (salary) are modelled by the membership functions depicted in Fig. 1. With the traditional approach, by applying Eq. (5) we obtain,

Table 1. Database sample

EID	Age	Seniority	Salary	Department
001	30	6	1,700	Admin
002	⊥	2	1,000	Admin
003	23	2	1,100	⊥
004	48	15	2,500	Finance
005	22	1	1,000	Admin
006	26	⊥	1,000	Admin
007	22	⊥	900	Finance
008	27	⊥	⊥	⊥
009	28	3	⊥	Finance
010	21	1	⊥	Admin

$$\begin{aligned}\mathcal{J}(S) &= \mu_{most}\left(\frac{1}{10}\sum_{j=1}^{10}\mu_{low}(Salary[r_j])\right)\\ &= \mu_{most}\left(\frac{0+1+0.8+0+1+1+1+0+0+0}{10}\right)\\ &= \mu_{most}(0.48) = 0.\end{aligned}$$

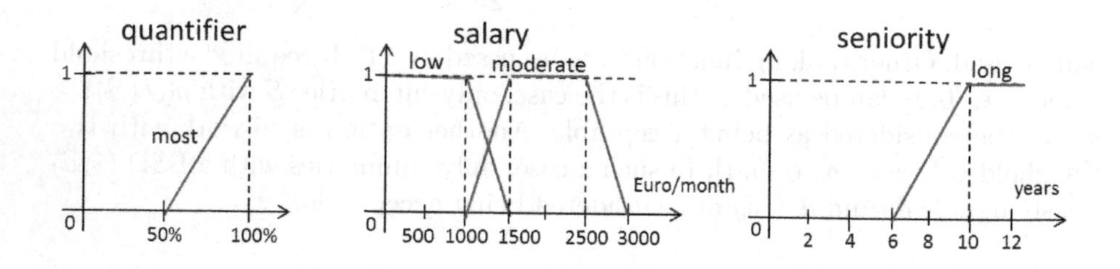

Fig. 1. Membership functions for the linguistic quantifier 'most' and the linguistic terms 'low' (salary), 'moderate' (salary) and 'long' (seniority)

Hereby, null-values as pseudo-values have an associated membership grade 0, i.e., $\mu_{low}(\perp) = 0$. However, by applying our novel technique, i.e., by using Eq. (7), we obtain

$$h^D_{Salary} = \frac{card(\{r|r \in D \wedge Salary[r] = \perp\})}{10} = \frac{3}{10} = 0.3 \text{ (cf. Eq. (6))}$$

and

$$D^*_{Salary} = \{r|r \in D \wedge Salary[r] \neq \perp\} = \{001, 002, 003, 004, 005, 006, 007\}$$

such that

$$\mathcal{J}(S) = (s, d)$$

with

$$\begin{aligned} s &= (1 - h^D_{Salary}) \cdot \left(\mu_{most} \left(\frac{1}{card(D^*_{Salary})} \sum_{r \in D^*_{Salary}} \mu_{low}(Salary[r]) \right) \right) \\ &= 0.7 \cdot \mu_{most} \left(\frac{0 + 1 + 0.8 + 0 + 1 + 1 + 1}{7} \right) \\ &= 0.7 \cdot \mu_{most}(0.69) = 0.27 \end{aligned}$$

and

$$\begin{aligned} d &= (1 - h^D_{Salary}) \cdot \left(1 - \mu_{most} \left(\frac{1}{card(D^*_{Salary})} \sum_{r \in D^*_{Salary}} \mu_{low}(Salary[r]) \right) \right) \\ &= 0.7 \cdot \left(1 - \mu_{most} \left(\frac{0 + 1 + 0.8 + 0 + 1 + 1 + 1}{7} \right) \right) \\ &= 0.7 \cdot (1 - \mu_{most}(0.69)) = 0.43. \end{aligned}$$

Hence, the hesitation of the validity of S is $h = 1 - s - d = 0.3$ and using Eq. (10), its overall quality can be considered as being

$$\rho((s, d)) = \frac{s + (1 - d)}{2} = 0.42.$$

When analysing the results of the old and new approaches, i.e., $\mathcal{J}(S) = 0$ vs. $\mathcal{J}(S) = (0.27, 0.43)$, we can observe that by neglecting the null-values for $Salary$, these are interpreted as not satisfying the summarizer low (salary) at all. This results in the too pessimistic assumption that the summary is not valid at all. However, the null-values represent missing data which cause hesitation and might cover cases that actually satisfy the summarizer. The BSD $(0.27, 0.43)$ reflects this hesitation of 0.3, caused by the 30% tuples that have a null-value for $Salary$. The satisfaction about the summary is still lower than the dissatisfaction.

Now, we consider the more complex linguistic summary

$S =$ Among all employees with long seniority, most have moderate salary.

The linguistic quantifier 'most', the linguistic qualifier 'long' (seniority) and the linguistic term 'moderate' (salary) are again modelled by the membership functions depicted in Fig. 1. Computing the validity of the summary description using the traditional approach, by applying Eq. (4) yields

$$\mathcal{J}(S) = \mu_{most}\left(\frac{\sum_{j=1}^{10} i(\mu_{long}(Seniority[r_j]), \mu_{moderate}(Salary[r_j]))}{\sum_{j=1}^{10} \mu_{long}(Seniority[r_j])}\right)$$
$$= \mu_{most}\left(\frac{\min(0,1)+\min(0,0)+\min(0,0.2)+\min(1,1)+6.\min(0,0)}{0+0+0+1+0+0+0+0+0+0}\right)$$
$$= \mu_{most}\left(\frac{1}{1}\right) = \mu_{most}(1) = 1.$$

Also here, null-values as pseudo-values have an associated membership grade 0, i.e., $\mu_{moderate}(\bot) = 0$ and $\mu_{long}(\bot) = 0$.

If we use Eq. (9) for applying our novel technique, we obtain

$$h^D_{Sal+Sen} = \frac{card(\{r|r \in D \wedge (Salary[r] = \bot \vee Seniority[r] = \bot)\})}{10}$$
$$= \frac{5}{10} = 0.5 \text{ (cf. Eq. (8))}$$

and

$$D^*_{Sal+Sen} = \{r|r \in D \wedge Salary[r] \neq \bot \wedge Seniority[r] \neq \bot\}$$
$$= \{001, 002, 003, 004, 005\}$$

such that

$$\mathcal{J}(S) = (s, d)$$

with

$$s = \left(1 - h^D_{Sal+Sen}\right)\cdot$$
$$\left(\mu_{most}\left(\frac{\sum_{r\in D^*_{Sal+Sen}} i(\mu_{long}(Seniority[r_j]), \mu_{moderate}(Salary[r_j]))}{\sum_{r\in D^*_{Sal+Sen}} \mu_{long}R(Seniority[r_j])}\right)\right)$$
$$= 0.5 \cdot \mu_{most}\left(\frac{\min(0,1)+\min(0,0)+\min(0,0.2)+\min(1,1)+\min(0,0)}{0+0+0+1+0}\right)$$
$$= 0.5 \cdot \mu_{most}(1) = 0.5$$

and

$$d = \left(1 - h^D_{Sal+Sen}\right)\cdot$$
$$\left(1 - \mu_{most}\left(\frac{\sum_{r\in D^*_{Sal+Sen}} i(\mu_{long}(Seniority[r_j]), \mu_{moderate}(Salary[r_j]))}{\sum_{r\in D^*_{Sal+Sen}} \mu_{long}R(Seniority[r_j])}\right)\right)$$
$$= 0.5 \cdot \left(1 - \mu_{most}\left(\frac{\min(0,1)+\min(0,0)+\min(0,0.2)+\min(1,1)+\min(0,0)}{0+0+0+1+0}\right)\right)$$
$$= 0.5 \cdot (1 - \mu_{most}(1)) = 0.$$

So, the hesitation of the validity of S is $h = 1 - s - d = 0.5$ and using Eq. (10), its overall quality can be considered as being

$$\rho((s,d)) = \frac{s + (1-d)}{2} = 0.75.$$

When comparing both the results of the old and new approaches, i.e., $\mathcal{J}(S) = 1$ vs. $\mathcal{J}(S) = (0.5, 0)$, we can now observe that neglecting the null-values for *Salary* and *Seniority*, results in a too optimistic assumption that the summary is completely valid. Null-values again cause hesitation and might this time cover cases that actually do not satisfy the summary. The BSD $(0.5, 0)$ reflects the hesitation of 0.5, caused by the 50% tuples that have a null-value for *Salary* or *Seniority*. The satisfaction about the summary is still higher than the dissatisfaction.

7 Conclusions

In this paper we propose a novel approach to cope with missing data when mining linguistic data(base) summaries. Hereby, we assumed that missing data cause hesitation about the validity of a summary. The approach concerns linguistic summaries that are specified using linguistic quantifiers and are of the general form

$$\mathcal{J}(\text{Among all } r \text{ with } R,\ Q \text{ are|have } P)$$

and computes and expresses the validity of such summaries by means of a bipolar satisfaction degree (BSD). The proposed technique is a generalisation of Yager's approach. Using BSDs has the advantage that the hesitation about the validity of a summary can be explicitly computed and expressed. This allows for a semantically richer feedback to users and applications. The added value of the approach has been demonstrated and discussed using illustrative examples.

Acknowledgements. Mateusz Dziedzic's contribution is supported by the Foundation for Polish Science under International PhD Projects in Intelligent Computing: A project financed by The European Union within the Innovative Economy Operational Programme (2007–2013) and European Regional Development Fund.

References

1. Castillo-Ortega, R., Marín, N., Sánchez, D.: A Fuzzy Approach to the Linguistic Summarization of Time Series. Multiple-Valued Logic and Soft Computing 17(2-3), 157–182 (2011)
2. Codd, E.F.: RM/T: Extending the Relational Model to capture more meaning. ACM Transactions on Database Systems 4(4) (1979)
3. Díaz-Hermida, F., Ramos-Soto, A., Bugarín, A.: On the role of fuzzy quantified statements in linguistic summarization of data. In: 11th International Conference on Intelligent Systems Design and Applications, ISDA 2011, pp. 166–171. IEEE Press (2011)

4. Dyreson, C.E.: A Bibliography on Uncertainty Management in Information Systems. In: Motro, A., Smets, P. (eds.) Uncertainty Management in Information Systems: From Needs to Solutions, pp. 415–458. Kluwer Academic Publishers, Boston (1997)
5. Kacprzyk, J., Yager, R.R., Zadrożny, S.: A fuzzy logic based approach to linguistic summaries of databases. International Journal of Applied Mathematics and Computer Science 10, 813–834 (2000)
6. Kacprzyk, J., Yager, R.R.: Linguistic summaries of data using fuzzy logic. International Journal of General Systems 30, 133–154 (2000)
7. Kacprzyk, J., Zadrożny, S.: Linguistic database summaries and their protoforms: toward natural language based knowledge discovery tools. Information Sciences 173, 281–304 (2005)
8. Kacprzyk, J., Wilbik, A., Zadrożny, S.: An approach to the linguistic summarization of time series using a fuzzy quantifier driven aggregation. International Journal of Intelligent Systems 25(5), 411–439 (2010)
9. Laurent, A.: A new approach for the generation of fuzzy summaries based on fuzzy multidimensional databases. Intelligent Data Analysis 7, 155–177 (2003)
10. Matthé, T., De Tré, G., Zadrożny, S., Kacprzyk, J., Bronselaer, A.: Bipolar Database Querying Using Bipolar Satisfaction Degrees. International Journal of Intelligent Systems 26(10), 890–910 (2011)
11. Matthé, T., De Tré, G.: Ranking of bipolar satisfaction degrees. Communications in Computer and Information Sciences 298(8), 461–470 (2012)
12. Niewiadomski, A., Ochelska, K., Szczepaniak, P.S.: Interval-valued linguistic summaries of databases. Control and Cybernetics 35(2), 415–443 (2006)
13. Pilarski, D.: Linguistic summarization of databases with quantirius: a reduction algorithm for generated summaries. International Journal of Uncertainty, Fuzziness and Knowledge-Based Systems 18(3), 305–331 (2010)
14. Wu, D., Mendel, J.M.: Linguistic Summarization Using IF-THEN Rules and Interval Type-2 Fuzzy Sets. IEEE Transactions on Fuzzy Systems 19(1), 136–151 (2011)
15. Yager, R.R.: A new approach to the summarization of data. Information Sciences 28, 69–86 (1982)
16. Yager, R.R., Ford, K., Cañas, A.J.: An Approach to the Linguistic Summary of Data. In: Bouchon-Meunier, B., Zadeh, L.A., Yager, R.R. (eds.) IPMU 1990. LNCS, vol. 521, pp. 456–468. Springer, Heidelberg (1991)
17. Yager, R.R., Petry, F.E.: A multicriteria approach to data summarization using concept ontologies. IEEE Transactions on Fuzzy Systems 14(6), 767–780 (2006)
18. Zadeh, L.A.: A computational approach to fuzzy quantifiers in natural languages. Computers & Mathematics with Applications 9, 149–184 (1983)

Evaluation of the Truth Value of Linguistic Summaries – Case with Non-monotonic Quantifiers

Anna Wilbik[1], Uzay Kaymak[1], James M. Keller[2], and Mihail Popescu[2]

[1] Information Systems
School of Industrial Engineering,
Eindhoven University of Technology,
Eindhoven, The Netherlands,
{A.M.Wilbik,U.Kaymak}@tue.nl
[2] Electrical and Computer Engineering,
University of Missouri,
Columbia, MO, USA
{kellerj,PopescuM}@missouri.edu

Abstract. In this paper we investigate linguistic summaries of the form "Q y's are P". We consider a case with non-monotonic quantifier, exemplified by *a few* or *about a half*. We propose a method for evaluating the truth value of such summaries.

Keywords: linguistic summaries, fuzzy logic, computing with words, truth value, non-monotonic quantifiers.

1 Introduction

With the advent of current technology more and more data are created and stored and need to be analyzed. However very often the amount of available data is beyond human cognitive capabilities and comprehension skills. Acknowledging this problem, creating summaries of data has been a goal of the artificial intelligence and computational intelligence community for many years. This was especially visible in the context of written text, but also other data like images and sensor data were summarized [1,2].

In this paper we are dealing with linguistic summaries of numerical data. This topic has been widely investigated, cf. [3,4,5,6]. For instance, Dubois and Prade [3] proposed representation and reasoning for gradual inference rules for linguistic summarization in the form "the more X is F, the more/ the less Y is G" that could summarize various relationships. Such rules expressed a progressive change of the degree to which the entity Y satisfies the gradual property G when the degree to which the entity X satisfies the gradual property F is modified. Rasmussen and Yager [5] discussed the benefit of using fuzzy sets in data summaries based on generalized association rules. Gradual functional dependences were investigated, and a query language called SummarySQL was

P. Angelov et al. (eds.), *Intelligent Systems'2014*,
Advances in Intelligent Systems and Computing 322, DOI: 10.1007/978-3-319-11313-5_7

proposed. Bosc et al. [7] discussed the use of fuzzy cardinalities for linguistic summarization. The SAINTETIQ model [4] provides the user synthetic views of groups of tuples over the database. In [8], the authors proposed a summarization procedure to describe long-term trends of change in human behavior, e.g., "the quality of the 'wake up' behavior has been decreasing in the last month" or "the quality of the 'morning routine' is constant but has been highly unstable in the last month."

In this paper we follow the approach of Yager [6], in the form "Q objects in Y are S,". This approach was considerably advanced and implemented by Kacprzyk [9], Kacprzyk and Yager [10], Kacprzyk et al. [11,12,13]. Those summaries have been applied in different areas, e.g. in financial data [14,15,16,17] and eldercare [2,18,19]. The biggest limitation of this approach is the quantifier, that needs to be normal and monotonically increasing.

In this paper we focus on the summaries with non-monotonic quantifiers, exemplified by *about a half* or *a few*, and propose a method for evaluating their truth value.

2 Background

Linguistic summaries of data are usually short sentences in natural language that condense the information carried in a large numeric data set. The linguistic summaries are linguistically quantified propositions and may be written using protoforms [20], such as a simple protoform

$$Q\ y's \text{ are } P \tag{1}$$

e.g. "*most* cars are *big*", and extended protoform

$$Q\ R\ y's \text{ are } P \tag{2}$$

e.g. "*most big* cars are *fast*".

Here, Q is the linguistic quantifier, P is the summarizer and R is the qualifier. Q, P and R are modeled by fuzzy sets over appropriate domains. It is assumed that the domain for the quantifier is unit interval, i.e., $\mu_Q : [0,1] \to [0,1]$. For every sentence the truth (validity), $\mathcal{T}$ of the summary is calculated. It corresponds directly to the truth value of quantified statements of (1) or (2). The truth may be calculated using either original Zadeh's calculus of quantified propositions (cf. [21]) or other methods. However all those methods require that the quantifier is normal and monotonically nondecreasing, such as *most* or *almost all*.

This condition excludes many quantifiers such as *a few*, *about a half* or *almost none*. In this paper we will propose how to evaluate the linguistic summaries with such non-monotonic quantifiers. Note that we still assume that the fuzzy sets used to model the quantifiers are normal. Moreover, we are considering only simple protoform type summaries.

For simplicity, we assume that all fuzzy sets are modeled with trapezoidal membership functions, as they are sufficient in most applications [22]. However,

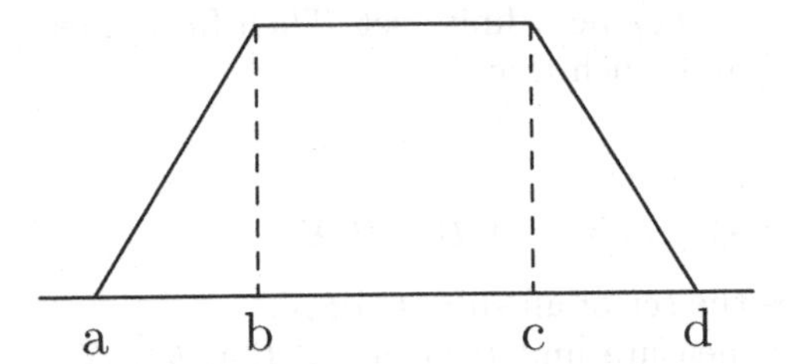

Fig. 1. Trapezoidal fuzzy membership function used in the numeric examples, denoted by Trap[a,b,c,d]

other types of membership functions can be also used. An example of such trapezoidal membership functions (Trap[a,b,c,d]) is shown in Fig 1.

The basic method of computing the truth value of linguistic summaries employs Zadeh's calculus of quantified propositions [21]. The truth value of a simple protoform summary is calculated as

$$\mathcal{T}(Qy's \text{ are } P) = \mu_Q\left(\frac{1}{n}\sum_{i=1}^{n}\mu_P(y_i)\right) \tag{3}$$

where μ_P and μ_Q are membership functions of P and Q, respectively. This method is fast, efficient and the results obtained generally confirm intuition. Unfortunately this is not always the case.

Example 1. Let us consider a tricky example: We have 100 balls in a box. All of them are big with the degree of 0.1. Let Q be *some*, defined as trapezoidal membership function Trap[0,0.1,1,1]. Consider a summary "*some* balls are *big*". Then according to the formula (3) the truth value is calculated as:

$$\mathcal{T}(\textit{some} \text{ balls are } \textit{big}) = \mu_{some}\left(\frac{1}{100}\sum_{i=1}^{100}0.1\right) = \mu_{some}(0.1) = 1 \tag{4}$$

This result is not exactly in line with our intuition since none of the balls is "really" big.

This divergence is caused by that fact that humans filter out objects with low membership values in question and Σ-count does not. With quantifiers such as *most* or *almost all*, that require the proportion of objects that are P to be high, this problem does not exist. Therefore to avoid such problems we should consider a method that not only counts the number of elements that are P but also allows to "filter out" objects with low values of μ_P.

Sugeno integral based aggregation seems to be a reasonable choice to address this problem, since it tries to find a balance between the quantity and quality of the elements.

Let $X = \{x_1, x_2, \ldots, x_n\}$ be a finite set. Then fuzzy measure [23] on X is a function $g : \mathcal{P}(X) \to [0, 1]$ such that

- $g(\varnothing) = 0$
- $g(X) = 1$
- If $A \subseteq B$ then $g(A) \leq g(B)$, $\forall A, B \in \mathcal{P}(X)$

where $\mathcal{P}(X)$ denotes the set of all subsets of X.

Let g be a fuzzy measure and h be a function $h : X \to [0, 1]$. Moreover, assume that $\{x_i\}$ are ordered so that $h(x_1) \geq h(x_2) \geq \ldots \geq h(x_n)$. Then the discrete Sugeno integral [24] of a function h with respect to g is a function $S_g : [0, 1]^n \to [0, 1]$ such that

$$S_g(h) = \max_{i=1,\ldots,n} [\min(h(x_i), g(A_i))] \tag{5}$$

where $A_i = \{x_1, x_2, \ldots, x_i\}$.

In our context of a linguistic summary "Q y's are P", X is the set of objects, i.e. $X = \{y_1, \ldots, y_n\}$. A fuzzy measure $g : \mathcal{P}(X) \to [0, 1]$ is defined as $g(A) = \mu_Q\left(\frac{|A|}{|X|}\right)$ where $|\cdot|$ denotes cardinality of the set, and μ_Q is the membership function of the quantifier, regular and monotonically non-decreasing. The partial support function $h : X \to [0, 1]$, is defined as $h(y_i) = \mu_P(y_i)$.

Then the truth value of the summary "Q y's are P" using the Sugeno integral [25,26] is expressed as

$$\mathcal{T}(Qy's \text{ are } P) = \max_{i=1,\ldots,n} \left[\min\left(\alpha, \mu_Q\left(\frac{|A_\alpha|}{n}\right)\right)\right] \tag{6}$$

where $|A_\alpha|$ is the number of objects with the degree of P bigger or equal to α.

Example 2. Consider the same settings as in example 1 and the same linguistic summary "*some* balls are *big*". Then according to the formula (6) we obtain

$$\begin{aligned}\mathcal{T}(\textit{some} \text{ balls are } \textit{big}) &= \max_{i=1,\ldots,100} \left[\min\left(\alpha, \mu_{some}\left(\frac{|A_\alpha|}{100}\right)\right)\right] \\ &= \min\left(0.1, \mu_{some}\left(\frac{100}{100}\right)\right) = 0.1\end{aligned}$$

This value is consistent with the human intuition.

In the subsequent sections the truth of the linguistic summary will be computed using the above described method with the Sugeno integral based aggregation.

3 Truth Value of Linguistic Summaries with Normal, Non-monotonic Quantifiers

Let us now consider Q defined as Trap[a,b,c,d] and $c < 1$. An example of such quantifier may be *about a half*. Consider a sentence "Q y's are P". It is not possible to compute the truth value of this sentence directly with any method shown above, since this quantifier does not fulfill the monotonicity condition.

Think now about two quantifiers $Q_1 = Trap[a, b, 1, 1]$ and $Q_2 = Trap[c, d, 1, 1]$, as in Fig. 2. They are normal and monotonic and non-decreasing quantifiers. Therefore we can compute the truth values of the summaries "Q_1 y's are P" and "Q_2 y's are P".

Note that $Q = Q_1 - Q_2$. Q_1 can be linguistically interpreted as *at least* Q. While Q_2 can be interepreted as *more than* Q. Hence the truth value of the summary "Q y's are P" may be calculated as:

$$\mathcal{T}(Q \text{ } y\text{'s are } P) = \mathcal{T}(Q_1 \text{ } y\text{'s are } P) - \mathcal{T}(Q_2 \text{ } y\text{'s are } P). \tag{7}$$

Linguistically we can interpret this as computing the truth value of the summary "*at least* Q are P and *no more than* Q are P".

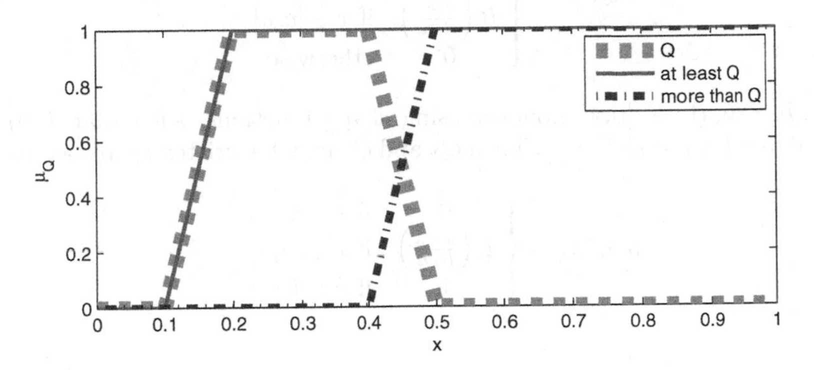

Fig. 2. Membership functions of Q, "*at least* Q" and "*no more than* Q"

Example 3. Consider similar settings as in the previous examples, namely 100 balls in a box. 85 of them are *big* with the degree of 0.0 and the other 15 are *big* with the degree of 0.99. Let Q be *a few*, defined as trapezoidal membership function Trap[0,0.1,0.2,0.3]. Consider a summary "*a few* balls are *big*". Then according to the (7) we need to use quentifiers Q_1=Trap[0, 0.1, 1, 1] and Q_2=Trap[0.2,0.3,1,1]. The truth value is calculated as:

$$\mathcal{T}(a\ few \text{ balls are } big) = \mathcal{T}(Q_1 \text{ balls are } big) - \mathcal{T}(Q_2 \text{ balls are } big) = \\ = 0.99 - 0 = 0.99$$

If 29 of the balls are *big* with the degree of 0.99 and the other 71 are not big at all (with the membership degree of 0.0), then the truth value is

$$\mathcal{T}(a\ few \text{ balls are } big) = \mathcal{T}(Q_1 \text{ balls are } big) - \mathcal{T}(Q_2 \text{ balls are } big) = \\ = 0.99 - 0.9 = 0.09$$

More examples are shown in the next section. However before we discuss them, let us make a few remarks.

First, this method can be also applied to other types of membership functions different than the trapezoidal ones, such as triangular, Gaussian and generalized bell. However the quantifiers we are investigating should be:

- normal, i.e., the height of the fuzzy set representing the quantifiers is equal to 1, and
- convex, i.e., for any $\lambda \in [0, 1]$, $\mu_Q(\lambda x_1 + (\lambda - 1)x_2) \geq min(\mu_Q(x_1) + \mu_Q(x_2))$.

More generally, we may use L-R fuzzy number to model the quantifiers with the membership function, for instance defined as

$$\mu_Q(x) = \begin{cases} L\left(\frac{x-a}{b-a}\right) & \text{if } x \in [a, b] \\ 1 & \text{if } x \in [b, c] \\ R\left(\frac{d-x}{d-c}\right) & \text{if } x \in [c, d] \\ 0 & \text{otherwise} \end{cases} \tag{8}$$

with $L, R : [0, 1] \to [0, 1]$ nondecreasing shape functions, such that $L(0) = R(0) = 0$ and $L(1) = R(1) = 1$. Then Q_1 and Q_2 may be written as, respectively:

$$\mu_{Q_1}(x) = \begin{cases} 0 & \text{if } x < a \\ L\left(\frac{x-a}{b-a}\right) & \text{if } x \in [a, b] \\ 1 & \text{if } x > b \end{cases} \tag{9}$$

and

$$\mu_{Q_1}(x) = \begin{cases} 0 & \text{if } x < c \\ R\left(\frac{x-c}{d-c}\right) & \text{if } x \in [c, d] \\ 1 & \text{if } x > d \end{cases} \tag{10}$$

and the truth value of "Q y's are P" is calculated as $\mathcal{T}(Q \text{ } y\text{'s are } P) = \mathcal{T}(Q_1 \text{ } y\text{'s are } P) - \mathcal{T}(Q_2 \text{ } y\text{'s are } P)$.

Secondly, this way of reasoning is also in line with computing the truth value of monotonically decreasing quantifiers such as *almost none*. In [27] the authors discussed some properties of the summaries. One of the properties is

$$\mathcal{T}(Q \text{ } y\text{'s are } P) = 1 - \mathcal{T}(\bar{Q} \text{ } y\text{'s are } P), \tag{11}$$

where $\bar{Q}$ is the negation of Q. This property holds also in our context.

4 Examples

The method was tested on several artificial data sets as well as on real data. In this section we will present the results shortly.

4.1 Artificial Examples

Consider similar settings as in the previous examples, namely 100 balls in a box. Let Q be *a few*, defined as trapezoidal membership function Trap[0,0.1,0.2,0.3].

Example 4. First let us assume we have 2 types of balls: *big* and not *big* (i.e., let assume *big* to the degree of 0.1). We vary the the number of balls that are *big* and the degree to which they are *big*. Figure 3 shows the effect of those two parameters on the truth value.

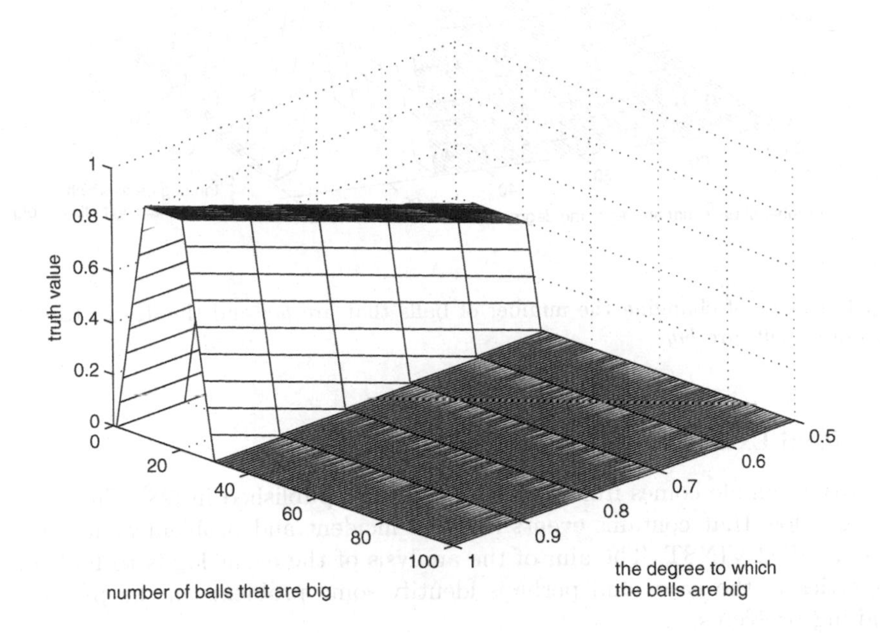

Fig. 3. Effects of changing the number of balls that are *big* and the degree to which they are *big*

In the case of the degree to which first group of balls is *big* is 1 and there are 10-20 such balls, truth value is very high. But when this degree decreases, also the truth value decreases, as it is expected, since the balls are not that *big* anymore.

Example 5. We again assume 2 types of balls: *big* to the degree 1 and not that *big*. This time we vary the number of balls that are *big* and the degree to which remaining balls are *big*. Figure 4 shows the effect of those two parameters on the truth value.

As the degree to which other balls are big increases, the truth value decreases, because now also the remaining balls are becoming bigger.

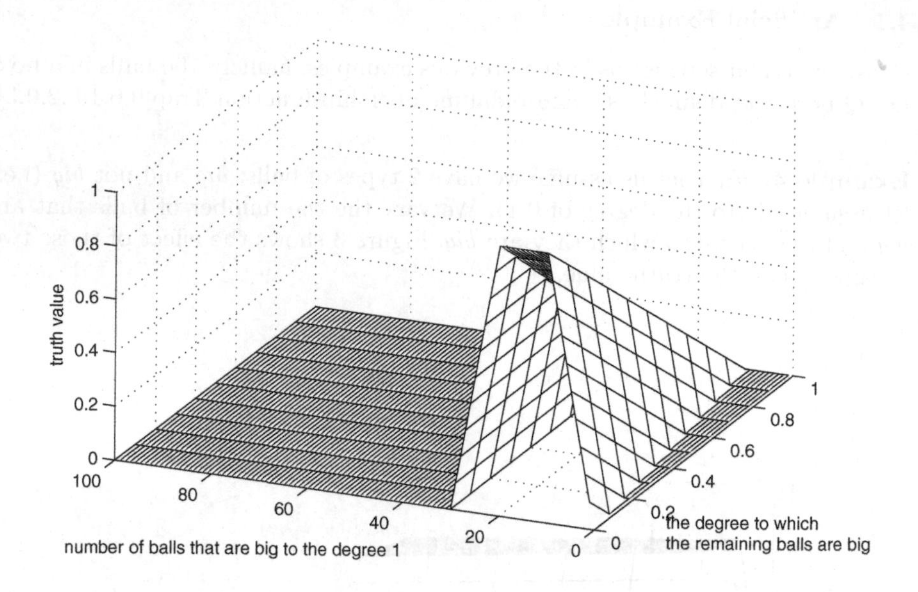

Fig. 4. Effects of changing the number of balls that are *big* and the degree to which remaining balls are *big*

4.2 Real Example

The real example comes from Volvo IT Belgium, published in [28]. The data is an event log, that contains events from an incident and problem management system called VINST. The aim of the analysis of the event log is to find some properties of the cases and perhaps identify some problems in the process of handling problems.

Generally all the problems are first registered, next they are solved by the help desk and closed. However if the help desk is unable to solve the problem, the problem is handed over to a support unit with more expertise, until is solved.

In this example we analyze only the sets that contain closed problems. There are 6660 events in the log that describe 1487 cases. Each event is described by several attributes such as problem number, its status and sub-status, time stamp and many others.

The cases in this event log had from 1 to 35 events. We define *small, moderate* and *large* number of recorded events, respectively as Trap[0,0,5,8], Trap[5,8,12,17], Trap[12,17,35,35] as in Fig. 5. The quantifiers *most* Trap[0.4, 0.6, 1, 1], *a few* Trap[0, 0.1, 0.2, 0.3] and *almost none* Trap[0,0, 0.05, 0.1] are shown in Figure 6 .

We can summarize this distribution as (with truth values in brackets):

- *Most* of the cases had a *small* number of recorded events. ($\mathcal{T} = 1$)
- *A few* cases had a *moderate* number of recorded events. ($\mathcal{T} = 0.81$)
- *Almost none* of the cases had a *large* number of recorded events. ($\mathcal{T} = 1$)

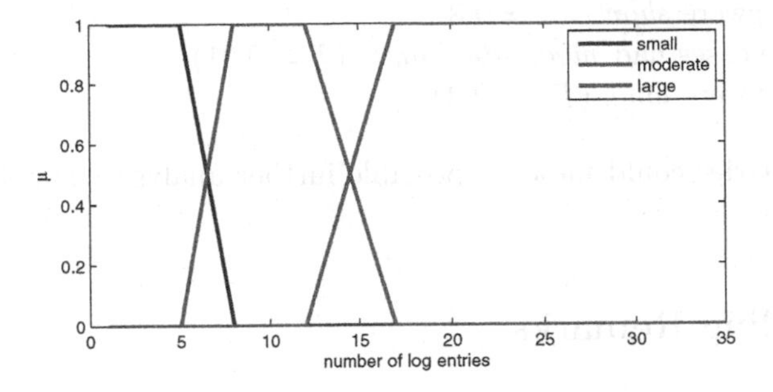

Fig. 5. The linguistic values used to describe the number of events in a case

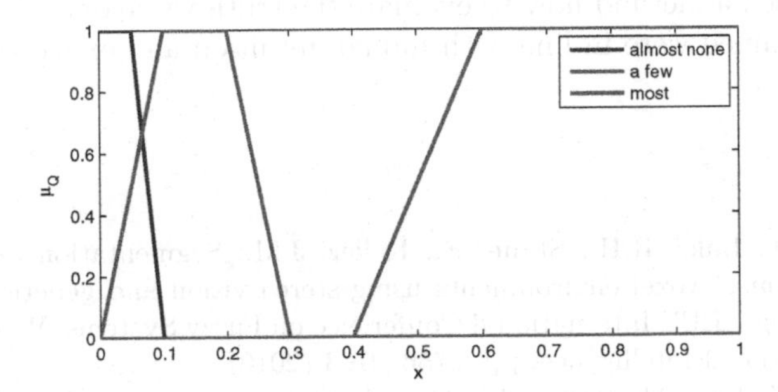

Fig. 6. The quantifiers used in this example

We can obtain similar summary for the time needed to close the case.

We defined *short* as Trap[0,0,10,20], *moderate length* as Trap[10,20,300,500] and *long* as Trap[300,500, ∞,∞], see Fig. 7. Quantifiers stayed the same.

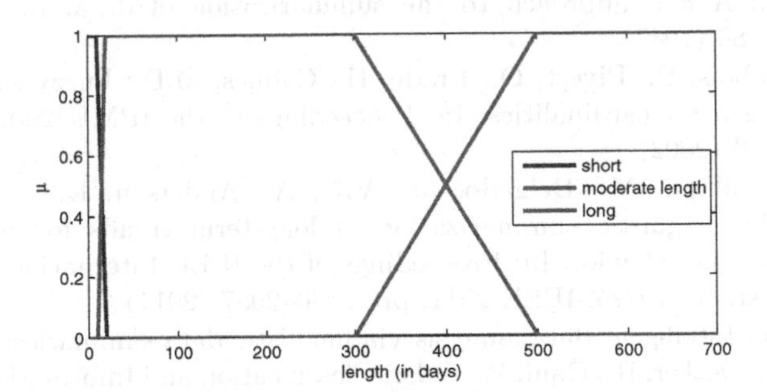

Fig. 7. The linguistic values used to describe the length

- *A few* cases were *short*. ($\mathcal{T} = 0.82$)
- *Most* of the cases had *moderate length*. ($\mathcal{T} = 0.91$)
- *A few* cases were *long*. ($\mathcal{T} = 0.94$)

Those summaries could indicate possible further analysis, namely of investigating long cases.

5 Concluding Remarks

In this paper we investigated linguistic summaries of the form "Q y's are P". We considered a case with non-monotonic quantifier, exemplified by *a few* or *about a half*. We proposed a method how to evaluate the truth value of such summaries. The results obtained were in line with human intuition and expectations.

References

1. Anderson, D., Luke, R.H., Stone, E., Keller, J.M.: Segmentation and linguistic summarization of voxel environments using stereo vision and genetic algorithms. In: Proceedings IEEE International Conference on Fuzzy Systems, World Congress on Computational Intelligence, pp. 2756–2763 (2010)
2. Wilbik, A., Keller, J.M., Alexander, G.L.: Linguistic summarization of sensor data for eldercare. In: Proceedings of the IEEE International Conference on Systems, Man, and Cybernetics (SMC 2011), pp. 2595–2599 (2011)
3. Dubois, D., Prade, H.: Gradual rules in approximate reasoning. Information Sciences 61, 103–122 (1992)
4. Raschia, G., Mouaddib, N.: SAINTETIQ: a fuzzy set-based approach to database summarization. Fuzzy Sets and Systems 129, 137–162 (2002)
5. Rasmussen, D., Yager, R.R.: Finding fuzzy and gradual functional dependencies with SummarySQL. Fuzzy Sets and Systems 106, 131–142 (1999)
6. Yager, R.R.: A new approach to the summarization of data. Information Sciences 28, 69–86 (1982)
7. Bosc, P., Dubois, D., Pivert, O., Prade, H., Calmes, M.D.: Fuzzy summarization of data using fuzzy cardinalities. In: Proceedings of the IPMU 2002 Conference, pp. 1553–1559 (2002)
8. Ros, M., Pegalajar, M., Delgado, M., Vila, A., Anderson, D.T., Keller, J.M., Popescu, M.: Linguistic summarization of long-term trends for understanding change in human behavior. In: Proceedings of the IEEE International Conference on Fuzzy Systems, FUZZ-IEEE 2011, pp. 2080–2087 (2011)
9. Kacprzyk, J.: Intelligent data analysis via linguistic data summaries: a fuzzy logic approach. In: Decker, R., Gaul, W. (eds.) Classification and Information Processing at the Turn of Millennium, pp. 153–161. Springer, Heidelberg (2000)
10. Kacprzyk, J., Yager, R.R.: Linguistic summaries of data using fuzzy logic. International Journal of General Systems 30, 33–154 (2001)
11. Kacprzyk, J., Yager, R.R., Zadrożny, S.: A fuzzy logic based approach to linguistic summaries of databases. International Journal of Applied Mathematics and Computer Science 10, 813–834 (2000)

12. Kacprzyk, J., Zadrożny, S.: Linguistic database summaries and their protoforms: toward natural language based knowledge discovery tools. Information Sciences 173, 281–304 (2005)
13. Kacprzyk, J., Zadrożny, S.: Fuzzy linguistic data summaries as a human consistent, user adaptable solution to data mining. In: Gabrys, B., Leiviska, K., Strackeljan, J. (eds.) Do Smart Adaptive Systems Exist?, pp. 321–339. Springer, Heidelberg (2005)
14. Kacprzyk, J., Wilbik, A., Zadrożny, S.: Linguistic summarization of time series using a fuzzy quantifier driven aggregation. Fuzzy Sets and Systems 159(12), 1485–1499 (2008)
15. Kacprzyk, J., Wilbik, A., Zadrożny, S.: An approach to the linguistic summarization of time series using a fuzzy quantifier driven aggregation. International Journal of Intelligent Systems 25(5), 411 (2010)
16. Castillo-Ortega, R., Marín, N., Sánchez, D.: Time series comparison using linguistic fuzzy techniques. In: Hüllermeier, E., Kruse, R., Hoffmann, F. (eds.) IPMU 2010. LNCS, vol. 6178, pp. 330–339. Springer, Heidelberg (2010)
17. Castillo-Ortega, R., Marín, N., Sánchez, D.: Linguistic local change comparison of time series. In: 2011 IEEE International Conference on Fuzzy Systems (FUZZ), pp. 2909–2915 (June 2011)
18. Wilbik, A., Keller, J.M.: A distance metric for a space of linguistic summaries. Fuzzy Sets and Systems 208, 79–94 (2012)
19. Wilbik, A., Keller, J.M., Bezdek, J.C.: Linguistic prototypes for data from eldercare residents. IEEE Transactions on Fuzzy Systems (2013) (in press)
20. Zadeh, L.A.: A prototype-centered approach to adding deduction capabilities to search engines – the concept of a protoform. In: Proceedings of the Annual Meeting of the North American Fuzzy Information Processing Society (NAFIPS 2002), pp. 523–525 (2002)
21. Zadeh, L.A.: Toward a theory of fuzzy information granulation and its centrality in human reasoning and fuzzy logic. Fuzzy Sets and Systems 9(2), 111–127 (1983)
22. Zadeh, L.A.: Computation with imprecise probabilities. In: Proceedings of the 12th International Conference Information Processing and Management of Uncertainty in Knowledge-based Systems (2008)
23. Grabisch, M., Murofushi, T., Sugeno, M. (eds.): Fuzzy Measures and Integrals. Theory and Applications. Physica Verlag, Heidelberg (2000)
24. Sugeno, M.: Theory of Fuzzy Integrals and Applications. PhD thesis, Tokyo Institute of Technology, Tokyo, Japan (1974)
25. Bosc, P., Lietard, L.: Monotonous quantifications and Sugeno fuzzy integrals. In: Proceedings of the 5th IPMU Conference, pp. 1281–1286 (1994)
26. Bosc, P., Lietard, L.: On the comparison of the Sugeno and the Choquet fuzzy integrals for the evaluation of quantified statements. In: Proceedings of EUFIT 1995, pp. 709–716 (1995)
27. Yager, R.R., Ford, K.M., Cañas, A.J.: An approach to the linguistic summarization of data. In: Bouchon-Meunier, B., Zadeh, L.A., Yager, R.R. (eds.) IPMU 1990. LNCS, vol. 521, pp. 456–468. Springer, Heidelberg (1991)
28. Steeman, W.: Volvo IT Belgium, closed cases event log, doi:10.4121/c2c3b154-ab26-4b31-a0e8-8f2350ddac11

Using Ant Colony Optimization and Genetic Algorithms for the Linguistic Summarization of Creep Data

Carlos A. Donis-Díaz[1], Rafael Bello[1], and Janusz Kacprzyk[2]

[1] Informatic Studies Center, Universidad Central Marta Abreu de Las Villas, Santa Clara, Cuba
{cadonis,rbellop}@uclv.edu.cu

[2] Systems Research Institute, Polish Academy of Sciences, Warsaw, Poland
Janusz.Kacprzyk@ibspan.waw.pl

Abstract. Some models using metaheuristics based in an "improvement of solutions" procedure, specifically Genetic Algorithms (GA), have been proposed previously to the linguistic summarization of numerical data (LDS). In the present work is proposed a new model for LDS based in Ant Colony Optimization (ACO), a metaheuristic that use a "construction of solution" procedure. Both models are compared in LDS over *creep* data. Results show how the ACO based model overcomes the measures of goodness of the final summary but fails to improve the results of the GA based model in relation to the diversity of the summary. Features of both models are considered to explain the results.

Keywords: Linguistic Data Summarization, Ant Colony Optimization, Genetic Algorithms, Fuzzy Logic, Creep rupture stress.

1 Introduction

Linguistic data summarization has been for a long time a subject of intensive research, and various tools and techniques from computational linguistics, natural language generation, etc. have been proposed. The use of the fuzzy logic with linguistic quantifiers is one of the most conceptually simple, developed and used approaches for the linguistic summarization of numerical data (LDS). The concept of a linguistic data summary, using fuzzy logic with linguistic quantifiers, which will be employed in this paper, was introduced by Yager [1], then considerably advanced in [2, 3] and presented in an implementable way in [4].

The process of generating linguistic data summaries for a given set of numerical data, usually a relational numerical database, can conveniently be represented as an optimization problem in which the best summaries from a large set of candidates are selected, and the basic objective function is assumed to be the truth degree of a linguistic summary that is equated with the degree of truth of a linguistically quantified proposition that is conceptually equivalent to the linguistic summary in question. Several works to deal with this problem have been developed [5-9]. Most of them obtain linguistic summaries by using heuristics based in an "improvement of solutions" procedure, specifically using evolutionary heuristics like Genetic Algorithms

P. Angelov et al. (eds.), *Intelligent Systems'2014*,
Advances in Intelligent Systems and Computing 322, DOI: 10.1007/978-3-319-11313-5_8

(GA). In a recent related work [10] has been proposed a hybrid model of GA with local search which improves the results obtained with the basic version of GA.

In the present work is proposed a different way of obtaining a linguistic summary by using a metaheuristic based in a "constructing of solutions" procedure, specifically an Ant Colony Optimization (ACO) based model. To the best of our knowledge, this metaheuristic has not been used previously for LDS although it has been used in problems with some commonalities like classification rules discovery and fuzzy rules learning. As final objective is compared the behavior of both metaheuristics when are applied in LDS over *creep* data.

2 Theoretical Background

This section presents the necessary theoretical precepts related to: the linguistic summarization of data, the hybrid GA proposed in [10] and used for comparison purposes in the present work, and finally some basics features of the ACO metaheuristic.

2.1 Linguistic Data Summarization Using Fuzzy Logic with Linguistic Quantifiers

In this paper is considered the linguistic data summarization approach proposed in [2].

Having: $Y=\{y_1, \ldots, y_n\}$ a set of objects in a database, e.g., the set of workers, and $A = \{A_1, \ldots, A_m\}$ a set of attributes (fuzzy variables) characterizing objects from Y, e.g., salary, age, etc. in a database D of workers, and $A_j(y_i)$ denotes the value of attribute A_j for object y_i. A linguistic summary from D consists of:

- a summarizer S, i.e. a linguistic expression composed by an attribute together with a linguistic value defined on the domain of attribute A_j (e.g. 'low salary');
- a quantifier Q (a linguistic quantifier), i.e. a fuzzy set with universe of discourse in the interval [0, 1] expressing a quantity in agreement, e.g. *most*;
- a qualifier R, i.e. a fuzzy filter determining a fuzzy subset of Y; can be composed for one or several linguistic expressions (e.g. 'young' for attribute 'age').
- a truth degree T (validity) of the summary, i.e. a number from the interval [0, 1] assessing the truth of the summary (e.g. 0.7); usually, only summaries with a high value of T are interesting;

Thus, linguistic summaries may be exemplified by

$$T\,(Most\ of\ young\ employees\ earn\ low\ salary) = 0.7$$

and their foundation is Zadeh's [11] linguistically quantified proposition corresponding to *QRy's are S*.

The truth value (T) may be calculated by using either original Zadeh's calculus of linguistically quantified statements [11] where a (proportional, nondecreasing) linguistic quantifier Q is assumed to be a fuzzy set in [0, 1] and the values of T are calculated as $T(QRy's\ are\ S) = \mu_Q(r)$ where

$$r = \frac{\Sigma_{i=1}^{n}(\mu_R(y_i) \wedge \mu_S(y_i))}{\Sigma_{i=1}^{n}\mu_R(y_i)} \quad (1)$$

Besides T (truth), other quality measures have been proposed to determine the quality of summaries. In [3] are described: the truth degree ($T1$) that correspond with the mentioned T, the degree of imprecision ($T2$), the degree of covering ($T3$) and the degree of appropriateness ($T4$).

In the present work is used the term proposition, to be more specific a linguistically quantified proposition, to refer a linguistic summary, and the term (linguistic) summary will be referred to a set of propositions. This is basically consistent with [2, 4] and in particular, the modern natural language generation (NLG) based approach.

2.2 A Hybrid Model of GA with Local Search for LDS

In this section are summarized the basic features of the hybrid model of GA with local search (HybridGA-LDS) proposed in [10] for LDS. It will be used for comparison purposes in this work.

Genetic Representation. Chromosomes in HybridGA-LDS model represent a whole linguistic summary (i.e. a set of linguistically quantified propositions) and each gene codifies just one of such propositions.

Fitness Function. The HybridGA-LDS model searches for a summary containing linguistic propositions with high values of quality (goodness) and high diversity between them. The fitness function to be maximized for a chromosome i is defined in the interval [0, 1] as $F_i = m_g G_i + m_d D_i$ where the term G_i and D_i represent the *Goodness* and *Diversity* of the summary respectively. *Goodness* is calculated as the mean value of the individual goodness $g_j = T_j \cdot St_j$ of propositions (genes) according to: $G_i = \frac{1}{n}\sum_{j=1}^{n} T_j \cdot St_j$ where T_j is the truth degree of the proposition j, St_j represents a called linguistic strength indicator and n is the total number of propositions in the summary. The term D_i expresses the degree of diversity between the propositions forming the summary; is calculate as $D_i = C_i / n$ where C_i represents the number of clusters of propositions existing in the summary.

Genetic Operators. The HybridGA-LDS model mixes the use of traditional operators like selection, crossover and mutation with two specifics operators proposed to improve the search.

The first proposed operator was the *Cleaning Operator*: this operator was introduced to "clean" those propositions inside a summary having no opportunities to evolve towards better solutions during the process due to the inexistence of cases in the data set to cover them. The operator substitutes the propositions with $T = 0$ by others randomly generated.

The second operator was the *Propositions Improver Operator*; it was introduced to overcome the weakness of basic operators (crossover and mutation) to improve the quality of individual propositions. Through the evolutionary process, the basic crossover operator improves the quality of the chromosomes (summary) as a whole but do not improve the quality of genes (propositions), i.e. improves the diversity degree of the summary but does not improves individual propositions. In other hand, the mutation operator does not guarantee the sufficient perturbation in the search to solve the above problem. The *Propositions Improver Operator* implements a randomly greedy search using a best first strategy based in six possible transformations of the linguistic proposition: four to modify the quantifier, one to modify the summarizer and one to modify the quantifier. Two parameters control the deep of the search: the length of the search (*ls*), (i.e. the total number of new considered propositions) and the maximum number of searches without improve the quality ($s_w i$).

2.3 Ant Colony Optimization

ACO is a metaheuristic inspired in the behavior of real ants to forage for food. This metaheuristic has been widely used in many optimization problems and fields including applications with some commonalities to the objective in this work like classification rules discovery [12, 13] and fuzzy rules learning [14]. The implementation of ACO used in the present work is Max-Min Ant System (MaxMin AS) [15] an extension of the Ant System (AS) [16] implementation in order to improve its performance. For an overview and recent reviews on ACO can be consulted [17, 18], following are presented some basics features of MaxMin AS:

- The pheromone update is applied offline and evaporated according to: $\tau_{(v_i,v_j)} = \rho \cdot \tau_{(v_i,v_j)}$ where $\tau_{(v_i,v_j)}$ is the pheromone value of arc a_{v_i,v_j} between nodes v_i and v_j; ρ is the persistence factor, a parameter defined by the user; *(1-ρ)* is the evaporation factor. After the evaporation, the pheromone is deposited on each arc corresponding with the solution of the best ant A_{best} as:

$$\tau_{(v_i,v_j)} = \tau_{(v_i,v_j)} + f(Q(A_{best})), \quad \forall a_{v_i,v_j} \in A_{best}$$

  where $f(Q(A_{best}))$ represents a function based on the quality of the solution in A_{best}. The ant that is permitted to add pheromone can be the ant with the best solution of the current iteration or the ant with the best global solution. Furthermore, it is common that ant solutions are improved by local searches before pheromone update.

- The possible values for pheromone trails are limited to the interval $[\tau_{min}, \tau_{max}]$.
- The initial pheromone trail of each arc is set to a high value.

3 ACO-LDS: A New Model for LDS

In this section is described the proposed ACO based model for searching linguistic data summaries (ACO-LDS). Similar to the HybridGA-LDS model, ACO-LDS aims to search not only propositions with good (possible, the best) qualities but also search propositions forming a good (possible, the best) summary in reference to the diversity among them. To meet this aim, an ant in ACO-LDS represents a whole summary and each iteration of the process discovers one summary, probably the best one between all iterations. Following subsections describes main aspects of the proposed model.

3.1 The General Algorithm of ACO-LDS

The high-level pseudo code of the algorithm is presented in Fig. 1.

```
Input: dataset
Output: best discovered summary
1. ComputeLocalHeuristicInformation()
2. InitPheromones()
3. gbSummary   = null             // Global best summary
4. currentIt = 1                  // Current iteration
5. while (currentIt < max.iter.) and (not stagnation) do
6.    ibSummary = null    // Iteration best Summary
7.    for  a=1 to colony size do
8.      summary_a  = null
9.      for i=1 to propositions per summary do
10.       proposition_i  = CreateProposition(dataset)
11.       summary_a = summary_a + proposition_i
12.     end for
13.     if Fitness(summary_a) > Fitness(ibSummary) then
14.       ibSummary = summary_a
15.     end if
16.   end for
17.   Improve(ibSummary)
18.   if Fitness(ibSummary) > Fitness(gbSummary) then
19.     gbSummary = ibSummary
20.   end if
21.   UpdatePheromone(gbSummary)
22.   currentIt = currentIt + 1
23. end while
24. return gbSummary
```

Fig. 1. High-level pseudocode for ACO-LDS

As mentioned before, ACO-LDS use the MaxMin AS implementation of ACO. The procedure starts computing the local heuristic information for each node in the graph and initializing the pheromones to a high value. Then each iteration of the algorithm (*while* loop) produces a summary that correspond to the best summary obtained from the construction process developed by the colony (*for* loop, lines 7 to 16). This best summary of the iteration (*ibSummary*) is improved by performing a local search on each of its propositions; this is a difference with most applications that use ACO for discovering classification or fuzzy rules where the local search is applied to all constructed solutions. The best global summary is updated with the best summary of the iteration if this latter has a better value of fitness. Finally, the pheromones are updated using the best global summary. The *while* loop iterates until the maximum number of iterations or the stagnation condition is reached. This latter condition occurs when more than 90 percent of propositions are stagnant. A proposition is stagnant if all nodes visited by the ant when constructing the proposition have the pheromone value equal to τ_{max} and the remaining nodes in the graph have τ_{min}.

3.2 ACO Representation for LDS

In ACO-LDS each ant represents/constructs/modifies a summary with a fixed number of linguistically quantified propositions. Considered propositions have the form:

$$<Q>(<a_1 = l_{1j}> \text{ and } [a_2 = l_{2j}] \text{ and } \ldots [a_i = l_{ij}] \text{ are/have/} \ldots <a_s = l_{sj}>) = <g>$$

where <Q> is the linguistic quantifier; terms $<a_i = l_{ij}>$ represents the linguistic expressions used in the qualifier being a_i the *i-th* fuzzy variable and l_{ij} the *j-th* linguistic term selected for a_i , observe that the qualifier require at least one linguistic expression; the summarizer is represented by only one linguistic expression $<a_s = l_{sj}>$ being a_s the fuzzy variable selected for that purpose and l_{sj} the linguistic term used for a_s; finally <g> represents the goodness (quality) of the proposition. The "*and*" operator is calculated as the minimum membership degree of both concatenated linguistic expressions (in general, is a *t-norm*).

The graph used by ants to construct a linguistic proposition is composed by nodes representing the possible linguistic expressions for the qualifier $<a_1 = l_{1j}>$ and the summarizer $<a_s = l_{sj}>$. In the graph will exist, for each fuzzy variable a_i, as many nodes as linguistics terms have been defined for a_i. To define an arcs between to nodes one rule apply: from the group of linguistic expressions corresponding to a fuzzy variable can only be selected one when construction a proposition, i.e. can not be established arcs between nodes representing linguistics expressions belonging to the same fuzzy variable.

3.3 Constructing a Proposition

When building a summary, the ants create a fixed number of propositions. In the construction process of one proposition (referred in line 10 of Figure 1), the ant selects linguistics expressions (nodes) for the summarizer and the qualifier in a tour through the graph. The ant start selecting only one node for the summarizer from the group of linguistic terms defined for it in the graph. Then the ant selects the first node for the qualifier guaranteeing that the subset of objects from the database meeting this partial qualifier is not null. The process continues adding nodes to the qualifier while the subsequent subsets contain one or more examples or all possible nodes have been added. Finally, the model selects the quantifier that better value of goodness (g) cause in the proposition.

Transition Rule. When selecting a node is important to note that arcs between nodes have not a special means. For the final proposition, the important thing is if a specific node is selected or not; the precedence relationship (that arcs represents) between two nodes is irrelevant in this case. This is why the pheromone is stored in the nodes and not in the arcs.

As ants have to construct several propositions, is necessary to keep different trails of pheromone, one for each proposition. To satisfy this condition in the pheromone matrix was included an additional index indicating the number of the proposition (*tour*) for which the trail is maintained.

In ACO-LDS is used a pseudo-random transition rule, similarly as does ACS. During the tour t, a node v_{ij} (representing the linguistic expression $<a_i = l_{ij}>$) is randomly selected using a probability distribution first calculated as:

if $q \leq q_0$

$$P_{v_{ij}} = \begin{cases} 1, & if\ v_{ij} = \arg\max\left\{(\alpha \cdot \tau_{(t,v_{ij})} + (1-\alpha)\eta_{(v_{ij})}), \quad \forall i \in N\right\} \\ 0, & in\ other\ case \end{cases}$$

else ($q > q_0$)

$$P_{v_{ij}} = \begin{cases} \dfrac{\alpha \cdot \tau_{(t,v_{ij})} + (1-\alpha)\eta_{(v_{ij})}}{\sum_{k=1}^{m}\sum_{j=1}^{n_k} \alpha \cdot \tau_{(t,v_{kj})} + (1-\alpha)\eta_{(v_{kj})}}, & \forall k \in N \\ 0, & in\ other\ case \end{cases}$$

where q_0 is a parameter in [0, 1] and q a random value in [0, 1], $\tau_{(t,v_{ij})}$ is the pheromone value accumulated by node v_{ij} in the trail of the t-th tour, $\eta_{(v_{ij})}$ is the local heuristic value for node v_{ij}, m is the number of attributes, n_k is the number of linguistic terms of attribute k, N represents the set of selectable attributes, i.e. attributes not yet used by the ant, α is a parameter to control de importance given to τ and η in the equation. Before to be used, pheromone and heuristic values are normalized in [0, 1].

The proposed model includes an extra heuristic called *Frequency of use* when selecting the next node v_{ij}. This heuristic aims to build the current proposition as different as possible in relation with the propositions previously added to the partial summary in construction, i.e. the *Frequency of use* contributes to increase the diversity of the summary. The heuristic calculates a term F_u using the number of times (u_{vij}) that v_{ij} has been used in the partial summary under construction: $F_u = 1 - (u_{v_{ij}} / p)^e$. Term p represents the amount of propositions added up to now to the partial summary and e is a parameter in [0, 1] to graduate the "power" of influence of F_u. Then the heuristic affects the final probability distribution according to: $P_{v_{ij}} = P_{v_{ij}} \cdot F_u$

Local Heuristics. For the transition rule, two local heuristics are used depending if the node to be selected is for the summarizer or for the qualifier. As known, the values for the local heuristics are calculated and stored as a pre-step (Line 1, Fig. 1).

For summarizer's nodes is used the degree of imprecision (ID) as local heuristic. This heuristic is calculated based in the degree of imprecision T2 mentioned in subsection 2.1 and proposed in [3]. This value depends only on the form of the summarizer and as in the present work, is considered the summarizer to has just one linguistic expression, the form of ID is: $ID = T2 = 1 - in(s)$ where in(s) defines the degree of fuzziness of the fuzzy set *s* defined for the fuzzy variable *S* as:

$in(s) = \frac{cardinality(x \in X_S : \mu_s(x) > 0)}{cardinality(X_S)}$ where X_S refers to the universe of discourse of the fuzzy variable *S* and $\mu_s(x)$ the membership degree of the element *x* to the fuzzy set *s*.

For qualifier's nodes is used a proposed *Relevance* degree (*RD*) as heuristic. The *Relevance* estimates the importance of a node v_{ij} for a given linguistic expression *s* in the summarizer by using the degree of covering (*T3*). *Relevance* is calculated as:

$$RD_{(v_{ij},s)} = \max(0, \quad T3_{(v_{ij} \wedge s)} - T3_{(s)}) \tag{2}$$

where $T3_{(v_{ij} \wedge s)}$ is calculated as in [3] and $T3_{(s)} = \frac{cardinality(\mu_s(y) > 0)}{number\ of\ examples}$. Equals values of $T3_{(v_{ij} \wedge s)}$ and $T3_{(s)}$ mean that "observing" node v_{ij} in the dataset has no influence to "observe" *s*, i.e. node v_{ij} is not relevant to *s*; in this case $RD_{(v_{ij},s)}$ is equal to cero. While higher the value of $RD_{(v_{ij},s)}$, more relevant is the node v_{ij} for the summarizer *s*. As this heuristic depends on previous selection of the summarizer, for each node should be calculated and stored many values as linguistic terms were defined for the fuzzy variable used in the summarizer.

Fitness Function. The fitness function for ACO-LDS was defined similar to that proposed in [10] and described in subsection 2.1 but incorporating other measures of quality in the calculus of goodness for individual propositions. The goodness *g* of a proposition *j* is calculated as: $g_j = w_1 T_1 St_j + w_2 T_2 + w_3 T_3 + w_4 T_4$, where $\sum_i w_i = 1$ and T_1, T_2, T_3, T_4 are obtained as proposed in [3].

Updating the Pheromone Trails. As defined for MaxMin AS the pheromone levels are bounded according to the interval [τ_{min}, τ_{max}]. In ACO-LDS is used an approach where limits are dynamically updated each time a new best solution is found, as detailed in [15]. The ant containing the best global solution is the only permitted to increases the pheromone level in nodes belonging to the solution. The general rule to calculate the new value of pheromone τ^{ts+1} for the tour *t* having a previous time stamp *ts* is expressed as

$$\tau_{(t,v_{ij})}^{ts+1} = \begin{cases} \max(\tau_{\min}, \rho \cdot \tau_{(t,v_{ij})}^{ts}), & \text{if } v_{ij} \notin gbSummary \\ \min(\tau_{\max}, \rho \cdot \tau_{(t,v_{ij})}^{ts} + \rho \cdot \tau_{(t,v_{ij})}^{ts} \cdot F_{gbSummary}), & \text{if } v_{ij} \in gbSummary \end{cases}$$

where τ^{ts} represents the pheromone value in the previous iteration and ρ is the pheromone persistent factor ($1 - \rho$ is the evaporation factor).

The local Search. ACO-LDS applies a local search to improve the propositions. This local search is similar to that used in [10] but do not include the transformation that modify the quantifier because in ACO-LDS, the quantifier is selected as the best one, i.e. the quantifier that better value of linguistic strength produce in the proposition is used. The local search is only applied to propositions of the best summary of the iteration (*ibSummary*); this approach permits to develop a deeper local search since the whole process do not increases the total number of considered propositions when compared with the approach (mostly used in works done so far using ACO in similar problems) where the local search is applied to all considered propositions.

4 Comparing GA and ACO Metaheuristics in LDS

HybridGA-LDS and ACO-LDS models were applied over *creep* data in the present work. The creep rupture stress (*creep*) is an important mechanical property considered in the design of new alloys. It measures the stress level in which a steel structure fails when exposed to quite aggressive conditions over long periods of time. The data and fuzzy modeling used was the same as employed in [10]. Is important to note that for *creep* problem, the propositions having *Most* or *Much* as quantifier are more interesting, that is why the parameter *St* (linguistic strength) was set in both model preferring these quantifiers by using the same values as in [10]. In order to achieve uniformity in the processing of both models, experiments were developed so that both considered the same total number of propositions (250 000, representing the 6.91E-15 percent from the total for this problem) when obtaining its results. Both models used the same fitness function as described in the present work; values for w_i were: w_1=0.4, w_2= 0.1, w_3=0.25, w_4=0.25. To get the results, ten runs of each model were made. The Wilcoxon's test and Monte Carlo's technique were used to compare the results pairs to pairs and to calculate a more precise signification of the differences respectively.

4.1 Results and Analysis

Results of experiments are presented in Table 1. Columns (2) to (4) present general quality measures of a summary: (2) Goodness, (3) Diversity, (4) number of proposition having the desired quantifier (*Most* or *Much*); its values represent the mean value from ten runs of each model. In turn, the quality measures of individual propositions (Column (6) to (11)) represent mean values from propositions composing the summaries obtained in all runs of models.

Table 1. Quality measures of summaries and propositions obtained by models

Model	Fitness	Goodness	Quantifier	Diversity	Quality measures of individuals propositions				
					T1	*T1·St*	*T2*	*T3*	*T4*
(1)	(5)	(2)	(4)	(3)	(6)	(7)	(8)	(9)	(10)
HybridGA-LDS	0.6931	0.5616	16.30	1.0000	0.9566	0.5157	0.8960	0.5287	0.5343
ACO-LDS	0.7359	0.6984	21.00	0.8233	0.9439	0.6638	0.8453	0.7357	0.6576

Let first observe the behavior of general quality measures in obtained summaries: ACO-LDS produces a better value of fitness (with significant difference) respect to the value of HybridGA-LDS model. This result is supported by the "Goodness" component of the fitness value ($Goodness_{HybridGA-LDS} < Goodness_{ACO-LDS}$) but not by the "Diversity" component ($Diversity_{HybridGA-LDS} > Diversity_{ACO-LDS}$); the differences are significant in both cases. Results for "Goodness" are a direct consequence of results obtained for quality measures of individuals propositions (Columns 6 to 11): except for *T2*, ACO-LDS improves or equals (in *T1*, the differences are not significant) the results obtained by HybridGA-LDS in each quality measure.

The analysis of *T1·St*, the most significant component when calculating the quality of a proposition, has great importance in the explanation of results; this parameter combines the truth degree of a proposition with its linguistic strength. Let start the analysis taking into account the relation *r* expressed in equation (1) and the calculus of *T3* proposed in [3]: observing the components and relations they use, could be expected that build a proposition with high *r* by using linguistic expressions with high *T3* in the qualifier is more probable that build a proposition with high *r* by using linguistic expressions with low *T3* in the qualifier, i.e., in the process of constructing a proposition, select nodes with high *T3* has a positive influence in obtaining a proposition with high *r*. Having into account that *Relevance degree* rewards to nodes with high *T3* (see equation 2) and the fact that high values of *r* produce high values of membership to the linguistic terms *Most* and *Much* of the quantifier (i.e., produce high values of *T1* (see the calculus of *T=T1*) in propositions having *Most* and *Much* as quantifiers) can be concluded that using the *Relevance degree* as local heuristic when constructing propositions in ACO-LDS stimulates the production of propositions with high degree of truth (*T1*) and having *Most* and *Much* as quantifier. This analysis explain the results obtained by ACO-LDS for *T1·St* since the parameter *St* was set in the present application precisely, to stimulate propositions having quantifiers like *Most* or *Much*. Values obtained by HybridGA-LDS for this component are lower despite values for *T1* are high; the main reason for this result is that the model generates linguistic terms for the qualifier without check any relation with the linguistic term generated for the summarizer, so the model can find propositions with high *T1* but do not having *Most* or *Much* as quantifier necessarily. Concluding this analysis can be established that in the search of propositions with high values of *T1* and having *Most* or *Much* as qualifier, using the approach of ACO-LDS that constructs the solutions and therefore permits the use of a local heuristic like the *Relevance* degree, has advantage over using the approach of HybridGA-LDS that improves the solutions by using the genetic operators. Column 4 shows the number of propositions with *Most* or *Much* as quantifiers, this values are consistent with previous analysis.

Results obtained for components *T3* and *T4* can be explained in a similar way since both are favored by the use of the *Relevance* heuristic, i.e. its features are considered in some way by the *Relevance* degree.

When analyzing values obtained for the *Diversity* component of the fitness (column 3) can be noted that despite using an additional local heuristic (*Frequency of use*) specifically designed to ensure diversity in the summary, ACO-LDS fails to improve the performance of HybridGA-LDS (differences in values are significant). In this sense can be highlighted the effectiveness of the crossover operator whose main function in HybridGA-LDS is to ensure diversity in the summary. Results of ACO-LDS are conditioned by the strict (reduced) pool of nodes (linguistics expressions) that imposes the *Relevance* heuristic when selecting nodes to construct the proposition.

4.2 Values of Parameters Used in Experiments

Table 2. Parameters and values used in the experimentation process

Param.	Description	Interval	Value
ρ	the persistent factor used in the pheromone updating rule, $(1-\rho)$ is the evaporation factor	[0 – 1]	0.7
α	controls the importance given to the pheromone τ and the heuristic η in the calculus of the probability selection of a node	[0 – 1]	0.5
q_0	used when constructing the proposition to determine if the next node, will be selected in a deterministic or stochastic way	[0 – 1]	0.8
e	graduates the influence of the frequency of use (of a node in a summary) when calculating the probability selection of a node	[0 – 1]	0.3
ls	length of the local search in ACO-LDS, specify the maximum total number of considered propositions for the local search		20
s_wi	maximum number of searches without improvement in the local search in ACO-LDS		15

Presented values were the final ones obtained during an experimentation process.

5 Conclusions

In the present work has been proposed a new model for LDS based in ACO. This model (ACO-LDS) overcomes measures of goodness of the final summary but fails to improve the diversity of the summary obtained by HybridGA-LDS. When searching linguistic summaries on *creep* data, good results obtained in ACO-LDS for goodness are influenced by the constructive procedure used in ACO which allows the use of a local heuristic as *Relevance* that selects the qualifier based in a previous selection of the summarizer. Respect to the degree of diversity in the final summary, has been shown how the crossover operator used in Hybrid-LDS result a more effective approach than that used in ACO-LDS to meet this requirement in the final summary. Concluding can be established that ACO-LDS do not overcome completely the results of Hybrid-GA. In future works will be mixed the best features of both models.

Acknowledgments. This research has been partially supported by the National Centre of Science under Grant No. UMO-2012/05/B/ST6/03068.

References

1. Yager, R.R.: A new approach to the summarization of data. Information Sciences 28, 69–86 (1982)
2. Kacprzyk, J.: Intelligent data analysis via linguistic data summaries: a fuzzy logic approach. In: Decker, R., Gaul, W. (eds.) Classification and Information Processing at the Turn of the Millennium, pp. 153–161. Springer, Heidelberg (2000)
3. Kacprzyk, J., Yager, R.R.: Linguistic summaries of data using fuzzy logic. International Journal of General Systems 30(2), 133–154 (2001)
4. Kacprzyk, J., Zadrożny, S.: Computing with words: towards a new generation of linguistic querying and summarization of databases. In: Sinčak, P., Vaščak, J. (eds.) Quo Vadis Computational Intelligence?, pp. 144–175. Physica-Verlag, Heidelberg (2000)
5. Castillo-Ortega, R., et al.: Linguistic Summarization of Time Series Data using Genetic Algorithms. In: 7th Conference of European Society for Fuzzy Logic and Technology - EUSFLAT 2011, Atlantis Press, Aix-les-Bains (2011)
6. Castillo-Ortega, R., et al.: A Multi-Objective Memetic Algorithm for the Linguistic Summarization of Time Series. In: 13th Annual Genetic and Evolutionary Computation Conference - GECCO 2011. ACM, Dublin (2011)
7. George, R., Srikanth, R.: Data summarization using genetic algorithms and fuzzy logic. In: Herrera, F., Verdegay, J.L. (eds.) Genetic Algorithms and Soft Computing, pp. 599–611. Physica-Verlag, Heidelberg (1996)
8. Kacprzyk, J., Wilbik, A., Zadrożny, S.: Using a Genetic Algorithm to Derive a Linguistic Summary of Trends in Numerical Time Series. In: International Symposium on Evolving Fuzzy Systems, Ambleside (2006)
9. Kacprzyk, J., Wilbik, A., Zadrożny, S.: Linguistic summarization of time series using a fuzzy quantifier driven aggregation. Fuzzy Sets and Systems 159(12), 1485–1499 (2008)
10. Donis-Diaz, C.A., et al.: A hybrid model of genetic algorithm with local search to discover linguistic data summaries from creep data. Expert System with Applications 41(4), 2035–2042 (2014)
11. Zadeh, L.: A computational approach to fuzzy quantifiers in natural languages. Computers and Mathematics with Applications 9, 149–184 (1983)
12. Parpinelli, R., Lopes, H., Freitas, A.: Data mining with an ant colony optimization algorithm. IEEE Transactions on Evolutionary Computation 6(4), 321–332 (2002)
13. Otero, F.B., Freitas, A., Johnson, C.G.: A New Sequential Covering Strategy for Inducing Classification Rules with Ant Colony Algorithms. IEEE Transactions on Evolutionary Computation 17(4), 64–76 (2013)
14. Alatas, B., Akin, E.: FCACO: Fuzzy Classification Rules Mining Algorithm with Ant Colony Optimization. In: Wang, L., Chen, K., S. Ong, Y. (eds.) ICNC 2005. LNCS, vol. 3612, pp. 787–797. Springer, Heidelberg (2005)
15. Stützle, T., Hoos, H.: MAX-MIN ant system. Future Generation Computer Systems 16(8), 889–914 (2000)
16. Dorigo, M., Colorni, A., Maniezzo, V.: The Ant System: Optimization by a colony of cooperating agents. IEEE Transactions on Systems, Man, and Cybernetics-Part B 26(1), 1–13 (1996)
17. Dorigo, M., Stützle, T.: Ant Colony Optimization: Overview and Recent Advances. In: Gendreau, M., Potvin, Y. (eds.) Handbook of Metaheuristics, pp. 227–263. Springer, New York (2010)
18. Dorigo, M., Birattari, M., Stützle, T.: Ant Colony Optimization- Artificial Ants as a Computational Intelligence. IEEE Computational Intelligence Magazine 1, 28–39 (2006)

Part III
Multicriteria Decision Making and Optimization

InterCriteria Decision Making Approach to EU Member States Competitiveness Analysis: Temporal and Threshold Analysis

Vassia Atanassova[1,3], Lyubka Doukovska[1], Deyan Mavrov[2], and Krassimir T. Atanassov[2,3]

[1] Intelligent Systems Department,
IICT – Bulgarian Academy of Sciences,
Acad. G. Bonchev Str., bl. 2, 1113 Sofia, Bulgaria
vassia.atanassova@gmail.com,
doukovska@iit.bas.bg

[2] Computer Systems and Technologies Department,
"Prof. Dr. Asen Zlatarov" University,
1 "Prof. Yakimov" Blvd., 8010 Burgas, Bulgaria
dg@mavrov.eu

[3] Bioinformatics and Mathematical Modelling Department,
IBPhBME – Bulgarian Academy of Sciences,
Acad. G. Bonchev Str., bl. 105, 1113 Sofia, Bulgaria
krat@bas.bg

Abstract. In this paper, we present some interesting findings from the application of our recently developed InterCriteria Decision Making (ICDM) approach to data extracted from the World Economic Forum's Global Competitiveness Reports for the years 2008–2009 to 2013–2014 for the current 28 Member States of the European Union. The developed approach which employs the apparatuses of index matrices and intuitionistic fuzzy sets is designed to produce from an existing index matrix with multiobject multicriteria evaluations a new index matrix that contains intuitionistic fuzzy pairs with the correlations revealed to exist in between the set of evaluation criteria, which are not obligatory there 'by design' of the WEF's methodology but exist due to the integral, organic nature of economic data. Here, we analyse the data from the six-year period within a reasonably chosen intervals for the thresholds of the intuitionistic fuzzy functions of membership and non-membership, and make a series of observations about the current trends in the factors of competitiveness of the European Union. The whole research and the conclusions derived are in line with WEF's address to state policy makers to identify and strengthen the transformative forces that will drive future economic growth.

Keywords: Global Competitiveness Index, Index matrix, InterCriteria decision making, Intuitionistic fuzzy sets, Multicriteria decision making.

1 Introduction

The present work contains a continuation of our recent research, started in [7], which aims at analyzing data about the performance of the 28 European Union Member

P. Angelov et al. (eds.), *Intelligent Systems'2014*,
Advances in Intelligent Systems and Computing 322, DOI: 10.1007/978-3-319-11313-5_9

States according to the Global Competitiveness Reports (GCRs) of the World Economic Forum (WEF), released in the period from 2008–2009 to 2013–2014. We use a recently developed multicriteria decision making method, based on intuitionistic fuzzy sets and index matrices, two mathematical formalisms proposed and significantly researched by Atanassov in a series of publications from 1980s to present day. As its title, InterCriteria Decision Making (ICDM, see [6]), suggests, the method aims at discovery of existing dependences *between* the evaluation criteria themselves.

The ICDM approach has been originally devised to reflect situations where some of the criteria come at a higher cost than others, for instance are harder, more expensive and/or more time consuming to measure or evaluate. Such criteria are generally considered unfavourable, hence if the method identifies certain level of correlation between such unfavourable criteria and others that are easier, cheaper or quicker to measure or evaluate, the unfavourable ones might be disregarded in the further decision making process.

In our work for applying ICDM to WEF GCR data [7] we have been interested to detect the eventual correlations between the 12 'pillars of country competitiveness', in order to outline fewer pillars on which policy makers should concentrate their efforts. Our motivation to conduct the analysis has been that it might be expected that improved country's performance against some pillars would positively affect the country's performance in the respective correlating ones. This is in line with WEF's address to state policy makers to identify and strengthen the transformative forces that will drive future economic growth of the countries, as formulated in the Preface of the latest Global Competitiveness Report 2013–2014, [8].

The twelve pillars in the WEF's methodology are grouped in three subindices:

- The first subindex 'Basic requirements' contains pillars 1–4: '1. Institutions', '2. Infrastructure', '3. Macroeconomic stability' and '4. Health and primary education', 25% weight for each pillar.
- The second subindex 'Efficiency enhancers' contains pillars 5–10: '5. Higher education and training', '6. Goods market efficiency', '7. Labor market efficiency', '8. Financial market sophistication', '9. Technological readiness' and '10. Market size', 17% weight for each pillar.
- The third subindex 'Innovation and sophistication factors' contains pillars 11–12: '11. Business sophistication' and '12. Innovation', 50% weight for each pillar.

On the basis of the evaluation of the countries according to these pillars and following a sophisticated methodology, WEF determines their 'stage of development', which is one of the five possible alternatives: '1. Factor driven', 'Transition 1–2', '2. Efficiency driven', 'Transition 2–3' or '3. Innovation driven'. From the 28 EU Member States, 19 are in stage '3. Innovation driven', 7 are in stage 'Transition 2–3', and 2 are in stage '2. Efficiency driven'.

In the first part of our research [7], we gave the comparison of the results of the ICDM for the two extreme years in the 6-year period, and discussed in more details the findings for the year 2013–2014. We showed the principle of gradual discovery of more correlations between the criteria by letting the two user defined thresholds involved in the ICDM approach change within the [0; 1]-interval. Example was given with a detailed description of the correlations in one partial case, and it was visually interpreted as a graph. Here, we will continue investigating the same selection of data, but we will further show how for each year, and for each pair of thresholds, the number of positive consonances for each of the twelve pillars change, and will accompany these observations with some initial conclusions.

This paper is organized as follows. In Section 2 the two basic mathematical concepts that we use – intuitionistic fuzzy sets and index matrices – are briefly presented and on this basis is described, the proposed method ICDM. Section 3 contains our results from applying the method to analysis of a selection of data about the performance of the currently 28 Member States of the EU during the last six years against the twelve pillars of competitiveness. We report the findings of our temporal and threshold analysis, and formulate our conclusions in the last Section 4.

2 Basic Concepts and Method

The presented multicriteria decision making method is based on two fundamental concepts: intuitionistic fuzzy sets and index matrices.

Intuitionistic fuzzy sets defined by Atanassov (cf. [1, 2, 4, 5]) represent an extension of the concept of fuzzy sets, as defined by Zadeh [9], exhibiting function $\mu_A(x)$ defining the membership of an element x to the set A, evaluated in the [0; 1]-interval. The difference between fuzzy sets and intuitionistic fuzzy sets (IFSs) is in the presence of a second function $\nu_A(x)$ defining the non-membership of the element x to the set A, where $\mu_A(x) \in [0; 1]$, $\nu_A(x) \in [0; 1]$, and moreover $(\mu_A(x) + \nu_A(x)) \in [0; 1]$.

The IFS itself is formally denoted by:

$$A = \{\langle x, \mu_A(x), \nu_A(x)\rangle \mid x \in E\}.$$

Comparison between elements of any two IFSs, say A and B, involves pairwise comparisons between their respective elements' degrees of membership and non-membership to both sets.

The second concept on which the proposed method relies is the concept of index matrix, a matrix which features two index sets. The theory behind the index matrices is described in [3]. Here we will start with the index matrix M with index sets with m rows $\{C_1, \ldots, C_m\}$ and n columns $\{O_1, \ldots, O_n\}$:

$$M = \begin{array}{c|ccccccc}
 & O_1 & \cdots & O_k & \cdots & O_l & \cdots & O_n \\ \hline
C_1 & a_{C_1,O_1} & \cdots & a_{C_1,O_k} & \cdots & a_{C_1,O_l} & \cdots & a_{C_1,O_n} \\
\vdots & \vdots & \ddots & \vdots & \ddots & \vdots & \ddots & \vdots \\
C_i & a_{C_i,O_1} & \cdots & a_{C_i,O_k} & \cdots & a_{C_i,O_l} & \cdots & a_{C_i,O_n} \\
\vdots & \vdots & \ddots & \vdots & \ddots & \vdots & \ddots & \vdots \\
C_j & a_{C_j,O_1} & \cdots & a_{C_j,O_k} & \cdots & a_{C_j,O_l} & \cdots & a_{C_j,O_n} \\
\vdots & \vdots & \ddots & \vdots & \ddots & \vdots & \ddots & \vdots \\
C_m & a_{C_m,O_1} & \cdots & a_{C_m,O_j} & \cdots & a_{C_m,O_l} & \cdots & a_{C_m,O_n}
\end{array},$$

where for every p, q ($1 \leq p \leq m$, $1 \leq q \leq n$), C_p is a criterion (in our case, one of the twelve pillars), O_q in an evaluated object (in our case, one of the 28 EU Member states), $a_{C_pO_q}$ is the evaluation of the q-th object against the p-th criterion, and it is defined as a real number or another object that is comparable according to relation R with all the rest elements of the index matrix M, so that for each i, j, k it holds the relation $R(a_{C_kO_i}, a_{C_kO_j})$. The relation R has dual relation $\overline{R}$, which is true in the cases when relation R is false, and vice versa.

For the needs of our decision making method, pairwise comparisons between every two different criteria are made along all evaluated objects. During the comparison, it is maintained one counter of the number of times when the relation R holds, and another counter for the dual relation.

Let $S^{\mu}_{k,l}$ be the number of cases where the relations $R(a_{C_kO_i}, a_{C_kO_j})$ and $R(a_{C_lO_i}, a_{C_lO_j})$ are simultaneously satisfied. Let also $S^{\nu}_{k,l}$ be the number of cases in which the relations $R(a_{C_kO_i}, a_{C_kO_j})$ and its dual $\overline{R}(a_{C_lO_i}, a_{C_lO_j})$ are simultaneously satisfied. As the total number of pairwise comparisons between the object is $n(n-1)/2$, it is seen that there hold the inequalities:

$$0 \leq S^{\mu}_{k,l} + S^{\nu}_{k,l} \leq \frac{n(n-1)}{2}.$$

For every k, l, such that $1 \leq k \leq l \leq m$, and for $n \geq 2$ two numbers are defined:

$$\mu_{C_k,C_l} = 2\frac{S^{\mu}_{k,l}}{n(n-1)}, \quad \nu_{C_k,C_l} = 2\frac{S^{\nu}_{k,l}}{n(n-1)}.$$

The pair constructed from these two numbers plays the role of the intuitionistic fuzzy evaluation of the relations that can be established between any two criteria C_k and C_l. In this way the index matrix M that relates evaluated objects with evaluating criteria can be transformed to another index matrix M^* that gives the relations among the criteria:

$$M^* = \begin{array}{c|ccc} & C_1 & \cdots & C_m \\ \hline C_1 & \langle \mu_{C_1,C_1}, \nu_{C_1,C_1} \rangle & \cdots & \langle \mu_{C_1,C_m}, \nu_{C_1,C_m} \rangle \\ \vdots & \vdots & \ddots & \vdots \\ C_m & \langle \mu_{C_m,C_1}, \nu_{C_1,C_m} \rangle & \cdots & \langle \mu_{C_m,C_m}, \nu_{C_m,C_m} \rangle \end{array}.$$

The final step of the algorithm is to determine the degrees of correlation between the criteria, depending on the user's choice of μ and ν. We call these correlations between the criteria: 'positive consonance', 'negative consonance' or 'dissonance'.

Let $\alpha, \beta \in [0; 1]$ be given, so that $\alpha + \beta \leq 1$. We say that criteria C_k and C_l are in:

- (α, β)-positive consonance, if $\mu_{C_k,C_l} > \alpha$ and $\nu_{C_k,C_l} < \beta$;
- (α, β)-negative consonance, if $\mu_{C_k,C_l} < \beta$ and $\nu_{C_k,C_l} > \alpha$;
- (α, β)-dissonance, otherwise.

Obviously, the larger α and/or the smaller β, the less number of criteria may be simultaneously connected with the relation of (α, β)-positive consonance. For practical purposes, it carries the most information when either the positive or the negative consonance is as large as possible, while the cases of dissonance are less informative and are skipped.

3 Main Results

We ran the described algorithm over a selection of data from last six WEF GCRs for the 28 (present) EU Member States from the period 2008–2009 to 2013–2014. The algorithm, as described in Section 2, produces the results with precision of 9 digits after the decimal point, however, we will use here precision of only 3 digits. From the six index matrices of data for 28 countries evaluated according to 12 pillars, we obtain six index matrices 12×12 for the intuitionistic fuzzy μ-function and six more index matrices 12×12 for the intuitionistic fuzzy ν-function. Obviously, in the first IM, along the main diagonal, $\mu_{C_k,C_k} = 1$, while in the second IM, along the main diagonal, $\nu_{C_k,C_k} = 0$, because naturally every pillar is in perfect positive consonance to itself. The sum of the values of the respective elements of the two IMs is generally:

$$0 \leq \mu_{C_k,C_l} + \nu_{C_k,C_l} \leq 1,$$

though in most cases it should be expected that this non-strict inequality is practically a strict one ($0 \leq \mu_{C_k,C_l} + \nu_{C_k,C_l} < 1$), thus leaving room for the complement to 1, that gives the measure of uncertainty.

Our aim will be to study how the positive consonance pairs of pillars behave over the considered 6-year period and for different runs of the values of the thresholds α, β. We focus on the positive consonances, although in a separate leg of research it might prove useful to focus on the negative ones, too. The study goes in two directions:

- how within a fixed year, changing the thresholds α, β changes the number of consonances formed for each of the twelve pillars, and
- how for a fixed pair of values of α, β these consonances change over time.

To start with, we will repeat from [7] the two 12×12 index matrices with the revealed intercriteria relations. Table 1 below gives the values of the intuitionistic fuzzy μ-function, and Table 2 gives the values of the intuitionistic fuzzy ν-function. Here, all cells are coloured in the greyscale, with the highest values coloured in the darkest shade of grey, while the lowest ones are coloured in white. Of course, every criteria perfectly correlates with itself, so for any i the value $\mu_{C_iC_i} = 1$, and $\nu_{C_iC_i} = \pi_{C_iC_i} = 0$. Also, the matrices are obviously symmetrical according to the main diagonals.

Table 1. Comparison of the calculated values of $\mu_{C_iC_j}$ for years 2008–2009 and 2013–2014

μ	1	2	3	4	5	6	7	8	9	10	11	12
1	1.000	0.844	0.685	0.757	0.788	0.833	0.603	0.828	0.823	0.497	0.794	0.802
2	0.844	1.000	0.627	0.751	0.749	0.743	0.529	0.741	0.775	0.582	0.831	0.807
3	0.685	0.627	1.000	0.616	0.638	0.664	0.653	0.648	0.693	0.434	0.651	0.667
4	0.757	0.751	0.616	1.000	0.780	0.720	0.550	0.704	0.725	0.524	0.765	0.772
5	0.788	0.749	0.638	0.780	1.000	0.746	0.622	0.728	0.757	0.558	0.767	0.796
6	0.833	0.743	0.664	0.720	0.746	1.000	0.627	0.817	0.802	0.505	0.786	0.765
7	0.603	0.529	0.653	0.550	0.622	0.627	1.000	0.664	0.611	0.389	0.563	0.590
8	0.828	0.741	0.648	0.704	0.728	0.817	0.664	1.000	0.820	0.476	0.733	0.751
9	0.823	0.775	0.693	0.725	0.757	0.802	0.611	0.820	1.000	0.548	0.817	0.815
10	0.497	0.582	0.434	0.524	0.558	0.505	0.389	0.476	0.548	1.000	0.648	0.601
11	0.794	0.831	0.651	0.765	0.767	0.786	0.563	0.733	0.817	0.648	1.000	0.860
12	0.802	0.807	0.667	0.772	0.796	0.765	0.590	0.751	0.815	0.601	0.860	1.000

μ	1	2	3	4	5	6	7	8	9	10	11	12
1	1.000	0.735	0.577	0.720	0.807	0.836	0.733	0.749	0.854	0.503	0.804	0.844
2	0.735	1.000	0.479	0.661	0.749	0.677	0.537	0.590	0.786	0.661	0.804	0.799
3	0.577	0.479	1.000	0.421	0.519	0.558	0.627	0.675	0.550	0.413	0.548	0.556
4	0.720	0.661	0.421	1.000	0.730	0.683	0.590	0.563	0.677	0.497	0.712	0.690
5	0.807	0.749	0.519	0.730	1.000	0.735	0.622	0.632	0.775	0.579	0.815	0.847
6	0.836	0.677	0.558	0.683	0.735	1.000	0.749	0.712	0.788	0.466	0.759	0.751
7	0.733	0.537	0.627	0.590	0.622	0.749	1.000	0.741	0.685	0.399	0.624	0.624
8	0.749	0.590	0.675	0.563	0.632	0.712	0.741	1.000	0.712	0.497	0.688	0.680
9	0.854	0.786	0.550	0.677	0.775	0.788	0.685	0.712	1.000	0.526	0.810	0.831
10	0.503	0.661	0.413	0.497	0.579	0.466	0.399	0.497	0.526	1.000	0.611	0.598
11	0.804	0.804	0.548	0.712	0.815	0.759	0.624	0.688	0.810	0.611	1.000	0.873
12	0.844	0.799	0.556	0.690	0.847	0.751	0.624	0.680	0.831	0.598	0.873	1.000

Table 2. Comparison of the calculated values of $\nu_{C_iC_j}$ for years 2008–2009 and 2013–2014

ν	1	2	3	4	5	6	7	8	9	10	11	12
1	0.000	0.114	0.241	0.140	0.140	0.077	0.275	0.116	0.116	0.458	0.148	0.127
2	0.114	0.000	0.304	0.156	0.190	0.167	0.365	0.220	0.180	0.384	0.127	0.138
3	0.241	0.304	0.000	0.265	0.265	0.209	0.204	0.270	0.225	0.495	0.270	0.241
4	0.140	0.156	0.265	0.000	0.108	0.140	0.294	0.201	0.169	0.381	0.138	0.111
5	0.140	0.190	0.265	0.108	0.000	0.135	0.233	0.198	0.164	0.378	0.156	0.130
6	0.077	0.167	0.209	0.140	0.135	0.000	0.209	0.090	0.095	0.397	0.114	0.127
7	0.275	0.365	0.204	0.294	0.233	0.209	0.000	0.212	0.259	0.497	0.315	0.265
8	0.116	0.220	0.270	0.201	0.198	0.090	0.212	0.000	0.132	0.476	0.217	0.196
9	0.116	0.180	0.225	0.169	0.164	0.095	0.259	0.132	0.000	0.399	0.122	0.116
10	0.458	0.384	0.495	0.381	0.378	0.397	0.497	0.476	0.399	0.000	0.307	0.336
11	0.148	0.127	0.270	0.138	0.156	0.114	0.315	0.217	0.122	0.307	0.000	0.079
12	0.127	0.138	0.241	0.111	0.130	0.127	0.265	0.196	0.116	0.336	0.079	0.000

ν	1	2	3	4	5	6	7	8	9	10	11	12
1	0.000	0.220	0.386	0.188	0.132	0.077	0.185	0.172	0.090	0.452	0.138	0.111
2	0.220	0.000	0.466	0.228	0.172	0.228	0.362	0.317	0.146	0.286	0.135	0.138
3	0.386	0.466	0.000	0.476	0.405	0.344	0.286	0.251	0.394	0.537	0.394	0.389
4	0.188	0.228	0.476	0.000	0.143	0.169	0.283	0.307	0.201	0.397	0.175	0.198
5	0.132	0.172	0.405	0.143	0.000	0.153	0.272	0.259	0.135	0.341	0.098	0.079
6	0.077	0.228	0.344	0.169	0.153	0.000	0.135	0.169	0.101	0.439	0.143	0.159
7	0.185	0.362	0.286	0.283	0.272	0.135	0.000	0.146	0.209	0.505	0.267	0.275
8	0.172	0.317	0.251	0.307	0.259	0.169	0.146	0.000	0.206	0.415	0.217	0.233
9	0.090	0.146	0.394	0.201	0.135	0.101	0.209	0.206	0.000	0.405	0.119	0.101
10	0.452	0.286	0.537	0.397	0.341	0.439	0.505	0.415	0.405	0.000	0.328	0.344
11	0.138	0.135	0.394	0.175	0.098	0.143	0.267	0.217	0.119	0.328	0.000	0.071
12	0.111	0.138	0.389	0.198	0.079	0.159	0.275	0.233	0.101	0.344	0.071	0.000

Case 1. Fixed years, changing thresholds α, β

In Tables 3–8, where α, β run from (0.7; 0.3) to (0.85; 0.15), for each pillar is given the number of the rest pillars, which it is in positive consonance with, as evaluated by $\mu_{C_k, C_l} > \alpha$ and $\nu_{C_k, C_l} < \beta$. We will note that we skip the columns for pillars '3. Macroeconomic stability' and '10. Market size', since both do not enter into positive consonance with any of the rest.

Table 3. Results for fixed year 2008–2009.

α	β	C_1	C_2	C_4	C_5	C_6	C_7	C_8	C_9	C_{11}	C_{12}	Rel	Uniq
0.85	0.15	0	0	0	0	0	0	0	0	1	1	1	2
0.825	0.175	3	2	0	0	1	0	1	0	2	1	5	6
0.8	0.2	5	3	0	0	3	0	3	5	3	4	13	7
0.775	0.225	7	4	1	3	4	0	3	6	5	5	19	9
0.75	0.25	8	8	5	7	6	0	4	7	7	8	30	9
0.725	0.275	8	8	5	8	7	0	7	7	8	8	33	9
0.7	0.3	8	8	8	8	8	0	8	8	8	8	36	9

Table 4. Results for fixed year 2009–2010.

α	β	C_1	C_2	C_4	C_5	C_6	C_7	C_8	C_9	C_{11}	C_{12}	Rel	Uniq
0.85	0.15	0	0	0	0	0	0	0	0	1	1	1	2
0.825	0.175	2	0	0	0	2	0	0	2	1	1	4	5
0.8	0.2	3	1	0	1	2	0	0	3	1	5	8	7
0.775	0.225	7	4	3	3	5	0	2	5	6	7	21	9
0.75	0.25	8	4	4	5	6	0	2	5	7	7	24	9
0.725	0.275	8	5	5	6	8	0	3	7	7	7	28	9
0.7	0.3	8	7	8	8	8	0	6	8	7	8	34	9

Table 5. Results for fixed year 2010–2011.

α	β	C_1	C_2	C_4	C_5	C_6	C_7	C_8	C_9	C_{11}	C_{12}	Rel	Uniq
0.85	0.15	0	0	0	0	0	0	0	0	1	1	1	2
0.825	0.175	3	0	0	0	1	0	0	1	1	2	4	5
0.8	0.2	5	2	0	2	1	0	0	2	3	5	10	7
0.775	0.225	5	3	0	2	4	0	0	5	5	6	15	7
0.75	0.25	6	4	1	3	5	0	0	5	6	6	18	8
0.725	0.275	6	4	3	5	6	0	0	6	7	7	22	8
0.7	0.3	8	5	5	7	7	0	2	6	7	7	27	9

Table 6. Results for fixed year 2011–2012.

α	β	C_1	C_2	C_4	C_5	C_6	C_7	C_8	C_9	C_{11}	C_{12}	Rel	Uniq
0.85	0.15	0	0	0	0	0	0	0	1	1	2	2	3
0.825	0.175	2	1	0	0	0	0	0	3	3	3	6	5
0.8	0.2	3	3	0	1	0	0	0	4	4	5	10	6
0.775	0.225	5	3	0	3	2	0	0	5	5	5	14	7
0.75	0.25	5	3	0	4	3	0	0	6	6	5	16	7
0.725	0.275	7	4	0	4	5	0	2	6	6	6	20	8
0.7	0.3	7	5	1	6	7	1	3	7	7	6	25	10

Table 7. Results for fixed year 2012–2013.

α	β	C_1	C_2	C_4	C_5	C_6	C_7	C_8	C_9	C_{11}	C_{12}	Rel	Uniq
0.85	0.15	0	0	0	0	0	0	0	1	1	2	2	3
0.825	0.175	3	0	0	0	1	0	0	3	2	4	7	5
0.8	0.2	5	1	0	2	2	0	0	4	4	4	11	7
0.775	0.225	5	3	0	4	2	0	0	6	5	5	15	7
0.75	0.25	5	3	0	4	3	0	0	6	6	5	16	7
0.725	0.275	8	5	1	7	6	2	3	6	6	6	25	10
0.7	0.3	9	5	6	7	8	3	4	8	7	7	32	10

Table 8. Results for fixed year 2013–2014.

α	β	C_1	C_2	C_4	C_5	C_6	C_7	C_8	C_9	C_{11}	C_{12}	Rel	Uniq
0.85	0.15	3	0	0	1	1	0	0	1	1	3	5	6
0.825	0.175	3	0	0	1	1	0	0	2	1	4	6	6
0.8	0.2	5	1	0	3	1	0	0	3	5	4	11	7
0.775	0.225	5	3	0	4	2	0	0	6	5	5	15	7
0.75	0.25	5	3	0	4	4	0	0	6	6	6	17	7
0.725	0.275	8	4	1	7	6	0	2	6	6	6	23	9
0.7	0.3	9	5	3	7	7	4	3	7	7	6	29	10

From the data in Tables 3–8, we can make some general observations concerning the thresholds α, β. These observations are needed for setting the general framework in which the findings of the present research can be usefully interpreted. Obviously, neither too few, nor too many correlating pillars would help yield an effective economic analysis.

- *Observation 1.* Three of the twelve pillars, namely, the basic requirement pillar '3. Macroeconomic stability' and the efficiency enhancer pillars '7. Labor market efficiency' and '10. Market size' tend to avoid positive consonances with the rest pillars in the WEF GCR methodology, which is especially well expressed for the 3rd and 10th pillars (for all studied years, the values of μ_{C_3,C_l} ranging from 0.648 to 0.693 and the maximal value of μ_{C_{10},C_l} ranging from 0.622 to 0.672). In general, it is worth analysing to what extent it is natural to have these pillars uncorrelated to the rest, or to what extent it results from particular governments' malperformance.
- *Observation 2.* Under a certain value for threshold α (respectively, above a certain value for threshold β), it is natural that all pillars start correlating, which is ineffective for the analysis. In the light of the Observation 1, we can safely focus only on the thresholds when the 9 out of 12 pillars start correlating, which is at $\alpha = 0.775$ for the first two years, and with α falling to around 0.725 in the next four years. Hence, it is interesting to analyse the data corresponding to larger values of threshold α.
- *Observation 3.* In the other extreme, analysing data for too few pillars being in positive consonance is not effective either. We observe that until 2013–2014, the number of unique correlating pillars when $\alpha = 0.85$ is 2 or 3 out of 12 (mainly '11. Business sophistication' and '12. Innovation', sometimes accompanied by the basic requirement pillar '1. Institutions'). The sudden number of six correlating pillars with as high threshold for α as 0.85 can be interpreted as a sign of raising mutual dependence of the different aspects of competitiveness, yet observations for next periods are needed for a more categorical conclusion.

Hence, it is most useful to focus the analysis in the range from (0.775; 0.225) to (0.825; 0.175). Besides the more general observations about the thresholds, we can also make some more specific observations about the pillars.

- *Observation 4.* The efficiency enhancer pillar '7. Labor market efficiency' starts correlating with the rest pillars only after year 2011–2012, and only in the low values of α from 0.7 to 0.75.
- *Observation 5.* The efficiency enhancer pillar '8. Financial market sophistication' in 2008–2009 and 2009–2010 was in consonance with other pillars as of $\alpha = 0.825$ or 0.775, respectively, after the α threshold for the 8th pillar falls down to $\alpha = 0.725$ and even 0.7, meaning that it becomes much weakly correlated.
- *Observation 6.* Pillars '11. Business sophistication' and '12. Innovation' have shown the greatest and most stable positive consonance between each other, as well as with the other pillars. This is not surprising, given that both pillars form with 50% weight each the third subindex 'Innovation and sophistication factors'.

Case 2. Fixed thresholds α, β, changing years

In the following temporal analysis (Tables 9–15) we compare how the pillars have correlated for the last six years at fixed values of the thresholds α, β.

Table 9. Results for fixed year $\alpha = 0.85$, $\beta = 0.15$.

Year	C_1	C_2	C_4	C_5	C_6	C_7	C_8	C_9	C_{11}	C_{12}	Rel	Uniq
2008–2009	0	0	0	0	0	0	0	0	1	1	1	2
2009–2010	0	0	0	0	0	0	0	0	1	1	1	2
2010–2011	0	0	0	0	0	0	0	0	1	1	1	2
2011–2012	0	0	0	0	0	0	0	1	1	2	2	3
2012–2013	0	0	0	0	0	0	0	1	1	2	2	3
2013–2014	3	0	0	1	1	0	0	1	1	3	5	6

Table 10. Results for fixed year $\alpha = 0.825$, $\beta = 0.175$.

Year	C_1	C_2	C_4	C_5	C_6	C_7	C_8	C_9	C_{11}	C_{12}	Rel	Uniq
2008–2009	3	2	0	0	1	0	1	0	2	1	5	6
2009–2010	2	0	0	0	2	0	0	2	1	1	4	5
2010–2011	3	0	0	0	1	0	0	1	1	2	4	5
2011–2012	2	1	0	0	0	0	0	3	3	3	6	5
2012–2013	3	0	0	0	1	0	0	3	2	4	7	5
2013–2014	3	0	0	1	1	0	0	2	1	4	6	6

Table 11. Results for fixed year $\alpha = 0.8$, $\beta = 0.2$.

Year	C_1	C_2	C_4	C_5	C_6	C_7	C_8	C_9	C_{11}	C_{12}	Rel	Uniq
2008–2009	5	3	0	0	3	0	3	5	3	4	13	7
2009–2010	3	1	0	1	2	0	0	3	1	5	8	7
2010–2011	5	2	0	2	1	0	0	2	3	5	10	7
2011–2012	3	3	0	1	0	0	0	4	4	5	10	6
2012–2013	5	1	0	2	2	0	0	4	4	4	11	7
2013–2014	5	1	0	3	1	0	0	3	5	4	11	7

Table 12. Results for fixed year $\alpha = 0.775, \beta = 0.225$.

Year	C_1	C_2	C_4	C_5	C_6	C_7	C_8	C_9	C_{11}	C_{12}	Rel	Uniq
2008–2009	7	4	1	3	4	0	3	6	5	5	19	9
2009–2010	7	4	3	3	5	0	2	5	6	7	21	9
2010–2011	5	3	0	2	4	0	0	5	5	6	15	7
2011–2012	5	3	0	3	2	0	0	5	5	5	14	7
2012–2013	5	3	0	4	2	0	0	6	5	5	15	7
2013–2014	5	3	0	4	2	0	0	6	5	5	15	7

Table 13. Results for fixed year $\alpha = 0.75, \beta = 0.25$

Year	C_1	C_2	C_4	C_5	C_6	C_7	C_8	C_9	C_{11}	C_{12}	Rel	Uniq
2008–2009	8	8	5	7	6	0	4	7	7	8	30	9
2009–2010	8	4	4	5	6	0	2	5	7	7	24	9
2010–2011	6	4	1	3	5	0	0	5	6	6	18	8
2011–2012	5	3	0	4	3	0	0	6	6	5	16	7
2012–2013	5	3	0	4	3	0	0	6	6	5	16	7
2013–2014	5	3	0	4	4	0	0	6	6	6	17	7

Table 14. Results for fixed year $\alpha = 0.725, \beta = 0.275$

Year	C_1	C_2	C_4	C_5	C_6	C_7	C_8	C_9	C_{11}	C_{12}	Rel	Uniq
2008–2009	8	8	5	8	7	0	7	7	8	8	33	9
2009–2010	8	5	5	6	8	0	3	7	7	7	28	9
2010–2011	6	4	3	5	6	0	0	6	7	7	22	8
2011–2012	7	4	0	4	5	0	2	6	6	6	20	8
2012–2013	8	5	1	7	6	2	3	6	6	6	25	10
2013–2014	8	4	1	7	6	0	2	6	6	6	23	9

Table 15. Results for fixed year $\alpha = 0.7, \beta = 0.3$

Year	C_1	C_2	C_4	C_5	C_6	C_7	C_8	C_9	C_{11}	C_{12}	Rel	Uniq
2008–2009	8	8	8	8	8	0	8	8	8	8	36	9
2009–2010	8	7	8	8	8	0	6	8	7	8	34	9
2010–2011	8	5	5	7	7	0	2	6	7	7	27	9
2011–2012	7	5	1	6	7	1	3	7	7	6	25	10
2012–2013	9	5	6	7	8	3	4	8	7	7	32	10
2013–2014	9	5	3	7	7	4	3	7	7	6	29	10

- *Observation 7.* We can observe that as of $\alpha = 0.85$, pillars 11 and 12 are again sustainably correlating during the whole period, and as of $\alpha = 0.825$ the basic requirement pillar '1. Institutions' also tends to enter in positive consonances, i.e. it is a factor of significant importance for the rest competitiveness pillars.
- *Observation 8.* With less but also visible importance are the efficiency enhancer pillars '9. Technological readiness' and '6. Goods market efficiency'.
- *Observation 9.* With $(\alpha; \beta)$ starting from (0.8; 0.2) down to (0.7; 0.3), we note that the number of positive consonances of the basic requirement pillar '2. Infrastructure' gradually reduces in time.

- *Observation 10.* We can also note that with the relatively high $\alpha > 0.75$, the efficiency enhancer pillar '8. Financial market sophistication' has been in positive consonances only in the period 2008–2010, and for smaller α it has been visibly less correlated after 2010, than it was before.
- *Observation 11.* A very similar to observation to the previous one can also be made for the basic requirement pillar '4. Health and primary education'. Compared to it, the efficiency enhancer pillar '5. Higher education and training' naturally shows higher levels of positive consonance with the rest competitiveness pillars for different runs of $(\alpha; \beta)$ throughout the whole period.
- *Observation 12.* The efficiency enhancer pillar '7. Labor market efficiency' has only started exhibiting for $\alpha > 0.75$ and only in the last year.

Beyond these observations, a more detailed and profound analysis of these data should be done by economists, taking into consideration various factors like the beginning of the world financial crisis, accession of new Member States to the European Union, certain changes in legislation, technological breakthroughs. It is also worth noting that in some pillars, like '4. Health and primary education' and '5. Higher education and training' effects should only be expected to occur after certain periods of time, which makes it necessary to continue the present research.

4 Conclusions

With the present temporal and threshold analysis of the WEF's Global Competitiveness Reports data for the EU Member States, we aim to continue and further elaborate on our analysis of the revealed relations between the twelve pillars of competitiveness. The observed positive consonances show certain changes and trends that may yield fruitful further analyses by interested economists. The presented approach for InterCriteria Decision Making also presents as a useful application of the theory of index matrices and of intuitionistic fuzzy sets.

The conclusions about how the competitiveness pillars correlate might help answer many questions about how European economies innovate, and would be useful for the EU Member States' national policy makers, in order to better identify and strengthen the transformative forces that drive the future economic growth of their countries. Despite that we have focused on data related to EU Member States, the same approach can be equally applied to other selections of countries and time periods, and analysing the differences with the hitherto presented results will be very informative and challenging.

Acknowledgements. The research work reported in the paper is partly supported by the project AComIn "Advanced Computing for Innovation", grant 316087, funded by the FP7 Capacity Programme (Research Potential of Convergence Regions).

References

1. Atanassov, K.: Intuitionistic fuzzy sets, VII ITKR's Session, Sofia (June 1983) (in Bulgarian)
2. Atanassov, K.: Intuitionistic fuzzy sets. Fuzzy Sets and Systems 20(1), 87–96 (1986)
3. Atanassov, K.: Generalized Nets. World Scientific, Singapore (1991)
4. Atanassov, K.: Intuitionistic Fuzzy Sets: Theory and Applications. Physica-Verlag, Heidelberg (1999)
5. Atanassov, K.: On Intuitionistic Fuzzy Sets Theory. Springer, Berlin (2012)
6. Atanassov, K., Mavrov, D., Atanassova, V.: InterCriteria decision making. A new approach for multicriteria decision making, based on index matrices and intuitionistic fuzzy sets. In: Proc. of 12th International Workshop on Intuitionistic Fuzzy Sets and Generalized Nets, Warsaw, Poland, October 11 (2013) (in press)
7. Atanassova, V., Doukovska, L., Atanassov, K., Mavrov, D.: InterCriteria Decision Making Approach to EU Member States Competitiveness Analysis. In: Proc. of 4th Int. Symp. on Business Modelling and Software Design, Luxembourg, June 24-26 (in press, 2014)
8. World Economic Forum. The Global Competitiveness Reports (2008-2013), `http://www.weforum.org/issues/global-competitiveness`
9. Zadeh, L.A.: Fuzzy Sets. Information and Control 8, 333–353 (1965)

InterCriteria Decision Making Approach to EU Member States Competitiveness Analysis: Trend Analysis

Vassia Atanassova[1], Lyubka Doukovska[2], Dimitar Karastoyanov[3], and František Čapkovič[4]

[1] Intelligent Systems Department, IICT – Bulgarian Academy of Sciences, Acad. G. Bonchev Str., bl. 2, 1113 Sofia, Bulgaria and Bioinformatics and Mathematical Modelling Dept., IBPhBME – Bulgarian Academy of Sciences, Acad. G. Bonchev Str., bl. 105, 1113 Sofia, Bulgaria
vassia.atanassova@gmail.com

[2] Intelligent Systems Department, IICT – Bulgarian Academy of Sciences, Acad. G. Bonchev Str., bl. 2, 1113 Sofia, Bulgaria
doukovska@iit.bas.bg

[3] Intelligent Systems Department, IICT – Bulgarian Academy of Sciences Acad. G. Bonchev Str., bl. 2, 1113 Sofia, Bulgaria
dkarast@iinf.bas.bg

[4] Institute of Informatics, Slovak Academy of Sciences, Dubravska cesta 9, 845 07 Bratislava, Slovak Republic
Frantisek.Capkovic@savba.sk

Abstract. In this paper, we continue our investigations of the newly developed InterCriteria Decision Making (ICDM) approach with considerations about the more appropriate choice of the employed intuitionistic fuzzy threshold values. In theoretical aspect, our aim is to identify the relations between the thresholds of inclusion of new elements to the set of strictly correlating criteria and the numbers of correlating pairs of criteria thus formed. We illustrate the findings with data extracted from the World Economic Forum's Global Competitiveness Reports for the years 2008–2009 to 2013–2014 for the current 28 Member States of the European Union. The study of the findings from the considered six-year period involves trend analysis and computation of two approximating functions: a linear function and a polynomial function of 6th order. The per-year trend analysis of each of the 12 criteria, called 'pillars of competitiveness' in the WEF's GCR methodology, gives an opportunity to prognosticate their values for the forthcoming year 2014–2015.

Keywords: Global Competitiveness Index, InterCriteria decision making, Intuitionistic fuzzy sets, Multicriteria decision making, Trend analysis.

1 Introduction

In a series of papers, we have started investigating the application of the newly proposed InterCriteria Decision Making (ICDM) approach, based on the concepts of

P. Angelov et al. (eds.), *Intelligent Systems'2014*,
Advances in Intelligent Systems and Computing 322, DOI: 10.1007/978-3-319-11313-5_10

intuitionistic fuzzy sets (see [1, 3, 4]) and index matrices (see [2]). ICDM aims to support a decision maker, who disposes of datasets of evaluations or measurements of multiple objects against multiple criteria, to more profoundly understand the nature of the utilized evaluation criteria, and discover some correlations existing among the criteria themselves. Theoretically, the ICDM approach has been presented in details in [5], and in [6, 7, 8] the approach was further discussed by the coauthors in the light of its application to data about EU Member States' competitiveness in the period 2008–2014, as obtained from the World Economic Forum's (WEF) annual Global Competitiveness Reports(GCRs), [9].

Shortly presented, in the ICDM approach, we have (at least one) matrix of evaluations or measurements of m evaluated objects against n evaluating criteria, and from these we obtain a respective $n \times n$ matrix giving the discovered correlations between the evaluating criteria in the form of intuitionistic fuzzy pairs, or, which is the same but more practical, two $n \times n$ matrices giving in separate views the membership-values (a μ-matrix) and the non-membership pairs (a ν-matrix). Once having these, we are interested to see which of the criteria are in *positive consonance* (situation of definitively correlating criteria), in *negative consonance* (situation of definitively non-correlating criteria), or in *dissonance* (situation of uncertainty, when no definitive conclusion can be made). In order to categorize all the values of the resultant $n(n-1)/2$ pairs of criteria, we need to define two thresholds, α and β, for the positive and for the negative consonance, respectively.

In [5, 7], where the emphasis was put on some other aspects of the ICDM's approach, we considered the rather simplistic case when the [0; 1]-limited threshold values α and β were changing with a predefined precision step and were always summing up to 1. Later, in [6], we notice that this approach is not enough discriminative and helpful for the decision maker, and reformulate the problem statement aiming to identify (shortlist) the k most strongly positively correlated criteria out of the totality of n disposable evaluation criteria. A general problem-independent algorithm for this shortlisting procedure is proposed there, and will be partially relied on here as well.

2 Data Presentation

The presented in [6] algorithm for identifying the values of threshold α under which a given element of the set of evaluating criteria starts entering in positive consonances with the rest criteria. The algorithm involves taking the maximal values of positive consonance per criterion (which in the terms of index matrices is the operation *max-row-aggregation*), sorting this list in descending order, and thus finding the ordering of the criteria by positive consonance.

In [6], we illustrated this algorithm using the datasets of EU Member States' competitiveness, as evaluated by the World Economic Forum. We made the calculations for both the positive and the negative consonance, in order to compare the results on this plane, as well as in time. We presented the results only for years 2008–2009 and 2013–2014, which are the two extreme years in the period we analyse from the publicly available GCRs, [9]. Here we will present the tables for all six years separately (Tables 1–6), as well as in aggregation (Table 7).

Table 1. Results for year 2008–2009

Number of correlating criteria	Number of pairs of correlating criteria	Criteria ordered by positive consonance	True when $\alpha \leq$
2	1	11	0.860
		12	
4	2	1	0.844
		2	
5	3	6	0.833
6	5	8	0.828
7	6	9	0.823
8	14	5	0.796
9	18	4	0.780
10	37	3	0.693
11	41	7	0.664
12	45	10	0.648

Number of correlating criteria	Number of pairs of correlating criteria	Criteria ordered by negative consonance	True when $\beta \geq$
2	1	1	0.077
		6	
4	2	11	0.079
		12	
5	3	8	0.09
6	4	9	0.095
8	5	4	0.108
		5	
9	8	2	0.114
11	35	3	0.204
		7	
12	54	10	0.307

Table 2. Results for year 2009–2010

Number of correlating criteria	Number of pairs of correlating criteria	Criteria ordered by positive consonance	True when $\alpha \leq$
2	1	1	0.856
		6	
4	2	11	0.852
		12	
5	3	9	0.849
7	8	2	0.807
		5	
8	16	8	0.783
9	18	4	0.778
10	36	7	0.693
11	37	3	0.690
12	52	10	0.622

Number of correlating criteria	Number of pairs of correlating criteria	Criteria ordered by negative consonance	True when $\beta \geq$
2	1	11	0.071
		12	
4	2	1	0.077
		6	
5	4	9	0.106
6	6	5	0.119
7	9	8	0.122
8	12	4	0.124
9	15	2	0.135
10	35	7	0.206
11	37	3	0.212
12	55	10	0.302

Table 3. Results for year 2010–2011

Number of correlating criteria	Number of pairs of correlating criteria	Criteria ordered by positive consonance	True when $\alpha \leq$
2	1	11	0.852
		12	
3	2	1	0.849
4	3	9	0.828
5	4	6	0.825
6	8	5	0.812

Number of correlating criteria	Number of pairs of correlating criteria	Criteria ordered by negative consonance	True when $\beta \geq$
2	1	11	0.087
		12	
3	2	1	0.095
4	3	6	0.103
5	5	9	0.106
6	6	4	0.114

Table 4. *(continued)*

7	9	2	0.810
8	18	4	0.751
9	24	8	0.720
10	31	7	0.690
11	34	3	0.683
12	42	10	0.653

7	7	5	0.116
8	10	2	0.127
9	28	8	0.185
10	29	3	0.190
11	30	7	0.198
12	54	10	0.294

Table 4. Results for year 2011–2012

Number of correlating criteria	Number of pairs of correlating criteria	Criteria ordered by positive consonance	True when $\alpha \leq$
2	1	11	0.870
		12	
3	2	9	0.854
4	3	1	0.841
5	6	2	0.831
6	10	5	0.807
7	11	6	0.796
8	19	8	0.738
9	24	7	0.706
10	25	4	0.704
11	29	3	0.685
12	35	10	0.672

Number of correlating criteria	Number of pairs of correlating criteria	Criteria ordered by negative consonance	True when $\beta \geq$
2	1	11	0.074
		12	
3	2	9	0.090
4	3	1	0.095
5	4	5	0.098
6	6	2	0.114
7	8	6	0.116
8	17	4	0.153
9	22	8	0.175
10	23	7	0.180
11	35	3	0.241
12	49	10	0.283

Table 5. Results for year 2012–2013

Number of correlating criteria	Number of pairs of correlating criteria	Criteria ordered by positive consonance	True when $\alpha \leq$
2	1	1	0.870
		9	
4	2	11	0.865
		12	
5	5	5	0.836
6	7	6	0.831
7	9	2	0.815
9	18	4	0.749
		8	
10	22	7	0.741
11	40	10	0.659
12	42	3	0.648

Number of correlating criteria	Number of pairs of correlating criteria	Criteria ordered by negative consonance	True when $\beta \geq$
2	1	1	0.071
		9	
3	2	6	0.074
5	3	11	0.077
		12	
6	6	5	0.090
7	10	2	0.111
8	11	4	0.114
10	25	7	0.153
		8	
11	42	3	0.267
12	46	10	0.286

Table 6. Results for year 2013–2014

Number of correlating criteria	Number of pairs of correlating criteria	Criteria ordered by positive consonance	True when $\alpha \leq$
2	1	11	0.873
		12	
4	2	1	0.854
		9	
5	3	5	0.847
6	11	2	0.804
7	13	6	0.788
9	20	7	0.749
		8	
10	25	4	0.730
11	37	3	0.675
12	39	10	0.661

Number of correlating criteria	Number of pairs of correlating criteria	Criteria ordered by negative consonance	True when $\beta \geq$
2	1	11	0.071
		12	
4	2	1	0.077
		6	
5	3	5	0.079
6	4	9	0.090
8	13	2	0.135
		7	
9	17	4	0.143
10	19	8	0.146
11	38	3	0.251
12	45	10	0.286

Table 7. Results for year 2008–2014

Number of correlating criteria	Number of pairs of correlating criteria	Criteria ordered by positive consonance	True when $\alpha \leq$
2	1	11	0.836
		12	
4	2	1	0.821
		6	
6	5	5	0.804
		9	
7	7	2	0.789
8	18	8	0.745
9	21	4	0.725
10	25	7	0.693
11	34	3	0.672
12	44	10	0.622

Number of correlating criteria	Number of pairs of correlating criteria	Criteria ordered by negative consonance	True when $\beta \geq$
2	1	11	0.091
		12	
4	2	1	0.092
		6	
5	3	5	0.109
6	4	9	0.124
7	8	4	0.139
8	9	2	0.144
9	18	8	0.163
10	26	7	0.190
11	34	3	0.239
12	48	10	0.306

While columns 1, 3 and 4 in each of the Tables 1–7 have been discussed and determined within the algorithm, we also support information in column 2 about the number of pairs formed by the so ordered correlating criteria. This separately determined, completely data-dependent, number gives information about the level of interconnectedness between the involved criteria, and is also considered by us to be useful to keep track of. From these 7 tables, we are interested to detect the dependences between the increase of the threshold values and the interconnectedness between the correlating criteria, and thus propose another way of determining the threshold values.

We are also interested to perform trend analysis of the threshold values for all the years from 2008–2009 to 2013–2014, and formulate a prognosis for year 2014–2015.

3 Main Results

Taking the data from Tables 1–6 from columns 2 and 4, we obtain six charts (Fig. 1–6), of the dependence between the number of pairs of correlating criteria, i.e. the interconnectedness (as plotted on the axis x), and the thresholds of inclusion of a new criterion to the set of correlating criteria (in descending order, as plotted on the axis y).

These figures allow us to make observations about the homogeneity of the positive consonances exhibited by the evaluation criteria, and thus more easily decide how to divide the set of criteria on strongly and weakly correlating ones. Finally, this helps us decide about the number k of the totality of n criteria, on which to focus our attention, as was formulated for parts of the problems ICDM solves, [6].

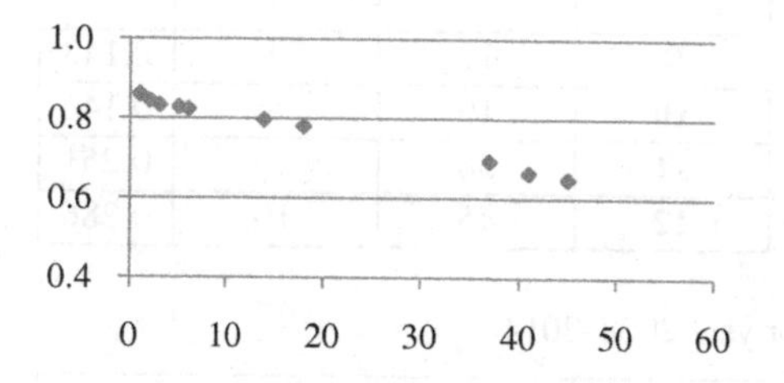

Fig. 1. Results for year 2008–2009.

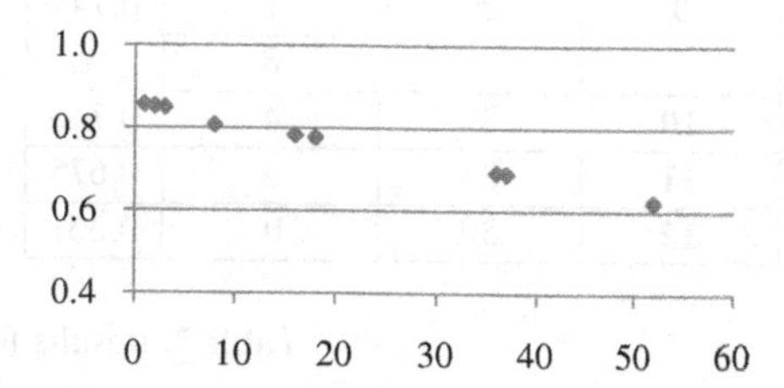

Fig. 2. Results for year 2009–2010.

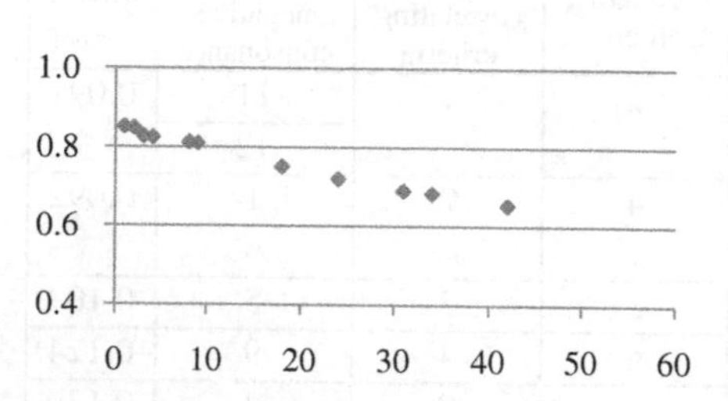

Fig. 3. Results for year 2010–2011.

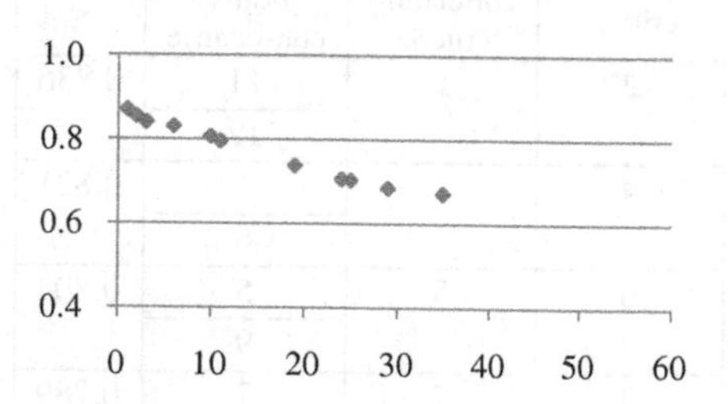

Fig. 4. Results for year 2011–2012.

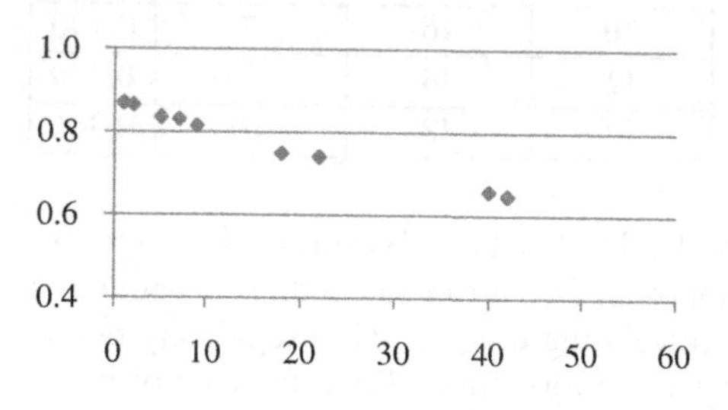

Fig. 5. Results for year 2012–2013.

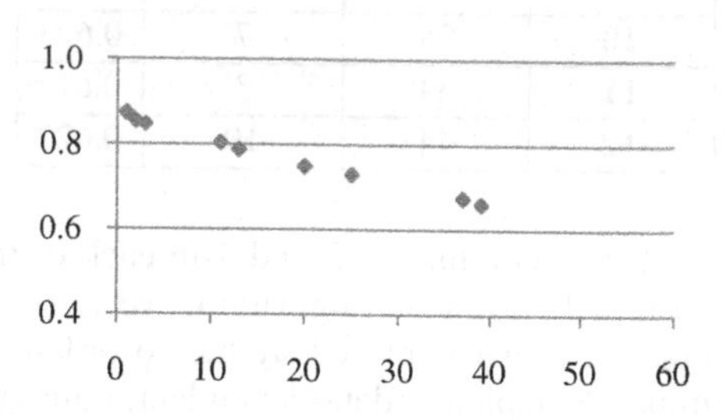

Fig. 6. Results for year 2013–2014.

On this basis we can conclude that for year 2013–2014, we can see that four groups of criteria are formed:

- five strongly correlating, with α varying from 0.847 to 0.873: '11. Business sophistication', '12. Innovation', '1. Institutions', '9. Technological readiness' and '5. Higher education and training';
- two weakly correlating, with α varying from 0.788 to 0.804: '2. Infrastructure' and '6. Goods market efficiency';
- three weakly non-correlating, with α varying from 0.730 to 0.749: '7. Labor market efficiency', '8. Financial market sophistication' and '4. Health and primary education'; and
- two strongly non-correlating, with α varying from 0.661 to 0.675: '3. Macroeconomic stability' and '10. Market size'.

With the data from this case, involving 12 criteria only, the decision maker can easily make the above observation without further calculations or application of other sophisticated methods. However, in cases involving a greater number of criteria, application of cluster analysis methods may prove unavoidable.

It is also interesting, to see how all these six charts combine in a single picture, as given in Fig. 7.

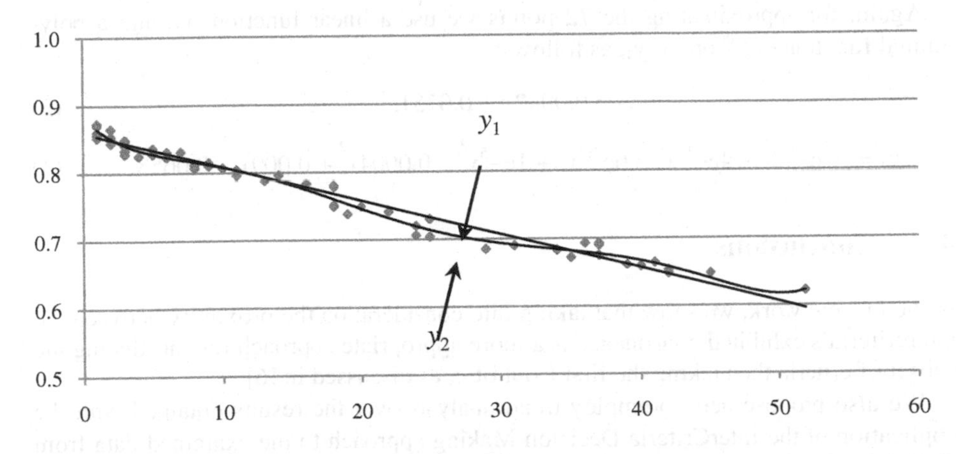

Fig. 7. Combined results for the years 2008–2014, based on dependencies between number of pairs in correlation (axis *x*) and level of positive consonance (axis *y*). Functions y_1 and y_2 approximate the set of plotted points.

We have further elaborated it by approximating the 72 points with a simple linear function, y_1, and with a polynomial function of 6th order, y_2. The form of both functions was produced with the aid of MS Excel, as follows:

$$y_1 = -0.0051x + 0.8591, \tag{1}$$

$$y_2 = 9e{-}10x^6 - 1e{-}7x^5 + 8e{-}6x^4 - 0.0002x^3 + 0.0028x^2 - 0.0203x + 0.8822. \tag{2}$$

Although we have already settled to give priority to the results related to positive consonance in this example, it is interesting to show also the last result, as obtained for the negative consonance. Skipping the charts for the six individual years, we show only the combined picture for negative consonance in the following Fig. 8.

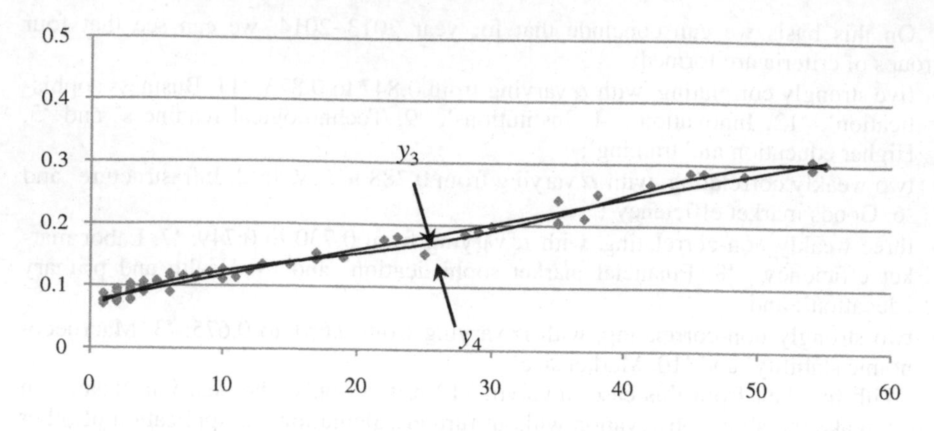

Fig. 8. Combined results for the years 2008–2014, based on dependencies between number of pairs in correlation (axis x) and level of negative consonance (axis y). Functions y_1 and y_2 approximate the set of plotted points.

Again, for approximating the 72 points we use a linear function, y_3, and a polynomial function of 6th order, y_4, as follows:

$$y_3 = 0.0042x + 0.0751, \tag{3}$$

$$y_4 = -3e{-}12x^6 - 4e{-}10x^5 - 6e{-}8x^4 + 1e{-}5x^3 - 0.0004x^2 + 0.009x + 0.0658. \tag{4}$$

4 Conclusions

In the present work, we show that taking into consideration the proximity between the intercriteria's exhibited consonance is a more appropriate approach for shortlisting the subset of criteria than taking the first k out of n, as discussed in [6].

We also propose here to employ trend analysis over the results obtained with the application of the InterCriteria Decision Making approach to the examined data from the Global Competitiveness Reports of the World Economic Forum. Further research in this direction has potential to reveal more knowledge about the nature of the evaluation criteria involved and their future development. Eventually, these results may help policy makers identify and strengthen the transformative forces that will drive future economic growth, [9].

Acknowledgements. The research work reported in the paper is partly supported by the project AComIn "Advanced Computing for Innovation", grant 316087, funded by the FP7 Capacity Programme (Research Potential of Convergence Regions).

References

1. Atanassov, K.: Intuitionistic fuzzy sets. Fuzzy Sets and Systems 20(1), 87–96 (1986)
2. Atanassov, K.: Generalized Nets. World Scientific, Singapore (1991)
3. Atanassov, K.: Intuitionistic Fuzzy Sets: Theory and Applications. Physica-Verlag, Heidelberg (1999)
4. Atanassov, K.: On Intuitionistic Fuzzy Sets Theory. Springer, Berlin (2012)
5. Atanassov, K., Mavrov, D., Atanassova, V.: InterCriteria decision making. A new approach for multicriteria decision making, based on index matrices and intuitionistic fuzzy sets. In: Proc. of 12th International Workshop on Intuitionistic Fuzzy Sets and Generalized Nets, Warsaw, Poland (October 11, 2013) (in Press)
6. Atanassova, V., Mavrov, D., Doukovska, L., Atanassov, K.: Discussion on the threshold values in the InterCriteria Decision Making Approach. Notes on Intuitionistic Fuzzy Sets 20(2), 94–99 (2014)
7. Atanassova, V., Doukovska, L., Atanassov, K., Mavrov, D.: InterCriteria Decision Making Approach to EU Member States Competitiveness Analysis. In: Proc. of 4th Int. Symp. on Business Modelling and Software Design, Luxembourg, June 24-26 (in press, 2014)
8. Atanassova, V., Doukovska, L., Atanassov, K., Mavrov, D.: InterCriteria Decision Making Approach to EU Member States Competitiveness Analysis: Temporal and Threshold Analysis. In: 7th Int. IEEE Conf. on Intelligent Systems, Warsaw, September 24-26 (submitted, 2014)
9. World Economic Forum (2008-2013). The Global Competitiveness Reports, `http://www.weforum.org/issues/global-competitiveness`

Computer-Based Support in Multicriteria Bargaining with Use of the Generalized Raiffa Solution Concept

Lech Kruś

Systems Research Institute, Polish Academy of Sciences,
Warsaw, Poland
krus@ibspan.waw.pl

Abstract. The paper deals with cooperation problems in the case of two parties having different sets of criteria measuring their payoffs. Using ideas of the game theory, a mathematical model describing multicriteria bargaining problem is formulated. In the paper an interactive procedure supporting multicriteria analysis and aiding consensus seeking is presented which can be implemented in a computer-based system. According to the procedure the system supports multicriteria analysis made by the parties and generates mediation proposals. The mediation proposals are derived on the basis of the original solution to the multicriteria problem. The solution expresses preferences of the decision makers. It generalizes the classic Raifa solution concept on the multicriteria case.

Keywords: computer-based intelligent systems, decision support, multicriteria analysis, cooperative games, mediation.

1 Introduction

A bargaining process is considered in the case of two parties, represented by decision makers discussing cooperation conditions to realize a joint project. The cooperation is possible if it is beneficial for both of them. It is assumed that each of the decision makers has his individual set of objectives which he would like to achieve. Achievements of the objectives are measured by given vectors of criteria, which are in general different for each decision maker. The criteria are conflicting in the case of the individual decision maker as well as between them. Each decision maker has also his individual preferences defined in the space of his criteria.

The bargaining process will succeed if the final cooperation conditions satisfy desirable benefits of each decision maker, measured by his criteria and valuated according to his individual preferences. Information about possibilities and preferences of each decision maker is confidential. In many situation, at beginning of the bargaining process, decision makers can not be conscious of their preferences if they have not enough information about the attainable payoffs.

The paper deals with computer intelligence problems related to construction of a computer-based system playing the role of a mediator in the bargaining

P. Angelov et al. (eds.), *Intelligent Systems'2014*,
Advances in Intelligent Systems and Computing 322, DOI: 10.1007/978-3-319-11313-5_11

process. The mediator in the negotiations is an impartial outsider who tries to aid the negotiators in their quest to find a compromise agreement. The mediator can help with the negotiation process, but he does not have the authority to dictate a solution.

A question arises: can a specially constructed computer-based system play a role of mediator and support the negotiation process? This question is discussed in the paper in the case of the mentioned decision situation formulated as the multicriteria bargaining problem. The problem is presented in Section 2. The multicriteria bargaining problem is a generalization of the bargaining problem formulated and discussed in the classic game theory by many researchers, including (Nash [9,10], Raiffa [12], Kalai and Smorodinsky [3], Roth [14], Thomson [15], Peters [11], Moulin [8]) and others. In these papers many different solution concepts have been proposed and analyzed. In the classic theory the decision makers are treated as players playing the bargaining game and it is assumed that each of them has explicitly given utility function measuring his payoffs. The solution is looked for in the space of the utilities. In the multicriteria problem considered in this paper, the payoff of each decision maker (player) is measured by a vector of criteria and we do not assume that his utility function is given explicitly. The solution is looked for in the space being the cartesian product of the multicriteria spaces of the players. The solution concepts proposed in the classic theory do not transfer in a simple way to the multicriteria case. Let us see that looking for a solution in the multicriteria bargaining problem we have to consider jointly two decision problems: the first – the solution should be related to the preferences of each of the players, and the second – the solution should fulfill fairness rules accepted by the players.

Section 3 presents formulation of the proposed solution concept to the multicriteria bargaining problem. It is based on the Raiffa concept generalized to the case of multicriteria payoffs of decision makers. It satisfies a set of axioms that can be considered as fair play rules to be accepted by rational decision makers. The solution can be constructed after the multicriteria analysis made by the decision makers. The multicriteria analysis is described in Section 4 with application of the reference point approach. A mediation procedure is presented in Section 5. According to the procedure the computer-based system plays a role of the mediator supporting the decision makers in reaching a consensus.

This paper continues the line of research dealing with multicriteria payoffs of players in bargaining, presented in the papers (Kruś and Bronisz [7], Kruś [6], [5], [4]).

2 Problem Formulation

A bargaining process is considered in the case of two decision makers negotiating conditions of possible cooperation.

Each decision maker i, $i = 1, 2$ has defined decision variables, denoted by a vector $x_i = (x_{i1}, x_{i2}, \ldots x_{ik^i})$, $x_i \in I\!R^{k^i}$, where k^i is a number of decision variables of the decision maker i, and $I\!R^{k^i}$ is a space of his decisions. Decision variables of all the decision makers are denoted by a vector $x = (x_1, x_2) \in I\!R^K$, $K = k^1 + k^2$, where $I\!R^K$ is the cartesian product of the decision spaces of the decision makers 1 and 2.

It is assumed that results of the cooperation are measured by a vector of criteria which are in general different for each decision maker. The criteria of the decision maker i, valuating his payoff are denoted by a vector $y_i = (y_{i1}, y_{i2}, \dots y_{im^i}) \in I\!R^{m^i}$, where m^i is a number of criteria of the decision maker i, and $I\!R^{m^i}$ is a space of his criteria. The criteria of all the decision makers are denoted by $y = (y_1, y_2) \in I\!R^M$, $M = m^1 + m^2$, where $I\!R^M$ is the cartesian product of the citeria spaces of all the decision makers.

We assume that a mathematical model is given describing payoffs of the decision makers. The payoffs are the result of decision variables undertaken. Formally we assume that the model is defined by a set of admissible decisions $X_0 \subset I\!R^K$, and by a mapping $F : I\!R^K \to I\!R^M$ from the decision space to the space of the criteria. A set of attainable payoffs, denoted by $Y_0 = F(X_0)$ is defined in the space of criteria of all decision makers. However each decision maker has access to information in his criteria space only. In the space of criteria of i-th decision maker a set of his attainable payoffs Y_{0i} can be defined. It is an intersection of the set Y_0. The set of attainable payoffs of each decision maker depends on his set of admissible decisions and on the set of admissible decisions of other decision maker.

A partial ordering is introduced in the criteria spaces. Let $I\!R^m$ denote a space of criteria for an arbitrary number m of criteria. Each of m criterions can be maximized or minimized. However, to simplify the notation and without loss of generality we assume that the decision makers maximize all their criteria.

Definition 1. *Let $z, y \in I\!R^m$, we say, that*
a vector z ***weakly dominates*** *y, and denote $z \geq y$, when $z_i \geq y_i$ for $i = 1, 2, \dots, m$,*
a vector z ***dominates*** *y, and denote $z > y$, when $z_i \geq y_i, z \neq y$ for $i = 1, 2 \dots, m$,*
a vector z ***strongly dominates*** *y, and denote $z \gg y$, when $z_i > y_i$ for $i = 1, 2 \dots, m$.*

Definition 2. *A vector $z \in I\!R^m$ is* ***weakly Pareto optimal*** *(weakly nondominated) in a set $Y \subset I\!R^m$ if $z \in Y$ and does not exist $y \in Y$ such, that $y \gg z$. A vector $z \in I\!R^m$ is* ***Pareto optimal*** *(nondominated) in a set $Y \subset I\!R^m$ if $z \in Y$ and does not exist $y \in Y$ such, that $y \geq z$.*

Definition 3. *A bargaining problem with multicriteria payoffs of decision makers (multicriteria bargaining problem) is formulated as a pair (S, d), where $d = (d_1, d_2) \in Y_0 \subset I\!R^M$ is a disagreement point, and S is an agreement set. The agreement set $S \subset Y_0 \subset I\!R^M$ is the subset of the set of the attainable payoffs dominating the disagreement point d. The agreement set defines payoffs attainable by all decision makers but under their unanimous agreement. If such agreement is not achieved, the payoffs of all decision makers are defined by the disagreement point d.*

The multicriteria bargaining problem is analyzed under the following general conditions:

C1 The agreement set S is compact and includes at least one point $y \in S$ such, that $y \gg d$.
C2 The agreement set S is comprehensive, i.e. for $y \in S$ if $z \leq y$ then $z \in S$.
C3 Let $\boldsymbol{M} = \{1, 2, \ldots M\}$. For any $y \in S$, let $P(S, y) = \{k \in \boldsymbol{M} : z \geq y, z_k > y_k$ for some $z \in S\}$. Then for any $y \in S$, there exists $z \in S$ such that $z \geq y, z_k > y_k$ for each $k \in P(S, y)$.

Let $\boldsymbol{B}$ denote class of all multicriteria bargaining problems satisfying the above conditions.

The conditions **C1** and **C2** are typically assumed in formulations of the bargaining problems. Additionally the convexity of the set S is usually assumed. The condition **C3** replaces the convexity assumption. The set S can be nonconvex but the set of Pareto optimal points in S should not contain any "holes". The condition **C3** is satisfied by any convex set.

We assume, that each decision maker i, $i = 1, 2$, defines the vector $d_i \in I\!R^{m^i}$ as his reservation point in his space of criteria. Every decision maker, negotiating possible cooperation, will not agree for payoffs decreasing any component of the vector. A decision maker can assume the reservation point as the status quo point. He can however analyze some alternative options to the negotiated agreement and he can define it on the basis of the BATNA concept presented in (Fisher, Ury [1]). The BATNA (abbreviation of **B**est **A**lternative **t**o **N**egotiated **A**greement) concept is frequently applied in processes of international negotiations in a prenegotiation step. According to the concept, each negotiating party should analyze possible alternatives to the negotiated agreement and select the best one according to its preferences. The best one is called as BATNA. It is the alternative that can be achieved by the party, if a consensus in the negotiations will not be obtained.

Let us see, that when looking for a solution to the multicriteria bargaining problem we have to consider two decision problems: the first one - the solution should be related to the preferences of all decision makers, the second - it should satisfy fairness rules accepted by the decision makers. The first problem relates to multicriteria decision making. The multicriteria decision support is proposed with application of the reference point method developed by A.P. Wierzbicki [16], [18]. In relation to the second problem, a solution concept to the multicriteria bargaining problem is proposed. The solution should have properties that could be accepted by rational decision makers.

Using the reference point approach each decision maker can explore Pareto optimal points in his objective space. Such Pareto optimal points derived for a given decision maker we call as individually nondominated. The individually nondominated points are generally not attainable. The individually nondominated point is an outcome that could be achieved by a decision maker if he would have full control of the moves of the other decision maker. We assume that each decision maker finishing such an exploration will select his preferred Pareto optimal outcome. A composition of the preferred Pareto optimal points of the both decision makers we call as an utopia point relative to their aspirations, and denote as RA utopia. The RA utopia point significantly differs from the ideal (utopia) point. The selected preferred Pareto optimal points of the decision makers as well as the RA utopia carry information about their preferences.

Definition 4. *For any multicriteria bargaining problem (S, d), a point $y^i \in S$ is* ***individually nondominated*** *by decision maker i, $i = 1, 2$, if $y^i \geq d$ and there is no $z \in S$ such that $z \geq d$, $z_i > y^i_i$. A point U^R is an* ***utopia point relative to the aspiration of decision makers*** *(RA utopia point) if for each decision maker $i = 1, 2$, there is an individually nondominated point y^i such that $U^R_i = y^i_i$.*

The set of all RA utopia points of the multicriteria bargaining problem (S, d) will be denoted $U(S, d)$.

Definition 5. *A solution to a multicriteria bargaining problem (S, d) is a function $F : \boldsymbol{B} \times I\!R^M \longrightarrow I\!R^M$ which associates to each problem (S, d) and each RA utopia point $U^R \in U(S, d)$, a point of S, denoted $f(S, d)$.*

3 Solution Concept

Let both decision makers have made multicriteria analysis in their spaces of criteria and have indicated their preferred nondominated payoffs $y^1, y^2 \in S$ defining the RA utopia point U^R. Let $U(S, d)$ be a set of all RA utopia points satisfying $U^R \gg d$. The following theorem can be proved.

Theorem 1. *There is a unique solution $y^R = f^R(S, d, U^R)$ to the multicriteria bargaining problem $(S, d) \in \boldsymbol{B}$ for the relative utopia U^R, satisfying the following axioms:*

(A1) *Weak Pareto-optimality*
$y^R = f^R(S, d, U^R)$ is weakly Pareto-optimal in the set S,

(A2) *Anonymity*
For any permutation on $\mathbf{M}$, π, let π^* denote the permutation on $I\!R^M$. Then $\pi^* f^R(S, d, U^R) = f^R(\pi^* S, \pi^* d, \pi^* U^R)$.

(A3) *Independence of equivalent preference representation (Independence of positive affine transformation of criteria)*
Let L be a positive affine mapping. Then $L f^N(S, d) = f^N(LS, Ld)$.

(A4) *Restricted monotonicity*
If $U^R \in U(S, d) \cap U(T, d)$ and $S \subseteq T$ then $f^R(S, d, U^R) \leq f^R(T, d, U^R)$.

The solution is defined by $f^R(S, d, U^R) = max_{\geq}\{y \in S : y = d + h(U^R - d)$ for some $h \in I\!R\}$.

Thus the solution presented in the theorem is a function which selects the unique outcome. The outcome maximizes improvements of the decision makers (players) payoffs in the direction of the relative utopia point U^R in comparison to the disagreement point d. The solution can be treated as a generalization of the Raiffa solution to the multicriteria bargaining problem. The Axioms 1, 2 and 3 are usually imposed on solutions of the classical bargaining problem, also in the case of the Raiffa solution [12] axiomatized by Kalai and Smorodinsky [3]. In the theory of the bargaining problem the decision makers are treated as players. Axiom A1 expresses a rational behavior of players. Axiom A2 demands that a solution does not depend on the order of players nor on the order of the criteria. Axiom A3 says that a solution does not depend on affine measure of any criterion. The Axioms A4 assures that all the players benefit, or at least not lose, from any enlargement of the agreement set, if the RA utopia does not change. The last axiom replaces the individual monotonicity axiom formulated by Kalai and Smorodinsky [3]. These four axioms can be considered as fair play rules imposed on the solution concepts. Let us see that the the solution is not a simple extension od the Raiffa concept. In the case of the Raiffa solution the cooperative outcome is derived as the maximal in S in the direction defined by the ideal point of S. In the presented concept the cooperative outcome is derived in the direction of the RA utopia point. The RA utopia point is defined by the preferred nondominated points selected independently by players in their criteria spaces. The presented generalized Raiffa solution concept expresses preferences of the players.

4 Multicriteria Analysis

The RA utopia point and the generalized Raiffa cooperative solution are constructed on the basis of the preferred nondominated outcomes in S, indicated by the decision makers. To indicate the preferred outcome consciously, each decision maker should make multicriteria analysis in his space of criteria. The analysis can be made applying the reference point approach developed by Wierzbicki [16], [18] with use of an order approximation functions. According the approach a decision maker assumes reference points in the space of his criteria and the system generates respective outcomes which are Pareto optimal in the set S. For some number of reference points assumed by the decision maker, a representation of the Pareto frontier of the set S can be obtained. Outcomes representing the Pareto frontier in the case of the i-th decision maker, $i = 1, 2$, can be derived by:

$$\max_{\substack{x \in X_0 \\ y_{3-i}=d_{3-i}}} [s(y_i(x), y_i^*)],$$

where: y_i^* is a reference point assumed by the decision maker i in the space $I\!R^{m^i}$, $y^i(x)$ defines the vector of criteria of the i-th decision maker, $i = 1, 2$ which are dependent by the model relations on the vector x of decision variables, $s(y, y*)$ is an order approximating achievement function. The function

$$s(y_i, y_i^*) = \min_{1 \le j \le m^i, i=1,2} [a_j(y_{ij} - y_{ij}^*)] + a_{m^i+1} \sum_{j=1,\ldots,m^i} a_j(y_{ij} - y_{ij}^*),$$

states an example of the achievement function suitable in this case, where $y_i^* \in \mathbb{R}^{m^i}$ is a reference point, a_j, $1 \leq j \leq m^i$, are scaling coefficients, and $a_{m^i+1} > 0$ is a relatively small number. The assumed reference points and the obtained Pareto outcomes are stored in a data base, so that a representation of the Pareto frontier of the set S can be derived and analyzed by the decision maker. The decision maker can select and indicate his preferred nondominated payoff.

5 Interactive Procedure Supporting Mediation Process

5.1 General Idea

The procedure has been proposed under inspiration of the Single Negotiation Text Procedure (SNTP) frequently applied in the international negotiations. The SNTP has been proposed by Roger Fisher during the Camp David negotiations to resolve an impasse which has occurred after several initial rounds of the positional negotiations (see Raiffa [13]). According to the procedure a negotiation process consists of a number of rounds. In each round a mediator prepares a package for the consideration of protagonists. Each package is meant as a single negotiation text to be criticized by protagonists then modified and remodified. Typically the negotiation process starts from the first single negotiation text which is far from the expectations of protagonists. The process is progressive for each of the protagonists.

In the considered case the role of mediator is played by a computer-based system. The procedure is realized in some number of rounds $t = 1, 2, ..., T$. In each round t:

- each decision maker makes independently interactive analysis of nondominated payoffs in his multicriteria space (the analysis is called further as unilateral) and indicates a preferred payoff.
- computer-based system collects the preferred payoffs indicated by the both decision makers, derives RA utopia point and generates on this basis a mediation proposal d^t,
- decision makers analyze the mediation proposal and correct the preferred payoffs, afterwards system derives next mediation proposal.

Each mediation proposals d^t is generated in the direction defined by the RA utopia point derived in the round t. Each decision maker can in each round reduce improvement of the payoffs (his own payoff and at the same time the payoff of the other decision maker). The process d^t is progressive.

5.2 Algorithm

Let $d^t \in S$ denote a vector of payoffs in round t for $t = 1, 2, \ldots$, and $d^0 = d$. Let $S^t = \{y : y \in S, y > d^{t-1}\}$.

Each decision maker i has the following parameters to control the process of multicriteria analysis and derivation of mediation proposal in the round t:

the reference points $y_i^{*t} \in I\!R^{m^i}$, the indicated preferred payoff nondominated in set S, and a parameter $\alpha_i^t \in (\delta, 1]$ called as confidence coefficient, where δ is a relatively small positive number $\delta > 0$.

Step 1. Set the number of round $t = 1$.

Step 2. The phase of unilateral analysis.

The system invites the decision makers $i = 1, 2$ to make independently multicriteria analysis of their nondominated payoffs in the multicriteria bargaining problem (d^{t-1}, S^t).

Step 2.1 The system presents to the decision maker i information about the ideal point I_i^t, and the status quo point d_i^{t-1} in the decision maker criteria space. The ideal point is derived as $I_i^t = (I_{i,1}^t, I_{i,2}^t, \ldots, I_{i,m_i}^t)$, where $I_{i,j}^t = \max y_{i,j} : y = (y_1, y_2) \in S^t \wedge y_{3-i} = d_{3-i}$.

Step 2.2 The decision maker i writes values of the components of his reference point y_{ij}^{*t}, $j = 1, 2, \ldots, m^i$.

Step 2.3 The system derives the nondominated solution in the set S, according to the reference point approach and stores the resulting payoff in a data base.

Step 2.4 The decision maker analyzes the generated nondominated payoff. He compares the payoff to other nondominated payoffs stored in the data base, obtained for other different reference points. If he has enough information to select the preferred payoff, he indicates it as $\widehat{y}_i$ and assumes a value for the confidence coefficient α_i^t. He signals finishing the unilateral analysis phase.

Step 2.5 Has the decision maker i finished unilateral analysis?

If no - go to Step 2.2, to generate next nondominated payoff.

If yes - system writes the preferred nondominated payoff indicated by the decision maker $\widehat{y}_i$ as well as the assumed value of the confidence coefficient α_i^t to the data base.

Step 3.

The system checks whether both decision makers have finished their unilateral analysis i.e. whether they have selected their preferred payoffs and have defined values of the confidence coefficients. If no - system waits as long as they will finish the unilateral analysis in Steps 2.1-2.5.

Step 4.

The system derives the individually nondominated points $y^1 = (\widehat{y}_1, d_2^{t-1})$, $y^2 = (d_1^{t-1}, \widehat{y}_2)$ and the RA utopia point U^{Rt}.

Step 5.

The system derives the mediation proposal $d^t = (d_1^t, d_2^t)$ at the round t,

$$d^t = d^{t-1} + \alpha^t [G^t - d^{t-1}],$$

where $G^t = f^R(S^t, d^{t-1}, U^{Rt}) = max_{\geq}\{y \in S^t : y = d^{t-1} + h(U^{Rt} - d^{t-1})$ for some $h \in I\!R\}$,
$\alpha^t = min\{\alpha_1^t, \alpha_2^t\}$, $0 < \rho < \alpha_i^t \leq 1$ for $i = 1, 2$.

Step 6.
The system presents the mediation proposal i.e. the proposed payoffs d_i^t to the decision makers $i = 1, 2$ respectively.
Step 7.
The system checks whether the mediation proposal of the round is Pareto optimal in the set S
If yes - end the procedure.
If no - set number of the next round $t = t + 1$ and go to Step 2.

In this algorithm a sequence of bargaining problems (S^t, d^{t-1}) is derived and analyzed. The decision makers make in each round independent analysis of non-dominated payoffs in the set S^t using the reference points. Then each of them selects his preferred payoff. This is made in Steps 2.1-2.5. The selected payoffs and confidence coefficients assumed by the decision makers are used by the system to derive a mediation proposal presented to the decision makers in the given round (Steps 4-7). The proposed construction of the mediation proposal assures that the proposal is consistent with the preferences of all decision makers in the given round. The decision makers, using the confidence coefficients, can inflow on the number of following rounds of the procedure. They can again analyze the Pareto optimal frontier of the set S in these following rounds and can correct the previously indicated preferences. The mediation proposal derived by the system according to the ideas of the generalized Raiffa cooperative solution, defines distribution of the cooperation benefits which fulfills axioms A1 - A4 describing fair play rules.

It can be proved that the sequence of the mediation proposals derived in the procedure converges to the Pareto optimal element in the set S.

The construction of the generalized Raiffa solution applied in the algorithm is illustrated in Fig. 1. In this example the decision maker 1 has two criteria $y_{1,1}$ and $y_{1,2}$ respectively, the decision maker 2 has only one criterion $y_{2,1}$. Let us assume that the decision maker 1 has made multicriteria analysis of attainable

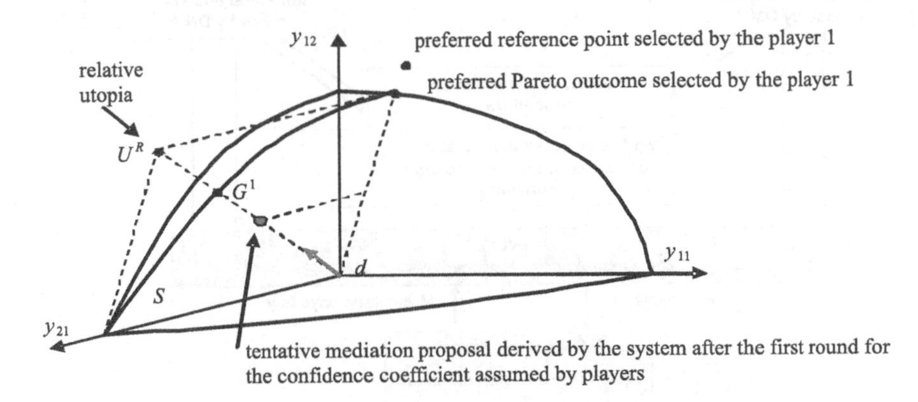

Fig. 1. The generalized Raiffa solution to the multicriteria bargaining problem, and the tentative mediation proposal after the first round of the procedure

Pareto outcomes in his space of criteria and has selected the preferred outcome presented in the figure. The preferred outcome of the decision maker 2 is the maximum value of $y_{2,1}$ in S. The RA utopia point is defined by the preferred outcomes of the decision makers. The solution G^1 maximizes improvements of the decision makers' payoffs in the direction of the relative utopia point U^R in comparison to the disagreement point d. The tentative mediation proposal is an outcome improving d in the direction of G^1, but limited by the assumed confidence coefficient.

6 Structure of the Computer-Based System

The proposed system includes a model representation, modules supporting unilateral analysis made by the decision makers, a module realizing the interactive mediation procedure, a module generating mediation proposals as well as modules including an optimization solver, respective data bases, and a graphical interface. A general structure of the system is presented in Fig. 2. The model describing the considered decision situation of the decision makers is the base for the decision analysis and support. The model is constructed with use of gathered information according to the rules of the system sciences. It includes the specification of decision variables, exogenous variables, output quantities, criteria, model relations. Values of the criteria of the decision makers are derived with use of the model for given assumed values of the decision variables. The criteria depend also on the exogenous variables representing quantities describing

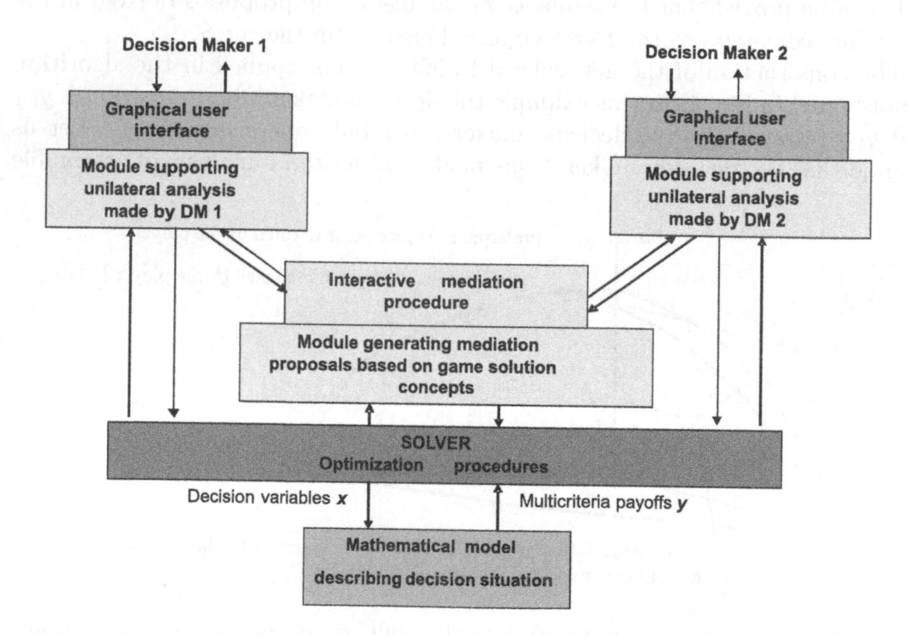

Fig. 2. The general structure of the computer-based system

external conditions, not dependent on the decision makers. Such variables are typically evaluated by experts in the forms of scenarios. It is assumed that the model parameters are properly identified and whole the model is verified and validated.

The module supporting unilateral analysis enables each decision maker to obtain independently information about possible multicriteria payoffs and look for the preferred option.

The system generates mediation proposals. The proposals are derived with use of the generalized Raiffa solution concept. They are generated and proposed to the decision makers according to the mediation procedure presented in Section 6.

The solver module includes respective optimization procedures. The procedures are utilized in the system in the modules supporting unilateral multicriteria analysis made by the decision makers and in the module generating the mediation proposals.

7 Conclusions

A mathematical background is presented for construction of a computer-based system, which can support decision makers in reaching a consensus in the case of the multicriteria bargaining problem. The problem describes decision situation of two decision makers negotiating conditions of possible cooperation. Each of them valuates effects of the cooperation by his own different set of criteria. The problem is defined by the disagreement point and the agreement set formulated in the space being the cartesian product of the criteria spaces of the decision makers. An interactive mediation procedure is proposed which can be implemented in the computer-based system. According to the procedure the decision makers look independently for the preferred variants of cooperation using reference point method. The preferred variants indicated by them are used to generate mediation proposals. The proposals are generated according to the preferences expressed by the decision makers. It can be shown that the sequence of the mediation proposals is convergent to a Pareto optimal outcome in the agreement set. The mediation proposals are derived on the basis of the proposed solution concept generalizing the Raiffa solution concept on the multicriteria case. The concept satisfies respective fair play rules to the multicriteria bargaining problem, that could be accepted by rational decision makers. The considered computer-based system supports multicriteria analysis made by the decision makers and plays a role of mediator generating mediation proposals.

References

1. Fisher, R., Ury, W.: Getting to Yes. Hougton Mifflin, Boston (1981)
2. Imai, H.: Individual Monotonicity and Lexicographical Maxmin Solution. Econometrica 51, 389–401 (1983)

3. Kalai, E., Smorodinsky, M.: Other Solutions to Nash's Bargaining Problem. Econometrica 43, 513–518 (1975)
4. Kruś, L.: Multicriteria Cooperative Decisions, Methods of Computer-based Support (in Polish: Wielokryterialne decyzje kooperacyjne, metody wspomagania komputerowego). Seria: Badania systemowe. Tom 70. Instytut Badań Systemowych PAN, Warszawa (2011)
5. Kruś, L.: Multicriteria Decision Support in Bargaining, a Problem of Players's Manipulations. In: Trzaskalik, T., Michnik, J. (eds.) Multiple Objective and Goal Programming Recent Developments, pp. 143–160. Physica Verlag, Heidelberg (2001)
6. Kruś, L.: Multicriteria Decision Support in Negotiations. Control and Cybernetics 25(6), 1245–1260 (1996)
7. Kruś, L., Bronisz, P.: Some New Results in Interactive Approach to Multicriteria Bargaining. In: Wierzbicki, A.P., et al. (eds.) User Oriented Methodology and Techniques of Decision Analysis. Lecture Notes in Economics and Mathematical Systems, vol. 397, pp. 21–34. Springer, Berlin (1993)
8. Moulin, H.: Axioms of Cooperative Decision Making. Cambridge University Press, Cambridge (1988)
9. Nash, J.F.: The Bargaining Problem. Econometrica 18, 155–162 (1950)
10. Nash, J.F.: Two-Person Cooperative Games. Econometrica 21, 129–140 (1953)
11. Peters, H.: Bargaining Game Theory. Ph.D. Thesis, Catholic University of Nijmegen, The Nederlands (1986)
12. Raiffa, H.: Arbitration Schemes for Generalized Two-Person Games. In: Annals of Mathematics Studies, Princeton, vol. 28, pp. 361–387 (1953)
13. Raiffa, H.: The Art and Science of Negotiations. Harvard Univ. Press, Cambridge (1982)
14. Roth, A.E.: Axiomatic Model of Bargaining. Lecture Notes in Economics and Mathematical Systems, vol. 170. Springer, Berlin (1979)
15. Thomson, W.: Two Characterization of the Raiffa Solution. Economic Letters 6, 225–231 (1980)
16. Wierzbicki, A.P.: On the Completeness and Constructiveness of Parametric Characterizations to Vector Optimization Problems. OR Spectrum 8, 73–87 (1986)
17. Wierzbicki, A.P., Kruś, L., Makowski, M.: The Role of Multi-Objective Optimization in Negotiation and Mediation Support. Theory and Decision 34, 201–214 (1993)
18. Wierzbicki, A.P., Makowski, M., Wessels, J.: Model-based Decision Support Methodology with Environmental Applications. Kluwer Academic Press, Dordrecht (2000)

Approach to Solve a Criteria Problem of the ABC Algorithm Used to the WBDP Multicriteria Optimization

Dawid Ewald[1], Jacek M. Czerniak[1], and Hubert Zarzycki[2]

[1] Casimir the Great University in Bydgoszcz,
Institute of Technology, ul. Chodkiewicza 30, 85-064 Bydgoszcz, Poland
{dawidewald,jczerniak}@ukw.edu.pl
[2] Wroclaw School of Applied Informatics "Horyzont",
ul. Wejherowska 28, 54-239 Wroclaw, Poland
hzarzycki@horyzont.eu

Abstract. This article describes the use of the bees algorithm for optimization of the multi-object structure of a welded beam. This problem has been used as a benchmark of the bees algorithm. This article presents an approach to improve the ABC algorithm criteria used for the WBDP multicriteria optimization. The contents of the article show the analysis of the problem, conducted based on the mathematical calculations of the beam dimensions. The further part constitutes a comparison of the received results with the results of the ABC operation. Based on the standards, an additional criterion was established, which verifies the correctness of the results.

Keywords: Bee Optimization, ABC, Fuzzy Bee, WBDP.

1 Introduction

During a global crisis the reduction of production costs is the primary goal of every manufacturer. Unfortunately, it is often very difficult to explicitly identify the aspects in which costs can be limited. Therefore, more and more often constructors are looking for software capable of structure optimization [1] [2]. Classic algorithms are very time-consuming and their results are not very satisfying. Currently, the field of artificial intelligence, dealing with swarm algorithms is developing very dynamically. This branch offers a number of optimization solutions which have been borrowed from nature. Man has noticed a long time ago that in nature numerous insects worked out very effective solutions for seeking optimum solutions. Ants, for example, optimize the distance they have to cover from the anthill to food [3]. An artificial bee colony or a so-called bee algorithm is also a solution observed in the nature and then implemented by man. It is an algorithm offering a multi-object optimization [4]. Thus, thanks to the simple optimization, it allows easy adjustment of its possibilities to numerous tasks.

P. Angelov et al. (eds.), *Intelligent Systems'2014*,
Advances in Intelligent Systems and Computing 322, DOI: 10.1007/978-3-319-11313-5_12

Combining the knowledge from the scope of IT with engineering and construction, we can modify the bee algorithm in such a manner that it can be applied in practice [5]. This article will discuss the issues of a multi-object structure optimization [6] [7], the so-called unilaterally fixed beam deflection [8], as well as fillet weld geometry. The bees algorithm prepared by its author [9] [10] and the version which was modified in order to enable practical application of the solution will be used for optimization.

2 Optimization Problem of a One-Side Fixed Beam Structure

The discussed case contains two elements. The first considered element is beam geometry [11], and in particular the dimensions of the cross-section which influences the deflection [12]. The second element of the discussed system is the weld geometry which depends upon the beam geometry. This system is presented in the Figure 1.

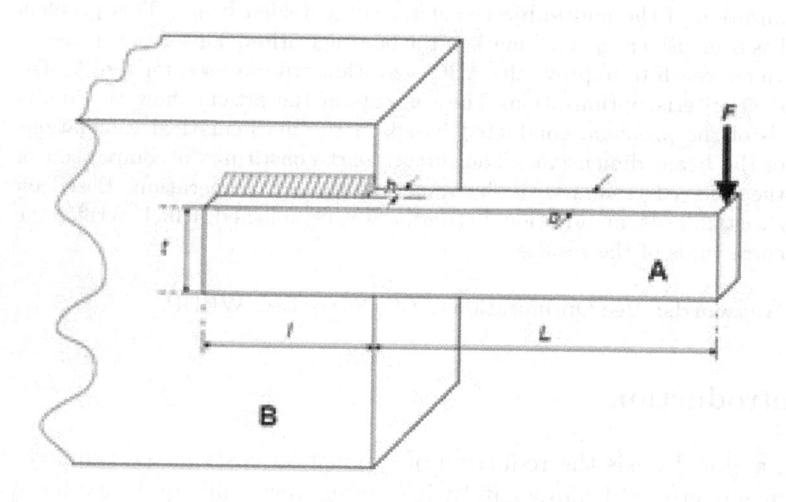

Fig. 1. Beam welding method

The entire structure will be subject to optimization with the use of two formulated functions [13]

$$Minf_1 = (1 + C_1)h^2l + C_2tb(L + l), \tag{1}$$

$$Minf_2 = \delta, \tag{2}$$

where

f_1- beam material and weld cost function,

f_2- beam deflection.

In order to compare the results, such parameters as beam length, weld material and the material the beam was made of, as well as the force, will remain the same as in the primary application:

C_1- unit cost of the weld $=6.3898 * 10^{-6}\$/mm^3$, $(0.10471\$/in3)$,

C_2- unit cost of the beam material $=6.9359 * 10^{-6}\$/mm^3$, $(0.04811\$/in3)$,

L- constant distance of the load from the fixation place = 356 mm 14 in,

h- width of the fillet weld,

l- weld length,

t- beam height,

b-beam width.

As we can see, the parameters h, l, t and b are variables determining the cost, as well as the deflection.

Beam deflection calculation analytic method. Deflection determination: δ_{dop}-207 MPa,.

L - 356 mm - 0.356 m,

P - 26689 N - 26.689 kN

b - beam width,

t - beam height.

b and t parameters are calculated based on the formula [14]:

$$W = \frac{bt^2}{6} \tag{3}$$

It should be remembered that in order to maintain an appropriate proportion of t to b, in order not to receive the result of the calculations in the form of an object which would resemble a board. Usually, for simplicity, it is assumed that t = 2b, therefore:

$$W = \frac{b(2b)^2}{6} = \frac{(2b)^3}{3} \tag{4}$$

$$W = \frac{M_{max}}{\delta_{dop}} \tag{5}$$

Where:

$M_{max} = P * L = 0.356 * 26.689 = 9,501kN * m$

Based on formula 5:

$$W = \frac{9.501}{207} * 10^3 = 45.899cm^3$$

Converting formula 4:

$$W = \frac{(2b)^3}{3} \Longrightarrow b = \sqrt[3]{\frac{3}{2}W} \tag{6}$$

Placing the values in formula 6, we receive:

$$b = \sqrt[3]{\frac{3}{2} * 45.898} = \sqrt[3]{68.847} = 4.098cm$$

After the calculations:
b - 4.098 cm,
t - 8.197 cm.

In order to calculate the deflection, the moment of inertia relative to the neutral axis is calculated, formula 7:

$$I = \frac{(bt)^3}{12} \tag{7}$$

Where:
b - beam width,
t - beam height.

Placing the values in formula 7, we receive:

$$I = \frac{4.098 * (8.197)^3}{12} = \frac{4.098 * 550.763}{12} = 188.085$$

The deflection is expressed in the formula:

$$f = \frac{P(l)^3}{3E * I} \tag{8}$$

Where:
P - applied force,
l - constant distance of the load from the fixation place,
E - Young module.

$$f = \frac{26.689 * (35.6)^3}{3 * 20500 * 188.085} = 0.104cm$$

The received result is the beam deflection with dimensions meeting the strength requirements. As it can be seen, there are vast numbers of b and t combinations.

A similar procedure must be applied to calculate the weld geometry. In order to do so, the admissible stress in welds must be calculated. This is necessary due to the decreased strength of the material in the area of the weld, which results from the local thermal stresses or weld incontinence. The value of admissible stresses is calculated from formula 9:

$$k' = k * x' \tag{9}$$

where:
k' - admissible weld stresses,
k - admissible welded material stresses,
x' - weld static strength coefficient specified based on table [12].

In the case in question, this coefficient amounts to 0.65, whereas k amounts to 93.8 MPa [15]. Having placed the values in the formula, we receive the following result: k'=60.97. The weld length is calculated based on formula 10, whereas it should be remembered that the minimum weld thickness amounts to 2.5 mm and maximum - 16 mm. It should also be noted that if the side of the weld is greater than 8 mm, the weld is made as a multi-run weld [16]. Therefore, it has been assumed that the side of a weld amounts to 8 mm:

$$L_S = \frac{P}{k' * (0.7 * h)} \tag{10}$$

Where:
P - force,
k' - admissible weld stresses,
h - height of the weld,
having placed the values in the formula, we get:

$$L_S = \frac{26689N}{60.97 * 8} = \frac{26689N}{487.76} = 54.71mm$$

As we can see, the length and width of the weld depends on the width of the beam.

3 Bees Algorithm

The bees algorithm was inspired by the intelligent behaviour of a swarm of bees, and later defined by Dervis Karaboga in 2005 [9] [14]. Due to the fact that it is not very complicated and in order to work it uses only controlling parameters, such as the maximum cycle number and colony size, it enables a simple implementation. ABC is an optimizing tool [17] [18], which action is based on a population where a single bee searches for places with food.

In a classic ABC model [19], a colony is composed of three bee groups, in the first group there are worker bees, in the second one - onlooker bees and in the third one - scout bees [13] [14]. It follows from the above that the number of worker bees is equal to the amount of available food. When a worker bee returns, the onlooker bees assess the amount of food it represents. The bee, whose food amount is lower than the greatest obtained becomes a scout bee and starts to search for a new source. In the ABC algorithm, the location of the food source constitutes a possible solution to the problem, whereas the amount of this food constitutes the quality of this solution [12]. The number of solutions in a given population is equal to the number of worker bees.

On the first stage of work of the algorithm, a random bee population is drawn. After this stage, other ones take place, which repeat in cycles. Worker bees check food sources in the given location and its neighbourhood, and then return to the beehive where the source quality is assessed. If the amount of food in a new place is higher than the current one, the bee remembers the new location. The main elements of the algorithm are presented below:

Drawing initial sources for all worker bees;
REPEAT
Each worker bee checks the remembered location and its neighbourhood and returns to the hive;

Each onlooker bee assesses and chooses one bee working based on the amount of food, after the selection it goes to the source;
The checked food sources are examined and replaced with new ones;
The best source is remembered;
UNTIL (meeting the requirements)

4 ABC – Application of the Algorithm to the Optimisation of the Structure

The solution to this problem based on the bees algorithm, which is available in literature, does not take into consideration the cross-section geometry of the beam. Such an action may lead to incorrect solutions, as a result of which we can get an object which is not a beam. Figure 2 presents an incorrect beam geometry generated during the operation of the program.

In order to avoid this, an additional parameter which guarantees correct beam geometry should be applied.

Introduction of some restrictions which verify the results obtained during the calculations guarantees that the final effect of algorithm performance will meet the expectations. The authors of this solution have prepared seven restrictions which, in their opinion, guarantee a correct result [16].

The first restriction is the fact that the maximum stresses used in the programme are smaller than the admissible shearing stresses of the weld. The second restriction guarantees that the maximum normal stress used during the work of the programme are smaller than the admissible normal stresses of the weld. The third restriction guarantees that the width of the weld is less than the thickness of the beam. The fourth and the fifth conditions guarantee that the width and length will not be negative values. Another restriction verifies whether the beam load is smaller than the admissible buckling load. The last restriction is to check whether the thickness of the weld is greater than the specified minimum.

5 Suggested Modification

As it can be seen on figure 2, the introduced restrictions do not guarantee a correct geometry. Supplementing the above restrictions additionally guarantees that

$$\frac{b}{t} \leq 15\sqrt{\frac{215}{fd}} \tag{11}$$

This restriction follows from a general assumption that the proportion of the beam height to its width must meet the criteria of local stability [20], which was presented in figure 3.

6 Summary

The solution of Karaboga et.al [9] [18] [13] [14] [21] presented in this paper concerning bee optimization, for optimization of the weld and beam structure

Fig. 2. Result of simulation of the received beam geometry performed in the SolidWorks programme

contains gross errors repeated in subsequent articles. The authors do not solve this issue in an appropriate manner, due to omission of the behaviour of the correct geometrical dimensions of the beam in the result. As a result, the received object is not a beam but a board, which can be seen on figure 2. As it follows from analytical calculations and numerical experiments, introduction of the improvement suggested in this paper considerably improves the results.

Table 1. Comparison of the performance of the algorithms

[cm]	ABC with 7 conditions	ABC with 8 conditions	Analytically calculated
h	0.522	1.872	0.8
l	8.815	4.724	5.471
t	22.953	8.196	8.197
b	0.522	4.097	4.098

Table 1 compares the results of ABCs performance before and after the introduction of the suggested improvement. Comparing the experimental results with the results of calculations, it can be explicitly seen that the introduced modification causes ABCs results to be similar to the calculations.

Introduction of an additional restriction described in the article guarantees maintenance of correct beam dimensions and allows applying the received results in practice. Introduction of an appropriate modification does not alter the manner of action of the algorithm itself, but only guarantees the selection of mechanically and actually appropriate solutions. This is not the only modification that can be introduced in order to make the obtained results more useful

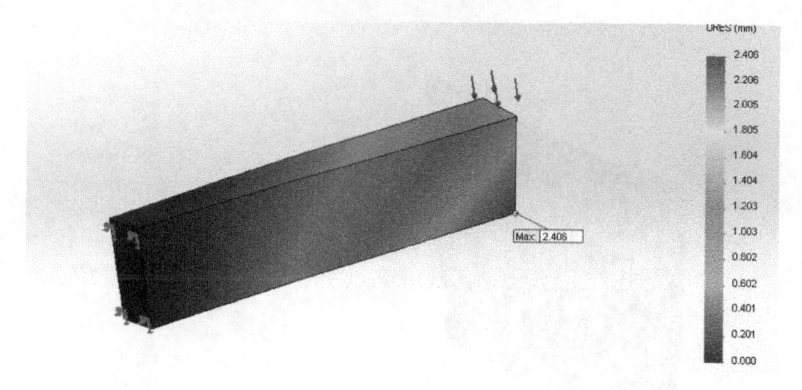

Fig. 3. Result of simulation of the received beam geometry after introduction of modifications, performed in the SolidWorks programme

for constructors. However, precise discussion of these issues goes far beyond the framework of this publication. What should be noted is the diligence in application of examples testing the optimization method, the so-called benchmarks. Even if the discussed problem of a beam with a rectangular cross section is a merely theoretical example, both diligence and rules applied in machine dynamics structure must be observed, which has been proved in this paper.

Acknowledgments. The authors would like to express their thanks to technical staff of AIRlab - Artificial Intelligence and Robotics Laboratory at Casimir the Great University in Bydgoszcz for their commitment and help during research and tests performed within this study.

References

1. Macko, M.: Economic-energetic analysis of multi edge comminution of polymer recyclates
2. Macko, M.: Metoda doboru rozdrabniaczy wielokrawędziowych do przeróbki materiałów polimerowych
3. Czerniak, J.M., Apiecionek, Ł., Zarzycki, H.: Application of ordered fuzzy numbers in a new OFNAnt algorithm based on ant colony optimization. In: Kozielski, S., Mrozek, D., Kasprowski, P., Małysiak-Mrozek, B., Kostrzewa, D. (eds.) BDAS 2014. CCIS, vol. 424, pp. 259–270. Springer, Heidelberg (2014)
4. Czerniak, J.: Evolutionary approach to data discretization for rough sets theory. Fundamenta Informaticae 92(1), 43–61 (2009)
5. Popielarski, W.: Bee algorithms in optimization of task scheduling for flow shop model. Studies and Materials in Applied Computer Science 2(2), 47–54 (2010)
6. Flizikowski, J., Macko, M., Czerniak, J., Mrozinski, A.: Implementation of genetic algorithms into development of mechatronic multi-edgeís grinder design. In: ASME 2011 International Mechanical Engineering Congress and Exposition, Denver, Colorado, USA, November 11-17. Dynamic Systems and Control; Mechatronics and Intelligent Machines, vol. 7, pp. 1227–1235. ASME (2011)

7. Faluyi, F., Arum, C.: Design optimization of plate girder using generalized reduced gradient and constrained artificial bee colony algorithms. International Journal of Emerging Technology and Advanced Engineering 2(7), 304–312 (2012)
8. Deb, K.: An efficient constraint-handling method for genetic algorithms. Computer Methods in Applied Mechanics and Engineering 186(0045-7825), 311–338 (2000)
9. Karaboga, D.: An idea based on honey bee swarm for numerical optimization. Engineering Faculty, Computer Engineering Department, Erciyes University, Turkey (2005)
10. Basturk, B., Karaboga, D.: An artificial bee colony (abc) algorithm for numeric function optimization. In: IEEE Swarm Intelligence Symposium, Indianapolis, Indiana, USA (2006)
11. Kumar, P., Pant, M., Singh, V.: Differential evolution with interpolation based mutation operators for engineering design optimization. Advances in Mechanical Engineering and its Applications 2(3), 221–231 (2012)
12. Anonymous: Values of the coefficients of the weld static strength, `http://www.pkm.edu.pl/index.php/polocenia-obl/spawane-obl/351-01050201`
13. Karaboga, D., Basturk, B.: On the performance of artificial bee colony (abc) algorithm. Applied Soft Computing 8(1), 687–697 (2008)
14. Karaboga, D., Basturk, B.: Artificial bee colony (abc) optimization algorithm for solving constrained optimization problems. In: Melin, P., Castillo, O., Aguilar, L.T., Kacprzyk, J., Pedrycz, W. (eds.) IFSA 2007. LNCS (LNAI), vol. 4529, pp. 789–798. Springer, Heidelberg (2007)
15. Anonymous: Admissible stresses in standard structural steels, `http://www.pkm.edu.pl/index.php/07/stale/64-07010202`
16. Pham, D., Ghanbarzadeh, A.: Multi-objective optimisation using the bees algorithm. In: 3rd International Virtual Conference on Intelligent Production Machines and Systems (IPROMS 2007), Whittles, Dunbeath, Scotland, vol. 242, pp. 111–116 (2007)
17. Tuba, M., Bacanin, N., Stanarevic, N.: Adjusted artificial bee colony (abc) algorithm for engineering problems. WSEAS Transactions on Computers 11(4), 111–120 (2012)
18. Karaboga, D., Basturk, B.: A powerful and efficient algorithm for numerical a powerful and efficient algorithm for numerical function optimization: artificial bee colony (abc) algorithm. Journal of Global Optimization 39, 459–471 (2007)
19. Tuba, M., Bacanin, N., Stanarevic, N.: Guided artificial bee colony algorithm. In: Proceedings of the 5th European Conference on European Computing Conference, ECC 2011, pp. 398–403 (2011)
20. PN-90/B03200 Steel structures. Static calculation and designing
21. Rekliatis, G., Ravindrab, A., Ragsdell, K.: Engineering Optimisation Methods and Applications. Wiley, New York (1983)
22. Czerniak, J., Angryk, R.: Heuristic algorithm for interpretation of multi-valued attributes in similarity-based fuzzy relational databases. International Journal of Approximate Reasoning 51(8), 1499–1502 (2010)
23. PN-EN 1990 Structure designing basics

Part IV
Issues in Intuitionistic Fuzzy Sets

Representation Theorem of General States on IF-sets

Jaroslav Považan

Faculty of Natural Sciences, Matej Bel University,
Tajovského 40, Banská Bystrica, Slovakia
jaroslav.povazan@umb.sk

Abstract. L. Ciungu and B. Riečan proved in [2] that any real state on IF-sets can be represented by integrals in sense that

$$m((\mu_A, \nu_A)) = \int \mu_A \mathrm{d}P + \alpha \left(1 - \int (\mu_A + \nu_A) \, \mathrm{d}Q\right).$$

However the formulation is unappropriate for general case with values from arbitrary Riesz space. This article shows that only small change in formulation make it appropriate for the general case.

Because Riesz spaces have similar structure it is natural question: Does this equality hold in general Riesz space? The answer is probably no because in general case it can be possible that there exist elements u, v of Riesz space such that $0 \leq u \leq v$, but there is no $\alpha \in \mathbb{R}$ such that $u = \alpha v$. However this inconvenience wanishes if we rewrite this by the folowing way:
For any real state on IF-set there exist measures P, Q such that

$$m((\mu_A, \nu_A)) = \int \mu_A \mathrm{d}P + \int (1 - \mu_A - \nu_A) \, \mathrm{d}Q.$$

This formulation is appropriate for general case.

Keywords: IF-sets, general state, Riesz space, general measure, Lukasiewicz operations, representation theorem, general integral.

1 Introduction

IF-sets are in some sense a generalization of fuzzy sets, which is a generalization of sets. Because of Kolmogorov definition of probability we can approach to probability by using strict analytically methods. Generalization of probability on sets are states on IF-sets. L. Ciungu and B. Riečan in [2] showed that one can use integrals only to describe or define state with real values. In this article we are going to generalize the notion of state in sense that it's values cannot be real numbers only, but range of states can be interval in arbitrary Riesz space. To realize this generalization we are going to use general integrals for which the range is interval in arbitrary Riesz space. Our results and prooves are very similar as real case described by L. Ciungu and B. Riečan.

P. Angelov et al. (eds.), *Intelligent Systems'2014*,
Advances in Intelligent Systems and Computing 322, DOI: 10.1007/978-3-319-11313-5_13

2 Representation Theorem in General Case

Definition 1 (Riesz space). *Ordered quadrupplet $(V, +, \cdot, \leq)$ is called Riesz space iff*

1. *$(V, +, \cdot)$ is vector (linear) space over real numbers.*
2. *$(V, \leq)$ is a lattice.*
3. *$(\forall a, b, c \in V)\,(a \leq b \Rightarrow a + c \leq b + c)$.*
4. *$(\forall a, b \in V)\left(\forall \alpha \in \mathbb{R}_0^+\right)(a \leq b \Rightarrow \alpha a \leq \alpha b)$.*

Definition 2 (Monotone limits). *Let $\{v_n\}_{n=1}^{\infty}$ be a sequence of elements of a Riesz space. We tell that v_n is monotonical converging to v iff it is satisfied one of the folowing conditions :*

1. $v_1 \leq v_2 \leq v_3 \leq \cdots \leq v_n \leq v_{n+1} \leq \cdots \;\&\; v = \bigvee_{n=1}^{\infty} v_n.$
2. $v_1 \geq v_2 \geq v_3 \geq \cdots \geq v_n \geq v_{n+1} \geq \cdots \;\&\; v = \bigwedge_{n=1}^{\infty} v_n.$

In the case 1 we write $v_n \nearrow v$ and in the case 2 we write $v_n \searrow v$.

Definition 3 (IF-subset). *By the notion of IF-subset of a set X we understand an ordered pair (f, g), such that*

$$f, g : X \to [0, 1] \;\&\; f + g \leq 1.$$

Definition 4 (Monotone convergence of IF-sets). *Let $\{(f_n, g_n)\}_{n=1}^{\infty}$ be a sequence of IF-sets. We tell that the sequence $\{(f_n, g_n)\}_{n=1}^{\infty}$ monotonically converges to IF-set (f, g) iff there is satisfied one of the folowing conditions :*

1. $f_n \nearrow f \;\&\; g_n \searrow g$.
2. $f_n \searrow f \;\&\; g_n \nearrow g$.

In the case 1 we write $(f_n, g_n) \nearrow (f, g)$ and in the case 2 we write $(f_n, g_n) \searrow (f, g)$.

Definition 5 (Lukasiewicz operations on IF-sets). *We define the folowing operations on IF-sets:*

- $(f, g) \oplus (h, k) = (\min(f + h, 1), \max(g + k - 1, 0))$.
- $(f, g) \odot (h, k) = (\max(f + h - 1, 0), \min(g + k, 1))$.

The operation $\oplus$ is called the Lukasiewicz sum, the operation $\odot$ is called the Lukasiewicz product.

Definition 6 (State). *A mapping $m : \mathcal{F} \to I$, where $I = [0, u]$ is an interval in a Riesz space, and $\mathcal{F}$ is a colection of IF-subsets of set Ω which is closed under Lukasiewicz operations and monotone limits, is called a state iff*

1. $m((0, 1)) = 0, m((1, 0)) = u$.

2. $(\forall (f,g), (h,k) \in \mathcal{F}) ((f,g) \odot (h,k) = (0,1) \Rightarrow$
$\Rightarrow m((f,g) \oplus (h,k)) = m((f,g)) + m((h,k)))$.
3. $(\forall \{(f_n, g_n)\}_{n=1}^{\infty}) (\forall (f,g) \in \mathcal{F})$
$((f_n, g_n) \nearrow (f,g) \Rightarrow m((f_n, g_n)) \nearrow m((f,g)))$.

Definition 7 (General measure). *Let V be a Riesz space , $\mathcal{S}$ be a $\sigma-$algebra of subsets of a set Ω. Then a set-mapping $\mu : \mathcal{S} \to \{v \in V; v \geq 0\}$ is called measure iff*
$\mu(A) = \bigvee_{n=1}^{\infty} \sum_{i=1}^{n} \mu(A_i)$, *whenever*
$A = \bigcup_{i=1}^{\infty} A_i \, (A_i \in \mathcal{S} \, (i = 1, 2, 3, \cdots), A_i \cap A_j = \emptyset \, (i \neq j))$.

Definition 8 (Nonnegative integrable function). *Let $f : \Omega \to \mathbb{R}_0^+$. Then f is called integrable iff $(\exists \alpha_i \in \mathbb{R}_0^+) (\exists A_i \in \mathcal{S})$, such that*

$$(\exists v \in V) \left(v = \bigvee_{n=1}^{\infty} \sum_{i=1}^{n} \alpha_i \mu(A_i) \right)$$

$$f(x) = \sum_{i=1}^{\infty} \alpha_i \chi_{A_i}(x)$$

Definition 9 (Integral of nonnegative integrable function). *We define the integral of an integrable function $f : \Omega \to \mathbb{R}_0^+$ by the equality*

$$\int_{\Omega} f \mathrm{d}\mu = \bigvee_{n=1}^{\infty} \sum_{i=1}^{n} \alpha_i \mu(A_i).$$

Definition 10 (Integrable function). *We tell that a function $h : \Omega \to \mathbb{R}$ is integrable iff there exist integrable $f, g : \Omega \to \mathbb{R}_0^+$ such that $h = f - g$.*

Definition 11 (Integral of integrable function). *We define the integral of an integrable function $h : \Omega \to \mathbb{R}$ by the equality*

$$\int_{\Omega} h \mathrm{d}\mu = \int_{\Omega} f \mathrm{d}\mu - \int_{\Omega} g \mathrm{d}\mu.$$

Theorem 1. *Let V be a Riesz space, $u \in V; 0 < u$ fixed positive element. Then for any state $m : \mathcal{F} \to [0, u]$ where $\mathcal{F}$ is set of IF sets (ordered pair of fuzzy sets f, g for which $f + g \leq 1$) closed under Lukasiewicz sum and product and monotone limits, there exist general measures $P, Q : \mathcal{S} \to [0, u]$ for which*

1.
$$P(\Omega) = u$$

2.
$$Q(\Omega) = m((0,0))$$

3.

$$m((\mu_A,\nu_A)) = \int\limits_{\Omega} (\mu_A)\,\mathrm{d}P + \int\limits_{\Omega} (1-\mu_A-\nu_A)\,\mathrm{d}Q.$$

Proof. Since $(f,g)=(f,1-f)\oplus(0,f+g)$ and $(f,1-f)\odot(0,f+g)=(0,1)$, there is

$$m(f,g)=m(f,1-f)+m(0,f+g).$$

Put $P(A)=m(\chi_A,1-\chi_A)$. Let $A,B\in\mathcal{S}; A\cap B=\emptyset$. Then

$$\begin{aligned} P(A)+P(B) &= m((\chi_A,1-\chi_A))+m((\chi_B,1-\chi_B)) = \\ &= m((\chi_A,1-\chi_A)\oplus(\chi_B,1-\chi_B)) = \\ &= m((\chi_A+\chi_B,1-\chi_A-\chi_B)) = \\ &= m((\chi_{A\cup B},1-\chi_{A\cup B})) = P(A\cup B) \end{aligned}$$

Let $A_n\in\mathcal{S}, A_n\nearrow A$. Then $\chi_{A_n}\nearrow\chi_A, 1-\chi_{A_n}\searrow 1-\chi_A$. Hence

$$P(A_n)=m((\chi_{A_n},1-\chi_{A_n}))\nearrow m((\chi_A,1-\chi_A))=P(A)$$

where for functions with values from Riesz space

$$f_n\nearrow f \Leftrightarrow f_1\le f_2\le\cdots\le f_n\le\cdots \text{ and } f=\bigvee_{i=1}^{\infty} f_i$$

Moreover

$$\begin{aligned} P(\Omega) &= m((\chi_\Omega,1-\chi_\Omega)) = m((1,0)) = u \\ P(\emptyset) &= m((\chi_\emptyset,1-\chi_\emptyset)) = m((0,1)) = 0 \end{aligned}$$

Now we prove by induction that for pairwise disjoint sets A_i and real numbers $0\le\alpha_i\le 1$ there holds equality

$$m\left(\left(\sum_{i=1}^{n}\alpha_i\chi_{A_i}1-\sum_{i=1}^{n}\alpha_i\chi_{A_i}\right)\right)=\sum_{i=1}^{n} m((\alpha_i\chi_{A_i},1-\alpha_i\chi_{A_i})).$$

For $n=1$ it holds trivially, for $n=2$ we have

$$\begin{aligned} & m\left((\alpha_1\chi_{A_1}+\alpha_2\chi_{A_2},1-\alpha_1\chi_{A_1}-\alpha_2\chi_{A_2})\right) = \\ &= m((\alpha_1\chi_{A_i},1-\alpha_1\chi_{A_1})\oplus(\alpha_2\chi_{A_2},1-\alpha_2\chi_{A_2})) = \\ & m((\alpha_1\chi_{A_1},1-\alpha_1\chi_{A_1}))+m((\alpha_2\chi_{A_2},1-\alpha_2\chi_{A_2})) \end{aligned}$$

Asume that equality holds for some $n \geq 2$. Then

$$m\left(\left(\sum_{i=1}^{n+1}\alpha_i\chi_{A_i}, 1-\sum_{i=1}^{n+1}\alpha_i\chi_{A_i}\right)\right) =$$

$$= m\left(\left(\sum_{i=1}^{n}\alpha_i\chi_{A_i} + \alpha_{n+1}\chi_{A_{n+1}}, 1-\sum_{i=1}^{n}\alpha_i\chi_{A_i} - \alpha_{n+1}\chi_{A_{n+1}}\right)\right) =$$

$$= m\left(\left(\sum_{i=1}^{n}\alpha_i\chi_{A_i}, 1-\sum_{i=1}^{n}\alpha_i\chi_{A_i}\right) \oplus \left(\alpha_{n+1}\chi_{A_{n+1}}, 1-\alpha_{n+1}\chi_{A_{n+1}}\right)\right) =$$

$$= m\left(\left(\sum_{i=1}^{n}\alpha_i\chi_{A_i}, 1-\sum_{i=1}^{n}\alpha_1\chi_{A_1}\right)\right) + m((\alpha_{n+1}\chi_{A_{n+1}}, 1-\alpha_{n+1}\chi_{A_{n+1}})) =$$

$$= \sum_{i=1}^{n} m((\alpha_i\chi_{A_i}, 1-\alpha_i\chi_{A_i})) + m((\alpha_{n+1}\chi_{n+1}, 1-\alpha_{n+1}\chi_{n+1})) =$$

$$= \sum_{i=1}^{n+1} m((\alpha_i\chi_i, 1-\alpha_i\chi_i))$$

Now we prove that for every $n \in \mathbb{N}$ there holds equality

$$nm\left(\left(\frac{\beta}{n}\chi_A, 1-\frac{\beta}{n}\chi_A\right)\right) = m((\beta\chi_A, 1-\beta\chi_A)).$$

for every $\beta \leq 1$

$$(n+1)\, m\left(\left(\frac{\beta}{n+1}\chi_A, 1-\frac{\beta}{n+1}\chi_A\right)\right) =$$

$$= nm\left(\left(\frac{n\beta}{n\,(n+1)}\chi_A, 1-\frac{n\beta}{n\,(n+1)}\chi_A\right)\right) +$$

$$+ m\left(\left(\frac{\beta}{n+1}\chi_A, 1-\frac{\beta}{n+1}\chi_A\right)\right) =$$

$$= m\left(\left(\frac{n}{n+1}\beta\chi_A, 1-\frac{n}{n+1}\beta\chi_A\right)\right) +$$

$$+ m\left(\left(\frac{\beta}{n+1}\chi_A, 1-\frac{\beta}{n+1}\chi_A\right)\right) =$$

$$= m((\beta\chi_A, 1-\beta\chi_A))$$

Similarly we prove that

$$m((n\beta\chi_A, 1-n\beta\chi_A)) = nm((\beta\chi_A, 1-\beta\chi_A))$$

whenever $n\beta \le 1$. For $n = 1$ it holds trivially. Asume that $(n+1)\beta \le 1$. Then $n\beta = (n+1)\beta - \beta \le 1 - \beta \le 1$ and

$$\begin{aligned}(n+1)\, m((\beta\chi_A, 1-\beta\chi_A)) &= \\ = nm((\beta\chi_A, 1-\beta\chi_A)) + m((\beta\chi_A, 1-\beta\chi_A)) &= \\ = m((n\beta\chi_A, 1-n\beta\chi_A)) + m((\beta\chi_A, 1-\beta\chi_A)) &= \\ = m(((n+1)\,\beta\chi_A, 1-(n+1)\,\beta\chi_A))&\end{aligned}$$

Hence the equality $m((\alpha\beta\chi_A, 1-\alpha\beta\chi_A)) = \alpha m((\beta\chi_A, 1-\beta\chi_A))$ holds for rational $0 \le \alpha \le 1$ and arbitrary $0 \le \beta \le 1$, especially for $\beta = 1$ For real $0 \le \alpha \le 1$ there exist rational $0 \le \alpha_i \le 1$ such that $\alpha_i \nearrow \alpha$. Then $\alpha_i\chi_A \nearrow \alpha\chi_A, 1-\alpha_i\chi_A \searrow 1-\alpha\chi_A$. Hence $m((\alpha\beta\chi_A, 1-\chi_A)) = \alpha m((\chi_A, 1-\beta\chi_A))$ for arbitrary real $0 \le \alpha \le 1$.

Let $f : \Omega \to [0,1]\,, f = \sum_{i=1}^{n} \alpha_i\chi_{A_i}, A_i \in \mathcal{S}, A_i \cap A_j = \emptyset\,(i \ne j)$. Then

$$\begin{aligned}m((f, 1-f)) &= \sum_{i=1}^{n} m((\alpha_i\chi_{A_i}, 1-\alpha_i\chi_{A_i})) = \\ &= \sum_{i=1}^{n} \alpha_i m((\chi_{A_1}, 1-\chi_{A_i})) = \\ &= \sum_{i=1}^{n} \alpha_i P(A_i) = \int_{\Omega} f \mathrm{d}P\end{aligned}$$

If $f_n \nearrow f, f_n$ simple, then $(f_n, 1-f_n) \nearrow (f, 1-f)$. Hence $m((f_n, 1-f_n)) \nearrow m((f, 1-f))$. Therefore $m((f, 1-f)) = \int_{\Omega} f \mathrm{d}P$ for measurable $f : \Omega \to [0,1]$.

Put

$$Q(A) = m((0, 1-\chi_A)).$$

Similarly as before it can be proved that $\sum_{i=1}^{n} m((0, 1-\alpha_i\chi_{A_i})) = m\left(\left(0, 1-\sum_{i=1}^{n} \alpha_i\chi_{A_i}\right)\right)$ and $m((0, 1-\alpha\chi_A)) = \alpha m((0, 1-\chi_A))$.

Let $f : \Omega \to [0,1]\,, f = \sum_{i=1}^{n} \alpha_i\chi_{A_i}, A_i \in \mathcal{S}, A_i \cap A_j = \emptyset\,(i \ne j)$. Then

$$\int_{\Omega} f \mathrm{d}Q = \sum_{i=1}^{n} \alpha_i Q(A_i) =$$

$$= \sum_{i=1}^{n} \alpha_i m((0, 1-\chi_{A_i})) =$$

$$= \sum_{i=1}^{n} m((0, 1 - \alpha_i \chi_{A_i})) = m\left(\left(0, 1 - \sum_{i=1}^{n} \alpha_i \chi_{A_i}\right)\right) =$$

$$= m((0, 1 - f))$$

Hence

$$m((0, f)) = \int_{\Omega} (1 - f)\, \mathrm{d}Q.$$

Therefore we get

$$m((\mu_A, \nu_A)) = m((\mu_A, 1 - \mu_A)) + m((0, \mu_A + \nu_A)) =$$

$$= \int_{\Omega} \mu_A \mathrm{d}P + \int_{\Omega} (1 - (\mu_A + \nu_A))\, \mathrm{d}Q = \int_{\Omega} \mu_A \mathrm{d}P + \int_{\Omega} (1 - \mu_A - \nu_A)\, \mathrm{d}Q.$$

3 Conclusions

Because of similar structure, real case and general case have similar rules and prooves. The main difference is in fact that for positive real numbers a, b exists real number α such that $a = \alpha \cdot b$. In general Riesz space this property need not to hold as we can check by taking Riesz space of functions with traditional operations. For this reason, the rule in real case must be simplified for general case. After this simplification we get general result that says that every general state has rule

$$m((\mu_A, \nu_A)) = \int_{\Omega} (\mu_A)\, \mathrm{d}P + \int_{\Omega} (1 - \mu_A - \nu_A)\, \mathrm{d}Q$$

for some general measures P, Q. Interesting question is: What about another operations instead of Lukasiewicz operations between IF-sets. This problem is (If I don't mistake) still open.

Acknowledgement. The support of the grant VEGA 1/0120/14 is kindly announced.

References

1. Riečan, B.: Analysis of Fuzzy Logic Models. In: Intelligent Systems. Intech, Croatia (2012)
2. Ciungu, L., Riečan, B.: General form of probabilities on IF-sets. In: Fuzzy Logic and Applications. Proc. WILF Palermo, pp. 101–107 (2009)
3. Riečan, B.: On a problem of Radko Mesiar: general form of a IF-probablities. Fuzzy Sets an Systems 152, 1485–1490 (2006)
4. Riečan, B., Neubrunn, T.: Integral, Measure and Ordering. Kluwer Academic Publishers, Dordrecht (1997)

On Finitely Additive IF-States

Beloslav Riečan

Faculty of Natural Sciences, Matej Bel University,
Tajovského 40, Banská Bystrica, Slovakia
and
Mathematical Institut, Slovak Academy of Sciences,
Štefánikova 49, Bratislava, Slovakia
Beloslav.Riecan@umb.sk

Abstract. It is well known that the set F of IF-sets can be embedded to an MV-algebra M. In the contribution to any finitely additive state m on F there is constructed a finitely additive state $\bar{m}$ on M which is an extension of m.

1 Introduction

It is well known that MV-algebras play a remarkable role in the multivalued logic as well as in the probability theory (see [9], [13]). On the other hand fuzzy sets opened some new possibilities for mathematical research with applications ([16], [17]). The present paper is concerned with their special case - intuitionistic fuzzy case (IF-sets, [1], [2]). Recall some results concerning probability on IF-sets (e.g. [5], [6], [8], [11], [15]). A review of some probability results is contained in [12]. It concerns with two objects. First by a representation of IF-states (= probability measures) by the help of classical real-valued probability measures ([3], [4], [10]). Secondly by embedding of IF-spaces to MV-algebras. The present paper contains a result in this direction.
In *Section 2* some basic facts about IF-sets and MV-algebras are contained. *Section 3* contains, using the paper [14], a new embedding proof for the embedding IF-states to MV-algebra states.

2 Finitely Additive IF-States

Any subset A of a given space Ω can be identified with its characteristic function

$$I_A : \Omega \to \{0, 1\}$$

where

$$I_A(\omega) = 1,$$

if $\omega \in A$,

$$I_A(\omega) = 0,$$

P. Angelov et al. (eds.), *Intelligent Systems'2014*,
Advances in Intelligent Systems and Computing 322, DOI: 10.1007/978-3-319-11313-5_14

if $\omega \notin A$. From the mathematical point of view a fuzzy set is a natural generalization of I_A (see [14], [15]). It is a function

$$\varphi_A : \Omega \to [0,1] \ .$$

Evidently any set (i.e. two-valued function on $I_A : \Omega \to \{0,1\}$) is a special case of a fuzzy set (multi-valued function), $\varphi_A : \Omega \to [0,1]$. There are many possibilities for characterizations of operations with sets (union $A \cup B$ and intersection $A \cap B$). We shall use so called Lukasiewicz characterization:

$$I_{A \cup B} = (I_A + I_B) \wedge 1,$$

$$I_{A \cap B} = (I_A + I_B - 1) \vee 0 \ .$$

(Here $(f \vee g)(\omega) = \max(f(\omega), g(\omega)), (f \wedge g)(\omega) = \min(f(\omega), g(\omega))$.) Hence if $\varphi_A, \varphi_B : \Omega \to [0,1]$ are fuzzy sets, then the union (disjunction φ_A or φ_B of corresponding assertions) can be defined by the formula

$$\varphi_A \oplus \varphi_B = (\varphi_A + \varphi_B) \wedge 1,$$

the intersection (conjunction φ_A and φ_B of corresponding assertions) can be defined by the formula

$$\varphi_A \odot \varphi_B = (\varphi_A + \varphi_B - 1) \vee 0 \ .$$

In the paper we shall work with the Atanassov generalization of the notion of fuzzy set so-called IF-set (see [1], [2]), what is a pair

$$A = (\mu_A, \nu_A) : \Omega \to [0,1] \times [0,1]$$

of fuzzy sets $\mu_A, \nu_A : \Omega \to [0,1]$, where

$$\mu_A + \mu_A \le 1 \ .$$

Evidently a fuzzy set $\varphi_A : \Omega \to [0,1]$ can be considered as an IF-set, where

$$\mu_A = \varphi_A : \Omega \to [0,1], \nu_A = 1 - \varphi_A : \Omega \to [0,1] \ .$$

Here we have

$$\mu_A + \nu_A = 1,$$

while generally it can be $\mu_A(\omega) + \nu_A(\omega) < 1$ for some $\omega \in \Omega$. Geometrically an IF-set can be regarded as a function $A : \Omega \to \Delta$ to the triangle

$$\Delta = \{(u,v) \in R^2 : 0 \le u, 0 \le v, u + v \le 1\} \ .$$

Fuzzy set can be considered as a mapping $\varphi_A : \Omega \to D$ to the segment

$$D = \{(u,v) \in R^2; u + v = 1, 0 \le u \le 1\}$$

and the classical set as a mapping $\psi : \Omega \to D_0$ from Ω to two-point set

$$D_0 = \{(0,1), (1,0)\} \ .$$

In the next definition we again use the Lukasiewicz operations.

Definition 1. *By an IF subset of a set Ω a pair $A = (\mu_A, \nu_A)$ of functions*

$$\mu_A : \Omega \to [0,1], \nu_A; \Omega \to [0,1]$$

is considered such that

$$\mu_A + \nu_A \leq 1 \ .$$

We call μ_A the membership function, ν_A the non membership function and

$$A \leq B \Longleftrightarrow \mu_A \leq \mu_B, \nu_A \geq \nu_B \ .$$

If $A = (\mu_A, \nu_A), B = (\mu_B, \nu_B)$ are two IF-sets, then we define

$$A \oplus B = ((\mu_A + \mu_B) \wedge 1, (\nu_A + \nu_B - 1) \vee 0),$$

$$A \odot B = ((\mu_A + \mu_B - 1) \vee 0, (\nu_A + \nu_B) \wedge 1),$$

$$\neg A = (1 - \mu_A, 1 - \nu_A) \ .$$

Denote by $\mathcal{F}$ a family of IF sets such that

$$A, B \in \mathcal{F} \Longrightarrow A \oplus B \in \mathcal{F}, A \odot B \in \mathcal{F}, \neg A \in \mathcal{F} \ .$$

Example 1. Let $\mathcal{F}$ be the set of all fuzzy subsets of a set Ω. If $f : \Omega \to [0,1]$ then we define

$$A = (f, 1 - f),$$

i.e. $\nu_A = 1 - \mu_A$.

Example 2. Let $(\Omega, \mathcal{S})$ be a measurable space, $\mathcal{S}$ a σ-algebra, $\mathcal{F}$ the family of all pairs such that $\mu_A : \Omega \to [0,1], \nu_A : \Omega \to [0,1]$ are measurable. Then $\mathcal{F}$ is closed under the operations $\oplus, \odot, \neg$.

Definition 2. *A mapping $m : \mathcal{F} \to [0,1]$ is called a finitely additive IF-state, if the following properties are satisfied:*

(1.1) $m((0,1)) = 0, \quad m((1,0)) = 1,$
(1.2) $A \odot B = (0,1) \Longrightarrow m(A \oplus B) = m(A) + m(B)$.

3 MV-Algebras

A prototype of an MV-algebra is the unit interval [0,1] with two binary operations

$$a \oplus b = (a + b) \wedge 1,$$

$$a \odot b = (a + b - 1) \vee 0,$$

and one unary operation

$$\neg a = 1 - a,$$

and the usual ordering. The operation $\oplus$ corresponds to the disjunction of statements (union of sets), $\odot$ corresponds to the conjunction of statements (product of sets), $a \to a'$ to the negation of a statement (complement of a set), $a \leq b$ to the implication of statements (inclusion of sets). Generally we shall use the Mundici characterization of MV-algebras [9]. It starts with the notion of an l-group. An l-group is an algebraic structure $(G, +, \leq)$, where $(G, +)$ is a commutative group, $(G, \leq)$ is a lattice, and the implication $a \leq b \Longrightarrow a + c \leq b + c$ holds. An MV-algebra is an algebraic structure

$$(M, 0, u, \neg, \oplus, \odot),$$

where 0 is the neutral element in G, u is a positive element, $M = \{x \in G; 0 \leq x \leq u\}, \neg : M \to M$ is the unary operation given by the equality

$$\neg x = u - x,$$

and $\oplus, \odot$ are two binary operations given by

$$a \oplus b = (a + b) \wedge u,$$

$$a \odot b = (a + b - u) \vee 0 \ .$$

Example 3. Consider $(R^2, +, \leq)$, where $(x_1, y_1) + (x_2, y_2) = (x_1 + x_2, y_1 + y_2 - 1)$ and $(x_1, y_1) \leq (x_2, y_2) \Longleftrightarrow x_1 \leq x_2, y_1 \geq y_2$. Then R^2 is an l-group. Put $u = (1, 0)$. Then

$$\mathcal{M} = \{(x, y) \in R^2; (0, 1) \leq (x, y) \leq u = (1, 0)\}$$

is an MV-algebra.

Definition 3. *Let $(\mathcal{M}, 0, u, \neg, \oplus, \odot)$ be an MV-algebra. Let $m : \mathcal{M} \to [0, 1]$ be a monotone mapinng (i.e. $a \leq b \Longrightarrow m(a) \leq m(b)$). We shall say that m is a finitely additive state, if the following properties are satisfied*

(2.1) $m(0) = 0, \quad m(u) = 1,$
(2.2) $a \odot b = 0 \Longrightarrow m(a \oplus b) = m(a) + m(b)$.

4 Extension Theorem

In the following text we shall consider a measurable space $(\Omega, \mathcal{S})$ with a σ-algebra $\mathcal{S}$. By $\mathcal{F}$ the family will be denoted of all pairs $A = (\mu_A, \nu_A)$ with $\mu_A : \Omega \to [0, 1], \nu_A : \Omega \to [0, 1]$ measurable and $\mu_A + \nu_A \leq 1$. By $\mathcal{M}$ the family fo all mappings $A = (\mu_A, \nu_A) : \Omega \to [0, 1]^2$ will be considered with $\mathcal{S}$-measurable mapings $\mu_A, \nu_A : \mathcal{S} \to [0, 1]$.
The aim of the section is the construction of an additive state $\bar{m} : \mathcal{M} \to [0, 1]$ being an extension of a given IF-additive state $m : \mathcal{F} \to [0, 1]$.

Proposition 1. *Let $A \in \mathcal{M}, A = (\mu_A, \nu_A)$. Define $A^0, A^1, A^2 : \Omega \to [0,1]^2$ by the following formulas*

$$A^0(\omega) = (\mu_A^0(\omega), \nu_A^0(\omega)) = (\mu_A(\omega), \nu_A(\omega)),$$

if $\mu_A(\omega) + \nu_A(\omega) \leq 1$,

$$A^0(\omega) = (0, 1),$$

if $\mu_A(\omega) + \nu_A(\omega) > 1$;

$$A^1(\omega) = (\mu_A^1(\omega), \nu_A^1(\omega)) = (0, 1)$$

if $\mu_A(\omega) + \nu_A(\omega) \leq 1$,

$$A^1(\omega) = (\mu_A(\omega), 1 - \mu_A(\omega)),$$

if $\mu_A(\omega) + \nu_A(\omega) > 1$;

$$A^2(\omega) = (\mu_A^2(\omega), \nu_A^2(\omega)) = (0, 1),$$

if $\mu_A(\omega) + \nu_A(\omega) \leq 1$,

$$A^2(\omega) = (0, 2 - \mu_A(\omega) - \nu_A(\omega)),$$

if $\mu_A(\omega) + \nu_A(\omega) > 1$.

Then $A^0, A^1, A^2 \in \mathcal{F}$.

Proof. Let $\Delta = \{(u, v) \in R^2; 0 \leq u, 0 \leq v, u + v \leq 1\}$.
If $\mu_A(\omega) + \nu_A(\omega) \leq 1$, then

$$\mu_A^0(\omega) + \nu_A^0(\omega) \leq 1,$$

$$\mu_A^1(\omega) + \nu_A^1(\omega) = 0 + 1 = 1,$$

$$\mu_A^2(\omega) + \nu_A^2(\omega) = 0 + 1 = 1\ .$$

If $\mu_A(\omega) + \nu_A(\omega) > 1$, then

$$\mu_A^0(\omega) + \nu_A^0(\omega) = 0 + 1 = 1,$$

$$\mu_A^1(\omega) + \nu_A^1(\omega) = \mu_A(\omega) + 1 - \nu_A(\omega) = 1,$$

$$\mu_A^2(\omega) + \nu_A^2(\omega) = 0 + 2 - \mu_A(\omega) - \nu_A(\omega) \leq 1\ .$$

We have seen that $\mu_A^0(\omega) + \nu_A^0(\omega) \leq 1, \mu_A^1(\omega) + \nu_A^1(\omega) \leq 1, \mu_A^2(\omega) + \nu_A^2(\omega) \leq 1$ for any $\omega \in \Omega$, hence $A^0, A^1, A^2 \in \mathcal{F}$. □

Definition 4. *Let $m : \mathcal{F} \to [0,1]$ be an IF-state, $A \in \mathcal{M}$. Define $\bar{m} : \mathcal{M} \to [0,1]$ by the formula*

$$\bar{m}(A) = m(A^0) + m(A^1) - m(A^2)\ .$$

Proposition 2. *$\bar{m} : \mathcal{M} \to [0,1]$ is an extension of $m : \mathcal{F} \to [0,1]$.*

Proof. If $A = (\mu_A, \nu_A) \in \mathcal{F}$, then $A^0 = (\mu_A, \nu_A), A^1 = (0,1), A^2 = (0,1)$, hence

$$\bar{m}(A) = m(A^0) + m((0,1)) - m((0,1)) = m(A) + 0 - 0 = m(A) \ .$$

□

Theorem 1. *Let $m : \mathcal{F} \to [0,1]$ be an additive IF-state. Then there exists exactly one MV-algebra finitely additive state $\bar{m} : \mathcal{M} \to [0,1]$ that is an extension of m.*

Proof. The uniqueness is evident.
Let $k : \mathcal{M} \to [0,1]$ be an MV-algebra additive state being an extension of $m, C \in \mathcal{M}$. Then

$$k(C) = k(C^0) + k(C^1) - k(C^2) = \bar{m}(C)$$

for any $C \in \mathcal{M}$.
Let $A, B \in \mathcal{M}, A \odot B = (0,1)$. We must prove the equality

$$\bar{m}(A \oplus B) = \bar{m}(A) + \bar{m}(B) \ .$$

Since $A \odot B = (0,1)$, we have

$$((\mu_A + \mu_B - 1) \vee 0, (\nu_A + \nu_B) \wedge 1) = (0,1),$$

hence

$$\mu_A + \mu_B \leq 1, \ \nu_A + \nu_B \geq 1 \ .$$

Therefore we have

$$A \oplus B = ((\mu_A + \mu_B) \wedge 1, (\nu_A + \nu_B - 1) \vee 0) = (\mu_A + \mu_B, \nu_A + \nu_B - 1) \ .$$

Since $A(\omega) \in \Delta$ for some ω, $A(\omega) \notin \Delta$ for other ω, we shall consider six possibilities:

$$C_1 = \{\omega \in \Omega; \mu_A(\omega) + \nu_A(\omega) > 1, \mu_B(\omega) + \nu_B(\omega) > 1\},$$

$$C_2 = \{\omega \in \Omega; \mu_A(\omega) + \nu_A(\omega) > 1, \mu_B(\omega) + \nu_B(\omega) \leq 1,$$
$$\mu_A(\omega) + \mu_B(\omega) + \nu_A(\omega) + \nu_B(\omega) - 1 > 1\},$$

$$C_3 = \{\omega \in \Omega; \mu_A(\omega) + \nu_A(\omega) \leq 1, \mu_B(\omega) + \nu_B(\omega) > 1,$$
$$\mu_A(\omega) + \mu_B(\omega) + \nu_A(\omega) + \nu_B(\omega) - 1 > 1\},$$

$$C_4 = \{\omega \in \Omega; \mu_A(\omega) + \nu_A(\omega) > 1, \mu_B(\omega) + \nu_B(\omega) \leq 1,$$
$$\mu_A(\omega) + \mu_B(\omega) + \nu_A(\omega) + \nu_B(\omega) - 1 \leq 1\},$$

$$C_5 = \{\omega \in \Omega; \mu_A(\omega) + \nu_A(\omega) \leq 1, \mu_B(\omega) + \nu_B(\omega) > 1,$$
$$\mu_A(\omega) + \mu_B(\omega) + \nu_A(\omega) + \nu_B(\omega) - 1 \leq 1\},$$

$$C_6 = \{\omega \in \Omega; \mu_A(\omega) + \nu_A(\omega) \le 1, \mu_B(\omega) + \nu_B(\omega) \le 1\} .$$

For any $D \in \mathcal{M}$ define

$$D_i = (\mu_D I_{C_i}, 1 - (1 - \nu_D) I_{C_i}) .$$

If $i \neq j$ then

$$D_i \odot D_j = ((\mu_D I_{C_i} + \mu_D I_{C_j} - 1) \vee 0, (1 - (1 - \nu_D) I_{C_i} + 1 - (1 - \nu_D) I_{C_j}) \wedge 1) = (0, 1),$$

$$D_i \oplus D_j = ((\mu_D I_{C_i} + \mu_D I_{C_j}) \wedge 1, (1 - (1 - \nu_D) I_{C_i} + 1 - (1 - \nu_D) I_{C_j} - 1) \vee 0) =$$
$$= (\mu_D I_{C_i \cup C_j}, 1 - (1 - \nu_D) I_{C_i \cup C_j}) .$$

Therefore

$$\bar{m}(D_i \oplus D_j) = \bar{m}(D_i) + \bar{m}(D_j),$$

and

$$\bar{m}(D) = \sum_{i=1}^{6} \bar{m}(D_i) .$$

Consider now e.g. $i = 1$. Put $\bar{\mu}_{D_1} = \mu_D I_{C_1}, \bar{\nu}_{D_1} = 1 - (1 - \nu_D) I_{C_1}$. Then

$$\bar{m}(A_1) = m(\bar{\mu}_{A_1}, 1 - \bar{\mu}_{A_1}) - m(0, 2 - \bar{\mu}_{A_1} - \bar{\nu}_{A_1}),$$

$$\bar{m}(B_1) = m(\bar{\mu}_{B_1}, 1 - \bar{\mu}_{B_1}) - m(0, 2 - \bar{\mu}_{B_1} - \bar{\nu}_{B_1}),$$

$$\bar{m}((A \oplus B)_1) = m(\bar{\mu}_{A_1} + \bar{\mu}_{B_1}, 1 - \bar{\mu}_{A_1} - \bar{\mu}_{B_1}) - m(0, 2 - \bar{\mu}_{A_1} - \bar{\mu}_{B_1} - (\bar{\nu}_{A_1} + \bar{\nu}_{B_1} - 1)),$$

$$\bar{m}(A_1) + \bar{m}(B_1) =$$
$$= \bar{m}(\bar{\mu}_{A_1} + \bar{\mu}_{B_1}, 1 - \bar{\mu}_{A_1} + 1 - \bar{\mu}_{B_1} - 1) - \bar{m}(0, 2 - \bar{\mu}_{A_1} - \bar{\nu}_{A_1} + 2 - \bar{\mu}_{B_1} - \bar{\nu}_{B_1} - 1) =$$
$$= \bar{m}(\bar{\mu}_{A_1} + \bar{\mu}_{B_1}, 1 - \bar{\mu}_{A_1} - \bar{\mu}_{B_1}) - \bar{m}(0, 3 - \bar{\mu}_{A_1} - \bar{\mu}_{B_1} - \bar{\nu}_{A_1} - \bar{\nu}_{B_1}) =$$
$$= \bar{m}((A \oplus B)_1) .$$

We have proved that

$$\bar{m}(A_1) + \bar{m}(B_1) = \bar{m}((A \oplus B)_1) .$$

Similarly it can be proved that

$$\bar{m}(A_i) + \bar{m}(B_i) = \bar{m}((A \oplus B)_i) .$$

for $i = 2, 3, 4, 5, 6$. Therefore

$$\bar{m}(A \oplus B) = \sum_{i=1}^{6} \bar{m}((A \oplus B)_i) = \sum_{i=1}^{6} \bar{m}(A_i) + \sum_{i=1}^{6} \bar{m}(B_i) =$$
$$= \bar{m}(A) + \bar{m}(B) .$$

□

5 Conclusion

A new proof has been presented concerning the embedding of IF-states to MV-algebra states. The result presents a new point of view to the problematic using the notion of finitely additive IF-state and its extension.

Acknowledgement. The support of the grant VEGA 1/0120/14 is kindly announced.

References

1. Atanassov, K.T.: Intuitionistic Fuzzy Sets: Theory and Applications. STUDFUZZ, vol. 35. Physica Verlag, Heidelberg (1999)
2. Atanassov, K.T.: On Intuitionistic Fuzzy Sets. Springer, Berlin (2012)
3. Ciungu, L., Riečan, B.: General form of probabilities on IF-sets. In: Fuzzy Logic and Applications. Proc. WILF Palermo, pp. 101–107 (2009)
4. Ciungu, L., Riečan, B.: Representation theorem for probabilities on IFS-events. Information Sciences 180, 793–798 (2010)
5. Grzegorzewski, P., Mrowka, E.: Probability of intuistionistic fuzzy events. In: Grzegorzewski, P., et al. (eds.) Soft Methods in Probability, Statistics and Data Analysis, pp. 105–115 (2002)
6. Lendelová, K.: Convergence of IF-observables. In: Issues in the Representation and Processing of Uncertain and Imprecise Information - Fuzzy Sets, Intuitionistic Fuzzy Sets, Generalized nets, and Related Topics, pp. 232–240. EXIT, Warsaw (2005)
7. Michalíková, A.: Absolute value and limit of the function defined on IF sets. Notes on Intuitionistic Fuzzy Sets 18, 8–15 (2012)
8. Michalíková, A.: The probability on square. Journal of Electrical Engeeniering 56(12/S), 21–22 (2005)
9. Mundici, D.: Interpretation of AFC* -algebras in Lukasiewicz sequential calculus. J. Funct. Anal. 65, 15–63 (1986)
10. Riečan, B.: On a problem of Radko Mesiar: general form of IF-probabilities. Fuzzy Sets and Systems 152, 1485–1490 (2006)
11. Riečan, B.: Probability theory on IF events. In: Aguzzoli, S., Ciabattoni, A., Gerla, B., Manara, C., Marra, V. (eds.) ManyVal 2006. LNCS (LNAI), vol. 4460, pp. 290–308. Springer, Heidelberg (2007)
12. Riečan, B.: Analysis of Fuzzy Logic Models. In: Koleshko, V.M. (ed.) Intelligent Systems, pp. 219–244. INTECH (2012)
13. Riečan, B., Mundici, D.: Probability in MV-algebras. In: Pap, E. (ed.) Handbook of Measure Theory II, pp. 869–910. Elsevier, Heidelberg (2002)
14. Skřivánek, V.: States on IF events. In: Klement, E.P., et al. (eds.) Twelfth Int. Conf. on Fuzzy Set Theory FSTA, Abstracts, Liptovský Ján, vol. 102 (2014)
15. Valenčáková, V.: A note on the conditional probability on IF-events. Math. Slovaca 59, 251–260 (2009)
16. Zadeh, L.A.: Fuzzy sets. Information and Control 8, 338–358 (1965)
17. Zadeh, L.A.: Probability measures on fuzzy sets. J. Math. Abal. Appl. 23, 421–427 (1968)

Embedding of IF-States to MV-Algebras

Beloslav Riečan

Faculty of Natural Sciences, Matej Bel University,
Tajovského 40, Banská Bystrica, Slovakia
and
Mathematical Institut, Slovak Academy of Sciences,
Štefánikova 49, Bratislava, Slovakia
Beloslav.Riecan@umb.sk

Abstract. In [13] any finitely additive IF-state has been embedded to some MV-algebra. Using this result we embedde any IF-state to an MV-σ-algebra.

1 Introduction

One of the most important results of the set theory in the second part of the 20th century there was the discovery of fuzzy sets ([13]). And it is interesting that one of the first direction in the theory there was the probability theory ([14]).The present paper is devoted to the intuitionistic fuzzy set theory ([1], [2]). Recall that there are some results devoted to the probability on IF-sets ([3], [4], [5], [6], [7], [10], [11]). One of the main result is the embedding of IF-sets to MV-algebras (for a review of probability on MV-algebras see [8], [9]). As in the classical theory also in IF-sets theory the probability is defined as an additive and continuous mapping (so-called state). In the present paper only the additive mappings are studied without any assumptions of continuity.
In *Section 2* basic information about finitely additive IF-states are presented. *Section 3* is devoted to the embedding theory of IF-sets into MV-algebras. *Section 4* contains the main result: the extension theorem of a finitely additive IF-state to a finitely additive state on the corresponding MV-algebra.

2 IF-sets and MV-algebras

Recall that an IF subset of a set Ω is a pair $A = (\mu_A, \nu_A)$ of functions

$$\mu_A : \Omega \to [0,1], \nu_A; \Omega \to [0,1]$$

such that

$$\mu_A + \nu_A \leq 1 \ .$$

We call μ_A the membership function, ν_A the non membership function and

$$A \leq B \Longleftrightarrow \mu_A \leq \mu_B, \nu_A \geq \nu_B \ .$$

P. Angelov et al. (eds.), *Intelligent Systems'2014*,
Advances in Intelligent Systems and Computing 322, DOI: 10.1007/978-3-319-11313-5_15

If $A = (\mu_A, \nu_A), B = (\mu_B, \nu_B)$ are two IF-sets, then we define

$$A \oplus B = ((\mu_A + \mu_B) \wedge 1, (\nu_A + \nu_B - 1) \vee 0),$$
$$A \odot B = ((\mu_A + \mu_B - 1) \vee 0, (\nu_A + \nu_B) \wedge 1),$$
$$\neg A = (1 - \mu_A, 1 - \nu_A) .$$

Denote by $\mathcal{F}$ a family of IF sets such that

$$A, B \in \mathcal{F} \Longrightarrow A \oplus B \in \mathcal{F}, A \odot B \in \mathcal{F}, \neg A \in \mathcal{F} .$$

Example 1. Let $\mathcal{F}$ be the set of all fuzzy subsets of a set Ω. If $f : \Omega \to [0,1]$ then we define

$$A = (f, 1 - f),$$

i.e. $\nu_A = 1 - \mu_A$.

Example 2. Let $(\Omega, \mathcal{S})$ be a measurable space, $\mathcal{S}$ a σ-algebra, $\mathcal{F}$ the family of all pairs such that $\mu_A : \Omega \to [0,1], \nu_A : \Omega \to [0,1]$ are measurable. Then $\mathcal{F}$ is closed under the operations $\oplus, \odot, \neg$.

Now have a look for MV-algebras. A prototype of an MV-algebra is the unit interval [0,1] with two binary operations

$$a \oplus b = (a + b) \wedge 1,$$

$$a \odot b = (a + b - 1) \vee 0,$$

one unary operation

$$\neg a = 1 - a,$$

and the usual ordering. The operation $\oplus$ corresponds to the disjunction of statements (union of sets), $\odot$ corresponds to the conjunction of statements (product of sets), $a \to a'$ to the negation of a statement (complement of a set), $a \leq b$ to the implication of statements (inclusion of sets).
Generally we shall use the Mundici characterization of MV-algebras [9]. It starts with the notion of an l-group. An l-group is an algebraic structure $(G, +, \leq)$, where $(G, +)$ is a commutative group, $(G, \leq)$ is a lattice, and the implication $a \leq b \Longrightarrow a + c \leq b + c$ holds. An MV-algebra is an algebraic structure

$$(M, 0, u, \neg, \oplus, \odot),$$

where 0 is the neutral element in G, u is a positive element, $M = \{x \in G; 0 \leq x \leq u\}, \neg : M \to M$ is the unary operation given by the equality

$$\neg x = u - x,$$

and $\oplus, \odot$ are two binary operations given by

$$a \oplus b = (a + b) \wedge u,$$

$$a \odot b = (a + b - u) \vee 0 .$$

Example 3. Consider $(R^2, +, \leq)$, where $(x_1, y_1) + (x_2, y_2) = (x_1 + x_2, y_1 + y_2 - 1)$ and $(x_1, y_1) \leq (x_2, y_2) \iff x_1 \leq x_2, y_1 \geq y_2$. Then R^2 is an l-group. Put $u = (1, 0)$. Then

$$\mathcal{M} = \{(x, y) \in R^2; (0, 1) \leq (x, y) \leq u = (1, 0)\}$$

is an MV-algebra.

Now we can present an embedding theorem for (finitely additive) IF-states.

Definition 1. *A mapping $m : \mathcal{F} \to [0, 1]$ is called a finitely additive IF-state, if the following properties are satisfied:*

(1.1) $m((0, 1)) = 0, \quad m((1, 0)) = 1,$
(1.2) $A \odot B = (0, 1) \Longrightarrow m(A \oplus B) = m(A) + m(B)$.

A finitely additive IF-state is an IF-state, if moreover

(1.3) $A_n \nearrow A \Longrightarrow m(A_n) \nearrow m(A)$.

Definition 2. *Let $(\mathcal{M}, 0, u, \neg, \oplus, \odot)$ be an MV-algebra. Let $m : \mathcal{M} \to [0, 1]$ be a monotone mapping (i.e. $a \leq b \Longrightarrow m(a) \leq m(b)$). We shall say that m is a finitely additive state, if the following properties are satisfied*

(2.1) $m(0) = 0, \quad m(u) = 1,$
(2.2) $a \odot b = 0 \Longrightarrow m(a \oplus b) = m(a) + m(b)$.

A finitely additive MV-state is an MV-state, if moreover

(2.3) $a_n \nearrow a \Longrightarrow m(a_n) \nearrow m(a)$.

3 Extension Theorem

In the following text we shall consider a measurable space $(\Omega, \mathcal{S})$ with a σ-algebra $\mathcal{S}$. By $\mathcal{F}$ the family will be denoted of all pairs $A = (\mu_A, \nu_A)$ with $\mu_A : \Omega \to [0, 1], \nu_A : \Omega \to [0, 1]$ measurable, and $\mu_A + \nu_A \leq 1$. By $\mathcal{M}$ the family fo all mappings $A = (\mu_A, \nu_A) : \Omega \to [0, 1]^2$ will be considered with $\mathcal{S}$-measurable mapings $\mu_A, \nu_A : \mathcal{S} \to [0, 1]$.

Proposition 1. *Let $A \in \mathcal{M}, A = (\mu_A, \nu_A)$. Define $A^0, A^1, A^2 : \Omega \to [0, 1]^2$ by the following formulas*

$$A^0(\omega) = (\mu_A^0(\omega), \nu_A^0(\omega)) = (\mu_A(\omega), \nu_A(\omega)),$$

if $\mu_A(\omega) + \nu_A(\omega) \leq 1$,

$$A^0(\omega) = (0, 1),$$

if $\mu_A(\omega) + \nu_A(\omega) > 1$;

$$A^1(\omega) = (\mu_A^1(\omega), \nu_A^1(\omega)) = (0, 1)$$

if $\mu_A(\omega) + \nu_A(\omega) \le 1$,

$$A^1(\omega) = (\mu_A(\omega), 1 - \mu_A(\omega)),$$

if $\mu_A(\omega) + \nu_A(\omega) > 1$;

$$A^2(\omega) = (\mu_A^2(\omega), \nu_A^2(\omega)) = (0, 1),$$

if $\mu_A(\omega) + \nu_A(\omega) \le 1$,

$$A^2(\omega) = (0, 2 - \mu_A(\omega) - \nu_A(\omega)),$$

if $\mu_A(\omega) + \nu_A(\omega) > 1$.
Then $A^0, A^1, A^2 \in \mathcal{F}$.

Proof. See [13] Proposition 1. □

Definition 3. *Let* $m : \mathcal{F} \to [0,1]$ *be an IF-state,* $A \in \mathcal{M}$*. Define* $\bar{m} : \mathcal{M} \to [0,1]$ *by the formula*

$$\bar{m}(A) = m(A^0) + m(A^1) - m(A^2) \ .$$

Proposition 2. $\bar{m} : \mathcal{M} \to [0,1]$ *is an extension of* $m : \mathcal{F} \to [0,1]$.

Proof. If $A = (\mu_A, \nu_A) \in \mathcal{F}$, then $A^0 = (\mu_A, \nu_A), A^1 = (0,1), A^2 = (0,1)$, hence

$$\bar{m}(A) = m(A^0) + m((0,1)) - m((0,1)) = m(A) + 0 - 0 = m(A) \ .$$

□

Theorem 1. *Let* $m : \mathcal{F} \to [0,1]$ *be an additive IF-state. Then there exists exactly one MV-algebra finitely additive state* $\bar{m} : \mathcal{M} \to [0,1]$ *that is an extension of* m.

Proof. See [13] Theorem 1. □

Theorem 2. *Let* $m : \mathcal{F} \to [0,1]$ *be an IF-state. Then there exist probabilities* $P, Q : \mathcal{S} \to [0,1]$ *and* $\alpha \in R$ *such that*

$$m(A) = \int_\Omega \mu_A dP + \alpha(1 - \int_\Omega \mu_A + \nu_A)dQ$$

for all $A \in \mathcal{F}$.

Proof. See [3], [4], [10], [12]. □

Theorem 3. *Let* $A_n = (\mu_n, \nu_n) \in \mathcal{F}(n = 1, 2, ...), A = (\mu, \nu) \in \mathcal{F}, \lim_{n\to\infty} \mu_n = \mu, \lim_{n\to\infty} \nu_n = \nu$*. Then*

$$\lim_{n\to\infty} m(A_n) = m(A) \ .$$

Proof. By Theorem 2.

$$m(A_n) = \int_\Omega \mu_n dP + \alpha(1 - \int_\Omega (\mu_n + \nu_n) dQ,$$

hence by the Lebesgue integration theorem

$$\lim_{n\to\infty} m(A_n) = \lim_{n\to\infty} \int_\Omega \mu_n dP + \alpha(1 - \lim_{n\to\infty} \int_\Omega (\mu_n + \nu_n) dQ) =$$

$$= \int_\Omega \mu dP + \alpha(1 - \int_\Omega (\mu + \nu) dQ) = m(A) \ .$$

□

The main result of the paper is contained in the following theorem.

Theorem 4. *Let $m : \mathcal{F} \to [0,1]$ be an IF-state. Then there exists exactly one extension $\bar{m} : \mathcal{M} \to [0,1]$ that is an MV-algebra state.*

Proof. Consider the mapping $\bar{m} : \mathcal{M} \to [0,1]$ described in Definition 3. We must prove that $\bar{m}$ is continuous. Let $A_n \in \mathcal{M}(n = 1,2,...)$, $A_n \nearrow A$. Put

$$B = \{\omega \in \Omega; \mu_A(\omega) + \nu_A(\omega) > 1\},$$

$$C = \{\omega \in \Omega; \mu_A(\omega) + \nu_A(\omega) = 1\},$$

$$D = \{\omega \in \Omega; \mu_A(\omega) + \nu_A(\omega) < 1\} \ .$$

Put further

$$B_n = (\mu_{A_n} I_B, 1 - (1 - \nu_{A_n}) I_B) = (\bar{\mu}_{B_n}, \bar{\nu}_{B_n}) \ .$$

For sufficiently large n

$$\bar{\mu}_{B_n} + \bar{\nu}_{B_n} > 1 \ .$$

Therefore

$$\bar{m}(B_n) = m(\bar{\mu}_{B_n}, 1 - \bar{\mu}_{B_n}) - m(0, 2 - \bar{\mu}_{B_n} - \bar{\mu}_{B_n}),$$

and

$$\lim_{n\to\infty} \bar{m}(B_n) = m(\bar{\mu}_B, 1 - \bar{\mu}_B) - m(0, 2 - \bar{\mu}_B - \bar{\mu}_B) = \bar{m}(B) \ .$$

Similarly

$$\lim_{n\to\infty} \bar{m}(C_n) = \bar{m}(C), \ \lim_{n\to\infty} \bar{m}(D_n) = \bar{m}(D),$$

and hence

$$\lim_{n\to\infty} \bar{m}(A_n) = \bar{m}(A) \ .$$

□

4 Conclusion

Let F be a set of IF-sets embedded to the MV-algebra M. In the paper we proved an extension theorem for a finitely additive F-valued IF-state to a finitely M-valued state. It follows that some results given for MV-algebras can be applied to families of IF-sets.

Acknowledgement. The support of the grant VEGA 1/0120/14 is kindly announced.

References

1. Atanassov, K.T.: Intuitionistic Fuzzy Sets: Theory and Applications. STUDFUZZ, vol. 35. Physica Verlag, Heidelberg (1999)
2. Atanassov, K.T.: On Intuitionistic Fuzzy Sets. Springer, Berlin (2012)
3. Ciungu, L., Riečan, B.: General form of probabilities on IF-sets. In: Fuzzy Logic and Applications. Proc. WILF Palermo, pp. 101–107 (2009)
4. Ciungu, L., Riečan, B.: Representation theorem for probabilities on IFS-events. Information Sciences 180, 793–798 (2010)
5. Grzegorzewski, P., Mrowka, E.: Probability of intuistionistic fuzzy events. In: Grzegorzewski, P., et al. (eds.) Soft Methods in Probability, Statistics and Data Analysis, pp. 105–115 (2002)
6. Lendelová, K.: Convergence of IF-observables. In: Issues in the Representation and Processing of Uncertain and Imprecise Information - Fuzzy Sets, Intuitionistic Fuzzy Sets, Generalized nets, and Related Topics, pp. 232–240. EXIT, Warsaw (2005)
7. Michalíková, A.: Absolute value and limit of the function defined on IF sets. Notes on Intuitionistic Fuzzy Sets 18, 8–15 (2012)
8. Michalíková, A.: The probability on square. Journal of Electrical Engeeniering 56(12/S), 21–22 (2005)
9. Mundici, D.: Interpretation of AFC* -algebras in Lukasiewicz sequential calculus. J. Funct. Anal. 65, 15–63 (1986)
10. Riečan, B.: On a problem of Radko Mesiar: general form of IF-probabilities. Fuzzy Sets and Systems 152, 1485–1490 (2006)
11. Riečan, B.: Probability theory on IF events. In: Aguzzoli, S., Ciabattoni, A., Gerla, B., Manara, C., Marra, V. (eds.) ManyVal 2006. LNCS (LNAI), vol. 4460, pp. 290–308. Springer, Heidelberg (2007)
12. Riečan, B.: Analysis of Fuzzy Logic Models. In: Koleshko, V.M. (ed.) Intelligent Systems, pp. 219–244. INTECH (2012)
13. Riečan, B.: On finitely additive IF-states. In: Intelligent Systems. IEEE (2014)
14. Riečan, B., Mundici, D.: Probability in MV-algebras. In: Pap, E. (ed.) Handbook of Measure Theory II, pp. 869–910. Elsevier, Heidelberg (2002)
15. Valenčáková, V.: A note on the conditional probability on IF-events. Math. Slovaca 59, 251–260 (2009)
16. Zadeh, L.A.: Fuzzy sets. Information and Control 8, 338–358 (1965)
17. Zadeh, L.A.: Probability measures on fuzzy sets. J. Math. Abal. Appl. 23, 421–427 (1968)

Definitive Integral on the Interval of IF Sets

Alžbeta Michalíková

Faculty of Natural Sciences,
Matej Bel University,
Tajovského 40,
Banská Bystrica,
Slovakia
Alzbeta.Michalikova@umb.sk

Abstract. In previous research the differential calculus for the functions defined on IF-sets was studied. Then there was the natural question if it is possible to define also definitive integral on this structure. In this paper the properties of definitive integral defined on IF-sets are studied.

1 Introduction

First publication about Intuitionistic fuzzy sets (IF sets in short) was publish in the year 1986 by professor Atanassov [1]. The set $A = \{\langle x, \mu_A(x), \nu_A(x)\rangle | x \in \Omega\}$ is an IF set if for each $x \in \Omega$ it holds

$$0 \leq \mu_A(x) + \nu_A(x) \leq 1 \ .$$

Function $\mu_A : \Omega \to [0, 1]$ is called membership function and the function $\nu_A : \Omega \to [0, 1]$ is called nonmembership function. For any IF set we will use following shorter notation $A = (\mu_A, \nu_A)$. Denote by $\mathcal{F}$ the family of all IF sets. On $\mathcal{F}$ we shall define two binary operations $\oplus, \odot$ and one unary operation $\neg$

$$A \oplus B = (\min(\mu_A + \mu_B, 1), \max(\nu_A + \nu_B - 1, 0))$$
$$A \odot B = (\max(\mu_A + \mu_B - 1, 0), \min(\nu_A + \nu_B, 1))$$
$$\neg A = (1 - \mu_A, 1 - \nu_A) \ .$$

A partial ordering on $\mathcal{F}$ is given by

$$A \leq B \Longleftrightarrow \mu_A \leq \mu_B, \nu_A \geq \nu_B \ .$$

Consider the set $A = (\mu_A, \nu_A)$, where $\mu_A, \nu_A : \Omega \to \mathbb{R}$. It is not difficult to construct an additive group $\mathcal{G} \supset \mathcal{F}$ with an ordering such that $\mathcal{G}$ is a lattice ordered group, also called ℓ-group, where

$$A + B = (\mu_A + \mu_B, \nu_A + \nu_B - 1)$$

with the neutral element $0 = (0_\Omega, 1_\Omega)$ and

$$A \leq B \Longleftrightarrow \mu_A \leq \mu_B, \nu_A \geq \nu_B \ .$$

P. Angelov et al. (eds.), *Intelligent Systems'2014*,
Advances in Intelligent Systems and Computing 322, DOI: 10.1007/978-3-319-11313-5_16

Lattice operations are given by

$$A \wedge B = (\mu_A \wedge \mu_B, \nu_A \vee \nu_B)$$
$$A \vee B = (\mu_A \vee \mu_B, \nu_A \wedge \nu_B) \ .$$

Evidently

$$A - B = (\mu_A - \mu_B, \nu_A - \nu_B + 1)$$

and

$$-A = (-\mu_A, 2 - \nu_A) \ .$$

The operations on $\mathcal{F}$ can be derived from operations on $\mathcal{G}$ if we use the unit $u = (1_\Omega, 0_\Omega)$. Then

$$A \oplus B = (A + B) \wedge u$$
$$A \odot B = (A + B - u) \vee 0$$
$$\neg A = u - A \ .$$

In our contribution also the following operations will be used

$$A.B = (\mu_A.\mu_B, \nu_A + \nu_B - \nu_A.\nu_B)$$

and if $\mu_B \neq 0, \nu_B \neq 1$ then

$$\frac{A}{B} = \left(\frac{\mu_A}{\mu_B}, 1 - \frac{1 - \nu_A}{1 - \nu_B}\right) .$$

Moreover

$$\sum_{i=1}^{n} C_i = \left(\sum_{i=1}^{n} \mu_{C_i}, \sum_{i=1}^{n} \nu_{C_i} - (n-1)\right) \ .$$

2 Definition of the Function and Its Properties

We will study the functions which are defined on the ℓ-group $\mathcal{G}$. From the previous text it follows that these results could be easily applied also for IF sets. The function on the ℓ-group $\mathcal{G}$ is defined by the following way

$$\tilde{f}(X) = (f(\mu_X), 1 - f(1 - \nu_X))$$

where $f : \mathbb{R} \to \mathbb{R}$ and $X \in \mathcal{G}$ [3].
In the first step we will look better on the domain of the function $\tilde{f}$. Since

$$\tilde{f}(X) = (f(\mu_X), 1 - f(1 - \nu_X))$$

where $f : \mathbb{R} \to \mathbb{R}$ then if the domain of the function $\tilde{f}$ is an interval $[A, B]$ then it must hold

$$A \leq B \Longleftrightarrow \mu_A \leq \mu_B \text{ and } \nu_A \geq \nu_B$$

and therefore

$$[\mu_A, \mu_B] \cup [\nu_B, \nu_A] \subset Dom f \ .$$

In the paper [4] there were studied the properties of absolute value of the function defined on the set $\mathcal{G}$. There was given the following definition

Definition 1. *Let $A \in \mathcal{G}$. Then the absolute value of A is defined by following formula*

$$|A| = (|\mu_A|, 1 - |1 - \nu_A|) \ .$$

In the papers [4] and [5] also the following properties of absolute valued were proved.

Lemma 1. *Let $|A|$ be the absolute value of A. Then $|A| = A$ for each $A \geq (0,1)$ and $|A| = -A$ for each $A < (0,1)$.*

Lemma 2. *For each $A, B \in \mathcal{G}$ it holds*

1. $|A.B| = |A|.|B|$
2. $|A - B| = |B - A|$
3. $|A + B| \leq |A| + |B|$
4. $|A - B| \geq |A| - |B|$
5. $||A| - |B|| \leq |A - B|$

Definition 2. *Let $A_0, A, \tilde{\delta} = (\delta, 1 - \delta)$, where $0 < \delta < 1$ be from the ℓ-group $\mathcal{G}$. A point A is in the $\tilde{\delta}$-neighborhood of a point A_0 if it holds*

$$|A - A_0| < \tilde{\delta} \ .$$

Definition 3. *Function $\tilde{f}$ is bounded on the interval $[A, B]$ if there exist such $H \in \mathcal{G}$ that for each $X \in [A, B]$ it holds*

$$|\tilde{f}(X)| \leq H \ .$$

In the paper [4] the definition of the limit of the function on ℓ-group $\mathcal{G}$ was given.

Definition 4. *Denote $\tilde{\varepsilon} = (\varepsilon, 1-\varepsilon)$ and $\tilde{\delta} = (\delta, 1-\delta)$ where $0 < \delta < 1, 0 < \varepsilon < 1$. Let $\tilde{f}$ be a function defined on the ℓ-group $\mathcal{G}$ and let $X_0, X, L, \tilde{\varepsilon}, \tilde{\delta}$ be from $\mathcal{G}$. For a function $\tilde{f}$ of a variable X defined on a $\tilde{\delta}$-neighborhood of a point X_0 except possibly the point X_0 itself, for each $\tilde{\varepsilon} > (0,1)$ there exists $\tilde{\delta} > (0,1)$ such that $|\tilde{f}(X) - L| < \tilde{\varepsilon}$ holds whenever $(0,1) < |X - X_0| < \tilde{\delta}$. Then we say that the function $\tilde{f}(X)$ tends to the limit L as X approaches X_0 and we write*

$$\lim_{X \to X_0} \tilde{f}(X) = L \ .$$

Definition 5. *The function $\tilde{f}$ is continuous at the point X_0 if it holds*

$$\lim_{X \to X_0} \tilde{f}(X) = f(X_0) \ .$$

Function $\tilde{f}$ is continuous if it is continuous at each point of its domain.

Remark 1. Function $\tilde{f} = (f(\mu_X), 1 - f(1 - \nu_X))$ is continuous on the interval $[A, B]$ if and only if f is continuous on the interval $[\mu_A, \mu_B]$ and at the same time on the interval $[\nu_B, \nu_A]$.

3 Definitive Integral

Lemma 3. *Let M_i, X_i $(i = 1, 2, \dots n)$ be elements from the set $\mathcal{G}$. Then it holds*

$$\sum_{i=1}^{n} M_i.(X_i - X_{i-1}) = \left(\sum_{i=1}^{n} \mu_{M_i}.(\mu_{X_i} - \mu_{X_{i-1}}), \sum_{i=1}^{n} (1 - \nu_{M_i}).(\nu_{X_i} - \nu_{X_{i-1}}) + 1 \right)$$

Proof.

$$X_i - X_{i-1} = (\mu_{X_i} - \mu_{X_{i-1}}, \nu_{X_i} - \nu_{X_{i-1}} + 1)$$

$$M_i.(X_i - X_{i-1}) =$$

$$= (\mu_{M_i}.(\mu_{X_i} - \mu_{X_{i-1}}), \nu_{M_i} + \nu_{X_i} - \nu_{X_{i-1}} + 1 - \nu_{M_i}.(\nu_{X_i} - \nu_{X_{i-1}} + 1)) =$$

$$= (\mu_{M_i}.(\mu_{X_i} - \mu_{X_{i-1}}), (1 - \nu_{M_i}).(\nu_{X_i} - \nu_{X_{i-1}}) + 1)$$

Therefore

$$\sum_{i=1}^{n} M_i.(X_i - X_{i-1}) =$$

$$= \left(\sum_{i=1}^{n} \mu_{M_i}.(\mu_{X_i} - \mu_{X_{i-1}}), \sum_{i=1}^{n} [(1 - \nu_{M_i}).(\nu_{X_i} - \nu_{X_{i-1}}) + 1] - (n - 1) \right) =$$

$$= \left(\sum_{i=1}^{n} \mu_{M_i}.(\mu_{X_i} - \mu_{X_{i-1}}), \sum_{i=1}^{n} [(1 - \nu_{M_i}).(\nu_{X_i} - \nu_{X_{i-1}})] + n - (n - 1) \right) =$$

$$= \left(\sum_{i=1}^{n} \mu_{M_i}.(\mu_{X_i} - \mu_{X_{i-1}}), \sum_{i=1}^{n} (1 - \nu_{M_i}).(\nu_{X_i} - \nu_{X_{i-1}}) + 1 \right) .$$

□

Definition 6. *We say that $\mathcal{D}$ is a partition of the interval $[A, B]$ if there exist finite sequence $X_0, X_1, \dots, X_n$ of the points from $\mathcal{G}$ such that it holds*

$$A = X_0 < X_1 < \dots < X_n = B .$$

Lemma 4. *Let $\mathcal{D}$ be a partition of the interval $[A, B]$. Then it holds*

$$\sum_{i=1}^{n} \mu_{X_i} - \mu_{X_{i-1}} = \mu_B - \mu_A$$

and

$$\sum_{i=1}^{n} \nu_{X_i} - \nu_{X_{i-1}} = \nu_B - \nu_A .$$

The proof of the Lemma 4 is obvious.

Theorem 1. *Let $\tilde{f}$ be continuous on $[A, B]$. There exists exactly one $I \in \mathcal{G}$ such that*

$$\underline{S}(\tilde{f}, \mathcal{D}) \leq I \leq \overline{S}(\tilde{f}, \mathcal{D})$$

for any partition $\mathcal{D}$ of interval $[A, B]$.

Proof. Of course the proof of this theorem has two parts. In the first part we will proved existence of the element I and in the second part we will proved the uniqueness.

Existence

Take fixed $\omega \in \Omega$ and a partition $\mathcal{D}$ of the interval $[A, B]$. Since f is continuous on $[\mu_{X_{i-1}}(\omega), \mu_{X_i}(\omega)]$ for each $i = 1, 2, \ldots n$ then f has the maximum $K_i(\omega)$ and also the minimum $k_i(\omega)$ on this interval. Similarly we can denote by $L_i(\omega)$ maximum and by $l_i(\omega)$ minimum of the function f on the interval $[1 - \nu_{X_{i-1}}(\omega), 1 - \nu_{X_i}(\omega)]$. Put

$$\begin{aligned} \mu_{M_i}(\omega) &= K_i(\omega), \\ \nu_{M_i}(\omega) &= 1 - L_i(\omega), \\ \mu_{m_i}(\omega) &= k_i(\omega), \\ \nu_{m_i}(\omega) &= 1 - l_i(\omega) \ . \end{aligned}$$

Then for $i = 1, 2, \ldots, n$ it holds

$$\mu_{m_i}(\omega) \leq f(\mu_X) \leq \mu_{M_i}(\omega)$$

and

$$1 - \nu_{M_i}(\omega) \leq 1 - f(1 - \nu_X) \leq 1 - \nu_{m_i}(\omega) \ .$$

Therefore for any ω there exists $K(\omega)$ such that

$$\sum_{i=1}^{n} \mu_{m_i}(\omega).\left(\mu_{X_i}(\omega) - \mu_{X_{i-1}}(\omega)\right) \leq K(\omega) \leq \sum_{i=1}^{n} \mu_{M_i}(\omega).\left(\mu_{X_i}(\omega) - \mu_{X_{i-1}}(\omega)\right) \ .$$

Similarly there exists $L(\omega)$ such that

$$\sum_{i=1}^{n} (1 - \nu_{M_i}(\omega)) . \left(\nu_{X_i}(\omega) - \nu_{X_{i-1}}(\omega)\right) + 1 \leq$$

$$\leq L(\omega) \leq \sum_{i=1}^{n} (1 - \nu_{m_i}(\omega)) . \left(\nu_{X_i}(\omega) - \nu_{X_{i-1}}(\omega)\right) + 1 \ .$$

Put

$$I(\omega) = (K(\omega), L(\omega)) \ .$$

Then

$$\sum_{i=1}^{n} m_i . (X_i - X_{i-1}) \leq I \leq \sum_{i=1}^{n} M_i . (X_i - X_{i-1}) \ .$$

Uniqueness
Let there exist $I_1 < I_2$ such that for any partition $\mathcal{D}$ of interval $[A, B]$ it holds

$$\underline{S}(\tilde{f}, \mathcal{D}) \leq I_1 < I_2 \leq \overline{S}(\tilde{f}, \mathcal{D}) \ .$$

Since $I_1 \neq I_2$ that there exists such ω that $\mu_{I_1}(\omega) < \mu_{I_2}(\omega)$ or $\nu_{I_1}(\omega) > \nu_{I_2}(\omega)$. Let $\mu_{I_1}(\omega) < \mu_{I_2}(\omega)$ and denote $\varepsilon = \mu_{I_2}(\omega) - \mu_{I_1}(\omega)$.
We know that in classical calculus holds following theorem: If f is continuous on an interval $[\alpha, \beta]$ then f is uniformly continuous on the interval $[\alpha, \beta]$. Therefore to each $\varepsilon > 0$ there exist such $\delta > 0$ that for each $u, v \in [\alpha, \beta]$ it holds

$$|u - v| < \delta \Rightarrow |f(u) - f(v)| < \frac{\varepsilon}{\beta - \alpha} \ .$$

Let $|\mu_{X_i}(\omega) - \mu_{X_{i-1}}(\omega)| < \delta$ for $i = 1, 2, \ldots n$ and

$$M_i(\omega) = \max\{f(\mu_x(\omega)), X_{i-1} \leq X \leq X_i\}$$
$$m_i(\omega) = \min\{f(\mu_x(\omega)), X_{i-1} \leq X \leq X_i\} \ .$$

Denote

$$M_i(\omega) = f(u_i), \quad m_i(\omega) = f(v_i)$$

then

$$u_i, v_i \in [\mu_{X_{i-1}}(\omega) - \mu_{X_i}(\omega)]$$

hence

$$|u_i - v_i| < \delta \ .$$

Therefore

$$M_i(\omega) - m_i(\omega) = f(u_i) - f(v_i) < \frac{\varepsilon}{\mu_{X_{i-1}}(\omega) - \mu_{X_i}(\omega)} \ .$$

Moreover

$$\sum_{i=1}^{n} (\mu_{X_i}(\omega) - \mu_{X_{i-1}}(\omega)) = \mu_B(\omega) - \mu_A(\omega) \ .$$

Then

$$\varepsilon = \mu_{I_2}(\omega) - \mu_{I_1}(\omega) \leq$$
$$\leq \sum_{i=1}^{n} M_i(\omega).(\mu_{X_i}(\omega) - \mu_{X_{i-1}}(\omega)) - \sum_{i=1}^{n} m_i(\omega).(\mu_{X_i}(\omega) - \mu_{X_{i-1}}(\omega)) =$$
$$= \sum_{i=1}^{n} (M_i(\omega) - m_i(\omega)) . \left(\sum_{i=1}^{n} (\mu_{X_i}(\omega) - \mu_{X_{i-1}}(\omega)) \right) <$$
$$< \frac{\varepsilon}{\mu_B(\omega) - \mu_A(\omega)} . (\mu_B(\omega) - \mu_A(\omega)) = \varepsilon \ .$$

Since the inequality $\varepsilon < \varepsilon$ is not correct, it must hold $I_1 = I_2$ and therefore there exists exactly one I such that

$$\underline{S}(\tilde{f}, \mathcal{D}) \leq I \leq \overline{S}(\tilde{f}, \mathcal{D}) \ .$$

□

Therefore we can use the notation

$$I = \int_A^B \tilde{f}(X) dX \ .$$

To prove another property of definitive integral we need the definition of the derivative of the function and also Lagrange mean valued theorem. These results were published in the papers [6] and [7].

Definition 7. *Let $X_0, H, \tilde{\delta} = (\delta, 1 - \delta)$ be from $\mathcal{G}$ and let $\tilde{f}$ be the function defined for such H that it hold $(0,1) < |H - X_0| < \tilde{\delta}$. Let*

$$\lim_{H \to (0,1)} \frac{\tilde{f}(X_0 + H) - \tilde{f}(X_0)}{H}$$

exist. Then this limit is the derivative of the function $\tilde{f}$ at the point X_0 and we can denote it as $\tilde{f}'(X_0)$.

Remark 2. Let for each $X \in [A, B]$ there exist $\tilde{f}'(X)$ then it holds

$$\tilde{f}'(X) = (f'(\mu_X), 1 - f'(1 - \nu_X)) \ .$$

Lagrange mean value theorem has the following form:

Theorem 2. *Let $\tilde{f}$ be continuous on $[A, B]$ and differentiable on (A, B). Then there exists $C \in (A, B)$ such that*

$$\tilde{f}(B) - \tilde{f}(A) = \tilde{f}'(C)(B - A) \ .$$

Now we can prove the property of definitive integral which in classical calculus is called the Newton-Leibniz formula.

Theorem 3. *Let $\tilde{F}'(X) = \tilde{f}(X)$ for any $X \in [A, B]$. Then*

$$\int_A^B \tilde{f}(X) dX = \tilde{F}(B) - \tilde{F}(A) \ .$$

Proof. Let $\mathcal{D}$ be any partition of interval $[A, B]$. By using the Lagrange mean valued theorem we get:

$$\tilde{F}(B) - \tilde{F}(A) = \sum_{i=1}^{n} \left(\tilde{F}(X_i) - \tilde{F}(X_{i-1}) \right) = \sum_{i=1}^{n} \tilde{F}'(C_i).(X_i - X_{i-1}) =$$

$$= \sum_{i=1}^{n} \tilde{f}(C_i).(X_i - X_{i-1}) \leq \sum_{i=1}^{n} M_i.(X_i - X_{i-1}) = \overline{S}(\tilde{f}, \mathcal{D}) \ .$$

Similarly

$$\underline{S}(\tilde{f}, \mathcal{D}) \leq \tilde{F}(B) - \tilde{F}(A) \ .$$

Therefore from Theorem 1 it follows

$$\int_A^B \tilde{f}(X)dX = \tilde{F}(B) - \tilde{F}(A) .$$

□

4 Conclusion

In the paper the definition of definitive integral defined on IF sets was given. It was shown that the definitive integral on interval could be constructing by using the partition of the IF integral. Then the validity of Newton-Leibniz formula was proved.

Acknowledgement. The support of the grant VEGA 1/0120/14 is kindly announced.

References

1. Atanassov, K.T.: Intuitionistic Fuzzy Sets. Fuzzy Sets and Systems 20(1), 87–96 (1986)
2. Atanassov, K.T.: Intuitionistic Fuzzy Sets. Springer, Berlin (1999)
3. Hollá, I., Riečan, B.: Elementary function on IF sets. In: Advances in Fuzzy Stes, Intuitionistic Fuzzy Sets, Generalized Nets and Related Topics. Foundations, vol. I, pp. 193–201. Academic Publishing House EXIT, Warszawa (2008)
4. Michalíková, A.: Absolute value and limit of the function defined on IF sets. In: Proceedings of the Sixteenth International Conference on Intuitionistic Fuzzy Sets, Sofia, Bulgaria, vol. 18(3), pp. 8–15 (2012)
5. Michalíková, A.: Some notes about boundaries on IF sets. In: New Trends in Fuzzy Sets, Intuitionistic Fuzzy Sets, Generalized nets and Related Topics, vol. I, pp. 105–113. Systems Research Institute, Polish Academy of Sciences, Poland (2013)
6. Michalíková, A.: The differential calculus on IF sets. In: International Conference on Fuzzy Systems, FUZZ-IEEE 2009. Proccedings [CD-ROM], Jeju Island, Korea, pp. 1393–1395 (2009)
7. Riečan, B.: On Lagrange mean value theorem for functions on Atanassov IF sets. In: Proceedings of the Eighth International Workshop on Intuitionistic Fuzzy Sets, Banská Bystrica, Slovakia, October 9, vol. 18(4), pp. 8–11 (2012)
8. Zadeh, L.A.: Fuzzy sets. Information and Control 8, 338–353 (1965)

Intuitionistic Fuzzy Tautology Definitions for the Validity of Intuitionistic Fuzzy Implications: An Experimental Study

Marcin Detyniecki[1,2,3], Marie-Jeanne Lesot[1,2], and Paul Moncuquet[1,2]

[1] CNRS, UMR 7606, LIP6, Paris, France
[2] Sorbonne Universités, UPMC Univ Paris 06, UMR 7606, LIP6, Paris, France
[3] Polish Academy of Sciences, IBS PAN, Warsaw, Poland

Abstract. The central issue of inference validity, that guarantees the correctness of reasoning and thus that of derived knowledge, depends on the definition of both implication operator and tautology. This paper studies the *modus ponens* validity in the case of Intuitionistic Fuzzy logic, in an experimental framework: considering 18 classical implication operators, it shows that validity usually does not hold for the classical definition of Intuitionistic Fuzzy tautology. It proposes two alternative, more constrained, tautology definitions, studying them with the same protocol, showing they make it possible to decrease the number of invalid implication operators.

Keywords: Intuitionistic Fuzzy logic, modus ponens, validity, tautology.

1 Introduction

The validity of *modus ponens* is a crucial aspect for any form of knowledge inference, as it guarantees the correctness of reasoning. In the general case, the validity property of an inference rule ensures that true premises always lead to a true conclusion. In other words, a valid inference is one in which the conclusion must be true, if the premise is.

In classical propositional logic, the *modus ponens* is indeed a valid rule of inference, as can be proven using Boolean logic definitions. It can be summarized as allowing to infer the conclusion C from a premise $P = \{A, A \rightarrow C\}$: if one asserts that P is *true*, one can infer that C is *true*.

Intuitionistic Fuzzy logic [2,1] introduces more expressivity, in a gradual logic framework: it allows not only to define degrees of truth, but also to simultaneously observe degrees of falsehood. It then proposes new definitions for implication operators and for tautology [2,3], that must be used to prove the validity of the *modus ponens*. Unfortunately, as observed in some particular cases by some authors [2,6,4], and thoroughly in this paper, the validity of the modus ponens is not guaranteed for most Intuitionistic Fuzzy implications.

We propose to interpret this observation as due to the fact that the classical Intuitionistic Fuzzy tautology [2] is too optimistic and we propose alternative,

P. Angelov et al. (eds.), *Intelligent Systems'2014*,
Advances in Intelligent Systems and Computing 322, DOI: 10.1007/978-3-319-11313-5_17

stricter, definitions. The latter also aim at proposing a more intuitive approach to tautology with respect to ignorance processing: through the expressivity allowed by the 2 degrees, of truth and falsehood, Intuitionistic Fuzzy logic makes it possible to model distinct levels of knowledge and in particular ignorance, as neither positive (i.e. formula with degree of truth greater than degree of falsehood) nor negative knowledge (reciprocally, formula with degree of truth lower than degree of falsehood). The proposed tautology definitions make it possible to express the notion of being "certainly true", whereas the classical definition considers ignorance as a tautology.

The paper is organized as follows: after recalling some notations and formal definitions in Section 2, it introduces in Section 3 the two new, more constrained, Intuitionistic Fuzzy tautology definitions. Section 4 presents the experimental study that aims at identifying counterexamples to possibly invalidate Intuitionistic Fuzzy implications, considering 18 operators from the state of the art. The modus ponens validity is experimentally checked for the 3 definitions of tautology, indicating that the more constrained alternatives indeed make it possible to decrease the number of invalid implication operators.

2 Usual Definitions in Intuitionistic Fuzzy logic

2.1 Intuitionistic Fuzzy Sets

An Intuitionistic Fuzzy set A defined over a universe X can be seen as an extension of a fuzzy set, characterized by both a membership function μ_A and a non-membership function ν_A. It is written $A = \{\langle x, \mu_A(x), \nu_A(x)\rangle \mid x \in X\}$. The membership and non-membership functions take their values in $[0, 1]$ and satisfy the condition that $\forall x \in X, \mu_A(x) + \nu_A(x) \leq 1$.

An Intuitionistic Fuzzy set can be represented on the complete lattice $(L, \leq_L)$ defined as follows:

Definition 1. *The* Intuitionistic Fuzzy lattice *is defined as*

$$L = \{(x_1, x_2) \in [0, 1]^2 \mid x_1 + x_2 \leq 1)\}$$

equipped with the comparison operator $\leq_L$ *defined as*

$$(x_1, x_2) \leq_L (y_1, y_2) \quad \textit{iff} \quad x_1 \leq y_1 \textit{ and } x_2 \geq y_2$$

Reference points of special interest in this lattice are the three extreme points denoted $1_L = (1, 0)$, $0_L = (0, 1)$ and $U_L = (0, 0)$.

2.2 Intuitionistic Fuzzy Set Operators

The set operations on Intuitionistic Fuzzy sets are written through extensions of the fuzzy set operations, we recall here the definition of Intuitionistic Fuzzy t-norms, used to compute the intersection:

Definition 2. *An* Intuitionistic Fuzzy t-norm $\mathcal{T}$ *is a mapping* $\mathcal{T} : L \times L \to L$ *that is commutative* ($\mathcal{T}(x,y) = \mathcal{T}(y,x)$)*, associative* ($\mathcal{T}(x,\mathcal{T}(y,z)) = \mathcal{T}(\mathcal{T}(x,y),z)$)*, monotonous* ($x \leq_L y$ *and* $x' \leq_L y' \Rightarrow \mathcal{T}(x,y) \leq_L \mathcal{T}(x',y')$) *and satisfies the boundary condition* $\mathcal{T}(x, 1_L) = x$.

Likewise, an *Intuitionistic Fuzzy t-conorm* $\mathcal{S}$ is defined as a mapping $\mathcal{S} : L \times L \to L$ that is commutative, associative, monotonous and satisfies the boundary condition $\mathcal{S}(x, 0_L) = x$.

Among the Intuitionistic Fuzzy t-norms, the category of t-representable t-norms is of special interest:

Definition 3. *A* t-representable Intuitionistic Fuzzy t-norm *is a t-norm* $\mathcal{T}$ *such that there exists a fuzzy t-norm* T *and a fuzzy t-conorm* S *such that*

$$\mathcal{T}((x_1,x_2),(y_1,y_2)) = (T(x_1,y_1), S(x_2,y_2))$$

2.3 Intuitionistic Fuzzy Implication Operators

General Definition. Intuitionistic Fuzzy logic extends fuzzy logic to process Intuitionistic Fuzzy sets and is based on extensions of the fuzzy operators, membership and non-membership degrees being interpreted as degrees of truth and falsehood respectively. Regarding the implication operators, that play a crucial role in the definition of any new logical framework, very few properties have been required for a mapping to be considered as such [2,5]. Formally

Definition 4. *An* Intuitionistic Fuzzy implication operator $\mathcal{I}$ *is a mapping* $\mathcal{I} : L \times L \to L$ *that*

- *satisfies the 4 boundary conditions:* $\mathcal{I}(0_L, 0_L) = 1_L$, $\mathcal{I}(0_L, 1_L) = 1_L$, $\mathcal{I}(1_L, 0_L) = 0_L$ *and* $\mathcal{I}(1_L, 1_L) = 1_L$
- *is decreasing in its first component:* $x \leq_L x' \Rightarrow \mathcal{I}(x,y) \geq_L \mathcal{I}(x',y)$
- *is increasing in its second component:* $y \leq_L y' \Rightarrow \mathcal{I}(x,y) \leq_L \mathcal{I}(x,y')$

Residuated Implication Operators. Intuitionistic Fuzzy residuated implication operators, also called $\mathcal{R}$-implicators, have been introduced by [4], as a direct extension of the fuzzy case: the inequality constraint between fuzzy degrees in $[0,1]$ is replaced by the lattice comparison operator between intuitionistic couples in L:

Definition 5. *Given an Intuitionistic Fuzzy t-norm* $\mathcal{T}$*, the* Intuitionistic Fuzzy residuated implication operator *is defined as the mapping* $\mathcal{I} : L \times L \to L$ *such that*

$$\forall (x,y) \in L^2, \quad \mathcal{I}(x,y) = \sup\{\gamma \in L \mid \mathcal{T}(x,\gamma) \leq_L y\}$$

The residuation makes it possible, based on different Intuitionistic Fuzzy break t-norms, to obtain a large number of implications operator.

Table 1. Considered implication operators. $sg(x) = \mathbb{1}_{x>0}$ and $\overline{sg}(x) = \mathbb{1}_{x\leq 0}$

Name	Value of the implication $\mathcal{I}((x_1, x_2), (y_1, y_2))$
Zadeh	$\langle \max(x_2, \min(x_1, y_1)), \min(x_1, y_2) \rangle$
Gaines-Rescher	$\langle 1 - sg(x_1 - y_1), y_2 \cdot sg(x_1 - y_1) \rangle$
Gödel	$\langle 1 - (1 - y_1) \cdot sg(x_1 - y_1), y_2 \cdot sg(x_1 - y_1) \rangle$
Kleene-Dienes	$\langle \max(x_2, y_1), \min(x_1, y_2) \rangle$
Łukasiewicz	$\langle \min(1, x_2 + y_1), \max(0, x_1 + y_2 - 1) \rangle$
Reichenbach	$\langle x_2 + x_1 y_1, x_1 y_2 \rangle$
Willmott	$\langle \min(\max(x_2, y_1), \max(x_1, x_2), \max(y_1, y_2)),$ $\max(\min(x_1, y_2), \min(x_1, x_2), \min(y_1, y_2)) \rangle$
Wu	$\langle 1 - (1 - \min(x_2, y_1)) \cdot sg(x_1 - y_1),$ $\max(x_1, y_2) \cdot sg(x_1 - y_1) \cdot sg(y_2 - x_2) \rangle$
Klir and Yuan 1	$\langle x_2 + x_1^2 y_1, x_1 x_2 + x_1^2 y_2 \rangle$
Klir and Yuan 2	$\langle y_1 \overline{sg}(1 - x_1) + sg(1 - x_1)(\overline{sg}(1 - y_1) + x_2 \cdot sg(1 - y_1)),$ $y_2 \overline{sg}(1 - x_1) + x_1 \cdot sg(1 - x_1) sg(1 - y_1) \rangle$
Atanassov 1	$\langle 1 - (1 - y_1) \cdot sg(x_1 - y_1), y_2 \cdot sg(x_1 - y_1) sg(y_2 - x_2) \rangle$
Atanassov 2	$\langle \max(x_2, y_1), 1 - \max(x_2, y_1) \rangle$
Atanassov and Kolev	$\langle x_2 + y_1 - x_2 y_1, x_1 y_2 \rangle$
Atanassov and Trifonov	$\langle 1 - (1 - y_1) sg(x_1 - y_1) - y_2 \overline{sg}(x_1 - y_1) sg(y_2 - x_2),$ $y_2 sg(y_2 - x_2) \rangle$
Atanassov 3	$\langle 1 - (1 - \min(x_2, y_1)) \cdot sg(sg(x_1, y_1) + sg(y_2, x_2))$ $- \min(x_2, y_1) \cdot sg(x_1 - y_1) \cdot sg(y_2 - x_2),$ $1 - (1 - \max(x_1, y_2)) \cdot sg(\overline{sg}(x_1 - y_1) + \overline{sg}(y_2 - x_2))$ $- \max(x_1, y_2) \cdot \overline{sg}(x_1 - y_1) \cdot \overline{sg}(y_2 - x_2) \rangle$
$\mathcal{R}$-Goguen	residuation of $\mathcal{T}(x, y) = (x_1 \cdot y_1, x_2 + y_2 - x_2 \cdot y_2)$
$\mathcal{R}$-Łukasiewicz	residuation of $\mathcal{T}(x, y) = (\max(x_1 + y_1 - 1, 0), \min(x_2 + y_2, 1))$
$\mathcal{R}$-Gödel	residuation of $\mathcal{T}(x, y) = (\min(x_1, y_1), \max(x_2, y_2))$

Considered Implication Operators. In this paper, we study 18 classical implication operators, listed in Table 1: the first 15 are the general operators listed in [3], the last 3 are residuated operators.

The Łukasiewicz residuated implication is obtained with the t-representable Intuitionistic Fuzzy t-norm based on the fuzzy t- norm and conorm of the same name; the Goguen operator is built with the probabilistic one. The Gödel residuated operator is obtained using Zadeh t-representable Intuitionistic Fuzzy t-norm and is not to be confused with the Gödel Intuitionistic Fuzzy implication introduced in [1]. The latter is also considered in this study, its definition is given in the third row of Table 1.

2.4 Intuitionistic Fuzzy Tautology

In formal logic, a tautology translates the idea of universal truth. In the Intuitionistic Fuzzy logic case, it has been expressed in the following definition [2], to which we will refer as *classical* throughout this paper: confusing a formula with its truth value,

Definition 6. $x = (x_1, x_2) \in L$ *is an* Intuitionistic Fuzzy tautology *iff* $x_1 \geq x_2$.

This definition imposes that the degree of truth is greater than the degree of falsehood for an intuitionistic couple to be considered as a tautology.

2.5 Validity

As mentioned in the introduction, the validity of the modus ponens can be characterized in terms of truth values. The transposition to the Intuitionistic Fuzzy framework is defined as [4]:

Definition 7. *The modus ponens is* valid *for an Intuitionistic Fuzzy implication $\mathcal{I}$ iff, if P is tautology and $\mathcal{I}(P, C)$ is a tautology, then C is a tautology.*

It can immediately be noticed that the validity of an implication operator directly depends upon the definition of the tautology. In the following we present some alternative definitions.

3 Proposed Alternative Intuitionistic Fuzzy Tautologies

Several authors [6,4,2] have noticed that for usual Intuitionistic Fuzzy implication operators the modus ponens is not valid according to the tautology definition given in Def. 6, as also observed and commented in Section 4. In this section, we propose and discuss two alternative definitions.

Besides the validity objective, the proposition of alternative tautology definitions also aims at better matching intuitive expectations: with the classical definition (Def. 6), the point $U_L = (0, 0)$ is considered as a tautology, which is counter-intuitive. Indeed, this value, for which both truth and falsehood degrees equal 0, is closer to the notion of "unknown", modelling total ignorance, rather than to the one of "true".

The two proposed tautology definitions attempt to overcome these flaws, by imposing additional requirements to exclude some cases from being considered as true.

The first one, which we propose to call *certain tautology*, requires a minimum knowledge level, through the addition of a constraint on the uncertainty degree, defined for $x = (x_1, x_2) \in L$ as $\pi(x) = 1 - x_1 - x_2$: it makes it possible to prevent an almost unknown situation to be considered as a tautology. Formally

Definition 8. $x = (x_1, x_2) \in L$ *is a* certain Intuitionistic Fuzzy tautology *iff* $x_1 \geq x_2$ *and* $\pi(x) = 1 - x_1 - x_2 \leq 0.5$.

The second tautology definition, inspired by the fuzzy case [7], is even stronger. It imposes the truth degree to be meaningful enough, instead of comparing it to the non-truth degree as is done in the usual definition:

Definition 9. $x = (x_1, x_2) \in L$ *is a* truth Intuitionistic Fuzzy tautology *iff* $x_1 \geq 0.5$.

It can be underlined that choosing 0.5 as threshold automatically guarantees that the truth degree is larger than the non-truth degree. Furthermore, it also guarantees that the uncertainty degree is lower than 0.5.

The three definitions thus satisfy an inclusion property: all truth tautologies (Def. 9) are certain tautologies (Def. 8) which, in turn, are classical tautologies (Def. 6). This also means that the truth tautology is stronger than the certain one which is stronger than the classical one.

4 Validity Experimental Study

4.1 Experimental protocol

The principle of the proposed experimental study consists in looking for counterexamples of the validity property, i.e. (x, y) couples in L^2 such that x and $\mathcal{I}(x, y)$ are tautologies but y is not: the protocol consists in first randomly generating $10,000$ (x, y) couples in L under a uniform distribution and keeping the ones for which x and $\mathcal{I}(x, y)$ are tautologies, which can be considered as candidate counterexamples. The candidates are then tested to check whether y is a tautology, leading to actual counterexamples. This protocol is applied to the 18 implication operators listed in Table 1, for the three considered tautology definitions.

Figures 1, 2 and 3, commented below, display the results graphically for all implications studied in this work, in the same order as their definition in Table 1. They show the identified actual counterexamples: each segment joins a point $x = (x_1, x_2)$ and a point $y = (y_1, y_2)$ that constitute a counterexample; for each segment, the conclusion y is represented by a black point. Valid cases are not represented.

Table 2 gives the corresponding numerical values in the form $a/b = c$: in each case a and b respectively indicate the number of actual and candidate counterexamples and c the quotient of these two quantities.

4.2 Validity Results

Classical Tautology Definition. It can be seen that in 14 cases out of 18, validity counterexamples are indeed found: this shows that, for most considered implication operators, the validity property does not hold.

A major difference can be observed among the 4 implication operators where no counterexample is identified, namely Willmott, Klir-Yuan 2, Atanassov 2 and Atanassov 3. Indeed, for the Klir-Yuan 2 operator, among the $10,000$ tested random intuitionistic couples, no one is such that x and $\mathcal{I}(x, y)$ both are tautologies. This corresponds to a very pessimistic behavior, which makes this operator unusable for any practical purpose. Quite the opposite, the Willmott operator maximizes the number of candidate counterexamples, as 2099 are considered as such; however not a single of these candidates turns out to be an actual counterexample, which constitutes an interesting property. The Atanassov 2 and Atanassov 3 operators have intermediary behaviors.

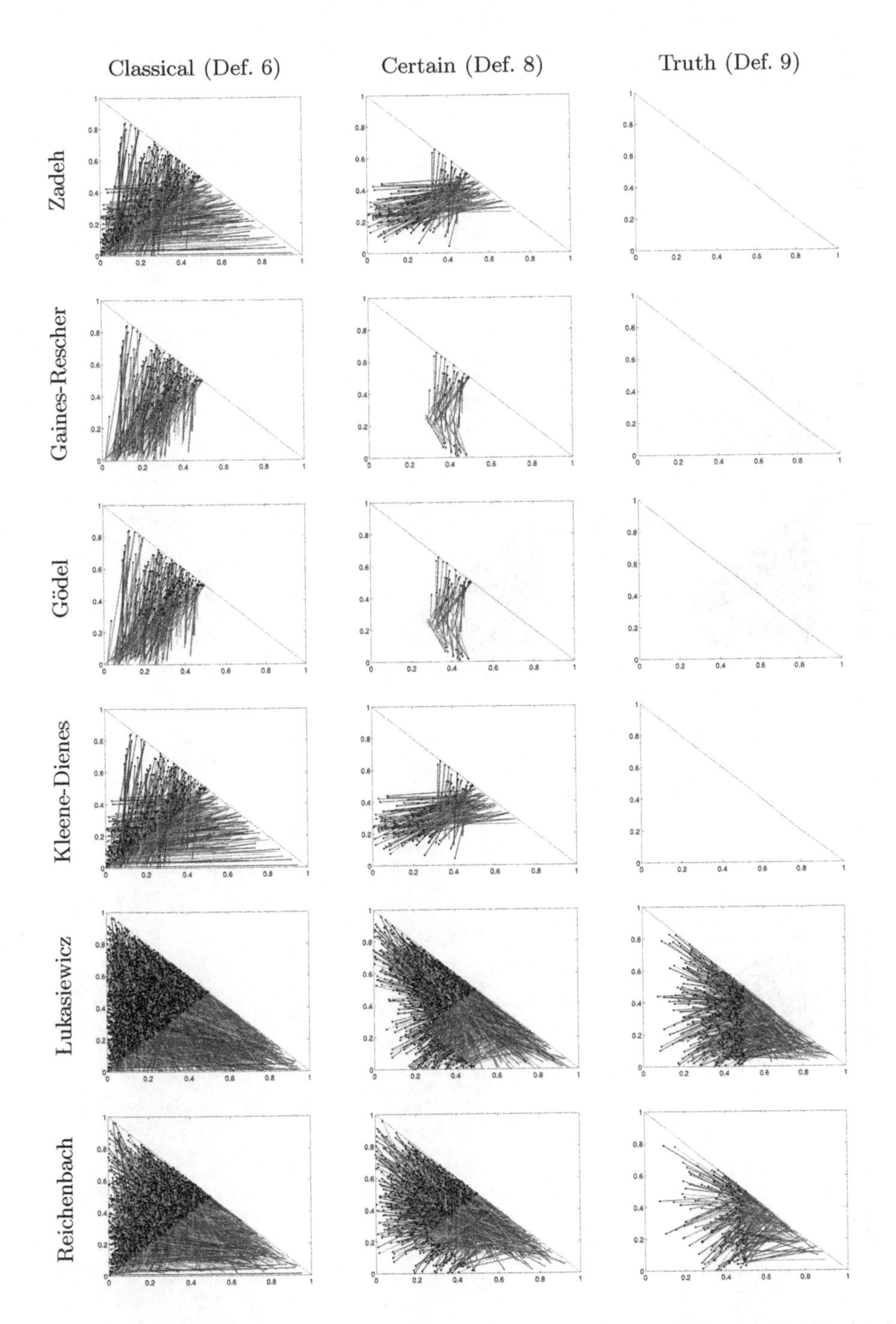

Fig. 1. Validity counterexamples for the first 6 implications and the 3 tautology definitions

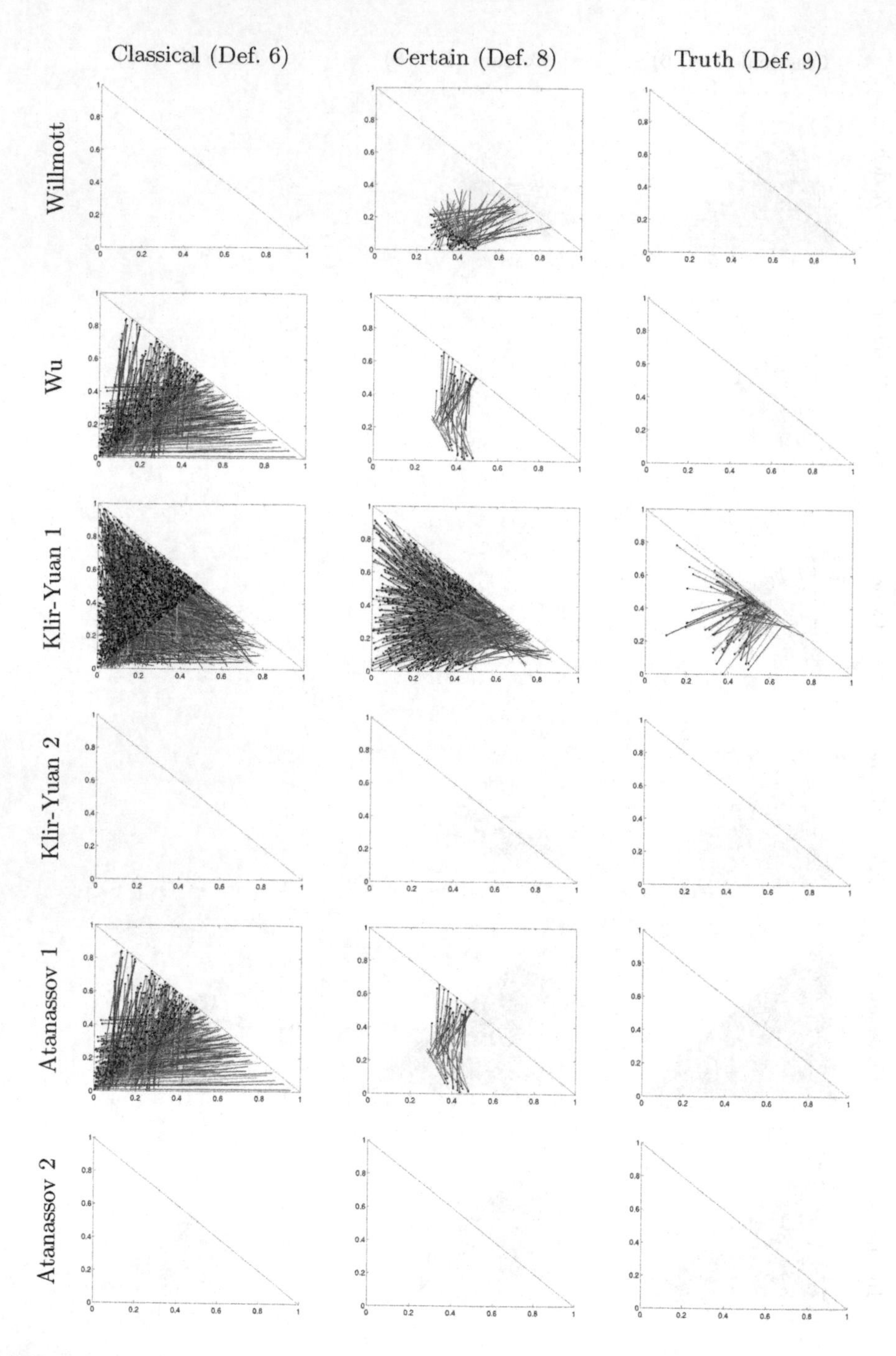

Fig. 2. Validity counterexamples for the next 6 implications and the 3 tautology definitions

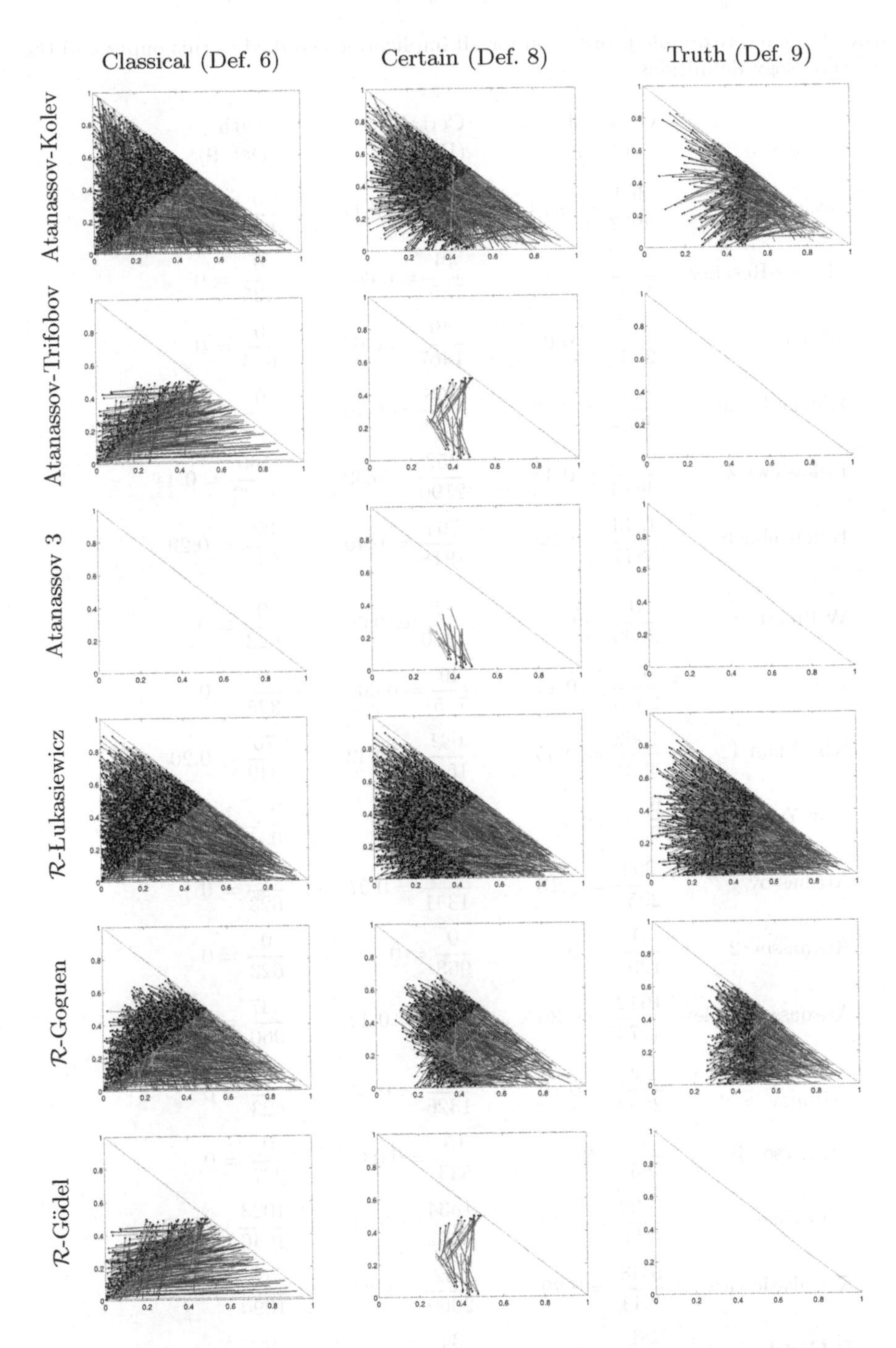

Fig. 3. Validity counterexamples for the last 3 and the 3 residuated implications and the 3 tautology definitions

Table 2. Counterexample proportion for all implications studied in this paper and the three tautology definitions

	Classical (Def. 6)	Certain (Def. 8)	Truth (Def. 9)
Zadeh	$\frac{394}{2957} = 0.13$	$\frac{158}{1444} = 0.11$	$\frac{0}{623} = 0$
Gaines-Rescher	$\frac{197}{1509} = 0.13$	$\frac{49}{785} = 0.06$	$\frac{0}{325} = 0$
Gödel	$\frac{197}{2760} = 0.07$	$\frac{49}{1467} = 0.03$	$\frac{0}{623} = 0$
Kleene-Dienes	$\frac{394}{2957} = 0.13$	$\frac{158}{1576} = 0.10$	$\frac{0}{623} = 0$
Łukasiewicz	$\frac{2047}{4610} = 0.44$	$\frac{852}{2190} = 0.39$	$\frac{500}{1123} = 0.44$
Reichenbach	$\frac{1314}{3877} = 0.34$	$\frac{764}{1918} = 0.40$	$\frac{199}{693} = 0.29$
Willmott	$\frac{0}{2099} = 0$	$\frac{81}{1450} = 0.06$	$\frac{0}{623} = 0$
Wu	$\frac{394}{2355} = 0.17$	$\frac{49}{785} = 0.06$	$\frac{0}{325} = 0$
Klir-Yuan 1	$\frac{1362}{3925} = 0.35$	$\frac{641}{1530} = 0.42$	$\frac{70}{349} = 0.20$
Klir-Yuan 2	$\frac{0}{0} = NaN$	$\frac{0}{0} = NaN$	$\frac{0}{0} = NaN$
Atanassov 1	$\frac{394}{2957} = 0.13$	$\frac{49}{1341} = 0.07$	$\frac{0}{623} = 0$
Atanassov 2	$\frac{0}{1297} = 0$	$\frac{0}{963} = 0$	$\frac{0}{623} = 0$
Atanassov-Kolev	$\frac{1612}{4175} = 0.39$	$\frac{1003}{2409} = 0.42$	$\frac{337}{960} = 0.35$
Atanassov-Trifonov	$\frac{289}{2852} = 0.10$	$\frac{34}{1326} = 0.03$	$\frac{0}{623} = 0$
Atanassov 3	$\frac{0}{685} = 0$	$\frac{15}{533} = 0.03$	$\frac{0}{227} = 0$
$\mathcal{R}$-Goguen	$\frac{1723}{4286} = 0.40$	$\frac{1634}{3052} = 0.54$	$\frac{1023}{1646} = 0.62$
$\mathcal{R}$-Łukasiewicz	$\frac{1048}{3611} = 0.29$	$\frac{652}{2060} = 0.32$	$\frac{467}{1090} = 0.43$
$\mathcal{R}$-Gödel	$\frac{289}{2851} = 0.10$	$\frac{34}{1325} = 0.03$	$\frac{0}{623} = 0$

Along the same lines, it can be observed that the invalid operators vary to a large extent regarding the number of candidate counterexamples. Some of these implications (Łukasiewicz, Reichenbach, Klir-Yuan 1, Atanassov-Kolev, $\mathcal{R}$-Goguen and $\mathcal{R}$-Łukasiewicz operators) appear to be very often true, generating more than 3,000 candidate counterexamples. However it appears that the latter very often turn out to be actual counterexamples, leading to proportion larger than 29% and up to 44% for the Łukasiewicz operator.

Certain Tautology Definition. It can be seen from the second column of the figures that, using the more constrained definition of certain tautology, in all cases but 2 (Willmott and Atanassov 3 operators), the number of counterexamples significantly decreases; it is also very low for the two exceptions.

Nevertheless, as some counterexamples are still identified, the certain tautology does not solve the validity issue: the 14 previous invalid operators still are invalid, the Willmott and Atanassov 3 operators are added.

Table 2 shows that although the number of counterexamples decreases, the number of candidates also decreases, leading to a proportion that does not significantly decreases and even increases in some cases.

Truth Tautology Definition. The third column in the figures shows that the truth tautology definition only leads to 6 invalid implication operators, namely Łukasiewicz , Reichenbach, Klir-Yuan 1, Atanassov-Kolev as well as the residuated Goguen and the residuated Łukasiewicz. For the last two, we believe that the problem may come from the residuation procedure; future work will focus on the latter. In all cases, the number of counterexamples decreases as compared to the certain tautology definition.

It can also be noticed that for all operators for which no counterexamples is found except Gaines-Rescher, Wu and Atanassov 3, the number of candidate counterexamples is actually the same, 623: the (x, y) couples such that x and $\mathcal{I}(x, y)$ as tautologies are the same whatever the implication is, which suggests that this definition of tautology is a robust one: this definition provides a stable compromise of what can be considered as true and can be implied as true.

Regarding the invalid operators, the graphics show areas where the modus ponens is not valid: for the classical tautology, all counterexample conclusions are placed above the diagonal, which is consistent with its definition. This area is extended when using the certain tautology and for the truth tautology, the false conclusions are placed on the left side of the graph.

5 Conclusion and Future Works

The experimental study conducted in this paper shows that, when considering the classical Intuitionistic Fuzzy tautology definition, among 18 classical Intuitionistic Fuzzy implication operators, 14 fail to satisfy the validity property that is crucial for knowledge inference.

We introduced a stricter tautology definition, called truth tautology, that appears to possess 3 essential properties: first, it overcomes one usual critic of the classical definition, namely the fact that ignorance is considered as a tautology, which is counter-intuitive and not satisfactory. Second, the carried out experiments suggest that under this definition, most implications satisfy the desired validity property. This makes it a candidate viable tautology in terms of knowledge inference and thus artificial intelligence in the Intuitionistic Fuzzy framework. Third, the result comparison across these non invalid implication operators, and more precisely the number of candidate counterexamples, suggests that the truth tautology provides a stable compromise of what can be considered as true and can be implied as true.

Further works will focus on a formal detailed analysis of the proposed tautology definition and in particular the mathematical proof of the validity property for implication operators. Another perspective aims at re-examining Intuitionistic Fuzzy results established with the classical tautology definition and replacing it with its truth variant. It can be expected that the results hold, since the truth version is included in the classical one.

References

1. Atanassov, K.: Intuitionistic fuzzy sets. STUDFUZZ, vol. 35. Springer, Heidelberg (1999)
2. Atanassov, K., Gargov, G.: Elements of intuitionistic fuzzy logic: part I. Fuzzy Sets and Systems 95(1), 39–52 (1998)
3. Atanassov, K.T.: On some intuitionistic fuzzy implications. Comptes Rendus de l'Académie Bulgare des Sciences 59(1), 19–24 (2006)
4. Cornelis, C., Deschrijver, G.: The compositional rule of inference in an intuitionistic fuzzy logic setting. In: Striegnitz, K. (ed.) Proceedings of the 13th European Summer School in Logic, Language and Information, ESSLLI 2001, pp. 83–94 (2001)
5. Cornelis, C., Deschrijver, G., Kerre, E.E.: Implication in intuitionistic fuzzy and interval-valued fuzzy set theory: construction, classification, application. International Journal of Approximate Reasoning 35(1), 55–95 (2004)
6. El-Hakeim, K., Zeyada, F.: Comments on some versions of intuitionistic fuzzy propositional calculus due to K. Atanassov and G. Gargov. Fuzzy Sets and Systems 110(3), 451–452 (2000)
7. Negoita, C.V., Ralescu, D.A.: Applications of fuzzy sets to systems analysis. Interdisciplinary Systems Research, vol. 11. Birkhaeuser, Basel (1975)

Short Remark on Fuzzy Sets, Interval Type-2 Fuzzy Sets, General Type-2 Fuzzy Sets and Intuitionistic Fuzzy Sets

Oscar Castillo[1], Patricia Melin[1],
Radoslav Tsvetkov[2], and Krassimir T. Atanassov[3]

[1] Tijuana Institute of Technology,
Division of Graduate Studies,
Tijuana, Mexico
{ocastillo,pmelin}@tectijuana.mx
[2] Technical University of Sofia,
Faculty of Applied Mathematics and Informatics,
Sofia, Bulgaria
rado_tzv@tu-sofia.bg
[3] IBPhBME - Bulgarian Academy of Sciences,
Bioinformatics and Mathematical Modelling Department,
Sofia, Bulgaria
krat@bas.bg

Abstract. In this paper, we introduce specific types of intuitionistic fuzzy sets, inspired by the multi-dimensional intuitionistic fuzzy sets and the General Type-2 fuzzy sets. The newly proposed sets extend the opportunities of the General Type-2 fuzzy sets when modelling of particular types of uncertainty. Short comparison between concepts of interval type-2 fuzzy sets and intuitionistic fuzzy sets is given.

Keywords: Fuzzy set, Interval type-2 fuzzy set, General type-2 fuzzy set, Intuitionistic fuzzy set.

1 Introduction

In 1965, when Lofti A. Zadeh first proposed Fuzzy Sets (FSs) [14] his vision was set on giving more control over decision making, with his Fuzzy Logic an immeasurable amount of decision making situations could be easily modeled whereas hard logic, true or false, could not. This opened a new era in decision making with FSs that have been evolving since its initial days, first starting out with the concept of a Type-1 Fuzzy Sets, then coming into an Interval Type-2 Fuzzy Sets (IT2FSs) [10] and finally arriving at the current state of advanced form of FS, which is a Generalized Type-2 Fuzzy Sets (GT2FSs) [12, 13, 15]. Another extension of the FSs are Intuitionistic Fuzzy Sets (IFSs) [1, 2].

Here, in continuation of [7], we discuss the interpretability of the FSs, IT2FSs and GT2FSs by IFSs and one of their extensions – the mMultidimensional IFSs (MDIFSs) [3, 4, 5, 6].

P. Angelov et al. (eds.), *Intelligent Systems'2014*,
Advances in Intelligent Systems and Computing 322, DOI: 10.1007/978-3-319-11313-5_18

2 Fuzzy Sets, Interval Type-2 Fuzzy Sets and General Type-2 Fuzzy Sets

A Type-1 Fuzzy Set A, denoted by $\mu_A(x)$ where $x \in X$, represented by

$$A = \{\langle x, \mu_A(x)\rangle \mid x \in X\},$$

which is a FS which takes on values between the interval [0, 1].

A type-2 fuzzy set A', is characterized by the membership function [8, 11]:

$$A' = \{((x,u), \mu_{A'}(x,u)) \mid x \in X, u \in J_x \subseteq [0,1]\} \tag{1}$$

in which $0 \le \mu_{A'}(x,u) \le 1$. Another expression for A' is

$$A' = \int_{x \in X} \int_{u \in J_x} \mu_{A'}(x,u) / (x,u) \tag{2}$$

for $J_x \subseteq [0, 1]$, where $\iint$ denotes the union over all admissible input variables x and u. For discrete universes of discourse $\int$ is replaced by $\sum$, [9].

A General Type-2 Fuzzy Set A' is described by

$$A' = \int_X \mu_{A'}(x) \; / x = \int_X \left[\int_{J_x} f_x(u)/u \right] / x,$$

where $J_x \subseteq [0,1]$, x is the partition of the primary membership function, and u is the partition of the secondary membership function.

3 Intuitionistic Fuzzy Sets, Multidimensional Intuitionistic Fuzzy Sets and Their Interpretations of the Fuzzy Sets, Interval Type-2 Fuzzy Sets and General Type-2 Fuzzy Sets

Let X be a fixed universe. The intuitionistic fuzzy set over X has the form

$$A = \{\langle x, \mu_A(x), \nu_A(x)\rangle \mid x \in X\},$$

where $\mu_A(x)$, $\nu_A(x)$ are degrees of membership and non-membership of elements $x \in X$ to a fixed set $A \subset X$, $0 \le \mu_A(x), \nu_A(x) \le 1$ and

$$0 \le \mu_A(x) + \nu_A(x) \le 1.$$

Let the set X be fixed. The Intuitionistic Fuzzy Multi-Dimensional Set (IFMDS) A in X, $Z_1, Z_2, \ldots, Z_n$ is an object of the form

$$A(Z_1, Z_2, \ldots, Z_n) = \{\langle x, \mu_A(x, z_1, z_2, \ldots, z_n), \nu_A(x, z_1, z_2, \ldots, z_n)\rangle \mid$$

$$\langle x, z_1, z_2, ..., z_n \rangle \in X \times Z_1 \times Z_2 \times ... \times Z_n\},$$

where: $0 \leq \mu_A(x, z_1, z_2, ..., z_n), \nu_A(x, z_1, z_2, ..., z_n) \leq 1$ are the degrees of membership and non-membership, respectively, of the elements $\langle x, z_1, z_2, ..., z_n \rangle \in X \times Z_1 \times Z_2 \times ... \times Z_n$ and

$$0 \leq \mu_A(x, z_1, z_2, ..., z_n) + \nu_A(x, z_1, z_2, ..., z_n) \leq 1.$$

Now, following the ideas for IFMDS, we can modify a given IF2-dimensional-S (IF2DS) to the next form:

$$A_1 = \{\langle x, \mu_{A_1}(x,u), \nu_{A_1}(x,u) \rangle \mid x \in X, u \in J_x \subseteq [0,1]\}, \tag{3}$$

where $\mu_{A_1}(x,u)$ and $\nu_{A_1}(x,u)$ are the degrees of membership and non-membership of $x \in X$ and $u \in J_x \subseteq [0,1]$ and

$$0 \leq \mu_{A_1}(x,u) + \nu_{A_1}(x,u) \leq 1.$$

For this set we construct the set

$$A_1^* = \left\{\left\langle z, \mu_{A_1^*}(z), \nu_{A_1^*}(z) \right\rangle \mid z \in E\right\},$$

that for the universe $E = \bigcup_{x \in X} (\{x\} \times J_x)$ and $J_x \in [0, 1]$ is an IFS. Obviously, for each fixed u, u being the second projection of $z \in E$, the new set is bijective one to set A_1, while for a fixed x, A_1 is a projection of the new set.

When $\nu_{A_1^*}(x,u) = \nu_x(u) = 1 - \mu_x(u) = 1 - \mu_{A_1^*}(x,u)$, we directly obtain an interpretation of the standard Type-1 Fuzzy Set for the case of universe E.

We must mention that A_1 can also be expressed as

$$A_1 = \int_{x \in X} \left(\int_{u \in J_x} \langle \mu_x(u), \nu_x(u) \rangle \right) / x \quad J_x \subseteq [0,1] \tag{4}$$

or

$$A_1 = \{\langle x, \bigcup_{u \in J_x} \langle \mu_x(u), \nu_x(u) \rangle \rangle \mid x \in X, J_x \subseteq [0,1]\}$$

When X is a discrete set we have that

$$A_1 = \sum_{x \in X} \left(\sum_{u \in J_x} \langle \mu_x(u), \nu_x(u) \rangle \right) / x .$$

We can get different sets by expressions

$$A_1 = \{\langle x, \underset{u \in J_x}{*} \langle \mu_x(u), \nu_x(u) \rangle \rangle \mid x \in X, J_x \subseteq [0,1]\}$$

where $*$ can denote the operation $\bigcup$, $\vee$, $\sum$ or another.

We can give the following definition as a generalization of IF2DS A_1. The set

$$A_2 = \{\langle x, \mu_{A_2}(x,u), \nu_{A_2}(x,w)\rangle \mid x \in X, u \in J_x \subseteq [0,1], w \in H_x \subseteq [0,1]\} \quad (5)$$

in which $\mu_{A_2}(x,u)$ and $\nu_{A_2}(x,w)$ are the degrees of membership and non-membership of $x \in X$ and $u \in J_x \subseteq [0,1]$ and

$$0 \leq \mu_{A_2}(x,u) + \nu_{A_2}(x,w) \leq 1.$$

As above, A_1 can also be expressed in A_2-forms by

$$A_2 = \int_{x \in X} \left(\int_{u \in J_x} \langle \mu_x(u), \int_{w \in H_x} \nu_x(w)\rangle\right) / x \quad J_x \subseteq [0,1], \mathrm{H}_x \subseteq [0,1] \quad (6)$$

and

$$A_2 = \{\langle x, \bigcup_{u \in J_x} \langle \mu_x(u), \bigcup_{w \in H_x} \nu_x(w)\rangle\rangle \mid x \in X, J_x \subseteq [0,1], \mathrm{H}_x \subseteq [0,1]\}.$$

When X is a discrete set we have that

$$A_2 = \sum_{x \in X} \left(\sum_{u \in J_x} \langle \mu_x(u), \sum_{w \in H_x} \nu_x(w)\rangle\right) / x.$$

We can get different sets by expressions

$$A_2 = \{\langle x, \mathop{*}_{u \in J_x} \langle \mu_x(u), \mathop{*}_{w \in H_x} \nu_x(w)\rangle\rangle \mid x \in X, J_x \subseteq [0,1], \mathrm{H}_x \subseteq [0,1]\}$$

where $*$ can denote the operation $\bigcup$, $\vee$, $\sum$ or another.

At each value of x, say $x = x'$, the 2-D plane whose axes are u and $\mu_{\tilde{\tilde{A}}}(x',u)$ is called a *vertical slice* of $\mu_{\tilde{\tilde{A}}}(x,u)$. At each value of x, say $x = x'$, 2-D plane whose axes are w and $\nu_{\tilde{\tilde{A}}}(x',w)$ is called a *vertical slice* of $\nu_{\tilde{\tilde{A}}}(x,w)$. A *secondary membership function* is a vertical slice of $\mu_{\tilde{\tilde{A}}}(x,u)$. A *secondary non-membership function* is a vertical slice of $\nu_{\tilde{\tilde{A}}}(x,w)$. It is $\langle \mu_{\tilde{\tilde{A}}}(x',u), \nu_{\tilde{\tilde{A}}}(x',w)\rangle$ for every $u \in J_{x'} \subseteq [0,1]$ and for every $w \in H_{x'} \subseteq [0,1]$, i.e.,

$$\langle \mu_{\tilde{\tilde{A}}}(x',u), \nu_{\tilde{\tilde{A}}}(x',w)\rangle \equiv \langle \mu_{\tilde{\tilde{A}}}(x'), \nu_{\tilde{\tilde{A}}}(x')\rangle$$
$$= \int_{u \in J_{x'}} \langle \mu_{x'}(u), \int_{w \in H_{x'}} \nu_{x'}(w)\rangle / \langle u,w\rangle \; J_{x'} \subseteq [0,1] \, H_{x'} \subseteq [0,1], \quad (7)$$

in which $0 \le \mu_{x'}(u) + \nu_{x'}(w) \le 1$. Because for every $x' \in X$, we drop the prime notation on $\mu_{\tilde{\tilde{A}}}(x')$, and refer to $\mu_{\tilde{\tilde{A}}}(x)$ as a secondary membership function, we drop the prime notation on $\nu_{\tilde{\tilde{A}}}(x')$, and refer to $\nu_{\tilde{\tilde{A}}}(x)$ as a secondary non-membership function; it is an intuitionistic fuzzy set, which we also refer to as a *secondary set.*

Based on the concept of secondary sets, we can reinterpret a restricted 2-dimensional intuitionistic fuzzy set as the union of all secondary sets, i.e., using (7), *we can re-express in a vertical-slice manner*, as

$$\tilde{\tilde{A}} = \{\langle x, \mu_{\tilde{\tilde{A}}}(x), \nu_{\tilde{\tilde{A}}}(x)\rangle \mid x \in X\} \tag{8}$$

or, as

$$\tilde{\tilde{A}} = \int_{x\in X} \langle \mu_{\tilde{\tilde{A}}}(x), \nu_{\tilde{\tilde{A}}}(x)\rangle / \langle u, w\rangle) / x = \int_{x\in X} (\int_{u\in J_x} \langle \mu_x(u), \int_{w\in H_x} \nu_x(w)\rangle / \langle u, w\rangle) / x \tag{9}$$

When X is a discrete set we have that

$$\tilde{\tilde{A}} = \sum_{x\in X} (\sum_{u\in J_x} \langle \mu_x(u), \sum_{w\in H_x} \nu_x(w)\rangle / \langle u, w\rangle) / x.$$

We can get different sets by expression

$$\tilde{\tilde{A}} = \{\langle x, \underset{u\in J_x}{*} \langle u, \mu_x(u), \underset{w\in H_x}{*} \langle w, \nu_x(w)\rangle\rangle\rangle \mid x \in X, J_x \subseteq [0,1], \mathrm{H}_x \subseteq [0,1]\},$$

where $*$ can denote the operation $\bigcup$, $\vee$, $\sum$ or another.

In the same case, the intuitionistic fuzzy set $\tilde{A}$ is characterized by the membership and non-membership functions, and has the form:

$$\tilde{A} = \{((x,u), \mu_{\tilde{A}}(x,u), \nu_{\tilde{A}}(x,u)) \mid x \in X, u \in J_x \subseteq [0,1]\}, \tag{10}$$

in which $0 \le \mu_{\tilde{A}}(x,u) \le 1$, $0 \le \nu_{\tilde{A}}(x,u) \le 1$, $0 \le \mu_{\tilde{A}}(x,u) + \nu_{\tilde{A}}(x,u) \le 1$. The analogue of the second expression (2) now is:

$$\tilde{A} = \int_{x\in X} \int_{u\in J_x} \langle \mu_{\tilde{A}}(x,u), \nu_{\tilde{A}}(x,u)\rangle / (x,u) \tag{11}$$

for $J_x \subseteq [0, 1]$, where, as above, $\iint$ denotes the union over all admissible input variables x and u. For discrete universes of discourse $\int$ is replaced by $\sum$.

Let X be a universe, $g : \mathrm{X} \to [0, 1]$ be an interval function and $X_g = \bigcup_{x\in X} x \times g(x)$.

Let us consider the IFS

$$A_g^* = \{\langle z, \mu_{\tilde{A}}(z), \nu_{\tilde{A}}(z)\rangle \mid z \in X_g\}.$$

Then any type-2 fuzzy set $\tilde{A}$ can be represented by the IFS A_g^* with a suitable choice of g.

Below, we will express the same set $\tilde{A}$ by using the following definition.

At each value of x, say $x = x'$, the 2-D plane whose axes are u and $\mu_{\tilde{A}}(x',u)$ is called a *vertical slice* of $\mu_{\tilde{A}}(x,u)$. At each value of x, say $x = x'$, 2-D plane whose axes are w and $\nu_{\tilde{A}}(x',u)$ is called a *vertical slice* of $\nu_{\tilde{A}}(x,u)$. A *secondary membership function* is a vertical slice of $\mu_{\tilde{A}}(x,u)$. A *secondary non-membership function* is a vertical slice of $\nu_{\tilde{A}}(x,u)$. It is $\langle \mu_{\tilde{A}}(x',u), \nu_{\tilde{A}}(x',u)\rangle$ for every $u \in J_{x'} \subseteq [0,1]$, i.e.,

$$\langle \mu_{\tilde{A}}(x',u), \nu_{\tilde{A}}(x',u)\rangle \equiv \langle \mu_{\tilde{A}}(x'), \nu_{\tilde{A}}(x')\rangle = \int_{u \in J_{x'}} \langle \mu_{x'}(u), \nu_{x'}(u)\rangle / u \; J_{x'} \subseteq [0,1], \tag{12}$$

in which $0 \le \mu_{x'}(u) + \nu_{x'}(u) \le 1$. Because for every $x' \in X$, we drop the prime notation on $\mu_{\tilde{A}}(x')$, and refer to $\mu_{\tilde{A}}(x)$ as a secondary membership function, we drop the prime notation on $\nu_{\tilde{A}}(x')$, and refer to $\nu_{\tilde{A}}(x)$ as a secondary non-membership function; it is an intuitionistic fuzzy set, which we also refer to as a *secondary set.*

Based on the concept of secondary sets, we can reinterpret a restricted 2-dimensional intuitionistic fuzzy set as the union of all secondary sets, i.e., using (12), *we can re-express in a vertical-slice manner*, as

$$\tilde{A} = \{\langle x, \mu_{\tilde{A}}(x), \nu_{\tilde{A}}(x)\rangle \mid x \in X\} \tag{13}$$

or, as

$$\tilde{A} = \int_{x \in X} \langle \mu_{\tilde{A}}(x), \nu_{\tilde{A}}(x)\rangle / u) / x = \int_{x \in X} \left(\int_{u \in J_x} \langle \mu_x(u), \nu_x(u)\rangle / u \right) / x \tag{14}$$

When X is a discrete set we have that

$$\tilde{A} = \sum_{x \in X} \left(\sum_{u \in J_x} \langle \mu_x(u), \nu_x(u)\rangle / u \right) / x .$$

We can get different sets by expression

$$\tilde{A} = \{\langle x, \underset{u \in J_x}{*} \langle u, \mu_x(u), \nu_x(u)\rangle\rangle \mid x \in X, J_x \subseteq [0,1]\},$$

where $*$ can denote the operation $\bigcup$, $\vee$, $\sum$ or another.

The set $\tilde{A}$ can be extended to the form of $\tilde{\tilde{A}}$. The intuitionistic fuzzy set $\tilde{\tilde{A}}$ is characterized by a membership and a non-membership function and has the form:

$$\tilde{\tilde{A}} = \left\{ \left\langle (x,u,w), \mu_{\tilde{\tilde{A}}}(x,u), \nu_{\tilde{\tilde{A}}}(x,w) \right\rangle \mid x \in X, u \in J_x \subseteq [0,1], w \in H_x \subseteq [0,1] \right\}, \quad (15)$$

in which $0 \le \mu_{\tilde{\tilde{A}}}(x,u) \le 1$, $0 \le \nu_{\tilde{\tilde{A}}}(x,w) \le 1$, $0 \le \mu_{\tilde{\tilde{A}}}(x,u) + \nu_{\tilde{\tilde{A}}}(x,w) \le 1$. The analogue of the second expression (2), now is:

$$\tilde{\tilde{A}} = \int_{x \in X} \int_{u \in J_x} \int_{w \in H_x} \langle \mu_{\tilde{A}}(x,u), \nu_{\tilde{A}}(x,w) \rangle / (x,u,w) \quad (16)$$

for $J_x \subseteq [0,\ 1]$ and $H_x \subseteq [0,\ 1]$, where, as above, $\iint$ denotes the union over all admissible input variables x and u. For discrete universes of discourse $\int$ is replaced by $\sum$.

Let X be a universe, $g : X \to [0,\ 1]$ be an interval function. Let X be a universe, $f : X \to [0,\ 1]$ be an interval function and

$$X_{\langle f,g \rangle} = \bigcup_{x \in X} (\{x\} \times f(x) \times g(x)).$$

We can also note that another form of expressing the IFS $\tilde{\tilde{A}}$ is the form:

$$A^*_{\langle f,g \rangle} = \{ \langle z, \mu_{\tilde{A}}(\mathrm{pr}_1\, z, \mathrm{pr}_2\, z), \nu_{\tilde{A}}(\mathrm{pr}_1\, z, \mathrm{pr}_3\, z) \rangle \mid z \in X_{\langle f,g \rangle} \}.$$

4 Conclusion

In conclusion, we mention that in future a detail comparison between both concepts will be prepared and relationships between the operations, relations and especially the operators will be compared.

References

1. Atanassov, K.: Intuitionistic Fuzzy Sets. STUDFUZZ, vol. 35. Springer, Heidelberg (1999)
2. Atanassov, K.: On Intuitionistic Fuzzy Sets Theory. STUDFUZZ, vol. 283. Springer, Heidelberg (2012)
3. Atanassov, K., Szmidt, E., Kacprzyk, J.: On intuitionistic fuzzy multi-dimensional sets. Issues in Intuitionistic Fuzzy Sets and Generalized Nets 7, 1–6 (2008)
4. Atanassov, K., Szmidt, E., Kacprzyk, J.: On intuitionistic fuzzy multi-dimensional sets. Part 3. In: Developments in Fuzzy Sets, Intuitionistic Fuzzy Sets, Generalized Nets and Related Topics. Foundations, vol. I, pp. 19–26. SRI Polish Academy of Sciences, Warsaw (2010)

5. Atanassov, K., Szmidt, E., Kacprzyk, J.: On intuitionistic fuzzy multi-dimensional sets. Part 4. Notes on Intuitionistic Fuzzy Sets 17(2), 1–7 (2011)
6. Atanassov, K., Szmidt, E., Kacprzyk, J., Rangasamy, P.: On intuitionistic fuzzy multi-dimensional sets. Part 2. In: Advances in Fuzzy Sets, Intuitionistic Fuzzy Sets, Generalized Nets and Related Topics. Foundations, vol. I, pp. 43–51. Academic Publishing House EXIT, Warszawa (2008)
7. Castillo, O., Melin, P., Tsvetkov, R., Atanassov, K.: Short remark on interval type-2 fuzzy sets and intuitionistic fuzzy sets. Notes on Intuitionistic Fuzzy Sets 20(2), 1–5 (2014)
8. Mendel, J.M.: Uncertain Rule-Based Fuzzy Logic Systems: Introduction and new directions. Prentice Hall, New Jersey (2001)
9. Mendel, J.M., Bob John, R.I.: Type-2 Fuzzy Sets Made Simple. IEEE Transactions on Fuzzy Systems 10, 117–127 (2002)
10. Mendel, J.M., Liu, X.: Simplified Interval Type-2 Fuzzy Logic Systems. IEEE Trans. Fuzzy Syst. 21(6), 1056–1069 (2013)
11. Mendel, J.M., Mouzouris, G.C.: Type-2 fuzzy logic systems. IEEE Transactions on Fuzzy Systems 7, 643–658 (1999)
12. Rizzi, A., Livi, L., Tahayori, H., Sadeghian, A.: Matching general type-2 fuzzy sets by comparing the vertical slices. In: 2013 Joint IFSA World Congress and NAFIPS Annual Meeting (IFSA/NAFIPS), pp. 866–871 (2013)
13. Sanchez, M.A., Castro, J.R., Castillo, O.: Formation of general type-2 Gaussian membership functions based on the information granule numerical evidence. In: 2013 IEEE Workshop on Hybrid Intelligent Models and Applications (HIMA), pp. 1–6 (2013)
14. Zadeh, L.A.: Fuzzy Sets. Inf. Control 8, 338–353 (1965)
15. Zhao, L., Li, Y., Li, Y.: Computing with words for discrete general type-2 fuzzy sets based on α plane. In: Proc. of 2013 IEEE International Conference on Vehicular Electronics and Safety, pp. 268–272 (2013)

Part V
Fuzzy Cognitive Maps and Applications

Integrated Approach for Developing Timed Fuzzy Cognitive Maps

Evangelia Bourgani[1,2], Chrysostomos D. Stylios[2], George Manis[1], and Voula C. Georgopoulos[3]

[1] Dept. of Computer Science & Engineering, University of Ioannina, Ioannina, Greece
ebourgani@gmail.com, manis@cs.uoi.gr

[2] Dept. of Computer Engineering, TEI of Epirus, Artas, Greece
stylios@teiep.gr

[3] School of Health and Welfare Professions, TEI of Western Greece, Patras, Greece
voula@teipat.gr

Abstract. Time is a basic aspect in any field, as factors evolve over time and influence the progression of any procedure. This work proposes an integrated approach to developing Timed Fuzzy Cognitive Maps (T-FCMs), an extension of FCMs that can handle uncertainty to infer a result. TFCMs take into consideration the time evolution of any procedure and permit the production of intermediate results and the influence of exterior parameters. It described the proposed method to develop T-FCM and then T-FCM is applied to develop a Medical Decision Support System.

Keywords: Fuzzy Cognitive Map, Time Evolution, Soft Computing, Medical Decision Support Systems.

1 Introduction

Nowadays systems are characterized by high complexity and it is essential to take into consideration various and complementary factors and reengineering alternatives before concluding to a final decision. Decision making needs exist in almost any field, while immediate and reliable decisions may be necessary in an emergent case. Decision Support Systems (DSS) can provide assistance during the process of decision making, which involves the comparison and selection of the best (or optimal) decision. Fuzzy Cognitive Maps (FCMs) have been used successfully to develop DSS because they can handle even incomplete or conflicting information.

In this work, we insert the time unit in FCMs for cases that time determines and changes the route of progression and so the final result. Intermediate results illustrate the progression of a case, while influences during the runtime can provoke concept triggering and/or deactivation, as well as changing the decision. An example from the medical domain illustrates the proposed procedure. Alternative models have been developed that insert temporal concepts into FCM. Hagiwara proposed the extended FCM (e-FCM) , introducing the need for non-linear weights, conditional weights and

P. Angelov et al. (eds.), *Intelligent Systems'2014*,
Advances in Intelligent Systems and Computing 322, DOI: 10.1007/978-3-319-11313-5_19

time delays weights. Park proposed the Fuzzy Time Cognitive Maps (FTCMs) with dummy nodes for showing time lags between nodes . Carvalho et al. proposed Rule Based FCM (RB-FCM) where a parameter called B-Time represents the highest level of temporal detail that a simulation can provide in the modeled system . Miao et al. introduced dynamic functions for the arcs to represent the dynamic and temporal effects of causal relationships . Zhong et al. proposed the Temporal FCMs (tFCM) for discrete linear temporal domain . But, some of them use only numerical weights, other fixed time lags or unique B-Time, while others are highly complicated.

2 Fuzzy Cognitive Maps

Fuzzy Cognitive Maps (FCMs) is a soft computing modeling technique for complex systems. FCMs introduced as an extension to Cognitive Maps . They are a graphical representation for the description and modeling of the behavior and operation of a system. FCM supports causal knowledge reasoning process and belongs to neuro-fuzzy systems that aim at solving decision making and modeling problems. FCM resembles human reasoning; it relies on the human expert knowledge for a domain, making associations along generalized relationships between domain descriptors, concepts and conclusions. It models any real world system as a collection of concepts and causal relations between concepts [8].

FCMs are dynamical, fuzzy signed directed graphs, permitting feedback, where the weighted edge w_{ij} from causal concept C_i to affected concept C_j describes the amount by which the first concept influences the latter (Fig.1). Experts design and develop the structure of the system and the nodes that represent the key factors of the system operation. They determine the way of network's interconnections, using linguistic variables to describe the relationships among concepts, then all variables are combined and weights are determined. Learning methods and historical data have also applied .

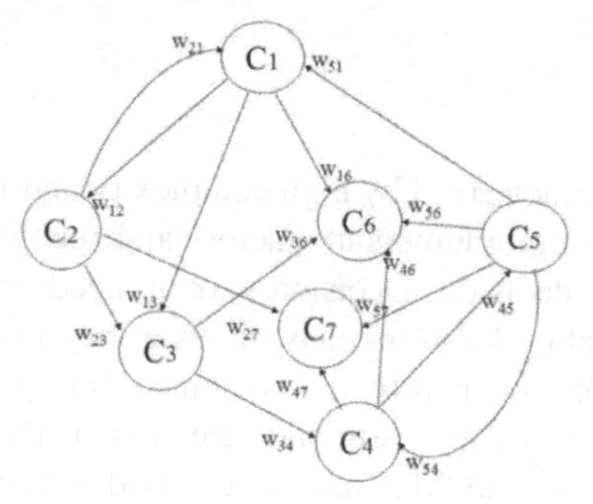

Fig. 1. The Fuzzy Cognitive Map model

Nodes of the graph are concepts, which correspond to variables, states, factors and other characteristics that are used in the model and describe the behavior of the system. A FCM is a conceptual network; the connection between the signed and weighted arcs represents the causal relationships. That is, if $w_{ji}>0$ then concept C_j increases C_i, if the weight is $w_{ji}<0$ then concept Cj decreases Ci, while if $w_{ji}=0$ concept C_j has no causal effect on C_i.

3 A Method for Timed FCM Development

One simulation approach to approximate real situations and check results could be by resembling the human way of thinking. It includes the (represented) knowledge (concepts-factors) and the progress of a case during the time, taking into account the influence of each concept either it is clearly related or not. However, the progress of a case may be related with many concepts from various fields and for every time units that determine the final decision. This is highly difficult for the human mind but decision making models permit easier and faster decisions and/or rest any doubts to estimations. The substantial elements that define and determine the problem are: knowledge and time. Knowledge can be inferred from literature and it is combined with all aspects that a decision maker has adopted from his/her experience, observing data, etc. Time insertion is imperative as it can determine the evolution of a case, influence and change the dependence and interconnections of the initial weights. Here, a methodology is proposed to suggest the concepts and the weights per time, simulate and take the final results for an under investigation case. The overall procedure is illustrated with a flow chart of Fig.2. Specifically, each step includes the following actions:

- *Concepts*

Step 1: Determination of concepts and categorizing as factor and decision concepts.
Step 2: Grouping factor-concepts into areas with respect to their origin (concepts that come from experts, literature, observation data, questionnaires etc.). So every aspect of a problem is taken into consideration, while every possibility, even the rarest can be included as the more diversity there is the more complete and closer to a real problem will be, taking every possible circumstance into consideration.
Step 3: After grouping, a credibility value can be assigned by the experts for each group, judging the source and its reliability.
Step 4: Experts define the time span. This time unit depends on the field and the represented knowledge. There are cases that a long-term supervision is needed or in others the time unit is shorter. The ideal time unit will be the exact time that an important change can affect and provoke a change to the weight dependencies.
Step 5: Additional parameters definition. These parameters are characteristics that are significant and basic to the progression of a case as they can change completely the route of progression and as a result the final decision. Literature and experts define these parameters and the direction of their influence to the overall procedure.

- *Weights per time unit*

Step 6: Weight definition. Experts define the weights that take various values according to each time unit. That is, weights of each interconnection for each time unit are defined linguistically, using either the conclusions of related literature, studies and/or other sources as is described to the following section. This means that each interconnection between concepts takes values proportionally to the time unit. Thus, the influence can be increased or alternatively decreased, giving a detailed illustration of the change direction for each time unit.

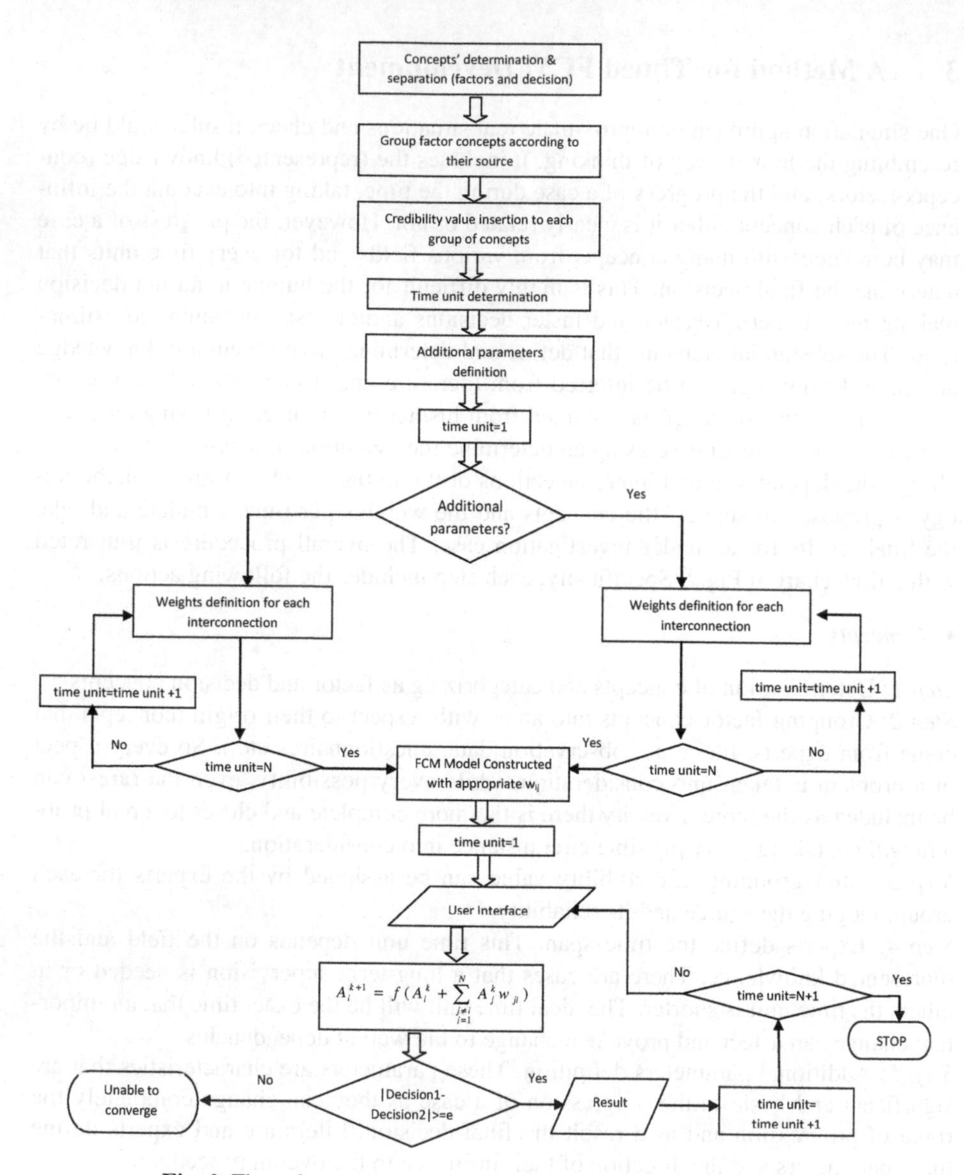

Fig. 2. Flow chart of the proposed method for developing T-FCMs

Step 7: Here the step 6 procedures are repeated until all the possible combinations are fulfilled and defined, for each additional parameter separately and combined with the other additional parameter(s), taking all the possible sub-cases into consideration. The result of this weight setting will be an array that each cell will be a matrix of the interconnections/dependencies to the correspondent time and the activated additional parameters. The final weight will represent the weight of a specific case (under the specific activated parameters and time unit), which depends both on the simulated time unit and the additional parameters. The final weight will be given by:

$$w_{\gamma,t,ij} = \left[w_{\gamma 1,t1,ij}, w_{\gamma 1,t2,ij} \dots w_{\gamma 1,tn,ij}, w_{\gamma 2,t1,ij}, w_{\gamma 1,t2,ij} \dots w_{\gamma n,t1,ij}, w_{\gamma n,t2,ij} \dots w_{\gamma n,tn,ij}\right] \quad (1)$$

where γ shows the activated state of additional parameters. Table 1 defines the possible states that the parameters can take. The index n of γ shows which weight combinations are going to be used during the specific runtime. In order to calculate the final value of $w_{\gamma,t,ij}$, a process is followed to define the remaining individual variables. Table 1 presents the $p_1, p_2, \dots p_n$ that are the additional characteristics that influence completely the overall procedure. These parameters take binary values, denoting whether this characteristic is activated or not, setting the weights accordingly. The indicator *t* shows the time interval that is running each time, while the indicators *ij* show the two concepts that influence each other. Thus, weights per time unit are defined, depend and described on these three parameters: γ, *t*, *ij*. The γ_0 state is the case that no additional parameter is activated and the procedure will follow the basic FCM without exterior changes influencing the overall simulation. The other states are possible individual parameters that their triggering can change the route of simulation.

Table 1. Matrix of states of the activated parameters

γ_n	Activated parameters ($p_1, p_2, \dots . p_n$)
γ_0	000….0
γ_1	100….0
γ_2	010….0
$\gamma_{\dots}$	111…..1

Subsequently, for each state (γ) the interconnections among concepts for each time unit should be defined. This information is illustrated in the variable $d_{\gamma,t,ij}$, where γ and *t* will determine the state and the time unit, respectively. Table2 represents the weight dependencies ($d_{\gamma,t,ij}$) for each state.

Table 2. Correspondent interconnections between time and additional parameters

γ_n / time unit	γ_0	γ_1	…	γ_n
t_1	$d_{\gamma 0,t1,ij}$	$d_{\gamma 1,t1,,ij}$	$d_{\gamma \dots,t1,ij}$	$d_{\gamma n,t1,ij}$
t_2	$d_{\gamma 0,t2,ij}$	$d_{\gamma 1,t2,ij}$	$d_{\gamma \dots,t2,ij}$	$d_{\gamma n,t2,ij}$
….	$d_{\gamma 0,t\dots,ij}$	$d_{\gamma 1,t\dots,ij}$	$d_{\gamma \dots,t\dots,ij}$	$d_{\gamma n,t\dots,ij}$
t_n	$d_{\gamma 0,tn,ij}$	$d_{\gamma 1,tn,ij}$	$d_{\gamma \dots,t,n,ij}$	$d_{\gamma n,tn,ij}$

Particularly, $d_{\gamma,t,ij}$ is an array whose dimensions are depending to the number of variables γ and *t*. Each cell of this array will be a matrix and the dimensions of this matrix will be defined as follows: if the number of concepts is m, then the $d_{\gamma,t,ij}$ will be an *mxm* matrix that will contain all the weights among the concepts, for the corresponding time unit and for each additional parameters combination (each state).

Table 3 contains the values of $d_{\gamma,t,ij}$ for a specific time unit (t_1). C_1 to C_n are the numbered concepts of the FCM model that characterize the simulated model and are determined in step 1.

Thus, $w_{\gamma,t,ij}$ refers to the value of weight of a particular case. These values will represent the direction of change compared to the originally constructed FCM (default FCM) with no additional parameters activated. For example, the weight $w_{\gamma1,t2,23}$ is referring to the weight between concepts C_2 and C_3 at time unit t_2 for the case when only p_1 parameter is activated. Table 3 contains just the values of γ_1 case and time unit 1 i.e. t_1. In the same way, the whole array should be completed.

Table 3. Weights for each interconnection and time unit t_1

	C_1	C_2	.....	C_n
C_1	$w_{\gamma1,t1,11}$	$w_{\gamma1,t1,12}$	.....	$w_{\gamma1,t1,1n}$
C_2	$w_{\gamma1,t1,21}$	$w_{\gamma1,t1,22}$	.....	$w_{\gamma1,t1,2n}$
....	....	....	.....	....
C_n	$w_{\gamma1,t1,n1}$	$w_{\gamma1,t1,n2}$	.....	$w_{\gamma1,t1,nn}$

- Simulation and Results

Step 8: Run per time unit giving the result for each time span. The time unit is characteristic for each application and field. Thus, the results per time unit are adaptable on each change and influence, reducing the possibility of an erroneous decision.
Step 9: The simulation will stop when the possible decisions will have sufficient difference among them. This difference depends on the nature of problem and the field that decision support system is used. In the case of no convergence, redesigning of the T-FCM model is needed.

3.1 Defining the interconnection dependencies (weights)

In order to define the weights between each interconnection and each time unit, experts need to recall the progression of the phenomenon during the time. They should define the initial weights, denoting which concepts' dependencies (weights) have lower or higher influence during the progression of the case. The direction of this change depends on the contribution of the additional parameters. During the time the interdependencies among some factors have different degrees of influence compared to others and this change depends on both time and the additional parameters; interconnections can become weaker or stronger, while some concepts may be deactivated and others activated.

Initially, experts determine the interconnections using fuzzy rules; a linguistic variable that describes the relationship between two concepts is inferred according to each expert who also determines the grade of causality between the two concepts. After the collecting and definition of concepts, experts are asked to propose the degree of influence among all the concepts, using IF- THEN rules. Their answers will be influenced by their knowledge, experience and etc.

These rules are according to the following statement and they will construct the weight matrix of the default FCM (without additional parameters): For time unit 0….N: "IF a {none, very-very low, very low, low, negatively medium, positively medium, high, very high, and very very high} change in concept value C_i occurs THEN a {none, very-very low, very low, low, negatively medium, positively medium, high, very high, and very very high} change in value C_j is caused THUS the influence of C_i to C_j is a T_{Cij}{influence}". In this way each influence is determined. The inferred fuzzy weights across experts are aggregated and deffuzzified giving an initial numerical weight in the interval [-1, 1]. Thus, the basic weights and weight matrix are defined.

However, the progression of a case depends on additional factors, which are different for each one case. These are defined from various sources as well. Taking these changes into account during the simulation process, experts need to define the direction of change for each time unit. Thus, for each interconnection influenced by the time and the additional parameters, the direction of change is determined leading to the appropriate weight change (step 6 and 7). Linguistic degrees are used to indicate the direction of change (such as none, very-very low, very low, low, negatively medium, positively medium, high, very high, and very very high). These degrees correspond to negative or positive direction. In this case the statement will be: For time unit 0 ....N: "IF (Additional Parameter 1=TRUE (AND/OR Additional Parameter 2...3 =TRUE) AND IF time has a { none, very-very low, very low, low, negatively medium, positively medium, high, very high, and very very high} change (increase or decrease) in concept value Ci THEN a { none, very-very low, very low, low, negatively medium, positively medium, high, very high, and very very high} change in value C_j is caused THUS the influence of C_i to C_j is a T'_{Cij}{influence}ELSE {keep the default value for the corresponding time unit (T_{Cij}{influence}).

The inferred new fuzzy weights, for each time unit and for each parameter are aggregated and defuzzified producing a new numerical weight in the interval [-1,1]. This procedure can better approximate real life situations, as the weights can change per time unit incorporating more elements and parameters, taking into consideration every possible situation, no matter how rare or underestimated it may be.

4 Time Varying Method for Different Fields

The described methodology is suitable for various fields and applications that time is substantial and influences the overall procedure. For every field, experts usually make decisions, control and check a process, predict the progression etc. But time should be taken into account for more realistic results. In medicine and business-management, for example, the lack of the concept of time, regarding the order and the reaction time of a change, may provoke important changes in patient state or influence the output in a strategic/economic problem. The proposed method is capable to give results for each time unit separately. This is highly important as some concepts may have been underestimated, eluded and/or emerged to a following stage.

This method lets the professionals to insert changes to the system, keeping values from the previous stages and observe the influence from the changes at the corresponding time unit, which can give a different result. Thus, expert is able to check and/or control the result provoking different changes to various time units, so different scenarios can be applied as the model follows the progression of a case and gives the possible results for long-term or short term decision cases.

Such systems that approximate real situation demands the incorporation of each possible situation coming from any related domain. Time is essential in real complex systems. Decision systems have to incorporate a high amount of data and information from interdisciplinary sources and, in addition, information may be vague, missing or not available. Such decision support complex systems involve inexact, uncertain, imprecise and ambiguous information [10]. These systems can provide assistance in crucial judgments, particularly for inexperienced professionals.

5 A Paradigm from Medicine: FCM-Medical Decision Support Tool for Differential Diagnosis

The aforementioned process can be applied to various fields. The following section describes the application of the proposed method for differential diagnosis in medicine. In medicine, professionals need to make immediate decisions, eliminating the possibility of error. Generally, medical decisions can be referred to diagnosis, treatment, prognosis etc. Some cases need a long-term while other short-term evaluation.

In this example, we apply the time varying FCM for differential diagnosis of two pulmonary diseases: Acute Bronchitis and Community-acquired pneumonia. Their symptoms are highly confused and usually their symptoms can lead to an error diagnosis [11]. This case needs short-term evaluation. The TFCM-DSS tool will lead to a daily evaluation, so that the professional will be able to judge and make a clearer decision. This is important for making immediate decisions, as when the patient visits the doctor, s/he has to decide according to the patent's symptoms and clinical examination [12],[13], [14]. For this case, an assistant tool would be valuable in order to prevent an undesirable complication. Thus, it can present a suggestion to the doctor to early estimate the patient's situation until the laboratory tests will be available. Besides, it gives to the opportunity to doctor to try various scenarios and to study how the additional parameters can influence the result. TFCM-DSS tool has the following characteristics:

Concepts: Concepts are retrieved from literature and experts. They represent the symptoms of these two diseases. They have been organized according to the nature of source; that is the activated concepts came from the patient or from the doctor. Table 4 contains these factor concepts and their separation. Doctor has a model that can handle factors from patients and from clinical findings of the under examination patient, giving the appropriate credibility to doctors' findings. The weights have a credibility value compared to those that came from patient's description of symptoms because patients' answers are influenced from fear, overestimation and/or underestimation of their symptoms. The time unit has been defined as one day, as short term progression of the disease is needed and the patient should follow a different prescription for each case. For these two diseases some additional parameters, which are individual characteristics of each patient, are defined. Table 5 contains these additional elements. Their activation in combination with the symptoms and clinical findings can lead to a different result. These parameters are important as they determine the result of the overall procedure. The age range 14-65 has been regarded as the firstly (default) constructed model with no additional parameter activated.

Table 4. Categorization of concepts

Factor Concepts		
According to patient'sanswers	**Accordingto doctor's clinical exam**	**Diagnosis Concepts**
C1:body temperature C2: cough C3:bloody sputum C4:chest pain C6:sore throat C7:nasal obstruction C8: malaise C9: rhinorrhea C10: dyspnea C15:sputum production	C5: hypoxemia C12: bronchical breath sound C13: crackles C14: beats/min C11: tachypnea C16: scatter rhonchi C17: wheezing C1:body temperature	C18:Community-acquired pneumonia C19: Acute Bronchitis

Weights per time unit: Experts define the weights, they take into consideration which concepts are influenced by time and give the proportional weight value, according to the additional parameters and the direction of weight change compared to the default one (with no additional parameters activated) (Table 3).

Simulation and Results: The simulation for each day will give a daily estimation of the disease progression, letting the doctor judge the results. In order to give the system a decision, the distance that has been determined for this case is ≥10%.

Table 5. Possible states for the medical decision support tool

γ_n	**Age:<14, Age:>65, Smoking, Co-morbidity**
γ_0	0000
γ_1	1000
...	...
γ_{15}	1111

Figure 3 illustrates the basic FCM model for the differential diagnosis of the two pulmonary diseases according to the described procedure. It should be underlined that some interconnections are omitted as the high complexity makes it unreadable.

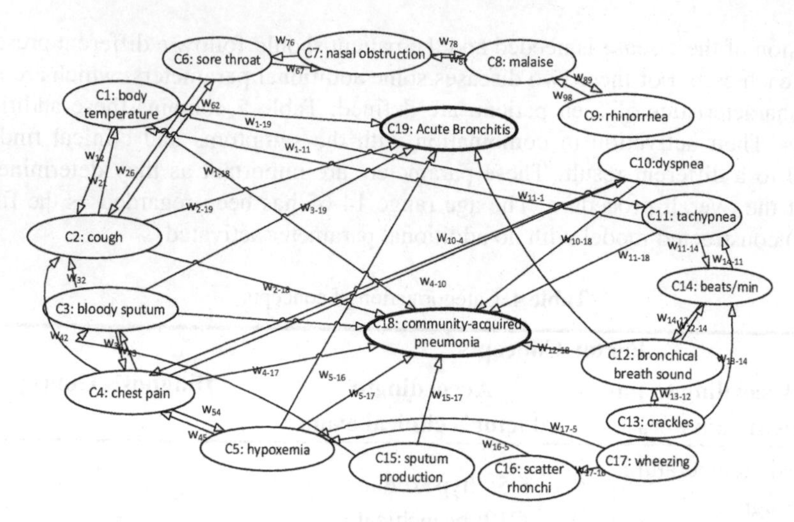

Fig. 3. T-FCM model for the pulmonary differential diagnosis

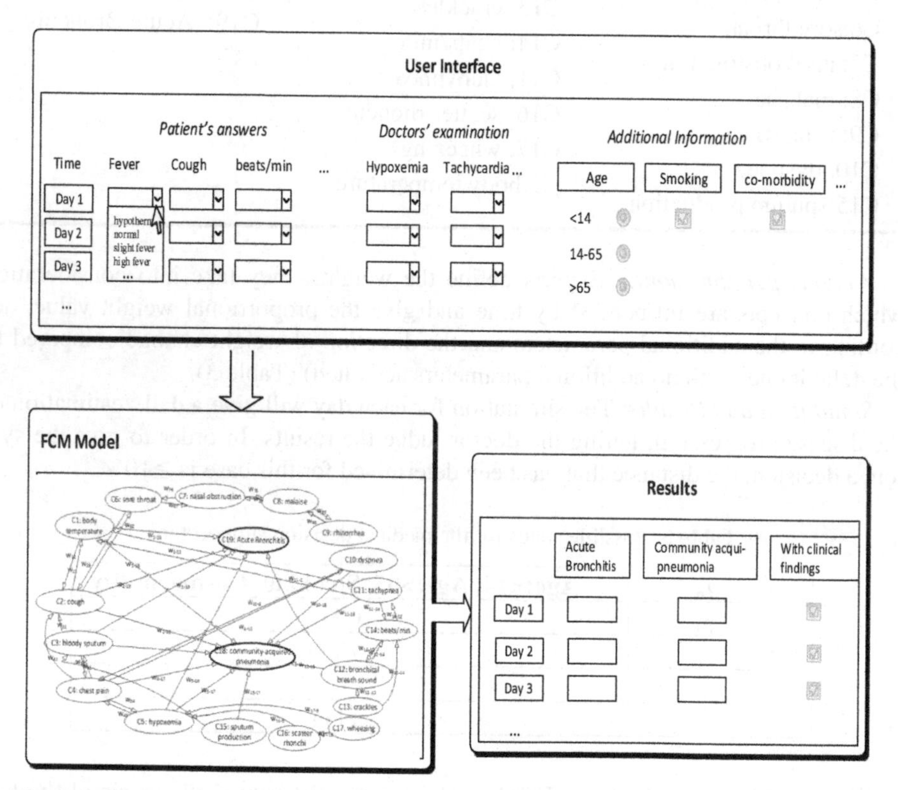

Fig. 4. An illustration of the FCM decision support system for the differential diagnosis

5.1 User Interface of the TFCM-Medical DSS Tool

The user interface contains all the possible situations that a doctor characterizes the patient's situation. The T-FCM takes into consideration all the possible degrees of a factor. For example, the body temperature factor-concept can be hypothermia (<35.6°), normal (35.7-37.3°), slight fever (37.4-37.8°), high fever (37.9-39°), very high fever (39.1-40.9°), hyperpyrexia (>41°). The T-FCM is able to model and handle various numbers of linguistic values and various types of values, as each concept has different characterization [15].

Fig. 4 illustrates a possible interface of the tool that the doctor would be able to set the appropriate values for each concept. The results for each simulation would appear in a window giving the decision for each day. Thus, the doctor will be able to make a decision for the patient, giving the appropriate prescription and/or laboratory tests.

6 Conclusions

The proposed method introduces the concept of time, which is essential for many problems to better approximate real situations. This procedure takes into consideration the evolution of a case, inserting additional parameters that can change the route of progression. These additional parameters are basic elements that make more particular and personalized a simulated case. The fact that concepts' values can change in proportion to the additional parameters' activation and/or deactivation which cause the change, make the model dynamic and adaptable to real situation problems. It takes each exterior factor into consideration and directs the evolution accordingly. Thus, it is suitable for observing the progression of a case and trying various scenarios by triggering and/or deactivating different concepts, achieving a better decision and better understanding of the evolution of a process. An example from the medical domain has been described for differential diagnosis among two diseases with many factors in common.

However, the process of constructing the T-FCM model is intensive for experts and other contributors, who need to determine the progression of the influenced by the time weights for each time unit. Thus, a more automated procedure is needed reducing the human intervention. The described tool is an assistant tool that will support the professional to have a general evaluation of a case progression, making faster decisions [16].

Acknowledgments. This research work was supported by the joint research project "Intelligent System for Automatic CardioTocoGraphic Data Analysis and Evaluation using State of the Art Computational Intelligence Techniques" by the programme "Greece-Czech Joint Research and Technology projects 2011-2013" of the General Secretariat for Research & Technology, Greek Ministry of Education and Religious Affairs, co-financed by Greece, National Strategic Reference Framework (NSRF) and the European Union, European Regional Development Fund.

References

1. Hagiwara, M.: Extended fuzzy cognitive maps. In: IEEE International Conference on Fuzzy Systems, San Diego, CA, pp. 795–801 (1992)
2. Park, K.S.: Fuzzy cognitive maps considering time relationships. Computer Studies 42, 157–168 (1995)
3. Carvalho, J.P., Tomé, J.A.: Rule Based Fuzzy Cognitive Maps- Fuzzy Causal Relations. In: Computational Intelligence for Modelling, Control and Automation (1999)
4. Miao, Y., Liu, Z.-Q., Siew, C.K., Miao, C.Y.: Dynamical Cognitive Network—an Extension of Fuzzy Cognitive Map. IEEE Transactions on Fuzzy Systems 9, 760–770 (2001)
5. Zhong, H., Miao, C., Shen, Z., Feng, Y.: Temporal Fuzzy Cognitive Maps. In: IEEE International Conference on Fuzzy Systems (FUZZ 2008), pp. 1830–1840 (2008)
6. Kosko, B.: Fuzzy Cognitive Maps. International Journal of Man-Machine Studies 24, 65–75 (1986)
7. Axelrod, R.: Structure of Decision: the cognitive maps of political elites. Princeton, NJ (1976)
8. Georgopoulos, V.C., Stylios, C.D.: Fuzzy Cognitive Map Decision Support System for Successful Triage to Reduce Unnecessary Emergency Romm Admissions for Elderly. In: Seising, R., Tabacchi, M. (eds.) Fuzziness and Medicine: Philosophical Reflections and Application Systems in Health Care. STUDFUZZ, vol. 302, pp. 415–436. Springer, Heidelberg (2013)
9. Papageorgiou, E.I., Stylios, C.D., Groumpos, P.: Active Hebbian Learning Algorithm to train Fuzzy Cognitive Maps. Int. Journal of Approximate Reasoning 37, 219–249 (2004)
10. Oja, E.: Simplified neuron model as a principal component analyzer. Journal of Mathematical Biology 16, 267–273 (1982)
11. Pulmonary Disorders. In: The Merck Manual of Diagnosis and Therapy, ch. 5, pp. 432–436. Merck Research Laboratories (2006)
12. Sprogar, M., Lenic, M., Alayon, S.: Evolution in medical decision making. Journal of Medical Systems 36, 479–489 (2002)
13. Georgopoulos, V., Stylios, C.: Complementary case-based reasoning and competitice Fuzzy Cognitive mapd for advanced medical decisions. Soft Computing 12(2), 191–199 (2008)
14. Bourgani, E., Stylios, C.D., Manis, G., Georgopoulos, V.: A study on Fuzzy Cognitive Map structures for Medical Decision Support Systems. In: 8th Conf. the European Society for Fuzzy Logic and Technology (EUSFLAT 2013), Milano, Italy, September 11-13, pp. 744–751 (2013)
15. Bourgani, E., Stylios, C.D., Manis, G., Georgopoulos, V.C.: Time Dependent Fuzzy Cognitive Maps for Medical Diagnosis. In: Likas, A., Blekas, K., Kalles, D. (eds.) SETN 2014. LNCS, vol. 8445, pp. 544–554. Springer, Heidelberg (2014)
16. Bourgani, E., Stylios, C.D., Manis, G., Georgopoulos, V.C.: Timed Fuzzy Cognitive Maps for supporting obstetricians' decisions. In: Lacković, I. (ed.) 6th European Conference of the International Federation for Medical and Biological Engineering. IFMBE Proceedings, vol. 45, pp. 753–756. Springer, Heidelberg (2015)

Linguistic Approach to Granular Cognitive Maps
User's Tool for Knowledge Accessing and Processing

Władyslaw Homenda[1] and Witold Pedrycz[2,3]

[1] Faculty of Mathematics and Information Science, Warsaw University of Technology, ul. Koszykowa 75, 00-662 Warszawa, Poland
[2] System Research Institute, Polish Academy of Sciences, ul. Newelska 6, 01-447 Warszawa, Poland
[3] Department of Electrical & Computer Engineering, University of Alberta, Edmonton, T6R 2G7 AB, Canada
homenda@mini.pw.edu.pl, wpedrycz@ualberta.ca

Abstract. In the paper we study an issue of building tools for accessing and processing knowledge. Such tools are fundamental facets of user friendly human-machine communication. An efficient understanding of data becomes of paramount relevance when dealing with a wealth of human-system interaction and communication. The underlying objective of this paper is to elaborate concepts of aforementioned tools based on automatic data understanding. The syntactic and semantic facets of data integrated in frames of granularity act as fundamental mechanism for data structuring in a form of granular cognitive maps. Linguistic approach to data structuring allows for establishing a suitable perspective at the problem at hand where knowledge structures need to be accessed and processed. Rather than embarking on the formal framework, our intent is to illustrate a realization of the paradigm in the realm of music information, its processing and understanding.

Keywords: granular cognitive maps, knowledge accessing, knowledge processing, linguistic approach, syntax v.s. semantics.

1 Introduction

In this paper we introduce the idea of linguistic approach of granular cognitive maps, which are the evident and inherent feature of intelligent human-machine communication. The idea is based on the syntactic and semantic characterization of data. Both syntactic and semantic facets of data are integrated in the frame of granular computing paradigm. Data granulation plays a central role in all processes of data understanding by facilitating establishing a suitable perspective at the problem at hand where knowledge structures need to be accessed and processed.

Intelligent human-machine as well as interpersonal communication is understood as a meaningful exchange of information. We assume that fundamental attributes of meaningful communication requires that parties participating in

P. Angelov et al. (eds.), *Intelligent Systems'2014*,
Advances in Intelligent Systems and Computing 322, DOI: 10.1007/978-3-319-11313-5_20

the communication (a) unambiguously decode passed information and (b) comprehend and reason about relationships between data in the same way. These attributes are contextually and/or implicitly obvious in the case of interpersonal communication. However, when human-machine communication is concerned, attributes of a meaningful communication usually are neither explicitly articulated, nor implicitly assumed. Human-centric communication requires at least that inanimate being (the machine) has (c) to use a map of concepts and relations between concepts the same as human and (d) to draw conclusions (reason) analogously as human. All these features are carried out in frames of linguistically accomplish granular cognitive maps.

In this paper we refer to fundamental ideas of syntax, semantics, granulation and understanding, which are described in c.f. [2].

A linguistic approach to granular cognitive maps is intended as a user's tool for accessing and processing knowledge in structured data spaces. It involves the paradigm of automatic data understanding, which is emanation of integrated syntactic description, semantic analysis and granular structuring as general methods. In our opinion, both automatic data understanding and linguistic interpretation of granular cognitive maps are the universal approaches to knowledge accessing and processing. However, details of such approaches depend on the application domain. Details of presented methodologies would differ in the domain of music and domain of geodesy. Both domains differ in basic symbols used, ways of assembling basic symbols, relations between syntactic and semantic items and methods of their structuring, operations specific for every domain (e.g. transposition is specific for music domain) etc. Therefore, details discussed in this paper may not be universally applicable in any domain. Anyway, there are common aspects, for instance, we have a languages of knowledge description (paginated music notations and geodesic maps), syntactic and semantic structures that would be integrated in frames of granular paradigm, common operations to be performed (e.g. selection and searching are common for both domains in example) etc.

An attempt to develop detailed methods of automatic data understanding, which would be domain independent, is not possible at current level of technology, if possible at all. This is not a place to open discussion on this opinion. Therefore, to give a justification, we refer to GPS paradigm (General Problem Solver), which was intensively studied in sixties of 20th century in frames of artificial intelligence domain and it finally failed. Our Therefore, we reflect meanings of automatic data understanding and knowledge accessing and processing in music data described in the language of paginated music notation.

Methods presented in the paper are illustrated with fragments of scores of pieces of classical music shown in Figure 1. Our discussion embraces only elements of these scores. Nevertheless, the methods illustrated with fragments of these pieces can be expanded on the entire pieces as well as on other scores. Alike, the study focused on spaces of music information oriented to paginated music notation may be adopted to other types of descriptions of music information, e.g. Music XML cf. [1,9], Braille Music cf. [7], MIDI cf. [8].

10. Polonaise

Polonaise from Sonata for Flute and Piano by Beethoven, the beginning excerpt

J. S. Bach, Suite (Ouverture) No. 3 in D major, BWV 1068, transcription for Piano by T. A. Johnson, the beginning excerpt of the second movement Air

Suite espagnole [Música impresa]. III, Sevilla: sevillanas

Isaac Albéniz

I. Albeniz, Suite espagnole, Sevilla: sevillanas, the beginning excerpt

Fig. 1. Scores - excerpts

2 Forming Accessible Knowledge Units

2.1 Syntactic Approach

The discussion on describing paginated music notation is based on common definition of grammars and of context-free grammars, c.f. [2]. We assume that the reader is familiar with the basic notions of mathematical linguistic.

Lexicon is the key concept of granulation of syntax. Lexicon is a set of language constructions, which describe objects and local and global structures of objects in the real world. Elements of lexicon are phrases generated in a grammar supplemented with parts of derivation tree build on them. These concepts are illustrated in Figure 2, c.f. also [2].

2.2 Valuation Relation

Lexicon elements describe objects in a real world. In the case of music information the real world includes sounds produced when a piece of music is being performed. For simplicity, we assume that sounds are of the form of notes described by three parameters: pitch, beginning time and duration of a note. Such the triple describing a sound/note roughly can be supplemented by other features, e.g. volume (loudness), articulation (legato, staccato, portato etc.).

Semantics, as interpreted in this paper, immerses the lexicon in the space of sounds/notes. Such the immersion is a relation defined in the Cartesian product of lexicon and the space of sounds/notes. Note that lexicon elements shown in Figure 2 a-d may define many objects in the world, c.f. [2].

Summarizing - semantics is a relation V in the lexicon L and the world W, called *valuation relation*:

$$V \subset L \times W$$

2.3 Granular Space Formation

The processing of music notation leads to extracting information from a score. A collection of data items is subjected to a process of mining conceptual entities of information. This process finally leads to an overall understanding of the processed score. The process of structuring the space of information fits the paradigms of granular computing and information granulation, c.f. [2].

Sizing Syntactic Granules. The meaning of granule size is defined accordingly to real application and should be consistent with commonsense and with the knowledge base. The intuition says that bigger the description, bigger the size of it. This intuition, compatible with commonsense, should be satisfied in sizing syntactic granules.

Let us recall that syntactic granules are simply elements of the lexicon. Therefore, we should describe size of lexicon elements. It can be done in terms of the number of nodes in the associated subtree. There are two aspects of lexicon elements affecting commonsense meaning of this concept of size. The first aspect is related to depth of the root of the subtree covering the phrase: deeper the root in the derivation tree, smaller the size. The second aspect is related to the length of path attached to the root of the subtree: shorter the length, smaller the size.

In the first case, for fixed derivation tree, i.e. for given piece of music and given grammar, height of the subtree is invertible to depth of its root. Let us recall that depth of a node in the tree is defined as length of the path from the root of the tree to this node. Of course, number of nodes depends increasingly on the height of the subtree: higher the subtree, higher the number of nodes in this subtree. Comparing indicated nodes in Figure 4 we can see that height and

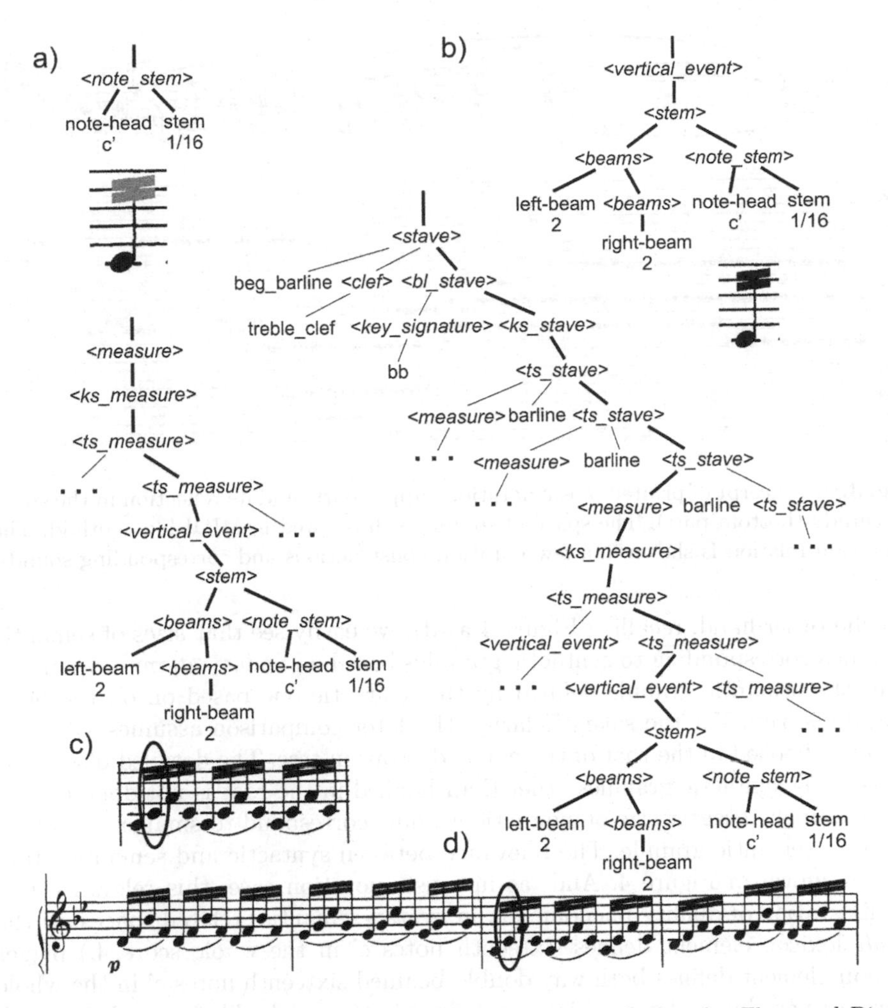

Fig. 2. A fragment of the derivation tree of *Polonaise from Sonata for Flute and Piano* by L. v. Beethoven. Derivation of the first beamed group in the second stave of the second system is expanded. Ellipsis indicate other parts of the score. The tree and right-beam, pitch for head and duration for stem.

size of subtrees hanged on *<stem>* nodes are the same while height and size of the subtree hanged on the *<ts_measure>* are bigger than former ones.

In the second case, dependency is quite obvious: longer path attached to the root of the subtree, more nodes we get. In Figure 4 height of subtrees (with attached path to the root) increase going from the case a) to the case d).

Sizing Semantic Granules. Size of semantic granule is estimated as a quantity of real world objects. Referring to granules outlined in Figure 2 and to discussion in Section 2.2, we see that sizes of semantic granules defined by lexicon elements a)-d) are reversible to sizes of corresponding syntactic granules.

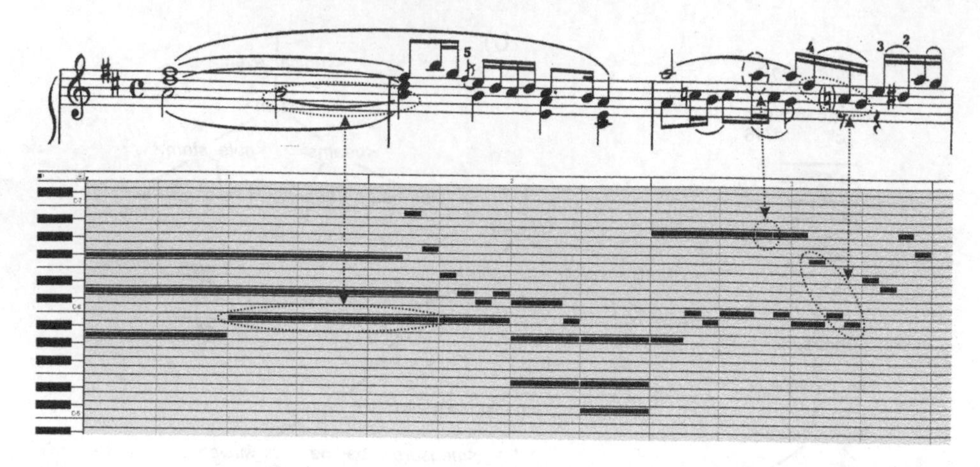

Fig. 3. An excerpt of printed music notation (upper part) and its reflection in the space of sounds (bottom part). The space of sounds is illustrated in MIDI-like method. The valuation relation is shown in a few notation constructions and corresponding sounds.

On the other hand, recalling Figure 4 a)-d), we easily see that sizes of semantic granules corresponding to syntactic granules based on nodes *<stem>* are smaller than the semantic granule defined by the syntactic one based on on the node *<ts_measure>*. For the sake of clarity, the later comparison assumes syntactic granules hooked in the root of the whole derivation tree. The detailed discussion concerning syntactic granules other than hooked in the root is not done here.

Amazingly, greater size of syntactic granule correspond to smaller size of respective semantic granule. The relevance between syntactic and semantic granules is shown in Figure 4. And, as in music notation case, this relevance is a manifestation of duality phenomenon in syntax-semantics related spaces: a) the *small* lexicon element defines sixteenth notes c' in the whole score, b) bigger lexicon element defines both way double beamed sixteenth notes c' in the whole score, c) next bigger lexicon element defines both way double beamed sixteenth notes c' in given place of any measure, d) next bigger lexicon element defines both way double beamed sixteenth notes c' in given place of the third measure of any stave, e) relevance of syntactic and semantic granules' pyramids.

3 Knowledge Accessing and Processing

Granular cognitive maps with their ability to recognize syntax structuring, semantic analysis and granulation of information are fundamental aspect of accessing and processing structures of knowledge. According to notes in Section 1, we do not attempt to develop a theoretical framework for automatic data understanding. Instead, we believe that case studies will be more informative in context of this paper. In subsequent sections we discuss examples of structural operations in the space of music information. Such operations are the best manifestation of what we mean as tools for knowledge accessing and processing.

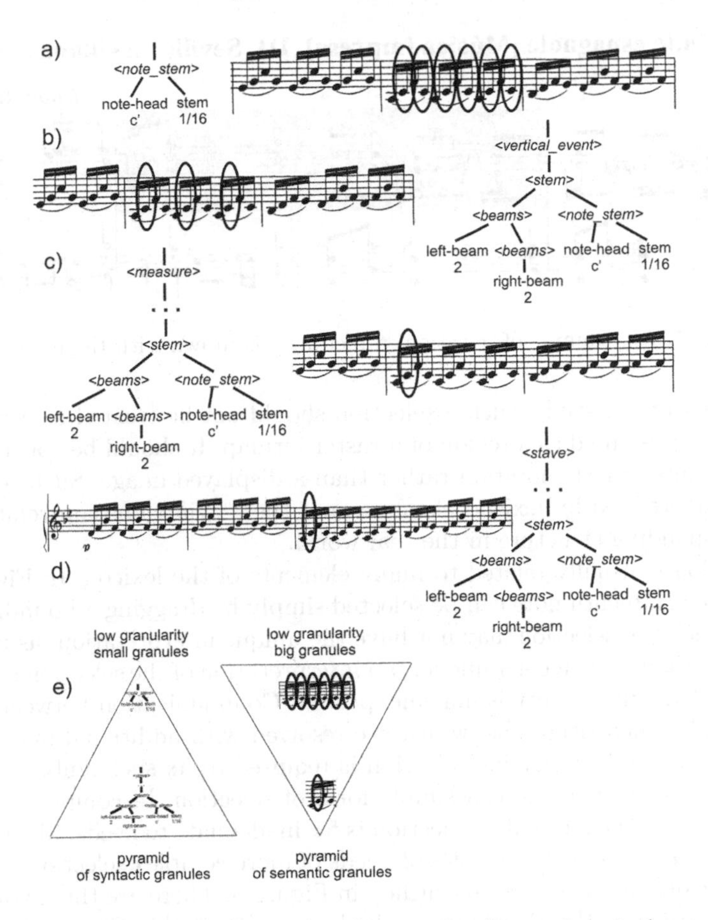

Fig. 4. Duality of syntax and semantics for music notation

3.1 Selecting

Among structural operations selection is the very fundamental one. Selection is the basis for other common structural operations like copying, finding, replacing etc. Selection is also the basis for domain specific operations. In case of paginated music notation, we can list such domain specific operations like transposition, conversion, harmonization, voice line identification etc.

This common operation in computer applications is usually performed on raw, low-level data. Examples of low level data selections are texts in text editors and regions of raster images (rectangles of pixels), c.f. [3,5]. An example of a raster image selection is shown in Figure 5: a rectangular region of the screen is selected and highlighted. This selection is interpreted as a part of a raster image and not the second and third measures of the upper stave of the paginated music notation. In this interpretation content of the image is of minor importance: it may be a part of printed music notation as well as a part of city map or a part of any other raster image.

Fig. 5. An example of selection: a rectangle is drawn with the mouse

In context of our study, such a selection should not be treated as a part of a screen and represented as a region of a raster bitmap. It should be considered as a part of printed music notation rather than a displayed image. Such selections can be characterized by lexicon elements and then - via valuation relation V - by a corresponding structure in the real world.

A selection is usually related to many elements of the lexicon. In Figure 4 d the indicated sixteenth note can be selected simply by dragging a bounding box. However, such the selection may not have the unique interpretation, as it is outlined in this Figure. As a consequence, an interpretation of the selection affects its semantics. This uncertainty is not unexpected. Communication between people may also raise such situations, which are resolved with additional information. Alike, an extra tool supporting selection is required to fix such ambiguities.

Mouse-dragged rectangle is a simple form of selection. In complex spaces of information such the method of selection is far inadequate to needs. Music spaces of information are perfect examples of needs of more complex selection methods. An example of such selections is clarified in Figure 6. There are three voice lines in the first stave of the Bach's score. In lower part of this Figure three voice lines are displayed in separate staves. It is clear that a rectangle dragging tool cannot select a voice line. This is why a more suitable tool should be used for such selections.

From human's perspective, a description in natural language would be the best way for describing such selection. For instance, the instruction *Select upper voice line in the upper stave of three beginning measures* clearly describes the selection. Analysis of such instructions opens a discussion on processing natural languages, which is not intended in this paper. Instead, for the sake of this study, we assume that instructions are formulated with simpler tools like mouse, keyboard and dialog boxes. For instance, we can use rectangle selection tool in the voice line mode to select voice lines.

3.2 Searching

Searching is an operation of locating object(s) matching a given pattern, i.e. locating instance(s) identical or similar to the pattern. In text editors obvious meaning of searching operation is finding instances of a given string and no analysis of information is done. Searching tools admit more sophisticated methods

Fig. 6. Identification of voice lines. Three voice lines of the upper stave of the upper system are split to three separate staves in bottom system.

of identification of instances. For example, regular expressions allow for finding strings of a language defined by such expression. Anyway, this is still the operation performed on strings and not on information brought by such strings. On the other hand, searching in raster bitmaps has no reasonable meaning unless content of the image is involved.

In context of this paper searching is performed on data structures rather than on such raw data as strings of symbols or region of rater bitmaps. The operation *Search* in a space of music information concerns a pattern, which is a part of space of music information. Searched pattern can be defined in different ways. Selection done as described in Section 3.1 is the simplest method for defining the search pattern.

Let us discuss the operation *Search* with the following example based on Figure 4 a-d. We assume that the *sixteenth c'* note is the pattern selected. We also assume that searching is accomplished along the piece with regard to pitch and duration. If the selection is interpreted as in the case a) of this Figure, i.e. *a sixteenth c'*, then any *sixteenth c'* note matches this pattern. This case is symbolically shown in Figure 4, case a), where any note c' matches the pattern. The pattern interpreted as in case b) of Figure 4 defines the sixteenth note c' beamed twice both ways. In this case a sixteenth note c' matches the pattern assuming that is beamed both ways. Case c) defines any sixteenth note c' beamed both ways and with given placement in a measure. Finally, in case d), only notes described in case c) belonging to the third measure of a stave are admitted.

Fig. 7. Illustration of searching. The pattern is in dotted ellipse, instances are dashed around. The upper part illustrates searching for exact matches (the same pitches, durations and time intervals between notes) without specific placement in the measure. The bottom part shows instances preserving pitch intervals, time intervals, durations and placement in the measure.

Considering another example based on Figure 7, assume that the first beamed group of four sixteenths of the first measure is selected as the searching pattern and that only minimal part of derivation tree creates the corresponding lexicon element, i.e. the lexicon element is hooked in its hang. This pattern matches next two beamed groups of sixteenths in the same measure and last two beamed groups of sixteenths in the fourth measure of the same stave, c.f. Figure 7. If the lexicon element is hooked in the node <*measure*>, c.f. 2, then the pattern *the first beamed group of four sixteenths in a measure* does not match any instance. However, if we admit notes to be moved up or down by the same interval, then the pattern matches the first beamed group of four sixteenths in the third measure: notes of the instance are moved down by three halftones (semitone) comparing to the pattern, c.f. lower part of Figure 7. Of course, durations and time successions are preserved in all instances of both cases.

3.3 Transposing

Transposition is a fundamental operation in music. Semantic interpretation of transposition is simple: it moves sounds/notes up or down by a number of halftones preserving intervals between sounds/notes. Human's communication often takes scales to define desired transposition. Transposition between different scales, e.g. from C major to B flat major, is easily expressed in terms of moving pitches by 3 halftones down. Syntactic interpretation of transposition is more complex due to irregularity of scales of paginated music notation. Therefore, automatic accomplishment of this operation is firmly based on semantic interpretation. Namely, a valuation relation V moves a given element of the lexicon to the world of sounds/notes, then the semantic image of the lexicon element is moved by the specified number of halftone and - finally - an inverse relation V^{-1} moves the semantic item back to the lexicon, c.f. Figure 8 for examples of transposition.

Fig. 8. J. S. Bach, Suite (Ouverture) No. 3 in D major, BWV 1068, transposition of voice lines: in middle part of the image the upper voice line is transposed by one octave up (12 halftones), the lower one - by one octave down; in bottom part of the image the middle voice line in the third measure is moved down by forth (seven halftones).

3.4 Converting

Conversions between different formats of music information is inherently based on semantic representation of music information. Formats of music representation are usually differently designed and direct conversion is highly difficult. Therefore a conversion from one format to another is usually done with a intermediate semantic analysis and representation of converted information. As mentioned above, firstly a valuation relation V_1 moves converted information to a world of objects (usually, to the world of sounds/notes) and then an inverse V_2^{-1} moves semantic image to target format, c.f. [4].

4 Conclusions

In this paper we introduced the idea of linguistic approach of granular cognitive maps as tools for knowledge accessing and processing. The idea stems from syntactic and semantic characterization of data and is soundly based on the paradigm of granular structuring of data and granular computing. We show the ways of employing syntactic structuring and semantic analysis in knowledge structuring and understanding data. The study is focused in the domain of music information as well as concepts are cast on the structured space of music information. Although we do not introduce a formal theory of automatic data understanding, the study points out the methodology of knowledge processing, which can be applicable to different domains.

The main objective of this paper, i.e. building tools for operating on knowledge structures, is illustrated by several representative examples. The discussion outlines the paradigm of linguistically designed granular cognitive maps as inherent feature of intelligent man-machine communication.

Future directions for the introduced ideas are both theoretical and applicational. The former ones include generalizing concepts to be more applicable in different domains. The later ones are intended to develop a music knowledge space founded on analysis of printed music notations, i.e. based on so called paper-to-computer technologies.

Acknowledgements. The research is supported by the National Science Center, grant No 2011/01/B/ST6/06478, decision no DEC-2011/01/B/ST6/06478.

References

1. Castan, G., Good, M., Roland, P.: Extensible Markup Language (XML) for Music Applications: An Introduction. In: Hewlett, W.B., Selfridge-Field, E. (eds.) The Virtual Score: Representation. The MIT Press (2001)
2. Homenda, W., Pedrycz, W.: Automatic Data Understanding, A Linguistic Tool for Granular Cognitive Maps Designing. In: Angelov, P., et al. (eds.) Intelligent Systems'2014. AISC, vol. 322, pp. 219–230. Springer, Heidelberg (2015)
3. Homenda, W., Rybnik, M.: Querying in Spaces of Music Information. In: Tang, Y., Huynh, V.-N., Lawry, J. (eds.) IUKM 2011. LNCS (LNAI), vol. 7027, pp. 243–255. Springer, Heidelberg (2011)
4. Homenda, W., Sitarek, T.: Notes on Automatic Music Conversions. In: Kryszkiewicz, M., Rybinski, H., Skowron, A., Raś, Z.W. (eds.) ISMIS 2011. LNCS, vol. 6804, pp. 533–542. Springer, Heidelberg (2011)
5. Homenda, W.: Automatic data understanding: a necessity of intelligent communication. In: Rutkowski, L., Scherer, R., Tadeusiewicz, R., Zadeh, L.A., Zurada, J.M. (eds.) ICAISC 2010, Part II. LNCS, vol. 6114, pp. 476–483. Springer, Heidelberg (2010)
6. Homenda, W.: Towards automation of data understanding: integration of syntax and semantics into granular structures of data. In: Proc. of the Fourth Int. Conf. on Modeling Decisions for Artificial Intelligence, MDAI 2007, pp. 134–145 (2007)
7. Krolick, B.: How to Read Braille Music, 2nd edn. Opus Technologies (1998)
8. MIDI 1.0, Detailed Specification, Document ver. 4.1.1 (February 1990)
9. MusicXML Version History (2004), http://www.recordare.com/dtds/versions.html

Automatic Data Understanding
A Linguistic Tool for Granular Cognitive Maps Designing

Władyslaw Homenda[1] and Witold Pedrycz[2,3]

[1] Faculty of Mathematics and Information Science, Warsaw University of Technology, ul. Koszykowa 75, 00-662 Warszawa, Poland
[2] System Research Institute, Polish Academy of Sciences, ul. Newelska 6, 01-447 Warszawa, Poland
[3] Department of Electrical & Computer Engineering, University of Alberta, Edmonton, T6R 2G7 AB, Canada
homenda@mini.pw.edu.pl, wpedrycz@ualberta.ca

Abstract. This study is concerned with an issue of automatic data understanding being treated as a fundamental tool used in cognitive maps creation. The underlying objective of this paper is to elaborate on a paradigm of automatic data understanding. We highlight the syntactic and semantic aspects of data. Granular Computing and information granules play a central role in all processes of data understanding by facilitating establishing a suitable perspective at the problem at hand where the data need to be looked at. Our intent is to illustrate a realization of the paradigm in the realm of music information, its processing and understanding, rather than embarking on the formal framework.

Keywords: granular cognitive maps, automatic data understanding, syntax vs. semantics.

1 Introduction

In this paper we introduce the paradigm of *automatic data understanding* to be used as a tool for cognitive map formation. The study concerns the domain of music information. Anyway, its adaptation to other domains is rather straightforward. The paradigm is firmly based on the syntactic and semantic characterization of data. Both syntactic and semantic facets of data are integrated in the frame of Granular Computing paradigm. Granulation of data plays a central role in all processes of data understanding by facilitating establishing a suitable perspective at the problem at hand where the data need to be looked at. Granulated data is intended to form a map of concepts and relations between concepts and to draw conclusions.

The term *cognitive maps* was used by E. Tollman in 1940s in studies on hidden learning process observed among vertebrate animals, cf. [11]. Experiments prove that data units gathered seemingly unwitting at a previous point of time could be efficiently processed in order to solve stimuli-triggered problem. These pieces of information residing in brain ordered in the cognitive map at the moment

P. Angelov et al. (eds.), *Intelligent Systems'2014*,
Advances in Intelligent Systems and Computing 322, DOI: 10.1007/978-3-319-11313-5_21

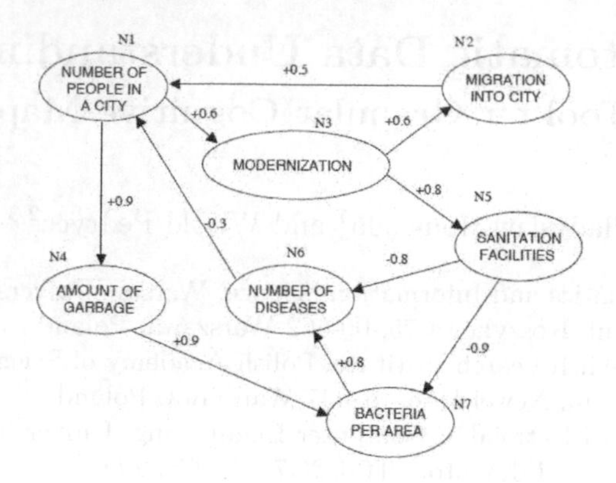

Fig. 1. An example of a fuzzy cognitive map (from [10])

of data processing are visualised and associated in order to increase chances of success. In 1970s R. Axelrod used cognitive maps to model causalities between economical and political phenomena, cf. [1]. In 1980s B. Kosko extended the concept of cognitive maps to fuzzy cognitive maps, cf. [6]. An example of a fuzzy cognitive map is given in Figure 1. Research record on fuzzy cognitive maps is long and we do not intend to go into details.

The concept of granular cognitive maps was raised by W. Pedrycz, cf. [9]. It is in its infant period. In this paper we attempt to give a view on nonstandard vision of granular data and cognitive maps. The vision is solely based on data structuring and leads to formation of granular cognitive maps, which relate concepts in terms of static knowledge structuring used in communication and reasoning rather than dynamic time causality.

In this paper we do not develop a formal and universal approach to automatic data understanding paradigm and its use for cognitive maps' formation. Instead, besides formulation of notions of automatic data understanding, granularity, semantics and syntax, we provide the case study based on the domain of music information. The case study shows how to cast these notion onto the domain. In the domain restricted area this methodology can be seen as a formal approach to this paradigm used as a tool for cognitive maps formation. Adaptation of the introduced ideas to other domains is straightforward.

The underlying objective of this paper is to elaborate on a paradigm of automatic data understanding and its usage in navigation inside structured bases of knowledge. We discuss integration of syntactic and semantic aspects of knowledge structuring integrated in frames of granularity paradigm. Finally, we come to granular cognitive maps formation as a frame covering discussed concepts.

The paper is structured as follows. In consecutive sections of Chapter 2 we provide the meaning of basic terms used in the paper: syntax, semantics, granulation etc. Chapter 3 covers discussion on syntactic structuring, which is nested in paginated music notation. Semantics and granulation reflected in paginated music notation is outlined in Chapter 4. An idea of cognitive map creation with automatic data understanding as a tool is presented in Chapter 5.

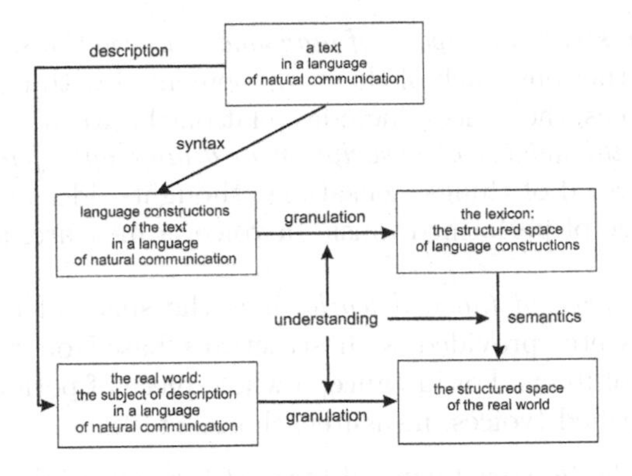

Fig. 2. The diagram of the paradigm of information understanding

2 Concepts

Data understanding requires data structuring. Data structures, if useful, are subjects of communication (information exchange), where information is understood as a description of data structure(s). Information exchange is accomplished in some language(s), which should be *understandable* to sides of communication. Sides of communication are human beings and/or computers. For instance, composer communicates musicians to perform in a given key using the key signature placed in music notation, teacher instructs students in a given natural language how to solve a problem. In all such cases information exchanged is expressed and/or described as a text in some language. Intelligent communication is just exchanging texts expressing information and/or describing data.

2.1 Syntax, Semantics, Granulation, Understanding

In Figure 2 we illustrate the meaning of the paradigm of understanding. This paradigm is built on the basis of syntactic structuring, semantic analysis and granulation. Below we briefly describe meaning of these terms referring either to common terms, or to respective sources. It is worth stressing that in general we consider known concepts and ideas integrating them into the novel approach.

We consider five entities in the diagram of understanding. These entities are commonly understood, or are described in further parts of this paper:

- *a text in a language of natural communication* as a whole. A whole score of paginated music notation or Braille music, a whole text (description, story, tale) in a natural language are examples of language constructions,
- *language constructions* of the text in a language of natural communication, i.e. a text as described in the above point. Parts and elements of whole scores (voices, measures, chords, notes) are examples of language constructions. Parts of sentences of a whole text in natural language are other examples. Language constructions make a base for lexicon,

- *the lexicon: the structured space of language constructions*, it is the space of language constructions, each of them supplemented with possible derivation (parsing) subtrees, the lexicon includes relations between items of this space,
- *the real world: the subject of description in a language of natural communication*, it is a world of things, sensations, thoughts, ideas etc. For instance, a part of a piece of performed music (a voice, a measure, a chord) that we can hear,
- *the structured space of the real world*, it is the space of things sensations, thoughts, ideas etc. provided with structures based on relations between items and sets of items. For instance, a whole piece of performed music with structures identified (voices, measures, chords etc.).

Syntax, i.e. *syntactic structuring* of texts of languages of natural communication (and also languages of formal communication) is the study on how words fit together structures describing a world of real objects, cf. [3]. Syntactic structuring is usually described by a kind of a grammar. Such a description can be given in a form of a set of rules or as a formal grammar. Our study is based on context-free grammars, cf. [3] for detailed justification of the choice of context-free methods.

Granulation is a concept of structuring of the space of language constructions and of the real world. Structuring the space of language constructions leads to the creation of the lexicon. The concept of lexicon is elaborated on in [4]. Language constructions supplemented by corresponding parts of derivation trees (parsing trees) create syntactic structures of the lexicon, cf. Figure 4. Creation of the lexicon on the basis of language constructions is an apparent application of the paradigm of granularity, cf. [7,8]. Structuring of the real world is what objectively exists, what is discovered and described by observers.

Semantics is a relation between items of lexicon and items of the structured space of the real world. In other words, semantics casts the lexicon onto the structured space of the real world. In general, such relations/casts are many-to-many dependencies. It may be many-to-one or one-to-one relation, i.e. it is a mapping form the lexicon to the structured space of the real world. Semantics concept is developed in Section 4.

Understanding is an ability to construct granular structures in both spaces: the lexicon and the real world and to construct semantics.

2.2 Generalization Versus Case Studies

The paradigm of automatic data understanding involves syntactic description, semantic analysis and granular structuring as general methods. However, we do not attempt to generalize these methods in a way that they would be directly applicable in any subject of communication. Instead, we provide a case study in a chosen domain. This approach would be adopted to other domains. In our opinion, the paradigm of automatic data understanding is the universal approach to intelligent human-machine communication. However, details of such approach depend on the application domain. Therefore, details discussed in this paper may not be universally applicable in all domains.

10. Polonaise
(from Sonata for Flute and Piano)
LUDWIG VAN BEETHOVEN (1770-1827)
Moderato

Fig. 3. Polonaise from Sonata for Flute and Piano by Beethoven, the beginning excerpt

2.3 The Case Study

In this paper, we analyze structured spaces of music data oriented to paginated music notation. Methods presented in the paper are illustrated with fragments of scores of pieces of classical music. Namely, we inspect *Polonaise from Sonata for Flute and Piano* by L. v. Beethoven, cf. Figure 3. Our discussion embraces only elements of the score. Nevertheless, the methods illustrated with an excerpt of this piece can be expanded on the entire pieces as well as on other scores. Alike, the study focused on spaces of music information oriented to paginated music notation may be adopted to other types of descriptions of music information, e.g. Music XML, Braille Music and MIDI.

3 Syntactic Approach

Syntactic approach is a crucial stage and a crucial problem in the wide spectrum of tasks as, for instance, pattern recognition, translation of programming languages, processing of natural languages, music processing, etc. By syntactic approach and syntactic methods we understand grammars, automata and algorithms used in processing languages. By analogy to the Chomsky's hierarchy of languages, syntactic approaches and methods can be categorized as regular, context-free and context-sensitive, cf. [5].

3.1 Context-Free Syntactic Description

The proposed approach to describing languages of natural communication will rely on the sensible application of context-free methods applied locally or on covering languages (i.e. generating all constructions of a language and constructions not belonging to it or incorrect constructions of the language). Moreover, we assume that the context-free methods will not be applied unfairly to generate incorrect constructions or constructions not belonging to them. These assumptions allow only for a raw approximation of languages of natural communication. Of course, such assumptions are real shortcomings comparing to an accurate description. The shortcomings must be solved by employing some other methods.

The Tool. The discussion on describing paginated music notation is based on common definition of grammars and of context-free grammars. Here, we only recall these basic notions assuming that the reader is familiar with them.

Let us recall that a system $G = (V, T, P, S)$ is a grammar, where: (a) V is a finite set of *variables* (called also *nonterminals*), (b) T is a finite set of terminal symbols (simply called *terminals*), (c) a nonterminal S is the initial symbol of the grammar and (d) P is a finite set of productions. A pair (α, β) of strings of nonterminals and terminals, usually denoted $\alpha \to \beta$, is a production assuming that the first element α of the pair is a nonempty string. Grammars having all productions with α being a nonterminal symbol are context-free grammars.

A derivation in a grammar is a finite sequence of strings of nonterminals and terminals such that: (a) the first string in this sequence is just the initial symbol of the grammar and (b) for any two consecutive strings in the sequence, the later one is obtained from the former one applying a production in the usual way, i.e. by replacing a substring of the former string equal to the left hand side of the production with the right hand side of the production. We say that the last element of the string is *derivable* in the grammar.

For a context-free grammar a derivation can be outlined in a form of derivation tree, i.e. (a) the root of the tree is labelled with the initial symbol of the grammar and (b) for any internal vertex labelled by the left side of a production, its children are labelled by symbols of the right side of the production.

The Application. Under the assumptions of Section 3.1, usage of a simplified context-free grammar for the purpose of syntactical structuring of paginated music notation is valid in practice. The grammar will be applied for analysis of constructions, which are assumed to be well grounded pieces of paginated music notation. Of course, such a grammar can neither be applied in checking correctness of constructions, nor in generation of such constructions.

Here we give a raw description of paginated music notation in a form of productions of a context-free grammar. The productions were constructed manually at the basis of observation of structures of paginated music notation. Paginated music notation is a collection of staves (staff lines) placed on pages. Every stave is surrounded by corresponding symbols placed on the stave and around it. However, the collection of staves has its own structure with staves grouped into higher-level units called systems. A raw description of the structure of music notation could be approximated by the set of context-free productions given below. Components of the grammar $G = (V, T, P, S)$ are as follows. The set of nonterminals includes all identifiers in triangle brackets printed in italic. The nonterminal *<score>* is the initial symbol of G. The set of terminal includes all non bracketed identifiers.

<score> → *<score_part>* *<score>* | *<score_part>*

<score_part> → *<page>* *<score_part>* | *<page>*

<page> → *<system>* *<page>* | *<system>*

<system> → *<stave>* *<system>* | *<stave>*

<system> → *<part_name>* *<stave>* *<system>* | *<part_name>* *<stave>*

<part_name> → Flute | Piano | *etc.*

<stave> → beginning-barline *<bl_stave>* | *<clef>* *<bl_stave>*

<stave> → beginning-barline *<clef>* *<bl_stave>* | *<bl_stave>*

<bl_stave> → *<key_signature>* *<ks_stave>* | *<ks_stave>*

<ks_stave> → *<time_signature>* *<ts_stave>* | *<ts_stave>*

<ts_stave> → *<measure>* *<barline>* *<ts_stave>* | *<measure>* *<barline>*

<measure> → *<change_k_sign.>* *<ks_measure>* | *<ks_measure>*

<ks_measure> → *<change_t_sign.>* *<ts_measure>* | *<ts_measure>*

<ts_measure> → *<vertical event>* *<ts_measure>* | *<vertical event>*

<vertical event> → *<stem>* *<vertical event>* | *<stem>*

<stem> → *<beams>* *<note_stem>* | *<flags>* *<note_stem>* | *<note_stem>*

<beams> → left-beam | right-beam | left-beam right-beam

<clef> → treble_clef | bass_clef

<flags> → flag *<flags>* | flag

<note_stem> → note-head *<note_stem>* | note-head

Music notation can be described by different grammars. Construction of such grammars may reflect various aspects of music notation, e.g. geometrical or logical structuring, cf. [2,3]. The above description is constructed from the point of view of geometrical properties of music notation. The nonterminals *<page>* and *<system>* define items of music notation strictly related to paginated music notation. The nonterminal *<vertical event>* defines items related to paginated music notation as well as a logical element of music information.

The grammar presented here may be expanded to a real grammar fully describing paginated music notation. Due to limitation of the paper, many details are skipped. For instance, pitch and duration of notes, articulation and ornamentation symbols, dynamics, placement of graphical elements on the page etc. are not outlined. On the other hand, expansion of the grammar is straightforward in its essence and could be done based on this study.

3.2 The Lexicon

Lexicon is the key concept of granulation of syntax. Lexicon is a set of language constructions, which describe objects and local and global structures of objects in the real world. Elements of lexicon are phrases generated in a grammar supplemented with parts of derivation tree build on them. Such a part of the derivation tree, which corresponds to the given phrase, should be the minimal one covering the phrase. This part has its upper vertex, which is its root. There is the path connecting the root of the subtree to the root of the whole derivation

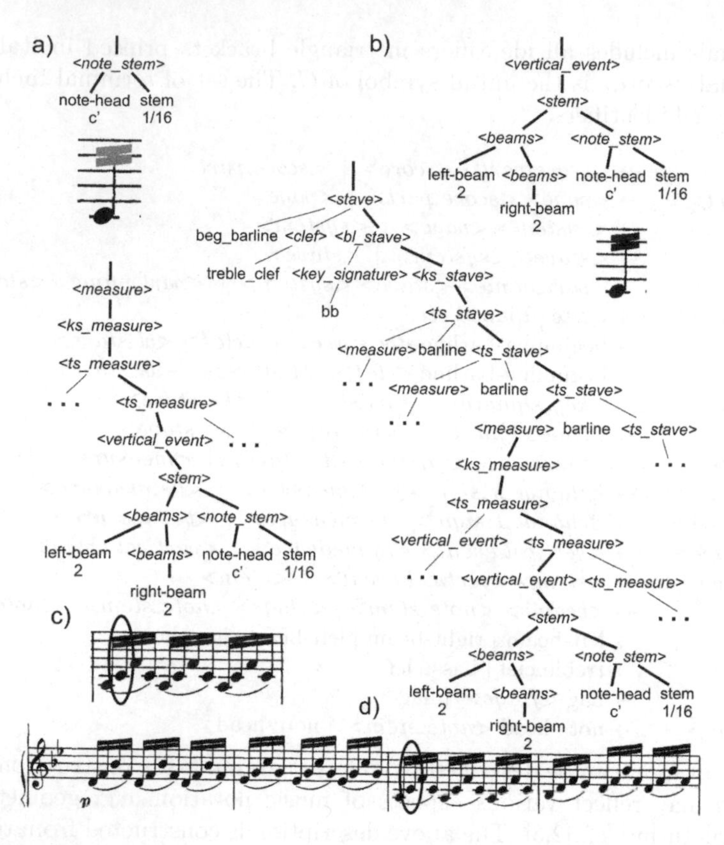

Fig. 4. Lexicon elements and their interpretation. Thick edges indicate the derivation tree of the Beethoven's score associated with the phrase (the sixteenth c' note). Extra nodes and edges are added for the sake of clarity.

tree. Extensions of the subtree along this path create more lexicon constructions based on the same phrase. Examples given in Figure 4 illustrate the concept of lexicon. All four elements of the lexicon are built on the same phrase of music notation, i.e. on the eight note. Part a) of the Figure shows the lexicon element (a sixteenth note c') with the minimal part of derivation tree. This lexicon element supplemented with two vertices and edges creates another lexicon element (a sixteenth note c' twice beamed both ways) indicated in part b). Parts c) and d) provide more detailed description of this note: the second note in a measure and the second note in the the third measure of the stave. These four lexicon elements are hanged to *<note_stem>* node. They are are hooked in *<note_stem>*, *<vertical_event>*, *<measure>*, *<stave>* nodes, respectively. Note that all these elements are built on the same phrase of music notation.

It is worth to recall that according to discussion in Section 3.1 the above grammar generates all valid paginated music notation as well as constructions not being correct music notations. However, we assume that only valid phrases of music notation will be subjected to analysis.

4 Semantics

The context-free grammar of music notation defined in Section 3.1 is a tool used to describe music notation rather than a subject of communication and understanding. Therefore, the syntactic approach to describing the music notation, as expressed in the form of this grammar, is a workout of a structured space of information. Such constructions describe music notation: notes, rests, vertical events, voice lines, measures, staves, systems, scores. On the other hand, music notation is context-sensitive, as noted in Section 3. For that reason, the study on automatic data understanding, if restricted to the pure syntactic approach based on context-free methods, will not give a complete perspective on the information being processed. So then, we consider switching to mutual utilization of syntactic and semantic methods, which leads to information granulation.

Lexicon elements describe objects in a real world. In the case of music information the real world includes sounds produced when a piece of music is being performed. For simplicity, we assume that sounds are of the form of notes described by three parameters: pitch, beginning time and duration of a note. Such the triple describing a sound/note roughly can be supplemented by other features, e.g. volume (loudness), articulation (legato, staccato, portato etc.)

Semantics, as interpreted in this paper, immerses the lexicon in the space of sounds/notes. Such the immersion is a relation defined in the Cartesian product of lexicon and the space of sounds/notes. Note that lexicon elements shown in Figure 4 may define many objects in the world. If the root of the lexicon element is hooked in the root of the derivation tree, ten such lexicon element defines unique object in the score. From geometrical perspective, semantics of the case a) defines all sixteenth notes placed at the second space of the stave. Considering logical perspective, since selected note is in scope of treble clef, semantics defines notes c' in the score, i.e. it defines one-lined c sixteenth notes having different beginning times. In case b) we have notes double beamed both ways. In case c) we have notes placed in the given context, i.e. context defined by position of the note in a measure. In case d) only third measure of a stave is considered. As mentioned, if the lexicon element begins with the root of the derivation tree, then semantics defines the unique element: the note with given beginning time and given pitch.

Summarizing - semantics is a relation V in the lexicon L and the world W, called *valuation relation*:

$$V \subset L \times W$$

The world W includes objects and sets of objects described in a given language. In fact, objects of the world W usually create meaningful sets and - moreover - create structures with internal relations between objects. Such structures are also described by constructions of the language. Therefore, the world W is a structured space of objects (or structured space of information). Namely, the world W includes notes/sounds related to each other according to a given piece of music.

5 Building Cognitive Maps

5.1 Granular Space

The processing of music notation, as described in Section 3, leads to extracting information from a score. A collection of data items is subjected to a process of mining conceptual entities of information. This process finally leads to an overall understanding of the processed score. The process of structuring the space of information fits the paradigms of granular computing and information granulation. Granular computing paradigm *raised from a need to comprehend the problem and provide a better insight into its essence rather than get buried in all unnecessary details. In this sense, granulation serves as an abstraction mechanism that reduces an entire conceptual burden. As a matter of fact, by changing the size of the information granules, we can hide or reveal a certain amount of details one intends to deal with during a certain design phase*, cf. [7].

The description of music notation as well as music notation itself could be innately subjected to the paradigm of granular computing elucidation. Information granules exhibit different levels of knowledge abstraction, what strictly corresponds to different levels of granularity. Depending upon the problem at hand, we usually group granules of similar size (i.e. similar granularity) together into a single layer. If more detailed (and computationally intensive) processing is required, smaller information granules are sought. Then, those granules are arranged in another layer. In total, the arrangement of this nature gives rise to the information pyramid. Information granularity implies the usage of various techniques that are relevant for the specific level of granularity.

5.2 Cognitive Maps Formation

Figure 5 present fragments of structured spaces of data. Such elements create concepts understandable by humans. The upper part of this Figure shows elements of lexicon, which are items of syntactic structuring of a given score. The second one shows semantic characterization of syntactic concepts. Lexicon items define concepts based on descriptions in a language of natural communication. They create a structure with dependencies between them, which is an image of humans perception. Lexicon items are reflected in a structured space of real world elements, phenomena, ideas etc.

Axelrod's and Kosko's ideas (cf. [1] and [6]) are adapted to create cognitive maps for granular structures of data. By analogy to Figure 1, granular entities, saying more precisely: syntactic and/or semantic entities, create a graph of connections. The cognitive map in Figure 1 models causalities between social and economical phenomena. Proposed granular cognitive maps model dependencies between granular structures of data. Among different types of dependencies the most important are inclusions, overlapping and similarities.

In Figure 5 only inclusion relation between units of structured syntactic and semantic spaces are delineated. Other types of relations also hold. As an example, overlapping relation can be easily distinguished in semantic space. One of

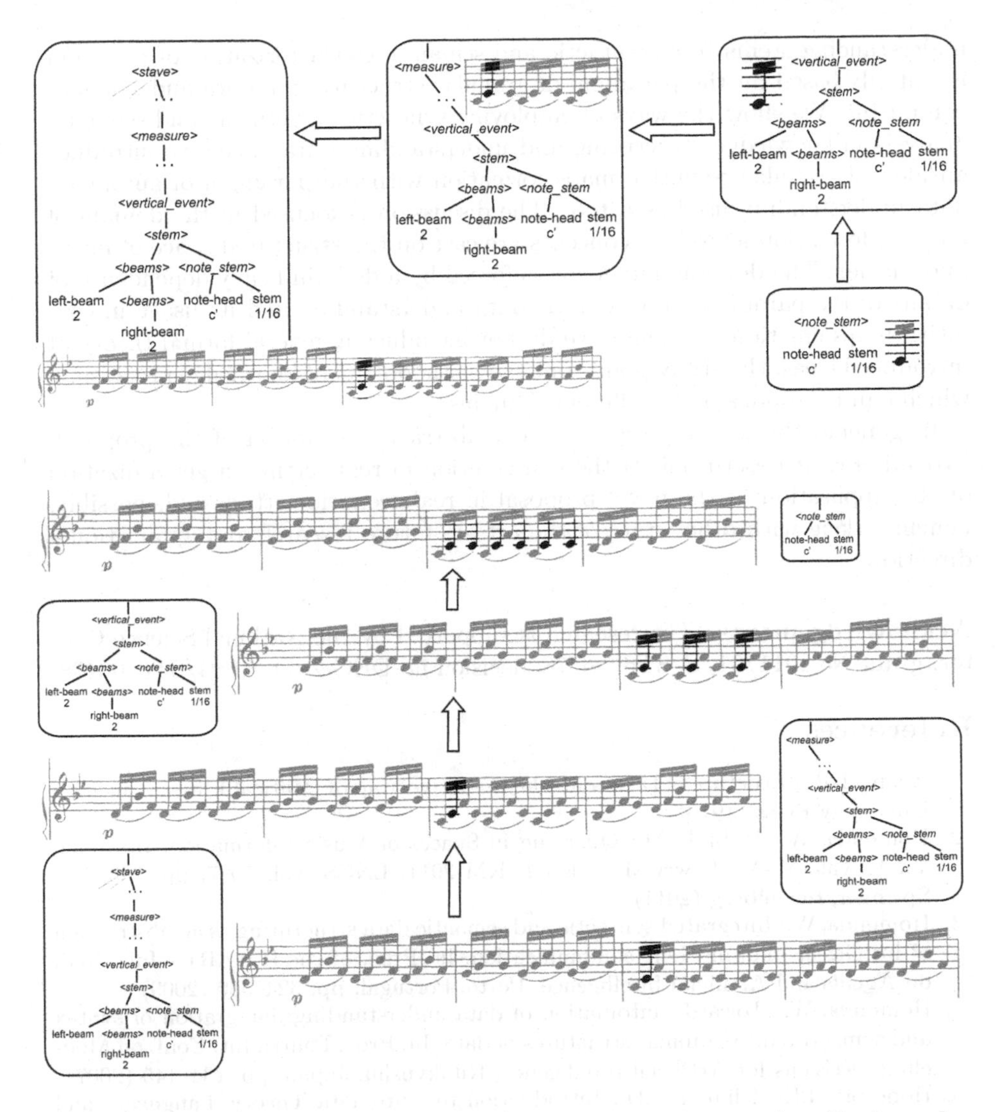

Fig. 5. Cognitive map formation by syntactic granulation - examples of lexicon elements (upper part) and their semantic reflection (lower part). Inclusion relation between these items are depicted by arrows. Note: transitive inclusions are not drawn.

the most important relation is similarity of units of both spaces: syntactic and semantic. A discussion on similarity relations is not in the scope of this paper, therefore we do not expose this topic.

6 Conclusions

In this paper we introduced the paradigm of automatic data understanding as a tool for cognitive maps creation. The paradigm of automatic data understanding stems from syntactic and semantic characterization of data and is soundly based on the paradigm of granular structuring of data and granular computing. We show the ways of employing syntactic structuring and semantic analysis in knowledge structuring and understanding data. Then we introduce an idea of granular cognitive maps formation with the paradigm of automatic data understanding used as a tool. The discussion is focused in the domain of music information as well as concepts are cast on the structured space of music information. The domain immersion is forced by a domain heavy dependence of details of the paradigm of automatic data understanding and its usage in cognitive maps creation. Although we do not introduce a general formal theory of introduced ideas, the study points out the methodology of knowledge processing, which can be applicable to different domains.

In general, the paper presents a very abstract description of the proposal. Providing clear description of the contribution in real setting, a generalization of the application of the novel proposal in real scenario setting and, possibly, benchmarking with other state of the art are future research and applicational direction.

Acknowledgements. The research is supported by the National Science Center, grant No 2011/01/B/ST6/06478, decision no DEC-2011/01/B/ST6/06478.

References

1. Axelrod, R.: Structure of Decision: The Cognitive Map of Political Elites. Princeton University Press (1976)
2. Homenda, W., Rybnik, M.: Querying in Spaces of Music Information. In: Tang, Y., Huynh, V.-N., Lawry, J. (eds.) IUKM 2011. LNCS, vol. 7027, pp. 243–255. Springer, Heidelberg (2011)
3. Homenda, W.: Integrated syntactic and semantic data structuring as an abstraction of intelligent man-machine communication. In: Proc. of the ICAART - Int. Conf. on Agents and Artificial Intelligence, Porto, Portugal, pp. 324–330 (2009)
4. Homenda, W.: Towards automation of data understanding: integration of syntax and semantics into granular structures of data. In: Proc. Fourth Int. Conf. on Modeling Decisions for Artificial Intelligence, Kitakyushu, Japan, pp. 134–145 (2007)
5. Hopcroft, J.E., Ullman, J.D.: Introduction to Automata Theory, Languages and Computation. Addison-Wesley Co. (1979, 2001)
6. Kosko, B.: Fuzzy cognitive maps. Int. J. Man Machine Studies 7 (1986)
7. Pedrycz, W.: Granular Computing: An introduction. In: Proc. of the Joint 9th IFSA World Congress and 20th NAFIPS, Vancouver, Canada (2001)
8. Pedrycz, W., Bargiela, A.: Granular clustering: A granular signature of data. IEEE Trans. Syst. Man And Cybernetics - Part B 32(2), 212–224 (2002)
9. Pedrycz, W.: The design of cognitive maps: A study in synergy of granular computing and evolutionary optimization. Expert Systems with Applications 37, 7288–7294 (2010)
10. Stach, W., Kurgan, L., Pedrycz, W., Reformat, M.: Genetic learning of fuzzy cognitive maps. Fuzzy Sets and Systems 153, 371–401 (2005)
11. Tolman, E.C.: Cognitive maps in rats and men. Psychological Review 55(4), 189–208 (1948)

Part VI
Issues in Logic and Artificial Intelligence

Fixed-Point Methods in Parametric Model Checking*

Michał Knapik[1] and Wojciech Penczek[1,2]

[1] Institute of Computer Science, PAS, Warsaw, Poland
{mknapik,penczek}@ipipan.waw.pl
[2] University of Natural Sciences and Humanities, II, Siedlce, Poland

Abstract. We present a general framework for the synthesis of the constraints under which the selected properties hold in a class of models with discrete transitions, together with Boolean encoding - based method of implementing the theory. We introduce notions of parametric image and preimage, and show how to use them to build fixed-point algorithms for parametric model checking of reachability and deadlock freedom. An outline of how the ideas shown in this paper were specialized for an extension of Computation Tree Logic is given together with some experimental results.

1 Introduction

The analysis of the existing software- and hardware systems is already a daunting task, and due to their increasing proliferation in everyday life will only become more difficult, and important. The systems employed in the critical areas such as avionics, security, and medical applications need to be thoroughly tested and verified, with the testing and verification present from the possibly earliest stages of design. While a battery of tests is often able to reveal some errors present in a design, the formal verification aims to mathematically *guarantee* that the abstraction of the system is compliant with its specification.

The classical model checking [7] is a simple, yet powerful methodology of systems analysis. In this approach, the behaviour of the system is presented as a mathematical model $\mathcal{M}$ with the intended level of granularity; a property to be verified (e.g., the lack of deadlocks) is then expressed as a formula ϕ of selected modal logic. The compliance of the model with the property is denoted as $\mathcal{M} \models \phi$. Symbolic model checking applies various methods for an efficient representation of the state-space, allowing to verify large systems with more than 10^{20} states [6]. Model checking needs a fairly complete description of the model and the property to be verified, which limits its applicability when some of the details are still unknown. Parametric model checking (also called parameter synthesis) is an extension of model checking that allows for the presence of free variables in the model and/or the formula; the goal is to describe all the valuations of the free variables under which ϕ holds in $\mathcal{M}$. In this way the parametric model

* Work partially funded by $DEC - 2012/07/N/ST6/03426$ NCN Preludium grant.

P. Angelov et al. (eds.), *Intelligent Systems'2014*,
Advances in Intelligent Systems and Computing 322, DOI: 10.1007/978-3-319-11313-5_22

checking tool can provide pointers to system designer or analyst, e.g., on how to instantiate the unknown variables.

In this work we do not focus on any logic from the menagerie of known modal logics. Instead, we propose a general framework that encompasses a range of state-based models with discrete transition guards and logics with model-checking procedures based on the computation of the (pre)image with respect to the transition relation.

1.1 Related Work

The problem of the synthesis of the set of assignments under which a given formula becomes true in a selected model was introduced in [1] in the context of a parametric extension of Linear Temporal Logic (LTL), called Parametric Temporal Logic (PLTL). This was later explored for other parametric extensions of LTL, and Computation Tree Logic (CTL) in [9,11]. The [9] paper seems to be especially important, as it shows how to extend standard fixed-point - based algorithms for model checking to encompass quantified parameters in Parametric Real Time CTL (PRTCTL). In the analysis of real-time systems, parametric versions of Timed CTL (PTCTL) interpreted over parametric extensions of timed automata appear in [4,5] and [24,25]. The first group of papers presents the techniques based on a translation of durations of runs of a timed automaton to a formula of Presburger arithmetic, the second group extends explicit-state methods to a parametric version of region graph. In [16,18,22] SAT-based Bounded Model Checking methods have been adapted to the task of RTCTL and PRTCTL verification and to a limited parameter synthesis for parametric reachability in a selected class of Petri Nets. In [8] the authors introduce algorithms for verification of specifications for software product lines expressed in Feature CTL. This work is closest to what we present in this paper, as it employs fixed-point algorithms implemented using Binary Decision Diagrams.

1.2 Paper Outline

In Section 2 we present the framework for parameter synthesis together with all the relevant notions. Section 3 contains a brief description of how the algorithms can be implemented using Boolean encodings. In Section 4 we sketch how the introduced theory can be specialized on an example of a selected modal logic PARCTL; we also cite some earlier experimental results that show the efficiency of our approach. The paper ends with Conclusions.

2 Framework for Parameter Synthesis

Let us start with establishing the computation model. Definition 1 allows to interpret many of the state-based models encountered in the theory and practice of model checking.

Definition 1. *Let X be a finite set of propositional variables (called* parameters*) and let $\mathcal{L}_X$ denote the set of propositional formulae over X. A $\mathcal{L}_X$*-labeled transition system *(also called* a model*) is a tuple $\mathcal{M} = (S, s^0, \rightarrow)$, where S is a finite set of states, $s^0 \in S$ is the* initial state, *and $\rightarrow \in S \times \mathcal{L}_X \times S$ is a* transition relation.

In what follows we fix a set of parameters X and $\mathcal{L}_X$-labeled transition system $\mathcal{M} = (S, s^0, \rightarrow)$. The set of all valuations of X is denoted by Ω. If $t = (s, g, s') \in \rightarrow$, then let $src(t) = s$, $trgt(t) = s'$ and $guard(t) = g$. We abbreviate $(s, g, s') \in \rightarrow$ as $s \xrightarrow{g} s'$. Consider a sequence of interleaved states and transitions $\pi = (s_0, t_0, s_1, t_1, \ldots)$. If the sequence is finite, then we assume that it ends with a state. By $|\pi|$ we denote the floor of the length of the sequence divided by two; the i–th state and transition of π are denoted by $\pi_i^S = s_i$ and $\pi_i^T = t_i$, respectively. Let $\upsilon \in \Omega$, then π is a υ-path if $s_i \xrightarrow{t_i} s_{i+1}$ and $\upsilon \models guard(t_i)$ for all $i < |\pi|$. The set of all υ-paths in $\mathcal{M}$ is denoted by $\Pi(\mathcal{M}, \upsilon)$ and the set of all υ-paths starting from $s \in S$ is denoted by $\Pi(s, \mathcal{M}, \upsilon)$; we omit the model symbol when it is evident from the context. One way of thinking about υ-paths is that a given valuation υ enables all and only those transitions whose guards evaluate to *true*.

Example 1. Consider the model in Fig. 1 with the set of parameters $X = \{x_1, x_2, x_3\}$. The loop $\pi = (s_0, x_1 \wedge \neg x_3, s_1, x_1 \vee x_2, s_2, x_1 \wedge \neg x_2, s_0, \ldots)$ is a υ-path iff $\upsilon(x_1) = true$ and $\upsilon(x_2) = \upsilon(x_3) = false$. In fact, this is the only maximal υ-path for this valuation, and this is the only valuation under which $\Pi(s_0, \upsilon)$ contains an infinite path.

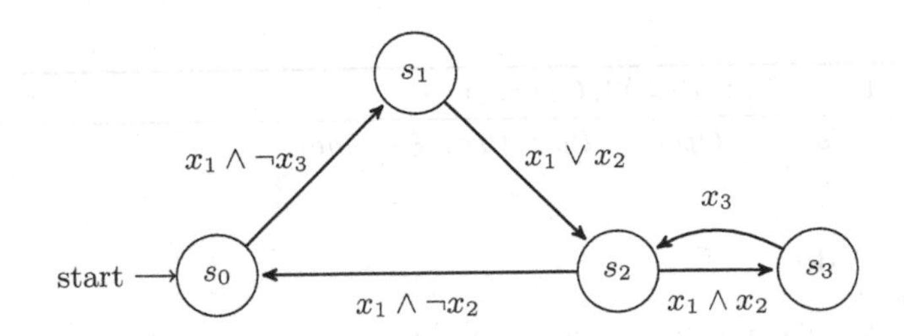

Fig. 1. A simple model.

A function $\xi : S \rightarrow 2^{\Omega}$ is called a *characterization*. The set of all characterizations is denoted by $Chars$. A characterization ξ assigns to an each state a set of valuations; intuitively, these are the constraints under which some given property holds. In what follows, for each $g \in \mathcal{L}_X$ by $[g]$ we denote the set of all valuations satisfying g.

Definition 2. *Let X be a set of parameters, and let $\mathcal{M} = (S, s_0, \rightarrow)$ be a $\mathcal{L}_X$-labeled transition system. We define the operators of* parametric image *and the* parametric preimage $Post, Pre \in Chars^{Chars}$ *with respect to* $\rightarrow$ *as follows. Let* $\xi \in Chars$*, then:*

$$Post(\xi)(s) = \bigcup_{t \in \rightarrow} \{[guard(t)] \cap \xi(src(t)) \mid trgt(t) = s\},$$

$$Pre(\xi)(s) = \bigcup_{t \in \rightarrow} \{[guard(t)] \cap \xi(trgt(t)) \mid src(t) = s\},$$

for all $s \in S$.

Let us assume that the characterization ξ defines the constraints under which some property in question holds. The set $Post(\xi)(s)$ consists of all the valuations that simultaneously enable the transition to s from some of its predecessors s' and ensure that the property ξ holds in s'. The set $Pre(\xi)(s)$ gathers all the valuations which simultaneously enable a transition to some successor s' of s and ensure that the property ξ holds in s'.

2.1 Reachability and Deadlock Freedom

The reachability analysis is arguably the most utilized formal methods-based task both in the research and the industry, employed in error detection [26], protocol analysis [12,19], planning [10], and many other areas. Algorithm 1 is a template that allows, depending on the passed operator *Oper* and initial property ξ, for the synthesis of constraints for reachability from the initial state (Lemma 1), reachability of a state satisfying a given property f (Lemma 2), and freedom from deadlocks in the model (Lemma 3).

Algorithm 1. $constrVals\,(\mathcal{M}, Oper, \xi)$

Input: $\mathcal{M} = (S, s^0, \rightarrow)$, $Oper \in \{Post, Pre\}$, $\xi \in Chars$

1: $f := \xi$
2: $h := \emptyset$
3: **while** $f \neq h$ **do**
4: $\quad h := f$
5: $\quad$ **for each** $s \in S$ **let** $f(s) := f(s) \cup Oper(f)(s)$
6: **end while**
7: **return** f

It is usually more feasible to use the parametric image (i.e., to test *forward reachability*) rather than the preimage operator (i.e., to test *backward reachability*) when synthesizing the constraints for the reachability from the initial state. This is due to the possible chance of pruning the state-space by removing the trajectories that start from the states unreachable from the initial one. The state $s \in S$ is *reachable* from the initial state s^0 under valuation $\upsilon \in \Omega$ iff there exists a υ-path $\pi \in \Pi(s^0, \upsilon)$ such that $\pi_i^S = s$ for some $i \in \mathbb{N}$.

Lemma 1. *Let $\mathcal{M} = (S, s^0, \rightarrow)$ be a model. Denote $\xi_I := \{(s^0, \Omega)\} \cup \{(s, \emptyset) \mid s \in S \wedge s \neq s^0\}$, then a state $s \in S$ is reachable from s^0 under $v \in \Omega$ iff $v \in constrVals(\mathcal{M}, Post, \xi_I)(s)$.*

Proof. For each $i \in \mathbb{N}$ by f^i denote the value of the f variable before entering the 3–6 loop. For a while replace the stop condition test of the loop with *true*. We prove that for each $i \in \mathbb{N}$ the set $f^i(s)$ consists of all the valuations v under which the state s can be reached in i or less steps from the initial state s^0, i.e., there exists $\pi \in \Pi(s^0, v)$ such that $\pi_j^S = s$ for some $j \leq i$. The proof follows by the induction with the trivial base case of f^0, as defined in Line 1. For the inductive step, notice that by the inductive assumption the state s is reachable under v in $i+1$ or less steps iff $v \in f^i(s)$ or if for some $\pi \in \Pi(s^0, v)$ we have $\pi_{j+1}^S = s$ and $v \in f^i(\pi_j^S)$ and $v \models \pi_j^T$ for some $j \leq i$. The latter is in turn equivalent to $v \in f^i(s) \cup Post(f^i)(s)$, i.e., $v \in f^{i+1}(s)$ (Line 5). Now it suffices to notice that for each of a finite number of $s \in S$ the sequence $(f^i(s))_{i \in \mathbb{N}}$ is monotonically increasing and consists of subsets of a finite set Ω, thus for some $k \in \mathbb{N}$ we have $f^{j+k} = f^k$ for all $j \in \mathbb{N}$. Therefore the fixed-point is reached, the loop stops, and $f = f^k$ is the correct characterization of the reachability from s^0 for $\mathcal{M}$. □

Let $\xi \in Chars$ be a characterization, $v \in \Omega$, and $s \in S$. We say that ξ is reachable under v from the state $s \in S$ iff there exists $\pi \in \Pi(s, v)$ such that $v \in \xi(\pi_i)$ for some $i < |\pi|$. To obtain the constraints for this type of reachability we employ the operator of the parametric preimage in Algorithm 1.

Lemma 2. *Let $\mathcal{M}$ be a model. A characterization $\xi \in Chars$ is reachable under $v \in \Omega$ from the state $s \in S$ iff $v \in constrVals(\mathcal{M}, Pre, \xi)(s)$.*

Proof. Again, for each $i \in \mathbb{N}$ by f^i we denote the value of the f variable before entering the loop 3–6. By the induction on $i \in \mathbb{N}$ we prove that for each $s \in S$ the set $f^i(s)$ consists of all the valuations under which for some $\pi \in \Pi(s, v)$ there exists $0 \leq j \leq i$ such that $v \in \xi(\pi_j)$. We omit the remaining details of the proof, as it follows very similarly to the proof of Lemma 1. □

One of the typical assumptions in the formal methods approach to systems analysis is that the model in question is free of deadlocks, i.e., every state reachable from the initial one has an enabled outgoing transition. A state $s \in S$ is in a *deadlock* under valuation $v \in \Omega$ iff there is no $s' \in S$, such that $s \xrightarrow{g} s'$ and $v \models g$, i.e., s has no successor under v. A state is *deadlock-free* under $v \in \Omega$ iff it is not in a deadlock. In what follows, let $\xi^\Omega \in Chars$ be such a characterization that $\xi^\Omega(s) = \Omega$ for all $s \in S$. For each $s \in S$ the set $Pre(\xi^\Omega)(s)$ consists of all those valuations under which there is an outgoing transition from s. This means that $Pre(\xi^\Omega) \in Chars$ is a characterization of *no-local-deadlock* property, what we want to obtain however, is the set of valuations under which no state in deadlock is reachable from a given state. Let $\xi \in Chars$ be a characterization, then we define its *complement* $\overline{\xi} \in Chars$ by $\overline{\xi}(s) := \Omega \setminus \xi(s)$ for all $s \in S$. Intuitively, $\overline{\xi}(s)$ gathers all those valuations under which the property ξ does not hold in s.

Lemma 3. *Let $\mathcal{M}$ be a model and $s \in S$. All states reachable from s under v are deadlock-free iff $v \in \overline{constrVals(\mathcal{M}, Pre, \overline{Pre(\xi^{\Omega})})(s)}$.*

Proof. Let $s \in S$, then the set $\overline{Pre(\xi^{\Omega})}(s)$ consists of the valuations under which s is in a deadlock, i.e., $\overline{Pre(\xi^{\Omega})} \in Chars$ is a characterization of a deadlock. By Lemma 2, the set $constrVals(\mathcal{M}, Pre, \overline{Pre(\xi^{\Omega})})(s)$ gathers all those valuations under which a state in a deadlock is reachable from s, thus the complement of the set consists of valuations under which all reachable states are deadlock-free. □

3 Boolean Encodings

The notions introduced in this paper are geared towards symbolic parameter synthesis. In contrast to the explicit approach, in symbolic verification and synthesis we do not deal with single states and transitions; instead - we manipulate sets of states, functions and relations [6]. In this section we show how to express the problem of parameter synthesis using operations on propositional formulae.

3.1 Boolean Encodings and Operations

Let $\mathcal{M} = (S, s^0, \rightarrow)$ be a $\mathcal{L}_X$-labelled transition system, and let V be a set of propositional variables such that V is disjoint with X and $|V| = \lceil log(|S|) \rceil$. Recall that $\mathcal{L}_V$ denotes the set of propositional formulae over V. It is easy to see that we can encode the set of states S using the variables from V, i.e., for each $s \in S$ we define $enc(s) \in \mathcal{L}_V$ such that $enc(s)$ is satisfiable and $\not\models enc(s) \wedge enc(s')$ for all $s, s' \in S, s \neq s'$. We also use an auxiliary set of propositional variables V', disjoint with V and X and such that $|V'| = |V|$. Assume that $V = \{v_1, \dots, v_k\}$ and $V' = \{v'_1, \dots, v'_k\}$; if $g \in \mathcal{L}_V$ then $g' \in \mathcal{L}_{V'}$ denotes the formula obtained by replacing in g each v_i with its primed counterpart. The transition relation of $\mathcal{M}$ is then encoded as: $enc(\rightarrow) = \bigvee_{t \in \rightarrow} enc(src(t)) \wedge guard(t) \wedge enc'(trgt(tr))$.

Example 2. Let us encode the transition relation of the model in Fig. 1. Denote $V = \{v_1, v_2\}$ and $X = \{x_1, x_2, x_3\}$. Let us put $enc(s_0) = v_1 \wedge v_2$, $enc(s_1) = v_1 \wedge \neg v_2$, $enc(s_2) = \neg v_1 \wedge v_2$, $enc(s_3) = \neg v_1 \wedge \neg v_2$, then: $enc(\rightarrow) = \big((v_1 \wedge v_2) \wedge (x_1 \wedge \neg x_3) \wedge (v'_1 \wedge \neg v'_2)\big) \vee \big((v_1 \wedge \neg v_2) \wedge (x_1 \vee x_2) \wedge (\neg v'_1 \wedge v'_2)\big) \vee \big((\neg v_1 \wedge v_2) \wedge (x_1 \wedge \neg x_2) \wedge (v'_1 \wedge v'_2)\big) \vee \big((\neg v_1 \wedge v_2) \wedge (x_1 \wedge x_2) \wedge (\neg v'_1 \wedge \neg v'_2)\big) \vee \big((\neg v_1 \wedge \neg v_2) \wedge (x_3) \wedge (\neg v'_1 \wedge v'_2)\big)$.

If $A \subseteq \Omega$ then let $enc(A) \in \mathcal{L}_X$ be such that $[enc(A)] = A$. Let $f \in Chars$ be a characterization, then we define its encoding as: $enc(f) = \bigvee_{s \in S} enc(s) \wedge enc(f(s))$. Let $\xi \in Chars$, then the operations of parametric image and preimage of ξ are performed symbolically as follows:

$$enc'(Post(\xi)) = \bigvee_{bf \in 2^V} (enc(\rightarrow) \wedge enc(\xi))[v_1 \leftarrow bf(v_1), \dots, v_k \leftarrow bf(v_k)],$$

$$enc(Pre(\xi)) = \bigvee_{bf \in 2^{V'}} (enc(\rightarrow) \wedge enc'(\xi))[v'_1 \leftarrow bf(v'_1), \dots, v'_k \leftarrow bf(v'_k)].$$

It is quite easy to see that the encoding of a join of two relations or characterizations is encoded as a disjunction of their encodings; the same follows for the meet, encoded as a conjunction. In order to implement the operation of complement, we make use of the following procedure.

Algorithm 2. *Complement(f)*

Input: $f \in Chars$; **Output:** $enc(\overline{f}) \in Chars$
1: **return** $\neg enc(f) \wedge \bigvee_{bf \in 2^{V'}} enc(f)[v'_1 \leftarrow bf(v'_1), \ldots, v'_k \leftarrow bf(v'_k)]$

The following lemma (first presented in [14]), applied to $F = enc(f)$ with f_i substituted with state encodings and g_i substituted with associated guards, proves that $enc(\overline{f}) = Complement(f)$.

Lemma 4. *Let $F = \bigvee_{i \in A} f_i \wedge g_i$, where A is a finite set of indices, and f_i, g_i are propositional formulae such that:*

- *sets of propositional variables present in f_i, g_i are disjoint,*
- *for all $i, j \in A$ such that $i \neq j$ we have that $\not\models f_i \wedge f_j$.*

Let $F' = \bigvee_{i \in A} f_i$, then $\neg F \wedge F' = \bigvee_{i \in A} f_i \wedge \neg g_i$.

Proof. Let $v \models \neg F \wedge F'$, then there exists exactly one $i \in A$ such that $v \models f_i$. On the other hand $v \not\models f_i \wedge g_i$, thus $v \models f_i \wedge \neg(f_i \wedge g_i)$. Now it suffices to notice that $f_i \wedge \neg(f_i \wedge g_i) \equiv f_i \wedge \neg g_i$. Now let $v \models \bigvee_{i \in A} f_i \wedge \neg g_i$. As previously, there exists exactly one $i \in A$ such that $v \models f_i \wedge \neg g_i$, and obviously $v \models F'$. It suffices to notice that if we had $v \models F$ then $v \models f_i \wedge g_i$ would hold. This is not possible, therefore $v \models \neg F$, and $v \models \neg F \wedge F'$. □

4 Application and Evaluation

In this section we present an outline of how the general ideas and models introduced so far have been specialized for a selected logic, providing parametric frameworks for action synthesis [17]. We have implemented the theory in stand-alone tool SPATULA [15,17]. The tool employs Reduced Ordered Binary Decision Diagrams (BDDs) CUDD package [23] for an efficient representation of transition relation and operations on Boolean encodings. To provide the means for the comparison with the non-parametric approach we also implemented the *naïve*, enumerative parameter synthesis. In the naïve approach, a tool simply iterates through all the possible substitutions of parameters and records all those for which the result yields true.

4.1 Preliminaries

In what follows, we introduce a parameterized logic that allows for the presence of the parameters (i.e., free variables) in the formulae. In order to deal with the formulae with multiple parameters, we need to extend the notion of a characterization. Let ϕ be a formula with parameters $Pars$ (i.e., a modal logic formula, interpreted in a $\mathcal{L}_X$-labeled transition system $\mathcal{M}$). Let $\upsilon : Pars \to 2^X$ be a valuation of $Pars$ (we use the symbol Ω_{Pars} to denote the set of all such valuations). In this work, we assign to $Pars$ the subsets of the set of allowed actions (subsection 4.2). By $\phi[\upsilon]$ we denote the result of substitution of the variables in ϕ in accordance to υ, and by $\mathcal{M}, s \models_\upsilon \phi$ we denote that the ground formula $\phi[\upsilon]$ holds in the state s of $\mathcal{M}$ (we usually omit the model symbol). The methods presented in the previous section can be extended to allow for an efficient construction, for each formula ϕ of a chosen logic, of the function $f_\phi : S \to 2^{\Omega_{Pars}}$ such that: $\upsilon \in f_\phi(s)$ iff $s \models_\upsilon \phi$, for all $s \in S$. We only present a brief explanation of the relevant extensions, referring to the earlier work for a detailed description.

Let $\mathcal{M}$ be a model with the set of states S, and let $\mathcal{PV}$ be a finite set of fresh (i.e., not appearing in the model) propositions. A function $\mathcal{V} : S \to 2^{\mathcal{PV}}$ is called a *labeling*. Intuitively, $\mathcal{V}(s)$ denotes the set of all the propositions from $\mathcal{PV}$ that are true in the state $s \in S$.

4.2 Parametric Action-Restricted CTL

Action-restricted CTL [20] (ARCTL) is a simple branching time logic with actions. The formulae of ARCTL limit the set of actions allowed along a given run. In [17] we have introduced a parametric extension of the logic, called PARCTL that allows for the presence of the variables in place of concrete sets of actions.

Definition 3 (PARCTL syntax). *Let Acts be a finite set of* actions, *and Pars be a finite set of* action variables. *The set of the formulae of Parametric Action-Restricted CTL is defined by the following grammar:*

$$\phi ::= p \mid \neg\phi \mid \phi \vee \phi \mid E_\alpha X\phi \mid E_\alpha G\phi \mid E_\alpha G^\omega \phi \mid E_\alpha(\phi\, U \phi),$$

where $p \in \mathcal{PV}$, $\alpha \in (2^{Acts} \cup Pars)$, *and* $Y \in Pars$.

The E path selector is read as *"there exists a path"*, and the superscript $_\alpha$ selects the actions allowed along a given run. The X modality stands for *"in a next state"*, the modalities G, G^ω stand for *"globally"* with the second one pertaining to the infinite paths only, and the U modality is read as *"until"*. In what follows, we also use the universal path selector A (read as *"for all paths"*) and the temporal modality F (standing for *"in future"*); both of these can be derived from the already introduced notions. As to give an example, $E_{\{\text{left, right}\}}G(E_{\{\text{forward}\}}F \text{ safe})$ is a formula without parameters that may be read as *"there exists a path over* left *and* right, *on which it holds globally that a state satisfying* safe *is reachable along some path over* forward*"*. A formula $E_Y G(E_Z F \text{ safe})$ contains two parameters we seek to valuate, i.e., for a given state $s \in S$ we wish to obtain all valuations υ such that $E_{\upsilon(Y)}G(E_{\upsilon(Z)}F \text{ safe})$ holds in s.

We refer to [17] for the full description of the semantics of the logic and the construction of the function f_ϕ for each $\phi \in$ PARCTL. In general, the implementation of the Boolean operations is rather straightforward, and the following equalities hold: $E_Y G^\omega \phi \equiv \phi \wedge E_Y X E_Y G^\omega \phi$, and $E_Y G \phi \equiv \phi \wedge (\neg E_Y X true \vee E_Y X E_Y G^\omega \phi)$, and $E_Y(\phi U \psi) \equiv \psi \vee (\phi \wedge E_Y X E_Y(\phi U \psi))$, where $\phi, \psi \in$ PARCTL, and $Y \in Pars$. These equivalences can be converted into fixed-point algorithms [17] based on the consecutive computation of $f_{E_Y X(\cdot)}$, which in turn can be obtained by a single application of parametric preimage operation with some additional variable relabeling.

4.3 Experimental Results

We have performed a batch of tests to establish the efficiency of our approach as compared to the naïve, enumerative one, as to our best knowledge there is no tool comparable to ours. In all the experiments, the timeout was set at 15 minutes.

In the first test, we used a version of Train-Gate-Controller, a classical benchmark [2] with the injected faulty behaviour, as inspired by [3]. The system consists of k trains and the controller that monitors the access to the tunnel. The safety requires that at most one train at a time enters the tunnel. In a non-faulty version of the model this is ensured by a simple protocol where a train gains an access if it reaches an agreement with the controller. In our version of the system, the (*red*/*green* light - based) communication between the selected faulty train and a controller can malfunction. We have tested the properties:

- $\psi_1 = A_Y G(\neg \bigvee_{1 \leq i < l \leq k}(in_i \wedge in_l)) \wedge \bigwedge_{1 \leq i \leq k} E_Y F in_i$, with the meaning that: *"it is not possible for any pair of trains to be in the tunnel at the same time, and each train will eventually be in the tunnel"*;
- $\psi_2 = E_Y F A_Y G((\bigwedge_i^k \neg in_i) \wedge green)$, with the meaning that: *"it is possible for the system to execute in such a way that at some state, in all the possible executions of the system, all the trains remain outside the tunnel while the controller remains in the* green *state"*.

As we aim to synthesize actions that are used for communication (via synchronized actions) between the participating entities, the results of the synthesis can be interpreted as finding the sets of messages that have to be turned off in order to provide the compliance with a specification.

In Table 1 we provide the results on the relative time speedup of the parametric approach versus the naïve one. As it can be clearly seen, the benefits of moving from an exponential number of simple tests operating on sets (enumerative synthesis) to a small number of complex tests operating on characteristic functions are very substantial. This is in line with the results reported in [8], where the relative speedup exceeded 750 for the emptiness and universality tests.

Table 1. Speedup for Faulty Train-Gate-Controller

Property	**Speedup** (naïve/parametric time)			
	2 trains	3 trains	4 trains	5 trains
ψ_1	76.0	463.59	4021.68	17378.02
ψ_2	48.96	276.01	703.97	1553.73

In the second test, we analyze a pipeline network inspired by [21]. The model in question consists of the chain of k nodes, each of which has two states: *in* and *out*. A node can synchronize via shared action with up to four other surrounding ones, depending on its position in the pipeline. The first node can be perceived as a Producer, and the last one as a Consumer, and each of the intermediate tokens can obtain an information token from one of its predecessors by firing a shared action and moving to the *in* state. In this way the token can be moved from the Producer to the Consumer through a series of intermediate nodes. We have tested the properties:

- $\phi_1 = A_Y F(\bigwedge_{1 \leq i \leq \lfloor \frac{k}{2} \rfloor} out_i \wedge \bigwedge_{\lceil \frac{k}{2} \rceil \leq j \leq k} in_j)$, describing unavoidability of a configuration in which the first half of the nodes is in *out*- and the other half is in *in* states;
- $\phi_2 = A_Y G A_Y F(\bigwedge_{1 \leq i \leq k} in_i)$, expressing that the configuration with all the nodes simultaneously in their *in* states appears infinitely often or ends a path;
- $\phi_3 = E_Y F A_Y G(\bigwedge_{1 \leq i \leq \lceil \frac{k}{2} \rceil} in_{2i-1} \wedge \bigwedge_{1 \leq i \leq \lfloor \frac{k}{2} \rfloor} out_{2i})$, describing that the configuration with the first half of the nodes such that the odd nodes are in their *in*- and the even are in their *out* states becomes persistent starting from some point in the future.

As previously, we present the relative speedup of our approach in Table 2. The relative efficiency of our approach is even more evident in this case, as we were able to verify formulae for which the naïve approach timed out within the set time limits.

Table 2. Speedup for Generic Pipeline Paradigm

Property	**Speedup** (naïve/parametric time)			
	7 processes	8 processes	9 processes	10 processes
ϕ_1	1402.60	4115.96	9171.02	22669.83
ϕ_2	1202.53	3265.79	8723.40	$> 12344.49^{\dagger}$
ϕ_3	2985.93	7979.04	18633.09	$> 34531.71^{\dagger}$

(† - the naïve approach exceeded set timeout of 15 minutes)

We refer to [17] for the detailed presentation and analysis of the experimental results, as well as for some applications of our tool to concurrent systems security.

5 Conclusions

In this work we presented a framework for parametric model checking for models with discrete transitions. We have shown how to synthesize constraints for reachability by means of computing the fixpoint of consecutively applied sequence of parametric image operations, and how to synthesize constraints for deadlock-freedom by means of computing the fixpoint of consecutively applied sequence of operations of parametric preimage. We have also outlined how to implement the presented theory using Boolean encodings (typically, BDDs), and how to extend it to properties expressed in selected modal logics on an example of PARCTL. The benefits of the approach are illustrated on two scalable benchmarks.

Acknowledgements. Michał Knapik is supported by the Foundation for Polish Science under Int. PhD Projects in Intelligent Computing. Project financed from the EU within the Innovative Economy OP 2007-2013 and ERDF.

References

1. Alur, R., Etessami, K., Torre, S.L., Peled, D.: Parametric Temporal Logic for "Model Measuring". ACM Trans. Comput. Log. 2(3), 388–407 (2001)
2. Alur, R., Henzinger, T., Vardi, M.: Parametric real-time reasoning. In: Proc. of the 25th Ann. Symp. on Theory of Computing (STOC 1993), pp. 592–601. ACM (1993)
3. Belardinelli, F., Jones, A.V., Lomuscio, A.: Model checking temporal-epistemic logic using alternating tree automata. Fundam. Inform. 112(1), 19–37 (2011)
4. Bozzelli, L., La Torre, S.: Decision problems for lower/upper bound parametric timed automata. Form. Methods Syst. Des. 35(2), 121–151 (2009)
5. Bruyère, V., Dall'Olio, E., Raskin, J.F.: Durations, parametric model-checking in timed automata with presburger arithmetic. In: Alt, H., Habib, M. (eds.) STACS 2003. LNCS, vol. 2607, pp. 687–698. Springer, Heidelberg (2003)
6. Burch, J.R., Clarke, E., McMillan, K.L., Dill, D.L., Hwang, L.J.: Symbolic model checking: 10^{20} states and beyond. Information and Computation 98(2), 142–170 (1990)
7. Clarke, E.M.: The birth of model checking. In: Grumberg, O., Veith, H. (eds.) 25 Years of Model Checking. LNCS, vol. 5000, pp. 1–26. Springer, Heidelberg (2008)
8. Classen, A., Heymans, P., Schobbens, P.Y., Legay, A.: Symbolic model checking of software product lines. In: Proc. of the 33rd Int. Conf. on Software Engineering, ICSE 2011, pp. 321–330. ACM, New York (2011)
9. Emerson, E.A., Trefler, R.: Parametric quantitative temporal reasoning. In: Proc. of the 14th Symp. on Logic in Computer Science (LICS 1999), pp. 336–343. IEEE Computer Society (July 1999)
10. Ghallab, M., Nau, D.S., Traverso, P.: Automated planning - theory and practice. Elsevier (2004)
11. Di Giampaolo, B., La Torre, S., Napoli, M.: Parametric metric interval temporal logic. In: Dediu, A.-H., Fernau, H., Martín-Vide, C. (eds.) LATA 2010. LNCS, vol. 6031, pp. 249–260. Springer, Heidelberg (2010)

12. Holzmann, G.J.: Protocol design: Redefining the state of the art. IEEE Software 9(1), 17–22 (1992)
13. Jensen, K., Donatelli, S., Koutny, M. (eds.): Transactions on Petri Nets and Other Models of Concurrency IV. LNCS, vol. 6550. Springer, Heidelberg (2010)
14. Jones, A.V., Knapik, M., Penczek, W., Lomuscio, A.: Parametric computation tree logic with knowledge. In: Proc. of the Int. Workshop on Concurrency, Specification and Programming (CS&P 2011), pp. 286–300. Białystok University of Technology (2011)
15. Knapik, M., https://michalknapik.github.io/spatula
16. Knapik, M., Penczek, W., Szreter, M., Pólrola, A.: Bounded parametric verification for distributed time Petri nets with discrete-time semantics. Fundam. Inform. 101(1-2), 9–27 (2010)
17. Knapik, M., Męski, A., Penczek, W.: Action synthesis for branching time logic: Theory and applications. In: Proc. of the 14th Int. Conf. on Application of Concurrency to System Design. IEEE Computer Society (to appear, 2014)
18. Knapik, M., Szreter, M., Penczek, W.: Bounded parametric model checking for elementary net systems. In: T. Petri Nets and Other Models of Concurrency [13], pp. 42–71
19. Lin, F.J., Chu, P.M., Liu, M.T.: Protocol verification using reachability analysis: The state space explosion problem and relief strategies. SIGCOMM Comput. Commun. Rev. 17(5), 126–135 (1987)
20. Pecheur, C., Raimondi, F.: Symbolic model checking of logics with actions. In: Edelkamp, S., Lomuscio, A. (eds.) MoChArt IV. LNCS (LNAI), vol. 4428, pp. 113–128. Springer, Heidelberg (2007)
21. Peled, D.: All From One, One For All: On Model Checking Using Representatives. In: Courcoubetis, C. (ed.) CAV 1993. LNCS, vol. 697, pp. 409–423. Springer, Heidelberg (1993)
22. Penczek, W., Pólrola, A., Zbrzezny, A.: Sat-based (parametric) reachability for a class of distributed time petri nets. In: T. Petri Nets and Other Models of Concurrency [13], pp. 72–97
23. Somenzi, F.: CUDD: CU decision diagram package - release 2.3.1., http://vlsi.colorado.edu/~fabio/CUDD/cuddIntro.html
24. Wang, F.: Parametric timing analysis for real-time systems. Inf. Comput. 130(2), 131–150 (1996)
25. Wang, F.: Parametric analysis of computer systems. Formal Methods in System Design 17(1), 39–60 (2000)
26. Xie, Y., Aiken, A.: Scalable error detection using boolean satisfiability. SIGPLAN Not. 40(1), 351–363 (2005)

Specialized vs. Multi-game Approaches to AI in Games

Maciej Świechowski[1] and Jacek Mańdziuk[2]

[1] Phd Studies at Systems Research Institute,
Polish Academy of Sciences, Warsaw, Poland
m.swiechowski@ibspan.waw.pl

[2] Faculty of Mathematics and Information Science,
Warsaw University of Technology, Warsaw, Poland
j.mandziuk@mini.pw.edu.pl

Abstract. In this work, we identify the main problems in which methodology of creating multi-game playing programs differs from single-game playing programs. The multi-game framework chosen in this comparison is General Game Playing, which was proposed at Stanford University in 2005, since it defines current state-of-the-art trends in the area. Based on the results from the International General Game Playing Competitions and additional results of our agent named MINI-Player we conclude on what defines a successful player. The most successful players have been using a minimal knowledge and a mechanism called Monte Carlo Tree-Search, which is simulation-based and self-improving over time.

Keywords: Games, Artificial Intelligence, General Game Playing, Heuristic Search.

1 Introduction

The focus of this study is to review some of the approaches to AI in games and elaborate on a shift from these approaches to multi-game playing. We emphasize the difference between those two approaches while focusing on the latter. Games have been a legitimate challenge of AI since the early days of the field. There have been some common traits formulated by various researches [25], [21], [19], such as deduction, reasoning, problem solving, intelligent search, knowledge representation, planning, learning, creativity, perception, motion (manipulation) and natural language processing. Most of them, especially the first seven, are part of intelligent game playing. Researchers started to work on AI in the so-called mind-games, often driven by the ultimate goal of getting closer towards general intelligence. The programs began to be more and more specialized. Despite displaying often a remarkable set of skills, their intelligence and usefulness of the applied methods in the real-world problems are questionable. A top level chess-playing computer program [10] can defeat any human in chess but cannot help to form a medical diagnosis or even play a simple Tic-Tac-Toe game. A lack for multi-game playing systems has emerged. There have been many attempts

P. Angelov et al. (eds.), *Intelligent Systems'2014*,
Advances in Intelligent Systems and Computing 322, DOI: 10.1007/978-3-319-11313-5_23

to bring back to life the early-day concepts of AI, such as SAL [9], Morph [16], Hoyle [5] and METAGAMER [26]. The most recent and probably the most well-designed is General Game Playing [8] which we introduce in the next section. In Section III we enumerate key differences to the methodology in both of the aforementioned disciplines. Next, we review a variety of heuristics applied to General Game Playing some of which are contributed by ours. Then, in section V, we provide some experimental results and conclude in section VI.

2 General Game Playing

General Game Playing (GGP) [8] is a realization of the multi-game concept and the current grand challenge of AI in games. The term was coined at Stanford University in 2005, together with specification of the international General Game Competition. The competition is held annually at the AAAI Conference (one exception is IJCAI in 2011) and works as an official world championships. The winners define the state-of-the-art solutions in the field. The GGP is about creating agents capable of playing games which can be defined in the Game Description Language (GDL) [17]. This declarative-logic language is a subset of Prolog. It is used to define rules of games as well as for communication between the agents and the Game Manager. The Game Manager is a communication hub between the players and a referee checking ensuring if the game is played legally. The communication, realized via HTTP, is also part of the GGP specification. General Game Playing revisits the early AI concepts. With programs tailored for playing specific games, most of the interesting analysis is done by the authors (or consulted experts). The goal of the GGP is to transfer this analysis to the programs, which start from scratch, without any knowledge. The programs are also autonomous i.e. no human intervention is possible.

3 General vs. Specialized Game Playing

In the previous section we emphasized the characteristic concepts of General Game Playing. There are various practical differences in research tools, key problems, implementation and the overall research mind-set, when comparing GGP and specialized game playing. We identify some of them in this section.

3.1 Rules Representation

In General Game Playing a rules interpretation system also known as an inference engine is required to be able to determine the game states and perform any kind of search. The GDL, which rules are represented in, is a declarative first-order logic language which contains no high level game-related logic. It means, that every concept present in games such as a piece, board, coordinate or adjacency of coordinates has to be defined from scratch within the language and there is no meta-information on what the concept is. Even mathematical formulas such as addition must be defined by logical rules for every arguments and results. Such an approach is very general but very costly at the same time in terms of computational efficiency. As a consequence, the first choice to make in GGP is an inference method. It affects the player in a way how fast it will be able to search the game tree and also what the fidelity of the simulation will be. The more control over the inference process the more game-playing algorithms can be put inside which use the internal/intermediate state of the process. The speed of a rules interpreter also affects the feasible game-tree search which can be applied. It also quite convincingly hampers the methods of computational intelligence to be used in GGP as long as the start and play clocks apply. There are a few approaches to the problem of rules interpretation. Some players convert rules directly to Prolog. FluxPlayer uses fluent calculus implemented in ECLiPSE Prolog [1]. Many agents use custom-made interpreters for the GDL. Another problem, apart from efficiency, arises with the rules representation. In General Game Playing, we cannot design an efficient representation of a state because we do not know dimensions of the state (state-space, possible elements and arities of the elements) in advance. We do not even know what defines the state exactly although a sensible approach is to approximate the state by taking all facts used by *init* and *next* relations. An optimized representation such as 0x88 and Zobrist's Hashing in chess, can be used for Transposition Tables [12] and serialization required for the communication in a distributed parallel system.

3.2 Access to Knowledge

In General Game Playing, there is no such thing as expert knowledge. Not only any human-intervention is illegal but also the games can be unknown beforehand or obfuscated so there would not exist an expert in most games. Moreover, the start clocks are usually set to from a couple of seconds to a couple of minutes, so there is no time for too complex methods. In specialized game playing lots of analysis can be done either by experts or by a long off-line tuning. In General Game Playing the agents either has to use robust simulation-based methods or discover the knowledge relatively quickly. It can be assumed that the discovered knowledge will not be very deep, but it may change in future when the GGP area becomes more advanced. The lack of expert knowledge is connected with the lack of databases of opening and endings. When a game is known, players quickly discover the reasonable ways how to start. Subsequent responses to the start moves are studied, players seek new moves and a database of opening grows. The endings are usually calculated using brute-force methods when the complexities of states become low enough to allow for this. Those two features are staples in single-game programs. Lastly, games in General Game Playing contain no interpretation of what particular predicates mean i.e. there is no connection to the game-specific objects. Even if there was such a connection defined in the rules (some kind of meta-information) both the designers of games and programs would have to predict names of all the possible game objects in advance. If the games' authors (or automatic generators) were not able to define custom types for the game objects, the GGP would not be general anymore. To sum it up, there are no predefined game-specific elements which agents could use in construction of evaluation functions. Every building block for the evaluation function has to be discovered at runtime and only with certain probability of being accurate.

3.3 Game Tree Search

The introduction of start and play clocks requires General Game Playing programs to use the so-called anytime game tree search methods i.e. which can be interrupted at any time and return the currently best action. Moreover, in GGP, a search method cannot use reaching a particular tree level as a stop condition because the required time cannot be predicted beforehand. The most popular search method in GGP is Monte Carlo Tree-Search (MCTS) [2] which will be presented in the next section. Lack of heuristics may limit possible usage of methods such as alpha-beta pruning or MTD [27]. The problem which does not exist in the specialized game-playing is that in multi-game playing the chosen tree search should be universally good at the cost of being suboptimal in some games.

3.4 Diversity of Games

There are many properties of games which can be took advantage of when designing an AI algorithm. Such properties include:

- Is the game zero-sum or not - zero-sum games can use the plain min-max idea. Moreover, there is no need to store all scores.
- Is the game co-operative or competitive - co-operation, to implement well, requires a completely different method.
- How many players there are - puzzles are inherently different whereas the goal of two-player games is usually more focused on beating an opponent. The game-tree structure can be also optimized for the number of players.
- Is the game turn-based or simultaneous - simultaneous games are usually more difficult. Action selection for turn-based games can be optimized.
- What is the Branching-Factor - a huge branching factor can render some game-tree search methods useless whereas a tiny branching-factor can make solving a game a viable attempt.
- Is there a board in the game - there have been various board-related concepts well-studied and established in games.

4 Heuristics in GGP

In this section we go through various heuristics applied in General Game Playing. We include both the related work and our contributions. We believe that the existence of such a rich variety of approaches shows that there is no single best solution to apply. Moreover, although the strongest programs use the Monte Carlo Tree-Search (MCTS) combined with the UCT [14] algorithm, which has become a *de-facto* standard, it is not clear whether with improvement in the heuristics field a different type of search will be a dominant one.

4.1 ClunePlayer

ClunePlayer [4] is the name of the first (2005) GGP Competition champion. It used the min-max search with an evaluation function. The key idea was an abstract model representing the so-called simplified game: $P : \Omega \rightarrow [0, 100]$ - a function which approximates the payoff $C : \Omega \rightarrow [-1, 1]$ - control: a relative mobility $T : \Omega \rightarrow [0, 1]$ - probability that a state is terminal $S_p : [1, \infty]$ - stability of the payoff $S_m : [1, \infty]$ - stability of the mobility; the Ω represents the set of legal states. The payoff function uses cardinality and distance between features as the building blocks. A feature is a GDL expression, which denotes a condition on a fact. Because its a condition and not the ground fact, it can contain uninstantiated variables. Candidate features are all conditions using facts defined in the GDL rules as well as new ones which are created by replacing variables with symbols of the respective domains. The cardinality means how many of facts fulfill the condition. The distances are used if a board relation and a next relation, which defines ordering of the board's coordinates, are detected semantically in the description using predefined template. Only stable features are used in the payoff function, where stability is computed based on the so-called adjacent variance (between the consecutive states) and total variance. The mobility is computed as follows:

$$C = \frac{Moves_a(\omega) - Moves_b(\omega)}{max(Moves_a(\omega) - Moves_b(\omega))} \tag{1}$$

where $Moves_a$ denotes the number of our player's legal moves. The formula as well as the whole approach is dedicated to two-player games only. The termination function is computed purely statistically using least-squares regression. The final formula for the evaluation function used in the min-max search combines five elements of the abstract game:

$$V = T * P + (1 - T) * [(50 + 50 * C) * S_c + S_p * P] \tag{2}$$

4.2 FluxPlayer

FluxPlayer [30] has been the second (2006) Competition champion. The key features of FluxPlayer are: using fuzzy logic to determine the degree to which any given state is terminal; identification of a few syntactic structures; fluent calculus implemented in a custom-made tool for reasoning in the GDL. In GDL, the rules and facts can be either true or false, there is no such concept as partial satisfaction. Moreover, the goal rule is defined to give a meaningful result only a in terminal state. However, if a rule is FALSE, it is convenient to have a mechanism of determining how close the rule is to TRUE. Moreover, it is useful to test the goal rule in non-terminal states because it is often a structural representation of the game objective. FluxPlayer is equipped with such a mechanism to reason about the terminal and goal rules. The procedure is recurrently applied to the GDL rules. When it reaches simple expressions (condition on facts) it returns 1 or 0 depending if the condition is true or not. More complex expressions can be

AND operators (between conditions in a rule), OR operator (defined explicitly between conditions or implicitly between implications) and distinct conditions. Distinct and negative conditions are ignored, whereas for AND and OR conjunctions, t-norms and s-norms are used respectively. The authors settled down with norms of the Yager family. Structures are identified using a syntactic templates. For example an order relation such as (*next* 1 2) is a binary, anti-symmetrical, injective and functional relation. The possible structures are step counter, order relation, board's pieces, quantities, control (turn) and board (ordered domains + pieces). The identified structures are used in a heuristic evaluation function which improves the basic fuzzy-logic goal/terminal satisfaction procedure. Where possible, complex expressions are replaced by the structures if the respective rule uses them. In general, the evaluation function still evaluates the degree of fulfillment for a terminal and goal states but in a more informed fashion. The search method used on top of this function is a modified iterative-deepening depth-first search with transposition tables. If a game is detected to be zero-sum and two-player, the alpha-beta pruning technique is used.

4.3 Boards Continued

Board detection is present in various General Game Playing approaches. One of them [11] is inspired by Hoyle [5] where game evaluators of two kind (structure and definition) are proposed following the idea of Hoyle's advisors. All of the game structure evaluators such as distance-initial, distance-to-target, count-pieces and occupied-columns are related to a two-dimensional board which is discovered using pattern detections algorithm. The novelty of the approach are smart algorithms used to sort GDL facts to store them in a consistent manner. Variance analysis helps to discover pieces which move around the board. A similar approach to this and FluxPlayer's [30] was proposed by [15]. The detection of boards and candidate features for an evaluation function is essentially similar but the application of the function is different. The authors introduce heuristics as linear mappings from the lowest to the highest goal value parametrized by the numeric value assigned to features (each one scaled to the [0,1] interval). The approach uses almost 200 of the so-called slave nodes from which each one is equipped with one heuristic. Another example exploring a board detection is [23] but this time boards of any shapes are taken into account. A distance between two board positions depends on which piece's movement we are interested in. The distance is equal to the number of steps needed to move a particular piece from one position to another. However, one of the problems of this approach is requirement to convert the GDL rules into a Disjunctive Normal Form (DNF) which is not always feasible as well as not taking certain constructions into account such as distinct keywords or negative conditions.

4.4 Monte Carlo Tree-Search Variations

The method was first proposed [3] as an approach to Go and was initially considered not serious. However, a confidence-based algorithm originating from gambling mathematics called Upper Confidence Bounds Applied for Trees (UCT) [14] made the MCTS the state-of-the-art approach both in Go and General Game Playing. The MCTS together with the UCT works as follows: Start from the root of a game-tree and traverse down until reaching a leaf node. During this traversal choose an action a^*, i.e. an edge in the tree, according to the following formula:

$$a^* = arg \max_{a \in A(s)} \left\{ Q(s,a) + C\sqrt{\frac{ln\,[N(s)]}{N(s,a))}} \right\} \tag{3}$$

where a - is an action; s - is the current state; $A(s)$ - is a set of actions available in state s; $Q(s,a)$ - is an assessment of performing action a in state s; $N(s)$ - is a number of previous visits of state s; $N(s,a)$ - is a number of times an action a has been sampled in state s; C - is a coefficient defining a degree to which the second component (exploration) is considered. The action which falls out of the tree and the subsequent state define a new node which is added. This phase is called expansions since the tree expands. Next, starting from that state a random simulation is performed until reaching a terminal state (a simulation phase). In general, players are simulated by making random actions but this is a room for potential algorithms altering that choice. We will present several of them. The last phase is called back-propagation in which the goals achieved by players in the simulation are used to update the statistics in the tree. If the MCTS visits a terminal state inside the selection phase then the simulation phase is skipped.

Monte Carlo Tree-Search is the method of choice of all the GGP Competition's winners since 2007: CadiaPlayer [6], Ary [22] and TurboTurtle. Our agent, called MINI-Player [32] also is based on this technique. Many enhancements to the plain algorithm have been proposed. The most widely used is History Heuristic [29]. History Heuristic was proposed initially to determine order of branches to be visited by alpha-beta algorithm in Chess. It was first used in General Game Playing either by CadiaPlayer [6] or FluxPlayer [30]. The key idea is that actions occur repeatedly in simulations and certain actions are universally strong i.e. independent of the state they are taken in. Examples of actions which are usually good are capturing of pieces or placing markers in the middle in the classic connect-n games. The first application of the history heuristic was to assign non-uniform probability of choosing an action during a Monte Carlo simulation. The process was modeled as a Gibbs Sampling and the historical average score defined a parameter Q of the Gibbs Distribution. Then CadiaPlayer simplify the way history heuristic was used. The authors called the method ϵ greedy which means that an action with the best historical score is chosen with probability equal to ϵ and otherwise a random move is chosen. MINI-Player uses a similar approach. An extension of History Heuristic is called N-grams [33] in which not just one historical action is taken into account but arbitrary longer sequences. A similar approach presented in [33] as well is Last-Good Reply Policy (LGRP). The idea is to store the best replies for particular moves, think of a counterattack. The method originates from Go [3]. MINI-Player uses a portfolio of heuristics called strategies [32] because they encapsulate the whole logic of how an action during a semi-random simulation is chosen. The strategies include

History Heuristic, Mobility, Approximate Goal Evaluation, Statistical Symbols Counting and Exploration. The first three are quite self-explanatory. The exploration heuristic is aimed at choosing states which are bring the highest possible difference to explore the state-space more efficiently. In order not to get stuck in a local minimum we maximize the difference between the next consecutive state and the most similar state to it among several last visited ones. The Statistical Symbol Counting is our approach for board-like heuristics where we count occurrences of symbols at certain positions in their respective arguments lists. We assign weights to that occurrences according to the correlation between their average count and the game outcome. An evaluation function is then constructed as a linear combination of the respective weights and occurrences. A detailed description is contained in [20].

4.5 Selected Other Concepts

A player called MAGICIAN [34] features a construction of a simulation-based evaluation function. The function is based on counting and weighing the GDL terms in a similar fashion to ClunePlayer's and FluxPlayer's. The process is split into five phases: initialization, generalization, specialization, selection and weighting. The generalization phase replaces symbols found in the GDL description by "?" which means any symbol. Then the specialization generates all possible combination of terms by instantiating symbols from their domains. This way, more terms can be detected as useful candidates for the evaluation function than in the previous approaches. Another novel part is the way how the evaluation function is combined with the UCT search. When a GGP game is simple enough the question arises whether it can be solved. One of the approach aimed at solving games is [13]. It attempts to instantiate all rules in a DNF form and then use a symbolic breadth-first search (BFS) algorithm to generate the whole game graph to solve the game. We found three works involving setting neural networks in the GGP scenario. In the first one [24] a constructed neural network works as a state evaluation function. A dependency graph of GDL rules is translated directly into a neural network by the $C-IL^2P$ [7] algorithm. The result is a neuro-fuzzy goal inference engine. Unfortunately, a fully-fledged player was not build on top of this method. Another approach is NeuroEvolution of Augmenting Topologies NEAT [28]. A player named nrng.hazel uses NEAT with a shallow min-max search. NEAT combines genetic algorithms with neural networks. Each genotype encodes a neural network whereas each neural network encodes a heuristic evaluation function. Two populations of neural networks co-evolve. The reproduction uses the so-called explicit fitness sharing in which individuals of one specie share the same fitness level. The last approach also employs co-evolution of populations represented by an ant colony system [31]. Each ant models a General Game Playing agent which performs cyclical simulations leaving pheromone. A simulation is driven by a local knowledge, common cultural knowledge and the pheromone. The local knowledge is a simple evaluation function being a linear combinations of GDL state terms. The pheromone is defined as a factor which alters the weights in the evaluation function. The more often a particular GDL term appears during a simulation the higher weight it gets. The global knowledge are sequences of states which appear in simulations of ants with the highest fitness level. The fitness level is computed in a tournament between populations where ants of one population play against random ants from the other population. It is proportional to the number of wins.

5 Results

Let us start from the results of the International General Game Playing Competitions. Most of the published work is expected to contain some positive results but both the choice of games and opponents can be arguable. The GGP Competitions use a decent variety of games and, moreover, there is no reason not to participate with a good player, so it is relatively safe to say that the winners are all around the most efficient and universal players. In the Table 1 we present the top 2 players from all the previous competitions and whether they are based on MCTS/UCT. In Table 2 we show the results of our own MINI-Player playing 270 matches against itself stripped down only to the basic MCTS/UCT player. Although the fully-fledged player using strategies to guide the Monte Carlo simulations fares better in 7 of 9 games, the advantage is not as big as it could be expected. In some games such as Pentago or Pilgrimage, the simplest approach is better while in Connect-4 Suicide the MINI-Player's winning margin is insignificant. Players based on neural networks have not been strong in General Game Playing so far. The same rule applies for other approaches in the spirit of the computational intelligence which are common in the non-GGP game systems [18]. Restrictive time limits and lack of knowledge transfer hinder any long-term learning.

Table 1. Results of MINI-Player vs. MINI-PlainUCT denoting a player without additional heuristics. A 95% confidence intervals are included in the square brackets.

Year	Winner	Runner-up	Winner MCTS?	Runner-up MCTS?
2005	ClunePlayer	Goblin	NO	NO
2006	FluxPlayer	ClunePlayer	NO	NO
2007	CadiaPlayer	ClunePlayer	YES	NO
2008	CadiaPlayer	ClunePlayer	YES	NO
2009	Ary	FluxPlayer	YES	NO
2010	Ary	Maligne	YES	YES
2011	TurboTurtle	CadiaPlayer	YES	YES
2012	CadiaPlayer	TurboTurtle	YES	YES
2013	TurboTurtle	CadiaPlayer	YES	YES

Table 2. Results of MINI-Player vs. MINI-PlainUCT denoting a player without additional heuristics. A 95% confidence intervals are included in the square brackets.

Game	MINI vs. MINI-PlainUCT
Connect4	61.33 [5.69]
Cephalopod Micro	59.33 [5.86]
Free for all 2P	75.33 [4.86]
Pentago	40.00 [5.84]
9 Board Tic-Tac-Toe	66.30 [5.42]
Connect4 Suicide	51.33 [5.72]
Checkers	79.33 [4.83]
Farming Quandries	66.67 [5.36]
Pilgrimage	39.33 [5.40]
Average	**59.88** [5.44]

6 Conclusions

We have shown a numerous approaches to General Game Playing. Most competitive players use the Monte Carlo Tree-Search as the baseline method with either none or only light-weight heuristics. Some other players use a min-max inspired depth search whereas some other use neural networks without any deep search. Since 2007, only simulation-based players have been the winners of the GGP Competition. Together with our results, we conclude that in General Game Playing, a too heavily marked heuristic bias is unfavorable on a large-enough variety of games. The MCTS approaches have advantage of *not being inaccurate* for any game. That means that while not being predefined for any class of games, the players will always converge (at some arbitrary speed) to the optimal play. For some games, the convergence rate may be very slow but it is still better than inaccurate heuristic which is a barrier preventing a player from being successful. Not suitable heuristics may even make the player going towards the wrong goals. We predict that a truly multi-game player should be either aheuristic, using statistical methods and crunching simulations efficiently or based on self-adapting heuristics. Self-adaptation means that heuristics can be based on any aspect of the played game (**variety**) and that only such heuristics are currently used which are suitable for the game (**accuracy**). Although the MCTS currently dominates, it is interesting to see which approach will the most be successful in the future competitions.

Acknowledgment. M. Świechowski was supported by the Foundation for Polish Science under International Projects in Intelligent Computing (MPD) and The European Union within the Innovative Economy Operational Programme and European Regional Development Fund. This research was partially funded by the National Science Centre in Poland, based on the decision DEC-2012/07/B/ST6/01527.

References

1. Apt, K.R., Wallace, M.: Constraint Logic Programming using Eclipse. Cambridge University Press, New York (2007)
2. Browne, C., Powley, E., Whitehouse, D., Lucas, S., Cowling, P., Rohlfshagen, P., Tavener, S., Perez, D., Samothrakis, S., Colton, S.: A Survey of Monte Carlo Tree Search Methods. IEEE Transactions on Computational Intelligence and AI in Games 4(1), 1–43 (2012)
3. Brügmann, B.: Monte Carlo Go. Technical report, Max Planck Institute for Physics (1993)
4. Clune, J.: Heuristic Evaluation Functions for General Game Playing. In: AAAI, pp. 1134–1139. AAAI Press (2007)
5. Epstein, S.: Toward an ideal trainer. Machine Learning 15(3), 251–277 (1994)
6. Finnsson, H., Björnsson, Y.: Simulation-based approach to general game playing. In: AAAI. AAAI Press (2008)
7. d'Avila Garcez, A.S., Gabbay, D.M., Broda, K.B.: Neural-Symbolic Learning System: Foundations and Applications. Springer-Verlag New York, Inc., Secaucus (2002)
8. Genesereth, M.R., Love, N., Pell, B.: General game playing: Overview of the aaai competition. AI Magazine 26(2), 62–72 (2005)
9. Gherrity, M.: A Game-learning Machine. PhD thesis, University of California at San Diego, La Jolla, CA, USA (1993) UMI Order No. GAX94-14755
10. Hsu, F.H.: Behind Deep Blue: Building the Computer that Defeated the World Chess Champion. Princeton University Press, Princeton (2002)
11. Kaiser, D.M.: Automatic feature extraction for autonomous general game playing agents. In: Proceedings of the 6th International Joint Conference on Autonomous Agents and Multiagent Systems, AAMAS 2007, pp. 93:1–93:7. ACM, New York (2007)
12. Kishimoto, A., Schaeffer, J.: Transposition table driven work scheduling in distributed game-tree search. In: Cohen, R., Spencer, B. (eds.) Canadian AI 2002. LNCS (LNAI), vol. 2338, pp. 56–68. Springer, Heidelberg (2002)
13. Kissmann, P., Edelkamp, S.: Gamer, a General Game Playing Agent. KI 25(1), 49–52 (2011)
14. Kocsis, L., Szepesvári, C.: Bandit based Monte-Carlo planning. In: Fürnkranz, J., Scheffer, T., Spiliopoulou, M. (eds.) ECML 2006. LNCS (LNAI), vol. 4212, pp. 282–293. Springer, Heidelberg (2006)
15. Kuhlmann, G., Dresner, K., Stone, P.: Automatic heuristic construction in a complete general game player. In: Proceedings of the Twenty-First National Conference on Artificial Intelligence, pp. 1457–1462 (July 2006)
16. Levinson, R.: Morph ii: A universal agent - progress report and proposal. Technical report, University of California at Santa Cruz, Santa Cruz, CA, USA (1994)
17. Love, N., Hinrichs, T., Haley, D., Schkufza, E., Genesereth, M.: General game playing: Game description language specication (March 2008)
18. Mańdziuk, J.: Computational Intelligence in Mind Games. In: Duch, W., Mańdziuk, J. (eds.) Challenges for Computational Intelligence. SCI, vol. 63, pp. 407–442. Springer, Heidelberg (2007)
19. Mańdziuk, J.: Knowledge-Free and Learning-Based Methods in Intelligent Game Playing. SCI, vol. 276. Springer, Heidelberg (2010)
20. Mańdziuk, J., Świechowski, M.: Generic heuristic approach to general game playing. In: Bieliková, M., Friedrich, G., Gottlob, G., Katzenbeisser, S., Turán, G. (eds.) SOFSEM 2012. LNCS, vol. 7147, pp. 649–660. Springer, Heidelberg (2012)

21. Mccarthy, J., Hayes, P.J.: Some philosophical problems from the standpoint of artificial intelligence. In: Machine Intelligence, pp. 463–502. Edinburgh University Press (1969)
22. Méhat, J., Cazenave, T.: Ary, a general game playing program. In: Board Games Studies Colloquium, Paris (2010)
23. Michulke, D., Schiffel, S.: Distance features for general game playing. In: Proceedings of the IJCAI 2011 Workshop on General Game Playing (GIGA 2011), pp. 7–14 (2011)
24. Michulke, D., Thielscher, M.: Neural networks for state evaluation in general game playing. In: Proceedings of the European Conference on Machine Learning (EMCL), pp. 95–110 (2009)
25. Nilsson, N.J.: Artificial Intelligence: A New Synthesis. Morgan Kaufmann Publishers Inc., San Francisco (1998)
26. Pell, B.: Metagame: A new challenge for games and learning. In: Programming in Artificial Intellegence: The Third Computer Olympiad, pp. 237–251. Ellis Horwood Limited (1992)
27. Plaat, A., Schaeffer, J., Pijls, W., de Bruin, A.: Best-first fixed-depth minimax algorithms. Artificial Intelligence 87(1-2), 255–293 (1996)
28. Reisinger, J., Bahceci, E., Karpov, I., Miikkulainen, R.: Coevolving strategies for general game playing. In: Proceedings of the IEEE Symposium on Computational Intelligence and Games, pp. 320–327. IEEE, Piscataway (2007)
29. Schaeffer, J.: The history heuristic and alpha-beta search enhancements in practice. IEEE Trans. Pattern Anal. Mach. Intell. 11(11), 1203–1212 (1989)
30. Schiffel, S., Thielscher, M.: Fluxplayer: A successful general game player. In: Proceedings of the 22nd AAAI Conference on Artificial Intelligence (AAAI 2007), pp. 1191–1196. AAAI Press (2007)
31. Sharma, S., Kobti, Z., Goodwin, S.: Coevolving intelligent game players in a cultural framework. In: Proceedings of the IEEE Congress on Evolutionary Computing, Special Session on Computational Intelligence in Games (2009)
32. Świechowski, M., Mańdziuk, J.: Self-Adaptation of Playing Strategies in General Game Playing. IEEE Transactions on Computational Intelligence and AI in Games (2014); Accepted for publication 20/06/2013. Available in Early Access, doi: 10.1109/TCIAIG.2013.2275163
33. Tak, M.J.W., Winands, M.H.M., Bjornsson, Y.: N-grams and the last-good-reply policy applied in general game playing. IEEE Transactions on Computational Intelligence and AI in Games 4(2), 73–83 (2012)
34. Walędzik, K., Mańdziuk, J.: An Automatically-Generated Evaluation Function in General Game Playing. IEEE Transactions on Computational Intelligence and AI in Games (2014); Accepted for publication 08/10/2013. Available in Early Access, doi: 10.1109/TCIAIG.2013.2286825

Part VII
Group Decisions, Consensus, Negotiations

Consensus Reaching Processes under Hesitant Linguistic Assessments

José Luis García-Lapresta[1], David Pérez-Román[2], and Edurne Falcó[3]

[1] PRESAD Research Group, IMUVA, Dept. de Economía Aplicada, Universidad de Valladolid, Spain
lapresta@eco.uva.es

[2] PRESAD Research Group, Dept. de Organización de Empresas y Comercialización e Investigación de Mercados, Universidad de Valladolid, Spain
david@emp.uva.es

[3] PRESAD Research Group, IMUVA, Dept. de Economía, Universidad Pública de Navarra, Spain
edurne.falco@unavarra.es

Abstract. In this paper, we introduce a flexible consensus reaching process when agents evaluate the alternatives through linguistic expressions formed by a linguistic term, when they are confident on their opinions, or by several consecutive linguistic terms, when they hesitate. Taking into account an appropriate metric on the set of linguistic expressions and an aggregation function, a degree of consensus is obtained for each alternative. An overall degree of consensus is obtained by combining the degrees of consensus on the alternatives by means of an aggregation function. If that overall degree of consensus reaches a previously fixed threshold, then a voting system is applied. Otherwise, a moderator initiates a consensus reaching process by inviting some agents to modify their assessments in order to increase the consensus.

1 Introduction

Usually, in group decision making problems, agents show their opinions on a set of alternatives and an aggregative procedure generates an outcome (a winning alternative, a set of winning alternatives, a ranking on the set of alternatives, etc.). However, this outcome could not represent the opinions of a large number of agents if the agreement among them is low. In these cases, the outcome is very sensitive to the aggregative procedure used for obtaining the outcome. On the contrary, when the agreement among agents is high, the outcome is not very different across proper aggregative procedures or voting systems.

For measuring the degree of consensus among a group of agents that provide their opinions on a set of alternatives, different proposals can be found in the literature (see Martínez-Panero [32] for an overview of different notions of consensus).

In the fuzzy framework, there exist a huge literature where some degrees of consensus have been defined (see Spillman *et al.* [44], Kacprzyk and Fedrizzi [27,28,29], Kacprzyk *et al.* [30], Fedrizzi *et al.* [15], Herrera *et al.* [24], Bordogna *et al.* [7], among others). A referenced survey can be found in Palomares *et al.* [35].

P. Angelov et al. (eds.), *Intelligent Systems'2014*,
Advances in Intelligent Systems and Computing 322, DOI: 10.1007/978-3-319-11313-5_24

More recently, the Social Choice Theory has been interested on how to measure consensus in different settings. First, the notion of *consensus measure* was introduced by Bosch [8] in the context of linear orders. Additionally, Bosch [8] and Alcalde-Unzu and Vorsatz [2] provided axiomatic characterizations of several consensus measures in the context of linear orders. García-Lapresta and Pérez-Román [20] extended that notion to the context of weak orders and they analyzed a class of consensus measures generated by distances. Alcantud *et al.* [3] provide axiomatic characterizations of some consensus measures in the setting of approval voting. In turn, Erdamar *et al.* [12] extended the notion of consensus measure to the preference-approval setting through different kinds of distances.

All the mentioned consensus analyses are static: once the agents show their opinions, a degree of consensus is obtained. In this paper the degree of consensus in a specific situation is only the starting point of a dynamic and iterative procedure that pursues to increase the agreement among agents when they evaluate the alternatives by means of hesitant linguistic assessments.

A consensus reaching process consists of several rounds where agents are invited to modify their opinions in order to increase the agreement. These rounds can be conducted by a human or virtual moderator (see Fedrizzi *et al.* [17,16], Saint and Lawson [43], Herrera *et al.* [25], Herrera-Viedma *et al.* [26], Eklund *et al.* [11], Martínez and Montero [31], Cabrerizo *et al.* [9,10], Pérez *et al.* [39,38], Mata *et al.* [33,34], Palomares *et al.* [37] and Palomares and Martínez [36], among others). Once a previously fixed consensus threshold is reached, a group decision making procedure can be carried out in order to rank the alternatives or to select a winning alternative.

In this paper, we have considered that agents evaluate the alternatives through linguistic expressions formed by a linguistic term, when they are confident on their opinions, or by several consecutive linguistic terms, when they hesitate. This approach is based on an adaptation of the *absolute order of magnitude spaces* introduced by Travé-Massuyès and Dague [46] and Travé-Massuyès and Piera [47]; more specifically in the extensions devised by Roselló *et al.* [40,41,42] (see also Falcó *et al.* [13,14] and García-Lapresta and Pérez-Román [21]).

It is important to note that there is empirical evidence showing that experts may hesitate when they assess alternatives through linguistic terms. For instance, in the tastings described in Agell *et al.* [1] and García-Lapresta *et al.* [19], 40% and 17.59%, respectively, of the assessments given by trained sensory panelists were linguistic expressions with two or more linguistic terms.

The rest of the paper is organized as follows. Section 2 includes the notation and some notions we need for developing the proposal. Section 3 is devoted to measure the consensus in each alternative and also the overall consensus; additionally, some properties of the corresponding degrees of consensus are shown. Section 4 contains our proposal of consensus reaching process in the framework of hesitant linguistic assessments. Finally, Section 5 includes some concluding remarks.

2 Preliminaries

First we introduce some pieces of notation. Vectors in $[0,1]^t$ will be denoted by means of boldface characters: $\boldsymbol{x} = (x_1,\ldots,x_t)$, $\mathbf{1} = (1,\ldots,1)$, $\mathbf{0} = (0,\ldots,0)$. For $x \in [0,1]$, we have $x \cdot \mathbf{1} = (x,\ldots,x)$. If $\boldsymbol{x},\boldsymbol{y} \in [0,1]^t$, with $\boldsymbol{x} \geq \boldsymbol{y}$ we mean $x_i \geq y_i$ for every $i \in \{1,\ldots,t\}$; by $\boldsymbol{x} > \boldsymbol{y}$ we mean $\boldsymbol{x} \geq \boldsymbol{y}$ and $\boldsymbol{x} \neq \boldsymbol{y}$. Given a set I, the cardinality of I is denoted by $\#I$. Given a permutation on $\{1,\ldots,t\}$, i.e., a bijection $\sigma : \{1,\ldots,t\} \longrightarrow \{1,\ldots,t\}$, and $\boldsymbol{x} \in [0,1]^t$, with $\boldsymbol{x}_\sigma$ we denote $(x_{\sigma(1)},\ldots,x_{\sigma(t)})$.

Let $A = \{1,\ldots,m\}$, with $m \geq 2$, be a set of agents and let $X = \{x_1,\ldots,x_n\}$, with $n \geq 2$, be the set of alternatives which have to be evaluated. Under total certainty, each agent assigns a linguistic term to every alternative within a linguistic ordered scale $L = \{l_1,\ldots,l_g\}$, where $l_1 < l_2 < \cdots < l_g$. The elements of L are linguistic terms as 'very bad', 'bad', 'acceptable', 'good' and 'very good'.

2.1 Linguistic Expressions

Taking into account the *absolute order of magnitude spaces* introduced by Travé-Massuyès and Piera [47], the *set of linguistic expressions* is defined as

$$\mathbb{L} = \{[l_h,l_k] \mid l_h,l_k \in L,\ 1 \leq h \leq k \leq g\},$$

where $[l_h,l_k] = \{l_h,l_{h+1},\ldots,l_k\}$. Since $[l_h,l_h] = \{l_h\}$, this linguistic expression can be replaced by the linguistic term l_h. In this way, $L \subset \mathbb{L}$.

The set of linguistic expressions can be represented by a graph $G_{\mathbb{L}}$. In the graph, the lowest layer represents the linguistic terms $l_h \in L \subset \mathbb{L}$, the second layer represents the linguistic expressions created by two consecutive linguistic terms $[l_h,l_{h+1}]$, the third layer represents the linguistic expressions generated by three consecutive linguistic terms $[l_h,l_{h+2}]$, and so on up to last layer where we represent the linguistic expression $[l_1,l_g]$. As a result, the higher an element is, the more imprecise it becomes.

The vertices in $G_{\mathbb{L}}$ are the elements of $\mathbb{L}$ and the edges $\mathscr{E} - \mathscr{F}$, where $\mathscr{E} = [l_h,l_k]$ and $\mathscr{F} = [l_h,l_{k+1}]$, or $\mathscr{E} = [l_h,l_k]$ and $\mathscr{F} = [l_{h+1},l_k]$. Fig. 3 shows the graph representation for $g = 5$.

The inverse of $\mathscr{E} = [l_h,l_k] \in \mathbb{L}$ is defined as $\mathscr{E}^{-1} = [l_{g+1-k},l_{g+1-h}]$.

A *profile* V is a matrix (v_i^a) consisting of m rows and n columns of linguistic expressions, where the element $v_i^a \in \mathbb{L}$ represents the linguistic assessment given by the agent $a \in A$ to the alternative $x_i \in X$. Then,

$$V = \begin{pmatrix} v_1^1 & \cdots & v_i^1 & \cdots & v_n^1 \\ \cdots & \cdots & \cdots & \cdots & \cdots \\ v_1^a & \cdots & v_i^a & \cdots & v_n^a \\ \cdots & \cdots & \cdots & \cdots & \cdots \\ v_1^m & \cdots & v_i^m & \cdots & v_n^m \end{pmatrix} = (v_i^a).$$

Given a profile $V = (v_i^a)$, its inverse is defined as $V^{-1} = (u_i^a)$ with $u_i^a = (v_i^a)^{-1}$. Given a permutation π on A and a profile $V = (v_i^a)$, the profile $V^\pi = (u_i^a)$ is defined as $u_i^a = v_i^{\pi(a)}$.

The set of linguistic expressions can be ranked in different ways. We now introduce a linear order on this set.

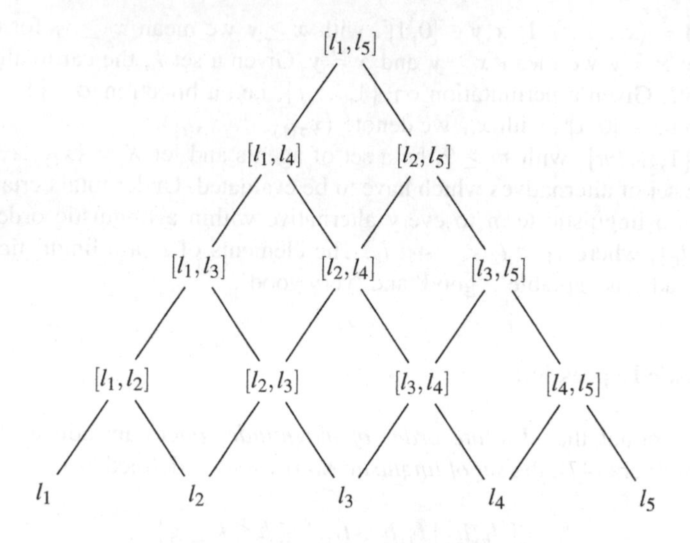

Fig. 1. Graph representation of the linguistic expressions for $g = 5$

Proposition 1. *The binary relation* $\succcurlyeq$ *on* $\mathbb{L}$ *defined as*

$$[l_h, l_k] \succcurlyeq [l_{h'}, l_{k'}] \Leftrightarrow \begin{cases} h+k > h'+k' \\ \text{or} \\ h+k = h'+k' \ \text{and} \ k-h \leq k'-h' \end{cases}$$

is a linear order, and it is called the canonical order on $\mathbb{L}$.

As usual in linear orders, $\mathscr{E} \succ \mathscr{F}$ means $\mathscr{E} \succcurlyeq \mathscr{F}$ and $\mathscr{E} \neq \mathscr{F}$.

Example 1. For $g = 5$, the elements of $\mathbb{L}$ are ordered as follows (see also Fig. 2):

$$l_5 \succ [l_4, l_5] \succ l_4 \succ [l_3, l_5] \succ [l_3, l_4] \succ [l_2, l_5] \succ l_3 \succ [l_2, l_4] \succ$$
$$\succ [l_1, l_5] \succ [l_2, l_3] \succ [l_1, l_4] \succ l_2 \succ [l_1, l_3] \succ [l_1, l_2] \succ l_1.$$

The geodesic distance[1] between two linguistic expressions $\mathscr{E}, \mathscr{F} \in \mathbb{L}$ is defined as the distance in the graph G between their associated vertices:

$$d_G\big([l_h, l_k], [l_{h'}, l_{k'}]\big) = |h - h'| + |k - k'|.$$

In Falcó *et al.* [13] the geodesic distance is modified for penalizing the imprecision by means two parameters, α and β. For simplicity, in this paper we only consider the first penalization through the parameter α.

Given $\mathscr{E} = [l_h, l_k] \in \mathbb{L}$, with $\#\mathscr{E}$ we denote the cardinality of $\mathscr{E}$, i.e., the number of linguistic terms in the interval $[l_h, l_k]$: $\#\mathscr{E} = k + 1 - h$.

[1] The geodesic distance between two vertices in a graph is the number of edges in one of the shortest paths connecting them.

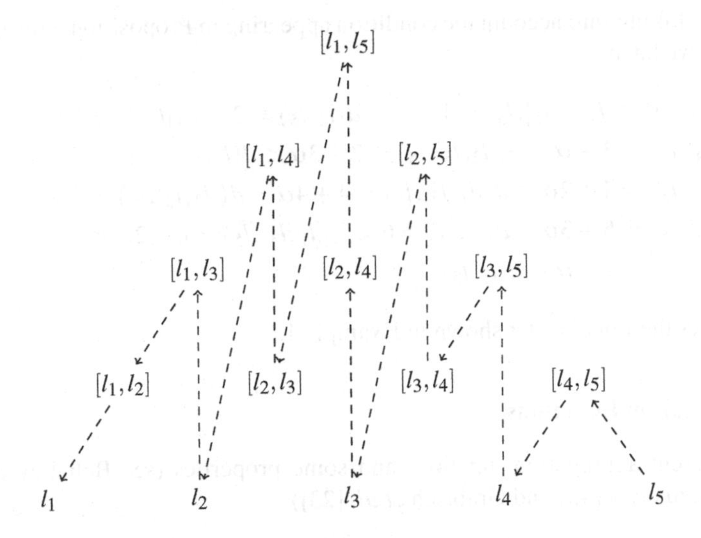

Fig. 2. Canonical order in $\mathbb{L}$ for $g = 5$

Proposition 2. *For every* $\alpha \geq 0$*, the function* $d : \mathbb{L}^2 \longrightarrow \mathbb{R}$*, defined as*

$$d(\mathscr{E}, \mathscr{F}) = d_G(\mathscr{E}, \mathscr{F}) + \alpha\, |\#\mathscr{E} - \#\mathscr{F}|,$$

i.e.,

$$d\big([l_h, l_k], [l_{h'}, l_{k'}]\big) = |h - h'| + |k - k'| + \alpha\, |h' - h + k - k'|.$$

is a metric, and it is called the metric associated with α.

For instance, $d(l_h, l_{h+1}) = 2$ and $d(l_h, [l_h, l_{h+1}]) = d(l_{h+1}, [l_h, l_{h+1}]) = 1 + \alpha$, for every $h \in \{1, \ldots, g-1\}$. In turn, if $g = 5$, we have $d(l_3, l_5) = 4$, $d([l_2, l_4], l_5) = 4 + 2\alpha$, and $d([l_1, l_5], l_5) = 4 + 4\alpha$.

The following result shows that the canonical order on the set of linguistic expressions can be obtained from the distance to the highest possible assessment, by considering appropriate values of the parameter $alpha$. This result is a direct consequence of Falcó *et al.* [13, Prop. 3].

Proposition 3. *Let* $d : \mathbb{L}^2 \longrightarrow \mathbb{R}$ *be the metric associated with* $\alpha \geq 0$*. The following statements are equivalent:*

1. $\forall \mathscr{E}, \mathscr{F} \in \mathbb{L} \quad \big(\mathscr{E} \succ \mathscr{F} \Leftrightarrow d(\mathscr{E}, l_g) < d(\mathscr{F}, l_g)\big)$.
2. $\alpha \in T_g$*, where* $T_g = \left(0, \frac{1}{g-2}\right)$*, if g is odd, and* $T_g = \left(0, \frac{1}{g-1}\right)$*, if g is even.*

Example 2. Taking into account the condition appearing in Proposition 3 for $g=5$, i.e., $0<\alpha<\frac{1}{3}$, we have

$$d(l_5,l_5)=0<d([l_4,l_5],l_5)=1+\alpha<d(l_4,l_5)=2<d([l_3,l_5],l_5)=2+2\alpha<$$
$$d([l_3,l_4],l_5)=3+\alpha<d([l_2,l_5],l_5)=3+3\alpha<d(l_3,l_5)=4<$$
$$d([l_2,l_4],l_5)=4+2\alpha<d([l_1,l_5],l_5)=4+4\alpha<d([l_2,l_3],l_5)=5+\alpha<$$
$$d([l_1,l_4],l_5)=5+3\alpha<d(l_2,l_5)=6<d([l_1,l_3],l_5)=6+2\alpha<$$
$$d([l_1,l_2],l_5)=7+\alpha<d(l_1,l_5)=8\,,$$

that generates the linear order shown in Example 1.

2.2 Aggregation Functions

We now present aggregation functions and some properties (see Beliakov *et al.* [6], Torra and Narukawa [45] and Grabisch *et al.* [23]).

Definition 1. *Let $A:[0,1]^t\longrightarrow\mathbb{R}$ be a function.*

1. *A is* idempotent *if for every $x\in[0,1]$ it holds $A(x\cdot\mathbf{1})=x$.*
2. *A is* symmetric *if for every permutation σ on $\{1,\dots,t\}$ and every $\mathbf{x}\in[0,1]^t$ it holds $A(\mathbf{x}_\sigma)=A(\mathbf{x})$.*
3. *A is* monotonic *if for all $\mathbf{x},\mathbf{y}\in[0,1]^t$ it holds $\mathbf{x}\geq\mathbf{y}\Rightarrow A(\mathbf{x})\geq A(\mathbf{y})$.*
4. *A is* compensative *if for every $\mathbf{x}\in[0,1]^t$ it holds*

$$\min\{x_1,\dots,x_t\}\leq A(\mathbf{x})\leq\max\{x_1,\dots,x_t\}.$$

Definition 2. *A function $A:[0,1]^t\longrightarrow[0,1]$ is called an* (t-ary) aggregation function *if it is monotonic and satisfies $A(\mathbf{0})=0$ and $A(\mathbf{1})=1$.*

It is easy to see that every idempotent aggregation function is compensative, and viceversa.

3 Measuring Consensus

For measuring the consensus of the agents over an alternative, we propose a method related to the Gini coefficient[2] [22] and the consensus measures introduced by García-Lapresta and Pérez-Román [20] in the context of weak orders.

Definition 3. *Let $d:\mathbb{L}^2\longrightarrow[0,1]$ be the metric associated with $\alpha\in T_g$ and $F:[0,1]^{m(m-1)/2}\longrightarrow[0,1]$ an aggregation function. Given a profile $V=(v_i^a)$, the* degree of consensus over the alternative $x_i\in X$ relative to d and F *is defined as*

$$C_{d,F}(V,x_i)=1-F\left(d\left(v_i^1,v_i^2\right),d\left(v_i^1,v_i^3\right),\dots,d\left(v_i^{m-1},v_i^m\right)\right).$$

[2] The Gini coefficient is a classical measure of statistical dispersion of income distributions that is based on the sum of absolute differences between all the pairs of individual incomes.

The following result includes some interesting properties of the proposed consensus measure. Normalization means that consensus is always between 0 and 1. Positiveness means that with more than two agents, the consensus is always positive. Anonymity requires symmetry with respect to agents. Unanimity means that the maximum consensus is only achieved when all opinions are the same. Reciprocity means that if all individual assessments are reversed, then the consensus does not change.

Proposition 4. *Let* $d : \mathbb{L}^2 \longrightarrow [0,1]$ *be the metric associated with* $\alpha \in T_g$ *and* $F : [0,1]^{m(m-1)/2} \longrightarrow [0,1]$ *an aggregation function. For every profile* $V = (v_i^a)$ *and every alternative* $x_i \in X$*, the following properties hold:*

1. *Normalization:* $C_{d,F}(V,x_i) \in [0,1]$.
2. *Positiveness: If* $\#A > 2$*, then* $C_{d,F}(V,x_i) > 0$.
3. *Anonymity: If* F *is symmetric, then* $C_{d,F}(V^\pi,x_i) = C_{d,F}(V,x_i)$ *for every permutation* π *on* A.
4. *Unanimity: If* $v_i^a = v_i^b$ *for all* $a,b \in A$*, then* $C_{d,F}(V,x_i) = 1$.
5. *Reciprocity:* $C_{d,F}(V^{-1},x_i) = C_{d,F}(V,x_i)$.

Once the degrees of consensus over the alternatives are known, it is interesting to define an overall degree of consensus. We define such overall degree as the outcome generated by an aggregation function to the degrees of consensus over the alternatives.

Definition 4. *Let* $d : \mathbb{L}^2 \longrightarrow [0,1]$ *be the metric associated with* $\alpha \in T_g$ *and two aggregation functions* $F : [0,1]^{m(m-1)/2} \longrightarrow [0,1]$ *and* $G : [0,1]^n \longrightarrow [0,1]$*. Given a profile* $V = (v_i^a)$*, the* overall degree of consensus relative to d, F and G *is defined as*

$$C_{d,F}^G(V) = G\left(C_{d,F}(V,x_1),\ldots,C_{d,F}(V,x_n)\right).$$

In the following result, some properties of the overall degree of consensus are shown. The meaning is similar to the corresponding properties appearing in Proposition 4. Neutrality means that all the alternatives are treated in an unbiased way. Compensativeness means that the overall degree of consensus is between the minimum and the maximum degrees of consensus over the alternatives.

Proposition 5. *Let* $d : \mathbb{L}^2 \longrightarrow [0,1]$ *be the metric associated with* $\alpha \in T_g$ *and two aggregation functions* $F : [0,1]^{m(m-1)/2} \longrightarrow [0,1]$ *and* $G : [0,1]^n \longrightarrow [0,1]$*. For every profile* $V = (v_i^a)$*, the following properties hold::*

1. *Normalization:* $C_{d,F}^G(V) \in [0,1]$.
2. *Anonymity: If* F *is symmetric, then* $C_{d,F}^G(V^\pi) = C_{d,F}^G(V)$ *for every permutation* π *on* A.
3. *Neutrality: If* G *is symmetric, then* $C_{d,F}^G(V_\sigma) = C_{d,F}^G(V)$ *for every permutation* σ *on* $\{1,\ldots,n\}$.
4. *Reciprocity:* $C_{d,F}^G(V^{-1}) = C_{d,F}^G(V)$.
5. *Compensativeness: If* G *is idempotent, then*

$$\min\{C_{d,F}(V,x_1),\ldots,C_{d,F}(V,x_n)\} \leq C_{d,F}^G(V) \leq \max\{C_{d,F}(V,x_1),\ldots,C_{d,F}(V,x_n)\}.$$

4 The Consensus Reaching Process

Given a profile $V = (v_i^a)$, with $\boldsymbol{v}_i$ we denote the assessments vector of alternative x_i, i.e., $\boldsymbol{v}_i = (v_i^1, \ldots, v_i^m) \in \mathbb{L}^m$; with $\overline{v}_i$ we denote the median of $\boldsymbol{v}_i$ with respect to the canonical order on $\mathbb{L}$, if m is odd, and the lower median if m is even, as in Majority Judgment (Balinski and Laraki [4,5]).

Before starting to ask the agents their opinions on the alternatives, a consensus threshold $\gamma \in (0,1]$ is fixed. If $C_{d,F}^G(V) \geq \gamma$, then a voting system is applied[3]. But if $C_{d,F}^G(V) < \gamma$, then the moderator initiates a reaching consensus process. In that case, the set of alternatives where the degrees of consensus are lower than the overall consensus is determined: $X^- = \left\{x_i \in X \mid C_{d,F}(V,x_i) < C_{d,F}^G(V)\right\}$.

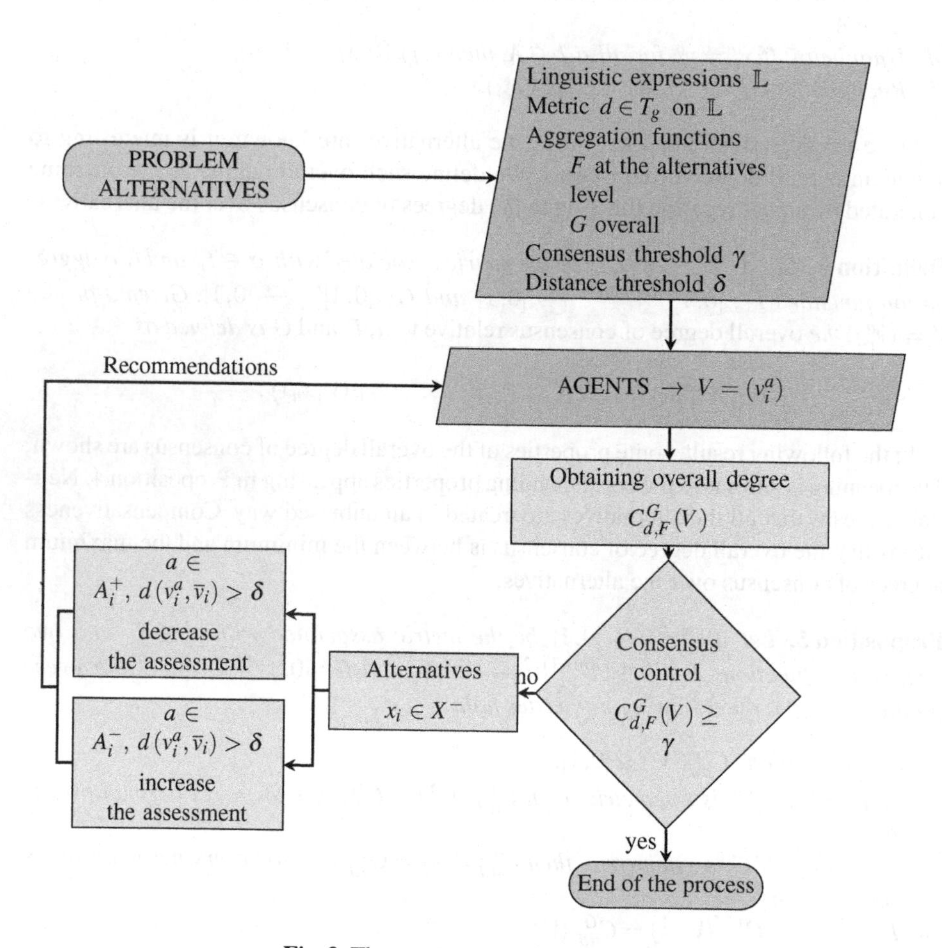

Fig. 3. The consensus reaching process

[3] In this setting, it would be suitable the voting system proposed by Falcó *et al.* [13].

For each $x_i \in X^-$, we define $A_i^+ = \{a \in A \mid v_i^a \succ \overline{v}_i\}$ and $A_i^- = \{a \in A \mid \overline{v}_i \succ v_i^a\}$. Given a distance threshold $\delta \in [0,1)$, if $a \in A_i^+$ and $d(v_i^a, \overline{v}_i) > \delta$, then agent a is invited to decrease the assessment v_i^a; if $a \in A_i^-$ and $d(v_i^a, \overline{v}_i) > \delta$, then agent a is invited to increase the assessment v_i^a.

Once the new profile is obtained, the degrees of consensus are calculated and the process is re-initiated. The number of rounds has to be fixed previously to the reaching consensus process. If the consensus threshold γ is reached before finishing the maximum number of rounds, then the voting system is applied. But if the rounds finish without reaching the consensus threshold, then it would be necessary to adopt a strategy (see Saint and Lawson [43]). One possibility is to exclude the agents whose overall contributions to consensus are negative, as in García-Lapresta [18].

A schematic view of the proposed consensus reaching process is represented in Fig. 3.

5 Concluding Remarks

In this paper, we have considered that agents evaluate the alternatives with linguistic terms, when they are confident on their opinions, or with linguistic expressions, when they hesitate. Taking into account the individual assessments provided by the agents, a voting system may be applied to make a collective decision. These assessments may diverge and then the collective decision could not be representative of the individual opinions. It is well-known that the outcomes provided by different voting systems can be totally different whenever the dispersion of the individual opinions is high. Therefore, it is convenient to decrease that dispersion if possible. In this way, consensus reaching processes are a fruitful tool for conducting collective decisions. These processes are iterative and they need the consensus to be measured in each stage. Thus, it is essential to devise appropriate consensus measures.

The degree of consensus over an alternative has been defined as one minus the dispersion of the individual assessments over that alternative. Such dispersion is based on the Gini coefficient through a parameterized family of metrics, where the parameter penalizes the imprecision of the individual opinions. The overall degree of consensus has been defined from the degrees of consensus over the alternatives through an appropriate aggregation function.

It is worth mentioning the good properties satisfied by the proposed consensus measures, specially when the aggregation functions involved in the process are symmetric.

An important feature of the proposed consensus reaching process is its flexibility. Specifically, the following elements are involved:

1. The set of linguistic terms and the corresponding set of linguistic expressions that agents may use for evaluating the alternatives.
2. The metric that calculates the distances among the individual assessments over each alternative.
3. The aggregation function that combines the mentioned distances for obtaining the degree of consensus over each alternative.
4. The aggregation function that merges the consensus over the alternatives for defining the overall consensus.

5. The consensus threshold that determines when the iterative process finishes.
6. The distance threshold that shows who are the agents that should be invited to modify their assessments.
7. The number of rounds for increasing the consensus among the agents.
8. The voting system that is applied to make the collective decision.

Acknowledgments. The financial support of the Spanish *Ministerio de Economía y Competitividad* (project ECO2012-32178) and *Consejería de Educación de la Junta de Castilla y León* (project VA066U13) are acknowledged.

References

1. Agell, N., Sánchez, G., Sánchez, M., Ruiz, F.J.: Selecting the best taste: a group decision-making application to chocolates design. In: Proceedings of the 2013 IFSA-NAFIPS Joint Congress, Edmonton, pp. 939–943 (2013)
2. Alcalde-Unzu, J., Vorsatz, M.: Measuring the cohesiveness of preferences: an axiomatic analysis. Social Choice and Welfare 41, 965–988 (2013)
3. Alcantud, J.C.R., de Andrés, R., Cascón, J.M.: On measures of cohesiveness under dichotomous opinions: some characterizations of Approval Consensus Measures. Information Sciences 240, 45–55 (2013)
4. Balinski, M., Laraki, R.: A theory of measuring, electing and ranking. Proceedings of the National Academy of Sciences of the United States of America 104, 8720–8725 (2007)
5. Balinski, M., Laraki, R.: Majority Judgment. Measuring, Ranking, and Electing. The MIT Press, Cambridge (2011)
6. Beliakov, G., Pradera, A., Calvo, T.: Aggregation Functions: A Guide for Practitioners. STUDFUZZ, vol. 221. Springer, Heidelberg (2007)
7. Bordogna, G., Fedrizzi, M., Pasi, G.: A linguistic modeling of consensus in group decision making based on OWA operators. IEEE Transactions on Systems, Man and Cybernetics, Part A: Systems and Humans 27, 126–133 (1997)
8. Bosch, R.: Characterizations of Voting Rules and Consensus Measures. Ph. D. Dissertation, Tilburg University (2005)
9. Cabrerizo, F.J., Moreno, J.M., Pérez, I.J., Herrera-Viedma, E.: Analyzing consensus approaches in fuzzy group decision making: advantages and drawbacks. Soft Computing 14, 451–463 (2010)
10. Cabrerizo, F.J., Pérez, I.J., Herrera-Viedma, E.: Managing the consensus in group decision making in an unbalanced fuzzy linguistic context with incomplete information. Knowledge-Based Systems 23, 169–181 (2010)
11. Eklund, P., Rusinowska, A., de Swart, H.: Consensus reaching in committees. European Journal of Operational Research 178, 185–193 (2007)
12. Erdamar, B., García-Lapresta, J.L., Pérez-Román, D., Sanver, M.R.: Measuring consensus in a preference-approval context. Information Fusion 17, 14–21 (2014)
13. Falcó, E., García-Lapresta, J.L., Roselló, L.: Aggregating imprecise linguistic expressions, in: P. Guo and W. Pedrycz (Eds.), Human-Centric Decision-Making Models for Social Sciences. Springer-Verlag, Berlin, pp. 97-113 (2014)
14. Falcó, E., García-Lapresta, J.L., Roselló, L.: Allowing agents to be imprecise: A proposal using multiple linguistic terms. Information Sciences 258, 249–265 (2014)
15. Fedrizzi, M., Kacprzyk, J., Nurmi, H.: Consensus degrees under fuzzy majorities and fuzzy preferences using OWA (ordered weighted average) operators. Control and Cybernetics 22, 71–80 (1993)

16. Fedrizzi, M., Kacprzyk, J., Owsińnski, J.W., Zadrozny, S.: Consensus reaching via a GDSS with fuzzy majority and clustering of preference profiles. Annals of Operations Research 51, 127–139 (1994)
17. Fedrizzi, M., Kacprzyk, J., Zadrozny, S.: An interactive multi-user decision support system for consensus reaching processes using fuzzy logic with linguistic quantifiers. Decision Support Systems 4, 313–327 (1988)
18. García-Lapresta, J.L.: Favoring consensus and penalizing disagreement in group decision making. Journal of Advanced Computational Intelligence and Intelligent Informatics 12(5), 416–421 (2008)
19. García-Lapresta, J.L., Aldavero, C., de Castro, S.: A linguistic approach to multi-criteria and multi-expert sensory analysis. In: Laurent, A., Strauss, O., Bouchon-Meunier, B., Yager, R.R. (eds.) IPMU 2014, Part II. CCIS, vol. 443, pp. 586–595. Springer, Heidelberg (2014)
20. García-Lapresta, J.L., Pérez-Román, D.: Measuring consensus in weak orders. In: Herrera-Viedma, E., García-Lapresta, J.L., Kacprzyk, J., Fedrizzi, M., Nurmi, H., Zadrożny, S. (eds.) Consensual Processes. STUDFUZZ, vol. 267, pp. 213–234. Springer, Heidelberg (2011)
21. García-Lapresta, J.L., Pérez-Román, D.: Consensus-based clustering under hesitant qualitative assessments. Fuzzy Sets and Systems (in press), doi:10.1016/j.fss, 05.0040165-0114
22. Gini, C.: Variabilità e Mutabilità. Tipografia di Paolo Cuppini, Bologna (1912)
23. Grabisch, M., Marichal, J.L., Mesiar, R., Pap, E.: Aggregation Functions. Cambridge University Press, Cambridge (2009)
24. Herrera, F., Herrera-Viedma, E., Verdegay, J.L.: A model of consensus in group decision making under linguistic assessments. Fuzzy Sets and Systems 78, 73–87 (1996)
25. Herrera, F., Herrera-Viedma, E., Verdegay, J.L.: Linguistic measures based on fuzzy coincidence for reaching consensus in group decision making. International Journal of Approximate Reasoning 16, 309–334 (1997)
26. Herrera-Viedma, E., Herrera, F., Chiclana, F.: A consensus model for multiperson decision making with different preference structures. IEEE Transactions on Systems, Man and Cybernetics - Part A: Systems and Humans 32, 394–402 (2002)
27. Kacprzyk, J., Fedrizzi, M.: 'Soft' consensus measures for monitoring real consensus reaching processes under fuzzy preferences. Control and Cybernetics 15, 309–323 (1986)
28. Kacprzyk, J., Fedrizzi, M.: A 'soft' measure of consensus in the setting of partial (fuzzy) preferences. European Journal of Operational Research 34, 315–325 (1988)
29. Kacprzyk, J., Fedrizzi, M.: A 'human-consistent' degree of consensus based on fuzzy logic with linguistic quantifiers. Mathematical Social Sciences 18, 275–290 (1989)
30. Kacprzyk, J., Fedrizzi, M., Nurmi, H.: Group decision making and consensus under fuzzy preferences and fuzzy majority. Fuzzy Sets and Systems 49, 21–31 (1992)
31. Martínez, L., Montero, J.: Challenges for improving consensus reaching process in collective decisions. New Mathematics and Natural Computation 3, 203–217 (2007)
32. Martínez-Panero, M.: Consensus perspectives: Glimpses into theoretical advances and applications. In: Herrera-Viedma, E., García-Lapresta, J.L., Kacprzyk, J., Fedrizzi, M., Nurmi, H., Zadrożny, S. (eds.) Consensual Processes. STUDFUZZ, vol. 267, pp. 179–193. Springer, Heidelberg (2011)
33. Mata, F., Martínez, L., Herrera-Viedma, E.: An adaptive consensus support model for group decision-making problems in a multigranular fuzzy linguistic context. IEEE Transactions on Fuzzy Systems 17, 279–290 (2009)
34. Mata, F., Pérez, L.G., Zhou, S.M., Chiclana, F.: Type-1 OWA methodology to consensus reaching processes in multi-granular linguistic contexts. Knowledge-Based Systems 58, 11–22 (2014)
35. Palomares, I., Estrella, F.J., Martínez, L., Herrera, F.: Consensus under a fuzzy context: taxonomy, analysis framework AFRYCA and experimental case of study. Information Fusion 20, 252–271 (2014)

36. Palomares, I., Martínez, L.: A semi-supervised multi-agent system model to support consensus reaching processes. IEEE Transactions on Fuzzy Systems (in press), doi:10.1109/TFUZZ.2013.2272588
37. Palomares, I., Martínez, L., Herrera, F.: A consensus model to detect and manage noncooperative behaviors in large-scale group decision making. IEEE Transactions on Fuzzy Systems 22(3), 516–530 (2014)
38. Pérez, I.J., Cabrerizo, F.J., Alonso, S., Herrera-Viedma, E.: A new consensus model for group decision making problems with non homogeneous experts. IEEE Transactions on Systems, Man, and Cybernetics: Systems 44, 494–498 (2014)
39. Pérez, I.J., Wikström, R., Mezei, J., Carlsson, C., Herrera-Viedma, E.: A new consensus model for group decision making using fuzzy ontology. Soft Computing 17, 1617–1627 (2013)
40. Roselló, L., Prats, F., Agell, N., Sánchez, M.: Measuring consensus in group decisions by means of qualitative reasoning. International Journal of Approximate Reasoning 51, 441–452 (2010)
41. Roselló, L., Prats, F., Agell, N., Sánchez, M.: A qualitative reasoning approach to measure consensus. In: Herrera-Viedma, E., García-Lapresta, J.L., Kacprzyk, J., Fedrizzi, M., Nurmi, H., Zadrożny, S. (eds.) Consensual Processes. STUDFUZZ, vol. 267, pp. 235–261. Springer, Heidelberg (2011)
42. Roselló, L., Sánchez, M., Agell, N., Prats, F., Mazaira, F.A.: Using consensus and between generalized multiattribute linguistic assessments for group decision-making. Information Fusion 17, 83–92 (2014)
43. Saint, S., Lawson, J.R.: Rules for Reaching Consensus. A Modern Approach to Decision Making. Jossey-Bass, San Francisco (1994)
44. Spillman, B., Bezdek, J., Spillman, R.: Development of an instrument for the dynamic measurement of consensus. Communication Monographs 46, 1–12 (1979)
45. Torra, V., Narukawa, Y.: Modeling Decisions: Information Fusion and Aggregation Operators. Springer, Berlin (2007)
46. Travé-Massuyès, L., Dague, P. (eds.): Modèles et Raisonnements Qualitatifs. Hermes Science, Paris (2003)
47. Travé-Massuyès, L., Piera, N.: The orders of magnitude models as qualitative algebras. In: Proceedings of the 11th International Joint Conference on Artificial Intelligence, Detroit (1989)

Consensus Modeling in Multiple Criteria Multi-expert Real Options-Based Valuation of Patents

Andrea Barbazza[1], Mikael Collan[2], Mario Fedrizzi[1,2], and Pasi Luukka[2]

[1] Department of Engineering, University of Trento, Italy
{andrea.barbazza,mario.fedrizzi}@unitn.it
[2] School of Business, Lappeenranta University of Technology, Finland
{mikael.collan,pasi.luukka}@lut.fi

Abstract. In this paper we introduce a decision system for supporting the ranking of patents carried on by a group of experts in a multiple criteria fuzzy environment. The process starts with the creation of three value scenarios for each considered patent by each expert which are then used for the construction of individual fuzzy pay-off distribution functions for the patent value, here represented with triangular fuzzy numbers. Then, for each expert a TOPSIS matrix is estimated assuming that the scores are linguistically expressed due to the vagueness of individual judgments. We assume that the criteria are represented by the first three possibilistic moments of the pay-off distribution function and by a set of "strategic" attributes that describe different relevant aspects of the patents under analysis. A novel consensus modeling mechanism is then introduced to determine a coalition of experts whose TOPSIS-based evaluations are close enough. Finally, the coalition-based group TOPSIS ranking of patents is determined.

Keywords: Patents, Fuzzy Pay-Off Method, TOPSIS, Group Consensus.

1 Introduction

Ranking and selection of patents is an important issue from the point of view of intellectual property (IPR) and represents a recurring task in companies that commonly have their IPR managers visit the patent and R&D portfolios at least once year, analyzing the composition of the portfolios and making decisions about the portfolio composition.

Commonly there are three main approaches for the valuation of patents, these are the "cost approach" the "market method" and the "income approach" also known as the discounted cash-flow method (DCF), e.g. see [1, 2]. It is important to note that patent analysis is a forward-looking exercise, this means that methods used in the valuation and analysis of patents should be able to take into consideration the estimation inaccuracy present in forward-looking estimation, as the estimation of future cash-flows for patents, since it is not realistic to expect anyone to be able to produce precise estimates for future cash-flows [3]. Using cash-flow scenarios is a widespread practice of modeling the inaccurate and uncertain future cash-flows, and it can also be

P. Angelov et al. (eds.), *Intelligent Systems'2014*,
Advances in Intelligent Systems and Computing 322, DOI: 10.1007/978-3-319-11313-5_25

applied to patent analysis [4]. Information to support the creation of cash-flow scenarios can come from systems specifically designed for supporting patent analysis, such as are presented in [5] and [6], or it can come directly from experts, most often from within the firm itself. Fuzzy logic is an established way to express imprecision precisely and as such is a usable tool also when patent cash-flows are considered. Fuzzy pay-off method, introduced in [7] and further presented in [8] is a tool for investment analysis and is based on using cash-flow scenarios to create an asset's pay-off distribution that is considered as a fuzzy number and is one of the most effective methods usable for the valuation of patents [9].

In our work, assuming that a group of experts (managers) is involved in the selection process and that several criteria should be considered for the ranking of patents, the group decision making process starts with the estimation of individual fuzzy pay-off distribution functions for each patent, here represented with triangular fuzzy numbers. Then, for each expert a TOPSIS matrix is constructed based on a set of criteria represented respectively by the first three possibilistic moments of the pay-off distribution function and by a bundle of "strategic" attributes that describe different relevant aspects of the patents under analysis. Here we assume that the scores in TOPSIS matrices are linguistically expressed due to the vagueness and imprecision of individual judgments. A new consensus modeling mechanism, based on a matrix congruence index and on a constrained optimization problem, is then introduced to determine a coalition of experts whose TOPSIS-based evaluations are similar. At the end of process, the TOPSIS matrices of the experts belonging to the coalition subset are pooled to find a unique group TOPSIS matrix used for calculating the final ranking of patents.

The reminder of the paper is organized as follows. In section 2 the Fuzzy Pay-Off Method (FPOM) is introduced. Section 3 is devoted to a short description of TOPSIS method and its extension to a fuzzy environment, starting from the assumption that individual scores and weights are linguistically expressed. In section 4 a new consensus modeling mechanism is introduced based on a matrix congruence index and on the solution of a constrained optimization aiming to find a consensual coalition of experts. Some conclusions in section 5 close the paper.

2 Cash-Flows Valuation via FPOM (Fuzzy Pay-Off Method)

The innovative feature of the FPOM, as fully described in [7] and in [8], concerns the inclusion of the inaccuracy and the imprecision inherent to the human estimation process by means of the fuzzy numbers. The starting point for the valuation of a generic investment project consists in estimating the future cash flows that it is able to generate, regarding both costs and revenues that will take place in the future, by means of cash flow scenarios. Basically, several experts from different company's areas are gathered and asked to generate forecasts for the future trends of the project's main drivers under different plausible market scenarios (usually three). Therefore the experts provide, for each year, three estimates of costs and benefits, thus simplifying their future levels to three different values contingent to three different scenarios that could occur. Typically, these scenarios are most likely (with higher probability to occur), best case and worst case (with lower probabilities to occur). Once the future cash flow forecasts are created, the next step consists in calculating the fuzzy Net Present Value (NPV) of the project, which will shape the fuzzy pay-off distribution. Clearly, in order to build the project's NPV, a discounting process has to take place, aiming to bring all the expected project's steams at present time. As a result, three NPVs are calculated, one for each scenario instance, and used to build the pay-off distribution of the project, which represents the set of all the possible project's outcomes. Since each one of the involved experts provides his/her own estimations about the future cash flows that the project being analysed is deemed able to generate, goes without saying that a number of distributions equal to the number of experts is created. Clearly, every distribution is affected by the expertise and the expectations of who has actually created it. The main idea proposed in the FPOM is that of looking at the pay-off distribution just obtained as a fuzzy number, which will be triangular if the scenarios are three, as in most cases. Therefore, the fuzzy pay-off is a fuzzy number representing the set of all the possible project's outcomes, in which the most likely scenario has fully degree of membership whereas the best case and worst case scenarios have membership degrees close to zero, thus looking more in detail, the fuzzy pay-off distribution should be represented as in Figure 1.

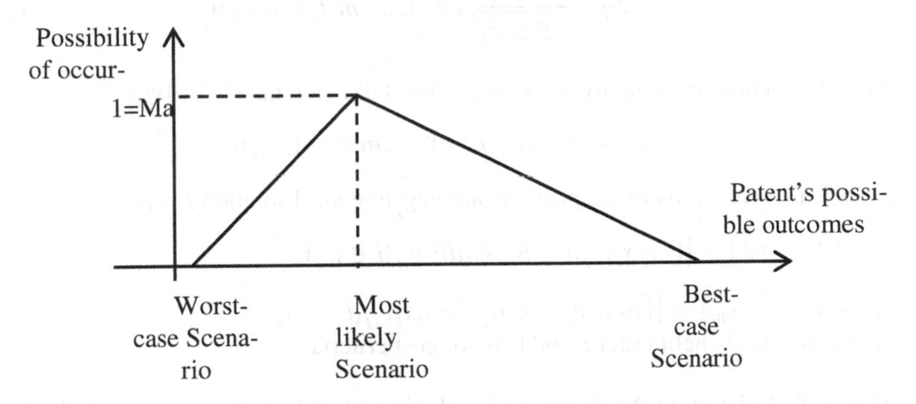

Fig. 1. Triangular fuzzy pay-off distribution

To summarize, the required steps in order to build the fuzzy pay-off distribution, as clearly stated in [2], can be summarized as follows:

1. Observing that the best guess scenario NPV is the most likely one and assigning it full degree membership in the set of expected outcomes;
2. Deciding that the maximum possible and the minimum possible scenarios' NPVs are the upper and lower bounds of the distribution–there is also a simplifying assumption, of bounding the distribution: to not consider values higher than the optimistic scenario and lower than the pessimistic scenario;
3. Assuming that the shape of the pay-off distribution is triangular.

3 TOPSIS-Based Individual Ranking in a Fuzzy Environment

In this section, for each expert a TOPSIS matrix is introduced assuming that the scores are linguistically expressed due to the vagueness of individual judgments. Here, the criteria are represented by the first three possibilistic moments of the pay-off distribution function (see [10, 11]) and by a set of "strategic" attributes that describe different relevant aspects of the patents under analysis.

TOPSIS (Technique for Order Preference by Similarity to an Ideal Solution) is a popular approach to multiple criteria decision making problems [12] which simultaneously considers the distances to the positive and negative ideal solution regarding each alternative and selects the most relative closeness to the ideal solution and the farthest one from the negative ideal solution.

Herewith we summarize the main steps of TOPSIS method starting from the construction of the matrix of scores X=[xij] where xij denotes score of the ith alternative with respect to the jth criterion, and can be summarized as follows:

Step 1: Calculation of normalized decision matrix Z=[zij]

$$z_{ij} = \frac{x_{ij}}{\sqrt{\sum_{j=1}^{m} x_{ij}^2}}, j = 1, \ldots, m, i = 1, \ldots, n \tag{1}$$

Step 2: Calculation of the weighted normalized decision matrix V=[vij]

$$v_{ij} = z_{ij}(\cdot)w_j, j = 1, \ldots, m, i = 1, \ldots, n \tag{2}$$

Step 3: Determination of the positive and negative ideal solution A+ and A-:

$$A^+ = \{v_1^+, \ldots, v_n^+\} = \{(max_j v_{ij} | i \in B), (min_j v_{ij} | i \in C)\} \tag{3}$$

$$A^- = \{v_1^-, \ldots, v_n^-\} = \{(min_j v_{ij} | i \in B), (max_j v_{ij} | i \in C)\}$$

where B is for benefit criteria and C is for cost criteria.

Step 4: Calculation of the distance of each alternative from the positive ideal solution and negative ideal solution:

$$d_i^+ = \sqrt{\sum_{j=1}^{m} (v_{ij} - v_j^+)^2}, \text{j=1,...,m} \tag{4}$$

$$d_i^- = \sqrt{\sum_{j=1}^{m} (v_{ij} - v_j^-)^2}, \text{j=1,...,m} \tag{5}$$

Step 5: Calculation of the relative closeness to the ideal solutions:

$$CC_i = \frac{d_i^-}{d_i^+ + d_i^-}, j = 1, \ldots, m \tag{6}$$

Step 6: Ranking of alternatives: The closer the CCi is to one implies the higher priority of the i:th alternative.

The fuzzy extension illustrated hereafter is based on [13] and aims to allow the experts to convey their expertise in the form of natural language, using the linguistic variables listed below. In order to make the information exploitable for the required calculations, such linguistic variables will then be expressed with the triangular fuzzy numbers shown alongside.

Table 1. Linguistic terms for rating alternatives and related fuzzy numbers

Linguistic terms	Triangular fuzzy numbers
Very Poor	(0, 0, 1)
Poor	(0, 1, 3)
Medium Poor	(1, 3, 5)
Fair	(3, 5, 7)
Medium Good	(5, 7, 9)
Good	(7, 9, 10)
Very Good	(9, 10, 10)

The triangular fuzzy numbers, used in order to express the experts' linguistic assessments, can be generally represented as follows:

$$\tilde{x}_{ij} = (a_{ij}, b_{ij}, c_{ij}) \quad (7)$$

denoting the rating of the i-th alternative with respect to the j-th selection criteria. The elements a_{ij}, b_{ij} and c_{ij} are, respectively, the left extreme, the centre and the right extreme of the triangular fuzzy number, as depicted below. The tilde simply indicates that we are dealing with a fuzzy number.

A procedure very similar to that just illustrated for the alternatives' ratings, is carried out with regard to the importance weights of the selection criteria. Once more, the experts are asked to convey personal opinions about something that is quite vague and that can be hardly represented by a punctual numerical value. The fuzzy extension of TOPSIS allows them to provide their personal valuations by means of linguistic variables, which, exactly as happens for the patents' ratings, will be subsequently expressed with triangular fuzzy numbers. The conversion from natural language into fuzzy numbers will take place according to the Table 2.

Table 2. Linguistic terms for the importance weights and related fuzzy numbers

Linguistic terms	Triangular fuzzy numbers
Very Low	(0, 0, 0.1)
Low	(0, 0.1, 0.3)
Medium Low	(0.1, 0.3, 0.5)
Medium	(0.3, 0.5, 0.7)
Medium High	(0.5, 0.7, 0.9)
High	(0.7, 0.9, 1.0)
Very High	(0.9, 1.0, 1.0)

Each expert thus provides an importance weight for each criterion in the form of natural variable, which will be subsequently expressed by means of a triangular fuzzy number indicated as $\tilde{w}_j$. As suggested in [13], the classical normalization process performed in the original formulation of TOPSIS is replaced by a simpler, linear scale transformation. Consequently, each element of the fuzzy evaluation matrix is normalized in the way described below, and denoted as $\tilde{r}_{ij}$:

$$\begin{cases} \tilde{r}_{ij} = \left(\frac{a_{ij}}{c_j^*}, \frac{b_{ij}}{c_j^*}, \frac{c_{ij}}{c_j^*}\right), & if\ j \in benefits \\ \tilde{r}_{ij} = \left(\frac{a_j^-}{c_{ij}}, \frac{a_j^-}{b_{ij}}, \frac{a_j^-}{a_{ij}}\right), & if\ j \in costs \end{cases} \tag{8}$$

Where

$c_j^* = \max_i c_{ij}$, if j is a criterion representing benefits;

$a_j^- = \min_i a_{ij}$, if j is a criterion representing costs;

Once that all the elements of the fuzzy evaluation matrix has been normalized in the way just exposed, they are gathered in the so-called normalized fuzzy evaluation matrix ($\tilde{R}$).

The next step of TOPSIS consists in weighting the normalized elements according to the importance of the selection criteria. This operation is rather straightforward ad performed in the following way:

$$\tilde{v}_{ij} = \tilde{r}_{ij} * \tilde{w}_j. \tag{9}$$

The weighted normalized fuzzy evaluation matrix ($\tilde{V}$) is obtained simply multiplying the normalized elements of the matrix $\tilde{R}$ by the weights of the corresponding selection criteria.

Since we have k experts, at the end of the process we obtain the following set of fuzzy TOPSIS matrices

$$V=\{\tilde{V}_1, \tilde{V}_2, \ldots, \tilde{V}_k\} \tag{10}$$

4 Consensus Modeling Mechanism

The problem we are going to address now is how to find a consensual TOPSIS ranking starting from the individual fuzzy TOPSIS matrices. The problem of consensus modeling under fuzziness has been studied from the early seventies and the interested reader can find surveys of the main theoretical and application oriented results obtained, among others, in [14, 15, 16, 17, 18]. Extensions to consensus modeling under linguistically expressed individual judgments have been developed in [19, 20, 21, 22, 23]. Some authors studied the problem of consensus under majority rules, see e. g. [24, 25, 26].

Our approach here fits into this last line of research and aims at finding a consensual coalition of experts acting as a steering committee in the overall ranking of patents. We start defuzzifying the matrices in the set V through the centroid method to obtain the crisp matrices denoted as V_1, V_2, ... , V_k. Then, a congruence relation is introduced to quantify the similarity between pairs (V_i, V_j) and the corresponding congruence indexes are computed as

$$\gamma_{ij}=\mathrm{tr}(V_iV_j^T)/\sqrt{tr(V_iV_i^T)}\sqrt{tr(V_jV_j^T)} \quad (11)$$

The distance between the defuzzified valuations of ith expert and jth expert is then defined as

$$\delta_{ij} = 1 - \gamma_{ij} \quad (12)$$

The distance δ_{ij} is here interpreted as a degree of agreement between the corresponding pair of experts. Accordingly, the overall degree of agreement between the group of experts is here represented as

$$D_w(d_1,\ldots,d_k)=SOWA_w(\delta_{12},\ldots,\delta_{k-1,k}) \quad (13)$$

where weights are taken in]0,1[and SOWA means Strict OWA.

It's easy to show that such a function satisfies reasonable properties like commutativity, idempotence, and strict monotonicity.

Following now an approach similar to the one adopted in [27] consensus is determined finding the coalition of experts solving the following constrained optimization problem

$$argmin\{D_w(e_{\pi_1},\ldots,e_{\pi_m}): \pi_1 < \cdots < \pi_m, \textstyle\sum_h w_h > \overline{w}\}, \quad (14)$$

i.e. the most consensual coalition is determined in such a way that the relative weight of the coalition is greater than a given threshold. Here the weights have been introduced to measure the different level of expertise of the decision makers. Then, the ranking of the patents is determined starting from the fuzzy TOPSIS matrix obtained through the aggregation of the fuzzy TOPSIS matrices (see) of the experts in the coalition.

As briefly outlined in the previous description of the method, TOPSIS exploits the distances from the so-called positive and negative ideal solutions in order to rank the alternatives under analysis. They represent, respectively, the best and the worst rating that the patents can obtain with regard to each one of the selection criteria. In the fuzzy-extended version of TOPSIS, these solutions are constituted of fuzzy numbers, and are quite easy to determine. In fact, the fuzzy numbers composing the weighted normalized fuzzy evaluation matrix can only range between a minimum value of 0 and a maximum value of 1. Consequently, it goes without saying that the FPIS is characterized by triangular fuzzy numbers whose elements are all 1s (the highest possible value) for the benefits attributes and all 0s (the lowest possible value) for the costs attributes. The FNIS is built in exactly the opposite way. In general, we have:

$$FPIS = (\tilde{v}_1^*, \tilde{v}_2^*, \ldots, \tilde{v}_n^*) \quad (15)$$

$$FNIS = (\tilde{v}_1^-, \tilde{v}_2^-, \ldots, \tilde{v}_n^-)$$

where

$$\tilde{v}_j^* = \begin{cases} (1,1,1), & If\ j\ is\ a\ benefit\ attribute \\ (0,0,0), & If\ j\ is\ a\ cost\ attribute \end{cases}$$
$$\tilde{v}_j^- = \begin{cases} (0,0,0), & If\ j\ is\ a\ benefit\ attribute \\ (1,1,1), & If\ j\ is\ a\ cost\ attribute \end{cases}$$

In order to rank the vying alternatives, TOPSIS requires the calculation of the distances of each alternative from what has been defined fuzzy positive ideal solution and fuzzy negative ideal solution, and the procedure used is the so called vertex method. According to such method, the distance between two generic triangular fuzzy numbers, $\tilde{m} = (m_1, m_2, m_3)$ and $\tilde{n} = (n_1, n_2, n_3)$, is calculated in the following way:

$$d(\tilde{m}, \tilde{n}) = \sqrt{\frac{1}{3}[(m_1 - n_1)^2 + (m_2 - n_2)^2 + (m_3 - n_3)^2]} \tag{16}$$

Thus, the distances from FPIS (d_i^*) and from FNIS (d_i^-) are calculated, for every patent, as follows:

$$d_i^* = \sum_{j=1}^{n} d(\tilde{v}_{ij}, \tilde{v}_j^*), \quad i = 1,2,\dots,m \tag{17}$$
$$d_i^- = \sum_{j=1}^{n} d(\tilde{v}_{ij}, \tilde{v}_j^-), \quad i = 1,2,\dots,m \tag{18}$$

At the end of this stage, each alternative will have a distance measure both from the FPIS and FNIS, expressed by a crisp number, the so called closeness coefficient, represented as follows:

$$0 \le \left[C_i = \frac{d_i^-}{d_i^* + d_i^-}\right] \le 1 \ , \ \forall_i = 1,2 \dots m \tag{19}$$

Consequently, the alternatives are decreasingly ranked according to their similarity closeness coefficients, starting from the highest values and ending with the lowest ones. The IPR managers can therefore use the ranking as decision support tool in identifying the most valuable patents that deserve to be funded.

5 Concluding Remarks

Drawing on the decision making framework introduced in [28], in this paper we have addressed the issue of designing a model based on Fuzzy Pay-Off Method and on fuzzy TOPSIS to support the ranking process of patents (R&D projects) as carried on by a group of experts (managers). The process starts with the creation of three value scenarios for each considered patent by each expert that are then used for the construction of individual fuzzy pay-off distribution functions for the patent value, here represented with triangular fuzzy numbers. Then, for each expert a fuzzy TOPSIS matrix is constructed assuming that the scores are linguistically expressed due to the vagueness of individual judgments and that the criteria are both quantitative and qualitative. A novel consensus modeling mechanism is then introduced to determine a coalition of experts whose TOPSIS-based evaluations are close enough. This coalition is entrusted with the task of drawing up the final ranking of patents. Future work will be focused firstly on numerical simulations using data from real contexts to test, also in comparison with the results obtained in [28], the usability of the model, and secondly on theoretical extensions which take account of the possibilistic risk aversion profiles of the experts.

References

1. Smith, G., Parr, R.: Valuation of Intellectual Property and Intangible Assets. John Wiley & Sons, New York (2000)
2. Reilly, R., Schweihs, R.: Valuing Intangible Assets. McGraw-Hill Professional Publishing, Blacklick (1998)
3. Karsak, E.E.: A Generalized Fuzzy Optimization Framework for R&D Project Selection Using Real Options Valuation. In: Gavrilova, M.L., Gervasi, O., Kumar, V., Tan, C.J.K., Taniar, D., Laganá, A., Mun, Y., Choo, H. (eds.) ICCSA 2006. LNCS, vol. 3982, pp. 918–927. Springer, Heidelberg (2006)
4. Collan, M., et al.: Numerical Patent Analysis with the Fuzzy Pay-Off Method: Valuing a Compound Real Option. In: IEEE 4th International Conference on Business Intelligence and Financial Engineering (BIFE 2011). Wuhan, PRC (2011)
5. Littman-Hillmer, G., Kuckartz, M.: SME Tailor-Designed Patent Portfolio Analysis. World Patent Information 31, 273–277 (2009)
6. Camus, C., Brancaleon, R.: Intellectual Assets Management: from Patents to Knowledge. World Patent Information 25, 155–159 (2003)
7. Collan, M., et al.: Fuzzy Pay-Off Method for Real Option Valuation. Journal of Applied Mathematics and Decision Systems (2009)
8. Collan, M.: The Pay-Off Method: Re-Inventing Investment Analysis. CreateSpace Inc., Charleston (2012)
9. Collan, M., Heikkilä, M.: Enhancing Patent Valuation with the Pay-Off Method. Journal of Intellectual Property Rights 16, 377–384 (2011)
10. Carlsson, C., Fullér, R.: On Possibilistic Mean Value and Variance of Fuzzy Numbers. Fuzzy Sets and Systems 122, 315–326 (2001)
11. Fuller, R., Majlender, P.: On Weighted Possibilistic Mean and Va-riance of Fuzzy Numbers. Fuzzy Sets and Systems 136, 363–374 (2003)
12. Hwang, C., Yoon, K.: Multiple Attributes Decision Making Methods and Applications. Springer, Heidelberg (1981)
13. Chen, C.-T.: Extensions of the TOPSIS for Group Decision Making under Fuzzy Environment. Fuzzy Sets and Systems 114, 1–9 (2000)
14. Kacprzyk, J., Nurmi, H., Fedrizzi, M. (eds.): Consensus under Fuzziness. International Series in Intelligent Technologies. Kluwer Academic Publishers, Dordrecht (1997)
15. Fedrizzi, M., Pasi, G.: Fuzzy Logic Approaches to Consensus Modeling in Group Decision Making. In: Ruan, D., Hardeman, F., Van Der Meer, K. (eds.) Intelligent Decision and Policy Making Support Systems. SCI, vol. 117, pp. 19–37. Springer, Heidelberg (2008)
16. Cabrerizo, F., Moreno, J., Perez, I., Herrera-Viedma, E.: Analyzing Consensus Approaches in Fuzzy Group Decision Making: Advantages and Drawbacks. Soft Computing 14, 451–463 (2010)

17. Herrera-Viedma, E., García-Lapresta, J.L., Kacprzyk, J., Fedrizzi, M., Nurmi, H., Zadrożny, S. (eds.): Consensual Processes. STUDFUZZ, vol. 267. Springer, Heidelberg (2011)
18. Herrera-Viedma, E., Cabrerizo, F., Kacprzyk, J., Pedrycz, W.: A Review of Soft Consensus Models in a Fuzzy Environment. Information Fusion 17, 4–13 (2014)
19. Mata, F., Martinez, L., Herrera-Viedma, E.: An Adaptive Consensus Support Model for Group Decision Making Problems in a Multigranular Fuzzy Linguistic Context. IEEE Transactions on Fuzzy Systems 17, 279–290 (2009)
20. Cabrerizo, F., Alonso, S., Herrera-Viedma, E.: A Consensus Model for Group Decision Making Problems with Unbalanced Fuzzy Linguistic Information. International Journal of Information Technology & Decision Making 8, 109–131 (2009)
21. Cabrerizo, F., Pérez, I., Herrera-Viedma, E.: Managing the Con-sensus in Group Decision Making in an Unbalanced Fuzzy Linguistic Context with Incomplete Information. Knowledge-Based Systems 23, 169–181 (2010)
22. Pérez, I., Cabrerizo, F., Herrera-Viedma, E.: Group Decision Making Problems in a Linguistic and Dynamic Context. Expert Systems With Applications 38, 1675–1688 (2011)
23. Tapia Garcia, J., del Moral, M., Martinez, M., Herrera-Viedma, E.: A Consensus Model for Group Decision Making Problems with Linguistic Interval Fuzzy Preference Relations. Expert Systems with Applications 39, 10022–10030 (2012)
24. Kacprzyk, J., Fedrizzi, M., Nurmi, H.: Group Decision Making and Consensus under Fuzzy Preferences and Fuzzy Majority. Fuzzy Sets and Systems 49, 21–31 (1992)
25. Fedrizzi, M., Kacprzyk, J., Owinsky, J., Zadrozny, S.: Consensus Reaching via a GDSS with Fuzzy Majority and Clustering of Preference Profiles. Annals of Operations Research 51, 127–139 (1994)
26. Pasi, G., Yager, R.: Modeling the Concept of Majority Opinion in Group Decision Making. Information Sciences 176, 330–414 (2006)
27. Brunelli, M., Fedrizzi, M., Fedrizzi, M.: OWA-Based Fuzzy *m*-ary Adjacency Relations in Social Network Analysis. In: Yager, R.R., Kacprzyk, J., Beliakov, G. (eds.) Recent Developments in the Ordered Weighted Averaging Operators: Theory and Practice. STUDFUZZ, vol. 265, pp. 255–267. Springer, Heidelberg (2011)
28. Collan, M., Fedrizzi, M., Luukka, P.: A Multi-Expert System for Ranking Patents: an Approach Based on Fuzzy Pay-Off Distributions and a TOPSIS-AHP Framework. Expert Systems with Applications 40, 4749–4759 (2013)

Modeling Different Advising Attitudes in a Consensus Focused Process of Group Decision Making

Dominika Gołuńska[1], Janusz Kacprzyk[2,3], and Enrique Herrera-Viedma[4]

[1] Department of Electrical and Computer Engineering Cracow, University of Technology Warszawska 24, 31-155 Cracow, Poland
[2] Systems Research Institute, Polish Academy of Sciences, Newelska 6, 01-447 Warsaw, Poland
kacprzyk@ibspan.waw.pl
[3] WIT - Warsaw School of Information Technology Newelska 6, 01-447 Warsaw, Poland
[4] DECSAI - University of Granada, 18071- Granada, Spain
viedma@decsai.ugr.es

Abstract. The main goal of this paper is to support reaching a consensus type solution in a group decision making problem. In this context, we present a new model which assists the support system which is strongly integrated with our consensus reaching model. It is based on a role of a moderator who helps agents (individuals), by rational argument, persuasion, etc. change their opinions and towards a higher agreement within the entire group. As in our previous works, the consensus degree determines the agreement among most of (important) agents as to most of (relevant) options. Information about the current state of agreement and the main obstacles in reaching consensus are provided in a human consistent, hence easy to use form, by linguistic data summaries that can be derived, for instance, via natural language generation (NLG). In this paper, we extend the consensus evaluation and reaching model by carrying out various consensus reaching scenarios depending on a context of the process and considering different natures of a group of agents involved. Though not each discussion requires the involvement of each member, which may be time consuming, it may often be necessary to avoid the accounting for group member's interests or emotional needs of agents. Therefore, to cope with these types of scenarios, we propose here an efficiently focused use of additional linguistic consensus indicators to provide the moderator with appropriate mechanisms for guiding the decision makers towards a higher degree of consensus. Finally, we present an illustrative implementation and numerical evaluation of various attitudes of the moderator's actions in the model proposed.

Keywords: consensus, consensus reaching process, group decision making, linguistic data summary, moderator, discussion, rule of influence.

1 Introduction

This paper deals with a relevant aspect of an important problem that is pertinent to various human activities, namely group decision making (GDM). Basically, in a GDM problem considered here, agents (individuals, decision makers, experts,...) openly reveal their opinions (preferences) as to the options considered [10]. The main

P. Angelov et al. (eds.), *Intelligent Systems'2014*,
Advances in Intelligent Systems and Computing 322, DOI: 10.1007/978-3-319-11313-5_26

goal of the group decision making process is to reach a consensus, i.e. a solution that an overwhelming majority (all, if possible) agents are willing to support. Many results of psychological investigations and also a real life experience clearly show that obtaining a solution without a sufficient agreement among the agents is not reasonable as it may be not accepted, and can lead to a solution with no chance for its practical implementation or even prevent the survival of the group in the long time period [1]. Thus, it is desirable that the agents carry out a consensus reaching process to achieve a sufficient agreement before proceeding to getting to a final solution.

Initially, the agents usually disagree in their opinions, i.e. they are far from consensus. Normally, the communication process between the agents involves a designated person who coordinates all group members and tries to convince other agents to change their preferences for the benefit of the group. This role in our consensus model is directly attributed to a moderator who runs the consensus reaching session until the group gets close enough to consensus, i.e. a sufficient agreement.

One of the problem in this field is to find the best indicators to make use of information from the agents' preferences and use them as an efficient tool to aid the moderator driving the group towards consensus.

In our previous works we applied a new concept of the consensus model, introduced by Kacprzyk and Fedrizzi [6], based on a "soft" degree of consensus under fuzzy preferences and a fuzzy majority. Basically, the degree of consensus was meant as the degree to which "most of the agents (individuals) agree in their preferences as to most of the options". We concentrated on the approach in which the agreement was obtained via a consensus reaching process run by the moderator [3]. Moreover, we extended this idea by employing linguistic data summaries to support the consensus reaching process, as proposed by Kacprzyk and Zadrożny [9].

In the present paper we further develop this approach and propose how to use a linguistic approach as an efficient mechanism to support the moderator's actions. We determine some rules to drive this part of the process in which the agents are expected to update their preferences in line with an advice given by the moderator. In other words, we propose various rules (tactics) of the moderators' persuasion, depending on the context of the process and considering different natures of the group of agents.

The next parts of this paper deal with the following. The basic concept of the new consensus reaching support system was precisely described in the authors' previous papers [2],[3],[4]. Nevertheless, the overall framework of this model and its essential elements are briefly mentioned in Section 2. An original extension of that proposed model, i.e. via the modeling of different advising attitudes, is presented in Section 3. An illustrative numerical example and final conclusions are given in Sections 4 and 5, respectively.

2 Preliminaries

We have a finite set of $n \geq 2$ options, $S = \{s_1, s_2, ..., s_n\}$ which are ranked using information given by a finite set of $m \geq 2$ agents (individuals), $E = \{e_1, e_2, ..., e_m\}$ who express their opinions as fuzzy preference relations.

Each agent e_k specifies his/her preferences as to the particular pairs of options in S. These testimonies are assumed to be an agent fuzzy preference relation R_k defined in $S \times S$ [8], characterized by its membership function:

$$\mu_{R_k} : S \times S \to [0,1], \tag{1}$$

such that the value $\mu_{R_k}(s_i, s_j) = r_{ij}^k$ is interpreted as a preference degree of option s_i over option s_j.

In this paper, we assume a (relatively) small group of homogeneous agents. Moreover, the feasible set of options is not very large. Thus, the agents can express their opinions about each pair of options and also the cardinality of S is small enough that allow us to represent agent fuzzy preference relations, R_k, by a $n \times n$ matrix $R_k = [r_{ij}^k]$, such that $r_{ij}^k = \mu_{R_k}(s_i, s_j)$, $i, j = 1,...,n; k = 1,....,m.$ We also do not get into any serious numerical complexity problem.

Initially, the agents usually disagree in their opinions. The overall goal of the group decision making is to attain the consensus among the group members, i.e. the highest acceptable degree of an agreement between themselves and as to the available options. We assume that the concept of a consensus is measurable via a degree of consensus. Specifically, we apply the concept of a "soft" degree of consensus proposed by Kacprzyk and Fedrizzi [6] that is, the degree to which: "most of agents agree in their preferences as to most of the options" (the degree of truth is obtained using Zadeh's calculus of linguistic quantified propositions [12]). The highest value, 1, corresponds to the unanimity of the group and the lowest value, 0, to the complete disagreement, with all partial values from [0,1]. The consensus is usually a prerequisite for being able to reach a good group decision solution.

If the degree of consensus is not sufficiently high and if the number of consensus reaching runs has not reached the time limit, some agents' preferences should be updated to possibly increase the degree of agreement. This means to go through several rounds of the process until the agent fuzzy preference relations become similar enough.

There are different approaches to the consensus reaching process. We apply the approach which is usually more promising in practice, namely the one run by a moderator [5]. It is defined as a dynamic and iterative group discussion process, coordinated by a moderator who helps agents bring their opinion closer by stimulating an exchange of information, suggesting arguments, convincing appropriate agents to change their preferences, focusing the discussion on issues which may resolve the conflict of opinions in the group, etc. For simplicity, in this paper we assume an ideal situation, i.e. that in each discussion round the agents agree on changes of their preferences following the advice given by the moderator.

A crucial component of the consensus reaching support system is a set of indicators which provide an additional knowledge of and insight into the current state of the group, i.e. how far the group is from the consensus, which options are the most preferable, whose preferences are in the highest agreement and disagreement with the rest of the group, etc. This information may greatly facilitate the work of the moderator by providing him/her some hints as to the promising changes of the agents' preferences in a further discussion.

2.1 Linguistic Consensus Indicators

We consider consensus indicators in a human consistent form of linguistic data summaries [7],[9], i.e. short sentences in natural language. Here, we briefly present a general concept of these additional measure, and technical details are presented in our previous works [2],[8].

- The *personal consensus degree* of each agent e_k, $PCD(e_k) \in [0,1]$, is defined as the degree of truth value of the following linguistically quantified statement:

  "Preferences of an agent e_k as to the most pairs of options are in agreement with the preference of most other agents"

The $PCD(e_k)$ takes values from 0 for an agent who is the most isolated with his opinion as to rest of the group; to 1 for an agent whose preferences are shared by most of other agents; through all intermediate values.

- *The response to omission* of each agent e_k, $RTO(e_k) \in [-1,1]$, is defined as a difference between the consensus degree calculated for the whole group and the consensus degree for the group without a given agent e_k.
  This measure allows to estimate the influence of a given agent on the agreement in the group. The range of values are from -1, for an absolutely negative influence, through 0 for a lack of effect, to +1 for a definite influence.
- The *option consensus degree* for each option s_i, $OCD(s_i) \in [0,1]$, is the degree of truth value of the statement:

  "Most pair of agents agree with their preferences with respect to the option s_i"

The $OCD(s_i)$ indicator may take values from 0 which means that preferences of most agents differ basically from a given option to 1 when there is an agreement among the agents as to this particular option. In other words, this indicator points out either the most controversial options or the most promising ones for a further discussion.

- The *detailed option consensus degree* for each pair of options (s_i, s_j), $DOCD(s_i, s_j) \in [0,1]$, is the degree of truth value of the following statement:

  "Preferences of most agents are in agreement with respect to the pair of options (s_i, s_j)"

This indicator makes it possible to compare the agreement of agents as to the particular pair of options.

- *Response to exclusion* of option $s_i \in S$, $RTE(s_i) \in [-1,1]$, is determined as a difference between the consensus degree calculated while considering the whole set of options and for the set without option s_i.

This measure provides an estimate of the influence of a given option on the consensus degree, and may be interpreted similarly as $RTO(e_k)$.

3 Advising Attitudes – Different Rules of Recommendation

Once the moderator evaluates the current degree of consensus and calculates proper additional indicators that are needed to support the consensus reaching process, it is necessary to find a tactic in the moderators' persuasion, rational argument, etc. for driving the group towards consensus in the further discussion.

There are several procedures that can be applied at this level of consensus reaching model. According to Perez et al. [11], these procedures may be classified along the following line: from a limited scope approach in which consensus is reached within a limited number of agents, to a more participatory process in which consensus is reached by all the agents involved.

The moderator may pursue various tactics, following different attitudes, during the modification of the agents' opinions. In this section, we propose the two basic ways of the opinions' changing process depending on two opposite standpoints of the moderator:

1. Optimistic
2. Pessimistic

This distinction may be helpful while dealing with various types of problems, and within a specified amount of time available for running the session among the agents [5]. In such a way, this feedback mechanism may be constitute a powerful tool in the consensus reaching process.

What matters here is that we need to devise easy-to-comprehend rules for each of the moderator's reasoning modes. To do so, the process of providing the advice is divided into two phases: the *identification phase* and the *recommendation phase*. During the *identification phase* we select those agents who should participate in the opinion change process, options which should be taken into account in the opinion change process, and finally particular pairs of options for which the assessments of the agents should be changed, whereas the *recommendation phase* is aimed at finding the direction of changes to be recommended in each case [11],[5]. Both phases differ with their assumptions according to the particular moderators' attitude.

3.1 Optimistic Attitude

In this scenario, the moderator gets the agents closer to the consensus by convincing the most consistent agents, while agents who are isolated in their opinion are omitted.

– *Identification of agents* : we identify the subgroup of agents who are the most consistent in their preferences. This set of agents is denoted as:

$$IND1 = \{e_k \in E \mid PCD(e_k) > \gamma \wedge RTO(e_k) \geq 0\}, \tag{2}$$

where γ is the minimum consistency level required for the agents who should participate in the opinion change process.

– *Identification of options:* we identify options with a positive influence on the degree of consensus. This set of options is denoted as:

$$OPT1 = \{s_i \in S \mid OCD(s_i) > \gamma \wedge RTE(s_i) \geq 0\}, \tag{3}$$

– *Identification of pairs of options*: we select the particular pairs of options (s_i, s_j) which should be considered during the opinion change process. This set of options is denoted as:

$$DOPT1 = \{(s_i, s_j) \mid DOCD(s_i, s_j) > \gamma \wedge s_i \in OPT1\} \tag{4}$$

– *Recommendation phase*: we should formulate a general direction of updating preferences of each agent $e_k \in IND1$ as to the this pair of options $(s_i, s_j) \in DOPT1$ with the maximum value of the $DOCD(s_i, s_j)$ indicator. The new value of preference should be the same as those represented by the agent with the maximum value of the $PCD(e_k)$ indicator. After this iteration the degree of consensus should be computed. If it is not high enough, then all indicators are calculated once again and we start another round of discussion.

3.2 Pessimistic Attitude

This attitude presupposes an equal participation of all decision makers during the consensus reaching process, in particular the participation of so-called "outsiders" who are isolated in their opinions from the rest of the group.

– *Identification of agents:* we identify the particular agent or the set of agents who are the most isolated with their opinion as to the rest of the group. This set is denoted as:

$$IND2 = \{e_k \in E \mid PCD(e_k) < \gamma \wedge RTO(e_k) < 0\}. \tag{5}$$

– *Identification of options:* we identify options which are the most promising direction of a further discussion. This set of options is denoted by:

$$OPT2 = \{s_i \in S \mid OCD(s_i) > \gamma \wedge RTE(s_i) \geq 0\}. \tag{6}$$

- *Identification of pairs of options*: we select the particular pairs of options (s_i, s_j) which should be considered during the opinion change process.

$$DOPT2 = \{(s_i, s_j) \mid DOCD(s_i, s_j) > \gamma \wedge s_i \in OPT2\} \quad (7)$$

- *Recommendation phase:* the preferences of one agent $e_k \in IND2$ with the minimum value of $PCD(e_k)$ should be changed with respect to the option $s_i \in OPT2$. In this approach there is no need to apply the identification of the particular pairs of options. The new value of the agents' preferences as to this option should be the same as those represented by the agent with the highest $PCD(e_k)$ value.

4 Numerical Example

To illustrate the functioning of the new model proposed, we will show an illustrative example. There is a group of 6 agents who are requested to provide testimonies about 4 options. The fuzzy preference relations of each agent (the left lower triangular part is omitted) are:

$$R_1 = \begin{pmatrix} 0.9 & 0.9 & 1 \\ & 0.8 & 0.7 \\ & & 0.7 \end{pmatrix} R_2 = \begin{pmatrix} 0.7 & 1 & 1 \\ & 0.8 & 0.9 \\ & & 0.5 \end{pmatrix} R_3 = \begin{pmatrix} 1 & 0.6 & 1 \\ & 0.0 & 0.4 \\ & & 1 \end{pmatrix}$$

$$R_4 = \begin{pmatrix} 0.9 & 0.8 & 0.6 \\ & 0.6 & 0.3 \\ & & 0.3 \end{pmatrix} R_5 = \begin{pmatrix} 0.9 & 0.5 & 1 \\ & 0.0 & 0.4 \\ & & 1 \end{pmatrix} R_6 = \begin{pmatrix} 0.6 & 1 & 0.7 \\ & 0.0 & 0.5 \\ & & 1 \end{pmatrix}$$

Initially, the degree of consensus equals 0,69 which is below the sufficient level of agreement within the group members. All of the consensus indicators, calculated after the first iteration of the process, are as follows:

Table 1. Values of PCD(e_k)

Personal Consensus Degree

Agent	Value
Agent 1	1
Agent 6	0.95
Agent 3	0.68
Agent 5	0.68
Agent 4	0.47
Agent 2	0.31

Table 2. Values of RTO(e_k)

Reaction To Omission

Omitted agent	Value
Agent 1	0.18
Agent 6	0.13
Agent 3	-0
Agent 5	-0
Agent 4	-0.11
Agent 2	-0.19

Table 3. Values of OCD(s_i)

Option Consensus Degree

Option	Value
Option 1	0.99
Option 2	0.71
Option 4	0.48
Option 3	0.33

Table 4. Values of RTE(s_i)

RTE

Omitted option	Value
Option 1	0.33
Option 2	0.12
Option 4	0.04
Option 3	-0.25

Table 5. Values of DOCD(s_i,s_j)

Detailed Option Consensus Degree

Options	Value
Options (1,2)	1
Options (1,4)	0.867
Options (2,4)	0.733
Options (1,3)	0.467
Options (3,4)	0.467
Options (2,3)	0.2

According to the rules proposed in Section 3, we followed both types of the moderator's attitudes: optimistic and pessimistic. It is clear that our main goal is to obtain the highest possible degree of consensus within the available time period. Nevertheless, to compare the efficiency of different types of the moderators' actions, we introduced one more indicator. When the particular fuzzy preference relation of some agent is replaced by one obtained by changing it by 0.1, it is assumed to be equal to the unit cost of change (i.e., 1). The results of the evaluation due to the above rule is shown in Table 6.

Table 6. A comparison of results for the optimistic and pessimistic moderator's attitude

	Optimistic	**Pessimistic**
Initial degree of consensus	0,69	0,69
Final degree of consensus	0,98	0,99
Number of rounds	3	3
Cost	14	10

5 Concluding Remarks

In this paper we proposed a new approach to an efficiency focused consensus reaching support run by the moderator. Specifically, we defined various tactics (attitudes) of the moderators' persuasion for driving the consensus reaching process towards the highest degree of agreement. We determined two kinds of rules representing the opposite attitudes of the moderators' actions depending on the problem considered or a specific nature of the group of agents. Results of an illustrative numerical example of a moderator run consensus reaching process, under the two moderator's attitudes mentioned, fully confirmed the effectiveness and efficiency of our proposed approach.

Acknowledgments. Dominika Gołuńska's contribution is supported by the Foundation for Polish Science under International PhD Projects in Intelligent Computing. Project financed from The European Union within the Innovative Economy Operational Programme 2007-2013 and European Regional Development Fund. Janusz Kacprzyk contribution is partially supported by the National Centre of Science grant UMO The research has been partially supported, first, by the National Centre of \science under Grant No. UMO-2012/05/B/ST6/03068. Enrique Herrera-Viedma's contribution is supported by the grants TIC-05299, TIC-5991, and TIN2010-17876.

References

1. Gołuńska, D., Kacprzyk, J.: The Conceptual Framework of Fairness in Consensus Reaching Process Under Fuzziness. In: Proceedings of the 2013 Joint IFSA World Congress NAFIPS Annual Meeting, Edmonton, Canada, June 24-28, pp. 1285–1290 (2013)

2. Gołuńska, D., Kacprzyk, J., Zadrożny, S.: A Model of Efficiency-Oriented Group Decision and Consensus Reaching Support System In a Fuzzy Environment, Communications in Computer and Information Science. In: 15th International Conference on Information Processing and Management of Uncertainty in Knowledge-Based Systems (in press)
3. Herrera-Viedma, E., Cabrerizo, F.J.: J, Kacprzyk, W, Pedrycz: A review of soft consensus models in a fuzzy environment. Information Fusion 17, 4–13 (2014)
4. Herrera-Viedma, E., García-Lapresta, J.L., Kacprzyk, J., Fedrizzi, M., Nurmi, H., Zadrożny, S. (eds.): Consensual Processes. STUDFUZZ, vol. 267. Springer, Heidelberg (2011)
5. Herrera-Viedma, E., Martinez, L., Mata, F., Chiclana, F.: A Consensus Support System Model for Group Decision-Making Problems With Multigranular Linguistic Preference relations. IEEE Tranactions on Fuzzy Sysems 13(5), 644–658 (2005)
6. Kacprzyk, J., Fedrizzi, M.: A 'soft' measure of consensus in the setting of partial (fuzzy) preferences. European Journal of Operational Research 34, 315–325 (1988)
7. Kacprzyk, J., Zadrożny, S.: Computing with words is an implementable paradigm: fuzzy queries, linguistic data summaries and natural language generation. IEEE Transactions on Fuzzy Systems 18(3), 461–472 (2010)
8. Kacprzyk, J., Zadrożny, S.: On the use of fuzzy majority for supporting consensus reaching under fuzziness. In: Proceedings of FUZZ-IEEE 1997 - Sixth IEEE International Conference on Fuzzy Systems, Barcelona, Spain, vol. 3, pp. 1683–1988 (1997)
9. Kacprzyk, J., Zadrożny, S.: Supporting consensus reaching in a group via fuzzy linguistic data summaries. In: IFSA 2005 World Congress - Fuzzy Logic, Soft Computing and Computational Intelligence, Beijing, China, pp. 1746–1751. Tsinghua University Press/Springer (2005)
10. Kacprzyk, J., Zadrożny, S., Fedrizzi, M., Nurmi, H.: On Group Decision Making, Consensus Reaching, Voting and Voting Paradoxes under Fuzzy Preferences and a Fuzzy Majority: A Survey and some Perspectives. In: Bustince, H., Herrera, F., Montero, J. (eds.) Fuzzy Sets and Their Extensions: Representations, Aggregation and Models. STUDFUZZ, vol. 220, pp. 263–295. Springer, Heidelberg (2008)
11. Perez, I.J., Wikström, R., Mezei, J., Carlsson, C., Anaya, K., Herrera-Viedma, E.: Linguistic Consensus Models Based on a Fuzzy Ontology. Procedia Computer Science 17, 498–505 (2013)
12. Zadeh, L.: A computational approach to fuzzy quantifiers in natural languages. Computers and Mathematics with Applications 9, 149–184 (1983)

Automated Negotiation with Multi-agent Systems in Business Processes

Manuella Kadar[1], Maria Muntean[1], Adina Cretan[2], and Ricardo Jardim-Gonçalves[3]

[1] Computer Science Department, "1 Decembrie 1918" University, Alba Iulia, Romania,
[2] Computer Science Department, "Nicolae Titulescu" University of Bucharest, Romania,
[3] CTS, Departamento de Engenharia Electrotecnica,
Universidade Nova de Lisboa, UNINOVA, Lisboa, Portugal
{Mkadar,mmuntean}@uab.ro, badina20@yahoo.com, rg@uninova.pt

Abstract. This paper presents the design and implementation of a collaborative multi-agent system for automated negotiation processes occurring in dynamic networked environments. The focus is given to a negotiation mechanism required at business level. The mechanism is achieved by software agents within a collaborative working environment in the special case of notified transient situations. Automated negotiation is one of the most common approaches used to make decisions and manage disputes between computational entities leading them to optimal agreements.

The proposed solution is based on a multi-agent system architecture that applies rule-based negotiation at business level. This architecture has been tested in the case of on-line auctions in distributed networks.

Keywords: multi-agent systems, automated negotiation, business processes, rule-based expert system.

1 Introduction

Traditional approaches for analysing and designing business processes are focused mainly on tangible static areas such as functional requirements, architecture, interfaces, databases, specifications and regulations. The constant and rapid evolution of technology has changed the static scene, bringing an increased computing capability and storage that lead to a gradual servitisation of the world. In the new digitised environment, businesses become more and more sets of services in which a strong conjunction of people, technology, and organisations creates the added value and high performance. Business processes are increasingly based on adaptive and flexible services offered by computational entities. Among such computational entities are the virtual intelligent agents. In business to business relationships an effective approach is to involve intelligent agents to perform negotiation tasks within intelligent frameworks on behalf of their owners [1], [2]. Automated negotiation is one of the most common approaches used to make decisions and manage disputes between computational entities leading them to optimal agreements. The uses of automated negotiating agents for business processes reduces significantly the negotiation time and thus

P. Angelov et al. (eds.), *Intelligent Systems'2014*,
Advances in Intelligent Systems and Computing 322, DOI: 10.1007/978-3-319-11313-5_27

provide an efficient way to deal with the limitations of the human relationships in such endeavours. Virtual agents are able to achieve and maintain interoperability between business players who have decided to collaborate and further on, they are able to provide the framework for common understanding between business players. This paper presents a holistic, integrated view of the automated negotiation framework, showing the involved partners and their processes as they are orchestrated in the system through techniques of business process design and agent engineering. In the last decades a number of industrial projects have investigated business process design as a way of closing the gap between analysis and development of complex solutions [3]. Many ideas from agent theory, such as roles, rules, and communication, can be found in the Business Process Modelling Notation (BPMN) [4], which has already led to some efforts of mapping BPMN diagrams to multi-agent systems [5].

Previous works of the authors have presented the design and implementation of a collaborative multi-agent system that applies rule-based negotiation at various organizational levels such as: data systems, ICT, production/workflows [6], [7], [8]. The proposed architecture has been tested in collaborative networked environment in which several entities, namely funding authority, beneficiary of funds, contractors, sub-contractor have been contractually engaged. This paper continues the previous developments and focuses on the processes occurring at business level. The proposed methodology integrates BPMN, ontology engineering, multi-agent system design and service engineering to obtain a holistic integrated view of the enterprise interoperability at business level. The multi - agent systems act in the backstage, helping to communicate results among different services and software components within the business process flows. Agent technology with the inclusion of rules emulates intelligence and provides flexible software systems. In this paper the focus is given to a negotiation mechanism required at business level. The mechanism is achieved by software agents within the collaborative working environment. Flexible business processes require the design of adapted and intelligent software components that may support effective communication, coordination and collaboration. The proposed solution is presented in the following sections.

Section 2 discusses automated negotiation within the collaborative networked environment, the multi-layered negotiation process and communication at business level. The discussion covers three aspects: (i) the abstract negotiation environment with Multi-Agent Systems (MAS), (ii) design details of the negotiation layers structured into general negotiation architecture, (iii) the negotiation process model. Section 3 provides a real life case study of rule based automated negotiation process achieved by software agents with implementation of the rule-based proposal acceptance and negotiation finalization. Section 4 presents results and discussions and section 5 draws conclusions and points to future work.

2 Automated Negotiation

In the context of the growing digital world, enterprises raise their levels of business complexity and maturity. In the meantime, the whole surrounding environment

(ICT, services, providers, partners, etc) is less concerned about syntax and connectivity and more about semantics, contents, and knowledge sharing, facts that tremendously influences the enterprise interoperability at business level. Kaddouci et al. [9] state that in any case of divergence, negotiation is the most appropriate method to solve conflicts. Usually, interoperability conflicts are solved via one party forcing the other to change, or by reaching consensus towards a midway solution. Software agents are considered a suitable technology for automating negotiation processes.

Multi-agent systems can be used in layered system architectures that are designed for business processes. Intelligent agents are required to coordinate decision –making at different layers (e.g., to achieve appropriate contractual terms and conditions in a complex organizational system). The Interactive Recommendation and Automated Negotiation Systems (IRANS) [10] use multiple criteria decision-making and automated negotiation. In addition, the system is based on the multi- agent system approach in which agents can make autonomous negotiation decisions [11]. Such agents are required to reach an agreement on the joint action or the joint decision to be accepted by all involved parties. Automated negotiation has been identified as one of the key mechanisms for efficient and effective cooperation of the computational entities leading them to optimal agreements [12]. A number of models have been proposed in the literature to support agents reaching mutually acceptable agreements in automated negotiations [13] [14]. In most real world environments, multiple aspects of negotiation typically need to be taken into consideration with agents dynamically entering and leaving the environment, while new issues are proposed and new requirements and constraints becoming available. In such cases, a negotiator's parameters and the dynamics of the interaction has to be changed. Hence, a more general model for automated negotiation is required to accommodate and facilitate agents with flexible and adaptive behaviours.

2.1 Business Process Design with Multi-agent Systems

Business process modelling notation may disclose information on business partners, credentials, roles, relationships that have been established in business processes but does not define how such information may be implemented by agents. Therefore, we propose a methodology in which BPMN is combined with techniques of ontology engineering, multi-agent system design and service engineering. Such methodology provides a holistic view of the entire distributed multi-agent system. BPMN is hereby embedded in the methodology but the aspects that cannot be represented by BPMN including (i) the declaration of data types, (b) design of agent organization and distribution and (iii) textual programming for low-level algorithms are modelled by techniques used in ontology engineering and agent engineering. The development process starts with use case analysis, then for each use case, a negotiation process diagram is created. That processes conduct to the creation of the negotiation role model. Finally, the agent organization model and core services are integrated and tested and the process may go into iteration until the system is deployed.

The negotiation process diagrams provide a view on the system as a whole that can be used as groundwork for the following implementation of (a) the agents' behavior, e.g. in the form of plans, rules, or services, and (b) the organizational model, e.g. roles, agents, and agent platforms. The derivation of agent behaviors from the process diagrams is highly dependent on how behaviors are implemented for a given agent framework, e.g. imperative or declarative, as a set of rules, goal-oriented, or as hybrid approaches. Generally speaking, for each pool in the different process diagrams, or more precisely, for each start event in the pools, one behavior is derived, covering the workflow (branching, communication, etc.) from that starting point on. For instance, a pool with a message start event can be translated to a script for the content of the workflow and a rule for starting that script when the given message arrives as shown in Fig.1.

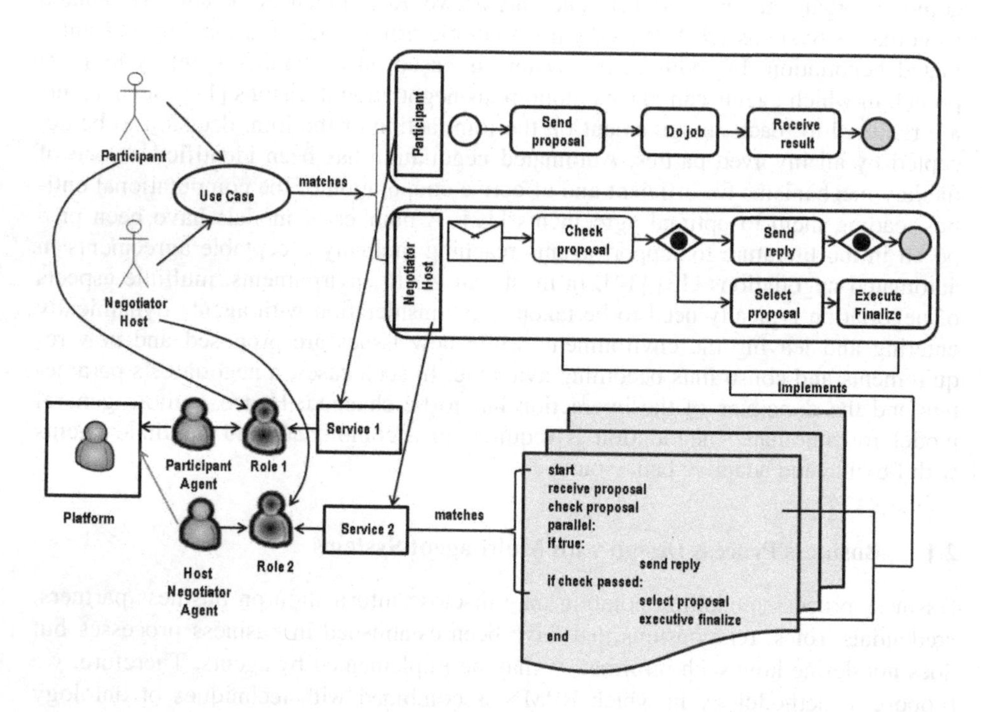

Fig. 1. Overview of models and their dependencies. Use Case, Negotiation Process Model, Negotiation Role Model, Implementation.

2.2 Abstract Negotiation Architecture with MAS

The general architecture of the negotiation system offers mechanisms to support negotiations in distributed environments. This architecture is structured in four main layers: Network of Hosts, Agent Oriented Platform, General Negotiation Protocol, and Negotiation Coordination Components consisted of Negotiation Rules, Negotiation Proposals, and Negotiation Finalization. The General Negotiation Protocol is obtained through ontology engineering, as well as the Negotiation Coordination Components that are presented in Fig. 2.

The Negotiation Coordination Components manage the coordination constraints among different negotiations which take place simultaneously. Such service components ensure that the communication process is shared by all negotiation partners. The Negotiation Coordination Components level allows the implementation of various types of appropriate behaviour in particular cases of negotiation, described through negotiation ontologies. Consequently, each component is considered to correspond to a particular negotiation type. Thus, different negotiation scenarios that are found in the lifecycles of the business to business negotiations can be modelled. The negotiation components have been described in detail in [6], [8].

The proposed infrastructure manages, in a decentralized manner, the coordination of multi-phase negotiations on a multi-attribute object and among a lot of participants.

In this approach, intelligent agents are regarded as computer systems situated in an agent oriented platform, being capable of autonomous action in the environment, in order to meet their design objectives [15]. The proposed model is an agent federation that is built on the main attributes of the agents, namely: autonomy, social ability, reactivity and pro-activeness. Agent federations share the common characteristic of a group of agents which have ceded some amount of autonomy to a single delegate which represents the group [16]. The Negotiator Host is the delegate and it is a distinguished agent member of the group, namely the Chief Negotiator. Other group members (Agent 01, Agent n) called Intermediary Agents interact with the Chief Negotiator developing various relationships and playing specific roles according to specified rules.

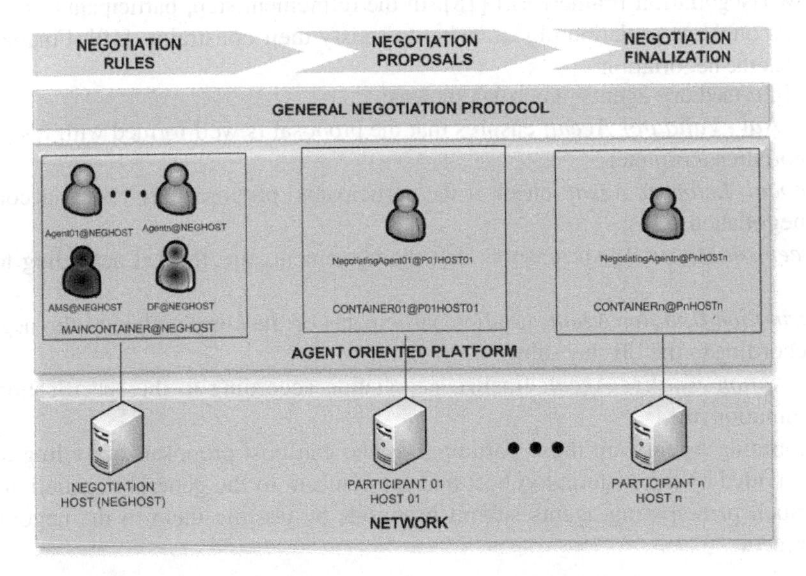

Fig. 2. Abstract architecture for automated negotiation with MAS

The Negotiation Host group is provided with a single, consistent interface. The Negotiator Host Agent is responsible for the creation and enforcement of rules governing participation, execution, resolution and finalization of a negotiation. Other group member agents, namely the Intermediary Agents receive messages from the Negotiator Host that may include skill descriptions, task requirements, status information and application-level data. Such information is communicated using general, declarative communication language which the team of agents and the Negotiator Host Agent understand. Implicit in this arrangement is that, while the intermediary agent must be able to interact with its local federation members (enterprise multi-layered MAS) [6], individual normal agents do not require a common language as they never directly interact. Such arrangement is particularly useful for integrating legacy or an otherwise heterogeneous group of agents. The intermediary agents from the Negotiation Host MAS assist the negotiations at a global level (negotiations with different participants on different jobs) and at a specific level (negotiation on the same job with different participants) handling all issues regarding its communication (synchronization among the Negotiating Agents of the several parties that are taking place in the negotiation process). During the negotiation the decisions are taken by the Negotiator Host, assisted by his Intermediary Team. For each negotiation, an Intermediary Agent manages one or more negotiation objects, one framework and the negotiation status represented as several graph structures. The Negotiator Host can specify some global parameters: duration; maximum number of messages to be exchanged; maximum number of candidates to be considered in the negotiation and involved in the business process; tactics; protocols for Intermediary Agent interactions. Each negotiation is organized in three main steps: initialization; refinement of the job under negotiation; and finalization [17]. The initialization step allows to define what has to be negotiated (Negotiation Object) and how (Negotiation Framework) [18]. In the refinement step, participants exchange proposals on the negotiation object trying to satisfy their constraints [19]. Finalization concludes the negotiation.

The Intermediary agents play roles such as:

Proposal - Validator Agent: ensures that the proposal is well formed with respect to the negotiation template;

Protocol- Enforcer Agent: check if the participants' proposals are posted according to the negotiation rules;

Agreement Maker Agent: ensures that the agreements are formed according to the rules

Information Updater Agent: notifies participants of the current state of the negotiation, according to the display rules.

Negotiation Finalize Agent: finalize negotiation according to the specifications of the termination rules.

Negotiating Agents are those participants who can post proposals according to the rules provided by the negotiation host and are conform to the general negotiation protocol. Such participating agents submit proposals by posting them to the negotiation environment.

2.3 Negotiation Process Model

The negotiation process is modelled using XPLORE, a middleware infrastructure for negotiation components [20]. According to Brandl et al. [20], each negotiation in

XPLORE is modelled as the collaborative construction of a negotiation graph among the negotiation participants. Each component has its own (partial) copy of the negotiation graph and expresses negotiation decisions manipulating that copy. For instance, a proposal made by the Negotiator Host Agent who initiated the negotiation is represented in the graph copy visible by the Outsrc component and the proposals made by possible Negotiating Agents (NA1, NA2) are represented in the graph copies visible through their Insrc components. In this way, the evolution of a negotiation in terms of proposals and counter-proposals is modelled by a bicolour graph in which white nodes represent *negotiation contexts*, and black nodes, represent *decision points* with multiple alternatives. Each context (white) node contains a parameter and a set of attributes with associated values. Parameter is the task to be negotiated (Negotiation Object) which is described in a time moment by a set of attributes that have to be negotiated depending on the specific information about the state of the negotiation at that node.

In the proposed study case, the unique parameter of the negotiation is a *construction job* and an attribute can be the *price* that can assume a range of possible values. Different branches of negotiation can be created in the negotiation graph, at the initiative either of the Negotiator Host Agent or the possible Negotiating Agents to explore alternatives (e.g., price under 100000 or over 100000), as shown in Fig. 3. The partners may then create new branches specifying different construction times (e.g., price over 100000 and construction time is greater than five months, or under 100000 and the construction time is less or equal to five months). The interaction specifying the construction time would occur in the context of one or the other branch created by the interaction concerning the price.

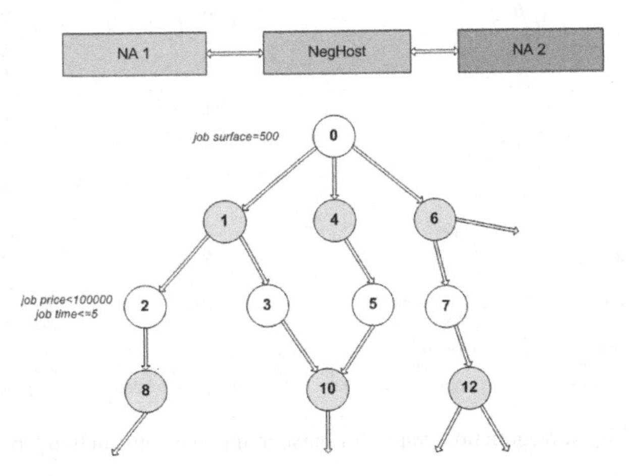

Fig. 3. Example of a negotiation graph with Negotiator Host and two Negotiating Agents

The purpose of the XPLORE Middleware is to allow the synchronization of different copies. Thus, the partners do not communicate directly but only via the operations they perform on their negotiation graph copies. These operations are calling the verbs of the XPLORE protocol [19]. For instance, the Negotiator Host Agent initiates the outsourcing of a construction job by communicating to the Information Updater Agent the attributes of the Negotiation Object to be used in the negotiation process with two negotiating agents, NA1 and NA2. Thus, the Negotiator Host Agent starts two bilateral negotiations with NA1 and NA2. These interactions correspond, at the negotiation component level, to the creation of two white nodes in the negotiation graph managed by the Outsrc component of NegHost and propagated as the creation of one white node in each negotiation graph managed by the Insrc components of NA1 and NA2 as in Fig. 4. Regarding the entire image of the negotiation graph, NA1 and NA2 have only a partial view of the complete negotiation graph (the one which is relative to the propositions sent to each of them by NegHost), while the initiator NegHost has a global view. When NA1 Negotiator and NA2 Negotiator receive these messages, they consult their own Manager within their own Enterprise Multilayer MAS and build the Negotiation Object and the Negotiation Framework that they will use to manage the negotiation with the initiator NegHost and generate different proposals. The proposals and the counter-proposals will be communicated through the Negotiation Component by opening nodes in the negotiation graphs of NegHost Outsrc component on one side, and in the negotiation graphs managed by the respective Insrc component of NA1 and NA2, on the other side. The negotiation goes on until NegHost decides to commit on the proposition made by NA1. It then closes the negotiation with an *select* act. This leads to the formulation of a contract with the complete instantiated Negotiation Object by the Middleware.

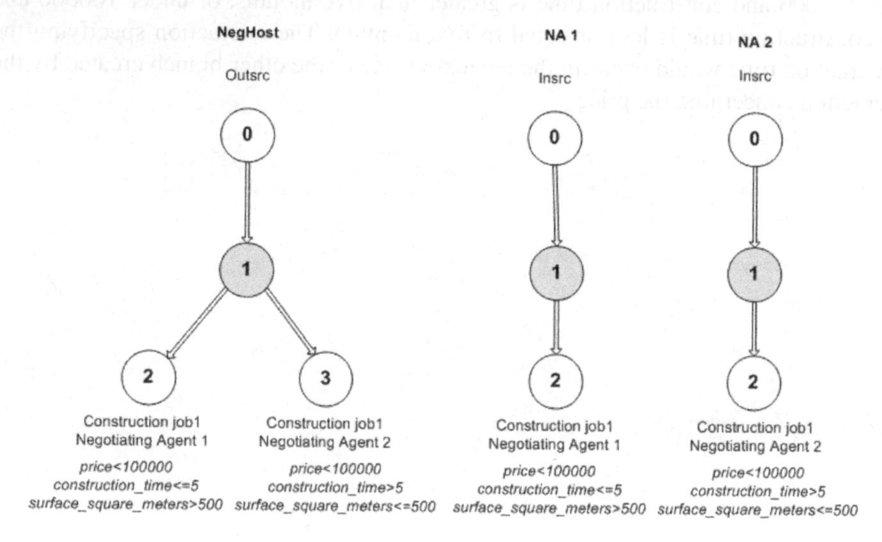

Fig. 4. Negotiation graphs for outsourcing of a construction job

3 Implementation of the Negotiation Process with MAS

3.1 Implementation of the Rule Based Negotiation

The experiment describes an online auction for a contract of road construction and uses rule-based negotiation. A rule-based system is a system that uses rules to derive conclusions from premises, where rules are IF-THEN statements. The IF part of a rule is called left-hand side (LHS), predicate, or premises; and the THEN part is the right hand side (RHS), actions, or conclusions [21].

The Negotiator Host decides to outsource a job for the Negotiating Agents. For this purpose, the Negotiator Host (NegHost) initiates the outsourcing by communicating to the Negotiator Agent 1 the attributes of the Negotiation Object and the Negotiation Framework. Further, the Negotiator Agents, the Coordination Components and the communication Middleware interact in order to establish a contract respecting the constraints defined by the NegHost.

For representing the "*select*" or "*reject*" proposals we used Java Expert System Shell templates [21]. The considered facts were the negotiation price, the construction time (in months) and the surface of the construction (in square meters). The negotiation template definition is presented below:

```
(deftemplate negotiation
  "Negotiation template"
  (slot price (type INTEGER))
  (slot construction_time_months (type INTEGER))
  (slot surface_square_meters (type INTEGER)))
```

Next, we asserted the facts with the values: 100000, 5 and 500. We introduced in the root of the negotiation graph the values set in the assert function. In this way we expressed the decision during the negotiation phase so that this was equal to the expressed property.

```
(assert negotiation (price 100000) (construction_time_months 5) (sur-
face_square_meters 500)))
```

The next rule fires if the negotiation price between the NegHost and one of its Negotiating Agents is greater than 100000 euro, and if the construction time is greater than five month and if the construction surface in less or equal to 500 square meters.

```
(defrule negotiation-protocol-reject
    "This is the negotiation protocol between NegHost and NA1, the
case: reject proposal"
    (negotiation {price > 100000})
    (negotiation {construction_time_months > 5})
    (negotiation {surface_square_meters <= 500})
    =>
    (printout t "Reject proposal" crlf))
```

If the price is less than 100000 euro, and the construction time is less or equal to 5 months and the construction surface is greater than 500 square meters, than the "select proposal" rule will be fired.

```
(defrule negotiation-protocol-select
    " This is the negotiation protocol between NegHost and NA1, the
case: select proposal"
    (negotiation {price < 100000})
    (negotiation {construction_time_months <= 5})
    (negotiation {surface_square_meters > 500})
    =>
    (printout t "Select proposal" crlf))
```

4 Results and Discussions

The automated negotiation environment was tested with agents that were designed with specific rule templates, where rules asserted information in their private fact bases. The agents responded to this information by sending messages according to the Send - Receive Protocol defined in the system. Once the system agent received the confirmation of starting a negotiation, an agent started to negotiate with his counterpart and automatically made decisions based on his negotiation parameters. The system also provided a functionality of viewing the communications between agents and the systems via a sniffer agent. Various types of rules and queries were used for representing valid, posted and active proposals, depending on the phase of the negotiation process.

Let us illustrate the above processes by a sample message exchange that occurs during a negotiation when the Negotiating Agent 1 encounters a technical problem and submit the proposal of restarting the construction in two months, proposal that is selected by the NegHost agent.

In this case, the message will have the following form:

```
(PROPOSE
:sender(agent-identifier :name na1@192.168.1.14:8888/JADE)
:receiver(set  agent-identifier :name negost@192.168.1.14:8888/JADE))
:content  "I propose to restart the construction in two months."
:X-JADE-real-sender rma@192.168.1.14:8888/JADE )
```

The NegHost agent analyses the proposal and decide to select it:

```
(SELECT-PROPOSAL
:sender(agent-identifier :name neghost@192.168.1.14:8888/JADE)
:receiver (set(agent-identifier :name na1@192.168.1.14:8888/JADE ))
:content  "I select your proposal."
:X-JADE-real-sender rma@192.168.1.14:8888/JADE)
```

As a result, the NegHost agent has to inform the other Negotiating Agents about the problem and negotiate with them the time needed in order to restart the activity.

```
(PROPOSE
:sender(agent-identifier :name neghost@192.168.1.14:8888/JADE)
:receiver(set(agent-identifier :name na2@192.168.1.14:8888/JADE))
:content  "I propose to restart the activity in two months."
:X-JADE-real-sender rma@192.168.1.14:8888/JADE )
(SELECT-PROPOSAL
:sender(agent-identifier :name na2@192.168.1.14:8888/JADE)
:receiver(set(agent-identifier :name neghost@192.168.1.14:8888/JADE))
:content  "I select your proposal."
:X-JADE-real-sender rma@192.168.1.14:8888/JADE)
(PROPOSE
:sender(agent-identifier :name neghost@192.168.1.14:8888/JADE)
:receiver(set( agent-identifier :name na3@192.168.1.14:8888/JADE))
:content  "I propose to restart the activity in two months."
:X-JADE-real-sender rma@192.168.1.14:8888/JADE )
```

```
(REJECT-PROPOSAL
 :sender(agent-identifier :name na3@192.168.1.14:8888/JADE)
 :receiver(set(agent-identifier :name neghost@192.168.1.14:8888/JADE))
 :content "In this case, I can restart the activity only in three
months."
 :X-JADE-real-sender rma@192.168.1.14:8888/JADE )
```

As one can see, the third Negotiating Agent considers that needs three months to restart the construction, and this proposal is rejected by the NegHost.

If the evaluated offer had a greater price than the maximum price considered, it was retracted from the fact base and a message was displayed.

The negotiation protocol was parameterised with specific rules taking into account the sustainable interoperability between agents.

The automated negotiation process has been achieved according to the designed workflow. Negotiation time is assessed and calculated for the entire process in various negotiation conditions for further optimisations. Here we have illustrated the negotiation process between the negotiator host and several participants that has been implemented in Java Agent Development Framework (JADE).

5 Conclusions and Future Work

This paper proposes a multi-agent system that adopts a rule-based approach to implement automated negotiations that support decision making in business processes. The developed intelligent system integrates responses from the collaborative networked environment and actively contributes to the negotiation process of various jobs and tasks in order to reach consensus at business level.

The main goal and originality of this work is represented by the modelling and implementation process of intelligent software components that support effective communication, coordination and collaboration at the enterprise's business level.

Future work outlooks: i) complete integration of the rule-based framework into the abstract negotiation architecture; ii) allowance of the logical specification of the rules in order to asses their correctness; iii) investigation of the e-effectiveness by system validation in several collaborative networked environments.

References

1. Luck, M., McBurney, P., Shehory, O., Willmott, S.: Agent Technology: Computing as Interaction (A Roadmap for Agent Based Computing). AgentLink (2005)
2. Endert, H., Hirsch, B., Küster, T., Albayrak, Ş.: Towards a Mapping from BPMN to Agents. In: Huang, J., Kowalczyk, R., Maamar, Z., Martin, D., Müller, I., Stoutenburg, S., Sycara, K. (eds.) SOCASE 2007. LNCS, vol. 4504, pp. 92–106. Springer, Heidelberg (2007)
3. Ahmad Decker, G., Puhlmann, F.: Extending BPMN for modeling complex choreographies. In: Proceedings of the COOPIS 2007, Vilamoura, Portugal, pp. 24–40 (2007)

4. Object Management Group: Business Process Modeling Notation, V1.1. Final Adopted Speci_cation formal/2008-01-17, OMG (January 2008), http://www.omg.rg/spec/BPMN/1.1/PDF
5. Endert, H.: Küster, T., Hirsch, B., Albayrak, S.: Mapping BPMN to agents: An analysis. In: Baldoni, M., Baroglio, C., Mascardi, V. (eds.) Agents, Web-Services, and Ontologies Integrated Methodologies, pp. 43–58 (2007)
6. Kadar, M., Cretan, A., Muntean, M., Jardim-Goncalves, R.: A Multi-Agent based Negotiation System for Re-establishing Enterprise Interoperability in Collaborative Networked Environments. In: Proceedings of 2013 UKSim 15th International Conference on Computer Modelling and Simulation, April 10-12, vol. 1, pp. 190–195. Cambridge University (2013) ISBN 978-0-7695-4994-1
7. Kadar, M., Muntean, M.: Intelligent and Collaborative Multi-Agent System to Generate Automated Negotiation for Sustainable Enterprise Interoperability. In: Jezic, G., Kusek, M., Lovrek, I., Howlett, R.J., Jain, L.C. (eds.) Agent and Multi-Agent Ssytems: Technologies and Applications, Chania, Greece. AISC, vol. 296, pp. 213–222. Springer, Heidelberg (2014)
8. Cretan, A., Coutinho, C., Bratu, B., Jardim-Goncalves, R.: A Framework for Sustainable Interoperability of Negotiation Processes. In: The 14th IFAC Symposium on Information Control Problems in Manufacturing (INCOM 2012), vol. 14(1), pp. 1258–1263. Elsevier (May 2012) ISBN: 978-3-902661-98-2, doi: 10.3182/20120523-3-RO-2023.00240
9. Kaddouci, A., Zgaya, H., Hammadi, S., Bretaudeau, F.: Multi-agents Based Protocols for Negotiation in a Crisis Management Supply Chain. In: 8th WSEAS International Conference on Computational Intelligence, Man-Machine Systems and Cybernetics (CIMMACS 2009), pp. 143–150 (2009)
10. Ahmadi, K.D., Charkari, N.M.: Multi Agent Based Interactive Recommendation and Automated Negotiation System in E-Commerce. In: 6th International Conference on Advanced Information Management and Service (IMS), pp. 279–284 (2010)
11. Ahmadi, K.D., Charkari, N.M., Enami, N.: E-Negotiation System Based on Intelligent Agents in B2C E-Commerce. Advances on Information Sciences and Service Sciences 3(2), 60–70 (2011)
12. Jennings, N.R., Faratin, P., Lomuscio, A.R., Parsons, S., Wooldridge, M., Sierra, C., et al.: Automated negotiation: Prospects, methods and challenges. International Journal of Group Decision and Negotiation 10(2), 199–215 (2001)
13. Fatima, S., Wooldridge, M., Jennings, N.R.: An agenda-based framework for multi-issue negotiation. Artificial Intelligence 152(1), 1–45 (2004)
14. Kowalczyk, R.: Fuzzy e-negotiation agents. Soft Computing 6(5), 337–347 (2002)
15. Wooldridge, M., Jennings, N.R.: Intelligent Agents: Theory and practice. The Knowledge Engineering review 10(2), 115–152 (1995)
16. Sycara, K., Dai, T.: Agent Reasoning in Negotiation. In: Handbook of Group Decision and Negotiation. Advances in Group Decision and Negotiation, Part 4, vol. 4, pp. 437–451 (2010)
17. Duan, L., Dogru, M., Ozen, U., Beck, J.: A negotiation framework for linked combinatorial optimization problems. Autonomous Agents and Multi-Agent Systems 25(1), 158–182 (2012)
18. Hu, J., Deng, L.: An Association Rule-Based Bilateral Multi-Issue Negotiation Model. In: Fourth International Symposium on Computational Intelligence and Design (ISCID), vol. 2, pp. 234–237 (2011)

19. Coutinho, C., Cretan, A., Jardim-Goncalves, R.: Negotiations framework for monitoring the sustainability of interoperability solutions. In: van Sinderen, M., Johnson, P., Xu, X., Doumeingts, G. (eds.) IWEI 2012. Lecture Notes in Business Information Processing, vol. 122, pp. 172–184. Springer, Heidelberg (2012)
20. Brandl, R., Andreoli, J.-M., Castellani, S.: Ubiquitous negotiation games: a case study. In: Proc. of DEXA "e-negotiations" Workshop, Prague, September 1-5 (2003)
21. Friedman-Hill, E.: Jess in Action - Rule-Based Systems in Java, Manning Publica-tions Co., ISBN 1-930110-89-8 (2003)

Part VIII
Issues in Granular Computing

Extended Index Matrices

Krassimir T. Atanassov

Dept. of Bioinformatics and Mathematical Modelling,
Institute of Biophysics and Biomedical Engineering,
Bulgarian Academy of Sciences,
105 Acad. G. Bonchev Str., 1113 Sofia, Bulgaria
and Intelligent Systems Laboratory,
Prof. Asen Zlatarov University,
1 Prof. Yakimov Blvd., 8010 Bourgas, Bulgaria
krat@bas.bg

Abstract. In this paper, an extension of the concept of an index matrix is introduced and some of its basic properties are studied. Operations, relations and operators are defined over the new object.

Keywords: Index matrix, Matrix, Operation, Relation.

1 Introduction

The concept of an Index Matrix (IM) was introduced in 1984 in [1,2]. During the next 25 years some of its properties were studied, but in general it was only used as an auxiliary tool for describing of other mathematical objects.

In the present paper, we introduce a new modification of the concept of an IM, that, probably, it will be the most general.

2 Definition of an Extended Index Matrix

We start with the basic definition of the concept of an IM in the extended case. After this, we discuss different particulal cases related to the forms of the IM-elements, that can be real numbers, elements of set $\{0, 1\}$, logical variables or predicates, intuitionistic fuzzy pair, functions and others. For each separate case, the basic definition will keep its form, but the forms of the operations and in some cases – of relations and operators over the new type of matrices, will be changed.

Let $\mathcal{I}$ be a fixed set of indices, $\mathcal{I}^n = \{\langle i_1, i_2, ..., i_n\rangle | (\forall j : 1 \leq j \leq n)(i_j \in \mathcal{I})\}$ and

$$\mathcal{I}^* = \bigcup_{1 \leq n \leq \infty} \mathcal{I}^n.$$

Let $\mathcal{X}$ be a fixed set of some objects. In the particular cases, they can be either real numbers, or only the numbers 0 or 1, or logical variables, propositions or predicates, etc.

P. Angelov et al. (eds.), *Intelligent Systems'2014*,
Advances in Intelligent Systems and Computing 322, DOI: 10.1007/978-3-319-11313-5_28

Let operations $\circ, * : \mathcal{X} \times \mathcal{X} \rightarrow \mathcal{X}$ be fixed.

An Extended IM (EIM) with index sets K and L ($K, L \subset \mathcal{I}^*$) and elements from set $\mathcal{X}$ is called the object:

$$[K, L, \{a_{k_i,l_j}\}] \equiv \begin{array}{c|ccccc} & l_1 & \dots & l_j & \dots & l_n \\ \hline k_1 & a_{k_1,l_1} & \dots & a_{k_1,l_j} & \dots & a_{k_1,l_n} \\ \vdots & \vdots & \dots & \vdots & \dots & \vdots \\ k_i & a_{k_i,l_1} & \dots & a_{k_i,l_j} & \dots & a_{k_i,l_n} \\ \vdots & \vdots & \dots & \vdots & \dots & \vdots \\ k_m & a_{k_m,l_1} & \dots & a_{k_m,l_j} & \dots & a_{k_m,l_n} \end{array},$$

where $K = \{k_1, k_2, ..., k_m\}$, $L = \{l_1, l_2, ..., l_n\}$, and for $1 \leq i \leq m$, and $1 \leq j \leq n : a_{k_i,l_j} \in \mathcal{X}$.

3 Operations over Extended Index Matrices

Let in this section, the sets $\mathcal{X}, \mathcal{Y}, \mathcal{Z}, \mathcal{U}$ be fixed. Let operations "$*$" and "$\circ$" be defined so that $* : \mathcal{X} \times \mathcal{Y} \rightarrow \mathcal{Z}$ and $\circ : \mathcal{Z} \times \mathcal{Z} \rightarrow \mathcal{U}$.

For the EIMs $A = [K, L, \{a_{k_i,l_j}\}]$ with $a_{k_i,l_j} \in \mathcal{X}$, $B = [P, Q, \{b_{p_r,q_s}\}]$ with $b_{p_r,q_s} \in \mathcal{Y}$, and $K, L, P, Q \subset \mathcal{I}$, operations that are analogous of the usual matrix operations of addition and multiplication and of IM-operations, are defined.

(1) Addition

$$A \oplus_{(\circ)} B = [K \cup P, L \cup Q, \{c_{t_u,v_w}\}],$$

where

$$c_{t_u,v_w} = \begin{cases} a_{k_i,l_j}, & \text{if } t_u = k_i \in K \text{ and } v_w = l_j \in L - Q \\ & \text{or } t_u = k_i \in K - P \text{ and } v_w = l_j \in L; \\ b_{p_r,q_s}, & \text{if } t_u = p_r \in P \text{ and } v_w = q_s \in Q - L \\ & \text{or } t_u = p_r \in P - K \text{ and } v_w = q_s \in Q; \\ a_{k_i,l_j} \circ b_{p_r,q_s}, & \text{if } t_u = k_i = p_r \in K \cap P \\ & \text{and } v_w = l_j = q_s \in L \cap Q \\ 0, & \text{otherwise} \end{cases}$$

We see that

- in the case of standard IM, $\mathcal{X} = \mathcal{Y} = \mathcal{R}$, where here and below, $\mathcal{R}$ is the set of the real numbers, operation "$*$" is the standard operation "$+$" or "$\times$" and obviously, $\mathcal{Z} = \mathcal{R}$ (cf. [4]);
- when $\mathcal{X} = \mathcal{Y} = \{0, 1\}$, operation "$*$" is "max" or "min", and $\mathcal{Z} = \mathcal{X}$ (cf. [4]);
- when $\mathcal{X} = \mathcal{Y}$ is a set of logical variables, sentences or predicates, then "$*$" is "$\vee$" or "$\wedge$" and $\mathcal{Z} = \mathcal{X}$ (cf. [4]);

- when $\mathcal{X} = \mathcal{Y} = \mathcal{L}^* \equiv \{\langle a, b\rangle | a, b, a + b \in [0, 1]\}$, then $\mathcal{Z} = \mathcal{X}$ and operation "$*$" is defined for the intuitionistic fuzzy pairs (see, [3,6,11])$\langle a, b\rangle$ and $\langle c, d\rangle$, by

$$\langle a, b\rangle * \langle c, d\rangle = \langle \max(a, c), \min(b, d)\rangle \tag{1}$$

or

$$\langle a, b\rangle * \langle c, d\rangle = \langle \min(a, c), \max(b, d)\rangle \tag{2}$$

(cf. [5]);
- when $\mathcal{X}$ is a set of real numbers and

$$\mathcal{Y} = \{f | f \text{ is of one-argument (let it always be } x) \text{ function and } f : \mathcal{X} \to \mathcal{X}\}, \tag{3}$$

then $\alpha * f = \alpha f$, i.e. $\mathcal{Z} = \mathcal{Y}$, but when

$$\mathcal{X} = \{f | f \text{ is of one-argument (let it always be } x) \text{ function and } f : \mathcal{Y} \to \mathcal{Y}\} \tag{4}$$

and Y – of real numberds, then $f * \alpha = f(\alpha)$, i.e. $\mathcal{Z} = \mathcal{X}$ (cf. [9]).

(2) Termwise multiplication

$$A \otimes_{(\circ)} B = [K \cap P, L \cap Q, \{c_{t_u, v_w}\}],$$

where $c_{t_u,v_w} = a_{k_i,l_j} \circ b_{p_r,q_s}$, for $t_u = k_i = p_r \in K \cap P$ and $v_w = l_j = q_s \in L \cap Q$.

The above discussion for the form of sets X, Y and Z here is the same. The basic difference between the first two operations is in the form of the indices of the resultant IM.

(3) Multiplication

$$A \odot_{(\circ,*)} B = [K \cup (P - L), Q \cup (L - P), \{c_{t_u, v_w}\}],$$

where

$$c_{t_u,v_w} = \begin{cases} a_{k_i,l_j}, & \text{if } t_u = k_i \in K \text{ and } v_w = l_j \in L - P - Q \\ b_{p_r,q_s}, & \text{if } t_u = p_r \in P - L - K \text{ and } v_w = q_s \in Q \\ \underset{l_j = p_r \in L \cap P}{\circ} a_{k_i,l_j} * b_{p_r,q_s}, & \text{if } t_u = k_i \in K \text{ and } v_w = q_s \in Q \\ 0, & \text{otherwise} \end{cases}.$$

Obviously, when $\circ$ is $+$, the symbol $\underset{j}{\circ}$ will coincide with $\sum_j$. We see immediately that

$$A^{2(\circ,*)} = A \odot_{(\circ,*)} A = [K \cup (K - L), L \cup (L - K), \{c_{t_u,v_w}\}] = [K, L, \{c_{t_u,v_w}\}]$$

where

$$c_{t_u,v_w} = \underset{l_j=p_r \in L\cap P}{\circ} a_{k_i,l_j} * b_{p_r,q_s}$$

and for $n \geq 2$:

$$A^{n+1_{(\circ,*)}} = A^{n_{(\circ,*)}} \odot_{(\circ,*)} A.$$

Now,

- in the case of standard IM, $\mathcal{X} = \mathcal{Y} = \mathcal{R}$, operation "$*$" is the standard operation "+" and operation "$\circ$" – standard operation ".", obviously, $\mathcal{Z} = \mathcal{R}$;
- when $\mathcal{X} = \mathcal{Y} = \{0,1\}$, operation "$*$" is "max" and "$\circ$" – "min", or opposite, "$*$" is "min" and "$\circ$" – "max", and $\mathcal{Z} = \mathcal{X}$;
- when $\mathcal{X} = \mathcal{Y}$ are a set of logical variables, sentences or predicates, then "$*$" is "$\vee$" and "$\circ$" – "$\wedge$", or vice versa, "$*$" is "$\wedge$" and "$\circ$" – "$\vee$", and $\mathcal{Z} = \mathcal{X}$;
- when $\mathcal{X} = \mathcal{Y} = \mathcal{L}^*$, then $\mathcal{Z} = \mathcal{X}$ and operation $*$ is defined for the intuitionistic fuzzy pairs $\langle a, b\rangle$ and $\langle c, d\rangle$, by (1) or (2);
- when at least one of the IMs has function-type components, this operation is not defined.

(4) Structural subtraction

$$A \ominus B = [K - P, L - Q, \{c_{t_u,v_w}\}],$$

where "–" is the set–theoretic difference operation and $c_{t_u,v_w} = a_{k_i,l_j}$, for $t_u = k_i \in K - P$ and $v_w = l_j \in L - Q$.

This operation is valid for each of the above discussed cases.

(5) Multiplication with a constant α

$$\alpha A = [K, L, \{\alpha a_{k_i,l_j}\}],$$

where α is a constant. Now,

- In the case of standard IM, i.e., $\mathcal{X} = \mathcal{R}$, $\alpha \in \mathcal{R}$;
- if $\mathcal{X} = \{0,1\}$, then $\alpha \in \{0,1\}$;
- when $\mathcal{X} = \mathcal{Y}$ is a set of logical variables, sentences or predicates, then α has sense only when it is an operation "negation";
- when $\mathcal{X} = \mathcal{Y} = \mathcal{L}^*$, then α has sense only when it is one of the following operations "negation", defined over IF pairs in [11].

(6) Termwise subtraction

$$A -_{(+)} B = A \oplus_{(+)} (-1)B.$$

For this operation,

- in the case of standard IM, $\mathcal{X} = \mathcal{Y} = \mathcal{R}$, operation "$*$" is the standard operatin "+" or "$\times$" and obviously, $\mathcal{Z} = \mathcal{R}$;
- when $\mathcal{X} = \mathcal{Y} = \{0,1\}$, operation "$*$" is "max" or "min", and $\mathcal{Z} = \mathcal{X}$;

- when $\mathcal{X} = \mathcal{Y}$ are a set of logical variables, sentences or predicates, then "$*$" is "$\vee$" or "$\wedge$" and $\mathcal{Z} = \mathcal{X}$;
- when $\mathcal{X} = \mathcal{Y} = \mathcal{L}^*$, then $\mathcal{Z} = \mathcal{X}$ and operation "$*$"' is defined for the IFPs $\langle a, b\rangle$ and $\langle c, d\rangle$.

In some cases, when we use $\mathcal{X} = \mathcal{Y} = \mathcal{R}$, it is suitable to define operation "$-$" by $A - B = [K \cup P, L \cup Q, \{c_{t_u,v_w}\}]$, where

$$c_{t_u,v_w} = \begin{cases} a_{k_i,l_j}, & \text{if } t_u = k_i \in K \text{ and } v_w = l_j \in L - Q \\ & \text{or } t_u = k_i \in K - P \text{ and } v_w = l_j \in L; \\ \max(0, a_{k_i,l_j} - b_{p_r,q_s}), & \text{if } t_u = k_i = p_r \in K \cap P \\ & \text{and } v_w = l_j = q_s \in L \cap Q \\ 0, & \text{otherwise} \end{cases}.$$

If $\circ : \mathcal{X} \times ... \times \mathcal{X} \to \mathcal{X}$, the aggregation operations have the forms

(7) $\circ$-row-aggregation

$$\rho_\circ(A, k_0) = \begin{array}{c|cccc} & l_1 & l_2 & \dots & l_n \\ \hline k_0 & \overset{m}{\underset{i=1}{\circ}} a_{k_i,l_1} & \overset{m}{\underset{i=1}{\circ}} a_{k_i,l_2} & \dots & \overset{m}{\underset{i=1}{\circ}} a_{k_i,l_n} \end{array},$$

(8) $\circ$-column-aggregation

$$\sigma_\circ(A, l_0) = \begin{array}{c|c} & l_0 \\ \hline k_1 & \overset{n}{\underset{j=1}{\circ}} a_{k_1,l_j} \\ k_2 & \overset{n}{\underset{j=1}{\circ}} a_{k_2,l_j} \\ \vdots & \\ k_m & \overset{n}{\underset{j=1}{\circ}} a_{k_m,l_j} \end{array},$$

where, as above, e.g., when $\circ$ is $+$, the symbol $\overset{m}{\underset{j=1}{\circ}}$ will coincide with $\sum_{j=1}^{m}$. Now,

- in the case of standard IM, $\mathcal{X} = \mathcal{Y} = \mathcal{R}$, operation "$*$" is "max", or "min", or "$\sum$", or "$\frac{1}{m}\sum$" and $\mathcal{Z} = \mathcal{R}$;
- when $\mathcal{X} = \mathcal{Y} = \{0, 1\}$, operation "$*$" is "max" or "min", and $\mathcal{Z} = \mathcal{X}$;
- when $\mathcal{X} = \mathcal{Y}$ is a set of logical variables, sentences or predicates, then "$*$" is "$\vee$" or "$\wedge$" and $\mathcal{Z} = \mathcal{X}$;
- when $\mathcal{X} = \mathcal{Y} = \mathcal{L}^*$, then $\mathcal{Z} = \mathcal{X}$ and operation "$*$" is defined for IFPs.

4 Relations over IMs

Let the two IMs $A = [K, L, \{a_{k,l}\}]$ and $B = [P, Q, \{b_{p,q}\}]$ be given, where $a_{k,l} \in \mathcal{X}$, $b_{p,q} \in \mathcal{Y}$, and $K, L, P, Q \subset \mathcal{I}$. Let R_s and R_n be a strict and a non-strict relations over $\mathcal{X} \times \mathcal{Y}$, respectively.

We shall introduce the following definitions, where $\subset$ and $\subseteq$ denote the relations *"strong inclusion"* and *"weak inclusion"*.

Definition 1: The strict relation "inclusion about dimension" is

$$A \subset_d B \text{ iff } (((K \subset P)\&(L \subset Q)) \vee ((K \subseteq P)\&(L \subset Q))$$
$$\vee((K \subset P)\&(L \subseteq Q))) \ \& \ (\forall k \in K)(\forall l \in L)(a_{k,l} = b_{k,l}).$$

Definition 2: The non-strict relation "inclusion about dimension" is

$$A \subseteq_d B \text{ iff } (K \subseteq P)\&(L \subseteq Q)\&(\forall k \in K)(\forall l \in L)(a_{k,l} = b_{k,l}).$$

Definition 3: The strict relation "inclusion about value" is

$$A \subset_v B \text{ iff } (K = P)\&(L = Q)\&(\forall k \in K)(\forall l \in L)(R_s(a_{k,l}, b_{k,l})).$$

Definition 4: The non-strict relation "inclusion about value" is

$$A \subseteq_v B \text{ iff } (K = P)\&(L = Q)\&(\forall k \in K)(\forall l \in L)(R_n(a_{k,l}, b_{k,l})).$$

Definition 5: The strict relation "inclusion" is

$$A \subset B \text{ iff } (((K \subset P) \ \& \ (L \subset Q)) \ \vee \ ((K \subseteq P) \ \& \ (L \subset Q))$$
$$\vee \ ((K \subset P) \ \& \ (L \subseteq Q))) \ \& \ (\forall k \in K)(\forall l \in L)(R_s(a_{k,l}, b_{k,l})).$$

Definition 6: The non-strict relation "inclusion" is

$$A \subseteq B \text{ iff } (K \subseteq P)\&(L \subseteq Q)\&(\forall k \in K)(\forall l \in L)(R_n(a_{k,l}, b_{k,l})).$$

It can be directly seen that for every two IMs A and B,

- if $A \subset_d B$, then $A \subseteq_d B$;
- if $A \subset_v B$, then $A \subseteq_v B$;
- if $A \subset B$, $A \subseteq_d B$, or $A \subseteq_v B$, then $A \subseteq B$;
- if $A \subset_d B$ or $A \subset_v B$, then $A \subseteq B$.

Operations "reduction", "projection" and "substitution" are the same, as in [4].

5 Hierarchical Operators over EIMs

In [4,7], two hierarchical operators are defined. They are applicable on EIM, when their elements are not only numbers, variables, etc, but when they also can be whole IMs.

Let A be an ordinary IM and let its element a_{k_f,e_g} be an IM by itself:

$$a_{k_f,l_g} = [P, Q, \{b_{p_r,q_s}\}],$$

where $K \cap P = L \cap Q = \emptyset$.

Here, we will introduce the first hierarchical operator:

$$A|(a_{k_f,l_g}) = [(K - \{k_f\}) \cup P, (L - \{l_g\}) \cup Q, \{c_{t_u,v_w}\}],$$

where

$$c_{t_u,v_w} = \begin{cases} a_{k_i,l_j}, \text{ if } t_u = k_i \in K - \{k_f\} \text{ and } v_w = l_j \in L - \{l_g\} \\ b_{p_r,q_s}, \text{ if } t_u = p_r \in P \text{ and } v_w = q_s \in Q \\ 0, \quad \text{otherwise} \end{cases}.$$

Let us assume that in the case when a_{k_f,l_g} is not an element of IM A, then $A|(a_{k_f,l_g}) = A$.

Let for $i = 1, 2, ..., m$:

$$a^i_{k_{i,f},l_{i,g}} = [P_i, Q_i, \{b^i_{p_{i,r},q_{i,s}}\}],$$

where for every i, j $(1 \leq i < j \leq m)$:

$$P_i \cap P_j = Q_i \cap Q_j = \emptyset,$$

$$P_i \cap K = Q_i \cap L = \emptyset.$$

Then, for $k_{1,f}, k_{2,f}, ..., k_{m,f} \in K$ and $l_{1,g}, l_{2,g}, ..., l_{m,g} \in L$:

$$A|(a^1_{k_{1,f},l_{1,g}}, a^2_{k_{2,f},l_{2,g}}, ..., a^m_{k_{m,f},l_{m,g}})$$

$$= (...((A|(a^1_{k_{1,f},l_{1,g}}))|(a^2_{k_{2,f},l_{2,g}}))...)|(a^m_{k_{m,f},l_{m,g}}).$$

Theorem 1. Let the IM A be given and let for $i = 1, 2$: $k_{1,f} \neq k_{2,f}$ and $l_{1,g} \neq l_{2,g}$ and

$$a^i_{k_{i,f},l_{i,g}} = [P_i, Q_i, \{b^i_{p_{i,r},q_{i,s}}\}],$$

where

$$P_1 \cap P_2 = Q_1 \cap Q_2 = \emptyset,$$

$$P_i \cap K = Q_i \cap L = \emptyset.$$

Then,

$$A|(a^1_{k_{1,f},l_{1,g}}, a^2_{k_{2,f},l_{2,g}}) = A|(a^2_{k_{2,f},l_{2,g}}, a^1_{k_{1,f},l_{1,g}}).$$

As it is mentioned in [7], the condition $k_{1,f} \neq k_{2,f}$ and $l_{1,g} \neq l_{2,g}$ is important (it was omitted in [4]), because if we have two elements of a given IM, that are IMs and that belong to one row or column, this will generate problems.

Let A and a_{k_f,l_g} be as above, let b_{m_d,n_e} be the element of the IM a_{k_f,l_g}, and let $b_{m_d,n_e} = [R, S, \{c_{t_u,v_w}\}]$, where

$$K \cap R = L \cap S = P \cap R = Q \cap S = K \cap P = L \cap Q = \emptyset.$$

Then,

$$(A|(a_{k_f,l_g}))|(b_{m_d,n_e})$$
$$= [(K - \{k_f\}) \cup (P - \{m_d\}) \cup R, (L - \{l_g\}) \cup (Q - \{n_e\} \cup S\,\{\alpha_{\beta_\gamma,\delta_\varepsilon}\}],$$

where

$$\alpha_{\beta_\gamma,\delta_\varepsilon} = \begin{cases} a_{k_i,l_j}, & \text{if } \beta_\gamma = k_i \in K - \{k_f\} \text{ and } \delta_\varepsilon = l_j \in L - \{l_g\} \\ b_{p_r,q_s}, & \text{if } \beta_\gamma = p_r \in P - \{m_d\} \text{ and } \delta_\varepsilon = q_s \in Q - \{n_e\} \\ c_{t_u,v_w}, & \text{if } \beta_\gamma = t_u \in R \text{ and } \delta_\varepsilon = v_w \in S \\ 0, & \text{otherwise} \end{cases}$$

Theorem 2. For the above IMs A, a_{k_f,l_g} and b_{m_d,n_e}

$$(A|(a_{k_f,l_g}))|(b_{m_d,n_e}) = A|((a_{k_f,l_g})|(b_{m_d,n_e})).$$

From the first definition of a hierarchical operator it follows that

$$A|(a_{k_f,l_g})$$

$$= \begin{array}{c|ccccccccc}
 & l_1 & \ldots & l_{g-1} & q_1 & \ldots & q_u & l_{g+1} & \ldots & l_n \\ \hline
k_1 & a_{k_1,l_1} & \ldots & a_{k_1,l_{g-1}} & 0 & \ldots & 0 & a_{k_1,l_{g+1}} & \ldots & a_{k_1,l_n} \\
\vdots & \vdots & \vdots & \vdots & \vdots & \vdots & \vdots & \vdots & \vdots & \vdots \\
k_{f-1} & a_{k_{f-1},l_1} & \ldots & a_{k_{f-1},l_{g-1}} & 0 & \ldots & 0 & a_{k_{f-1},l_{g+1}} & \ldots & a_{k_{f-1},l_n} \\
p_1 & 0 & \ldots & 0 & b_{p_1,q_1} & \ldots & b_{p_1,q_v} & 0 & \ldots & 0 \\
\vdots & \vdots & \vdots & \vdots & \vdots & \vdots & \vdots & \vdots & \vdots & \vdots \\
p_u & 0 & \ldots & 0 & b_{p_u,q_1} & \ldots & b_{p_u,q_v} & 0 & \ldots & 0 \\
k_{f+1} & a_{k_{f+1},l_1} & \ldots & a_{k_{f+1},l_{g-1}} & 0 & \ldots & 0 & a_{k_{f+1},l_{g+1}} & \ldots & a_{k_{f+1},l_n} \\
\vdots & \vdots & \vdots & \vdots & \vdots & \vdots & \vdots & \vdots & \vdots & \vdots \\
k_m & a_{k_m,l_1} & \ldots & a_{k_m,l_{g-1}} & 0 & \ldots & 0 & a_{k_m,l_{g+1}} & \ldots & a_{k_m,l_n}
\end{array}.$$

From this form of the IM $A|(a_{k_f,l_g})$ we see that for the hierarchical operator the following equality holds.

Theorem 3. Let $A = [K, L, \{a_{k_i,l_j}\}]$ be an IM and let $a_{k_f,l_g} = [P, Q, \{b_{p_r,q_s}\}]$ be its element. Then

$$A|(a_{k_f,l_g}) = (A \ominus [\{k_f\}, \{l_g\}, \{0\}]) \oplus a_{k_f,l_g}.$$

We see that the elements $a_{k_f,l_1}, a_{k_f,l_2}, ..., a_{k_f,l_{g-1}}, a_{k_f,l_{g+1}}, ..., a_{k_f,l_n}$ in the IM A now are replaced by "0". Therefore, as a result of this operator, information is being lost.

Below, we modify the first hierarchical operator, so that all the information from the IMs, participating in it, be kept. The new – second – form of this operator for the above defined IM A and its fixed element a_{k_f,l_g}, is

$$A|^*(a_{k_f,l_g})$$

$$= \begin{array}{c|cccccccc} & l_1 & \dots & l_{g-1} & q_1 & \dots & q_u & l_{g+1} & \dots & l_n \\ \hline k_1 & a_{k_1,l_1} & \dots & a_{k_1,l_{g-1}} & a_{k_1,l_g} & \dots & a_{k_1,l_g} & a_{k_1,l_{g+1}} & \dots & a_{k_1,l_n} \\ \vdots & \vdots & \vdots & \vdots & \vdots & \vdots & \vdots & \vdots & \vdots & \vdots \\ k_{f-1} & a_{k_{f-1},l_1} & \dots & a_{k_{f-1},l_{g-1}} & a_{k_{f-1},l_g} & \dots & a_{k_{f-1},l_g} & a_{k_{f-1},l_{g+1}} & \dots & a_{k_{f-1},l_n} \\ p_1 & a_{k_f,l_1} & \dots & a_{k_f,l_{g-1}} & b_{p_1,q_1} & \dots & b_{p_1,q_v} & a_{k_f,l_{g+1}} & \dots & a_{k_f,l_n} \\ \vdots & \vdots & \vdots & \vdots & \vdots & \vdots & \vdots & \vdots & \vdots & \vdots \\ p_u & a_{k_f,l_1} & \dots & a_{k_f,l_{g-1}} & b_{p_u,q_1} & \dots & b_{p_u,q_v} & a_{k_f,l_{g+1}} & \dots & a_{k_f,l_n} \\ k_{f+1} & a_{k_{f+1},l_1} & \dots & a_{k_{f+1},l_{g-1}} & a_{k_{f+1},l_g} & \dots & a_{k_{f+1},l_g} & a_{k_{f+1},l_{g+1}} & \dots & a_{k_{f+1},l_n} \\ \vdots & \vdots & \vdots & \vdots & \vdots & \vdots & \vdots & \vdots & \vdots & \vdots \\ k_m & a_{k_m,l_1} & \dots & a_{k_m,l_{g-1}} & a_{k_m,l_g} & \dots & a_{k_m,l_g} & a_{k_m,l_{g+1}} & \dots & a_{k_m,l_n} \end{array}.$$

Now, the following assertion is valid.

Theorem 4. Let $A = [K, L, \{a_{k_i,l_j}\}]$ be an IM and let $a_{k_f,l_g} = [P, Q, \{b_{p_r,q_s}\}]$ be its element. Then

$$A|^*(a_{k_f,l_g})$$

$$= (A \ominus [\{k_f\}, \{l_g\}, \{0\}]) \oplus a_{k_f,l_g} \oplus [P, L - \{l_g\}, \{c_{x,l_j}\}] \oplus [K - \{k_f\}, Q, \{d_{k_i,y}\}],$$

where for each $t \in P$ and for each $l_j \in L - \{l_g\}$, $c_{x,l_j} = a_{k_f,l_j}$ and for each $k_i \in K - \{k_f\}$ and for each $y \in Q$, $d_{k_i,y} = a_{k_i,l_g}$.

We can give other representations of IMs $A|(a_{k_f,l_g})$ and $A|^*(a_{k_f,l_g})$, using other operations defined over IMs.

The following equalities are valid.

Theorem 5. Let $A = [K, L, \{a_{k_i,l_j}\}]$ be an IM and let $a_{k_f,l_g} = [P, Q, \{b_{p_r,q_s}\}]$ be its element. Then

$$A|(a_{k_f,l_g}) = pr_{K-\{k_f\},L-\{l_g\}} A \oplus a_{k_f,l_g},$$

$$A|^*(a_{k_f,l_g})$$

$$= pr_{K-\{k_f\},L-\{l_g\}} A \oplus a_{k_f,l_g} \oplus [P, L - \{l_g\}, \{c_{x,l_j}\}] \oplus [K - \{k_f\}, Q, \{d_{k_i,y}\}],$$

where for each $x \in P$ and for each $l_j \in L - \{l_g\}$, $c_{x,l_j} = a_{k_f,l_j}$ and for each $k_i \in K - \{k_f\}$ and for each $y \in Q$, $d_{k_i,y} = a_{k_i,l_g}$.

Theorem 6. Let $A = [K, L, \{a_{k_i,l_j}\}]$ be an IM and let $a_{k_f,l_g} = [P, Q, \{b_{p_r,q_s}\}]$ be its element. Then

$$A|(a_{k_f,l_g}) = A_{(k_f,l_g)} \oplus a_{k_f,l_g},$$

$$A|^*(a_{k_f,l_g}) = A_{(k_f,l_g)} \oplus a_{k_f,l_g} \oplus [P, L - \{l_g\}, \{c_{x,l_j}\}] \oplus [K - \{k_f\}, Q, \{d_{k_i,y}\}],$$

where for each $x \in P$ and for each $l_j \in L - \{l_g\}$, $c_{x,l_j} = a_{k_f,l_j}$ and for each $k_i \in K - \{k_f\}$ and for each $y \in Q$, $d_{k_i,y} = a_{k_i,l_g}$.

Now, we can see that the newly introduced types of IMs, namely, IFIMs, EIFIMs, TIFIMs and ETIFIMs can be represented as EIMs, too. Indeed, if we put

$$\overline{\mathcal{I}} = \mathcal{I} \times [0,1] \times [0,1]$$

and

$$\overline{\mathcal{X}} = \mathcal{X} \times [0,1] \times [0,1],$$

then we directly see that the IFIMs and EIFIMs can be represented as EIMs, while, for the sets

$$\overline{\overline{\mathcal{I}}} = \mathcal{I} \times [0,1] \times [0,1] \times \mathcal{T}$$

and

$$\overline{\overline{\mathcal{X}}} = \mathcal{X} \times [0,1] \times [0,1] \times \mathcal{T},$$

the TIFIMs and ETIFIMs can be represented as sets of EIMs.

6 New Operations over EIMs

Now, we introduce some new (non-standard) operations over EIMs.

Let index set $\mathcal{I}$ and set $\mathcal{X}$ be fixed and let the EIMs $A_1, A_2, ..., A_n$ over both sets be given.

Let for s $(1 \leq s \leq n)$:

$$(\forall p, q)(1 \leq p < q \leq n)(K^p \cap K^q = L^p \cap L^q = \emptyset)$$

and

$$A_s = [K^s, L^s, \{a^s_{k_i, l_j}\}] = \begin{array}{c|ccccc} & l_{s,1} & \ldots & l_{s,j} & \ldots & l_{s,n_s} \\ \hline k_{s,1} & a_{k_{s,1},l_{s,1}} & \ldots & a_{k_{s,1},l_{s,j}} & \ldots & a_{k_{s,1},l_{s,n_s}} \\ \vdots & \vdots & \ldots & \vdots & \ldots & \vdots \\ k_{s,i} & a_{k_{s,i},l_{s,1}} & \ldots & a_{k_{s,i},l_{s,j}} & \ldots & a_{k_{s,i},l_{s,n_s}} \\ \vdots & \vdots & \ldots & \vdots & \ldots & \vdots \\ k_{s,m} & a_{k_{s,m},l_{s,1}} & \ldots & a_{k_{s,m},l_{s,j}} & \ldots & a_{k_{s,m},l_{s,n_s}} \end{array}.$$

The first new operation, that we call “composition” is defined by

$$\flat\{A_s | 1 \leq s \leq n\} = [\bigcup_{s=1}^{n} K^s, \bigcup_{s=1}^{n} L^s, \{\langle c_{1,t_{1,u},v_{1,w}}, c_{2,t_{2,u},v_{2,w}}, ..., c_{n,t_{n,u},v_{n,w}} \rangle\}],$$

where for r $(1 \leq r \leq n)$:

$$c_{r,t_u,v_w} = \begin{cases} a_{r,k_i,l_j}, \text{ if } t_u = k_i \in K^r \text{ and } v_w = l_j \in L^r \\ \bot, \quad\quad \text{otherwise} \end{cases}$$

Therefore, it is composed of a new EIM on the basis of n EIMs. The new EIM contains n-dimensional vectors as elements. By this reason, we define function *dim*, giving the dimensionality of the elements of the EIM A, i.e., for the above EIM, the equality $dim(A) = n$ holds.

The second new operator, that we call "automatic reduction" is defined for a given EIM A by

$$@(A) = [P, Q, \{b_{p_r,q_s}\}],$$

where $P \subseteq K, Q \subseteq L$ are index sets with the following property:

$$(\forall k \in K - P)(\forall l \in L)(a_{k_i,l_j} = \perp) \ \& \ (\forall k \in K)(\forall l \in L - Q)(a_{k_i,l_j} = \perp)$$

$$\& \ (\forall p_r = a_i \in P)(\forall q_s = b_j \in Q)(b_{p_r,q_s} = a_{k_i,l_j}).$$

For example, if

$$A = \begin{array}{c|cccc} & d & e & f & g \\ \hline a & 1 & 2 & \perp & 3 \\ b & \perp & \perp & \perp & \perp \\ c & 4 & 5 & \perp & \perp \end{array},$$

then

$$@A = \begin{array}{c|ccc} & d & e & g \\ \hline a & 1 & 2 & 3 \\ c & 4 & 5 & \perp \end{array}.$$

Let $\mathcal{X}$ be a set of n-dimensional vectors and A be an EIM with elements from set $\mathcal{X}$. Then

$$Pr_s(A) = \begin{cases} I_\emptyset, \text{ if } s \leq 0 \text{ or } s > n \\ \\ A_s, \text{ otherwise} \end{cases},$$

where $A_s = [K^s, L^s, \{a^s_{k_i,l_j}\}]$ and $a^s_{k_i,l_j}$ is the s-th component of vector $\langle a^1_{k_i,l_j}, a^2_{k_i,l_j}, ..., a^n_{k_i,l_j}\rangle$ that is an element of A.

Now, we can define an operation, that is in some sense opposite of operation $\flat$. It has the form

$$\sharp(A) = \{@Pr_s(A) | 1 \leq s \leq n\}.$$

We give an example. Let the EIMs A_1, A_2 have the forms

$$A_1 = \begin{array}{c|cccc} & d & e & f & g \\ \hline a & 1 & 2 & \perp & 3 \\ b & 4 & \perp & 5 & \perp \\ c & 6 & 7 & \perp & 8 \end{array}, \qquad A_2 = \begin{array}{c|ccc} & d & i & f \\ \hline a & 11 & \perp & 12 \\ c & \perp & 13 & 14 \\ h & 15 & \perp & \perp \end{array}.$$

Then

$$A = \flat\{A_1, A_2\} = \begin{array}{c|ccccc} & d & e & f & g & i \\ \hline a & \langle 1, 11\rangle & \langle 2, \perp\rangle & \langle \perp, 12\rangle & \langle 3, \perp\rangle & \langle \perp, \perp\rangle \\ b & \langle 4, \perp\rangle & \langle \perp, \perp\rangle & \langle 5, \perp\rangle & \langle \perp, \perp\rangle & \langle \perp, \perp\rangle \\ c & \langle 6, \perp\rangle & \langle 7, \perp\rangle & \langle \perp, 14\rangle & \langle 8, \perp\rangle & \langle \perp, 13\rangle \\ h & \langle \perp, 15\rangle & \langle \perp, \perp\rangle & \langle \perp, \perp\rangle & \langle \perp, \perp\rangle & \langle \perp, \perp\rangle \end{array}.$$

On the other hand,

$$\sharp(A) = \{@Pr_1(A), @Pr_2(A)\}$$

$$= \left\{ @\left(\begin{array}{c|ccccc} & d & e & f & g & i \\ \hline a & 1 & 2 & \bot & 3 & \bot \\ b & 4 & \bot & 5 & \bot & \bot \\ c & 6 & 7 & \bot & 8 & \bot \\ h & \bot & \bot & \bot & \bot & \bot \end{array} \right), @\left(\begin{array}{c|ccccc} & d & e & f & g & i \\ \hline a & 11 & \bot & 12 & \bot & \bot \\ b & \bot & \bot & \bot & \bot & \bot \\ c & \bot & \bot & 14 & \bot & 13 \\ h & 15 & \bot & \bot & \bot & \bot \end{array} \right) \right\}$$

$$= \left\{ \begin{array}{c|cccc} & d & e & f & g \\ \hline a & 1 & 2 & \bot & 3 \\ b & 4 & \bot & 5 & \bot \\ c & 6 & 7 & \bot & 8 \end{array}, \begin{array}{c|ccc} & d & f & i \\ \hline a & 11 & 12 & \bot \\ c & \bot & 14 & 13 \\ h & 15 & \bot & \bot \end{array} \right\}.$$

7 Transposed EIM

Let the EIM A be given as above. Then its Transposed EIM has the form

$$A' = [L, K, \{a_{l_j,k_i}\}] = \begin{array}{c|ccccc} & k_1 & \dots & k_i & \dots & k_m \\ \hline l_1 & a_{l_1,k_1} & \dots & a_{l_1,k_i} & \dots & a_{l_1,k_m} \\ \vdots & \vdots & \dots & \vdots & \dots & \vdots \\ l_j & a_{l_j,k_1} & a_{l_j,k_i} & \dots & a_{l_j,k_n} & \\ \vdots & \vdots & \dots & \vdots & \dots & \vdots \\ l_n & a_{l_n,k_1} & \dots & a_{l_n,k_i} & \dots & a_{l_n,k_m} \end{array}.$$

The EIMs $A \odot_{(\circ,*)} A'$ and $A' \odot_{(\circ,*)} A$ are square EIMs. For example, if

$$A = \begin{array}{c|cc} & c & d \\ \hline a & \alpha & \beta \\ b & \gamma & \delta \\ e & \varepsilon & \zeta \end{array},$$

then

$$A' = \begin{array}{c|ccc} & a & b & c \\ \hline d & \alpha & \gamma & \varepsilon \\ e & \beta & \delta & \eta \end{array}$$

and

$$A \odot_{(+,\times)} A' = \begin{array}{c|ccc} & a & b & c \\ \hline a & \alpha^2 + \beta^2 & \alpha\gamma + \beta\delta & \alpha\varepsilon + \beta\zeta \\ b & \alpha\gamma + \beta\delta & \gamma^2 & \gamma\varepsilon + \delta\zeta \\ c & \alpha\varepsilon + \beta\zeta & \gamma\varepsilon + \delta\zeta & \varepsilon^2 + \zeta^2 \end{array},$$

while

$$A' \odot_{(+,\times)} A = \begin{array}{c|cc} & d & e \\ \hline d & \alpha^2 + \gamma^2 + \varepsilon^2 & \alpha\beta + \gamma\delta + \varepsilon\zeta \\ b & \alpha\beta + \gamma\delta + \varepsilon\zeta & \beta^2 + \delta^2 + \zeta^2 \end{array}.$$

8 Conclusion

In next research some application of the newly defined EIMs will be discussed. They can be used as a tool for representing of the rest IM-modifications.

Acknowledgment. The present research is supported by Bulgarian Science Fund, Grant Ref. No. DFNI-I-01/0006.

References

1. Atanassov, K.: Conditions in generalized nets. In: Proc. of the XIII Spring Conf. of the Union of Bulg. Math. Sunny Beach, pp. 219–226 (April 1984)
2. Atanassov, K.: Generalized index matrices. Comptes rendus de l'Academie Bulgare des Sciences 40(11), 15–18 (1987)
3. Atanassov, K.: Intuitionistic Fuzzy Sets. STUDFUZZ, vol. 35. Springer, Heidelberg (1999)
4. Atanassov, K.: On index matrices. Part 1: Standard cases. Advanced Studies in Contemporary Mathematics 20(2), 291–302 (2010)
5. Atanassov, K.: On index matrices, Part 2: Intuitionistic fuzzy case. Proceedings of the Jangjeon Mathematical Society 13(2), 121–126 (2010)
6. Atanassov, K.: On Intuitionistic Fuzzy Sets Theory. Springer, Berlin (2012)
7. Atanassov, K.: On index matrices, Part 3: On the hierarchical operation over index matrices. Advanced Studies in Contemporary Mathematics 23(2), 225–231 (2013)
8. Atanassov, K.: On extended intuitionistic fuzzy index matrices. Notes on Intuitionistic Fuzzy Sets 19(4), 27–41 (2013)
9. Atanassov, K.: Index matrices with function-type of elements. Int. J. Information Models and Analyses 3(2), 103–112 (2014)
10. Atanassov, K., Sotirova, E., Bureva, V.: On index matrices. Part 4: New operations over index matrices. Advanced Studies in Contemporary Mathematics 23(3), 547–552 (2013)
11. Atanassov, K., Szmidt, E., Kacprzyk, J.: On intuitionistic fuzzy pairs. Notes on Intuitionistic Fuzzy Sets 19(3), 1–13 (2013)
12. Mink, H.: Permanents. Addison-Wesley, Reading (1978)

Some Remarks on the Fuzzy Linguistic Model Based on Discrete Fuzzy Numbers

Enrique Herrera-Viedma[1], Juan Vicente Riera[2], Sebastià Massanet[2], and Joan Torrens[2]

[1] Department of Computer Science and Artificial Intelligence,
University of Granada Granada, Spain
viedma@decsai.ugr.es

[2] University of Balearic Islands,
Ctra. de Valldemossa, Km.7.5, 07122 Palma de Mallorca, Spain
{jvicente.riera,s.massanet,jts224}@uib.es

Abstract. In this article, some possible interpretations of the computational model based on discrete fuzzy numbers are given. In particular, some advantages of this model based on the aggregation process as well as on a greater flexibilization of the linguistic expressions are analysed. Finally, a fuzzy decision making model based on this kind on fuzzy subsets is proposed.

Keywords: Discrete fuzzy numbers, subjective evaluation, hesitant fuzzy linguistic term set, multicriteria decision making problem.

1 Introduction

The presence of uncertainty is a common feature that characterizes a wide range of real problems related to decision making. Frequently, the uncertainty appears when we use assessments not necessary quantitative but we deal with terms or qualitative information when making the decision. For this reason, the fuzzy linguistic approximations have emerged as a tool which allow to properly handle the qualitative information. In this sense, we highlight the *symbolic linguistic model* based on ordinal scales (wherein an order is considered among the different linguistic labels) [1,2]; *the linguistic 2-tuples model* [3], which introduces the symbolic translation to the linguistic representation; *the linguistic model based on type-2 fuzzy sets representation* [4], which represents the semantics of the linguistic terms using type-2 fuzzy membership functions or *the proportional 2-tuple model* [5], which extends the 2-tuple model by using two linguistic terms with their proportion to model the information; *the linguistic model based on PSO and granular computing of linguistic information* [6], which proposes to model the linguistic information like expressed in terms of information granules defined as sets, among many others.

A common feature of many of these models is that experts should express their valuations choosing a single linguistic level associated with the linguistic variable. This kind of information is usually interpreted using a linguistic scale like this,

$$\mathcal{L} = \{\text{EB,VB,B,F,G,VG,EG}\} \tag{1}$$

where the linguistic terms correspond to the expressions *"Extremely Bad"*, *"Very Bad"*, *"Bad"*, *"Fair"*, *"Good"*, *"Very Good"* and *"Extremely Good"*, respectively. However

P. Angelov et al. (eds.), *Intelligent Systems'2014*,
Advances in Intelligent Systems and Computing 322, DOI: 10.1007/978-3-319-11313-5_29

in many cases the experts' opinions do not correspond exactly to a particular linguistic term. On the contrary, expressions like *"better than good"* or *"fair to very good"* or, even more complex ones, are commonly used by experts in order to make their opinions. To avoid this drawback some other approaches are appeared in very recent papers. In this way, different authors [7,8] have proposed a computational linguistic model based on hesitant fuzzy linguistic term sets, allowing that the expert could consider of several possible linguistic values or richer expressions than a single term for an indicator, alternative, variable, etc; increasing the richness of linguistic elicitation based on fuzzy linguistic approach. On the other hand, the authors [9,10,11,12,13] have considered another linguistic computational model based on discrete fuzzy numbers [14], which allows to interpret the qualitative information in a more flexible way and also enables to aggregate the information they expressed using this type of fuzzy subsets directly. The main idea lying in this approach is that any linguistic scale like (1) can be identified with the finite chain $L_n = \{0, 1, \ldots, n\}$ (with $n = 6$ in the concrete example). Moreover, with this identification any discrete fuzzy number with support an interval contained in L_n can be interpreted as a fuzzy subset on $\mathscr{L}$, and so, as a possible flexibilization of the linguistic terms and expressions.

Thus, in this article we want to make a step further in the study of this model based on discrete fuzzy numbers. To do this, we will deeply investigate possible interpretations of this computational model as well as their advantages. As we will see, some of these advantages are related to the existence of aggregation functions on the set of discrete fuzzy numbers which allow us to aggregate the opinions without any transformation or loss of information, and also to the possibility of a greater flexibilization of the opinions given by the experts.

The paper is organized as follows. In Section 2, we make a brief review of discrete fuzzy numbers. In Section 3, we explain the main characteristics of the fuzzy linguistic model based on discrete fuzzy numbers and we define the so-called comparative subjective evaluations. In Section 4, we propose a multi-criteria decision making approach based on these particular fuzzy subsets. The last section is devoted to give some conclusions and the future work that we want to develop.

2 Preliminaries

In this section, we recall some definitions and results about discrete aggregation functions and discrete fuzzy numbers which will be used later. More details on discrete aggregation functions can be found in [15].

Aggregation functions can be defined in any bounded partially ordered set (poset) [16]. An important case is when we take as poset a finite chain L_n with $n+1$ elements. In such a framework, only the number of elements is relevant (see [15]) and so the simplest finite chain, that is $L_n = \{0, 1, \ldots, n\}$, is usually considered. Several researchers have developed an extensive study of aggregation functions on L_n, usually called *discrete aggregation functions* [15,17,18,19,20].

Next, we will recall some concepts on discrete fuzzy numbers. In this sense, by a fuzzy subset of $\mathbb{R}$, we mean a function $A : \mathbb{R} \to [0,1]$. For each fuzzy subset A, let $A^{\alpha} = \{x \in \mathbb{R} : A(x) \geq \alpha\}$ for any $\alpha \in (0,1]$ be its α-level set (or α-cut). By $supp(A)$, we mean the support of A, i.e. the set $\{x \in \mathbb{R} : A(x) > 0\}$. By A^0, we mean the closure of $supp(A)$.

Definition 1. *[14] A fuzzy subset A of $\mathbb{R}$ with membership mapping $A : \mathbb{R} \to [0,1]$ is called a* discrete fuzzy number *if its support is finite, i.e., there exist $x_1, \ldots, x_n \in \mathbb{R}$ with $x_1 < x_2 < \ldots < x_n$ such that $supp(A) = \{x_1, \ldots, x_n\}$, and there are natural numbers s,t with $1 \leq s \leq t \leq n$ such that:*

1. *$A(x_i)=1$ for any natural number i with $s \leq i \leq t$ (core)*
2. *$A(x_i) \leq A(x_j)$ for each natural number i,j with $1 \leq i \leq j \leq s$.*
3. *$A(x_i) \geq A(x_j)$ for each natural number i,j with $t \leq i \leq j \leq n$.*

If the fuzzy subset A is a discrete fuzzy number then the support of A coincides with its closure, i.e. $supp(A) = A^0$.

From now on, we will denote the set of discrete fuzzy numbers using the abbreviation DFN and *dfn* will denote a discrete fuzzy number. Also, we will denote by $\mathscr{A}_1^{L_n}$ the set of all discrete fuzzy numbers whose support is a subinterval of the finite chain L_n. Moreover, note that for any $A \in \mathscr{A}_1^{L_n}$, not only $supp(A)$ is an interval of L_n, but also any α-level set. So, let $A, B \in \mathscr{A}_1^{L_n}$ be two discrete fuzzy numbers, and we will denote by $A^\alpha = [x_1^\alpha, x_p^\alpha]$, $B^\alpha = [y_1^\alpha, y_k^\alpha]$ the α-level cuts for A and B, respectively.

The following result holds for $\mathscr{A}_1^{L_n}$, but is not true for the set of discrete fuzzy numbers in general (see [21]).

Theorem 1. *[21] The triplet $(\mathscr{A}_1^{L_n}, MIN, MAX)$ is a bounded distributive lattice where $1_n \in \mathscr{A}_1^{L_n}$ (the unique discrete fuzzy number whose support is the singleton $\{n\}$) and $1_0 \in \mathscr{A}_1^{L_n}$ (the unique discrete fuzzy number whose support is the singleton $\{0\}$) are the maximum and the minimum, respectively, and where $MIN(A,B)$ and $MAX(A,B)$ are the discrete fuzzy numbers belonging to the set $\mathscr{A}_1^{L_n}$ such that they have the sets*

$$\begin{aligned} MIN(A,B)^\alpha &= \{z \in L_n \mid \min(x_1^\alpha, y_1^\alpha) \leq z \leq \min(x_p^\alpha, y_k^\alpha)\}, \\ MAX(A,B)^\alpha &= \{z \in L_n \mid \max(x_1^\alpha, y_1^\alpha) \leq z \leq \max(x_p^\alpha, y_k^\alpha)\} \end{aligned} \tag{2}$$

as α-cuts respectively for each $\alpha \in [0,1]$ and $A, B \in \mathscr{A}_1^{L_n}$.

Remark 1. [21] Using these operations, we can define a partial order on $\mathscr{A}_1^{L_n}$ in the usual way: $A \preceq B$ if and only if $MIN(A,B) = A$.

Aggregation functions defined on L_n have been extended to the bounded lattice $\mathscr{A}_1^{L_n}$ (see for instance [9,12]) according to the next result.

Theorem 2. *[9,12] Let consider a binary aggregation function F on the finite chain L_n. The binary operation on $\mathscr{A}_1^{L_n}$ defined as follows*

$$\begin{aligned} \mathscr{F} : \mathscr{A}_1^{L_n} \times \mathscr{A}_1^{L_n} &\longrightarrow \mathscr{A}_1^{L_n} \\ (A,B) &\longmapsto \mathscr{F}(A,B) \end{aligned}$$

being $\mathscr{F}(A,B)$ the discrete fuzzy number whose α-cuts are the sets

$$\{z \in L_n \mid \min F(A^\alpha, B^\alpha) \leq z \leq \max F(A^\alpha, B^\alpha)\}$$

for each $\alpha \in [0,1]$ is an aggregation function on $\mathscr{A}_1^{L_n}$. This function will be called the extension of the discrete aggregation function F to $\mathscr{A}_1^{L_n}$. *In particular, if F is a t-norm, a t-conorm, an uninorm or a nullnorm, its extension $\mathscr{F}$, so is. In addition, if F is a compensatory aggregation function in L_n, so is its extension $\mathscr{F}$.*

3 Linguistic Model Based on Discrete Fuzzy Numbers

In this section, some remarks on the fuzzy linguistic model based on discrete fuzzy numbers whose support is an interval of the finite chain $L_n = \{0,1,\ldots,n\}$, presented in [11], are given. In particular, we will show how to define some comparative linguistic expressions in this framework, which extend the comparative linguistic expressions based on the hesitant fuzzy linguistic term sets (HFLTS) model [8].

First of all, note that we can consider a bijective mapping between the ordinal scale $\mathfrak{L} = \{s_0,\ldots,s_n\}$ and the finite chain L_n which keeps the original order. Furthermore, each normal discrete convex fuzzy subset defined on the ordinal scale $\mathfrak{L}$ can be considered like a discrete fuzzy number belonging to $\mathscr{A}_1^{L_n}$, and vice-versa.

For example, consider the linguistic scale

$$\mathfrak{L} = \{EB,VB,B,F,G,VG,EG\} \tag{3}$$

where the letters refer to the linguistic terms Extremely Bad, Very Bad, Bad, Fair, Good, Very Good and Extremely Good and they are listed in an increasing order:

$$EB \prec VB \prec B \prec F \prec G \prec VG \prec EG$$

and the finite chain L_6.

Thus, the discrete fuzzy number $A = \{0.6/2, 0.7/3, 1/4, 0.8/5\} \in \mathscr{A}_1^{L_6}$ can be also expressed as $A = \{0.6/B, 0.7/F, 1/G, 0.8/VG\}$. Note that these fuzzy subsets can be interpreted as a possible flexibilization of the linguistic labels (see Figure 1).

Definition 2. *Let $L_n = \{0,\ldots,n\}$ be a finite chain. We call a subjective evaluation to each discrete fuzzy number belonging to the partially ordered set $\mathscr{A}_1^{L_n}$.*

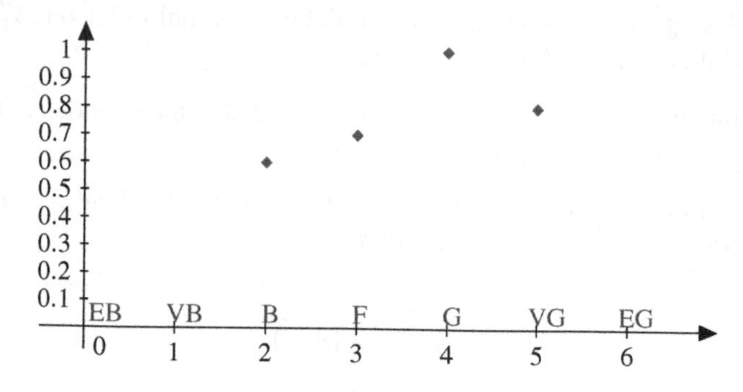

Fig. 1. Graphical representation of the discrete fuzzy number A which can be interpreted as a possible flexibilization of the linguistic expression "Good"

According to the previous comments, a subjective evaluation can be interpreted equivalently like a normal convex fuzzy subset defined on the ordinal scale $\mathfrak{L}$.

Moreover, let us denote by $S = \{s_0, \dots, s_n\}$ a general linguistic ordinal scale like $\mathfrak{L}$. In this general setting, typical linguistic expressions like the ones commented in the introduction can be expressed as follows:

$$\begin{array}{ll} (\text{label } s_i) & = \{s_i\}, \\ (\text{less than } s_i) & = \{s_j \mid s_j \in S \text{ and } s_j \leq s_i\}, \\ (\text{greater than } s_i) & = \{s_j \mid s_j \in S \text{ and } s_j \geq s_i\}, \\ (\text{between } s_i \text{ and } s_j) & = \{s_k \mid s_k \in S \text{ and } s_i \leq s_k \leq s_j\}, \end{array} \tag{4}$$

for all $0 \leq i, j \leq n$.

Note that these expressions can be also interpreted in our context as discrete fuzzy numbers $A \in \mathscr{A}_1^{L_n}$ such that, both, $core(A)$ and $supp(A)$ coincide with the corresponding interval.

In this way, we will have:

$$\begin{array}{rl} \text{Label } s_i \equiv 1_{s_i} & = \text{the only discrete fuzzy number } A \text{ with } supp(A) = \{s_i\}, \\ \text{Between } s_i \text{ and } s_j & \equiv \{A \in \mathscr{A}_1^{L_n} \mid core(A) = supp(A) = [s_i, s_j]\} \\ \text{Worse than } s_i & \equiv \{A \in \mathscr{A}_1^{L_n} \mid core(A) = supp(A) = [s_0, s_{i-1}]\} \\ \text{Better than } s_i & \equiv \{A \in \mathscr{A}_1^{L_n} \mid core(A) = supp(A) = [s_{i+1}, s_n]\} \end{array} \tag{5}$$

See for instance, Figure 2-(a), where the subjective evaluation "between Bad and Good" is depicted.

It is clear that, with this equivalence, any manipulation that we can do with the linguistic expressions as in (4) can also be done with the corresponding discrete fuzzy numbers described in (5).

Moreover, with this approach we have the following advantages:

First Advantage. An interesting aspect about the linguistic interpretation based on subjective evaluations is that it does not need making previous transformations when we wish to aggregate the information. Each subjective evaluation considered by an expert is directly interpreted as a discrete fuzzy number of the bounded set $\mathscr{A}_1^{L_n}$ and by this reason we can handle this information directly according to Theorem 2.

Second Advantage. This approach allows more flexibilization of the linguistic term sets. As we have mentioned above, Figure 1 represents a flexibilization of the linguistic label *Good* (G). Similarly, it is possible to define different *flexibilizations* as the linguistic expressions generated by using production rules of the context-free grammar G_H such as it is presented in (4).

In this sense, we define the following linguistic expressions interpreted as subjective evaluations:

$$\begin{array}{rl} \text{Between } s_i \text{ and } s_j & = \{A \in \mathscr{A}_1^{L_n} \mid core(A) = [s_i, s_j]\} \\ \text{Worse than } s_i & = \{A \in \mathscr{A}_1^{L_n} \mid core(A) = [s_0, s_{i-1}]\} \\ \text{Better than } s_i & = \{A \in \mathscr{A}_1^{L_n} \mid core(A) = [s_{i+1}, s_n]\} \end{array} \tag{6}$$

for all $0 \leq i, j \leq n$.

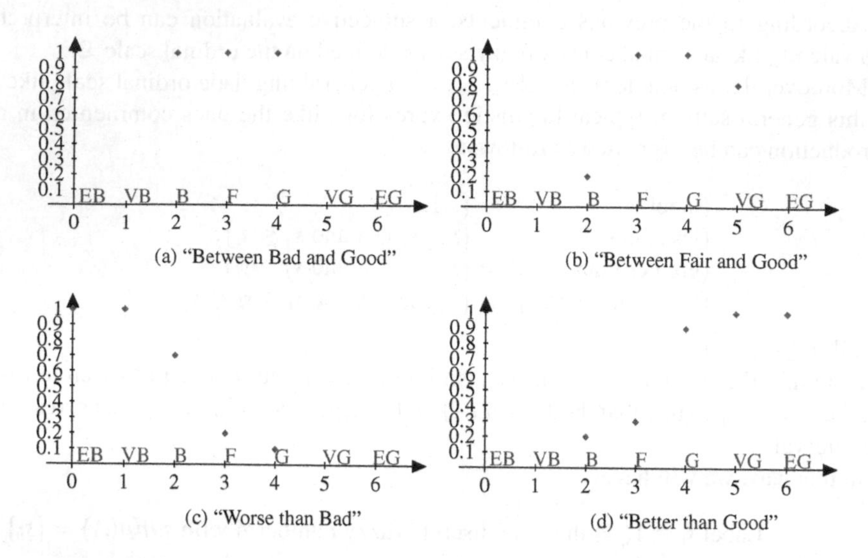

(a) "Between Bad and Good" (b) "Between Fair and Good"

(c) "Worse than Bad" (d) "Better than Good"

Fig. 2. Graphical representations of some comparative subjective evaluations

Thus, discrete fuzzy numbers $A \in \mathscr{A}_1^{L_n}$ with $core(A) = [s_i, s_j]$, but with a different support, can be interpreted as flexibilizations of the subjective evaluation "between s_i and s_j", and similarly with the other expressions in (5). In addition, the above linguistic expressions are not unique and allow us to describe in a more flexible way the different interpretations that may have the experts when performing a qualitative assessment. Thus, the flexibility that may give an expert assigning values outside the core will depend on the security he has about his evaluation. When the expert does not hesitate about the election of the set of possible linguistic labels, we would be on the case that the subjective evaluation satisfies that its support coincides with its core. In this last case, we would obtain a coincidence between the interpretation presented in [7] and the purpose of this work (see Figure 2-(a)). However, Figure 2-(d) can be seen as an interpretation of the linguistic expression "*Better than Good*" where the expert does not reject the possibility to assess some positive values outside of core, allowing a greater flexibility of the linguistic expression and therefore his assessment (see the example below).

Example 1. Figures 2-(b), (c) and (d) represent the subjective evaluations

$$A = \{0.2/B, 1/F, 1/G, 0.8/VG\},$$
$$B = \{1/EB, 1/VB, 0.7/B, 0.2/F, 0.1/G\},$$
$$C = \{0.2/B, 0.3/F, 0.9/G, 1/VG, 1/EG\},$$

that interpret the linguistic expressions *between Fair and Good*, *Worse than Bad* and *Better than Good*, respectively.

4 Multicriteria Linguistic Decision-Making Problem with Linguistic Expressions Based on Subjective Evaluations

In [7,8] a linguistic computational model for multi-criteria decision making based on symbolic computational models that allows to the experts to express its assessments through comparative linguistic expressions or simple linguistic terms is presented . Similarly, in this section we present a model based on subjective evaluations.

Following the ideas presented in [8], our proposed model has the following phases:

1. *Transformation phase*: Fixed a linguistic scale $\mathscr{L}$, each expert assesses using subjective evaluations defined on this scale.
2. *Aggregation of the assessments represented by subjective evaluations*: The assessments represented by subjective evaluations are aggregated using the operators obtained according to Theorem 2.
3. *Exploitation phase:* A method is chosen to sort the subjective evaluations and it is used to select the best alternative.

In this section we wish to give an illustrative example of multi-criteria decision making following the three phases explained previously.

4.1 Transformation Phase

Suppose that we have $X = \{x_1, x_2, x_3, x_4\}$ a set of alternatives and $C = \{c_1, c_2, c_3\}$ a set of criteria . Let $\mathscr{L} = \{\text{EB,VB,B,F,G,VG,EG}\}$ be the linguistic scale considered in (1). Table 1 contains the assessments that are provided for the decision problem.

Table 1. Linguistic assessments for the decision problem

	c_1	c_2	c_3
x_1	Between VB and F	Between G and VG	G
x_2	Between B and F	G	Worse than B
x_3	Better than G	Between VB and B	Better than G
x_4	B	Between B and G	G

According to Table 1, let us suppose that the expert provides these linguistic assessments through the following discrete fuzzy numbers:

- For the alternative x_1 and criteria c_1,c_2 and c_3 we have respectively:

$$Between\,VB\,and\,F = \{0.5/0, 1/1, 1/2, 1/3, 0.6/4\},$$
$$Between\,G\,and\,VG = \{0.5/2, 0.8/3, 1/4, 1/5, 0.7/6\},$$
$$G = \{0.6/2, 0.7/3, 1/4, 0.8/5\}.$$

- For the alternative x_2 and criteria c_1,c_2 and c_3 we have respectively:

$$Between\,B\,and\,F = \{0.6/0, 0.7/1, 1/2, 1/3, 0.8/4\},$$
$$G = \{0.8/3, 1/4, 0.7/5, 0.6/6\},$$
$$Worse\,than\,B = \{1/0, 1/1, 0.8/2, 0.7/3\}.$$

- For the alternative x_3 and criteria c_1,c_2 and c_3 we have respectively:

$$Better\ than\ G = \{0.6/3, 0.8/4, 1/5, 1/6\},$$
$$Between\ VB\ and\ B = \{0.8/0, 1/1, 1/2, 0.7/3\},$$
$$Better\ than\ G = \{0.7/3, 0.9/4, 1/5, 1/6\}.$$

- For the alternative x_4 and criteria c_1,c_2 and c_3 we have respectively:

$$B = \{0.6/0, 0.7/1, 1/2, 0.7/3\},$$
$$Between\ B\ and\ G = \{0.6/1, 1/2, 1/3, 1/4, 0.7/5\},$$
$$G = \{0.7/3, 1/4, 0.8/5\}.$$

4.2 Aggregation of the Assessments Represented by Subjective Evaluations

The next step is to aggregate the assessments expressed by the previous subjective evaluations. As result of such fusion we obtain another subjective evaluation which represents the collective preference.

In order to show the potential of our aggregation method using subjective evaluations we consider the interesting family of discrete idempotent uninorms which depend on a parameter $e \in \{1,\dots,n-1\}$, the neutral element of the uninorm [22]. Moreover, we study the behaviour of the method according to the value of the chosen parameter. Thus, if we get the extension of the idempotent uninorm

$$U(x,y) = \begin{cases} \max(x,y), & \text{if } x,y \in [e,n], \\ \min(x,y), & \text{otherwise.} \end{cases} \quad (7)$$

taking as a particular case $e = 3$, we obtain that the aggregations for each alternative are:

- For the alternative x_1: $A_{x_1} = \{0.5/0, 1/1, 1/2, 1/3, 1/4, 1/5, 0.7/6\}$.
- For the alternative x_2: $A_{x_2} = \{1/0, 1/1, 0.8/2, 0.7/3, 0.7/4, 0.7/5, 0.6/6\}$.
- For the alternative x_3: $A_{x_3} = \{0.8/0, 1/1, 1/2, 0.7/3, 0.7/4, 0.7/5, 0.7/6\}$.
- For the alternative x_4: $A_{x_4} = \{0.6/0, 0.7/1, 1/2, 0.7/3, 0.7/4, 0.7/5\}$.

Instead of taking as neutral element of the uninorm $e = 3$, we could also have taken any value $e \in \{1,2,3,4,5\}$. The choice of the neutral element is crucial for the behaviour of the aggregation phase. Values close to 1 would give us more optimistic subjective evaluations, while greater values would rise on less optimistic subjective evaluations. The structure of the idempotent uninorms is the main reason of this behaviour. As lower is the neutral element, greater is the region where the uninorm acts as the maximum and consequently, between a good evaluation and a bad one, the output will be the good one determining an optimistic behaviour.

4.3 Exploitation Phase

Previously we dealt with the aggregation phase in order to obtain a preference value for each possible alternative based on the subjective evaluations given by the experts. Now, we need to choose the best one, i.e., to exploit the collective linguistic preference (see [23,24]). To do that, we will use the ranking method proposed by L. Chen and H. Lu in [25]. This method fits perfectly to our framework and it was already considered in [11]. It is based on the concepts of left and right dominance of one fuzzy number over the other one and on one parameter $\beta \in [0,1]$, called the index of optimism (see [25] for the full details of the method). The larger the index of optimism β is, more important is the right dominance. Herein, the index of optimism is used to reflect a decision maker's degree of optimism. A more optimistic decision maker generally will take a larger value of the index, for example, a situation in which $\beta = 1$ (or 0) represents an optimistic (pessimistic) decision maker's perspective, and only right (left) dominance is considered.

Finally, in this phase, using this method, we will arrange the above subjective evaluations to decide which is the best alternative among the considered ones. Taking into account the aggregated subjective evaluations obtained in Section 4.2 by means of the idempotent uninorm with $e = 3$, the method decides that x_4 is the best alternative when $\beta < 0.1$, x_1 when $\beta > 0.1$ and evaluates equally x_1 and x_4 when $\beta = 0.1$. Furthermore, in Figure 3, we show the ranking of the alternatives depending on the choice of the neutral element of the idempotent uninorm and the value of β.

Table 2. Notations used for the rankings of the alternatives in Figure 3

Symbol	Ranking	Symbol	Ranking	Symbol	Ranking
$\star_1$	$x_4 > x_1 > x_3 > x_2$	$\star_5$	$x_1 > x_3 > x_4 > x_2$	$\star_9$	$x_3 > x_1 > x_4 > x_2$
$\star_2$	$x_1 = x_4 > x_3 > x_2$	$\star_6$	$x_1 > x_3 > x_2 = x_4$	$\star_{10}$	$x_3 > x_1 > x_2 = x_4$
$\star_3$	$x_1 > x_4 > x_3 > x_2$	$\star_7$	$x_1 > x_3 > x_2 > x_4$	$\star_{11}$	$x_3 > x_1 > x_2 > x_4$
$\star_4$	$x_1 > x_3 = x_4 > x_2$	$\star_8$	$x_1 = x_3 > x_4 > x_2$		

At a first glance, we can observe that the alternatives are ranked in a different way depending on the parameter of the function and the value of β. In particular, x_1, x_3 and x_4 are assessed as the best alternative for some particular values of the parameters e and β, while x_2 is never ranked as the first option since its three evaluations are negative. In addition, note that the results are coherent with the evaluations given by the expert and the behaviour of the parameter. In Table 1 which is included at the beginning of Section 4.1, we can see that x_1 and x_3 have some worse evaluation than the three obtained by x_4 while they have some other better evaluations. Setting a lower neutral element, the uninorm acts more as the maximum and therefore, we promote the positive evaluations instead of the negative ones resulting on a better ranking for x_1 and x_3 with respect to the rankings obtained for greater values of the neutral element.

The β value also takes into account these evaluations. For instance, in the uninorm case with $e = 1$, if we consider $\beta > \frac{9}{11}$, we get that x_3 is the best alternative. For lower values of β, which are more pessimistic values, the best alternative is x_1 due to the bad

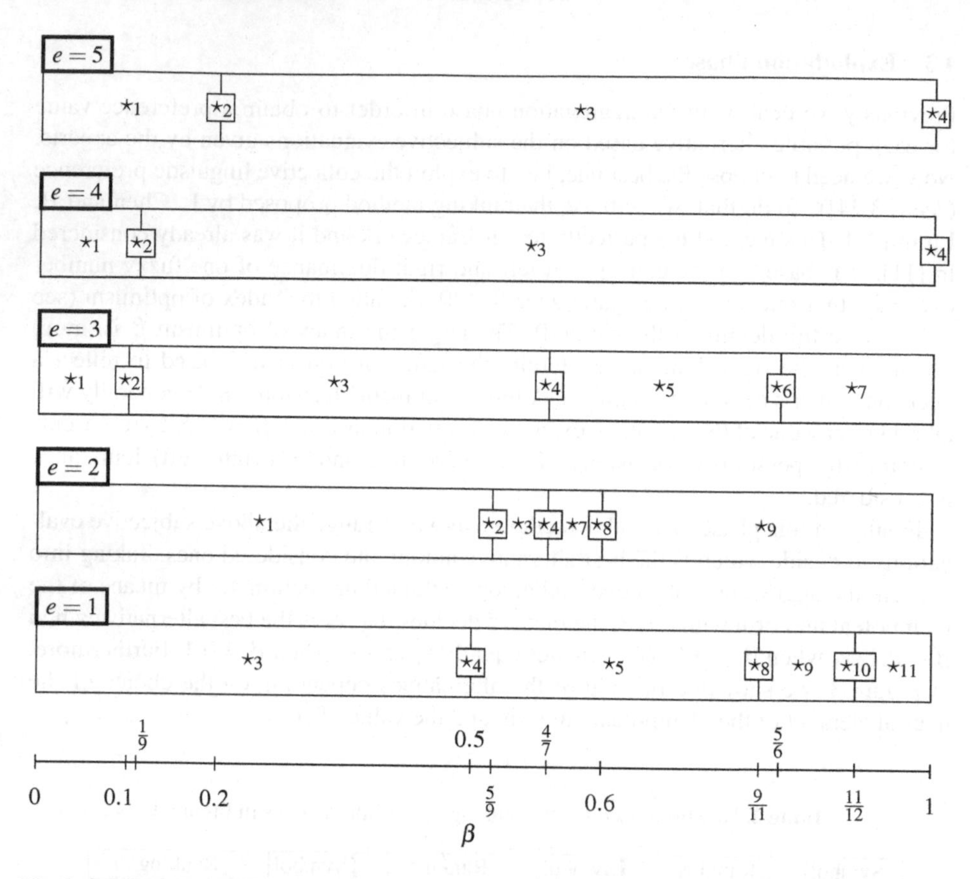

Fig. 3. Ranking of the alternatives according to the discrete idempotent uninorm with different values of the neutral element e and the parameter $\beta \in [0, 1]$. See Table 2 for the notations used to represent the rankings

evaluation of x_3 given by one of the experts. So, to sum up, the parameter values of the aggregation functions allow to design a computational linguistic model based on discrete fuzzy numbers meeting the needs and desires of the person who is performing this multicriteria decision making problem.

5 Conclusions and Future Work

In this paper, we have analysed the linguistic computational model based on discrete fuzzy numbers. We have defined comparative subjective evaluations in this framework which subsume the hesitant fuzzy linguistic term sets of the model recently published in [7]. Since the hesitant fuzzy linguistic term sets can be interpreted as particular cases of discrete fuzzy numbers whose support coincides with their core, with the appropriate aggregation functions and exploitation method, both methods agree in their final decision. However, we should point out that the model based on discrete fuzzy numbers presents some advantages with respect to the other model. Particularly, in the linguistic computational model based on discrete fuzzy numbers, the experts can evaluate the alternatives using different linguistic scales with a greater flexibility. In addition, there is no need of any transformation since the aggregation functions on the set of discrete fuzzy numbers are well-defined and studied.

In the future, we think to compare this model with other important linguistic computational models existing in the literature, like the hesitant fuzzy linguistic term sets model [7], the unbalanced linguistic computational model [26] and the linguistic computational model based on granular computing [6].

Acknowledgments. This paper has been partially supported by the Spanish Grants MTM2009-10320, MTM2009-10962 and TIN2013-42795-P with FEDER support also with the nancing of FEDER funds in FUZZYLING-II Project TIN2010-17876, Andalusian Excellence Projects TIC-05299 and TIC-5991.

References

1. Herrera, F., Herrera-Viedma, E., Verdegay, J.L.: Direct approach processes in group decision making using linguistic OWA operators. Fuzzy Sets and Systems 79, 175–190 (1994)
2. Herrera, F., Alonso, S., Chiclana, F., Herrera-Viedma, E.: Computing with words in decision making: Foundations, trends and prospects. Fuzzy Optimization and Decision Making 8(4), 337–364 (2009)
3. Herrera, F., Martínez, L.: A 2-tuple fuzzy linguistic representation model for computing with words. IEEE Transactions on Fuzzy Systems 8(6), 746–752 (2000)
4. Türksen, I.: Type 2 representation and reasoning for CWW. Fuzzy Sets and Systems 127(1), 17–36 (2002)
5. Wang, J.-H., Hao, J.: A new version of 2-tuple fuzzy linguistic representation model for computing with words. IEEE Transactions on Fuzzy Systems 14(3), 435–445 (2006)
6. Cabrerizo, F.J., Herrera-Viedma, E., Pedrycz, W.: A method based on PSO and granular computing of linguistic information to solve group decision making problems defined in heterogeneous contexts. European Journal of Operational Research 230(3), 624–633 (2013)
7. Rodríguez, R., Martínez, L., Herrera, F.: Hesitant fuzzy linguistic term sets for decision making. IEEE Transactions on Fuzzy Systems 20(1), 109–119 (2012)
8. Rodríguez, R., Martínez, L., Herrera, F.: A group decision making model dealing with comparative linguistic expressions based on hesitant fuzzy linguistic term sets. Information Sciences 241, 28–42 (2013)
9. Casasnovas, J., Riera, J.V.: Extension of discrete t-norms and t-conorms to discrete fuzzy numbers. Fuzzy Sets and Systems 167(1), 65–81 (2011)
10. Casasnovas, J., Riera, J.V.: Weighted means of subjective evaluations. In: Seising, R., Sanz González, V. (eds.) Soft Computing in Humanities and Social Sciences. STUDFUZZ, vol. 273, pp. 331–354. Springer, Heidelberg (2012)
11. Massanet, S., Riera, J.V., Torrens, J., Herrera-Viedma, E.: A new linguistic computational model based on discrete fuzzy numbers for computing with words. Information Sciences 258, 277–290 (2014)
12. Riera, J.V., Torrens, J.: Aggregation of subjective evaluations based on discrete fuzzy numbers. Fuzzy Sets and Systems 191, 21–40 (2012)
13. Riera, J.V., Torrens, J.: Aggregation functions on the set of discrete fuzzy numbers defined from a pair of discrete aggregations. Fuzzy Sets and Systems 241, 76–93 (2014)

14. Voxman, W.: Canonical representations of discrete fuzzy numbers. Fuzzy Sets and Systems 118(3), 457–466 (2001)
15. Mayor, G., Torrens, J.: Triangular norms in discrete settings. In: Logical, Algebraic, Analytic, and Probabilistic Aspects of Triangular Norms, pp. 189–230. Elsevier (2005)
16. De Baets, B., Mesiar, R.: Triangular norms on product lattices. Fuzzy Sets and Systems 104, 61–75 (1999)
17. Mas, M., Monserrat, M., Torrens, J.: Kernel aggregation functions on finite scales, Constructions from their marginals. Fuzzy Sets and Systems 241, 27–40 (2014)
18. Godo, L., Torra, V.: On aggregation operators for ordinal qualitative information. IEEE Transactions on Fuzzy Systems 8, 143–154 (2000)
19. Mas, M., Mayor, G., Torrens, J.: t-Operators and uninorms on a finite totally ordered set. International Journal of Intelligent Systems 14, 909–922 (1999)
20. Yager, R.R.: Aggregation of ordinal information. Fuzzy Optimization and Decision Making 6, 199–219 (2007)
21. Casasnovas, J., Riera, J.V.: Lattice properties of discrete fuzzy numbers under extended min and max. In: Proceedings of IFSA-EUSFLAT-2009, Lisbon (2009)
22. De Baets, B., Fodor, J.C., Ruiz-Aguilera, D., Torrens, J.: Idempotent uninorms on Finite Ordinal Scales. International Journal of Uncertainty, Fuzziness and Knowledge-Based Systems 17(1), 1–14 (2009)
23. Herrera, F., Herrera-Viedma, E., Martínez, L.: A fusion approach for managing multi-granularity linguistic term sets in decision making. Fuzzy Sets and Systems 114, 43–58 (2000)
24. Alonso, S., Pérez, I.J., Cabrerizo, F.J., Herrera-Viedma, E.: A linguistic consensus model for Web 2.0 communities. Applied Soft Computing 13(1), 149–157 (2013)
25. Chen, L.H., Lu, H.W.: An approximate approach for ranking fuzzy numbers based on left and right dominance. Computers and Mathematics with Applications 41(12), 1589–1602 (2001)
26. Herrera, F., Herrera-Viedma, E., Martínez, L.: A fuzzy linguistic methodology to deal with unbalanced linguistic term sets. IEEE Transactions on Fuzzy Systems 16(2), 354–370 (2008)

Differences between Moore and RDM Interval Arithmetic

Marek Landowski

Maritime University of Szczecin, Waly Chrobrego 1-2, 70-500 Szczecin, Poland
m.landowski@am.szczecin.pl

Abstract. The uncertainty theory solves problems with uncertain data. Often to perform arithmetic operations on uncertain data, the calculations on intervals are necessary. Interval arithmetic uses traditional mathematics in the calculations on intervals. There are many methods that solve the problems of uncertain data presented in the form of intervals, each of them can give in some cases different results. The most known arithmetic, often used by scientists in calculations is Moore interval arithmetic. The article presents a comparison of Moore interval arithmetic and multidimensional RDM interval arithmetic. Also, in both Moore and RDM arithmetic the basic operations and their properties are described. Solved examples show that the results obtained using the RDM arithmetic are multidimensional while Moore arithmetic gives one-dimensional solution.

Keywords: interval arithmetic, uncertainty theory, fuzzy arithmetic, granular computing, computing with words.

1 Introduction

Interval arithmetic was deemed as necessary with the development of the theory of uncertainty [1]. It was realized that the use of uncertain parameters and uncertain data is very important for the description of reality in the form of a mathematical model. Interval arithmetic is used in scientific fields such as uncertainty theory [1], grey systems [4], granular computing [8], fuzzy systems [3,7], to determine the uncertain data and modeling of uncertain systems. The most common and most frequently used interval arithmetic is Moore arithmetic [5,6,8]. A number of limitations and the drawbacks has been found in the Moore interval arithmetic [2,9,14] such as: the excess width effect problem, dependency problem, difficulties of solving even simplest equation problem, interval equation's right-hand side problem, absurd solutions and request to introduce negative entropy into the system problem. In Moore arithmetic basic operations on intervals $A = [\underline{a}, \overline{a}]$ and $B = [\underline{b}, \overline{b}]$ are realized by formulas (1).

$$\begin{aligned} [\underline{a}, \overline{a}] + [\underline{b}, \overline{b}] &= [\underline{a} + \underline{b}, \overline{a} + \overline{b}] \\ [\underline{a}, \overline{a}] - [\underline{b}, \overline{b}] &= [\underline{a} - \overline{b}, \overline{a} - \underline{b}] \\ [\underline{a}, \overline{a}] \cdot [\underline{b}, \overline{b}] &= [\min(\underline{a}\underline{b}, \underline{a}\overline{b}, \overline{a}\underline{b}, \overline{a}\overline{b}), \max(\underline{a}\underline{b}, \underline{a}\overline{b}, \overline{a}\underline{b}, \overline{a}\overline{b})] \\ [\underline{a}, \overline{a}] / [\underline{b}, \overline{b}] &= [\underline{a}, \overline{a}] \cdot [1/\overline{b}, 1/\underline{b}] \ \text{ if } 0 \notin [\underline{b}, \overline{b}] \end{aligned} \tag{1}$$

P. Angelov et al. (eds.), *Intelligent Systems'2014*,
Advances in Intelligent Systems and Computing 322, DOI: 10.1007/978-3-319-11313-5_30

The alternative for Moore arithmetic can be multidimesional RDM interval arithmetic. The idea of multidimensional RDM arithemtic was developed by A. Piegat [10,11,12,13]. Abbreviation RDM stands for Relative Distance Measure where given value x from interval $X = [\underline{x}, \overline{x}]$ is described using RDM variable α_x, $\alpha_x \in [0,1]$, as shown in (2).

$$x = \underline{x} + \alpha_x(\overline{x} - \underline{x}) \quad (2)$$

In notation RDM the interval $X = [\underline{x}, \overline{x}]$ is described in the form (3).

$$X = \{x : x = \underline{x} + \alpha_x(\overline{x} - \underline{x}),\ \alpha_x \in [0,1]\} \quad (3)$$

The RDM variable α_x gives possibility to obtain any value between left border $\underline{x}$ and right border $\overline{x}$ of interval X. For $\alpha_x = 0$ the value from interval X equals $\underline{x}$ and the variable $\alpha_x = 1$ gives $\overline{x}$. Lets consider value $x \in [2,4]$, in RDM notation the value x is written as $x = 2 + 2\alpha_x$, where $\alpha_x \in [0,1]$. Fig. 1 shows the interval $X = [\underline{x}, \overline{x}]$ and the meaning of the RDM variable α_x in case $\underline{x} \leq \overline{x}$.

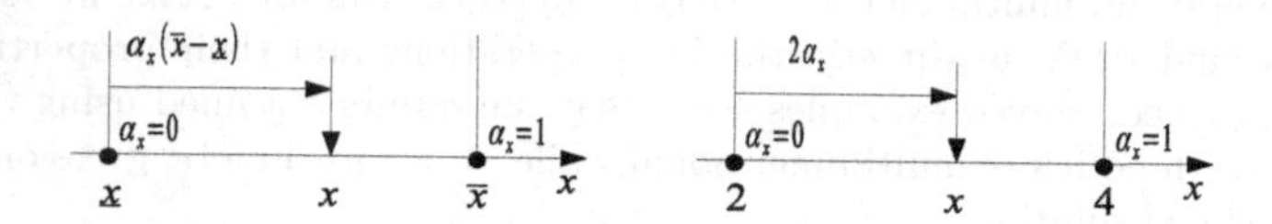

Fig. 1. The interval $X = [\underline{x}, \overline{x}]$ and the meaning of the RDM variable $\alpha_x \in [0,1]$, $\underline{x} \leq \overline{x}$

2 Operations in RDM Interval Arithmetic

In RDM arithmetic the following operations are definied: addition, subtraction, multiplication and division. Depending on the numbers of variables in the calculations, the obtained solution is in multidimensional space, in Moore arithmetic the solution are in 1-dimensional space.

Let X and Y are two intervals: $X = [\underline{x}, \overline{x}] = \{x : x = \underline{x} + \alpha_x(\overline{x} - \underline{x}),\ \alpha_x \in [0,1]\}$ and $Y = [\underline{y}, \overline{y}] = \{y : y = \underline{y} + \alpha_y(\overline{y} - \underline{y}),\ \alpha_y \in [0,1]\}$.

Addition in RDM

$$X + Y = \{x + y : x + y = \underline{x} + \alpha_x(\overline{x} - \underline{x}) + \underline{y} + \alpha_y(\overline{y} - \underline{y}),\ \alpha_x, \alpha_y \in [0,1]\} \quad (4)$$

Subtraction in RDM

$$X - Y = \{x - y : x - y = \underline{x} + \alpha_x(\overline{x} - \underline{x}) - \underline{y} - \alpha_y(\overline{y} - \underline{y}),\ \alpha_x, \alpha_y \in [0,1]\} \quad (5)$$

Multiplication in RDM

$$X \cdot Y = \{xy : xy = [\underline{x} + \alpha_x(\overline{x} - \underline{x})] \cdot [\underline{y} + \alpha_y(\overline{y} - \underline{y})],\ \alpha_x, \alpha_y \in [0,1]\} \quad (6)$$

Division in RDM

$$X/Y = \{x/y : x/y = [\underline{x} + \alpha_x (\overline{x} - \underline{x})] \,/\, [\underline{y} + \alpha_y (\overline{y} - \underline{y})], \alpha_x, \alpha_y \in [0,1]\}, \text{if } 0 \notin Y \quad (7)$$

For intervals $X = [\underline{x}, \overline{x}]$ and $Y = [\underline{y}, \overline{y}]$ and the base operations $* \in \{+, -, \cdot, /\}$ span is an interval defined as (8), operation / is defined only if $0 \notin Y$.

$$s(X * Y) = [\min\{X * Y\}, \max\{X * Y\}] \quad (8)$$

Example 1. To show multidimensionality of solution in RDM arithmetic and 1-dimensional solution in Moore arithmetic, we will consider operations such as addition, subtraction, multiplication and division of two intervals $A = [1,3]$ and $B = [3,4]$.

The first solution in 1-dimension space by Moore arithmetic will be presented by equations (9).

$$\begin{aligned} A + B &= [1,2] + [3,4] = [4,6] \\ A - B &= [1,2] - [3,4] = [-3,-1] \\ A \cdot B &= [1,2] \cdot [3,4] = [3,8] \\ A/B &= [1,2]/[3,4] = [1,2] \cdot [1/4, 1/3] = [1/4, 2/3] \end{aligned} \quad (9)$$

To find solutions by RDM arithmetic we should write intervals in RDM notation using RDM variable α_a and α_b, where $\alpha_a \in [0,1]$ and $\alpha_b \in [0,1]$, formula (10).

$$\begin{aligned} A &= [1,2] = \{a : a = 1 + \alpha_a,\ \alpha_a \in [0,1]\} \\ B &= [3,4] = \{b : b = 3 + \alpha_b,\ \alpha_b \in [0,1]\} \end{aligned} \quad (10)$$

Obtained solutions are presented in equations (11).

$$\begin{aligned} A + B &= \{a + b : a + b = 4 + \alpha_a + \alpha_b,\ \alpha_a, \alpha_b \in [0,1]\} \\ A - B &= \{a - b : a - b = -2 + \alpha_a - \alpha_b,\ \alpha_a, \alpha_b \in [0,1]\} \\ A \cdot B &= \{ab : ab = 3 + 3\alpha_a + \alpha_b + \alpha_a\alpha_b,\ \alpha_a, \alpha_b \in [0,1]\} \\ A/B &= \{a/b : a/b = (1 + \alpha_a)/(3 + \alpha_b),\ \alpha_a, \alpha_b \in [0,1]\} \end{aligned} \quad (11)$$

To show the illustration of solution obtained by RDM arithmetic we should find border values of the results, Table 1.

Illustration of the solutions obtained by RDM arithmetic presents Fig. 2.

Spans of the 3-dimensional solutions (12), (13), (14) and (15) are the same as intervals calculated by Moore arithmetic (9).

$$s(A + B) = \left[\min_{\substack{\alpha_a \in [0,1] \\ \alpha_b \in [0,1]}} (4 + \alpha_a + \alpha_b),\ \max_{\substack{\alpha_a \in [0,1] \\ \alpha_b \in [0,1]}} (4 + \alpha_a + \alpha_b) \right] = [4,6] \quad (12)$$

$$s(A - B) = \left[\min_{\substack{\alpha_a \in [0,1] \\ \alpha_b \in [0,1]}} (-2 + \alpha_a - \alpha_b),\ \max_{\substack{\alpha_a \in [0,1] \\ \alpha_b \in [0,1]}} (-2 + \alpha_a - \alpha_b) \right] = [-3,-1] \quad (13)$$

Table 1. Results of the basic operations for two intervals $A = [\underline{a}, \overline{a}] = [1, 2]$ and $B = [\underline{b}, \overline{b}] = [3, 4]$ for border values of RDM-variables $\alpha_a \in [0, 1]$ and $\alpha_b \in [0, 1]$

α_a	0	0	1	1
a	1	1	2	2
α_b	0	1	0	1
b	3	4	3	4
$a+b$	$\underline{a}+\underline{b}$	$\underline{a}+\overline{b}$	$\overline{a}+\underline{b}$	$\overline{a}+\overline{b}$
	4	5	5	6
$a-b$	$\underline{a}-\underline{b}$	$\underline{a}-\overline{b}$	$\overline{a}-\underline{b}$	$\overline{a}-\overline{b}$
	-2	-3	-1	-2
ab	$\underline{a}\underline{b}$	$\underline{a}\overline{b}$	$\overline{a}\underline{b}$	$\overline{a}\overline{b}$
	3	4	6	8
a/b	$\underline{a}/\underline{b}$	$\underline{a}/\overline{b}$	$\overline{a}/\underline{b}$	$\overline{a}/\overline{b}$
	1/3	1/4	2/3	1/2

$$s(AB) = \left[\min_{\substack{\alpha_a \in [0,1] \\ \alpha_b \in [0,1]}} (3 + 3\alpha_a + \alpha_b + \alpha_a\alpha_b), \max_{\substack{\alpha_a \in [0,1] \\ \alpha_b \in [0,1]}} (3 + 3\alpha_a + \alpha_b + \alpha_a\alpha_b) \right] = [3, 8] \tag{14}$$

$$s(A/B) = \left[\min_{\substack{\alpha_a \in [0,1] \\ \alpha_b \in [0,1]}} [(1 + \alpha_a)/(3 + \alpha_b)], \max_{\substack{\alpha_a \in [0,1] \\ \alpha_b \in [0,1]}} [(1 + \alpha_a)/(3 + \alpha_b)] \right] = [1/4, 2/3] \tag{15}$$

As it can be seen in Fig. 2 spans are only partial information pieces about full 3-dimensional result granules.

3 Properties of RDM and Moore Interval Arithmetic

Commutativity

Both Moore arithmetic and RDM arithnetic are commutative. For any intervals X and Y equations (16) and (17) are true.

$$X + Y = Y + X \tag{16}$$

$$X \cdot Y = Y \cdot X \tag{17}$$

Associativity

Also it is easy to show that both interval addition and multiplication in Moore arithmetic and RDM arithnetic are associative. For any intervals X, Y and Z there are true equations (18) and (19).

$$X + (Y + Z) = (X + Y) + Z \tag{18}$$

$$X \cdot (Y \cdot Z) = (X \cdot Y) \cdot Z \tag{19}$$

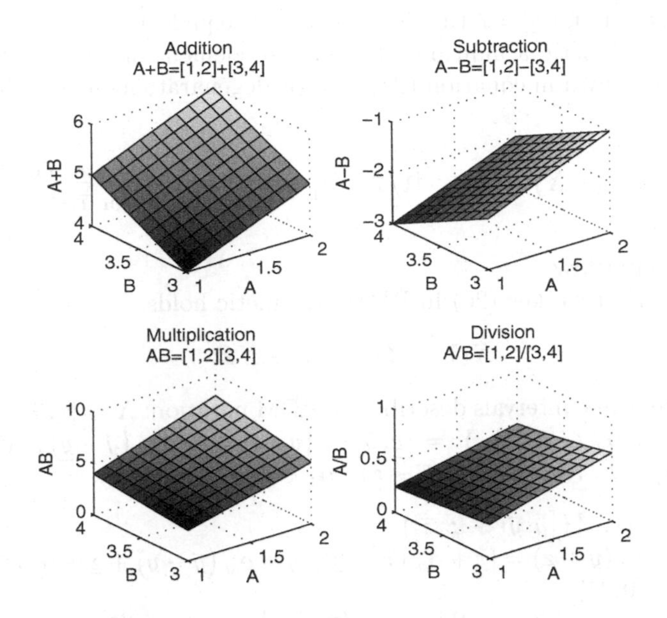

Fig. 2. Example of addition, subtraction, multiplication and division in RDM arithmetic of two intervals $A = [1, 2]$ and $B = [3, 4]$, 3-dimensional result

Neutral elements of addition and multiplication

In the conventional and RDM arithmetic there exist additive and multiplicative neutral elements such as degragative intervals 0 and 1 for any interval X, as shown in equations (20) and (21).

$$X + 0 = 0 + X = X \tag{20}$$

$$X \cdot 1 = 1 \cdot X = X \tag{21}$$

Inverse elements

In RDM arithmetic $-X = -[\underline{x}, \overline{x}] = \{-x : -x = -\underline{x} - \alpha_x (\overline{x} - \underline{x}),\ \alpha_x \in [0, 1]\}$, is an additive inverse element of interval $X = [\underline{x}, \overline{x}] = \{x : x = \underline{x} + \alpha_x (\overline{x} - \underline{x}),\ \alpha_x \in [0, 1]\}$, so:

$$X - X = \{x - x : x - x = \underline{x} + \alpha_x (\overline{x} - \underline{x}) - \underline{x} - \alpha_x (\overline{x} - \underline{x}),\ \alpha_x \in [0, 1]\} = 0. \tag{22}$$

An multiplicative inverse element of $X = \{x : x = \underline{x} - \alpha_x (\overline{x} - \underline{x}),\ \alpha_x \in [0, 1]\}$, if $0 \notin X$, in RDM arithmetic is $X = \{x : x = \underline{x} - \alpha_x (\overline{x} - \underline{x}),\ \alpha_x \in [0, 1]\}$:

$$X/X = \{x/x : x/x = [\underline{x} + \alpha_x (\overline{x} - \underline{x})] / [\underline{x} - \alpha_x (\overline{x} - \underline{x})],\ \alpha_x \in [0, 1]\} = 1. \tag{23}$$

In Moore arithmetic an additive inverse element for interval X does not exist, as equation (24) shows.

$$X - X = [\underline{x}, \overline{x}] - [\underline{x}, \overline{x}] = [\underline{x}, \overline{x}] + [-\overline{x}, -\underline{x}] = [\underline{x} - \overline{x}, \overline{x} - \underline{x}] \tag{24}$$

Equation (24) only for $\underline{x} = \overline{x}$ (if width of X is 0) equals $[0, 0]$.

In Moore arithmetic a multiplicative inverse element of interval X, $0 \notin X$, does not exist, shown in equation (25), except degenarate intervals where width equals zero.

$$X/X = X \cdot (1/X) = [\underline{x}, \overline{x}] \cdot [1/\overline{x}, 1/\underline{x}] = \begin{cases} [\underline{x}/\overline{x}, \overline{x}/\underline{x}] \text{ for } \underline{x} > 0 \\ [\overline{x}/\underline{x}, \underline{x}/\overline{x}] \text{ for } \overline{x} < 0 \end{cases} \tag{25}$$

Subdistributive law

The subdistributive law (26) in RDM arithmetic holds,

$$X(Y + Z) = XY + XZ. \tag{26}$$

Proof. For any three intervals described in RDM notation: $X = [\underline{x}, \overline{x}] = \{x : x = \underline{x} + \alpha_x (\overline{x} - \underline{x}),\ \alpha_x \in [0, 1]\}$, $Y = [\underline{y}, \overline{y}] = \{y : y = \underline{y} + \alpha_y (\overline{y} - \underline{y}),\ \alpha_y \in [0, 1]\}$ and $Z = [\underline{z}, \overline{z}] = \{z : z = \underline{z} + \alpha_z (\overline{z} - \underline{z}),\ \alpha_z \in [0, 1]\}$, we have:

$$\begin{aligned} &X(Y + Z) = [\underline{x}, \overline{x}] \left([\underline{y}, \overline{y}] + [\underline{z}, \overline{z}]\right) \\ &= \{x(y + z) : x(y + z) = [\underline{x} + \alpha_x (\overline{x} - \underline{x})] \left[\underline{y} + \alpha_y (\overline{y} - \underline{y}) + \underline{z} + \alpha_z (\overline{z} - \underline{z})\right], \\ &\alpha_x, \alpha_y, \alpha_z \in [0, 1]\} \\ &= \{xy : xy = [\underline{x} + \alpha_x (\overline{x} - \underline{x})] \left[\underline{y} + \alpha_y (\overline{y} - \underline{y})\right],\ \alpha_x, \alpha_y \in [0, 1]\} \\ &+ \{xz : xz = [\underline{x} + \alpha_x (\overline{x} - \underline{x})] [\underline{z} + \alpha_z (\overline{z} - \underline{z})],\ \alpha_x, \alpha_z \in [0, 1]\} \\ &= [\underline{x}, \overline{x}] [\underline{y}, \overline{y}] + [\underline{x}, \overline{x}] [\underline{z}, \overline{z}] = XY + XZ. \end{aligned}$$

□

In Moore arithmetic the subdistributive law holds only in the form (27).

$$X(Y + Z) \subseteq XY + XZ. \tag{27}$$

Cancellation law

The cancellation law for addition of intervals (28) holds for both Moore and RDM arithmetic.

$$X + Z = Y + Z \Rightarrow X = Y \tag{28}$$

Proof. Concerns RDM arithmetic. For any intervals $X = \{x : x = \underline{x} + \alpha_x(\overline{x} - \underline{x}),\ \alpha_x \in [0, 1]\}$, $Y = \{y : y = \underline{y} + \alpha_y (\overline{y} - \underline{y}),\ \alpha_y \in [0, 1]\}$ and $Z = \{z : z = \underline{z} + \alpha_z (\overline{z} - \underline{z}),\ \alpha_z \in [0, 1]\}$ in RDM notation using inverse element of interval Z and associativity we have:

$$X + Z = Y + Z$$

$$[\underline{x}, \overline{x}] + [\underline{z}, \overline{z}] = [\underline{y}, \overline{y}] + [\underline{z}, \overline{z}]$$

$$\begin{aligned} &\{x + z : x + z = \underline{x} + \alpha_x (\overline{x} - \underline{x}) + \underline{z} + \alpha_z (\overline{z} - \underline{z}), \alpha_x, \alpha_z \in [0, 1]\} \\ &= \{y + z : y + z = \underline{y} + \alpha_y (\overline{y} - \underline{y}) + \underline{z} + \alpha_z (\overline{z} - \underline{z}), \alpha_y, \alpha_z \in [0, 1]\} \end{aligned} \tag{29}$$

Adding an inverse interval $-Z = \{-z : -z = -\underline{z} - \alpha_z (\overline{z} - \underline{z}),\ \alpha_z \in [0, 1]\}$, to both sides of equation (29), we obtain:

$$\{x : x = \underline{x} + \alpha_x (\overline{x} - \underline{x}), \alpha_x \in [0, 1]\} = \{y : y = \underline{y} + \alpha_y (\overline{y} - \underline{y}), \alpha_y \in [0, 1]\}$$

$$[\underline{x}, \overline{x}] = [\underline{y}, \overline{y}]$$
$$X = Y$$

□

Example 2 shows that multiplicative cancellation does not hold in interval arithmetic, from equation $XZ = YZ$ we cannot imply $X = Y$.

Example 2. Let us give three intervals: $X = [1, 3]$, $Y = [2, 3]$ and $Z = [-1, 1]$. Using Moore arithmetic values of multiplication XZ and YZ are equal (30), but X and Y are different intervals.

$$\begin{aligned} XZ &= [1,3][-1,1] = [-3,3] \\ YZ &= [2,3][-1,1] = [-3,3] \end{aligned} \tag{30}$$

To find a product by RDM arithmetic we write intervals X, Y and Z in RDM notation with RDM variable (31).

$$\begin{aligned} X &= [1,3] = \{x : x = 1 + 2\alpha_x,\ \alpha_x \in [0,1]\} \\ Y &= [2,3] = \{y : y = 2 + \alpha_y,\ \alpha_y \in [0,1]\} \\ Z &= [-1,1] = \{z : z = -1 + 2\alpha_z,\ \alpha_z \in [0,1]\} \end{aligned} \tag{31}$$

The solutions with RDM variable α_x, α_y and α_z are presented in (32).

$$\begin{aligned} XZ &= [1,3][-1,1] = \{xz : xz = (1+2\alpha_x)(-1+2\alpha_z), \alpha_x, \alpha_z \in [0,1]\} \\ YZ &= [2,3][-1,1] = \{yz : yz = (2+\alpha_y)(-1+2\alpha_z),\ \alpha_y, \alpha_z \in [0,1]\} \end{aligned} \tag{32}$$

To find graphical illustration of solution, the border values should be computed, Table 2 and Table 3.

Table 2. Multiplication results of fwo intervals $XZ = [\underline{x}, \overline{x}]\,[\underline{z}, \overline{z}] = [1,3][-1,1]$ for border values of RDM-variables $\alpha_x \in [0,1]$ and $\alpha_z \in [0,1]$

α_x	0	0	1	1
x	1	3	1	3
α_z	0	1	0	1
z	-1	1	-1	1
xz	$\underline{x}\underline{z}$	$\underline{x}\overline{z}$	$\overline{x}\underline{z}$	$\overline{x}\overline{z}$
	-1	1	-3	3

Fig. 3 and Fig. 4 show fully 3-dimensional results of interval multiplication XZ and YZ, the span of solution in both cases are equal $[-3, 3]$, but the solution surfaces are different.

Comparing solutions obtained by Moore and RDM arithmetic we see that in Moore arithmetic multiplication interval $Z = [-1, 1]$ by different intervals $X = [1, 3]$ or $Y = [2, 3]$ ($X \neq Y$) gives the same results $XZ = YZ$ and the differences in multiplication XZ and YZ are not noticeable. Analizing the solution obtained by RDM arithmetic Fig. 3 and Fig. 4 show that the results of multiplication

Table 3. Multiplication results of fwo intervals $YZ = [\underline{y}, \overline{y}]\ [\underline{z}, \overline{z}] = [2,3][-1,1]$ for border values of RDM-variables $\alpha_y \in [0,1]$ and $\alpha_z \in [0,1]$

α_y	0	0	1	1
y	2	2	3	3
α_z	0	1	0	1
z	-1	1	-1	1
YZ	$\underline{y}\underline{z}$	$\underline{y}\overline{z}$	$\overline{y}\underline{z}$	$\overline{y}\overline{z}$
	-2	2	-3	3

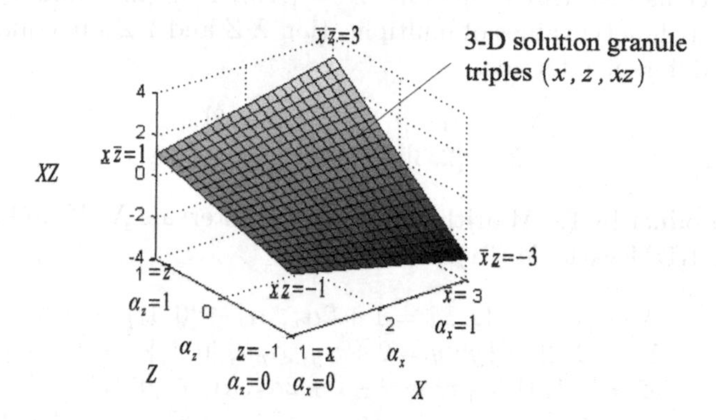

Fig. 3. 3-dimensional result of interval multiplication $XZ = [1,3][-1,1]$ with use of RDM interval arithmetic with span $[-3,3]$

intervals XZ and YZ are different, $XZ \neq YZ$. The span $[-3,3]$ are equal in both multiplications but the surfaces of solutions have different shapes.

Example 2 also shows that Moore arithmetic does not give a full solution, the solution obtained by Moore method is 1-dimensional and describes only the span. Example 3 shows that the results of operation made by Moore arithmetic depend on the form of equation.

Example 3. Let us consider the results of Moore and RDM interval arithmetic for nonlinear equation (33) where $A = [0,2]$.

$$C = A - A^2 \tag{33}$$

Equation (33) can take a form (34) and (35).

$$C = A(1 - A) \tag{34}$$

$$C = (A - 1) + (1 - A)(1 + A) \tag{35}$$

Calculating value C from equation (33), (34) and (35) for $A = [0,2]$ using Moore arithmetic we obtain different results (36), (37) and (38).

$$C_1 = [0,2] - ([0,2])^2 = [-4,2] \tag{36}$$

$$C_2 = [0,2](1 - [0,2]) = [-2,2] \tag{37}$$

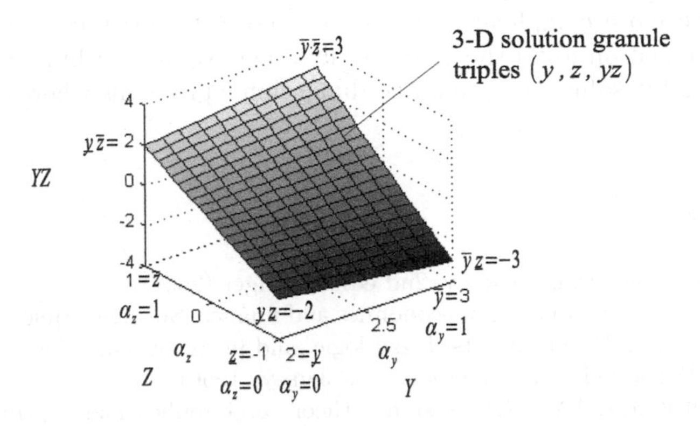

Fig. 4. 3-dimensional result of interval multiplication $YZ = [2,3][-1,1]$ with use of RDM interval arithmetic with span $[-3,3]$

$$C_3 = ([0,2]-1) + (1-[0,2])(1+[0,2]) = [-4,4] \tag{38}$$

Which solution is correct?

Solving equations (33), (34) and (35) by RDM arithmetic the inetrval $A = [0,2]$ in notation RDM takes the form (39).

$$A = [0,2] = \{a : a = 2\alpha_a,\ \alpha_a \in [0,1]\} \tag{39}$$

The solution obtained by RDM arithmetic for different forms of the equation (33) gives the same results (40), (41) and (42).

$$C = A - A^2 = \{c : c = 2\alpha_a - 4\alpha_a^2,\ \alpha_a \in [0,1]\} \tag{40}$$

$$\begin{aligned} C &= A(1-A) = \{c : c = 2\alpha_a(1-2\alpha_a), \alpha_a \in [0,1]\} \\ &= \{c : c = 2\alpha_a - 4\alpha_a^2, \alpha_a \in [0,1]\} \end{aligned} \tag{41}$$

$$\begin{aligned} C &= (A-1) + (1-A)(1+A) \\ &= \{c : c = (2\alpha_a - 1) + (1-2\alpha_a)(1+2\alpha_a), \alpha_a \in [0,1]\} \\ &= \{c : c = 2\alpha_a - 4\alpha_a^2, \alpha_a \in [0,1]\} \end{aligned} \tag{42}$$

The solution calculated by RDM arithmetic has only one RDM variable $\alpha_a \in [0,1]$ so is 1-dimensional and has the form (43).

$$C = \left[\min_{\alpha_a \in [0,1]}(2\alpha_a - 4\alpha_a^2),\ \max_{\alpha_a \in [0,1]}(2\alpha_a - 4\alpha_a^2)\right] = [-2, 1/4] \tag{43}$$

4 Conclusions

The paper compares the Moore and RDM interval arithmetic. The results obtained with Moore arithmetic are one-dimensional, the RDM arithmetic gives a multidimensional solution. In some cases the solutions in Moore arithmetic depend on the form of the equation, so it suggests that Moore arithmetic cannot correctly solve more complicated problems. The RDM arithmetic for different forms of the equation gives the same results. The Moore arithmetic gives only the span, not a full solution, except one-dimensional problems where the solution is an interval.

References

1. Boading, L.: Uncertainty theory, 2nd edn. Springer (2007)
2. Dymova, L.: Soft computing in economics and finance. Springer, Heidelberg (2011)
3. Klir, G.J., Yuan, B.: Fuzzy sets, fuzzy logic, and fuzzy systems. Selected paper by L. Zadeh, World Scientic, Singapor, New Jersey (1996)
4. Liu, S., Lin Forest, J.Y.: Grey systems, theory and applications. Springer, Heidelberg (2010)
5. Moore, R.E.: Interval analysis. Prentice Hall, Englewood Cliffs (1966)
6. Moore, R.E., Kearfott, R.B., Cloud, J.M.: Introduction to interval analysis. SIAM, Philadelphia (2009)
7. Pedrycz, W., Gomide, F.: Fuzzy systems engineering. Wiley, Hoboken (2007)
8. Pedrycz, W., Skowron, A., Kreinovich, V. (eds.): Handbook of granular computing. Wiley, Chichester (2008)
9. Piegat, A.: On practical problems with the explanation of the difference between possibility and probability. Control and Cybernetics 34(2) (2005)
10. Piegat, A., Landowski, M.: Is the conventional interval-arithmetic correct? Journal of Theoretical and Applied Computer Science 6(2), 27–44 (2012)
11. Piegat, A., Landowski, M.: Multidimensional approach to interval uncertainty calculations. In: Atanassov, K.T., et al. (eds.) New Trends in Fuzzy Sets, Intuitionistic: Fuzzy Sets, Generalized Nets and Related Topics, Volume II: Applications, Warsaw, Poland. IBS PAN - SRI PAS, Warsaw, pp. 137–151 (2013)
12. Piegat, A., Landowski, M.: Two Interpretations of Multidimensional RDM Interval Arithmetic - Multiplication and Division. International Journal of Fuzzy Systems 15(4), 488–496 (2013)
13. Piegat, A., Landowski, M.: Correctness-checking of uncertain-equation solutions on example of the interval-modal method. Paper presented on Twelfth International Workshop on Intuitionistic Fuzzy Sets and Generalized Nets, Warsaw, Poland (October 11, 2013)
14. Sevastjanov, P., Dymova, L., Bartosiewicz, L.: A framework for rule-base evidential reasoning in the interval settings applied to diagnosing type 2 diabets. Expert Systems with Applications 39, 4190–4200 (2012)

Interval-Valued Fuzzy Preference Relations and Their Properties

Urszula Bentkowska

University of Rzeszów, Department of Mathematics and Natural Sciences,
Al. Rejtana 16 C, 35-959 Rzeszów, Poland
ududziak@ur.edu.pl

Abstract. In the paper properties of interval-valued fuzzy preference relations are considered and preservation of a preference property by some operations, including lattice operations, the converse and the complement relations are studied. The concept of a preference relation presented here is a generalization of the concept of crisp preference relations. Moreover, weak properties of interval-valued fuzzy relations, namely reflexivity, irreflexivity, connectedness, asymmetry, antisymmetry, transitivity, and moderate transitivity are defined. Furthermore, the assumptions under which interval-valued fuzzy preference relations fulfil the mentioned properties are proposed.

Keywords: interval-valued relations, preference property, properties of interval-valued fuzzy relations, weak transitivity, moderate transitivity.

1 Introduction

Interval-valued fuzzy relations were introduced in [27] as a generalization of the concept of a fuzzy relation [25]. Fuzzy sets and relations have applications in diverse types of areas: data bases, pattern recognition, neural networks, fuzzy modeling, economy, medicine, multicriteria decision making. Similarly, interval-valued fuzzy sets are widely applied, for example in: group decision making where a solution from the individual preferences over some set of options should be derived [9], [24], in classification where the use of interval-valued fuzzy sets has allowed to improve the performance of some of the state-of-the-art algorithms for fuzzy rule based classification systems [20], [21], in applications to image processing [1], [8].

Preference relations appear in choice and utility theories. In economics, utility functions represent preferences over some set of goods and services. Choice theory is a framework for understanding and formally modeling social and economic behavior. Regarding preference relations the use of intervals allows more flexibility, since they allow to take into account possible lack of precision or certainty in the evaluation of the preferences. Interval-valued fuzzy relations are suitable in those setting where the expert has problems to provide a precise values for his/her judgement. In this sense, and as long as preference is given by a number to which a meaning has been attached (e.g. cost) intervals may

P. Angelov et al. (eds.), *Intelligent Systems'2014*,
Advances in Intelligent Systems and Computing 322, DOI: 10.1007/978-3-319-11313-5_31

provide a solution for those situations where that number cannot be precisely determined. However, note that the problem of measuring intensity of preference is very hard, and the literature about it is very scarce. This is a very deep issue and we do not consider it in the present work.

Interval-valued fuzzy preference relations were applied for example to goal programming [22]. Many other papers are devoted to the topic of preference relations and diverse problems are considered. For example interval-valued hesitant preference relations were introduced for decision making problems [10]. Consistency of preference relations in group decision making is an important subject and for interval-valued fuzzy relations it was discussed for example in [14], [16], [23].

Moreover, fuzzy preference relations could be understood as fuzzy order relations as introduced in [26] and studied in ([4],[5]). In this case completeness means $\max(R(x,y), R(y,x)) = 1$. The second way of considering fuzzy preferences is where completeness means $R(x,y) + R(y,x) = 1$, which is called reciprocity property and may be treated as a generalization of completeness of crisp preference relations [13]. We follow the first way and for interval-valued fuzzy relations we consider the adequate properties.

Here, properties of interval-valued fuzzy preference relations in the context of preservation of these properties by interval-valued operations are studied. Next, weak properties for interval-valued fuzzy relations are introduced: reflexivity, irreflexivity, connectedness, asymmetry, symmetry, transitivity and moderate transitivity. Moreover, suitable assumptions for interval-valued fuzzy preference relations to fulfil the considered properties are proposed.

Firstly, some concepts and results useful in further considerations are recalled (section 2). Next, results connected with the preservation of a preference property by the basic operations are given (section 3). Finally, weak properties of interval-valued fuzzy relations and moderate transitivity are defined. It is studied when such properties are fulfilled by interval-valued fuzzy preference relations (section 4).

2 Preliminary Notions

First we recall the notion of the lattice operations and the order in the family of interval-valued fuzzy relations. Let X, Y, Z be non-empty sets and $Int([0,1]) = \{[x_1, x_2] : x_1, x_2 \in [0,1], x_1 \leqslant x_2\}$. Interval-valued fuzzy relations reflect the idea that membership grades are often not precise and the intervals represent such uncertainty.

Definition 1 (cf. [19], [27]). *An interval-valued fuzzy relation R between universes X, Y is a mapping $R : X \times Y \to Int([0,1])$ such that*

$$R(x,y) = [\underline{R}(x,y), \overline{R}(x,y)] \in Int([0,1]), \tag{1}$$

for all pairs $(x,y) \in (X \times Y)$. The class of all interval-valued fuzzy relations between universes X, Y will be denoted by $IVFR(X \times Y)$ or $IVFR(X)$ for $X = Y$.

The boundary elements in $IVFR(X \times Y)$ are $\mathbf{1} = [1,1]$ and $\mathbf{0} = [0,0]$. Let $S, R \in IVFR(X \times Y)$. Then for every $(x,y) \in (X \times Y)$ we have

$$S(x,y) \leqslant R(x,y) \Leftrightarrow \underline{S}(x,y) \leqslant \underline{R}(x,y), \overline{S}(x,y) \leqslant \overline{R}(x,y), \tag{2}$$

$$(S \vee R)(x,y) = [\max(\underline{S}(x,y), \underline{R}(x,y)), \max(\overline{S}(x,y), \overline{R}(x,y))], \tag{3}$$

$$(S \wedge R)(x,y) = [\min(\underline{S}(x,y), \underline{R}(x,y)), \min(\overline{S}(x,y), \overline{R}(x,y))], \tag{4}$$

where operations $\vee$ and $\wedge$ are the supremum and the infimum in $IVFR(X \times Y)$, respectively. Similarly for arbitrary set $T \neq \emptyset$

$$(\bigvee_{t \in T} R_t)(x,y) = [\sup_{t \in T} \underline{R_t}(x,y), \sup_{t \in T} \overline{R_t}(x,y)], \tag{5}$$

$$(\bigwedge_{t \in T} R_t)(x,y) = [\inf_{t \in T} \underline{R_t}(x,y), \inf_{t \in T} \overline{R_t}(x,y)]. \tag{6}$$

The pair $(IVFR(X \times Y), \leqslant)$ is a partially ordered set. As a result, the family $(IVFR(X \times Y), \vee, \wedge)$ is a lattice (for the notion of a lattice and other related concepts see [3]) which is a consequence of the fact that $([0,1], \max, \min)$ is a lattice. The lattice $IVFR(X \times Y)$ is complete. This fact follows from the notion of the supremum $\bigvee$ and the infimum $\bigwedge$ and from the fact that the values of fuzzy relations are from the interval $[0,1]$ which, with the operations maximum and minimum, forms a complete lattice. As a result $(IVFR(X \times Y), \vee, \wedge)$ is a complete, infinitely distributive lattice. Let us now recall the notion of the composition.

Definition 2 (cf. [7], [15]). *Let $S \in IVFR(X \times Y)$, $R \in IVFR(Y \times Z)$. The* $\sup - \min$ *composition of the relations S and R is called the relation $S \circ R \in IVFR(X \times Z)$,*

$$(S \circ R)(x,z) = [(\underline{S} \circ \underline{R})(x,z), (\overline{S} \circ \overline{R})(x,z)],$$

where

$$(\underline{S} \circ \underline{R})(x,z) = \sup_{y \in Y} \min(\underline{S}(x,y), \underline{R}(y,z)),$$

$$(\overline{S} \circ \overline{R})(x,z) = \sup_{y \in Y} \min(\overline{S}(x,y), \overline{R}(y,z))$$

and $(\underline{S} \circ \underline{R})(x,z) \leqslant (\overline{S} \circ \overline{R})(x,z)$.

The structure $(IVFR(X), \circ)$ is a semigroup (see [17]), thus we can consider the powers of its elements, i.e. relations R^n for $R \in IVFR(X)$, $n \in \mathbb{N}$.

Definition 3 ([7]). *A power of relation $R \in IVFR(X)$ is defined in the following way*

$$R^1 = R, \; R^{n+1} = R^n \circ R, \; n \in \mathbb{N}.$$

We do not consider here other composition types in the family $IVFR(X)$, in which max and min are replaced with adequate operations ([6], p. 295).

Definition 4 ([7]). *For arbitrary $R \in IVFR(X \times Y)$ we may consider:*
- *the converse relation: R^{-1}, where $R^{-1}(x,y) = [\underline{R}(y,x), \overline{R}(y,x)]$,*
- *the complement: $R' = 1 - R$, where $R'(x,y) = [1 - \overline{R}(x,y), 1 - \underline{R}(x,y)]$,*
- *the dual relation: R^d, where $R^d(x,y) = [1 - \overline{R}(y,x), 1 - \underline{R}(y,x)]$.*

As a result $R^d = (R')^{-1} = (R^{-1})'$.

If we consider decision making problems in the interval-valued fuzzy environment we deal with the finite set of alternatives $X = \{x_1, \ldots, x_n\}$ and an expert who needs to provide his/her preference information over alternatives. There are the following possibilities: the decision maker can provide clear judgement whether or not he/she prefers the alternative x_i over x_j, the alternatives are indifferent to them or the alternatives are incomparable.

In the fuzzy setting there are possible two approaches. The first approach where $R(x,y) = 1$ means full strict preference of x over y, and is equivalent to $R(y,x) = 0$ which expresses full negative preference, and suggest indifference be modelled by $R(x,y) = R(y,x) = 0.5$, and more generally the property $R(x,y) + R(y,x) = 1$ is assumed. This property generalizes completeness, and $R(x,y) > 0.5$ expresses a degree of strict preference, so antisymmetry means $R(x,y) = 0.5 \Rightarrow x = y$. In the second approach $R(x,y)$ evaluates weak preference, completeness means $\max(R(x,y), R(y,x)) = 1$ and indifference is when $R(x,y) = R(y,x) = 1$ while incomparability is when $R(x,y) = R(y,x) = 0$. As a result this convention allows a direct extension of crisp preference relations to the fuzzy ones. Conventions of crisp relations are obtained when restricting to Boolean values (see [13]).

In this paper we follow the second convention for interval-valued fuzzy relations (the results on the first convention for interval-valued fuzzy relations are discussed in [2]). In the sequel, we will consider a preference relation on a finite set $X = \{x_1, \ldots, x_n\}$. In this situation interval-valued fuzzy relations may be represented by matrices.

Definition 5. *Let card $X = n$. An interval-valued fuzzy relation R on the set X is represented by a matrix $R = (r_{ij})_{n \times n}$ with $r_{ij} = [\underline{R}(i,j), \overline{R}(i,j)]$ for all $i, j = 1, ..., n$, where $r_{ij} \in Int([0,1])$ and $r_{ij} \vee r_{ji} = [1,1]$.*

The element r_{ij} indicates the interval-valued degree or intensity of the alternative x_i over alternative x_j and $\underline{R}(i,j)$, $\overline{R}(i,j)$ are the lower and upper limits of r_{ij}, respectively. Moreover, $r_{ij} = r_{ji} = [1,1]$ indicates indifference between x_i and x_j (and obviously by Definition 5 one has $r_{ii} = [1,1]$), $r_{ij} = r_{ji} = [0,0]$ indicates incomparability between x_i and x_j and r_{ij} indicates that x_i is weakly preferred to x_j. If $\overline{R}(i,j) = \underline{R}(i,j) = r_{ij}$ for $i, j = 1, ..., n$, then an interval-valued fuzzy relation reduces to a fuzzy relation. Furthermore, by the fact that for interval-valued fuzzy preference relation we have $r_{ij} \in Int([0,1])$ and $r_{ij} \vee r_{ji} = [1,1]$, it follows that $r_{ij} = [1,1]$ or $r_{ji} = [1,1]$ for all $i, j = 1, ..., n$,.

3 Basic Operations on Preference Relations

In this section, we will consider interval-valued fuzzy relations on a finite set $X = \{x_1, ..., x_n\}$ and some fundamental operations on these relations in the context of preservation of the preference property by these operations.

Proposition 1. *Let $R \in IVFR(X)$. If R is an interval-valued fuzzy preference relation, then R^{-1} is an interval-valued fuzzy preference relation.*

Proof. Let $R \in IVFR(X)$ be an interval-valued fuzzy preference relation, $i, j \in \{1, ..., n\}$. Let us denote $S = R^{-1}$, $S = [s_{ij}]$. Then by commutativity of maximum

$$s_{ij} \vee s_{ji} = [\max(\underline{S}(i,j), \underline{S}(j,i)), \max(\overline{S}(i,j), \overline{S}(j,i))] =$$

$$[\max(\underline{R}(j,i), \underline{R}(i,j)), \max(\overline{R}(j,i), \overline{R}(i,j))] = [1,1].$$

As a result $S = R^{-1}$ is an interval-valued fuzzy preference relation.

Proposition 2. *Let $R \in IVFR(X)$. If R is an interval-valued fuzzy preference relation and $r_{ij} = [0,0]$ or $r_{ji} = [0,0]$ for $i, j \in \{1, ..., n\}$, then R' and R^d are interval-valued fuzzy preference relations.*

Proof. Let $R \in IVFR(X)$ be an interval-valued fuzzy preference relation, $r_{ij} = [0,0]$, $i, j \in \{1, ..., n\}$. Let us denote $T = R'$, $T = [t_{ij}]$. Then

$$t_{ij} \vee t_{ji} = [\max(1 - \overline{R}(i,j), 1 - \overline{R}(j,i)), \max(1 - \underline{R}(i,j), 1 - \underline{R}(j,i))] =$$

$$[1 - \min(\overline{R}(i,j), \overline{R}(j,i)), 1 - \min(\underline{R}(i,j), \underline{R}(j,i))] = [1,1].$$

This means that $T = R'$ is an interval-valued fuzzy preference relation and by the property $R^d = (R')^{-1} = (R^{-1})'$ and Proposition 1 we see that R^d is also an interval-valued fuzzy preference relation.

Now we will consider preservation of preference property by the lattice operations in the family $IVFR(X)$.

Proposition 3. *Let $R, S \in IVFR(X)$, be interval-valued fuzzy preference relations. Then $R \vee S$ is an interval-valued fuzzy preference relation.*

Proof. Let $R, S \in IVFR(X)$ be interval-valued fuzzy preference relations, $i, j \in \{1, ..., n\}$. Let us denote $R = [r_{ij}]$, $S = [s_{ij}]$. Then by commutativity and associativity of maximum and by assumptions on R and S we get

$$(r \vee s)_{ij} \vee (r \vee s)_{ji} = [1,1].$$

As a result $R \vee S$ is an interval-valued fuzzy preference relation.

Example 1. Let card $X = 2$ and $R, S \in IVFR(X)$ be preference relations represented by the matrices:

$$R = \begin{bmatrix} [1,1] & [1,1] \\ [0.4,0.6] & [1,1] \end{bmatrix}, S = \begin{bmatrix} [1,1] & [0.3,0.7] \\ [1,1] & [1,1] \end{bmatrix}.$$

Then we obtain

$$R \wedge S = \begin{bmatrix} [1,1] & [0.3,0.7] \\ [0.4,0.6] & [1,1] \end{bmatrix},$$

We see that relation $R \wedge S$ is not a preference relation.

By definition of an interval-valued fuzzy preference relations and definition of $\wedge$ operation it is easy to see that

Proposition 4. *Let $R, S \in IVFR(X)$ be interval-valued fuzzy preference relations. If for all $i, j = 1, ...n$, $r_{ij} = [1,1] \Leftrightarrow s_{ij} = [1,1]$, then $R \wedge S$ is an interval-valued fuzzy preference relation.*

4 Properties of Interval-Valued Fuzzy Preference Relations

In this section we consider some properties of interval-valued fuzzy relations in the context of preference relations. We will pay our attention to the transitivity property which is crucial for preference relations since it is the most important property guaranteing consistency of a preference relation. Since transitivity is so important we examine three types of this property (for other types of transitivity see [14]).

Now, we recall the "standard" properties of interval-valued fuzzy relations which are direct generalizations of well-known fuzzy relation properties and we define weak versions of these properties (for transitivity there is additionally given property of moderate transitivity). We follow the concept of such properties given by Drewniak [12] for fuzzy relations. Other types of transitivity properties for fuzzy setting and fuzzy preference relations are discussed for example in [11]. In the sequel, to simplify the notations we will use the symbols: $\vee$ and $\wedge$ instead of max and min, respectively.

Definition 6 ([18]). *An interval-valued fuzzy relation $R \in IVFR(X)$ is called:*

- *reflexive if*

$$\forall x \in X : R(x,x) = [1,1], \tag{7}$$

- *irreflexive if*

$$\forall x \in X : R(x,x) = [0,0], \tag{8}$$

- *symmetric if*

$$\forall x, y \in X : R(x,y) = R(y,x), \tag{9}$$

- *asymmetric if*

$$\forall x, y \in X : R(x,y) \wedge R(y,x) = [0,0], \tag{10}$$

- *antisymmetric if*

$$\forall x, y \in X, x \neq y : R(x, y) \wedge R(y, x) = [0, 0], \tag{11}$$

- *totally connected if*

$$\forall x, y \in X : R(x, y) \vee R(y, x) = [1, 1], \tag{12}$$

- *connected if*

$$\forall x, y \in X, x \neq y : R(x, y) \vee R(y, x) = [1, 1], \tag{13}$$

- *transitive if*

$$\forall x, y, z \in X : R(x, z) \geqslant R(x, y) \wedge R(y, z). \tag{14}$$

Corollary 1 (cf. [2]). *Let $R \in IVFR(X)$. R is transitive if and only if $R^2 \leqslant R$.*

The transitivity has the following interpretation for the preference relation: the interval value r_{ij}, representing the preference between the alternatives x_i and x_j, should be equal or greater than the minimum of interval values r_{ik} and r_{kj} between the alternatives x_i, x_k and x_k, x_j.

Definition 7. *An interval-valued fuzzy relation $R \in IVFR(X)$ is called:*

- *weakly reflexive if*

$$\forall x \in X : R(x, x) > [0, 0], \tag{15}$$

- *weakly irreflexive if*

$$\forall x \in X : R(x, x) < [1, 1], \tag{16}$$

- *weakly symmetric if*

$$\forall x, y \in X : R(x, y) = [1, 1] \Leftrightarrow R(y, x) = [1, 1], \tag{17}$$

- *weakly asymmetric if*

$$\forall x, y \in X : R(x, y) \wedge R(y, x) < [1, 1], \tag{18}$$

- *weakly antisymmetric if*

$$\forall x, y \in X, x \neq y : R(x, y) \wedge R(y, x) < [1, 1], \tag{19}$$

- *totally weakly connected if*

$$\forall x, y \in X : R(x, y) \vee R(y, x) > [0, 0], \tag{20}$$

- *weakly connected if*

$$\forall x, y \in X, x \neq y : R(x, y) \vee R(y, x) > [0, 0], \tag{21}$$

- *weakly transitive if*

$$\forall x, y, z \in X : R(x, y) \wedge R(y, z) > [0, 0] \Rightarrow R(x, z) > [0, 0], \tag{22}$$

- *moderately transitive if*

$$\forall x, y, z \in X : R(x, y) \wedge R(y, z) = [1, 1] \Rightarrow R(x, z) = [1, 1]. \tag{23}$$

By definition of weak and moderate transitivity and definition of the composition of interval-valued fuzzy relations we obtain characterizations of moderate and weak transitivity.

Lemma 1. *Let $R \in IVFR(X)$ be an interval-valued fuzzy relation. Relation R is weakly transitive if and only if*

$$\forall x, z \in X : R^2(x,z) > [0,0] \Rightarrow R(x,z) > [0,0]. \tag{24}$$

Relation R is moderately transitive if and only if

$$\forall x, z \in X : R^2(x,z) = [1,1] \Rightarrow R(x,z) = [1,1]. \tag{25}$$

Proof. If R is weakly transitive, then by (22), definition of the order and by applying the tautologies for quantifiers we obtain

$$\forall x, y, z \in X : \underline{R}(x,y) \wedge \underline{R}(y,z) > 0 \Rightarrow \underline{R}(x,z) > 0$$

and

$$\forall x, y, z \in X : \overline{R}(x,y) \wedge \overline{R}(y,z) > 0 \Rightarrow \overline{R}(x,z) > 0.$$

As a result

$$\forall x, z \in X (\forall y \in X : \underline{R}(x,y) \wedge \underline{R}(y,z) > 0 \Rightarrow \underline{R}(x,z) > 0)$$

and

$$\forall x, z \in X (\forall y \in X : \overline{R}(x,y) \wedge \overline{R}(y,z) > 0 \Rightarrow \overline{R}(x,z) > 0).$$

This implies

$$\forall x, z \in X : \sup_{y \in X} (\underline{R}(x,y) \wedge \underline{R}(y,z)) > 0 \Rightarrow \underline{R}(x,z) > 0 \tag{26}$$

and

$$\forall x, z \in X : \sup_{y \in X} (\overline{R}(x,y) \wedge \overline{R}(y,z)) > 0 \Rightarrow \overline{R}(x,z) > 0, \tag{27}$$

so by the definition of composition we get (24).

Let us assume that condition (24) is fulfilled which is equivalent to conditions (26) and (27). We will show that R is weakly transitive. Let $x, y, z \in X$ and the antecedent in (22) be fulfilled. As a result we have $\underline{R}(x,y) \wedge \underline{R}(y,z) > 0$ and $\overline{R}(x,y) \wedge \overline{R}(y,z) > 0$. By definition of supremum we obtain

$$\sup_{y \in X} (\underline{R}(x,y) \wedge \underline{R}(y,z)) \geqslant \underline{R}(x,y) \wedge \underline{R}(y,z) > 0$$

and

$$\sup_{y \in X} (\overline{R}(x,y) \wedge \overline{R}(y,z)) \geqslant \overline{R}(x,y) \wedge \overline{R}(y,z) > 0.$$

By (26), (27) we have $\underline{R}(x,z) > 0$ and $\overline{R}(x,z) > 0$. By definition of the interval-valued fuzzy relation this finishes the proof. The justification for the property of moderate transitivity is analogous.

Properties from Definition 7 in some cases may be more useful for fuzzy preference relations than the ones which are direct generalizations, to the interval-valued fuzzy environment, of the properties introduced by Zadeh (cf. [25]) because they are weaker versions of the properties from Definition 6, i.e. if R has the given property from Definition 6, then it is the appropriate weak property from Definition 7. Here, we have two types of transitivity - weak and moderate transitivity, which are weaker properties than transitivity but they are not comparable with one another.

Corollary 2. *Let $R \in IVFR(X)$ be an interval-valued fuzzy relation. If R is transitive, then it is weakly and moderately transitive.*

Proof. Let $R \in IVFR(X)$ be transitive, $x, z \in X$. If $R^2(x,z) > [0,0]$, then by transitivity $[0,0] < R^2(x,z) \leq R(x,z)$, so $R(x,z) > [0,0]$ and R is weakly transitive. If $R^2(x,z) = [1,1]$, then by transitivity $[1,1] = R^2(x,z) \leq R(x,z)$ and definition of interval valued fuzzy relations we have $R(x,z) = [1,1]$. It means that R is moderately transitive.

Example 2. Properties if weak and moderate transitivity are not comparable. Let $card\ X = 3$ and relations $R, S \in IVFR(X)$ be given by the matrices:

$$R = \begin{bmatrix} [1,1] & [1,1] & [0.2,0.5] \\ [0.6,0.8] & [1,1] & [1,1] \\ [1,1] & [0.2,0.7] & [1,1] \end{bmatrix}, \quad S = \begin{bmatrix} [1,1] & [1,1] & [1,1] \\ [0.4,0.6] & [1,1] & [1,1] \\ [0.2,0.5] & [0,0] & [1,1] \end{bmatrix}.$$

Then $T = R^2 \equiv [1,1]$, so for example $t_{32} = [1,1]$, but $r_{32} = [0.2,0.7]$ and by Lemma 1 it means that R is not moderately transitive but it is weakly transitive. Moreover, we have

$$S^2 = \begin{bmatrix} [1,1] & [1,1] & [1,1] \\ [0.4,0.6] & [1,1] & [1,1] \\ [0.2,0.5] & [0.2,0.5] & [1,1] \end{bmatrix},$$

so $W = S^2$ and by Lemma 1 relation S is moderately transitive but it is not weakly transitive ($w_{32} > [0,0]$, but $s_{32} = [0,0]$).

Now, we will check under which assumptions an interval-valued fuzzy preference relation has the given property. Directly by the definition of an interval-valued fuzzy preference relation we obtain

Corollary 3. *Each interval-valued fuzzy preference relation is weakly reflexive, reflexive, weakly (totally) connected, (totally) connected but it cannot be neither weakly irreflexive nor irreflexive.*

Corollary 4. *Let $card\ X = n$, $R \in IVFR(X)$ be an interval-valued fuzzy preference relation. If $r_{ij} = [1,1]$ for all $i, j \in \{1, ..., n\}$, then R weakly symmetric and symmetric.*

Corollary 5. *Let $card\ X = n$, $R \in IVFR(X)$ be an interval-valued fuzzy preference relation. If for all $i, j \in \{1, ..., n\}$ we have $\min(\overline{R}(i,j), \overline{R}(j,i)) = 0$, then R is asymmetric (antisymmetric). If for all $i, j \in \{1, ..., n\}$ we have $\min(\overline{R}(i,j), \overline{R}(j,i)) < 1$, then R is weakly asymmetric (weakly antisymmetric).*

Proof. Let $i, j \in \{1, ..., n\}$. If we assume that $\min(\overline{R}(i,j), \overline{R}(j,i)) = 0$, then also $\min(\underline{R}(i,j), \underline{R}(j,i)) = 0$, so R is asymmetric and antisymmetric. The proof for weak asymmetry and weak antisymmetry is analogous.

Proposition 5. *Let card $X = n$, $R \in IVFR(X)$ be an interval-valued fuzzy preference relation. If*

$$\forall i, j \in \{1, ..., n\} : r_{ij} = [1, 1] \Leftrightarrow i \geq j \tag{28}$$

(or

$$\forall i, j \in \{1, ..., n\} : r_{ij} = [1, 1] \Leftrightarrow i \leq j), \tag{29}$$

then R is moderately transitive.

Proof. Let $i, j, k \in \{1, ..., n\}$. We will show that for a preference relation the implication $r_{ij} \wedge r_{jk} = [1, 1] \Rightarrow r_{ik} = [1, 1]$ holds. There are the following cases: $1^0)\ i > k > j$, $2^0)\ i > j > k$, $3^0)\ j > i > k$, $4^0)\ k > j > i$, $5^0)\ k > i > j$, $6^0)\ j > k > i$.
In the first three cases the consequence in the given implication is true, so the implication is true. In the remaining cases, the antecedent of the given implication is false, so the implication is true. In a similar way we may show the property assuming the condition (29).

Example 3. Condition given in (28) is only the sufficient one for a relation R to be moderately transitive. Let *card* $X = 3$ and relation $R \in IVFR(X)$ with the determined R^2 be given by the matrices:

$$R = \begin{bmatrix} [1,1] & [1,1] & [1,1] \\ [0.5, 0.8] & [1,1] & [1,1] \\ [0.2, 0.6] & [1,1] & [1,1] \end{bmatrix}, \quad R^2 = \begin{bmatrix} [1,1] & [1,1] & [1,1] \\ [0.5, 0.8] & [1,1] & [1,1] \\ [0.5, 0.8] & [1,1] & [1,1] \end{bmatrix}.$$

Then relation R does not fulfil (28) however by Lemma 1 it is moderately transitive.

It is easy to see that the following property holds for interval-valued fuzzy relations (including the ones having preference property).

Proposition 6. *Let card $X = n$, $R \in IVFR(X)$ be an interval-valued fuzzy relation. If for all $i, j \in \{1, \ldots, n\}$ we have $r_{ij} > [0, 0]$, then $R \in IVFR(X)$ is weakly transitive.*

5 Conclusion

In this paper we studied properties of interval-valued fuzzy preference relations in the context of preservation of this property by the basic interval-valued operations. We also introduced the weak properties and moderate transitivity for interval-valued fuzzy relations and we investigated fulfilment of these properties by interval-valued fuzzy preference relations. These properties may be important because of its possible applications for preference procedure.

Acknowledgments. This work was partially supported by the Center for Innovation and Transfer of Natural Sciences and Engineering Knowledge in Rzeszów.

References

1. Barrenechea, E., Bustince, H., De Baets, B., Lopez-Molina, C.: Construction of interval-valued fuzzy relations with application to the generation of fuzzy edge images. IEEE Transactions on Fuzzy Systems 19(5), 819–830 (2011)
2. Bentkowska, U., Bustince, H., Pękala, B.: Some properties of interval-valued fuzzy relations. Fuzzy Sets and Systems (submitted
3. Birkhoff, G.: Lattice Theory. AMS Coll. Publ. 25, Providence (1967)
4. Bodenhofer, U.: Representations and constructions of similarity-based fuzzy orderings. Fuzzy Sets and Systems 137, 113–136 (2003)
5. Bodenhofer, U., De Baets, B., Fodor, J.: A compendium of fuzzy weak orders: Representations and constructions. Fuzzy Sets and Systems 158, 811–829 (2007)
6. Bustince, H., Burillo, P.: Structures on intuitionistic fuzzy relations. Fuzzy Sets and Systems 78, 293–303 (1996)
7. Bustince, H., Burillo, P.: Mathematical Analysis of Interval-Valued Fuzzy Relations: Application to Approximate Reasoning. Fuzzy Sets and Systems 113, 205–219 (2000)
8. Bustince, H., Barrenechea, E., Pagola, M., Fernandez, J.: Interval-valued fuzzy sets constructed from matrices: application to edge detection. Fuzzy Sets and Systems 160(13), 1819–1840 (2009)
9. Chen, H., Zhou, L.: An approach to group decision making with interval fuzzy preference relations based on induced generalized continuous ordered weighted averaging operator. Expert Systems with Applications 38, 13432–13440 (2011)
10. Chen, N., Xu, Z., Xia, M.: Interval-valued hesitant preference relations and their applications to group decision making. Knowledge-Based Systems 37, 528–540 (2013)
11. Chiclana, F., Herrera-Viedma, E., Alonso, S., Pereira, R.A.M.: Preferences and consistency issues in group decision making. In: Bustince, H., et al. (eds.) Fuzzy Sets and Their Extensions: Representation, Aggregation and Models. STUDFUZZ, vol. 220, pp. 219–237. Springer, Berlin (2008)
12. Drewniak, J.: Fuzzy Relation Calculus. Silesian University, Katowice (1989)
13. Dubois, D.: The role of fuzzy sets in decision sciences: Old techniques and new directions. Fuzzy Sets and Systems 184(1), 3–28 (2011)
14. Genç, S., Boran, F.E., Akay, D., Xu, Z.: Interval multiplicative transitivity for consistency, missing values and priority weights of interval fuzzy preference relations. Information Sciences 180, 4877–4891 (2010)
15. Goguen, A.: L-fuzzy sets. Journal of Mathematical Analysis and Applications 18, 145–174 (1967)
16. Liu, F.: Acceptable consistency analysis of interval reciprocal comparison matrices. Fuzzy Sets and Systems 160, 2686–2700 (2009)
17. Pękala, B.: Operations on Interval Matrices. In: Kryszkiewicz, M., Peters, J.F., Rybiński, H., Skowron, A. (eds.) RSEISP 2007. LNCS (LNAI), vol. 4585, pp. 613–621. Springer, Heidelberg (2007)
18. Pękala, B.: Properties of Interval-Valued Fuzzy Relations and Atanassovs Operators. In: Atanassov, K.T., et al. (eds.) Developments in Fuzzy Sets, Intuitionistic Fuzzy Sets, Generalized Nets and related Topics, vol, vol. 1, pp. 153–166. EXIT, Warszawa (2010)

19. Sambuc, R.: Fonctions ϕ-floues: Application á l'aide au diagnostic en pathologie thyroidienne. Ph.D. Thesis. Université de Marseille, France (1975) (in French)
20. Sanz, J., Fernandez, A., Bustince, H., Herrera, F.: Improving the performance of fuzzy rule-based classification systems with interval-valued fuzzy sets and genetic amplitude tuning. Information Sciences 180(19), 3674–3685 (2010)
21. Sanz, J., Fernandez, A., Bustince, H., Herrera, F.: A genetic tuning to improve the performance of fuzzy rule-based classification systems with intervalvalued fuzzy sets: degree of ignorance and lateral position. International Journal of Approximate Reasoning 52(6), 751–766 (2011)
22. Wang, Z.-J., Li, K.W.: Goal programming approaches to deriving interval weights based on interval fuzzy preference relations. Information Sciences 193, 180–198 (2012)
23. Xu, Z.: Consistency of interval fuzzy preference relations in group decision making. Applied Soft Computing 11, 3898–3909 (2011)
24. Yager, R.R., Xu, Z.: Intuitionistic and interval-valued intuitionistic fuzzy preference relations and their measures of similarity for the evaluation of agreement within a group. Fuzzy Optimization and Decision Making 8, 123–139 (2009)
25. Zadeh, L.A.: Fuzzy sets. Information and Control 8, 338–353 (1965)
26. Zadeh, L.A.: Similarity relations and fuzzy orderings. Information Sciences 3, 177–200 (1971)
27. Zadeh, L.A.: The Concept of a Linguistic Variable and its Application to Approximate Reasoning-I. Information Sciences 8, 199–249 (1975)

Combining Uncertainty and Vagueness in Time Intervals

Christophe Billiet and Guy De Tré

Department of Telecommunications and Information Processing, Ghent University, Sint-Pietersnieuwstraat 41, B-9000, Ghent, Belgium
{Christophe.Billiet,Guy.DeTre}@UGent.be

Abstract. Database systems contain data representing properties of real-life objects or concepts. Many of these data represent time indications and such time indications are often subject to imperfections. Although several existing proposals deal with the modeling of uncertainty or vagueness in time indications in database systems, only a few of them summarily examine the interpretation and semantics of such imperfections. The work presented in this paper starts at a more thorough examination of the semantics and modeling of uncertainty or vagueness in time intervals in database systems and presents methods to model combinations of uncertainty and vagueness in time intervals in database systems, based on examinations of their requisite interpretations.

1 Introduction

Information systems in general and database systems (DBS) in specific usually have two main purposes: to preserve data, information or knowledge concerning real-life objects or concepts and to allow humans to retrieve these data, information or knowledge. In order to do this, DBS usually dispose of data representing properties of these real-life objects or concepts. Through such data (a.o.), DBS model these real-life objects or concepts [2], [14]. For example, patient DBS may model patients' medical state by containing data representing these patients' heart rate, blood pressure, ...

Many real-life objects or concepts are somehow time-related. Thus, DBS often dispose of data representing time indications [1], [11], [15], [17]. Time indications modeled by DBS refer to parts of time and generally take the form of either *time intervals* [3], [10] or *instants* [3], [10]. A time interval can intuitively be seen as a continuous period of time, whereas an instant can intuitively be seen as an infinitesimally short 'moment' in time [3], [10]. For example, patient DBS may need to model patients' medical history by containing data representing the time intervals during which the patients suffered from certain diseases or data representing the instants when the patients took medication. Because time intervals in DBS are studied more exhaustively in literature [1], [2], [5], [8], [12], [14], [15], [16], [17] than instants in DBS [7], [8] and because an instant can be seen (and modeled) as a time interval, the work presented in this paper only considers time intervals in DBS.

P. Angelov et al. (eds.), *Intelligent Systems'2014*,
Advances in Intelligent Systems and Computing 322, DOI: 10.1007/978-3-319-11313-5_32

Usually, a lot of the data disposed of by DBS are created by humans or represent information or knowledge created by humans. Such data represent properties which can be the result of measurements, estimations, derivations, ... However, human-made data, information and knowledge can be prone to imperfections, due to the inherent imperfect nature of humans and human reasoning or imperfections in measuring equipment [1], [2], [5], [6], [7], [8], [9], [11], [12], [13], [14], [15]. For example, a medical professional may need to estimate the exact moment in time when a patient was given a certain medicine, as nobody remembers exactly, or the period of time during which a patient suffered from a certain disease, fully knowning that the emergence and the disappearance of the disease are gradual. Obviously, the degree to which data, information or knowledge preserved in an DBS present a correct reflection of reality and thus are any useful, is heavily influenced by the degree to which the imperfections in these data, information or knowledge can be defined and described in the DBS. For example, if a patient DBS only allows to specify one single exact date on which a patient started suffering from a disease, medical professionals will never be able to make the information modeled by this DBS reflect the gradual emergence and disappearance of a disease over the period of several days. Thus, the authors of this paper strongly believe that modeling imperfections in data, information or knowledge to incorporate such imperfections into DBS is highly preferable to only allowing perfect, but inherently wrong data, information or knowledge into DBS. Hence, the work presented in this paper considers modeling time intervals subject to imperfection in DBS.

The two most common types of imperfections to which time intervals in DBS could be subject, are [2], [5], [8], [11], [12], [14], [15], [16], [17]:

- **Uncertainty.** A time interval is subject to *uncertainty* when it is known that a precise time interval is intended, but it is somehow uncertain which interval this is. For example, surgery on a patient may have started exactly at 7h42 and ended exactly at 11h46, but a secretary, not knowning this, might incorporate in the patient DBS that surgery started 'around 8' and ended 'around 12'. Exact, precise starting and ending instants exist, but the secretary doesn't know them.
- **Vagueness/imprecision.** A time interval is subject to *imprecision* or *vagueness* when it is certainly known exactly which time interval is intended, but not every instant belongs to this time interval to the same extent. For example: it could be known for sure that a patient showed no symptoms of a disease up to 8h12 and all the possible symptoms of this disease starting from 10h11. In this case, the patient is considered ill with this disease starting at 8h12, but in the period between 8h12 and 10h11, the patient isn't considered to be fully sick yet. At every instant between 8h12 and 10h11, he is only ill with the disease to some degree. When such imprecision is described or indicated using linguistic terms, it is called *vagueness.*

Up to now, many proposals have considered the modeling of uncertainty [1], [2], [4], [15], [14] or vagueness [17] in time intervals modeled by DBS. Surprisingly, to the knowledge of the authors, only very few proposals examine the interpretation and semantics of imperfections in time intervals modeled by DBS and all proposals doing this, do it in a very summary way. Even more surprisingly, to the knowledge of the authors, no proposal has ever considered the modeling of combinations of both uncertainty and vagueness in time intervals modeled by DBS. However, the authors deem it likely that situations exist where it is necessary to be able to model such combinations. For example: different sources may disagree about the time interval subject to vagueness during which a patient became increasingly ill with a disease, resulting in uncertainty about which time interval is the correct one.

In order to respond to such situations, the work presented in this paper makes three contributions:

1. Both the modeling and semantics of uncertainty and vagueness in time intervals modeled by DBS are clearly explained. This is done in section 3.
2. The semantics of possible combinations of uncertainty and vagueness in time intervals modeled by DBS are explored and based on this exploration, a technique to model these combinations will be presented. This is done in section 4.
3. It is the opinion of the authors that (at least) one combination could semantically be reduced to a more usable form. The main arguments for this opinion and a proposed reduction strategy will be introduced in section 5.

First, however, section 2 presents and describes a few necessary concepts.

2 Preliminaries

2.1 Time as Perceived by Database Systems

Time is usually perceived by DBS as a linear given that flows at a constant pace and in one direction (from the past to the future). Therefore, time is often thought of in the context of DBS as following an axis (a *time axis*), where the points on this axis are infinitesimally short 'moments' in time [2], [8], [11], [12], [14], [15], [17]. Thus, time is often seen as a totally ordered set of such points (the *time axis*), which are called *instants* [3], [10].

Definition 1. Instant [3], [10]
An instant *is a time point on an underlying time axis.*

For example, the point on an unspecified time axis described by the words 'March 6th at 19h exactly' is an instant.

Using two of its instants, an interval subset of a time axis can be defined. Such interval subset is called a *time interval* [3], [10].

Definition 2. Time interval [3], [10]
In the presented work, a subset of all instants of a time axis between two given instants of this time axis is called a time interval

For example, the subset of an unspecified time axis described by the words 'from March 6th at 19h exactly until March 7th at noon exactly' is a time interval, bounded by the instants described by the words 'March 6th at 19h exactly' and 'March 7th at noon exactly'.

In theory, time intervals can be closed, halfopen or open intervals. An instant itself is in fact a closed time interval of which the bounding instants coincide. The work presented in this paper deals with imperfection in closed time intervals.

Definition 3. Duration [3], [10]
A duration *is an amount of time with known length, but no specific starting or ending instants.*

A time interval is bounded by two instants, whereas a duration is not. For example, the amount of time described by the words 'a month' is a duration: its length is given, but its bounding instants are not.

2.2 Time as Modeled by Database Systems

As many properties of real-life objects or concepts are time-related, many DBS model time intervals or instants using *time models*. However, as DBS are usually finitely precise and instants have an infinitesimally short duration, time models used by DBS usually model an underlying time axis using *chronons*.

Definition 4. Chronon [3], [10]
In a data model, a chronon *is a non-decomposable time interval of some fixed, minimal duration.*

A time model used by an DBS usually disposes of a datatype used to represent time indications. Such datatype usually has a time domain at it's disposal, which is a set of values allowed by the datatype. For example, a time model's datatype could use $\mathbb{Z}$ as time domain.

Usually, time models used by DBS model a time axis using a sequence of consecutive chronons. Every element of the time model's time domain then corresponds to exactly one chronon, the ordering of the consecutive time domain elements reflecting the temporal ordering of these consecutive chronons. These chronons are the smallest time intervals an DBS using the time model can distinguish, which is generally why they are used [11], [15].

An instant is usually modeled by a time model as a single element of a time domain, corresponding to the single chronon containing the instant somehow. A time interval can be modeled as an interval in the time domain, corresponding to a set of consecutive chronons called a *chronon interval.*

Definition 5. Chronon interval
In a data model, a chronon interval *is a set of (one or more) consecutive chronons, used to represent a time interval.*

An example is in order here. Consider an DBS with a time model modeling a time axis, where this time axis is a totally ordered set containing all instants in the first week of the year 2014. The time domain $\mathbb{E}$ used by this model could for example be the interval $\mathbb{E} = [1, 7] \subset \mathbb{N}$. Every different element e of $\mathbb{E}$ could now correspond to a different chronon c, which have a duration of one day. For example: element 1 corresponds to the first day of 2014, element 2 to the second, etc. A (closed) time interval starting half an hour before the start of the second day of 2014 and ending two hours before the end of the second day of 2014 can now be mapped to a chronon interval consisting of only the chronons corresponding to the first and second day of 2014, and thus be represented by time domain interval $[1, 2]$.

The work presented in this paper deals with imperfection in both closed time intervals and closed intervals in time domains (which are called closed *time domain intervals* in this paper). Approaches to determine chronons or chronon intervals or elements of or intervals in a time domain based on corresponding instants or time intervals are not dealth with in this paper.

3 Uncertainty and Vagueness in Time (Domain) Intervals

In sections 3.1 and 3.2, the semantics and interpretations of respectively uncertainty and vagueness in time (domain) intervals are clearly explained and the ways to model these are described. For reasons of clarity, an interval which is not subject to any imperfections will be called a *regular* interval in this paper.

3.1 Uncertainty in Time (Domain) Intervals: Semantics and Modeling

Although other sources may exist, uncertainty in time indications modeled by DBS is usually caused by a (partial) lack of knowledge: the exact part of time intended by the time indication exists, but uncertainty exists concerning exactly which part of time is intended, because knowledge about the intended part of time is incomplete. This is mainly because DBS model time concerning real-life objects or concepts, and in most cases, temporal aspects of real-life objects or concepts can simply not be perfectly and completely controlled. For example, a medieval document may be written on one specific day in history, but it might be unknown exactly which day this was, resulting in uncertainty about the exact day intended.

Therefore, the work presented in this paper only considers time (domain) intervals subject to uncertainty caused by a (partial) lack of knowledge. Confidence in the context of such uncertainty should be modeled using possibility theory [1], [2], [4], [7], [8], [9], [15]. In the work presented in this paper, possibility is always interpreted as plausibility, given all available knowledge.

The work presented in this paper will introduce the modeling of uncertainty in time (domain) intervals using the concept of *ill-known intervals* (IKI). However, before introducing IKI, the concepts of *possibilistic variables* and *ill-known values* (IKV) should be introduced [1], [2], [4], [14], [15]:

Definition 6. Possibilistic variable

A possibilistic variable X on a universe U is a variable taking exactly one value in U, but for which this value is (partially) unknown. The possibility distribution π_X on U models the available knowledge about the value that X takes: for each $u \in U$, $\pi_X(u)$ represents the possibility that X takes the value u.

Consider a universe U and a possibilistic variable X on U, defined and described by its possibility distribution π_X on U. For every $u \in U$, $\pi_X(u)$ now represents the possibility that X takes the value u. The interpretation is that $\pi_X(u)$ represents how plausible it is that X takes the value u, given all available knowledge about the value X is intended to take, fully understanding that it is (partially) unknown which value X is intended to take.

Definition 7. Ill-known value

Consider a set U containing single values (and not collections of values). When a possibilistic variable $X_{\tilde{v}}$ is defined on U, the unique value $X_{\tilde{v}}$ takes, which is (partially) unknown, will be a single value in U and is called an ill-known value *(in U).*

Consider a set U of single values and an IKV $\tilde{v}$ in U defined and described by a possibilistic variable $X_{\tilde{v}}$ on U, which is defined and described by its possibility distribution $\pi_{X_{\tilde{v}}}$ on U. The IKV $\tilde{v}$ now intends to be a single value in U. However, due to a (partial) lack of knowledge about the exact value $\tilde{v}$ is intended to be, it is not certain exactly which value $\tilde{v}$ is. The degree of possibility $\pi_{X_{\tilde{v}}}(u)$ of an arbitrary element u of U is now interpreted as the degree of plausibility that $X_{\tilde{v}}$ takes value u and thus that $\tilde{v}$ is u, given all available knowledge about the value $\tilde{v}$ is intended to be, fully understanding that it is (partially) unknown which value $\tilde{v}$ is intended to be.

Definition 8. Ill-known interval

Consider a totally ordered set U containing single values (and not collections of values) and its powerset $\wp(U)$. Now consider the subset $\wp_I(U)$ of $\wp(U)$ and let this subset contain every element of $\wp(U)$ that is an interval, but no other elements. When a possibilistic variable $X_{\tilde{I}}$ is defined on this $\wp_I(U)$, the unique value $X_{\tilde{I}}$ takes, which is (partially) unknown, will be a regular interval in U and is called an ill-known interval *(in U).*

Consider a totally ordered set U of single values and an IKI $\tilde{I}$ in U defined and described by a possibilistic variable $X_{\tilde{I}}$ on U, which is defined and described by its possibility distribution $\pi_{X_{\tilde{I}}}$ on U. The IKI $\tilde{I}$ now intends to be a single regular interval in U. However, due to a (partial) lack of knowledge about the exact interval $\tilde{I}$ is intended to be, it is uncertain exactly which regular interval $\tilde{I}$ is. The degree of possibility $\pi_{X_{\tilde{I}}}(J)$ of an arbitrary regular interval $J \subseteq U$ is now interpreted as the degree of plausibility that $X_{\tilde{I}}$ takes value $J \in \wp_I(U)$ and thus that $\tilde{I}$ is J, given all available knowledge about the regular interval $\tilde{I}$ is intended to be, fully understanding that it is (partially) unknown which value $\tilde{I}$ is intended to be.

In the work presented in this paper, an IKI in a time axis is called an *Ill-known Time Interval* (IKTI), while an IKI in a time domain is called an *Ill-known Time Domain Interval* (IKTDI).

In conclusion, it should be clear that the occurrence of uncertainty in time (domain) intervals only implies that the time (domain) interval intended is not exactly known, as some of the knowledge to determine this is lacking. The occurrence of uncertainty in time (domain) intervals does *not* imply that the intended time (domain) interval isn't a precise one.

3.2 Vagueness in Time (Domain) Intervals: Semantics and Modeling

As mentioned before, many time intervals modeled in DBS correspond to real-life objects or concepts, which could be (to some degree) defined by humans. Often, such objects or concepts may have a gradual nature: at some instants, they are in existence to a higher extent than at other instants. At some instants, they are fully non-existing, at others they are fully in existence and at yet others, they only exist to some (and not full) extent. For example, the 'industrial revolution' is a concept defined by humans based on the developmental state of a part of the world. It took place during the 18th and 19th century, so one could correspond it to the time interval containing every instant between the first instant of the 18th century and the last instant of the 19th century. However, as the developmental state of the world gradually changed (and the Industrial Revolution thus gradually came into existence) between 1750 and 1800 and the inverse took place during the middle of the 19th century, the instants belonging to these parts of time belong to the time interval corresponding to the industrial revolution to a lesser extent than the instants belonging to the part of time between 1800 and the middle of the 19th century, when the industrial revolution was in full existence.

The time intervals corresponding to such 'gradual' objects or concepts contain every instant which corresponds to the object or concept to some degree. However, some instants fully are moments in time when the object or concept is in existence and thus fully belong to its corresponding time interval and some instants merely partially correspond to moments in time when the object or concept is in existence and thus merely partially belong to its corresponding time interval.

As a result of the gradualness of such object or concept, (parts of) the time interval corresponding to it can only be approximately described. When such a description is given using precise boundaries, the time interval is subject to imprecision. For example: the industrial revolution gradually came into existence between 1750 and 1800. Thus, a part of the time interval during which the industrial revolution was in existence is approximately described using the precise boundaries 'between 1750 and 1800'. When such a description is given using linguistic terminology, the time interval is subject to vagueness. For example: the industrial revolution gradually faded out during the middle of the 19th century. Thus, a part of the time interval during which the industrial revolution was in existence is approximately described using the linguistic description 'the middle of the 19th century'.

The work presented in this paper will introduce the modeling of imprecision or vagueness in time (domain) intervals using the concept of *fuzzy time (domain) intervals* (FT(D)I).

Definition 9. Fuzzy Time Interval
Consider a time axis T, which is a totally ordered set of instants. A fuzzy time interval *$\tilde{V}$ (in T) is a fuzzy subset of T of which the support is an interval in T and which has a clearly conjunctive interpretation.*

Definition 10. Fuzzy Time Domain Interval
Consider a time domain D, which is a totally ordered set. A fuzzy time domain interval *$\tilde{W}$ (in D) is a fuzzy subset of D of which the support is an interval in D and which has a clearly conjunctive interpretation.*

The imprecision or vagueness in a time (domain) interval corresponding to a real-life object or concept can now be modeled by allowing the time (domain) interval to be a FT(D)I. Given a time axis T and a FTI $\tilde{V}$ in T defined by its membership function $\mu_{\tilde{V}}$, the membership degree $\mu_{\tilde{V}}(u)$ of instant $u \in T$ now expresses to what extent u belongs to $\tilde{V}$. Such a membership degree can be interpreted in two ways.

- **Similarity**. Here, instants at which the real-life object or concept fully exist are given the highest possible membership degrees. The membership degree of any other instant then quantifies the extent to which that instant is similar to the aforementioned instants, where two instants are more similar when their extents to which the real-life object or concept exists at them are more similar.
- **Truth or preference**. Here, an FTI corresponding to a real-life object or concept is seen as a set of all instants at which the real-life object or concept existed. The membership degree of an instant is seen as reflecting the extent to which the person(s) that defined the FTI find(s) it true or preferred that the real-life object or concept exists at this instant.

In conclusion, it should be clear that the occurrence of imprecision or vagueness in time (domain) intervals only implies that not every instant belongs to the time (domain) interval to the same degree. However, it is perfectly and certainly known which instants belong to the time (domain) interval to which degree.

4 Combining Uncertainty and Vagueness in Time (Domain) Intervals

In this section, two types of situations are presented, in which it appears necessary to model combinations of uncertainty and vagueness in time (domain) intervals. Both situation types are accompanied by explanatory examples. For every situation type, a technique for modeling the imperfection in the time (domain) intervals is presented.

4.1 Uncertainty about a Time (Domain) Interval Subject to Vagueness: Semantics and Modeling

In this type of situation, it is perfectly known that a real-life object or concept corresponds to a single time (domain) interval subject to vagueness, however, it is (partially) unknown exactly which time (domain) interval subject to which vagueness this is. As a result, there exists uncertainty concerning which time (domain) interval subject to vagueness corresponds to the object or concept.

Consider a situation in which both a nurse and a doctor examine a patient, searching for the evolution of a disease in the patient. The doctor finds no signs of the disease in the patient up to instant $t_{d,1}$, all signs of full disease starting from instant $t_{d,2}$ and ending at instant $t_{d,3}$, and no signs anymore starting from instant $t_{d,4}$, where $t_{d,1}$ is earlier than $t_{d,2}$, which is earlier than $t_{d,3}$, which is earlier than $t_{d,4}$. The doctor concludes that the time interval during which the patient is ill is the interval starting at $t_{d,1}$ and ending at $t_{d,4}$ and that the parts of the time interval between $t_{d,1}$ and $t_{d,2}$, resp. $t_{d,3}$ and $t_{d,4}$ are subject to imprecision. The nurse now finds the same general disease evolution, but concludes that the time interval during which the patient is ill is the interval starting at $t_{n,1}$ and ending at $t_{n,4}$ and that the parts of the time interval between $t_{n,1}$ and $t_{n,2}$, resp. $t_{n,3}$ and $t_{n,4}$ are subject to imprecision, where $t_{n,1}$ is earlier than $t_{n,2}$, which is earlier than $t_{n,3}$, which is earlier than $t_{n,4}$. Thus, there is uncertainty about the time interval during which the patient is ill, which is a time interval subject to imprecision.

In the work presented in this paper, it is suggested to model all imperfection in such types of situations by making the time (domain) interval under consideration an *ill-known FT(D)I* (IKFT(D)I).

Definition 11. Ill-known Fuzzy Time Interval

Consider a time axis T, which is a totally ordered set of instants. Consider the fuzzy powerset $\tilde{\wp}(T)$ of T and let this set contain every fuzzy subset of T and no other elements. Now consider the subset $\tilde{\wp}_{\tilde{V}}(T)$ of $\tilde{\wp}(T)$ and let this set contain every fuzzy time interval in T and no other elements. When a possibilistic variable $X_{\tilde{\tilde{V}}}$ is defined on this $\tilde{\wp}_{\tilde{V}}(T)$, the unique value $X_{\tilde{\tilde{V}}}$ takes, which is (partially) unknown, will be a fuzzy time interval in T and is called an ill-known fuzzy time interval *(in T).*

Definition 12. Ill-known Fuzzy Time Domain Interval

Consider a time domain D, which is a totally ordered set. Consider the fuzzy powerset $\tilde{\wp}(D)$ of D and let this set contain every fuzzy subset of D and no other elements. Now consider the subset $\tilde{\wp}_{\tilde{V}}(D)$ of $\tilde{\wp}(D)$ and let this set contain every fuzzy time domain interval in D and no other elements. When a possibilistic variable $X_{\tilde{\tilde{V}}}$ is defined on this $\tilde{\wp}_{\tilde{V}}(D)$, the unique value $X_{\tilde{\tilde{V}}}$ takes, which is (partially) unknown, will be a fuzzy time domain interval in T and is called an ill-known fuzzy time domain interval *(in D).*

An IKFT(D)I now intends to be an FT(D)I, thus, a time (domain) interval subject to vagueness or imprecision. This is modeled through the use of the set of every existing FT(D)I in the definition of the IKFT(D)I, where the FT(D)I inherently model vagueness or imprecision. However, due to a (partial) lack of knowledge, there exists uncertainty about exactly which FT(D)I is the one intended by the IKFT(D)I. Confidence in the context of this uncertainty is modeled using the possibilistic variable on the set of existing FT(D)I.

4.2 Vagueness about Time (Domain) Intervals Subject to Uncertainty: Semantics and Modeling

In this type of situation, there is no intention to represent a single time (domain) interval. Instead, the intention is to represent a set of time (domain) intervals, all of which may belong to this set to different degrees. However, as a result of a (partial) lack of knowledge, it is not known exactly which time (domain) interval is intended by every element of the aforementioned set. As a result, in this type of situation, the intention is to represent a set of time (domain) intervals, for some of which there exists uncertainty about exactly which time (domain) interval is intended.

Consider a hospital where a secretary attempts to model the surgery scheduling preferences of a surgeon. Consider the surgeon telling her he can't perform surgery before 10 o'clock because of a meeting ending at 10, which could run late up to 15 minutes. After the meeting, he could perform surgery, but only mildly prefers to do so. From around 12 o'clock until around 13 o'clock (15 minutes of delay could occur) is his lunch break and after that, he prefers most to perform surgery. Thus, the surgeon has given the secretary two intervals in time during which he prefers (more or less) to perform surgery, but due to possible delay of other obligations, it is not known exactly which time intervals are intended.

In the work presented in this paper, it is suggested to model all imperfection in such types of situations by making the time (domain) interval under consideration a *fuzzy IKT(D)I* (FIKT(D)I).

Definition 13. Fuzzy Ill-known Interval
Consider a totally ordered set U containing single values (and not collections of values). Now consider the subset $\tilde{\wp}_{\tilde{I}}(U)$ of U and let this set contain every ill-known interval in U, but no other elements. A fuzzy ill-known interval *(in U) is a fuzzy subset of $\tilde{\wp}_{\tilde{I}}(U)$ which has a clearly conjunctive interpretation.*

A fuzzy ill-known interval (FIKI) in a time axis is now called a *Fuzzy Ill-known Time Interval* (FIKTI), whereas a FIKI in a time domain is called a *Fuzzy Ill-known Time Domain Interval* (FIKTDI).

A FIKT(D)I now models a set of time (domain) intervals, where uncertainty about exactly which time (domain) intervals are intended is modeled through the use of IKT(D)I. However, different IKT(D)I may have different membership degrees in the set, representing the different extent to which they belong to the set, modeling imprecision or vagueness about this set of IKT(D)I. Such membership degrees can now again be interpreted as degrees of similarity or degrees of truth or preference.

5 A Reducible Combination

Consider a type of situation where one wants to model uncertainty about a time (domain) interval subject to uncertainty. In theory, it would be possible to model this as an IKV in a set of IKI in a time axis or time domain. Semantically, this construction intends to be an IKT(D)I, however, there is uncertainty about exactly which IKT(D)I is intended. Moreover, every IKT(D)I intends to be a regular time (domain) interval, however, there is uncertainty about exactly which interval is intended. Overall, this construction intends to be a regular time (domain) interval, however, there is uncertainty about exactly which interval is intended. Thus, it is the opinion of the authors that situations of this type could be modeled using a single IKT(D)I.

Consider a totally ordered set U containing single values (and not collections of values). Now consider the subset $\tilde{\wp}_{\tilde{I}}(U)$ of U and let this set contain every IKI in U, but no other elements. Now consider an IKV $\tilde{\tilde{I}}$ in $\tilde{\wp}_{\tilde{I}}(U)$ defined by the possibility distribution $\pi_{X_{\tilde{\tilde{I}}}}$ of its possibilistic variable. Now consider an arbitrary element $\tilde{I}_i$ of $\tilde{\wp}_{\tilde{I}}(U)$, defined by the possibility distribution $\pi_{X_{\tilde{I}_i}}$ of its possibilistic variable. For an arbitrary regular interval I in U, $\pi_{X_{\tilde{I}_i}}(I)$ now expresses the possibility that I is the regular interval intended by $\tilde{I}_i$, whereas $\pi_{X_{\tilde{\tilde{I}}}}(\tilde{I}_i)$ expresses the possibility that the interval intended by $\tilde{\tilde{I}}$ is the interval intended by $\tilde{I}_i$. Following the rules of possibility theory [7], the possibility $\pi(I)$ that the interval intended by $\tilde{\tilde{I}}$ is the interval I should thus be calculated as follows.

$$\pi(I) = \max_{\tilde{I}_i \in \tilde{\wp}_{\tilde{I}}(U)} (\min(\pi_{\tilde{I}_i}(I), \pi_{\tilde{\tilde{I}}}(\tilde{I}_i)))$$

Thus, a construction as introduced above can semantically be reduced to a single IKI.

6 Conclusions and Future Work

In the work presented in this paper, the semantics and interpretation of uncertainty and vagueness in time (domain) intervals in database systems are thoroughly examined. Based on the semantics and interpretations of combinations of such uncertainty and vagueness, methods to model such combinations are introduced. A reducible combination is identified and described. Future work should either focus on the remaining combination or on the temporal relationships between such combinations and regular time intervals. As stressed throughout the paper, interpretation and semantics should never be downplayed.

References

1. Billiet, C., Pons Frias, J.E., Pons, O., De Tré, G.: Bipolarity in the Querying of Temporal Databases. In: Atanassov, K.T., Homenda, W., Hryniewicz, O., Kacprzyk, J., Krawczak, M., Nahorski, Z., Szmidt, E., Zadrozny, S. (eds.) New trends in fuzzy sets, intuitionistic fuzzy sets, generalized nets and related topics volume II: applications, New trends edn., pp. 21–37. SRI PAS/IBS PAN (2013)
2. Billiet, C., Pons Frias, J.E., Pons Capote, O., De Tré, G.: A Comparison of Approaches to Model Uncertainty in Time Intervals. In: Pasi, G., Montero, J., Ciucci, D. (eds.) Advances in Intelligent Systems Research, pp. 626–633. Atlantis Press, Milano (2013)
3. Jensen, C.S., et al.: The consensus glossary of temporal database concepts - february 1998 version. In: Etzion, O., Jajodia, S., Sripada, S. (eds.) Dagstuhl Seminar 1997. LNCS, vol. 1399, pp. 367–405. Springer, Heidelberg (1998)
4. Bronselaer, A., Pons, J.E., De Tré, G., Pons, O.: Possibilistic evaluation of sets. International Journal of Uncertainty, Fuzziness and Knowledge-Based Systems 21(3), 325–346 (2013)
5. Chountas, P., Petrounias, I.: Modelling and Representation of Uncertain Temporal Information. Requirements Engineering 5(3), 144–156 (2000)
6. De Tré, G., Zadrozny, S., Bronselaer, A.J.: Handling Bipolarity in Elementary Queries to Possibilistic Databases. IEEE Transactions on Fuzzy Systems 18(3), 599–612 (2010)
7. Dubois, D., Hadjali, A., Prade, H.: A Possibility Theory-Based Approach to the Handling of Uncertain Relations between Temporal Points. International Journal of Intelligent Systems 22(2), 157–179 (2007)
8. Dubois, D., Prade, H.: Processing Fuzzy Temporal Knowledge. IEEE Transactions on Systems, Man and Cybernetics 19(4), 729–744 (1989)
9. Dubois, D., Prade, H.: The Three Semantics of Fuzzy Sets. Fuzzy Sets and Systems 90(2), 141–150 (1997)
10. Dyreson, C.E., et al.: A Consensus Glossary of Temporal Database Concepts. SIGMOD Record 23(1), 52–64 (1994)
11. Dyreson, C.E., Snodgrass, R.T.: Supporting Valid-Time Indeterminacy. ACM Transactions on Database Systems 23(1), 1–57 (1998)
12. Garrido, C., Marín, N., Pons, O.: Fuzzy Intervals To Represent Fuzzy Valid Time in a Temporal Relational Database. International Journal of Uncertainty, Fuzziness and Knowledge-Based Systems 17(Suppl.1), 173–192 (2009)
13. Kacprzyk, J., Zadrozny, S.: Computing with Words in Intelligent Database Querying: Standalone and Internet-based Applications. Information Sciences 134(1-4), 71–109 (2001)
14. Pons, J.E., Marín, N., Pons, O., Billiet, C., Tré, G.D.: A Relational Model for the Possibilistic Valid-time Approach. International Journal of Computational Intelligence Systems 5(6), 1068–1088 (2012)
15. Pons Frias, J.E., Billiet, C., Pons, O., De Tré, G.: Aspects of Dealing with Imperfect Data in Temporal Databases. In: Pivert, O., Zadrozny, S. (eds.) Flexible Approaches in Data, Information and Knowledge Management. SCI, vol. 497, pp. 189–220. Springer, Heidelberg (2013)
16. Qiang, Y., Asmussen, K., Delafontaine, M., Stichelbaut, B., De Tré, G., De Maeyer, P., Van De Weghe, N.: Handling Imperfect Time Intervals in a Two-Dimensional Space. Control and Cybernetics 39(4), 983–1010 (2010)
17. Schockaert, S., Cock, M.D.: Temporal reasoning about fuzzy intervals. Artificial Intelligence 172, 1158–1193 (2008)

Equality in Approximate Tolerance Geometry

Gwendolin Wilke

Institute for Information Science,
University of Applied Sciences and Arts Northwestern Switzerland
Riggenbachstr. 16, 4600 Olten, Switzerland
gwendolin.wilke@fhnw.ch
http://www.fhnw.ch/personen/gwendolin-wilke

Abstract. The framework of Approximate Tolerance Geometry (ATG) has been proposed in [1] as an approach to handling large and heterogeneous imperfections in geometric data in vector-based geographic information systems. Here, different types of positional error can often only be subsumed as possibilistic location constraints. The application of the ATG framework to a classical geometry provides a calculus for the propagation of this error type in geometric reasoning. As a first step towards an implementation of an ATG geometry, the paper applies the framework to the geometric equality relation. It thereby lays the basis for the application of ATG to the other axioms of classical geometry.

Keywords: geographic information system, fuzzy geometry, uncertain points, fuzzy logic with evaluated syntax, rational Pavelka logic, Łukasiewicz logic, fuzzy similarity relation, fuzzy equivalence relation.

1 Introduction

Vector-based geographic information systems (GIS) are used to store, analyze and visualize geographical data. Entities in the world, such as buildings or parcel boundaries, are represented by simple geometric figures such as points, lines and polygons. To construct these figures from measured or derived coordinate points (e.g. from surveying measurements or satellite imagery), Euclidean or projective geometry is used. Even though geometric calculi can only handle exact points and lines, the various kinds of imperfections present in geographic coordinate data (such as statistical or possibilistic imperfections, fuzzyness or vagueness) are usually manageable as long as they are small. This has been the case until recently with big mapping agencies acting as the main data providers and having high standards in quality control. Yet, with the proliferation of mobile computing and the development of smaller and cheaper global positioning system (GPS) sensors in the last decade, a paradigm change took place: Now, ordinary citizens can volunteer in collecting geographic information. As a consequence, the sources of geographic information have become highly inhomogeneous, and the degree of imperfection of the collected coordinate data is often very high [2]. When integrated in a GIS, the data and their imperfections can exhibit a high degree

P. Angelov et al. (eds.), *Intelligent Systems'2014*,
Advances in Intelligent Systems and Computing 322, DOI: 10.1007/978-3-319-11313-5_33

of variability, and the geometric algorithms may produce highly erroneous results that cannot always be mitigated with existing heuristic algorithms.

To meet this challenge, we proposed *Approximate Tolerance Geometry (ATG)* [1,3]: ATG is a framework for fuzzifying classical geometries that allows for interpreting the geometric primitives *point* and *line* by "approximate" points and lines (cf. figure 1). It assumes that the exact coordinates of a point are not known, but instead, a region can be given that specifies the set of its *possible* positions (i.e., a possibilistic location constraint). Possibilistic location constraints for coordinate points subsume a lot of imperfections in positional data that occur in GIS, and in the GIS literature, they are often referred to as *points/lines with positional tolerance* (cf., e.g., [4]). As a consequence of the possibilistic interpretation of primitive geometric objects, ATG interprets primitive geometric relations (such as equality, incidence or betweenness) by *possible* relations. E.g., instead of stating that a point *lies on* a line, an ATG geometry can only state that a point with tolerance *possibly lies on* a line with tolerance. The result of applying the ATG framework to the axioms of a classical geometry is a fuzzy set of axioms, and a fuzzy deduction apparatus allows for deriving fuzzy geometric statements from the axioms and from given facts. The fuzzy membership degree is interpreted as the degree of similarity of a geometric statement to the truth, and its value depends on the characteristics of the given geometric configuration at hand. Figure 1 illustrates an example: Here, the statement "the approximate point P approximately lies on the approximate line L" is not fully true, but almost true: The possibilistic location constraints P and L do not overlap, hence it is not fully true that P *possibly lies on* L. Yet, we can say that it is almost true, since L is very close to P. ATG is based on fuzzy logic with evaluated syntax [5,6]. Its graduated deduction apparatus can be used to propagate similarity to the truth. Since deduction is sound by design, an ATG geometry treats imperfections (in the form of possibilistic location constraints) as an intrinsic part of geometry, and geometric reasoning is automatically sound. ATG can also be seen as an error propagation calculus.

In the present paper, we apply the ATG framework to the equality axioms: Equality is the most basic geometric relation, and all other geometric axioms rely on it. We do this in two steps: First, we show that not all of the equality axioms are fully true in the possibilistic interpretation of ATG. More specifically, reflexivity and symmetry hold with truthlikeness degree 1, while transitivity is not only not fully true, but not even close to the truth. We show that the underlying reason for this result is the additional degree of freedom introduced

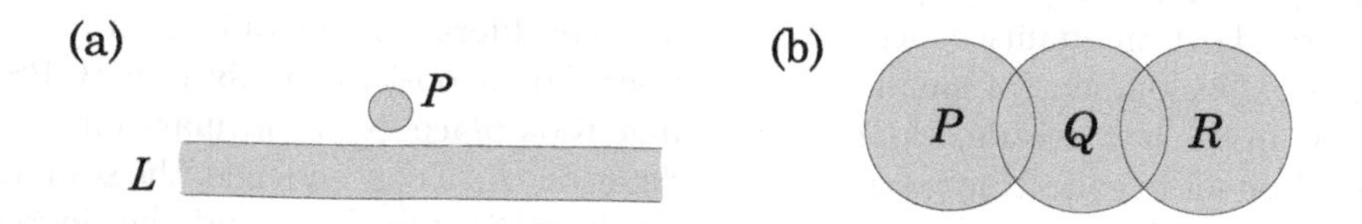

Fig. 1. (a) Approximate point and line. (b) Approximate equality is not transitive.

to geometry by the fact that the possibilistic location constraints have size. This is in contrast to classical geometry, where points are interpreted by extensionless coordinate points. In a second step, following Gerla [7], we address this issue by generalizing the transitivity axiom so that size is accounted for in the form of an additional geometric predicate we call *exactness* predicate. The resulting generalized transitivity axiom has a truthlikeness degree of 1 and is consistent with the classical axiom for extensionless points.

The paper is structured as follows: Section 2 gives a brief account of related work. Section 3 provides the basic definitions of fuzzy logic with evaluated syntax that are needed in the paper. Section 4 introduces the ATG framework, and sections 5 and 6 apply the framework to the equality axioms. Section 7 concludes the paper and gives an outlook to future work.

2 Related Work

The treatment of positional tolerance of coordinate points and derived geometric figures has a long standing history in the GIS community. It has been first addressed by J. Perkal [8], who defined the so-called epsilon-band model, where every geometric feature is associated with a zone of width epsilon around it. The type of imperfection associated with this zone can vary, and a huge amount of literature has been sparked off by Perkal's work (e.g. [9,10,11,12]). Today, the research concentrates mainly on statistical treatment of positional error, for which Gaussian error propagation can be utilized. Other approaches to the representation of and reasoning with imperfections in positional information of point and line features that integrate different types of positional error are, e.g., [13,14,15,16,17,18,4,19,20,21]. Most of them do not account for generalized equality relations that are not in general transitive and do not guarantee that reasoning is sound. In the field of digital geometries much work has been done (cf., e.g., [22,23]). Yet, for GIS it is necessary to provide a more general approach, where points and lines with tolerance are not restricted to the constraints imposed by the structure of the digital plane. Region-based geometries are logical theories that use *region* as a primitive object to axiomatize classical geometries with exact points and lines. In contrast, ATG uses a region primitive to axiomatize classical geometries with *in*exact points and lines. The present work is built on the work of F. Roberts' *tolerance geometry* [24] and M. Katz' *inexact geometry* [25], who use Łukasiewicz fuzzy predicate logic to fuzzify geometry in a similar way, but apply it only to one dimensional geometry. Łukasiewicz fuzzy predicate logic is the basis for fuzzy logic with evaluated syntax, which did not exist at the time. While Łukasiewicz fuzzy predicate logic uses classical deduction, fuzzy logic with evaluated syntax has graduated deduction, which allows for propagating truthlikeness and spatial error.

3 Fuzzy Logic with Evaluated Syntax

Fuzzy logic with evaluated syntax (FL_{ev}) is a generalization of Łukasiewicz Rational Pavelka Logic ($RPL\forall$), which is in turn a generalization of Łukasiewicz fuzzy predicate logic (Ł$\forall$), cf. [26,5]. The Łukasiewicz t-norm is given by $a \otimes b = max\{a+b-1, 0\}$, its residuum is $(a \Rightarrow b) = min\{b-a+1, 1\}$, and the negation is $\neg a = (a \Rightarrow 0) = 1 - a$. The universal quantifier in Ł$\forall$ is interpreted by the infimum, the existential quantifier is interpreted by the supremum. The following definitions are taken from [5] and [6]:

An *evaluated formula* or *signed formula* $(\varphi; a)$ in FL_{ev} is a formula φ in Ł$\forall$, together with an element $a \in [0, 1]$. Here, a does not belong to the language, but is a formal evaluation of the syntax. a is called the *syntactic evaluation* or *sign* of φ. The intended meaning of a signed formula$(\varphi; a)$ is that the truth value of φ is greater or equal than a.A *fuzzy theory* τ in Ł$\forall$ is a fuzzy subset $\tau : F \to [0, 1]$ of the set F of formulas of Ł$\forall$. The intended meaning of a fuzzy set of axioms is that some of the axioms are "not fully convincing".

4 The ATG Framework

The ATG framework consists of the following constituents:

- An interpretation of points and lines by possibilistic location constraints in a classical interpretation domain, e.g. the Euclidean plane.
- An interpretation of primitive geometric relations that is consistent with the above interpretation of primitive objects.
- A semantic similarity measure that quantifies the similarity of a classical geometric statements to the truth when interpreted by the possibilistic interpretation referred to above. The similarity measure is called *truthlikeness measure*. Due to the spatiality of geometric configurations, ATG measures truthlikeness by spatial nearness, i.e. by an inverse spatial distance.
- Fuzzy logic with evaluated syntax, which is used as a similarity logic [27]. Every classical geometric statement is assigned a truthlikeness degree in the ATG-interpretation as a *sign* (i.e. syntactic evaluation). Together, they constitute a signed formula. The truthlikeness degree indicates the degree of fuzzy membership of the classical statement to the approximate tolerance geometry. The graduated deduction apparatus of fuzzy logic with evaluated syntax allows for calculating the truthlikeness degree of derived geometric statements from the truthlikeness degree of the given ones.

Applying the ATG framework to a classical geometry, e.g., to Euclidean geometry, requires taking the following steps:

1. Given an axiomatization of the geometry, quantify for every axiom its similarity to the truth in the above possibilistic interpretation.
2. Assign the resulting truthlikeness degree as a fuzzy membership degree to the respective axiom. The result is a fuzzy set of axioms $\{(\varphi_i; a_i)\}$, where every classical axiom φ_i is paired with a sign a_i that indicates the degree of similarity of the classical statement to the truth in the ATG interpretation.

3. Check if the truthlikeness degrees of all axioms are greater than zero. Only if this is the case, the fuzzification of the classical geometry has been successful: If the truthlikeness degree of an axiom is zero, it is not only not true, but not even close to the truth, and only absolutely false statements can be derived by graduated deduction in fuzzy logic with evaluated syntax.
4. If an axiom has truthlikeness degree zero, check if the axiom can be generalized, such that the generalized axiom has positive truthlikeness degree and at the same time is consistent with classical geometry in the limit case, (i.e., in the case where location constraints have size zero).

In the following, we define the possibilistic interpretation of geometric primitives in ATG. In this paper, we restrict our considerations to points and equality of points. Lines and equality of lines can be defined analogously by utilizing the dual space, cf. [3,1]. The following definitions are taken from [1]: Given a metric space (X, d_X) and an induced metric topology τ_{d_X} on X, we say that $P \subseteq X$ is an *approximate point in* X, if P is either a τ_{d_X}-topological neighborhood of a point $p \in X$ or a singleton, $P = \{p\}$. P is interpreted as a set of possible locations of the exact point p. We denote the set of approximate points in X by $\mathcal{P}_X$ and call $s_{(d_X)}(P) = \sup\{d_X(p, \bar{p}) | p, \bar{p} \in P\}$ the *size* of the approximate point P. We call $x_{(d_X)}(p) = 1 - s_{(d_X)}(p)$ the *exactness* of P. Since the intended application space of an ATG geometry is a geographic map section embedded in the Euclidean plane, we assume in the remainder of the paper that X is a connected subset of $\mathbb{R}^2$ with the Euclidean metric d. We also assume that X itself is a neighbourhood in the Euclidean plane, and that size of X is normalized, i.e., that $s(X) = 1$.

Geometric relations between approximate points cannot be recognized with certainty, and they must be interpreted by *possible* relations between exact points. We interpret the equality predicate by the Boolean relation $e_{\mathbb{B}}(P, Q) = (P \cap Q \neq \varnothing)$ and call it *equality with tolerance.* If $e_{\mathbb{B}}(P, Q) = 1$, the location constraints P, Q overlap, and the corresponding exact points p and q are *possibly* equal. Equality with tolerance is the overlap relation, and it is reflexive and symmetric. Yet, notice that it is not transitive in general (cf. figure 1b on page 366).

To quantify the similarity of a geometric statement to the truth, ATG defines a truthlikeness measure for atomic statements and derives the truthlikeness degree of compound statements using the truth functional semantic of Łukasiewicz fuzzy logic. Atomic statements involve no logical operators or quantifiers, but only a single predicate. For the equality predicate, ATG measures the truthlikness of the atomic statement $e_{\mathbb{B}}(P, Q) = 1$ by an inverse spatial distance measure, namely by $e_{(d_X)}(P, Q) := 1 - d_{(d_X)}(P, Q)$. Here, $d_{(d_X)}(P, Q) = \inf\{d_X(\bar{p}, \bar{q}) | \bar{p} \in P, \bar{q} \in Q\}$ is the set distance of P and Q in X, where the domain X has size 1. The relation $e_{(d_X)}$ is called *approximate equality,* and $e_{(d_X)}(P, Q)$ measures the similarity of $e_{\mathbb{B}}(P, Q) = 1$ to the truth. The fuzzy relation $e_{(d_X)}$ and the Boolean relation $e_{\mathbb{B}}$ coincide at the value 1.

5 Equality in Approximate Tolerance Geometry

Equality, or, more generally, equivalence is a reflexive, symmetric and transitive relation with the following axiomatization:

(E1) $\forall \mathtt{p}.\, [\mathtt{E}(\mathtt{p},\mathtt{p})]$,
(E2) $\forall \mathtt{p},\mathtt{q}.\, [\mathtt{E}(\mathtt{p},\mathtt{q}) \rightarrow \mathtt{E}(\mathtt{q},\mathtt{p})]$,
(E3) $\forall \mathtt{p},\mathtt{q},\mathtt{r}.\, [\mathtt{E}(\mathtt{p},\mathtt{q})\ \&\ \mathtt{E}(\mathtt{q},\mathtt{r}) \rightarrow \mathtt{E}(\mathtt{p},\mathtt{r})]$.

To apply the ATG framework to the axioms (E1)-(E3), we need to quantify their similarity to the truth in the possibilistic interpretation given in section 4, i.e., we need to derive their truthlikeness degrees $\sigma_{(E1)}$-$\sigma_{(E3)}$. To do this, we interpret the equality predicate E, in (E1)-(E3) by the approximate equality relation $e_{(d_X)}$. Using the Łukasiewicz semantic of logical operators and quantifiers, we can write the truthlikeness degrees $\sigma_{(E1)}$-$\sigma_{(E3)}$ of the axioms (E1)-(E3) as follows:

$$\sigma_{(E1)} := \inf_{P \in \mathcal{P}_X} \left[e_{(d_X)}(P,P) \right],$$
$$\sigma_{(E2)} := \inf_{P,Q \in \mathcal{P}_X} \left[e_{(d_X)}(P,Q) \;\Rightarrow\; e_{(d_X)}(Q,P) \right],$$
$$\sigma_{(E3)} := \inf_{P,Q,R \in \mathcal{P}_X} \left[e_{(d_X)}(P,Q) \;\otimes\; e_{(d_X)}(Q,R) \;\Rightarrow\; e_{(d_X)}(P,R) \right].$$

To find the actual numerical values of $\sigma_{(E1)}$-$\sigma_{(E3)}$, we check the definition of $e_{(d_X)}$: Since $e_{(d_X)}$ is symmetric, $\sigma_{(E1)} = 1$. Since $e_{(d_X)}$ is also reflexive, it follows with the definition of the residuated implication, $(a \Rightarrow b) = min\{b - a + 1, 1\}$, that $\sigma_{(E2)} = 1$ holds. Yet, since $e_{\mathbb{B}}$ is not always transitive, $e_{(d_X)}$ cannot always be transitive either, and $\sigma_{(EP3)} < 1$. More specifically, $\sigma_{(EP3)} = 0$ holds. To see this, remember that all exact points p are also approximate points P, with the identification $P = \{p\} \equiv p$. Since X has size 1, two exact points $\bar{p}, \bar{r} \in X$ exist that have maximal distance, $d_X(\bar{p},\bar{r}) = 1$. If we set $\bar{P} = \{\bar{p}\}, \bar{R} = \{\bar{r}\}$ and $\bar{Q} := X \in \mathcal{P}_X$, we have a triple of approximate points, for which approximate equality has truthlikness degree of zero: $e_{(d_X)}(\bar{P},\bar{R}) = 1 - d_{d_X}(\bar{P},\bar{R}) = 1 - d_X(\bar{p},\bar{r}) = 0$. Consequently, the infimum over all triples is also zero, and $\sigma_{(E3)} = 0$. I.e., application of ATG to the equality axioms (E1)-(E3) yields the following set of signed axioms in fuzzy logic with evaluated syntax, FL_{ev}:

(E1$_{ev}$) $(\forall \mathtt{p}.\, \mathtt{E}(\mathtt{p},\mathtt{p});\quad 1)$,
(E2$_{ev}$) $(\forall \mathtt{p},\mathtt{q}.\, [\mathtt{E}(\mathtt{p},\mathtt{q}) \rightarrow \mathtt{E}(\mathtt{q},\mathtt{p})];\quad 1)$,
(E3$_{ev}$) $(\forall \mathtt{p},\mathtt{q},\mathtt{r}.\, [\mathtt{E}(\mathtt{p},\mathtt{q}) \wedge \mathtt{E}(\mathtt{q},\mathtt{r}) \rightarrow \mathtt{E}(\mathtt{p},\mathtt{r})].\quad 0)$,

In other words, the transitivity axiom is not only not true, but not even close to being true (i.e., absolutely false). In the following subsection, we show that the underlying reason for this result is that the generalization of ideal points by approximate points introduces an additional degree of freedom to geometry, namely the size of approximate points. This additional degree of freedom is not accounted for in the classical axioms. Using the size parameter, we show that a generalization of the transitivity axiom exists that is consistent with the classical axiom and has truthlikeness degree 1.

6 A Generalization of the Equality Axioms for ATG

Following the approach proposed by Gerla [7], subsection 371 shows that the classical transitivity axiom can be generalized by adding an exactness predicate X that is interpreted by an inverse size measure. Subsection 372 shows that the generalized axiom is consistent with the classical axiom (i.e., they coincide for exact points) and that it has a truthlikeness degree of 1.

6.1 Extensive Pseudometrics and Approximate Fuzzy Equivalences

To generalize the concept of a pseudometric space so that size and a partial order of objects are accounted for, Gerla [7] introduced the concept of a *pointless pseudometric space* or *ppm-space*. In the present work, the order relation is not needed, but the size function s is. For ease of terminology, we define an *extensive pseudometric space* $(\mathcal{X}, d, s)$ as a ppm-space where the partial order relation coincides with set equality:

Definition 1. *An* extensive pseudometric space *or* epm-space *is a triple $(\mathcal{X}, d, s)$, where $\mathcal{X}$ is a set, $d : \mathcal{X} \times \mathcal{X} \to \mathbb{R}^+ = [0, \infty)$, and $s : \mathcal{X} \to [0, \infty]$ are functions, such that for every $A, B, C \in \mathcal{X}$ the following axioms hold:*

$$d(A, A) = 0, \tag{1}$$

$$d(A, B) = d(B, A), \tag{2}$$

$$d(A, B) + s(B) + d(B, C) \geq d(A, C). \tag{3}$$

Given s, we refer to d as an extensive distance. *We refer to inequality (3) as the* extensive triangle inequality.

The set $\mathcal{P}_X$ of approximate points is a class of nonempty subsets of a metric space, and $(\mathcal{P}_X, d_{(d_X)}, s_{(d_X)})$ is a canonical example of an epm-space. Figure 2 on page 371 illustrates the extensive triangle inequality for approximate points P, Q, R: Here, $d_{(d_X)}(P, Q) + s_{(d_X)}(Q) + d_{(d_X)}(Q, R) \geq d_{(d_X)}(P, R)$ holds.

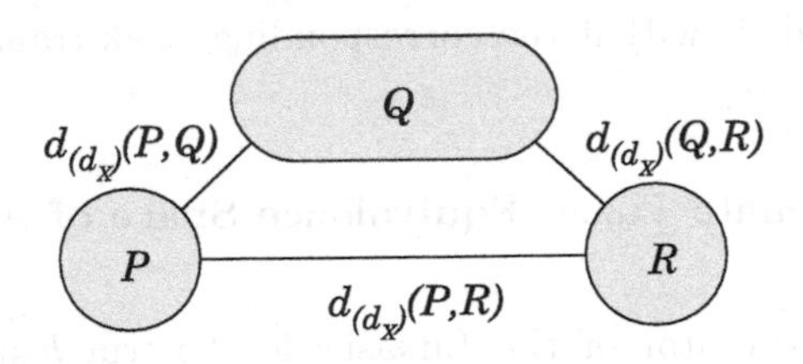

Fig. 2. The extensive triangle inequality

Gerla [7] shows that ppm-spaces (and therefore also epm-spaces) are dual to so-called *approximate fuzzy $*$-equivalence spaces*, as long as $*$ is an Archimedian t-norm (i.e., $\lim_{n\to\infty} a^n = 0$ for all $a \in [0, 1]$). An approximate fuzzy *-equivalence space is defined as follows:

Definition 2. [7] *An* approximate fuzzy $*$-equivalence space *is a set* $\mathcal{X}$*, together with two functions* $e : \mathcal{X} \times \mathcal{X} \to [0,1]$ *and* $x : \mathcal{X} \to [0,1]$*, such that for* $A, B, C \in \mathcal{X}$*,*

$$e(A, A) = 1, \tag{4}$$
$$e(A, B) = e(B, A), \tag{5}$$
$$e(A, B) * x(B) * e(B, C) \leq e(A, C) \tag{6}$$

holds. Given x*, we call* e *an* approximate fuzzy $*$-equivalence relation. (6) *is called* weak transitivity.

Approximate fuzzy $*$-equivalence relations are also called *approximate fuzzy* $*$*-similarity relations.* Notice that the weak transitivity axiom (6) is a generalization of transitivity [7]: Using the adjunction property, $(a \Rightarrow b) \geq c$ iff $a * c \leq b$, (6) can be written as $[e(A, B) * x(B) * e(B, C) \Rightarrow e(A, C)] \geq 1$, which is equivalent to $[e(A, B) * x(B) * e(B, C) \Rightarrow e(A, C)] = 1$. (6) coincides with transitivity whenever $x(B) = 1$.

The duality theorem states:

Theorem 1. [7] *Let* h *be the additive generator of an Archimedean t-norm* $*$*,* $h^{[-1]}$*its pseudoinverse, and* $(\mathcal{X}, d, s)$ *a pointless pseudometric space. Then the triple* $(\mathcal{X}, e, x)$*, with*

$$e(A, B) = h^{[-1]}\left(d(A, B)\right) \in [0, 1], \tag{7}$$
$$x(A) = h^{[-1]}(s(A)) \in [0, 1], \tag{8}$$

is an approximate fuzzy $*$*-equivalence space. Conversely, let* $(\mathcal{X}, e, t)$ *be a fuzzy* $*$*-equivalence space. Then* $(\mathcal{X}, d, s)$*, with*

$$d(A, B) = h\left(e(A, B)\right) \in [0, 1], \tag{9}$$
$$s(A) = h(x(A)) \in [0, 1], \tag{10}$$

is a pointless pseudometric space.

In the following subsection, we apply the duality theorem to the the epm-space $(\mathcal{P}_X, d_{(d_X)}, s_{(d_X)})$ and show that the corresponding weak transitivity axiom has a truthlikeness degree of 1.

6.2 The Approximate Fuzzy Equivalence Space of Approximate Points

With the additive generator of the Łukasiewicz t-norm $h_{\otimes}(a) = 1 - a$ and its pseudoinverse $h_{\otimes}^{[-1]}(a) = 1 - \min\{a, 1\}$ Theorem 1 yields the approximate $\otimes$-equivalence space $\left(\mathcal{P}_X, \max\left\{1 - d_{(d_X)}, 0\right\}, \max\left\{1 - s_{(d_X)}, 0\right\}\right)$. More specifically, since $d_{(d_X)}$ and $s_{(d_X)}$ are normalized to [0,1] we get the approximate $\otimes$-equivalence space $(\mathcal{P}_X, e_{(d_X)}, x_{(d_X)})$, where

$$e_{(d_X)} = 1 - d_{(d_X)}, \tag{11}$$
$$x_{(d_X)} = 1 - s_{(d_X)}. \tag{12}$$

Corollary 1. *Let $\mathcal{P}_X$ denote the set of approximate points in the normalized metric space (X, d_X), let $e_{(d_X)}$ and $x_{(d_X)}$ be defined as in section 4. Then the triple $(\mathcal{P}_X, e_{(d_X)}, x_{(d_X)})$ is an approximate fuzzy $\otimes$-equivalence space w.r.t. the Łukasiewicz t-norm $\otimes$.*

Using the Łukasiewicz interpretation of the universal quantifier by the infimum, the axioms of $(\mathcal{P}_X, e_{(d_X)}, x_{(d_X)})$ can be written as follows:

$$\sigma_{(E1)} = \left[\inf_{P \in \mathcal{P}_X} \left[e_{(d_X)}(P,P)\right]\right] = 1,$$

$$\sigma_{(E2)} = \left[\inf_{P,Q \in \mathcal{P}_X} \left[e_{(d_X)}(P,Q) \Rightarrow e_{(d_X)}(Q,P)\right]\right] = 1,$$

$$\sigma_{(E3x)} := \left[\inf_{P,Q,R \in \mathcal{P}_X} \left[e_{(d_X)}(P,Q) \otimes x_{(d_X)}(Q) \otimes e_{(d_X)}(Q,R) \Rightarrow e_{(d_X)}(P,R)\right]\right] = 1.$$

If we introduce an additional predicate X to the language, the axioms can also be written as signed formulas in FL_{ev} with signs $\sigma_{(E1)} = \sigma_{(E2)} = \sigma_{(E3x)} = 1$:

(E1) $(\forall \mathtt{p}.\ \mathtt{E}(\mathtt{p},\mathtt{p});$ $\quad 1)$,
(E2) $(\forall \mathtt{p},\mathtt{q}.\ [\mathtt{E}(\mathtt{p},\mathtt{q}) \to \mathtt{E}(\mathtt{q},\mathtt{p})];$ $\quad 1)$,
(E3x) $(\forall \mathtt{p},\mathtt{q},\mathtt{r}.\ [\mathtt{E}(\mathtt{p},\mathtt{q}) \wedge \mathtt{X}(\mathtt{q}) \wedge \mathtt{E}(\mathtt{q},\mathtt{r}) \to \mathtt{E}(\mathtt{p},\mathtt{r})].$ $\quad 1)$,

The possibilistic interpretation of X(P) is the exactness measure $x_{(d_X)}(P) = 1 - s_{(s_X)}(P)$ of the approximate point P, and we call X the *exactness* predicate. It is inverse to the size $s_{(d_X)}(P)$ of P. (E3x) is a generalization of (E3) in ATG that is consistent with the the classic axiom: If an approximate point $Q = \{q\}$ has size zero (i.e., Q is exact), then $s_{(d_X)}(Q) = 0$ and $x_{(d_X)}(Q) = 1$. 1 is the neutral element of $\otimes$, and (E3x) collapses to (E3). Consequently, $(E1) - (E3x)$ is a vlaid axiomatization of approximate equality in ATG.

Remark 1. For any specific triple of approximate points P, Q, R, the weak transitivity formula $e(P,Q) * x(Q) * e(Q,R) \leq e(P,R)$ can also be written as signed transitivity, with sign $x(Q)$: $(e(P,Q) * e(Q,R) \Rightarrow e(P,R); x(Q))$[1]. In other words, the exactness degree $x(Q)$ is a *local* measure of the truthlikeness of transitivity. We have shown in section 5, that *globally*, a triple $\bar{P}, \bar{Q}, \bar{R}$ always exists with $x(\bar{Q}) = 0$ yielding the sign $\sigma_{(E3)} = 0$ for the transitivity axiom. Weak transitivity incorporates the local parameter of exactness in the transitivity formula itself and thus allows for restoring a positive sign globally. Following Gerla [7], it can be shown that the exactness measure $x_{(d_X)}$ is the most accurate local transitivity measure that is compatible with the approximate fuzzy similarity structure of $(\mathcal{P}_X, e_{(d_X)}, x_{(d_X)})$.

[1] In FL_ev the meaning of a signed formula $(\varphi; a)$ is that the truth(-likeness) degree of φ is not lower than a. I.e., a is a lower bound of the formula's truthlikeness. This is indeed the case here, because, with the adjunction property, we can write the weak transitivity formula $e(P,Q) * x(Q) * e(Q,R) \leq e(P,R)$ as $[e(P,Q) * e(Q,R) \Rightarrow e(P,R)] \geq x(Q)$, and $x(Q)$ is a lower bound to the truthlikeness of transitivity.

7 Conclusions and Future Work

We have successfully applied the ATG framework to the equality axioms, thereby laying the basis for the application of ATG to other axioms of classical geometries, such as Euclid's First Postulate. More specifically, we showed in a first step that a straight forward interpretation of the equality axioms by approximate points and approximate equality does not yield a usable fuzzy axiom set in ATG, because transitivity has truthlikness degree 0. Following Gerla [7], we showed in a second step that a generalization of the classical transitivity axiom can be given that has truthlikeness degree 1 and that is consistent with classical transitivity for the limit case of exact points. We discussed that the necessity to generalize the transitivity axiom arises from the fact that approximate points have size. This can be seen by the duality of approximate fuzzy equivalence spaces with extended pseudometric spaces. It is in contrast to the classical interpretation of geometry, where points are exact.

Due to size restrictions, the paper considered only approximate points. It can be shown that all produced results also hold for approximate lines as defined in the ATG framework in [1]. We also could not discuss here the fact that an alternative to the proposed generalization of the transitivity axiom exists, which consists in slightly changing the interpretation of the point primitive in ATG: If we introduce a maximum and minimum allowable size of approximate points in the possibilistic interpretation, the sign of the transitivity axiom does not vanish, but has a positive value. Furthermore, the results of the paper can be used to illustrate how the truthlikness degree of deduced geometric statements can be derived from the axioms using the graduated deduction apparatus of fuzzy logic with evaluated syntax. Since truthlikness in ATG is inverse to a spatial distance measure, truthlikeness propagation is at the same time a propagation of spatial error, and it can be illustrated how ATG can be used as an error propagation calculus in GIS.

As a next step of future work, we will apply the ATG framework to Euclid's First Postulate, which states that any two points can be connected by exactly one line. The topic has been partially discussed in preliminary work, cf. [28,29], but for a veristic instead of a possibilistic interpretation of approximate points and lines, and without an embedding in the rigorous framework of ATG.

Acknowledgment. The work is part of the author's PhD thesis and was produced while having been employed as a research assistant at the Research Group Geoinformation at the Vienna University of Technology. It was supported by Scholarships of the Austrian Marshallplan Foundation and the Vienna University of Technology. The author would like to thank Prof. Andrew U. Frank, Dr. John Stell and Prof. Lotfi Zadeh for their continuous support and inspiration.

References

1. Wilke, G.: Fuzzy logic with evaluated syntax for sound geometric reasoning with geographic information. In: Proceedings of the Annual Meeting of the North American Fuzzy Information Processing Society, NAFIPS 2014 (2014)
2. Goodchild, M.F.: Citizens as voluntary sensors: Spatial data infrastructure in the world of web 2.0. International Journal of Spatial Data Infrastructures Research 2, 24–32 (2007)
3. Wilke, G.: Towards approximate tolerance geometry for gis - a framework for formalizing sound geometric reasoning under positional tolerance. Ph.D. dissertation, Vienna University of Technology (2012)
4. Pullar, D.V.: Consequences of using a tolerance paradigm in spatial overlay. In: Proceedings of the Auto-Carto 11, pp. 288–296 (1993)
5. Novak, V., Perfilieva, I., Mockor, J.: Mathematical Principles of Fuzzy Logic. Kluwer Academic Publishers (1999)
6. Gerla, G.: Fuzzy Logic - Mathematical Tools for Approximate Reasoning. In: Trends in Logic, vol. 11. Kluwer Acadeic Publishers (2001)
7. Gerla, G.: Approximate similarities and poincare paradox. Notre Dame Journal of Formal Logic 49(2), 203–226 (2008)
8. Perkal, J.: On epsilon length. Bulletin de l'Academie Polonaise des Sciences 4, 399–403 (1956)
9. Shi, W.: A generic statistical approach for modelling error of geometric features in gis. International Journal of Geographical Information Science 12(2), 131–143 (1998)
10. Buyong, T.B., Frank, A.U., Kuhn, W.: A conceptual model of measurement-based multipurpose cadastral systems. Journal of the Urban and Regional Information Systems Association URISA 3(2), 35–49 (1991)
11. Shi, W., Cheung, C.K., Zhu, C.: Modelling error propagation in vector-based buffer analysis. International Journal of Geographical Information Science 17(3), 251–271 (2003)
12. Leung, Y., Ma, J.-H., Goodchild, M.F.: A general framework for error analysis in measurement-based gis, parts 1-4. Journal of Geographical Systems 6(4), 323–428 (2004)
13. Dougenik, J.A.: Whirlpool: A geometric processor for polygon coverage data. In: Proceddings AUTOCARTO, vol. 4, pp. 304–311 (1979)
14. Beard, K.M.: Zipping: New software for merging map sheets. In: Proceedings of the American Congress on Surveying and Mapping, vol. 1, pp. 153–161 (1986)
15. Chrisman, N.R.: The accuracy of map overlays: A reassessment. In: Landscape and Urban Planning, vol. 14, pp. 427–439 (1987), `http://www.sciencedirect.com/science/article/B6V91-46MN01M-1X/2/297d236facc206cdf028eedd653eb36d`
16. Beard, K.M., Chrisman, N.R.: Zipper: A localized approach to edgematching. Cartography and Geographic Information Science 15(2), 163–172 (1988)
17. Chrisman, N., Dougenik, J.A., White, D.: Lessons for the design of polygon overlay processing from the odyssey whirlpool algorithm. In: Bresnahan, P., Corwin, E., Cowen, D. (eds.) Proceedings of the 5th International Symposium on Spatial Data Handling, vol. 2, pp. 401–410. IGU Commission of GIS (1992)
18. Zhang, G., Tulip, J.: An algorithm for the avoidance of sliver polygons and clusters of points in spatial overlay. In: Proceedings of the 4th Inthernational Symposium on Spatial Data Handling (1990)

19. Harvey, F., Vauglin, F.: Geometric match processing: Applying multiple tolerances. In: Proceedings of 7th International Symposium on Spatial Data Handling (1996)
20. Abdelmoty, A.I., Jones, C.B.: Towards maintaining consistency of spatial databases. In: Proceedings of the Sixth International Conference on Information and Knowledge Management, CIKM 1997, pp. 293–300. ACM, New York (1997), http://doi.acm.org/10.1145/266714.266913
21. Clementini, E.: A model for uncertain lines. Journal of Visual Languages & Computing 16(4), 271–288 (2005)
22. Salesin, D., Stolfi, J., Guibas, L.: Epsilon geometry: building robust algorithms from imprecise computations. In: SCG 1989: Proceedings of the Fifth Annual Symposium on Computational Geometry, pp. 208–217. ACM, New York (1989)
23. Veregin, H.: Data quality parameters. In: Longley, P.A., Goodchild, M.F., Maguire, D.J., Rhind, D.W. (eds.) Geographical Information Systems, 2nd edn., vol. 1&2, pp. 177–189. John Wiley & Sons (1999)
24. Roberts, F.S.: Tolerance geometry. NDJFAM 14(1), 68–76 (1973)
25. Katz, M.: Inexact geometry. Notre Dame Journal of Formal Logic 21(3), 521–535 (1980)
26. Hájek, P.: Fuzzy logic. In: Zalta, E.N. (ed.) The Stanford Encyclopedia of Philosophy, Fall 2010 edn. (2010), http://plato.stanford.edu/archives/fall2010/entries/logic-fuzzy/
27. Godo, L., Rodriguez, R.O.: Logical approaches to fuzzy similarity-based reasoning: an overview, vol. 504, pp. 75–128. Springer (2008)
28. Wilke, G., Frank, A.: On equality of lines with positional uncertainty. In: Proceedings of the Sixth International Conference on Geomgraphic Information Science (2010)
29. Wilke, G., Frank, A.: Tolerance geometry - euclids first postulate for points and lines with extension. In: Proceedings of the ACM SIGSPATIAL 2010, San Jose, California, USA (2010)

Estimators of the Relations of: Equivalence, Tolerance and Preference on the Basis of Pairwise Comparisons with Random Errors

Leszek Klukowski

Systems Research Institute Polish Academy of Sciences,
Newelska 6, 01-447 Warsaw, Poland
Leszek.Klukowski@ibspan.waw.pl

Abstract. The paper presents the estimators of three relations: equivalence, tolerance and preference in a finite set on the basis of multiple pairwise comparisons, disturbed by random errors; they have been developed by the author. The estimators can rest on: binary (qualitative), multivalent (quantitative) and combined comparisons. The estimates are obtained on the basis of discrete programming problems. They require weak assumptions about distributions of comparisons errors, especially allow non-zero expected values. The estimators have good statistical properties, in particular are consistent. The estimates can be verified using statistical tests. The paper summarizes briefly the results obtained lastly by the author.

Keywords: estimators of the relations of equivalence, tolerance and preference, binary and multivalent comparisons with random errors, nearest adjoining order principle.

1 Introduction

Estimation of the relations of: equivalence, tolerance, or preference, on the basis of multiple pairwise comparisons with random errors, is aimed at detection of an actual structure of data. It also provides the properties of estimates, i.e.: consistency, efficiency, distributions of errors, etc. The estimates can be verified using statistical tests - the assumptions about: comparison errors, type of the relation (e.g. equivalence or tolerance) and existence of a relation (or randomness of comparisons).

The approach proposed rests on a statistical paradigm-is a simple nonparametric M-estimator. Its idea is: to determine the relation form, which minimizes the inconsistencies (differences) with a sample, i.e. multiple pairwise comparisons. It is similar to the concept of the nearest adjoining order (NAO-Slater, 1961). The comparisons can be obtained with the use of statistical tests, experts' opinions or other procedures, prone to generating random errors. The estimates are obtained on the basis of discrete optimization problems.

P. Angelov et al. (eds.), *Intelligent Systems'2014*,
Advances in Intelligent Systems and Computing 322, DOI: 10.1007/978-3-319-11313-5_34

The approach presented here is a contribution of the author to the subject. The main components comprise: weak assumptions about distributions of comparison errors, definition of two types of estimators for two types of data - binary (qualitative) and multivalent (quantitative), properties of the estimators and tests for verification of estimates. The features provides the extension of the application sphere of pairwise techniques.

The assumptions about distributions of comparison errors are weaker than those commonly used in the literature (David, 1988) and are satisfied in the case of each rational scientific investigation. The estimators can be also applied in the case of unknown distributions of comparison errors.

The properties of the estimators have been obtained on the basis of: • properties of random variables expressing sum of absolute differences between the actual relation and pairwise comparisons (determined by the author), • the probabilistic inequalities (Hoeffding 1963, Chebyshev), • properties of order statistics (David, 1970). The main property is consistency, i.e. convergence of estimates to actual relation for the number of independent comparisons of each pair approaching infinity. The analytical properties of the estimators has been complemented with the use of a simulation experiment (Klukowski 2011 Chap. 9).

The literature on pairwise comparisons with random errors concerns mainly ranking problems-classical results are presented in: David (1988), Bradley (1976, 1984), Brunk (1960), Flinger and Verducci (1993). The basic methods are based on the linear model and the combinatorial models (David, 1988 Chap. 2, 4).

The literature concerning classification methods (overlapping and non-overlapping), based on pairs of elements, is extremely extensive (see e.g. Gordon 1999, Hand 1986, Kaufman, Rousseeuv 1990, Hastie, Tibshirani, Friedman 2002, Kohonen 1995). However, it should be emphasized that existing approaches do not cover entirely the problems presented in the work.

The paper consists with 5 sections. The second section presents main ideas of estimation and general form of the estimators. The next section presents properties of estimators obtained by the author. The forth section discusses briefly optimization algorithms, which can be applied for determining of estimates. Last section summarizes results of the author in the area under consideration and shows problems for further researches.

2 Estimation of the Relations-Main Ideas

2.1 Definitions, Notations and Formulation of the Estimation Problems

The problem of estimation of relation on the basis of pairwise comparisons can be stated as follows. We are given a finite set of elements $\mathbf{X} = \{x_1, ..., x_m\}$ $(3 \leq m < \infty)$. It is assumed that there exists in the set $\mathbf{X}$: the equivalence relation $R^{(e)}$ (reflexive, transitive, symmetric), or the tolerance relation $R^{(\tau)}$ (reflexive, symmetric), or the preference relation $R^{(p)}$ (alternative of the equivalence relation and strict preference relation). Each relation generates some family of subsets $\chi_1^{(l)^*}, ..., \chi_n^{(l)^*}$ $(l \in \{p, e, \tau\}; n \geq 2)$ The equivalence relation generates the family $\chi_1^{(e)^*}, ..., \chi_n^{(e)^*}$ having the following properties:

$$\bigcup_{q=1}^{n} \chi_q^{(e)^*} = \mathbf{X}, \quad (1)$$

$$\chi_r^{(e)^*} \cap \chi_s^{(e)^*} = \{\mathbf{0}\}, \quad (2)$$

where: **0**-the empty set,

$$x_i, x_j \in \chi_r^{(e)^*} \equiv x_i, x_j - \text{equivalent elements}, \quad (3)$$

$$(x_i \in \chi_r^{(e)^*}) \wedge (x_j \in \chi_s^{(e)^*}) \equiv x_i, x_j - \text{non} - \text{equivalent elements for i} \neq \text{j}, \text{r} \neq \text{s}. \quad (4)$$

The tolerance relation generates the family $\chi_1^{(\tau)^*} \cap \chi_n^{(\tau)^*}$ with the property (1), and: $\exists r, s (r \neq s)$ such that $\chi_r^{(\tau)^*} \cap \chi_s^{(\tau)^*} \neq \{\mathbf{0}\}$,

$$x_i, x_j \in \chi_r^{(\tau)^*} \equiv x_i, x_j - \text{equivalent elements}, \quad (5)$$

$$(x_i \in \chi_r^{(\tau)^*}) \wedge (x_j \in \chi_s^{(\tau)^*}) \equiv x_i, x_j - \text{non} - \text{equivalent elements for} \quad (6)$$

$$i \neq j, (x_i, x_j) \notin \chi_r^{(\tau)^*} \cap \chi_s^{(\tau)^*},$$

each subset $\chi_r^{(\tau)^*} (1 \leq r \leq n)$ includes an element x_i such that $x_i \notin \chi_s^{(\tau)^*} (s \neq r)$. (7)

The preference relation generates the family $\chi_1^{(p)^*}, ..., \chi_n^{(p)^*}$ with the properties (1), (2) and the property:

$$(x_i \in \chi_r^{(p)^*}) \wedge (x_j \in \chi_s^{(p)^*}) \equiv x_i \text{ is preferred to } x_j \text{ for } r < s. \quad (8)$$

The relations defined by the relationships (1) - (8) can be determined, alternatively, by the values $T_v^{(l)}(x_i, x_j)$ $(x_i, x_j) \in \mathbf{X} \times \mathbf{X}$; $l \in \{p, e, \tau\}$, $v \in \{b, \mu\}$ symbols b, μ denote-respectively-the binary and multivalent comparisons), defined as follows:

$$T_b^{(e)}(x_i, x_j) = \begin{cases} 0 \; \textit{if exists } r \textit{ such that } (x_i, x_j) \in \chi_r^{(e)^*}, \\ 1 \; \textit{otherwise}; \end{cases} \quad (9)$$

• the values $T_b^{(e)}(x_i, x_j)$, describing the equivalence relation, assuming binary values, expresses the fact if a pair (x_i, x_j) belongs to a common subset or not;

$$T_b^{(\tau)}(x_i, x_j) = \begin{cases} 0 \; \textit{if exists } r, s (r = s \textit{ not excluded}) \textit{ such that } (x_i, x_j) \in \chi_r^{(\tau)^*} \cap \chi_s^{(\tau)^*}, \\ 1 \; \textit{otherwise}; \end{cases} \quad (10)$$

• the values $T_b^{(\tau)}(x_i, x_j)$, describing the tolerance relation, assuming binary values, expresses the fact if a pair (x_i, x_j) belongs to any conjunction of subsets (also to the same subset) or not; the condition (7) guarantees uniqueness of the description;

$$T_{\mu}^{(\tau)}(x_i, x_j) = \#(\Omega_i^* \cap \Omega_j^*) \tag{11}$$

where: Ω_l^* - the set of the form $\Omega_l^* = \{| \ x_i \in \chi_s^{(\tau)^*}\}$, $\#(\Xi)$ - the number of elements of the set Ξ;

• the values $T_{\mu}^{(\tau)}(x_i, x_j)$, describing the tolerance relation, assuming multivalent values, expresses the number of subsets of conjunction including both elements; condition (7) guarantees the uniqueness of the description;

$$T_b^{(p)}(x_i, x_j) = \begin{cases} 0 \ \textit{if there} \ \exists \ r \ \textit{such that} \ (x_i, x_j) \in \chi_r^{(p)^*}, \\ -1 \ \textit{if} \ x_i \in \ \chi_r^{(p)^*}, x_j \in \chi_s^{(p)^*} \ \textit{and} \ r < s; \\ 1 \ \textit{if} \ x_i \in \ \chi_r^{(p)^*}, x_j \in \chi_s^{(p)^*} \ \textit{and} \ r > s; \end{cases} \tag{12}$$

• the values $T_b^{(p)}(x_i, x_j)$, describing the preference relation, assuming binary values, expresses the direction of preference in a pair or the equivalence of its elements;

$$T_{\mu}^{(p)}(x_i, x_j) = d_{ij} \Leftrightarrow x_i \in \chi_r^{(p)^*}, \ x_j \in \chi_s^{(p)^*}, \ d_{ij} = r - s; \tag{13}$$

• the values $T_{\mu}^{(p)}(x_i, x_j)$, describing the preference relation, assuming multivalent values, expresses the difference of ranks of elements x_i and x_j.

2.2 Assumptions about Pairwise Comparisons

The relation $\chi_1^{(l)^*}, ..., \chi_n^{(l)^*}$ is to be determined (estimated) on the basis of $N (N \geq 1)$ comparisons of each pair $(x_i, x_j) \in \mathbf{X} \times \mathbf{X}$; any comparison $g_k^{(l)^*}(x_i, x_j)$ evaluates the actual value of $T_v^{(l)^*}(x_i, x_j)$ and can be disturbed by a random error. The following assumptions are made:

A1. The relation type (equivalence or tolerance or preference) is known, the number of subsets n - unknown.

A2. The probabilities of errors $g_{vk}^{(l)^*}(x_i, x_j) - T_v^{(l)^*}(x_i, x_j)$ ($l \in \{e, \tau, p\}$; $v \in \{b, \mu\}$; $k = 1, ..., N$) have to satisfy the following assumptions:

$$P(g_{bk}^{(l)}(x_i, x_j) - T_b^{(l)}(x_i, x_j) = 0 | T_b^{(l)}(x_i, x_j) = \kappa_{bij}^{(l)}) \geq 1 - \delta$$
$$(\kappa_{bij}^{(l)} \in \{-1, 0, 1\}, \ \delta \in (0, \tfrac{1}{2})), \tag{14}$$

$$\sum \sum\nolimits_{r \leq 0} \ P(g_{\mu k}^{(l)}(x_i, x_j) - T_{\mu}^{(l)}(x_i, x_j) = r | T_{\mu}^{(l)}(x_i, x_j) = \kappa_{\mu ij}^{(l)}) > \tfrac{1}{2}$$
$$(\kappa_{\mu ij}^{(l)} \in \{0, \pm 1, ..., \pm m\}, \ r - \text{zero or an integer number}), \tag{15}$$

$$\sum_{r\geq 0} P(g_{\mu k}^{(l)}(x_i,x_j) - T_{\mu}^{(l)}(x_i,x_j) = -r | T_{\mu}^{(l)}(x_i,x_j) = \kappa_{\mu ij}^{(l)}) > \tfrac{1}{2}$$
$$(\kappa_{\mu ij}^{(l)} \in \{0, \pm 1, ..., \pm m\}, r - \text{zero or an integer number}), \qquad (16)$$

$$P(g_{\mu k}^{(l)}(x_i,x_j) - T_{\mu}^{(l)}(x_i,x_j) = r) \geq P(g_{\mu k}^{(l)}(x_i,x_j) - T_{\mu}^{(l)}(x_i,x_j) = r+1 |$$
$$T_{\mu}^{(l)}(x_i,x_j) = \kappa_{\mu ij}^{(l)})(\kappa_{\mu ij}^{(l)} \in \{0, ..., m\}, r \geq 0), \qquad (17)$$

$$P(g_{\mu k}^{(l)}(x_i,x_j) - T_{\mu}^{(l)}(x_i,x_j) = r) \geq P(g_{\mu k}^{(l)}(x_i,x_j) - T_{\mu}^{(l)}(x_i,x_j) = r-1 |$$
$$T_{\mu}^{(l)}(x_i,x_j) = \kappa_{\mu ij}^{(l)})(\kappa_{\mu ij}^{(l)} \in \{0, ..., m\}, r \leq 0), \qquad (18)$$

$$\sum_{r} P(g_{vk}^{(l)}(x_i,x_j) - T_{v}^{(l)}(x_i,x_j) = r) = 1, \qquad (19)$$

A3. The comparisons $g_{vk}^{(l)}(x_i,x_j)$ ($l \in \{e, \tau, p\}$; $v \in \{b, \mu\}$; $(x_i,x_j) \in \mathbf{X} \times \mathbf{X}$; $k = 1, ..., N$) are independent random variables.

The assumption A3 makes it possible to determine the probability functions of the estimators proposed. However, determination of the exact distribution, in an analytic way, is complicated and, in practice, unrealizable. The main properties of the estimators, especially consistency, are valid without the assumption. The assumption A3 can be relaxed in the following way: the comparisons $g_{vk}^{(l)}(x_i,x_j)$ and $g_{vl}^{(l)}(x_r,x_s)$ ($l \neq k$; $r \neq i,j$; $s \neq i,j$), i.e. including different elements, have to be independent. In the case of the preference relation with equivalent elements, the condition (14) can be relaxed to the form (15)-(16). The assumptions A2-A3 reflect the following properties of distributions of comparisons errors:

- the probability of correct comparison is greater than of the incorrect one - in the case of binary comparisons (inequality (14));
- zero is the median of each distribution of comparison error (inequalities (14)-(16));
- zero is the mode of each distribution of comparison error (inequalities (14)-(18));
- the set of all comparisons comprises the realizations of independent random variables;
- the expected value of any comparison error can differ from zero.

2.3 The Form of Estimators

Two kinds of estimators are examined. The first one is based on the total sum of absolute differences between relation and comparisons (denoted $\hat{\chi}_1^{(l)}, ..., \hat{\chi}_{\hat{n}}^{(l)}$

or $\hat{T}_v^{(l)}(x_i, x_j)$, $(x_i, x_j) \in \mathbf{X} \times \mathbf{X}))$. The estimate is obtained on the basis of the discrete minimization problem:

$$\min_{\chi_1^{(l)},\dots,\chi_r^{(l)} \in F_{\mathbf{X}}^{(l)}} \left\{ \sum_{<i,j> \in R_m} \sum_{k=1}^{N} |g_{vk}^{(l)}(x_i, x_j) - t_v^{(l)}(x_i, x_j)| \right\}, \tag{20}$$

where:
$F_{\mathbf{X}}^{(l)}$ - the feasible set, i.e. the family of all relations $\chi_1^{(l)}, \dots, \chi_r^{(l)}$ of l-th type in the set $\mathbf{X}$,
$t_v^{(l)}(x_i, x_j)$ - the values describing any relation $\{\chi_1^{(l)}, \dots, \chi_r^{(l)}\}$ of l-th type, from feasible set,
R_m - the set of the form $R_m = \{< i, j > | 1 \leq i, j \leq m; j > i\}$
(symbol $g_{vk}^{(l)}(x_i, x_j)$ is used for random variables and realizations).

In the case of preference relation (binary comparisons) the following transformation can be applied:

$$\theta(g_{bk}^{(p)}(x_i, x_j) - t_b^{(p)}(x_i, x_j)) \begin{cases} 0 \ \mathit{if} \ g_{bk}^{(p)}(x_i, x_j) = t_b^{(p)}(x_i, x_j); \\ 1 \ \mathit{if} \ g_{bk}^{(p)}(x_i, x_j) \neq t_b^{(p)}(x_i, x_j). \end{cases} \tag{21}$$

The criterion function (20) with the use of (21), expresses the number of differences between the comparisons and values $T_b^{(p)}(x_i, x_j)$. It is simpler from the computational point of view, because the variables $\theta(g_{vk}^{(l)}(x_i, x_j) - t_{vb}^{(l)}(x_i, x_j))$ assume binary values (zero or one), while the differences $|g_{vk}^{(l)}(x_i, x_j) - t_{vb}^{(l)}(x_i, x_j)|$ assume values from the set $\{0, 1, 2\}$. The properties of both approaches are similar.

The second kind of estimate is based on medians from comparisons of each pair and denoted $\hat{\chi}_1^{(l)}, \dots, \hat{\chi}_r^{(l)}$ (or $\hat{T}_v^{(l)}(x_i, x_j)$); it is obtained on the basis of the following minimization problem:

$$\min_{\chi_1^{(l)},\dots,\chi_r^{(l)} \in F_X} \left\{ \sum_{<i,j> \in R_m} |g_v^{(l,me)}(x_i, x_j) - t_v^{(l)}(x_i, x_j)| \right\}, \tag{22}$$

where:
$g_v^{(l,me)}(x_i, x_j)$- the sample median in the set $g_{v,1}^{(l)}(x_i, x_j), \dots, g_{vN}^{(l)}$.

The number of estimates, resulting from the criterion functions (20), (22) can exceed one; the unique estimate can be determined in a random way or as a result of verification. Multiple estimates can appear also in other methods (see David 1988, Chap. 2). The minimal values of the criterion functions (20), (22) are equal zero. The assumptions A1-A3 allow for inference about distributions of errors of estimators. Let us discuss firstly the estimator based on of the criterion (20). For each relation type one can determine a finite set including all possible realizations of comparisons $g_{vk}^{(l)}(x_i, x_j), (l \in \{e, \tau, p\}, v \in \{b, \mu\}, k = 1, \dots, N; < i, j > \in R_m)\}$ and the probability of each realization. The use of the criterion (20) determines: the estimate, its probability and estimation error. The error has the form: $\{\hat{T}_v^{(l)}(x_i, x_j) - T_v^{(l)}(x_i, x_j);\ <i,j> \in R_m\}$, i.e. is a multidimensional random variable.

The estimate with the value of error equal zero is the errorless estimate. The probability of such error can be determined in the analytic way - as a sum of probabilities of all realizations of comparisons indicating the errorless estimate. The probabilities of errors $\{\hat{T}_v^{(l)}(x_i, x_j) - T_v^{(l)}(x_i, x_j);\ <i,j> \in R_m\}$, which assume values different than zero, can be determined in a similar way; such the errors and their probabilities determine the distribution function of estimation error. Determination of the probability function in the analytic way is complicated - even for moderate m. Therefore, simulation approach has to be used instead of the analytic. Similar considerations apply for the criterion (22).

3 Properties of Estimators

The analytical properties of the estimators, obtained by the author, have mainly asymptotic character, i.e. they apply to the case $N \to \infty$. They are based on the random variables: $\sum_{R_m} \sum_k |g_{vk}^{(l)}(x_i, x_j) - T_v^{(l)}(x_i, x_j)|$ and $\sum_{R_m} |g_v^{(l,me)}(x_i, x_j) - T_v^{(l)}(x_i, x_j)|$. The following results have been obtained:

(i) the expected values $E(\sum_{R_m} \sum_k |g_{vk}^{(l)}(x_i, x_j) - T_v^{(l)}(x_i, x_j)|)$, corresponding to the actual relation, are lower than the expected values $E(\sum_{R_m} \sum_k |g_{vk}^{(l)}(x_i, x_j) - \tilde{T}_v^{(l)}(x_i, x_j)|)$ corresponding to any other relation (denoted by $\tilde{T}_v^{(l)}(x_i, x_j)$), i.e.:
$$E(\sum_{R_m} \sum_k |g_{vk}^{(l)}(x_i, x_j) - T_v^{(l)}(x_i, x_j)|) < E(\sum_{R_m} \sum_k |g_{vk}^{(l)}(x_i, x_j) - \tilde{T}_v^{(l)}(x_i, x_j)|);$$
it is valid also for the median case: $E(\sum_{R_m} |g_v^{(l,me)}(x_i, x_j) - T_v^{(l)}(x_i, x_j)|) < E(\sum_{R_m} |g_v^{(l,me)}(x_i, x_j) - \tilde{T}_v^{(l)}(x_i, x_j)|)$;

(ii) the variances of the random variables divided by the number of comparisons N, in the case of sum of differences, converge to zero for $N \to \infty$, i.e.: $\lim_{N\to\infty} Var(\frac{1}{N} \sum_{R_m} \sum_k |g_{vk}^{(l)}(x_i, x_j) - T_v^{(l)}(x_i, x_j)|) = 0$, $\lim_{N\to\infty} Var(\frac{1}{N} \sum_{R_m} \sum_k |g_{vk}^{(l)}(x_i, x_j) - \tilde{T}_v^{(l)}(x_i, x_j)|) = 0$; the similar property is true for the median case $\lim_{N\to\infty} Var(\sum_{R_m} |g_v^{(l,me)}(x_i, x_j) - T_v^{(l)}(x_i, x_j)|) = 0$, $\lim_{N\to\infty} Var(\sum_{R_m} |g_v^{(l,me)}(x_i, x_j) - \tilde{T}_v^{(l)}(x_i, x_j)|) = 1$;

(iii) the probability of the event that the variable expressing differences, corresponding to actual relation, assumes a value lower than the variable corresponding to any relation other than actual converges to one for $N \to \infty$, i.e.: $\lim_{N\to\infty} P(\sum_{R_m} \sum_k |g_{vk}^{(l)}(x_i, x_j) - T_v^{(l)}(x_i, x_j)| < \sum_{R_m} \sum_k |g_{vk}^{(l)}(x_i, x_j) - \tilde{T}_v^{(l)}(x_i, x_j)|) = 1$; the similar property is valid in the median case: $\lim_{N\to\infty} P(\sum_{R_m} |g_v^{(l,me)}(x_i, x_j) - T_v^{(l)}(x_i, x_j)| < \sum_{R_m} |g_v^{(l,me)}(x_i, x_j) - \tilde{T}_v^{(l)}(x_i, x_j)|) = 1$. The speed of convergence (obtained on the basis of Hoeffding inequality for the first estimator and properties of the sample median for the median estimator) guarantees good efficiency of the estimates in both cases.

The facts (i) - (iii) guarantee good statistical properties of the estimators: consistency and fast convergence to actual relation.

Let us illustrate these general facts by the simplest case, i.e. equivalence relation and the estimator based on the criterion (20). The differences between any comparison $g_{bk}^{(e)}(x_i, x_j)$ and the value $T_b^{(e)}(x_i, x_j)$ assume the form:

$$U_{bk}^{(e)^*}(x_i, x_j) = \begin{cases} 0 \text{ if } g_{bk}^{(e)}(x_i, x_j) = T_b^{(e)}(x_i, x_j); T_b^{(e)}(x_i, x_j) = 0; \\ 1 \text{ if } g_{bk}^{(e)}(x_i, x_j) \neq T_b^{(e)}(x_i, x_j); T_b^{(e)}(x_i, x_j) = 0. \end{cases} \quad (23)$$

$$V_{bk}^{(e)^*}(x_i, x_j) = \begin{cases} 0 \text{ if } g_{bk}^{(e)}(x_i, x_j) = T_b^{(e)}(x_i, x_j); T_b^{(e)}(x_i, x_j) = 1; \\ 1 \text{ if } g_{bk}^{(e)}(x_i, x_j) \neq T_b^{(e)}(x_i, x_j); T_b^{(e)}(x_i, x_j) = 1. \end{cases} \quad (24)$$

The sum of differences assumes, for any $k (1 \leq k \leq N)$, the form:

$$\sum_{<i,j> \in I^{(e)^*}} U_{bk}^{(e)^*}(x_i, x_j) + \sum_{<i,j> \in J^{(e)^*}} V_{bk}^{(e)^*}(x_i, x_j), \quad (25)$$

where:

$I^{(e)^*}$ - the set of pairs $\{<i,j> | T_b^{(e)^*}(x_i, x_j) = 0\}$,
$J^{(e)^*}$ - the set of pairs $\{<i,j> | T_b^{(e)^*}(x_i, x_j) = 1\}$.

The total sum of the differences between the relation form and the comparisons is equal:

$$W_{bN}^{(e)^*} = \sum_{k=1}^{N} (\sum_{<i,j> \in I^{(e)^*}} U_{bk}^{(e)^*}(x_i, x_j) + \sum_{<i,j> \in I^{(e)^*}} V_{bk}^{(e)^*}(x_i, x_j)). \quad (26)$$

Under the assumptions A1, A2, A3, the expected values of the variables $U_{bk}^{(e)^*}(x_i, x_j)$, $V_{bk}^{(e)^*}(x_i, x_j)$ satisfy the inequalities: $E(U_{bk}^{(e)^*}(x_i, x_j)) \leq \delta$, $E(V_{bk}^{(e)^*}(x_i, x_j)) \leq \delta$. Therefore, the expected value of the variable $W_{bN}^{(e)^*}$ satisfies the inequality $E(V_{bk}^{(e)^*}) \leq \frac{Nm(m-1)}{2}\delta$. Assumptions A1-A3 allow for determining the variance $Var(W_{bN}^{(e)^*})$; its value is finite and satisfies the inequality $Var(W_{bN}^{(e)^*}) \leq \frac{Nm(m-1)}{2}\delta(1-\delta)$. Obviously:

$$E(\frac{1}{N} W_{bN}^{(e)^*}) \leq \frac{m(m-1)}{2}\delta, \quad (27)$$

$$\lim_{N \to \infty} Var(\frac{1}{N} W_{bN}^{(e)^*}) = 0. \quad (28)$$

Let us consider any relation $\tilde{\chi}_1^{(e)}, ..., \tilde{\chi}_{\tilde{n}}^{(e)}$ different than $\chi_1^{(e)^*}, ..., \chi_n^{(e)^*}$; this means that there exist pairs (x_i, x_j) such, that $\tilde{T}_b^{(e)} \neq T_b^{(e)}(x_i, x_j)$. Define the random variables $\tilde{U}_{bk}^{(e)}(x_i, x_j)$, $\tilde{V}_{bk}^{(e)}(x_i, x_j)$, $\tilde{W}_b^{(e)}(x_i, x_j)$ corresponding to the values $\tilde{T}_b^{(e)}(x_i, x_j)$:

$$\tilde{U}_{bk}^{(e)}(x_i, x_j) = \begin{cases} 0 \text{ if } g_{bk}^{(e)}(x_i, x_j) = \tilde{T}_b^{(e)}(x_i, x_j); \tilde{T}_b^{(e)}(x_i, x_j) = 0; \\ 1 \text{ if } g_{bk}^{(e)}(x_i, x_j) \neq \tilde{T}_b^{(e)}(x_i, x_j); \tilde{T}_b^{(e)}(x_i, x_j) = 0, \end{cases}$$

$$\tilde{V}_{bk}^{(e)}(x_i, x_j) = \begin{cases} 0 \text{ if } g_{bk}^{(e)}(x_i, x_j) = \tilde{T}_b^{(e)}(x_i, x_j); \tilde{T}_b^{(e)}(x_i, x_j) = 1; \\ 1 \text{ if } g_{bk}^{(e)}(x_i, x_j) \neq \tilde{T}_b^{(e)}(x_i, x_j); \tilde{T}_b^{(e)}(x_i, x_j) = 1, \end{cases} \quad (29)$$

$$\tilde{W}_{bN}^{(e)^*} = \sum_{k=1}^{N} \left(\sum_{\tilde{I}^{(e)}} \tilde{U}_{bk}^{(e)}(x_i, x_j) + \sum_{\tilde{J}^{(e)}} \tilde{V}_{bk}^{(e)}(x_i, x_j) \right). \quad (30)$$

where:
$\tilde{I}^{(e)}$ - the set of pairs $\{< i, j > | \tilde{T}_b^{(e)^*}(x_i, x_j) = 0\}$,
$\tilde{J}^{(e)}$ - the set of pairs $\{< i, j > | \tilde{T}_b^{(e)^*}(x_i, x_j) = 1\}$.

The expected values $E(\tilde{U}_{bk}^{(e)}(x_i, x_j))$, $E(\tilde{V}_{bk}^{(e)}(x_i, x_j))$ satisfy the relationships:

$$E(\tilde{U}_{bk}^{(e)}(x_i, x_j)) = 0 * P(g_{bk}^{(e)}(x_i, x_j)) = 0 |\ T_b^{(e)^*}(x_i, x_j) = 1) + 1 * P(g_{bk}^{(e)}(x_i, x_j)) = 1 |\ T_b^{(e)^*}(x_i, x_j) = 1) \geq 1 - \delta, \quad (31)$$

$$E(\tilde{V}_{bk}^{(e)}(x_i, x_j)) = 0 * P(g_{bk}^{(e)}(x_i, x_j)) = 0 |\ T_b^{(e)^*}(x_i, x_j) = 1) + 1 * P(g_{bk}^{(e)}(x_i, x_j)) = 1 |\ T_b^{(e)^*}(x_i, x_j) = 1) \geq 1 - \delta, \quad (32)$$

and:

$$E(\frac{1}{N}\tilde{W}_{bN}^{(e)}) = \frac{1}{N}\left(\sum_{k=1}^{N}\left(\sum_{\tilde{I}^{(e)}} E(\tilde{U}_{bk}^{(e)}(x_i, x_j)) + \sum_{\tilde{J}^{(e)}} E(\tilde{V}_{bk}^{(e)}(x_i, x_j))\right)\right) > \frac{m(m-1)}{2}\delta. \quad (33)$$

The formulae (26) and (33) indicate that the expected value $E(\frac{1}{N}W_{bN}^{(e)^*})$, corresponding to the actual relation $\chi_1^{(e)^*}, ..., \chi_n^{(e)^*}$, is lower than the expected value $E(\frac{1}{N}\tilde{W}_{bN}^{(e)})$, corresponding to any other relation $\tilde{\chi}_1^{(e)^*}, ..., \tilde{\chi}_{\tilde{n}}^{(e)^*}$. The variances of both variables converge to zero for $N \to \infty$. The variables $U_{bk}^{(e)^*}(x_i, x_j)$, $V_{bk}^{(e)^*}(x_i, x_j)$ assume values $|g_{bk}^{(e)^*}(x_i, x_j) - T_b^{(e)^*}(x_i, x_j)|$, used in the criterion function (20). Moreover, it can be also shown on the basis of Hoeffding (1963) inequality, that:

$$P(W_{bN}^{(e)^*} < \tilde{W}_{bN}^{(e)}) \geq 1 - \exp\{-2N(\frac{1}{2} - \delta)^2\}. \quad (34)$$

The above facts indicate that the estimator $\hat{\chi}_1^{(e)}, ..., \hat{\chi}_{\hat{n}}^{(e)}$, minimizing the number of inconsistencies with comparisons, guarantees the errorless estimate for

$N \to \infty$. The inequality (34) indicates the influence of δ and N on the precision of the estimator. The properties of the median estimator are based on the fact that the random variables $\frac{1}{N}\sum_{k=1}^{N} U_{bk}^{(e)^*}(x_i, x_j)$ and $\sum_{k=1}^{N} V_{bk}^{(e)^*}(x_i, x_j)$ converge, with probability one, to a limit equal or lower than δ, as $N \to \infty$. Therefore, the median $g_b^{(e,me)}(x_i, x_j)$ converges to the actual value $T_b^{(e)}(x_i, x_j)$ (David 1970). As a result, minimization of (22) guarantees that the estimate $\hat{\chi}_1^{(e)}, ..., \hat{\chi}_{\hat{n}}^{(e)}$ converges to $\chi_1^{(e)^*}, ..., \chi_n^{(e)^*}$. Moreover, it can be shown (see Klukowski, 1994) that:

$$P(W_{bN}^{(e,me)^*} < \tilde{W}_{bN}^{(e,me)}) \geq 1 - 2\exp\{-2N(\frac{1}{2} - \delta)^2\}. \quad (35)$$

The inequality (35), weaker than (34), gives some evaluation of convergence of the median estimator. The evaluation of the error of the estimator has been determined, in the case of the preference relation, with the use of simulation approach (see Klukowski 2011, Chap. 9). The above results are valid also in the case of the tolerance and preference relations, with binary comparisons.

The case of multivalent comparisons (tolerance and preference relations) can be analyzed in a similar way. The properties (i)-(iii) hold true, as well, but the inequality (34), proved on the basis of Hoeffding inequalities, becames more complicated; the subtrahend assumes the form $-2\exp\{-2N\tilde{\theta}^{-2}\}$, where $\tilde{\theta}$- constant depending on $\tilde{T}_\mu^{(l)}(x_i, x_j)(l \in \{p, \tau\})$. The probability corresponding to median estimator also converges to one, as , but this fact is proved on the basis of distribution of sample median (see David 1970). The details about these estimators are presented in Klukowski 2011, Chap. 6 and 8.

4 Solving of Optimization Problems

Minimization of the functions (20), (22) is, in general, not an easy problem, because of the dimension of the feasible set. Currently, the algorithms are available only for the equivalence and preference relations and single binary comparisons (see Hansen P., et al 1994, David, 1988, Chapt. 2); they refer to the dynamic programming or branch-and-bound algorithms, some of them can be applied for known n. The algorithms are efficient for the moderate number of elements m, i.e. about 50. In the case of large m, the problems can be solved with the use of heuristic algorithms: genetic (Falkenauer, 1998), artificial neural networks, random search (Ripley, 2006), swarm intelligence (Abraham and Grosan, 2006), etc.

In the case of multivalent comparisons the exact algorithms are not available now. The problems with moderate number of elements m, i.e. 3-12, can be solved with the use of complete enumeration. Problems with higher number of elements can be solved using heuristic algorithms, mentioned above. It is obvious that the estimators based on multivalent comparisons require more computations than those based on binary comparisons.

5 Summary - Achievements of the Work and Further Researches

The work presents the synthesis of the results, of the author (Klukowski 1990, 1994, 2000, 2011), concerning estimation of three relations-equivalence, tolerance, and preference-on the basis of pairwise comparisons with random errors. The following new results should be emphasized.

1^0. Two types of data have been taken into account: binary and multivalent. Binary data reflect qualitative features of the compared elements, e.g. direction of preference in a pair, while multivalent data-quantitative features, e.g. the distance between elements - difference of ranks.

2^0. The assumptions about the comparison errors are weaker than those commonly used in the literature, especially: expected values of comparison errors can differ from zero, distributions of comparison errors may be unknown, comparisons including the same element cannot be independent. Therefore, the estimators can be used in the cases, when the existing algorithms are not applicable and can produce incorrect results.

3^0. Two estimators have been examined; the first one is based on the sum of absolute differences between the relation form and comparisons, the second is based on differences between the relation form and medians from comparisons of each pair. The estimators have the form of solutions of optimization problems. The efficiency of the first estimator is better, but involves higher cost of computations. The median estimator requires a lower amount of computations, in the case of application of optimization algorithms; however, it is more robust (robustness is important feature especially in the case of multivalent comparisons).

4^0. The analytical properties of the estimators have been complemented with the simulation experiment (Klukowski 2011). It indicates good efficiency of the estimators, especially those based on multivalent comparisons, and confirms "advantage" of precision the estimator based on total sum of absolute differences.

5^0. The properties of estimates can be thoroughly verified using statistical tests (known and developed by the author); verification comprises the assumptions as to the comparison errors and existence of the relation.

6^0. The approach proposed allows for combining of comparisons obtained from different sources, e.g. statistical tests, experts, neural networks. It is also possible to combine binary and multivalent data.

7^0. The basis for estimates - solutions of the problems (20) and (22) - can be obtained with the use of: complete enumeration or discrete algorithms or heuristic algorithms-depending on number of m.

8^0. The approach can be applied to more complex structures of data, e.g. trees .

References

Abraham, A., Grosan, C. (eds.): Swarm Intelligence in Data Mining. Springer (2006)

Brunk, H.D.: Mathematical models for ranking from paired comparisons. JASA Vol 55, 503–520 (1960)

Bradley, R.A.: Science, statistics and paired comparisons. Biometrics vol 32, 213–232 (1976)
Bradley, R.A.: Paired comparisons: some basic procedures and examples. In: Krishnaiah, P.R., Se, P.K. (eds.) Handbook of Statistics, vol. 4, pp. 299–326. North-Holland, Amsterdam (1984)
David, H.A.: Order Statistics. J. Wiley (1970)
David, H.A.: The Method of Paired Comparisons, 2nd edn. Griffin, London (1988)
Falkenauer, E.: Genetics Algorithms and Grouping Data. J. Wiley (1998)
Flinger, Verducci (eds.): Probability Models and Statistical Analyses for Ranking Data. Springer, Heidelberg (1993)
Gordon, A.D.: Classification, 2nd edn. Chapman&Hall/CRC (1999)
Hand, D.J.: Discrimination and Classification. J. Wiley (1986)
Hansen, P., Jaumard, B., Sanlaville, E.: Partitioning Problems in Cluster Analysis: A Review of Mathematical Programming Approaches. In: Studies in Classification, Data Analysis, and Knowledge Organization, Springer (1994)
Hastie, T., Tibshirani, R., Friedman, J.: The Elements of Statistical Learning. Data Mining, Inference and Prediction. Springer (2002)
Hoeffding, W.: Probability inequalities for sums of bounded random variables. JASA 58, 13–30 (1963)
Kaufman, L., Rousseeuw, P.J.: Findings Groups in Data: An Introduction to Cluster Analysis. J. Wiley (1990)
Klukowski, L.: Algorithm for classification of samples in the case of unknown number variables generating them, Przeglad Statystyczny XXXVII, pp. 167–177 (1990) (in Polish)
Klukowski, L.: Some probabilistic properties of the nearest adjoining order method and its extensions. Annals of Operational Research 51, 241–261 (1994)
Klukowski, L.: The nearest adjoining order method for pairwise comparisons in the form of difference of ranks. Annals of Operations Research vol 97, 357–378 (2000)
Klukowski, L.: Methods of Estimation of Relations of: Equivalence, Tolerance, and Preference in a Finite Set. IBS PAN, Warsaw. Systems Research, vol. 69 (2011)
Kohonen, T.: Self-Organizing Maps. Springer (1995)
Ripley, B.D.: Stochastic Simulation. J. Wiley (2006)
Slater, P.: Inconsistencies in a schedule of paired comparisons. Biometrika 48, 303–312 (1961)

Part IX
Multiagent Systems

An Intelligent Architecture for Autonomous Virtual Agents Inspired by Onboard Autonomy

Kaveh Hassani and Won-Sook Lee

School of Electrical Engineering and Computer Science, University of Ottawa, Ottawa, Canada
{kaveh.hassani,wslee}@uottawa.ca

Abstract. Intelligent virtual agents function in dynamic, uncertain, and uncontrolled environments, and animating them is a chaotic and error-prone task which demands high-level behavioral controllers to be able to adapt to failures at lower levels of the system. On the other hand, the conditions in which space robotic systems such as spacecraft and rovers operate, inspire by necessity, the development of robust and adaptive control software. In this paper, we propose a generic architecture for developing autonomous virtual agents that let them to illustrate robust deliberative and reactive behaviors, concurrently. This architecture is inspired by onboard autonomous frameworks utilized in interplanetary missions. The proposed architecture is implemented within a discrete-event simulated world to evaluate its deliberative and reactive behaviors. Evaluation results suggest that the architecture supports both behaviors, consistently.

Keywords: Virtual agents, Autonomous systems, Cognitive architectures.

1 Introduction

Embodied agents have been extensively investigated in both robotics and virtual environments. In latter field, they are referred as intelligent virtual agents (IVA). Believability is a critical aspect of an IVA which can be augmented by surface realization and intelligent behavior. Although advances in computer graphics has led to realistic surface realization, yet the IVA's behavior is monotonous due to its scripted nature. Intelligent behavior emerges from cognitive characteristics such as recognition, decision making, perception, situation assessment, prediction, problem solving, planning, reasoning, belief maintenance, execution, interaction and communication, reflection, and learning [1].

Traditional IVA architecture follows a dualist perspective which decomposes the agent to a mind and a body. Mind as an abstract layer provides the agent with cognitive functionalities. It receives perceptions from the body, makes decisions, and sends the decisions in terms of abstract actions to the body. The body as an embodied layer animates the received actions within the virtual environment and provides the mind with perceptions acquired from its virtual sensors. Continues interaction between mind and body forms a closed perception-cognition-action loop. However, some recent studies challenge the strict separation between mind and body [2].

P. Angelov et al. (eds.), *Intelligent Systems'2014*,
Advances in Intelligent Systems and Computing 322, DOI: 10.1007/978-3-319-11313-5_35

In terms of AI, Russell and Norving [3] utilized the notation of rational agents, and categorized them to simple reflex, model-based reflex, goal-based, and utility-based agents. In robotics literature, agents are classified to cognitive, behavioral and hybrid agents. Wooldridge [4] categorized intelligent agents to logic-based, reactive, belief-desire-intention (BDI) and layered agents. Among them, BDI agent has been widely utilized as an IVA architecture. Logic-based agents exploit symbolic logic deductions and cannot handle uncertainties. Cognitive agents such as BDI provide deliberative decision making capabilities for long temporal horizons. However, they cannot react to the situations which need instant responds. Furthermore, knowledge representation is a main challenge in these architectures. Reactive agents (i.e. behavior-based agents in robotics literature) couple the control and decision making mechanisms to the current local sensory information to provide real-time reactions. Although this approach minimizes the complexity of the representational structures and provides quick responses to dynamic environments, it is not scalable and suffers from the lack of reasoning capabilities and task-oriented behaviors. Hybrid agents have a layered structure. Layers function in different abstraction and operational frequency levels, and thus let the agent to combine reactive and deliberative behaviors.

An IVA functions in a dynamic, uncertain, and uncontrolled environment, and animating it is a chaotic and error-prone task which demands high-level behavioral controllers to be able to adapt to failure at lower levels of the system (e.g. when a navigation system fails to direct a walking agent to a desired waypoint). The conditions in which space robotic systems such as satellites, spacecraft and rovers operate, inspire by necessity, the development of robust and adaptive control software. These autonomous systems which have been successfully employed by NASA and ESA[1] can achieve mission goals and handle unpredicted situations, autonomously [5].

According to Muscettola et al. [6], challenges of developing agent architectures for onboard autonomy in space missions are driven by four major characteristics of the spacecraft as follows. First, the spacecraft must perform autonomous operations for long periods of time without human guidance. Second, the performed operations must guarantee success, given tight deadlines and resource constraints. Third, due to high cost of the spacecraft, its operations require high reliability. Fourth, spacecraft operation involves concurrent activities among a set of tightly coupled subsystems.

In this paper, we adopt Remote Agent (RA) architecture [6] (i.e. developed by NASA to autonomously control the DS-1 spacecraft as part of New Millennium Deep Space Mission-1 to flyby an asteroid) to develop a generic architecture for IVA development. Furthermore, we utilize a fuzzy ontology as the agent's knowledge representation scheme. Adopting RA architecture reinforces IVA by a reliable and intelligent platform that has already shown to be successful in complex inter-planetary missions. Moreover, embedded fuzzy ontology lets IVA to acquire knowledge from environmental uncertainties and construct a proper belief model. This paper is organized as follows: in section 2 an overview of related works is presented. In section 3, we describe our proposed architecture. In section 4, experimental results and evaluations are discussed. Finally, section 5 concludes the paper.

---

[1] European space agency.

2 Related Works

According to Langley, Laird and Rogers [1] "A cognitive architecture specifies the underlying infrastructure for an intelligent system". During the last decades, several cognitive architectures have been proposed. ACT-R [7] cognitive framework emphasizes human psychological verisimilitude. Soar [8] as a rule-based cognitive architecture formulates the tasks as attempts to achieve goals. ICARUS [9] model designed for embodied agents emphasizes perception and action over abstract problem solving. SASE [10] is based on Markov decision processes, and utilizes the concept of autonomous mental development. PRODIGY [11], DUAL [12] and Polyscheme [13] are other examples of cognitive architectures. Probably, BDI [14] is the most representative model of cognitive agents. It triggers behaviors driven by conceptual models of intentions and goals in complex dynamic scenarios. BBSOAA [15] is an extension of BDI architecture that enhances the knowledge representation and inference capabilities, and is suitable for simulating virtual humans. Although BDI-inspired architectures such as IRMA [16] support long term behaviors, their current implementations are whether hardware-based or logic-based. More information regarding cognitive architectures can be found in [1].

A few IVA architectures concern with software engineering issues. As an instance, CAA [17] is a generic object-oriented architecture that supports context-sensitive behaviors. Additionally, A few architectures such as CMION [18] are developed based on the notion of migrating agent which refers to the ability of an agent to morph from one form of embodiment to another. A few studies such as embodied cognition model [19] challenge the strict separation between mind and body. This model embeds a secondary control loop, subconscious mind, into the body layer. Moreover, some research works emphasize on machine learning techniques to enhance the robustness. Reinforcement learning for behavioral animation [20] and FALCON-X [21], an IVA learning architecture that utilizes self-organizing neural model, are examples of these studies. OML [22] is an agent architecture for virtual environments equipped with neural network-based learning mechanism. In this model, a sensory neuron represents an object, and a motor neuron represents an action. An alternative paradigm for developing IVA is to employ a middleware to integrate existing multi-agent systems such as 2APL [23], GOAL [24], Jadex [25] and Jason [26] with existing game engines. This systematic approach benefits from reusability and rapid prototyping characteristics. As an example, CIGA [27] is a middleware that amalgamates an arbitrary multi-agent system with a game engine by employing domain ontology.

A few IVA architectures, similar to behavioral robotic frameworks, investigate the behavioral organization and action selection. SAIBA [28] is a popular framework that defines a pipeline for abstract behavior generation. It consists of intent planner, behavior planner and behavior realizer. Thalamus [29] framework adds a perceptual loop to SAIBA to let the embodied agent to perform continuous interaction. AATP [30] is a coupled planning and execution architecture for action selection in cognitive IVAs. Neural-dynamic architecture [31] utilizes a dynamic neural field to describe and learn the behavioral state of the system, which in turn, enables the agent to select the appropriate action sequence regarding its environment.

Ultimately, layered architectures (i.e. hybrid models) perform deliberative and reactive operations, simultaneously. COGNITIVA [32] is a reactive-deliberative agent architecture that consists of reactive, deliberative and social layers. In those situations that there is no time for planning, the reactive layer reacts to the situation. Otherwise, the architecture generates goals, plans sequence of actions to reach those goals, schedules the actions, and executes them. Hybrid architectures are widely utilized in space robotic systems as well [5]. RA [6] is a hybrid architecture tested on deep space-1. It is designed to provide reliable autonomy for extended periods. IDEA [33] is a multi-agent architecture that supports distributed autonomy by separating the layers of architecture to independent agents. IDEA has been successfully evaluated on K9 rover. MDS [34] is a hybrid software framework that emphasizes state estimation and control whereas TITAN [34] emphasizes model-based programming. LAAS [35] and Claraty [36] are other examples of hybrid architectures utilized in space missions. The modern space systems including satellite systems (e.g. EO-1 and Techsat-21) and interplanetary missions (e.g. DS-1) exploit hybrid architectures. We adopt RA framework as a reliable architecture to design a generic architecture for autonomous virtual agents that consistently supports deliberative and reactive behaviors.

3 Autonomy for Virtual Agents

RA [6] architecture provides the spacecraft with onboard autonomy and is developed as a hybrid platform with three operational layers including: deliberative planner-scheduler (PS), reactive executive, and mode identification and recovery system (MIR). PS determines the optimal execution sequence of actions in a way that spacecraft can reach its predefined mission goals. Also, it schedules the start time of the actions. Reactive executive receives scheduled actions from PS, and decomposes them to sub-actions understandable by flight software. Flight software is an interface between RA and spacecraft hardware, and consists of collection of software packages such as motor controllers managed by RA. Moreover, executive monitors the execution process to detect the inconsistencies in plans. MIR consists of mode identification (MI) and mode recovery (MR) units. MI transfers the low-level sensor data to high-level perceptions and provides its upper levels with the current system configuration. Ultimately, MR provides the system with error detection and recovery services.

Schematic of our proposed architecture, inspired by RA, is shown in Figure 1. It consists of two layers including cognitive and executive layers. Furthermore, it utilizes a middleware as an interface between abstract agent and its embodied counterpart animated by a game engine. In this architecture, the components are placed in their corresponding layers regarding their operational frequency and abstraction level. The cognitive layer is responsible for providing cognitive functionalities whereas the executive layer is responsible for executing the decisions made by cognitive layer and providing the cognitive layer with high level feedbacks. Cognitive layer functions in low frequency and high level knowledge representation, and plans for long temporal horizons whereas executive layer functions in high frequency and deals with the current situations in a reactive and soft real-time manner.

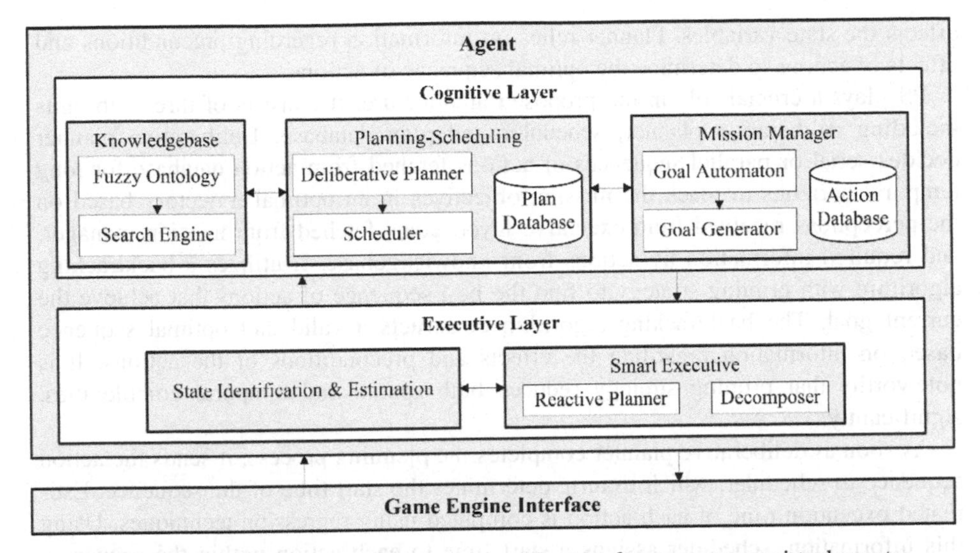

Fig. 1. Schematic of the proposed generic architecture for autonomous virtual agents

3.1 Cognitive Layer

Cognitive layer provides IVA with autonomy, and consists of three components including mission manager (MM), planning-scheduling (PS) and knowledgebase (KB). MM contains the agent's goals and feasible actions. It consists of three sub-units including goal automaton, goal generator, and action database. Goal automaton keeps a network of predefined goals, and is defined as a DFA (deterministic finite automaton) $A=<Q,\Sigma,\sigma,q,F>$ where Q denotes a set of goals, Σ is the evaluation signal indicating whether the current goal has been achieved, σ is the transition function (i.e. $\sigma:Q\times\Sigma\rightarrow Q$) which determines the priority of goals, $q\in Q$ determines the initial goal, and $F\subseteq Q$ is the set of final goals. Structurally, goal automaton is a graph whose nodes present the goals, and edges determine the satisfaction criteria of corresponding goals.

Goal generator functions as the transition function of the goal network. In each time step, it evaluates current goal and received perceptions in order to determine whether the current goal is satisfied. If so, it transforms the goal state to a new goal within the goal automaton, and sends the new goal to PS, so that it can plan new sequence of actions. In case that the goal generator detects the current goal is not satisfied, it keeps the current goal as mission objective.

Regarding the physical constraints of controlled system, there is a limited set of valid actions that agent can execute. These feasible actions are stored in action database. An action is a high-level abstract activity that encapsulates a few low-level sub-actions and consists of a unique identifier, some preconditions and effects, estimated execution duration, set of sub-actions, and their execution timeline. This abstraction scheme dramatically reduces the complexity of planning and scheduling processes. Preconditions determine the constraints on state variables which must be satisfied in order to an action can be executed. Effects determine how the execution of an action affects the state variables. Planner relies on information regarding preconditions and effects of actions to determine the optimal sequence of actions.

PS plays a crucial role in the proposed architecture. It consists of three sub-units including deliberative planner, scheduler and plan database. Deliberative planner decides serial or parallel sequences of actions fetched from action database for long temporal horizons to reach the mission objectives in an optimal trajectory based on the perceptions received from executive layer, goals fetched from mission manager, and required information by actions from knowledgebase. It utilizes a backtracking algorithm with pruning strategy to find the best sequence of actions that achieve the current goal. The backtracking algorithm constructs a valid and optimal sequence based on information regarding the effects and preconditions of the actions. It is noteworthy that pruning strategy reduces both spatial and temporal complexities, significantly.

As soon as deliberative planner completes the planning process, it sends the action sequence to scheduler, which in turn, determines the start time of the sequence. Estimated execution time of each action is computed using regression techniques. Using this information, scheduler assigns a start time to each action within the sequence. Then, the planned and scheduled action sequence is inserted into the plan database. In each time step, this temporal database retrieves actions regarding their start time and sends them to the executive layer, which in turn, executes them.

KB component as a profound memory provides the agent with knowledge acquired from perception sequence. Essentially, a knowledgebase consists of a set of sentences that claim something about the world, an updating mechanism, and a knowledge extraction engine [1]. Our KB consists of two sub-modules: fuzzy ontology and search engine. Fuzzy ontology represents the concepts, objects, features and their relations based on the agent's perceptional history. The ontology can be constructed either in design-time to keep the built-in knowledge, or in run-time to automatically capture the knowledge, or in a hybrid manner. It utilizes a maintainer as an updating mechanism that receives current perceptions from the executive layer and compares them with the knowledge represented in the ontology. Based on this comparison, it may decide to insert new concepts, objects or relations, update them, or even prune the ontology to omit the redundancies or inconsistencies. It is noteworthy that extending the ontology with fuzzy theory enables the agent to model both internal and external uncertainties. We utilize the fuzzy ontology proposed in [37] to design the agent's knowledgebase. Search engine receives queries from PS and searches the ontology by applying iterative first depth search. Then, it returns the resultant knowledge to PS.

3.2 Executive Layer

Executive layer executes the decisions made by cognitive layer, monitors the execution process, and provides the cognitive layer with high-level feedbacks. As illustrated in Figure 1, this layer consists of two main components including state identification and estimation unit, and smart executive. The first component, state identification and estimation unit is responsible for providing the framework with perceptions and estimations. It receives the sensory data from the game engine interface and maps it to the formal knowledge representation used by cognitive and executive layers. In other words, it converts data acquired from IVA's virtual sensors to the perceptions cognoscible by the agent. In order to complete this task, it utilizes Kalman filters for data assimilation and fuzzifiers for data conceptualization. Moreover, it can exploit variety of software libraries to perform specialized data processing activities such as automatic speech recognition, image processing, etc. Therefore, state identification and estimation unit enables the agent to deal with a variety of sensory data acquired from different sensory channels.

Smart executive is responsible for executing sequences of planned actions within plan database, and monitoring the execution process in order to prevent inconsistencies. It consists of two sub-components including decomposer and reactive planner. Decomposer fetches the scheduled action sequences from the plan database, assigns a software thread to each of the retrieved actions, and starts the threads according to the schedule. Using this approach, agent can perform parallel plan execution. As aforementioned, each action is an abstract activity that embodies a set of low-level activities (i.e. sub-actions). In the beginning of execution of an action, it invokes its corresponding sub-actions according to a predefined timeline. This timeline is a built-in knowledge defined by system experts. Execution of each sub-action results in an activity in embodied layer (i.e. agent's avatar). Thus, using this hierarchal scheme, abstract decisions are mapped to physical manipulations and actuations within the virtual environment. Furthermore, decomposer can employ specialized software libraries to provide the actions with required facilities such as text-to-speech engine. Additionally, decomposer monitors the execution to prevent inconsistencies. It compares the current states of the system with the expected states predicted by state estimation, and in case of any irregularities, it halts the inconsistent thread and sends a signal to the reactive planner so that it can take a proper action to eliminate the inconsistency.

Reactive planner is an event-oriented planner that reacts to the unpredicted situations. It exploits a greedy algorithm to choose an action that can handle the unpredicted situation with minimum cost. The cost is computed based on the number of actions required to be performed in order to return the agent's configuration to the planned trajectory. Therefore, in case of unpredicted situation, reactive planner switches from monitoring mode to reacting mode and executes minimal number of activities to solve the inconsistency. It is noteworthy that in those situations that reactive planner executes an unplanned action, it informs PS regarding the divergence in plan sequence, which in turn, investigates whether the current sequence can still reach the current goal. If the goal is not reachable, it repairs the sequence using a backtracking algorithm, and informs the decomposer to fetch the repaired sequence to execute.

Ultimately, game engine interface as a middleware provides an interface between the abstract agent (i.e. mind) and its embodied counterpart (i.e. body). It directly maps the sub-actions to physical activities within the virtual environment. Moreover, it transfers the virtual sensory data from the avatar to its controlling layer. Additionally, it performs a few pre-processing steps such as encoding and decoding data to facilitate the coordination between the abstract and animated layers. It is noteworthy that our proposed architecture is body-independent in a sense that it can adapt to a given body by embedding proper knowledge within its mission manager.

4 Experimental Results

Reliability of our proposed architecture is partially supported by evaluation results of its predecessors operating in inter-planetary missions. However, for further evaluations, we design a discrete event-based 2D world-- the world of circles. In this virtual world, we define three different types of circles regarding their colors (i.e. red, green and blue). These circles may either pop into the world randomly or in a predefined order. The only constraint is that they can only enter the world from a fixed position called the origin. In case of random entrance, a circle with random color enters the world from the origin point. Otherwise, the order of circles is announced before their arrival. There are three non-stationary target zones in the world, each corresponding to a particular color. Also, a few non-stationary obstacles are embedded in the world. The agent's goal in this simulation is to pick up a circle from the origin and move it to one of the target zones regarding the color similarity while avoiding the obstacles (e.g. blue circles to blue target zone). It is noteworthy that because the random circles can pop into the world among the predefined sequence of circles, and both targets and obstacles are non-stationary, the world provides proper characteristics to evaluate both reactive and deliberative behaviors.

We applied our proposed architecture to the world of circles by implementing an agent based on our architecture to move the circles to their target zones while avoiding the obstacles. The agent and environment are implemented in C#.Net programming language. A sample scene of the simulation is shown in Figure 2. The gray rectangles present obstacles whereas blue, purple and green rectangles present the target zones. In this simulation, user can determine the number of circles, their colors, and the randomness level (i.e. randomness of 0.1 means one out of every ten circles pops into the world randomly following a uniform distribution). The agent's KB contains the positions of the origin, obstacles and target zones as variables. Its action database contains the following action:

```
Action identifier: Move (Circle)
Preconditions:
           Status: Idle
          #Circles>0
Effects:
           #Circles--
Estimated execution time: #Steps
Sub-Actions:
           Status: Moving
           Fetch(Circle);
           Loop: FindPath(Circle);
           Move(Circle);
          If(Failure)
             FindTarget(Circle); Update(KB);Goto Loop;
          Else
             Status: Idle
```

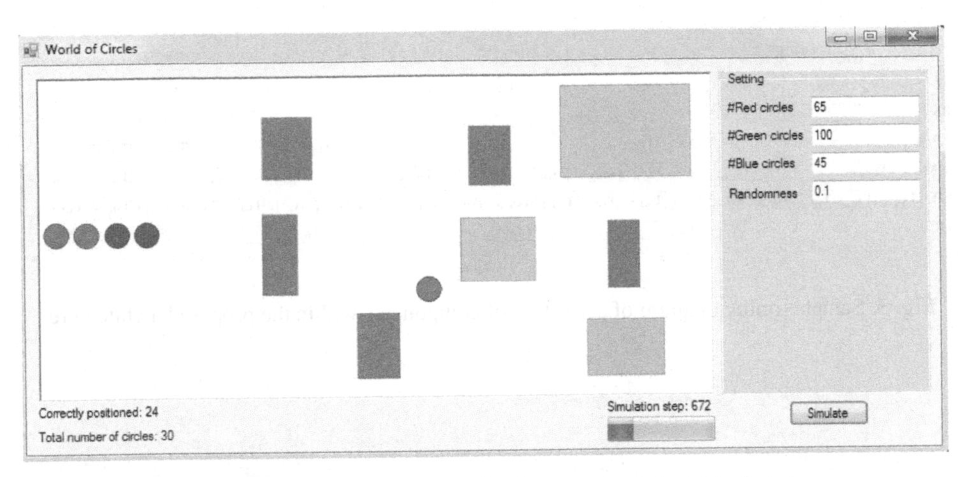

Fig. 2. Snapshot of discrete event implementation of world of circles

Preconditions of this action state that the agent must be idle and there must be new circles in queue. This action first moves the agent to the origin position and then finds an optimal path from origin point to the corresponding target zone using A^* algorithm based on current values of position variables. Then, it moves the front circle in queue to its target zone. If it succeeds, the action is completed. Otherwise, it finds the new position of the target, updates its knowledgebase and repeats the path planning process. As soon as the agent gets information regarding the arrival of new sequence, its deliberative planner plans paths for those circles whose arrival is announced. However, it is possible that a random circle arrives among the predetermined circles. In this case, the reactive planner activates, plans the proper path for the random circle, and then informs the PS. Then, the scheduler reschedules the remaining actions.

In order to facilitate the evaluations, we employ timing diagrams of simulation process to represent the activation sequences of components. A sample timing diagram of the simulations is shown in Figure 3 in which vertical axis indicates the components (i.e. FO: fuzzy ontology, SE: search engine, PS: planner-scheduler, RP: reactive planner, MM: mission manager, SIE: state identification and estimation, DE: decomposer). As shown in Figure 3, components function in different frequencies. As an example, while agent is executing the plans from $T_{PS:1}$, a random circles pops in and reactive planner reacts to the situation in $T_{RP:1}$.

We run the simulations for 1100 times. After each 100 simulations, we increase the randomness by 0.1. Also, we use 200 circles with uniform random colors in each trial. The length of a sequences is selected by a random uniform distribution in range of [5,25]. Simulation results are shown in Figure 4. As illustrated, by increasing the randomness of the simulation, the number of activations of deliberative planner decreases whereas this number increases for reactive planner. Furthermore, even when there is no randomness in the simulation, the reactive planner activates when two consecutive sequences appear with a small delay. In this case, deliberative planner does not have enough time to plan for all the circles, thus agent detects them as unplanned circles, and activates the reactive planner.

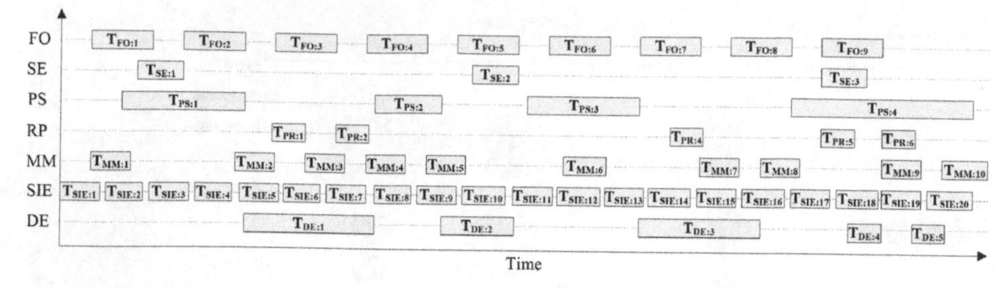

Fig. 3. Sample timing diagram of activation of components within the proposed architecture

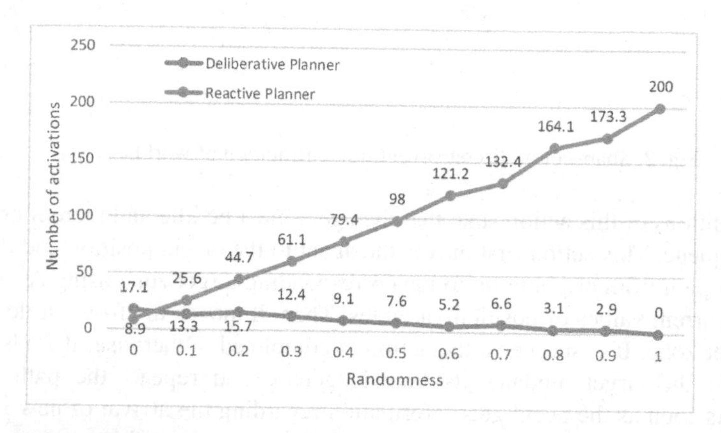

Fig. 4. Average number of reactive and deliberative behaviors in response to changes in randomness of the simulations

5 Conclusion

In this paper, we introduced a generic architecture for autonomous virtual agents inspired by onboard autonomy utilized in inter-planetary space missions. Our proposed architecture provides the agent with concurrent deliberative and reactive behaviors. Furthermore, it equips the agent with necessary components for autonomy such as fuzzy knowledgebase and smart executive. In order to validate our proposed architecture, we implemented a discrete event simulated world. The evaluation results of applying our agent architecture on this world suggest that the architecture is valid and consistent, and is able to handle deliberative and reactive functionalities, simultaneously. Moreover, it can properly support the required parallelism among the processes. As future works, we are planning to apply our agent architecture to a complex game scenario and investigate it is capability in playing the game autonomously. Moreover, we are planning to exploit reinforcement learning so that agent can learn its feasible actions, independently. Ultimately, we are planning to utilize a self-organizing neuro-fuzzy architecture to learn the ontology automatically from perception sequence. Using these two learning schemes renders the need for expert's knowledge obsolete.

References

1. Langley, P., Laird, J.E., Rogers, S.: Cognitive Architectures: Research Issues and Challenges. Cogn. Syst. Res. 10, 141–160 (2009)
2. Ribeiro, T., Vala, M., Paiva, A.: Censys: A Model for Distributed Embodied Cognition. In: Aylett, R., Krenn, B., Pelachaud, C., Shimodaira, H. (eds.) IVA 2013. LNCS, vol. 8108, pp. 58–67. Springer, Heidelberg (2013)
3. Russell, S., Norving, P.: Artificial Intelligence: A Modern Approach. Prentice Hall, New Jersey (2010)
4. Wooldridge, M.: Intelligent Agents: The Key Concepts. In: Marik, V., Stepankova, O., Krautwurmova, H., Luck, M. (eds.) Multi-Agent Systems and Applications II, pp. 3–43. Springer, Heidelberg (2002)
5. Hassani, K., Lee., W.-S.: A Software-in-the-Loop Simulation of an Intelligent Micro-Satellite within a Virtual Environment. In: IEEE International Conference on Computational Intelligence and Virtual Environments for Measurement Systems and Applications, pp. 31–36. IEEE Press (2013)
6. Muscettola, N., Nayak, P., Pell, B., Williams, B.: Remote Agent: To Boldly Go Where no AI System has Gone Before. Artificial Intell. 103, 5–48 (1998)
7. Anderson, J., Bothell, D., Byrne, M., Douglass, S., Lebiere, C., Qin, Y.: An Integrated Theory of the Mind. Psych. Rev. 111, 1036–1060 (2004)
8. Laird, J., Newell, A., Rosenbloom, P.: Soar: An Architecture for General Intelligence. Artificial Intell. 33, 1–64 (1987)
9. Langley, P., Choi, D.: A Unified Cognitive Architecture for Physical Agents. In: 21st National Conference on Artificial Intelligence, pp. 1469–1474. AAAI Press (2006)
10. Weng, J.: Developmental Robotics: Theory and Experiments. Int. J. Humanoid Robot. 1, 199–236 (2004)
11. Carbonell, J., Etzioni, O., Gil, Y., Joseph, R., Knoblock, C., Minton, S., Veloso, M.: PRODIGY: An Integrated Architecture for Planning and Learning. In: Lehn, K. (ed.) Architectures for Intelligence, pp. 51–55. ACM Press (1991)
12. Kokinov, B.: The DUAL Cognitive Architecture: A Hybrid Multi-Agent Approach. In: 11th European Conference of Artificial Intelligence, ECAI, pp. 203–207 (1994)
13. Cassimatis, N., Nicholas, L.: Polyscheme: A Cognitive Architecture for Integrating Multiple Representation and Inference Schemes. MIT Ph.D. Dissertation (2002)
14. Rao, A., Georgeff., M.: BDI-agents: From Theory to Practice. In: 1st International Conference on Multi-agent Systems, ICMAS, pp. 312–319 (1995)
15. Liu, J., Lu., Y.: Agent Architecture Suitable for Simulation of Virtual Human Intelligence. In: 6th World Congress on Intelligent Control and Automation, pp. 2521–2525. IEEE Press (2006)
16. Bratman, M., Israel, D., Pollack, M.: Plans and Resource-Bounded Practical Reasoning. Comput. Intell. 4, 349–355 (1988)
17. Kim, I.-C.: CAA: A Context-Sensitive Agent Architecture for Dynamic Virtual Environments. In: Panayiotopoulos, T., Gratch, J., Aylett, R.S., Ballin, D., Olivier, P., Rist, T. (eds.) IVA 2005. LNCS (LNAI), vol. 3661, pp. 146–151. Springer, Heidelberg (2005)
18. Kriegel, M., Aylett, R., Cuba, P., Vala, M., Paiva, A.: Robots Meet IVAs: A Mind-Body Interface for Migrating Artificial Intelligent Agents. In: Vilhjálmsson, H.H., Kopp, S., Marsella, S., Thórisson, K.R. (eds.) IVA 2011. LNCS, vol. 6895, pp. 282–295. Springer, Heidelberg (2011)
19. Vala, M., Ribeiro, T., Paiva, A.: A model for embodied cognition in autonomous agents. In: Nakano, Y., Neff, M., Paiva, A., Walker, M. (eds.) IVA 2012. LNCS, vol. 7502, pp. 505–507. Springer, Heidelberg (2012)

20. Conde, T., Tambellini, W., Thalmann, D.: Behavioral Animation of Autonomous Virtual Agents Helped by Reinforcement Learning. In: Rist, T., Aylett, R.S., Ballin, D., Rickel, J. (eds.) IVA 2003. LNCS (LNAI), vol. 2792, pp. 175–180. Springer, Heidelberg (2003)
21. Kang, Y., Tan, A.-H.: Self-organizing Cognitive Models for Virtual Agents. In: Aylett, R., Krenn, B., Pelachaud, C., Shimodaira, H. (eds.) IVA 2013. LNCS, vol. 8108, pp. 29–43. Springer, Heidelberg (2013)
22. Starzyk, J.A., Graham, J., Puzio, L.: Simulation of a Motivated Learning Agent. In: Papadopoulos, H., Andreou, A.S., Iliadis, L., Maglogiannis, I. (eds.) AIAI 2013. IFIP AICT, vol. 412, pp. 205–214. Springer, Heidelberg (2013)
23. Dastani, M.: 2APL: A Practical Agent Programming Language. Auton. Agents and Multi-Agent Sys. 16, 214–248 (2008)
24. Hindriks, K.: Programming Rational Agents in GOAL. In: Seghrouchni, A., Dix, J., Dastani, M., Bordini, R. (eds.) Multi-Agent Programming: Languages, Tools and Applications, pp. 119–157. Springer (2009)
25. Pokahr, A., Braubach, L., Lamersdorf, W.: Jadex: A BDI Reasoning Engine. In: Bordini, R., Dastani, M., Dix, J., Seghrouchni, A. (eds.) Multi-Agent Programming: Languages, Platforms and Applications, pp. 149–174. Springer (2005)
26. Bordini, R., Hübner., J., Wooldridge, M.: Programming Multi-Agent Systems in AgentSpeak Using Jason. John Wiley (2007)
27. van Oijen, J., Vanhée, L., Dignum, F.: CIGA: A middleware for intelligent agents in virtual environments. In: Beer, M., Brom, C., Dignum, F., Soo, V.-W. (eds.) AEGS 2011. LNCS, vol. 7471, pp. 22–37. Springer, Heidelberg (2012)
28. Kopp, S., et al.: Towards a common framework for multimodal generation: The behavior markup language. In: Gratch, J., Young, M., Aylett, R.S., Ballin, D., Olivier, P. (eds.) IVA 2006. LNCS (LNAI), vol. 4133, pp. 205–217. Springer, Heidelberg (2006)
29. Ribeiro, T., Vala, M., Paiva, A.: Thalamus: Closing the mind-body loop in interactive embodied characters. In: Nakano, Y., Neff, M., Paiva, A., Walker, M. (eds.) IVA 2012. LNCS, vol. 7502, pp. 189–195. Springer, Heidelberg (2012)
30. Edward, L., Lourdeaux, D., Barthès, J.-P.: An action selection architecture for autonomous virtual agents. In: Nguyen, N.T., Katarzyniak, R.P., Janiak, A. (eds.) New Challenges in Computational Collective Intelligence. SCI, vol. 244, pp. 269–280. Springer, Heidelberg (2009)
31. Sandamirskaya, Y.: Richtert. M., Schoner, G.: A Neural-Dynamic Architecture for Behavioral Organization of an Embodied Agent. In: IEEE International Conference on Development and Learning, pp.1–7. IEEE Press (2011)
32. Spinola, J., Imbert, R.: A Cognitive Social Agent Architecture for Cooperation in Social Simulations. In: Nakano, Y., Neff, M., Paiva, A., Walker, M. (eds.) IVA 2012. LNCS, vol. 7502, pp. 311–318. Springer, Heidelberg (2012)
33. Muscettola, N., Dorais, G., Fry, C., Levinson, R., Plaunt, C.: IDEA: Planning at the Core of Autonomous Reactive Agents. In: 3rd International NASA Workshop on Planning and Scheduling for Space. NASA (2002)
34. Horvath, G., Ingham, M., Chung, S., Martin, O.: Practical Application of Model-Based Programming and State-Based Architecture to Space Missions. In: 2nd IEEE Conference on Space Mission Challenges for Information Technology, pp. 80–88. IEEE Press (2006)
35. Alami, R., Chautila, R., Fleury, S., Ghallab, M., Ingrand, F.: Architecture for Autonomy. Int. J. Robot. Res. 17, 315–337 (1998)
36. Nesnas, I., Simmons, R., Gaines, D., Kunz, C., Calderon, A., Estlin, T., Madison, R., Guineau, J., McHenry, M., Shu, I., Apfelbaum, D.: CLARAty: Challenges and Steps toward Reusable Robotic Software. Int. J. Advance Robot. Sys. 3, 23–30 (2006)
37. Hassani, K., Nahvi, A., Ahmadi, A.: Architectural Design and Implementation of Intelligent Embodied Conversational Agents Using Fuzzy Knowledgebase. J. Intell. Fuzzy Sys. 25, 811–823 (2013)

A Decentralized Multi-agent Approach to Job Scheduling in Cloud Environment

Jakub Gąsior[1] and Franciszek Seredyński[2]

[1] Systems Research Institute, Polish Academy of Sciences, Warsaw, Poland
[2] Cardinal Stefan Wyszyński University, Warsaw, Poland

Abstract. Paper proposes a novel solution to a job scheduling problem in the Cloud Computing systems. The goal of this scheme is allocating a limited quantity of resources to a specific number of jobs minimizing their execution failure probability and completion time. It employs the Pareto dominance concept implemented at the client level. To select the best scheduling strategies from the Pareto frontier and construct a global scheduling solution we develop decision-making mechanisms based on the game-theoretic model of Spatial Prisoner's Dilemma and realized by selfish agents operating in the two-dimensional Cellular Automata space. Their behavior is conditioned by objectives of the various entities involved in the scheduling process and driven towards a Nash equilibrium solution by the employed social welfare criteria. The related results show the effectiveness and scalability of this scheme in the presence of a large number of jobs and resources involved in the scheduling process.

Keywords: Multiobjective optimization, Genetic algorithm, Risk resilience.

1 Introduction

The increasing demand for computational resources has led to an introduction of a new type of heterogeneous, distributed computing platform, where customers do not own any part of the infrastructure. In Cloud Computing (CC) systems resources such as computational nodes and storage space are provided in the form of virtual machines (VM) to the clients. A scheduler has to decide how to allocate these virtual machines in order to guarantee a reasonable level of security-aware computation while maintaining efficient resource management of the CC environment.

Obviously, the conflict between achieving good performance, in terms of the job completion time, and providing high security-assurance introduces new challenges in CC scheduling. The aim of our study is to propose an efficient algorithm which effectively handles such a multi-criteria parallel job scheduling problem, taking into account not only the job completion time but also the security constraints inherent in the CC system.

We propose a solution to the issue of reliable and secure computation in CC, where independent brokering agents take scheduling decisions based only on

P. Angelov et al. (eds.), *Intelligent Systems'2014*,
Advances in Intelligent Systems and Computing 322, DOI: 10.1007/978-3-319-11313-5_36

locally available information. Each agent tries to allocate a batch of jobs from the associated user in a pure selfish way, without considering actions of the other brokers and assuming total availability of system resources. A method employed for the scheduling purposes is the Multiobjective Genetic Algorithm (MOGA) optimization resulting in a Pareto-based evaluation, necessary to find the best scheduling strategy accommodating often conflicting optimization objectives.

In order to make a dynamic choice among the Pareto set of the proposed solutions we develop a distributed scheduling scheme employing a non-cooperative game-theoretic approach, in particular a spatially generalized Prisoner's Dilemma (SPD) model [9]. The whole process is realized in the two-dimensional Cellular Automata (CA) space where individual agents are mapped onto a regular square lattice. Main issues that are addressed here are: a) incorporating the global goal of the system into the local interests of all agents participating in the game, and b) such a formulation of the local interaction rules that will allow to achieve those interests.

The remainder of this paper is organized as follows. In Section 2, we present the works related to the security-aware scheduling in Grid and Cloud computing systems. In Section 3, we describe the Cloud system model. Section 4 briefs the proposed agent-based game-theoretic scheduling scheme and its application. Section 5 demonstrates the performance metrics, the input parameters and experimental results. Finally, Section 6 concludes the paper.

2 State of the Art

The integration of the security mechanisms with the scheduling algorithms can be perceived as one of the most important issues in Cloud scheduling. Due to the NP-hardness [5] of the job scheduling problem, finding the exact solutions to solve the large-scale task scheduling problems in the dynamic environments is not always feasible. Therefore the approximation methods providing a near optimal solution are some of the most promising approaches.

In [3] authors proposed a solution capable of meeting diverse security requirements of multiple users simultaneously, while minimizing the number of jobs aborted due to resource limitations. Insecure conditions in on-line job scheduling in Grids caused by the software vulnerabilities were also analyzed in [11, 13] by considering the heterogeneity of the fault-tolerant mechanisms in the security-assured job scheduling.

These works proved that (meta)heuristics can be useful in the design of effective schedulers mapping a large number of jobs to the available resources. Several works proposed game theoretic models to solve job scheduling problems using the concept of Nash equilibrium [1]. The convergence time to such equilibria for several selfish scheduling problem variants was considered in [4], while in [6] four genetic-based hybrid meta-heuristics have been proposed and evaluated as a non-cooperative, non-zero sum game of the Grid users in order to address the security requirements. Finally, the concept of collocation games have been applied to Cloud computing in [8], while an utility-driven solution for optimal resource allocation was developed in [7]. In particular, a selfish scheduling model was analyzed in [10], proposing allocation scheme for federated Cloud organizations based on independent and competing agents.

3 Cloud Model

Our model is based on the architecture introduced in [12]. A system consists of a set of geographically distributed Cloud nodes $M_1, M_2, ..., M_m$, which are connected to each other via a wide area network. Each node M_i is described by a parameter m_i, which denotes the number of identical processors P_i and its computational power s_i. Figure 1(a) depicts an exemplary set of parallel machines in the CC system.

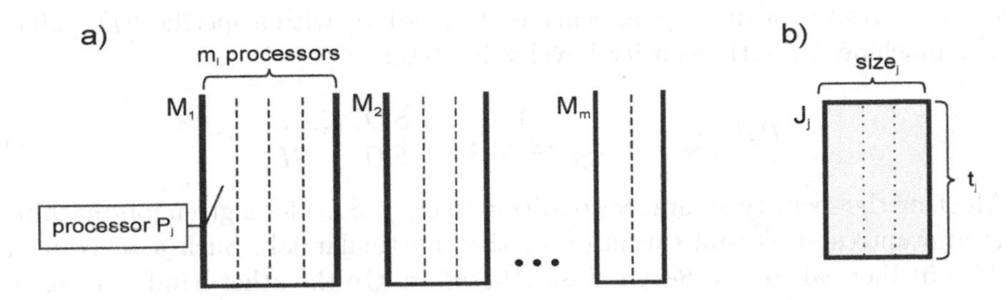

Fig. 1. Example of the Cloud Computing system. A set of parallel machines (a) and the multi-threaded job model (b).

Users $(U_1, U_2, ..., U_n)$ submit their jobs to the system, expecting their completion before a required deadline. Job (denoted as J_k^j) is jth job produced by user U_k. J_k stands for the set of all jobs produced by user U_k, while $n_k = |J_k|$ is the number of such jobs. Each job has varied parameters defined as a tuple $< r_k^j, size_k^j, t_k^j, d_k^j >$, specifying its release date $r_k^j = 0$; processor requirements $size_k^j$; execution requirements t_k^j defined by a number of operations and a deadline d_k^j.

We assume that job J_k^j can only run on machine M_i if $size_k^j \leq m_i$ holds, that is, we do not allow multi-site execution and co-allocation of processors from different machines. Let $p_k^{i,j} = \frac{t_k^j}{s_i}$ define job's J_k^j execution time on machine M_i. Further, $W_k^{i,j} = p_k^{i,j} * size_k^j$ denotes the work of job J_k^j, also called its *area* or *resource consumption*. Figure 1(b) shows an example of the multi-threaded job model. We assume that the scheduler is clairvoyant and working in off-line mode. Jobs are scheduled independently, and there is no communication between them. All jobs and resources are available from time zero.

3.1 Security Model

We consider a security-driven scheduling scheme, to address the reliability issues in a computational Cloud environment and apply a modified version of the approach presented in [11] to match job's security requirements submitted by the user with security index defined for each Cloud site. During a job submission, users define a *Security Demand* (SD) for each job dependent on data sensitivity, access control or required level of authentication. On the other hand, the defense capability of a resource can be attributed to the available intrusion detection mechanisms or its response capacity. This capability is modeled by a *Security Level* (SL) factor, evaluating the risk existing in the allocation of a submitted job to a specific machine.

The SD is a real fraction in the range [0,1] with 0 representing the lowest and 1 the highest security requirement. The SL is in the same range with 0 for the most risky resource site and 1 for a risk-free or fully trusted site. We define a job execution model as a function of the difference between the job security demand and site trust. The Equation 1 expresses the probability of a *Successful Job Completion* regarding the allocation of the job J_j with a specific SD_j value, to the machine M_i with Security Level value SL_i:

$$P^{i,j}_{Success} = \begin{cases} 1, & SD_j \leq SL_i, \\ exp^{-(SD_j - SL_i)}, & SD_j > SL_i. \end{cases} \tag{1}$$

Meeting the security assurance condition ($SD_j \leq SL_i$) for a given job-machine pair guarantees successful execution of that particular job. Such a scheduling will be further called as a *Secure Task Allocation*. On the other hand, successful execution of the job assigned to machine without meeting this condition ($SD_j > SL_i$), will be dependent on the calculated probability and further referred to as a *Risky Task Allocation*.

3.2 Problem Formulation

A typical way of assessing system's performance is measuring the completion time of submitted jobs. There exist various levels of aggregation, on which the performance can be measured, e.g., the level of individual jobs or the level of the complete system. In our work we evaluate the performance of individual users submitting their jobs to the Cloud. Let us denote S_k as a user's U_k schedule. The completion time of jobs on machine M_i in the schedule S_k is denoted by $C_i(S_k)$. Two different objectives are considered in this work:

- The minimization of *Makespan*, $E(C^k_{max})$, which means the estimated duration of all the user's U_k jobs. We consider minimization of the time $C_i(S_k)$ on each machine M_i over the system in such a way that the *Makespan* is defined as: $E(C^k_{max}) = max_i\{C_i(S_k)\}$;
- The maximization of the *Security Assurance Level*, $\overline{P^k_{Success}}$, defined as an average successful job completion probability of each user's U_k job allocation in the schedule S_k.

Therefore, the Multiobjective Optimization Problem (MOP) can be formulated as follows:

$$Min(E(C^k_{max}), 1 - \overline{P^k_{Success}}). \tag{2}$$

We assume that there is no centralized control and the assignment of the resources available within the Cloud is governed exclusively by the brokers assigned to individual Cloud users submitting their jobs to the Cloud. To develop a truly distributed multiobjective scheduling algorithm we propose a three-stage procedure. At the first stage, a Pareto front is calculated for each user U_k submitting its batch of jobs J_k to the Cloud. The Pareto front is calculated under an assumption that all cloud resources belong exclusively to the user U_k using the NSGA-II algorithm [2].

At the second stage some solutions (scheduling strategies) from the Pareto front are selected according to the Pareto *Selection Policies* characterizing user's preferences, and these solutions will be subsequently used by a broker B_k representing user's U_k interests. In the third stage all brokers employ previously selected scheduling strategies in an iterated, non-cooperative Spatial Prisoner's Dilemma game [9] realized in a two-dimensional cellular automata (CA) space, where they try to find scheduling strategies providing compromise global scheduling solution (Nash equilibrium point) in their selfish attempts to obtain Cloud resources. These three stages will be further explained in the subsequent sections.

4 Multiobjective Scheduling Framework

Genetic Algorithms (GAs) are meta-heuristics mimicking the process of natural selection by applying a set of genetic operators on the population of candidate solutions. In order to formulate our problem without overriding previously defined constraints, we propose the following encoding scheme for the MOGA individuals. In the case of job scheduling, each chromosome represents a schedule allocating a batch of user's U_k jobs on a group of machines.

We use the structure in which each gene consists of a pair of values (J_k^j, M_m), indicating that job J_k^j is assigned to machine M_m, where j is the index of a job in user's U_k batch of jobs J_k and m is the index of the machine. The execution order of jobs allocated to the same machine is given by the positions of the corresponding genes in the chromosome on First-Come-First-Serve basis.

As was stated before, we apply in this work the Pareto optimality approach to solve the Multiobjective Optimization Problem (Equation 2). In particular, we use the second version of the Nondominated Sorting Genetic Algorithm (NSGA-II) resulting in the Pareto frontier of nondominated scheduling solutions as depicted in Figure 2. The overall structure of the NSGA-II can be found in [2].

4.1 Selection of the Pareto-efficient Strategy

We wish to further restrict the solution search space to the Pareto-efficient strategies, therefore it is necessary to implement additional mechanisms limiting a set of possible scheduling strategies from the Pareto frontier. We developed multiple *Selection Polices* aiming to achieve this goal by specifying the basic requirements of the Cloud user:

- **Maximum Reliability policy** selecting a strategy from Pareto frontier yielding the maximum Security Assurance Level;
- **Minimum Cost policy** selecting a strategy from Pareto frontier yielding the minimum fees for the Cloud provider;
- **Minimum Cost with Deadline policy** selecting a strategy from Pareto frontier yielding the minimum fees for the Cloud provider while meeting the deadline required by the user;
- **Optimum policy** selecting a strategy from Pareto frontier optimizing all above criteria.

We depict in Figure 2 an exemplary Pareto frontier with highlighted strategies corresponding to the aforementioned *Selection Polices*. Each agents selects its scheduling strategy S_k in a pure selfish way according to one of the *Selection Polices*. Agents aim at designing the best schedule minimizing their own *Utility*, that is one using the most powerful and reliable machines and minimizing the associated costs. Because agents are not isolated, their actions influence and are influenced by the actions of other brokers. Accordingly, they must be forced to interact in order to generate a global schedule in which all the users have their jobs processed on the Cloud. This can be achieved by incorporating the global goal of the system into the local interests of all agents and such a formulation of the local interaction rules that will allow to achieve those interests.

With this goal in mind, we formulate the *Utility* function $\Xi_k(S_k)$ of an agent B_k selecting a scheduling strategy S_k as follows:

$$\Xi_k(S_k) = \sum_{(i,j)\in S_k} [\frac{p_k^{i,j} * Cost_k^{i,j}}{P_{Success}^{i,j}}], \tag{3}$$

where:

- *Scheduling Strategy* (S_k) defines a complete set of associations among the jobs belonging to a specific user U_k and the available machines. Each job J_k^j to be completed on a machine M_i requires a processing time defined as $p_k^{i,j} = \frac{t_k^j}{s_i}$;
- *Scheduling Cost* ($Cost_k^{i,j}$) denotes the fee for a Cloud provider, dependent on the type of allocation defined previously in Section 3.1. For simplicity's sake, we assume that *Risky Task Allocation* costs $1\frac{Unit(s)}{Second}$, while *Secure Task Allocation* costs $5\frac{Unit(s)}{Second}$ in Cloud renting fees. *Units* represent stipulated currency used by the agents to make conscious financial decisions regarding their scheduling choice;
- *Job Completion Probability* ($P_{Success}^{i,j}$) associated with each job allocation in a schedule S_k denotes the confidence level that the job will be completed without interruptions and is a direct result of the *Security Demand* and *Security Level* factors describing job and machine, respectively.

By gathering information about the component strategies proposed by other participating agents, each broker B_k is able to construct a tentative global

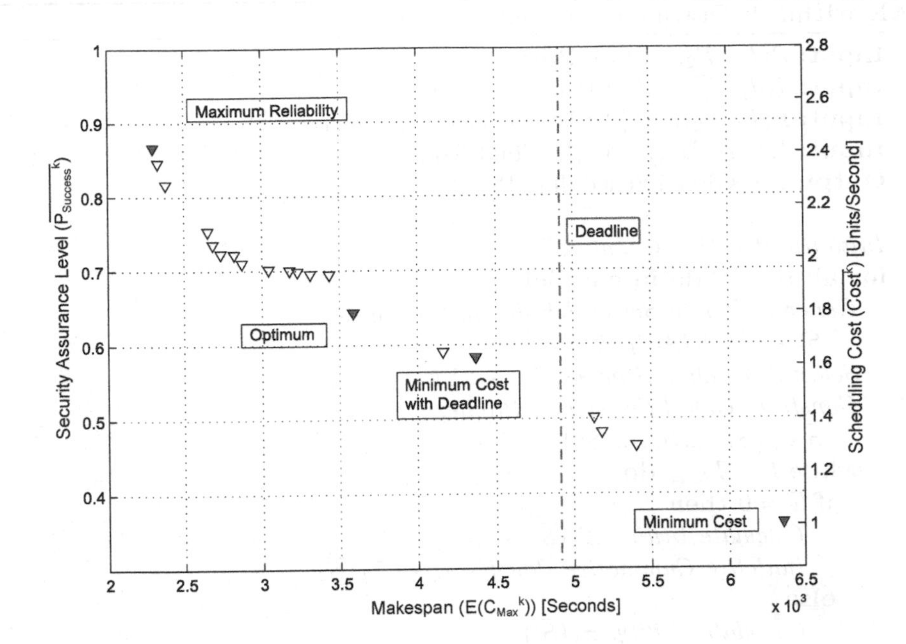

Fig. 2. Pareto Frontier generated by the Agent B_k assigned to User U_k consisting of the viable, non-dominated scheduling solutions ($S_k \in S_k^*$). Exemplary strategies selected by the static Pareto *Selection Polices* are marked as red triangles.

scheduling profile $\overline{S} = (S_1, ..., S_k, ..., S_n)$ containing all the solutions proposed by agents operating in the Cloud. Such a global schedule $\overline{S}$, referred to as a *Global Scheduling Profile*, represents a combination of allocation strategies for all the agents and hence defines a specific state of the scheduling game.

Due to lack of any central coordination mechanism, such a global schedule resulting from the above interactions may not efficiently share the available Cloud resources. Thus, the goal of the scheduler is transforming this global scheduling solution $\overline{S}$ into a pure Nash equilibrium [9], through a sequence of steps, where each agent, starting from the initial allocation strategy, proposes a solution that minimizes his local *Utility* function. Such an equilibrium defines a fundamental point of stability within the system, because no broker can unilaterally perform any action to improve its social welfare.

4.2 The Game-Theoretic Scheduling Scheme

The whole distributed scheduling process is realized using a modified version of the Prisoner's Dilemma (PD) model [9] and we present a simplified pseudo-code of the proposed scheduling scheme in Algorithm 1. We employ a variation of Prisoner's Dilemma game working in the two-dimensional Cellular Automata space with the *von Neumann* neighborhood ($z = 4$ neighbor cells: the cell above

Algorithm 1. Distributed Scheduling Algorithm

Input: $B(B_1, B_2, ..., B_n)$: Brokers
Input: $J(J_1, J_2, ..., J_n)$: Batches of Jobs
Input: $P(P_1, P_2, ..., P_n)$: Pareto Selection Policies
Input: $M(M_1, M_2, ..., M_m)$: Cloud Nodes
Output: $\overline{S}$: Global Scheduling Profile

Initialize Iteration Counter, $T \leftarrow 0$
for all $B_k \in B$ **do in parallel**
  Generate Pareto Set of Scheduling Strategies,
    $S_k^* \leftarrow MOGA(J_k, M)$
  Select Scheduling Strategy, $S_k \leftarrow P_k(S_k^*)$
  Construct Global Scheduling Profile,
    $\overline{S} \leftarrow \{S_1, ..., S_k, ..., S_n\}$
  while $T < T_{Max}$ **do**
    if $T = 0$ **then**
      Calculate Utility, $\Xi_k(\overline{S})$
      Initialize Cumulative Payoff, $G_k \leftarrow \Xi_k(\overline{S})$
    else
      Calculate Utility, $\Xi_k(\overline{S'})$
      Calculate Utility Gain, $\Gamma_k(\overline{S}, \overline{S'}) \leftarrow \Xi_k(\overline{S}) - \Xi_k(\overline{S'})$
      Calculate Cumulative Payoff, $G_k \leftarrow \sum_{i=1}^{z} G_k^i$
    end if
    Update Pareto Selection Policy, $P'_k \leftarrow P_z$
    Update Scheduling Strategy, $S'_k \leftarrow P'_k(S_k^*)$
    Reconstruct Global Scheduling Profile, $\overline{S'} \leftarrow \{S'_1, ..., S'_k, ..., S'_n\}$
    Update Iteration Counter, $T \leftarrow T + 1$
  end while
end for
return *Global Scheduling Profile,* $\overline{S} \leftarrow \overline{S'}$

and below, right and left from a given cell), where agents are mapped onto a regular square lattice.

We assign randomly to each agent B_k, located at the cell, a behavioral variable P_k specifying one of the available *Selection Policies.* Agents can cope with potential allocation conflicts and unsatisfactory *Utility* scores by choosing new *Selection Policy* P'_k and selecting a corresponding scheduling strategy S'_k from the Pareto Frontier. Clearly, each change in the job allocation for an individual agent in our competitive resource allocation game may create new conflicts and influence the *Utilities* of several other agents, thus implying reconstruction of the *Global Scheduling Profile* $\overline{S'}$ and recalculation of the *Utility* function $\Xi_k(\overline{S'})$. Potential gain in the agent's B_k utility due to modification of the scheduling strategy is calculated as follows:

$$\Gamma_k(\overline{S}, \overline{S'}) = \Xi_k(\overline{S}) - \Xi_k(\overline{S'}). \quad (4)$$

The metric Γ_k is further used to determine the agent's payoff in the games against each of its neighbors. Every agent plays with z agents belonging to his local neighborhood getting a payoff calculated according to the formula presented in Table 1. The cumulative payoff, G_k, of each individual is determined by summing the acquired payoffs. In the next round, all individuals update their *Selection Polices* by ranking agents from a local neighborhood in an ascending order of their cumulative payoffs $G_1 \leq G_2 \leq ... \leq G_z$ and adopting the Pareto *Selection Policy* of the top scoring player B_z.

Table 1. The Prisoner's Dilemma Payoff Matrix

	Change strategy	Keep strategy
Change strategy	$\frac{\Gamma}{2}, \frac{\Gamma}{2}$	Γ, 0
Keep strategy	0, Γ	0, 0

It is clear from Equation 4 that the only way for an agent to improve his social welfare is to modify the Pareto *Selection Policy* in such a way that a new value of the *Utility* function will be lower than the last one. This encourages agents to experiment with other scheduling strategies and reduces the chances of stagnation, where agents are content with the first scheduling choice they made and do not try to further improve their social welfare.

An equilibrium is reached when none of the participating brokers is interested in changing (otherwise its *Utility* will increase) its own job allocation strategy. This is experienced when no further improvement can achieved in the acquired *Utility* values and a valid global problem solution $\overline{S}$ is obtained.

5 Experimental Analysis and Performance Evaluation

Our experiments employed 16 independent users competing for a limited number of Cloud resources. Each user submitted to the system a randomly generated *Bag of Tasks* containing $n_k = 1000$ job instances. Jobs were then scheduled within $m = 8, 16, 32, 64$ nodes by independent brokering agents assigned to individual users. Game was conducted on 4 x 4 size square lattice. Initial Pareto *Selection Polices* were equally and randomly distributed between the participating players. The maximum number of game iterations has been fixed at a value of $T_{Max} = 200$ steps. Table 2 summarizes key simulation parameters used in the experiment. To evaluate the scheduling performance, we have used the following metrics:

– ***Total Makespan:*** the total running time of all jobs submitted to the Cloud for the *Global Scheduling Profile*, $\overline{S}$, defined as $max\{E(C^k_{max}), k = 1, 2, ..., n\}$;
– ***Scheduling Success Rate:*** the percentage of jobs successfully completed in the system for the *Global Scheduling Profile*, $\overline{S}$.

5.1 Results and Discussion

The experiment was conducted in order to compare the results obtained by the proposed solution with several static Pareto *Selection Polices* defined in Section 4.1. We conducted five different experiments: four static experiments where each agent was employing the same *Selection Policy* from the available set (Maximum Reliability, Minimum Cost, Minimum Cost with Deadline and Optimum) throughout the whole scheduling cycle. No changes in the scheduling strategy were allowed. In the last experiment, our proposed solution (denoted further as *SPD-NSGA-II*) employing dynamic strategy modification in the course of the Spatial Prisoner's Dilemma game was evaluated. The simulation results are given in Figure 3 for each proposed performance metric.

Table 2. Simulation Parameter Settings

System Parameters	**Value setting**
Average number of cores ($\overline{m_i}$)	6
Average processor speed ($\overline{s_i}$)	2
Node Security Level (SL)	0.3 - 1.0
Job Parameters	**Value setting**
Average number of threads ($\overline{size_j^k}$)	4
Average execution time of a job ($\overline{t_j^k}$)	5
Job Security Demand (SD)	0.6 - 0.9

Let us start with a discussion of the results achieved by the static Pareto *Selection Polices.* Not surprisingly, *Maximum Reliability* policy achieves the best *Makespan* performance of all compared static *Polices*, due to allocation of jobs to the most reliable resources, regardless of the scheduling cost to the user. As can be seen in Figure 3(b) it also results in one of the highest *Scheduling Success Rates.* On the other hand, job allocations selected by the *Minimum Cost* and *Minimum Cost with Deadline* policies result in a rather poor performance. The obvious reason is the assignment of jobs to resources offering the lowest *Cost*, regardless of their overall reliability which results in higher probability of failures and more frequent rescheduling events.

It is clear, however, that the proposed scheduling scheme (*SPD-NSGA-II*) clearly outperforms the static Pareto *Selection Polices. Polices* focusing on similar objectives are simply not capable of achieving compromise solution with competing Cloud's users. Their similar goals and optimization criteria result in allocating tasks to the same pool of machines which leads to the overall congestion and load imbalance, and in the effect, inferior scheduling performance.

Achieved results validate the feasibility of the proposed competitive approach applied in the distributed scheduling scheme. It can lead to a satisfactory solutions both in terms of quality and scalability and provide a substantial performance gain in terms of job completion time, security assurance and scheduling costs. When the job scheduling problem scales towards large Cloud environments with a lot of computational nodes and users to be served, a fully distributed scheduling scheme exhibits an indispensable benefit of breaking down a single scheduling problem into smaller, more manageable pieces.

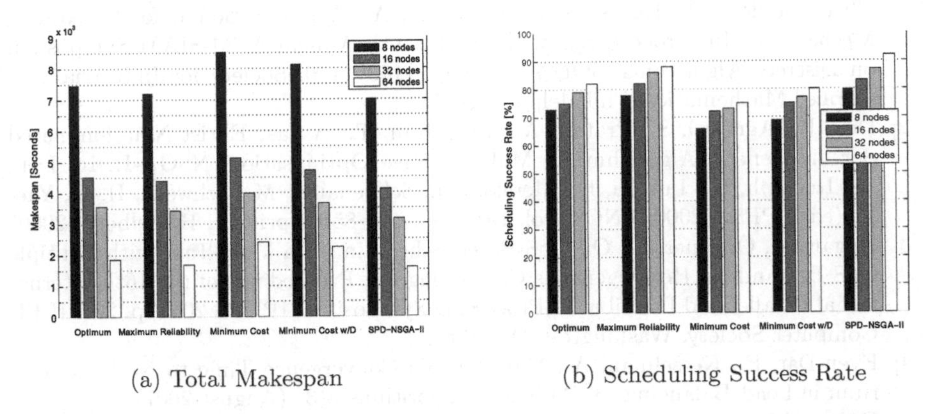

(a) Total Makespan (b) Scheduling Success Rate

Fig. 3. Performance results of conducted experiments for a total of 16000 jobs scheduled within 8, 16, 32 and 64 computational nodes by 16 independent agents. (a) Total Makespan. (b) Scheduling Success Rate.

6 Conclusions and Future Work

Security-driven job scheduling is crucial to achieving high performance in the Cloud computing environment. However, existing scheduling algorithms largely ignore the security induced risks involved in dispatching jobs to untrustworthy resources. The paper proposes a new agent-based game-theoretic scheme for scheduling jobs within Cloud infrastructure. It combines the paradigm of MOGA-based scheduling with a game-theoretic model of the Spatial Prisoner's Dilemma game.

It incorporates security-awareness into scheduling process and aims to minimize both job completion time and possible security risks. Due to its very nature, it is capable of exploring and exploiting the whole range of solution search space. By employing non-cooperative agents, we are able to use the competition among the entities involved to converge towards Nash equilibrium. It allows to account for often contradicting interests of the clients within the Cloud, without the need of any centralized control.

Results of conducted simulations proved the potential of the presented approach. Our future research will be oriented on implementing dynamic load balancing and rescheduling techniques allowing to react proactively to any changes that can occur in the on-line Cloud models.

Acknowledgment. The first author was supported by the Foundation for Polish Science under International PhD Projects in Intelligent Computing. Project financed from The European Union within the Innovative Economy Operational Programme 2007-2013 and European Regional Development Fund (ERDF).

References

[1] Christodoulou, G., Koutsoupias, E., Vidali, A.: A Lower Bound for Scheduling Mechanisms. In: Proceedings of the Eighteenth Annual ACM-SIAM Symposium on Discrete Algorithms, SODA 2007, pp. 1163–1170. Society for Industrial and Applied Mathematics, Philadelphia (2007)

[2] Deb, K., Agrawal, S., Pratap, A., Meyarivan, T.: A Fast Elitist Non-dominated Sorting Genetic Algorithm for Multiobjective Optimisation: NSGA-II. In: Deb, K., Rudolph, G., Lutton, E., Merelo, J.J., Schoenauer, M., Schwefel, H.-P., Yao, X. (eds.) PPSN 2000. LNCS, vol. 1917, pp. 849–858. Springer, Heidelberg (2000)

[3] Dogan, A., Özgüner, F.: On QoS-Based Scheduling of a Meta-Task with Multiple QoS Demands in Heterogeneous Computing. In: Proceedings of the 16th International Parallel and Distributed Processing Symposium, IPDPS 2002, p. 227. IEEE Computer Society, Washington, DC (2002)

[4] Even-Dar, E., Kesselman, A., Mansour, Y.: Convergence Time to Nash Equilibrium in Load Balancing. ACM Trans. Algorithms 3(3) (August 2007)

[5] Garey, M.R., Johnson, D.S.: Computers and Intractability: A Guide to the Theory of NP-Completeness. W. H. Freeman & Co., New York (1990)

[6] Kolodziej, J., Xhafa, F.: Meeting Security and User Behavior Requirements in Grid Scheduling. Simulation Modelling Practice and Theory 19(1), 213–226 (2011)

[7] Li, Z.-J., Cheng, C.-T., Huang, F.-X.: Utility-Driven Solution for Optimal Resource Allocation in Computational Grid. Comput. Lang. Syst. Struct. 35(4), 406–421 (2009)

[8] Londoño, J., Bestavros, A., Teng, S.-H.: Colocation Games And Their Application to Distributed Resource Management. In: Proceedings of the 2009 Conference on Hot Topics in Cloud Computing, HotCloud 2009. USENIX Association, Berkeley (2009)

[9] Nowak, M.A., May, R.M.: Evolutionary Games and Spatial Chaos. Nature 359, 826 (1992)

[10] Palmieri, F., Buonanno, L., Venticinque, S., Aversa, R., Di Martino, B.: A Distributed Scheduling Framework Based on Selfish Autonomous Agents for Federated Cloud Environments. Future Gener. Comput. Syst. 29(6), 1461–1472 (2013)

[11] Song, S., Hwang, K., Kwok, Y.-K.: Risk-Resilient Heuristics and Genetic Algorithms for Security-Assured Grid Job Scheduling. IEEE Trans. Comput. 55(6), 703–719 (2006)

[12] Tchernykh, A., Schwiegelshohn, U., Yahyapour, R., Kuzjurin, N.: Online Hierarchical Job Scheduling on Grids with Admissible Allocation. J. of Scheduling 13(5), 545–552 (2010)

[13] Wu, C.-C., Sun, R.-Y.: An Integrated Security-Aware Job Scheduling Strategy For Large-Scale Computational Grids. Future Gener. Comput. Syst. 26(2), 198–206 (2010)

Model Checking Properties of Multi-agent Systems with Imperfect Information and Imperfect Recall

Jerzy Pilecki[1], Marek A. Bednarczyk[2,3], and Wojciech Jamroga[4,5]

[1] Systems Research Institute, Polish Academy of Sciences, Warsaw, Poland
[2] Polish-Japanese Institute of Information Technology, Warsaw, Poland
[3] Institute of Computer Science, Polish Academy of Sciences, Warsaw, Poland
[4] Computer Science and Communications Research Unit & Interdisciplinary Centre for Security, Reliability and Trust, University of Luxembourg, Luxembourg
[5] Department of Informatics, Clausthal University of Technology, Germany

Abstract. The problem of practical model checking Alternating-time Temporal Logic (ATL) formulae under imperfect information and imperfect recall is considered. This is done by synthesis and subsequent verification of *strategies*, until a good one is found. To reduce the complexity of the problem we define an equivalence relation on strategies. Then an algorithm for model checking a class of modal properties with a single coalitional modality is presented, which utilises the observation that there is no need to verify more than one strategy from an equivalence class. The experimental results of the approach are also discussed.

Keywords: ATL, model checking, imperfect information and imperfect recall.

1 Introduction

A large number of temporal logics for reasoning about properties of multi-agent systems has been proposed in recent years. These contributions include theoretical results concerning the complexity of model checking. However, there is relatively little research on practical model checking algorithms.

This article deals with model checking for a subset of formulae of alternating-time temporal logic with imperfect information and imperfect recall (ATL_{ir}). The problem is known to be hard, both computationally and conceptually. We present some observations that lead to a reduction of the search space for certain instances of the problem. In consequence, we hope that a reduction by a significant factor is possible for many practical instances.

The main contribution of this paper is an algorithm that utilizes these observations. We provide some experimental results obtained from a software implementation of this algorithm. The results we have obtained indicate that there exists the possibility of greatly decreasing the search space of the ATL_{ir} model checking problem. This lead to a noteworthy reduction in computation time for the problems we have tested.

P. Angelov et al. (eds.), *Intelligent Systems'2014*,
Advances in Intelligent Systems and Computing 322, DOI: 10.1007/978-3-319-11313-5_37

2 Preliminaries

We begin by presenting the syntax and semantics of alternating-time temporal logic, as well as defining the model checking problem formally.

2.1 ATL: Syntax and Semantics

In [2,3] a logic was proposed for reasoning about abilities of agents in multi-agent systems. Alternating-time temporal logic (ATL) is interpreted over concurrent game structures.

Concurrent Game Structures. Quoting from [2,3], a concurrent game structure is a tuple $S = \langle \Sigma, Q, \Pi, \pi, d, \delta \rangle$, where:

- Σ is a finite set of players (also called agents), $\Sigma = \{1, \ldots, k\}$
- Q is a finite set of states
- Π is a finite set of propositions
- for each state $q \in Q$ the set $\pi(q) \subseteq \Pi$ denotes propositions true at q (function π is called the labeling function)
- for each player $a \in \{1, \ldots, k\}$ and each state $q \in Q$, we have the natural number $d_a(q) \geq 1$ of moves available at state q to player a. We identify the moves of player a at state q with the numbers $1, \ldots, d_a(q)$. For each state $q \in Q$, a *move vector* at q is a tuple $\langle j_1, \ldots, j_k \rangle$ such that $1 \leq j_a \leq d_a(q)$ for each player a. Given a state $q \in Q$, we write $D(q)$ for the set $\{1, \ldots, d_1(q)\} \times \ldots \times \{1, \ldots, d_k(q)\}$ of move vectors. The function D is called *move function* or *protocol*.
- For each state $q \in Q$ and each move vector $\langle j_1, \ldots, j_k \rangle \in D(q)$, a state $\delta(q, j_1, \ldots, j_k) \in Q$ that results from state q if every player $a \in \{1, \ldots, k\}$ chooses move j_a. The function δ is called *transition function*.

ATL Syntax. An ATL formula is one of the following:

- p, for propositions $p \in \Pi$
- $\neg\varphi$ or $\varphi_1 \vee \varphi_2$, where $\varphi, \varphi_1, \varphi_2$ are ATL formulas
- $\langle\langle A \rangle\rangle X \varphi$, $\langle\langle A \rangle\rangle G \varphi$, $\langle\langle A \rangle\rangle \varphi_1 U \varphi_2$

Above $A \subseteq \{1, \ldots, k\}$ ranges over *coalitions*, i.e., subsets of players. Additional Boolean connectives can be defined from $\neg, \vee$ in the usual way. The future temporal operator $\langle\langle A \rangle\rangle F \varphi$ can be defined as $\langle\langle A \rangle\rangle true U \varphi$.

ATL Semantics. The meaning of ATL formulae of type $\langle\langle A \rangle\rangle \varphi$ can be defined informally as: *coalition A has a strategy to enforce φ.*

The notion of strategy was defined in [2,3] as follows: A *strategy* for player $a \in \Sigma$ is a function f_a that maps every nonempty finite sequence of states $\lambda \in Q^+$ to a natural number such that if the last state of λ is q, then $f_a(\lambda) \leq d_a(q)$.

Thus, the strategy f_a determines for every finite prefix λ of a computation a move $f_a(\lambda)$ for player a. Each strategy f_a for player a induces a set of computations that player a can enforce. Given a state $q \in Q$, a set $A \subseteq \{1, \ldots, k\}$ of players, and a set $F_A = \{f_a | \, a \in A\}$ of strategies, one for each player in A, we define the *outcomes* of F_A from q to be the set $out(q, F_A)$ of q-computations that the players in A enforce when they follow the strategies in F_A.

The fact that a state q in structure S satisfies formula φ is denoted by $S, q \models \varphi$. When S is obvious, we omit it and write $q \models \varphi$. The satisfaction relation $\models$ has been defined for all states $q \in S$, inductively as follows:

- $q \models p$, for proposition $p \in \Pi$, iff $p \in \pi(q)$
- $q \models \neg\varphi$ iff $q \not\models \varphi$
- $q \models \varphi_1 \vee \varphi_2$ iff $q \models \varphi_1$ or $q \models \varphi_2$
- $q \models \langle\langle A\rangle\rangle X\varphi$ iff there exists a set F_A of strategies, one for each player in A, such that for all computations $\lambda \in out(q, F_A)$, we have $\lambda[1] \models \varphi$.
- $q \models \langle\langle A\rangle\rangle G\varphi$ iff there exists a set F_A of strategies, one for each player in A, such that for all computations $\lambda \in out(q, F_A)$, and all positions $i \geq 0$, we have $\lambda[i] \models \varphi$.
- $q \models \langle\langle A\rangle\rangle \varphi_1 U \varphi_2$ iff there exists a set F_A of strategies, one for each player in A, such that for all computations $\lambda \in out(q, F_A)$, there exists a position $i \geq 0$ such that $\lambda[i] \models \varphi_2$ and for all positions $0 \leq j < i$, we have $\lambda[j] \models \varphi_1$.

2.2 ATL$_{ir}$ and iCGS

In [5,10,1,7], it was pointed out that the assumption that agents know the entire state of the system at each step of its execution is often unrealistic. Similarly, agents being able to record the entire history of their information was shown to be unlikely. In [10] the author proposed to distinguish between *perfect information* and *imperfect information* as well as between *perfect recall* and *imperfect recall.* Concurrent game structures were extended to *imperfect information concurrent game structures* via addition of an indistinguishability relation $\sim : \Sigma \to 2^{Q\times Q}$ (where Σ is the set of all k players). $\sim$ is a family of equivalence relations $\sim_a$ (one for each agent a), that indicate which states are indistinguishable from the point of view of this agent. $\sim_a$ determines equivalence classes $[q]_a$ on Q. Q_a is the set of such classes $[q]_a$. From the point of view of agent a, the total number of unique states the agent is able to distinguish is $|Q_a|$.

The composition of two binary choices led to the definition of four types of strategies. Strategies were defined as a function of an agents view of the past V to a choice C ($f_a : V \to C$). By changing the definition of V, different strategies semantics are achieved:

1. IR (perfect information and perfect recall) strategies. The view of the past $V = Q^+$ is the sequence of all steps of the execution of the system.
2. Ir (perfect information and imperfect recall) strategies. The view of the past $V = Q$ is limited to the current state of the system.
3. iR (imperfect information and perfect recall) strategies. The view of the past $V = Q_a^+$ is the sequence of all states of the system observed by the agent.

4. ir (imperfect information and imperfect recall) strategies. The view of the past $V = Q_a$ is limited to the agent's view of the current state.

We write ATL_{XY} to denote a model where strategies are assumed to be of type XY ($X \in \{I, i\}, Y \in \{R, r\}$), e.g. ATL_{ir} is a model where imperfect information and imperfect recall strategies are considered.

Given a formula $\langle\langle A \rangle\rangle \varphi$, we require ATL_{ir} strategies for A to be *uniform*, i.e. to specify the same choices in indistinguishable states. This ensures that the choice of an action for an agent does not depend on information that is not accessible to this agent.

2.3 The Model Checking Problem

The model checking problem is usually defined as follows. Given

- a model M,
- a formula φ,
- a set of distinguished *initial states* Q_0 (often assumed to be a singleton),

determine whether $\forall_{q \in Q_0}\ M, q \models \varphi$.

It was argued in [10,7,8] that agents $a \in A$ should have a strategy in ATL_{iY} variants not only for the initial state, but also for all indistinguishable states ($[initialState]_a$ for each a). It was argued that otherwise the agents would be given perfect information on the initial state.

We would like to comment on this observation. One can easily imagine a model where the constant initial state is unknown to the checked coalition, however the agents are able to test strategies a finite number of times. Equally, they could just guess a strategy or would be provided a strategy by an external entity. Such a strategy, obtained in one of such ways, could still be a viable and executable strategy. Therefore it can be claimed agents do have a viable strategy even if such a strategy is only viable in the initial state of the model.

In [2,3] it was shown that ATL_{IR} model checking can be performed with the CTL fixpoint model checking algorithm, when using a modified pre-image auxiliary function. In [10] a proof is given that checking a formula with a single coalitional modality for an ATL_{ir} model is *NP*-complete. Model checking of ATL_{ir} turns out to be slightly harder, namely Δ_2^P-complete [10,6]. This is mainly because fixpoint characterizations of strategic modalities do *not* hold under imperfect information [4], and hence purely incremental synthesis of winning strategies is not possible for ATL_{ir}.

3 ATL_{ir} Model Checking

In [10], the procedure for performing ATL_{ir} model checking for a single coalitional modality formula $\langle\langle B \rangle\rangle \varphi$ is described as a two-step algorithm:

1. Guess nondeterministically a strategy.
2. Perform CTL model checking of $A\varphi$ in a model that follows this strategy.

The problem of performing ATL_{ir} model checking for the full semantics of ATL can be reduced to performing the above procedure for each coalitional subformula (at most linearly many calls of this procedure).

3.1 Determinization

In order to construct a deterministic and complete version of the ATL_{ir} model checking algorithm, the following brute force procedure for determinization of *NP* problems would seem sufficient:

1. Generate a new potential solution that has not been checked before. Verify if that potential solution is valid.
2. If this solution is not valid, check if there exists any previously unchecked solution.
   (a) If all solutions have been checked, terminate with the answer *NO*.
   (b) Otherwise go to 1.
3. If this solution is valid, terminate with the answer *YES*.

Obviously, such a determinization is a heuristic, since acceptable computation time can be achieved only for problems where a solution exists. However it is hoped that in a large number of classes of models where a solution exists, a heuristic ATL_{ir} model checking algorithm may yield satisfactory results despite the theoretical computational hardness of the problem. This has been experimentally observed for several classes of computationally hard problems, most notably SAT.

Limiting the Search Space of Solutions. In case of ATL_{ir} model checking, the solutions are strategies that a coalition can utilise to enforce a property. Since the space of solutions is computationally large, it is crucial for such an algorithm to limit the searched space of solutions to the maximum possible extent in order to achieve acceptable computation time. We have limited the search space of solutions by identifying equivalences between solutions.

In order to describe the space of searched strategies, some definitions and observations are needed. Please note the symbols defined in this chapter will be used later for the remainder of the chapter.

Definition 1. *For a model M and strategy S for coalition A, we define a* trimmed model M_S *as a restriction of model M, where agents from coalition A have their choices restricted to S. We have the set Q of states reachable in M and the set Q_S of states reachable in M_S.*

Observation 1. Obviously, $Q_S \subseteq Q$. We believe that in many practical cases not only $Q \setminus Q_S \neq \emptyset$ holds for many a strategy, but in fact $|Q_S| \ll |Q|$.

Definition 2. *We define the* proper domain *of strategy S as Q_S.*

Observation 2. Many authors define the domain of a strategy as Q. The main issue is that the assignment of actions for states in $Q \setminus Q_S$ does not have any significance, because those states are never reached with strategy S. We note that strategies S_1, S_2 that assign identical actions in the same *proper domain* $Q_{S_1} = Q_{S_2}$ can be considered *equal*, regardless of actions assigned in $Q \setminus Q_{S_1}$:

$$\forall_{S_1,S_2}(Q_{S_1} = Q_{S_2} \wedge \forall_{q \in Q_{S_1}} S_1(q) = S_2(q) \iff M_{S_1} = M_{S_2})$$

Definition 3. *We define a* proper strategy *as a strategy S that has its domain limited to Q_S.*

Observation 3. Having shown that only proper strategies are worth considering, we can significantly limit the searched strategy space by treating all strategies equal to S as a single proper strategy. This single proper strategy can be viewed a representative of a equivalence class of strategies. The size of each such equivalence class can be described as

$$\prod_{a \in A} \prod_{[q]_a \in [Q \setminus Q_S]_a} Act([q]_a)$$

where $Act([q]_a)$ denotes the number of actions available for agent a in the equivalence class of states $[q]_a$.

Definition 4. *We define a imperfect information and imperfect recall* partial strategy *for agent a in model M as a piecewise-defined function from equivalence classes of states to sets of actions. This function is defined by two pieces (subfunctions), an* explicit strategy *and an* implicit strategy.

The domain Q_S of a partial strategy S is a sum of two domains, the *explicit domain* (the domain of the explicit strategy) and the *implicit domain* (the domain of the implicit strategy).

Definition 5. *We define an* explicit strategy *as a function from equivalence classes of states to sets of actions containing one and only one action. Those actions are required to be allowed for those equivalence classes of states in the original model M.*

Definition 6. *We define an* implicit strategy *as a function from equivalence classes of states to sets of actions, containing possibly more than one action. This function is equivalent to the protocol present in the original model M.*

Definition 7. *Since the domain of a strategy is discrete, we define the* size *of a strategy as the number of equivalence classes of states contained in the domain. For partial strategies the size will be understood as the size of the explicit strategy.*

Definition 8. *With such a definition, we define an* empty strategy *as a partial strategy with explicit strategy of size 0 (zero).*

Definition 9. *With such a definition, we define a* complete strategy *as a partial strategy with implicit strategy of size 0 (zero).*

Observation 4. In the original model M, a protocol determines a set of actions allowed for an agent in any state. A strategy is a possible restriction enforced upon this protocol. An empty strategy is just a strategy that restricts nothing, for an empty strategy ES the trimmed model $M_{ES} \equiv M$. A complete strategy on the other hand assigns a single action for any state in Q_S. All other partial strategies have some explicit assignments for some states, and some implicit (resulting from the protocol in the original model M).

We consider the *ir* setting default and therefore such strategies will be called just *partial strategies* or even *strategies.*

Example 1. Consider a model with 2 states $Q = \{color, noColor\}$, with a single agent with 2 actions $Act = \{push, wait\}$. The protocol in the model permits the execution of both actions in both states. For the sake of the example we assume $Q_S = Q$ for any strategy S.

An empty strategy ES is equivalent to that protocol, i.e. it permits the execution of both actions in both states as well. More formally, the explicit strategy is of size 0 and the implicit strategy is the protocol of the original model M.

An example complete strategy CS defined in the following way: $CS(x) =$ {push if x = color, wait if x = noColor} assigns a single action for all states in Q_S, leaving the implicit strategy empty (of size 0).

An example partial strategy PS defined in the explicit part in the following way: $PS_{explicit}(x) =$ {push if x = color} must have the implicit domain cover the rest of states in Q_S, and therefore the implicit strategy is: $PS_{implicit}(x) =$ {{push, wait} if x = noColor}.

Observation 5. The strategies we have described above are *local* strategies, they apply to a single agent. The strategies for a coalition are easily constructed as a tuple of such individual strategies.

Definition 10. *We define a partial strategy S_A for coalition $A = \{a_1, \ldots, a_n\}$ in model M as a tuple of local partial strategies of agents: $S_A = (S_{a_1}, S_{a_2}, \ldots, S_{a_n})$*

Observation 6. As a results of the above definition we have the coalitional explicit (implicit) strategy defined as a tuple of the explicit (implicit respectively) individual strategies. This extends to domains as well.

Definition 11. *We define an* empty strategy *for coalition A as a tuple of local strategies for all agents $a \in A$ where each local strategy is empty.*

Definition 12. *We define a* complete strategy *for coalition A as a tuple of local strategies for all agents $a \in A$ where each local strategy is complete.*

4 The SMC Model-Checker

4.1 Introduction

The project with the "working title" SMC is a software program capable of performing model checking.

4.2 Capabilities

The current version of SMC is capable of performing $\langle\langle\Gamma\rangle\rangle$-ATL model checking. The formulae of $\langle\langle\Gamma\rangle\rangle$-ATL are restricted to contain at most a single coalitional modality. The verification is performed under the imperfect recall and imperfect information setting.

4.3 The Model Checking Algorithm

SMC utilises a custom algorithm for model checking $\langle\langle\Gamma\rangle\rangle$-ATL formulae.

We present the most interesting fragments of this algorithm by starting with a general description, in order to provide a more detailed description of the most important step.

High-level description of the algorithm

We present a generalised description of the algorithm, in order to provide further details in the following sections.

1. For formula of type $\langle\langle B\rangle\rangle\varphi$, synthesize a previously unverified strategy S to be verified.
2. Perform CTL-style model checking of $A\varphi$ in the trimmed model M_S (model M restricted to follow S).
3. If the verification yielded a *valid* result for S, return the answer *true* and the strategy S, then terminate.
4. If all strategies have been verified, return the information that no strategy was found.
5. If the verification yielded a *invalid* result for S, return to start.

We observe that step 1. is of most significance in this algorithm, as step 2. can be performed with the well-known CTL fix-point model checking algorithms with a modified pre-image auxiliary function. Steps 3. - 5. are simple binary decision steps.

High-level description of the strategy synthesis step

In the previous section we have provided a description of the steps of the algorithm. Now we present a more detailed description of the first step (strategy synthesis step).

1. Start with an empty partial strategy and with the initial state.
2. In a loop, generate potential partial strategies by fixing actions for newly discovered states that do not have already fixed actions. These newly discovered states are required to be reachable with the employment of this strategy.
3. Continue the above step until a valid strategy is found or all strategies have been explored.

Further considerations concerning this algorithm are presented in the conclusions section.

5 Experimental Results

We present results of our model-checker SMC for a scalable model *Castles*.

5.1 Castles — A Running Example

Te model consists of an agent called Environment, that keeps track of the health points of 3 castles. Health points (HP) are meant to represent the condition of the castle, 0 HP denotes a castle is *defeated*. Additionally, there is a number of agents called *Workers*, that work for the benefit of a castle.

An instance with 1 worker assigned with the first castle, 3 workers assigned to the second and 4 to the third castle will be denoted by 9 $(1, 3, 4)$.

Workers are able to:

1. attack a castle they do not work for
2. defend the castle they do work for
3. do nothing

Doing nothing is the only action allowed once the castle this agent is working for is defeated.

Any agent can not defend its castle twice in a row, it must wait 1 step before being able to defend again.

Castles get damaged if the number of attackers is greater than the number of defenders. (Example: castle 3 is attacked by 2 agents, it loses 2 HP if not defended, or it loses 1 HP if defended by a single agent).

Here we present experimental results for the following formula:
$\langle\langle c12 \rangle\rangle$G(! castle1Defeated)

The semantics of this formula is that agents working for Castle 1 and agents working for Castle 2 have a strategy to prevent Castle 1 from ever being defeated, no matter what the other agents do.

This formula was *true* in the models we have tested. It is important to note partial strategies verification was disabled for the tests, only complete strategies were verified.

5.2 Performance Results

We present performance results for a class of models of various size. Results where computation exceeded 180 seconds are not presented. The columns in the table below should be interpreted in the following way (from left to right):

- the scalability factor N, the total number of agents is indicated first (including agent Environment), followed by the number of agents working for Castle 1, 2, 3 (in that order)
- total "wall clock" time taken by the model checking algorithm (input parsing time is deemed irrelevant and not accounted), in milliseconds
- "wall clock" time taken by the first step of the algorithm (strategy synthesis), in milliseconds

- "wall clock" time taken by the second step of the algorithm (CTL verification), in milliseconds
- peak memory usage observed during the execution of the program (the default Java Virtual Machine makes it hard to determine real maximum usage, as memory is freed nondeterministically), in megabytes

N	Total time (ms)	1. step (ms)	2. step (ms)	Peak memory (MB)
4 (1 1 1)	107	99	7	10
5 (2 1 1)	190	187	2	22
6 (2 2 1)	623	615	5	35
7 (2 2 2)	11 867	3 151	8 714	314
8 (3 2 2)	150 077	28 869	121 205	665

5.3 The Number of Strategies Considered

The number of considered strategies is of even greater interest to us, as we consider only proper strategies. The columns in the table below should be interpreted in the following way (from left to right):

- the scalability factor N, the total number of agents is indicated first, followed by the number of agents working for Castle 1, 2, 3 (in that order)
- the number of potential strategies
- the number of proper strategies
- the number of strategies verified until a valid one was found or all proper strategies were verified

Please note some results for potential and proper strategies were omitted as the computation of these values tends to be very time-consuming. It is hoped the initial results are enough to be somewhat representative.

N	Potential strategies	Proper strategies	Tested strategies
4 (1 1 1)	429981696	283	1
5 (2 1 1)	8916100448256	16616	1
6 (2 2 1)	?	?	1
7 (2 2 2)	?	?	5
8 (3 2 2)	?	?	5

5.4 Comparison to MCMAS

At the moment, the only tool we know of that is capable of performing ATL_{ir} model checking (with uniform strategies) is an experimental version of MCMAS [9], not yet released publicly at the time of writing. We have been able to compare the results of our SMC model-checker and those of an experimental version of MCMAS kindly provided to us by the authors.

We were unable to obtain any results from MCMAS for the formula presented above due to exceedingly long computation time, therefore we have used the following two formulae for comparison tests:

(a) $\langle\langle c12\rangle\rangle$F(castle3Defeated) (*true* in the models we have tested)
(b) $\langle\langle w12\rangle\rangle$F(allDefeated) (*false* in the models we have tested)

The semantics of the first formula is that agents working for Castle 1 and agents working for Castle 2 have a strategy to enforce Castle 3 become defeated, not matter what the other agents do.

The semantics of the second formula is that agents Worker 1 and Worker 2 have a strategy to enforce all castles become defeated, not matter what the other agents do.

N	Formula	MCMAS tested strategies	SMC tested strategies
4 (1 1 1)	a	$\approx 20\,000$	1
5 (2 1 1)	a	$> 540\,000$ (interrupted)	1
4 (1 1 1)	b	$\approx 20\,000$	283

N	Formula	MCMAS execution time	SMC execution time
4 (1 1 1)	a	72 s	0.1 s
5 (2 1 1)	a	$>$ 30 minutes (interrupted)	0.2 s
4 (1 1 1)	b	78 s	5.4 s

6 Conclusions

Our results indicate that while the ratio of potential to proper strategies can be very small for small models, it can quickly become overwhelmingly large. This observation was universally true in the several classes of models we have tested, not only the class of models presented in this article. This is a strong argument to consider only proper strategies.

The structure of the algorithm resembles bounded model checking (BMC). It generates strategies of bounded explicit domain size. The domain is discrete, we are generating strategies for explicit strategy size $k = 0, 1, 2, \ldots$.

We observe similar results to those observed in BMC algorithms. Most often the explicit strategy size of a valid strategy is smaller than the entire domain of the model for this strategy. Furthermore, when a solution exists it can often be found *quickly*. The number of examined proper strategies before finding a valid one is often significantly smaller than the number of all considered proper strategies.

One important difference with respect to BMC, is that the algorithm is complete, i.e. it searches all proper strategies.

Our approach enables the capability of constructing strategies of limited explicit strategy size. It should be noted that partial strategies will be constructed no matter what by the algorithm, but it is possible not to subject them to verification. In such a case only complete strategies will be verified, but partial strategies will still be constructed, as they serve as an refinement source for further partial and eventually complete strategies.

A proper partial strategy can be described by the explicit strategy part. This part can often be of small size. Example benefits of smaller size of the descriptions are improved readability for humans and reduced memory/processing requirements for computers.

The problem itself remains computationally difficult, but our algorithm enables a potential speedup on two levels. Firstly, it considers only proper strategies. Secondly, if a valid strategy exists, it often finds a strategy after a number of attempts vastly smaller than the number of all proper strategies. The first speedup yielded a reduction of the search space by order of 10^8 for one of the presented problems. The second speedup yielded a solution after only a single attempt for a problem where 16616 proper strategies exist. As a result, the strategy verification sub-routine was called only once, yielding a 16616-fold speedup.

Acknowledgements. This contribution has been supported by the Foundation for Polish Science under International PhD Projects in Intelligent Computing; project financed from The European Union within the Innovative Economy Operational Programme 2007-2013 and European Regional Development Fund. Wojciech Jamroga acknowledges the support of the FNR (National Research Fund) Luxembourg under project GALOT – INTER/DFG/12/06. We gratefully acknowledge the help of Hongyang Qu and Alessio Lomuscio who kindly provided us with the latest experimental version of MCMAS, as well as MCMAS model generators.

References

1. Ågotnes, T.: A note on syntactic characterization of incomplete information in ATEL. In: Procedings of the Workshop on Knowledge and Games, Liverpool, pp. 34–42 (2004)
2. Alur, R., Henzinger, T.A., Kupferman, O.: Alternating-time Temporal Logic. In: Proceedings of the 38th Annual Symposium on Foundations of Computer Science (FOCS), pp. 100–109. IEEE Computer Society Press (1997)
3. Alur, R., Henzinger, T.A., Kupferman, O.: Alternating-time Temporal Logic. Journal of the ACM 49, 672–713 (2002)
4. Bulling, N., Jamroga, W.: Comparing variants of strategic ability. Journal of Autonomous Agents and Multi-Agent Systems 28(3), 474–518 (2014)
5. Jamroga, W.: Some remarks on alternating temporal epistemic logic. In: Dunin-Keplicz, B., Verbrugge, R. (eds.) Proceedings of Formal Approaches to Multi-Agent Systems (FAMAS 2003), pp. 133–140 (2003)
6. Jamroga, W., Dix, J.: Model checking ATL_{ir} is indeed Δ_2^P-complete. In: Proceedings of EUMAS 2006 (2006)
7. Jamroga, W., van der Hoek, W.: Agents that know how to play. Fundamenta Informaticae 63(2-3), 185–219 (2004)
8. Jonker, G.: Feasible strategies in Alternating-time Temporal Epistemic Logic, master thesis, University of Utrecht (2003)
9. Lomuscio, A., Qu, H., Raimondi, F.: MCMAS: A model checker for the verification of multi-agent systems. In: Bouajjani, A., Maler, O. (eds.) CAV 2009. LNCS, vol. 5643, pp. 682–688. Springer, Heidelberg (2009)
10. Schobbens, P.Y.: Alternating-time logic with imperfect recall. Electronic Notes in Theoretical Computer Science 85(2), 82–93 (2004)

AjTempura – First Software Prototype of C3A Model

Vladimir Valkanov[1], Asya Stoyanova-Doycheva[1], Emil Doychev[1],
Stanimir Stoyanov[1], Ivan Popchev[2], and Irina Radeva[2]

[1] University of Plovdiv, Bulgaria
{stani,astoyanova}@uni-plovdiv.bg,
{vvalkanov,e.doychev}@uni-plovdiv.net
[2] Bulgarian Academy of Science, Bulgaria
{ipopchev,iradeva}@iit.bas.bg

Abstract. The paper provides a general description of a model for context-aware agent architecture (C3A) and first steps in AjTempura creation vie C3A model. The approach adopts the definition of context and context-awareness given by Dey. The C3A model aims at creating of smart virtual spaces. The applicability of the model is demonstrated by development of an agent-oriented application.

Keywords: Context-aware architecture, Formal models, Intelligent agents, Agent-Oriented architectures, Tempura.

2010 Mathematics Subject Classification: 68T42 Agent technology

1 Introduction

One of the main characteristics of the modern systems today (i.e. eLearning) is the 'anytime-anywhere-anyhow' delivery of electronic content, personalized and customized for each individual user. To satisfy this requirements new types of context-aware and adaptive software architectures are needed, which are enabled to sense aspects of the environment and use the information to adapt their behaviour in response to changing situation. From a network of computer networks the Internet has transformed into the Internet of Things [1], where people and everyday objects can be assigned identifiers. These identifiers can be created and managed by computers. Within the infrastructure of the current Web, the new Symantec Web [18] is rising making all sorts of informational resources addressable by the existing identification protocol URI. A model for development of context-aware software architectures, known as Context-Aware Agent Architecture (C3A), is presented in the publication. There are a lot of definition of context-awareness, first attempts to give satisfactory definition are related to application of mobile devices and reading users' position. In the specialized literature it is presented as the term context-aware computing is used to describe software possible to render an account of users' location [10] or as the ability of a program or device to sense various states of its environment and itself [4]. We adopt Dey's context definition for developing C3A model. Dey [3] criticizes the

P. Angelov et al. (eds.), *Intelligent Systems'2014*,
Advances in Intelligent Systems and Computing 322, DOI: 10.1007/978-3-319-11313-5_38

above definitions in two ways: context if defined by examples, i.e. by enumeration of various cases, and context is defined by synonyms, mainly as environments or situations. In his opinion, there definitions should bother development of context-aware applications. In this sense he proposes a more common definition whereby context is any information that can be used to characterize the situation of an entry. An entry can be assumed with a person, a place, or an object that is considered relevant to the interaction between a user and an application, including themselves [10]. In this point of view if a system which provides relevant information and/or services to the user, where relevancy depends on the user's task, is called context-aware system. An implementation of the C3A in real software demonstrating the features of the model is also presented in this paper.

2 C3A Model

The model aims to propose a framework which can be used for implementation of context-aware applications for various domains. A context-aware architecture includes autonomous intelligent components which can: operate in a changing environment; detect, identify and localize changes (events) in the environment; initiate corresponding compensatory actions depending on the types of changes (events). A compensatory action is defined as an activity (functionality) fired as a reaction to the occurrence of some event (change in the environment). Two basic compensatory actions are: *Adaptation* – before the information resources (electronic services, electronic content) are provided, they can be modified so that they reflect as completely as possible the specifics of the change or event; *Personalization* – before the information resources (electronic services, electronic content) are provided, they can be modified so that they reflect as completely as possible the users' desires, intentions, background, etc.

Using C3A a formal presentation is proposed which reflects the spatial and temporal aspects of: virtual space structure and building components; relations between the components; operations allowed in the space; control of the processes taking place in the space.

In the model, the fundamental notion is this of ***smart space***, defined as

$SmartSpace = \bigcup_{i=1}^{p} SP_i$, where there exists a set $SubSpaces = \{SP_i \mid i = 1, \ldots, n\}$ of subspaces and $p \leq n$.

For two subspaces $SP_i \in SubSpaces$ and $SP_j \in SubSpaces$ we introduce the following definitions:

- *Empty subspace*, if $SP_i = \varnothing$;
- *Overlapping subspaces*, if $SP_i \cap SP_j \neq \varnothing$;
- *Identical (or completely overlapping) subspaces*, if $SP_i = SP_j$;
- *Disjunctive subspaces*, if $SP_i \cap SP_j = \varnothing$.

In the smart space, ***services*** will be provided through autonomous intelligent components, known as *agents*. The agents are with 'limited rationality', which in this case means:

- They operate with limited capacity (resources) in order to plan, predict, make choices and execute actions;
- They have partial (local) control and impact on the smart space.

Due to the limited rationality, we would assume that:

- The set of services, provided in the space, alters dynamically;
- There is a *minimal functionality* (minimal set of services), that the space cannot operate without.

Let $Agents = \{a_i / i = 1, \ldots, m\}$ be the set of all agents potentially operating in the smart space. Due to the limited rationality we assume that each agent operates in its own subspace (known as *agent's range*) operating as agent's environment. Agents can operate and impact only within their own range. In this sense, the overall smart space (noted as $SmartSpace^A$) is built as union of agents' ranges $R(a_i)$. Furthermore, let $Services = \{s_i / i = 1, \ldots, k\}$ be the set of services provided in the $SmartSpace^A$.

Various types of agents and services are distinguished in the $SmartSpace^A$. The set of agents is decomposed as $Agents = PA \cup OA$, so that $PA \cap OA = \emptyset$. Respectively, the set of services can be decomposed as $Services = F \cup AF \cup SF$, i.e. as a union of three mutually disjunctive subsets ($F \cap AF = \emptyset$, $F \cap SF = \emptyset$ and $AF \cap SF = \emptyset$). This decomposition characterizes the following types of services:

- F is the minimal functionality of SmartSpace;
- AF includes the compensatory actions;
- $SF = \{generate, remove, selfremove\}$ includes three special operations which are used to create and remove operational agents. For instance, the generate operation is defined as $generate : PA \times E \rightarrow OA$.

Here, the two types of agents are presented shortly. PA includes agents, known as *persistent agents,* which are always available in the space. These agents are used to:

- Provide the minimal functionality of the space;
- Identify and localize the events which take place in the space;
- Generate operative agents;
- Remove operative agents.

The persistent agents' behavior is defined as the next genetic function

$$Beh_{pa} : PA \rightarrow 2^F \cup \{generate, remove\},$$

where $\{generate, remove\} \subset SF$.

OA includes agents, called *operative agents,* generated dynamically by the persistent agents. An operative agent provides some kind of compensatory actions as reaction to the events identified in the range of the generating persistent agent. After compensatory actions completing, the operative agent will be removed (or self-removed).

The operative agents' behavior is defined as the below genetic function

$$Beh_{oa} : OA \rightarrow 2^{AF} \cup \{selfremove\},$$

where $\{selfremove\} \subset SF$

The events which take place in the space are introduced as $Events = \{e_i \,/\, i = 1, \ldots, e\}$. *Location* and *time of occurrence* are the two basic features of the events. C3A examines the locations of the events' occurrence as unique subspaces, denoted as $R(e_i)$, so that $SmartSpace^E$ presents the overall SmartSpace in terms of the locations of expected events' occurrences. The events are always viewed as *atomic primitives,* and they can be combined to make more complex structures such as situations, scenarios, intervals. An interval is the sequence $EInt = [e_1, \ldots, e_{pr-1}, e_{pr}, e_{pr+1}, \ldots, e_l]$ of events ordered in time where an event e_{pr} can be identified at any time. Event e_{pr} is interpreted as the *present*; the subinterval $[e_1, \ldots, e_{pr-1}]$ is the *past*; respectively, $[e_{pr+1}, \ldots, e_l]$ is the *future*.

As already pointed out, the persistent agents perform two different functions. First, they provide the minimal functionality in the SmartSpace. Second, they can identify the occurrence of certain events in its ranges and dynamically generate appropriate operative agents, respectively, which in turn execute the necessary compensatory actions. In the model, this is formally presented, as follows:

$$\forall a_i \in PA\ (\exists\, a_j \in OA, \exists\, e^* \in E : generate(a_i, e^*) = a_j),$$

under the condition that $R(a_i) \equiv R(e^*)$.

The general lifecycle of a context-aware agent-based software architecture compatible with C3A is shown (in pseudocode) in Fig. 1. Presented in time, the operation of the architecture looks like a pulsating core (implementing the minimal functionality of the SmartSpace) which periodically 'inflates' in different directions (depending on the changes in the environment) and again 'deflates' to its usual size (minimal functionality).

```
repeat
        running(a1), running(a2), ..., running(ap);
        anytime ∀ ai {
                if ( (∃ek ∈ R(ek)) ∧ (R(ek) = R(ai)) )
                then {
                        aj ← GENERATE (aj , ek );
                        send(ai, aj, REQUEST (ek ));
                        when INFORM (aj , ai , 'done' )
                        then ( REMOVE( ai , aj ) ∨ SELFREMOVE(aj) ) }
                endif
        endanytime
forever
ai ∈ PA, i = 1, ..., p; aj ∈ OA; ek ∈ E.
```

Fig. 1. C3A Architecture Lifecycle

3 AjTempura as a C3A Application

Application of C3A model for development of real software will be demonstrated in this section. E-Learning environments providing electronic educational services are becoming an integral part of modern education. In the Faculty of Mathematics and Informatics at the University of Plovdiv, an infrastructure is being developed, known as Distributed e-Learning Center (DeLC), as a response to the need for supporting learning using modern information and communication technologies [14,11,13]. The center aims to provide adaptive and personalized e-learning services and teaching content located on physically separated servers. DeLC is a dynamic network structure consisting of [12]:

- Nodes – these operate as repositories of services and content;
- Relations – these specify various kinds of dependencies arising during an education experience.

The nodes can operate independently or dynamically link to each other, making complex virtual structures, called educational clusters. In the current DeLC versions, two educational clusters are built. The first cluster called MyDeLC is used to organize and perform e-learning by allowing fixed access to the services and electronic content via a specialized educational portal [15,16]. The second cluster, called InfoStation cluster offers a three-layer architecture which allows mobile access to services and informational resources via intelligent wireless access points (called Information Stations - ISs), situated around the university building [17,7,6].

In 2013 year we have started a new project that aims at the transformation of DeLC in a new infrastructure called Virtual eLearning Space (VeLSpace). In this space, the context-aware provision education services and teaching content will be supported by autonomous intelligent agents with 'limited rationality', i.e. changes can be identified only locally within the ranges of the agents positioned there. Each event causes a change in the local state of the range. In this way the state of the overall VeLSpace (global state) depends on the local states ordered in time. In order to manage the global state a suitable formalism, known as Interval Temporal Logic (ITL), was chosen as a theoretical model. Interval Temporal Logic [9] is a kind of temporal logic for describing time-dependent processes. Tempura interpreter [8] is an executable subset of ITL which uses the ITL syntax to identify time as a finite consequence of states. The existing version of Tempura was created in the C language [5] and repeatedly amended and expanded with new functionalities. Since the VeLSpace is built by separate intelligent autonomous components where the electronic services are equipped with their own operational agents, we created a new agent-oriented version of Tempura, called AjTempura, which can be easily integrated in VeLSpace while maintaining the space's homogeneity.

The transformation of the original Tempura interpreter was an iterative hand-made process which consist three basic steps. The software result of each step was tested and his run-time was compared with the original Tempura interpreter. Firstly we made a direct translation from C to Java code without changing the imperative structure of the interpreter. Second step was a refactoring process which aims at creating an object-oriented version of Tempura called jTempura [19]. The final step was creating an agent-oriented version AjTempura. The AjTempura architecture is an implementation of the C3A model which is shortly described below.

The minimal functionality of AjTempura is implemented through two persistent agents - IOAgent and TempuraAgent. Both agents listen to the occurrence of events in VeLSpace. According to the kind of the identified event, an operative agent will be generated. This new agent has to analyze the event in more details and to initiate a corresponding compensatory action. AjTempura is implemented in development environment JADE [2]. The operating of the architecture is presented in the next diagrams.

Besides the two persistent agents, another one, known as Sniffer, is of key importance for tracking communication in agents. Its role is to provide a simple visual means for presenting the consecutive exchange of messages between the agents in a system. In its nature, the Sniffer agent draws diagrams similar to the Sequence diagrams in the UML language. The starting condition of this diagram, at the AjTempura start-up, can be seen in Fig. 2. In the left part of the figure, there is a list of the existing agents, among which there is a representative of the persistent IOAgent and a service DF agent. In accordance with AjTempura's lifecycle, the IOAgent starts sending periodic queries to DF for the presence of agents from the TempuraAgent type. The messages are sent every second until the reception of a response which contains a list of the present agents from the wanted type.

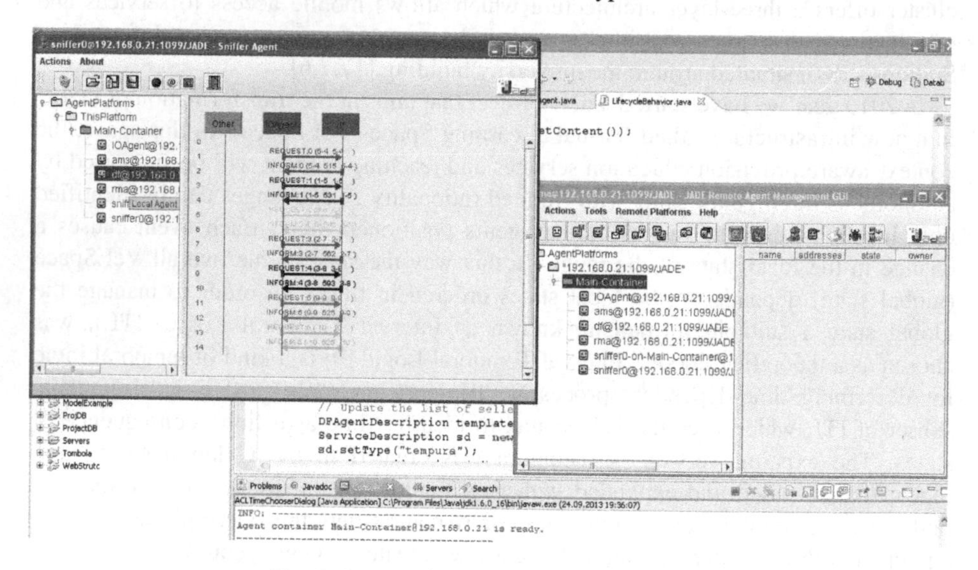

Fig. 2. Starting situation of the Sniffer agent

At the appearance of an agent from the TempuraAgent type in the DF list, an exchange of messages between this agent and IOAgent immediately takes place, which leads to the generation of an operative interpreting agent. The appearance of the necessary interpreting agent marks the start of the exchange of ITL sequences between this agent and IOAgent until the IOAgent goals are met. The communication between the entire set of agents can be seen in the following state of the diagram, generated by the Sniffer agent (Fig. 3).

As a result from the appearance of the TempuraAgent and its communication with IOAgent, a new operative agent IA_1 has been generated. Its name is unique, and each following similar agent will be generated with a consecutive number. The operative agents exist until the IOAgent sends a 'done' message, i.e. confirmation for completion of the compensation action. In this case, this means that there are no more ITL sequences for processing and the IA_1 is redundant. After that the operative agents can be removed or self-removed (Fig. 4).

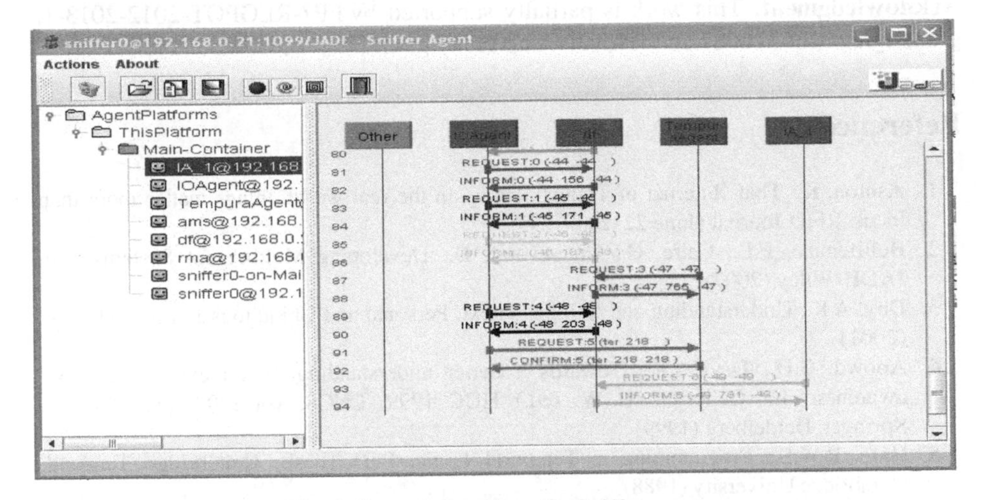

Fig. 3. Second view from the Sniffer agent

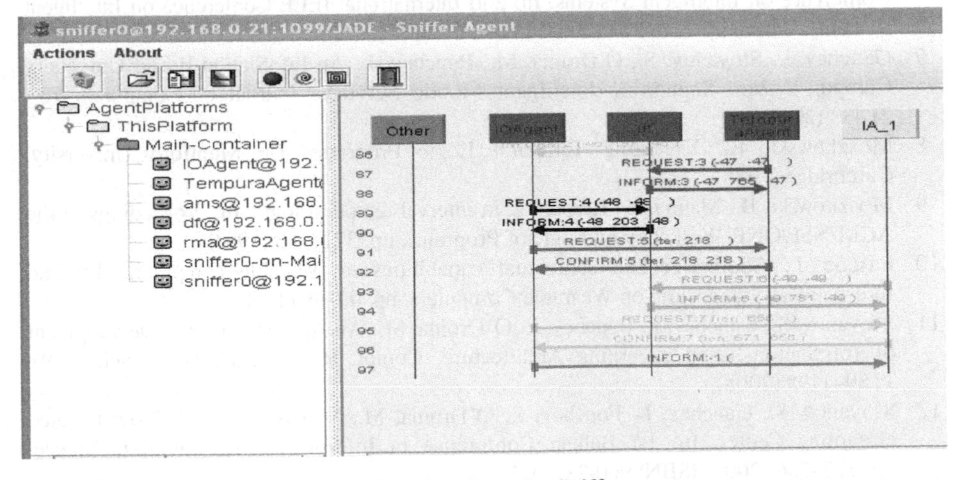

Fig. 4. Third view from the Sniffer agent

4 Conclusion

The current state of C3A model is presented in this paper. The research continues to expand and refine the existing version of the model. There are various issues being tackled. To name a few, for example, conflict resolving in overlapping or identical subspaces, saving traces of the removed operative agents as subintervals (the past), formalizing the interfaces between agents and education services, formalizing the notion of adaptation and personalization. The model extensions will be prototyped and examined in the VeLSpace.

Acknowledgment. This work is partially supported by FP7-REGPOT-2012-2013-1, grant agreement 316087 and Plovdiv University grant NI13-FMI-02.

References

1. Ashton, K.: That 'Internet of Things' Thing, in the real world, things matter more than ideas. RFID Journal (June 22, 2009)
2. Bellifemine, F.L., Caire, G., Greenwood, D.: Developing Multi-Agent Systems with JADE. Wiley (2007)
3. Dey, A.K.: Understanding and Using Context. Personal and Ubiquitous Journal 5(1), 4–7 (2001)
4. Abowd, G.D., Dey, A.K.: Towards a better understanding of context and context-awareness. In: Gellersen, H.-W. (ed.) HUC 1999. LNCS, vol. 1707, pp. 304–307. Springer, Heidelberg (1999)
5. Hale, R.W.S.: Programming in Temporal Logic. PhD Thesis. Crambridge, England. Cambridge University (1988)
6. Ganchev, I., Stoyanov, S., O'Droma, M., Popchev, I.: Enhancement of International IEEE Conference on Intelligent Systems. In: 2nd International IEEE Conference on Intelligent Systems, Varna, pp. 359–364 (2004) ISBN: 0-7803-8278-1
7. Ganchev, I., Stoyanov, S., O'Droma, M., Popchev, I.: An InfoStation-Based University Campus System Supporting Intelligent Mobile Services. Journal of Computers 2(3), 21–33 (2007)
8. Moszkowski, B.: Executing Temporal Logic Programs. De Montford University, Cambridge (1985)
9. Moszkowski, B., Manna, Z.: Reasoning in interval temporal logic. In: Proceedings of the ACM/NSF/ONR Workshop on Logic of Programs, pp. 371–383 (1984)
10. Pascoe, J.: Adding generic contextual capabilities to wearable computers. In: 2nd International Symposium on Wearable Computers, pp. 92–99 (1998)
11. Stoyanov, S., Ganchev, I., Popchev, I., O'Droma, M.: An Approach for the Development of InfoStation-Based eLearning Architecture. Compt. Rend. Acad. Bulg. Sci., 62(9), 1189–1198 (2008)
12. Stoyanov, S., Ganchev, I., Popchev, I., O'Droma, M., Venkov, R.: DeLC -Distributed eLearning Center. In: 1st Balkan Conference in Informatics, Thessaloniki, Greece, pp. 327–336 (2003) ISBN: 960-287-045-1
13. Stoyanov, S., Popchev, I., Doychev, E., Mitev, D., Valkanov, V., Stoyanova-Doycheva, A., Valkanov, V., Mitev, I.: Educational portal. Cybernetics and Information Technologies (CIT) 10(3), 49–69 (2010)

14. Stoyanov, S., Popchev, I., Ganchev, S., O'droma, M.: From CBT to e-Learning. Information Technologies and Control 4, 2–10 (2005)
15. Stoyanov, S., Stoyanova-Doycheva, A., Popchev, I., Sandalski, M.: ReLE - A refactoring Supporting Tool. Compt. Rend. Acad. Bulg. Sci. 64(7), 1017–1026 (2011)
16. Stoyanov, S., Ganchev, I., Popchev, I., Dimitrov, I.: Request Globalization in an InfoStation Network. Compt. Rend. Bulg. Acad. Sci. 63(6), 901–908 (2010)
17. Stoyanov, S., Ganchev, I., O'Droma, M., Zedan, H., Meere, D., Valkanova, V.: Semantic Multi-Agent mLearning System. In: Kone, M.T., Orgun, M.A., Elci, A. (eds.) Semantic Agent Systems. SCI, vol. 344, pp. 243–272. Springer, Heidelberg (2011)
18. Berners Lee, T., Handler, J., Lassila, O.: The Semantic Web. Scientific American 284, 34–43 (2001)
19. Вълканов, В.: Контекстно-ориентирано управление на електронни услуги. София, България: Академично издателство "Проф. Марин Дринов" (2013) ISBN 978-954-322-701-3

Part X
Metaheuristics and Applications

A Neutral Mutation Operator in Grammatical Evolution

Christian Oesch and Dietmar Maringer

Faculty of Business and Economics,
University of Basel, Switzerland

Abstract. In this paper we propose a Neutral Mutation Operator (NMO) for Grammatical Evolution (GE). This novel operator is inspired by GE's ability to create genetic diversity without causing changes in the phenotype. Neutral mutation happens naturally in the algorithm; however, forcing such changes increases success rates in symbolic regression problems profoundly with very low additional CPU and memory cost. By exploiting the genotype-phenotype mapping, this additional mutation operator allows the algorithm to explore the search space more efficiently by keeping constant genetic diversity in the population which increases the mutation potential. The NMO can be applied in combination with any other genetic operator or even different search algorithms (e.g. Differential Evolution or Particle Swarm Optimization) for GE and works especially well in small populations and larger problems.

Keywords: Grammatical Evolution, Neutral Evolution, Genetic Operator, Genotype-Phenotype Mapping.

1 Introduction

Grammatical Evolution (GE) is a form of Evolutionary Computation to perform symbolic regressions. It is inspired by the biological process of deriving proteins from DNA. By using a problem specific Backus-Naur Form grammar, phenotypic expressions are represented on a numerical string. This genotype-phenotype mapping allows the inclusion of domain knowledge and permits the use of a variety of search algorithms. GE has been applied in many domains such as automatic music composition [1], biochemical network models [4], and discovering market index trading rules [5].

As with all such population-based approaches, the algorithms success will depend on its ability to balance exploration of the global search space and exploitation of locality. When a population reaches a local optimum in the fitness landscape, diversity drops and the population enters evolutionary stasis. This will decrease the chances of finding the global optimum. Furthermore, over the course of generations, selection pressure decreases genetic diversity, which leads to a decline in the rate of discovery of new solutions. Methods that increase diversity without disruptive effects can have a powerful influence on the final outcome of the experiment. Neutral Mutations are a possible solution to this problem.

P. Angelov et al. (eds.), *Intelligent Systems'2014*,
Advances in Intelligent Systems and Computing 322, DOI: 10.1007/978-3-319-11313-5_39

In biology, neutral mutations are random changes in the genome sequence which do not affect the phenotype of an individual [2]. This means the changes are momentarily neither beneficial nor destructive for the individual's fitness. But, the induced genetic diversity increases the exploration potential of the population. Through the distinctive genotype-phenotype mapping in GE, neutral mutations can arise naturally. However, maximizing the frequency of occurrence will increase the performance of the method.

We propose a novel Neutral Mutation Operator (NMO) to achieve constant genetic diversity over the course of generations. Our experiments show that success rates in symbolic regression problems can increase up to 250% with low additional CPU and memory cost. The NMO can be used in combination with any other genetic operator. Additionally, the principle can even be included in other search methods for Grammatical Evolution such as Differential Evolution or Particle Swarm Optimization.

The next section gives a short introduction to GE and its genotype-phenotype mapping and explains the grammar used in this study. Section 3 describes the proposed NMO and section 4 presents the results of the experiments which compare the performance of GE with and without the NMO. Section 5 concludes this study and sets the stage for further research.

2 Grammatical Evolution

2.1 Background

Grammatical Evolution as introduced by [7], is a grammar-based form of Genetic Programming. It uses a combination of ideas from molecular biology, natural selection and grammar representations to evolve symbolic solutions to a predefined problem. This allows for the automatic creation and execution of computer programs without human intervention. GE employs a linear string of numerical values which act as genes on a chromosome, called the genome. By following a Backus-Naur-Form (BNF) grammar definition a mapping process derives a set of rules from the genes on the genome creating an executable program, the so called phenotype. The grammar ensures closure for the rule set as defined by [3]. The requirement of closure demands that for every genetic operation such as crossover or mutation a valid function set is defined. In plain words this means that all the possible solutions of the grammar are executable.

2.2 The Natural Analogy

GE is strongly inspired by the natural system to generate proteins from genetic code [6]. In natural organisms the genetic instructions are mostly encoded in the DNA (deoxyribonucleic acid). The DNA contains a sequence of nucleotides which are made up of four different acid pairs. These nucleotides encode the information to construct amino acids. In the process to develop the proteins from the DNA it is first transcribed into RNA. The RNA is very similar to the

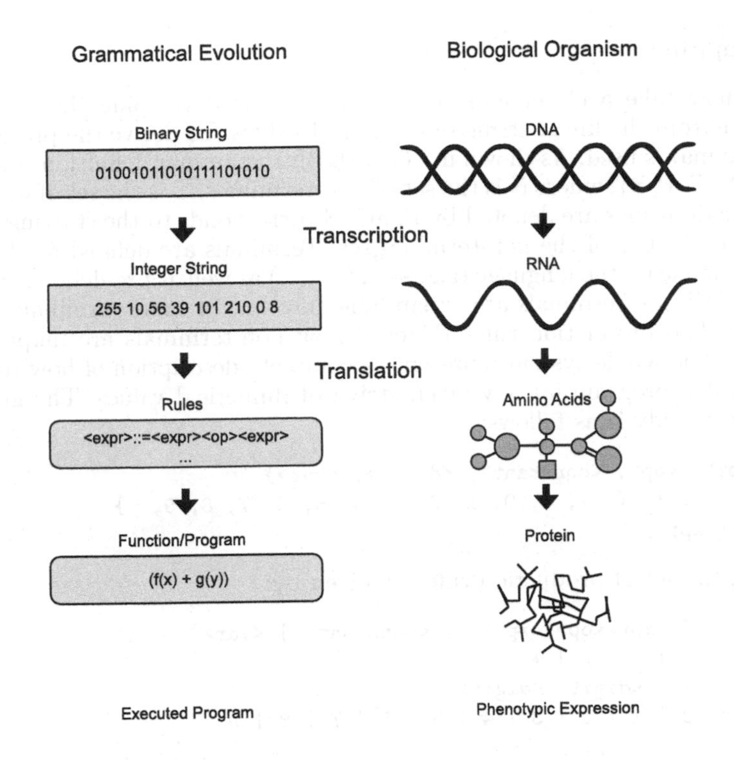

Fig. 1. A simple comparison of the biological system and the GE algorithm

DNA but acts as an intermediary in the construction of the proteins. To build an amino acid, three nucleotides, called a codon, are interpreted together. When the amino acids are put together, they form a protein. This protein drives together with other factors (such as the environment) a phenotypic effect [6]. This effect can be anything from the height of an individual to it's eye color.

In Figure 1 a simplified graphical comparison of the process of constructing proteins in organic cells and the process used by GE is shown. The DNA of an organism is represented by the binary string making up the numerical values in GE. The transcription can also be identified in both systems: the binary representation of GE is rendered into an integer string and the DNA string of the organism is translated into the RNA molecule. The third step derives the different rules for the algorithm in GE and the amino acids are built in the biological organism. The last step produces the final program in the algorithm and the protein in the natural process.

This decoupling of the genotype from the phenotype makes it possible to apply any search method to the linear string representation without having to control for the underlying mapping system. For example an evolutionary algorithm can be used to create new solutions by mutation and crossover of the linear numerical string.

2.3 Mapping

We will now take a closer look at the process for developing the executable phenotype from the linear string of numerical values. To derive the phenotype a BNF grammar is used. As shown in [6] each BNF grammar definition consists of the set {N, T, P, S}, where N is the set of non-terminals, T is the set of terminals, the production rules are denoted by P, and S corresponds to the starting symbol which is a member of the non-terminal set. Terminals are defined as the items which can make up the language (e.g. +, -, *, /, ...) as well as pre-defined variables (x, y, ...). All non-terminals are expandable into one or more terminals or non-terminals. The production rules P detail how non-terminals are mapped into terminals. The whole system represents a complete description of how to create an executable program for any given string of numerical values. The grammar used in this study is as follows:

```
N = {<expr>, <op>, <constant>, <digit>, <var>}
T = {+, -, /, *, (, ), x, 0, 1, 2, 3, 4, 5, 6, 7, 8, 9, .}
S =      {<expr>}.
```

And P, the set of production rules, is given by:

```
<expr>::=     (<expr><op><expr>) | <constant> | <var>
<op>::=       + | - | / | *
<constant>::=    <digit>.<digit>
<digit>::=   0 | 1 | 2 | 3 | 4 | 5 | 6 | 7 | 8 | 9
<var>::=     x
```

To ensure that no illegal mathematical expressions occur the division operator / is defined to be fail-save and will return the numerical value 0 if a division by 0 occurs. These production rules can now be used to develop a phenotypic expression from the genome. To do this with the grammar as defined above we start with the starting symbol <expr> which gives us 3 options to choose from. Either the <expr> is expanded into <expr><op><expr>, <constant> or directly into a variable <var>. To decide which non-terminal to choose a modulo operation is performed on the first codon of the genome:

$$o = c\%n, \tag{1}$$

where o is the numerical index of the option to choose, c is the value of the current codon, n is the number of available choices, and % is the modulo operator which returns the remainder of an integer division. To further expand the phenotype the process is repeated for each non-terminal by using the next codon on the string to decide between the options. If the process runs out of codons to choose from it is common to either expand the genome or to wrap back to the first value on the string and continue from there. This is called wrapping.

This describes the whole process of mapping a genotype to a phenotype. One can now easily apply a variety of search mechanisms to find new solutions to a given problem. In the next section we will describe the novelty mutation operator which can be used in combination with all of the search mechanisms applicable to GE.

3 The Neutral Mutation Operator

With standard mutation and crossover the population converges quickly to a homogenic gene pool because it is very hard to find new solutions which will survive the high fitness threshold in late generations. Neutral Mutation allows the genes to be changed without any changes in the fitness of the individuals.

Standard Point Mutation lets the algorithm draw a random number from the whole search space Ω and is done to just one single codon on the gene. This is done to explore the search space and heighten diversity while keeping disruptive changes at a minimum. In contrast to standard point mutation, the Neutral Mutation Operator acts on every codon of an individual in every generation. With the NMO we can change each codon on the gene since we restrict the possible mutations to the sub-set $\Omega_r \subseteq \Omega$ of the original search space which does not cause any phenotypic changes. To achieve Neutral Mutation, whenever the algorithm conducts the genotype-phenotype mapping the number m of rules a codon chooses from is saved on a second integer string of the same length as the codon. Without an upper or lower limit Ω_r, the space of Neutral Mutations, can be defined as follows:

$$\Omega_r := c + \rho * m, \tag{2}$$

where c is the numerical value of the codon and ρ is any integer number. In practice, one usually specifies a maximum and minimum value for the codons to take. To perform the actual Neutral Mutation the operator randomly draws a number for each codon c_i by

$$c_i = \text{floor}(u * m/n_i) * n_i + r_i, \tag{3}$$

where u is a uniformly distributed random variable between 0 and 1, m is the maximum value a codon can take, n_i again is the number of choices the codon choses from, and r is the remainder of the modulo operation i.e. the choice index of the current solution. This mutation will not change the phenotype of the individual in the current generation, but rather creates the possibility for future large changes in the solution when standard mutation occurs.

Figure 2 shows an example of this process. On the very top the original genome can be seen. In this snapshot the mapping is currently at the first codon containing the value 24 and the grammar implies two options: choosing an operation or a single operator. The number of options, two, is then saved to a second string of integers. The second codon decides between four different options $(+, -, *, /)$ and again the number of options is saved to the second string. One can see that there is a range of possible values for every codon which do not change the current phenotype, the sub-space Ω_r. We can use this information to change the current genome and thus introduce additional diversity. A pitfall of the method is that wrapping should not occur since the mutations would be highly disturbing if a codon encodes decisions with different number of choices. However, this is easily avoided by variable or larger genome lengths.

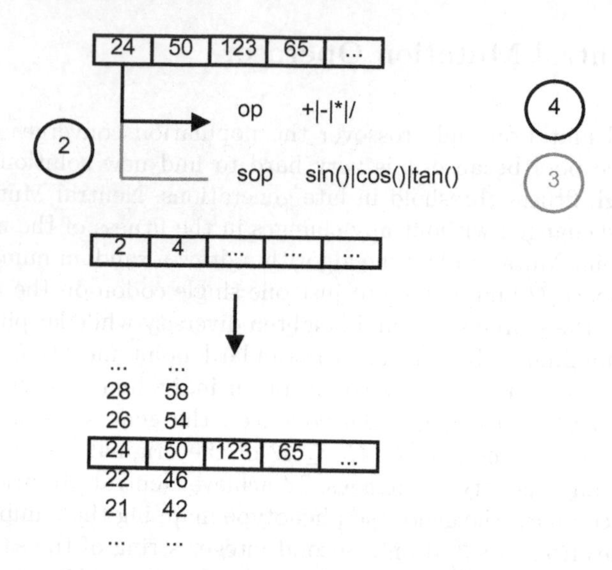

Fig. 2. An example of the process to derive the restricted search space of Neutral Mutation. The top shows the genome and the different choices to choose from. The number of choices is then saved in a second integer string. The restricted search space Ω_r then contains numbers which are linear combinations of the original codons with added or subtracted multiples of the number of choices.

4 Experimental Study

4.1 Setup

To compare the standard evolutionary algorithm and the extended version including the Neutral Mutation Operator we compare the performances on symbolic regression problems for different population sizes. The grammar is as defined in section 2. It is notable that while the target functions include power functions the grammar only includes a multiplication operator to get to the same result. The fitness measure is the sum of the squared deviations of 100 fitness cases from $x = 1, ..., x = 100$. The target functions are polynomials of different degrees including one constant:

1. $x + 2.5x^2 + x^3 + x^4$,
2. $x + 2.5x^2 + x^3 + x^4 + x^5$,
3. $x + 2.5x^2 + x^3 + x^4 + x^5 + x^6$,
4. $x + 2.5x^2 + x^3 + x^4 + x^5 + x^6 + x^7$.

Table 1 lists the parameters of the experiments. Initialization was done through ramped half-and-half and 1000 runs were performed for each setting. We chose to look at the results of the NMO for different population sizes: 250, 500 and 1000. The evolutionary standard parameters are all close to what has been used in previous literature. To ensure no wrapping events occur during the simulations the genome length has been set to 512.

Table 1. Settings

Parameter	Value
Number of runs	1000
Generations	200
Population size	250/500/1000
Crossover probability	0.8
Mutation probability	0.065
Truncation selection	0.6
Initialization	Ramped Half-and-Half
Initial max. depth	8
Max. depth	15
Genome length	512
Successful run	Fitness value < 0.1

4.2 Performance

Table 2 reports the success rates after 200 generations. The increase in success rates vary from 23.5% for the simplest problem with a mid-sized population to 270.0% with a small population and the hardest problem. Since the NMO can be applied in combination with any other evolutionary operators further improvements can be expected in combination with other advanced evolutionary operators.

Table 2. Success rates

		Problem			
Pop. Size		**1**	**2**	**3**	**4**
	NMO	35.1%	16.3%	6.6%	3.7%
250	no NMO	26.8%	11.6%	3.2%	1.0%
	Increase	**31.0%**	**40.5%**	**106.3%**	**270.0%**
	NMO	50.0%	31.4%	17.1%	9.0%
500	no NMO	40.5%	18.3%	9.6%	5.2%
	Increase	**23.5%**	**71.6%**	**78.1%**	**73.1%**
	NMO	71.7%	44.6%	32.1%	20.1%
1000	no NMO	53.5%	32.2%	21.0%	12.3%
	Increase	**34.0%**	**38.5%**	**52.9%**	**63.4%**

Figure 3 shows the performance comparison of GE with and without the proposed NMO. The graphs show the percentage of successful runs after a given amount of generations. The columns differ in population size and the rows differ in the problem size. With the addition of the NMO the standard approach is dominated in nearly all scenarios over the full generation span and dominates the standard approach in all scenarios after around 100 generations. Especially

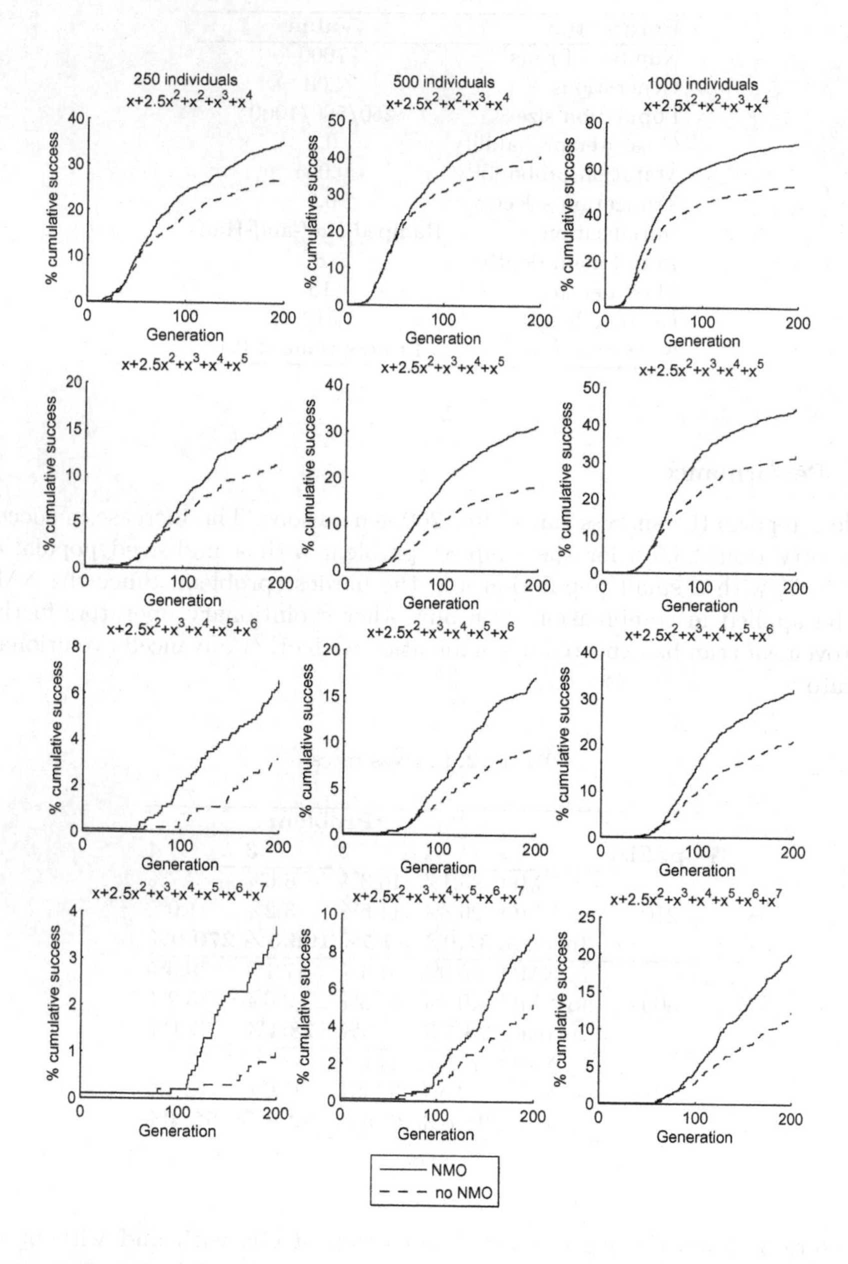

Fig. 3. Cumulative success for symbolic regressions with and without the NMO and different population sizes

in later generations the advantage of using the NMO to increase genetic diversity becomes obvious.

Figure 4 shows parameter variations for the mutation probability for problem 2 and a population size of 500. The statistics are shown for 200 runs. To make sure that the NMO does not only perform better because of an implicit higher mutation rate, we plotted it against different mutation probabilities. Increasing the mutation probability to 10% does indeed bring the results closer together, but they are still dominated by the setup where the mutation rate is 6.5% and the NMO is included. We think that this can be attributed to the fact that the higher mutation rate becomes too destructive and consumes the gains from the additional diversity. The NMO on the other hand only increases diversity without the damaging effects.

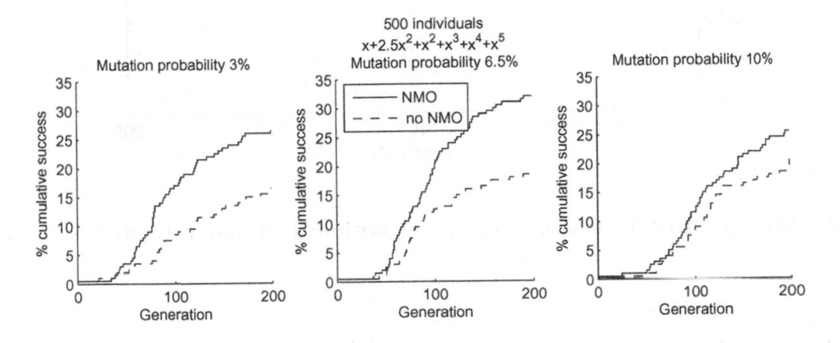

Fig. 4. Cumulative success for a simple symbolic regression with and without the NMO and different mutation probabilities

4.3 Diversity

Figure 5 shows a simple genetic diversity measure for experiments with and without the NMO over the generations. The diversity is measured by

$$d = 1/1 + \sum_i \sum_j \sum_k^{i \neq k} \left\{ \begin{matrix} \frac{1}{a} & \text{if} \quad c_{i,j} = c_{k,j} \\ 0 & \text{otherwise} \end{matrix} \right\}, \quad (4)$$

where again $c_{i,j}$ is the j-th codon of individual i, and $a = p^2 * q$ is the product of the squared population size p and the length of the genome q. This measure is 1 if all individuals differ in every codon and 0.5 if all individuals are genetically identical. As expected, the unconditional genetic diversity is nearly constant over time when the Neutral Mutation Operator is applied. This is one of the interesting properties of the proposed operator. It exploits the fact that genetic and phenotypic diversity don't need to behave in the same way. While the phenotypes converge and lead the search into the right direction i.e. exploit the search space, the genotypes keep constant diversity and allow the algorithm

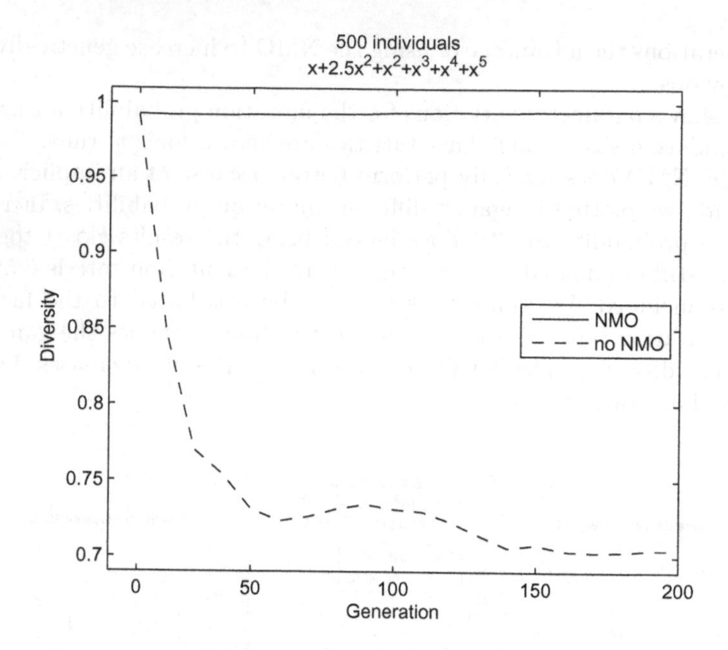

Fig. 5. Average diversity measure for the algorithm with and without NMO for 20 runs

to explore further away from current solutions by increasing the mutation potential. Especially in late generations of smaller populations where genetic diversity drops after the first generations, the advantage of the constant genetic diversity becomes apparent.

5 Conclusion

In this paper we propose a novel Neutral Mutation Operator which adds large performance gains to standard Grammatical Evolution without much additional computation costs. By restricting the search space to a sub-set which does not affect the phenotype we can constantly perform Neutral Mutations on every codon and keep the genetic diversity constant over the course of generations. These mutations do not create immediate changes in the phenotype but rather increase the potential for standard mutations and crossover in later generations and so exploit the genotype-phenotype decoupling inherent to GE. This highlights another advantage of algorithms which differentiate between genetic and phenotypic expressions. The proposed operator is tested on symbolic regressions with increasing difficulty. The experimental study shows increases in performance between 23-250%. This performance increase is mainly attributable to the constant genetic diversity and seems to be stronger in harder problems with smaller population sizes.

While this is only a small study with a very limited test-scope, the operator seems to be highly promising and further studies with different problem sets and in combination with other algorithm configurations should be performed. It should be also very interesting to compare different structures of the grammar since the performance can be expected to be dependent on the diversity of the number of choices in the production rules.

References

1. de la Puente, A.O., Alfonso, R.S., Moreno, M.A.: Automatic composition of music by means of grammatical evolution. ACM SIGAPL APL Quote Quad 32(4), 148–155 (2002)
2. Kimura, M.: The neutral theory of molecular evolution. Cambridge University Press (1984)
3. Koza, J.R.: Genetic Programming: on the programming of computers by means of natural selection. A Bradford Book. The MIT Press (1992)
4. Moore, J.H., Hahn, L.W.: Petri net modeling of high-order genetic systems using grammatical evolution. BioSystems 72(1), 177–186 (2003)
5. O'Neill, M., Brabazon, A., Ryan, C., Collins, J.J.: Evolving Market Index Trading Rules Using Grammatical Evolution. In: Boers, E.J.W., Gottlieb, J., Lanzi, P.L., Smith, R.E., Cagnoni, S., Hart, E., Raidl, G.R., Tijink, H. (eds.) EvoWorkshop 2001. LNCS, vol. 2037, pp. 343–352. Springer, Heidelberg (2001)
6. O'Neill, M., Ryan, C.: Grammatical evolution. IEEE Transactions on Evolutionary Computation 5(4), 349–358 (2001)
7. Ryan, C., Collins, J.J., Neill, M.O.: Grammatical evolution: Evolving programs for an arbitrary language. In: Banzhaf, W., Poli, R., Schoenauer, M., Fogarty, T.C. (eds.) EuroGP 1998. LNCS, vol. 1391, pp. 83–95. Springer, Heidelberg (1998)

Study of Flower Pollination Algorithm for Continuous Optimization

Szymon Łukasik[1,2] and Piotr A. Kowalski[1,2]

[1] Systems Research Institute, Polish Academy of Sciences,
ul. Newelska 6, 01-447 Warsaw, Poland
[2] Department of Automatic Control and Information Technology,
Cracow University of Technology,
ul. Warszawska 24, 31-155 Cracow, Poland
{slukasik,pakowal}@ibspan.waw.pl

Abstract. Modern optimization has in its disposal an immense variety of heuristic algorithms which can effectively deal with both continuous and combinatorial optimization problems. Recent years brought in this area fast development of unconventional methods inspired by phenomena found in nature. Flower Pollination Algorithm based on pollination mechanisms of flowering plants constitutes an example of such technique. The paper presents first a detailed description of this algorithm. Then results of experimental study of its properties for selected benchmark continuous optimization problems are given. Finally, the performance the algorithm is discussed, predominantly in comparison with the well-known Particle Swarm Optimization Algorithm.

Keywords: flower pollination algorithm, metaheuristics, optimization, nature-inspired algorithms.

1 Introduction

Inspiration from nature is a feature found in many traditional [1,2] or newly developed metaheuristics. Among the examples of mechanisms mimicked by recent unconventional heuristic algorithms one can name, e.g. bioluminescent communication of fireflies [3], social spider's cooperative schemes [4] or swarm behavior of krill herds [5]. This paper studies an approach based on flower pollination, namely Flower Polination Algorithm (FPA) proposed in 2012 by Xin-She Yang [6].

The goal of the paper is to evaluate FPA performance and study its properties in continuous optimization. Up to date the only contributions in this area have been made by creator of the algorithm. They establish the basic form of FPA [6] and describe its multi-objective variant [11]. Here we try to supplement those studies, e.g. by more detailed analysis of parameters and evaluation of algorithm's performance with regards to standard variant of Particle Swarm Optimization (PSO) [12].

The paper is organized as follows. First, it provides a foundation upon the algorithm was constructed along with its formal description. Then, it describes the

P. Angelov et al. (eds.), *Intelligent Systems'2014*,
Advances in Intelligent Systems and Computing 322, DOI: 10.1007/978-3-319-11313-5_40

results of numerical experiments, devoted to exploring possible parameter values. Algorithm's performance in relation to PSO method is also studied therein. Finally, the last part of the contribution contains some concluding remarks on FPA and proposals for its further development.

2 FPA Formulation

Pollination itself as a natural phenomenon constitutes a process of transferring pollen grains from the stamens, the flower parts that produce them, to the ovule-bearing organs or to the ovules (seed precursors) themselves [7]. It is crucial for fertilization and reproduction of flowering plants also known as the angiosperms – the largest and most conspicuous group of modern plants [8]. Two types of pollination can be distinguished with regards to the methods of pollen transfer. First, abiotic pollination that does not involve using other organisms and employs wind, water or gravity as pollination mediators. Second, biotic pollination, which is the dominant one, requires pollinators, i.e. organisms, predominantly insects, that carry or move the pollen grains. Evolutionary, pollination by insects probably occurred in primitive seed plants, with reliance on other means being a relatively recent development [7]. Pollination can be accomplished by cross-pollination or by self-pollination. Self-pollination occurs when pollen from one flower pollinates the same flower or other flowers of the same individual [9]. In contrast to that, cross-pollination happens when pollen is delivered to a flower from a different plant. About 70% of the present day angiosperms are cross-pollinated while the rest are self-pollinated [10]. To sum up, it is worth to mention that nowadays in nature all kind of pollination modes can be found, with some plants being able to effectively employ each and every one of them. Consequently both biotic and cross-pollination as well as abiotic and self-polination strategies were translated into the optimization domain and included in novel unconventional metaheuristic of FPA.

Pollination process constitutes a set of complex mechanisms crucial to the success of plants reproductive strategies. To build an effective algorithm mimicking those mechanisms several simplifications in understanding their fundamental aspects have to be made. Firstly, global, i.e. long-distance pollination should occur for biotic and cross-pollination pollen transfers. At the same time abiotic and self-pollination are to be associated with local reproductive strategies. Plants can employ both pollination schemes, however attractiveness of selected plant should influence the tendency of individual pollinators to exclusively visit it. The strength of this attraction is referred to as flower constancy.

When moving to the optimization domain pollination concept can be understood as follows. A single flower or pollen gamete will constitute a solution of the optimization problem, with the whole flower population being actually used. Their constancy will be understood as solution fitness. Pollen will be transferred in the course of two operations used interchangeably, that is: global and local pollination. The first one employs pollinators to carry pollen to long distances towards individual characterized by higher fitness. Local pollination on the other hand occurs within limited range of individual flower thanks to pollination mediators like wind or water. FPA for continuous optimization built on assumptions presented above will be more strictly formulated in the subsequent part of this Section.

The general goal of continuous optimization is to find x^* which satisfies:

$$f(x^*) = \min_{x \in S} f(x), \tag{1}$$

where $S \subset R^N$, and $f(x)$ constitutes solution's x cost function value. Therefore actual task of the optimizer is to find argument minimizing f.

To solve the optimization problem (1) the population of M individuals will be used. It is represented by a set of N-dimensional vectors – equivalent to individuals' positions – within the iteration k denoted by:

$$x_1(k), x_2(k), ..., x_M(k). \tag{2}$$

The best position found by given individual m prior to iteration k is given by $x_m^*(k)$ with cost/fitness function value $f(x_m^*(k))$. At the same time:

$$x^*(k) = \arg \min_{m=1,\ldots,M} f(x_m(k)), \tag{3}$$

or

$$x^*(k) = \arg \max_{m=1,\ldots,M} f(x_m(k)), \tag{4}$$

corresponds to the best solution found by the algorithm in its k iterations, with $f(x^*(k))$ representing its related cost (3) or fitness (4) function value. The formula used depends on type of optimization task (minimization or maximization of function f).

Flower Pollination Algorithm's formal description using aforementioned notation will be given below (as Algorithm 1). It can be seen that global pollination occurs with probability p defined by so called switch probability. If this phase is omitted local pollination takes place instead. The first one constitutes of pollinator's movement towards best solution $x^*(k)$ found by the algorithm, with s representing N-dimensional step vector following a Lévy distribution :

$$L(s) \sim \frac{\lambda \Gamma(\lambda) \sin(\pi\lambda/2)}{\pi s^{1+\lambda}}, \ (s \gg s_0 > 0), \tag{5}$$

with Γ being the standard gamma function and parameters $\lambda = 1.5, s_0 = 0.1$ as suggested by Yang [11]. Practical method for drawing step sizes s following this distribution by means of Mantegna algorithm are given in [13]. Local pollination includes two randomly selected members of the population and is performed via movement towards them, with randomly selected step size ϵ. Finally, the algorithm is terminated when number of iteration k reaches predetermined limit defined by K.

Next Section of the paper will present results of conducted tests which were aimed at exploring the performance of the Flower Pollination Algorithm in the form being presented here extensively.

Algorithm 1. Flower Pollination Algorithm

```
1: k ← 1 {initialization}
2: f(x*(0)) ← ∞
3: for m = 1 to M do
4:    Generate_Solution(x_m(k))
5: end for
6: {find best}
7: for m = 1 to M do
8:    f(x_m(k)) ← Evaluate_quality(x_m(k))
9:    if f(x_m(k)) < f(x*(k − 1)) then
10:      x*(k) ← x_m(k)
11:   else
12:      x*(k) ← x*(k − 1)
13:   end if
14: end for
15: {main loop}
16: repeat
17:    for m = 1 to M do
18:       if Real_Rand_in_(0, 1) < p then
19:          {Global pollination}
20:          s ← Levy(s_0, γ)
21:          x_trial ← x_m(k) + s(x*(k) − x_m(k))
22:       else
23:          {Local pollination}
24:          ε ← Real_Rand_in_(0, 1)
25:          r, q ← Integer_Rand_in(1, M)
26:          x_trial ← x_m(k) + ε(x_q(k) − x_r(k))
27:       end if
28:       {Check if new solution better}
29:       f(x_trial) ← Evaluate_quality(x_trial)
30:       if f(x_trial) < f(x_m(k)) then
31:          x_m(k) ← x_trial
32:          f(x_m(k)) ← f(x_trial)
33:       end if
34:    end for
35:    {find best and copy population}
36:    for m = 1 to M do
37:       if f(x_m(k)) < f(x*(k − 1)) then
38:          x*(k) ← x_m(k)
39:       else
40:          x*(k) ← x*(k − 1)
41:       end if
42:       f(x(k + 1)) ← f(x_k)
43:       x(k + 1) ← x(k)
44:    end for
45:    f(x*(k + 1)) ← f(x*k)
46:    x*(k + 1) ← x*(k)
47:    stop_condition ← Check_stop_condition()
48:    k ← k + 1
49: until stop_condition = false
50: return f(x*(k)), x*(k), k
```

3 Experimental Results

For computational experiments examining properties of the FPA set of benchmark problems already considered in CEC'13 competition was used [14]. Table 1 presents those functions along with their mathematical expressions, dimensionality and optimum values f^*.

Table 1. Benchmark functions used for experimental studies

f Name	Expression	Feasible bounds	N	f^*
f_1 Sphere	$f_1(x) = \sum_{i=1}^{N} z_i^2 + f_1^*$ $z = x - o$	$[-100, 100]^N$	10	-1400
f_2 Rotated Bent Cigar	$f_2(x) = z_1^2 + 10^6 \sum_{i=2}^{N} z_i^2 + f_2^*$ $z = M_2 T_{asy}^{0.5}(M_1(x - o))$	$[-100, 100]^N$	10	-1000
f_3 Different Powers	$f_3(x) = \sqrt{\sum_{i=1}^{N} \lvert z_i\rvert^{2+4\frac{i-1}{N-1}}} + f_3^*$ $z = x - o$	$[-100, 100]^N$	10	-1000
f_4 Rotated Rastrigin	$f_4(x) = \sum_{i=1}^{N} (z_i^2 - 10\cos(2\pi z_i) + 10) + f_4^*$ $z = M_1 \Lambda^{10} M_2 T_{asy}^{0.2}(T_{osz}(M_1 \frac{5.12(x-o)}{100}))$	$[-100, 100]^N$	10	-300
f_5 Schwefel	$f_5(x) = 418.9829N \sum_{i=1}^{N} g(z_i) + f_5^*$ $z = \Lambda^{10}(\frac{1000(x-o)}{100}) + 4.209687462275036 * 10^2$ $g(z_i) = z_i \sin(\lvert z_i\rvert^{1/2})$	$[-100, 100]^N$	10	-100
f_6 Rotated Katsuura	$f_6(x) = \frac{10}{N^2} \prod_{i=1}^{N} (1 + i \sum_{j=1}^{32} \frac{\lvert 2^j z_i - round(2^j z_i)\rvert}{2^j})^{\frac{10}{N^{1.2}}} - \frac{10}{N^2} + f_6^*$ $z = M_2 \Lambda^{100}(M_1 \frac{5(x-o)}{100})$	$[-100, 100]^N$	10	200

Symbols:
$o = x^*$ the shifted global optimum , which is randomly distributed in $[-80, 80]^N$,
M_1, M_2 - orthogonal (rotation) matrix generated from standard normally distributed entries by Gram-Schmidt orthonormalization.
Λ^α - a diagonal matrix in N dimensions with the ith diagonal element as $\lambda_{ii} = \alpha^{\frac{i-1}{2(N-1)}}$ for $i = 1, 2, ..., N$.
T_{asy}^{β} - if $x_i > 0, x_i = x_i^{1+\beta\frac{i-1}{N-1}\sqrt{x_i}}$ for $i = 1, 2, ..., N$.
T_{osz} - for $x_i = sign(x_i)\exp(\hat{x}_i + 0.049(sin(c_1\hat{x}_i) + sin(c_2\hat{x}_i)))$ for $i = 1, 2, ..., N$.
where:
$\hat{x}_i = log(\lvert x_i\rvert)$ for $x_i \neq 0$, otherwise $\hat{x}_i = 0$,
$c_1 = 10$ if $x_i > 0$, otherwise $c_1 = 5.5$,
$c_2 = 7.9$ if $x_i > 0$, otherwise $c_2 = 3.1$.

The experiments were conducted for fixed number of iterations $K = 10000$ $(1000 * N)$, and 30 trials for each function. As a performance measure mean optimization error $\overline{E}(K)$ was used (with $E(K) = \lvert f(x^*(K)) - f^*\rvert$) along with its standard deviation $\sigma_{E(K)}$. We have also studied mean execution time $\bar{t}$ in seconds needed for one algorithm's run.

First the influence of population size M was under investigation. We tested FPA with $M = \{10, 20, 30, ..., 100\}$ and $p = 0.8$ as suggested in [6]. Obviously, as the algorithm is characterized by $O(M)$ complexity, it is recommended to limit population size if satisfactory outcomes are achieved for smaller M. It is the case for unimodal functions f_1, f_2 and f_3 where acceptable results were obtained for 20 individuals. For multimodal functions it is recommended to further increase population size, e.g. to 80 as significant improvement in optimization performance is observed. It is illustrated by mean optimization error shown for f_5 and f_6 on Fig. 1.

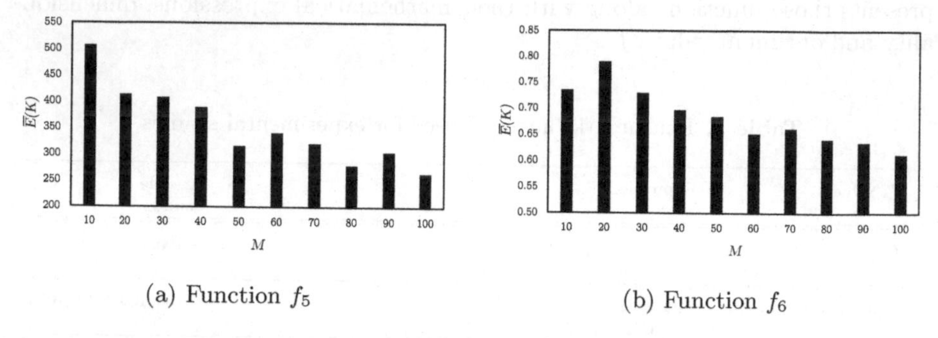

(a) Function f_5

(b) Function f_6

Fig. 1. Mean error values for varying population size

In the subsequent phase of numerical experiments we used $M = 20$ for f_1, f_2, f_3 and f_4 and $M = 80$ for f_5 and f_6. During these tests suggested value for switch probability p were under investigation. Most significant findings were summarized in Table 2. It presents values of p which ensure minimum error value in single run. It was established that for unimodal functions FPA provided optimum values regardless of p being used. It was noted however that higher value of switch probability is recommended, as computational cost of the algorithm is lower in this case. For multimodal problems best results can be achieved for $p \in [0.5, 0.8]$.

Table 2. Switch probability values ensuring $min(E(K))$

Function	Minimum error for
f_1	p={0.1,...,0.9}
f_2	p={0.3,...,0.8}
f_3	p={0.1,...,0.9}
f_4	p=0.8
f_5	p=0.5
f_6	p=0.7

Finally, experimental evaluation of algorithm's performance in reference to the standard Particle Swarm Optimization was performed. PSO variant with constriction factor χ was used [2]. For parameter values $\chi = 0.72984$ and $c_1 = c_2 = 2.05$ were chosen [15]. Experiments were executed for the same population sizes and switch probabilities established in previous tests. Mean error, its deviation and minimum value along with execution time were under investigation. Summary of obtained results is provided in Table 3.

Table 3. Performance comparison of Flower Pollination Algorithm and Particle Swarm Optimization for selected benchmark functions

Function	Algorithm	$\overline{E}(K)$	$\sigma_{E(K)}$	$\min E(K)$	$t[s]$
f_1	FPA	0.00	0.00	0.00	3.69
	PSO	0.00	0.00	0.00	2.88
f_2	FPA	1.51	2.36	0.00	3.78
	PSO	3288692	7359309	0.16	3.03
f_3	FPA	0.00	0.00	0.00	3.72
	PSO	0.00	0.00	0.00	2.91
f_4	FPA	12.55	6.58	2.98	3.84
	PSO	13.69	5.27	3.98	3.19
f_5	FPA	227.78	70.78	39.02	18.78
	PSO	130.30	80.71	0.31	7.96
f_6	FPA	0.59	0.16	0.16	18.02
	PSO	0.42	0.13	0.19	11.02

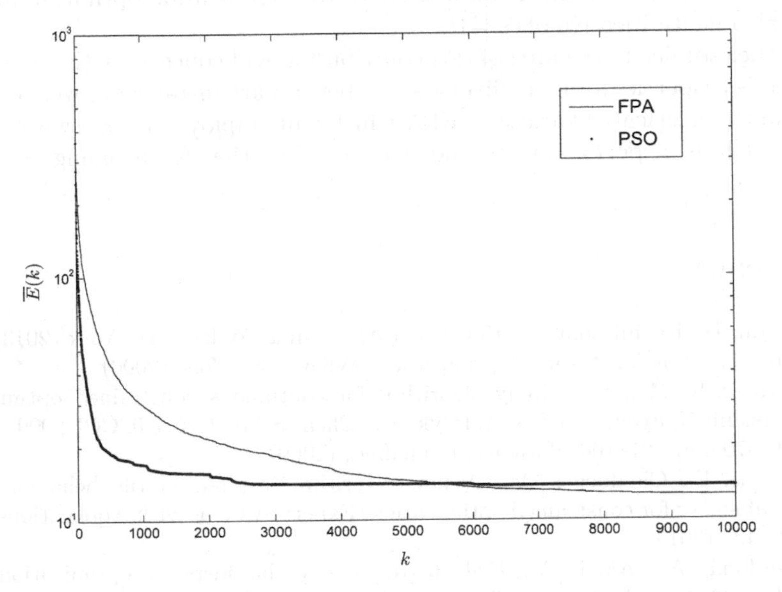

Fig. 2. Average error values during 10000 iterations for PSO and FPA (function f_4, logarithmic scale)

It can be seen that FPA and PSO exhibit similar performance for selected benchmark functions. The algorithm studied in this paper performs better for f_2 and f_4. Choice between PSO and FPA is not so obvious for f_1, f_3 and f_6 where indistinguishable performance differences were observed. PSO in turn would be an apparent choice in case of f_5 function.

To illustrate the dynamics of both algorithms average error values obtained for f_4 function during 10000 iterations were enclosed on Figure 2. It can be seen that in this case PSO converges faster in the first phase of the optimization process. It stucks in the local minimum however and is outperformed by FPA in the latter part of the process.

4 Conclusion

The paper examined novel nature-inspired metaheuristics of Flower Pollination Algorithm. Besides its detailed description, experimental evaluation of its sensitivity to parameter values and performance with regards to the Particle Swarm Optimization were under consideration.

We found that the initial variant of the algorithm – as introduced by Yang – proved to be quite competitive even in comparison to the well-studied PSO. Mechanisms of local and global exploration embedded within the algorithm are complementary and allow the algorithm to effectively search the domain of optimization problem at hand. What is more FPA is characterized by small number of parameters which make it an attractive tool for real-life optimization problems, also multiobjective ones [11].

Further studies in the area of this contribution will concern, both studying algorithm's properties for more diverse set of benchmark instances as well as assessing more sophisticated variant of FPA which will employ variable switch probability. Those aspects are to be covered by the forthcoming follow-up paper [16].

References

1. Simon, D.: Evolutionary Optimization Algorithms. Wiley, New York (2013)
2. Clerc, M.: Particle Swarm Optimization. Wiley, New York (2006)
3. Łukasik, S., Żak, S.: Firefly algorithm for continuous constrained optimization tasks. In: Nguyen, N.T., Kowalczyk, R., Chen, S.-M. (eds.) ICCCI 2009. LNCS, vol. 5796, pp. 97–106. Springer, Heidelberg (2009)
4. Cuevas, E., Cienfuegos, M.: A new algorithm inspired in the behavior of the social-spider for constrained optimization. Expert Systems with Applications 41(2), 412–425 (2014)
5. Gandomi, A., Alavi, A.: Krill herd: A new bio-inspired optimization algorithm. Communications in Nonlinear Science and Numerical Simulation 17(12), 4831–4845 (2012)
6. Yang, X.-S.: Flower pollination algorithm for global optimization. In: Durand-Lose, J., Jonoska, N. (eds.) UCNC 2012. LNCS, vol. 7445, pp. 240–249. Springer, Heidelberg (2012)
7. Encyclopædia Britannica Online: Pollination (2014) (Online; accessed June 27, 2014)
8. Takhtajan, A.: Flowering Plants. Springer, New York (2009)
9. Cronk, J., Fennessy, M., Siobhan, M.: Wetland plants: biology and ecology. CRC Press/Lewis Publishers, Boca Raton (2001)

10. Maiti, R., Satya, P., Rajkumar, D., Ramaswam, A.: Crop plant anatomy. CABI, Wallingford (2012)
11. Yang, X.S., Karamanoglu, M., He, X.: Multi-objective flower algorithm for optimization. Procedia Computer Science 18, 861–868 (2013)
12. Kennedy, J., Eberhart, R.: Particle Swarm Optimization. In: Proceedings of IEEE International Conference on Neural Networks, vol. IV, pp. 1942–1948 (1995)
13. Yang, X.: Nature-Inspired Optimization Algorithms. Elsevier, London (2014)
14. Liang, J., Qu, B., Suganthan, P., Hernandez-Diaz, A.: Problem Definitions and Evaluation Criteria for the CEC 2013 Special Session and Competition on Real-Parameter Optimization. Technical Report 201212, Computational Intelligence Laboratory, Zhengzhou University, Zhengzhou China and Technical Report, Nanyang Technological University, Singapore (2013)
15. Bratton, D., Kennedy, J.: Defining a standard for particle swarm optimization. In: Swarm Intelligence Symposium, SIS 2007, pp. 120–127. IEEE (2007)
16. Łukasik, S., Kowalski, P.: Flower Pollination Algorithm – facts, conjectures and improvements (in preparation, 2014)

Search Space Reduction in the Combinatorial Multi-agent Genetic Algorithms

Łukasz Chomątek and Danuta Zakrzewska

Lodz University of Technology,
Institute of Information Technology,
Lodz, Poland
{lukasz.chomatek,danuta.zakrzewska}@p.lodz.pl

Abstract. Genetic algorithms are widely used for solving the optimization problems. However combinatorial problems are usually hard to solve using genetic algorithms as the chromosomes are very long, what causes the increase of the computational complexity of such solutions. In the previous research, authors proposed the method of efficient search space reduction, which was applied to lower the complexity of random searches. In the current work, a modification of the multi-agent genetic algorithm and its application for solving the single source shortest path problem are proposed. Presented approach is compared to the former implementation of the genetic algorithm. Investigations showed that the genetic diversity of the population in multi-agent genetic algorithm can be successfully measured, which allows to identify the premature convergence.

Keywords: genetic algorithms, multi-agent systems, search space reduction, shortest path problem.

1 Introduction

Nowadays proper route planning plays a big role in both commercial and private transport due to the increasing number of vehicles on the roads. Finding good itineraries became simple because of the growing availability of the both: online and standalone services offering such a feature. Modern algorithms are usually based on the hierarchical division of the road network and some kind of the graph pre-processing. Routing based on the hierarchical division of the road network is based on the actual behaviour of drivers who are supposed to plan a long trip. At first they try to reach a big road and follow it to get near to their destination. When they are close enough, they choose minor roads to reach the planned place.

In all the countries a certain division of the roads into classes is provided by the government (ie. in Poland there are 7 classes of roads). Such a division is not sufficient for the algorithms used to find optimal route between points on the map, as these algorithms usually require all vertices on the highest hierarchy level to belong to the one connected component. However in many countries highway network consists of parts which are connected by smaller roads. To mitigate this problem an artificial hierarchical division of the road network should be prepared.

P. Angelov et al. (eds.), *Intelligent Systems'2014*,
Advances in Intelligent Systems and Computing 322, DOI: 10.1007/978-3-319-11313-5_41

Algorithms for finding the optimal path in the road network graph which are based on the artificial hierarchical division of the road network consist of two phases : pre-processing and querying phase. The assumptions are that the first phase can take a long but acceptable amount of time, while queries should be realized faster than in the classical algorithms. When the hierarchies are prepared, any of the known algorithms for finding the optimal path can be applied to find actual route. For the reduced number of the road segments on the higher hierarchy levels, the querying algorithm performs better as the search space is smaller. In this work authors discuss the possibility of the application of genetic algorithms to find optimal route in the limited search space. The research has shown that such algorithm performs better than known solutions in this area.

The paper is organized as follows. In the next section related work concerning application of multi-agent technology in optimization problems as well as search space reduction in the shortest path problem is presented. Then proposed methodology and the algorithm are described. The following section is devoted to experiments and their results, taking into account converegence of the algorithm and its performance. Finally some conclusions and remarks on future work are depicted.

2 Related Work

2.1 Multi-agent Systems in Optimization Problems

Multi-agent approach to solve optimization problems has been the matter of interest of researchers for several years. As the main advantages of agent technology Shi and Wang mentioned its strong reliability and high efficiency, features which arise due to its good autonomy, collaboration, self-organizing and self learning abilities [1]. Barbati et al. stated that agent based models and classical heuristics have complementary characteristics. As one of the integration approaches they mentioned embedding optimizing behaviours in physical agents, by translating search algorithms into agents behaviour [2].

Multi-agent technology was very often integrated with evolutionary algorithms. Liu et al. [3] stated that both agents and evolutionary algorithms have a high potential in solving complex and ill-defined problems. They introduced multi-agent evolutionary algorithm for constraint satisfaction problem solving, where agents live in a lattice environment, in which they can sense and act. They concluded that the model of the agent lattice is closer to the real evolutionary model in nature than the population in traditional evolutionary algorithms [3]. Shi and Wang introduced a multi-objective genetic algorithm based on self-adaptive agent, with evolution parameters adjusted adaptively in the evolutionary process [1].

Zhong et al. showed the good performance of multi-agent genetic algorithm for solving global numerical optimization problems [4]. They introduced a multi-agent genetic algorithm MAGA, similar to cellular genetic algorithms [5], in which agents live in a lattice environment and represent candidates for solutions of optimization problems. In spite of other cellular genetic algorithms in MAGA each individual is considered as an agent, which has its own purpose and behaviors [4].

Multi-agent technology was used for solving Vehicle Routing Problem for many years [6]. The broad review of the applications of agent based models for solving optimization problems including vehicle routing problem is presented in [2]. Genetic algorithms have been considered as search strategies for VHR solving by some researchers. Baker and Ayechew considered the straightforward genetic algorithm and its hybrid, which performs well competitively with tabu search and simulated annealing in terms of solution time and quality [7]. Masum et al., in turn, added some heuristic during crossover or mutation for tuning the system to obtain better results. They concluded that genetic algorithms provide an effective approach to solve problems where an exact method cannot be applied [8].

2.2 Search Space Reduction in the Shortest Path Problem

The methods of search space reduction in the Shortest Path Problem can be divided into three groups:

- pruning of nodes,
- decomposition of the problem,
- hierarchical division of edges.

The pruning methods applied for the graphs, where some coordinates can be added to their nodes, are based on the observations that in the real world examples one doesn't have to analyse the whole graph, but only a certain region containing the origin and the destination node. Such an attempt was made by Karimi [9] and Fu [10], where the search space was limited to a rectangle with specified dimensions.

Nicholson [11] proposed a method with two independent searches for each query. One of them was a classical search from source to destination node; the second one was carried from destination to source. Direction of graph edges in case of backward search is opposite to the real one. To complete the query, at least one node must be visited by both forward and backward searches.

Algorithms that belong to the third group are inspired by the observation of real drivers behaviour. When one has to take a long trip, he usually tries to reach the major road and get as close as possible to the destination. When he is close to the destination, he chooses smaller roads to reach his goal. When driver is travelling on big roads he is not considering leaving them while he is far from the destination - actually this means the reduction of the search space by pruning the edges. Administrative division of the roads into categories is not sufficient for such algorithms, as the roads on the certain level often are not a connected component. To overcome this problem, several authors [12], [13] proposed their own hierarchical division. This significantly reduced the querying time, but the hierarchy-building process was a time consuming task.

Fig. 1. Example of the hierarchical division

In [15] authors proposed the algorithm that combines all three methods and can be applied for graphs where nodes are associated with some coordinates. At first a road network is divided into sectors. For each sector an artificial hierarchical division is performed to reduce the search space. During the query execution, a bidirectional searches are started between the borders of sectors.

3 Proposed Methodology

In the proposed approach agents that represent the solutions are organized in a grid. The grid has a fixed size during the simulation process. Agents can communicate with their neighbours in case of crossover. Agents that are located at the border of the grid, can also communicate with agents located on its opposite side. In each iteration we gather information about the best individual in the grid, however the best individual is not directly copied to the next iteration.

In general, the algorithm is divided into four steps (Fig. 2). In the first step we construct an initial population of the individuals. In each cell of the grid we locate a single agent. The next steps are repeated several times to reach the specified iteration number. In the second step we calculate fitness of each individual as well as the diversity measures which help us to identify the convergence of the algorithm. After this step, we perform the crossover between neighbouring agents. The last step is a mutation which is performed on each individual, with certain probability. The details of the algorithm are described in the following subsections.

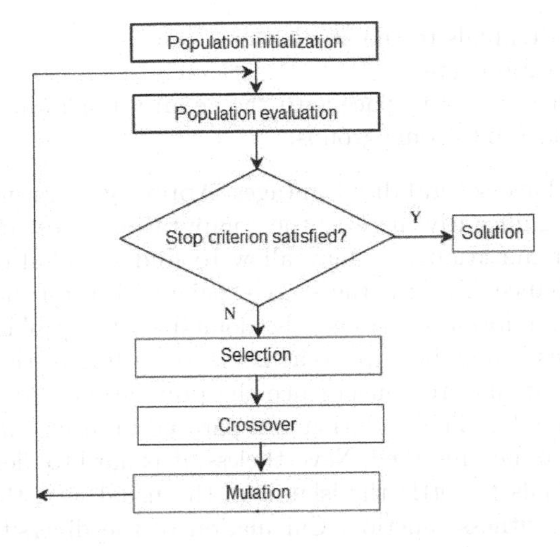

Fig. 2. General schema of the genetic algorithm

3.1 Initial Population

Initial population consists of randomly generated individuals that represent paths between source and destination nodes. The former research revealed that the proposed method of the reduction of the search space can significantly fasten the individual construction process. We do not apply any heuristic rules during the random searches. The only one rule is known from the hierarchical search. According to that rule the search front is not allowed to choose lower hierarchy levels. We use bidirectional search in each sector between source and the destination node. We stop the construction process when in each considered sector at least one node is visited by forward and backward search.

3.2 Fitness Function

Let $G = (V, E)$ be a graph, where V and E represent its nodes and edges, respectively. Let $w > 0$ be a travelling time associated with the edge $e \in E$. Moreover we denote a path between nodes s and t as $p(s,t) = \langle s, v_1, v_2, \dots v_n, t \rangle$, where $v_1, v_2, \dots v_n$ denote nodes. In the path, all the neighbouring nodes are connected by the edge. In our experiments we consider a fitness of each individual as a sum of travelling times of all edges that belong to the path, which is represented by that individual.

3.3 Diversity Measures

One of the biggest problems in application of the genetic algorithms is the premature convergence. When the search space has a local minima, the population may stuck in one of them. To avoid this problem, in the classical genetic algorithms, one can use the following techniques:

- apply more individuals to the population,
- use higher mutation rate,
- manually limit the search space with the penalty function,
- divide population into some groups.

These techniques have several disadvantages. Applying more individuals to the population may significantly increase the computational cost of the algorithm. The use of higher mutation rate may allow to find a global optimum, but it will be done by chance. To limit the search space with the penalty function we have to gather the information about the domain of the problem. In the combinatorial problems it may be hard to achieve. According to the last mentioned way to avoid the premature convergence the population is divided into some groups (named islands). The evolution is separated on each island, so that the genetic diversity can be preserved. Nevertheless, it is hard to identify genetically diversified individuals to settle the island. In the usual way, the only criterion is the value of the fitness function. Calculation of the diversity between each pair of individuals, especially in the combinatorial problems, where the length of the chromosome can be very large, is a time consuming task, especially for the prolific populations.

To avoid the premature convergence in the agent based genetic algorithms we consider using of three measures of the genetic diversity:

$$\frac{\sum_{x=1}^{xSize}\sum_{y=1}^{ySize}\sum_{x_n,y_n}\left(\frac{g(f(x,y),f(x_n,y_n))}{g_b(f(x,y),f(x_n,y_n))}\right)}{xSize*ySize} \tag{1}$$

$$\frac{\sum_{x=1}^{xSize}\sum_{y=1}^{ySize}\sum_{x_n,y_n} max(g(f(x,y),f(x_n,y_n))}{xSize*ySize} \tag{2}$$

$$\frac{\sum_{x=1}^{xSize}\sum_{y=1}^{ySize}\sum_{x_n,y_n} min(g(f(x,y),f(x_n,y_n))}{xSize*ySize} \tag{3}$$

where:

- $xSize$, $ySize$ - the dimensions of the grid,
- x_n, y_n - the coordinates of the neighbouring agents
- $f(x,y)$ - value of the fitness function for the agent located in the position (x,y) in the grid

The functions g and g_b are given by (4) and (5) respectively.

$$g(a,c) = \begin{cases} c-a & for\ a<c, \\ 0 & otherwise \end{cases} \tag{4}$$

$$g_b(a,c) = \begin{cases} 1 & for\ a<c, \\ 0 & otherwise. \end{cases} \tag{5}$$

All of the population diversity measures specified by (1 - 3) are the global measures calculated for the whole population - we sum the partial values calculated for each cell. According to (1) we calculate the average fitness difference between each cell and those cells in the neighbourhood that has a better fitness values. The second and third measures, given by (2) and (3) respectively, specify average maximum or minimum difference between the fitness value of each cell and its neighbouring cells which have better fitness. Function g returns the difference of the fitness values, if the fitness value of the neighbouring cell is better than in the current cell. Function g_b is applied to calculate the number of cells in the neighbourhood of better fitness with respect to the fitness of the current cell.

3.4 Fair and Selfish Crossover

For our experiments we implemented a single point crossover. Such an operation can be applied, if at least one node that belongs to both paths is represented by the crossed individuals. At first, we identify such nodes and randomly choose one of them. Next, we substitute the part of the path that is placed after this point between the individuals. For given paths p_1 and p_2, let v be a such a vertex that $v \in p_1$ and $v \in p_2$. Let us assume, that paths look as follows:

- $p_1 = \{s, v_{a1}, v_{a2}, \ldots v_{ak}, v, v_{al}, \ldots v_n, t\}$
- $p_2 = \{s, v_{b1}, v_{b2}, \ldots v_{bm}, v, v_{bn}, \ldots v_{bp}, t\}$

After the crossover operation, the paths will take the form:

- $p_1' = \{s, v_{b1}, v_{b2}, \ldots v_{bm}, v, v_{al}, \ldots v_n, t\}$
- $p_2 = \{s, v_{a1}, v_{a2}, \ldots v_{ak}, v, v_{bn}, \ldots v_{bp}, t\}$

We consider two methods of choosing the mating agents. In the first method, named *fair crossover*, the second parent is chosen from all the neighbouring agents. In the second method, denoted as *selfish crossover*, only agents with the better fitness are taken into account.

3.5 Mutation

The mutation operator is applied with the certain probability to each individual in the population. If the mutation occurs, two nodes on the path represented by the individual are chosen and a new partial path is calculated randomly. We do not reset the path even if after mutation the value of the fitness function for the mutated individual is worse than before.

4 Experiment Results and Discussion

The research was carried out on the map of the city of Lodz, Poland, containing about 7000 nodes and 17000 edges. Some of the nodes do not represent crossroads but are placed on the map to preserve geographical coordinates of the roads.

Table 1. Convergence of the proposed algorithm

Max iter	Size X	Size Y	mut. Prob.	Result	PC
20	5	6	10%	621	no
20	5	6	5%	635	no
20	5	4	15%	621	no
20	5	4	10%	633	no
20	5	4	5%	1049	yes
20	4	3	15%	1231	yes
25	4	4	15%	627	no
25	4	4	10%	642	no

We have chosen two nodes on the map which represented the starting and the finishing point on the route that was supposed to be found. The first node was located in the suburbs and the second one was placed in the opposite side of the city. As the simulation environment for finding the optimal routes, we used Matsim[1] integrated with our algorithms. To check the correctness of our method, we performed four experiments. Af first we examined the convergence of the algorithm by comparing its results to those obtained by use of Dijkstra's algorithm [14]. In the second experiment we discussed efficiency of the algorithm against the Simple Genetic Algorithm. The third and the fourth experiment aimed at checking the correctness of the diversity measures mentioned in the previous section and results obtained by the use of the selfish crossover.

4.1 Convergence of the Algorithm

During the first experiment we examined the convergence of the algorithm by multiple runs of the same query. We have taken into account the following:

- the maximum number of iterations,
- size of the grid, in which the agents reside,
- mutation probability,
- obtained results,
- observation of the premature convergence.

The results of our research are gathered in the Table 1. In the last column we presented information if we have observed the premature convergence for the specified algorithm's parameters. All of the results were calculated with the fair crossover applied. The minimum cost of the path calculated by the deterministic methods is 605.

Conducted research has shown that the algorithm converges to the near optimal solution in almost all experiments. When the dimensions of the gird were too small, we have observed the premature convergence. It can be explained in two ways: at first, when the population is small, the genetic diversity is not sufficient to properly explore the search space. Moreover, in a proposed approach agents

[1] www.matsim.org

Table 2. Comparison of the proposed approach and SGA

ItCount	POP SIZE		mut prob	RESULT		PC	
	Grid	SGA		Grid	SGA	Grid	SGA
20	30	30	15%	616	631	no	no
20	30	30	5%	635	614	no	no
20	20	20	15%	621	618	no	no
20	20	20	10%	633	800	no	yes
20	20	20	5%	1049	760	yes	yes
20	12	12	15%	1231	1051	yes	yes
25	16	16	15%	627	608	no	no
25	16	16	10%	642	759	no	yes

can only exchange their genetic material with neighbours, so that in case of the small grid, it is not a better solution. The second observation is that one can try to improve the results by the use of higher mutation rate. Our suggestion is to make some experiments to choose optimal dimensions of the grid, while keeping constantly low mutation probability.

4.2 Comparison to the Simple Genetic Algorithm

After checking the convergence of the algorithm, the results were compared with those obtained by our implementation of the Simple Genetic Algorithm. Implementation details of the SGA are desribed in [15]. During the experiment, such parameters as the mutation probability, size of the population and the maximum iteration number were the same for the both versions of the algorithm. In the multi-agent genetic algorithm we used a fair version of the crossover operator. We performed several runs of the both of the algorithms on the same query as was specified in the former experiment.

The results, gathered in the Table 2, show the comparison of the algorithms. In the last four columns actual cost of the path and the information if the premature convergence was observed are presented. It can be easily noticed, that both of the algorithms yield the similar results, but for a small number of individuals in the population, the SGA algorithm converges too fast more often than the algorithm based on the proposed approach. One can conclude that the application of the grid version of a genetic algorithm allowed us to use smaller populations without stuck in the local minima.

4.3 Measuring Diversity of the Population

The third experiment aimed at finding out, whether proposed measures of the population's diversity are correctly built. For the same nodes as in the former experiments we investigated values of the diversity measures for two different algorithm configurations. In both cases, the experiment was conducted on the 4x4 grid of agents. The number of iteration was set to 25. The mutation probability was 0.05 in the first attempt and 0.15 in the second one. Results of the

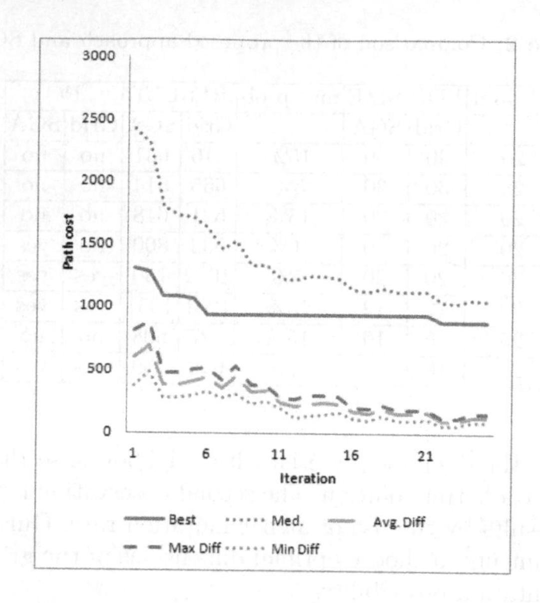

Fig. 3. Diversity measures for small mutation

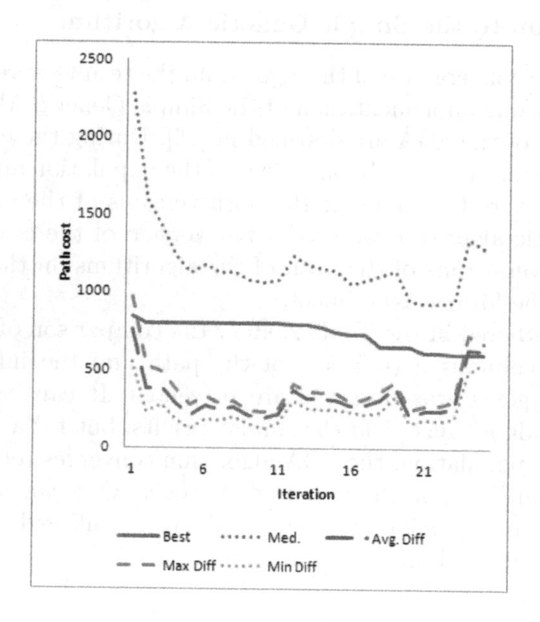

Fig. 4. Diversity measures for more frequent mutation

first and the second run are presented in the Fig.3 and Fig.4, respectively. On both figures we show the average and the best fitness value in the population as well as values of the three considered diversity measures given by (1 - 3) denoted as *Avg. Diff*, *Max Diff* and *Min Diff* respectively.

Table 3. Comparison of selfish & fair crossover operators

Grid size		mut prob	RESULT		PC	
X	Y		fair	selfish	fair	selfish
5	6	15%	617	607	no	no
5	6	5%	615	870	no	yes
5	4	15%	632	622	no	no
5	4	5%	625	810	no	yes

For the low mutation rate all of the diversity measures quickly accomplished the low values, what showed that the algorithm stuck in some local minimum. When the mutation probability was higher, the values of the diversity measures were higher too. Higher values of the diversity measures allows us to expect that the better solution can still be found.

4.4 Selfish Crossover

In the last experiment we compared the results obtained by use of the different implementations of the crossover operator. We performed 25 iterations of the algorithm for each run. In the Table 3 we present values of the fitness function of the best individual after the specified number of iterations and the information if we observed the premature convergence. As one can see, the selfish crossover performs worse than the fair implementation. Such a situation happens as the selfish crossover is performed less often, because the individuals can only be mated if almost one neighbour has a better value of the fitness function with respect to the current individual. Moreover, till the next iteration, the genes of the selfish individual can be reproduced by its weak neighbour. Finally, good individuals can be slightly genetically diversified, so such a rule can lead to the premature convergence.

5 Conclusions and Future Work

In the paper the genetic approach to find optimal route in the limited search space is considered. Conducted research has shown that the multi-agent based genetic algorithm can be successfully applied to the combinatorial problem of finding the optimal path between two nodes in the graph. What is more, such an algorithm performs better than the classical Simple Genetic Algorithm taking into account the quality of its convergence. The search space reduction obtained by combining pruning, reduction number of edges, and search decomposition allow to use the algorithm for complex graphs. Considered measures of the genetic diversity allows to discover whether the algorithm can find a better solution during the next iterations. Finally, it occurs that the application of the selfish crossover is not effective.

In our future work we would like to focus on the intelligent way of recovering the genetic diversity of the population while diversity measures have sufficiently small values. What is more, we would like to propose several new diversity measures. This improvements will allow us to utilize developed algorithms for the solution of the multi-criteria optimal path problem.

Acknowledgment. Development of the application for handling the simulation data was financed from the project "Platforma Informatyczna TEWI" funded from the European Union Innovative Economy, grant no. POIG.02.03.00-00-028/09.

References

1. Shi, L.S., Wang, H.H.: Multi-Agent Genetic Algorithm Based on Self-Adaptive Operator. In: 2012 Fifth International Conference on Intelligent Networks and Intelligent Systems, pp. 105–108 (2012)
2. Barbati, M., Bruno, G., Genovese, A.: Applications of agent-based models for optimization problems: A literature review. Experts Systems with Applications 39, 6020–6028 (2012)
3. Liu, J., Zhong, W., Jiao, L.: A Multiagent Evolutionary Algorithm for Constraint Satisfaction Problems. IEEE Transactions on Systems, Man and Cybernetics-Part B: Cybernetics 36(1), 54–73 (2006)
4. Zhong, W., Liu, J., Xue, M., Jiao, L.: A Multiagent Genetic Algorithm for Global Numerical Optimization. IEEE Transactions on Systems, Man and Cybernetics-Part B: Cybernetics 34(2), 54–73 (2004)
5. Whitley, D.: Cellular Genetic Algorithms. In: 5th International Conference on Genetic Algorithms, San Mateo, CA (1993)
6. Thangiah, S.M., Shmygelska, O., Mennell, W.: An Agent Architecture for Vehicle Routing Problems. In: SAC 2001, Las Vegas, NV, pp. 517–521 (2001)
7. Baker, B.M., Ayechew, M.A.: A Genetic Algorithm for the Vehicle Routing Problem. Computers & Operations Research 30, 787–800 (2003)
8. Masum, A.K.M., Shahjalal, M., Faruque, F., Sarker, I.H.: Solving the Vehicle Routing Problem Using Genetic Algorithm. International Journal of Advanced Computer Science and Applications 2(7), 126–131 (2011)
9. Karimi, H.A.: Real-time Optimal Route Computation: A Heuristic Approach. ITS Journal 3(2), 111–127 (1996)
10. Fu, L.: Real-Time Vehicle Routing and Scheduling in Dynamic and Stochastic Traffic Networks, PhD thesis, University of Alberta (1996)
11. Nicholson, J.A.T.: Finding the Shortest Route between Two Points in a Network. The Computer Journal 9(3), 275–280 (1966)
12. Sanders, P., Schultes, D.: Highway Hierarchies Hasten Exact Shortest Path Queries. In: Brodal, G.S., Leonardi, S. (eds.) ESA 2005. LNCS, vol. 3669, pp. 568–579. Springer, Heidelberg (2005)
13. Schultes, D., Sanders, P.: Dynamic Highway-Node Routing. In: Demetrescu, C. (ed.) WEA 2007. LNCS, vol. 4525, pp. 66–79. Springer, Heidelberg (2007)
14. Dijkstra, E.W.: A Note on Two Problems in Connexion with Graphs. Numerische Mathematik 1, 269–271 (1959)
15. Chomatek, L.: Parallel Hierarchies for Solving Single Source Shortest Path Problem. Journal of Applied Computer Science 21(1), 25–39 (2013)

Experimental Study of Selected Parameters of the Krill Herd Algorithm

Piotr A. Kowalski[1,2] and Szymon Łukasik[1,2]

[1] Systems Research Institute, Polish Academy of Sciences,
ul. Newelska 6, PL-01-447 Warsaw, Poland
{pakowal,slukasik}@ibspan.waw.pl

[2] Department of Automatic Control and Information Technology,
Cracow University of Technology,
ul. Warszawska 24, PL-31-155 Cracow, Poland
{pkowal,szymonl}@pk.edu.pl

Abstract. The Krill Herd Algorithm is the latest heuristic technique to be applied in deriving best solution within various optimization tasks. While there has been a few scientific papers written about this algorithm, none of these have described how its numerous basic parameters impact upon the quality of selected solutions. This paper is intended to contribute towards improving the aforementioned situation, by examining empirically the influence of two parameters of the Krill Herd Algorithm, notably, maximum induced speed and inertia weight. These parameters are related to the effect of the herd movement as induced by individual members. In this paper, the results of a study – based on certain examples obtained from the CEC13 competition – are being presented. They appear to show a relation between these selected two parameters and the convergence of the algorithm for particular benchmark problems. Finally, some concluding remarks, based on the performed numerical studies, are provided.

Keywords: Krill Herd Algorithm, Biologically Inspired Algorithm, Optimization, Metaheuristic.

1 Introduction

The optimization problems [5] are encountered when deriving solutions to many engineering issues [15]. Optimization can be considered as choosing of the best possible use of limited resources (time, money, etc.) while attempting to bring about certain goals. In achieving this, the optimization problem can be written in a formal way. Let us introduce the 'cost function' K:

$$K : A \longmapsto \mathbb{R}, \tag{1}$$

(in some instances, this function can be also called the 'fitness function') where $A \subset \mathbb{R}^n$. The optimization task consists of finding the value $x^* \in A$, such that for every $x \in A$ the following relationship is true:

$$K(x) \geq K(x^*). \tag{2}$$

P. Angelov et al. (eds.), *Intelligent Systems'2014*,
Advances in Intelligent Systems and Computing 322, DOI: 10.1007/978-3-319-11313-5_42

Although the optimization problem can be easily defined and described, determining its solution is already a very difficult issue. To find solutions for this problem, certain optimization algorithms are commonly used. Generally, these are placed within two classes: deterministic and stochastic algorithms.

Deterministic algorithms operate according to strictly defined procedures, and the results obtained through employing them are repeatable. A simple and well-known example of this group is the numerical algorithm called 'Newton's Method' [5]. This can easily be used to find the minimum of the function. Assuming no change in the input data, the intermediate results and the final result obtained in subsequent runs of the algorithm will be identical.

On the other hand, we have the stochastic algorithms. These have a random factor and each execution of some algorithm belonging to this class implies that results will vary. Typically this group of algorithms allows users to obtain solution in a short time, but very often it constitutes only an approximation of optimum value. Among stochastic algorithms that can be distinguished there exists a group of procedures inspired by Nature. These are divided into three main streams: evolutionary algorithms, artificial immune systems, and swarm intelligence.

Evolutionary Algorithms [1] are based on biological evolution, and they are built upon conceptualising a population that adapts to its surrounding world by using mechanisms such as selection, reproduction, mutation and crossover. There are many implementations of the evolutionary approach. Among these we can distinguish Genetic Algorithms, Genetic Programming, and Evolutionary Strategies [9]. In each of those, a population is created which evolves in successive iterations towards the best solution.

Artificial Immune Systems [2] are adaptive systems inspired by the results of studies on the immune systems of living organisms. Their main task is to detect anomalies. Examples of such systems are the Negative Selection Procedure and the Clonal Selection Procedure.

Swarm intelligence, also known as the 'intelligent group' [4], is a decentralized self-organizing system which is formed by a population consisting of a number of individuals called 'agents'. In this approach, there is no commanding authority that determines the way the agents behave. Instead, individual 'animals' roam the space of proposed solutions, while following few simple rules. The most well-known implementations of this are: Ant Colony Optimization [3], Bee Colony Algorithm [8], Particle Swarm Optimization [14], Glowworm Swarm Optimization [13], and Firefly Algorithm [12]. In 2012, A. H. Gandomi and A. H. Alavi contributed to these by developing Krill Herd Algorithm (KHA) [6].

The goal of the paper is to reveal the impact of selected parameters in KHA, to the quality and convergence speed of an optimization procedure. Presented tests empirically provide suggested strategies of selecting two parameters within the Krill Herd Algorithm: maximum induced speed and inertia weight. These quantities are related to the part of the herd movement that is induced by other swarm members. Parameter choice is based upon numerical simulation results which indicate solution's quality: average error value with its standard deviation, the best result obtained by the swarm, and, additionally, the speed of its convergence. Finally, to increase the quality of solutions, a procedure for decreasing the value of the inertia weight parameter has also been proposed.

2 Krill Herd Algorithm

Antarctic krill (Euphausia superba) is a species of krill found in the Southern Ocean. An adult body has a length of 6 cm, and its weight comes about 2 grams. This species feeds on phytoplankton. A characteristic feature of krill, and the inspiration of the present algorithm [6,16], is the ability for individual krill to be moulded within a large herd that is even hundreds of meters in length.

Algorithm 1. Krill Herd Algorithm

1: $k \leftarrow 1$ {initialization}
2: Initialize parameters (D^{max}, N^{max}, etc.)
3: **for** $i = 1$ to M **do**
4:     Generate Solution ($x_i(k)$)
5:     {evaluate and update best solutions}
6:     $K(x_i(0)) \leftarrow$ Evaluate quality($x_i(0)$)
7: **end for**
8: $x^* \leftarrow$ Save best individual $x^*(0)$
9: {main loop}
10: **repeat**
11:     sort population of krills
12:     **for** $i = 1$ to M **do**
13:         Perform motion calculation and genetic operators:
14:         $N_i \leftarrow$ Motion induced by other individuals
15:         $F_i \leftarrow$ Foraging activity
16:         $D_i \leftarrow$ Random diffusion
17:         Crossover
18:         Mutation
19:         {update krill position}
20:         Update Solution ($x_m(k)$)
21:         {evaluate and update best solutions}
22:         $K(x_i(k)) \leftarrow$ Evaluate quality($x_i(k)$)
23:     **end for**
24:     $x^* \leftarrow$ Save best individual $x^*(k)$
25:     stop condition $\leftarrow$ Check stop condition ()
26:     $k \leftarrow k + 1$
27: **until** stop condition = **false**
28: **return** $K(x^*(k))$, $x^*(k)$, k

The Krill Herd procedure is shown in Algorithm 1. Operation of the algorithm starts with the initialization of data structures, i.e. describing individuals, as well as the whole population. Initializing data structure representing a krill, means situating it in a certain place (at the 'solution space') by giving it a set of coordinates. For this purpose, it is recommended to employ random number generation according to a uniform distribution. Like other algorithms inspired by Nature, each individual represents one possible solution of the problem under consideration.

After the initialization phase of the algorithm, it follows into a series of iterations. The first step of each is to calculate fitness function value for each individual of the population. This is equivalent to the calculation of optimized functions with arguments which are the coordinates of the krill's position. Then, for each individual of the population, the vector which indicates the displacement in the solution space is calculated. The movement of krill is described by an equation dependent on three factors:

$$\frac{dX_i}{dt} = N_i + F_i + D_i, \tag{3}$$

where N_i refers to the movement induced by the influence of other krills (subsection 2.1), F_i is a movement associated with the search for food (subsection 2.2), and D_i constitutes a random diffusion factor (subsection 2.3).

In order to improve efficiency, the algorithm includes two genetic operators: crossover and mutation (subsection 2.4). This phase is optional, implementation of these operators can be completely omitted or only one of them can be employed. As shown by certain preliminary tests [6] the Krill Herd Algorithm achieves the best results when only the crossover operator is implemented.

The next step is to update the position of krills in the solution space, in accordance with the designated movement vectors (3). In the next iteration, based on these values, the new value of the fitness function will be calculated.

A termination of the algorithm can be implemented in many ways. One possibility is to stop it when the solution achieves or is below predetermined cost function value. This approach is used where the main purpose is accuracy. The second way is to terminate the procedure upon reaching a predetermined number of iterations. This method is applicable in situations where the solution must be achieved in a fixed time. The most common method is a combination of both conditions.

2.1 Motion Induced by Other Individuals

The movement stimulated by the presence of other individuals (krills) is defined as:

$$N_i = N^{max} \alpha_i + \omega N_i^{old}, \tag{4}$$

where

$$\alpha_i = \alpha_i^{local} + \alpha_i^{target}. \tag{5}$$

In the above formulas N^{max} constitutes the maximum speed induced by the presence of other krills, and it is determined experimentally (in paper [6] expressing this value as 0.01 was recommended – this recommendation rate is based on the maximum induced speed of the krill herd [7]. Furthermore, the parameter ω describes the inertia weight of the induced motion, and takes the value from the interval $[0, 1]$. What is more, N_i^{old} is the last iteration value limiting this motion. Moreover, α_i^{local} denotes the local effect provided by neighbouring krill, and α_i^{target} constitutes a target direction effect based on the position of the best individuals. The neighbours effect, connected with attractiveness, can be described as:

$$\alpha_i^{local} = \sum_{j=1}^{M_{neighbours}} \hat{K}_{i,j} \hat{X}_{i,j}. \tag{6}$$

Where the variable $\hat{K}_{i,j}$ and vector $\hat{X}_{i,j}$ are defined as follows:

$$\hat{X}_{i,j} = \frac{X_i - X_j}{\|X_i - X_j\| + \epsilon}, \tag{7}$$

$$\hat{K}_{i,j} = \frac{K_i - K_j}{K^{worst} - K^{best}}. \tag{8}$$

Here K^{worst} and K^{best} are the worst and best values of the fitness function within a population (the wording "best" means the smallest value of $K(X_i)$, taking into account the fact that a minimum of a function (1) is to be located). In addition the variable K_i in equation (8) represents the fitness function's value of the i-th krill, and the K_j j-th neighbour for $j = 1, 2, ..., M_{neighbours}$. Moreover a notation $M_{neighbours}$ expresses the number of krill neighbours. In order to avoid zero coming about within the equation (7), in the denominator, a factor ϵ is added, which is a small positive number. Finally, the neighbours of the krill are those individuals that are in range, i.e. in the area of a circle centred at the position of the i-th krill, and which has a radius equal to:

$$d_{s,i} = \frac{1}{5N} \sum_{j=1}^{M} \|X_j - X_i\|, \tag{9}$$

where M represents the total number of individuals in a population.

The movement of krill is also dependent on the best individual location. This factor determines the value α_i^{target} in following way:

$$\alpha_i^{target} = \gamma^{best} \hat{K}_{i,best} \hat{X}_{i,best}, \tag{10}$$

where γ^{best} is the ratio of the individual impact with the best fitness function value for the i-th krill:

$$\gamma^{best} = 2(\xi + \frac{k}{I_{max}}). \tag{11}$$

In the previous equation, ξ is a random number coming from the interval $[0, 1]$, while k denotes the number of the current iteration, and I_{max} indicates a maximum number of iterations.

2.2 Foraging Activity

The movement associated with the search for food (foraging) depends on two components. The first one is the current food location and the second points out the previous location of food intake. For the i-th individual, this component of movement is defined as:

$$F_i = V_f \beta_i + \omega_f F_i^{old}, \tag{12}$$

where

$$\beta_i = \beta_i^{food} + \beta_i^{best}. \tag{13}$$

Furthermore, quantity V_f describes the speed of searching for food, and has been selected empirically. Its recommended value is 0.02. The notation ω_f denotes the inertia weight of the foraging motion in range $[-1, 1]$. The location of food is the quantity that for KHA is defined on the basis of the distribution of the fitness function. It is given by following equation:

$$X^{food} = \frac{\sum_{i=1}^{M} K_i^{-1} X_i}{\sum_{i=1}^{M} K_i^{-1}}. \tag{14}$$

The previous equation participates in determining a quantity which is a measure of the impact of food on the i-th krill:

$$\beta_i^{food} = C^{food} \hat{K}_{i,food} \hat{X}_{i,food}, \tag{15}$$

where C^{food} represents a coefficient determining the influence of the impact of food location on the i-th krill

$$C^{food} = 2(1 - \frac{k}{I_{max}}). \tag{16}$$

In this algorithm, the best individual and its position are also included in accordance with the following formula:

$$\beta_i^{best} = \hat{K}_{i,best} \hat{X}_{i,best}. \tag{17}$$

2.3 Physical Diffusion

Physical diffusion is a random process, and this vector is defined as follows:

$$D_i = D^{max}(1 - \frac{k}{I_{max}})\delta, \tag{18}$$

where $D^{max} \in [0.002, 0.01]$ is the maximal diffusion speed, while denoted as δ represents the random directional vector, and its elements belong within the interval $[-1, 1]$.

The previously described factors affect that the krill additionally changes its position to some extent randomly over time.
Finally, its location at time $t + \Delta t$ is determined as follows:

$$X_i(t + \Delta t) = X_i(t) + \Delta t \frac{dX_i}{dt}, \tag{19}$$

wherein Δt is the scaling factor for the speed of the search of the solution space, and is defined as:

$$\Delta t = C_t \sum_{j=1}^{N} (UB_j - LB_j). \tag{20}$$

In the above formula, the condition $C_t \in [0, 2]$ is fulfilled, while N is search space dimensionality. In addition, UB_j and LB_j are the upper and lower limitations of j-th coordinates of vector X.

2.4 Genetic Operators

The final stage of the main iteration in KHA is the use of genetic operators [17]. The crossover results in a change of the m-th coordinate of i-th krill as shown below by the formula:

$$x_{i,m} = \begin{cases} x_{r,m} \text{ for } \gamma \leq Cr \\ x_{i,m} \text{ for } \gamma > Cr \end{cases} \tag{21}$$

where $Cr = 0.2\hat{K}_{i,best}$; $r \in \{1, 2, ..., i-1, i+1, ..., N\}$ denotes a random index, and γ is a random number from the interval $[0, 1)$ generated according to the uniform distribution. In this approach the crossover operator is acting on a single individual.

The mutation modifies the m-th coordinate of the i-th krill in accordance with the formula:

$$x_{i,m} = \begin{cases} x_{gbest,m} + \mu(x_{p,m} - x_{q,m}) & \text{for } \gamma \leq Mu \\ x_{i,m} & \text{for } \gamma > Mu \end{cases} \tag{22}$$

wherein $Mu = 0.05/\hat{K}_{i,best}$; $p, q \in \{1, 2, ..., i-1, i+1, ..., K\}$ and $\mu \in [0, 1)$.

Additional information about the Krill Herd Algorithm in [16,18] could be found.

3 Numerical Studies

In this section, the performance of the KHA metaheuristic for different sets of selected parameters connected with motion is verified – by running a series of experiments related to various global numerical optimization tasks. One of the main goals in presenting these numerical experiments is to show the dynamics of an optimization process based on KHA for diverse sets of the algorithm's parameters. In doing so, the minimal values of an error, as well as a set of mean values with their standard deviations were derived in order to perform such a comparison.

For numerical experiments, we applied a selected set of benchmark functions considered in the CEC'13 competition [11]. Table 1 revels a list of functions with their formulas, dimensionality, feasible bounds and optimal values. From the set of benchmark tasks, six functions were selected, the first three having

Table 1. Benchmark functions used for experimental study

K	Name	Expression	Feasible bounds	N	$K(x^*)$
$K1$	Sphere	$K_1(x) = \sum_{i=1}^{N} z_i^2 + K_1^*$ $z = x - o$	$[-100, 100]^N$	10	-1400
$K2$	Rotated Bent Cigar	$K_2(x) = z_1^2 + 10^6 \sum_{i=2}^{N} z_i^2 + K_2^*$ $z = M_2 T_{asy}^{0.5}(M_1(x - o))$	$[-100, 100]^N$	10	-1000
$K3$	Different Powers	$K_3(x) = \sqrt{\sum_{i=1}^{N} \lvert z_i \rvert^{2+4\frac{i-1}{N-1}}} + K_3^*$ $z = x - o$	$[-100, 100]^N$	10	-1000
$K4$	Rotated Rastrigin	$K_4(x) = \sum_{i=1}^{N} (z_i^2 - 10\cos(2\pi z_i) + 10) + K_4^*$ $z = M_1 \Lambda^{10} M_2 T_{asy}^{0.2}(T_{osz}(M_1 \frac{5.12(x-o)}{100}))$	$[-100, 100]^N$	10	-300
$K5$	Schwefel	$K_5(x) = 418.9829N \sum_{i=1}^{N} g(z_i) + K_5^*$ $z = \Lambda^{10}(\frac{1000(x-o)}{100}) + 4.209687462275036e+002$ $g(z_i) = z_i \sin(\lvert z_i \rvert^{1/2})$	$[-100, 100]^N$	10	-100
$K6$	Rotated Katsuura	$K_6(x) = \frac{10}{N^2} \prod_{i=1}^{N} (1 + i \sum_{j=1}^{32} \frac{\lvert 2^j z_i - round(2^j z_i) \rvert}{2^j})^{\frac{10}{N^{1.2}}} - \frac{10}{N^2} +$ K_6^* $z = M_2 \Lambda^{100}(M_1 \frac{5(x-o)}{100})$	$[-100, 100]^N$	10	200

an unimodal character, while the others are of a multimodal type. Due to text limitation, a description of the symbols used in Table 1 can be found in [11].

The experiments were carried out for a fixed number of iterations $I_{max} = 10000$ $(1000N)$, and 30 trials were undertaken for each benchmark function. In all numerical experiments the population size M = 40 was used. What is more KHA was configured with the parameters given in section 2. However, this does not apply to the parameters of the algorithm described in subsection 2.1, which are the subject of this study.

As a performance measure the mean optimization error $\hat{E}_{mean}$ was used (with $\hat{E}(k) = |K(x^*(x)) - K^*|$) along with its standard deviation $\sigma_{\hat{E}}$. In addition, as a measure of the optimization algorithm quality, a minimal error is also reported.

3.1 Selection of N^{max} Parameter

The first set of tests was associated with the selection of the N^{max} parameter. This can be defined as maximum speed induced by the presence of other individuals. In order to investigate the best value of this parameter, the value of N^{max} was set in following order $\{0.005, 0.010, 0.015, 0.020, 0.025\}$. Moreover, in these tests, the decreasing character of the parameter w was proposed according to the following formula:

$$w = 0.1 + 0.8 * \left(1 - \frac{k}{I_{Max}}\right). \tag{23}$$

The results of this investigation can be seen in Table 2, as well as in the exemplary diagrams (Figures 1 and 2). These show the first 200 iterations of the algorithm's convergence for the unimodal (K1) and multimodal (K4) functions, respectively.

In interpreting the presented results, the following remarks can be concluded. If we optimize the function of the unimodal character, the smallest parameter value of the N^{max} should be chosen. In addition, for the K1 and K3 functions,

Table 2. Optimisation results for benchmark functions with varying N^{max}

		N^{max} 0.005	0.010	0.015	0.020	0.025
K1	$\hat{E}_{min}$	0.00000	0.00000	0.00000	0.00000	0.00000
	$\hat{E}_{mean}$	0.00000	0.00000	0.00000	0.00001	0.00001
	$\sigma_{\hat{E}}$	0.00000	0.00000	0.00000	0.00000	0.00000
K2	$\hat{E}_{min}$	4.993E+03	1.124E+03	7.949E+02	2.750E+02	1.961E+02
	$\hat{E}_{mean}$	1.617E+08	2.797E+07	1.246E+07	5.227E+07	2.008E+07
	$\sigma_{\hat{E}}$	5.436E+08	7.600E+07	3.279E+07	9.625E+07	3.836E+07
K3	$\hat{E}_{min}$	0.00005	0.00002	0.00003	0.00006	0.00004
	$\hat{E}_{mean}$	0.00011	0.00008	0.00009	0.00011	0.00013
	$\sigma_{\hat{E}}$	0.00004	0.00004	0.00004	0.00005	0.00005
K4	$\hat{E}_{min}$	14.92439	9.94961	9.94960	4.97481	8.95464
	$\hat{E}_{mean}$	47.19417	38.13998	31.44063	25.30509	29.94820
	$\sigma_{\hat{E}}$	17.08109	16.87140	15.33291	12.15950	13.50465
K5	$\hat{E}_{min}$	269.95241	309.41975	496.45608	285.41008	367.79078
	$\hat{E}_{mean}$	992.74120	953.26188	846.03055	786.37048	775.80619
	$\sigma_{\hat{E}}$	346.93644	288.11901	270.35735	365.50859	277.75687
K6	$\hat{E}_{min}$	0.02427	0.03769	0.02559	0.05169	0.03126
	$\hat{E}_{mean}$	0.14897	0.13162	0.12815	0.15088	0.15652
	$\sigma_{\hat{E}}$	0.10647	0.08539	0.08918	0.08900	0.09986

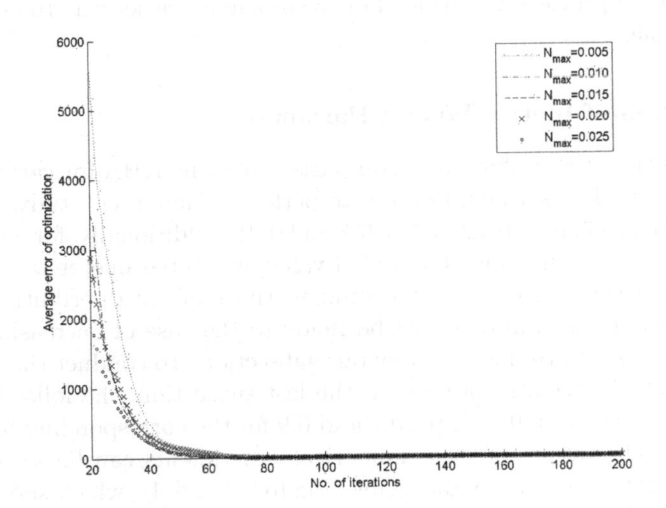

Fig. 1. Convergence of KHA with varying N^{max} for the benchmark function K1

the best results for N^{max} belong to the interval range [0.005, 0.01]. On the other hand, it can be seen that an increase in of this parameter results in a faster convergence of the optimization algorithm. For the functions of a multimodal character (K4-K6), a higher value of this parameter (even to 0.025) is recommended. This is due to the presence of numerous local minima. Thus increasing the value of this parameter allows for a stronger influence upon the solutions

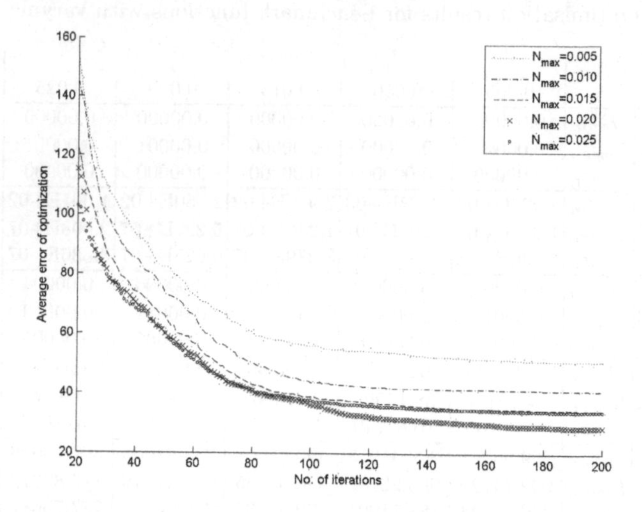

Fig. 2. Convergence of KHA with varying N^{max} for the benchmark function K4

derived from the previous iteration. This results in an easier way to escape from a local extreme.

3.2 Selection of Inertia Weight Parameter

In this subsection, the performance comparison of using KHA for varying inertia weight parameter is described. In order to perform a numerical study, a parameter w was taken as being $\{0.1, 0.3, 0.5, 0.7 \text{ and } 0.9\}$. Additionally, for comparison, a variant is described in which the added value of this parameter was generated randomly from the range $[0, 1]$ according to the uniform distribution. In this situation, a comparison also should be made to the case of decreasing the parameter that was reported in the previous subsection. To conduct the numerical tests, based on the results obtained in the last subsection, the following values of N^{max} parameter $0.01, 0.01, 0.01, 0.7$ and 0.9 for the corresponding function K were set out. The results of these numerical experiments can be seen in Table 3, as well as in the exemplary diagrams (Figures 3 and 4), which show the first 200 iterations of algorithm convergence for unimodal (K1) and multimodal (K4) functions, respectively.

In this quantitative comparison, both means with standard deviations and best obtained values of error were reported. The results show that here as well, two cases should be considered. An example of the former are the unimodal functions $K1 - K3$. For these instances, the smaller value of the parameter w is preferred. In this research, the best results were obtained for $w = 0.1$. The exception was the K2 function, here, the best results were obtained by way of the random inertia weight parameter. The second case incorporates examples of multimodal functions ($K4 - K6$). In this situation, we recommend a set up of

Table 3. Optimisation results for benchmark functions with varying inertia weight w

		w 0.100	0.300	0.500	0.700	0.900	rand
K1	$\hat{E}_{min}$	0.00000	0.00000	0.00000	0.00007	0.03219	0.00001
	$\hat{E}_{mean}$	0.00000	0.00001	0.00002	0.00019	0.07729	0.00002
	$\sigma_{\hat{E}}$	0.00000	0.00000	0.00001	0.00007	0.02606	0.00001
K2	$\hat{E}_{min}$	6.348E+03	6.757E+02	1.911E+02	8.015E+02	5.140E+04	2.915E+01
	$\hat{E}_{mean}$	1.234E+08	2.750E+07	2.133E+07	7.599E+07	3.060E+07	2.206E+07
	$\sigma_{\hat{E}}$	2.584E+08	6.249E+07	7.178E+07	1.794E+08	5.036E+07	6.098E+07
K3	$\hat{E}_{min}$	0.00001	0.00009	0.00026	0.00056	0.04050	0.00028
	$\hat{E}_{mean}$	0.00003	0.00024	0.00087	0.00175	0.09250	0.00050
	$\sigma_{\hat{E}}$	0.00001	0.00008	0.00038	0.00100	0.02360	0.00017
K4	$\hat{E}_{min}$	50.74270	41.78823	28.85386	42.78334	9.05741	23.87924
	$\hat{E}_{mean}$	123.30676	115.57938	91.40320	82.58270	33.56695	100.25815
	$\sigma_{\hat{E}}$	37.03286	46.96247	42.16632	29.73748	17.12788	38.81281
K5	$\hat{E}_{min}$	388.25117	283.30526	467.88315	390.25919	173.34292	374.17893
	$\hat{E}_{mean}$	795.00710	810.78165	808.17313	962.54859	467.21808	955.54884
	$\sigma_{\hat{E}}$	319.50109	276.70611	200.73907	274.54883	139.34147	279.48430
K6	$\hat{E}_{min}$	0.08638	0.02508	0.01941	0.08221	0.60140	0.05913
	$\hat{E}_{mean}$	0.26328	0.17722	0.16912	0.26530	0.96770	0.26897
	$\sigma_{\hat{E}}$	0.16465	0.12818	0.10473	0.13189	0.19718	0.17963

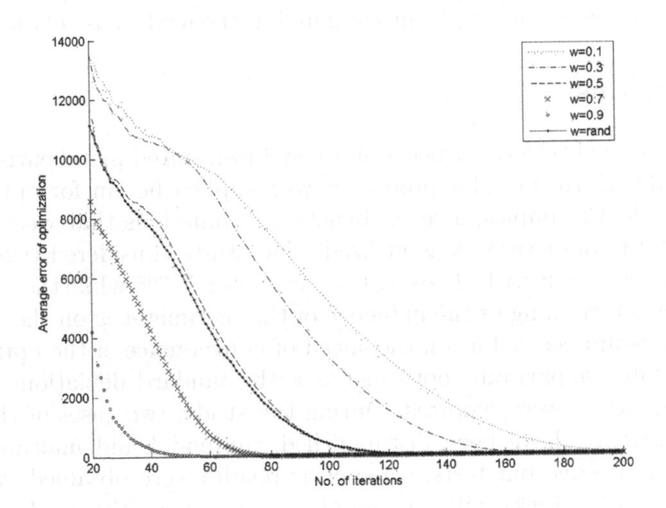

Fig. 3. Convergence of KHA with varying w for the benchmark function K1

$w = 0.9$. However, if we look at the results of the best value of cost function, then the best solution seems to be to adopt a variable parameter w derived by way of formula (23). Based on the presented Figures 3 and 4, we can easily conclude that the increase of the parameter w influences a faster convergence. Finally, it is also worth to note that KHA was found to underperform for some of examined

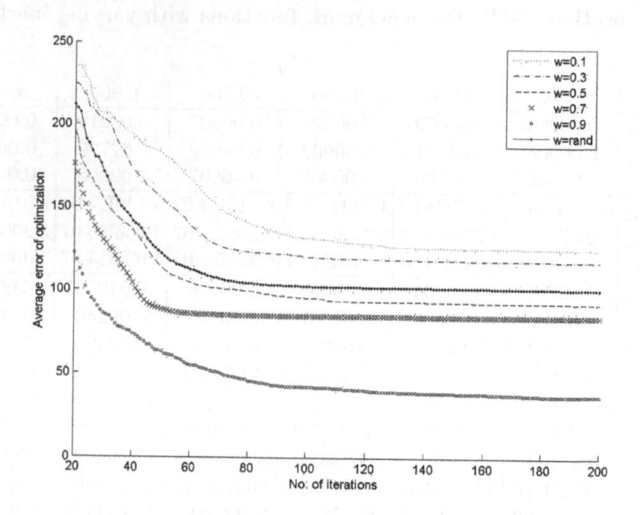

Fig. 4. Convergence of KHA with varying w for the benchmark function K4

benchmark instances (K2 in particular) - regardless of selected parameters set. This phenomenon is planned to be investigated in the next phase of our research.

4 Conclusions

The paper examined selected aspects of a new bio-inspired metaheuristic called the Krill Herd Algorithm. This procedure was applied herein for optimization tasks. As practical examples, a set of benchmark functions that were obtained from the CEC13 competition was utilized. The study considered two types of tests. The first one dwelt with choosing the parameter N^{max} while the second one concerned the determining of the influence of the parameter w on the quality of the obtained results, as well as on the speed of convergence of the optimization algorithm. In all comparisons, both means with standard deviations and best obtained error values were reported. During the study, two cases of the nature of the cost functions have been distinguished: unimodal and multimodal. On the basis of the carried-out tests, some clear results were obtained. We found that for optimization tasks with unimodal cost functions, the smaller value of the parameter N^{max} and w is highly recommended. Otherwise, for describing the multimodal character of cost function, we recommend higher values of these parameters or the application of the decreasing parameter w. This also results in a faster convergence of the optimization algorithm.

Finally, it should be noted that the performed experiments do not include a full analysis of the KHA parameters. The obtained results may, however, provide a starting point for further research and practical applications. In the near future, it is planned to publish research containing a full empirical analysis of the KHA procedure [10].

References

1. Ashlock, D.: Evolutionary Computation for Modeling and Optimization. Springer (2006)
2. Castro, L.N., Timmis, J.: Artificial Immune Systems: A New Computational Intelligence Approach. Springer (2002)
3. Dorigo, M., Birattari, M., Stutzle, T.: Ant colony optimization. IEEE Computational Intelligence Magazine 1, 28–39 (2006)
4. Engelbrecht, A.: Fundamentals of Computational Swarm Intelligence. John Wiley & Sons (2005)
5. Fletcher, R.: Practical Methods of Optimization. John Wiley & Sons (2000)
6. Gandomi, A.H., Alavi, A.H.: Krill herd: A new bio-inspired optimization algorithm. Communications in Nonlinear Science and Numerical Simulation 17, 4831–4845 (2012)
7. Hofmann, E.E., Haskell, A.G.E., Klinck, J., Lascara, M.: Lagrangian modelling studies of Antarctic krill (Euphasia superba) swarm formation. ICES J. Mar. Sci. 61, 617–631 (2004)
8. Karaboga, D., Basturk, B.: On the performance of artificial bee colony (ABC) algorithm. Applied Soft Computing 8, 687–697 (2008)
9. Kowalski, P.A.: Evolutionary Strategy for the Fuzzy Flip-Flop Neural Networks Supervised Learning Procedure. In: Rutkowski, L., Korytkowski, M., Scherer, R., Tadeusiewicz, R., Zadeh, L.A., Zurada, J.M. (eds.) ICAISC 2013, Part I. LNCS, vol. 7894, pp. 294–305. Springer, Heidelberg (2013)
10. Kowalski, P.A., Łukasik, S.: Properties and experimental evaluation of Krill Head Algorithm (in preparation, 2014)
11. Liang, J., Qu, B., Suganthan, P., Hernandez-Diaz, A.: Problem Definitions and Evaluation Criteria for the CEC 2013 Special Session and Competition on Real-Parameter Optimization, technical Report 201212, Computational Intelligence Laboratory, Zhengzhou University, Zhengzhou China and Technical Report, Nanyang Technological University, Singapore (2013)
12. Łukasik, S., Żak, S.: Firefly Algorithm for Continuous Constrained Optimization Tasks. In: Nguyen, N.T., Kowalczyk, R., Chen, S.-M. (eds.) ICCCI 2009. LNCS, vol. 5796, pp. 97–106. Springer, Heidelberg (2009)
13. Łukasik, S., Kowalski, P.A.: Fully Informed Swarm Optimization Algorithms: Basic Concepts, Variants and Experimental Evaluation. In: 2014 Federated Conference on Computer Science and Information Systems, FedCSIS (in print, 2014)
14. Poli, R., Kennedy, J., Blackwell, T.: Particle swarm optimization. Swarm Intell. 1, 33–57 (2007)
15. Rao, S.: Engineering Optimization: Theory and Practice. John Wiley & Sons (2007)
16. Wang, G.G., Gandomi, A.H., Alavi, A.H., Hao, G.S.: Hybrid krill herd algorithm with differential evolution for global numerical optimization. Neural Comput. & Applic., 1–12 (2013)
17. Wang, G., Guo, L., Wang, H., Duan, H., Liu, L., Li, J.: Incorporating mutation scheme into krill herd algorithm for global numerical optimization. Neural Comput. & Applic. 24, 853–871 (2014)
18. Wang, G.G., Gandomi, A.H., Alavi, A.H.: Stud krill herd algorithm. Neurocomputing 128, 363–370 (2014)

Hybrid Cuckoo Search-Based Algorithms for Business Process Mining

Viorica R. Chifu, Cristina Bianca Pop, Ioan Salomie, Emil St. Chifu, Victor Rad, and Marcel Antal

Tehnical University of Cluj-Napoca, Computer Science Department, Baritiu 26-28, Cluj-Napoca, Romania
{viorica.chifu,cristina.pop,ioan.salomie, emil.chifu,marcel.antal}@cs.utcluj.ro

Abstract. In this paper, we analyze the impact of hybridization on the Cuckoo Search algorithm as applied in the context of business process mining. Thus, we propose six hybrid variants for the algorithm, as obtained by combining the Cuckoo Search algorithm with genetic, Simulated Annealing, and Tabu Search-based components. These components are integrated into the Cuckoo Search algorithm at the steps that correspond to generating the new business process models. The hybrid algorithm variants proposed have been comparatively evaluated on a set of event logs of different complexities. Our experimental results obtained have been compared with the ones as provided by the state of the art Genetic Miner algorithm.

Keywords: process mining, cuckoo search, simulated annealing, tabu search, genetic algorithm.

1 Introduction

Process mining is a rapidly-developing research area in computer science, aiming to discover business process models based on the information stored in event logs. Due to the large number of causality relations established between the activities in an event log, the discovery of the optimal business process model that is complete and precise according to the event log can be considered as an optimization problem. Such an optimization problem can be solved satisfactorily by applying heuristic algorithms. The heuristic algorithms are inspired from various sources, many of them coming from the natural behavior of swarms/groups of animals or from the physical characteristics of the evolutionary natural phenomena.

Recently, the idea of hybridization has become a challenging research area aiming to improve the performance of the existing metaheuristics by combining them with other algorithmic components (i.e. integer programming, other metaheuristics etc.) [11]. Many approaches focus on hybridizing metaheuristics with components from other metaheuristics. More specifically, such approaches integrate trajectory-based metaheuristic components into population-based metaheuristics. This hybridization is motivated by the fact that population-based

P. Angelov et al. (eds.), *Intelligent Systems'2014*,
Advances in Intelligent Systems and Computing 322, DOI: 10.1007/978-3-319-11313-5_43

metaheuristics can find, by exploration, promising search space areas which are then exploited, by using trajectory-based metaheuristics, in order to find the best solutions [11].

This paper presents how the Cuckoo Search algorithm [6] can be applied to discover business process models as starting from event logs. In addition, the paper discusses different variants of hybridizing the Cuckoo Search algorithm with genetic operators, Tabu Search and Simulated Annealing, aiming to improve the algorithm performance in terms of fitness values. The hybridization components have been used into the Cuckoo Search to implement the Levy Flight procedure and the procedure for generating new business process models that will replace the worst ones. We obtained several hybrid variants for the Cuckoo Search-based algorithm, and we integrated them as plugins in the PROM framework [13]. We evaluated the hybrid algorithm variants on a set of event logs provided by [12]. Additionally, we analyzed our experimental results by comparison with the results provided by Genetic Miner [1].

The paper is structured as follows. Section 2 presents related work. Section 3 introduces the hybrid Cuckoo Search-based model for business process mining, while Section 4 presents the hybridization variants for the Cuckoo Search algorithm. Section 5 evaluates the performance of the hybrid algorithm variants and compares our experimental results with the ones obtained by running Genetic Miner on the same event logs. The paper ends with conclusions.

2 Related Work

This section reviews some of the most relevant business process mining approaches available in the research literature.

The α algorithm [2] extracts business process models from event logs based on the ordering relations between the activities recorded in the log. The algorithm is able to extract the following types of ordering relations: *follows*, *causal*, *parallel*, and *unrelated*. The main disadvantage of the α algorithm is that it works only on noise-free and complete event logs [1].

The Heuristic Miner [3], which is an extension of the α algorithm, addresses the problem of processing noisy event logs by considering the frequency of the *follows* relation between activities.

In [1], the business process mining problem is modeled as an optimization problem that is solved by using a genetic-based algorithm. In this approach, an individual is represented as a causal matrix of a candidate business process model. Crossover and mutation operators are considered in order to alter the structure of causal matrices, and a completeness and preciseness-based fitness function is used to evaluate the quality of a business process model.

In [7], a simulated annealing-based process mining method is proposed. Similar to [1], a candidate process model is represented as a causal matrix which additionally stores the *parallel/select* relation among subsets in the input/output sets. Simulated annealing is used to identify basic and non-free choice structures, invisible tasks, and duplicate tasks. A business process model is evaluated in terms of completeness and preciseness, similar to [1].

In [4] the authors present a business process mining approach which combines a genetic programming algorithm with a directed graph structure used for representing a business process model. The genetic programming algorithm consists in four steps. In the first step the dependency relations between the activities are extracted from the event log, while in the second step the initial population of individuals is generated based on the dependency relationships identified. In the third step the fitness of each individual is computed by using a fitness function as defined in [1]. In the fourth step, the crossover and mutation operators are applied to create the next population.

In [5] an Ant Colony Optimization - based approach for discovering business process models from event logs is used. In this approach the business process models are represented as directed graphs, where the nodes are the activities from the event log and the edges are the process flows between the activities. The novelty of this work is twofold. First it applies the Ant Colony Optimization algorithm on the graph structure to extract the optimal or near optimal business process model. Second, it defines a tentative moving procedure which avoids the stagnation of the Ant Colony Optimization algorithm into a local optimum. The tentative moving procedure consists of moving an ant with a probability p. If the probability p is low, then the ant will be placed in a terminal node, otherwise, if the ant gets stuck (i.e. the ant hasn't moved during a predefined consecutive number of iterations) it will be relocated into another node.

3 Hybrid Cuckoo Search-Based Model for Business Process Mining

This section presents the Hybrid Cuckoo Search-based model which will be used to obtain several hybridization variants for the classical Cuckoo Search algorithm. The hybrid model consists of the Cuckoo Search core component and the hybridization components.

3.1 Cuckoo Search Core Component

The Cuckoo Search core component is defined by mapping the concepts from the Cuckoo Search Algorithm onto the concepts of business process mining (see Table 1).

In our approach, a business process model (BPM) is defined as a causal matrix [1] which is formally represented as:

$$CM = \{(In(a_1), Out(a_1)), (In(a_2), Out(a_2)), , (In(a_n), Out(a_n))\} \quad (1)$$

where a_i is an activity from the event log, $In(a_i)$ is the set of activity subsets that directly precede the activity a_i, $Out(a_i)$ is the set of activity subsets that directly follow the activity a_i, and n is the total number of activities from the event log considered.

Table 1. Mapping Cuckoo Search concepts to business process mining concepts

Cuckoo Search concepts	Business Process Mining concepts
Cuckoo	Agent with a business process model associated
Nest with egg	Container with a business process model
	model by applying a modification strategy based on genetic operations, Simulated Annealing or Tabu Search
Laying an egg	Replacing a container's business process model with the one associated to an agent
Nest destruction	Discard of a business process model within a container
Nest replacement	Replacement of the discarded business process model with a new business process model

To evaluate a business process model, the following fitness function (which takes values in the range $(-\infty, 1)$) is used [1]:

$$FDGA(CM) = F(L, CM, CMS) - \gamma * PF_{folding}(CM) \quad (2)$$

where L is the event log given, CMS is a set of causal matrices representing multiple process models derived from the event log and containing CM, γ is a penalty weight, $PF_{folding}(CM)$ is a component penalizing solutions having lowest fitness values, and $F(L, CM, CMS)$ evaluates the completeness and preciseness of a causal net by using the formula below [1]:

$$F(L, CM, CMS) = PF_{complete}(L, CM) - k * PF_{precise}(L, CM, CMS) \quad (3)$$

3.2 Hybridization Components

This sub-section presents the hybridization components that will be injected into the Cuckoo Search core component in order to obtain several hybridization variants for the classical Cuckoo Search algorithm.

Genetic Component. The genetic component uses a genetic mutation operator as defined in [1] so as to induce diversity in the population by introducing new genetic material. This is achieved through adding, removing or reorganizing causality relations in a BPM. Mutation is actually a simple strategy for modifying the business process model, which pushes the evolutionary direction of the population towards exploration.

Simulated Annealing Component. This component uses a Simulated Annealing based strategy for modifying a BPM. Simulated Annealing is a nature-inspired trajectory-based metaheuristic which is capable of escaping from local optima values. This metaheuristic focuses on accepting new solutions under certain conditions. The modification strategy implemented in the Simulated Annealing component consists in the following steps:

- A new BPM is generated in the neighborhood of the BPM which is currently being exploited. This is done by randomly generating a new BPM and then performing a crossover operation between the current BPM and the randomly-generated one. Out of the two business process models thus resulted, the BPM having higher fitness value is retained and further used during the algorithm.
- If the new model has greater fitness than the current model, then the current BPM is always replaced by the newly-generated model. Otherwise, the current BPM is only sometimes replaced by the newly-generated model, with the following probability [10]:

$$P = e^{\frac{-(FDGA(CM') - FDGA(CM))}{T}} \quad (4)$$

  where CM is the causal matrix associated to the current BPM, CM' is the causal matrix of the newly generated BPM, $FDGA$ is the fitness function (see Formula 2) and T is the temperature.
- The temperature is decreased at each step in order to lower the probability of a worse model being accepted later in the search process.

Tabu Search Component. This component uses a Tabu Search based strategy for modifying a BPM. Tabu Search [8] is a trajectory-based metaheuristic which uses short-term and long-term memory structures to guide the search for an optimal or near-optimal solution for an optimization problem. The modification strategy implemented in the Tabu Search component consists in the following steps:

- A new BPM is generated in the neighborhood of the currently processed model similar to the strategy for neighborhood generation used in the Simulated Annealing component.
- The newly generated BPM is searched for in the short-term memory and if the model has already been visited (i.e. the model is in the short-term memory), then the previous step is re-executed in order to attempt generating another model. If the model has not been visited, then the next step is performed.
- The currently generated BPM is placed in the short-term memory after calculating its fitness function. Thus, the model is marked as having been already visited, and it is added into a search trace.

Finally, once all the iterations have been completed, the search trace is scanned and the BPM with the maximum fitness among the visited models is returned.

4 Hybrid Cuckoo Search-Based Algorithms for Business Process Mining

This section presents the hybrid variants for the Cuckoo Search-based algorithm, which we obtained by plugging the hybridization components presented in Section 3.2 into the Cuckoo Search core component. In what follows, we first present the Cuckoo Search-based algorithm for business process mining [6] that corresponds to the Cuckoo Search core component. We then mark the algorithm steps in which the hybridization components are plugged in (see Algorithm 1). The algorithm takes as inputs the following parameters: (*i*) L - the event log from which we want to mine a process model, (*ii*) n - maximum number of generations, (*iii*) cn - number of cuckoos, (*iv*) nn - number of nests, (*v*) f - threshold for the fitness of an acceptable model, and (*vi*) el - proportion of worst models to discard at each step. The algorithm returns either the best business process model identified in the predefined number of iterations, or the business process model with a fitness equal or higher than the threshold for the fitness of an acceptable model.

Algorithm 1: Cuckoo Search-based Algorithm for Business Process Mining

```
1  Input: L, n, cn, nn, f, el
2  Output: cm_opt
3  begin
4     CuckooSet = Create_BPM_Set(L, cn)
5     NestSet = Create_BPM_Set(L, nn)
6     while( (i ≤ n) and (MaxFitness(CuckooSet, NestSet) < f)) do
7        Cuckoo = Get_Random_Model(Cuckoo_Set)
8        Nest = Get_Random_Model(NestSet)
9        [ Cuckoo = Perform_Levy_Flight(Cuckoo) ]
10       if (Fitness(Cuckoo) > Fitness(Nest)) then
11          Nest = Copy(Cuckoo)
12       end if
13       CuckooSet = Sort_Set(CuckooSet)
14       NestSet = Sort_Set(NestSet)
15       [ NestSet = Replace_Worst_Models(NestSet, el) ]
16    return Max_Fitness_Model(CuckooSet ∪ NestSet)
17 end
```

The algorithm consists of an initialization stage and an iterative stage. In the initialization stage, the sets of cuckoos and nests are initialized with candidate business process models based on the event log given as input.

In the iterative stage, the following steps are executed iteratively until the stopping condition is satisfied:

- Randomly select one BPM from the set of cuckoos and one from the set of nests. These two models will be used in the next step for the Levy flight and for the attempt to "lay an egg".
- The cuckoo model selected at the previous step undergoes a modification which is akin to the concept of a Levy flight within the solution space. This is the first step at which hybridization takes place by injecting one of the three

hybridization components defined in Section 3.2. Once this cuckoo process model has been modified, its fitness is recalculated and it is compared to the fitness of the model contained within the nest selected at the previous step. If the fitness of the model associated to the cuckoo is higher than the fitness of the model in the nest, then the model associated to the cuckoo is copied into the nest.
- The two sets of business process models, the one associated to the cuckoos and the one corresponding to the nests, are sorted in descending order of the fitness function value of the process model.
- The worst business process models in the set of nests may be replaced with newly generated models. The chance for this to happen mirrors the probability with which cuckoo eggs are discovered by the host bird in the nature; this probability is used to determine the fraction of models that have to be replaced. The current step is also seen as a hybridization point, because it involves creating a new set of models. This set contains some of the worst extant models as well as new models generated by using the modification strategy of the chosen hybridization component. Once the models are created, their fitness is calculated and they are added back into the set of nests.

By employing the hybridization components presented in Section 3.2, we propose 6 different hybrid variants for the Cuckoo Search-based algorithm for business process mining. The CSM algorithm variant uses mutation to perform the Levy flight, without applying any optimization on the new process models which are used to replace the discarded unfit models at each step. The CSMM algorithm variant uses mutation both for Levy flight and for optimizing the process of generating new models to replace the worst ones. The CSA algorithm variant uses Simulated Annealing for Levy flight, without applying optimizations in the process of generating new models to replace the worst ones at each step. The CSMA algorithm variant employes mutation to perform the Levy flight, as well as Simulated Annealing to induce modifications on the process models added to the population as substitutes for the solutions discarded at each step of the algorithm. The CST algorithm variant applies Tabu Search for Levy flights and does not perform any operations for generating new models to replace the worst ones. The CSTM algorithm variant employes Tabu Search for performing Levy flights and it uses mutation for generating new models to replace the worst ones.

5 Performance Evaluation

We evaluated our hybrid variants proposed for the Cuckoo Search-based algorithm, by implementing them as plugins for the PROM process mining framework [13]. The hybrid variants of the Cuckoo Search-based algorithm have been evaluated on a set of event logs provided in [12]. The performance of the hybrid algorithm variants depends on the following adjustable parameters: number of cuckoos, number of nests, percent of nests to be destroyed, mutation rate, cooling factor, and short-term memory size. Thus, we first performed a set of preliminary experiments aiming to identify the optimal values for these parameters.

For the parameters number of cuckoos, number of nests, and percent of nests to be destroyed, we have observed that the values 15, 15, and 25% respectively, as recommended in [6], indeed lead to the best results. In the case of mutation rate, we have varied its values in the range [0.1, 1] with an increment of 0.1, and we have observed that a value of 0.5 provides the best results. The cooling factor has been ranged from 0.2 to 0.8 with an increment of 0.2, and the value of 0.4 led to the best results. The values chosen for the short-term memory size are taken from roundings of the arithmetic progression which starts with initial value 2.5 and ratio 2.5, ending up to a value of 20. Experiments have shown that a value of 10 for the short-term memory size provides the best results.

In Table 2 we illustrate the experimental results obtained when running the hybrid variants of the Cuckoo Search-based algorithm (using the optimal values for the adjustable parameters) as compared to the Genetic Miner [1]. The input for all these experiments is an event log consisting of 132 process instances and 1642 events. As stopping condition for all the algorithms, we have used a predefined number of iterations and a specified threshold for the fitness value (i.e. 0.8 fitness value).

Table 2. Experimental results for the event log considered

Algorithm variant	Maximum fitness	Time(min:sec)
CSM	0.731	0:45
CSMM	0.672	0:47
CSA	0.72	5:34
CSMA	0.731	8:42
CST	0.810	2:08
CSTM	0.813	3:11
Genetic	0.735	7:04

From the experimental results presented in Table 2 it follows that the CST and CSTM algorithm variants provide the best results in terms of fitness value. Compared with Genetic Miner, the CST and CSTM algorithm variants provide better results in terms of both fitness value and execution time. To illustrate how the values of the maximum fitness and average fitness vary during the execution of the algorithm, we have plotted these values in Figures 1 to 6 for all the hybrid variants of the Cuckoo Search algorithm. As Figure 1 shows, in the case of the CSM algorithm, the average fitness of the population decreases sharply over time, with mutation inducing a too large amount of diversity in the population. The maximum fitness over the population remains relatively constant over a large portion of the execution of this algorithm variant.

Despite the short execution time (see Table 2), the CSMM algorithm variant is one of the hybridizations that perform poorly over the course of the given

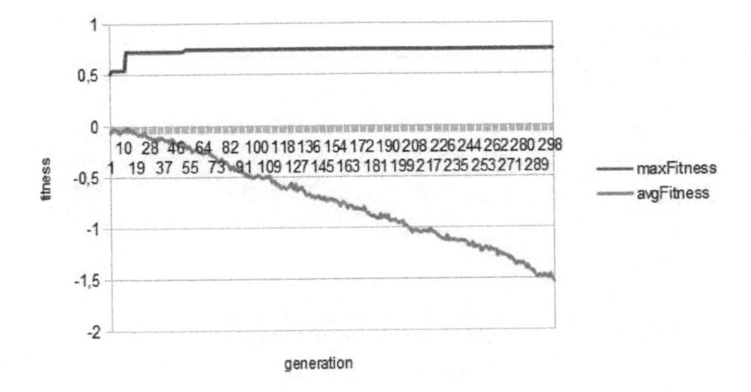

Fig. 1. Evolution of maximum fitness and average fitness for the CSM algorithm variant

number of generations. Inducing diversity in the population by random exploration is shown not to be a relatively viable solution, as the average fitness of the population decreases gradually over time, and more quickly than for the other solutions. The maximum fitness also increases more slowly, and it remains stuck in constant plateaus over a long series of iterations.

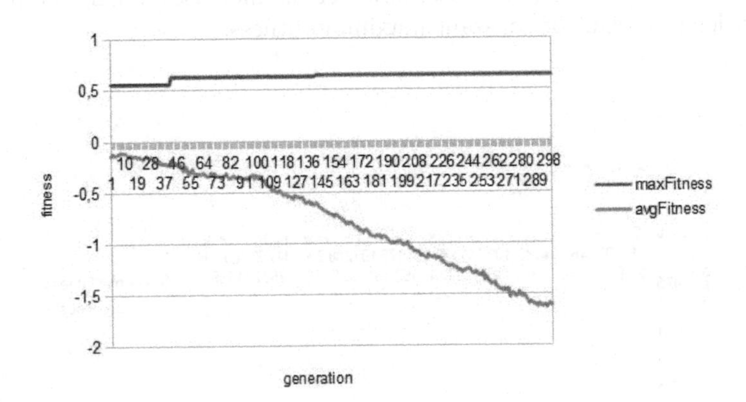

Fig. 2. Evolution of maximum fitness and average fitness for the CSMM algorithm variant

In the case of the CSA algorithm variant, the difference is visible on the evolution of the algorithm as compared to the previous variants. The average fitness of the population remains positive for most of the algorithm's running time, and it generally increases from beginning to end. At the start of the algorithm, we notice that the maximum fitness over the population has taken a dive. This can be attributed to the solution acceptance tactic of the Simulated Annealing

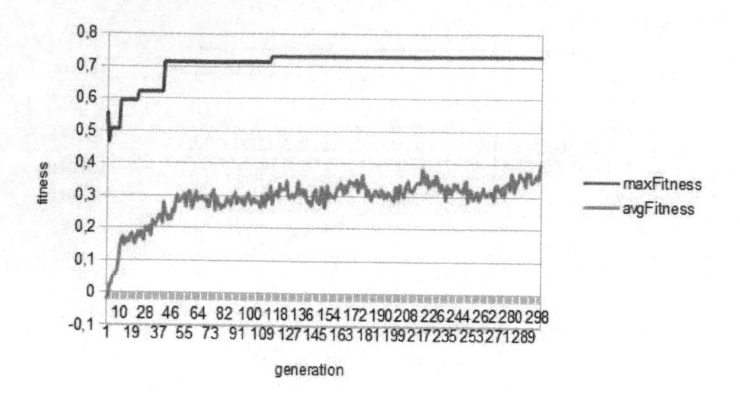

Fig. 3. Evolution of maximum fitness and average fitness for the CSA algorithm variant

algorithm, and models which undergo modifications are replaced with models that are not as fit, in order to escape from local extremes in the search space.

In the case of the CSMA algorithm variant, the mutation component induces a too large amount of diversity in the population, with the average fitness of the population decreasing gradually over time. However, a small increase is noticed in the average fitness at the start of the execution. The maximum fitness also experiences a more gradual increase; however the increases mostly happen after relatively long plateaus of constant maximum fitness.

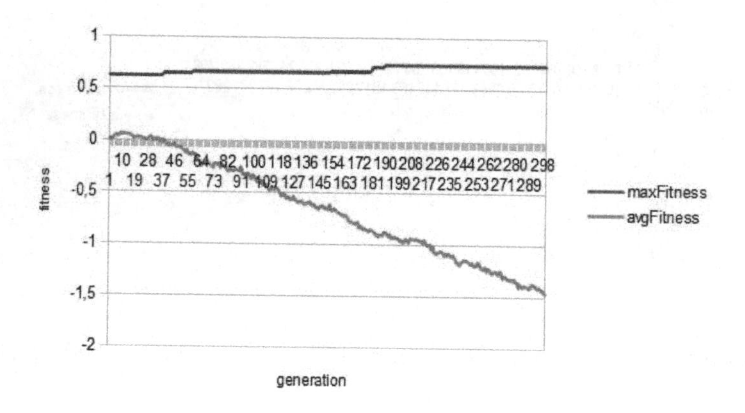

Fig. 4. Evolution of maximum fitness and average fitness for the CSMA algorithm variant

The CST algorithm variant terminates before the maximum number of iterations is reached, proving the efficiency of using Tabu Search as the hybridization component involved in Levy flight. The algorithm ends with the fitness value of the best individual above 0.8, while the average fitness of the population increases gradually over the course of the algorithm's execution.

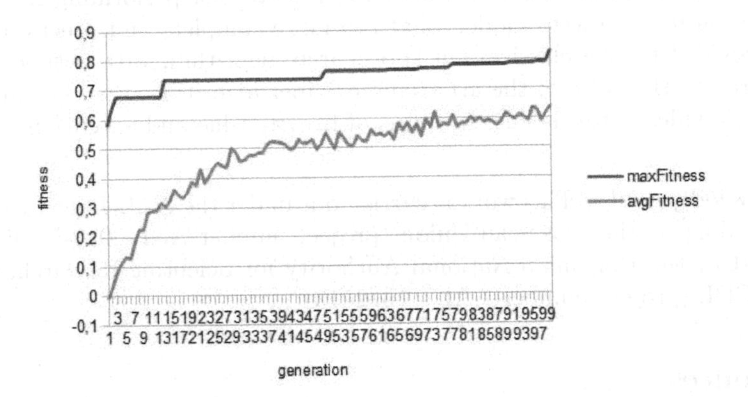

Fig. 5. Evolution of maximum fitness and average fitness for the CST algorithm variant

In the case of the CSTM algorithm variant, the mutation component once again decreases the average fitness of the population over the course of the execution. Moreover, the maximum fitness overcomes the acceptable fitness threshold after a larger number of iterations than the previous variants. The climb towards the maximum fitness value is also done more slowly, in smaller increments.

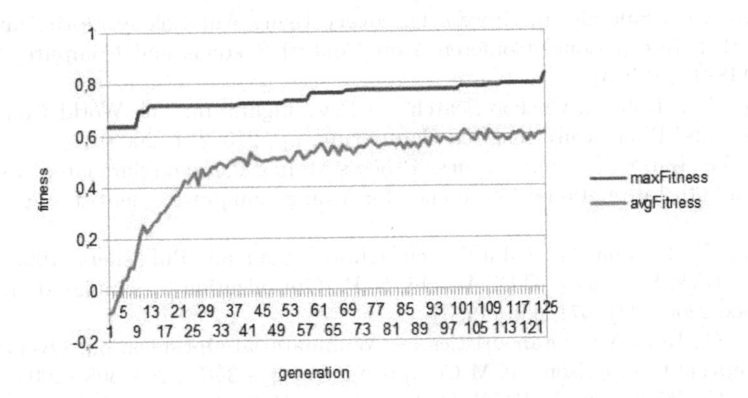

Fig. 6. Evolution of maximum fitness and average fitness for the CSTM algorithm variant

6 Conclusions and Future Work

This paper presented six hybridization variants for the Cuckoo Search algorithm in the context of business process mining. These variants have been obtained by hybridizing the Cuckoo Search with principles from genetic algorithms, Simulated Annealing, and Tabu Search. Experimental results have demonstrated that the CST algorithm variant which uses Tabu Search for performing Levy flights provides business process models that are more complete and precise than the ones provided by the classical Cuckoo Search algorithm, with a time penalty. Compared to the state of the art Genetic Miner algorithm, the CST algorithm variant provides better results in terms of fitness value and execution time.

Acknowledgments. This work is carried out under the AAL Joint Programme with funding by the European Union (project number AAL-2012-5-195) and is suported by the Romanian National Authority for Scientific Research, CCCDI UEFISCDI, (project number AAL - 16/2013).

References

1. Alves de Medeiros, A.K.: Genetic Process Mining, PhD Thesis (2006)
2. van der Aalst, W.M.P., Weijters, A.J.M.M., Maruster, L.: Workflow Mining: Discovering Process Models from Event Logs. IEEE Transactions on Knowledge and Data Engineering 16(9), 1128–1142 (2004)
3. Weijters, A.J.M.M., van der Aalst, W.M.P.: Rediscovering Workflow Models from Event-Based Data Using Little Thumb. Integrated Computer-Aided Engineering 10(2), 151–162 (2003)
4. Turner, C.J., Tiwari, A., Mehnen, J.: A Genetic Programming Approach to Business Process Mining. In: The 10th Annual Conference on Genetic and Evolutionary Computation, pp. 1307–1314 (2008)
5. Chinces, D., Salomie, I.: Process Discovery Using Ant Colony Optimization. In: The 19th International Conference on Control Systems and Computer Science, pp. 448–454 (2013)
6. Yang, X.S., Deb, S.: Cuckoo Search via Levy flights. In: The World Congress on Nature and Biologically Inspired Computing, pp. 210–214 (2009)
7. Song, W., Liu, S., Liu, Q.: Business Process Mining Based on Simulated Annealing. In: The 9th International Conference for Young Computer Scientists, pp. 725–730 (2008)
8. Glover, F., Laguna, M.: Tabu Search. Kluwer Academic Publishers (1997)
9. Kirkpatrick, S., Gelatt, C.D., Vecchi, M.P.: Optimization by simulated annealing. Science 220(4598), 671–680 (1983)
10. Blum, C., Roli, A.: Metaheuristics in Combinatorial Optimization: Overview and Conceptual Comparison. ACM Computing Surveys 35(3), 268–308 (2003)
11. Blum, C., Puchinger, J., Raidl, G.R., Roli, A.: Hybrid metaheuristics in combinatorial optimization: A survey. Applied Soft Computing 11, 4135–4151 (2011)
12. `http://www.processmining.org/`
13. ProM framework, `http://www.processmining.org/prom/start`

Direct Particle Swarm Repetitive Controller with Time-Distributed Calculations for Real Time Implementation

Piotr Biernat, Bartlomiej Ufnalski, and Lech M. Grzesiak

Institute of Control and Industrial Electronics,
Faculty of Electrical Engineering, Warsaw University of Technology,
75 Koszykowa Str., Warsaw 00-662, Poland
{piotr.biernat,bartlomiej.ufnalski,lech.grzesiak}@ee.pw.edu.pl
http://www.ee.pw.edu.pl

Abstract. In this paper, real-time implementation of recently developed direct particle swarm controller for repetitive process is presented. The proposed controller solves the dynamic optimization problem of shaping the control signal in the voltage source inverter. The challenges in real time implementation come from limited sampling period to evaluate a candidate solution. In this paper, the solution of PDPSRC time-distributed calculation is presented. This method can be implemented using digital signal controllers (DSC).

Keywords: particle swarm optimization (PSO), time-distributed calculations, real-time implementation.

1 Introduction

The particle swarm optimization is a population-based evolutionary algorithm (EA) modelled after biological swarming such as bird flocking, fish schooling, herding of land animals or collective behavior of insects. This method is usually applied to find the best solution of static, non-changing problems. However, many real world problems are dynamic i.e. variable load of the inverter. In this case, the PSO algorithm has to track the progression of the function optimum through the space as closely as possible. In [1] the plug-in direct particle swarm repetitive controller (PDPSRC) for the sine-wave constant-amplitude constant-frequency voltage source inverter (CACF VSI) is presented. To find the best control signal, the swarm itself is a repetitive controller cooperating in parallel with the controller shaping the signal behaviour along the pass (in the p-direction). The algorithm maintains a swarm of individuals called particles, where each particle stores all samples of the control signal along the pass. During the exploration process, all particles are rated according to user-defined cost function. The detailed explanation of the proposed algorithm is presented in [8].

In the basic control algorithm [2], all the samples have to be remembered and rated in the end of pass or trial. Although the calculations to update one sample of the particle are not computational burden, updating all samples in all particles in off-the-shelf

P. Angelov et al. (eds.), *Intelligent Systems'2014*,
Advances in Intelligent Systems and Computing 322, DOI: 10.1007/978-3-319-11313-5_44

industrial microcontrollers, most of the time cannot be completed within one sampling period.

The real-time implementation in microcontrollers of the PSO algorithm [3, 4] is much more complicated compared to a simulation model [5, 6]. There issues that need to be considered such as limited calculation time (due to sampling time) and size of memory. In view of these drawbacks, this paper introduces PDPSRC time-distribution calculation which prevent from conducting all calculations in one sample time. The main idea is to split up processes that can be calculated incrementally, e.g. calculating mean squared error (MSE), and do not require knowledge of a whole signal. The accuracy of the proposed code has been verified in a numerical simulation.

2 Particle Swarm Optimization Time-Distributed Calculations

In a particle swarm optimization, each particle represents a possible solution. Particles travel through search space in cooperation with other particles and are rated by the user defined cost function. The position of the particle is then determined based on the best solution found for the particle itself and the best solution in its surroundings.

In a conventional approach, updates of all particles are conducted after the last sample of the swarm based on gathered particles' response data. Therefore, the real-time system requires huge computing power. However, instead of using more powerful hardware, the PDPSRC algorithm should be organized in such a way that the calculation will be performed in every sample time.

The major problem is how to organise the algorithm computational burden variables such as fitness function (calculated for a whole particle). The proposed solution is to divide bigger problems into smaller ones calculated in every sample time. The flow chart consisting of time-distributed PDPSRC calculations is presented in Fig. 1. The colours green, yellow and red indicate respectively calculations performed in each sample time, after each particle and after a whole swarm.

The particles start with some random position and random speed. Next, the control signal is applied to the plant and the response is gathered. For each particle, the algorithm has to calculate their fitness:

- $VE_j(p)$ – error component, difference between the reference voltage and filter capacitor voltage for *j-th* particle *p-th* sample,
- $D_j(p)$ – control dynamics component of *j-th* particle *p-th* sample,
- $Fitness_j$ – fitness function of *j-th* particle

The pseudo-code is presented in Algorithm 1.

Algorithm 1: Calculation of fitness function

```
VEj=VEj(p-1)+(u^err)^2
Dj=Dj(p-1)+(qj(p)-qj(p-1))^2
if (end of particle){
VEj=VEj*δ1
Dj=Dj*δ2
Fitnessj= VEj+Dj*β}
```

The first two functions, $VE_j(p)$ and $D_j(p)$, are calculated by incrementing them in each sample time and β is the penalty factor for control dynamics. After the last sample of the particles, the fitness function is calculated by summing them. Next, the coefficients are set to zero for the next particle. Factor δ_1 is the reciprocal of the number of samples in a single particle and δ_2 is also the reciprocal of the number of particle samples, but without the first one.

A diversity [7] calculation of the swarm can also be distributed in every single sample time. First, the algorithm has to find the maximum and minimum of the particles' position.

Algorithm 2: Searching maximum and minimum of particles position.

```
if (swarm_j(p)>div_max(p))
    div_max(p)= swarm_j(p);

if (swarm_j(p)<div_min(p))
    div_min(p)= swarm_j(p);
```

Each *p-th* sample in all of the particles is compared to find the highest and lowest position. After every swarm iteration the diversity is calculated in every sample time, based on previous iterations *div_min* and *div_max* variables.

Algorithm 3: Calculation of diversity and choosing the repel or attract mode

```
swarm_div=0.5*(div_max(p)-div_min(p);
if(swarm_div<diversity_limit)
    swarm_dir=-1.0;
else(swarm_div>=diversity_limit)
    swarm_dir=1.0;
```

The result is then compared by the user defined diversity threshold and switching between attract (*swarm_dir=1)* and repel *(swarm_dir=-1)* modes.

After the last sample of the particle, the algorithm has to check if the actual cost function is better than the corresponding personal best fitness. If the proposed solution is better the table consisting of the best solutions is updated, if not the actual solution evaporate at the constant rate ρ.

Algorithm 4: Updating new personal best solution or evaporation actual solution

```
if (end of particle)
  if (fitness_j is less then pbest_FF* )
    Pbest_table.update();
    pbest_FF = fitness_j
  else
    pbest_FF = FF_j*
  end
end
```

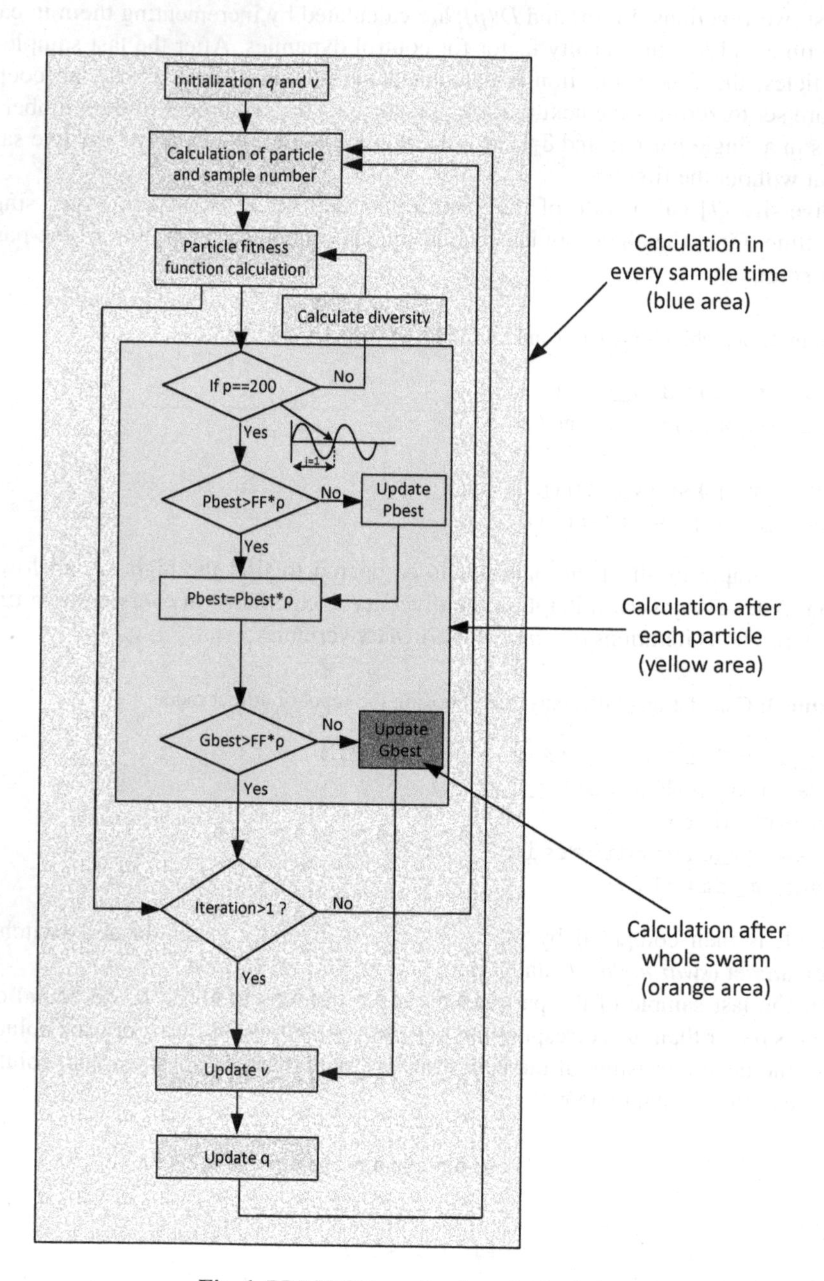

Fig. 1. PDPSRC time-distributed calculations

In the same sampling period the algorithm is checking if the current cost function value is better than the best solution found so far in the given swarm.

Algorithm 5: Checking if actual solution is better than previous one

```
if (fitness_j < Gbest_FF)
    Gbest=pbest_j;
    Gbest_FF= fitness_j;
end
if (last sample of swarm)
Gbest_table.update();
end
```

Data stored in global best solutions should be updated after the last sample of the swarm in synchronous mode.

In the first loop, the algorithm only tests the proposed solutions and gathers the required data: fitness function, personal best solutions, global best solution and swarm diversity. From the second loop, the speed and position update rules mechanism is applied. The position of the particles is then determined based on the best solution for the particle itself and its surroundings. Updated speed for the corresponding particle matrix element is calculated in every single sample period. The new velocity and position is calculated as follows

Algorithm 6: Particles position and velocity with clamping updates

```
if(second loop or more){
  update velocity v_j(p+1)=eq.(1)
  velocity clamping v_j(p)=V_clmp(p)
  update position q_j(p+1)=q_j(p)+v_j(p)
}
```

The new velocity is calculated by:

$$v_j^{p+1} = c_1 v_j^p + c_2 r_q \delta\left(q_j^{pbest} - q_j^p\right) + c_3 r_g \delta\left(q_j^{gbest} - q_j^p\right) \tag{1}$$

where: c_1, c_2 and c_3 are the inertia, cognitive and social coefficients, and r_q, r_g are random numbers between 0 and 1. Random numbers are uniformly distributed in the unit interval and are generated in each sample for each particle in every iteration. There are many different solutions for PSO randomisation like: the particle-wise randomization, the dimension-wise randomization, the list-based PSO, the list-based PSO with a memory-cheap table [8].

The algorithm of tracking the minimum of the cost function is then repeated until the system is in operation. A graph that represents the distribution of the calculation in time is presented in Fig. 1. The green tags represents the calculation carried out in every algorithm entrance to PWM interruption, the yellow ones constitute the calculations after every particle, and red ones after each trial.

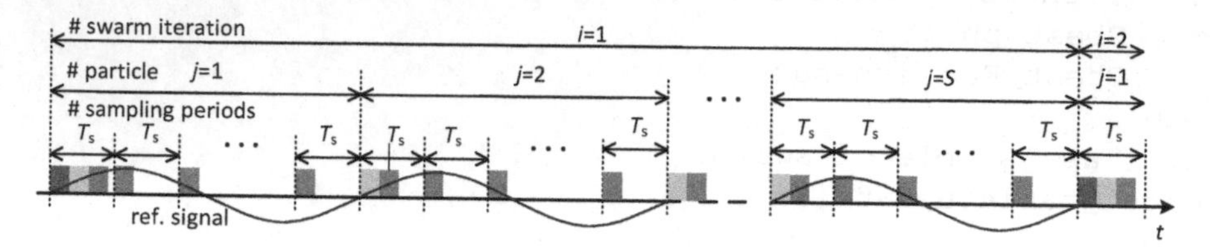

Fig. 2. Time-distributed calculations for real time implementation

In table 1 the parts of the PDPSRC are presented with explanation in which moment they can be calculated.

Table 1. Time-distribution of PDPSRC algorithm

Function	Fig. 1 marker	Time-distributed
Fitness function	green	Yes (calculated in each sample)
Swarm diversity	green	Yes (calculated in each sample)
Velocity update	green	Yes (calculated in each sample)
Position update	green	Yes (calculated in each sample)
Find personal best	yellow	Partly (calculated after each particle)
Personal best update	yellow	Partly (calculated after each particle)
Find global best	yellow	Partly (checking after each particle)
Global best update (synchronous mode)	red	No (updated after whole swarm)

3 Model of the Plant and Numerical Experiment Results

The time-distributed PSO code has been tested using the simulation model in Matlab/Simulink and Plecs toolbox [9] presented in Fig. 2. A plug-in direct particle swarm repetitive controller has been used as a k-direction controller (pass to pass) and it is required to support it with a controller in the p-direction (sample to sample). As a non-repetitive controller (p-direction), the full-state feedback has been implemented to increase damping in the plant.

In the simulation model the PWM voltage source inverter is approximated using a controlled voltage source with the LC filter on the output. Selected parameters of the model and swarm have been given in Table 3 and 4.

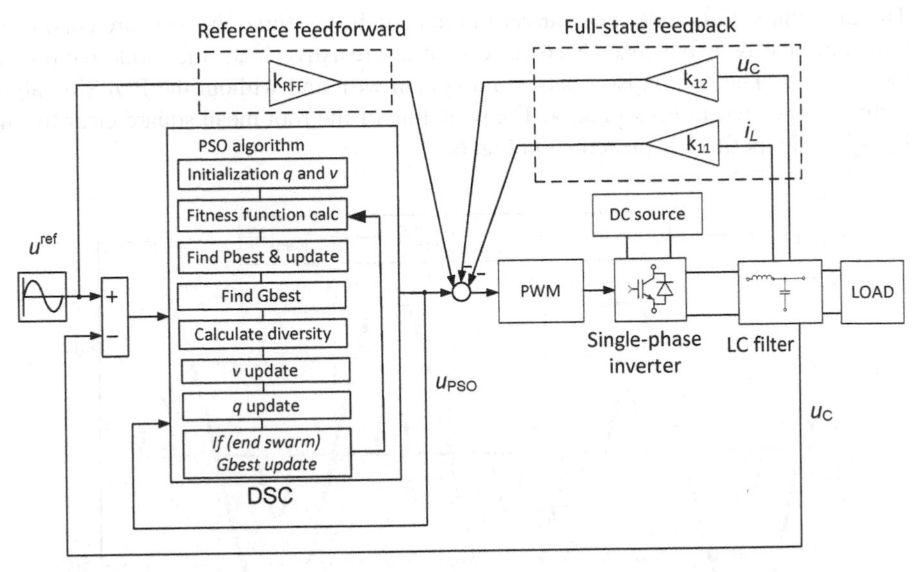

Fig. 3. Schematic diagram of the proposed repetitive control system

Table 2. Parameters of the simulation model

Parameter	Symbol	Value
DC-link Voltage	U_{DC}	450 V_{DC}
Reference voltage	U^{ref}	325 V
Reference frequency	f^{ref}	50 Hz
Filter inductance	L_f	300 μH
Filter capacitance	C_f	160 μF
Switching frequency	f_{sw}	10 kHz
Load power	-	ca. 5kW

Table 3. Parameters of the swarm

Dimensionality of the problem	α	200
Number of particles	S	25
Evaporation constant	ρ	1.1
Penalty factor	β	0.25
Velocity clamping level	V_{clmp}	9.0

The behaviour of the plant with full state feedback and without PSO controller is depicted in Fig. 3. The simulation model has been tested under two kinds of loads:

1. Resistive load (ca. 3kW), applied for first 150s,
2. Diode rectifier (ca. 2kW), applied from 150s to 500s.

The algorithm starts with the near zero initial condition. First, the tests are conducted with only the resistive load. Next instead of the resistive load, the diode rectifier is switched on. The comparison between a system with and without the PDPSRC algorithm is presented in Fig. 3 and 4. The evolution of the root mean square error for the PSO-based repetitive is presented in Fig. 6.

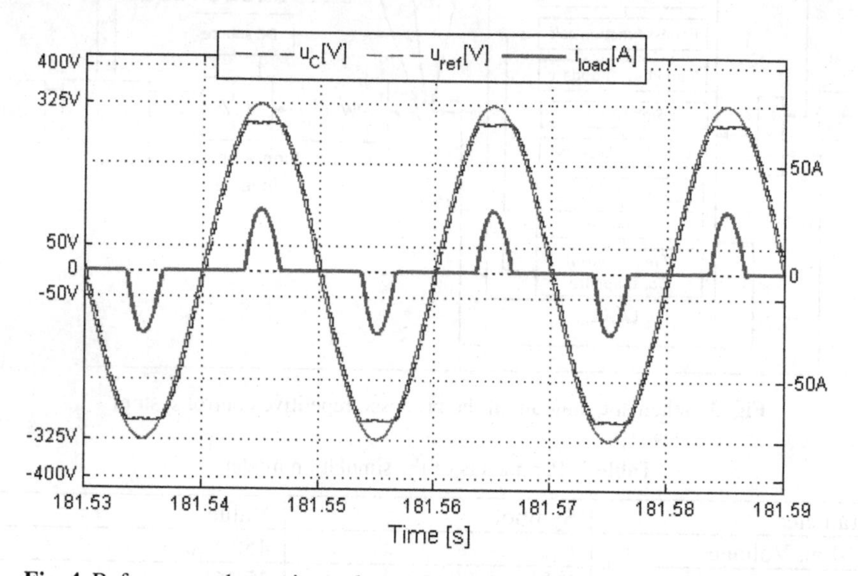

Fig. 4. Reference and capacitor voltage without PDPSRC (only FSF is implemented)

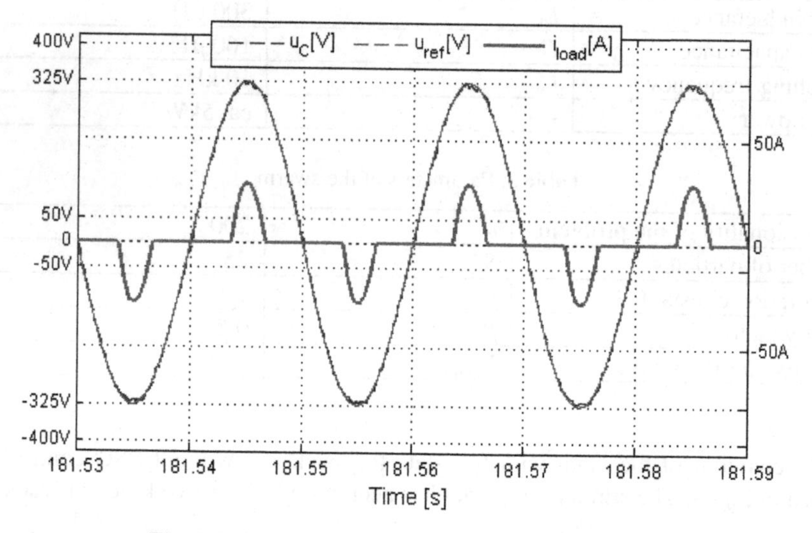

Fig. 5. Reference and capacitor voltage with PDPSRC

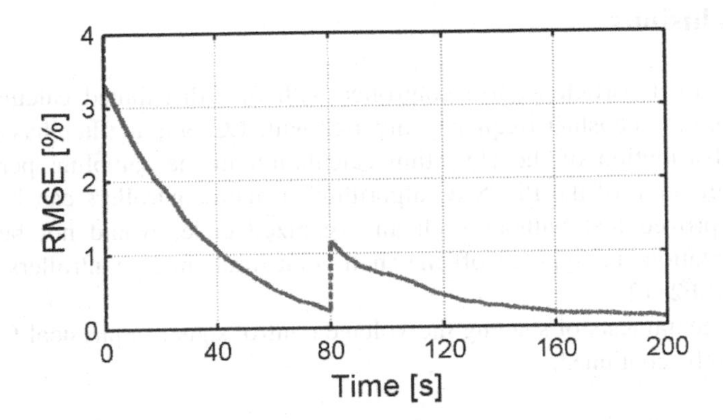

Fig. 6. Evolution of the root mean square error for the PSO-based repetitive controller (in percent of reference voltage)

The PSPSRC time–distributed calculations reduce the computational burden of the algorithm. Implementation tests on the TMS320F2812 controller using code composer studio [10] show that the time required for calculations in one sample period does not exceed 20μs (green+yellow+red tags). Therefore, time-distributed calculation makes it possible to implement the algorithm in off-the-shelf industrial microcontrollers. During the implementation procedure, an additional problem shows up. This problem is connected with the controller's memory size, which should be large enough to store about 15000 samples (Table 4).

Table 4. Data need to be stored in PDPSRC (for sampling time 10kHz, reference frequency 50Hz, swarm size 25 particles)

Parameter	Type	Number of samples
Swarm position	Float	5000
Swarm velocity	Float	5000
Personal best	Float	5000
Global best	Float	200
Diversity	Float	400
Others	Float + integer	ca. 100
	TOTAL	ca. 15700

For example the digital signal controller eZdspTMS320F2812 has a 18k-word internal memory and additional 64K words off-chip SRAM memory. The float type variable needs 4 bytes of memory space. Thus, for all variables the required space must be at least 60kB. Therefore, to solve this issue the controller external memory must be used, on which the swarm samples will be stored.

4 Conclusions

The plug-in direct particle swarm controller with time-distributed calculations for constant-amplitude constant-frequency inverter with LC output filter has been presented. By distribution of the algorithm calculation in the sampling period, real-time implementation of the PDPSRC algorithm in microcontrollers can be realized. It has been proved that without such an organized code, would not be possible its implementation in typical off-the-shelf industrial microcontrollers such as eZdspTMS320F2812.

An experimental tests of shaping the voltage control signal in physical CACF VSI are planned to be continued.

Acknowledgements. The research work was supported by the statutory funds of the Electrical Drive Division for 2014.

References

1. Ufnalski, B., Grzesiak, L.M.: A plug-in direct particle swarm repetitive controller for a single-phase inverter. Electrical Review (Przeglad Elektrotechniczny, paper in English, open access at http://pe.org.pl), 1–6 (2014)
2. Eberhart, R.C., Shi, Y., Kennedy, J.: Swarm Intelligence (The Morgan Kaufmann Series in Evolutionary Computation), 1st edn. Morgan Kaufmann Publishers (April 2001)
3. Liu, W., Chung, I., Cartes, D., Leng, S.: Real-Time Particle Swarm Optimization based Current Harmonic Cancellation based PMSM Parameter Identification. In: IEEE Power & Energy Society General Meeting, PES 2009 (2009)
4. Liu, W., Liu, L., Cartes, D.: Efforts on Real-Time Implementation of PSO. In: 2008 IEEE Power and Energy Society General Meeting - Conversion and Delivery of Electrical Energy in the 21st Century (2008)
5. Huang, J., Li, H., Guo, J.: Application of PSO in the improved real-time evolution. In: 11th International Conference on Hybrid Intelligent Systems (2011)
6. Liu, L., Cartes, D.: Real time implementation of particle swarm optimization based model parameter identification and an application example. in: IEEE Congress on Evolutionary Computation (CEC 2008) (2008)
7. Zhan, Z., Zhang, J., Shi, Y.: Experimental study on PSO Diversity. In: Third International Workshop on Advanced Computational Intelligence (2010)
8. Ufnalski, B., Grzesiak, L.M.: A comparative investigation on different randomness schemes in the particle-swarm-based repetitive controller for the sine-wave inverter. In: 7th International Conference on Intelligent Systems IS 2014 (2014)
9. Plecs simulation software for Power Electronics, ver.3.5.3, http://www.plexim.com/
10. Code composer studio – an integrated development environment, ver. 3.3, http://www.ti.com/tool/ccstudio

Artificial Fish Swarm Algorithm for Energy-Efficient Routing Technique

Asmaa Osama Helmy[1], Shaimaa Ahmed[1], and Aboul Ella Hassenian[2,3]

[1] Faculty of Engineering, Zagazig University,Zagazig, Egypt
[2] Faculty of Computers and Information, Cairo University, Egypt
[3] Scientific Research Group in Egypt (SRGE), Egypt
http://www.egyptscience.net

Abstract. Wireless Sensor Network consists of an enormous number of small disposable sensors which have limited energy. The sensor nodes equipped with limited power sources. Therefore, efficiently utilizing sensor nodes energy can maintain a prolonged network lifetime. This paper proposes an optimized hierarchical routing technique which aims to reduce the energy consumption and prolong network lifetime. In this technique, the selection of optimal cluster heads (CHs) locations is based on Artificial Fish Swarm Algorithm (AFSA). Various behaviors in AFSA such as preying, swarming, and following are applied to select the best locations of CHs. A fitness function is used to compare between these behaviors to select the best CHs. The model developed is simulated in MATLAB. Simulation results show the stability and efficiency of the proposed technique. The results are obtained in terms of number of alive nodes and the energy residual mean value after some communication rounds. To prove the AFSA efficiency of energy consumption, we have compared it to LEACH and PSO. Simulation results show that the proposed method outperforms both LEACH and PSO in terms of first node die (FND) round, total data received by base station, network lifetime, and energy consume per round.

1 Introduction

A Wireless Sensor Network (WSN) is a collection of sensor nodes that have some properties like low cost, low power, limited network lifetime etc. These sensor nodes are deployed to the region of interest (Area monitoring, Air pollution monitoring, Forest fire detection, Water quality monitoring etc.) for gathering data, each sensor collects data from the monitored area then it routs data back to the base station (BS). Initially WSN was developed for military and civilians purpose, now it is extended to wide range of applications such as Disaster management and Habitat monitoring [1], Target tracking and Security management [2], Medical and health care [3], home automation and traffic control [4–6], machine failure diagnosis and energy management [7], Rescue missions, climate change, earthquake warning and monitoring the enemy territory [4]. Due to this large number of applications, WSN has been identified as one of the pioneer technique in 21st century.

P. Angelov et al. (eds.), *Intelligent Systems'2014*,
Advances in Intelligent Systems and Computing 322, DOI: 10.1007/978-3-319-11313-5_45

For such applications, it is clear that routing protocols are required in order to insure an efficient communication by transmitting data between sensor nodes and the base stations. This has led to quite a number of different protocols operating in different layers of the network, with the goal of allowing the network working autonomously for as long as possible while maintaining data channels and network processing to provide the application's required quality of service QoS.

Many routing, power management, and data dissemination protocols have been specifically designed for WSNs where energy awareness is an essential design issue. Classifies routing in WSNs into the following three categories:

- Network structure (flat-based routing, hierarchical-based routing, and location-based routing).
- Protocol operation (multipath-based, query-based, negotiation-based, QoS-based, or coherent-based routing techniques).
- Spin, Directed diffusion, Pegasis, Qos based, Coherent based.

The most energy aware WSN routing techniques are based in hierarchical routing. The main aim of hierarchical routing is to efficiently maintain the energy consumption of sensor nodes by involving them in multi-hop communication within a particular cluster and by performing data aggregation and fusion in order to decrease the number of transmitted messages to the sink. Cluster formation is typically based on the energy reserve of sensors and sensor's proximity to the cluster head. However, most hierarchical techniques use single-hop routing where each node can transmit directly to the cluster-head and the sink. Therefore, it is not applicable to networks deployed in large regions. Furthermore, the idea of dynamic clustering brings extra overhead, e.g. head changes, advertisements etc., which may diminish the gain in energy consumption. One of the applied methods for clustering enhancement is the use of the swarm intelligence algorithms such as particle swarm optimization [9], and artificial fish swarm algorithm [10].

Swarm Intelligence is a modern method employed in optimization problems. It is based on the en masse movement of living animals like birds, fishes, ants and other social animals. Migration, seeking for food and fighting with enemies are social behaviors of animals. Artificial fish swarm algorithm (AFSA) was presented by Li Xiao Lei in 2002 [9]. This algorithm is a technique based on swarm behaviors that was inspired from social behaviors of fish swarm in nature. AFSA works based on population, random search, and behaviorism. This algorithm has been used in optimization applications, such as clustering [11], machine learning [12], PID control [13], data mining [14], image segmentation [4]. This algorithm is one of the best approaches of the Swarm Intelligence method with considerable advantages like high convergence speed, flexibility, error tolerance and high accuracy [8, 16]. In this paper, we optimize the cluster head selection by using AFSA. In cluster routing protocol, energy consumption is concentrated on CHs. The disadvantages of LEACH protocol are mainly the using of the local information and the randomness in selecting the clusters number and the initial clustering center. Our main contribution in this paper is proposing an optimized hierarchical routing technique which aims to reduce the energy consumption and prolong network lifetime. In this technique, the selection of optimal cluster heads (CHs) locations is based on AFSA.

The remainder of the paper is organized as follows: Section 2 depicts the related work. In section 3, the proposed algorithm is presented. Section 4 studies the experiments and analyzes the results. Section 5 concludes the paper.

2 Related Works

Hierarchical routing in WSN involves the arrangement of clusters in form of hierarchy when sending information from the sensor nodes to the base station. Hierarchical routing efficiently reduces energy consumption by employing multi-hop communication for a specific cluster and thus performing aggregation of data and fusion in a way that decreases the number of data carried across the network to the sink. Cluster formation is based on residual energy in the sensor nodes and election of a CH. A very good example of hierarchical routing protocol is low-energy adaptive clustering hierarchy (LEACH) [17]. The LEACH approach involves formation of clusters of sensor nodes centered on the received signal quality and the use of a local CH as a router to the BS. Clustering is a method by which sensor nodes are hierarchically organized on the basis of their relative proximity to each other. Clustering of sensor nodes helps to compress the routing table such that the discovery mode between sensor nodes is done more easily.

Threshold sensitive energy efficient sensor network protocol (TEEN) [18] is a hierarchical protocol. It is useful for time-critical applications in which the network operates in a reactive way. Closer nodes form clusters and elect a cluster head. Each cluster head is responsible for directly sending the data to the sink. After the clusters are formed, the cluster head broadcasts two thresholds to the nodes. Adaptive Periodic Threshold-sensitive Energy Efficient sensor Network protocol (APTEEN) [19] is an extension to TEEN. The main features of these protocols are that it combines proactive and reactive policies and modification of parameters that allow better control in the cluster heads. APTEEN supports three different query types: historical, to analyze past data values; one-time, to take a snapshot view of the network; and persistent to monitor an event for a period of time. The experiments have demonstrated that APTEEN's performance is between LEACH and TEEN in terms of energy dissipation and network lifetime. TEEN gives the best performance since it decreases the number of transmissions. The main drawbacks of the two approaches are the overhead and complexity of forming clusters in multiple levels, implementing threshold-based functions and dealing with attribute-based naming of queries.

In recent years, insect sensory systems have been inspirational to new communications and computing models like bio-inspired routing. It is due to their ability to support features like autonomous, and self-organized adaptive communication systems for pervasive environments like WSN and mobile ad hoc networks. Biological synchronization phenomena have great potential to enable distributed and scalable synchronization algorithms for WSN [20]. In [21], an adaptation of

Ant Colony Optimization (ACO) technique is demonstrated for network routing. This approach belongs to the class of routing algorithms inspired by the behavior of the ant colonies in locating and storing food. In Bio-inspired routing [22], The most popular ACO (Ant Colony Optimization) is a colony of artificial ants is used to construct solutions guided by the pheromone trails and heuristic information they are not strong or very intelligent; but they successfully make the colony a highly organized society. Swarms are useful in many optimization problems. A swarm of agents is used in a stochastic algorithm to obtain near optimum solutions to complex, non-linear optimization problems [23].

Minimum Ant-based Data Fusion Tree (MADFT) [24] is a sink selection heuristic routing algorithm .It is based on ACO for gathering correlated data in WSN. It first assigns ants to source nodes. Then, the route is constructed by one of the ants in which other ants search the nearest point of previous discovered route. The chosen formula is Probability function composed of pheromones and costs in order to find the minimum total cost path. MADFT not only optimizes over both the transmission and fusion costs, but also adopts ant colony system to achieve the optimal solution. Swarm Intelligence Optimization Based Routing Algorithm [25] works with the objective to balance global energy consumption and avoiding some node's premature energy exhausting. The algorithm chooses the nodes with less pheromone as next hop, taking less hop numbers into consideration. The algorithm is different from traditional ant colony algorithms. It is better than the directed diffusion routing protocol both in end-to-end delay and global energy balance. It can effectively balance the global energy consumption and prolong the network lifetime.

3 The Proposed System

The proposed hierarchical routing protocol is based on the principle of clustering algorithm. In cluster routing protocol, energy consumption is concentrated on CHs. The cluster heads have to collect the data from other neighboring nodes. After data aggregation in CHs, it has to forward the aggregated information to the base station. This section describes in details the proposed algorithm where the CHs selection is optimized by using $AFSA$. In the proposed technique, clusters are formed geographically. Geographical formation of cluster sizes is based on equal segmentation of area space, depending on the case being considered. Due to draining activities being constraint on a cluster head during data aggregation and transfer phase, the cluster head is rotated among the sensor nodes of each cluster at every transmission round. A completely new estimation of energy is carried out at the beginning of every transmission round to elect a new CH for the cluster and thereby energy wastage is being reduce to its minimum, and utilization of each nodes energy is being maximized to ensure a prolong network lifetime.

3.1 CH Election Process

After the cluster formation phase, the CH election phase proceeds. The selection of CHs within each cluster formed is carried out by electing a node that require less transmission energy (to BS or to the next hop CH nearer to the BS) to be the CH for a particular transmission round. This is achieved by applying Prey, Swarm, and Following behaviors. Each behavior provides a set of CHs. The fitness equation is used on the three sets to select the optimal CHs set. The smallest fitness represents combination among the others. Then the base station broadcasts the optimized CHs candidates to the network. The hierarchical routing technique consists of four main stages:

- Geographical formation of cluster.
- Selection of cluster heads in each cluster formed by using AFSA (subsection 3.2).
- Data aggregation phase which involves the gathering of collected data by the cluster head from the sensor nodes within its cluster.
- Data transmission phase which involves the transfer of all data from the nearest cluster head(s) to the BS.

3.2 CHs Selection by Using AFSA

The flow process of the optimal CHs selection by using $AFSA$ is described in Algorithm (1).

Algorithm 1. CHs selection by using AFSA

1: Select *ntree*, the number of trees to grow, and *mtry*, a number no larger than the number of variables.
2: Simulation parameters setting, M represent number of clusters, *none* is number of total nodes, v is the visual distance δ is the crowd factor where $(0 < \delta < 1)$ and the number of rounds, and the initial CHs set.
3: Applying AFSA behaviors: The AFSA behaviors; swarming, preying and following are applied. Each behavior produces a set of CHs.
4: Choose the optimal CHs set
5: Calculate fitness for each behavior's CHs set (3.2).
6: Choose the behavior set that gives the minimum fitness
7: 5: broadcast cluster heads candidates Base station broadcasts the optimal CHS candidates to the network.

3.3 Evaluating the AFSA Behaviors

Each $AFSA$ behavior outputs a set of CHs. To choose the best CHs set the fitness function is used. The fitness function is described in Equ. (1). The smallest fitness represents the best CHs set among others.

$$fintness = \alpha f_1 + (1 - \alpha) f_2 \tag{1}$$

Where f_1 and f_2 are defined as follows:

$$f_1 = \sum_{i=1}^{N} E(n_i) / \sum_{m=1}^{M} E(CH_m) \quad (2)$$

$$f_2 = \max \frac{\sum_{i \in m} d(n_1, CH_m)}{|C_m|} \quad (3)$$

where f_1 is the energy representation part, and it is equal to the sum of all member node energy $E(n_i)$ (not including CH) divided by the sum of all CH energy $E(CH_m)$. Referring to Equ.(2) f_2 represents the density and it is equal to cluster with highest average distance between CH and joined member nodes $d(n_i, CH_m)$ divided by the total member nodes in the same cluster $|C_m|$ as shown in equ.(3).

4 Simulation and Results

[H] The proposed algorithm discussed above is implemented using MATLAB 7.0 on HP Core2Duo laptop, 2.5GHZ, and 3GB RAM. To evaluate its performance, a total number of 250 nodes were randomly deployed within a space region on $300mx300m$. Table (1) shows the parameters setting, and Figure (1) illustrates the simulated environment of the 250 nodes we deployed. The coordinates of X and Y are measured in meters.

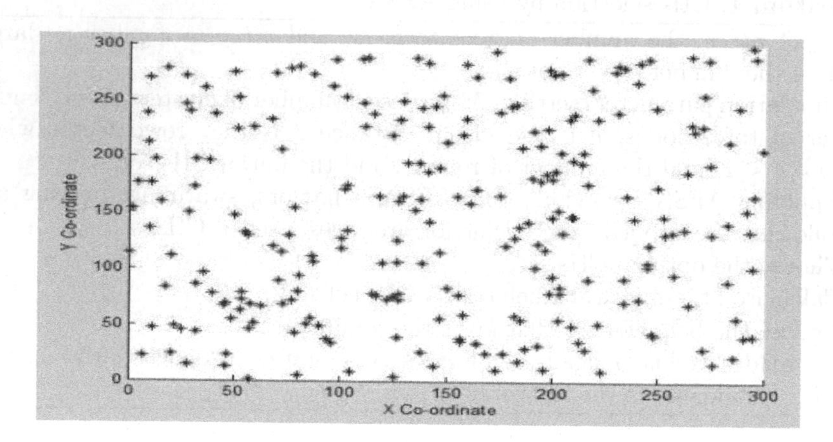

Fig. 1. 250 nodes deployed randomly in a geographical location of $300 \times 300m^2$

4.1 Radio Energy Model

The expended energy during transmission and reception for bits data to a distance between transmitter and receiver is given by equ. (4). Figure (2) shows

Table 1. Simulation and setting parameters

Parameters	Values
Simulation Rounds	400
Network size	300*300
Number of Nodes	250
Node distribution	Nodes are uniformly distributed
Data Packet size	4000 bits
BS position	(100,100)
Initial energy of node	0.2 joule
Energy dissipation	10 pJ per bit
Energy for Transmission	50 pj per bit
Energy for Reception	50 pj perbit
Energy for data aggregation	5nj per bit
Visual	2.5
Step	0.05
Iteration No.	200
Compression coefficient	0.05

the first order radio energy model. In this radio energy model, sensor nodes are assumed to have the ability to control the transmission power. It use minimum transmit power to transmit the data to its destination with acceptable signal to noise ratio (SNR).

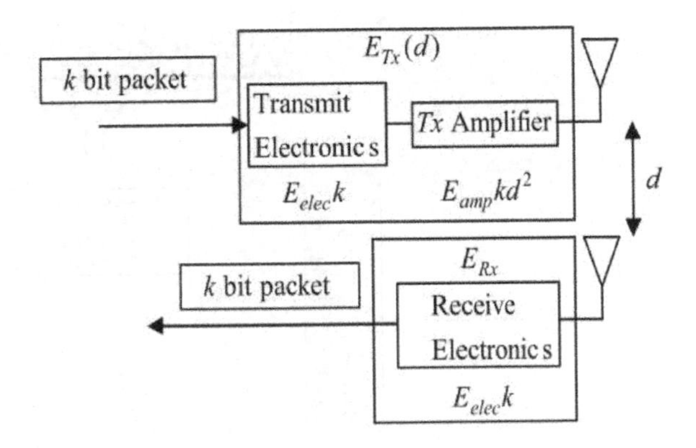

Fig. 2. Radio energy model

The first order radio model offers an evaluation of energy consumed when transmission or reception is made by a sensor node at each cycle. The radio has a power control to expend minimum energy required to reach the intended recipients. Mathematically, when a k-bit message is transmitted through a distance, d, required energy can be expressed as stated in the equation:

$$E_{TX} = E_{elect}.k + E_{amp}.d^2.k \tag{4}$$

Likewise, the energy consumed at the reception is illustrated as shown in equ. (5):

$$E_{Rx} = E_{elect}.k \tag{5}$$

Where ETx_{elect} is the energy dissipated per bit at transmitter, ERx_{elect} is the energy dissipated per bit at receiver, E_{amp} is the amplification factor, E_{elect} is the cost of circuit energy when transmitting or receiving one bit of data, E_{amp} is the amplifier coefficient, K is the number of transmitted data bits, and d is the distance between a sensor node and its respective cluster head or between a CH to another cluster head nearer to the BS or between CH and BS. The first order radio equation was used to verify the operation of our proposed protocol, assuming that the radio channel being symmetric.

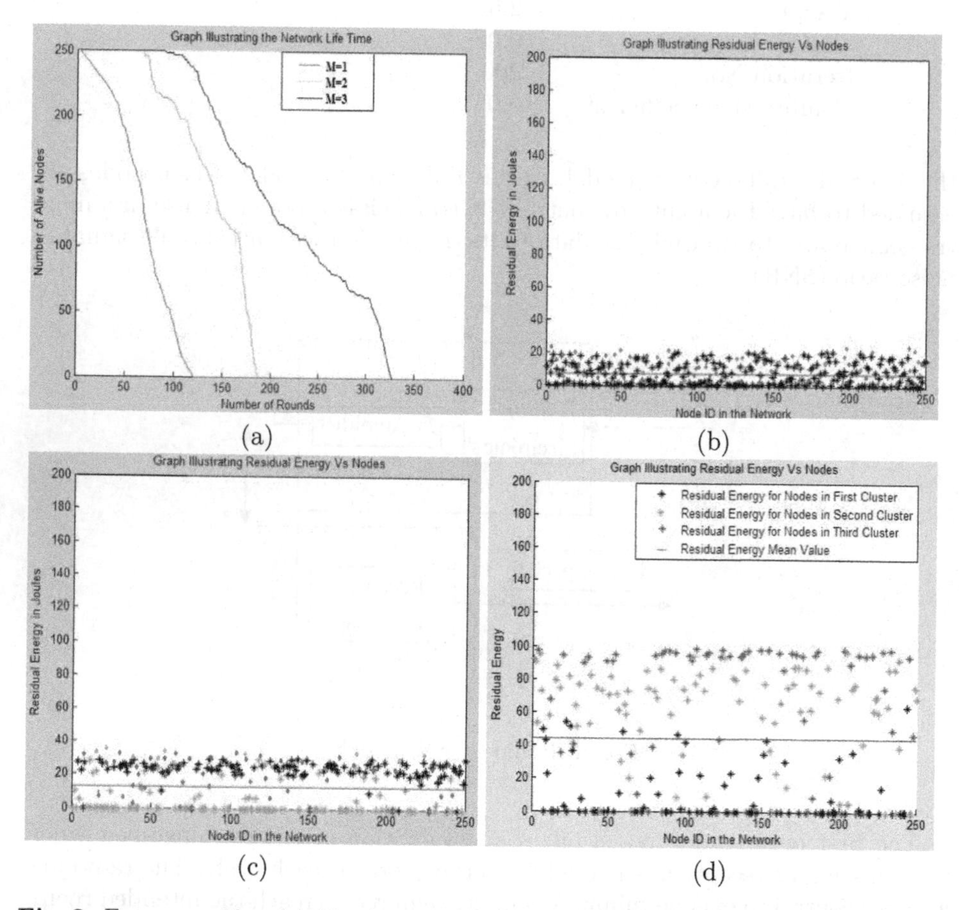

Fig. 3. Energy residue (a) Number of alive nodes for different clusters No.(M)s, (b) Nodes energy residue in LEACH technique different clusters No.(M)s, (c) Nodes energy residue in PSO technique for 400 rounds simulation, (d)Nodes energy residue in proposed technique for 400 rounds simulation

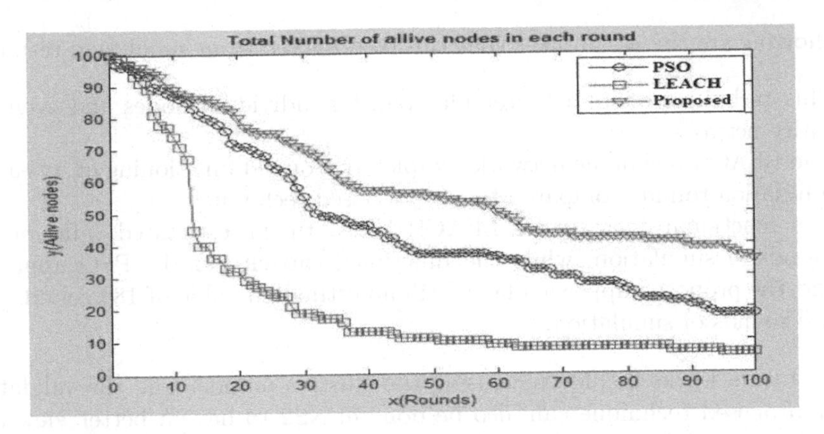

Fig. 4. Number of alive nodes in LEACH, PSO, and the proposed technique at different rounds

4.2 Experimental Results

Experiment-I analyses the performance of the proposed algorithm in terms of number of alive nodes at different clusters number and the residual energy at the network nodes. It is observed from Figure 3(a) that the network lifetime increased to a certain length in the three cluster formation scenario (M=3). With this increase, the WSN's lifetime was further prolonged when compared to the two cluster formation (M=2) and the non hierarchical technique (M=1).

Experiment-II compares the performance of the proposed technique to those of LEACH, and PSO techniques. Figure (3) shows that the mean residual energy value of all the sensor nodes of our proposed method is higher than that in cases of LEACH and PSO.

Figure (4) shows the result of alive nodes after each round for smaller number of nodes (100). The results of graphs clearly indicate that the number of active nodes in the proposed technique is greater than LEACH and PSO. In HEERP the simulation is run for 100 rounds.

5 Conclusion and Future Works

This paper proposes an optimized hierarchical routing technique which aims to reduce the energy consumption and prolong network lifetime. In this technique, the selection of optimal cluster heads (CHs) locations is based on AFSA. Applying AFSA into base station (BS) finds the best CHs formation by considering three parameters which are energy concentration, distance from BS, and centrality of sensor nodes selection. To prove the AFSA efficiency of energy consumption, we have compared it to both LEACH [17] and PSO [9]. Simulation results demonstrate that AFSA achieves better network lifetime over LEACH and PSO.

Following are the advantages that can be observed from simulation results:

- This technique offers a better life span for individual nodes and even the entire network.
- The LEACH technique network completely stopped functioning at an earlier simulation rounds compared to our proposed technique.
- The functional capacity for LEACH lasted till an estimated value of 120 rounds of simulation, while the functional capacity of the PSO approach and the proposed approach lasted till an estimated value of 180 rounds and 330 rounds of simulation.

The future works include increasing the clusters number and the validation of our proposed technique can also be done in NS2 to have a better view and understanding of the result analysis. Furthermore, in routing of packets in wireless sensor networks, some sensor node energy is wasted on relaying others data instead of using its energy own data. Therefore, limiting the energy of the nodes that has little or no energy is another area of future research. One can also take account the communication cost and its impact on network lifetime.

References

1. Akyildiz, I.F., Su, W., Sankarasubramaniam, Y., Cayirci, E.: Wireless sensor networks: A survey. Computer Networks 38(4), 393–422 (2002)
2. Estrin, D., Culler, D., Pister, K., Sukhatme, G.: Connecting the physical world with pervasive networks. IEEE Pervasive Computing, 59–69 (January-March 2002)
3. El-said, S.A., Hassanien, A.E.: Artificial Eye Vision Using Wireless Sensor Networks. In: Wireless Sensor Networks: Theory and Applications. CRC Press, Taylor and Francis Group (January 2013)
4. Culler, D., Estrin, D., Strivastava, M.: Overview of Sensor Networks. IEEE Computer Society 37(8), 41–49 (2004)
5. Liao, W., Chang, K., Kedia, S.: An Object Tracking Scheme for Wireless Sensor Networks using Data Mining Mechanism. In: Proceedings of the Network Operations and Management Symposium, Maui, HI, USA, pp. 526–529 (2012)
6. Ye, W., Heidemann, J., Estrin, D.: An energy-efficient MAC protocol for wireless sensor networks. In: Proceedings of the 21st Annual Joint Conference of the IEEE Computer and Communications Societies, vol. 3, pp. 1567–1576 (2002)
7. Wood, A., Stankovic, J., Virone, G., Selavo, L., Zhimin, H., Qiuhua, C., Thao, D., Yafeng, W., Lei, F., Stoleru, R.: Context-Aware wireless sensor networks for assisted living and residential monitoring. Network 22, 26–33 (2008)
8. Fathy, M.E., Hussein, A.S., Tolba, M.F.: Fundamental matrix estimation: a study of error criteria. Pattern Recognition Letters 32(2), 383–391 (2011)
9. Siew, Z.W., Wong, C.H., Chin, C.S., Kiring, A., Teo, K.T.K.: Cluster Heads Distribution of Wireless Sensor Networks via Adaptive Particle Swarm Optimization. In: Fourth International Conference on Computational Intelligence, Communication Systems and Networks, pp. 78–83 (2012)
10. Li, L.X., Shao, Z.J., Qian, J.X.: An Optimizing Method Based on Autonomous Animate: Fish Swarm Algorithm. In: Proceeding of System Engineering Theory and Practice, pp. 32–38 (2002)

11. Xiao, L.: A Clustering Algorithm Based on Artificial Fish school. In: 2nd International Conference on Computer Engineering and Technology, Chengdu, pp. 766–769 (2010)
12. Yazdani, D., Golyari, S., Meybodi, M.R.: A New Hybrid Algorithm for Optimization Based on Artificial Fish Swarm Algorithm and Cellular Learning Automata. In: 5th International Symposium on Telecommunication (IST), Tehran, pp. 932–937 (2010)
13. Luo, Y., Zhang, J., Li, X.: The Optimization of PID Controller Parameters Based on Artificial Fish Swarm Algorithm. In: IEEE International Conference on Automation and Logistics, Jinan, pp. 1058–1062 (2007)
14. Zhang, M., Shao, C., Li, M., Sun, J.: Mining Classification Rule with Artificial Fish Swarm. In: 6th World Congress on Intelligent Control and Automation, Dalian, pp. 5877–5881 (2006)
15. Li, C.X., Ying, Z., JunTao, S., Qing, S.J.: Method of Image Segmentation Based on Fuzzy C-means Clustering Algorithm and Artificial Fish Swarm Algorithm. In: International Conference on Intelligent Computing and Integrated Systems (ICISS), Guilin (2010)
16. Neshat, M., Adeli, A., Sepidnam, G., Sargolzaei, M., Toosi, A.N.: A Review of Artificial Fish Swarm Optimization Methods and Applications. International Journal on Smart Sensing and Intelligent Systems 5(1) (2012)
17. Heinzelman, W., Chandrakasan, A., Balakrishnan, H.: Energy-efficient communication protocol for wireless sensor networks. In: The Proceeding of the Hawaii International Conference System Sciences, Hawaii (January 2000)
18. Manjeshwar, A., Agrawal, D.: TEEN: a Routing Protocol for Enhanced Efficient in Wireless Sensor Networks. In: Proceedings of the 15th International Parallel and Distributed Processing Symposium, San Francisco, pp. 2009–2015 (April 2001)
19. Manjeshwar, A., Agrawal, D.P.: APTEEN: a hybrid protocol for efficient routing and comprehensive information retrieval in wireless sensor networks. In: Proceedings of the 2nd International Workshop on Parallel and Distributed Computing Issues in Wireless Networks and Mobile Computing, Ft. Lauderdale, FL (April 2002)
20. Atakan, B., Akan, O.B., Tugcu, T.: Bio-inspired Communications in Wireless. In: Guide to Wireless Sensor Networks, ch. 26, pp. 659–687. Springer-Verlag London Limited (2009)
21. Batra, N., Jain, A., Dhiman, S.: An Optimized Energy Efficient Routing Algorithm For Wireless Sensor Network. International Journal of Innovative Technology & Creative Engineering 1(5) (2011) ISSN: 2045-8711
22. Krings, A.W., Sam Ma, Z.: Bio-Inspired Computing and Communication in Wireless Ad Hoc and Sensor Networks. Ad Hoc Networks 7(4), 742–755 (2009)
23. Selvakennedy, S., Sinnappan, S., Shang, Y.: A biologically-inspired clustering protocol for wireless sensor networks. Computer Communications 30, 2786–2801 (2007)
24. Juan, L., Chen, S., Chao, Z.: Ant System Based Anycast Routing in Wireless Sensor Networks. In: International Conference on Wireless Communications, Networking and Mobile Computing, pp. 2420–2423 (2007)
25. Wang, C., Lin, Q.: Swarm intelligence optimization based routing algorithm for Wireless Sensor Networks. In: Proceedings of International Conference on Neural Networks and Signal Processing, pp. 136–141 (2008)

Part XI
Issues in Data Analysis and Data Mining

An Intelligent Flexible Querying Approach for Temporal Databases

Aymen Gammoudi[1,2], Allel Hadjali[2], and Boutheina Ben Yaghlane[3]

[1] ISGT, University of Tunis,
92, Boulevard 9 Avril 1938, 1007 Tunis, Tunisia
aymen.gammoudi@ensma.fr
[2] LIAS/ENSMA, 1 Avenue Clement Ader,
86960 Futuroscope Cedex, France
allel.hadjali@ensma.fr
[3] IHECT, University of Carthage,
IHEC-Carthage Presidence, 2016 Tunis, Tunisia
boutheina.yaghlane@ihec.rnu.tn

Abstract. Time is a crucial dimension in many application domains. This paper proposes an intelligent approach to querying temporal databases using fuzzy temporal criteria. Relying on fuzzy temporal Allen relations, a particular class of criteria are studied. First, a query language that supports flexible temporal query is discussed. Then, the architecture and the interface of the system developed are explicitly described. To evaluate intelligently temporal queries, our system is endowed with some reasoning capabilities.

Keywords: Temporal databases, temporal flexible queries, fuzzy Allen relations, fuzzy comparators.

1 Introduction

Temporal databases manage some temporal aspects (time, date ...) of the data they contain. In such databases, time-related attributes are not generally treated like other attributes, they although describe the properties of the same objects. This is because time-related attributes are considered to have an impact on the consistency of the set modeled by the database objects. Temporal information can often be expressed in imprecise/fuzzy manner. For example, periods of global revolutions are characterized by beginnings and endings naturally gradual and ill-defined (such as "well after early 20" or "to late 30"). Unfortunately, and to the best of our knowledge, there is not much work devoted to querying/handling fuzzy/imprecise information in a temporal databases context.

On the contrary, in Artificial Intelligence field, several works exist to represent and handle imprecise or uncertain information in temporal reasoning (see [4][11]). As mentioned above, only very few studies have considered the issue of modeling and handling flexible queries over regular/fuzzy temporal databases. Billiet et al. [1] have proposed an approach that integrates bipolar classifications to determine

P. Angelov et al. (eds.), *Intelligent Systems'2014*,
Advances in Intelligent Systems and Computing 322, DOI: 10.1007/978-3-319-11313-5_46

the degree of satisfaction of records by using both positive and negative imprecise and possibly temporal preferences. But this approach is still unable to model complex temporal relationships and not applied in historical temporal databases (for instance, the user may request one time period but reject a part of this period, when specifying the valid time constraint in the query). Deng et al. [7] have proposed a temporal extension to an extended ERT model to handle fuzzy numbers. They have specified a fuzzy temporal query which is an extension from the TQuel language and they have introduced the concepts of fuzzy temporal in specification expressions, selection, join and projection.

Tudorie et al. [2] have proposed a fuzzy model for vague temporal terms and their implication in queries' evaluation. Unfortunately, this approach does not allow to model a large number of temporal terms (such as: just after and much before). In [6], Galindo and Medina have proposed an extension of temporal fuzzy comparators and have introduced the notion of dates in Relational Databases (RDB) by adding two extra precise attributes on dates (VST, VET). Comparators proposed by Galindo and Medina do not treat all fuzzy temporal relations that may occur as well as combinations thereof. Note that some comparators might not be in full agreement with the intuitive semantics underlying the notion of the temporal relations that refer to (for instance, the relation [*a*, *a*'] Fuzzy_Overlaps [*b*, *b*'] cover the classical relation overlap and also the classical relation during, i.e., [*a*, *a*'] is included in [*b*, *b*']). Moreover, Galindo and Medinaś approach cannot support some sophisticated queries that need a step of reasoning before processing.

This paper is an attempt to filling the gap that exists in the field of querying temporal databases by means of flexible queries (involving fuzzy temporal criteria) that faithfully express users' needs. It relies on fuzzy temporal Allen relations and allows covering multiple and diverse fuzzy temporal relations (such as approximate equality between two temporal entities, more or less start and more or less finish relations) are supported by our approach, while they are untreated by Galindo and Medinaś approach. In addition, our proposal exhibits some reasoning on the basis of fuzzy Allen relations.

In section 2, a background on particular fuzzy comparators and fuzzy Allen relations is provided. Section 3 gives a critical review on the approach of Galindo and Medina that is the most closest to our proposal. Section 4 describes our approach for modeling and handling fuzzy temporal queries. An illustrated example is detailed in Section 5. Section 6 presents the interface of the system developed. Section 7 concludes the paper.

2 Background

The purpose of this background is twofold. We begin by recalling the fuzzy comparators modeled by approximate equalities or graded strict inequalities. Then, we present a fuzzy extension of temporal Allen relations[1]. This section is mainly browsed from [4].

---

[1] Allen approach [5] is a well-known and a powerful formalism for expressing and reasoning on relations between interval-based periods.

2.1 Fuzzy Comparators

We recall fuzzy comparators defined in terms of difference of values and the inference rules that express the basic properties of such comparators.

Approximate Equalities and Graded Inequalities. *An approximate equality* between two values, here representing dates, modeled by a fuzzy relation E with membership function μ_E (E stands for "equal"), can be defined in terms of a distance such as the absolute value of the difference. Namely,

$$\mu_E(x, y) = \mu_L(|x - y|)$$

For simplicity, fuzzy sets and fuzzy relations are assumed to be defined on the real line. Approximate equality can be represented by
$\forall x, y \in \mathbb{R}$
$\mu_E(x, y) = \mu_L(|x - y|) = max(0, min(1, \frac{\delta+\varepsilon-|x-y|}{\varepsilon})) =$

$$\begin{cases} 1 & if\,|x - y| \leq \delta \\ 0 & if\,|x - y| > \delta + \varepsilon \\ \frac{\delta+\varepsilon-|x-y|}{\varepsilon} & otherwise \end{cases}$$

where δ and ε are respectively positive and strictly positive parameters which affect the approximate equality. With the following intended meaning: the possible values of the difference a - b are restricted by the fuzzy set L=$(-\delta - \varepsilon, -\delta, \delta, \delta + \varepsilon)$ [2]. In particular a E(0) b means $a = b$. Similarly, a more or less strong inequality can be modeled by a fuzzy relation G (G stands for "greater"), of the form

$$\mu_G(x, y) = \mu_K(x - y)$$

We assume $\rho > 0$, i.e. G more demanding than the idea of "strictly greater" or "clearly greater". K=$(\lambda, \lambda + \rho, +\infty, +\infty)$ is a fuzzy interval which gathers all the values equal to or greater than a value fuzzily located between λ and $\lambda + \rho$. K is thus a fuzzy set of positive values with an increasing membership function. See Figure 1. According to the values of parameters $\lambda + \rho$, the modality, which indicates how much larger than b is a, may be linguistically labeled by "Slightly", "moderately", "much". In a given context. G(0) stands for '>'. The relation "smaller than" may be graded as well. Note that a G(K) $c \Leftrightarrow a$ - $c \in$ K $\Leftrightarrow c$ - $a \in K^{ant} \Leftrightarrow c\ S(K^{ant})a$ where S stands for smaller, and K^{ant} is the antonym of K defined by $\mu_{K^{ant}}(d) = \mu_K(-d)$. Thus, if a is much greater than c, c is much smaller than a. In the following we take
$\forall x, y \in \mathbb{R}$
$\mu_G(x, y) = \mu_K(x - y) = max(0, min(1, \frac{x-y-\lambda}{\rho})) =$

$$\begin{cases} 1 & if\,x > y + \lambda + \rho \\ 0 & if\,x \leq y + \lambda \\ \frac{x-y-\lambda}{\rho} & otherwise \end{cases}$$

[2] With the semantics A = (a, b, c, d), [a, d] (resp. [b, c]) is the core (resp. support).

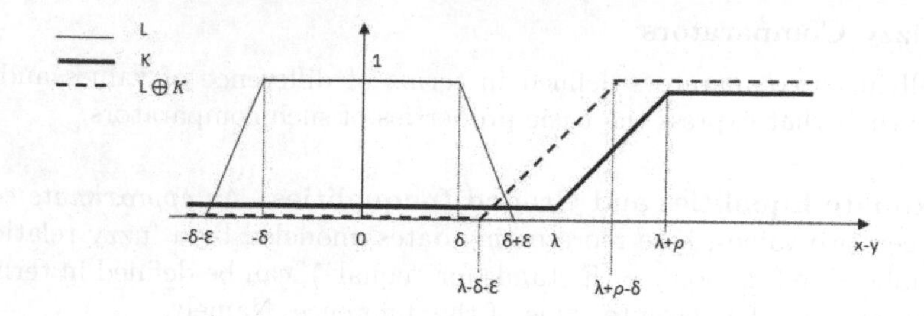

Fig. 1. Modeling "approximate equality" and "graded strict inequality"

Fuzzy Parameterized Inference Rules, taking advantage of the fact that the composition of the fuzzy relations E(L) and G(K) reduces to simple arithmetic operations on the fuzzy parameters K and L underlying the semantics of E and G. Fuzzily parameterized inference rules can be obtained. A set of inference rules describing the behaviors of the fuzzy comparators E and G have been established in [3]. For instance, we have $a\ E(L)\ b \wedge b\ G(K)\ c \Rightarrow a\ G(K \oplus L)\ c$, where $\oplus$ stands for the fuzzy addition.

2.2 Graded Allen Relations

Using the fuzzy parameterized comparators E(L) and G(K), fuzzy counterparts of Allen relations have been established [4]. The idea is that the relations which can hold between the endpoints of the intervals we consider may not be described in precise terms. For instance, we want to speak in terms of approximate equality (in the sense of E) rather in terms of precise equality in order to not introduce a brutal discontinuity between the case of a *"perfect"* meet relation and the case of a before relation when the upper bound of the first interval is close to the lower bound of the second interval. Then, in approximate terms, only two distinct relations may hold between two dates *a* and *b*. Indeed, a date *a* can be *"approximately equal"* to a date *b* in the sense of E(L), or *a* can be *"clearly different from"* *b* in the sense of not E(L). This last relation corresponds to *"much larger"* in the sense of G(K) or *"much smaller"* in the sense of S(K^{ant}). Let A = [*a*, *a'*] and B = [*b*, *b'*] be two time intervals, Table 1 summarizes the fuzzy Allen relations established (with L=$(-\delta-\varepsilon, -\delta, \delta, \delta+\varepsilon)$ and $L^c_+ = (\delta, \delta+\varepsilon, +\infty, +\infty)$).

In the modeling proposed above, a fuzzy Allen relation covers several situations corresponding to different ordinary Allen relations; for instance, fuzz-meets(L) covers the ordinary "meet" situation as well as situations as "slightly before" or "slight overlap". However here, the fuzzy parameter L controls to what extent we can shift from the ordinary "meet" situation, and provides a basis for the semantics of what "slightly" means in the above expressions. In the same way, we can see that F-equals(L) can cover the ordinary situation expressed by "slightly contains" or "slightly during".

Table 1. Fuzzy Allen Relations

Fuzzy Allen Relation	Definition	Label
A *F-before(L)* B B *F-after(L)* A	b $G(L_+^c)$ a'	*fb*(L) *fa*(L)
A *F-meets(L)* B B *F-met by(L)* A	a' E(L) b	*fm*(L) *fmi*(L)
A *F-overlaps(L)* B B *F-overlapped by(L)* A	b $G(L_+^c)$ a ∧ a' $G(L_+^c)$ b ∧ b' $G(L_+^c)$ a'	*fo*(L) *foi*(L)
A *F-during(L)* B B *F-contains(L)* A	a $G(L_+^c)$ b ∧ b' $G(L_+^c)$a'	*fd*(L) *fdi*(L)
A *F-starts(L)* B B *F-started by(L)* A	a E(L) b ∧ b' $G(L_+^c)$ a'	*fs*(L) *fsi*(L)
A *F-finishes(L)* B B *F-finished by(L)* A	a' E(L) b' ∧ a $G(L_+^c)$ b	*ff*(L) *ffi*(L)
A *F-equals(L)* B B *F-equals(L)* A	a E(L) b ∧ b' $E(L)$ a'	*fe*(L) *fe*(L)

3 Galindo and Medinaś Approach

Galindo and Medina [6] have proposed an extension of SQL (named FSQL) to allow expressing flexible queries. To do this, they extended the SELECT command. The main extensions added to the SQL command are:

- Linguistic labels: each label has a trapezoidal possibility distribution.
- Fuzzy comparators: in addition of common comparators (=, > ...), FSQL includes fuzzy comparators. For instance:
  Fuzzy Equal : CDEG(A FEQ B)= $\sup_{u \in U}$ {min(A(u), B(u))}
  Fuzzy Greater : CDEG(A FGT B)=

$$\begin{cases} 1 & if \gamma_A \geq \delta_B \\ \frac{\delta_A - \gamma_B}{(\delta_B - \gamma_B) - (\gamma_A - \delta_A)} & if \gamma_A < \delta_B \; y \; \delta_A > \gamma_B \\ 0 & in\, other\, case (\delta_A \leq \gamma_B) \end{cases}$$

  where U denotes the underlying domain or Universe, A=$[\alpha_A, \beta_A, \gamma_A, \delta_A]$, B=$[\alpha_B, \beta_B, \gamma_B, \delta_B]$, CDEG: a function, used in the select list, that shows a column with the fulfillment degree of the query condition for the specified attribute.
- Fulfillment thresholds (γ) for each condition. Which can be established with the format: <condition> THOLD γ, indicating that the condition must be fulfilled with a minimum degree of γ.

Galindo and Medina have also developed a tool *FSQL Server* PL/SQL to get the answers to FSQL queries and work with crisp or fuzzy databases. To incorporate temporal dates in the RDB (Relational DataBase), Galindo and Medina have added two additional attributes whose data type is one of the types of predefined time: VST (Valid Start Time) and VET (Valid End Time).

Table 2. Fuzzy comparators

Expression		Equivalence
$[t.VST, t.VET]$ F_INCLUDES	$[T1, T2]$	T1 FGEQ t.VST AND T2 FLEQ t.VET
$[t.VST, t.VET]$ F_INCLUDED_IN	$[T1, T2]$	T1 FLEQ t.VST AND T2 FGEQ t.VET
$[t.VST, t.VET]$ F_OVERLAPS	$[T1, T2]$	T1 FLEQ t.VET AND T2 FGEQ t.VST
$[t.VST, t.VET]$ F_BEFORE	$[T1, T2]$	T1 FGEQ t.VET
$[t.VST, t.VET]$ F_AFTER	$[T1, T2]$	T2 FLEQ t.VST
$[t.VST, t.VET]$ F_XBEFORE	$[T1, T2]$	T1 FGT t.VET
$[t.VST, t.VET]$ F_XAFTER	$[T1, T2]$	T2 FLT t.VST
$[t.VST, t.VET]$F_MUCH_BEFORE	$[T1, T2]$	T1 MGT t.VET
$[t.VST, t.VET]$ F_MUCH_AFTER	$[T1, T2]$	T2 MLT t.VST

These attributes in a tuple t express that its information is valid in the real world only during the time period $[t.VST, t.VET]$. So to select all versions of tuple that were valid on certain point of time T or were valid on a certain period of time $[T1, T2]$, the specified point of time or the time period may be fuzzy or crisp values and they are compared with the period of validity of each version of tuple t: $[t.VST, t.VET]$. Table 2 presents the extension of crisp comparators to possibility/necessity comparators.

Let us take an example from [6]: "retrieve all tuple versions that were valid (at any point) during around 1994 (in minimum degree 0.6)". Galindo and Medina transform this request into another FSQL query, through two stages : i)make fuzzy time (during 1994) at a specific interval ((1994, 01, 01), (1994, 12, 31)), ii) choose a fuzzy comparator, among those defined in Table 2, which responds to the request. The condition of the query will then write: [t.VST, t.VET] F-OVERLAPS [#(1994,1,1), #(1994,12,31)] THOLD 0.6. Unfortunately, Galindo and Medina's fuzzy temporal comparators for solving flexible queries, are unable to handle some kind of queries. For example:

- Select the tuples valid only from the end of [T1, T2] and does not exceed the end of the period [T3, T4].
- Select the valid tuples just at the beginning of the past two years or during the two months of the year.

Consequently Galindo and Medina's approach neglects the accuracy and precision of equality at the bounds of the intervals to compare. On the other hand, Galindo and Medina's approach is incapable of reasoning on fuzzy temporal relations to answer some complex user query. While our proposal can exhibit some aspect of reasoning on the basis of the introduced fuzzy Allen relations. For instance, if we have $A\ fb(L_1)\ B$ and $B\ fb(L_2)\ C$ then $A\ fb(L_1 \oplus L_2)\ C$.

Let us take the following example "select events that have their place just before the start of the Egyptian revolution, knowing that the end of the Tunisian revolution presents the beginning of the Egyptian revolution". Let us denote by

A (resp. B) the period related to the beginning of Egyptian revolution (resp. the end of Tunisian revolution). To answer the query at hand, it suffices to find periods of events (denoted by C) that have place just before A. To this end, on can first query the database to find the event such that B fmi(L) (i.e., F-met-by) C and then use the inference rule B fmi(L1) C and B fb(L2) (i.e., F-before) A to derive the result tuples that satisfy C fb(L) A. Hence, to solve this request a step of reasoning is required.

4 The Approach Proposed

4.1 Modeling and Language

The language, TSQLf, that supports our approach is an extension of SQLf language [10] by adding the dimension of time. The evaluation of flexible temporal queries expressed in TSQLf consists in two major steps: 1) conversion of the query in a standard SELECT query based on crisp criteria of research, 2) computing the satisfaction degree of the criteria for each candidate tuple. The structure of the three clauses: "select", "from" and "where" of the basic block SQL are preserved in TSQLf. Clause "from" does not change and the differences are mainly in two points: i) the calibration result can lead to a number of desired responses (denoted n) or a qualitative threshold (denoted by s), or both and ii) the nature of the conditions permitted. Accordingly, in TSQLf the formulation of the basic block is given by the following expression:
Select [distinct] [n — s — n, s] attributes
From relations
Where *Temp-cond-fuzzy*
Where "***Temp-cond-fuzzy***" may contain Boolean conditions and fuzzy temporal Allen relations. In the clause [Where] may appear various types of temporal fuzzy Allen relations or a combination of these, as well as join conditions.

Our approach allows to improve Galindo and Medina's approach. The idea is to use the principle of [6] to incorporate two additional attributes (VST and VET) and leverage fuzzy temporal Allen relations to better meet the needs of users expressing in TSQLf (see Table 3, with $t = [VST, VET]$ and $T = [T_1, T_2]$).

4.2 System Architecture

In this section, we present the architecture of the system developed to process an TSQLf query. The architecture presented in Figure 2 describes the different modules necessary to processing an TSQLf query. In the *Interface* module the user enters a gradual temporal query using a graphical interface, the latter gives him/her in the first place the possibility of choosing attributes, tables and build fuzzy temporal conditions, and second the system asks the user to identify the validity interval and the tolerance interval for each fuzzy temporal condition (i.e., the fuzzy parameter L). Thereafter, the system passes the request to the *Interpreting Query* module which transforms the request into a crisp query using (i) intervals of validity already introduced in the previous module, (ii) VST

Table 3. Temporal Fuzzy Allen Relations

Temporal Fuzzy Allen Relation	Definition	Label
t *F-before(L)* T T *F-after(L)* t	T1 $G(L_+^c)$ VET	*fb*(L) *fa*(L)
t *F-meets(L)* T T *F-met by(L)* t	VET E(L) T1	*fm*(L) *fmi*(L)
t *F-overlaps(L)* T T *F-overlapped by(L)* t	T1 $G(L_+^c)$ VST $\wedge$ VET $G(L_+^c)$ T1 $\wedge$ T2 $G(L_+^c)$ VET	*fo*(L) *foi*(L)
t *F-during(L)* T T *F-contains(L)* t	VST $G(L_+^c)$ T1 $\wedge$ T2 $G(L_+^c)$ VET	*fd*(L) *fdi*(L)
t *F-starts(L)* T T *F-started by(L)* t	VST E(L) T1 $\wedge$ T2 $G(L_+^c)$ VET	*fs*(L) *fsi*(L)
t *F-finishes(L)* T T *F-finished by(L)* t	VET E(L) T2 $\wedge$ VST $G(L_+^c)$ T1	*ff*(L) *ffi*(L)
t *F-equals(L)* T T *F-equals(L)* t	VST E(L) T1 $\wedge$ T2 $E(L)$ VET	*fe*(L) *fe*(L)

and VET attribute database and (iii) a fuzzy Allen relation that responds to the query. The result of this module is an TSQLf query to be sent to the management system database in order to select the attributes that meet the fuzzy temporal query criteria. The last module, *Computing Satisfaction Degree*, ensures the calculation of satisfaction degree of the TSQLf query at hand. The returned results are attached with a degree of satisfaction and then displayed on the user interface. In case where the TSQLf query requires a phase of reasoning, the system calls the module *Reasoning Step*. This module leverages the inference machinery of fuzzy Allen relations.

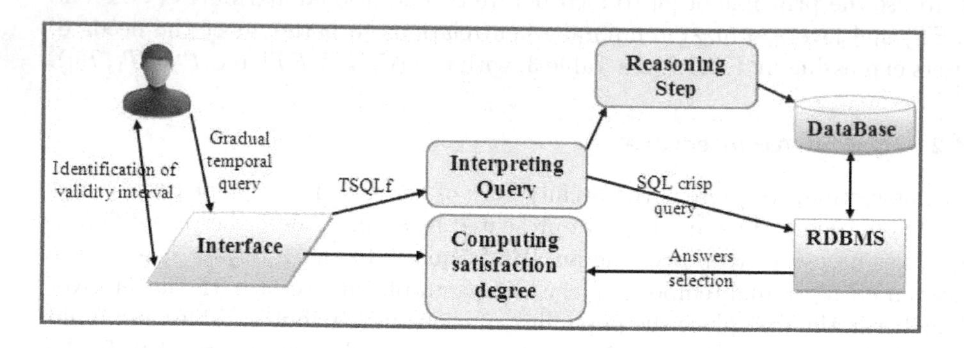

Fig. 2. Architecture of TSQLf system

5 An Illustrative Example

To better explain our proposal we present below an example from Archeology field. Consider the Archeology table (see Table 4) presenting the material remains from prehistoric times. The table schema is *Archeology* (Code_Ar#, Name_Ar, Location, Date_Discovery, VST_Dc, VET_Dc, Date_Dated, VST_Dd, VET_Dd). Where VST_Dc means the start validity date of Date_ Discovery, VET_Dc means the end validity date of Date_ Discovery, VST_Dd means the start validity date of Date_Dated and VET_Dd means the end validity date of Date_Dated.

Table 4. The Archeology table

Code_Ar	Name_Ar	Location	Date_ Discovery	VST_Dc	VET_Dc	Date_Dated	VST_Dd	VET_Dd
A011	Pyramid of six meters in height	Lima	Recently	10/07/2013	31/07/2013	5000 years ago	2987 BC	2988 BC
A015	Cone rocks	Lake Tiberias	In 2003	05/02/2003	08/08/2003	2050 years ago	37 BC	38 BC
A120	Church	Island of the City	Little ago	20/06/2013	25/07/2013	End 158	08/09/158	18/12/158
A002	Chanel monumental	Narbanne	In 2010	05/03/2010	10/11/2010	2455 years ago	440 BC	442 BC
A020	Hunting weapon	South Africa	Early 2009	03/01/2009	05/04/2009	500000 years ago	497985 BC	497987 BC
A042	Mammoth Skeleton	Seine-et-Marne	Summer 2013	05/06/2013	01/09/2013	Before beginning 159	15/11/189	13/12/189
A075	Sort of mini-dinosaur	Africa	October 2012	02/10/2012	31/10/2012	Before the end of 260	20/09/260	15/12/260
A101	Indian prints	Brazil	Before the end of November 2011	18/11/2011	30/11/2011	More than 3000 years BC	3000 BC	3002 BC
A111	Corps soldiers Allemends	Carspach	Before the end of 2011	12/10/2011	15/12/2011	In 1918	10/07/1918	15/12/1918
A224	Statues banned by the Nazis	Berlin	Beginning in November 2010	01/11/2010	10/11/2010	During the second war mandial	01/09/1939	02/09/1945

Given the following queries:

Q1: "Show archaeological discoveries that took place well after medium 2011 and just at end July 2013 (with a minimum 0.8 degree)."
TSQLf Query: Select 0.8 Nom_Ar, Lieu From Archeology Where [t.VST_Dc, t.VET_Dc] *FT_Contains* [01/08/2011, 30/07/2013] ;
The result of this query is the tuples A011, A120 and A075 (with a degree = 1) and A042 (with a degree = 0.9).
Q2: "Show places of archaeological discoveries that took place well after the end of 2008 and dated just after 440 years BC (with a minimum 0.7 degree)."
TSQLf Query: Select 0.7 Lieu From Archeology Where [t.VST_Dc, t.VET_Dc] *FT_After* [01/01/2009, 31/12/2009] and [t.VST_Dd, t.VET_Dd] *FT_Started by* [440 BC, 450 BC].
The result of this query is the tuple A002 (with a degree = 1).

Q3: "Show Names archaeological held just before or with the Date_Dc argholo-gie A011 (with a minimum 0.6 degree)" where the validity interval of archeology A011 is [10/07/2013, 31/07/2013].
TSQLf Query: Select 0.6 Nom_Ar From Archeology Where [t.VST_Dc, t.VET_Dc] *FT_Before* [10/07/2013, 31/07/2013] *or* [t.VST_Dc, t.VET_Dc] *FT_Overlaps* [10/07/2013, 31/07/2013].
The result of this query is the tuples A120 (with a degree = 1), A075 (with a degree = 0.9), A101 and A111 (with a degree = 0.8), A002 (with a degree = 0.7) A015 and A224 (with a degree = 0.6).

6 System Interface

The definition of intuitive and user-friendly interfaces that help to make gradual temporal queries is of great interest for unfamiliar users with the concepts of fuzzy set theory. This can improve the use of TSQLf as a language of query expression. The tool we have developed acts as an JAVA interface with the Oracle DBMS and generates TSQLf queries directly executable by calls to functions and PL/SQL bloc stored. The interface is connected to the database so as to store the tables in the field study that incorporates fuzzy temporal aspects. In this way, it is possible to add for each table, which contains a crisp/fuzzy temporal attribute, two specific temporal attributes (VST, VET). These two attributes express the time interval validity for the temporal attribute. Our approach supports fuzzy temporal Allen relations which compares two well-defined time intervals ([T1, T2] and [t.VET, t.VET]). As an TSQLf query involves a fuzzy temporal condition, the developing tool allows, first, to accept the definition of fuzzy temporal condition (Figure 3-left part), and second, the definition of validity interval and tolerance interval for each fuzzy temporal conditions (Figure 3-right part). After the above step, the generated TQSLf query is submitted to the database to select rows that match its fuzzy temporal criteria. Finally, a degree of satisfaction is calculated and assigned to each selected line (Figure 4).

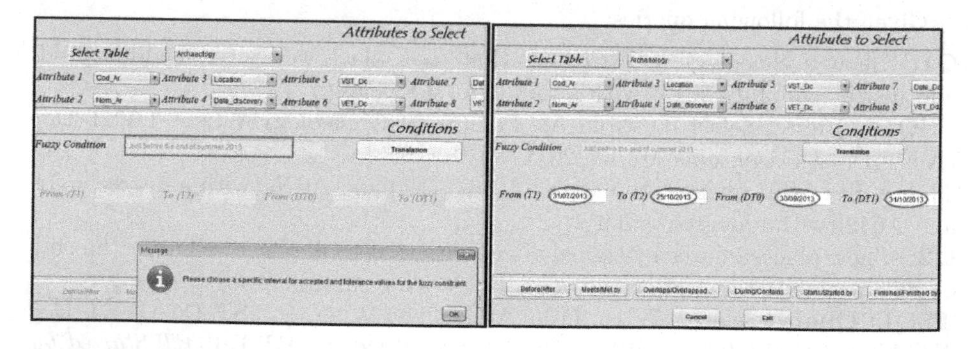

Fig. 3. Definition of fuzzy temporal conditions (left) and acceptable interval values(right)

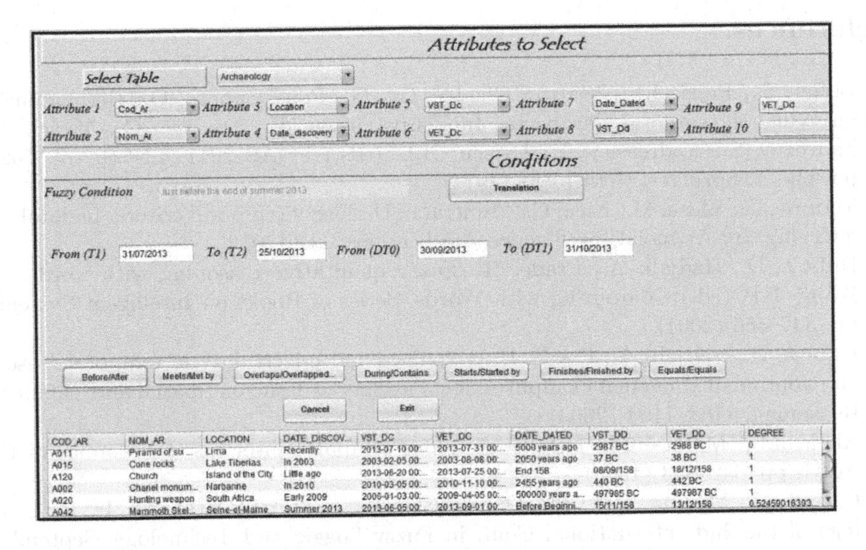

Fig. 4. Result of an TSQLf query

Let us note that some methods have been offered previously to define vague terms and how to handle fuzzy database queries. In [8], Goncalves and Tineo defined commands to add predicates (*create fuzzy predicate*) and fuzzy comparators (*create comparator*) to SQLf, but the major disadvantage of this method is its difficulty for users novice. Another solution, proposed in [9], is to include definitions of fuzzy predicates in the query itself. This solution can lead to complex queries. In [12], the authors suggest using the notion of fuzzy order to specify fuzzy predicates defined on numeric fields. They set three operators, namely *is at least*, *is at most* and *belongs to*. This method allows to specify only atomic vague terms. All methods reviewed above treat digital or categorical vague terms, and no solution has proposed to address temporal terms. The interface we developed, provides a user-friendly and simple method (Figure 3) to define fuzzy temporal terms. It displays the TSQLf query results with their satisfaction degrees (Figure 4) to user.

7 Conclusion

In this paper, we have put the first foundations for building an intelligent querying system to temporal databases. The key concept of our approach is the fuzzy Allen temporal relations to express some particular vague temporal criteria. We have provided a user-friendly interface to allow users to submit their fuzzy temporal query. The system developed make use of a step of reasoning to solve sophisticated queries expressing complex users' needs. In the future, we plan to investigate time intervals with fuzzy bounds, and to apply our approach in on real-life data.

References

1. Billiet, C., Pons, J.E., Matthé, T., De Tré, G., Pons Capote, O.: Bipolar fuzzy querying of temporal databases. In: Christiansen, H., De Tré, G., Yazici, A., Zadrozny, S., Andreasen, T., Larsen, H.L. (eds.) FQAS 2011. LNCS, vol. 7022, pp. 60–71. Springer, Heidelberg (2011)
2. Tudorie, C., Vlase, M., Nica, C., Muntranu, D.: Fuzzy temporal criteria in database querying. In: Artificial Intelligence Applications, 112 (2012)
3. Dubois, D., Hadjali, A., Prade, H.: Fuzzy qualitative reasoning with words. In: Wang, P.P. (ed.) Computing with Words. Series of Books on Intelligent Systems, pp. 347–366 (2001)
4. Dubois, D., Hadjali, A., Prade, H.: Fuzziness and uncertainty in temporal reasoning. Journal of Universal Computer Science, Special Issue on Spatial and Temporal Reasoning, 1168–1194 (2003)
5. Allen, J.F.: Maintaining knowledge about temporal intervals. In: Comm. of the ACM, pp. 832–843 (1983)
6. Galindo, J., Medina, J.M.: Ftsql2: Fuzzy time in relational databases. In: Proceedings of the 2nd International Conf. in Fuzzy Loggic and Technology, September 5-7 (2001)
7. Deng, L., Liang, Z., Zhang, Y.: A fuzzy temporal model and query language for fter databases. In: Eighth International Conference on Intelligent Systems Design and Applications, pp. 77-82 (2008)
8. Goncalves, M., Tineo, L.: Sqlf3: An extension of sqlf with sql features. In: Proc. of FUZZ-IEEE.01, pp. 477–480 (2001)
9. Pivert, O., Smits, G.: On fuzzy preference queries explicitly handling satisfaction levels. In: Information Science, pp. 341–350 (2012)
10. Bosc, P., Pivert, O.: SQLf: A relational database language for fuzzy querying. IEEE Transactions on Fuzzy Systems 3(1), 1 (1995)
11. Schockaert, S., De Cock, M.: Temporal reasoning about fuzzy intervals. In: Artificial Intelligence, pp. 1158–1193 (2008)
12. Bodenhofer, U., Kung, J.: Fuzzy orderings in flexible query answering systems. In: Soft Comput., pp. 512–522 (2004)

Effective Outlier Detection Technique with Adaptive Choice of Input Parameters

Agnieszka Duraj and Danuta Zakrzewska

Institute of Information Technology,
Lodz University of Technology,
ul. Wolczanska 215, 90-924 Lodz, Poland

Abstract. Detection of outliers can identify defects, remove impurities in the data and what is the most important it supports the decision-making processes. In the paper an outlier detection method based on simultaneous indication of outliers by a group of algorithms is proposed. Three well known algorithms: DBSCAN, CLARANS and COF are considered. They are used simultaneously with iteratively chosen input parameters, which finally guarantee stabilization of the number of detected outliers. The research is based on data retrieved from the Internet service allegro.pl, where comments in online auctions are considered as outliers.

Keywords: detection of outliers, detection of exceptions, data mining, online transactions.

1 Introduction

Outlier analysis is a very useful tool for finding out data, which significantly differ from the others collected in the same dataset. There exist many areas, where detection of outliers is necessary. As the most significant there should be mentioned: finance, medicine, biomedicine, insurance, various branches of industry including telecommunications or electronic commerce. In the paper we will focus on the last one. We will investigate using of outlier analysis for detecting negative comments in online auctions.

There exist many definitions of outlier concepts [1], [2], [3]. As outliers we will consider points belonging to class A but actually situated inside class B so the true (veridical) classification of the points is surprising to the observer [3].

Online auction sites contain sellers and buyers comments, which can be used in reputation management systems. From among comments negative ones can be considered as outliers. They are mainly connected with law violations. According to the police (*www.policja.pl*) usually law violation takes place when seller does not send the goods after receiving the money - the so-called non-delivery; or seller sends the item which is different than described or damaged; or there has been a withdrawal of payment or fictitious payment; or seller sells stolen goods. Finding out exceptions, which occur among comments may help in evaluating

P. Angelov et al. (eds.), *Intelligent Systems'2014*,
Advances in Intelligent Systems and Computing 322, DOI: 10.1007/978-3-319-11313-5_47

sellers. Thus including outlier detection techniques will support the performance of reputation systems. However to obtain the required results, effective methods of outlier detection are necessary.

Outlier detection algorithms create a model of the normal patterns in the data and then compute an outlier score of a given data point on the basis of the deviations from these patterns [4]. However score values, for which points should be indicated as outliers depend on the data character and input algorithm parameters. Automation of the process, would allow to use analytical techniques without user interference. In the paper improving the performance of outlier detection technique is considered. We propose the methodology, which enables automatic selection of input parameters, what in turn ensures the required effectiveness of finding outliers. Investigated approach is used for detecting negative comments of sellers in auction sites and thus to classify dishonest traders. The proposed method is validated by experiments conducted on data retrieved from the Internet service allegro.pl.

The paper is organized as follows. Related work concerning outlier detection as well as reputation services is presented in Section 2. In the following section the proposed methodology is described. Next the results of the experiments are presented. Finally concluding remarks as well as the directions of future research are depicted.

2 Related Work

There exist three basic approaches to the problem of detecting outliers ([5], [6], [7], [8], [9], [10]). The first one assumes that the errors and defects are separated from the correct data. In the second approach the presence of the both: normal and abnormal data in the database is supposed. The classifier learns the model and qualifies new patterns to the relevant classes. The last approach assumes the existence of only the correct data or a very small number of cases with abnormal patterns.

The methods used to detect outliers originate from methodologies of statistics and machine learning. For example, Barnett [2], Rousseeuw[11] describe and analyze an extensive family of statistical techniques for determining exceptions. Other examines the broad family of neural networks methods. In the above mentioned technologies the division into methods based on distance and density can also be applied.

Seller ratings can be considered as measures of their honesty. In [12], [13], [14], [15], one can find different approaches to seller reputation measuring. For example, in [16] computing the indirect propagated reputation is proposed, in [17] the so-called specially crafted comments were constructed with a view to performing the so-called coalition attacks. Douceur [14] has taken into account the increasing undeserved reputation of cheaters, and Morzy [18] proposed the method for the determination of classified or intentionally omitted comments. Assessment of the level of trust and mistrust between any pairs of participants on the basis of a small number of public comments was proposed by Guhy [15] and the definition of the seller reputation based on social networks was considered by [19].

From among the other works, there should be mentioned the papers [20], [21], [22], where the authors concentrated mainly on the methods of detecting shill bidding. Trevathan [21] applied statistical methods, Xu [23] considered using of Markov models. Application of Bayesian probability was discussed in [24].

3 Methodology

3.1 A Concept

Automation of the process of outlier detection is difficult to perform, because of the significant dependence of the existing techniques on input parameters and data characteristics. Therefore user interference is necessary to obtain satisfying results. In the presented approach we propose to avoid this problem by combining different outlier analysis techniques. The review of clustering algorithms presented in [25] showed that combinations of algorithms performed better in outlier detection than each of them independently. Accordingly, the concept of the proposed methodology is based on the heuristics that *simultaneous iterative application of different approaches enables to choose effectively input parameters and thus to improve the performance of outlier detection.* In the presented approach, three outlier detection algorithms DBSCAN, CLARANS and COF are considered. Good performance of all of them were described respectively in [26], [8],[10]. Additionally, each of the algorithms represents a different group of outlier detection methods. However both of them: DBSCAN and CLARANS are clustering techniques, but the first one is based on the density, while the second algorithm consists in partitioning. COF algorithm is dedicated to outlier analysis and is based on the both: distance and density. The proposed algorithm consists of the following steps, as illustrated in Fig. 1:

- **Data Preparation.** For the certain time period, n online auction sellers is selected from an auction site and statistics are calculated. Vendor dataset is created, it contains all kind of comments, including positive, negative or neutral ones; prices of sold items and respective statistical values , taking into account percentage of negative and positive comments, standard deviations as well as averages.
- **Outlier Detection.** the input parameters for each algorithm are introduced, and outliers are detected by using all the considered techniques separately.
- **Verification.** If the results obtained by the algorithms are different, the respective input parameters are changed and the previous step is repeated.
- **Final Results.** The algorithm terminates, when the results obtained by different outlier detection techniques are consistent with the assumed accuracy (we will further suppose that the difference between the number of obtained outliers cannot be greater than 10).

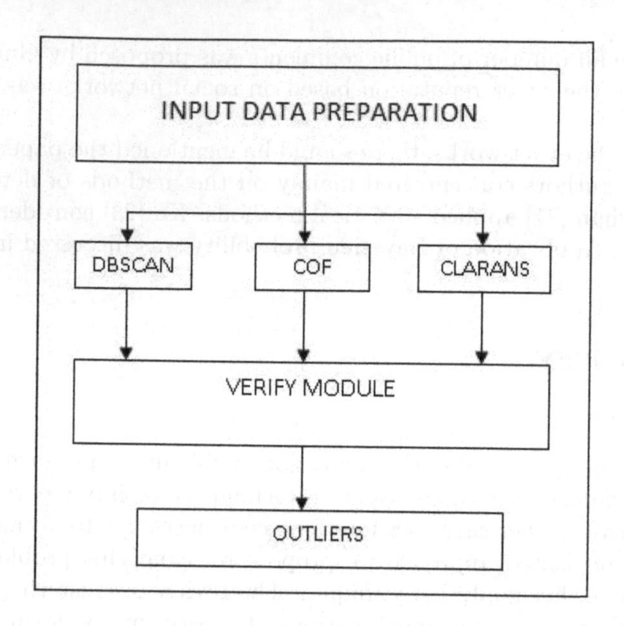

Fig. 1. Simplified diagram of the proposed methodology

3.2 DBSCAN

The Density Based Spatial Clustering of Applications with Noise - DBSCAN is the most popular technique belonging to the group of algorithms based on density [26]. As one of advantages of the algorithm, there should be mentioned its feature that in order to detect groups of any given shape or noise, the algorithm reviews the dataset only once. However the performance of the algorithm can be very good, but it significantly depends on the choice of input parameters, which are not easy to determine. DBSCAN requires two input parameters:

- ε - defines the maximum radius of neighborhood for a given object, all objects being in this area are called ε -neighbors;
- $MinPts$ - the minimum number of points required to form a cluster, a point is called the core point, if in the neighborhood of the given point there are at least MinPts objects.

DBSCAN defines a cluster as a maximal set of so called density-connected points. Cluster points are divided into core points (the ones inside the cluster) and border points. Every object not contained in any cluster is considered to be an outlier. The algorithm starts with any point, p, and finds out all points density-reachable [26] from p with respect to parameters: ε and $MinPts$. If p is a core point, the new cluster is created. If it is a border point, no points are density-reachable from p. As the main advantages of DBSCAN Ester et al. [26] mentioned the possibility it offers to discover clusters of arbitrary shape, and its outlier detection ability, as well as the minimal number of input parameters. However, the last two are strictly connected with the density of objects and difficult to determine. In many cases even experimental choice of input parameters does not guarantee obtaining satisfying results (see [28] for example).

3.3 CLARANS

CLARANS (Clustering Large Applications Based on Randomized Search) is the combination of PAM (Partitioning Around Medoids) and CLARA (Clustering for Large Applications). It searches the graph randomly to find points located centrally in the group. The algorithm requires the following initial parameters:

- $Maxneighbor$ - the maximum number of neighbors of the node to be checked.
- *numlocal* - the maximum number of local minima information on which can be gathered
- $numberofclasses$ - the number of classes which are considered to be possible
- *minimum* - the number of local minima to be considered before the end of the algorithm

The operation of CLARANS starts with selecting a node randomly. It is followed by checking the neighbors of the selected node and when a better neighbor is found (based on the cost difference of two nodes) there is a change of the node and the algorithm continues until the parameter $Maxneighbor$ is reached. If the algorithm does not obtain the improvement of the neighboring nodes in relation to the starting node, the current node is declared to be a local minimum, and the algorithm starts searching for a new local minimum. After gathering information about the specified number of local minima (*numlocal*), the algorithm returns the best of these local values as the medoid of the group. The declared values of the parameters *maxneighbor* and *numlocal* are not intuitive. In many studies, the authors indicate that the smaller the value of the parameter *maxneighbor*, the worse the quality of the group. The higher the value of the parameter *maxneighbor*, the quality of the obtained groups is the more similar to the results of PAM partitioning algorithm. $Numlocal$ value greater than 2 does not improve the quality of detected local minima [27].

3.4 COF Algorithm

COF (Connectivity-based Outlier Factor) algorithm was introduced in [10],as a modification of the LOF algorithm [5]. COF is based on the distance between objects, and takes into account the density of objects in the set. The main idea of the COF algorithm [10] consists in determining the so-called COF isolation coefficient for each object in a data set. This factor determines how much the given object is isolated from the whole dataset. Analysis of the COF values divides the data into a set of proper objects and a set of outliers. COF isolation factor is defined according to the formula:

$$COF_k(p) = \frac{|N_k(p)| * A(p)}{\Sigma_{o \epsilon N_k(p)} A(o)} \quad (1)$$

where $N_k(p)$is the k-nearest neighborhood of the point p and $A(p)$ means the average chaining distance of the point p to the points of the neighborhood. COF algorithm requires the following input parameters:

- k - the number of objects of the nearest neighborhood,
- Pr - threshold of isolation,

In the first step the algorithm determines the COF coefficient for each object p taking into account its k- nearest neighborhood, called $COF_k(p)$. Then, outliers are selected, that is, data that meets the following inequality:

$$COF_k(p) > P_r \quad (2)$$

The advantage of COF is that it is able to detect outliers from the data of different densities, and exceptions that are close to areas of low density, which makes them difficult to detect for other algorithms (especially those based only on density).

4 Experiment Results and Discussion

Experiments aimed at evaluating the performance of the proposed methodology. They were conducted by using data retrieved from the auction portal allegro.pl. Selection of the appropriate parameter set is necessary to construct a model of detection exceptions. During the preparation of the sets used in the present study, Poland's largest auction portal allegro.pl was taken into account. The sets are prepared in *.csv format. The algorithm is implemented in Java (JRE). Attributes include, among others, the number of positive opinions and negative ratings, average, standard deviation. The number of negative comments was most important in the collection. The studies adopted that:

- The users called positive are users whose percentage of positive comments in relation to the negative comments was close to 100%.
- The users called negative are users of the service, for whom the number of positive comments in relation to the negative comments was small or amounted to 0. Negative users are those whose per cent of positive comments is relatively high (nearly 98%) and yet there is a very large number of negative comments (over 1000 for a single period of bidding.)

Of course, the issue considered in the paper can be reduced to the problem of two classes. One class is the correct patterns the so-called positive users. The second class is the 'negative' users. In our experiments, we treat the detected negative users as the exceptions being sought. However, research does not apply only to detecting negative users as exceptions. It is mainly based on the automatic selection of input parameters for the three selected exceptions detection

Table 1. The outliers detected by the DBSCAN algorithm

ε	$MinPts$	**The number of detected outliers**
5	5	1197
10	5	506
15	5	294
20	5	211
25	5	151
5	3	989
10	3	398
15	3	205
20	3	148
25	3	113

Table 2. The outliers detected by the CLARANS algorithm

$numberofclasses$	$numlocal$	$minimum$	$Maxneighbor$	**Outliers**
60	60	10	3	1400
100	60	10	3	1205
30	30	10	3	195
60	30	10	3	618
100	50	10	3	1208
100	60	5	3	1300

algorithms. The classification of a user to the "positive" or "negative" group of users simultaneously by the three selected algorithms gives greater (or even 100%) certainty of classification accuracy.

In the first step, all the considered methods were used separately. Obtained results found out by DBSCAN , CLARANS and COF were presented respectively in Tables 1, 2, 3 . It can be easily noticed that the best results were obtained with the DBSCAN and COF algorithms. We pondered the number of $n = 1500$ auction sellers. The considered dataset contained 1,300 users for whom the percentage of positive comments in relation to the negative ones was equal almost to 100%. The remaining 200 records consisted of service users, for whom the number of positive comments in relation to the negative ones was close to 0. Experiments were conducted to verify the effectiveness of considered outlier detection methodology in finding out the improper auction transactions and, especially dishonest vendors.

The experiments for DBSCAN algorithm started from the initial values $\varepsilon = 1$ and $MinPTS = 1$, where the ε value belonged to the interval $\varepsilon \in \{1, 2, ..., 30\}$, and $MinPTS \in \{1, 2, ...10\}$. For COF initial accuracy threshold was $\Pr = 85$, and the number of nearest neighbor $k = 1$, $Pr \in \{85, ..., 99\}$, $k \in \{1, ...10\}$. For CLARANS $Maxneighbor \in \{10, 20, ..., 100\}$, $numlocal \in \{10, ...100\}$, $maxneighbor \in \{1, 2, .., 5\}$, $minimum \in \{1, ...10\}$. The changing of the parameter was by one value. DBSCAN algorithm, at fixed initial values of the $\varepsilon = 15$ and $MinPts = 3$, correctly detected 200 negative comments, and 5 comments defined

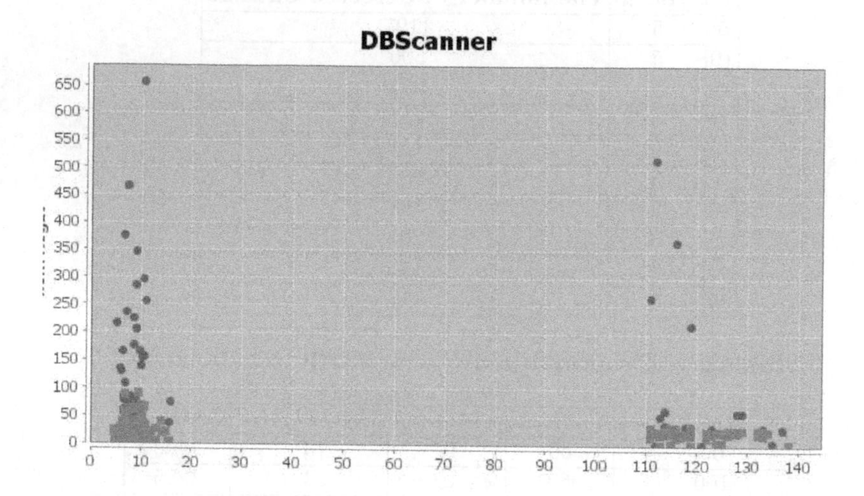

Fig. 2. The result outliers detection - DBSCAN algorithm

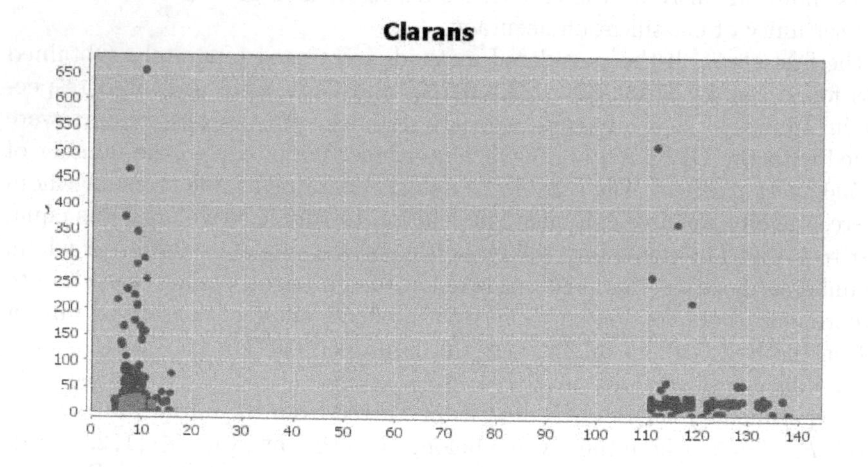

Fig. 3. The result outliers detection - CLARANS algorithm

as neutral ones. Increasing value of ε from 15 to 20 worsened the performance of outlier detection. The algorithm found 148 of the 200 exceptions contained in the considered dataset. This means that more than 25% of all the negative comments were not detected, and that they were classified to the group of positive comments. Increasing the value of ε leads to assignment of positive comments to the group of exceptions. The outliers detected as negative comments together with positive comments are illustrated in (Fig. 2) (negative comments - exceptions are marked in blue, positive comments marked in red).

Considering detection of outliers by using COF algorithm, the threshold values from among the numbers 90, ..., 99 were considered and the number of objects $k = 3, 5$ were taken into account. Analysis of the obtained results showed that the threshold value greater than 95 did not result in improving the accuracy of exception detection. The best results were obtained for $k = 3$ and $Pr = 95$. COF algorithm found 206 exceptions (200 negative comments, 6 comments defined as neutral ones). In the case of the CLARANS algorithm, the statement that the small value of the parameter $Maxneighbor$ deteriorates the quality of the groups was confirmed. CLARANS detected 195 outliers out of 200 only for the initial parameter values equal respectively: $numberofclasses = 30$, $numlocal = 30$, $minimum = 10$; $Maxneighbor = 3$. The outliers detected by CLARANS algorithm are illustrated in Fig. 3. The results of the algorithm CLARANS are not satisfactory, and, what is more, they are not repeatable. Due to random selection of the initial location, the algorithm may find different outliers each time it is started from the beginning even in situations when input parameters are of the same values. Contrarily to CLARANS, DBSCAN and COF detected exceptions with high accuracy and in a short time. For comparison, run times of all the algorithms are presented in Table 4.

In the second step of the experiments,the performance of the proposed approach was analysed. Firstly all the changes of input parameters were identified and compared for all the algorithms. Analysis showed that the change of the input parameters took place most often for the CLARANS algorithm. For DBSCAN total changes in input parameters took place from 1 to 3 times , for COF from 1 to 4 times, while for CLARANS from 5 to 15 times. During experiments input parameters were not changed if the difference in the number of the detected outliers was smaller than 10. For example, the resulting number of exceptions detected by the DBSSCAN algorithm was 205 , by COF was 206, and by CLARANS was 195 after respectively 1, 3 and 7 changes of input parameters (iterations) . DBSCAN and COF quickly obtained convergence in the number of detected outliers, after maximum of 4 iterations. However CLARANS required very careful analysis before changing input parameters. On average, difference in the numbers of outliers detected by all the parameters was smaller than 10 only after 10 to 15 changes of CLARANS input parameters. Then the results for CLARANS were similar to the ones obtained by DBSCAN and COF algorithms. However due to the random selection of the starting point in CLARANS, repeatability is impossible.

Table 3. The outliers detected by the COF algorithm

Pr	k	**The number of detected outliers**
90	5	320
93	5	310
95	5	310
97	5	310
98	5	310
99	5	310
90	3	290
93	3	208
95	3	206
97	3	206
98	3	206
99	3	206
90	10	170
95	10	170
98	10	170

Table 4. Run time of the algorithms

Algorithm	**Time of operation** $[s]$
DBSCAN	0.287007645
COF	0.29218723
CLARANS	13.185476419

5 Conclusions

In the paper the approach of simultaneous using of different outlier detection algorithms in order to find out the best possible input parameters is considered. Experiments showed that simultaneous monitoring of the number of detected outliers for the three algorithms makes possible not only to set the input parameters properly for each of them, but also to get accurate and reliable results.

Additionally, it was noted that the majority of false detections were obtained by the CLARANS algorithm. The convergence of results between DBSCAN and COF algorithms occurred very quickly, usually in the second iteration. CLARANS achieved similar results after 10 to 15 changes in input parameters. Random selection of the starting point of the CLARANS algorithm, is the reason why the results are not repeatable. Time of obtaining the response and the number of errors suggest considering inclusion of the other algorithm to the proposed approach.

As demonstrated in the paper, the proposed new solution accelerates setting of initial parameters for the considered algorithms as well as verifies the validity of the detected exceptions. This may constitute the basis for the further strategy of inference and decision-making processes. Future research will be related to increasing the number of algorithms involved in detecting exceptions and replacing

CLARANS algorithm by another one and thus improving the effectiveness of the proposed method.

References

1. Aggarwal, C.C., Yu, P.S.: Outlier Detection for High Dimensional Data. In: Proceedings of the ACM SIGMOD Conference (2001)
2. Barnett, V., Lewis, T.: Outliers in Statistical Data, 3 edn. John Wiley and Sons (1994)
3. John, G.H.: Robust Decision Trees: Removing Outliers from Databases. In: Proceedings of the First International Conference on Knowledge Discovery and Data Mining, Menlo Park, CA, pp. 174–179. AAAI Press (1995)
4. Aggarwal, C.C.: Outlier Analysis. Springer Science+Business Media, New York (2013)
5. He, Z., Xu, X., Deng, S.: Discovering Cluster-Based Local Outliers. Pattern Recognition Letters 24(9-10), 1641–1650 (2003)
6. Hodge, V.J., Austin, J.: A Survey of Outlier Detection Methodologies. Artificial Intelligence Review 22, 85–126 (2004)
7. Jin, W., Tung, A.K., Han, J.: Mining Top-n Local Outliers in Large Data Bases. In: Proceedings of International Conference on Knowledge Discovery in Data Bases, pp. 293–298 (2002)
8. Knorr, E.M., Ng, R.T., Tucakov, V.: Distance-Based Outliers: Algorithms and Applications. The International Journal on Very Large Data Bases 8, 237–253 (2000)
9. Ramaswamy, S., Rastogi, R., Shim, K.: Efficient Algorithms for Mining Outliers from Large Data Sets. In: Preceedings of ACM SIGMOD International Conference on Management of Data, pp. 427–438 (2000)
10. Tang, J., Chen, Z., Wai-chee Fu, A., Cheung, D.: A Robust Detection Scheme for Large Data Sets. In: 6th Pacific-Asia Conf. on Knowledge Discovery and Data Miting
11. Rousseeuw, P., Leroy, A.: Robust Regression and Outlier Detection, 3rd edn. John Wiley and Sons (1996)
12. Chae, M., Shim, S., Cho, H., Lee, B.: Empirical analysis of online auction fraud: Credit card phantom transactions. Expert Systems with Applications 37, 2991–2999 (2010)
13. Chau, D.H., Pandit, S., Faloutsos, C.: Detecting Fraudulent Personalities in Networks of Online Auctioneers. In: Fürnkranz, J., Scheffer, T., Spiliopoulou, M. (eds.) PKDD 2006. LNCS (LNAI), vol. 4213, pp. 103–114. Springer, Heidelberg (2006)
14. Douceur, J.R.: The Sybil Attack. In: Druschel, P., Kaashoek, M.F., Rowstron, A. (eds.) IPTPS 2002. LNCS, vol. 2429, p. 251. Springer, Heidelberg (2002)
15. Guha, R., Kumar, R., Raghavan, P., Tomkins, A.: Propagation of trust and distrust. In: 13th Int. Conf. on World Wide Web WWW4 (2004)
16. Mui, L.: Computational models of trust and reputation: Agents, evolutionary games, and social networks. Technical Report, MIT (2003)
17. Melnik, M.I., Alm, J.: Does a Seller as eCommerce Reputation Matter? Evidence from eBay Auctions. Journal of Industrial Economics (2002)
18. Morzy, M.: Density-based measure of reputation of sellers in online auctions. In: 9th 1st ADBIS Workshop on Data Mining and Knowledge Discovery, ADMKD 2005 (2005)

19. Morzy, M., Wierzbicki, A.: The Sound of Silence: Mining Implicit Feedbacks to Compute Reputation. In: Spirakis, P.G., Mavronicolas, M., Kontogiannis, S.C. (eds.) WINE 2006. LNCS, vol. 4286, pp. 365–376. Springer, Heidelberg (2006)
20. Dong, F., Shatz, S.M., Xu, H.: Reasoning under Uncertainty for Shill Detection in Online Auctions Using Dempster-Shafer Theory. International Journal of Software Engineering and Knowledge Engineering 20, 943–973 (2010)
21. Trevathan, J., Read, W.: Detecting Shill Bidding in Online English Auctions. In: Gupta, M., Sharman, R. (eds.) Handbook of Research on Social and Organizational Liabilities in Information Security. Information Science Reference, pp. 446–470 (2009)
22. Wang, J., Chiu, C.Q.: Detecting online auction inflated-reputation behaviors using social network analysis. In: Proceedings of North American Association for Computational Social and Organizational Science, Notre Dame, Indiana (2005)
23. Xu, H., Cheng, Y.: Model Checking Bidding Behaviors in Internet Concurrent Auctions. International Journal of Computer Systems Science And Engineering 22, 179–191 (2007)
24. Goel, A., Xu, H., Shatz, S.M.: A Multi-State Bayesian Network for Shill Verification in Online Auctions. In: Proceedings of the 22nd International conference on Software Engineering And Knowledge Engineering (SEKE 202010), Redwood City, San Francisco Bay, CA, USA, pp. 279–285 (2010)
25. Zakrzewska, D.: Cluster analysis in personalized E-learning systems. In: Nguyen, N.T., Szczerbicki, E. (eds.) Intelligent Systems for Knowledge Management. SCI, vol. 252, pp. 229–250. Springer, Heidelberg (2009)
26. Ester, M., Kriegel, H.-P., Sander, J., Xu, X.: A density-based algorithm for discovering clusters in large spatial databases with noise. In: Simoudis, E., Han, J., Fayyad, U.M. (eds.) Proceedings of the Second International Conference on Knowledge Discovery and Data Mining, Portland, OR, pp. 226–231. AAAI, Menlo Park (1996)
27. Ng, R.T., Han, J.: Efficient and effective clustering methods for spatial data mining. In: Proc. of 20th International Conference on Very Large Data Bases, pp. 144–155 (1994)
28. Zakrzewska, D.: Validation of clustering techniques for student grouping in intelligent e-learning systems. In: Jozefczyk, J., Orski, D. (eds.) Knowledge-Based Intelligent System Advancements: Systemic and Cybernetic Approaches, pp. 232–251. IGI Global (2011)

Data Quality Improvement by Constrained Splitting

Antoon Bronselaer and Guy De Tré

Dept. of Telecommunications and Information Processing,
Ghent University, Sint-Pietersnieuwstraat 41, B-9000 Ghent, Belgium
{antoon.bronselaer,guy.detre}@UGent.be

Abstract. In the setting of relational databases, the schema of the database provides a context in which the data should be interpreted. As a consequence, the quality of a relational database depends strongly on the assumption that data fits this context description. In this paper, we investigate the case where the information provided by an attribute value exceeds the framework provided by the schema. It is shown that such an information overflow can have two orthogonal causes: (i) data about multiple attributes are jointly stored as one attribute and (ii) data about multiple tuples are jointly stored as one tuple. Needless to say, such erroneous information storage deteriorates the quality of the database. In this paper, it is investigated how data quality can be improved by a split operator. The major difficulty hereby is to take into account the constraints that are present in a relational database. A generic algorithm is provided and tested on the well-know Cora dataset.

Keywords: Data Quality, Databases, Multi-Valued Data.

1 Introduction

The continuously increasing importance of information gained from data has resulted in a growing interest in the improvement of data quality [1]. In [1], data quality is treated as a multi-dimensional concept with dimensions such as completeness, consistency and correctness. In light of this view on data quality, the relational model [2] for databases has the potential to guarantee high data quality by specification of *constraints*. Indeed, completeness of data can be partially or fully enforced by specification of NOT NULL constraints; consistency can be enforced by means of CHECK constraints and duplicate data can be avoided to a large extent by usage of KEY and/or UNIQUE constraints.

Despite these possibilities in the relational model, it can not be ignored that many relational databases deal with serious data quality issues [1]. The reason for that is twofold. On the one hand, the design of the database and thus also the specification of constraints, is itself a data quality issue. Needless to say that a poor database design results in poor quality of the data. In many cases, there is an insufficient usage of constraints to provide high quality standards. On the other hand, a data storage system can only partially deal with faulty user

P. Angelov et al. (eds.), *Intelligent Systems'2014*,
Advances in Intelligent Systems and Computing 322, DOI: 10.1007/978-3-319-11313-5_48

input. To illustrate this point better, consider the problem of duplicate data [3], which deals with the fact that two or more tuples describe the same real-world entity. Suppose a relation that stores persons by means of name, first name and birth date. The requirement that the combination of these attributes must be unique, can be enforced by means of a UNIQUE constraint. The definition of this constraint could aid in the avoidance of duplicate entries. However, a simple typographical error in the name or first name of a person, can result in the presence of duplicate data. The key problem is hereby that the database system can not verify whether a given value is actually correct.

The particular data quality problem studied in this paper, deals with the fact that the database schema of a relational database provides a context in which the data should be interpreted. In a way, the schema gives meaning to the data. A direct consequence thereof is that the quality of a relational database strongly depends on the assumption that the meaning of data in the database perfectly matches this context description. A common problem in relational databases is that the information given by an attribute value *exceeds* the context provided by the schema. In the scope of this paper, this problem is referred to as *information overflow*[1] and a clear distinction is made between two orthogonal cases: *horizontal* information overflow and *vertical* information.

id	name	street	number
1	John	Main Street 156	-
2	Jane	Broadway 1000	-

Fig. 1. An example of horizontal overflow

In the case of horizontal information overflow, data about multiple attributes are jointly stored as one attribute. An example thereof that often occurs in practice, is observed when storing both street and house number in an attribute named "street" (Figure 1). In the case of vertical information overflow, data about multiple tuples are jointly stored as one tuple. An example thereof is shown in Figure 2, where multiple artists are stored jointly in one tuple. In this paper, we will focus on the latter case of vertical information overflow. A solution for this problem is proposed by means of a vertical split-replicate operator. The major difficulty of using such a split-replication operator, is to take into account the *constraints* that are present in a relational database.

The remainder of the paper is structured as follows. In Section 2, the necessary preliminary concepts are introduced. In Section 3, the relevant literature is reviewed. In Section 4, the split-replication operator is defined and discussed and its usage in the constrained context of a relational database is studied. A generic algorithm that models constrained split-replications is provided. This algorithm

[1] The dual case of information underflow is not considered here.

id	artist
1	Jimmy Hendrix,Keith Richards,Jimmy Page

Fig. 2. An example of vertical overflow

is tested in Section 5. Finally, in Section 6 the most important contributions of this paper are summarized.

2 Preliminaries

In the remainder of this paper, the relational model is assumed as the data storage model [2]. We shall briefly recapitulate the most important concepts of this model. Let $\mathcal{A}$ be a countable set of attributes. A *relational schema* $\mathcal{R}$ is defined by a non-empty and finite subset of $\mathcal{A}$. For each attribute $a \in \mathcal{A}$, let $\text{dom}(a)$ denote the domain of a. Having a relational schema $\mathcal{R} = \{a_1, ..., a_k\}$, a *relation* R over $\mathcal{R}$ is defined by $R \subseteq \text{dom}(a_1) \times ... \times \text{dom}(a_k)$. Each element of a relation R with schema $\mathcal{R}$ is called a tuple over $\mathcal{R}$. In the remainder, we shall denote an arbitrary tuple by t. For simplicity, we shall denote the combined universe $\text{dom}(a_1) \times ... \times \text{dom}(a_k)$ as $\text{dom}(\mathcal{R})$. A *database schema* $\mathcal{D}$ is defined by a non empty and finite set of relational schemata, i.e. $\mathcal{D} = \{\mathcal{R}_1, ..., \mathcal{R}_n\}$ and a database D over $\mathcal{D}$ is defined as a set of relations $\{R_1, ..., R_n\}$ where R_i is a relation over $\mathcal{R}_i$, for any i.

If R is a relation with schema $\mathcal{R}$ and $A \subseteq \mathcal{R}$, then the *projection* of R over A is defined by a relation $R[A]$ with schema A that is obtained by taking the tuples in R and retaining only the attribute values in A. Projected tuples are denoted as $t[A]$ (and $t[a]$ in the case where $A = \{a\}$).

For a schema $\mathcal{R}$, a *key* is a set of attributes $K \subseteq \mathcal{R}$ such that K is unique and irreducible. A set of attributes K is called unique if:

$$\forall R : \forall (t_1, t_2) \in R^2 : t_1[K] = t_2[K] \Rightarrow t_1 = t_2. \quad (1)$$

A unique set of attributes K is irreducible if no strict subset of K is unique. As multiple keys may exist for a schema $\mathcal{R}$, the relational model dictates that one key is selected which is called the *primary key.*

In the relational model, relationships between relations are made possible through the use of *foreign keys*. Let $\mathcal{R}_1$ and $\mathcal{R}_2$ be two schemata. FK is a foreign key of $\mathcal{R}_2$ that refers to $\mathcal{R}_1$ if $\mathcal{R}_1$ has a key K such that there exists a one-to-one mapping $g \subseteq \text{FK} \times K$ satisfying:

$$\forall (a_1, a_2) \in g : \text{dom}(a_1) = \text{dom}(a_2). \quad (2)$$

In addition, relations R_1 (with schema $\mathcal{R}_1$) and R_2 (with schema $\mathcal{R}_2$) must satisfy:

$$\forall t_2 \in R_2 : \exists t_1 \in R_1 : t_2[\text{FK}] = t_1[K]. \quad (3)$$

This last constraint is called *referential integrity.* Abiteboul remarks that referential integrity can be modeled as an inclusion dependency [4]:

$$R_2[\text{FK}] \subseteq R_1[K] \tag{4}$$

which provides us with a simple notation to characterize referential integrity between R_2 and R_1. If for some reason we wish to omit the key and foreign key, we will denote the existence of a relationship between R_2 and R_1 as $R_2 \triangleright R_1$, where the symbol $\triangleright$ should be read and interpreted as "refers to". For the sake of simplicity and without limiting the generality, it will be assumed that foreign keys always refer to the *primary key* of another relational schema. Note that, if FK is a key of $\mathcal{R}_2$, then each tuple in $\mathcal{R}_1$ is related to at most one tuple in $\mathcal{R}_2$. In this case, the relationship between $\mathcal{R}_1$ and $\mathcal{R}_2$ is called a one-to-one relationship. However, if FK is not a key of $\mathcal{R}_2$, then each tuple in $\mathcal{R}_1$ can be related to a *set* of tuples in $\mathcal{R}_2$. In this case, the relationship between $\mathcal{R}_1$ and $\mathcal{R}_2$ is called a one-to-many relationship.

3 Related Work

The literature on information extraction comprises many interesting works that deal with the problem of horizontal overflow. More specifically, these techniques focuss on the transformation of text into (a part of) a record. In [5], NoDose is presented as a semi-automatical, rule-based tool for extracting structured and semi-structured information from textual documents. A similar idea is developed in [6], but with the novelty that rule generation is based on regular expressions. In [7], Inductive Logic Programming techniques are used to infer transformation rules. In [8] and [9], transformation rules are modeled as a Hidden Markov Model (HMM) and learned by means of an annotated training set. A text is then transformed by classifying each word from the text as being part of one of the attributes.

Up till now, little to no studies have been made regarding vertical information overflow. The major reason therefore is because splitting a string into a bag of substrings is a straightforward preprocessing operation in the field of Information Retrieval and is considered as trivial. However, when horizontal splitting must be applied in the constrained context of a relational database, the problem of tuple replication does require a more in-depth study. This paper aims at filling this gap in the current state of the literature.

4 Constrained Splitting

4.1 Problem Statement

In the context of the relational model [2], the problem studied in this paper can be formalized as follows. Consider a database $D = \{R_1, ..., R_n\}$ with database schema $\mathcal{D} = \{\mathcal{R}_1, ..., \mathcal{R}_n\}$. Consider a schema $\mathcal{R}$ and consider an attribute $a^+ \in$

$\mathcal{R}$, called the overflow attribute. For simplicity, it is assumed in the scope of this paper that $\text{dom}(a^+)$ is the set of character strings, denoted as $\mathcal{S}$.

It is said that a^+ is a *vertical overflow attribute* for a tuple $t \in R$ if $t[a^+]$ represents multiple values of a^+. This problem can be solved by *splitting* the value $t[a^+]$ into a bag of values V and to replicate t for $|V|$ times, such that each replica (including t) holds one of the values from V. The key difficulty with such tuple replication is to deal with *constraints* that may be defined on attributes in $\mathcal{R}$. It is hereby assumed that the database before splitting does not violate any of the constraints defined. In the scope of this paper, we limit ourselves, for reasons of simplicity, to unique constraints and referential constraints.

4.2 A Constrained Splitting Approach

The basic operator from which the solution in this paper departs, is called a split operator and is defined as follows.

Definition 1 (Split operator). *Let $\mathcal{R}$ be a schema. A split operator for an attribute $a \in \mathcal{R}$ is defined as:*

$$\Theta_a : \text{dom}(a) \rightarrow \mathcal{M}(\text{dom}(a)) \tag{5}$$

where $\mathcal{M}(.)$ denotes the set of all multisets.

A split operator transforms a textual string into a multiset of substrings. In order to use a split operator in solving vertical overflow, it is required to replicate tuples in the sense that there must be as much tuples as there are values in the result of the split operator. Therefore, a replication operator is defined as follows.

Definition 2 (Replication operator). *Let $\mathcal{R}$ be a schema. A replication operator for $\mathcal{R}$ and an attribute $a \in \mathcal{R}$ is defined as:*

$$\lambda : \text{dom}(\mathcal{R}) \times \text{dom}(a) \rightarrow \mathcal{M}(\text{dom}(\mathcal{R})) \tag{6}$$

where $\mathcal{M}(.)$ denotes the set of all multisets and such that:

$$\forall t \in R : \forall a \in \mathcal{R} : \lambda(t,a)[\mathcal{R} \ominus a] = t[\mathcal{R} \ominus a] \tag{7}$$

where $\ominus$ denotes the set difference operator.

Definition 2 states that a replication operator replicates a tuple t for several times such that the replicated tuples only differ from each other in their value for attribute a. We can now combine a split operator and a replication operator into a so called *split-replication* operator as follows. First, we can require that the result of replication contains exactly as much tuples as there are splitted values. Second, we can require that each tuple in the replication result, contains a value that is splitted by Θ_a. Both requirements are reflected in the following constraint:

$$\forall t \in R : \forall a \in \mathcal{R} : \lambda(t,a)[\![a]\!] = \Theta_a\left(t[a]\right). \tag{8}$$

Hereby, $\lambda(t,a)[\![a]\!]$ denotes the multi-projection of $\lambda(t,a)$ over a, which yields a multiset instead of a set. This constraint states that the projection of the replication over attribute a is exactly the result of splitting the value $t[a]$. This combination of splitting and replication models a solution for vertical overflow. More specifically, if t is a tuple with vertical overflow for attribute a^+, then we can solve this by replicating t such that (8) holds for attribute a^+. Tuple t can then be replaced with it's replications. However, before this replacement can be done, it must be verified whether or not there are constraints defined over $\mathcal{R}$. As mentioned in the previous, we restrict ourselves to unique constraints and referential constraints.

Unique Constraints. Let us first focus on unique constraints. A unique constraint is hereby modeled by a set of attributes $U \subseteq \mathcal{R}$ and implies that, for each relation R and for each tuple $t \in R$, $t[U]$ is unique in R. In the context of split-replication, two distinct cases can occur: either a^+ is part of the unique constraint or not.

In the first case, we have that $a^+ \notin U$. Because of the assumption that the database is in a valid state before split-replication, it follows that the unique constraint is only violated in the scope of $\lambda(t,a)$. The purpose of the replication is to create a new tuple for each value in $\Theta_{a^+}(t[a^+])$. Therefore, the breach of the constraint can be avoided by providing a so-called U-generator that provides unique values for U in the context of the relation R. Such a U-generator is easy to construct if U comprises for example an artificial key, but can be more difficult if U is a (combination of) text-valued attribute(s).

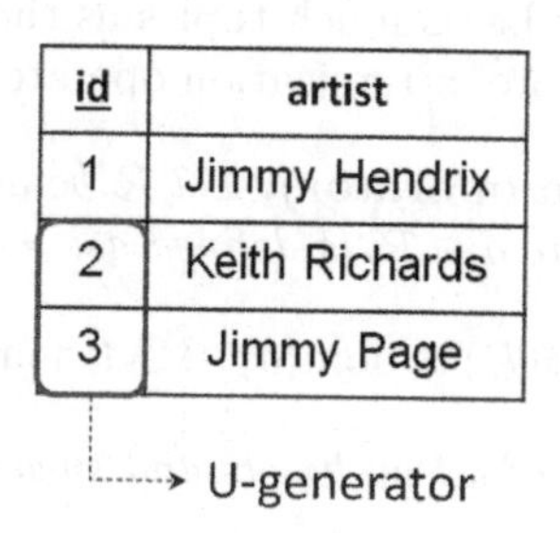

Fig. 3. Result of split-replication on 'artist' in relation from Figure 2

An example of this scenario is observed in Figure 2, where attribute 'artist' is a vertical overflow attribute and where there is a unique constraint on attribute 'id' because of the fact that it is a key attribute. We thus have that $U = \{\text{id}\}$. Suppose that attribute 'artist' is splitted by using the comma character as a separator symbol. The replication of the tuple shown in Figure 2 yields three tuples (i.e., one for each splitted value). In order to deal with the unique constraint on attribute 'id', a generator of unique values can be deployed. In this case, a U-generator is obtained by:

$$\max(R[\text{id}]) + 1. \tag{9}$$

The resulting relation after split-replication on attribute 'artist' is shown in Figure 3, where it can be seen that the relation has now three tuples and a unique value for 'id'. The quality of this relation is improved in the sense that attribute 'artist' now contains values that correspond to the frame of the schema of this relation.

In the second case, we have that $a^+ \in U$. In this case, the unique constraint can be violated in the scope of the entire relation. The purpose of replication is now to create a new tuple for each value in $\Theta_{a^+}(t[a^+])$ that leads to a unique value for attributes U. Therefore all tuples in $R \ominus t \oplus \lambda(t, a)$ with the same value for U, are combined. Such tuple combination can be done by use of a fusion function [10], [11] F which is defined by:

$$\mathrm{F} : \mathcal{M}(\mathrm{dom}(\mathcal{R})) \rightarrow \mathrm{dom}(\mathcal{R}) \tag{10}$$

A fusion function maps a bag of tuples on to one tuple and thus combines the information of multiple tuples in one tuple.

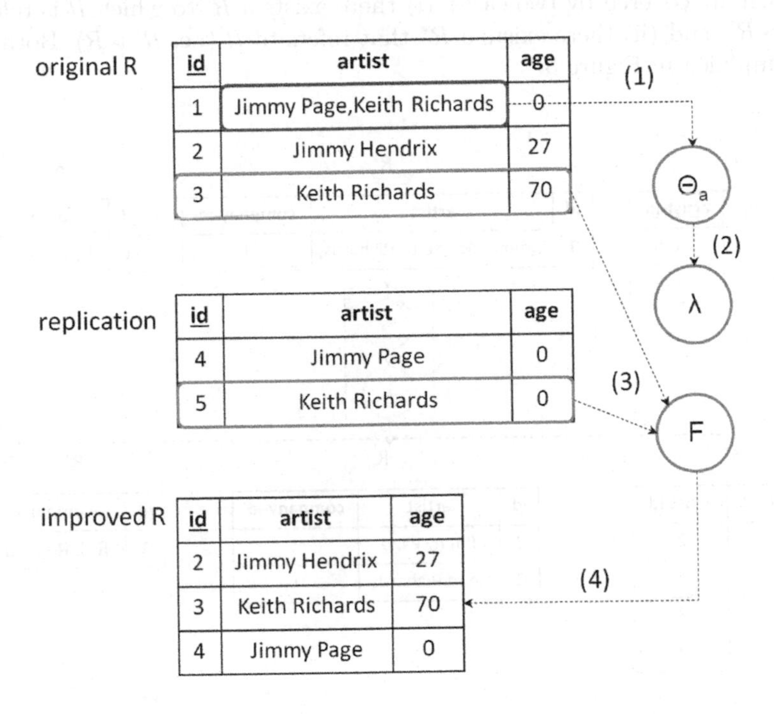

Fig. 4. An example of vertical overflow under unique constraints

As an example of this case, consider the original relation shown in Figure 4 and assume that there are two unique constraints: one specified by $U_1 = \{\text{'id'}\}$ and one specified by $U_2 = \{\text{'artist'}\}$. If we use a split operator with the comma character as separator to resolve the vertical overflow on 'artist', the first tuple will yield two values: "Jimmy Page" and "Keith Richards" (Figure 4, step 1). In the construction of the replication $\lambda(t, a)$, the unique constraint on attribute 'id' is taken into account by generating new unique values (Figure 4, step 2). The replication is then inspected for duplicate values within the replication and the original relation excluding t. It is observed that value "Keith Richards" occurs twice for attribute 'artist'. Because there is a unique constraint on this attribute, the tuples with an equal value for this attribute are fused (Figure 4, step 3). In this example, the attribute values that were present in the non-replicated tuple are preserved. Finally, tuple t is removed from R and the replicated tuples are inserted, taking into account the unique constraints.

A generalization of this second case is obtained when equality of values for attribute a^+ is weakened to *equivalence*. Two character strings could be equivalent because they represent the same value, but differ from each other by a typographical error. In that case, equivalence could be detected by using string similarity metrics [12], [13], [14].

Referential Constraints. Let us now consider referential constraints, which in their turn are covered by two cases: (i) there exists a R' to which R is referring (i.e. $R \triangleright R'$) and (ii) there exists a R^* that refers to R (i.e. $R^* \triangleright R$). Both cases are exemplified in Figure 5.

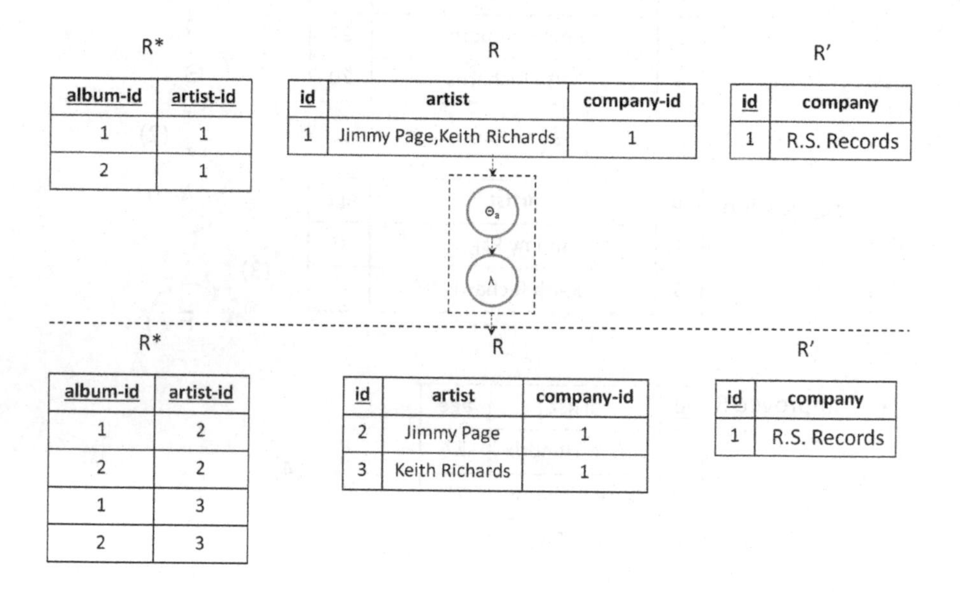

Fig. 5. An example of vertical overflow with referential constraints

The case where we have that $R \triangleright R'$, is treated as follows. If the relationship is of type *one-to-many*, then R contains a foreign key FK that is not part of a unique constraint. In that case, the replication operator will replicate references for each new tuple. If the relationship is of type *one-to-one*, then R contains a foreign key FK that must be unique. Resolving such unique constraints during split-replication, is explained in the previous. Assuming that the foreign key is not a^+, we should provide a generator for the foreign key. However, due to the fact that the generated values should refer to values in R', the construction of such a generator is virtually impossible. In addition, the one-to-one reference from R to R' indicates that each tuple in R semantically belongs to a tuple in R'. The replication of tuples in R without proper replication of tuples in R' is therefore a deterioration of the quality of the database. For that reason, split-replication should not be allowed on relations that refer to another relation through a one-to-one relationship.

The case where we have that $R^* \triangleright R$, is treated as follows. Regardless of the type of the relationship, the tuple $t \in R$ that is the subject of split-replication, corresponds to a *set* of tuples in R^*. Replications of the tuple should therefore also refer to this exact set of tuples. If a unique constraint U is resolved such that $a^+ \in U$, then there may exist multiple tuples that need to be fused. The corresponding sets of tuples in R are then combined by taking the union of these sets.

Algorithm 1. ConstrainedSplitReplication (CSR)

Require: R, Θ_{a^+}
1: **for all** $t \in R$ **do**
2: $T \leftarrow \lambda(t, a)$ conditioned by (8)
3: **for all** U **where** $a^+ \notin U$ **do**
4: $\forall t' \in T : t'[U] \leftarrow$ **generator.next**(U)
5: **end for**
6: **for all** U **where** $a^+ \in U$ **do**
7: **for all** $x \in (R \cup T \ominus t)\,[U]$ **do**
8: **fuse**$(\mathrm{F}, x, (R \cup T \ominus t))$
9: **end for**
10: **end for**
11: **for all** $R^* \triangleright R$ **do**
12: **replicate**$(\{t^* | t^* \in R^* \wedge t^*[\mathrm{FK}] = t[K]\})$
13: **end for**
14: **end for**

The observations made in the previous are summarized in pseudo-code (Algorithm 1). First, for each tuple, the split-replication operator is applied on a^+. Next, unique constraints that do not involve a^+, are taken care of by means of an appropriate U-generator. Fusion operators are used to handle unique constraints that do involve a^+ and finally, the replication operation is propagated to relations R^* that satisfy $R^* \triangleright R$.

5 Experimental Evaluation

The CSR algorithm is implemented in Java and tested on a dual core Intel machine (2.67 GHz) with 4 Gb of RAM memory. The dataset used is the Cora dataset in XML format as presented in [15]. This dataset describes articles from the Cora database by means of a title, venue and authors. The dataset is transformed into a relational format with three relations as shown in Figure 6.

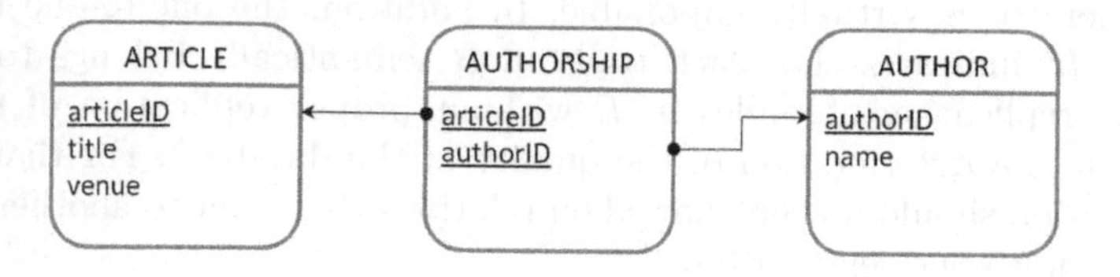

Fig. 6. Database schema

It is observed that attribute 'name' in the relation 'author' is a vertical overflow attribute. An extract of the relation 'author' is shown in Figure 7.

6	O. Inganas and M.R. Andersson
7	C. Ray Asfahl
8	Steve Benford and Lennart E. Fahlen
	...

Fig. 7. Extract from relation 'author'

Relation 'author' contains 710 tuples whereas 'authorship' contains 4571 tuples. When using the string ' and ' as delimiter to split attribute 'name', it is found that 184 tuples (about 26%) contain multiple values for attribute 'name'. Our implementation of the CSR algorithm generates an SQL script that models the required modifications to the relation in which the split-replication is performed and, if required, to linked relations. In the case of the Cora dataset, a script of 1848 statements is generated, of which 658 have impact on relation 'author' and 1190 have impact on relation 'authorship'. After execution of the script, relation 'author' contains 894 tuples and relation 'authorship' contains 5215 tuples.

6 Conclusion

The quality of a relational database depends strongly on the assumption that data fits the context implied by the schema. In this paper, we have investigated the case where information provided by an attribute value exceeds the frame provided by the schema. More specifically, a distinction has been made between horizontal and vertical information overflow. Whereas the case of horizontal overflow has been investigated widely, the case of vertical overflow has not yet reached this point. Therefore, a solution in the form of a split-replication operator has been proposed. The major difficulty is hereby to take into account constraints that hold on the database. The algorithm has been used to improve the quality of the Cora dataset.

References

1. Batini, C., Scannapieca, M.: Data quality: concepts, methodologies and techniques. Springer (2006)
2. Codd, E.F.: A relational model of data for large shared data banks. Communications of the ACM 13(6), 377–387 (1970)
3. Fellegi, I., Sunter, A.: A theory for record linkage. American Statistical Association Journal 64(328), 1183–1210 (1969)
4. Abiteboul, S., Hull, R., Vianu, V.: Foundations of Databases. Addison-Wesley (1995)
5. Adelberg, B.: Nodose: A tool for semi-automatically extracting structured and semistructured data from text documents. In: Proceedings of the SIGMOD Conference, pp. 283–294 (1998)
6. Soderland, S.: Learning information extraction rules for semi-structured and free text. Machine Learning 34, 1–44 (1999)
7. Califf, M.E., Mooney, R.: Relational learning of pattern-match rules for information extraction. In: Proceedings of the Sixteenth National Conference on Artificial Intelligence (AAAI 1999), pp. 328–334 (1999)
8. Borkar, V., Deshmukh, K., Sarawagi, S.: Automatically extracting structure from free text addresses. IEEE Data Engineering Bulletin 23(4), 27–32 (2000)
9. Freitag, D., McCallum, A.: Information extraction with HMM structures learned by stochastic optimization. In: Proceedings of the AAAI 2000, pp. 584–589 (1999)
10. Bronselaer, A., De Tré, G.: Aspects of object merging. In: Proceedings of the NAFIPS Conference, Toronto, Canada, pp. 27–32 (July 2010)
11. Bronselaer, A., Van Britsom, D., De Tré, G.: A framework for multiset merging. Fuzzy Sets and Systems 191, 1–20 (2012)
12. Levenstein, V.: Binary codes capable of correcting deletions, insertions and reversals. Physics Doklady 10(8), 707–710 (1966)
13. Jaro, M.: Unimatch: A record linkage system: User's manual. Technical report, US Bureau of the Census (1976)
14. Bronselaer, A., De Tré, G.: A possibilistic approach on string comparison. IEEE Transactions on Fuzzy Systems 17(1), 208–223 (2009)
15. Weis, M., Naumann, F., Brosy, F.: A duplicate detection benchmark for xml (and relational) data. In: Proceedings of the SIGMOD International Workshop on Information Quality for Information Systems (IQIS), pp. 1–9 (2005)

Auditing-as-a-Service for Cloud Storage

Alshaimaa Abo-alian, N.L. Badr, and M.F. Tolba

Faculty of Information and Computer Sciences, Ain shams University, Cairo, Egypt
{shimo.fcis83,dr.nagwabadr,fahmytolba}@gmail.com

Abstract. Cloud Storage Service (CSS) is a vital service of cloud computing which relieves the burden of storage management, cost and maintenance. However, Cloud storage introduces new security and privacy challenges that make data owners worry about their data. It is essential to have an auditing service to verify the integrity of outsourced data and to prove to data owners that their data is correctly stored in the Cloud. Recently, many researchers have focused on validating the integrity of outsourced data and proposed various schemes to audit the data stored on CSS. However, most of those schemes deal with static and single copy data files and do not consider data dynamic operations on replicated data. Furthermore, they do not have the facility to repair corrupt data. In this paper, we address these challenging issues and propose a public auditing scheme for multiple-copy outsourced data in CSS. Our scheme achieves better reliability, availability, and scalability by supporting replication and data recovery.

Keywords: cloud storage, auditing, probabilistic encryption, fountain codes, cryptographic hash algorithms.

1 Introduction

Recently, cloud storage service offers attractive features such as massive scalability, elasticity, reliability, pay as you go, and self-provisioning. On the other hand, it is susceptible to security and privacy threats. For example, data could be lost in the cloud because of cloud outages [18] and the cloud service providers may choose to hide data loss and claim that the data is still correctly stored in the cloud. In addition, the cloud service providers may be dishonest and they may discard the data which has not been accessed or rarely accessed or maintain fewer replicas than what is paid for to save the storage space in order to increase the profit margin by reducing cost.

Since users may not retain a local copy of outsourced data and may not trust the cloud service provider, It is a significant aspect for the cloud service provider to provide data security practices to convince data owners that their outsourced data is correct and safe. Thus, many researchers have focused on checking the integrity of outsourced data and proposed various schemes and protocols to audit the data stored on CSS.

Any system model of auditing scheme consists of three entities as mentioned in [4]:

P. Angelov et al. (eds.), *Intelligent Systems'2014*,
Advances in Intelligent Systems and Computing 322, DOI: 10.1007/978-3-319-11313-5_49

1. **Data Owner:** an entity, which has large data files to be stored in the cloud and can be either individual consumers or organizations.
2. **Cloud Storage Server (CSS):** an entity, which is managed by Cloud Service Provider (CSP), has significant storage space and computation resource to maintain the clients' data.
3. **Third Party Auditor or Verifier (TPA):** an entity, which has expertise and capabilities to check the integrity of data stored on CSS.

In view of the verifier role in the model, all auditing schemes fall into two classes: private auditing and pubic auditing [7]. In private auditing, only Data owner who can audit CSS to verify the correctness of outsourced data [8]. Unfortunately, private auditing schemes have two drawbacks: (a) They impose an online burden on the data owner to verify data integrity and (b) Data owner must have huge computational resources for auditing. In Public auditing or Third party auditing, Data owners can delegate the auditing task to an independent third party auditor (TPA), without dedication of their computation resources [8]. However, pubic auditing schemes should ensure that the privacy of the verified data is maintained against disclosure by the TPA.

Several auditing schemes such as [2,3,4], [9,10,11,12] were proposed under different cryptographic assumptions. Most of these schemes [12,13] deal with integrity verifications and do not support data recovery in case of data corruption. Some schemes [5], [9], [11] deal only with archival static data files and does not consider dynamic operations such as insert, delete and update. Whereas many schemes support only private auditing such as [3], [10], [12].

In this paper, we propose a public and privacy- preserving auditing scheme for single-copy and multiple-copy, moreover for dynamic data files. We improve the CSP's efficiency and achieve better reliability, availability, and scalability by supporting replication and data recovery.

The rest of the paper is organized as follows. Section 2 overviews related work. In section 3, we provide the detailed description of our auditing scheme. Then, we illustrate performance analysis of our scheme in section 4. Finally, we conclude in section 5.

2 Related Work

For verifying the integrity of single copy data outsourced in the cloud storage, Jun Liu et al. [4] considered the security problem of the auditing protocol proposed by Wang et al. [13] in the signature generation phase which allows the CSP to cheat by using blocks from different files during verification. Therefore, they presented a secure public auditing protocol based on the homomorphic hash function and BLS short signature scheme, which supports for public verifiability, data dynamics and privacy preserving. However, their protocol suffers from massive computation and communication costs.

Ren et al. [6] proposed a privacy-preserving public auditing scheme using random masking and homomorphic linear authenticators (HLAs) [1]. Their auditing scheme also supports data dynamics using Merkle Hash Tree (MHT). In addition, it enables the auditor to perform audits for multiple users simultaneously and efficiently. Unfortunately, their scheme is vulnerable to the TPA offline guessing attack.

Considering replicated data stored in multiple servers, Barsoum and Hasan [11] proposed two dynamic multi-copy provable data possession schemes: tree-based and map-based dynamic multi-copy provable data possession (TB-DMCPDP and MB-DMCPDP, respectively). These schemes prevent the CSP from cheating and using less storage by maintaining fewer copies through using the diffusion property of AES encryption scheme. The notable limitations of both schemes are high computation and communication costs. Besides, The replica number should be known to the authorized users to be able to generate the original file.

Etemad and Kupcu [3] proposed a distributed and replicated DPDP (DR- DPDP) which provides transparent distribution and replication of user data over multiple servers where the cloud storage provider (CSP) may hide its internal structure from the client. They use persistent rank-based authenticated skip lists to make data dynamics more efficient. Their scheme supports dynamic version control to enable accessing old values of updated data. On the other hand, DR-DPDP has three noteworthy disadvantages: First, it only supports private auditing. Second, it does not support recovery of corrupted data. Finally, the organizer looks like a central entity which may get overloaded and can cause a bottleneck.

3 Proposed Scheme

In this section, we first state some definitions applied in the design of the auditing scheme. Then, we describe the algorithms and the detailed phases of the auditing scheme for cloud storage.

3.1 Notations and Preliminaries

In this sub-section, we list some notations and define some preliminaries used in the proposed scheme.

- F is a data file to be outsourced and consists of a finite ordered set of m blocks, i.e. $F = \{b_1, b_2, ..., b_m\}$.
- $e : \mathbb{G}_1 \times \mathbb{G}_2 \rightarrow \mathbb{G}_T$ is a bilinear pairing; where $\mathbb{G}_1$, $\mathbb{G}_2$ and $\mathbb{G}_T$ be three multiplicative groups.
- *Paillier Encryption*: Paillier cryptosystem [14] is a probabilistic encryption scheme which creates different ciphertexts each time the same message is encrypted using the same key. Using a public key (N, g), a message m is encrypted to a ciphertext ct using equation (1):

$$ct = g^m x^N \bmod N. \quad (1)$$

Using a secret key λ, Plaintext can be decrypted as:

$$m = L(ct^{\lambda} \bmod N^2) * (L(g^{\lambda} \bmod N^2))^{-1} \bmod N. \quad (2)$$

Where :

- p, q are two prime numbers, N = p * q.
- λ = LCM (p-1, q-1).
- g is random number such that its order is a multiple of N; g $\in \mathbb{Z}_N$.
- x is a random number and x $\in \mathbb{Z}_N$.
- L(u) = (u-1)/N.

– *Raptor codes:* A Raptor code [15] is a fountain code that encodes a message of k symbols into a limitless sequence of encoding symbols such that knowledge of any k or more encoding symbols allows the message to be recovered with some non-zero probability. A Raptor code [16] is specified by parameters (k, C, Ω (x)), where C is the (n, k) erasure correcting block code, called the *pre-code*, and Ω (*x*)) is the generator polynomial of the degree distribution of the LT code.

$$\Omega\,(x) = \sum_{i=1}^{k} \Omega_i x^i \tag{3}$$

where Ω_i is the probability that the degree of an output node is *i*.

3.2 Proposed Model

The proposed scheme consists of nine polynomial time algorithms as shown in Figure 1: Key Generation (KeyGen), Probabilistic encryption and replica generation (ReplicaGen), File Encoding (RaptorEncode), Hashing and Tag Generation (TagGen), Challenge Generation, Proof Generation, Proof Verification, Data Recovery and Data Modification.

- **Key Generation (KeyGen):** This algorithm is executed by the data owner. It takes as input security parameter 1^λ and its outputs: private key *sk* and public key *pk* for block tag generation, Hash secret key sk_h, and pseudorandom function key Key_{PRF} for replica generation.
- **Replica Generation (ReplicaGen):** This algorithm is executed by the data owner if s/he chooses multiple replica version. The number of replicas *r* and the file *F* are taken as input and generates *r* unique differentiable replicas $\{F_i\}_{1\leq i\leq r}$. This algorithm is run only once. Unique copies of each file block of file *F* is created by encrypting it using a probabilistic encryption scheme. We utilize Paillier encryption scheme [14] for replica generation because it is semantically secure and has efficient encryption complexity.
- **File Encoding (RaptorEncode):** This algorithm is executed by the data owner in order to support data recovery when s/he outsources single-copy data file. RaptorEncode algorithm takes Key_{PRF}, outputs encoded file F', and works as follows: F = {b_1,, …, b_k}, b_i is s bits, is encoded by an erasure code (pre-code) to obtain F'= {y_1,…,y_k}. Then, Choose a random s × s binary matrix A = $[A_1, \ldots, A_s]^T$ where each A_l is an s –bit vector. For each $1\leq i\leq n$, Create authenticators $\delta_1,\ldots,\delta_n$ as:

$$\delta_i = PRF_{\,Key_{PRF}}(i) \oplus y_i A \tag{4}$$

Finally, F' = $\{y_1,\ldots,y_k, \delta_1,\ldots, \delta_n\}$ is the encoded file. For each encoded block, a coding vector Δ_j is attached where each bit represents whether the corresponding original block is combined into F' or not.

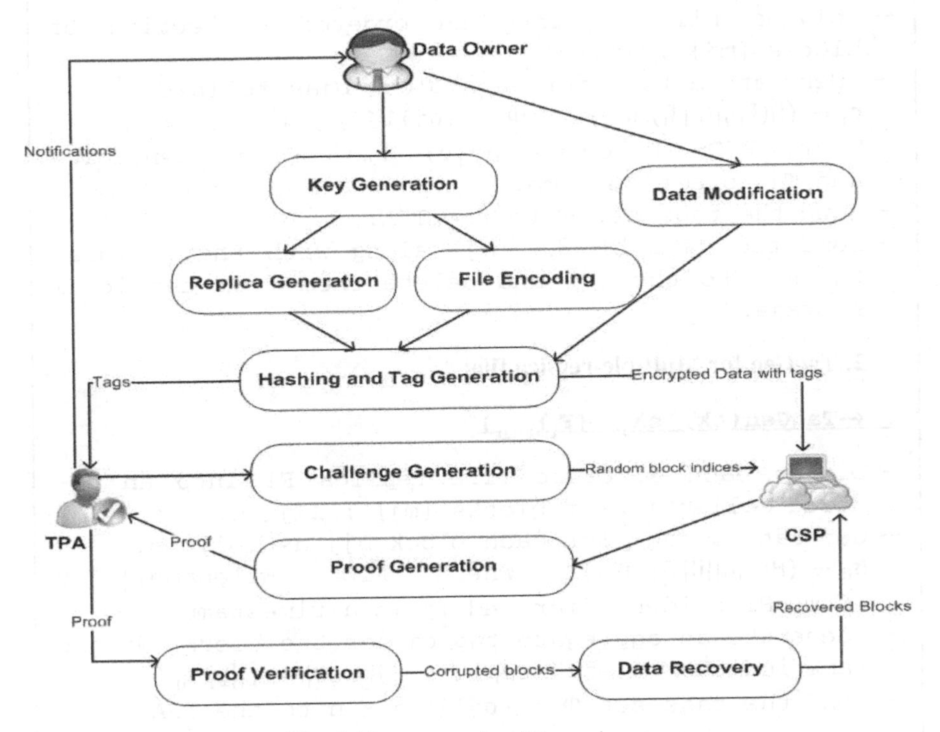

Fig. 1. The proposed auditing scheme

- **Hashing and Tag Generation (TagGen):** This algorithm is run by the data owner. It takes the private key sk, the secret hash key sk_h, and the unique differentiable file replicas $\{F_i\}_{1\leq i\leq r}$ or the encoded file F' as inputs. Its output is the tags set $\Phi = \{\sigma_j\}_{1\leq j\leq n}$ which is an ordered collection of tags for the data blocks. Figure 2 illustrates a detailed description of TagGen algorithm. It is valuable to note that we embed the file identifier F_{ID} into the block tag to prevent the CSP from cheating and using data blocks from different files and passing the audit. Embedding a time-stamp of each data block T_j into the block tag to authenticate the tag and maintain the block versions. We utilize the BLS tag generation due to its homomorphic verifiable property which aggregate the signatures of distinct blocks into a single short one and verify it at one time, and thus reduce storage overhead and communication costs for challenge and response messages.

1. TagGen for Single-Copy file:

Φ ←TagGen(sk, sk_h, F')

- Divide File F' into an ordered collection of blocks {mj} ; 1≤j≤ n.
- Generate a tag for each block bjas follows: $\sigma_j \leftarrow (H(F_{ID}||j||T_j).u^{m_j})^{\alpha}$, Φ = {σj}1≤ j ≤ n
  Where FID= Filename||n||u; i.e. File identifier and Tj is a timestamp.
- Send the tags set Φ to the TPA.
- Send the data blocks {mj} along with their signatures to the CSP and delete them from the local storage.

2. TagGen for Multiple-replica file:

Φ ←TagGen(sk, sk_h, {F_i}_{1≤i≤r})

- Divide each distinct file replica Fi into an ordered collection of blocks {mj} ; 1≤j≤ n.
- Generate a tag for each block bjj as follows: $\sigma_{ij} \leftarrow (H(F_{ID}||j||T_j).u^{m_{ij}})^{\alpha}$ Where FID= Filename||n||u ;i.e. File identifier and Tj is a timestamp.
- Generate an aggregate tag σj for the blocks at the same indices in each replica Fj as$\sigma_j = \prod_{i=1}^{r} \sigma_{ij}$
- Send the tags set Φ = {σj}1≤ j ≤ n to the TPA.

- Send the data blocks {mj} along with their signatures Φ to the CSP and delete them from the local storage.

Fig. 2. TagGen algorithm

- **Challenge Generation:** In this algorithm, the TPA challenges the CSP to verify the integrity of all outsourced replicas. The TPA sends c (number of blocks to be challenged; $1 \leq c \leq n$) and two distinct PRF keys at each challenge: k_1 and k_2.The PRF keyed with k_1 is used to generate c random indices which the file blocks that the CSP should use to prove the integrity. The PRF keyed with k_2 is used to generate y_j random values that are associated with each random index j and used by the CSP while generating the response. Then, the challenge set $Q = \{(j, y_j)\}$ of pairs of random indices and values is generated at the CSP.
- **Proof Generation:** This algorithm is run by the CSP, upon receiving the challenge set Q, to generate a proof that it is still correctly storing all replicas. The CSP computes

$$\sigma = \prod_{(j,y_j)} (\sigma_j)^{y_j}, \ \mu = \sum_{i=1}^{r} \left(\sum_{(j,y_j)} y_j \cdot m_{ij} \right) \quad (5)$$

The CSP sends the proof $\mathbb{P} = \{\sigma, \mu\}$ to the TPA.

- **Proof Verification:** This is run by the TPA. It takes as input the public key *pk*, the challenge set *Q*, and the proof $\mathbb{P}$ returned from the CSP, The TPA checks the following verification equation:

$$e(\sigma, g) ?= e([\prod_{(j,y_j)} H(F_{ID} \parallel j \parallel T_j)^{y_j}]^r . \mu, \mathcal{V}) \quad (6)$$

and outputs TRUE if the verification equation passed, or FALSE otherwise.

- **Data Recovery:** To repair a corruption on the ℓ- storage server, the TPA uses the corresponding coding vectors $\{\Delta_{\ell j}\}_{1 \le j \le n}$ to generate the encoded blocks $\{m_{\ell j}\}_{1 \le j \le n}$. $m_{\ell j}$is generated by the XOR combination of $|\Delta_{\ell j}|$ original blocks as

$$m_{\ell j} = M_{\ell 1} \oplus \dots \oplus M_{\ell |\Delta_{\ell j}|} \quad (7)$$

- **Data Modification:** To support efficient dynamic operations, we utilize a map-version table which is an authenticated data structure stored on the TPA to validate the data dynamics on all file replicas. The map-version table consists of four columns: Index (j) , block number (B_j), version number (V_j) and timestamp (T_j). The index denotes the current block number of the data block m_j. B_j denotes the original block number of the data block m_j. V_j denotes the current version number of the data block m_j which is increased by 1 each time the data block is modified. T_j denotes the timestamp used for generating the tag. The dynamic operations in our proposed scheme are preformed via a request in the form Modify(j, Op, m_j^*) where Op is the dynamic operation; i.e. 0 for deletion, 1 for insertion , and 2 for update. $m_j^*$is the new block value.

4 Performance Analysis

In this section we evaluate the performance of our proposed scheme as we list its features in Table 2 and make a comparison of our scheme and the state of the art (refer to Table 1 for notations). Let r, n, k denote the number of replicas, the number of blocks per replica and the number of sectors per block (in case of block fragmentation), respectively. s denotes the block size. c denotes the number of challenged blocks.

Table 1. Notation of cryptographic operations.

Notation	Cryptographic operation
MUL	Multiplication in group $\mathbb{G}$
ADD	Addition in group $\mathbb{G}$
EXP	Exponentiation in group $\mathbb{G}$
H	Hashing into group $\mathbb{G}$
Pairing	Bilinear pairing ; e(u, v)

5 Experiments and Discussion

Our experiments are conducted using MATLAB and Java on a system with an Intel Core i5 processor running at 2.2 GHz and 4 GB RAM running Windows 7. In our implementation, we use Java Pairing-Based Cryptography (JPBC) library version 2.0.0. To achieve 80-bit security parameter, the elliptic curve group we work on has a 160-bit group order and the size of modulus N is 1024 bits. All files used in the experiments are downloaded from the Human Genome Project at NCBI [17]. All results are the averages of 20 trials.

Table 2. Comparison between our proposed scheme and the-state-of-the-art

Scheme	Public Audit	Dynamic data	Replication	Data Recovery	Computational complexity		Storage overhead		Communication Compl.
					CSP	Auditor	CSP	Auditor	
[4]	√	√	×	×	H+ O(s)[MUL+EXP]	O(c) [MUL+ H+ EXP] + 2 *Pairing*	O((1+1/k) n)	O(1)	O(1)
[5]	√	×	×	√	O(c+s)[MUL+ EXP]	O(c) [MUL+ EXP] + + 4 *Pairing*	O(2nk)	O(1)	O(c)
[6]	√	√	×	×	O(c) [MUL+ EXP] +H + 1 *Pairing*	O(c) [MUL+ H+ EXP]+ + 2 *Pairing*	O((1+1/k) n)	O(1+1/k)	O(kc)
[3]	×	√	√	×	O(1+ log nr) [EXP+ MUL]	O(1+ log nr) [EXP+ MUL]	O(log nr)	O(1)	O(log nr)
[11]	√	√	√	×	O(c) [EXP+MUL + ADD	O(c)[H+ MUL+ EXP] + O(k) (MUL+ EXP) + O(r)ADD + 2 *Pairing*	O(2n)	O(2n)	O(kr)
Proposed Scheme (Single Copy)	√	√	×	√	O(c) [MUL+ EXP]	O(c) [EXP+ MUL+H] + O(s) [MUL + EXP]	O(log n)	O(1)	O(log n)
Proposed Scheme (Multi-replica)	√	√	√	√	O(c) [EXP+ MUL] + O(r) ADD	O(cr) [EXP+ MUL+H] +O(s) [MUL+ EXP]	O(log nr)	O(log n)	O(log nr)

Figure 3 illustrates the computational time for our proposed scheme and for the [11] scheme using different number of replicas, 1 GB file, 16 KB block size, corruption rate = 1% and detection probability = 99.99%.

The CSP computational time of our proposed scheme, as shown figure 3(a), is about 5 times faster than that of [11] due to the tag aggregation and fast decryption of the Paillier scheme. Although the TPA computational time of our scheme, as shown figure 3(b), is faster than that of [11] for small number of replicas, the TPA computational time of [11] is independent on number of replicas and performs efficiently for larger number of replicas.

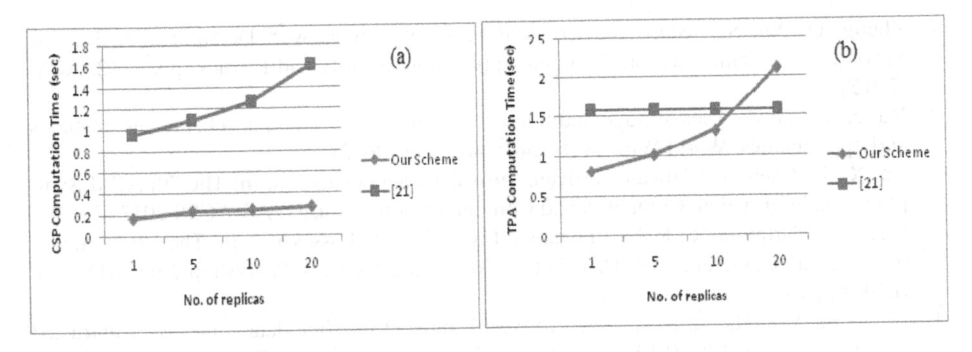

Fig. 3. The computational time of our scheme and [11]

6 Conclusion

In this paper, we propose a dynamic public auditing scheme for verifying the integrity of replicated data in cloud storage. We utilize the homomorphic BLS tags and cryptographic SHA-256 algorithm to guarantee that the scheme is privacy-preserving. We exploit Paillier probabilistic encryption scheme for efficient replica differentiation to prevent the CSP from cheating and maintaining fewer replicas than what is paid for. To achieve efficient data dynamics, we utilize the map-version table which improves the computation time of dynamic operations and the proof generation. we take benefit of the efficient encoding and decoding of Raptor codes to support data recovery. The performance analysis show that the proposed scheme is complete, provably secure, and efficiently comparable to the already existing schemes.

References

1. Ateniese, G., Kamara, S., Katz, J.: Proofs of Storage from Homomorphic Identification Protocols. In: Matsui, M. (ed.) ASIACRYPT 2009. LNCS, vol. 5912, pp. 319–333. Springer, Heidelberg (2009)
2. Zhang, Y., Blanton, M.: Efficient dynamic provable possession of remote data via balanced update trees. In: the 8th ACM SIGSAC Symposium on Information, Computer and Communications Security (ASIA CCS 2013), pp. 183–194. ACM, New York (2013)
3. Etemad, M., Küpçü, A.: Transparent, Distributed, and Replicated Dynamic Provable Data Possession. In: Jacobson, M., Locasto, M., Mohassel, P., Safavi-Naini, R. (eds.) ACNS 2013. LNCS, vol. 7954, pp. 1–18. Springer, Heidelberg (2013)
4. Liu, H., Zhang, P., Lun, J.: Public Data Integrity Verification for Secure Cloud Storage. Journal of Networks 8(2), 373–380 (2013)
5. Yuan, J., Yu, S.: Proof of retrievability with public verifiability and constant communication cost in cloud. In: The 2013 International ACM Workshop on Security in Cloud Computing, pp. 19–26. ACM (2013)
6. Wang, C., Chow, S.S.M., Wang, Q., Ren, K., Lou, W.: Privacy-Preserving Public Auditing for Secure Cloud Storage. IEEE Transactions on Computers 62(2), 362–375 (2013)

7. Zheng, Q., Xu, S.: Secure and Efficient Proof of Storage with Deduplication. In: The Second ACM Conference on Data and Application Security and Privacy, pp. 1–12. ACM (2012)
8. Yang, K., Jia, X.: Data storage auditing service in cloud computing: challenges, methods and opportunities. World Wide Web 15(4), 409–428 (2012)
9. Jia, X., Ee-Chein, C.: Towards efficient provable data possession. In: The 7th ACM Sympoium on Information, Computer, and Communications Security, ASIACSS 2012 (2012)
10. Chen, B., Curtmola, R.: Robust Dynamic Provable Data Possession. In: The 32nd International IEEE Conference on Distributed Computing Systems Workshops, pp. 515–525. IEEE (2012)
11. Barsoum, A.F., Hasan, M.A.: Integrity verification of multiple data copies over untrusted cloud servers. In: 12th IEEE/ACM International Symposium on Cluster, Cloud and Grid Computing (ccgrid 2012), pp. 829–834. IEEE Computer Society (2012)
12. Liu, F., Gu, D., Lu, H.: An improved dynamic provable data possession model. In: The 2012 IEEE International Conference on Cloud Computing and Intelligence Systems (CCIS), pp. 290–295. IEEE (2011)
13. Wang, Q., Wang, C., Ren, K., Lou, W., Li, J.: Enabling Public Auditability and Data Dynamics for Storage Security in Cloud Computing. IEEE Transactions on Parallel And Distributed Systems 22(5), 847–859 (2011)
14. Paillier, P.: Public-Key Cryptosystems Based on Composite Degree Residuosity Classes. In: Stern, J. (ed.) EUROCRYPT 1999. LNCS, vol. 1592, pp. 223–238. Springer, Heidelberg (1999)
15. Shokrollahi, A.: Raptor Codes. IEEE Transactions on Information Theory 52(6), 2551–2567 (2006)
16. Ho, T.: Summary of Raptor Codes. Scientific Commons (2003)
17. National Center for Biotechnology Information, `http://www.ncbi.nlm.nih.gov/genome/genomes/51`
18. Raphael, J.R.: The worst cloud outages of 2013, part 2 (2013), `http://www.infoworld.com/slideshow/133109/the-worst-cloud-outages-of-2013-part-2-232912`

Data Driven XPath Generation

Robin De Mol, Antoon Bronselaer, Joachim Nielandt, and Guy De Tré

Dept. of Telecommunications and Information Processing,
Ghent University, Sint-Pietersnieuwstraat 41, B-9000 Ghent, Belgium
{robin.demol,antoon.bronselaer,joachim.nielandt,guy.detre}@UGent.be

Abstract. The XPath query language offers a standard for information extraction from HTML documents. Therefore, the DOM tree representation is typically used, which models the hierarchical structure of the document. One of the key aspects of HTML is the separation of data and the structure that is used to represent it. A consequence thereof is that data extraction algorithms usually fail to identify data if the structure of a document is changed. In this paper, it is investigated how a set of tabular oriented XPath queries can be adapted in such a way it deals with modifications in the DOM tree of an HTML document. The basic idea is hereby that if data has already been extracted in the past, it could be used to reconstruct XPath queries that retrieve the data from a different DOM tree. Experimental results show the accuracy of our method.

Keywords: XPath Generation, Data Driven, HTML.

1 Introduction

With big data a hot topic at the moment, data extraction has been gaining a lot of attention over the last few years. In that field semi-structured data sources, such as HTML, present a large challenge due to their property of separating data from the structure that presents it. Semi-structured documents preserve a notion of hierarchy and dependencies between values but there is no tabular schema behind the structure. It is the hierarchy itself that supposedly provides metadata on the data.

Current data extraction tools, often referred to as wrappers, usually heavily rely on XPaths to identify where data of interest can be found. However, if the structure of a document is changed, the XPaths to the data change with it, which creates the need for a reconfiguration of the data extraction tool. With our approach we aim to improve existing tools by providing the functionality to automatically reconstruct XPaths to previously identified data.

First, we will extract data using an XPath based data extractor. Second, we will use this data to attempt to reconstruct the XPaths used to configure the data extractor.

The remainder of the paper is structured as follows. In Section 2, we explain why our approach is beneficial for extraction tools. In Section 3, we indicate how our methods can be placed in current technology. Section 4 provides some preliminaries for our work. In Section 5, we elaborate on our technique, illustrated

P. Angelov et al. (eds.), *Intelligent Systems'2014*,
Advances in Intelligent Systems and Computing 322, DOI: 10.1007/978-3-319-11313-5_50

by an example that will be used throughout the rest of the work. In Section 6, we discuss how we can use data to construct XPaths. In sections 7 and 8, we explain how we can use these XPaths to configure a new wrapper. In section 9 we present a use case to measure the accuracy of our research. In Section 10, we indicate what might still improve our work and finally, in Section 11 we present our final conclusions.

2 Problem Statement

XPath based data extraction tools show some shortcomings. One of their most important issues lies in the semi-structured nature of HTML. Because data and structure are separated, a purely structural approach to retrieve data carries inherent risk, because there is no way to verify if the given XPaths extract the correct data. When websites are generated through templates, generally it is true that similar data will be found in similar nodes, resulting in similar XPaths. However, because of a new visual update to the website, the template that generates the structure can be changed while the data remains the same, as to present it in a different fashion. Then, it can happen that the configured XPaths no longer point to data, or worse, point to the wrong data.

Therefore, a more intelligent approach is needed for data extraction from semi-structured documents, based on both structure and data. Under the assumption we can detect changes in the structure of an HTML document, we propose a technique to automatically reconstruct XPaths to the data of interest.

3 Related Work

The internet has been profiled as the largest source of data available. Therefore, there has been a lot of research effort into the creation of data extraction tools [1,2]. These are classified based on the degree of automation.

The most commonly represented tools are semi-automated, relying on a manual training phase after which the crawler can continue to run automatically [3,4,5,6]. Alternatively, there are fully automated techniques [7,8,9], which require a single URL as input and aim to return a set of objects extracted from the content. Though these tools are considered to be easier to use, their output generally requires more attention than that of semi-automated solutions. For the latter, the manual training phase often allows semantic annotation that is used to post-process the data, which is lost when using the former.

Due to the semi-structured nature of HTML, existing systems additionally almost exclusively use a structural approach. Most of these rely on the DOM tree representation, also called tag tree, of the document [4,5,7,8,9,10,11], though some also look at the way the data is visualized [3,12] using optical recognition software. Such techniques often focus on frequency and pattern analysis of subtrees to estimate where data of interest can be found.

On the topic of wrapper breakage, which occurs when a wrapper can no longer retrieve data because the HTML document is changed structurally, there has been less research [13,14,15,16,17]. These approaches discuss ways to identify data in labeled documents, but do not offer a solution for text analysis or entity recognition. Furthermore, these techniques often use a different than XPath based (re)configuration, reintroducing the need for another manual configuration.

Our approach can be classified as a semi-automated and part structural, part data driven technique for web data extraction. The novel factor of our methodology is the combination of both structural and data driven aspects, providing a strong, yet highly automated solution to reconfigure XPath based data extraction tools.

4 Preliminaries

4.1 XPath

XML Paths (XPaths) are used to uniquely identify one or more elements in XML documents. Due to the fact that HTML and XML are both based on Standard General Markup Language (SGML), they are highly similar. Therefore, XPaths can and have also be used to identify elements in HTML documents.

XPaths structurally heavily resemble resource locators from UNIX-based file systems and hence consist of a sequence of steps separated by forward slashes. XPaths can be absolute or relative. Absolute XPaths always starts with a forward slash and are evaluated starting from the root element, a required element in all SGML documents. Relative XPaths assume a context, usually an internal element, and describe a path starting from there.

XPaths can be used to navigate through HTML documents by executing them in sequence. A new navigation is always indicated by an absolute path. Subsequent XPath queries are relative to the results of the previous one. After each XPath query, the elements resulting from the navigation are returned and are used as a context for the next query.

In contrast to UNIX-based paths, XPaths offer additional functionality in the form of predicates, indicated by square brackets. Such predicates can be used to select specific elements from the results of an XPath query. Predicates can contain wildcards to point to multiple elements simultaneously.

To illustrate some of the possibilities, a small example is provided. Below is one of the numerous alternative ways to list the authors of this paper in HTML, followed by some simple XPaths that can be used to select data from it.

```
<html>
    <h1>Authors</h1>
    <p>Robin De Mol</p>
    <p>Antoon Bronselaer</p>
</html>
```

```
/html[1]/p[1]  → selects author Robin De Mol
/html[1]/p     → selects all authors by omitting the index predicate
/html[1]/*     → selects all child elements of the html element
```

XPaths can be used in a broader context with more flexibility but this functionality lies outside the scope of this paper.

4.2 DOM

The Document Object Model (DOM) is a tree-based representation of an HTML document's structure consisting of nodes. Each node has at least a type and possibly a name or value.

The type, which can be `element` for elements and `text` for data, among others, indicates how the node should be interpreted. In case of an element node, the name represents the kind of element, such as `p`, `a` and `html`. In case of a text node, the name is "#text"[1].

The value contains the actual data in case of a text node and equals null in case of an element node. Nodes of the element type can contain any number of child nodes. Getting the text component of any node concatenates its own value with the text component of all its descendants.

Below is the DOM tree for the HTML document from the previous example.

```
{type: element, name: html, value: null}
   {type: element, name: h1, value: null}
      {type: text, name: null, value: Author list}
   {type: element, name: p, value: null}
      {type: text, name: null, value: Robin De Mol}
   {type: element, name: p, value: null}
      {type: text, name: null, value: Antoon Bronselaer}
```

5 General Approach

It is assumed here that XPath queries are used to map HTML documents onto a relational database. Therefore, we have chosen to model our data in a tabular way. Hence, we assume that the data we want to extract consists of multiple records. Each record is a collection of one or more attributes. Among them we choose one attribute to be the key attribute. This attribute needs to be present for each record as it is used to identify it in the document and in the database. The other attributes are called relative attributes. If the HTML document is generated by a template, we assume that each relative attribute can always be found at the same relative location to the key data for each record. We use this assumption to model concepts like "X can be found below Y" or "A is located to the right of B".

Throughout the remainder of this work, we will use a case study of a bookstore: www.eci.be. We identify books as records. For each book, the title is the key attribute and the author is a relative attribute. Other candidates for relative attributes are the publisher, amount of pages, price and description, among others.

[1] `http://docs.oracle.com/javase/6/docs/api/org/w3c/dom/Node.html`

Fig. 1. Example data from http://www.eci.be/boeken/roman-literatuur/literatuur. Key attributes (titles) are marked in green, relative attributes (authors) in yellow.

First, we extract a data set of books from http://www.eci.be/boeken/roman-literatuur/literatuur using an XPath based data extraction tool. For this, we identify the attributes of interest, in our case title and author, and configure one XPath per attribute. For the key attribute, the title, we specify one absolute XPath which evaluates all book titles on the document. For the relative attribute, the author, we specify one relative XPath, which tells the wrapper where the author can be found, given the title. Our goal is to reconstruct these XPaths using our approach. First, we investigate how we can reconstruct the XPath for the key attribute, the title. Then, we look at how we can calculate the XPath for the relative attribute, the author.

To find the XPath that matches all book titles, we first select some sample books from the data set. We then search the DOM tree of the HTML document for titles of these samples. From this inspection, we will retrieve the XPaths to all elements containing title data from the samples. We then cluster, align and merge this information with the intent to extrapolate the XPaths from the samples into a single XPath that matches all book titles in the document, as is explained below.

Next, we turn our attention towards the relative attribute "author". For each sample of which the title was matched, we calculate the relative XPath to the nearest element containing author data. Under the assumption the HTML document is generated by a template, we expect this relative path will be the same for each book. However, we foresee the possibility some might differ for unknown reasons. In order to retain only one XPath we use a majority voting system, choosing the relative path that is found the most.

Finally, we end up with a single XPath per attribute: an absolute one for the titles and a relative one for the authors. These can be interpreted as follows: the absolute XPath indicates where all book titles can be found in the document, and the relative XPath specifies where the author of a book can be found with respect to its title.

When comparing these XPaths to the ones we used for the initial configuration, we should find that they are equal.

6 DOM Inspection

To retrieve the XPath to an element containing specific data, we crawl through the DOM tree recursively, inspecting each node we pass through. The algorithm is initialized at the root node, which is equivalent to the `html` element. At each call, we check if the node is a text node. In case it is, we verify whether its value matches the data we're searching for. In case it isn't, we descend deeper into the DOM tree by calling the function on each child of the current node.

Each time we can match the value of a text node to the data we're trying to locate, we add the XPath to that element to a list of results. At the end of the algorithm, we retrieve a list of XPaths to all elements that contain the data.

This algorithm depends on a function that performs text-based entity comparison. Its implementation has a big impact on the accuracy of the results. In its simplest form, this could be a case and diacritic insensitive string comparison, but for names of people this method should be extended to be more flexible. The exact implementation of this method justifies more research, as explained in 10. In our experiments, we have used a case and diacritic insensitive regex to match the wrapped data.

7 Key Data

Given the XPaths to the key data we want to combine them in such a manner that we can use them to extrapolate where the rest of the records in the document can be found. Because the DOM tree inspection algorithm does not guarantee that only one XPath is returned per sample, we need to analyse the results before we can extrapolate them. If multiple XPaths are returned for a single sample, it means the data was found multiple times on the page. Likely at least one of the matches will be part of a list containing all the records we aim to extract. To identify which XPaths point to the elements in this list and which are false positives, we apply a clustering, alignment and merging technique. Because the clustering itself depends on alignment and merging, we explain it last.

7.1 Aligning and Merging

Aligning and merging a set of XPaths combines them into a single XPath. The resulting XPath has the characteristic that it matches at least all elements the original XPaths match. Ideally, the merged XPath matches more elements, so it can be used to extrapolate information from some samples.

First, we align the XPaths according to an n-dimensional implementation of an adaptation to the Levenshtein algorithm to align lists [18]. The alignment will produce the optimal arrangement for all XPaths simultaneously resulting in the lowest merge cost.

Afterwards the XPaths can be merged into a minimal enclosing XPath by inserting wildcards for the differences in the aligned paths. Iterating over the items of the now equally long XPaths, we introduce a wildcard if at least one item differs from the others. A wildcard can be inserted for either tagname, index or both.

```
Unaligned:
    /html[1]/body[1]/div[1]/div[1]/ol[1]/li[1]/div[2]/h1[1]/a[1]
    /body[1]/div[1]/div[2]/ol[1]/li[1]/div[2]/h2[1]/a[1]
    /html[1]/body[1]/div[1]/div[3]/ol[1]/li[1]/div[2]/h3[1]

Aligned:
    /html[1]/body[1]/div[1]/div[1]/ol[1]/li[1]/div[2]/h1[1]/a[1]
            /body[1]/div[1]/div[2]/ol[1]/li[1]/div[2]/h2[1]/a[1]
    /html[1]/body[1]/div[1]/div[3]/ol[1]/li[1]/div[2]/h3[1]

Merged:
    */body[1]/div[1]/div/ol[1]/li[1]/div[2]/*[1]/*
```

Fig. 2. Example of aligning and merging several XPaths

The example in Figure 2 shows how three similar XPaths are aligned and then merged. Through the alignment, the amount of wildcards to be inserted during merging is minimal. The merging step illustrates a wildcard insertion for index, tagname and both. An index wildcard is inserted by omitting the index specifier. A tagname wildcard is inserted by replacing the tag by an asterisk.

7.2 Clustering

The purpose of clustering is to identify groups of similar XPaths. Therefore we iteratively assign each XPath to a cluster based on an adaptation to the Levenshtein distance measure for lists [19]. XPaths are essentially ordered lists of items, identified by a tagname and index. Two items are considered equal if they have both the same tagname and index. The distance from an XPath to a cluster is calculated using the centroid of the cluster. The centroid is calculated by aligning and merging all XPaths in the cluster.

During the clustering, we assign each XPath to its closest cluster, where closest cluster is defined as the cluster with centroid with the lowest Levenshtein-list distance to the XPath to assign. If the closest cluster has a distance higher than a certain threshold or no such cluster is found, the XPath is placed into a new cluster.

At the end of the clustering, no XPaths in one cluster should have a distance to each other greater than the threshold. This means a cluster contains a set of nodes that are similar. The following example shows how an XPath is assigned to a cluster.

```
XPath to assign:
    /body[1]/ul[2]/li[5]/div[1]/span[1]/span[3]/a[1]
  cluster 1: (distance = 1)
    /body[1]/ul[2]/li[1]/div[1]/span[1]/span[3]/a[1]
    /body[1]/ul[2]/li[3]/div[1]/span[1]/span[3]/a[1]
  cluster 2: (distance = 5)
    /body[1]/div[1]/div[2]/div][4]/div[2]/a[1]/span[1]

After assignment:
  cluster 1:
    /body[1]/ul[2]/li[1]/div[1]/span[1]/span[3]/a[1]
    /body[1]/ul[2]/li[3]/div[1]/span[1]/span[3]/a[1]
    /body[1]/ul[2]/li[5]/div[1]/span[1]/span[3]/a[1]
  cluster 2:
    /body[1]/div[1]/div[2]/div][4]/div[2]/a[1]/span[1]
```

Fig. 3. An example showing how an XPath is assigned to a cluster

Before we analyse relative data, we choose one cluster to continue with. Generally, the largest cluster is representative for the data we are searching. In case there are multiple clusters with the same size, a tie-breaker is needed. As a tie-breaker, we evaluate the centroids of all equally large clusters and count how many elements are matched. We continue by selecting the cluster that matches the most elements, assuming the bulk of the page contains data and the other clusters, matching less elements, contain outliers, which can often be found in a side menu. In case the amount of matched elements are equal again, a random cluster is selected. The selected cluster is called the dominant cluster.

8 Relative Data

For each sample record for which we have an XPath in the dominant cluster, we verify if there are relative attributes. For each relative attribute, we use the DOM tree inspection algorithm to find the XPaths to where the data can be found on the document. We select the shortest relative XPath, which means the element in closest proximity of the key data, as each item in a relative path means navigating one hop in the DOM tree.

Ideally, and under the assumption of template based website generation, all relative paths for a relative attribute will be the same. However, we foresee some records may be divergent and choose for a majority voting system for each relative attribute to select the most represented relative path. Consider the following example where two books have the same relative path to their author and there is one outlier. The majority voting system will select the most frequent relative path as the XPath for the relative attribute author:

9 Experiments

To test our approach, we have set up two experiments. For each of them, we used the same data set which we retrieved from the Eci book store website[2]. From there, we have extracted over 1000 books using a data extraction tool based on manually configured XPaths. Each page in the domain contains information on 12 books.

First, we verify whether or not we can reconstruct the initially configured XPaths by testing our method on the same website. We discern two scenarios: one where we provide well-chosen sample books of which we know they can be found on the page, and one where we use the entire data set.

Second, we apply our approach on a different website[3], from De Standaard Boekhandel. We have manually verified this page contains books from the dataset, so we know this website can serve as an example of a changed HTML document, where the structure is altered but the data is the same. Each page in this domain contains information on 16 books.

In both experiments we measure how many books are automatically discovered using the reconstructed XPaths.

9.1 Unchanged HTML

Scenario 1: In the first scenario, we select well-chosen sample books to verify if our DOM tree inspection and clustering, merging and alignment techniques work properly. A well-chosen sample is a book of which we know in advance it can be found on the page. Because we test our approach on the same website we used to extract the data set from, we expect the reconstructed XPaths to be identical to the originals. We count the amount of books identified by the reconstructed XPaths and list the percentage of the expected books that are found.

Table 1. The amount of books found in the original document based on the amount of well-chosen samples used

Amount of samples	Books found	% found
1	1	8
2	12	100
4	12	100
8	12	100
12	12	100

We identify that we need at least two samples to find all books. This is because we need more than one sample to extrapolate. Furthermore, these results

[2] `http://www.eci.be/boeken/roman-literatuur/literatuur`

[3] `http://www.standaardboekhandel.be/`

show that our approach produces the correct results. Moreover, no more than 2 well-chosen samples are needed to reconstruct the original XPaths with perfect accuracy.

Scenario 2: To simulate a more realistic scenario where it is not possible to identify well-chosen samples, we run the same experiment using the entire data set. We find the results are the same: all 12 books on the page are found and the original XPaths are reconstructed.

9.2 Changed HTML

We verify if our approach is able to construct XPaths to known data in an altered HTML document. We test our method using the entire data set on an alternate website, after manually verifying there are at least two books on the page that are also in our data set. We also evaluate the computed XPaths to verify all books are returned. In the results below, we have shortened some XPaths by omitting the part that is equal among all of them to better highlight their discrepancies.

```
Cluster 1 (dominant):
    /body[1]/ ... /ul[2]/li[1]/div[1]/span[1]/span[3]/a[1]
    /body[1]/ ... /ul[2]/li[3]/div[1]/span[1]/span[3]/a[1]
    /body[1]/ ... /ul[2]/li[5]/div[1]/span[1]/span[3]/a[1]
    /body[1]/ ... /ul[2]/li[11]/div[1]/span[1]/span[3]/a[1]
centroid: //body[1]/ ... /ul[2]/li/div[1]/span[1]/span[3]/a[1]

Cluster 2:
    /body[1]/div[1]/ ... /div[2]/div][4]/div[2]/a[1]/span[1]
    /body[1]/div[1]/ ... /div[2]/div][2]/div[2]/a[1]/span[1]
centroid: //body[1]/div[1]/ ... /div[2]/div/div[2]/a[1]/span[1]
```

Fig. 4. The results of the clustering, aligning and merging of the XPaths of matched book titles on an alternate HTML document

Fig. 5. Part of the output of the automatically reconfigured data extraction tool, which now highlights the titles (green) and authors (yellow) of all books based on the reconstructed XPaths, without manual reconfiguration, for the website of De Standaard Boekhandel

We notice there are 16 books on the website. Of all samples, only 4 books were matched. We manually verify these are the only 4 books that are also present in the data set. Of those 4, 2 were found multiple times on the page. Indeed, there is a sidebox where bestsellers are listed and those 2 books are among them. However, the dominant cluster leads to the correct list of data on the website. Its centroid serves an the absolute XPath for all book titles. Furthermore, all matched books return the same calculated relative path for the author. Reconfiguring the data extraction tool with the newly constructed absolute XPath for the titles and relative XPath for the authors, we see all books are returned, as shown in Figure 5.

10 Future Work

An important part of the DOM tree inspection is the function that does text-based entity comparison. The difficulty of the decision whether or not a text field contains information on an entity we're interested in should reflect the complexity of the human decision making process. Concluding if two strings of text represent the same data often results in a certain degree of hesitation. This is usually expressed in linguistic terms such as "probably" and "likely". To translate this into numbers we turn to soft computing, which allows us to model such terms using special functions. We can then add an uncertainty model to the results.

11 Conclusion

In this work, we have provided a way to extend XPath-based data extraction tools with a method to automatically (re)construct XPaths based on data. This creates the possibility to create smarter data extraction tools that use both structure and data to extract information. First, data is gathered from HTML document A and stored as a set of records, each consisting of a set of attributes, of which one is a mandatory key attribute. Second, this data is used to search HTML document B for nodes containing information, which are represented by their XPaths. These XPaths are then clustered, aligned and merged. The new XPaths then tell us where the information on records can be found in document B. Our approach succeeds in identifying the records by locating the key data and generates XPaths for all attributes with high accuracy.

References

1. Laender, A.H., Ribeiro-Neto, B.A., da Silva, A.S., Teixeira, J.S.: A brief survey of web data extraction tools. ACM Sigmod Record 31(2), 84–93 (2002)
2. Chang, C.H., Kayed, M., Girgis, M.R., Shaalan, K.F.: A survey of web information extraction systems. IEEE Transactions on Knowledge and Data Engineering 18(10), 1411–1428 (2006)

3. Liu, W., Meng, X., Meng, W.: Vide: A vision-based approach for deep web data extraction. IEEE Transactions on Knowledge and Data Engineering 22(3), 447–460 (2010)
4. Sugibuchi, T., Tanaka, Y.: Interactive web-wrapper construction for extracting relational information from web documents. In: Special Interest Tracks and Posters of the 14th International Conference on World Wide Web, WWW 2005, pp. 968–969. ACM, New York (2005)
5. Myllymaki, J.: Effective web data extraction with standard xml technologies. Computer Networks 39(5), 635–644 (2002)
6. Pan, A., Raposo, J., Álvarez, M., Hidalgo, J., Viña, Á.: Semi-automatic wrapper generation for commercial web sources. In: Engineering Information Systems in the Internet Context, pp. 265–283. Springer (2002)
7. Buttler, D., Liu, L., Pu, C.: A fully automated object extraction system for the world wide web. In: 21st International Conference on Distributed Computing Systems, pp. 361–370. IEEE (2001)
8. Crescenzi, V., Mecca, G., Merialdo, P., et al.: Roadrunner: Towards automatic data extraction from large web sites. VLDB 1, 109–118 (2001)
9. Reis, D.: d.C., Golgher, P.B., Silva, A., Laender, A.: Automatic web news extraction using tree edit distance. In: Proceedings of the 13th International Conference on World Wide Web, pp. 502–511. ACM (2004)
10. Zhai, Y., Liu, B.: Web data extraction based on partial tree alignment. In: Proceedings of the 14th International Conference on World Wide Web, pp. 76–85. ACM (2005)
11. Liu, L., Pu, C., Han, W.: Xwrap: An xml-enabled wrapper construction system for web information sources. In: Proceedings of the16th International Conference on Data Engineering, pp. 611–621. IEEE (2000)
12. Zhu, J., Nie, Z., Wen, J.R., Zhang, B., Ma, W.Y.: Simultaneous record detection and attribute labeling in web data extraction. In: Proceedings of the 12th ACM SIGKDD International Conference on Knowledge Discovery and Data Mining, pp. 494–503. ACM (2006)
13. Carlson, A., Schafer, C.: Bootstrapping information extraction from semi-structured web pages. In: Daelemans, W., Goethals, B., Morik, K. (eds.) ECML PKDD 2008, Part I. LNCS (LNAI), vol. 5211, pp. 195–210. Springer, Heidelberg (2008)
14. Dalvi, N., Bohannon, P., Sha, F.: Robust web extraction: An approach based on a probabilistic tree-edit model. In: Proceedings of the 2009 ACM SIGMOD International Conference on Management of Data, SIGMOD 2009, pp. 335–348. ACM, New York (2009)
15. Dalvi, N., Kumar, R., Soliman, M.: Automatic wrappers for large scale web extraction. Proc. VLDB Endow. 4(4), 219–230 (2011)
16. Hao, Q., Cai, R., Pang, Y., Zhang, L.: From one tree to a forest: A unified solution for structured web data extraction. In: Proceedings of the 34th International ACM SIGIR Conference on Research and Development in Information Retrieval, SIGIR 2011, pp. 775–784. ACM, New York (2011)
17. Lerman, K., Minton, S., Knoblock, C.: Wrapper maintenance: A machine learning approach. J. Artif. Intell. Res (JAIR) 18, 149–181 (2003)
18. Gusfield, D.: Algorithms on Strings, Trees and Sequences. Cambridge University Press (1997)
19. Levenshtein, V.: Binary codes capable of correcting deletions, insertions and reversals. Physics Doklady 10(8), 707–710 (1966)

Selection of Semantical Mapping of Attribute Values for Data Integration

Marcin Szymczak[1,2], Antoon Bronselaer[2], Sławomir Zadrożny[1], and Guy De Tré[2]

[1] Systems Research Institute, Polish Academy of Sciences, Newelska 6, 01-447 Warsaw, Poland
{szymczak,zadrozny}@ibspan.waw.pl

[2] Department of Telecommunications and Information Processing, University Ghent, St-Pietersnieuwstraat 41, 9000 Ghent, Belgium
{antoon.bronselaer,guy.detre}@ugent.be

Abstract. Useful information is often scattered over multiple sources. Therefore, automatic data integration that guarantees high data quality is extremely important. One of the crucial operations in data integration from different sources is the detection of different representations of the same piece of information (called *coreferent* data) and translation to a common, unified representation. That translation is also known as *value mapping*. However, values mappings are often not explicit i.e. the specific value may be mapped to more than one value. In this paper, we investigate automatic selection method which reduces the set of one-to-many mappings to the set of one-to-one mappings for attributes whose domains are partially ordered and where the given order relation reflects a notion of generality.

Keywords: selection of mappings, objects mapping, integration.

1 Introduction

The proper integration of objects from different data collections is extremely important to preserve high data quality in a database. The data integration process consists of two steps. The first step, reconciling structural heterogeneity of data by mapping schema elements across the data sources is known as the *schema matching* problem [11,15,8,4,14]. The second step resolves semantic heterogeneity of data by mapping data instances across the datasets and is known as the *object mapping* problem [1,5,9,7,6,3,16]. Moreover, mappings of objects are often not explicit, i.e., the specific object may be mapped to more than one object what yields one-to-many mappings.

In this paper, we present a novel approach for a specific part of the object mapping problem, namely we study automatic selection method which reduces the set of one-to-many mappings to the set of one-to-one mappings for attributes whose domains are partially ordered and where the given partial order relation reflects a notion of generality. We make the following assumptions. First of all,

P. Angelov et al. (eds.), *Intelligent Systems'2014*,
Advances in Intelligent Systems and Computing 322, DOI: 10.1007/978-3-319-11313-5_51

the schema matching between input datasets is established. Next, the schema of each considered dataset contains an attribute which satisfies the above condition, i.e. there exist a partial order relation defined on its domain and such an order relation is known. Finally, one-to-many mappings for values of considered attributes are generated in advance by our approach which is presented in [13]. The approach is based on textual descriptions of the considered values and information retrieval techniques. These descriptions are constructed automatically from the values of other attributes (such as `name`) or from an external source (e.g. World Wide Web) and compared by a method which estimates the possibility that two values are coreferent.

Table 1. Example of objects extracted from dataset S

Id	Name	Lon.	Lat.	Category
1	Kinepolis Gent	3.730169	51.040456	movie theater
2	Borluut Bed Breakfast	3.657992	51.018882	lodging
3	Carlton Hotel	3.713951	51.036280	lodging

As a running example in this contribution, we consider two datasets that both contain representations of some points of interest (POIs). POIs are geographical annotations of a map that pinpoint locations of specific interest. Each object is characterized by at least four attributes: `name`, `longitude`, `latitude` and `category`. The attribute `name` represents the name of the specific POI, the `longitude` and `latitude` give the geographic coordinates of the place, and the `category` specifies the type or function of the location. The content of the first dataset is illustrated by Table 1 which contains objects extracted from a Google database[1], called the source S, for short, with a known partial order relation on the domain of the `category` attribute. Figure 1 presents a part of this relation. The second dataset contains objects extracted from the RouteYou dataset[2], called the target T, also with a known partial order relation on the domain of the `category` attribute (Figure 2).

Figure 3 presents some of the one-to-many mappings between categories from the source and the target which are presented in Figure 1 and Figure 2 respectively [13]. It means that each category from the source is mapped to at least one category from the target and it is indicated by the dotted arrows, e.g., the mapping of *lodging* and *Hotel*, *lodging* and *Guest Room*, *lodging* and *Accommodation*, etc.

Let us consider a data integration scenario in which objects from the dataset S have to be merged with objects from the dataset T. Values of attributes as `name`, `longitude` and `latitude` might be stored in the target dataset T without any additional processing assuming that there is no coreference between imported

[1] Google, `http://maps.google.com`

[2] RouteYou, `http://www.routeyou.com/`

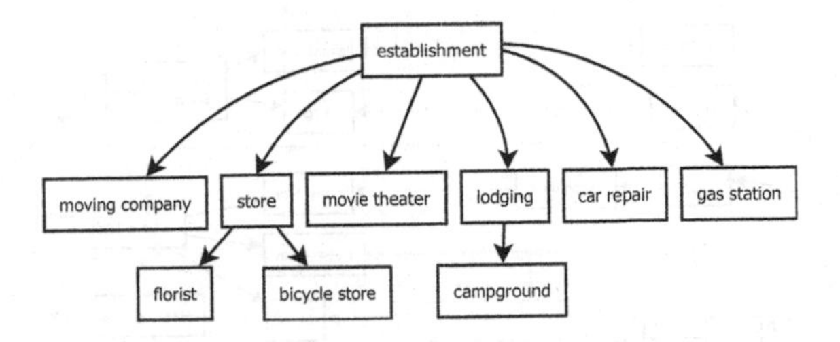

Fig. 1. A part of the hierarchy of the values of the category attribute from S

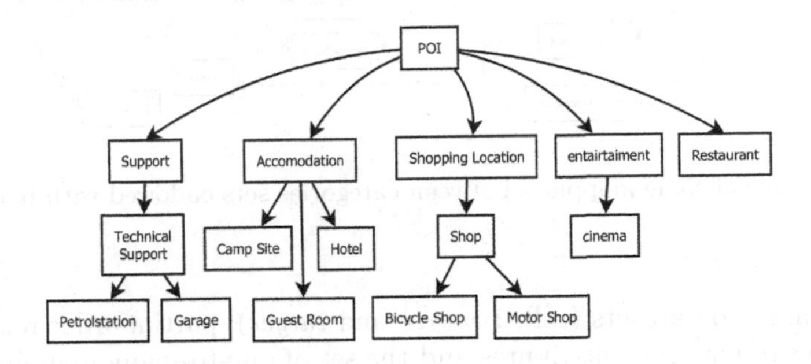

Fig. 2. A part of the hierarchy of the values of the category attribute from T

values and values of objects in the dataset T for the corresponding attributes. However, importing of values of attributes such as `category`, representing information on the type of a point of interest, is less trivial as they may refer to the same concepts presented in different ways in both datasets. For instance, the concept *accommodation* is represented by the category *lodging* in the dataset S and by the categories *Accommodation*, *Hotel*, *Guest Room*, etc. in the dataset T. In this case, following intuition, there can be established a mapping between concepts on the same level of abstraction: *lodging* and *Accommodation*. However, for specific objects in the source dataset, e.g. *Carlton Hotel* with the category *lodging* in Table 1, there may exist a mapping which better expresses the category of this specific object, e.g. *Hotel*. Therefore, for successful data integration, it may be crucial to select proper mappings of values of the category attribute in the datasets S and T. Thus, such mappings help to maintain consistency and decrease the number of duplicates in the integrated dataset which has extreme influence on data quality. This in turn decreases the cost of a database maintenance. To sum up, a number of problems occur that deserve attention:

- How to select the proper mapping for each object?
- How to select the proper mapping for each category?

With respect to the problems given above, the following important contributions are made by this paper. The automatic mapping selection algorithm is proposed for attributes domains of which are endowed with partial order relations that reflect a notion of generality (called *category* attributes). The input for this

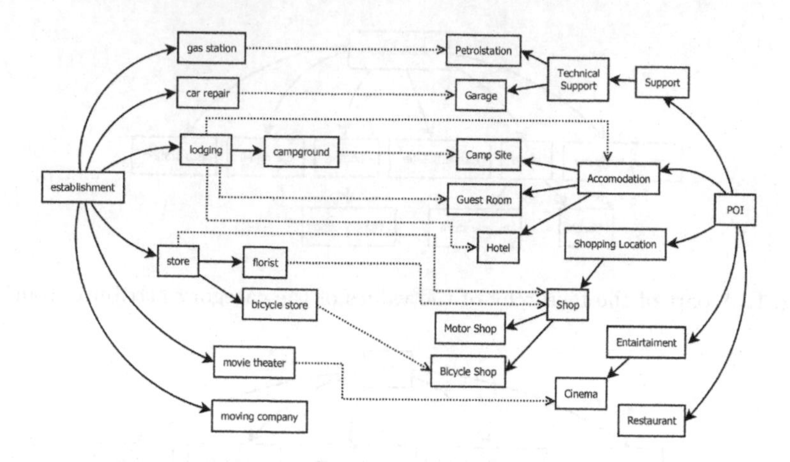

Fig. 3. Example of some mappings between categories sets endowed with partial order relations

algorithm are two datasets (called *source* and *target*), partial order relations on the domains of matching attributes and the set of one-to-many mappings which is produced by our method [13]. The algorithm returns a set of one-to-one mappings between values of the category attribute from the source and the target datasets. In other words, the category of each object from the source is mapped to exactly one category from the target.

The remainder of this paper is structured as follows. In Section 2, an overview of work related to the topic of this paper is provided. Next, in Section 3 some preliminary concepts are introduced that serve as a theoretical foundation of this paper. In Section 4, the proposed algorithm is described. The detailed description of applied mapping selection heuristics is contained in Section 5. Next, Section 6 presents our experimental results. Finally, we conclude and point our future work in Section 7.

2 Related Work

The problem of objects mappings has been studied in different contexts by many authors as, e.g. record linkage [6], duplicate detection [1] or data integration [9]. Most of the previous works assume that data values of each pair of corresponding attributes are drawn from the same domain or at least they bear some similarity that can be measured using a kind of the distance (e.g. edit distance, Jaccard). Some approaches are based on statistical information processing [7,3]. For instance, Kang et al. in [7] use a statistical model which captures the co-occurrence of values of all attributes characterizing datasets. Next, constructed models are aligned assuming various matchings between the values of a given attribute in both datasets. The alignment with the minimum distance between thus aligned models is returned as the mapping. Moreover, in [16] domain independent string transformations are proposed to compare syntactically object's shared attributes. The established mappings depend on the mapping rules which are determined by a mapping learner and supervised by the user. In contrast, in [5], selection of mappings based on non-overlapping correlated attributes using a combination of profilers which contain the specific knowledge about what constitutes a typical concept. To the best of our knowledge, none of the mentioned approaches rely on an order relation to select automatically the proper mapping for attributes values with different domains. In this paper, a partial order relation is used for the first time in the context of object mapping.

3 Preliminaries

Within the scope of this paper it is assumed that entities from the real world are described as *objects* which are characterized by a number of *attributes* $a \in A$.

A *schema* of a given dataset is identified with the set of attributes A. For each attribute $a \in A$, let $dom(a)$ denote the domain of a (the set of possible values for attribute a) and $dom'(a)$ denote the domain of a comprising values of a actually present in the dataset.

Problem Definition

Two datasets are considered. The source dataset over the schema $A_S = \{a_1^S, \ldots, a_n^S\}$ is denoted as S, while the target dataset over the schema $A_T = \{a_1^T, \ldots, a_m^T\}$ is denoted as T. We assume that the schema matching is known:

$$f : A_S \rightarrow A_T \tag{1}$$

Moreover, at least one *category* attribute $a_C^S \in A_S$ is distinguished with values $c^S \in dom(a_C^S)$ (called *categories* for short), and similarly $a_C^T \in A_T$ with values $c^T \in dom(a_C^T)$. The category attribute of an object is an ordinal attribute [12] of which the values (categories) are partially ordered by means of a generalization/specialization relation. These values are a upper semilattice, because it is a partially ordered set that has a least upper bound (supremum) for any nonempty finite subset, i.e., for any values (categories) of the category attribute exist supremum. Besides that, the set of one-to-many categories mappings is defined as follows.

Definition 1 (One-To-Many Categories Mapping). *A relation $R_{M1:m} \subseteq dom'(a_C^S) \times dom(a_C^T)$ is a one-to-many categories mapping if it maps one category $c^S \in dom'(a_C^S)$ to a nonempty subset of categories $\{c_1^T, ..., c_i^T\} \subseteq dom(a_C^T)$, called the candidate categories set, representing the coreferent categories of a pair of corresponding attributes $a_C^S \in A_S$ and $a_C^T \in A_T$, where $f(a_C^S) = a_C^T$. This mapping is denoted as:*

$$R_{M1:m} : dom'(a_C^S) \longrightarrow_{1:m} co\_dom(a_C^T) \tag{2}$$

where $dom'(a_C^S)$ is a domain of the attribute a_C^S comprising values of a_C^S actually present in the dataset S and $co\_dom(a_C^T)$ is a set of the subsets of $dom(a_C^T)$

comprising values of a_C^T. *Thus, one-to-many categories mapping* $R_{M1:m}$ *defines for each element* $c^S \in dom'(a_C^S)$ *a set of pairs* $\{(c^S, c_1^T), ..., (c^S, c_i^T)\}$.

The desired result of our approach are one-to-one mappings which are defined below.

Definition 2 (One-To-One Categories Mapping). *An injective function:*

$$R_{M1:1} : dom'(a_C^S) \longrightarrow_{1:1} range(a_C^T) \tag{3}$$

is referred to as a one-to-one categories mapping. It maps one category $c^S \in dom'(a_C^S)$ *to exactly one category* $c^T \in dom(a_C^T)$ *representing its coreferent category.*

In our approach we use PTVs to express the confidence (certainty) in the validity of the produced mappings. Hereby, a PTV is a normalized possibility distribution [17] defined over the set of Boolean values $\mathbb{B}$ [10]. A PTV expresses the uncertainty about the Boolean value of a *proposition* p. In the context considered here, the propositions p of interest are of the form:

$$p = c^S \text{ and } c^T \text{ are coreferent}$$

where c^S and c^T are two given categories, i.e., values of the category attributes from two datasets under consideration.

Let P denote a set of all propositions under consideration. Then each $p \in P$ can be associated with a PTV denoted $\tilde{p} = \{(T, \mu_{\tilde{p}}(T)), (F, \mu_{\tilde{p}}(F))\}$, where $\mu_{\tilde{p}}(T)$ represents the possibility that p is true and $\mu_{\tilde{p}}(F)$ denotes the possibility that p is false. The domain of all possibilistic truth values is denoted $\mathcal{F}(\mathbb{B})$, i.e., is the fuzzy power set of (normalised) fuzzy sets over $\mathbb{B}$.

Let us define the order relation $\geq$ on the set $\mathcal{F}(\mathbb{B})$ by:

$$\tilde{p} \geq \tilde{q} \Longleftrightarrow \begin{cases} \mu_{\tilde{p}}(F) \leq \mu_{\tilde{q}}(F), \mu_{\tilde{p}}(T) = \mu_{\tilde{q}}(T) = 1 \\ \mu_{\tilde{q}}(T) \leq \mu_{\tilde{p}}(T), \quad \text{else.} \end{cases} \tag{4}$$

4 Mappings Selection

Equivalent and Non-equivalent Mappings

Figure 3 presents mappings from the input mappings set which is obtained by our method [13]. For instance, *Movie theater* is mapped to *Cinema* or *car repair* to *Garage*. These mappings are valid in both directions, i.e., from the dataset S to the dataset T and inversely, because these categories represent coreferent concepts on the same level of abstraction. Thus, these mappings are called *equivalent* mappings.

In contrast, mappings such as the one between concepts *lodging* and *Hotel* are asymmetric, in a sense. On the one hand, not each *lodging* is a *Hotel*. Therefore *lodging* should be mapped to a more general concept than *Hotel*. On the other hand, each *Hotel* is a *lodging*. These categories describe different levels of abstraction, they are not equivalent, i.e. *lodging* is more general concept than *Hotel* and *Hotel* is a specialization of *lodging*. Therefore, these mappings are called *non-equivalent* mappings which are further divided into two subclasses. The first one, called *generalized* mappings, contains mappings in which the target category is a generalization of the source category and it is a valid mapping but on a different level of abstraction. In contrast to that, a mapping of which the target category is a specialization of the source category is called *specialized* mapping and it may be an invalid mapping; however, between those categories there still exists a strong semantical relation.

Due to the above described conditions, the direction of mapping has to be considered during the data integration. In this paper we consider directional mappings from the source dataset S to the target dataset T.

Algorithm

The Category Selection Algorithm (CSA, Algorithm 1) selects for each object's $o \in S$ category $c^S \in dom'(a_C^S)$ exactly one category $c^T \in dom(a_C^T)$ using the order relation R_T defined on the domain of the category attribute a_C^T from the target dataset T. Therefore, the inputs for the algorithm are the source and target datasets (S and T respectively), the order relation R_T and the set of one-to-many mappings $R_{M1:m}$. The objective of our algorithm is to select as many as possible true and, if possible, equivalent one-to-one mappings $R_{M1:1}$ for each object from the source S.

In the first step (lines 1-3 in Algorithm 1), for each category c_S from the source the method selectDefaultMapping selects *the best possible* one-to-one mappings (called *default* mappings) from the set of one-to-many mappings $R_{M1:m}$ using the heuristics which are described in Section 5. *The best possible* mapping means here that the desired mapping is an equivalent mapping if there exists one, or non-equivalent but the most specific (supremum) otherwise. The detection if a mapping is equivalent is based on the proposed heuristics. One of them is based on the intuition that the most certain mapping may be equivalent. The others are based on the following idea. For instance, suppose that the category *lodging* is mapped to *Hotel*, *Camp Site* and *Guest Room* in Figure 3. Following intuition we know that the considered concept cannot be at the same time *Hotel*, *Camp Site* and *Guest Room*. Therefore, our heuristics uses a partial order relation to select a concept which generalizes all these categories, in this case it is *Accommodation*.

In the second step (lines 4-10 in Algorithm 1), for each object's category from the source the *possible most specific* category is selected, in the sense that the algorithm tries to find first of all a mapping specific for the considered object (lines 5-6 in Algorithm 1), called a *specific* mapping. For instance, let us consider object *Carlton Hotel* with the category *lodging* in Table 1. In the first step the selectDefaultMapping method returns equivalent default mapping (*lodging*, *Accommodation*). However, in the partial order relation of the target categories (R_T), there exists a specific category (*Hotel*) for this object and in fact this category is contained in the name of considered object. This allows to create the mapping (*lodging,Hotel*) for the object. The high precision of this mapping

Algorithm 1. CategorySelectionAlgorithm

Require: Dataset S, Dataset T, Order Relation R_T, Mappings $R_{M1:m}$
Ensure: Mappings $R_{M1:1}$, Dataset S^*
1: **for all** $c^S \in dom'(a_C^S)$ **do**
2: $R_{M1:1} \leftarrow$ selectDefaultMapping(c^S, $R_{M1:m}$, R_T)
3: **end for**
4: **for all** $o \in S$ **do**
5: **if** existObjectSpecificCategoryMapping($o.name$, $dom(a_C^T)$) **then**
6: $o.c^S \leftarrow$ getObjectSpecificCategory(o, $dom(a_C^T)$)
7: **else if** existDefaultCategoryMapping($o.c^S$, $R_{M1:1}$) **then**
8: $o.c^S \leftarrow$ getDefaultCategory($o.c^S$, $R_{M1:1}$)
9: **end if**
10: **end for**

is ensured by selecting only that category which is in the relation with the target category of the default mapping for the considered object's category. For example, for the considered object's category *lodging* the target category of the default mapping is *Accommodation* which is in the relation R_T with the detected category *Hotel*. Moreover, if there exist more than one category mapping for the particular object then the selection mappings methods from the first step are applied.

If there does not exist a specific mapping for the category of the particular object from the source, the algorithm checks if there exists a default mapping for that category (line 7 in Algorithm 1). If so, then the considered category is mapped according to the established default mapping in the first step (line 8 in Algorithm 1). Finally, the considered object from the source is mapped to exactly one category from the target and can be later on imported to the target dataset.

It should be clear that the proper mapping for any object is not always an equivalent mapping of objects categories (e.g., the default mapping between *lodging* and *Accommodation*). In some cases it is possible to establish a mapping which is more appropriate to the particular object, e.g., the specific mapping between *lodging* and *Hotel*.

5 Selection Heuristics

The set of different heuristics are considered to establish *the best possible* one-to-one category mappings $R_{M1:1}$ from the set of one-to-many categories mappings $R_{M1:m}$ using the partial order relation R_T.

5.1 Uncertainty Based Mapping Selection

This mapping selection method chooses the mapping for the source category based only on the uncertainty about the mapping which is expressed by a PTV.

It applies the order relation of PTVs (Equation 4) which helps to select the mapping with the lowest uncertainty. In case of indistinguishable mappings, i.e. with equal PTVs, a mapping is selected randomly.

5.2 The Simple Balanced Mapping Selection

That method is based on the balanced selection originaly applied for data fusion [2] which uses a partial order relation, in our case R_T. The balanced selection is the most specific category (supremum) from the candidate categories provided that the selected category is comparable to all other candidate categories (that category is called *total*), otherwise the balanced selection is unspecified. As a consequence, the heuristics returns a mapping with supremum: the most specific *total* category if it exists or with a common more general category of candidate categories from R_T otherwise. For instance, consider again the category *lodging* from the source and assume that it is mapped only to the following candidate categories from the target in Figure 3:

- *Accommodation* and *POI.* They are total so the balanced selection is the most specific category, *Accommodation*
- *Hotel, Guest Room, Accommodation* and *POI.* Then the balanced selection is the category *Accommodation,* because it is the most specific total category
- *Hotel* and *Guest Room.* Then the balanced selection is unspecified because these categories are not total. Thus, the *lodging* will be mapped to *Accommodation* because it is the common more general category of candidate categories in the order relation R_T

5.3 The Subset Balanced Mappings Selection

This method works like the heuristics in Subsection 5.2, but with one difference which may help to avoid the situation where the answer of the balanced selection is unspecified, i.e., if no category in the candidate set is total. Namely, the candidate category set is splitted into subsets and a subset with the largest cardinality of which categories may be fused using the balanced selection [2] is used to select the proper mapping by the Simple Balanced Mappings Selection method in Subsection 5.2. However, if a subset with the largest cardinality is not unique then a random subset is chosen. The criterion to split the candidate set into subsets is that a subset have to contain at least one category which is total and the remaining categories in a subset have to be related to the total category(ies) or the heuristics returns a mapping with a common more general category of candidate categories from R_T otherwise. This method reduces the set of candidate categories which helps to select a mapping which points to the category different than the most general category in the partial order relation.

For instance, consider again the category *lodging* from the source and assume that it is mapped only to the following candidate categories from the target: *Restaurant, Accommodation, Guest Room* and *Hotel.* For those candidate categories the balanced selection is unspecified because there does not exist any total category based on the order relation in Figure 3. However, the candidate set can be splitted into two subsets which contain total categories. The first contains only *Restaurant*, while the second consists of the other categories. Following in the rules of this heuristics the second subset is selected. Finally, the balanced selection selects the most specific total category: *Accommodation.*

6 Evaluation

The evaluation of our algorithm is conducted based on two real-world datasets. The source dataset contains 15886 objects and 98 categories for which a partial order is available. This dataset is extracted from the Google Maps database by the Google Places API[6]. The object's categories from the source dataset are mapped to the categories taxonomy from the target dataset which is shared by RouteYou[7]. The taxonomy consists of 502 categories. The input set of 215 mappings is known and generated by mapping methods which are presented in [13].

The effectiveness of our method is tested using three measures: recall, precision and FMeasure. Precision is a fraction of detected true mappings among all detected mappings, recall is a fraction of detected true mappings among all true mappings, and FMeasure is a harmonic mean of this two measures, calculated as follows:

$$\text{FMeasure} = 2 * \frac{precision * recall}{precision + recall} \quad (5)$$

The set of true coreferent equivalent and generalized non-equivalent one-to-one category mappings and object's specific mappings has been determined manually by the authors.

Table 2. Results of the evaluation: Precision, Recall and FMeasure. H1 is the simple balanced mapping selection, H2 is the subset balanced mappings selection, H3 is uncertainty based mappings selection heuristic. Column Not Root contains the percent of objects from the source which are not mapped to the most general category.

H	Equivalent			Equivalent & Generalized			
	Precision	Recall	FMeasure	Precision	Recall	FMeasure	Not Root
H1	0.68	0.51	0.58	0.82	0.33	0.47	0.09
H2	0.57	0.58	0.57	0.70	0.37	0.49	0.65
H3	0.61	0.62	0.62	0.72	0.38	0.50	0.65

[6] Google Places, `http://developers.google.com/places/`
[7] RouteYou, `http://routeyou.com/`

First of all the object's specific category mappings which are established by method in line 6 of Algorithm 1 are evaluated. This method mapped 196 objects with precision equal 0.84 which are only 1.2% of all objects from the source. However, each of these objects is mapped to the category which is specific for the considered object, i.e. a hotel with the category *lodging* is mapped to *Hotel* or a car repair company with the category *establishment* is mapped to *Garage*, etc. The rest of the input objects are mapped using the default category mapping selection heuristics from Section 4 (line 8 in Algorithm 1).

Each category from the source is mapped at least to the most general category from the target by Definition 1, in our case it is *POI*. However, the objective of our approach is to establish as many as possible equivalent or generalized mappings different from the mapping to *POI*. Thus, we do not consider mappings to *POI* in the following evaluation. Table 2 presents the precision, recall and FMeasure for the heuristics used in our experiments.

On the one hand, the simple balanced mapping selection method has higher precision for equivalent as well as for equivalent with generalized mappings than other heuristics. On the other hand, it has lower recall than the other methods. As consequence, only 9% objects from the source are not mapped to the most general category. In contrast, the two other heuristics established mappings which allow to map more than 65% objects from the source to the category different than *POI*. Because of that and the fact the desired mapping is not a mapping to the most general category the two others heuristics are definitely more useful in the data integration.

7 Conclusion

In the approach reported in this paper we employ the partial order relation that reflects a notion of generality in data integration to develop an effective algorithm for selecting the proper semantic mappings of attributes values. On the one hand, an input set of one-to-many mappings is reduced to a set of one-to-one mappings by proposed heuristics which select default mappings between concepts on the same level of abstraction if there exists one, or non-equivalent but the most specific (supremum) otherwise. On the other hand, a mapping which better fits to the particular object is suggested if there exists one. The experimental evaluation of our method shows the benefits from using the partial order relation for the proper mappings selection. In the future, we intend to consider an extended set of heuristics to select the proper mappings.

Acknowledgments. This contribution is supported by the Foundation for Polish Science under International PhD Projects in Intelligent Computing. Project financed from The European Union within the Innovative Economy Operational Programme 2007-2013 and European Regional Development Fund.

References

1. Ananthakrishna, R., Chaudhuri, S., Ganti, V.: Eliminating fuzzy duplicates in data warehouses. In: Proceedings of the 28th International Conference on Very Large Data Bases (2002)
2. Bronselaer, A., Szymczak, M., Zadrożny, S., De Tré, G.: Dynamical order construction in data fusion. Submitted for review in VLDB Journal (2014)
3. Cohen, W.W.: Integration of heterogeneous databases without common domains using queries based on textual similarity. In: Proceedings of the 1998 ACM SIGMOD International Conference on Management of Data, SIGMOD 1998, pp. 201–212. ACM, New York (1998)
4. Do, H., Rahm, E.: Coma - a system for flexible combination of schema matching approaches. In: Proceedings of the 28th International Conference on Very Large Data Bases, pp. 610–621 (2002)
5. Doan, A., Lu, Y., Lee, Y., Han, J.: Object matching for information integration: A profiler-based approach. In: Proceedings of the IJCAI 2003 Workshop on Information Integration on the Web, pp. 53–58 (2003)
6. Fellegi, I.P., Sunter, A.B.: A theory for record linkage. Journal of the American Statistical Association 64, 1183–1210 (1969)
7. Kang, J., Lee, D., Mitra, P.: Identifying value mappings for data integration: An unsupervised approach. In: Ngu, A.H.H., Kitsuregawa, M., Neuhold, E.J., Chung, J.-Y., Sheng, Q.Z. (eds.) WISE 2005. LNCS, vol. 3806, pp. 544–551. Springer, Heidelberg (2005)
8. Madhavan, J., Bernstein, P.A., Rahm, E.: Generic schema matching with Cupid. In: Proceedings of the 27th International Conference on Very Large Data Bases, VLDB 2001, pp. 49–58. Morgan Kaufmann Publishers Inc., San Francisco (2001)
9. Naumann, F., Bilke, A., Bleiholder, J., Weis, M.: Data fusion in three steps: Resolving inconsistencies at schema-, tuple-, and value-level. Bulletin of The Technical Committee on Data Engineering, 21–31 (2006)
10. Prade, H.: Possibility sets, fuzzy sets and their relation to Lukasiewicz logic. In: Proceedings 12th Int. Symp. on Multiple-Valued Logic, pp. 223–227 (1982)
11. Rahm, E., Bernstein, P.A.: A survey of approaches to automatic schema matching. The VLDB Journal 10(4), 334–350 (2001)
12. Stevens, S.S.: On the theory of scales of measurement. Science 103(2684), 677–680 (1946)
13. Szymczak, M., Bronselaer, A., Zadrożny, S., De Tré, G.: Semantical mappings of attribute values for data integration. In: Proceedings of the 2014 NAFIPS Annual Meeting, pp. 1–8. IEEE (2014)
14. Szymczak, M., Koepke, J.: Matching methods for semantic annotationbased xml document transformations. In: New Developments in Fuzzy Sets, Intuitionistic Fuzzy Sets, Generalized Nets and Related Topics. Applications, pp. 297–308. SRI PAS (2012)
15. Szymczak, M., Zadrożny, S., De Tré, G.: Coreference detection in XML metadata. In: Proceedings of the 2013 Joint IFSA World Congress NAFIPS Annual Meeting, pp. 1354–1359 (2013)
16. Tejada, S., Knoblock, C.A., Minton, S.: Learning domain-independent string transformation weights for high accuracy object identification. In: Proceedings of the Eighth ACM SIGKDD International Conference on Knowledge Discovery and Data Mining, KDD 2002, pp. 350–359. ACM, New York (2002)
17. Zadeh, L.A.: Fuzzy sets as a basis for a theory of possibility. Fuzzy Sets Syst. 100, 9–34 (1999)

A Knowledge-Driven Tool for Automatic Activity Dataset Annotation*

Gorka Azkune[1], Aitor Almeida[1], Diego López-de-Ipiña[1], and Liming Chen[2]

[1] Deusto Institute of Technology – DeustoTech, University of Deusto, Bilbao, Spain
{gorka.azkune, aitor.almeida,dipina}@deusto.es
[2] De Montfort University, Leicester, United Kingdom
liming.chen@dmu.ac.uk

Abstract. Human activity recognition has become a very important research topic, due to its multiple applications in areas such as pervasive computing, surveillance, context-aware computing, ambient assistive living or social robotics. For activity recognition approaches to be properly developed and tested, annotated datasets are a key resource. However, few research works deal with activity annotation methods. In this paper, we describe a knowledge-driven approach to annotate activity datasets automatically. Minimal activity models have to be provided to the tool, which uses a novel algorithm to annotate datasets. Minimal activity models specify action patterns. Those actions are directly linked to sensor activations, which can appear in the dataset in varied orders and with interleaved actions that are not in the pattern itself. The presented algorithm finds those patterns and annotates activities accordingly. Obtained results confirm the reliability and robustness of the approach in several experiments involving noisy and changing activity executions.

Keywords: Activity Recognition, Knowledge-Driven, Activity Annotation.

1 Introduction

Being able to recognize what human beings are doing in their daily life can open the door to a lot of possibilities in diverse areas. Technology is becoming more and more human-centred in order to provide personalized services. In other words, technological services have to adapt to human users and not the other way around.

Following that trend, human activity recognition becomes a natural enabler to such adaptive technologies. Let us consider the case where an intelligent assistive system has to promote a healthy lifestyle. For such a system it is mandatory to recognize the activities carried out by human users in order to analyse what kinds of recommendations would be better to improve the health of users. Many

* This work was supported by the Spanish Government under the TIN2013-47152-C3-3-R (FRASEWARE) project.

P. Angelov et al. (eds.), *Intelligent Systems'2014*,
Advances in Intelligent Systems and Computing 322, DOI: 10.1007/978-3-319-11313-5_52

other examples can be found in domains like pervasive and mobile computing, surveillance-based security, context-aware computing or social robotics.

The literature shows multiple ways of approaching activity recognition, based on the sensors to be used and techniques adopted. A good survey by Chen et al. can be found in [1]. There are two main currents inside the activity recognition scientific community: the data-driven current and the knowledge-driven current. It turns out that for both activity recognition currents annotated datasets are needed to ensure a successful development and/or evaluation.

However, the scientific community has already identified a lack of work in methods to get annotated datasets for activity recognition, as noted by Rashidi and Cook in [11]: *"An aspect of activity recognition that has been greatly underexplored is the method used to annotate sample data that the scientist can use to train the activity model"*.

This paper tackles the problem of annotating activity datasets, adopting a knowledge-driven approach. The basic idea is that domain experts have important knowledge about how activities are carried out by users. However, that knowledge is not complete in the sense that every user has different ways to perform activities. The software tool presented in this paper uses expert knowledge to build minimal activity models that contain for each activity, those concepts that always appear in it. For example, to make a coffee every user will use coffee and a liquid container to pour and drink that coffee. Additional actions may be executed by several users, such as adding milk and sugar, but no user will make a coffee without a container and coffee.

A novel algorithm that uses those minimal activity models to annotate activity datasets is presented in this paper, which is called SA^3 (*Semantic Activity Annotation Algorithm*). The input of SA^3 is a time-stamped sensor activation dataset. First, sensor activations are transformed to actions. For example, if a contact sensor installed in a coffee-cup is triggered, the algorithm transforms it to the action *hasContainer(cup)*. Minimal activity models are expressed as a sequence of actions to perform activities. In the second step, the algorithm iterates through the actions dataset and finds all the occurrences of minimal activity models. In this step the algorithm finds all action patterns regardless of the order of actions and the occurrence of any interleaved action. In the third step the set of non-overlapping activity patterns is found, which is used to annotate the original dataset.

The three-step annotation algorithm has been extensively tested. To make those tests more reliable, a synthetic dataset generator has been developed. The synthetic dataset generator allows simulating an arbitrary number of days for a user, specifying the activities performed, several ways to perform those activities and sensor noise. The tool generates a proper ground-truth to test SA^3 in many diverse ways and scenarios. The results obtained are very encouraging.

The paper is structured as follows: in Section 2, the literature about activity dataset annotation will be analysed. In Section 3, the details of the annotation algorithm will be described. Afterwards, Section 4 will explain briefly the usage of the synthetic dataset generator and present all the experiments performed and results obtained. Those results will be discussed in Section 5, to finish the paper with some conclusions and future work in Section 6.

2 Related Work

Dataset annotation for activity recognition is an under-explored area. Most of the researchers rely on manually annotated datasets. Many research works show experimental methodologies where participants have to manually annotate the activities they are performing. Good examples can be found in [12], [10] and [7]. Nevertheless, such system does not allow participants to behave naturally, interfering the experiment itself. In addition, correct annotation cannot be assumed when working with cognitively impaired people.

There are some other research works where the experimenters prepare a script with the activities that participants should perform, as can be seen in [2], [8], [3] and [4]. The problem of such an approach is again that participants cannot behave naturally. In the case of [2], participants cannot even decide how to perform the activities. Those situations are far from being real-world situations.

To finish up with manual annotation methods, Wren et al. show in [13] experiments where an expert had to go through raw sensor data to find activities and annotate them. In general, such an approach is very time consuming and results depend on the ability of the expert.

Although not many, there are some alternative methods to manual annotation. For instance, Kasteren and Noulas present in [6] a novel method that implies the use of a bluetooth headset equipped with speakers in order to capture the voice of the participant. While performing an activity, the participant has to name the activity itself. Speech recognition algorithms are used to automatically annotate the activity. This method is much more comfortable than manual annotation from the point of view of the participant. However, it is invasive and presents the same problems with cognitively impaired people.

A different approach is presented in [5] by Huynh et al. They provide three annotation methods. The first one is a mobile phone application that asks several questions while the participant is performing activities. Depending on the answers, the system decides the label that corresponds to the ongoing activity. The second method is a hand-written diary, i.e. manual annotation. And the third system uses the mobile phone again, but in a different way. Another application has been developed to take pictures regularly. Those pictures are used by experts to annotate the activities. Authors admit that after several experiments where participants could choose among the three options, the vast majority tend to use the manual annotation system.

Finally, Rashidi and Cook show in [11] an activity recognition approach that does not need any annotated dataset. They use an unsupervised approach that discovers frequent patterns in raw sensor data. Models are trained to be able to recognize the extracted patterns, so they do not require any annotated activity. However, the activity models they generate do not have any semantic meaning. Hence, a manual process should be run afterwards to understand activity models and assign a semantic label to them. In that sense, the work of Rashidi and

Cook cannot be considered a dataset annotator. What they present is an activity recognition system that overcomes the problem of not having an annotated activity dataset.

The work presented in this paper is novel respect to the state of the art, since an algorithm which uses minimal expert knowledge is developed to automatically annotate activity datasets. Notice that our work cannot be considered as an activity recognition approach, since it works offline, once a whole dataset of activities has been collected. In that sense, the requirements and constraints of activity annotation are softer than real-time activity recognition.

3 The Approach to Automatic Annotation

The annotation algorithm presented in this paper is inspired in knowledge-driven activity recognition methods, specially in [2]. Three important concepts from that work are extracted and used for the presented approach:

- **Domain knowledge allows recognizing activities reliably:** in contrast with data-driven techniques, which use intensively annotated datasets in order to learn activity models for recognition, knowledge-driven techniques use expert domain knowledge to model activities. Those models can be reliably used for activity recognition
- **Sensor activations are linked to user-object interactions:** if a contact sensor installed in a cup changes its state, the object is assumed to be used for an activity. Sensors' information is modelled to allow transformations from sensor activations to user-object interactions or **actions**. Following with the contact sensor of the cup, whenever an activation is received, it is interpreted as an action, which can be named as *hasContainer(cup)*
- **Activities can be modelled as sequences of actions:** even though richer information can be provided as location and time, to keep activity models as simple as possible, we only consider necessary actions. As such, the *MakeCoffee* activity can be modelled as the action sequence {*hasContainer(x)*, *hasCoffee(y)*}.

Real-time activity recognition needs complete activity models to have a reliable recognition performance. However, providing complete activity models is very complicated, since each user executes varied action sequences to perform the same activity. For example, to make a coffee, some users may use milk and sugar, while others may add only some cream. But making coffee will always imply using coffee and having a container, i.e. the action sequence {*hasContainer(x)*, *hasCoffee(y)*}. This prior knowledge is used in SA^3 for activity annotation.

Activity annotation and recognition is not the same thing. Activity annotation can make use of the whole dataset offline, with no time restrictions. On the other hand, activity recognition is required to work while activities are being performed, so only past sensor activations can be used. This key difference makes feasible using incomplete activity models to annotate activities, in contrast with the real-time recognition problem.

3.1 Basic Concepts

- **Sensor activations (SA):** it is assumed that environments are equipped with sensors that can collect information about user-object interactions. Whenever those sensors change their states, a sensor activation is collected. Sensor activations are then composed by a sensor identifier and a time-stamp
$$SA = \{\textit{time-stamp}, \textit{sensorID}\}$$
- **Sensor dataset:** a time-ordered sequence of sensor activations
- **Actions:** actions are the primitives of activities and are directly linked to sensor activations. The link between sensor activations and actions is manually established. Tables are used where several sensor activations are linked to an action. For example, *cupObjSensor* and *glassObjSensor* are linked to the action *hasContainer*. A sensor activation can only be linked to one single action. The transformation function is defined as:
$$Trans(SA) : SA \rightarrow Action$$
- **Action dataset:** a time-ordered sequence of actions
- **Minimal activity models:** activity models are sequences of actions defined by a domain expert. Minimal activity models refer to the minimal number of necessary actions to perform an activity. The objective of such models is to represent incomplete but generic activity models which provide enough information to detect activities. Minimal activity models also have an estimation of the maximum duration of the activity based on a heuristic
$$Activity_n = \{action_a, action_b, \ldots, max\_duration\}$$

3.2 SA^3: Semantic Activity Annotation Algorithm

Having as inputs a sensor dataset, sensor-action transformation functions and minimal activity models, the objective of SA^3 is to generate an annotated sensor dataset, where each sensor activation will have an activity label or the special label *None*, which represents lack of known activity. Single-user single-activity scenarios are considered, so no interleaved activities will be in the sensor dataset. Considering those constraints, a three-step algorithm has been designed:

1. **Sensor-action transformation step:** the first step takes as input the sensor dataset and transformation functions to generate an action dataset. For each sensor activation in the sensor dataset, the transformation function is applied to obtain the corresponding action. In consequence, a sensor sequence such as:
$$\{cupObjSensor, whiteSugarSensor, skimmedMilkSensor\}$$
will be transformed to:
$$\{hasContainer(cup), hasFlavour(\textit{white-sugar}), hasMilk(\textit{skimmed-milk})\}$$
Since only actions and not objects are relevant for activity annotation, *hasContainer(cup)* will be used as *hasContainer*

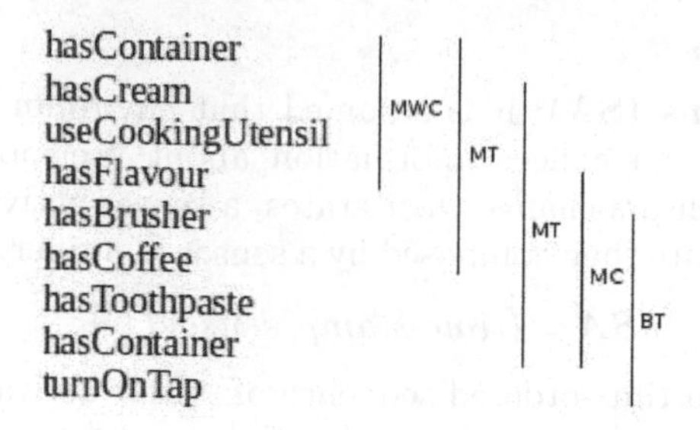

Fig. 1. Illustrative example of the output of the step 2 of SA^3. MWC refers to *MakeWhippedCream*, MT to *MakeTiramisu*, MC to *MakeCoffee* and BT to *BrushTeeth*

2. **Activity sequence finding step:** all the occurrences of minimal activity models are found iterating on the action dataset. For an action pertaining to a minimal activity model following actions are searched. Those actions are considered to describe the activity if two criteria are fulfilled:
   (a) *Completion criterion*: the action sequence has to contain all the actions of the corresponding minimal activity model
   (b) *Duration criterion*: the duration of the action sequence has to be smaller than the duration estimation of the corresponding minimal activity model

   Depending on the activity models, the actions dataset and noise levels, detected activities can overlap each other. Here is an illustrative example, where time-stamps are ignored since duration criterion is satisfied. Imagine we define minimal activity models for *MakeCoffee, MakeTiramisu, MakeWhippedCream* and *BrushTeeth,* where:

$$\begin{aligned} MakeCoffee &= \{hasCoffee, hasContainer, hasFlavour\} \\ MakeTiramisu &= \{hasCream, hasContainer, hasCoffee\} \\ MakeWhippedCream &= \{hasFlavour, hasContainer, hasCream\} \\ BrushTeeth &= \{hasBrusher, hasToothPaste, turnOnTap\} \end{aligned}$$

   Let us consider the following action sequence from the action dataset:

$$\{hasContainer, hasCream, useCookingUtensil, hasFlavour, hasBrusher, hasCoffee, hasToothPaste, hasContainer, turnOnTap\}$$

   Figure 1 shows all the activities found applying the completion and duration criterion. Activities overlap each other, because there are several actions that belong to several activities. Notice also that there are some actions that are not in any activity model, which is totally feasible for our approach.
3. **Correct activity sequence fitting step:** having the overlapping activities, the objective of this step is to find the maximum number of activities that do not overlap. This heuristic is derived from the fact that only none interleaved activities are considered. Applying it, the example of Figure 1 can be solved

appropriately. The solution found by the algorithm is that there are only two activities in that action sequence:

$$MakeWhippedCream = \{hasContainer, hasCream, useCookingUtensil, hasFlavour\}$$
$$BrushTeeth = \{hasBrusher, hasCoffee, hasToothPaste, hasContainer, turnOnTap\}$$

Hence, *hasCoffee* is probably due to a faulty sensor activation and *hasContainer* may refer to a glass used in the bathroom to rinse the mouth.

This three-step algorithm is depicted as pseudo-code in Algorithm 1. It has been designed to work with noisy sensor activations, varying order for activity executions and sensor activations that do not belong to any activity model. That flexibility allows using the tool in many different datasets. Section 4 contains many cases to show the performance of the algorithm in several demanding situations.

Algorithm 1. SA^3 algorithm for semantic activity annotation

```
Require: sensor_dataset, transformation_function, minimal_activity_models
Ensure: annotated_dataset
  action_dataset ← apply_transform_function(sensor_dataset, transformation_function)
  for all action ∈ action_dataset do
    if action ∈ minimal_activity_models then
      activities ← obtain_activities(action, minimal_activity_models)
    end if
    for all activity ∈ activities do
      // Use duration and completion criteria
      detected_activities ← find_proper_activities(minimal_activity_models)
    end for
  end for
  annotated_dataset ← find_non_overlapping_activities(detetected_activities)
  return annotated_dataset
```

4 Evaluation

4.1 Synthetic Dataset Generator

To evaluate the SA^3 algorithm, a ground-truth was required. For that purpose, a synthetic dataset generator algorithm was developed, inspired by the tool described by Okeyo et al. in [9]. The tool offers the following functionalities:

- Define activation patterns: activation patterns are sensor activation sequences linked to an activity. Each activity can have different activation patterns with associated probabilities. Typical time intervals between two successive activations are also provided. Synthetic dataset generator uses Gaussian time intervals, simulating activation patterns more realistically

– Define activity patterns: activity patterns specify sequences of activities that occur in a given time interval. Time intervals between two successive activities are provided. Activity patterns may model that between 7 AM and 9 AM a person usually makes coffee and brushes teeth. Additionally, alterations can also be modelled, where an occurrence probability is assigned to an activity for a given interval. This allows us to model the probability of a person watching TV between 6 PM and 8 PM
– Sensor noise: two kinds of noise can be modelled: (i) missing sensors, where a sensor that should be activated fails and (ii) positive noise, where a sensor that should not be activated, suddenly gets activated. For all sensors, both noise modalities can be provided with probabilistic models

Activation patterns contain different ways of performing activities, while activity patterns capture typical activities performed by a user everyday. Synthetic dataset generator uses probabilistic mechanisms to simulate sensor activations for a given number of days. Generated datasets are already labelled to work as ground-truth.

4.2 Experimental Set-Up and Results

Four experimental set-ups were prepared to evaluate the SA^3 algorithm:

Ideal Scenario. There is no noise in this set-up. Seven activities are defined, with two variations for each activity. Three typical days are modelled through activity patterns. Generated datasets contain around 900 sensor activations. Five tests that simulate 30 days each, were run using five datasets generated from the same script. Overall results for the seven activities are depicted in Table 1. Success rate for all activities is 100%. No false positives or negatives were detected.

Table 1. Results for *Ideal scenario*

Activity	Instances	Correctly annotated	Total annotated
MakeChocolate	150	150	150
WatchTelevision	133	133	133
BrushTeeth	411	411	411
WashHands	150	150	150
MakePasta	178	178	178
ReadBook	130	130	130
MakeCoffee	67	67	67

Missing Sensor Noise Scenario. The same seven activities are used in this set-up. Three typical days are modelled and 60 days are simulated. A typical dataset contains around 1900 sensor activations. Two activities, *MakeChocolate* and *BrushTeeth*, have a probability of being performed with a missing sensor activation which is used in their respective minimal activity models. Five experiments are run, where failing probabilities for both sensors are set to 0.01, 0.02, 0.05, 0.07 and 0.1, to see the behaviour of SA^3 in presence of increasing noise. Results show that the performance for the rest of activities is the same as the previous experiment. However, as noise increases, correctly annotated activities for *MakeChocolate* and *BrushTeeth* decrease, as it can be seen in Table 2.

Table 2. Results for *Missing sensor noise scenario*

Activity	Missing sensor probability				
	0.01	0.02	0.05	0.07	0.1
MakeChocolate	59/60	58/60	58/60	54/60	53/60
BrushTeeth	162/163	158/161	146/159	150/161	145/157

Positive Sensor Noise Scenario. The same set-up as previous scenario, but containing positive sensor noise and no missing activations. Some sensor activations that are linked to minimal activity models are assigned a probability to generate faulty activations in any time of the day. Some other faulty sensors that are not linked to activity models are also used. To sum up, 7 out of 9 noisy actions are used in minimal activity models. Once again, five experiments are run, where activation probability is increased. Results are shown in Table 3, where for each activity and noise probability, three parameters are given: instances of that activity in the dataset, correctly annotated activities and total annotated activities.

Table 3. Results for *Positive sensor noise scenario*; I: instances; C: correctly annotated; T: total annotated

Activity	Positive sensor probability														
	0.05			0.1			0.15			0.2			0.25		
	I	C	T	I	C	T	I	C	T	I	C	T	I	C	T
MakeChocolate	60	60	60	60	58	59	60	59	59	60	59	60	60	59	59
WatchTelevision	55	55	55	54	53	54	54	54	54	47	47	47	60	60	60
BrushTeeth	162	162	162	162	161	161	162	162	162	161	160	160	168	166	166
WashHands	60	60	60	60	60	60	60	60	60	60	60	60	60	60	60
MakePasta	66	66	66	65	65	65	64	63	64	63	63	63	77	76	77
ReadBook	55	55	55	49	49	50	50	50	51	52	52	53	50	50	50
MakeCoffee	24	24	24	23	23	24	22	22	26	22	22	32	29	29	51

Challenging Activity Models Scenario. In this scenario, seven activities are used again, but three of them contain the same actions to be described. Concretely:

$$MakeCoffee = \{hasContainer, hasCoffee, hasFlavour\}$$
$$MakeTiramisu = \{hasContainer, hasCoffee, hasCream\}$$
$$MakeWhippedCream = \{hasFlavour, hasCream, hasContainer\}$$

The other four activities remain the same as in previous examples. The aim of the current set-up is to test the validity of the algorithm with activities that contain very similar action sequences. Five experiments were run in ideal conditions, where each experiment simulates 30 days. Combined results are depicted in Table 4. For all activities, 100% success rate has been achieved, with 0 false positives and false negatives.

Table 4. Results for *Challenging activity models scenario*

Activity	Instances	Correctly annotated	Total annotated
MakeCoffee	222	222	222
MakeTiramisu	120	120	120
MakeWhippedCream	72	72	72
WashHands	150	150	150
BrushTeeth	420	420	420
ReadBook	127	127	127
WatchTelevision	151	151	151

5 Discussion

Results shown in Section 4 show that SA^3 can handle varying order of actions in activity performance without any problem. All four set-ups contain different activation patterns, where actions are executed in different orders. For example, for the activity *MakeChocolate*, two activation patterns are considered: (i) {*hasContainer*, *hasMilk*, *useCookingAppliance*, *hasChocolate*} and (ii) {*useCookingAppliance*, *useCookingUtensil*, *hasMilk*, *hasChocolate*, *hasContainer*}. To define the minimal activity model, only two actions are used: *hasContainer* and *hasChocolate*. As the noiseless scenario shows, the success rate of the algorithm for that activity is 100%. Notice that activation patterns used for each activity contain more actions that are not used in minimal activity models. Hence, SA^3 can also handle non-considered actions.

As far as noisy scenarios regard, results suggest that missing sensor activations cannot be properly handled by the algorithm, when those activations are linked to actions that are used in minimal activity models. The non-captured activities are proportional to the missing activation probability of those actions. That is due to the completeness criterion used in the second step of SA^3, as explained in Section 3. We believe this error can be mitigated, using completion metrics, but as minimal activity models contain very few actions (2-3), those metrics do not seem very useful. Moreover, user-object interaction recognition information gathered by Chen et al. in [2], show that contact sensors, for instance, have a missing probability of around 2%. Hence, due to the nature of minimal activity models and small missing errors, developing special mechanisms that deal with missing sensor activations is not crucial.

For positive noise, the scenario is different. SA^3 has shown to perform reliably even with high levels of noise. For example, even having certain sensors with faulty activation probabilities of 0.25 per hour, average success rate was 99%. Results show an interesting case: *MakeCoffee* activity shows an increasing rate of false positives, scaling up to 76% (look at Table 3). However, none other activities show this behaviour. That is due to the model used for *MakeCoffee*, which only contains *hasContainer* and *hasCoffee* actions. Both actions are involved in the noise generation, hence the probability of getting false activation patterns of those two actions is very high. Overall, results obtained in noisy scenarios can be considered as very good.

Finally, a test involving challenging activity models showed that as far as activity models are unique, SA^3 does not have any problem finding them. A success rate of 100% was achieved under those conditions, without any noise.

In conclusion, considering usual noise levels described in [2], where missing sensor activation rates are around 6% (including poorly behaving pressure sensors), SA^3 showed to be able to annotate very accurately the activities in a time-stamped sensor dataset.

6 Conclusions and Future Work

A novel algorithm for activity dataset annotation was presented in this paper. SA^3 is a knowledge-driven algorithm, which uses sensor-action transformations and minimal activity models as prior expert domain knowledge, to annotate time-stamped sensor datasets.

The results of the experiments described in Section 4 and discussed in Section 5, show that SA^3 can handle varying order of sensors, incomplete activity models and several forms of noise. The main weakness of the algorithm appears when missing sensor activations occur for actions that are used to describe an activity. However, due to the low missing activation rates reported by previous research, the problem does not compromise the performance of the algorithm.

We believe that SA^3 can be a very useful tool for those working on activity recognition. Activity annotation is accurate, as far as sensor-action transformations and minimal activity models can be defined. SA^3 can be a good complement to a human annotator, who only would have to review the results obtained by the automatic tool. In the experiments we carried out, it was easy to identify some activities that were not captured by SA^3. It was also easy to discard false positive activities emerging from sensor noise. From our experience, SA^3 makes activity annotation much easier and faster.

As future work, it would be very useful to test SA^3 with publicly available real-world activity datasets. Additionally, further research can be done to take advantage of frequent behaviours of users to mitigate more the sensor noise, specially for missing activations. The idea would be to add another step in the algorithm in order to find frequent activity sequences. Afterwards, that information could be used to analyse actions around activities that are in frequent activity patterns. That way, even having noisy scenarios, better results could be obtained for activity annotation.

References

1. Chen, L., Hoey, J., Nugent, C.D., Cook, D.J., Yu, Z.: Sensor-based activity recognition. IEEE Transactions on Systems, Man, and Cybernetics-Part C 42(6), 790–808 (2012)
2. Chen, L., Nugent, C.D., Wang, H.: A knowledge-driven approach to activity recognition in smart homes. IEEE Transactions on Knowledge and Data Engineering 24(6), 961–974 (2012)
3. Cook, D.J., Schmitter-Edgecombe, M.: Assessing the quality of activities in a smart environment. Methods of Information in Medicine 48(5), 480 (2009)
4. Gu, T., Wu, Z., Tao, X., Pung, H.K., Lu, J.: epsicar: An emerging patterns based approach to sequential, interleaved and concurrent activity recognition. In: IEEE (ed.) IEEE International Conference on Pervasive Computing and Communications (PerCom), pp. 1–9 (2009)
5. Huynh, T., Fritz, M., Schiele, B.: Discovery of activity patterns using topic models. In: Proceedings of the 10th International Conference on Ubiquitous Computing, pp. 10–19. ACM (2008)
6. Van Kasteren, T., Noulas, A.: Accurate activity recognition in a home setting. In: Proceedings of the 10th International Conference on Ubiquitous Computing, pp. 1–9 (2008)
7. Liao, L., Fox, D., Kautz, H.: Location-based activity recognition. Advances in Neural Information Processing Systems 18, 787 (2006)
8. Maurer, U., Smailagic, A.: Activity recognition and monitoring using multiple sensors on different body positions. In: Wearable and Implantable Body Sensor Networks, pp. 4–116 (2006)
9. Okeyo, G., Chen, L., Wang, H., Sterritt, R.: Dynamic sensor data segmentation for real-time knowledge-driven activity recognition. Pervasive and Mobile Computing (December 2012)
10. Philipose, M., Fishkin, K.P.: Inferring activities from interactions with objects. Pervasive Computing 3(4), 50–57 (2004)
11. Rashidi, P., Cook, D.J.: Discovering activities to recognize and track in a smart environment. IEEE Transactions on Knowledge and Data Engineering 23(4), 527–539 (2011)
12. Tapia, E.M., Intille, S.S., Larson, K.: Activity Recognition in the Home Using Simple and Ubiquitous Sensors. In: Ferscha, A., Mattern, F. (eds.) PERVASIVE 2004. LNCS, vol. 3001, pp. 158–175. Springer, Heidelberg (2004)
13. Wren, C.R., Tapia, E.M.: Toward scalable activity recognition for sensor networks. In: Hazas, M., Krumm, J., Strang, T. (eds.) LoCA 2006. LNCS, vol. 3987, pp. 168–185. Springer, Heidelberg (2006)

Multimodal Statement Networks for Organic Rankine Cycle Diagnostics – Use Case of Diagnostic Modeling

Tomasz Rogala and Marcin Amarowicz

Institute of Fundamentals of Machinery Design, Silesian University of Technology, Konarskiego 18A, Gliwice, Poland
{tomasz.rogala,marcin.amarowicz}@polsl.pl

Abstract. The paper shows an example of application of modeling process of diagnostic knowledge with the use of multimodal statement networks. The goal is to present generic approach to modeling complex diagnostic domains by many independent experts. As an object cogeneration power plant with Organic Rankine Cycle is presented. Implementation of diagnostic model was performed using *REx* system that enables among other things knowledge management and the application and preliminary evaluation of diagnostic knowledge of gathered knowledge for the purpose of its further incorporation to diagnostic systems.

Keywords: multimodal statement networks, *REx* system, diagnostic modeling, knowledge integration, Organic Rankine Cycle.

1 Introduction

Contemporary information systems process a large amount of data, which often are approximate, inaccurate and ambiguous. Knowledge-based systems for diagnostics of complex technical objects are a typical example. Knowledge base of these systems requires defining comprehensive knowledge from many domains which needs to gather its from many independent sources, e.g. from different domain experts. Integration of knowledge from many sources, ensuring adequate semantic consistency and taking into account the uncertain nature of the gathered knowledge, requires development of appropriate tools for this purpose.

The paper shows an example of modeling process of diagnostic knowledge with the use of multimodal statement networks [2], [5] in the *REx* environment [3]. The main objective of this paper is to make the approach to diagnostic modeling clear to potential domain experts who could be interested in building diagnostic knowledge for complex technical objects in *REx* environment. The *REx* environment enables among other things knowledge management and the application and preliminary evaluation of diagnostic knowledge of gathered knowledge for the purpose of its further incorporation to diagnostic systems.

P. Angelov et al. (eds.), *Intelligent Systems'2014*,
Advances in Intelligent Systems and Computing 322, DOI: 10.1007/978-3-319-11313-5_53

2 Multimodal Statement Networks and *REx* Environment

During the inference process about technical state of an object more than once is required accurate representation of relationships between the selected quantitative variables which was recorded on the considered object. In many cases, this task may be very difficult to achieve and sometimes even impossible. Therefore, it is necessary to conversion from quantitative to qualitative representation. One of the possibilities are graphical models which can be built based on a set of statements.

Each statements s can be recorded as a pair $s =< c(s), b(s) >$, where $c(s)$ is called the statement content and $b(s)$ is assigned to its a certain value. Statement content is a sentence that declares about observed facts or expresses some opinion [2]. A some value, usually one of the element of set {*True, False*}, can be assigned to this content. This value indicates about truthfulness of considered statement. One example of the statement content can be the following sentence: *Broken tooth of active gearwheel of gear no. PZ02.* The logic value *True* can be assigned to this content.

Each statement network can be recorded as a pair $< V, E >$, where V is a set of nodes (which are represented by statements) of this network and E is set of directed edges which joining selected nodes. Relationships between particular nodes of considered network can express different aspects of knowledge, applied for different conditions and assumptions and at the presence of different contexts. An attempt to take into account all possible dependencies on a single, global statement network can lead to too much its growth. Considerable difficulty in the analysis of such network may be a result of this approach. As an alternative solution multimodal statement networks can be used. There are defined as $< V, E, \Gamma >$, where Γ is a set of layers such network. Each layer is defined as $\Gamma_i =< V_i, E_i >$, wherein $V_i \subset V$, $E_i \subset E$. A characteristic feature of the multimodal statement networks is occurrence of common nodes between the selected layers. Final values of such nodes are determined using selected methods of aggregation. Relationships between particular nodes of such statement network can belong to different classes. Solutions based on conditional probability expressed by Bayesian equation are commonly used [6], [4]. These solutions are also called as belief networks. For the description of the required dependencies it is also possible to use intuitionistic logic [1]. All of the mentioned solutions can be used as layers of multimodal statement networks for the purpose of diagnostic knowledge modeling.

3 Description of Studied Example

3.1 Object Description

The installation of cogeneration system with Rankine cycle (shown in Figure 1) consists of four circuits: bio-waste combustion system and other low enthalpy sources, thermal oil cycle which receives heat from the boiler and delivers it to the Rankine cycle, Organic Rankine cycle and cooling cycle which transfer some part of heat energy from Rankine cycle. Proper conduct of the process in the Rankine cycle is determined by conserving appropriate thermodynamic parameter values of organic working fluids at the selected points of the cycle. For example, thermodynamic parameters of working fluid at the output of evaporator should meet the values for superheated vapour. The conserving of these values has an significant influence on not only proper operation of the process and its efficiency, but also extends the time of the life of selected system components.

Typical conduct of process in Rankine cycle during nominal operating conditions is as follows. The working fluid in the evaporator is heated up to a temperature at which it becomes superheated vapour. It results from exchange of heat energy between thermal oil cycle and Organic Rankine cycle. The superheated vapour expands in the turbine and generates kinetic energy which drives the generator and results in electricity production. Expanded working fluid delivers its own relatively hight heat energy to the working fluid on the high pressure site of Rankine cycle. The wet vapour of working fluid at the low pressure site exchanges the rest of its thermal energy with cooling fluid via condenser. The low pressure of working fluid is next raised by a pump. Before the next cycle, the fluid, now on the high pressure site, is preheated by regenerator.

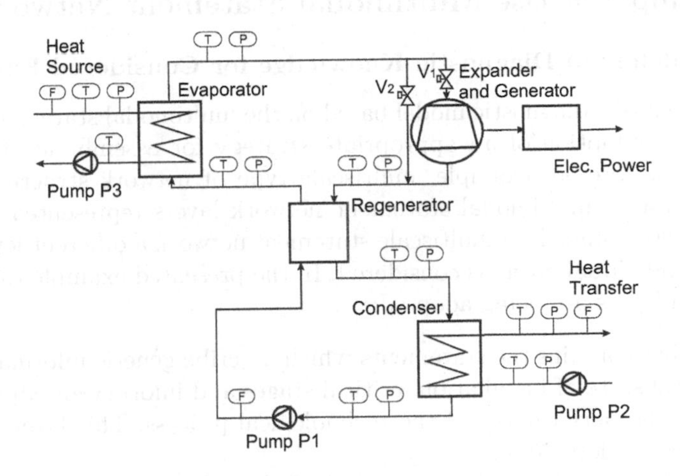

Fig. 1. Part of schema of cogeneration installation with Organic Rankine Cycle. T - temperature measurement, P - pressure measurement, F - mass rate flow measurement.

3.2 Considered Faults

The goal of technical diagnostics is to recognize faults or malfunctions on the base of all available information on the considered object. Conclusions resulting from the inference process can be, among others, done on the basis of the measurement of observed variables and known definitions of faults and malfunctions. Hence, knowledge on the relationships between the observed values which covariate with changes of technical states named symptoms and changes of technical states are important. Information about the symptoms and technical states can be stored in the *REx* system using statements whose values can be represented by different type of degree of truthfulness of the statement content. The degree of belief in the truthfulness of statements based on Bayes theorem is an example. For the considered example the following division of statements has been introduced and stored in the thesaurus of *REx* system:

- statements which relate to the potential faults of components and which were defined on the basis of available diagnostic knowledge on possible component faults e.g. *Fouling in the evaporator on the working fluid site.*
- statements concerned with possible critical states (e.g. pointed out by producers of installation components as important factors which have an impact on developing faults and malfunctions as well as other threads), e.g. *Maximum working pressure of evaporator at the Rankine cycle site is exceeded.*
- statements with are assertion about control errors, which contents were defined on the base of knowledge about proper process conduction e.g. *Flow rate through the pump P1 is less then the recommended flow rate.*

4 Example of Use Multimodal Statement Network

4.1 Modeling of Diagnostic Knowledge for Considered Example

Development of a diagnostic model based on the multimodal statement networks requires the adoption of an appropriate strategy for assembling of individual layers. In the present example, multiscale type of network structure which is a special case of multimodal statement network layers represented by a belief networks was assumed. In multiscale statement network a different level of granularity of individual layers is considered. In the presented example the following division on layers are being adopted:

- layer that contains the statements which describe generic information about technical state of installation, critical stages and information about correctness of the performance of the technological process. This layer is depicted in Figure 2 (left side),
- layers that contain the statements which describe possible malfunctions and faults, control errors and critical states. This division was introduced because of the need to examine not only the faults or malfunctions of individual components, but also the need to take into account the different situations that maintained for a long period of time may lead to failure or insufficient energy efficiency of the system. An example of layer that is connected with control errors is shown in Figure 2 (right side),
- layers that extend the existing layers of faults and malfunctions of individual components (evaporator, turbine, pump etc.), or some elements of these components (e.g. layers connected with pump).

Division of considered example of multimodal model on individual layers is shown in Figure 3.

In the case of layers that are represented by belief networks, for each of their nodes that represent selected statements the conditional probability table is necessary to define. Required number of parameters that describes this table depends on Cartesian product of the number of available variants of considered statement and sum of all variants of the statements values represented by the parent nodes.

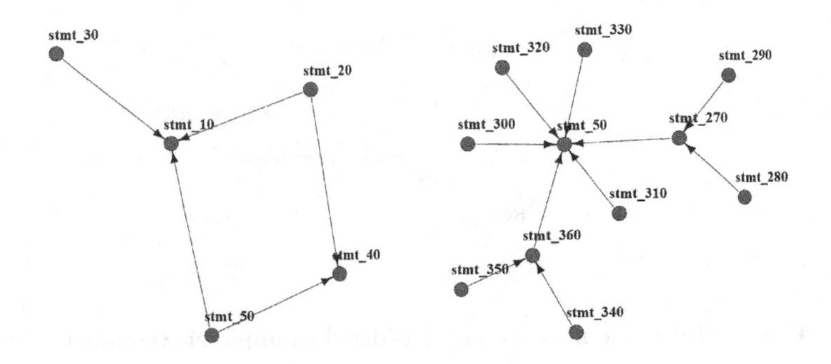

Fig. 2. Generic layer of multimodal model for considered ORC installation (left side) and layer connected with control errors of ORC system (right side)

4.2 Use Case

The use of multimodal statement network that layers are represented by belief networks enables to obtain relevant values for particular statements when information on observed variables about object is available. Values of selected statements are determined based on the known measurement values. The process of inference computes marginal probability distribution for each node [6] and generate values of individual statements. Values of common nodes that exist in many layers are determined based on appropriate method of aggregation. This method may also take into account weights of individual layers of the multimodal model. Weight of these layers can be determined based on e.g. experiences of authors in considered domain.

Three scenarios of use cases were elaborated:

- Scenario I: It was assumed that in object does not exist any malfunctions or faults, control unit doest not have any control errors and any critical state doest not exist as well. The status *Fixed* is assigned to these statements values which values are known, statements which values are not known, are determined in inference process and have the status *Calculated*, and finally another statements, that exist in many nodes and values of these statements are determined by an aggregation have the status *Aggregated*.

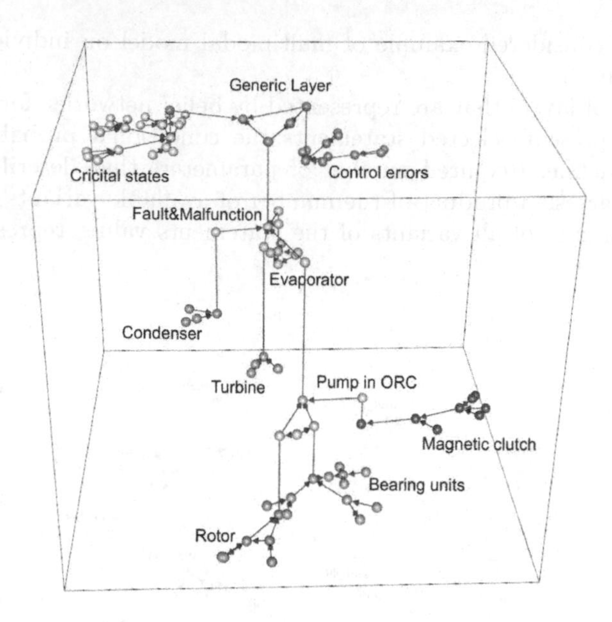

Fig. 3. Multiscale network for considered example of Rankine Cycle

- Scenario II: Sudden reduction of thermal power delivered to the Rankine system has been observed. After a while it causes a decrease in the temperature of the working fluid at the output of the evaporator. Hence, the value of statement has changed its value *stmt_310 Thermodynamic parameters of working fluid at the output of evaporator does not exceed the point of saturated vapour* = {*yes*}. The supervised control system has computed new optimal operating point of Rankine cycle and new set-point values are delivered to the local control system which reduce the flow rate in the Rankine cycle. Assuming that the decrease in delivered power was significant, the action of the control system may not be sufficient to compensate changes of thermal power input. This results in a decrease in system efficiency and statement *stmt_40* is set to new value *Decrease of ORC system efficiency* = {*yes*}.
- Scenario III: This time sudden increase of thermal power delivered to the Rankine system has been observed. The control system accelerates the input shaft of the pump P1 in order to decrease the temperature of working fluid at the output of the evaporator. Because of too high rate of acceleration the limited value of torque at input shaft of pump P1 was exceeded and the magnetic drive was partially decoupled from the impeller (*stmt_630 Upper limit of torque at the input shaft of pump P1 is exceeded*)= {*yes*}. It caused a pressure drop in the system and the following statements changed its values *stmt_350 Turbine is underloaded*= {*yes*} and *stmt_40 Decrease of ORC system efficiency*= {*yes*}.

Values of selected statements for considered scenarios are shown in Table 1. Presented results allow to preliminary evaluation of implemented knowledge. This can be done by taking advantage to conduct inference process in bidirectional way. For example, changing the values of nodes that represent possible faults or critical states allows to obtain most probable explanations for symptoms. These activities can be very helpful during the validation process of designed model.

Table 1. Values of selected statements for considered scenarios

ID	Content	Status	Scen.1 Val.(Yes)	Scen.2 Val.(Yes)	Scen.3 Val.(Yes)
stmt_10	ORC system does not work properly	Calc.	0.01	0.63	0.63
stmt_20	Installation component is faulty	Agg.	0.02	0.27	0.27
stmt_30	Critical state occurred	Agg.	0.16	0.16	0.16
stmt_40	Decrease of ORC system efficiency	Fixed	0	1	1
stmt_50	Control error occurred	Agg.	0.05	0.73	0.73
stmt_270	Pump does not work properly	Calc.	0.5	0.5	0.5
stmt_280	Flow rate through the pump P1 is less then the recommended flow rate	Fixed	0	0	0
stmt_300	Thermodynamic parameters of working fluid at the output of condenser does not exceed the point of saturated liquid	Fixed	0	0	0
stmt_310	Thermodynamic parameters of working fluid at the output of evaporator does not exceed the point of saturated vapour	Fixed	0	1	0
stmt_320	Temperature of working fluid at the output of evaporator has exceeded the limit value	Fixed	0	0	0
stmt_330	Temperature of working fluid at the output of condenser has exceeded the limit value	Fixed	0	0	0
stmt_340	Turbine is overloaded	Fixed	0	0	0
stmt_350	Turbine is underloaded	Fixed	0	0	1
stmt_360	Turbine does not work properly	Calc.	0	0	1
stmt_630	Upper limit of torque at the input shaft of pump P1 is exceeded	Fixed	0	0	1
stmt_620	Magnetic clutch is decoupled	Calc.	0.2	0.2	0.71
stmt_550	Magnetic clutch is faulty	Agg.	0.1	0.1	0.63
stmt_400	Pump is faulty	Agg.	0.11	0.11	0.19

5 Summary

The paper presents an example of application of multimodal statement networks for the purposes of diagnostic knowledge modeling complex technical object that is part of the installation of cogeneration unit with Organic Rankine Cycle. Recalled example has been developed using the *REx* system, which is available as the R package at `http://ipkm.polsl.pl/index.php?n=Projekty.Rex`. System *REx* support the process of building diagnostic models designed by many independent experts. This example should allow to explain the generic approach to diagnostic modeling using multimodal statement networks. In the given example, individual layers of multimodal statement networks were represented by using belief networks. In order to be able to take into account any partial contradiction in the incorporated knowledge further work on the development of *REx* system functionality are planed. One of the is the implementation of intuitionistic statement networks [1].

Acknowledgments. Described herein are selected results of study, supported partly from scientific from the budget of Research Task No. 4 entitled *Development of integrated technologies to manufacture fuels and energy from biomass, agricultural waste and others* implemented under The National Centre for Research and Development (NCBiR) in Poland and ENERGA SA strategic program of scientific research and development entitled *Advanced technologies of generating energy.*

References

1. Cholewa, W.: Intuitionistic notice boards for expert systems. In: Gruca, A., Czachórski, T., Kozielski, S. (eds.) Man-Machine Interactions 3. AISC, vol. 242, pp. 337–344. Springer, Heidelberg (2014)
2. Cholewa, W.: Multimodal Statements Networks for Diagnostic Applications. In: Sas, P., Bergen, B. (eds.) Proceedings of the International Conference on Noise and Virbation Engineering ISMA 2010, September 20-22, pp. 817–830. Katholique Universiteit Lueven, Lueven (2010)
3. Cholewa, W., Amarowicz, M., Chrzanowski, P., Rogala, T.: Development Environment for Diagnostic Multimodal Statement Networks. In: Key Engineering Materials, Smart Diagnostics V, vol. 588, pp. 74–83 (2014)
4. Cowell, R.G., Dawid, P.A., Lauritzen, S.L., Spiegelhalter, D.J.: Probabilistic Networks and Expert Systems. Springer, New York (2003)
5. Heath, L., Sioson, A.: Multimodal Networks: Structure and Operations. IEEE/ACM Trans. Computational Biology and Bioninformatics 6(2), 321–332 (2009)
6. Nielsen, T.D., Jensen, F.V.: Bayesian Networks and Decision Graphs. Springer, Berlin (1997)

Gradual Forgetting Operator in Intuitionistic Statement Networks

Wojciech Cholewa*

Silesian University of Technology,
ul. Konarskiego 18a, 44-100 Gliwice, Poland
wojciech.cholewa@polsl.pl

Abstract. The investigated intuitionistic statement networks facilitate designing and development of complex expert systems. Within these networks, statements consist of contents and values. Statement values are based on a concept introduced in intuitionistic fuzzy sets, i.e. they contain independent belief about validity and nonvalidity of information presented by statement contents. The relationships between statements are modeled in the form of a set of necessary as well as sufficient conditions, and these conditions are considered as corresponding inequalities between statement values. The paper introduces gradual forgetting operator of statement values. The networks with such an operator may be used as models of dynamic objects. They allow for non-monotonic reasoning required for monitoring and diagnostic systems.

Keywords: expert system, intuitionistic statement network, forgetting operator, non-monotonic reasoning.

1 Motivation

The main task of technical diagnostics is to determine a condition, or changes in the condition, of an object on the basis of all available information within this object. Such a task may be aided by expert systems. Needs arising from development of diagnostic experts systems became a founding ground for actions that resulted in the proposed intuitionistic statement networks. The diagnostic expert systems considered connecting results of independently realized individual procedures of knowledge acquisition as well as methods of identification of conditional inconsistencies that occur in the resultant knowledge base. These tasks could not be realized directly with a use of commonly available methods.

One finds a great variety of the currently available expert systems and their detailed overview would go beyond the purpose of the presented paper. Nevertheless, it is essential to note that before research into new classes of such systems can proceed, the corresponding needs must be meticulously analysed.

* Described herein are selected results of study, supported partly from the budget of Research Task No. 4 implemented under The National Centre for Research and Development in Poland and ENERGA SA Strategic Program of Scientific Research and Development entitled *Advanced Technologies of Generating Energy*.

P. Angelov et al. (eds.), *Intelligent Systems'2014*,
Advances in Intelligent Systems and Computing 322, DOI: 10.1007/978-3-319-11313-5_54

Such an analysis should indicate those features of the available systems that require further modifications.

While taking a decision, the user should know whether the result of the inference process is conditioned by direct results of both observation and measurements, or whether it originates from the default data whose occurrence is expected for an average object, however, not necessarily for the analysed object in question. The decisions should base on a belief that there are indications of their validity, i.e. plausibility, advisability, legality. Systems basing on intuitionistic logic are considered to be an interesting example of inference systems that enable collection of judgments. In this logic, propositional calculus allows for a syntax transformation that ensures its justification. This differentiates intuitionistic logic from classical logic whose propositional calculus maintains the truthfulness of transformed syntax.

This paper not only addresses a possibility of using intuitionistic statement networks to represent knowledge in diagnostic expert systems, but also emphasizes the advantages of such networks. Records of analyzed data in a form of statements allow for their consideration along with the degree of granulation. Resigning from using too detailed data has a positive effect onto stability of data-based inference process, and enables acquisition of more precise results of this process. A particular attention to the need of maintaining a rational precision degree, i.e. adequate data granulation, is crucial for technical diagnostics-related applications with exceptionally large data streams.

A content which should appear as a declarative sentence is the main element of the statement. Any information concerning validity or lack of validity of the statement content is recorded as an intuitionistic value. These values are assigned to a statement content. A possibility of indicating a partial validity or lack of validity of the statement contents allows for development of both approximate and partially uncertain models.

Knowledge of the analyzed field is recorded as a set of relations between statements. These relations occur in a form of necessary and/or sufficient conditions. It was proven that sets of such relations may be considered as sets of inequalities referring to intuitionistic values.

One of the particularly interesting feature of introduced statement networks includes a possibility of a process of forgetting of beliefs concerning acknowledgment or lack of acknowledgment of statement contents, which enables non-monotonous inference being a requirement for diagnostic systems.

2 Statements

Knowledge acquisition for the needs of diagnostic systems may be realized as a result of applied processes of machine learning as well as as a result of cooperation with domain experts. In each case, a need to define a degree of generality of considered data arises. Not only a considerably high, but also significantly low degree of generality of data may hinder or even block entirely a possible definition of a sustainable solution. An optimization should result in assuming appropriate granularity of information represented by the data.

A number of formalisms relating to possible methods of information granulation is currently well known [2]. Statements, most suitable for practical uses, constitute a kind of informational granules that facilitate development and construction of knowledge models in a form of statement networks [4].

The noun *statement* has several meanings. Here, the statement is information on recognition of proposition resulting from observed facts or representing an opinion. The statement s consists of a content $c(s)$ and value $b(s)$

$$s = \langle\, c(s), b(s)\, \rangle. \tag{1}$$

The content $c(s)$ of a statement s is presented as an indicative sentence, i.e. a sentence which may be either entirely true or entirely false. On the basis of the content only, one may not conclude whether the statement has been recognized or not. The information on recognition of the statement is represented as a value $b(s)$ of the statement s, independently of its content $c(s)$.

It is very important that not only data on the object design and working conditions, but also results of inferring on the object state may appear in a form of statements. This, in turn, allows for an inference process carried out in a set of statements in which diagnostic knowledge is represented by relations occurring between the statement values.

A value of a statement constitutes a carrier of information on validity or the lack of validity of the statement content. The boolean type is the fundamental type of a statement value. It consists of logical values *true* and *false* which are frequently encoded as numerical values, 1 and 0, respectively. If the value of a statement is not equal to one of these values, then it must be equal to the other value. It means the boolean type comply with a so-called law of excluded middle.

The boolean type of statement values facilitates construction of accurate inference systems. Due to restricted quality of available knowledge and available information, a use of approximate inference systems is frequently required. A degree of belief about the validity of a statement content is an example of a value of statements considered as nodes of belief networks which are also referred to as bayesian networks [5]. Not only a certainty factor or subjective probability of a statement truthfulness or degree of belief in the truth of a statement, but also a degree of belonging to a fuzzy or rough set of statements recognized as true, and a large number of similar concepts, may become a value of a statement. A common feature that is shared by the enumerated types of statement values is that these values are represented in a form of a point value by a one number only.

In practical applications of expert systems one needs to determine statement values in a manner that allows differentiation between the following groups of statements

- statements that are not recognized as either valid or nonvalid,
- statements for which the numbers of for and against arguments on their recognition are equal.

The expected manner of representing a statement value is hard to introduce if a statement value is written down as a single number. For this purpose it is essential to consider a statement value defined in a form of pairs of numbers. An example of such a value may include a pair consisting of a necessity degree and possibility degree originating from modal logic. Another concept of a statement value which is worthy of attention is a definition of this value (2) derived from the theory of intuitionistic fuzzy sets (IFS) [1].

An intuitionistic value $b(s)$ of statement s, also referred to as the statement value, is represented by the ordered pair

$$b(s) = \langle p(s), n(s) \rangle \quad \text{for} \quad p(s), n(s) \in [0,1] \tag{2}$$

where $p(s)$ is a degree of validity (recognition, justification, truthfulness, acceptance) and $n(s)$ is a degree of nonvalidity (lack of justification, lack of truthfulness) of statement s. These values are also referred to as evaluations of positive and negative information. Within IFS it is assumed that

$$p(s) + n(s) \in [0,1], \tag{3}$$

and it is not assumed that

$$p(s) + n(s) = 1.0 . \tag{4}$$

Considering (2), the value $b(s)$ of an unrecognized statement s, i.e. a value of the statement not recognized as valid or nonvalid, takes the following form

$$b(s) = \langle 0, 0 \rangle \tag{5}$$

Values $p(s)$ and $n(s)$ allow for defining a hesitation margin $h(s)$ for statement s which is also known as a degree of nondetermination

$$h(s) = 1 - p(s) - n(s) \tag{6}$$

where (3) results in

$$h(s) \in [0,1] \tag{7}$$

The literature on IFS uses notation $\mu(s)$, $\nu(s)$, $\pi(s)$ instead of $p(s)$, $n(s)$, $h(s)$.

The statement values (2) may be determined directly with respect to the assumption (3). They may also result from inference on the basis of other statements in accordance with an assumed knowledge model. A potential imperfection of the applied knowledge model and inconsistency of the studied set of statement values may lead to unexpected results of the inference process manifested as disagreements and contradictions in the set of selected statement values which may not satisfy restrictions (3). Such possibility leads to an extension of the definition (6) of hesitation margin, with respect to (7)

$$h(s) = \begin{cases} 1 - p(s) - n(s) & \text{if } p(s) + n(s) \leq 1 \\ 0 & \text{if } p(s) + n(s) > 1 \end{cases} \tag{8}$$

In addition a disagreement level $d(s)$ for values $p(s)$ and $n(s)$ of statement s not satisfying restriction (3) is also introduced

$$d(s) = \begin{cases} 0 & \text{if } p(s) + n(s) \leq 1 \\ p(s) + n(s) - 1 & \text{if } p(s) + n(s) > 1 \end{cases} \tag{9}$$

where

$$d(s) \in [\,0\,,1\,]\,. \tag{10}$$

Occurrence of positive values $d(s)$ implies contradiction in either the applied knowledge model or in the assumed statement values.

3 Intuitionistic Statement Network

Elements which statement networks consist of are statements $s \in S$ considered as pairs (1) of both contents $c(s)$ and values $b(s)$ of these statements. The state of statement networks is determined by a set of statement values

$$b(S) = \{\, b(s) : s \in S\} \tag{11}$$

The essence of statement networks is that a change in a value of a particular statement occurring in the network may induce changes in values of other statements.

For a given set $S = \{s\}$ of statements s (1) a statement network may be represented as a graph G

$$G = \langle\, C, V, E, A\,\rangle \tag{12}$$

with a subgraph H defining a constant structure of a statement network and its elements

$$H = \langle\, C, V, E\,\rangle\,, \tag{13}$$

and a subgraph D defining interrelations of actual statement values

$$D = \langle\, V, A\,\rangle\,, \tag{14}$$

where:

C - a set of vertices representing statement contents

$$C = \{c(s) : s \in S\} \tag{15}$$

V - a set of vertices representing values of degrees of validity and nonvalidity,

E - a set of edges connecting vertices $c \in C$ which represent statement contents with vertices $v \in V$ which represent degrees of validity or nonvalidity, where

$$E = E_p \cup E_n \tag{16}$$

E_p - a set of p-edges $e_p(s) = \langle c(s), p(s)\rangle$ connecting contents $c(s)$ of statements s with values of degrees of their validity $p(s)$,

E_n - a set of n-edges $e_n(s) = \langle c(s), n(s) \rangle$ connecting contents $c(s)$ of statements s with values of validity degrees of their contradictions, i.e. with values of degrees of their nonvalidity $n(s)$,

A - a set of directed edges (arcs) defining relations between values of degrees of validity (or nonvalidity).

The vertex $v \in V$ representing the value of degree of validity or nonvalidity may be connected by means of edges $e \in E$ with numerous vertices $c \in C$ representing statement contents, which allows for, inter alia, concurrent considering of equivalent or contradictory statement contents. Each vertex $c \in C$ may be connected with no more than one p-edge and one n-edge.

Each edge $a = \langle s_1, s_2 \rangle \in A$ of the graph G defines a true implicational relationship $s_1 \to s_2$ between statements s_1 and s_2

$$(a = \langle s_1, s_2 \rangle \in A) \quad \Rightarrow \quad (s_1 \to s_2) \tag{17}$$

This relationship proves that the antecedent, i.e. statement s_1, is a sufficient condition for the consequent, i.e. statement s_2. The sufficiency condition denotes that recognition of the validity of statement s_1 is always accompanied by recognition of validity of statement s_2.

The necessary and sufficient conditions can be modelled by means of inequalities [3]. For considered intuitionistic values of a pair of statements s_1 and s_2 connected by an edge $a = \langle s_1, s_2 \rangle$ and for the selected value $b(s)$ the following sufficiency condition is obtained

$$(s_1 \to s_2,\, b(s_1) = \langle p(s_1), n(s_1) \rangle) \quad \Rightarrow \quad (p(s_2) \geq p(s_1)) \tag{18}$$

Furthermore, for the given value $b(s_2)$ the following necessity condition is obtained

$$(s_1 \to s_2,\, b(s_2) = \langle p(s_2), n(s_2) \rangle) \quad \Rightarrow \quad (n(s_1) \geq n(s_2)) \tag{19}$$

One considers, in the similar fashion, a sufficient condition including one consequent s and a disjunction of a number of antecedents $s_1, s_2, \cdots$

$$((\vee_i s_i) \to s) \quad \Rightarrow \quad p(s) \geq \max_i p(s_i) \tag{20}$$

as well as conjunctive of a number of antecedents $s_1, s_2, \cdots$

$$((\wedge_i s_i) \to s) \quad \Rightarrow \quad p(s) \geq \min_i p(s_i) \tag{21}$$

In a set of vertices representing values of degrees of validity and nonvalidity one can distinguish between three subsets

$$V = V_{or} \cup V_{and} \cup V_{fix}\,, \tag{22}$$

where

V_{or} - a set of vertices whose values are defined according to (20), when they appear as consequents of sufficient conditions,

V_{and} - a set of vertices whose values are defined according to (21), when they appear as consequents of sufficient conditions,

V_{fix} - a set of vertices whose values are arbitrarily acknowledged or are defined by means of measurement sets or other systems.

4 Gradual Forgetting Operator

A statement network represents knowledge in a form of graph G (12). While applying the statement network, statement values serve as variable elements of the graph. The remaining elements, i.e. statement contents and graph edges, are not subject to change.

In a number of applications, one can assume that values $v \in V_{fix}$ are constant. Solving of a statement network consists in application of known values of vertices $v \in V_{fix}$ as well as relations $a \in A$ according to (14) in order to determine unknown values of vertices $v \in V \backslash V_{fix}$. A simple algorithm consists in assuming initial zero values of degrees of validity for $v \in V \backslash V_{fix}$, and, subsequently, in iterative increasing selected degrees of validity that fulfils relations $a \in A$. The implemented process of inference is a monotonic one.

For dynamic objects or processes, the values of statements representing observation results of an object are subjects to changes. A statement network which is studied as a model of a dynamic object or process should adapt to changing operation environment due to which values $v \in V_{fix}$ of selected statements may be subject to changes. It is essential to emphasize that updates of values $v \in V_{fix}$ may be performed in an asynchronous manner. One of the approaches to solve a network, not satisfactory though, is to temporarily freeze the state of the statement network, and, at the same time, to look for solutions of such snapshots of the network. When searching for an alternative method it was determined that a process enabling simulation of gradual forgetting of statement values was required.

In general forgetting refers to loss of information already stored in the variable elements of the network. It should be a context-dependent gradual process. It was assumed that the forgetting process should offer a possibility of individual selection of a rate of forgetting for particular statements. Implementation of forgetting process when the values of statements are represented as single numbers (such as a degree of truthfulness) constitutes a difficult task, however, the use of intuitionistic values facilitates introduction of such a process. If intuitionistic values, i.e. independent ratings of a belief for and against validity of the statement contents, are available, then the process of forgetting can be defined as an even decrease of these values. The range of such decreased values shall include value 0 (zero), as a limit meaning that no knowledge is available, i.e. that all has been forgotten.

The gradual forgetting operator should be applied to all values $v \in V$, i.e. to both values of primary statements $v \in V_{fix}$ and values of secondary statements $v \in V \backslash V_{fix}$. The values of these degrees are studied as values dependent on a discrete time t, i.e.

$$v_i(t),\ v_i(t+1),\ \cdots, \tag{23}$$

The simplest process of forgetting may be considered in the following form:

$$v_i(t+1) := q_i\ v_i(t)\ ;\quad 0 < q_i \leq 1\,, \tag{24}$$

where the introduced constants q_i assigned separately to all vertices $v_i \in V$ allow for tuning of dynamic properties of the network.

The statement network integrity should be subject to verification during the system development phase as well as application phase. It is assumed that potentially contradicting elements of the network should be detected during network operation, basing on the disagreement level (9)

$$d(s) > 0\,, \tag{25}$$

as elements that are conditionally contradicting, i.e. contradicting for a specific case only.

5 Conclusions

Intuitionistic statement networks may be applied, among others, in diagnostic expert systems. Values of statements in such networks are based on a general concept used for intuitionistic fuzzy sets. They are assigned to statement contents and consist of independently determined degree of validity as well as degree of nonvalidity of a statement content. The domain knowledge is represented in a form of a set of necessary and sufficient conditions occurring between the statements. While the statement network is being solved, these conditions are considered as corresponding inequalities between statement values.

In a number of applications, one can not assume that values of statements are constant. The changes in the values depend on operation of considered object and may occur asynchronously. Intuitionistic statement networks with the introduced gradual forgetting operator may be used as models of dynamic objects or processes. They may be applied, among others, to diagnostic expert systems.

References

1. Atanassov, K.T.: Intuitionistic Fuzzy Sets: Theory and Applications. STUDFUZZ, vol. 35. Springer, Heidelberg (1999)
2. Bargiela, A., Pedrycz, W.: Granular Computing. An Introduction. Kluwer Academic Publishers, Boston (2003)
3. Cholewa, W.: Expert systems in technical diagnostics. In: Korbicz, J., Kościelny, J., Kowalczuk, Z., Cholewa, W. (eds.) Fault Diagnosis: Models, Artificial Intelligence, Applications, pp. 591–674. Springer, Heidelberg (2004)
4. Cholewa, W.: Mechanical analogy of statement networks. International Journal of Applied Mathematics and Computer Science 18(4), 477–486 (2008)
5. Jensen, V.J.: Bayesian Networks and Decision Graphs. Springer, New York (2002)

Part XII
Issues in Generalized Nets

A Generalized Net Model Based on Fast Learning Algorithm of Unsupervised Art2 Neural Network

Todor Petkov and Sotir Sotirov

Intelligent Systems Laboratory,
University of Prof. d-r Assen Zlatarov, Bourgas, Bulgaria
{todor_petkov,ssotirov}@btu.bg

Abstract. In this paper the fast learning algorithm of unsupervised adaptive resonance theory ART2 neural network is described. At the beginning of the process the algorithm is illustrated step by step by mathematical formulas and it is shown how individual vector changes its values during the training. The network supports clustering by using competitive learning, normalization and suppression of the noise. At the end of the process we have stable recognition clusters with values according to the vectors.

The learning process algorithm is presented by a Generalized net model.

Keywords: Generalized Nets, Neural Networks, Adaptive Resonance Theory.

1 Introduction

The Adaptive Resonance Theory (ART) [3, 4] is unsupervised neural network devised by Stephen Grossberg in 1976. In this work a generalized net model [9, 10, 11] based on fast learning [5] algorithm of ART2 [2, 3, 4] neural network [8] is presented. ART2 is designed to preform operation over input vectors of continuous values or noisy binary input vectors. In general the network consists of two layers composed of fully connected neurons with a set of weights also known as bottom-up and top-down, and one orienting sub system (Fig.1). First layer (F1) is composed of three sub layers of neurons, each sub layer supports combination of vector normalization and noise suppression. Second layer (F2) is a competitive one and the neuron with maximum value will win therefore the layer weights will learn according to the vector. The orienting sub system takes design whether or not the neuron with maximum value responds to the criteria of insert.

The fast learning algorithm according to [6, 7] can be expressed by the following steps:

Step 0. Initialize the parameters: a, b, Θ, c, d, e, α, δ, EP, l
where:

- a, b – fixed weights in the $F1$ layer;
- Θ - noise suppression parameter;
- c – fixed weight used in testing for reset;
- d – activation of winning $F2$ unit;

P. Angelov et al. (eds.), *Intelligent Systems'2014*,
Advances in Intelligent Systems and Computing 322, DOI: 10.1007/978-3-319-11313-5_55

- e – small parameter using preventing division by zero;
- S-matrix with input vectors ($s_1, s_2, \ldots, s_n$);
- n – number of the neurons on the input layer;
- m – number of the neurons in the second layer;
- α – learning rate;
- ρ – vigilance threshold;
- b_j – initial bottom-up weights;
- t_j– initial top-down weights;
- EP – number of epochs;
- l – number of loops in $F1$ layer.

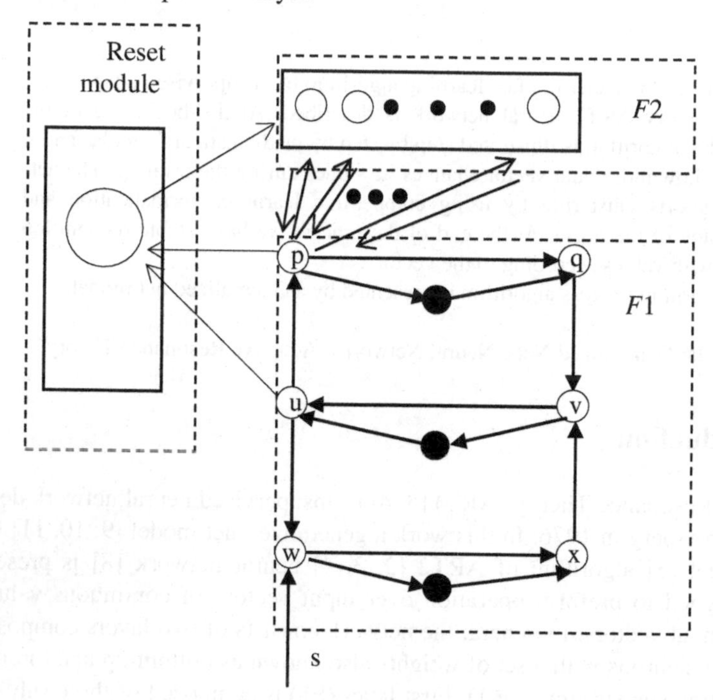

Fig. 1. ART 2 Neural network

Step 1. Do Steps 2-12 $N...EP$ times.
Step 2. For each input vector “s“ do steps 3-11.
Step 3. Update $F1$ unit activation “l=1”:

$$u_i = 0,\ x_i = \frac{w_i}{e + \|w\|}, q_i = 0$$

$$w_i = s_i,\ \ p_i = 0, v_i = f(x)_i$$

The activation function is

$$f(x)=\begin{cases} x \text{ if } x \geq \theta \\ 0 \text{ if } x < \theta \end{cases}$$

Update F_1 activations again "$l=2$"

$$u_i = \frac{v_i}{e+\|v\|},\quad x_i = \frac{w_i}{e+\|w\|},\quad q_i = \frac{p_i}{e+\|p\|}$$

$$w_i = s_i + a * u_i,\quad p_i = u_i,\quad v_i = f(x)_i + bf(q)_i$$

Step 4. Compute the signals to $F2$ units:

$$y_j = \sum_j b_j * p_i$$

Step 5 While reset is true, do steps 6-7

Step 6 Find $F2$ unit with the largest signal. (Define J such that $y_J \geq y_j$ for j=1, 2…m.)

Step 7 Check for reset:

$$u_i = \frac{v_i}{e+\|v\|},\quad r_i = \frac{u_i + cp_i}{e+\|u\|+c\|p\|},\quad p_i = u_i + dt_{Ji}$$

If

$$\|r\| < p - e$$

then y_J = -1 (inhibit J)

(Reset is true; repeat step 5)

If

$$\|r\| \geq p - e$$

Reset is false; proceed to step 8

Step 8 Do steps 9-11 N_IT times

(Perform the specified number of learning iterations.)

Step 9 Update weights for the winning unit

$$t_{Ji} = \alpha du_i + \{1 + \alpha d(d-1)\}t_{Ji}$$

$$b_{iJ} = \alpha du_i + \{1 + \alpha d(d-1)\}b_{iJ}$$

Step 10 Update $F1$ activations

$$u_i = \frac{v_i}{e+\|v\|},\quad x_i = \frac{w_i}{e+\|w\|},\quad q_i = \frac{p_i}{e+\|p\|}$$

$$w_i = s_i + a * u_i,\quad p_i = u_i,\quad v_i = f(x)_i + bf(q)_i$$

Step 11 Test stopping conditions for weight updates

$$N\_IT = t_{Ji} = b_{iJ}$$

Step 12 Test stopping condition for number of epochs.

2 GN-Model

Initially the following tokens enter the generalized net [1]:

In place L_1 – α token with initial characteristic

$$x_0^{\alpha} = "\langle s_1, s_2, s_3, ..., s_k \rangle",$$

where k is the number of input vectors.

In place L_2 – β token with initial characteristics

$$x_0^{\beta} = "\langle a, b, c, e, d, n, m, \rho, \alpha, \theta, EP, l \rangle",$$

Generalized net (Fig. 2) is presented by a set of transitions:

$$A = \{ Z_1, Z_2, Z_3, Z_4, Z_5, Z_6, Z_7 \},$$

where transitions describe the following processes:

- Z_1="Extraction of a vector from the matrix";
- Z_2="Calculation values of the vector";
- Z_3="Normalization of the data";
- Z_4="Calculation of resonance state";
- Z_5="Suppression of the noise";
- Z_6="Determination of a winning neuron";
- Z_7="Determination of the weights";

The GN-Model consists of eight transitions with the following descriptions:

$$Z_1 = \langle \{L_1, L_2, L_5, L_{22}\}, \{L_3, L_4, L_5\}, R_1 \vee (L_5, L_{22}, \wedge (L_1, L_2)) \rangle$$

$$R_1 = \begin{array}{c|ccc} & L_3 & L_4 & L_5 \\ \hline L_1 & False & False & True \\ L_2 & False & False & True \\ L_5 & W_{5,3} & W_{5,4} & True \\ L_{22} & False & False & True \end{array},$$

where:

$W_{5,3} = W_{5,4}$ = "There is an active signal from place L_{22}";

The token β that enters place L_5 from place L_2 unites with token α'(from place L_1) in one δ token with characteristic

$$x_{cu}^{\delta} =<" pr_1 x_0^{\beta}; pr_2 x_0^{\beta}; pr_3 x_0^{\beta}; pr_4 x_0^{\beta}; pr_5 x_0^{\beta}; pr_6 x_0^{\beta}; pr_7 x_0^{\beta}; pr_8 x_0^{\beta};$$

$$pr_9 x_0^{\beta}; pr_{10} x_0^{\beta}; pr_{11} x_0^{\beta}; pr_{12} x_0^{\beta}; pr_1 x_0^{\alpha}, pr_2 x_0^{\alpha}, ..., pr_k x_0^{\alpha} ">;$$

The token δ' enters place L_3 from place L_5 with characteristic

$$x_{cu}^{\delta'} =<" pr_1 x_0^{\beta}; pr_2 x_0^{\beta}; pr_3 x_0^{\beta}; pr_4 x_0^{\beta}; pr_5 x_0^{\beta}; pr_6 x_0^{\beta};$$

$$pr_7 x_0^{\beta}; pr_8 x_0^{\beta}; pr_9 x_0^{\beta}; pr_{10} x_0^{\beta}; pr_{12} x_{cu}^{\beta}; pr_i x_0^{\alpha} ">;$$

where $i \in [1:k]$ represent current number of iteration.

The token α' enters place L_4 from place L_5 with characteristic

$$x_{cu}^{\alpha'} =<" pr_i x_0^{\alpha} ">;$$

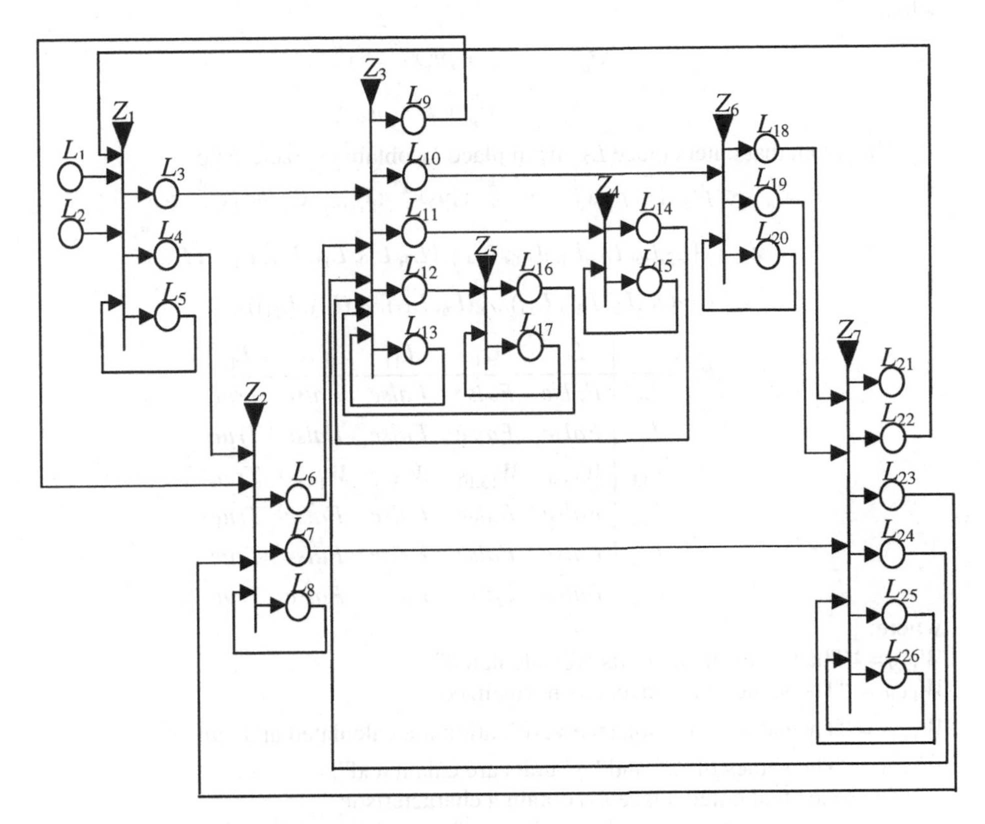

Fig. 2. Generalized net model based on fast learning algorithm

$$Z_2 = \langle \{L_4, L_8, L_9, L_{23}\}, \{L_6, L_7, L_8\}, R_2, \vee (L_8, \wedge (L_4, L_9, L_{23}))\rangle$$

$R_2 =$

	L_6	L_7	L_8
L_4	*False*	*False*	*True*
L_8	$W_{8,6}$	$W_{8,7}$	*True* ,
L_9	*False*	*False*	*True*
L_{23}	*False*	*False*	*True*

where:

$W_{8,6}$= "The values of p and w units are calculated and $pr_{12}x_{cu}^{\beta} = 1$";

$W_{8,7}$ = "The values of p units are calculated and $pr_{12}x_{cu}^{\beta} > 1$".

The token that enters place L_8 obtain a characteristic

$$x_{cu}^{\delta^{IV}} = <" pr_1 x_0^{\beta}; pr_2 x_0^{\beta}; pr_4 x_0^{\beta}; pr_5 x_0^{\beta}; pr_6 x_0^{\beta}; pr_7 x_0^{\beta}; pr_9 x_0^{\beta}; pr_{12} x_{cu}^{\beta}; pr_i x_0^{\alpha}; x_{cu(it)}^{\varepsilon}; \eta; \zeta ">;$$

The token that enters place L_6 from place L_8 obtain a characteristic

$$x_{cu}^{\gamma} = "\eta, \zeta",$$

where

$$x^{\zeta}_{cu(it)} = "x^{\alpha'}_{cu} + pr_1 x_0^{\delta^{IV}} * x^{\varepsilon'}_{cu}",$$

$$x^{\eta}_{cu(it)} = "x^{\varepsilon'}_{cu} + \kappa' * pr_5 x_0^{\delta''}$$

The token that enters place L_7 from place L_8 obtain characteristic

$$x^{\delta^V}_{cu} = <" pr_5 x_0^{\delta^{IV}}; pr_6 x_0^{\delta^{IV}}; pr_7 x_0^{\delta^{IV}}; pr_9 x_0^{\delta^{IV}}; x^{\eta}_{cu(it)}; x^{\alpha'}_{cu}; pr_{12} x^{\delta^{IV}}_{cu} ">.$$

$$Z_3 = \langle \{L_3, L_6, L_{13}, L_{14}, L_{16}, L_{24}\}, \{L_9, L_{10}, L_{11}, L_{12}, L_{13}\}, R_3,$$
$$\vee(\wedge(L_3, L_{16}, L_{24}), \wedge(L_6, L_{13}), \wedge(L_{13}, L_{14}))\rangle,$$

$R_3 =$

	L_9	L_{10}	L_{11}	L_{12}	L_{13}
L_3	*False*	*False*	*False*	*False*	*True*
L_6	*False*	*False*	*False*	*False*	*True*
L_{13}	$W_{13,9}$	$W_{13,10}$	$W_{13,11}$	$W_{13,12}$	*True*
L_{14}	*False*	*False*	*False*	*False*	*True*
L_{16}	*False*	*False*	*False*	*False*	*True*
L_{24}	*False*	*False*	*False*	*False*	*True*

where:
$W_{13,9}$ = "The values of "*u*" units are calculated";
$W_{13,10}$ = "The value of resonance is normalized";
$W_{13,11}$ = "The values for resonance verification are calculated and $pr_{12} x^{\beta}_{cu} > 1$; ";
$W_{13,12}$ = "The values of "*x*" and "*q*" units are calculated".

The token that enters place L_{13} obtain a characteristic

$$x^{\delta''}_{cu} = < pr_1 x_0^{\beta}; pr_2 x_0^{\beta}; pr_3 x_0^{\beta}; pr_4 x_0^{\beta}; pr_5 x_0^{\beta}; pr_6 x_0^{\beta};$$
$$pr_7 x_0^{\beta}; pr_8 x_0^{\beta}; pr_9 x_0^{\beta}; pr_{10} x_0^{\beta}; pr_{12} x^{\beta}_{cu}; pr_i x_0^{\alpha}; x^{\varepsilon'}_{cu}; \gamma; \sigma">,$$

where

$$x^{\varepsilon'}_{cu} = \frac{x^{\varepsilon}_{cu}}{pr_4 x_0^{\delta''} + \| x^{\varepsilon}_{cu} \|},$$

and

$$pr_{12} x^{\beta}_{cu} = cu + 1.$$

The token δ" enters place L_9 from place L_{13} with a characteristic

$$x^{\delta'''}_{cu} = <" pr_1 x_0^{\beta}; pr_2 x_0^{\beta}; pr_4 x_0^{\beta}; pr_5 x_0^{\beta}; pr_6 x_0^{\beta}; pr_7 x_0^{\beta}; pr_9 x_0^{\beta}; pr_{12} x^{\beta}_{cu}; x^{\varepsilon'}_{cu(it)} ">,$$

where
it – current iteration of *F*1 layer.

The token σ that enters place L_{13} obtain a characteristic

$$x^{\varepsilon'}_{cu(it)} = "it".$$

The token γ' enters place L_{11} from place L_{13} with characteristic

$$x^{\gamma'}_{cu} = " pr_1 x^{\gamma}_{cu}, \varepsilon', \varepsilon'', \eta', pr_3 x_0^{\delta''}, pr_4 x_0^{\delta''} "$$

where

$$x_{cu}^{\varepsilon''} = \sqrt{\sum_{i=1}^{n} x_1^2 + x_2^2 + ... x_n^2},$$

and

$$x_{cu}^{\eta'} = \sqrt{\sum_{i=1}^{n} x_1^2 + x_2^2 + ... x_n^2},$$

where x – different elements of the vector s_i.

The token μ'' enters place L_{10} from place L_{13} with a characteristic

$$x_{cu}^{\mu''} = "\mu', pr_8 x_0^{\delta''}, pr_4 x_0^{\delta''}",$$

where

$$x_{cu}^{\mu'} = \sqrt{\sum_{i=1}^{n} \mu_1^2 + \mu_2^2 + ... \mu_n^2},$$

The token γ'' enters place L_{12} from place L_{13} with a characteristic

$$x_{cu}^{\gamma''} = \eta'', \zeta', pr_2 x_0^{\delta''}, pr_6 x_0^{\delta''}, pr_{10} x_0^{\delta''},$$

where

$$x_{cu}^{\zeta'} = \frac{pr_2 x_{cu}^{\gamma}}{pr_4 x_0^{\delta''} + \| pr_2 x_{cu}^{\gamma} \|},$$

and

$$x_{cu}^{\eta''} = \frac{pr_1 x_{cu}^{\gamma}}{pr_4 x_0^{\delta''} + \| pr_1 x_{cu}^{\gamma} \|}.$$

$$Z_4 = \langle \{L_{11}, L_{15}\}, \{L_{14}, L_{15}\}, R_4, \wedge(L_{11}, L_{15}) \rangle$$

$R_4 =$

	L_{14}	L_{15}
L_{11}	*False*	*True*,
L_{15}	$W_{15,14}$	*True*

where:
$W_{15,14}$ = "Resonance state is calculated".

The token that enters place L_{15} from place L_{11} does not obtain a new characteristic.

The token that enters place L_{14} from place L_{15} obtains a characteristic

$$x_{cu}^{\mu} = \frac{pr_2 x_{cu}^{\gamma'} + pr_5 x_0^{\gamma'} * pr_1 x_{cu}^{\gamma'}}{pr_6 x_0^{\gamma'} + pr_3 x_{cu}^{\gamma'} + pr_5 x_0^{\gamma'} * pr_4 x_{cu}^{\gamma'}}$$

$$Z_5 = <\{L_{12}, L_{17}\}, \{L_{16}, L_{17}\}, R_5, \wedge(L_{12}, L_{17})>,$$

$R_5 =$

	L_{16}	L_{17}
L_{12}	*False*	*True*,
L_{17}	$W_{17,16}$	*True*

where:
$W_{17,16}$ = "Noise suppression is determined".

The token that enters place L_{17} obtain a characteristic

$$x_{cu}^{\gamma^{IV}} = "x_{cu}^{\gamma''}, x_{cu}^{\gamma'''}";$$

where

$$x_{cu}^{\gamma'''} = f(pr_2 x_{cu}^{\gamma''}) + pr_3 x_0^{\gamma''} * f(pr_1 x_{cu}^{\gamma''});$$

where

$$f(x) = \begin{cases} x & if \quad x \geq pr_5 x_0^{\gamma''} \\ 0 & if \quad x < pr_5 x_0^{\gamma''} \end{cases}$$

The token that enters place L_{16} from place L_{17} obtain a characteristic

$$x_{cu}^{\varepsilon} = "pr_2 x_{cu}^{\gamma^{IV}}".$$

$$Z_6 = \langle \{L_{10}, L_{20}\}, \{L_{18}, L_{19}, L_{20}\}, R_6, \wedge (L_{10}, L_{20}) \rangle,$$

$$R_6 = \begin{array}{c|ccc} & L_{18} & L_{19} & L_{20} \\ \hline L_{10} & False & False & True \\ L_{20} & W_{20,18} & W_{20,19} & True \end{array},$$

where:

$W_{20,18}$= "$pr_1 x_{cu}^{\mu''} \geq pr_8 x_0^{\beta} - pr_4 x_0^{\beta}$";

$W_{20,19}$ = „$\neg W_{20,18}$";

The token that enters place L_{20} obtain a characteristic

$$x_{cu}^{\mu'''} = pr_2 x_0^{\mu''}; pr_3 x_0^{\mu''}.$$

The token that enters place L_{18} from place L_{20} obtain a characteristic

$$x_{cu}^{\xi} = "true".$$

The token that enters place L_{19} from place L_{20} obtain a characteristic

$$x_{cu}^{o} = "false".$$

$$Z_7 = \langle \{L_7, L_{18}, L_{19}, L_{25}, L_{26}\}, \{L_{21}, L_{22}, L_{23}, L_{24}, L_{25}, L_{26}\}, R_7, \vee (\wedge L_{25}(L_7, L_{18}, L_{26})(\wedge \ L_{26}(L_7, L_{19}))\rangle$$

$$R_7 = \begin{array}{c|cccccc} & L_{21} & L_{22} & L_{23} & L_{24} & L_{25} & L_{26} \\ \hline L_7 & False & False & False & False & W_{7,25} & W_{7,26} \\ L_{18} & False & False & False & False & True & False \\ L_{19} & False & False & False & False & False & True \\ L_{25} & False & False & False & W_{25,24} & True & W_{25,26} \\ L_{26} & W_{26,21} & W_{26,22} & W_{26,23} & False & W_{26,25} & True \end{array},$$

where:

$W_{7,25}$ = "There is active σ token ";

$W_{7,26}$ = " $pr_7 x_{cu}^{\delta^V} = 2$ ";

$W_{25,24}$ = "The token σ is activated";

$W_{25,26}$ = "Update bottom-up and top-down weights";

$W_{26,21}$ = "$j>m$";

$W_{26,22}$ = "Request for the next input vector";
$W_{26,23}$ = "Top – down weights are determined";
$W_{26,25}$ = "Bottom-up and top-down weights are determined".

The token that enters place L_{26} obtain a characteristic

$$\delta^{VI} =" pr_5 x_0^{\delta^{IV}}, pr_6 x_0^{\delta^{IV}}, pr_7 x_0^{\delta^{IV}}, x_{cu}^{\eta}, x_{cu}^{\alpha'}, x_{cu}^{\theta}, x_{cu}^{\kappa}, x_{cu}^{\iota}, x_{cu}^{\kappa'}, x_{cu}^{\theta'}",$$

where

$$x_{cu}^{\theta} =" \frac{1}{(1 - pr_1 x_0^{\delta^{VI}}) * \sqrt{pr_2 x_0^{\delta^{VI}}}} "$$

$$x_{cu}^{\kappa} ="(t_j)",$$

$$x_{cu}^{\iota} =" \sum_{1}^{pr_3 x_0^{\delta^{VI}}} pr_4 x_{cu}^{\delta^{VI}} * x_{cu}^{\theta}",$$

$$x_{cu}^{\kappa'} ="\max(x_{cu}^{\iota}(\kappa)_j)",$$

$$x_{cu}^{\theta'} ="\max(x_{cu}^{\iota}(\theta)_j)",$$

where $j \in [1,2,...pr_3 x_0^{\delta^{VI}}]$ – represents current neuron with a maximum value.

The tokens $x_{cu}^{\theta}, x_{cu}^{\kappa}$ update their values from token λ that origins from place L_{25}.

The token o from place L_{18} enters place L_{26} and obtain a characteristic

$$x_{cu}^{\iota} =" next \max(j)",$$

The token that enters place L_{21} from place L_{26} obtain a characteristic

$$x_{cu}^{\alpha'} =" reject \quad set",$$

The token that enters place L_{22} from place L_{26} obtain a characteristic

$$x_{cu}^{\varsigma} = ("next \quad vector"),$$

The token κ' that enters place L_{23} from place L_{26} does not obtain a new characteristic.

The token that enters place L$_{25}$ from place L_{26} obtain a characteristic

$$x_{cu}^{\rho} =" x_{cu}^{\kappa'}, x_{cu}^{\theta'}",$$

The token that enters place L_{25} obtain a characteristic

$$x_{cu}^{\rho'} =" pr_1 x_{cu}^{\rho}, pr_2 x_{cu}^{\rho}, pr_5 x_0^{\delta^{IV}}, pr_9 x_0^{\delta^{IV}}, (x_{cu}^{\eta})_{it}, x_{cu}^{\sigma}, x_{cu}^{\kappa''}, x_{cu}^{\theta''}",$$

where

$$x_{cu}^{\kappa''} = pr_4 x_0^{\rho'} * pr_3 x_0^{\rho'} * pr_5 x_0^{\rho'} + (1 + pr_5 x_0^{\rho'} * pr_3 x_0^{\rho'} (pr_3 x_0^{\rho'} - 1) * pr_1 x_0^{\rho'})$$

$$x_{cu}^{\theta''} = pr_4 x_0^{\rho'} * pr_3 x_0^{\rho'} * pr_5 x_0^{\rho'} + (1 + pr_4 x_0^{\rho'} * pr_3 x_0^{\rho'} (pr_3 x_0^{\rho'} - 1) * pr_2 x_0^{\rho'}),$$

The token that enters place L_{24} from place L_{25} obtain a characteristic

$$x_{cu}^{\sigma} =" \kappa'' \neq \theta''",$$

The token that enters place L_{26} from place L_{25} obtain a characteristic

$$x_{cu}^{\lambda} =" \kappa'', \theta''".$$

3 Conclusion

The process of fast learning algorithm of ART2 neural network was presented. It was shown how each vector changes its values during the network training process. The first layer normalizes the vector values and suppresses the noise. Therefore when it gets to the second layer the values of the vector will be stable and consequently the clusters learned will also be stable. In that way the network can learn and recognize a lots of different problems, that cannot be solved by others methods.

Acknowledgment. The authors are grateful for the support provided by the project DFNI-I-01/0006 "Simulating the behavior of forest and field fires", funded by the National Science Fund, Bulgarian Ministry of Education, Youth and Science.

References

1. Atanassov, K.: Generalized nets. World Scientific, Singapore (1991)
2. Ben, K., van der Smagt, P.: An introduction to Neural Networks, 8th edn., ch. 6. University of Amsterdam (1996)
3. Carpenter, G., Grossberg, S.: The ART of adaptive pattern recognition by a self-organizing neural network. Computer 21(3), 77–88 (1988)
4. Grossberg, S., Carpenter, G.: ART2: Self-organization of stable category recognition codes of analog input patterns. Boston University, Center for Adaptive Systems 26(23), 4919–4930 (1987)
5. Grossberg, S.: Adaptive pattern classification and universal recoding. II. Feedback, expectation, olfacation, and illusions. Bioi. Cybemet. 23, 187–202 (1976)
6. Sivanandam, S., Deepa, S.: Introduction to Neural Networks using matlab 6.0
7. Fausett, L.: Fundamentals of Neural Networks; Architecture algorithms and applications (1993)
8. McCulloch, W., Pitts, W.: A logical calculus of the ideas immanent in nervous activity. Bulletin of Mathematical Biophysics (1943)
9. Petkov, T., Sotirov, S.: Generalized net model of the ART1 Neural Network. Annual of "Informatics" Section Union of Scientists in Bulgaria 6, 32–38 (2013)
10. Petkov, T., Sotirov, S.: Bio-inspired Artificial Intelligence: Generalized Net Model of the Cognitive and Neural Algorithm for Adaptive Resonance Theory 1. Int. J. Bioautomation 17(4), 207–216 (2013)
11. Petkov, T., Surchev, S., Sotirov, S.: Generalized net model of the forest-fire detection with ART2 neural network. In: International Workshop on Generalized Nets, IWGN 2013 (2013)

Intuitionistic Fuzzy Evaluation of the Behavior of Tokens in Generalized Nets

Velin Andonov[1] and Anthony Shannon[2]

[1] Institute of Biophysics and Biomedical Engineering,
Bulgarian Academy of Sciences,
Acad. G. Bonchev Str., bl. 105, 1113 Sofia, Bulgaria
velin_andonov@yahoo.com

[2] Faculty of Engineering & IT,
University of Technology,
Sydney, NSW 2007, Australia
tshannon38@gmail.com, Anthony.Shannon@uts.edu.au

Abstract. Two methods for evaluation of the behavior of tokens in Generalized Nets (GNs) are discussed. In the ordinary GNs the evaluations are based on determining whether the characteristics of the tokens meet a predefined criterion. It is shown that in Generalized Nets with Characteristics of the Places (GNCP) the evaluations of the tokens can also be obtained on the basis of the characteristics of the places. The evaluations are obtained in the form of Intuitionistic Fuzzy Pairs (IFPs). Modification of a given GN model is proposed which allows for the evaluation of tokens on the basis of the characteristics of the places to be obtained during the functioning of the net. The modified GN preserves the functioning and the results of the work of the given net.

Keywords: generalized nets, evaluation of tokens, intuitionistic fuzzy pairs.

1 Short Remark on Generalized Nets

Generalized Nets (GNs) (see [6,4,7]) are extentions of Petri Nets (see [10]). They were defined in 1982 and up to now more than 800 papers on the theory and applications of GNs have been published. Formally, every transition is described by a seven-tuple:

$$Z = \langle L', L'', t_1, t_2, r, M, \square \rangle,$$

where:

(a) L' and L'' are finite, non-empty sets of places (the transition's input and output places, respectively);

(b) t_1 is the current time-moment of the transition's firing;

(c) t_2 is the current value of the duration of its active state;

(d) r is the transition's *condition* determining which tokens will be transferred from the transition's inputs to its outputs. Parameter r has the form of an Index Matrix (IM);

P. Angelov et al. (eds.), *Intelligent Systems'2014*,
Advances in Intelligent Systems and Computing 322, DOI: 10.1007/978-3-319-11313-5_56

(e) M is an IM of the capacities of transition's arcs;

(f) $\square$ is called transition type and it is an object having a form similar to a Boolean expression. It may contain as variables the symbols that serve as labels for transition's input places, and it is an expression constructed of variables and the Boolean connectives.

The ordered four-tuple

$$E = \langle\langle A, \pi_A, \pi_L, c, f, \theta_1, \theta_2\rangle, \langle K, \pi_K, \theta_K\rangle, \langle T, t^0, t^*\rangle, \langle X, \Phi, b\rangle\rangle$$

is called a *Generalized Net* if:

(a) A is a set of transitions (see above);

(b) π_A is a function giving the priorities of the transitions, i.e., $\pi_A : A \to \mathcal{N}$;

(c) π_L is a function giving the priorities of the places, i.e., $\pi_L : L \to \mathcal{N}$, where

$$L = pr_1 A \cup pr_2 A$$

and obviously, L is the set of all GN-places;

(d) c is a function giving the capacities of the places, i.e., $c : L \to \mathcal{N}$;

(e) f is a function that calculates the truth values of the predicates of the transition's conditions;

(f) θ_1 is a function giving the next time-moment, for which a given transition Z can be activated, i.e., $\theta_1(t) = t'$, where $pr_3 Z = t, t' \in [T, T+t^*]$ and $t \leq t'$; the value of this function is calculated at the moment when the transition terminates its functioning;

(g) θ_2 is a function giving the duration of the active state of a given transition Z, i.e., $\theta_2(t) = t'$, where $pr_4 Z = t \in [T, T+t^*]$ and $t' \geq 0$; the value of this function is calculated at the moment when the transition starts functioning;

(h) K is the set of the GN's tokens. In some cases, it is convenient to consider this set in the form

$$K = \bigcup_{l \in Q^I} K_l,$$

where K_l is the set of tokens which enter the net from place l, and Q^I is the set of all input places of the net;

(i) π_K is a function giving the priorities of the tokens, i.e., $\pi_K : K \to \mathcal{N}$;

(j) θ_K is a function giving the time-moment when a given token can enter the net, i.e., $\theta_K(\alpha) = t$, where $\alpha \in K$ and $t \in [T, T+t^*]$;

(k) T is the time-moment when the GN starts functioning; this moment is determined with respect to a fixed (global) time-scale;

(l) t^0 is an elementary time-step, related to the fixed (global) time-scale;

(m) t^* is the duration of the GN functioning;

(n) X is a function which assigns initial characteristics to every token when it enters input place of the net;

(o) Φ is a characteristic function that assigns new characteristics to every token when it makes a transfer from an input to an output place of a given transition;

(p) b is a function giving the maximum number of characteristics a given token can receive, i.e., $b : K \to N$.

For the algorithms of transition and GN functioning the reader can refer to [7,9].

Many extensions of GNs have been defined through the years and all are proved to be conservative extensions of the standard GNs, i.e. the functioning and the results of the work of every net from the extensions can be represented by a standard GN. Here we shall briefly mention one of the most recent GN extensions which is relevant to the present paper — Generalized Nets with Characteristics of the Places (GNCP). The concept of GNCP is defined in [1] and the algorithm for transition functioning is proposed in [3]. In GNCP the places can also receive characteristics during the functioning of the net when tokens enter them. A GNCP G is the ordered four-tuple

$$G = \langle\langle A, \pi_A, \pi_L, c, f, \theta_1, \theta_2\rangle, \langle K, \pi_K, \theta_K\rangle, \langle T, t^0, t^*\rangle, \langle X, Y, \Phi, \Psi, b\rangle\rangle , \quad (1)$$

where all components with the exception of the characteristic functions Y and Ψ have the same meaning as in the standard GNs. Here Y assigns initial characteristics to some of the the places and Ψ assigns characteristics to some places when tokens enter them. In the graphical representation of the net, the places which can receive characteristics are marked with two concentric circles.

Operations and relations over transitions and GNs are defined in [6] and many assertions about them are proved. Of these, related to the present paper is the relation $\subset_*$. Let

$$E_i = \langle\langle A_i, \pi_A^i, \pi_L^i, c^i, f^i, \theta_1^i, \theta_2^i\rangle, \langle K_i, \pi_K^i, \theta_K^i\rangle, \langle T_i, t_i^0, t_i^*\rangle, \langle X_i, Y_i, \Phi_i, \Psi_i, b_i\rangle\rangle .$$

Definition 1. *For every two GNCP E_1 and E_2:*
$E_1 \subset_* E_2$ *iff* $(\forall Z_1 \in A_1)(\exists Z_2 \in A_2)(Z_1 \subset Z_2)\&(\pi_A^1 = \pi_A^2|E_1)\&(\pi_L^1 = \pi_L^2|E_1)\&(c^1 = c^2|\overline{E_1})\&(f^1 = f^2|E_1)\&(\theta_1^1 = \theta_1^2|E_1)\&(\theta_2^1 = \theta_2^2|E_1)\&(K_1 \subset K_2)\&(\pi_K^1 = \pi_K^2|E_1)\&(\theta_K^2 = \theta_K^2|E_1)\&(T_2 \le T_1 \le T_1 + t_1^* \le T_2 + t_2^*)\&(t_1^0 = t_2^0)\&(X_1 = X_2|E_1)\&(Y_1 = Y_2|E_1)\&(\Phi_1 = \Phi_2|E_1)\&(\Psi_1 = \Psi_2|E_1)\&(b_1 \le b_2|E_1)$.

2 Evaluation of Tokens Based on Their Characteristics

The results of the work of the net are stored in the form of characteristics of the tokens. For arbitrary token α by $\overline{x}^\alpha = \langle x_0^\alpha, x_1^\alpha, ..., x_{fin}^\alpha\rangle$ we denote the vector of all characteristics obtained by the token during its transfer in the net. A way to evaluate the tokens with respect to a given criterion is proposed in [2]. In the simplest case when all characteristics of the tokens belong to one type, for instance they are all real number, let the criterion be such that all characteristics less than a given threshold T are "bad" (i.e. they do not meet the criterion) and all characteristics greater or equal to T are "good" (i.e. they meet the criterion). Using the indicator function

$$I^\alpha(x_i^\alpha) = \begin{cases} 0, & \text{if} x_i^\alpha < T \\ 1, & \text{if} x_i^\alpha \ge T \end{cases} \quad (2)$$

an evaluation of the token α with respect to the criterion can be obtained through the function

$$\mu_\alpha = \frac{\sum_{i=0}^{fin} I^\alpha(x_i^\alpha)}{fin+1}. \quad (3)$$

In this special case, μ_α is a fuzzy membership function (see [11]).

In the definition of GN no restriction is imposed on the type of characteristics that a token can receive, i.e. we may have tokens all characteristics of which belong to one type, and tokens the characteristics of which are of different types. For example, if the vector of characteristics of the token is $\langle 10, 12, \text{"}red\text{"}, 23, -10\rangle$ we cannot determine whether the characteristic *"red"* satisfies a criterion of evaluation with regard to numerical characteristics. Also, the criterion for evaluation of the token may be such that we have two disjoint sets Ξ^α and Δ^α the first of which is the set of all possible "bad" characteristics, i.e. they do not satisfy the criterion, and the second is set of all possible "good" characteristics, i.e. they satisfy the criterion. Using the indicator functions of these sets:

$$I^\alpha_\Delta(x_i^\alpha) = \begin{cases} 0, & \text{if } x_i^\alpha \notin \Delta^\alpha \\ 1, & \text{if } x_i^\alpha \in \Delta^\alpha \end{cases}, \quad (4)$$

$$I^\alpha_\Xi(x_i^\alpha) = \begin{cases} 0, & \text{if} x_i^\alpha \notin \Xi^\alpha \\ 1, & \text{if} x_i^\alpha \in \Xi^\alpha \end{cases}, \quad (5)$$

the token α can be evaluated with the pair $\langle \mu_\alpha, \nu_\alpha\rangle$ where

$$\mu_\alpha = \frac{\sum_{i=0}^{fin} I^\alpha_\Delta(x_i^\alpha)}{fin+1}, \quad (6)$$

$$\nu_\alpha = \frac{\sum_{i=0}^{fin} I^\alpha_\Xi(x_i^\alpha)}{fin+1}. \quad (7)$$

Obviously, $\mu_\alpha, \nu_\alpha \in [0,1]$ and $\mu_\alpha + \nu_\alpha \leq 1$. In this case $\langle \mu_\alpha, \nu_\alpha\rangle$ is an intuitionistic fuzzy pair (see [8]). The number $\pi_\alpha = 1 - \mu_\alpha - \nu_\alpha$ is the degree of indeterminacy in intuitionistic fuzzy sense. As it is pointed out in [2], the sources of indeterminacy of the evaluation are two:

- indeterminacy due to the criterion;
- indeterminacy due to the GN model.

In formulae (3), (6) and (7) the time moments when the characteristics are assigned to the token are not taken into account. However, depending on the model, we may want the more recently obtained characteristics to have greater impact on the evaluation. This can be achieved through the introduction of weights. Let the vector with characteristics of the token α obtained up to the current time moment be $\langle x_0^\alpha, x_1^\alpha, ..., x_k^\alpha\rangle$. The evaluation with weights is given by

$$\mu_\alpha^w = \frac{\sum_{i=0}^{k} w_i^\alpha I_\Delta^\alpha(x_i^\alpha)}{k+1}, \tag{8}$$

$$\nu_\alpha^w = \frac{\sum_{i=0}^{k} w_i^\alpha I_\Xi^\alpha(x_i^\alpha)}{k+1}, \tag{9}$$

where w_i^α is the weight of the i-th characteristic of the token and $w_i^\alpha \in [0,1]$ for $i = 0, 1, ..., k$. The additional condition which the weights must satisfy is $w_i^\alpha < w_{i+1}^\alpha$. The use of weights for the evaluation allows to detect tendencies in the behavior of the tokens. As an example, a simple way to choose the weights is to use the vector $\langle \frac{1}{k+1}, \frac{2}{k+1}, ..., \frac{k}{k+1}, 1 \rangle$. In this case (8) and (9) become

$$\mu_\alpha^w = \frac{\sum_{i=0}^{k} \frac{i+1}{k+1} I_\Delta^\alpha(x_i^\alpha)}{k+1}, \tag{10}$$

$$\nu_\alpha^w = \frac{\sum_{i=0}^{k} \frac{i+1}{k+1} I_\Xi^\alpha(x_i^\alpha)}{k+1}. \tag{11}$$

In the case of Intuitionistic Fuzzy Generalized Net of type 1 (IFGN1) (see [5]) the function f evaluates the truth values of predicates in the form of IFP $\langle \mu(r_{i,j}), \nu(r_{i,j}) \rangle$, where $r_{i,j}$ is the predicate corresponding to the i-th input and j-th output place of the transition. The tokens in IFGN1 are transferred from input to output place if one of the following conditions holds:

C1 $\mu(r_{i,j}) = 1, \nu(r_{i,j}) = 0$ (the case of ordinary GN)

C2 $\mu(r_{i,j}) > \frac{1}{2}$ $(> \nu(r_{i,j}))$

C3 $\mu(r_{i,j}) \geq \frac{1}{2}$ $(\geq \nu(r_{i,j}))$

C4 $\mu(r_{i,j}) > \nu(r_{i,j})$

C5 $\mu(r_{i,j}) \geq \nu(r_{i,j})$

C6 $\mu(r_{i,j}) > 0$

C7 $\nu(r_{i,j}) < 1$, i.e. at least $\pi(r_{i,j}) > 0$, where $\pi(r_{i,j}) = 1 - \mu(r_{i,j}) - \nu(r_{i,j})$ is the degree of uncertainty (indeterminancy).

The condition for transfer of the tokens is chosen for every transition before the firing of the net. To evaluate the tokens in IFGN1 we can use as weights the degrees of truth of the predicates corresponding to the characteristics.

The tokens in GNs do not keep all obtained characteristics during their stay in the net. A token α keeps only the last $b(\alpha)$ characteristics. In [2] an extension of a given GN model is proposed which allows us to preserve all characteristics of the tokens that are object of evaluation and also to obtain the evaluations during the functioning of the net. On the basis of this extension, in the next section we propose a modification of a given GNCP aimed at evaluation of tokens on the basis of the characteristics of the places.

3 Evaluation of Tokens in GNCP Based on the Characteristics of the Places

The characteristics assigned to the places in GNCP can be used to evaluate the tokens. This is so because the characteristics are assigned to the places as a result of the transfer of tokens and they provide indirect information about the behaviour of the token. Let $\langle l_0, l_1, ..., l_k\rangle$ be the places through which token α has consecutively passed and as a result characteristic has been assigned to them. Let $\langle \psi^{\alpha}_{l_0}, \psi^{\alpha}_{l_1}, ..., \psi^{\alpha}_{l_k}\rangle$ be the characteristics obtained by the places upon the entering of the token. Let Δ^{l_j} be the set of all possible "good" characteristics, i.e. that satisfy the criterion for evaluation and Ξ^{l_j} be the set of all "bad" characteristics, i.e. those that do not satisfy the criterion for evaluation of place l_j. Using the indicator functions of these two sets:

$$I^{l_j}_{\Delta}(\psi^{\alpha}_{l_j}) = \begin{cases} 0, & \text{if } \psi^{\alpha}_{l_j} \notin \Delta^{l_j} \\ 1, & \text{if } \psi^{\alpha}_{l_j} \in \Delta^{l_j} \end{cases}, \quad (12)$$

$$I^{l_j}_{\Xi}(\psi^{\alpha}_{l_j}) = \begin{cases} 0, & \text{if} \psi^{\alpha}_{l_j} \notin \Xi^{l_j} \\ 1, & \text{if} \psi^{\alpha}_{l_j} \in \Xi^{l_j} \end{cases}, \quad (13)$$

we evaluate the token with the pair $\langle \mu^l_{\alpha}, \nu^l_{\alpha}\rangle$ where

$$\mu^l_{\alpha} = \frac{\sum_{j=0}^{k} I^{l_j}_{\Delta}(\psi^{\alpha}_{l_j})}{k+1}, \quad (14)$$

$$\nu^l_{\alpha} = \frac{\sum_{j=0}^{k} I^{l_j}_{\Xi}(\psi^{\alpha}_{l_j})}{k+1}. \quad (15)$$

In analogy to the evaluation of the tokens on the basis of their characteristics, it is reasonable to use weights for the characteristics of the places. If $\langle w_0, w_1, ..., w_k\rangle$ is the vector of weights where $w_j \in [0,1]$ for $j = 0, 1, ..., k$ then

$$\mu^{l,w}_{\alpha} = \frac{\sum_{j=0}^{k} w_j I^{l_j}_{\Delta}(\psi^{\alpha}_{l_j})}{k+1}, \quad (16)$$

$$\nu^{l,w}_{\alpha} = \frac{\sum_{j=0}^{k} w_j I^{l_j}_{\Xi}(\psi^{\alpha}_{l_j})}{k+1}. \quad (17)$$

To give more significance to the newly obtained characteristics the weights can be chosen so that $w_j < w_{j+1}$ for $j = 0, 1, ..., k-1$. For example, $w_j = \frac{1+j}{k+1}$ for $j = 0, 1, ..., k$. The weights can also be given for every place before the start of the functioning of the net to give more significance to some of the places with respect to other.

With some minor changes the extended GN model proposed in [2] can be used to collect the characteristics assigned to the places as a result of the transfer of the tokens which are object of evaluation and also to obtain the evaluations of the tokens during the functioning of the net. Let

$$E = \langle\langle A, \pi_A, \pi_L, c, f, \theta_1, \theta_2\rangle, \langle K, \pi_K, \theta_K\rangle, \langle T, t^0, t^*\rangle, \langle X, Y, \Phi, \Psi, b\rangle\rangle$$

be a GNCP with graphical representation in Fig. 1.

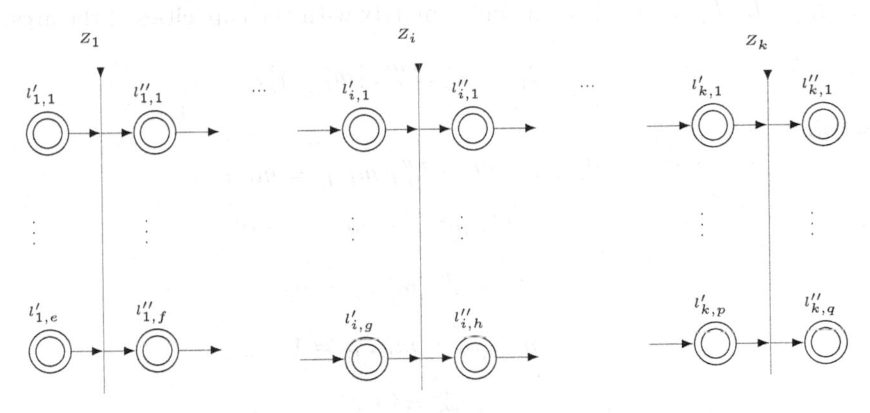

Fig. 1. Graphical representation of the GNCP E

The two concentric circles denote that the places can obtain characteristics. We consider that $|A| = k$. We shall construct a new GNCP E^* based on E which allows us to collect the characteristics of the places and obtain evaluation of tokens during the functioning of the net. In comparison to the modification proposed in [2], here we shall modify only those transitions of E for which at least one of the output places can obtain characteristics. To every transition $Z_i = \langle L'_i, L''_i, t^i_1, t^i_2, r_i, M_i, \Box_i\rangle$ where $1 \leq i \leq k$ for which at least one of the output places can receive characteristics we add two more places l^*_i and l^{**}_i the first of which is output for the transition while the second is both input and output. The so constructed new transition we denote by $Z^*_i = \langle L'^*_i, L''^*_i, t^i_1, t^i_2, r^*_i, M^*_i, \Box^*_i\rangle$, where

$$L'^*_i = L'_i \cup \{l^{**}_i\},$$
$$L''^*_i = L''_i \cup \{l^*_i, l^{**}_i\}.$$

If $r_i = [L'_i, L''_i, \{r_{l_s,l_t}\}]$ is the index matrix of transition's conditions, then

$$r^*_i = [L'^*_i, L''^*_i, \{r^*_{l_s,l_t}\}],$$

where

$$(\forall l_s \in L'_i)(\forall l_t \in L''_i)(r^*_{l_s,l_t} = r_{l_s,l_t});$$

$$(\forall l_s \in L'_i)(r^*_{l_s,l^*_i} = r^*_{l_s,l^{**}_i} = \text{“}false\text{”});$$

$$(\forall l_t \in L''_i)(r^*_{l^{**}_i,l_t} = \text{“}false\text{”});$$

$r^*_{l^{**}_i,l^{**}_i}$ = *“at least one token which is object of evaluation has been tranferred to output place of the transition and a characteristic has been assigned to the output place as a result of the transfer”*;

$$r^*_{l^{**}_i,l^*_i} = r^*_{l^{**}_i,l^{**}_i}.$$

If $M_i = [L'_i, L''_i, \{m_{l_s,l_t}\}]$ is the index matrix with the capacities of the arcs, then

$$M^*_i = [L'^*_i, L''^*_i, \{m^*_{l_s,l_t}\}],$$

where

$$(\forall l_s \in L'_i)(\forall l_t \in L''_i)(m^*_{l_s,l_t} = m_{l_s,l_t});$$

$$(\forall l_s \in L'_i)(m^*_{l_s,l^*_i} = m^*_{l_s,l^{**}_i} = 0);$$

$$(\forall l_t \in L''_i)(m^*_{l^{**}_i,l_t} = 0);$$

$$m^*_{l^{**}_i,l^{**}_i} = m^*_{l^{**}_i,l^*_i} = 1.$$

$$\Box^*_i = \Box_i.$$

Let A' be the set of all transitions obtained from transitions of E through the procedure described above and $A'' \subset A$ be the set of those transitions of E for which the above modification is not applied because their output places do not obtain characteristics. We shall denote by S the set of the indices of the modified transitions.

The modified net E^* will also have a transition Z_{ev} with no analogue in E.

$$Z_{ev} = \langle L'_{ev}, L''_{ev}, t_1^{ev}, t_2^{ev}, r_{ev}, M_{ev}, \Box_{ev}\rangle,$$

where

$$L'_{ev} = \{l^*_n | Z^*_n \in A'\} \cup \{l_{ev}, l_{cr}\},$$

$$L''_{ev} = \{l_{ev}, l_{cr}\},$$

$$r_{ev} = \begin{array}{c|cc} & l_{ev} & l_{cr} \\ \hline l^*_1 & true & false \\ \vdots & \vdots & \vdots \\ l^*_k & true & false \\ l_{ev} & true & false \\ l_{cr} & false & true \end{array},$$

$$M_{ev} = \begin{array}{c|cc} & l_{ev} & l_{cr} \\ \hline l_1^* & 1 & 0 \\ \vdots & \vdots & \vdots \\ l_k^* & 1 & 0 \\ l_{ev} & 1 & 0 \\ l_{cr} & 0 & 1 \end{array},$$

$$\square_{ev} = \wedge(\vee(l_1^*, l_2^*, ..., l_k^*), l_{cr}).$$

The graphical representation of the extended GNCP is in Fig. 2.

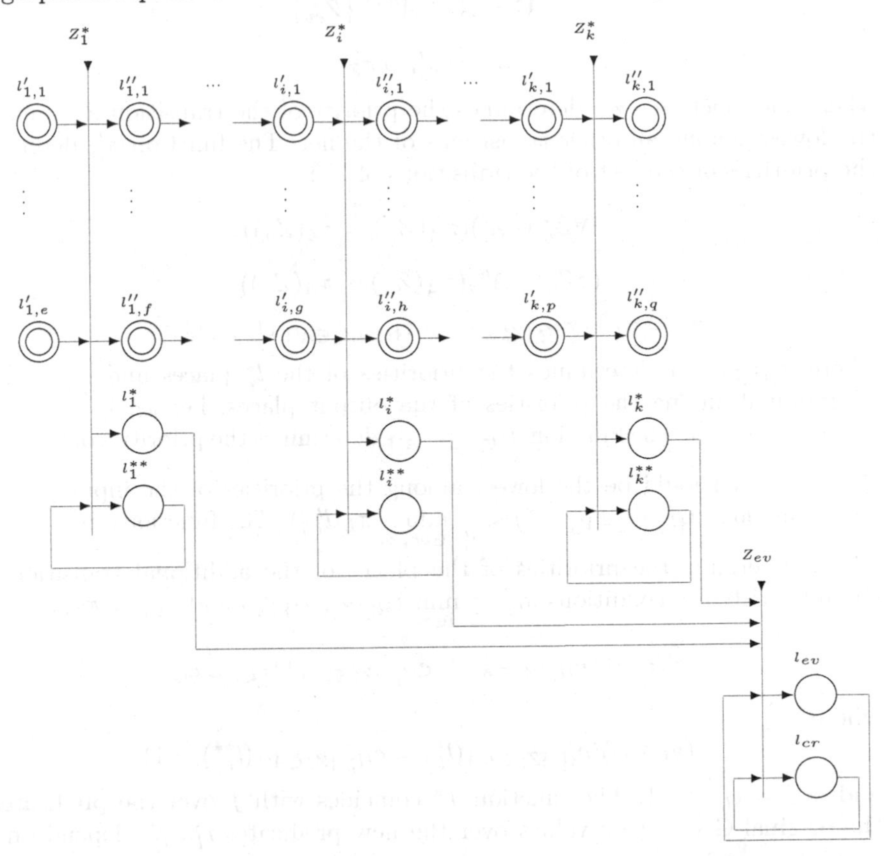

Fig. 2. Graphical representation of the extended GNCP

In each of the additional places l_i^{**} a token α_i^* stays in the initial time moment without characteristic. These tokens shall be used to store the characteristics obtained by the places when a token which is object of evaluation enters them. In place l_{cr} a token β stays in the initial time moment with characteristic a list with the tokens which are to be evaluated and the criteria for each place which can obtain characteristics in the form

"tokens to be evaluated, $\langle l_{i_1}$, criterion for $l_{i_1}\rangle, ..., \langle l_{i_j}$, criterion for $l_{i_j}\rangle$" .

Token α^* stays in place l_{ev} in the initial time moment without initial characteristic. This token will be used to preserve the characteristics of the places related to the tokens which are object of evaluation. Also, the evaluations of the tokens in the form of IFPs will be obtained as characteristics of α^*. Let

$$E^* = \langle\langle A^*, \pi_A^*, \pi_L^*, c^*, f^*, \theta_1^*, \theta_2^*\rangle, \langle K^*, \pi_K^*, \theta_K^*, \rangle, \langle T, t^0, t^{**}\rangle, \langle X^*, Y, \Phi^*, \Psi, b^*\rangle\rangle,$$

where the set of transitions of E^* consists of all modified transitions of E, all transitions of E which remain the same and the transition Z_{ev}, i.e.

$$A^* = A' \cup A'' \cup \{Z_{ev}\}.$$

$$\pi_A^* = \pi_A' \cup \pi_{Z_{ev}},$$

where the function $\pi_{Z_{ev}}$ determines the priority of the transition Z_{ev} and it is the lowest among all other transitions of the net. The function π_A' determines the priorities of the rest of the transitions of E^*:

$$(\forall Z_n^* \in A')(\pi_A'(Z_n^*) = \pi_A(Z_n)),$$

$$(\forall Z_n \in A'')(\pi_A'(Z_n) = \pi_A(Z_n)).$$

$$\pi_L^* = \pi_L \cup \pi_{\{l_n^* | Z_n^* \in A'\}} \cup \pi_{\{l_n^{**} | Z_n^* \in A'\}} \cup \pi_{l_{ev}} \cup \pi_{l_{cr}},$$

where $\pi_{\{l_n^* | Z_n^* \in A'\}}$ determines the priorities of the l_i^* places and they should be minimal among the priorities of the output places, i.e. $\pi_{\{l_n^* | Z_n^* \in A'\}}(l_i^*) < \min_{l''_{i,j} \in pr_2 Z_i} \pi_L(l''_{i,j})$. The function $\pi_{\{l_n^{**} | Z_n^* \in A'\}}$ determines the priorities of the places l_i^{**} and they should be the lowest among the priorities of the input places of the transition: $\pi_{\{l_n^{**} | Z_n^* \in A'\}}(l_i^{**}) < \min_{l'_{i,j} \in pr_1 Z_i} \pi_L(l'_{i,j})$. The functions $\pi_{l_{ev}}$ and $\pi_{l_{cr}}$ which determine the priorities of the places of the additional transition Z_{ev} should satisfy the conditions $\pi_{l_{ev}} < \min_{i \in S} \pi_{\{l_n^* | Z_n^* \in A'\}}(l_i^*)$ and $\pi_{l_{cr}} < \pi_{l_{ev}}$.

$$c^* = c \cup c_{\{l_n^* | Z_n^* \in A'\}} \cup c_{\{l_n^{**} | Z_n^* \in A'\}} \cup c_{l_{cr}} \cup c_{l_{ev}},$$

where

$$(\forall i \in S)(c_{\{l_n^* | Z_n^* \in A'\}}(l_i^*) = c_{\{l_n^{**} | Z_n^* \in A'\}}(l_i^{**}) = 1)$$

and $c_{l_{cr}} = c_{l_{ev}} = 1$. The function f^* coincides with f over the predicates of the original GN and its values over the new predicates $r^*_{l_i^{**}, l_i^{**}}$ depend on the concrete model.

The moment of time when E^* starts functioning and the elementary time step are the same as in E. The duration of the net functioning is exactly one time step longer than that of E:

$$t^{**} = t^* + 1.$$

The reason for this is that the evaluation of the tokens is obtained on the next time step after their transfer. The extra time step allows us to include in the evaluation the characteristics obtained at time $T + t^0.t^*$.

$$K^* = K \cup \{\alpha_n^* | n \in S\} \cup \{\alpha^*, \beta\},$$

$$\pi_K^* = \pi_K \cup \pi_{\{\alpha_n^*|n \in S\}} \cup \pi_{\alpha^*} \cup \pi_\beta ,$$

where the function $\pi_{\{\alpha_n^*|n \in S\}}$ determines the priorities of the tokens in places l_i^{**} for $i \in S$. These priorities have no effect on the functioning of E^*. One way to assign priorities to the α_i^* tokens is by using the priorities of the transitions, i.e. $\pi_{\{\alpha_n^*|n \in S\}}(\alpha_i^*) = \pi_A(Z_i)$, $\forall i \in S$. The priorities of the tokens α^* and β also do not have effect on the functioning of the net.

When the truth value of the predicate $r^*_{l_i^{**},l_i^*}$ is "true" the α_i^* in place l_i^{**} splits into two tokens — the original which remains in place l_i^{**} and a new one $\alpha_i^{*'}$ which enters place l_i^* where it obtains the characteristics of the output places (one or more) to which the tokens which are object of evaluation have been transferred.

$$\Phi^* = \Phi \cup \Phi_{\{l_n^*|Z_n^* \in A'\}} \cup \Phi_{\{l_n^{**}|Z_n^* \in A'\}} \cup \Phi_{l_{ev}} \cup \Phi_{l_{cr}} ,$$

where

$$\Phi_{\{l_n^{**}|Z_n^* \in A'\}}(\alpha_i^*) = \text{"}\emptyset\text{"},$$

$$\Phi_{\{l_n^*|Z_n^* \in A'\}}(\alpha_i^{*'}) = \text{"}\langle\langle\alpha_p, \Psi_{l''_{i,t_1}}\rangle, \langle\alpha_q, \Psi_{l''_{i,t_2}}\rangle, ..., \langle\alpha_t, \Psi_{l''_{i,t_s}}\rangle\rangle\text{"} .$$

Here by $\langle\alpha_p, \Psi_{l''_{i,t_1}}\rangle$,$\langle\alpha_q, \Phi_{l''_{i,t_2}}(\alpha_q)\rangle$,...,$\langle\alpha_t, \Psi_{l''_{i,t_s}}\rangle$ we denote the tokens which are being evaluated and which have been transferred to the output places of the transition and the characteristics assigned to the places. The tokens from places l_i^* enter place l_{ev} where they merge into a token α^* with characteristic a list of the tokens which are object of evaluation together with their evaluation according to the criteria kept as characteristic of token β which loops in l_{cr}:

$$\Phi_{l_{ev}}(\alpha^*) = \text{"}\langle\alpha_1, \overline{\psi}_{0,cu}^{\alpha_1}, \langle\mu_{\alpha_1}^l, \nu_{\alpha_1}^l\rangle\rangle, ..., \langle\alpha_j, \overline{\psi}_{0,cu}^{\alpha_j}, \langle\mu_{\alpha_j}^l, \nu_{\alpha_j}^l\rangle\rangle\text{"} ,$$

where $\langle\mu_{\alpha_i}^l, \nu_{\alpha_i}^l\rangle$ is the IFP for the token α_i for $i = 1, 2, ..., j$ and j is the number of the tokens which are object of evaluation. By $\overline{\psi}_{0,cu}^{\alpha_i}$ for $i = 1, 2, ..., j$ we denote the vector of all characteristics obtained by the places into which token α_i has been transferred up to the current time moment.

$$b^* = b \cup b_{\{\alpha_n^*|n \in S\}} \cup b_{\alpha^*} \cup b_\beta ,$$

where $b_{\{\alpha_n^*|n \in S\}}$ determines the number of characteristics that the α_i^* tokens can keep and for $1 \leq i \leq k$ we have $b_{\{\alpha_n^*|n \in S\}}(\alpha_i^*) = 1$. Functions b_{α^*} and b_β determine the number of characteristics that the tokens α^* and β respectively can keep: $b_{\alpha^*}(\alpha^*) = \infty$ and $b_\beta(\beta) = 1$.

The following theorem specifies the relation between the given GNCP E and the modified E^*.

Theorem 1. $E \subset_* E^*$.

4 Conclusion and Future Work

We have discussed two ways for evaluation of tokens — one based on their characteristics in ordinary GNs, and one based on the characteristics of the places in GNCP. Based on the modification suggested in [2] an extension of a given net is proposed that allows for the evaluation of tokens through the characteristics obtained by the places in GNCP. The extended GNCP preserves the functioning and the results of work of the given net. In future, we shall discuss the possibility to use the proposed modifications for the evaluation of the behavior of places, transitions and the net. Also, parallel evaluation of tokens and places on the basis of the characteristics of the tokens and on the basis of the characteristics of the places would allow for easier detection of problems related to the functioning of the net.

References

1. Andonov, V., Atanassov, K.: Generalized nets with characteristics of the places. Compt. Rend. Acad. Bulg. Sci. 66(12), 1673–1680 (2013)
2. Andonov, V.: Intuitionistic fuzzy evaluation of tokens in generalized nets based on their characteristics. Notes on Intuitionistic Fuzzy Sets 20(2), 109–118 (2014)
3. Andonov, V.: Reduced generalized nets with characteristics of the places. ITA 2013 – ITHEA ISS Joint International Events on Informatics, Winter Session, December 18-19, Sofia, Bulgaria (2013) (in press)
4. Atanassov, K.: Applications of Generalized Nets. World Scientific Publ. Co., Singapore (1993)
5. Atanassov, K., Dimitrov, D., Atanassova, V.: Algorithms for Tokens Transfer in Diferent Types of Intuitionistic Fuzzy Generalized Nets. Cybernetics and Information Technologies 10(4), 22–35 (2010)
6. Atanassov, K.: Generalized Nets. World Scientific, Singapore (1991)
7. Atanassov, K.: On Generalized Nets Theory. Prof. M. Drinov Academic Publ. House, Sofia (2007)
8. Atanassov, K., Szmidt, E., Kacprzyk, J.: On intuitionistic fuzzy pairs. Notes on Intuitionistic Fuzzy Sets 19(3), 1–13 (2013)
9. Dimitrov, D.: Optimized Algorithm for Tokens Transfer in Generalized Nets. In: Recent Advances in Fuzzy Sets, Intuitionistic Fuzzy Sets, Generalized Nets and Related Topics, vol. 1, pp. 63–68. SRI PAS, Warsaw (2010)
10. Murata, T.: Petri Nets: Properties, Analysis and Applications. Proceedings of the IEEE 77(4), 541–580 (1989)
11. Zadeh, L.A.: Fuzzy sets. Information and Control 8, 338–353 (1965)

Generalized Net Description of Essential Attributes Generator in SEAA Method

Maciej Krawczak[1,2] and Grażyna Szkatuła[1]

[1] Systems Research Institute, Polish Academy of Sciences, Warsaw, Poland
[2] Warsaw School of Information Technology,
Newelska 6, Warsaw, Poland
{krawczak,szkatulg}@ibspan.waw.pl

Abstract. The paper considers the generalized net description of essential attributes generator which is one of the main part of SEAA method developed for dimensionality reduction of time series. SEAA method (*Symbolic Essential Attributes Approximation*) (Krawczak and Szkatuła, 2014) was developed to reduce the dimensionality of multidimensional time series by generating a new nominal representation of the original data series. The approach is based on the concept of data series envelopes and essential attributes obtained by a multilayer neural network. The considered neural network architecture is based on Cybenko's theorem and consists of two three-layer neural networks. In this paper the generalized net description of this part of SEAA method is developed in order to show the beauty of generalized nets.

Keywords: Modeling, Generalized nets, Data series, Dimensionality reduction, Essential Attributes Approximation.

1 Introduction

The idea of generalized nets was invented in 1982 by Krassimir Attanasov (1987, 1991 and 1997). Generalized nets were proposed as a new definition of nets for modelling dynamic systems, and in several papers the theory of generalized nets was developed by Atanassov (1987, 1991, 2007 and 2010) and Krawczak (2003, 2013) in order to show advantage of this kind of modeling compare to Petri nets. Up to now more than 1000 scientific works related to generalized nets were published, an early survey of theoretical as well as practical scientific works was published by Radeva, Krawczak and Choy (2002).

Nowadays we often face a problem with huge amount of data represented by data series or time series. One of a way to overcome such problem is to reduce dimensionality of time series, which become a very crucial in many areas, for example in medicine, finance, industry, climate. One of the most effective method to reduce dimensionality of time series is SEAA method developed by Krawczak and Szkatuła (2014), which based on previous works the authors (2008, 2010a,b,c, 2011, 2012a,b,c).

P. Angelov et al. (eds.), *Intelligent Systems'2014*,
Advances in Intelligent Systems and Computing 322, DOI: 10.1007/978-3-319-11313-5_57

The method consists of three main parts:

- generation of so-called envelopes of time series,
- generating of essential attributes,
- generating of symbolic (nominal) representation of time series.

This paper is devoted to the second part of SEAA method, and especially to generating essential attributes by feedforward neural network and to show its generalized net description. Generalized nets can be used to find logistic model of each step of neural network designed for generating of essential attributes.

2 Essentaial Attribues Generation

Within this part of SEAA method Krawczak and Szkatuła (2014) proposed to use a feedforward neural network as auto-associative memory where the output of hidden layer constitute a new representation of time series called essential attributes and the number of them E must be adjusted, where $E << \left\lfloor \frac{M}{m} \right\rfloor$. This part of SEAA methodology is based on Cybenko's theorem (1989) as well as nonlinear principle component analysis (Oja, 1992) and auto-associative neural networks.

Selection of the proper values of the number of essential attributes E is of crucial importance because their values have strong influence on quality of new data series representations. It worth to say that original data as well as the envelopes data representation have a form of time series, and the order of samples is natural and of principal importance. The essential attributes generated by the neural network have a set form and the order of elements is not important.

In order to generate essential attributes we use an auto-associative feedforward neural network with five layers, including the input layer. The inputs of the neural network are described by envelopes

$$[y_1(n), y_2(n), \ldots, y_{\left\lfloor \frac{M}{m} \right\rfloor}(n)], \quad n = 1, \ldots, N \tag{1}$$

of dimensionality $\left\lfloor \frac{M}{m} \right\rfloor$, while the given outputs of the designed neural network are described by

$$[\hat{y}_1(n), \hat{y}_2(n), \ldots, \hat{y}_{\left\lfloor \frac{M}{m} \right\rfloor}(n)] \tag{2}$$

The designed neural network consists of two three-layer joint networks where the middle layer is responsible for generation of essential attributes denoted as follows

$$\{b_1'(n), b_2'(n), \ldots, b_E'(n)\}, n = 1, \ldots, N \tag{3}$$

where it assumed that $E \ll \left\lfloor \frac{M}{m} \right\rfloor$. The architecture of the neural network is shown in Fig. 1

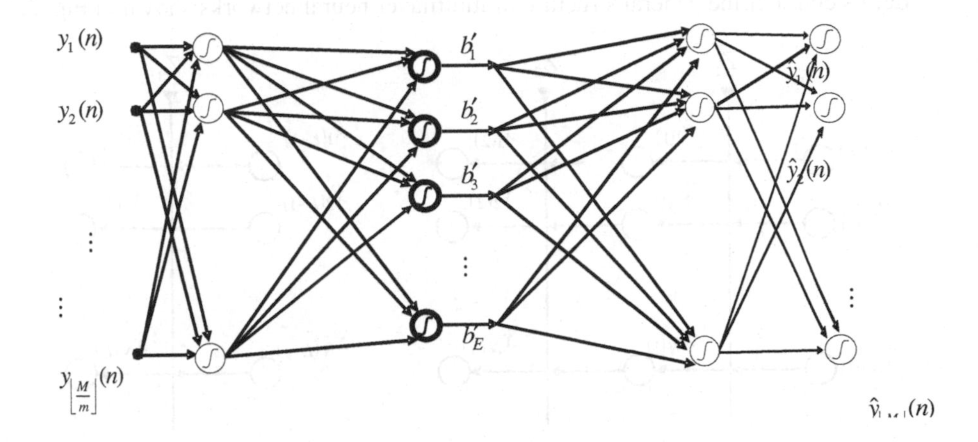

Fig. 1. Neural network generating essential attributes

In the next section we will describe three generalized net models which illustrate the essential attributes generation.

3 Generalized Net Description of Essential Attributes Generator

It should be said that the conception of generalized nets was described by Krassimir Atanassov in 1982 for describing and analyzing various kinds of discrete events systems. Compare to Petri nets (e.g. Petri 1962, 1980) generalized nets have some new extra elements. This section is based on books by Atanassov (1991, 1992, 1997, 1998, 2007) as well as a book by Krawczak (2003, 2013) and by Krawczak et al. (2010).

We will consider the case without aggregation, it means that each separated neuron is treated as a subsystem, the considered neural network depictured in Fig.1.

The considered neural network consists of the following number of subsystems (neurons) $NL = \sum_{l=1}^{L} N(l)$ described by the activation function as follows

$$x_{pj(l)} = f\left(net_{pj(l)}\right), \quad \text{where} \quad net_{pj(l)} = \sum_{i(l-1)=1}^{N(l-1)} w_{i(l-1)\,j(l)}\, x_{pi(l-1)} \tag{4}$$

while $x_{pi(l-1)}$ denotes the output of the i-th neuron with respect to the pattern p, $p=1,2,...,P$, and the weight $w_{i(l-1)j(l)}$ connects the $i(l-1)$-th neuron from the $(l-1)$-st layer with the $j(l)$-th from the l-th layer, $j(l)=1,2,...,N(l)$, $l=1,2,...,L$.

Let us consider the general structure of multilayer neural networks shown in Fig. 2.

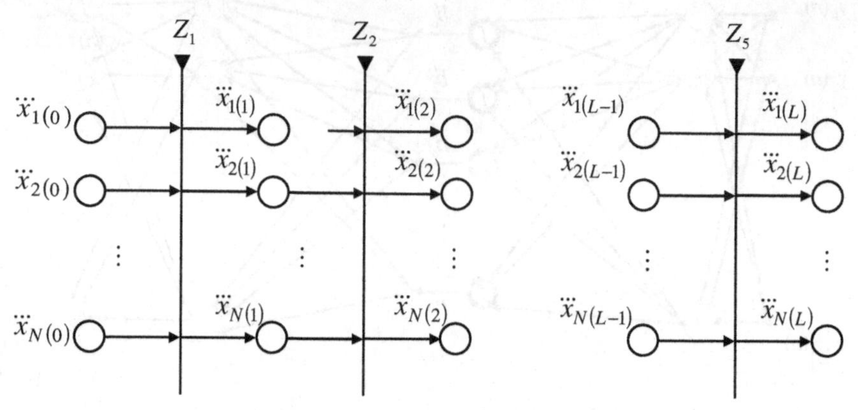

Fig. 2. The generalized net description of neural network simulation

The generalized net from Fig. 2, modelling of the multilayer neural network from Fig. 1, consists of a set of 5 transitions, each transition has the following form

$$Z_l = \left\langle \left\{ \dddot{x}_{1(l-1)}, \dddot{x}_{2(l-1)}, ..., \dddot{x}_{N(l-1)} \right\}, \left\{ \dddot{x}_{1(l)}, \dddot{x}_{2(l)}, ..., \dddot{x}_{N(l)} \right\}, \tau_l, \tau'_l, r_l, M_l, \Box_l \right\rangle \quad (5)$$

for $l=1,2,...,5$,

where $\left\{ \dddot{x}_{1(l-1)}, \dddot{x}_{2(l-1)}, ..., \dddot{x}_{N(l-1)} \right\}$ is the set of input places of the l-th transition,

$\left\{ \dddot{x}_{1(l)}, \dddot{x}_{2(l)}, ..., \dddot{x}_{N(l)} \right\}$ is the set of output places of the l-th transition,

τ_l is the time when the l-th transition is fired out,

τ'_l is the duration time of firing of the l-th transition,

r_l denotes the l-th transition condition determining the transfer of tokens from the transition's inputs $\left\{ \dddot{x}_{1(l-1)}, \dddot{x}_{2(l-1)}, ..., \dddot{x}_{N(l-1)} \right\}$ to its outputs $\left\{ \dddot{x}_{1(l)}, \dddot{x}_{2(l)}, ..., \dddot{x}_{N(l)} \right\}$, and has the index matrix form, see Table 1.

Table 1. Index matrix r_l

	$\dddot{x}_{1(l)}$	$\dddot{x}_{2(l)}$	..	$\dddot{x}_{N(l)}$
$\dddot{x}_{1(l-1)}$	*true*	*true*	..	*true*
$\dddot{x}_{2(l-1)}$	*true*	*true*	..	*true*
$\vdots$	$\vdots$	$\vdots$	..	$\vdots$
$\dddot{x}_{N(l-1)}$	*true*	*true*	..	*true*

where the value true indicates that the tokens representing the neurons can be transferred from the $i(l-1)$ -st input place to the $j(l)$ -th output place, $i(l-1)=1,2,\ldots,N(l-1)$, $j(l)=1,2,\ldots,N(l)$,

M_l indicates an index matrix describing the capacities of transition's arcs, see Table 2.

Table 2. Index matrix M_l

	$\dddot{x}_{1(l)}$	$\dddot{x}_{2(l)}$	$\ldots$	$\dddot{x}_{N(l)}$
$\dddot{x}_{1(l-1)}$	1	1	$\ldots$	1
$\dddot{x}_{2(l-1)}$	1	1	$\ldots$	1
$\vdots$	$\vdots$	$\vdots$	$\ldots$	$\vdots$
$\dddot{x}_{N(l-1)}$	1	1	$\ldots$	1

$\square_l$ has a form of Boolean expression $\wedge\left(\dddot{x}_{1(l-1)},\dddot{x}_{2(l-1)},\ldots,\dddot{x}_{N(l-1)}\right)$ and stipulates that each input place $\dddot{x}_{i(l-1)}$, $i(l-1)=1,2,\ldots,N(l-1)$, must contain a token that will be transferred to the l -th transition.

The generalized net describing the considered neural network simulation process (Fig. 2) has the following form:

$$GN=\left\langle \langle A,\pi_A,\pi_X,c,g,\Theta_1,\Theta_2\rangle, \langle K,\pi_k,\Theta_K\rangle, \langle T,t^0,t^*\rangle, \langle Y,\Phi,b\rangle \right\rangle \tag{6}$$

where $A=\{Z_1,Z_2,\ldots,Z_5\}$ is the set of transitions,

π_A is a function classifying the transitions,

π_X is a function describing the priorities of the places in the following way:

$$pr_1\{Z_1,Z_2,\ldots,Z_5\}= \left\{\dddot{x}_{1(0)},\dddot{x}_{2(0)},\ldots,\dddot{x}_{N(0)},\ \dddot{x}_{1(1)},\dddot{x}_{2(1)},\ldots,\dddot{x}_{N(1)},\ldots,\ \dddot{x}_{1(L-1)},\dddot{x}_{2(L-1)},\ldots,\dddot{x}_{N(L-1)}\right\} \tag{7}$$

$$pr_2\{Z_1,Z_2,\ldots,Z_5\}= \left\{\dddot{x}_{1(1)},\dddot{x}_{2(1)},\ldots,\dddot{x}_{N(1)},\ \dddot{x}_{1(2)},\dddot{x}_{2(2)},\ldots,\dddot{x}_{N(2)},\ldots,\ \dddot{x}_{1(L)},\dddot{x}_{2(L)},\ldots,\dddot{x}_{N(L)}\right\} \tag{8}$$

$$pr_1A\cup pr_2A= \left\{\dddot{x}_{1(0)},\dddot{x}_{2(0)},\ldots,\dddot{x}_{N(0)},\ \dddot{x}_{1(1)},\dddot{x}_{2(1)},\ldots,\dddot{x}_{N(1)},\ \dddot{x}_{1(2)},\dddot{x}_{2(2)},\ldots,\dddot{x}_{N(2)},\ldots \ldots,\ \dddot{x}_{1(L)},\dddot{x}_{2(L)},\ldots,\dddot{x}_{N(L)}\right\} \tag{9}$$

c is a function describing the capacities of the places; of the places; in our case the capacity function has the form:

$$c\left(x_{i(l)}\right)=1\ ,\qquad \text{for } i(l)=1,2,\ldots,N(l),\ l=0,1,2,\ldots,5\ , \tag{10}$$

g is a function that calculates the truth values of the predicates of the transition conditions, here the function is constant

$$g\left(r_{l,i(l-1)j(l)}\right)=true \quad (11)$$

Θ_1 is a function yielding the next time-moment when the transitions can be again activated,

Θ_2 is a function giving the duration of activity of a given transition Z_l,

$$\Theta_2(t_l)=t_l'',\ l=1,2,...,5 \quad (12)$$

$$t_l''=pr_4 Z_l=\tau_l' \quad (13)$$

K is the set of tokens entering the generalized net, in the considered case there are $N(0)=\left\lfloor \frac{M}{m} \right\rfloor$ input places and each place contains one token

$$K=\left\{\alpha_{1(0)},\alpha_{2(0)},...,\alpha_{N(0)}\right\} \quad (14)$$

π_K is a function describing the priorities of the tokens, here all tokens have the same priorities, and it will be denoted by "*",

Θ_K is a function giving the time-moment when a given token can enter the net,

T is the time when the generalized net starts functioning,

t^0 is an elementary time-step, here this parameter is not used and is denoted by *,

t^* determines the duration of the generalized net functioning,

Y denotes the set of all the initial characteristics of the tokens

$$Y=\left\{y\left(\alpha_{1(0)}\right),y\left(\alpha_{2(0)}\right),...,y\left(\alpha_{N(0)}\right)\right\} \quad (15)$$

$$y\left(\alpha_{i(0)}\right)=\left\langle NN1,N(0),N(1),imX_{i(0)},imW_{i(0)},F_{(1)},imout_{i(0)}\right\rangle \quad (16)$$

is the initial characteristic of the token $\alpha_{i(0)}$ that enters the place $\dddot{x}_{i(0)}$, $i(0)=1,2,...,N(0)$,

where $NN1$ the neural network identifier,

$N(0)$ number of input places to the net as well as to the transition Z_1 (equal to the number of inputs to the neural network),

$N(1)$ number of the output places of the transition Z_1,

$$imX_{i(0)}=\left[0,...,0,x_{i(0)},0,...,0\right]^T \quad (17)$$

is the index matrix, indicating the inputs to the network, of dimension $N(0)\times 1$ in which all elements are equal 0 except for the element $i(0)$ whose value is equal $x_{i(0)}$ (the $i(0)$-th input of the neural network),

$$imW_{i(0)} = \begin{array}{c|c|c|c|c} & \dddot{x}_{1(1)} & \dddot{x}_{2(1)} & \cdots & \dddot{x}_{N(1)} \\ \hline \dddot{x}_{i(0)} & w_{i(0)1(1)} & w_{i(0)2(1)} & \cdots & w_{i(0)N(1)} \end{array} \tag{18}$$

has a form of an index matrix and denotes the weights connecting the $i(0)$-th input with all neurons allocated to the 1-st layer,

$$F_{(1)} = \left[f_{1(1)}\left(\sum_{i(0)=1}^{N(0)} x_{i(0)}\, w_{i(0)1(1)}\right), f_{2(1)}\left(\sum_{i(0)=1}^{N(0)} x_{i(0)}\, w_{i(0)2(1)}\right), \ldots, f_{N(1)}\left(\sum_{i(0)=1}^{N(0)} x_{i(0)}\, w_{i(0)N(1)}\right)\right]^T \tag{19}$$

denotes a vector of the activation functions of the neurons associated with the 1-st layer,

$$imout_{i(0)} = \begin{array}{c|c|c|c|c} & \dddot{x}_{1(1)} & \dddot{x}_{2(1)} & \cdots & \dddot{x}_{N(1)} \\ \hline \dddot{x}_{i(0)} & x_{i(0)}\, w_{i(0)1(1)} & x_{i(0)}\, w_{i(0)2(1)} & \cdots & x_{i(0)}\, w_{i(0)N(1)} \end{array} \tag{20}$$

is an index matrix describing the signal outgoing from the $i(0)$-th input place $i(0) = 1, 2, \ldots, N(0)$, to all output places of the Z_1 transition,

Φ is a characteristic function that generates the new characteristics of the new tokens after passing the transition; for the transition Z_l, $l = 1, 2, \ldots, L$, there are $N(l-1)$ input places $\{\dddot{x}_{1(l-1)}, \dddot{x}_{2(l-1)}, \ldots, \dddot{x}_{N(l-1)}\}$ and with each place there is associated a single token $\alpha_{i(l-1)}$, $i(l-1) = 1, 2, \ldots, N(l-1)$, having the characteristic

$$y(\alpha_{i(l-1)}) = \langle NN1, N(l-1), N(l), imX_{i(l-1)}, imW_{i(l-1)}, F_{(l)}, imout_{i(l-1)} \rangle \tag{21}$$

where $NN1$ is the neural network identifier,

$N(l-1)$ is number of input places to the net as well as to the transition Z_l,

$N(l)$ is number of the output places of the transition,

$$imX_{i(l-1)} = [0, \ldots, 0, x_{i(l-1)}, 0, \ldots, 0]^T \tag{22}$$

is the index matrix of dimension $N(l-1) \times 1$ in which all elements are equal 0 except the element $i(l-1)$ whose value is equal $x_{i(l-1)}$ - the $i(l-1)$-st input value associated with the Z_l transition,

$$imW_{i(l-1)} = \begin{array}{c|c|c|c|c} & \dddot{x}_{1(1)} & \dddot{x}_{2(1)} & \cdots & \dddot{x}_{N(1)} \\ \hline \dddot{x}_{i(l-1)} & w_{i(l-1)1(l)} & w_{i(l-1)2(l)} & \cdots & w_{i(l-1)N(l)} \end{array} \tag{23}$$

is an index matrix describing the weight connection between the $i(l-1)$-st input places with all output places of the Z_l transition,

$$F_{(l)} = \left[f_{1(l)}\left(\sum_{i(l-1)=1}^{N(l-1)} x_{i(l)}\, w_{i(l-1)1(l)}\right), f_{2(l)}\left(\sum_{i(l-1)=1}^{N(l-1)} x_{i(l-1)}\, w_{i(l-1)2(l)}\right), \ldots, f_{N(l)}\left(\sum_{i(l-1)=1}^{N(l-1)} x_{i(l-1)}\, w_{i(l-1)N(l)}\right)\right]^T \quad (24)$$

is a vector of the activation functions of the neurons associated with the l-th layer of the neural network,

$$imout_{i(l-1)} = \begin{array}{c|c|c|c|c} & \dddot{x}_{1(l)} & \dddot{x}_{2(l)} & \cdots & \dddot{x}_{N(l)} \\ \hline \dddot{x}_{i(l-1)} & x_{i(l-1)}\, w_{i(l-1)1(l)} & x_{i(l-1)}\, w_{i(l-1)2(l)} & \cdots & x_{i(l-1)}\, w_{i(l-1)N(l)} \end{array} \quad (25)$$

is an index matrix describing the signals outgoing from the $i(l)$-th input place $i(l)=1,2,\ldots,N(l)$, to all output places of the Z_l transition,

b is a function describing the maximum number of characteristics a given token can receive,

$$b(\alpha_{j(l)})=1, \quad \text{for} \quad j(l)=1,2,\ldots,N(l),\ l=1,2,\ldots,5, \quad (26)$$

Due to the above considerations the transitions now can be rewritten in the following form

$$Z_l = \langle \{\dddot{x}_{1(l-1)}, \dddot{x}_{2(l-1)},\ldots, \dddot{x}_{N(l-1)}\}, \{\dddot{x}_{1(l)}, \dddot{x}_{2(l)},\ldots, \dddot{x}_{N(l)}\}, \tau_l, \tau'_l, *, *, \square_l \rangle \quad (27)$$

for $l=1,2,\ldots,5$.

The *reduced form* of the generalized net describing the neural network has the following form:

$$GN1 = \langle\, \langle A, *, \pi_X, c, *, \Theta_1, \Theta_2 \rangle, \langle K, *, \Theta_K \rangle, \langle T, *, t^* \rangle, \langle Y, \Phi, b \rangle \,\rangle \quad (28)$$

where $A=\{Z_1, Z_2,\ldots, Z_L\}$ is a set of transitions,

π_X is a function describing the priorities of the places,

c is a function describing the capacities of the places, i.e. $c(x_{i(l)})=1$, $i(l)=1,2,\ldots,N(l),\ l=0,1,2,\ldots,L$,

Θ_1 is a function yielding the next time-moment when the transitions can be again activated, $\Theta_1(t_l)=t'_l,\ l=1,2,\ldots,L$,

Θ_2 is a function giving the duration of activity of the transition Z_l $\Theta_2(t_l)=t''_l$, $l=1,2,\ldots,L$,

$K=\{\alpha_{1(0)}, \alpha_{2(0)},\ldots, \alpha_{N(0)}\}$ is the set of tokens entering the generalized net,

$\Theta_K = T$ for all tokens entering the net and at this moment the net starts to function,

t^* determines the duration of the generalized net's functioning,
Y denotes the set of all initial characteristics of the tokens described by (27)

$$Y = \{ y(\alpha_{1(0)}), y(\alpha_{2(0)}), ..., y(\alpha_{N(0)}) \} \tag{29}$$

where

$$y(\alpha_{i(0)}) = \langle NN1, N(0), N(1), x_{i(0)}, W_{i(0)}, F_{(1)}, out_{i(0)} \rangle, \; i(0) = 1, 2, ..., N(0), \tag{30}$$

Φ is a characteristic function that generates the new characteristics of the new tokens after passing the transition,
$b(\alpha_{j(l)}) = 1$, for $j(l) = 1, 2, ..., N(l), l = 1, 2, ..., 5$, is a function describing the number of characteristics memorized by each token.

Such generalized nets with some elements missing (the elements not being valid) are called reduced generalized nets (Atanassov 1991, 1992, 2007). In the above version of the generalized nets representation of the simulation process of multilayer neural network we preserve the parallelism of computation.

The generalized nets theory, as described by Atanassov (1991, 1992, 2007) contains many different operations and relations over the transitions, tokens as well as over the characteristics of tokens.

Here we would like to recall a different model of multilayer neural network applied for generating essential attributes. Let us consider the structure of the multilayer neural network shown in Fig. 1, and aggregate the neurons allocated within each layer l, $l = 1, 2, ..., 5$, and we aggregate all input tokens into only one token.

The generalized net representation of the neural network consists of 5 transitions. Each transition Z_l, $l = 1, 2, ..., 5$, has the following aggregated input place

$$\dddot{X}_{(l-1)} = \{ \dddot{x}_{1(l-1)}, \dddot{x}_{2(l-1)}, ..., \dddot{x}_{N(l-1)} \} \tag{31}$$

and the aggregated output place

$$\dddot{X}_{(l)} = \{ \dddot{x}_{1(l)}, \dddot{x}_{2(l)}, ..., \dddot{x}_{N(l)} \}. \tag{32}$$

Now the structure aggregated neural network is depicted in Fig. 3.

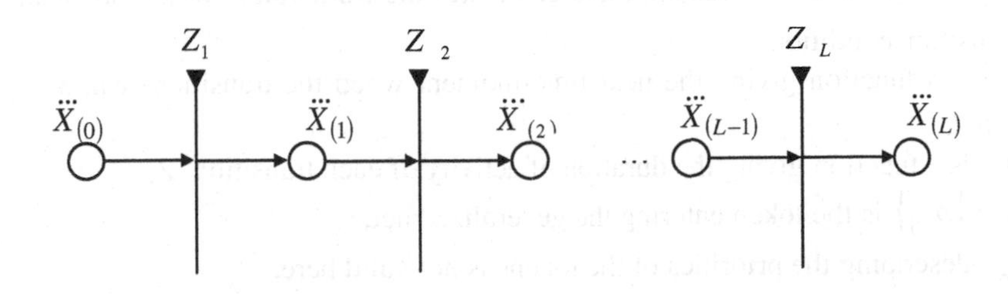

Fig. 3. The generalized net aggregated representation of the feedforward neural network

It is assumed that for each transition Z_l, $l = 1, 2, ..., 5$, there is a single input place as well as a single output place, additionally let us assume that each place contains only one token.

Each transition Z_l, $l = 1, 2, ..., 5$, has the following form

$$Z_l = \left\langle \left\{ \dddot{X}_{(l-1)} \right\}, \left\{ \dddot{X}_{(l)} \right\}, \tau_l, \tau'_l, r_l, M_l, \Box_l \right\rangle \quad (33)$$

where $\left\{ \dddot{X}_{(l-1)} \right\}$ is the input place of the l-th transition,

$\left\{ \dddot{X}_{(l)} \right\}$ is the output place of the l-th transition,

τ_l is a time when the l-th transition is fired out,

τ'_l is the duration of activity of the l-th transition,

r_l denotes the l-th transition condition of the $\alpha_{(l-1)}$ token from the transition's input place $\left\{ \dddot{X}_{(l-1)} \right\}$ to its output place $\left\{ \dddot{X}_{(l)} \right\}$, and it has a simple form: $r_l = true$ (i.e. the input token can be transferred to the output place without any condition),

$M_l = 1$ indicates that only one token can be transferred by arcs in the same time,

$\Box_l$ is not valid due to the existence of only one token in any place.

In this way the transitions Z_l, , now have the following reduced form

$$Z_l = \left\langle \left\{ \dddot{X}_{(l-1)} \right\}, \left\{ \dddot{X}_{(l)} \right\}, \tau_l, \tau'_l \right\rangle, \quad l = 1, 2, ..., L \quad (34)$$

The new aggregated generalized net has the following form:

$$GN2 = \left\langle \langle A, \pi_A, \pi_X, c, g, \Theta_1, \Theta_2 \rangle, \langle K, \pi_k, \Theta_K \rangle, \langle T, t^0, t^* \rangle, \langle Y, \Phi, b \rangle \right\rangle \quad (35)$$

The elements are described as follows:

$A = \{Z_1, Z_2, ..., Z_5\}$ is a set of transitions,

π_A is not valid here,

π_X is a function describing the priorities of the places,

$c(x_{i(l)}) = 1$ is a function describing the capacities of the places,

$g(r_{l,i(l-1)j(l)}) = true$ is a function that calculates the truth values of the predicates of the transition conditions,

Θ_1 is a function giving the next time-moment when the transitions can be again activated,

Θ_2 is a function giving the duration of activity of each transition Z_l.

$K = \{\alpha_{(0)}\}$ is the token entering the generalized net,

π_K describing the priorities of the tokens is not valid here,

$\Theta_K = T$ is the time when the token $\alpha_{(0)}$ enters the net,

T the time when the generalized net starts functioning,

t^0 describing the elementary time-step (not valid here),

t^* describing the time when the generalized net is functioning,

$Y = \{ y(\alpha_{(0)}) \}$ is the initial characteristic of the token which enters the place $\dddot{X}_{(0)}$,

Φ is a function generating the new characteristics of the new token after passing the transition Z_l, $l = 1, 2, ..., 4$; the input place $\dddot{X}_{(l-1)}$ has a token $\alpha_{(l-1)}$,

$b(\alpha_{(l)}) = 1$ for $l = 1, 2, ..., 5$.

The reduced form of the generalized net, describing the simulation process of the aggregated neural network has the following form:

$$GN2 = \langle \langle A, *, \pi_X, c, *, \Theta_1, \Theta_2 \rangle, \langle K, *, * \rangle, \langle T, *, t^* \rangle, \langle Y, \Phi, 1 \rangle \rangle \quad (36)$$

with the elements specified in detail above.

4 Conclusions

The generalized net representations described in this paper illustrate the possibilities of application of the generalized net methodology to model dynamic systems like neural networks. Here we showed only the simplest forms of the generalized net representations. Using generalized net model we are able to analyze functioning of dynamic systems in details. In order to apply more advanced generalized nets theory we must specify the detailed forms of transitions. Some specialized operators can be also applied, here we can constructed a generalized net with a structure similar to the structure of a given neural network. (or system of neural networks).

In this paper we modeled feedforward neural network called also autoassociative memory which was used as crucial part of SEAA method for reducing dimensionality of time series. W showed two different, but in some sense similar, models. The first is very similar to the network architecture, while the second one is with aggregated tokens. Due to restrictive space of this paper we were not able to describe the parameters in details and therefore our description is unfortunately only rough.

References

1. Atanassov, K.: Generalized nets. World Scientific, Singapore (1991)
2. Atanassov, K.: Generalized Nets and Systems Theory. Prof. M. Drinov. Academic Publishing House, Sofia (1997)
3. Atanassov, K.: Generalized Index Matrices. Competes Rendus de l'Academie Bulgare des Sciences 40(11), 15–18 (1987)
4. Cybenko, G.: Approximations by superpositions of sigmoidal functions. Mathematics of Control, Signals, and Systems 2(4), 303–314 (1989)
5. Krawczak, M.: Multilayer Neural Systems and Generalized Net Models. Ac. Publ. House EXIT, Warsaw (2003)

6. Krawczak, M.: Multilayer Neural Networks – Generalized Net perspective. Springer (2013)
7. Krawczak, M., Szkatuła, G.: On decision rules application to time series classification. In: Atanassov, K.T., et al. (eds.) Advances in Fuzzy Sets, Intuitionistic Fuzzy Sets, Generalized Nets and Related Topics, Ac. Publ. House EXIT (2008)
8. Krawczak, M., Szkatuła, G.: Time series envelopes for classification. In: IEEE Intelligent Systems Conference, London, July 7-9 (2010a)
9. Krawczak, M., Szkatuła, G.: On time series envelopes for classification problems. In: Atanassov, K.T., et al. (eds.) Developments in Fuzzy Sets, Intuitionistic Fuzzy Sets, Generalized Nets and Related Topics. II. SRI PAS, Warsaw (2010b)
10. Krawczak, M., Szkatuła, G.: Dimensionality reduction for time series. Case Studies of the Polish Association of Knowledge 31, 32–45 (2010c)
11. Krawczak, M., Szkatuła, G.: A hybrid approach for dimension reduction in classification. Control and Cybernetics 40(2), 527–552 (2011)
12. Krawczak, M., Szkatuła, G.z.: A clustering algorithm based on distinguishability for nominal attributes. In: Rutkowski, L., Korytkowski, M., Scherer, R., Tadeusiewicz, R., Zadeh, L.A., Zurada, J.M. (eds.) ICAISC 2012, Part II. LNCS, vol. 7268, pp. 120–127. Springer, Heidelberg (2012a)
13. Krawczak, M., Szkatuła, G.: Dimension reduction of time series for the Clustering Problem. In: Atanassov, K.T., Homenda, W., Hryniewicz, O., Kacprzyk, J., Krawczak, M., Nahorski, Z., Szmidt, E., Zadrożny, S. (eds.) New Developments in Fuzzy Sets, Intuitionistic Fuzzy Sets, Generalized Nets and Related Topics. II: Applications, pp. 101–110. SRI PAS, Warsaw (2012b)
14. Krawczak, M., Szkatuła, G.: Nominal Time Series Representation for the Clustering Problem. In: IEEE 6th International Conference Intelligent Systems, Sofia, pp. 182–187 (2012c)
15. Krawczak, M., Szkatuła, G.: An approach to dimensionality reduction in time series. Information Sciences 260, 15–36 (2014)
16. Oja, E.: Principal components, minor components and linear neural networks. Neural Networks 5, 927–935 (1992)
17. Radeva, V., Krawczak, M., Choy, E.: Review and Bibliography on Generalized Nets Theory and Applications. Advanced Studies in Contemporary Mathematics 4(2), 173–199 (2002)

Development of Generalized Net for Testing of Different Mathematical Models of *E. coli* Cultivation Process

Dimitar Dimitrov[1] and Olympia Roeva[2]

[1] Faculty of Mathematics and Informatics, Sofia University,
Sofia University St. Kliment Ohridski,
5 James Bourchier Str., Sofia, Bulgaria
dgdimitrov@fmi.uni-sofia.bg
[2] Bioinformatics and Mathematical Modelling Department,
IBPhBME – Bulgarian Academy of Sciences,
Acad. G. Bonchev Str., bl. 105, 1113 Sofia, Bulgaria
olympia@biomed.bas.bg

Abstract. The present paper proposes a developed generalized net model that compares different mathematical models of the process of *E. coli* fed-batch cultivation. A system of four ordinary differential equations describes the main variables of the considered cultivation process, namely, biomass, substrate and acetate. For the purposes of model simulation, we use the software package GN Lite. The GN model compares the simulated performance of the proposed set of mathematical models and selects the best performing one one the basis of a predefined criterion. During the simulation, GN Lite calls the Matlab software environment to solve the process model, presented as a system of nonlinear differential equations, and plots the dynamics of the main process variables' of the model that has been computed to perform best.

Keywords: Generalized net, GN Lite, GNTicker, GN IDE, *E. coli*, Fed-batch cultivation, Mathematical models.

1 Introduction

Cultivation of recombinant micro-organisms e.g. *E. coli*, in many cases is the only economical way to produce pharmaceutic biochemicals such as interleukins, insulin, interferons, enzymes and growth factors. To maximize the volumetric productivities of bacterial cultures it is important to grow *E. coli* to high cell concentration. This sets the challenge to model and optimize fed-batch cultivation processes.

The problem of model parameter estimation refers to situations where decisions have to be made depending on multiple conflicting criteria, oftentimes involving imprecision or uncertainty. A wide range of optimization methods are available for solving such problems, each featuring specific properties making it appropriate for specific cases. Choosing an optimization method very much depends on the nature of the problem and the availability of appropriate software that suits the problem statement.

P. Angelov et al. (eds.), *Intelligent Systems'2014*,
Advances in Intelligent Systems and Computing 322, DOI: 10.1007/978-3-319-11313-5_58

Up to now Generalized nets (GNs) are used as a tool for modelling of parallel processes in many areas as economics, transport, medicine, computer technologies etc. [7, 8, 11]. The theory of the GNs [9, 10, 11, 13, 14, 15, 16] proved to be quite successful when applied to the description of the functioning of expert systems, machine learning and different technological processes. Recently attempts have been made to describe various models, optimization procedures, control loops considering bioprocesses, based on GNs [1, 17, 19, 22].

In this paper the apparatus of GNs is used to describe a GN model for comparison of different mathematical models of an *E. coli* fed-batch cultivation process. Proposed here GN model is an extension of presented in Dimitrov and Roeva [2]. The new model is more complex including equation for acetate dynamics and thus two more model parameters. The GN model is simulated using GN Lite – a software package for modeling and simulation with GNs [25]. GN Lite implements the current state of the most aspects of GN theory, compared to other software tools for GNs [6]. Its graphical environment has user-friendly interface which assists the user through the whole modeling process [3, 4]. Some of the features of GN Lite are the visual editing and simulation of GN models, the usage of XML format to define the GN models, the support of several programming language for predicates and characteristic functions, as well as the supported integration of Matlab code into GN models.

2 GN Lite Software

GN Lite is a component oriented software product for running simulations of generalized net models. It has a client-server architecture and its core is the software interpreter of GN models GNTicker [24], which has most aspects of GN theory implemented. The interpreter on the server side uses optimized versions of the algorithms for transition and GN functioning [24] and is the first GN-related software that makes use of the XML format (XGN) for the model definitions. Compared to previous implementations of GN simulators, this approach has multiple advantages like the ensured platform independence, not restricting developers of new GN applications to particular operating systems or programming languages. The popularity of XML has also been taken as considerations due to the wide availability of free software tools for XML manipulation. Applications on the client side, such as visual environments for graphic editing and user input, communicate with the simulation core via TCP/IP connection and the specially designed The GNTicker Trace Protocol (GNTP).

Together with GNTicker, the GNTicker Characteristic Function Language (GNTCFL) programming language has been developed. It is a weakly-typed procedural language which bears syntactic resemblance to the LISP language. The language can code the GN transition predicates, characteristic functions, rules for tokens splitting and merging, as well as additional utility functions. Moreover, GNTicker comes with a number of standard primitive functions that occur in many models. GNTicker has been integrated to call functions written in Matlab environment, thus enabling the performance of complex computations, like solving differential equations, during the GN simulation [5].

GNLite features an integrated modeling environment, called GN IDE, written in Java, consisting of graphical user interface providing facilities for visual building of GN models, running simulations and tracing their execution, exporting models to various file formats, and many more. GN IDE can be run on any platform with Java Runtime Edition (JRE) installed, as well as in any browser using Java Web Start.

Developing new GN-related applications has been made easier due to the scalable, component-oriented design that allows reuse of available components. Some examples are GNProfy and GNvis [8]. New functionalities can be easily implemented over the top of the existing ones, and even main components such as the interpreter and the visual environment can be replaced with third-party products. On Fig. 2 is presented the main part of the GN model and its corresponding XGN file [4].

So far, the operating systems, GN Lite has been tested on, are Ubuntu Linux and Microsoft Windows (32- and 64-bit versions).

3 *E. coli* Fed-Batch Cultivation Process Model

The rates of cell growth, sugar consumption and product accumulation in the considered fed-batch cultivation could be described according to the mass balance as:

$$\frac{dX}{dt} = \mu X - \frac{F}{V} X \tag{1}$$

$$\frac{dS}{dt} = -q_S X + \frac{F}{V}(S_{in} - S) \tag{2}$$

$$\frac{dA}{dt} = -q_A A + \frac{F}{V} A \tag{3}$$

$$\frac{dV}{dt} = F \tag{4}$$

where $\mu = \mu_{\max} \frac{S}{S + k_S}$, $q_s = \frac{1}{Y_{S/X}} \mu_{\max} \frac{S}{S + k_S}$, $q_A = \frac{1}{Y_{A/X}} \mu_{\max} \frac{A}{A + k_A}$ and X – biomass concentration, [g·l^{-1}]; S – substrate (glucose) concentration, [g·l^{-1}]; A – acetate concentration, [g·l^{-1}]; F – influent flow rate, [h^{-1}]; V – bioreactor volume, [l]; S_{in} – influent glucose concentration, [g·l^{-1}]; μ – specific growth rate, [h^{-1}]; q_s, q_A – specific rates, [g·g^{-1}·h^{-1}]; μ_{max} – maximum specific growth rate, [h^{-1}]; $Y_{S/X}$, $Y_{A/X}$ – yield coefficients, [g·g^{-1}]; k_s, k_A – saturation constants, [g·l^{-1}].

We are solving the model parameter estimation problem, using real experimental data of the *E. coli* MC4110 fed-batch cultivation process. The initial process parameters are presented in Table 1. We have taken into consideration several different optimization techniques, namely, Genetic algorithms (GA), Tabu search (TS) and Firefly algorithm (FA), advancing some previous authors' research with mathematical models based on GA [21], TS [12] and FA [18, 20].

```
<?xml version="1.0" ?>
<gn xmlns="http://www.clbme.bas.bg/GN" name="SimpleGN" time="256">
  <transitions>
      <transition id="Z1" positionX="100" positionY="100" sizeY="100">
        <inputs>
          <input ref="L1">
            <arc>
              <point positionX="50" positionY="150"/>
              <point positionX="100" positionY="150"/>
            </arc>
          </input>
        </inputs>
        <outputs>
          <output ref="L2">
            <arc>
              <point positionX="100" positionY="150"/>
              <point positionX="150" positionY="150"/>
            </arc>
          </output>
        </outputs>
        <predicates>
          <predicate input="L1" output="L2">true</predicate>
        </predicates>
      </transition>
    </transitions>
    <places>
      <place id="L1" positionX="50" positionY="150"/>
      <place id="L2" positionX="150" positionY="150"/>
    </places>
    <tokens>
      <token id="alpha1" host="L1">
          <char name="Default" type="double" history="1">1</char>
      </token>
    </tokens>
    <functions/>
  </gn>
```

Fig. 1. Simple SN model

Table 1. Initial process parameters

Process parameters	Value
t_0, h	6.68
$X(0)$, g/l	1.25
$A(0)$, g/l	0.03
$S(0)$, g/l	0.81
$V(0)$, l	1.35
S_{in}, g/l	100

In the current leg of research, a parameter identification of model (1)-(4) has been done with the application of TS, thus elaborating on the model presented in [12]. We consider the following ranges of the model parameters:

$\mu_{max} \in [0, 0.9]$, $k_s \in [0, 2]$, $Y_{S/X}$, $Y_{A/X} \in [0, 1]$.

The following optimization techniques are applied to the mathematical model (1)-(4) in the present research, assigned with the following values of the parameters:

- Genetic algorithm: μ_{max} = 0.52 h^{-1}, k_s = 0.023 g/l, k_A = 0.59 g/l, $1/Y_{S/X}$ = 0.5, $1/Y_{A/X}$ = 0.013 [21];
- Tabu search: μ_{max} = 0.49 h^{-1}, k_s = 0.03 g/l, $1/Y_{S/X}$ = 0.5 according [12], and k_A = 0.46, $1/Y_{A/X}$ = 0.01 identified in this research, with TS used in [12];
- Firefly algorithm: μ_{max} = 0.46 h^{-1}, k_s = 0.013 g/l, k_A = 0.18 g/l, $1/Y_{S/X}$ = 0.5, $1/Y_{A/X}$ = 0.03 [18];
- Test model: μ_{max} = 2.4 h^{-1}, k_s = 6 g/l, k_A = 0.18 g/l, $1/Y_{S/X}$ = 5, $1/Y_{A/X}$ = 0.03.

The fourth set of model parameters are defined in a way to test if the considered parameters belong to the defined range.

The model choice is done on the basis of the evaluation of a distance measure J between experimental and model predicted values of state variables, as denoted by the vector y:

$$J = \sum_{i=1}^{n}\sum_{j=1}^{m}\left\{\left[y_{exp}(i) - y_{mod}(i)\right]_j\right\}^2 \rightarrow min \qquad (5)$$

where J *is* optimization criterion; n – number of measurements for each state variable; m – number of state variables; y_{exp} – experimental data vector; y_{mod} – model predicted data vector.

4 Developed Generalized Net Model

In Fig. 2 the graphical structure of the constructed GN is shown. Here is its formal definition.

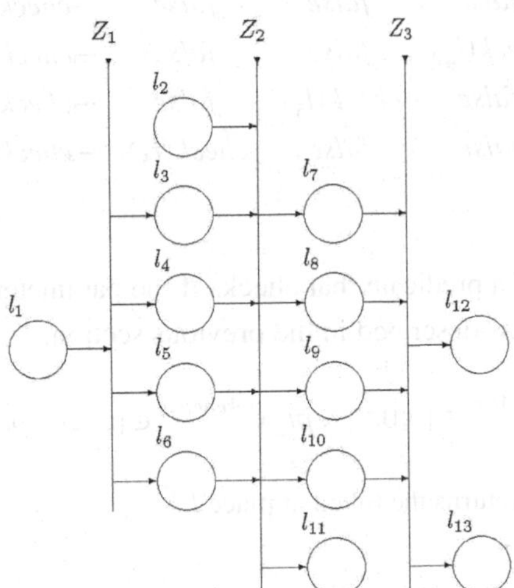

Fig. 2. Graphical structure of the GN model

$$E = \langle\langle\{Z_1, Z_2, Z_3\}\rangle, \langle\{\alpha, \beta\}\rangle, \langle\{x_0^{\alpha}, x_0^{\beta}\}\rangle, \Phi\rangle,$$

where the three transitions are described consequently, as follows

$$Z_1 = \langle\{l_1\}, \{l_3, l_4, l_5, l_6\}, \begin{array}{c|cccc} & l_3 & l_4 & l_5 & l_6 \\ \hline l_1 & true & true & true & true \end{array}, l_1\rangle,$$

Token α enters place l_1 with four sets of parameter values as initial characteristic: $(\mu_{max_j}, k_{s_j}, k_{A_j}, 1/Y_{S/X_j}, 1/Y_{A/X_j})$.

Token β enters place l_2 with values for $t, S_{in}, V_0, X_0, S_0, A_0, F, X_d, S_d$ and A_d. Characteristic function Φ is defined as follows:

After the token α is split into four tokens, $\Phi_{l_i}, 3 \le i \le 6$ keeps only (i – 2)-th set of parameter values. Token β is split into four tokens too. $\Phi_{l_i}, 7 \le i \le 10$ merges α and β tokens in its corresponding position. Then it calls a Matlab function named *simul* with the following declaration:

$$function[J, X_{mod}, S_{mod}, A_{mod}] = simul(t, \mu_{max}, k_s, k_A, 1/Y_{S/X}, \\ 1/Y_{A/X}, S_{in}, V_0, X_0, S_0, A_0, F, X_d, S_d, A_d).$$

$$Z_2 = \langle\{l_2, l_3, l_4, l_5, l_6\}, \{l_7, l_8, l_9, l_{10}, l_{11}\},$$

$$\begin{array}{c|ccccc} & l_7 & l_8 & l_9 & l_{10} & l_{11} \\ \hline l_2 & check(l_3) & check(l_4) & check(l_5) & check(l_6) & false \\ l_3 & check(l_3) & false & false & false & \neg check(l_3) \\ l_4 & false & check(l_4) & false & false & \neg check(l_4) \\ l_5 & false & false & check(l_5) & false & \neg check(l_5) \\ l_6 & false & false & false & check(l_6) & \neg check(l_6) \end{array},$$

$$\wedge(l_2, l_3, l_4, l_5, l_6)\rangle,$$

where $check(l)$ is a predicate that checks if the parameters in the token at the given place l are correct, as described in the previous section:

$$check(l) = pr_1 x^{tokenAt(l)} \in [0, 0.9] \wedge pr_2 x^{tokenAt(l)} \in [0, 2] \wedge pr_3 x^{tokenAt(l)} \in [0, 1],$$

where $tokenAt(l)$ returns the token at place l.

$$Z_3 = \langle \{l_7, l_8, l_9, l_{10}\}, \{l_{12}, l_{13}\},$$

$$\begin{array}{c|cc} & l_{12} & l_{13} \\ \hline l_7 & isMinError(l_7) & \neg isMinError(l_7) \\ l_8 & isMinError(l_8) & \neg isMinError(l_8) \\ l_9 & isMinError(l_9) & \neg isMinError(l_9) \\ l_{10} & isMinError(l_{10}) & \neg isMinError(l_{10}) \end{array},$$

$$\vee(l_7, l_8, l_9, l_{10})\rangle,$$

where $isMinError(l)$ is a predicate that checks whether the error characteristic of the token at the given place l is minimal among the characteristics of the four tokens.

A screenshot of GN IDE of presented in this paper GN model is shown in Fig. 3. Fig. 4 shows a fragment of the source code used for GN model functions definition.

Token splitting and merging must be enabled for the net, i.e. operators $D(2,1)$ and $D(2,4)$ must be defined for E.

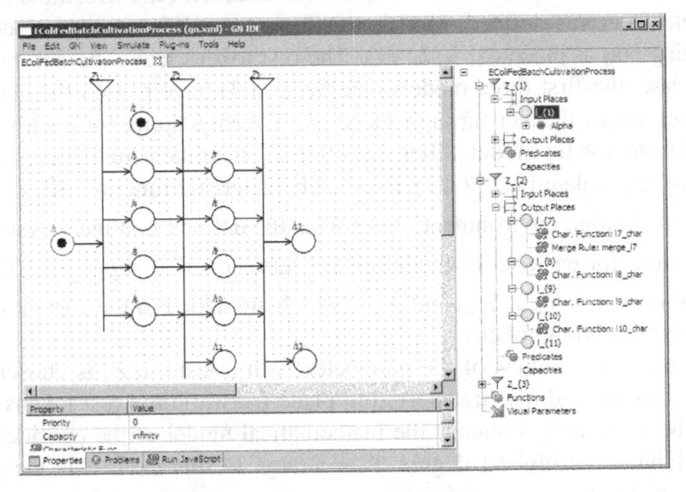

Fig. 3. Screenshot of GN IDE

5 Simulation Results

For the simulation of the presented here model, is used the GN Lite software package, run on Windows 32-bit operating system. A token α enters the GN model, having as an initial characteristic the four mathematical models. In the first transition, token α splits into four tokens corresponding to each model. In the simulation, each new token is given a distinctive color. The main initial characteristics of the *E. coli* fed-batch cultivation process are defined in token β that enters the net via the input place l_2.

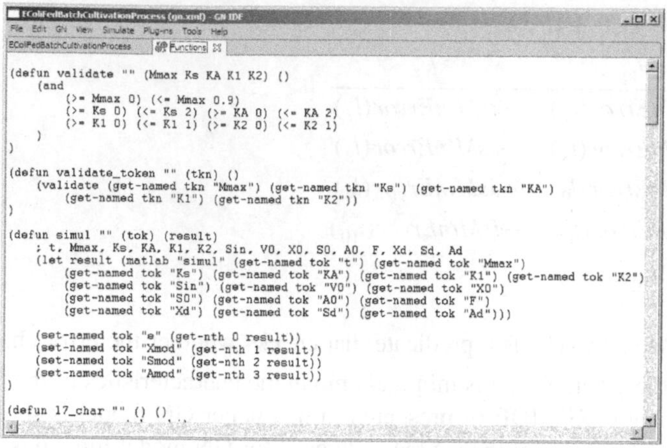

ECoilFedBatchCultivationProcess (gn.xml) - GN IDE
File Edit GN View Simulate Plug-ins Tools Help
ECoilFedBatchCultivationProcess | Functions

```
(defun validate "" (Mmax Ks KA K1 K2) ()
    (and
        (>= Mmax 0) (<= Mmax 0.9)
        (>= Ks 0) (<= Ks 2) (>= KA 0) (<= KA 2)
        (>= K1 0) (<= K1 1) (>= K2 0) (<= K2 1)
    )
)

(defun validate_token "" (tkn) ()
    (validate (get-named tkn "Mmax") (get-named tkn "Ks") (get-named tkn "KA")
        (get-named tkn "K1") (get-named tkn "K2"))
)

(defun simul "" (tok) (result)
    ; t, Mmax, Ks, KA, K1, K2, Sin, V0, X0, S0, A0, F, Xd, Sd, Ad
    (let result (matlab "simul" (get-named tok "t") (get-named tok "Mmax")
        (get-named tok "Ks") (get-named tok "KA") (get-named tok "K1") (get-named tok "K2")
        (get-named tok "Sin") (get-named tok "V0") (get-named tok "X0")
        (get-named tok "S0") (get-named tok "A0") (get-named tok "F")
        (get-named tok "Xd") (get-named tok "Sd") (get-named tok "Ad")))

    (set-named tok "e" (get-nth 0 result))
    (set-named tok "Xmod" (get-nth 1 result))
    (set-named tok "Smod" (get-nth 2 result))
    (set-named tok "Amod" (get-nth 3 result))
)

(defun 17_char "" () ()
```

Fig. 4. Fragment of the source code

In the next transition, Z_2, the four models are checked for correctness according to the parameter intervals defined. Models that have passed the check are moved to the corresponding output place (l_7, l_8, l_9, l_{10}). The rest tokens leave the net through place l_{11}. In this case, the first three models (see Section 3) have passed the check and the fourth model leaves the net through place l_{11}. Token β splits for each output place corresponding to a valid model. After transition Z_2, the tokens in places l_7 to l_{10} merge together. For each token in $l_i, 7 \le i \le 10$, the characteristic function Φ calls function *simul* from the Matlab environment, that solves the defined nonlinear system of differential equations for each set of model parameters, i.e. $x^{\alpha_{i-2}} = pr_{i-2}x^{\alpha}$. The function returns four values: J, X_{mod}, S_{mod} and A_{mod}, which are added to the current characteristics of the corresponding tokens.

Depending on the values of the parameter J, in transition Z_3 is chosen the token with minimal error, and it is transferred to place l_{12}, while the rest tokens proceed to place l_{13}. The token in l_{12} contains the mathematical model of the considered process that has exhibited the highest degree of accuracy. In this case, it is the first model (obtained using GA):

$$\mu_{max} = 0.52 \text{ h}^{-1},\ k_s = 0.023 \text{ g/l},\ k_A = 0.59 \text{ g/l},\ 1/Y_{S/X} = 0.5,\ 1/Y_{A/X} = 0.013 \text{ [17]}.$$

For every moment of the simulation, the user can use the GN IDE functionalities to inspect the current value of each token. Technically, this is done by rolling the mouse over the given token or by using the tree view in the right side of the application window. Fig. 5 shows the initial characteristics of the selected token β.

Also, GN IDE gives possibility for plotting chart diagrams, showing the values of a selected set of characteristics, as illustrated on Figs. 6-8, where the dynamics of the main process variables (biomass, substrate and acetate) are shown. Data from real experiments is compared to the data, as predicted by the model. The presented graphical results show a very good correlation between the real and modelled process.

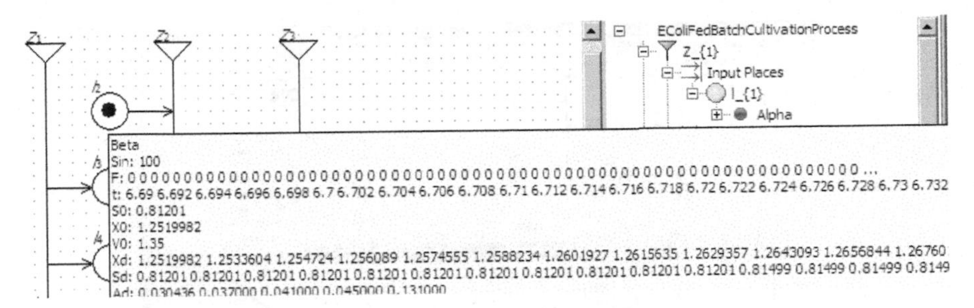

Fig. 5. Initial characteristics of the pointed token

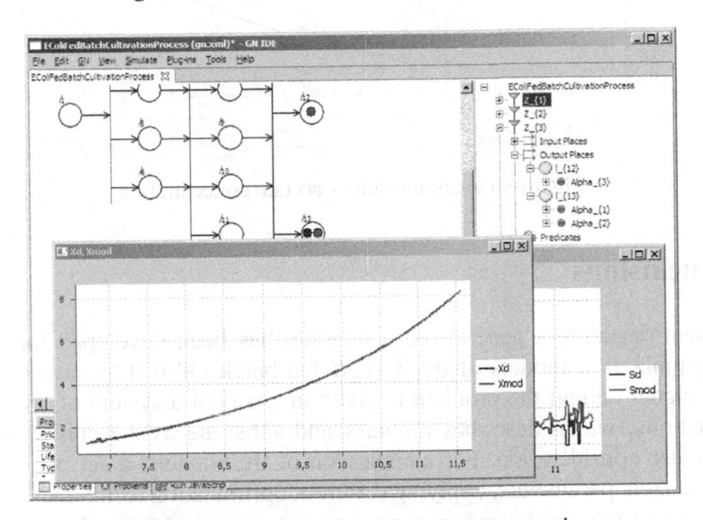

Fig. 6. Simulation results - biomass concentration

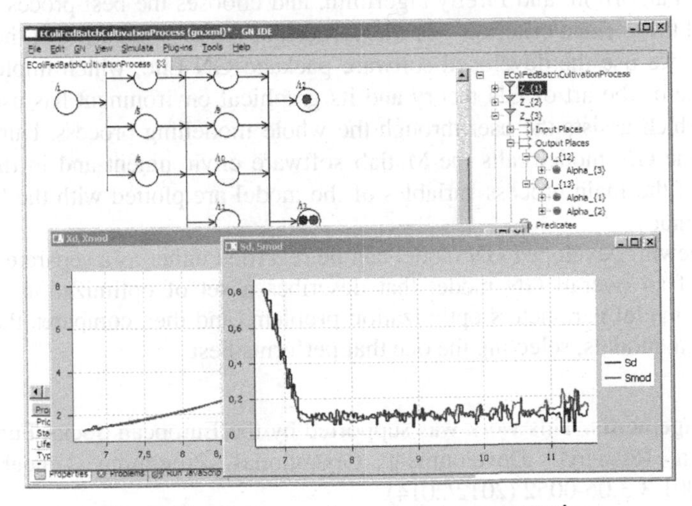

Fig. 7. Simulation results - substrate concentration

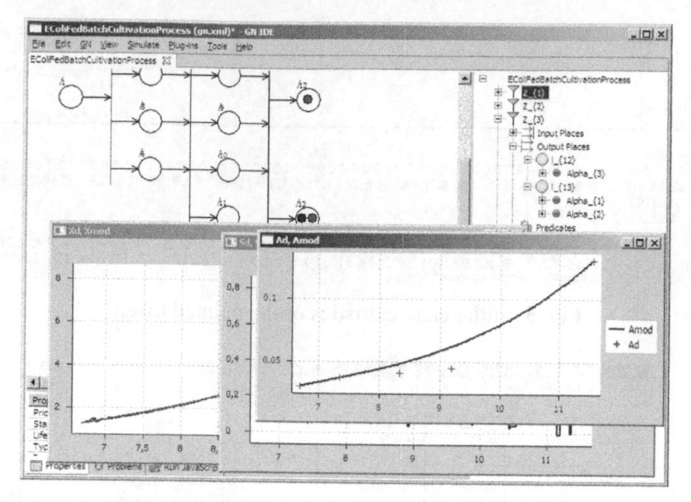

Fig. 8. Simulation results - acetate concentration

6 Conclusions

In the present research, a generalized net model has been developed for comparing different optimization models of the *E. coli* fed-batch cultivation process. The mathematical model behind the problem is given in terms of a system of ordinary differential equations, which describes biomass and substrate as the cultivation process variables under consideration. For comparison of the models, a set of resulting mathematical models parameters, applying different optimization techniques, is used.

The presented GN model compares the models obtained by a Genetic algorithm, Tabu search algorithm and Firefly algorithm, and chooses the best process model on the basis of the defined criterion. For the purpose of model simulation of the proposed GN model, we use the developed software package GN Lite, which implements the current state-of-the-art of GN theory and its graphical environment has user-friendly interface which assists the user through the whole modelling process. During the simulation, the GN model calls the Matlab software environment and in the end, the dynamics of the main process variables of the model are plotted with the highest degree of accuracy.

The herewith developed GN model can be regarded either as a separate module or as a part of an overall GN model that describes a set of optimization algorithms, solving the model parameters optimization problem, and then compares the obtained mathematical models, selecting the one that performs best.

Acknowledgements. This work was supported by the European Social Fund through the Human Resource Development Operational Programme under contract BG051PO001-3.3.06-0052 (2012/2014).

References

1. Shannon, A., Roeva, O., Pencheva, T., Atanassov, K.: Generalized Nets Modelling of Biotechnological Processes, Prof. M. Drinov Academic Publishing House, Sofia (2004)
2. Dimitrov, D.G., Roeva, O.: Comparison of Different Mathematical Models of an *E. coli* Fed-batch Cultivation Process Using Generalized Net Model. In: Proccedings of the 13th Int. Workshop on Generalized Nets, London, UK, October 29, pp. 15–23 (2012)
3. Dimitrov, D.G.: A Graphical Environment for Modeling and Simulation with Generalized Nets. Annual of "Informatics" Section, Union of Scientists in Bulgaria 3, 51–66 (2010) (in Bulgarian)
4. Dimitrov, D.G.: GN IDE - A Software Tool for Simulation with Generalized Nets. In: Proc. of the Tenth Int. Workshop on Generalized Nets, Sofia, pp. 70–75 (2009)
5. Dimitrov, D.G.: Integration of the Generalized Nets Simulator GNTicker with MATLAB. Annual of "Informatics" Section Union of Scientists in Bulgaria 4, 49–52 (2011) (in Bulgarian)
6. Dimitrov, D.G.: Software Products Implementing Generalized Nets. Annual of "Informatics" Section, Union of Scientists in Bulgaria 3, 37–50 (2010) (in Bulgarian)
7. Aladjov, H., Atanassov, K.: A Generalized Net for Genetic Algorithms Learning. In: Proc. of the XXX Spring Conference of the Union of Bulgarian Mathematicians, Borovets, Bulgaria, pp. 242–248 (2001)
8. Atanassov, K., Aladjov, H.: Generalized Nets in Artificial Intelligence. Generalized Nets and Machine Learning, vol. 2. Prof. M. Drinov Academic Publishing House, Sofia (2000)
9. Atanassov, K.: Generalized Nets and Systems Theory. Prof. M. Drinov Academic Publishing House, Sofia (1997)
10. Atanassov, K.: Generalized Nets. World Scientific, Singapore (1991)
11. Atanassov, K.: On Generalized Nets Theory. Prof. M. Drinov. Academic Publishing House, Sofia (2007)
12. Kosev, K., Trenkova, T., Roeva, O.: Tabu Search for Parameter Identification of an Fermentation Process Model. J. of Int. Scientific Publication: Materials, Methods & Technologies 6(2), 457–464 (2012)
13. Borisova, L., Krawczak, M., Zadrozny, S.: Modelling the Process of Introducing New Product to the Market via Generalized Nets. In: Atanassov, K., Kacprzyk, J., Krawczak, M. (eds.) Issues in Intuitionistic Fuzzy Sets and Generalized Nets, vol. 4, pp. 89–93. Wydawnictwo WSISiZ, Warszawa (2007)
14. Krawczak, M.: A Novel Modeling Methodology: Generalized Nets. In: Cader, A., Rutkowski, L., Tadeusiewicz, R., Zurada, J. (eds.) Artificial Intelligence and Soft Computing, pp. 1160–1168. Publishing House EXIT, Warsaw (2006)
15. Krawczak, M.: Algebraic Aspects of Generalized Nets, Development of Methods and Technologies of Informatics for Process Modeling and Management. In: Studzinski, J., Drelichowski, L., Hryniewicz, O. (eds.) Systems Research Institute, pp. 37–47. Polish Academy of Sciences, Warsaw (2006)
16. Krawczak, M.: Generalized Net Models of MLNN Learning Algorithms. In: Duch, W., Kacprzyk, J., Oja, E., Zadrożny, S. (eds.) ICANN 2005. LNCS, vol. 3697, pp. 25–30. Springer, Heidelberg (2005)
17. Roeva, O., Pencheva, T.: Generalized Net for Control of Temperature in Fermentation Processes. Issues in Intuitionistic Fuzzy Sets and Generalized Nets, Wydawnictwo WSISiZ, Warszawa 4, 49–58 (2007)

18. Roeva, O., Trenkova, T.: Genetic Algorithms and Firefly Algorithms for Non-linear Bioprocess Model Parameters Identification. In: Proccedings of the 4th Int. Joint Conference on Computational Intelligence (ECTA), Barcelona, Spain, October 5-7, pp. 164–169 (2012)
19. Roeva, O.: Generalized Net Model of Oxygen Control System using Intuitionistic Fuzzy Logic. In: Proccedings of the 1st Int. Workshop on Intuitionistic Fuzzy Sets, Generalized Nets and Knowledge Engineering. University of Westminister, London, September 6-7, pp. 49–55 (2006)
20. Roeva, O.: Optimization of *E. coli* Cultivation Model Parameters using Firefly Algorithm. Int. J. Bioautomation 16(1), 23–32 (2012)
21. Roeva, O.: Parameter Estimation of a Monod-type Model based on Genetic Algorithms and Sensitivity Analysis. In: Lirkov, I., Margenov, S., Waśniewski, J. (eds.) LSSC 2007. LNCS, vol. 4818, pp. 601–608. Springer, Heidelberg (2008)
22. Roeva, O., Pencheva, T., Tzonkov, S.: Generalized Net for Proportional-Integral-Derivative Controller. In: Book Series "Challenging Problems of Sciences" – Computer Sciences, pp. 241–247 (2008)
23. Chountas, P., Kolev, B., Tasseva, V., Atanassov, K.: Generalized Net Model for Binary Operations Over Intuitionistic Fuzzy OLAP Cubes. In: Proccedings of the Eighth Int. Workshop on Generalized Nets, Sofia, pp. 66–72 (2007)
24. Trifonov, T., Georgiev, K.: GNTicker – A Software Tool for Efficient Interpretation of Generalized Net Models. Issues in Intuitionistic Fuzzy Sets and Generalized Nets, Wydawnictwo WSISiZ, Warsaw 3, 71–78 (2006)
25. Trifonov, T., Georgiev, K., Atanassov, K.: Software for Modelling with Generalised Nets. Issues in Intuitionistic Fuzzy Sets and Generalized Nets, Wydawnictwo WSISiZ, Warsaw 6, 36–42 (2008)

Part XIII
Neural Networks, Modeling and Learning

An Approach to RBF Initialization with Feature Selection

Ireneusz Czarnowski and Piotr Jędrzejowicz

Department of Information Systems, Gdynia Maritime University
Morska 83, 81-225 Gdynia, Poland
{irek,pj}@am.gdynia.pl

Abstract. The paper focuses on a radial basis function network initialization. An application of the agent-based population learning algorithm to set RBF networks main parameters including number and locations of centroids is discussed. The main contribution of the paper is proposing and evaluating an agent-based approach to determine unique subset of features independently for each hidden unit. Two versions of the proposed algorithm for selecting values of the RBF networks parameters are considered. The approach is validated experimentally. Advantages and main features of the PLA-based RBF designs are discussed basing on results of the computational experiment.

Keywords: RBF networks, feature selection, population learning algorithm, multi-agent systems.

1 Introduction

Radial basis function networks (RBFNs), introduced by Bromhead and Lowe [1], are a popular type of the feedforward networks. RBF networks are used to obtain solutions to a variety of different problems and they provide accurate generalization property in a wide range of applications. They are known to serve well as an universal approximation tool, similarly to the multilayer perceptrons (MLPs). However, radial basis function networks usually achieve faster convergence since only one layer of weights is required [2].

An RBF network is constructed from only one hidden layer. RBF uses different activation functions at each node. The activation functions in the RBF nodes compute a distance between the input examples and the centers (centroids). Each hidden unit in the RBFN represents a particular point in the space of the input instances. The output of the hidden unit depends on the distance between the processed instances and the particular point in the input space of instances. The distance is calculated as a value of the activation function. Next, the distance is transformed into a similarity measure that is produced through a nonlinear transformation carried-out by the output function. The output function of the hidden unit is called the radial basis function. Each particular point in the space of the input instances, is called an initial seed point, prototype, centroid or kernel of the basis function.

Performance of the RBF network depends on numerous factors. The basic problem when designing an RBFN is to select an appropriate number of radial basis functions,

P. Angelov et al. (eds.), *Intelligent Systems'2014*,
Advances in Intelligent Systems and Computing 322, DOI: 10.1007/978-3-319-11313-5_59

i.e. a number of the hidden units. Deciding on this number results in fixing the number of clusters and their centroids. Another factor influencing the quality of the RBFN involves parameters describing the radial basis functions, such as a shape parameter of the radial basis function. Another problem concerns connection weights as well as the method used for learning the input-output mapping. Both have a direct influence on the RBFN performance, and both, traditionally, need to be somehow set. Hence, designing a RBFN is neither simple nor straightforward task

In general, designing the RBFN involves two stages: initialization and training. At the first stage the RBF parameters, including the number of centroids with their locations and the number of the radial basis functions with their parameters, need to be calculated or somehow induced. It is a very important stage from the point of view of achieving a good approximation and stability by the RBF network. At the second stage weights used to form a linear combination of the outputs of the hidden neurons are estimated. Also connection weights as well as the method used for learning the input-output mapping have a direct influence on the RBFN performance.

Different approaches to setting the RBF initialization parameters have been proposed in the literature. However, none of the approaches proposed, so far, for RBF network design can be considered as a superior and guaranteeing optimal results in terms of the learning error reduction or increased efficiency of the learning process. It can be observed that search for effective methods of designing RBF networks is far from being closed and a quest for robust and efficient approaches to RBFN construction and learning is still a lively and exciting field of research.

The paper focuses on a radial basis function network initialization. An application of agent-based population learning algorithm to determining RBF networks parameters including numbers and locations of centroids is discussed. The main contribution of the paper is proposing and evaluating an agent-based approach to determine unique subset of features independently for each hidden unit.

It is well known that through feature selection the dimension of data is decreasing which may result in classification accuracy improvement and savings of the computation time. In the proposed approach, it is assumed that kernels of the basis functions can be selected using an unique set of features where not all features are used as inputs of the hidden units. Thus, different feature masks are used for different radial basis functions and that, finally, means that the heterogeneous activation functions are introduced (see, for example, [3], where the problem of the heterogeneous RBFNs has been discussed). The assumption is based on the fact that "*a feature may have different power in discriminating classes, and thus a subset of features may be selected to discriminate one classes from other classes*" [4]. Following the above observation our approach aims at verifying whether introduction of the heterogeneous kernels can improve RBFN performance. Two versions of the proposed agent-based population learning algorithm to determining RBF networks parameters are considered. Within the first version, the agent-based population learning algorithm is used to select prototypes and to determine the best set of features for each class subset using the class-dependent feature selection procedure. In the second variant, the agent-based algorithm locates prototypes and determines the independent set of features for each cluster using the cluster-dependent feature selection procedure. To validate the approach, an extensive computational experiment has been carried out using several benchmark datasets from UCI repository [5].

The paper is organized as follows. Section 2 is dedicated to short description of the RBF networks. First the initialization of the RBF networks is discussed. Second, the approaches to increasing RBFNs performance are described. This is followed by some details of data processing procedure within RBFNs. The idea of the agent-based population learning algorithm is presented in Section 3. Section 4 provides details on the computational experiment setup and discusses its results. Finally, the last section contains conclusions and suggestions for future research.

2 RBF Neural Network

2.1 RBF Network Initialization

The RBF network initialization is a process, where the set of parameters, including the number of centroids with their locations and the number of the radial basis functions with their parameters, need to be calculated or somehow induced. It is a very important stage from the point of view of achieving a good approximation by the RBF network under development.

Different approaches to learning the RBF parameters have been proposed in the literature. Among classical methods used to RBFN initialization one should mention clustering techniques, such as the vector quantization or input-output clustering. Besides clustering methods, the support vector machine or the orthogonal forward selection approaches are used. In [6] several strategies for RBFN initialization were reviewed. They include random selection, clustering [7], sequential growing [8], [9], [10], systematic seeding and editing techniques [11], for determining centroids.

In [12] a similarity-based approach as an editing technique has been proposed for cluster initialization. The approach produces the clusters using the procedure based on the similarity coefficients calculated in accordance with the scheme proposed in [13]. Clusters contain instances with identical similarity coefficient and the number of clusters is determined by value distribution of this coefficient. Thus the clusters are initialized automatically, which also means that the number of radial basis function is initialized automatically. Under this procedure all the required operations including data transformation, calculation of the similarity coefficient values and vector mapping are carried-out automatically. Next from thus obtained clusters of instances centroids are selected. The similarity-based approach is also used to RBFN initialization within the approach proposed in this paper.

2.2 Data Dimensionality Reduction for RBF

Data dimensionality reduction techniques aim at decreasing the quantity of information required to learn a good quality classifiers, increasing learning efficiency and improving predictive performance. Data dimensionality reduction is also called editing, prototype selection, condensing, filtering, etc. It should be noted however, that the scope of the process depends on the object of reduction. Data reduction can be achieved by selection of instances, by selection of features or by simultaneous reduction in both dimensions.

Data reduction is an active research area in pattern recognition, machine learning, data mining, computer vision, bioinformatics, and etc. [14].

An interested example of implementation of data dimensionality reduction for computer vision has been discussed in [15]. It has been shown that it is generally difficult to meet cluster suitability criteria when different features have different meanings for given domain. According to this a dedicated approach to irrelevant feature selection, for specific regions in a video scene, based on clustering technique has been proposed. It has been shown that quality of video analysis can be increased by cluster modeling using heterogeneous set of features. In other words, the quality of video analysis and predictive performance have been increased by reducing data dimensionality via selection of heterogeneous set of features for produced clusters, i.e. by cluster-dependent feature selection.

In general, two type of feature selection tasks can be investigated: class-dependent and class-independent. The main difference between class dependent and independent approach is whether to use unique subsets for each class or uniform features for all classes [4], [16]. It is worth noting, that the individual discriminatory capability of each feature is often ignored. This should not be the case, since the individual discriminatory capability of each feature is a very important information hidden in the data [4]. Cluster-dependent feature selection is a special case of the class-dependent selection.

An example of the class-independent feature selection for RBFN has been investigated in [17]. In [16] a class-dependent feature selection for discriminating all considered classes by selecting heterogeneous subset of features for each class, has been proposed. The approach has been applied to solve the cancer biomarker discovery problem. It has been shown that the class-dependent feature selection can help select the most distinguishing feature sets for classifying different cancers. The results shown that the class-dependent approaches can effectively identify biomarkers related to each cancer type and improve classification accuracy as compared to class independent feature selection methods.

In [18] multiple MLP classifiers based on the class-dependent feature selection have been proposed. The main idea was to construct multiple MLP, for each class, and each one MLP classifier with feature inputs selected individually for a given class. In [4] and [19] a RBF classifier with class-dependent features has been proposed. For different groups of hidden units corresponding to different classes in the RBF neural network, different feature subsets have been selected as inputs. In this implementation, feature subsets have been produced by a generic algorithm. In [18] it was also concluded that a single RBF network is more suitable than multiple MLPs for the multi-class problem with selection of the class-dependent features.

It should be concluded that RBFNs based on class-dependent feature selection leads to a heterogeneous networks. Different feature masks are used for different kernel functions, thus each Gaussian kernel function of the RBF neural network is active for a subset of patterns. This results in obtaining a heterogeneous activation functions. It also means, that heterogeneous network uses different type of decisions at each stage, enabling discovery of the most appropriate bias for the data. Thus, according to [19], such heterogeneous network can be classified as a heterogeneous system.

2.3 RBF Network Processing

The output of the RBF network is a linear combination of the outputs of the hidden units, i.e. a linear combination of the nonlinear radial basis function generating approximation of the unknown function. In case of the classification problems the output value is produced using the sigmoid function with a linear combination of the outputs of the hidden units as an argument. In general, the RBFN output function has the following form:

$$f(x) = \sum_{i=1}^{M} w_i G_i(x, c_i, p_i) = \sum_{i=1}^{M} w_i G_i(\|x - c_i\|, p_i),$$

where M defines the number of hidden neurons, G_i denotes a radial basis function associated with *i-th* hidden neuron, p_i is a vector of parameters, which can include the location of centroids, dispersion or other parameters describing the radial function, x is an n-dimensional input instance ($x \in \Re^n$), $\{c_i \in \Re^n : i = 1,\dots, M\}$ are cluster's centers, $\{w_i \in \Re^n : i = 1,\dots, M\}$ represents the output layer weights and $\|\cdot\|$ is the Euclidean norm.

In a conventional RBF neural network, it is also true that all features are used as inputs for activation function. In this paper, a novel RBF classifier with feature selection for each initialized clusters is proposed. In the proposed RBF classifier hidden units are mapped on clusters that are obtained from different sets of features through transformation from original set of input features.

According to the proposed RBF initialization based on feature selection the RBF output function has the following form:

$$f(x) = \begin{cases} \sum_{i=1}^{M} w_i G_i(\|x \cdot g_i - c_i \cdot g_i\|, p_i), \text{ for the cluster-dependent feature selection} \\ \sum_{i=1}^{K} \sum_{j=1}^{k_i} w_{ij} G_{ij}(\|x \cdot g_i - c_i \cdot g_i\|, p_{ij}), \text{ for the class-dependent feature selection} \end{cases}$$

where g_i is the feature mask of i-th hidden neuron or class i, K is the number of classes, k_i is the number of hidden units serving i-th class.

3 Proposed Approach

The proposed approach to RBF initialization with feature selection aims to select both the optimal prototypes and feature sets for each hidden units. Each center c_i corresponding to hidden unit i is allowed to have own set of local features as a subset of original set of input features.

Selecting an optimal feature subset from a high dimensional space and the prototype selection are known to belong to the NP-hard problem class [20], [21]. From practical point of view, it means that the both problems are solved using approximation algorithms, which is a standard approach for solving the NP-hard optimization problems.

In this paper an agent-based population learning algorithm is used to optimize parameters in initialization stage of the RBFN design process. The main author's hypothesis is that the RBF network initialization using an agent-based population learning algorithm is an effective approach assuring, on average, better performance than the existing methods.

3.1 Agent-Based Population Learning Algorithm

In [22] it has been shown that agent-based population learning search can be used as a robust and powerful optimizing technique. In the agent-based population learning implementation both - optimization and improvement procedures are executed by a set of agents cooperating and exchanging information within an asynchronous team of agents (A-Team). The A-Team concept was originally introduced in [23].

Concept of the A-Team was motivated by several approaches like blackboard systems and evolutionary algorithms, which have proven to be able to successfully solve some difficult combinatorial optimization problems. Within an A-Team agents achieve an implicit cooperation by sharing a population of solutions, to the problem to be solved.

An A-Team can be also defined as a set of agents and a set of memories, forming a network in which every agent remains in a closed loop. Each agent possesses some problem-solving skills and each memory contains a population of temporary solutions to the problem at hand. It also means that such an architecture can deal with several searches conducted in parallel. In each iteration of the process of searching for the best solution agents cooperate to construct, find and improve solutions which are read from the shared, common memory.

In the discussed population-based multi-agent approach multiple agents search for the best solution using local search heuristics and population based methods. The best solution is selected from the population of potential solutions which are kept in the common memory. Specialized agents try to improve solutions from the common memory by changing values of the decision variables. All agents can work asynchronously and in parallel. During their work agents cooperate to construct, find and improve solutions which are read from the shared common memory. Their interactions provide for the required adaptation capabilities and for the evolution of the population of potential solutions.

Main functionality of the agent-based population learning approach includes organizing and conducting the process of search for the best solution. It involves a sequence of the following steps:

- Generation of the initial population of solutions to be stored in the common memory.
- Activation of optimizing agents which execute some solution improvement algorithms applied to solutions drawn from the common memory and, subsequently, store them back after the attempted improvement in accordance with the user defined replacement strategy.
- Continuation of the reading-improving-replacing cycle until a stopping criterion is met. Such a criterion can be defined either or both as a predefined number of iterations or a limiting time period during which optimizing agents do not manage to improve the current best solution. After computation has been stopped the best solution achieved so far is accepted as the final one.

3.2 Approach to the RBFN Initialization

The main goal of the proposed approach is to find the optimal set of RBF network parameters with respect to:

- Producing clusters and determining their centroids,
- Determining feature sets for each hidden units.

Under the proposed approach clusters are produced at the first stage of the initialization process. They are generated using the procedure based on the similarity coefficients calculated in accordance with the scheme proposed in [14] and described in subsection 2.1.

Next, for each clusters an independent heterogeneous subset of features and centroids are selected. An agent-based algorithm with a dedicated set of agents is used to locate centroids within clusters and feature selection.

Thus, in the proposed approach, an A-Team consists of agents which execute the improvement procedures and cooperate with a view to solve the RBF network initialization problem.

Most important assumptions behind the proposed approach, can be summarized as follows:

- Shared memory of the A-Team is used to store a population of solutions to the RBFN training problem.
- A solution is represented by a string consisting of two parts. The first contains integers representing numbers of instances selected as centroids. The length of the first part of the string is equal to the number of clusters (i.e. the number of hidden units). The second – the numbers of features chosen to represent the clusters.
- The initial population is generated randomly.
- Initially, potential solutions are generated through randomly selecting exactly one single centroid from each of the considered clusters and through randomly selection of the feature numbers.
- Each solution from the population is evaluated in terms of the RBFN performance and the value of its fitness is calculated. The evaluation is carried out by estimating classification accuracy or error approximation of the RBFN, which is initialized using prototypes and trained using the backpropagation algorithm.

To solve the RBFN training problem two groups of optimizing agents have been proposed. The first group includes agents executing procedures for centroid selection. To this end the following procedures have been implemented:

- Local search with the tabu list for prototype selection – this procedure modifies a solution by replacing a randomly selected reference instance with some other randomly chosen reference instance thus far not included within the improved solution. The modification takes place providing the replacement move is not on the tabu list. After the modification, the move is placed on the tabu list and remains there for a given number of iterations.
- Simple local search – this procedure modifies the current solution either by removing the randomly selected reference instance or by adding some other randomly selected reference instance thus far not included within the improved solution.

The second group of optimizing agents includes a local search procedure with tabu list for feature selection. This procedure modifies a solution by replacing a randomly selected feature with some other randomly selected feature thus far not included within the current solution, providing the move is tabu active and not on tabu list. It procedure has been introduced before in [14].

In each of the above cases solutions that are forwarded to the optimizing agents for improvement are selected from the population using in accordance with the so-called working strategy. For the presented algorithm each optimizing agent receives a solution drawn at random from the population of solutions. A returning individual replaces randomly selected individual in the common memory. In each cases the modified solution replaces the current one if it is evaluated as a better one. Evaluation of the solution is carried-out by estimating classification accuracy or error approximation of the respective RBFN.

4 Computational Experiment

This section contains results of several computational experiments carried-out with a view to evaluate the performance of the proposed approach. In particular, the reported experiments aimed at evaluating quality of the RBF-based classifiers initialized using the proposed approach. Experiments aimed at answering the question whether the proposed agent-based approach to RBF network initialization denoted further on as *ABRBFN_FS* performs better than classical methods of RBFN initialization (without feature selection)?

Two versions of the proposed approach have been validated. The first one, denoted as *ABRBFN_FS 1*, uses the class-dependent feature selection procedure. The second version, denoted as *ABRBFN_FS 2*, uses the cluster-dependent feature selection procedure. Here, the experiment aimed at answering the question which type of feature selection assures better results.

In the reported experiments the following RBFN approaches have been compared:

- The proposed similarity-based clustering approach with the agent-based population learning algorithm used to locate prototypes for each of the Gaussian kernels within the obtained clusters – *sim-ABRBFN*.
- The k-means clustering with the agent-based population learning algorithm used to locate prototypes (in this case at the first stage the k-means clustering has been implemented and next, from thus obtained clusters, the prototypes have been selected using the agent-based population learning algorithm) – denoted as *k-meansABRBFN*.
- The k-means algorithm used to locate centroids for each of the Gaussian kernels (in this case at the first stage the k-means clustering has been implemented and the cluster centers have been used as prototypes) – denoted as k-*meansRBFN*.
- The random search for kernel selection – denoted as *randomRBFN*.

Evaluation of the proposed approaches and performance comparisons are based on the classification and the regression problems. In both cases the proposed algorithms have been applied to solve respective problems using several benchmark datasets obtained from the UCI Machine Learning Repository [5]. Basic characteristics of these datasets are shown in Table 1.

Each benchmark problem has been solved 50 times, and the experiment plan involved 10 repetitions of the 10-cross-validation scheme. The reported values of the goal function have been averaged over all runs. The goal function, in case of the classification problems, was the correct classification ratio – accuracy (Acc). The overall goal function for regression problems was the mean squared error (MSE) calculated as the approximation error over the test set.

For algorithms based on k-*means* and random initialization the number of prototypes has been set in a way assuring comparability with respect to the number of prototypes produced by the similarity-based procedure.

During the experiment population size for each investigated A-Team architecture was set to 60. The process of searching for the best solution has been stopped either after 100 iterations or after there has been no improvement of the current best solution for one minute of computation. The number of epoch for backpropagation cycles has been set to 500. Values of these parameters have been set arbitrarily.

Table 1. Datasets used in the experiment

Dataset	Type of problem	Number of instances	Number of attributes	Number of classes	Best reported results
Forest Fires	Regression	517	12	-	-
Housing	Regression	506	14	-	-
WBC	Classification	699	9	2	97.5% [25] (Acc.)
ACredit	Classification	690	15	2	86.9% [25] (Acc.)
GCredit	Classification	999	21		77.47% [26] (Acc.)
Sonar	Classification	208	60	2	97.1% [25] (Acc.)
Diabetes	Classification	768	9	2	77.34% [25] (Acc.)

The Gaussian function has been used for computation within the RBF hidden units. The dispersion of the Radial function has been calculated as a double value of minimum distance between basis functions [1].

The proposed A-Team has been implemented using the middleware environment called JABAT [22].

Table 2 shows mean values of the classification accuracy of the classifiers (Acc) and the means squared error of the function approximation models (MSE) obtained using the RBFN architecture initialized by the proposed approach. Table 2 also contains results obtained by the RBF network initialized using the set of prototypes produced through clustering based on the k-*means* algorithm and through random selection.

From Table 2, one can observe that the best results have been obtained by the *ABRBFN_FS_2* using the cluster-dependent feature selection procedure. It can be also observed that, in general, the competitive results have been obtained by the

ABRBF_FS algorithm. The *ABRBFN_FS* can be considered as superior to the other methods including MLP, Multiple linear regression, SVR/SVM and C4.5. This statement is supported by the fact that in three cases the proposed algorithm (its considered versions) have been able to provide better generalization ability as compared to other methods.

Table 2. RBFN results obtained for different initialization schemes compared with performance of some different competitive models

Problem / Algorithm	Forest fires	Housing	WBC	ACredit	GCredit	Sonar	Diabetes
	MSE		Acc. (%)				
ABRBF_FS 1	2.08	**34.17**	95.02	83.4	70.08	81.71	**76.8**
ABRBF_FS 2	**1.98**	34.71	**96.11**	**86.13**	**71.31**	**83.14**	75.1
sim- ABRBFN	2.15	35.24	94.56	84.56	70.01	82.09	**77.6**
k-*meansABRBFN*	2.29	35.87	95,83	84.16	70.07	81.15	73.69
k-*meansRBFN*	2.21	36.4	93.9	82.03	68.3	78.62	70.42
randomRBFN	3.41	47.84	84.92	77.5	67.2	72.79	62.15
Neural network – MLP	2.11[24]	40.62[24]	96.7[25]	84.6[25]	**77.2[25]**	**84.5**[25]	70.5[25]
Multiple linear regression	2.38[24]	36.26[24]	-	-	-	-	-
SVR/SVM	**1.97**[24]	44.91[24]	**96.9**[25]	84.8[25]	-	76.9[25]	75.3[25]
C 4.5	-	-	94.7[25]	85.5[25]	70.5	76.9[25]	73.82[26]

5 Conclusions

The paper investigates an influence of feature selection on the quality and performance of RBF networks. Two feature selection approaches have been used, i.e. class-dependent feature selection and cluster-dependent feature selection. The paper proposes an integration of feature selection and prototype selection with an agent-based population learning algorithm used to locate prototypes within obtained clusters and determine unique subset of features independently for each hidden unit. The RBFN performance using the proposed approach is analyzed and experimentally evaluated. It can be observed that in most cases our approach outperforms other methods used for designing RBFNs.

The important property of the proposed initialization technique is that the number of clusters is determined on the fly. The role of the applied agent-based population learning algorithm is to select the appropriate prototypes and reduce the dimensional space. In the reported computational experiment the proposed algorithm outperforms other techniques for RBF initialization.

Future research will focus on identifying the influence of different feature selection on the RBFN quality with a view to further increase effectiveness of the approach.

References

1. Broomhead, D.S., Lowe, D.: Multivariable Functional Interpolation and Adaptive Networks. Complex Systems 2, 321–355 (1988)
2. Gao, H., Feng, B., Zhu, L.: Training RBF Neural Network with Hybrid Particle Swarm Optimization. In: Wang, J., Yi, Z., Żurada, J.M., Lu, B.-L., Yin, H. (eds.) ISNN 2006. LNCS, vol. 3971, pp. 577–583. Springer, Heidelberg (2006)
3. Wilson, D.R., Martinez, T.R.: Heterogeneous Radial Basis Function Networks. In: Proceedings of the International Conference on Neural Networks (ICNN 1996), vol. 2, pp. 1263–1267 (1996)
4. Wang, L., Fu, X.: A Rule Extraction System with Class-Dependent Features. In: Ghosh, A., Jain, L.C. (eds.) Evolutionary Computation in Data Mining. STUDFUZZ, vol. 163, pp. 79–99. Springer, Heidelberg (2005)
5. Asuncion, A., Newman, D.J.: UCI Machine Learning Repository. University of California, School of Information and Computer Science, Irvine (2007), http://www.ics.uci.edu/~mlearn/MLRepository.html (accessed June 24, 2009)
6. Krishnaiah, P.R., Kanal, L.N.: Handbook of Statistics 2: Classification, Pattern Recognition and Reduction of Dimensionality. North Holland, Amsterdam (1982)
7. Qasem, S.N., Shamsuddin, S.M.H.: Radial Basis Function Network Based on Multi-Objective Particle Swarm Optimization. In: Proceeding of the 6th International Symposium on Mechatronics and its Applications (ISMA 2009), Sharjah, UAE, March 24-26 (2009)
8. Huang, G.-B., Saratchandra, P., Sundararajan, N.: A Generalized Growing and Pruning RBF(GGAP-RBF) Neural Network for Function Approximation. IEEE Transactions on Neural Networks 16(1), 57–67 (2005)
9. Ros, F., Pintore, M., Chretie, J.R.: Automatic Design of Growing Radial Basis Function Neural Networks Based on Neighboorhood Concepts. Chemometrics and Intelligent Laboratory Systems 87, 231–240 (2007)
10. Wei, L.Y., Sundararajan, N., Saratchandran, P.: Performance Evaluation of a Sequential Minimal Radial Basis Function (RBF) Neural Network Learning Algorithm. IEEE Transactions on Neural Networks 9, 308–318 (1998)
11. Wilson, D.R., Martinez, T.R.: Reduction Techniques for Instance-based Learning Algorithm. Machine Learning 33(3), 257–286 (2000)
12. Czarnowski, I., Jędrzejowicz, P.: Agent-Based Approach to the Design of RBF Networks. Cybernetics and Systems 44(2-3), 155–172 (2013)
13. Czarnowski, I.: Cluster-based Instance Selection for Machine Classification. Knowledge and Information Systems 30(1), 113–133 (2012)
14. Czarnowski, I.: Distributed Learning with Data Reduction. In: Nguyen, N.T. (ed.) TCCI IV 2011. LNCS, vol. 6660, pp. 3–121. Springer, Heidelberg (2011)
15. Harit, G., Chaudhury, S.: Clustering in video data: Dealing with heterogeneous semantics of features. Journal Pattern Recognition 39(5), 789–811 (2006)
16. Zhu, W., Dickerson, J.A.: A novel class dependent feature selection methods for cancer biomarker discovery. Computers in Biology and Medicine 47, 66–75 (2014)

17. Novakovic, J.: Wrapper approach for feature selections in RBF network classifier. Theory and Applications of Mathematics & Computer Science 1(2), 31–41 (2011)
18. Oh, I.S., Lee, J.S., Suen, C.Y.: Analysis of class separation and combination of class-dependent features for handwriting recognition. IEEE Trans. PAMI 21, 1089–1094 (1999)
19. Duch, W., Grąbczewski, K.: Heterogeneous adaptive systems. In: IEEE World Congress on Computational Intelligence, Honolulu, pp. 524–529 (2002)
20. Hamo, Y., Markovitch, S.: The COMPSET Algorithm for Subset Selection. In: Proceedings of The Nineteenth International Joint Conference for Artificial Intelligence, Edinburgh, Scotland, pp. 728–733 (2005)
21. Kohavi, R., John, G.H.: Wrappers for Feature Subset Selection. Artificial Intelligence 97(1-2), 273–324 (1997)
22. Barbucha, D., Czarnowski, I., Jędrzejowicz, P., Ratajczak-Ropel, E., Wierzbowska, I.: e-JABAT - An Implementation of the Web-Based A-Team. In: Nguyen, N.T., Jain, I.C. (eds.) Intel. Agents in the Evol. of Web & Appl. SCI, vol. 167, pp. 57–86. Springer, Berlin (2009)
23. Talukdar, S., Baerentzen, L., Gove, A., de Souza, P.: Asynchronous Teams: Co-operation Schemes for Autonomous, Computer-Based Agents. Technical Report EDRC 18-59-96, Carnegie Mellon University, Pittsburgh (1996)
24. Zhang, D., Tian, Y., Zhang, P.: Kernel-based Nonparametric Regression Metod. In: Proceedings of the IEEE/WIC/ACM International Conference on Web Intelligence and Intelligent Agent Technology, pp. 410–413 (2008)
25. Datasets used for classification: comparison of results. In. directory of data sets, `http://www.is.umk.pl/projects/datasets.html` (accessed September 1, 2009)
26. Jędrzejowicz, J., Jędrzejowicz, P.: Cellular GEP-Induced Classifiers. In: Pan, J.-S., Chen, S.-M., Nguyen, N.T. (eds.) ICCCI 2010, Part I. LNCS, vol. 6421, pp. 343–352. Springer, Heidelberg (2010)

Artificial Neural Network Ensembles in Hybrid Modelling of Activated Sludge Plant

Jukka Keskitalo and Kauko Leiviskä*

University of Oulu, Control Engineering, P.O Box 4300,
FIN-90014 Oulun yliopisto, Finland
kauko.leiviska@oulu.fi

Abstract. Combining first-principles knowledge in the form of a mechanistic model with artificial neural networks (ANN) into so-called hybrid models is an attractive approach to improve models for biological wastewater treatment. Although neural networks have been proven to be an effective method in learning nonlinear input-output mappings, their generalization ability is of concern. Generalization ability of ANNs has been improved by combining several ANNs into ensembles, where each ANN provides a solution to the same problem, and the solutions are combined into a single ensemble output. In this paper, a parallel hybrid modeling approach was developed where ANN ensembles were used to improve Activated Sludge Model (ASM) predictions in modeling a pulp mill wastewater treatment plant. This approach was successful in improving the generalization ability of the ASM in modeling three different pollutants. The best performing methods were the cross-validation ensemble and bagging, and in calculating the ensemble output, simple averaging outperformed stacking. It was also shown that the generalization performance was improved by adding more members in the ensemble.

Keywords: Activated sludge model, Hybrid model, Artificial neural networks, Neural network ensemble.

1 Introduction

Biological wastewater treatment has been extensively used for decades in the treatment of municipal and industrial wastewaters. Activated sludge process (ASP) in its numerous different configurations is the most common way in this case (Gernaey et al., 2004). Despite its long history of application, operation of the activated sludge process is not without problems. Industrial wastewater treatment plants (WWTP), especially in pulp and paper mills, have operational problems resulting in incidental discharges to receiving waterways. The composition and flow rates of industrial wastewater fluctuate depending on the production schedules of the upstream processes. Some of the wastewater components are even toxic for the microorganisms in biological WWTPs. Reasons for the occasionally deteriorating treatment perfor-

* Corresponding author.

P. Angelov et al. (eds.), *Intelligent Systems'2014*,
Advances in Intelligent Systems and Computing 322, DOI: 10.1007/978-3-319-11313-5_60

mance are often unknown, and therefore it is difficult to find suitable corrective control actions. As the limitations for effluent discharges are getting more stringent, there is a need to achieve better control of the process. Development of process models for the WWTPs is a possible method for achieving better understanding and control of the process (Gernaey et al., 2004).

Both mechanistic models based on first-principles knowledge and data-driven black-box models have been developed for ASPs. The mechanistic Activated Sludge Models (ASM) published by the IWA Task Group on Mathematical Modelling for Design and Operation of Biological Wastewater Treatment (Henze et al., 2000) are by far the most widely used models for ASPs. The ASMs have been applied to modeling both municipal (Brun et al., 2002; Nuhoglu et al., 2005; Koch et al., 2000) and industrial wastewater treatment (Barañao and Hall, 2004; Keskitalo et al., 2010; Hulle and Vanrolleghem, 2004), although they were primarily developed for municipal WWTPs (Gernaey et al., 2004). The ASM1 has also been modified for pulp and paper mill wastewaters (Lindblom et al., 2004). The modified ASM1 was applied to modeling a full-scale pulp mill WWTP by Keskitalo et al. (2010). It was noticed, however, that the full-scale process had non-modeled phenomena. In a long-term simulation, there were periods during which the model could not predict the treatment performance accurately (Keskitalo et al., 2010).

Black-box approaches for modeling activated sludge process were developed concurrently with the ASM development. Black-box modeling of ASPs was motivated by the development of new modeling methods such as artificial neural networks (ANN). Problems with mechanistic models, e.g. poor structural and practical identifiability, together with calibration only for certain operating conditions, and the lack of information on sludge settling properties, also generated interest in applying black-box methods. Black-box methods applied to modeling ASPs include ANNs (Gontarski et al., 2000), multivariate statistical methods (Mujunen et al., 1998), neuro-fuzzy modeling (Wan et al., 2011) and combinations of the methods (Oliveira-Esquerre et al., 2002). However, black-box modeling of ASPs has problems of its own. The accuracy of purely data-based models is often poor due to the limitations of available process data. The black-box approach has also been criticized for the lack of physical interpretation of parameters and the poor extrapolation capability of the models (Lee et al., 2005).

Combining first-principles knowledge in the form of deterministic model equations with neural networks into so-called hybrid models is a promising approach to overcome the problems in modeling biological wastewater treatment. The available a priori knowledge is provided in the mechanistic model while the neural network accounts for the unknown parameters or non-modeled dynamics. Hybrid models have been applied in serial and parallel configurations. In the serial configuration, the neural network estimates the unknown parameters of the mechanistic model (Psichogios and Ungar, 1992). The neural network in the parallel configuration estimates the difference between the mechanistic model outputs and the real process values (Lee et al., 2005). The parallel hybrid model output is obtained by summing the outputs of the mechanistic model and the neural network.

Studies have shown that using the combined output of multiple learning machines in an approach called ensemble model results in improved accuracy over a single model (Brown, 2010). The ensemble approach, where each ANN provides a solution to the same problem, and the solutions are combined into a single ensemble output, has become popular in improving generalization ability of ANNs. Applications of ANN ensembles for regression problems include the prediction of the water discharge to a river (Ahmad and Zhang, 2005), the estimation of the ice thickness on lakes (Zaier et al., 2010) and use in a hybrid model structure for the prediction of leaching rate (Hu et al., 2011). A number of methods exist for both generating individual ensemble members and combining the outputs. Performance of the ensemble methods depends on the data and the applied learning machine (Graczyk et al., 2010; Opitz and Maclin, 1999; West et al., 2005). Therefore, it is difficult to decide the best ensemble method prior to the application.

A parallel hybrid modeling approach is developed in this paper where ANN ensembles are used to improve ASM predictions. Popular methods for ensemble generation and output combination are compared. The developed ASM – ANN ensemble hybrid is applied for modeling a pulp mill WWTP. The generalization ability of the developed models is evaluated on an independent testing data set which has not been used for calibration or training the models. The good generalization ability is necessary for applying the models in practical applications, where the models are expected to stay accurate for long periods after calibration.

This paper is structured as follows. Section 2 introduces ensemble approaches including the model structures for individual ensemble members, the methods for generating ensembles members and the methods for combining the outputs of the ensemble members. The pulp mill WWTP and the hybrid modeling approach are described in Section 3. Section 4 describes the experiments conducted to evaluate the performance of the ensemble methods and Section 5 includes results and discussion. Section 6 is the conclusion.

2 Materials and Methods

2.1 Process and Data

The WWTP under study is an aerobic activated sludge plant designed for the removal of suspended solids and organic carbonaceous material from bleached sulfate pulp mill wastewater. Simplified diagram of the process is presented in Fig. 1. Wastewater is treated in the primary sedimentation before the biological treatment. The primary treatment stages also include pH control, cooling and nitrogen dosing. The biological treatment consists of an aeration basin and a secondary clarifier. In the aeration basin, microorganisms oxidize carbonaceous material for producing energy and cell growth. Sludge from the aeration basin is settled in the secondary clarifier. Clarified effluent of the secondary clarifier is discharged to the sea and settled sludge is returned to the aeration basin to maintain sufficient solids concentration.

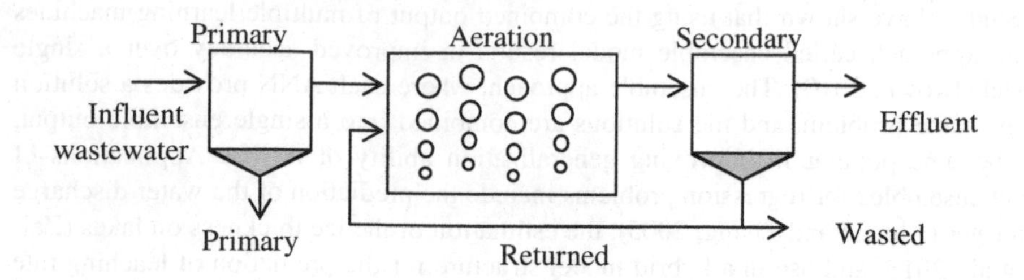

Fig. 1. Simplified diagram of the activated sludge process

The process data is extracted from the pulp mill databases. The primary clarifier is not included in this study. Therefore the influent data is from the aeration basin influent. Effluent data refers to the secondary clarifier effluent discharged into the sea. The process data covers the period from 2 February 2010 to 31 March 2011 and it has been split into training and testing data sets of equal size.

2.2 Mechanistic Model

When applying the Activated Sludge Models, the aeration basin of the activated sludge process is modeled as a continuous stirred-tank reactor (CSTR) or as a series of CSTRs. The ASMs describe the reactions of the wastewater components, for which mass balances are written as ordinary differential equations. The mechanistic model is a modified version of the original Activated Sludge Model No. 1 (ASM1) (Henze et al., 2000) modified by Lindblom et al. (2004) for modeling the biological treatment of nutrient deficient wastewaters. The original ASM1 was developed to describe simultaneous removal of organic nitrogen and carbon. Concentration of nitrogen in pulp mill wastewaters is, however, low and the biological nitrogen removal has been omitted from the modified ASM. Limitations on heterotrophic biomass growth resulting from nitrogen- and phosphorus-deficient conditions have also been considered in the modified model in the pulp mill case (Keskitalo et al., 2010).

Parameters of the ASMs are not universal and most applications require adjustment of at least some of the parameters. Calibration of the ASMs consists of choosing a subset of model parameters and identifying values for parameters in this subset. The subset is often chosen based on practical identifiability analysis (Weijers and Vanrolleghem, 1997; Brun et al., 2002). Parameter values in the subset can be adjusted either manually or by an optimization algorithm. In this paper, the calibrated parameter values from Keskitalo and Leiviskä (2011) are used. There, practical identifiability of the model has been analyzed following the methodology of Brun et al. (2002) and evolutionary optimizers have been used to identify the model parameters.

2.3 Hybrid Mechanistic-ANN Ensemble Modeling of the Activated Sludge Plant

In the parallel hybrid model developed in this paper, a black-box model is used in parallel with the mechanistic model in order to reduce the error of the mechanistic model predictions. The modified ASM1 presented in the previous section is used as the mechanistic model. The parallel hybrid model is later called an ensemble hybrid model (EHM) when the black-box model is an ensemble of models. The term a single hybrid model (SHM) is used when the black-box model is a single model. The parallel hybrid model structure is shown in Fig. 2. Parallel hybrid model utilizes the information available in the residuals of the mechanistic model (Lee et al., 2005). The residuals can be expected to contain non-modeled dynamics, i.e. effects of disturbances and variables not included in the mechanistic model

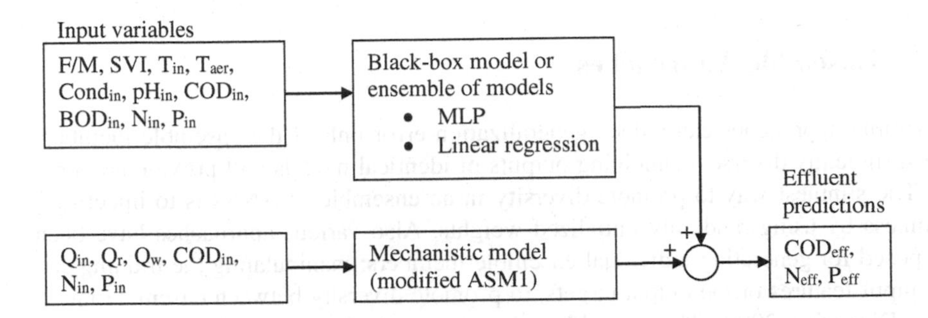

Fig. 2. Parallel hybrid model structure

Inputs in both the mechanistic and the black-box models are influent concentrations of COD (COD_{in}), total nitrogen (N_{in}) and total phosphorus (P_{in}). Flow rates of influent wastewater (Q_{in}), return sludge (Q_r) and wasted sludge (Q_w) are fed to the mechanistic model. Food to microorganism ratio (F/M), aeration basin temperature (T_{aer}), SVI, and influent wastewater temperature (T_{in}), pH (pH_{in}), conductivity ($Cond_{in}$) and BOD concentration (BOD_{in}) are used as inputs in the black-box model. The mechanistic model outputs considered in this paper are total COD, soluble N and soluble P. The black-box model outputs are correspondingly the residuals of total COD, soluble N and soluble P. The parallel hybrid model outputs are obtained by summing the outputs from the mechanistic model and the black-box model.

Two black-box modeling methods are used in this paper: multiple linear perceptron (MLP) and multiple linear regression (MLR). Three black-box models are tested: (1) one MLP network for estimating residuals of all three outputs (denoted by 1xMLP in the sequel) (2) a separate MLP network for each output (3xMLP) and (3) a multiple linear regression model (MLR) for estimating all outputs. 1xMLP and 3xMLP can consist of a single network (SHM) or several networks forming a network ensemble (EHM). The MLR model is used in comparison with the MLP models to evaluate the benefits of more complex model structures capable of estimating nonlinear functions.

In this case, the comparison between different models is based on the performance criterion, which is the weighted sum of the normalized root mean square errors (RMSE) of the COD, N and P predictions:

$$J_i = \sum_{j=1}^{3} w_j \frac{RMSE_{i,j} - RMSE_j^{\min}}{RMSE_j^{\max} - RMSE_j^{\min}}$$

where J_i is the performance criterion value for the ensemble method i, w_j are the weight factors, $RMSE_{i,j}$ is the RMSE value of predictions of the jth output in the method i and $RMSE_j^{\min/\max}$ are the minimum and maximum RMSE values of all the models in the prediction of the jth output.

3 Ensemble Approaches

Ensemble approaches can reduce generalization error only if the ensemble members are sufficiently diverse. Combining outputs of identical models will provide no benefit. The simplest way to promote diversity in an ensemble of ANNs is to inject randomness by using randomly initialized weights. Also various approaches have been proposed for generating individual ensemble members: manipulating the training set, the input features or the output targets, to promote diversity between ensemble members (Dietterich, 2000). The ensemble generation methods used in this paper are based on manipulating the training set, namely bagging, boosting and cross-validation ensemble.

Bagging and *boosting* are popular ensemble generation methods which create diversity by manipulating the training data sets. The bagging (bootstrap aggregating) algorithm (Breiman, 1996a) creates varied but overlapping data sets by drawing samples randomly and with replacement from the original training set. Each time a sample is drawn from the original training set of *N* samples, each sample may be picked with the equal probability of 1/*N*. Data set constructed with bagging may have some samples appearing multiple times while others are left out. The data set generation is repeated for each member of the ensemble. Bagging gives the most benefit when the base model is unstable, i.e. small changes in the training examples can produce large changes in the model parameters and behavior (Breiman, 1996a; Barutçuoğlu and Alpaydin, 2003; Brown, 2010)

Boosting methods produce ensemble members in sequence, where each new model has higher probability of being trained on samples which were not predicted well by the previous model. Each sample in the training set is assigned a probability of being included when choosing samples for training the ensemble members. The probabilities are updated based on the models accuracy of estimation. Several different boosting algorithms have been published since Schapire (1990) proposed the original idea. While the research on boosting has been mostly focused on classification problems, few boosting algorithms are available also for regression tasks (Freund and Schapire, 1997; Drucker, 1997; Shrestha and Solomatine, 2006). Drucker's boosting algorithm for regression tasks (Drucker, 1997) is used in this paper as it allows an efficient and simple implementation.

The cross-validation ensemble (Krogh and Vedelsby, 1995) is another method which resamples the training set to advance diversity. This method has also been called cross-validated committees (Parmanto et al., 1996, cited in Dietterich, 2000) and cross-validation aggregating (Barutçuoğlu and Alpaydin, 2003). The training set is randomly separated into K sets of equal or almost equal size. For each K members of the ensemble, one of the K sets can be used to estimate the generalization error while the remaining sets are used in training. In this paper, an independent test data set is used for evaluating generalization ability. The samples left out from the training data are used for early stopping in ANN training.

Another approach for promoting diversity in ensemble generation is the manipulation of input variables available in training of the individual members (Dietterich, 2000). Methods using manipulation of input variables are the random subspace method for constructing decision tree based classifier (Ho, 1998), extension of the random subspace method for regression problems (Rooney et al., 2004) and random forests (Breiman, 2001). In the method developed in this paper, each member in an ensemble is trained with a different random subset of the input variables. This method has the potential to avoid overfitting while promoting diversity. Diversity in ensembles was further enhanced by combining randomization of the number of neurons in the MLP hidden layer with the random selection of input variables. This method is later called randomization ensemble.

After individual ensemble members have been created using one the methods described before the second step is to combine the outputs of the individual members to produce the ensemble output. A commonly used and simple approach for ensemble combination is computing the mean of individual ensemble member outputs. As an alternative to the simple mean, the weighted mean can be used. Stacking (Wolpert, 1992; Breinan, 1996b) was proposed to improve generalization ability of ensembles. Stacking utilizes cross-validation data and applies non-negativity constraints when determining weights in the linear combination of ensemble outputs.

4 Results and Discussion

4.1 Modeling the Pulp Mill WWTP

Comparison of performance criterion values for the EHMs and the SHMs are presented in Table 1. Smaller performance criterion values indicate better generalization performance. The results are presented separately for three different model structures, as explained before. The results in Table 1 are the best results found for each method when different parameter combinations were tested. Parameters were typical ANN parameters: network size and architecture.

From Table 1 it can be seen, that bagging and cross-validation ensembles have the best generalization performance. All ensemble methods except boosting produce better generalization performance than the SHM. Performance criterion values for bagging and cross-validation ensemble are almost equal. Stacking decreases generalization performance of all ensemble generation methods except boosting when compared to simple averaging. 3xMLP was the best performing and linear regression was the worst of the three different model structures.

Table 1. Best mean performance criterion values on the testing data for each ensemble method in the parallel hybrid structure. The table is for testing data.

Black-box method in the parallel hybrid	1xMLP	3xMLP	MLR
Single black-box model	0.0769	0.0686	0.0753
Bagging ensemble	0.0377	0.0303	0.0753
Bagging and stacking ensemble	0.072	0.0606	0.0776
Cross-validation ensemble	0.0364	0.0304	0.0764
Cross-validation ensemble and stacking	0.0567	0.0507	0.0756
Boosting ensemble	0.0888	0.0806	0.1627
Boosting and stacking ensemble	0.0871	0.0712	0.0909
Randomization ensemble	0.0445	0.0436	0.0639
Randomization ensemble and stacking	0.0626	0.0673	0.0656

Table 2 shows the results for the modified ASM1 without the hybrid structure, the best performing EHM and the best performing SHM. Significant improvements of generalization performance were achieved by using either the EHM or the SHM, with the EHM being slightly better. The RMSE reduction of COD predictions for the testing data was larger than for the training data. For P predictions, similar RMSE reduction was achieved. The RMSE reduction of N predictions for the testing data was significantly smaller than for the training data.

Table 2. RMSE of ASM without hybrid structure, the best performing EHM and the best performing SHM for each component in the outflow.

	Training data, RMSE			Testing data, RMSE		
	COD	Tot N	Tot P	COD	Tot N	Tot P
Best ensemble parallel hybrid (bagging, 3xMLP)	224.3896	0.3607	0.1077	127.7105	0.2311	0.1594
Best single black-box model parallel hybrid (3xMLP)	240.4771	0.4436	0.1278	131.0139	0.2406	0.1741
Modified ASM1, without hybrid structure	276.6356	0.5859	0.1787	188.8622	0.2508	0.2621

Boosting and stacking were proposed as improvements to the early ensemble approaches such as bagging. There are numerous studies reporting improved performance by using boosting or stacking (e.g. Dietterich, 2000; Drucker, 1997; Shrestha and Solomatine, 2006; Zaier et al., 2010). In this study, the boosting method had the worst generalization ability of the evaluated ensemble generation methods and using stacking instead of simple averaging decreased generalization performance. However, the results obtained in this study are not unique. Bagging has been found to perform better than boosting also by Barutçuoğlu and Alpaydin (2003) and Pardo et al. (2010). Boosting has been found to be sensitive to noise (Opitz and Maclin, 1999; West et al., 2005), which may partly explain the results obtained in this study. The contradicting results suggest that no single ensemble method has the best performance in all applications.

Generalization performance of the EHMs with MLPs as the ensemble members was better than with MLR models for all the applied ensemble methods. This suggests that the relationships between input and output variables are nonlinear, and simple linear models are insufficient for the task. On the other hand, the success of the ensemble approaches has been found to depend on sufficiently diverse ensemble members. All the MLPs used in this study had randomly initialized weights, which contributed to the diversity of the ensembles.

Simulation results of the best EHM and the ASM without hybrid structure are shown in Figs. 3-5. The ensemble method in the simulations is bagging with 3xMLP. Fig, 3 has the results for effluent COD, Fig. 4 for effluent N and Fig. 5 for effluent P. The split between training and testing data is indicated by the dotted vertical line in Figs. 3-5. Training data is on the left side of the line and testing data on the right. In Table 2, it was shown that the RMSE reduction of N predictions for the testing data was significantly smaller than for the training data. Fig. 4 shows that N concentration in effluent was rather constant in the testing data, and the ASM without hybrid structure was already capable of predicting the correct level. Therefore, there was not as much room for improvement in the testing data as there was in the training data.

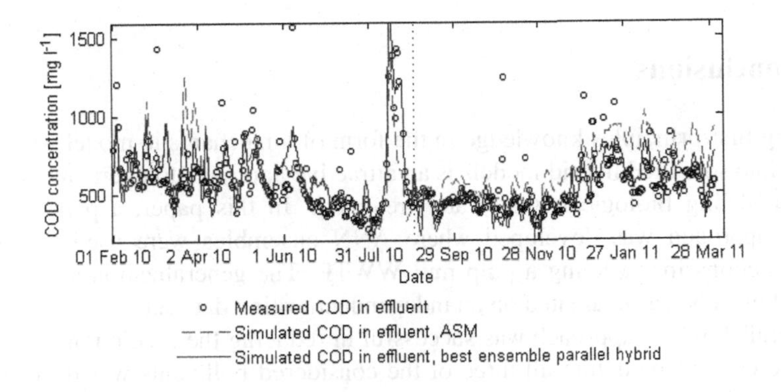

Fig. 3. The simulated and measured COD concentrations in the pulp mill WWTP effluent. Simulation results are shown for the best EHM and the ASM without the hybrid structure.

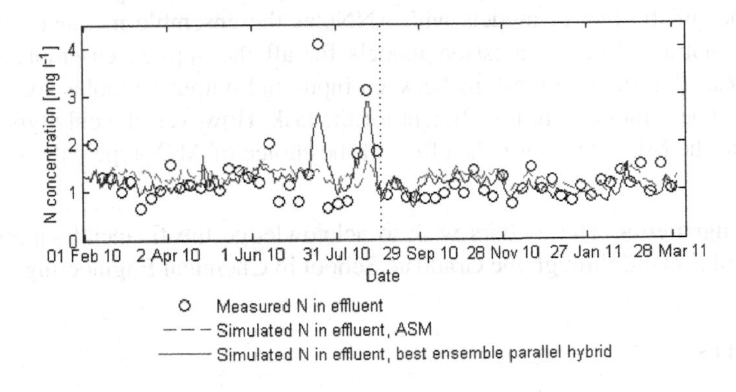

Fig. 4. The simulated and measured N concentrations in the pulp mill WWTP effluent. Simulation results are shown for the best EHM and the ASM without the hybrid structure.

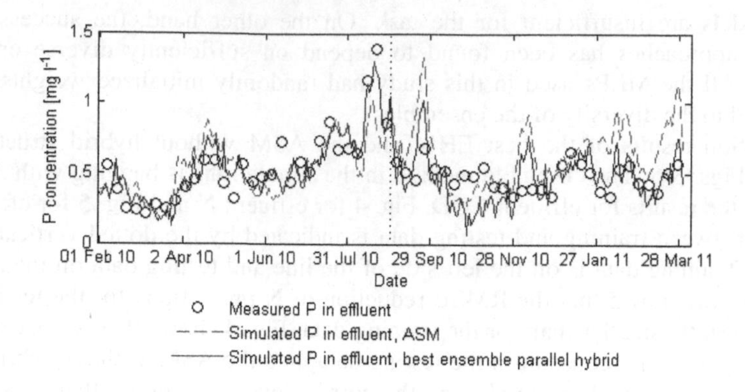

Fig. 5. The simulated and measured P concentrations in the pulp mill WWTP effluent. Simulation results are shown for the best EHM and the ASM without the hybrid structure.

5 Conclusions

Combining first-principles knowledge in the form of a mechanistic model with neural networks into so-called hybrid models is an attractive approach to overcome the problems in modeling biological wastewater treatment. In this paper, a parallel hybrid modeling approach was developed where ANN ensembles were used to improve ASM predictions in modeling a pulp mill WWTP. The generalization ability of the developed models was evaluated on an independent testing data set

The parallel hybrid approach was successful in reducing the prediction error of the ASM. The error in modeling all three of the considered pollutants was decreased by using the parallel hybrid approach. The best performing ensemble generation methods were the cross-validation ensemble and bagging, and in combining the ensemble output, simple averaging outperformed stacking. The best model structure as an individual ensemble member was a separate MLP for each of the outputs. Generalization performance of the hybrid models with ANNs as the ensemble members was better than with multiple linear regression models for all the applied ensemble methods. This suggests that the relationships between input and output variables are nonlinear, and simple linear models are insufficient for the task. However, the only type of ANN applied was the MLP. In future, the effect of the choice of ANN type will be studied.

Acknowledgements. The authors wish to acknowledge the financial support of the Academy of Finland through the Graduate School in Chemical Engineering.

References

1. Ahmad, Z., Zhang, J.: Combination of multiple neural networks using data fusion techniques for enhanced nonlinear process modelling. Computers and Chemical Engineering 30, 295–308 (2005)

2. Barañao, P.A., Hall, E.R.: Modelling carbon oxidation in pulp mill activated sludge systems: calibration of Activated Sludge Model 3. Water Science and Technology 50, 1–10 (2004)
3. Barutçuoğlu, Z., Alpaydin, E.: A comparison of model aggregation methods for regression. In: Kaynak, O., Alpaydın, E., Oja, E., Xu, L. (eds.) ICANN/ICONIP 2003. LNCS, vol. 2714, pp. 76–83. Springer, Heidelberg (2003)
4. Breiman, L.: Bagging predictors. Machine Learning 24, 123–140 (1996a)
5. Breiman, L.: Stacked regressions. Machine Learning 24, 49–64 (1996)
6. Breiman, L.: Random forests. Machine Learning 45, 5–32 (2001)
7. Brown, G.: Ensemble learning. In: Sammut, C., Webb, G.I. (eds.) Encyclopedia of Machine Learning, pp. 312–320. Springer US (2010)
8. Brun, R., Kühni, M., Siegrist, H., Gujer, W., Reichert, P.: Practical identifiability of ASM2d parameters – systematic selection and tuning of parameter subsets. Water Research 36, 4113–4127 (2002)
9. Dietterich, T.G.: Ensemble methods in machine learning. In: Kittler, J., Roli, F. (eds.) MCS 2000. LNCS, vol. 1857, pp. 1–15. Springer, Heidelberg (2000)
10. Drucker, H.: Improving regressors using boosting techniques. In: Proceedings of the Fourteenth International Conference on Machine Learning, pp. 107–115 (1997)
11. Freund, Y., Schapire, R.E.: A decision-theoretic generalization of on-line learning and an application to boosting. Journal of Computer and System Sciences 55, 119–139 (1997)
12. Gernaey, K., van Loosdrecht, M., Henze, M., Lind, M., Jørgensen, B.: Activated sludge wastewater treatment plant modelling and simulation: state of the art. Environmental Modelling & Software 19, 763–783 (2004)
13. Gontarski, C.A., Rodrigues, P.R., Mori, M., Prenem, L.F.: Simulation of an industrial wastewater treatment plant using artificial neural networks. Computers and Chemical Engineering 24, 1719–1723 (2000)
14. Graczyk, M., Lasota, T., Trawiński, B., Trawiński, K.: Comparison of bagging, boosting and stacking ensembles applied to real estate appraisal. In: Nguyen, N.T., Le, M.T., Świątek, J. (eds.) ACIIDS 2010. LNCS (LNAI), vol. 5991, pp. 340–350. Springer, Heidelberg (2010)
15. Haykin, S.: Neural networks: a comprehensive foundation. Prentice Hall (1999)
16. Henze, M., Gujer, W., Mino, T., Van Loosdrecht, M.C.M.: Activated Sludge Models ASM1, ASM2, ASM2d and ASM3. Scientific and Technical Report No 9. IWA (2000)
17. Ho, T.K.: The random subspace method for constructing decision forests. IEEE Transactions on Pattern Analysis and Machine Intelligence 20, 832–844 (1998)
18. Hu, G., Mao, Z., He, D., Yang, F.: Hybrid modeling for the prediction of leaching rate in leaching process based on negative correlation learning bagging ensemble algorithm. Computers and Chemical Engineering 35, 2618–2631 (2011)
19. Hulle, S., Vanrolleghem, P.A.: Modelling and optimisation of a chemical industry wastewater treatment plant subjected to varying production schedules. Journal of Chemical Technology and Biotechnology 79, 1084–1091 (2004)
20. Keskitalo, J., la Cour Jansen, J., Leiviskä, K.: Calibration and validation of a modified ASM1 using long-term simulation of a full-scale pulp mill wastewater treatment plant. Environmental Technology 31, 555–566 (2010)
21. Keskitalo, J., Leiviskä, K.: Application of evolutionary optimisers in data-based calibration of activated sludge models. Expert Systems with Applications 39(7), 6609–6617 (2011), http://dx.doi.org/10.1016/j.eswa.2011.12.041

22. Koch, G., Kühni, M., Gujer, W., Siegrist, H.: Calibration and validation of Activated Sludge Model No. 3 for Swiss municipal wastewater. Water Research 34, 3580–3590 (2000)
23. Krogh, A., Vedelsby, J.: Neural network ensembles, cross-validation, and active learning. In: Tesauro, G., Touretzky, D.S., Leen, T.K. (eds.) Advances in Neural Information Processing Systems 7, pp. 231–238. MIT Press (1995)
24. Lee, D.S., Vanrolleghem, P.A., Park, J.M.: Parallel hybrid modeling methods for a full-scale wastewater treatment plant. Journal of Biotechnology 115, 317–328 (2005)
25. Lindblom, E., Rosen, C., Vanrolleghem, P.A., Olsson, L.-E., Jeppsson, U.: Modelling a nutrient deficient wastewater treatment process. In: 4th IWA World Water Conference, Marrakech, Morocco (2004)
26. Mujunen, S., Minkkinen, P., Teppola, P., Wirkkala, R.: Modeling of activated sludge plants treatment efficiency with PLSR: a process analytical case study. Chemometrics and Intelligent Laboratory Systems 41, 83–94 (1998)
27. Nuhoglu, A., Keskinler, B., Yildiz, E.: Mathematical modelling of the activated sludge process – the Erzincan case. Process Biochemistry 40, 2467–2473 (2005)
28. Oliveira-Esquerre, K.P., Mori, M., Bruns, R.E.: Simulation of an industrial wastewater treatment plant using artificial neural networks and principal component analysis. Brazilian Journal of Chemical Engineering 19, 365–370 (2002)
29. Opitz, D., Maclin, R.: Popular ensemble methods: an empirical study. Journal of Artificial Intelligence Research 11, 169–198 (1999)
30. Pardo, C., Rodríguez, J.J., García-Osorio, C., Maudes, J.: An empirical study of multilayer perceptron ensembles for regression tasks. In: García-Pedrajas, N., Herrera, F., Fyfe, C., Benítez, J.M., Ali, M. (eds.) IEA/AIE 2010, Part II. LNCS, vol. 6097, pp. 106–115. Springer, Heidelberg (2010)
31. Psichogios, D., Ungar, L.: A hybrid neural network-first principles approach to process modeling. AIChE Journal 38, 1499–1511 (1992)
32. Rooney, N., Patterson, D., Tsymbal, A., Anand, S.: Random subspacing for regression ensembles. In: Proceedings of the Seventeenth International Florida Artificial Intelligence Research Society Conference, pp. 532–537 (2004)
33. Schapire, R.E.: The strength of weak learnability. Machine Learning 5, 197–227 (1990)
34. Shrestha, D.L., Solomatine, D.P.: Experiments with AdaBoost.RT, an improved boosting scheme for regression. Neural Computation 18, 1678–1710 (2006)
35. Wan, J., Huang, M., Ma, Y., Guo, W., Wang, Y., Zhang, H., Li, W., Sun, X.: Prediction of effluent quality of a paper mill wastewater treatment using an adaptive network-based fuzzy inference system. Applied Soft Computing 11, 3238–3246 (2011)
36. Weijers, S.R., Vanrolleghem, P.A.: A procedure for selecting best identifiable parameters in calibrating Activated Sludge Model no.1 to full-scale plant data. Water Science and Technology 36, 69–79 (1997)
37. West, D., Dellana, S., Qian, J.: Neural network ensemble strategies for financial decision applications. Computers & Operations Research 32, 2543–2559 (2005)
38. Wolpert, D.: Stacked generalization. Neural Networks 5, 241–259 (1992)
39. Zaier, I., Shu, C., Ouarda, T., Seidou, O., Chebana, F.: Estimation of ice thickness on lakes using artificial neural network ensembles. Journal of Hydrology 383, 330–340 (2010)

Implicit GPC Based on Semi Fuzzy Neural Network Model

Margarita Terziyska[1], Lyubka Doukovska[1], and Michail Petrov[2]

[1] Institutes of Information and Communication Technologies,
Bulgarian Academy of Sciences,
Acad. G. Bonchev st., block 2, floor 5, 1113 Sofia, Bulgaria
terziyska@dir.bg, l.doukovska@mail.bg
[2] Technical University-Sofia, branch Plovdiv,
25, Ts. Dustabanov st., 4000 Plovdiv, Bulgaria
mpetrov@tu-plovdiv.bg

Abstract. The model in Model Predictive Control (MPC) takes the central place. Therefore, it is very important to find a predictive model that effectively describes the behavior of the system and can easily be incorporated into MPC algorithm. In this paper it is presented implicit Generalized Predictive Controller (GPC) based on Semi Fuzzy Neural Network (SFNN) model. This kind of model works with reduced number of the fuzzy rules and respectively has low computational burden, which make it suitable for real-time applications like predictive controllers. Firstly, to demonstrate the potentials of the SFNN model test experiments with two benchmark chaotic systems - Mackey-Glass and Rossler chaotic time series are studied. After that, the SFNN model is incorporated in GPC and its efficiency is tested by simulation experiments in MATLAB environment to control a Continuous Stirred Tank Reactor (CSTR).

Keywords: Semi Fuzzy Neural Network, Fuzzy-Neural Models, Nonlinear Identification, Takagi-Sugeno fuzzy inference, NARX model, chaotic time series, Generalized Predictive Controller, Levenberg-Marguard Optimization Algorithm, CSTR.

1 Introduction

Model Predictive Control (MPC) is an advanced control methodology that originates in the late seventies, very quickly became popular and nowadays MPC is a well-known, classical control method. This is due to its ability to deal with input and output constraints while it can be applied to multivariable process control.

MPC algorithms based on linear models have been successfully used for years in numerous industrial applications [1]. It is provided a review [2] of the most commonly used methods that have been embedded in industrial model predictive control. Since, the nature of processes is inherently nonlinear and this implies the use of nonlinear models and respectively Nonlinear Model Predictive Control (NMPC) algorithms.

P. Angelov et al. (eds.), *Intelligent Systems'2014*,
Advances in Intelligent Systems and Computing 322, DOI: 10.1007/978-3-319-11313-5_61

NMPC is a variant of MPC that is characterized by the use of nonlinear system models in the prediction. The success of NMPC algorithm depends critically on the mathematical model. It is very important to find a predictive model that effectively describes the nonlinear behavior of the system and can easily be incorporated into NMPC algorithm. One possibility is to use first principle models and the other is to use empirical or black–box models (e.g. neural networks, fuzzy models, polynomial models, Volterra series models). The way for selection of a nonlinear model for NMPC is detailed described in [3].

This work presents a Nonlinear Generalized Predictive Control (NGPC) algorithm based on a Semi Fuzzy Neural Network (SFNN) model with a second order Takagi-Sugeno inference mechanism. Both neural networks and fuzzy logic are universal estimators. The fusion of the fuzzy logic with the neural networks allows combining the learning and computational ability of neural networks with the human like IF-THEN thinking and reasoning of the fuzzy system. This could be compared with the human brain [4] – neural networks concentrate on the structure of human brain, i.e., on the "hardware", whereas fuzzy logic system concentrate on "software". Combining neural networks and fuzzy systems in one unified framework has become popular in the last few years.

A lot of architectures have been proposed in the literature that combines fuzzy logic and neural networks. Some of the most popular are ANFIS [5], DENFIS [6]. They are all composed of a set of if-then rules. In principle, the number of fuzzy rules dependent exponentially on the number of inputs and membership functions. If **n** is number of inputs and **m** is number of membership functions, then the required number of fuzzy rules is $\mathbf{m}^{\mathbf{n}}$. The huge number of rules leads to determination of a large number of parameters during the learning procedure, which makes the neural fuzzy system difficult to implement.

In order to reduce the number of fuzzy rules without accuracy losses different fuzzy clustering approaches can be used, such as fuzzy C-means clustering [7], K-means clustering [8]. Evolving fuzzy systems [6], [9], [10] are used to evolve clustering and formed dynamically bases on fuzzy rules that had been created during the past learning process.

Another possibility to reduce the number of fuzzy rules gives self-constructing and self-organizing fuzzy neural network structures [11], [12]. In this type of structures, during the training procedure, inactive rules are dropped, which consequently leads to a reduction in the number of trained parameters.

In order to deal with the rule-explosion problem, hierarchical fuzzy neural networks could be used. But the normal learning method in these structures is very complex [13]. A method that compresses a fuzzy system with an arbitrarily large number of rules into a smaller fuzzy system by removing the redundancy in the fuzzy rule base is presented in [14]. Review of the most of existing rule base reduction methods for fuzzy systems, summary of their attributes and introduction of advanced techniques for formal presentation of fuzzy systems based on Boolean matrices and binary relations, which facilitate the overall management of complexity, is made in [15].

In this paper it is presented a SFNN model that works with a reduced number of fuzzy rules. Firstly, to demonstrate the potentials of the SFNN model test experiments with two benchmark chaotic systems - Mackey-Glass and Rossler chaotic time series are studied. After that, the SFNN model is incorporated in GPC and its efficiency is tested by simulation experiments in MATLAB environment to control a CSTR.

2 Classical Fuzzy Neural Network

In this section it is described so-called Classical Fuzzy Neural Network (CFNN).It is so named to distinguish it from the proposed in the next section SFNN. The structure of CFNN is shown on figure 1.

Layer 1: This layer accepts the input variables and then nodes in this layer only transmit the input values to the next layer directly.

Layer 2: Each node in this layer implements the fuzzyfication via a Gaussian membership function:

$$\mu_{Xp,m}^{(n)} = \exp\frac{-(x_p - c_{Xp,m})^2}{2\sigma_{Xp,m}^2} \tag{1}$$

where X_p are the input value, $c_{Xp,m}$ and $\sigma_{Xp,m}$ are the center and the standard deviation of the Gaussian membership function.

Layer 3: This layer is a kind of rules generator as it formed the fuzzy logic rules. Their number depends on the number of inputs **n** and the number of their fuzzy sets **m**, and is calculated according to the expression $N=m^n$. In this layer, each node represents a fuzzy rule in following form:

$$R^{(i)} : if\ x_1\ is\ \tilde{A}_1^{(i)}\ and\ x_p\ is\ \tilde{A}_p^{(i)}\ then\ f_y^{(i)}(k) \tag{2}$$

$$\begin{aligned} f_y^{(i)}(k) &= a_1^{(i)} y(k-1) + a_2^{(i)} y(k-2) + \cdots + a_{ny}^{(i)} y(k-n_y) \\ &+ b_1^{(i)} u(k) + b_2^{(i)} u(k-1) + \cdots + b_{nu}^{(i)} u(k-n_u) + c_0^{(i)} \end{aligned} \tag{3}$$

Layer 4: In the fourth layer it is realized implication operation:

$$\mu_{yq}^{(n)}(k+j) = \mu_{x_1,m}^{(n)}(k+j) * \mu_{x_2,m}^{(n)}(k+j) * \ldots * \mu_{x_p,m}^{(n)}(k+j) \tag{4}$$

Layer 5: In the fifth last layer it is taken the final decision which consists in determining the value of the model output by the expression:

$$\hat{y}(k+j) = \frac{\sum_{i=1}^{q} f_y^{(i)}(k+j)\mu_y^{(i)}(k+j)}{\sum_{i=1}^{q} \mu_y^{(i)}(k+j)} \tag{5}$$

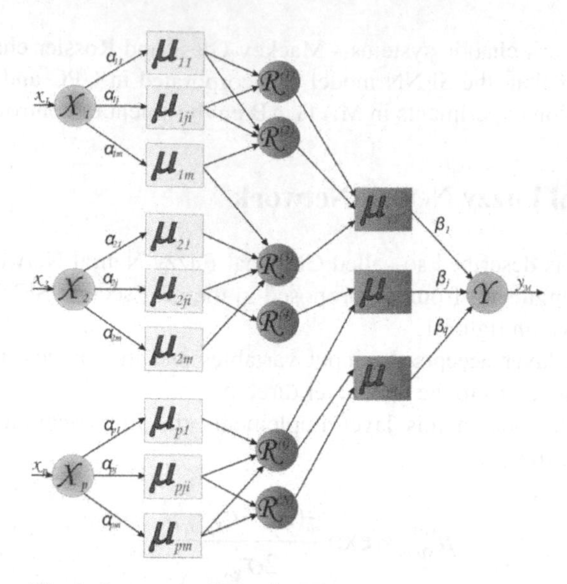

Fig. 1. Structure of Classical Fuzzy Neural Network

The learning algorithm for the CFNN model is very simple. It is based on minimization of an instant error measurement function between the measured real plant output and the model output, calculated by the CFNN:

$$E(k) = \frac{(y(k) - \hat{y}(k))^2}{2} \tag{6}$$

where y(k) denotes the measured real plant output and ŷ(k) is the CFNN model output.

During the gradient learning procedure it is needed to adjusted two groups of parameters in CFNN rules and they are: premise and consequent parameters in the fuzzy rules. The consequent parameters are the coefficients a_1, a_2...a_{ny}, b_1, b_2...b_{nu} in the Sugeno function f_y and they are calculated at first step by the following equations:

$$\beta_{ij}(k+1) = \beta_{ij}(k) + \eta(y(k) - y_M(k))\bar{\mu}_y^{(j)}(k)x_i(k) \tag{7}$$

$$\beta_{0j}(k+1) = \beta_{0j}(k) + \eta(y(k) - y_M(k))\bar{\mu}_y^{(j)}(k) \tag{8}$$

where η is the learning rate and β_{ij} is an adjustable i^{-th} coefficient (a_i or b_i) in the Sugeno function of the j^{-th} activated rule.

The premise parameters are the centre c_{ij} and the deviation σ_{ij} of a Gaussian fuzzy set. They can be calculated at second step using the following equations:

$$c_{ij}(k+1) = c_{ij}(k) + \eta(y - y_M)\bar{\mu}_y^{(j)}(k)[f_y^{(i)} - \hat{y}(k)]\frac{[x_i(k) - c_{ij}(k)]}{c_{ij}^2(k)} \tag{9}$$

$$\sigma_{ij}(k+1) = \sigma_{ij}(k) + \eta(y - y_M)\bar{\mu}_y^{(j)}(k)[f_y^{(i)} - \hat{y}(k)]\frac{[x_i(k) - \sigma_{ij}(k)]^3}{\sigma_{ij}^2(k)} \tag{10}$$

3 Semi Fuzzy Neural Network

The structure of the proposed SFNN model is shown on figure 2. SFNN model is a modification of the CFNN model described in previous section. Actually, the SFNN model is also five-layer architecture with the Takagi-Sugeno inference mechanism. However, in SFNN model a part of input signals are not fuzzyfied, but they come with their real values, weighted by the appropriate coefficient, into the third layer (fuzzy rules layer), i.e. directly into the THEN part of the functions of Sugeno. Thus, on the one hand it is reduced the number of fuzzy rules with which the model is working and on the other hand – it is reduced the number of the non-linear parameters that must be determined during learning procedure. Furthermore, SFNN model is similar in structure to a linear model, therefore, it can also be called semi- or quasi-linear. This ensure reduced computational burden during the optimization procedure for the calculation of the optimal value of the control action, when the SFNN model is used as a part of a NMPC algorithm.

In this paper on the basis on SFNN model it is realized a NARX model. Two types of SFNN models are used during the simulation experiments. The first one is a case in which the values of vector-regressor **y**(y(k-1), y(k-2)) are fuzzyfied inputs and the values of vector-regressor **u** (u(k), u(k-1)) are nonfuzzyfied inputs – SFNN Type I. The second one is an opposite – the values of vector-regressor **y**(y(k-1), y(k-2)) are nonfuzzyfied inputs and the values of vector-regressor **u** (u(k), u(k-1))are fuzzyfied inputs – SFNN Type II .

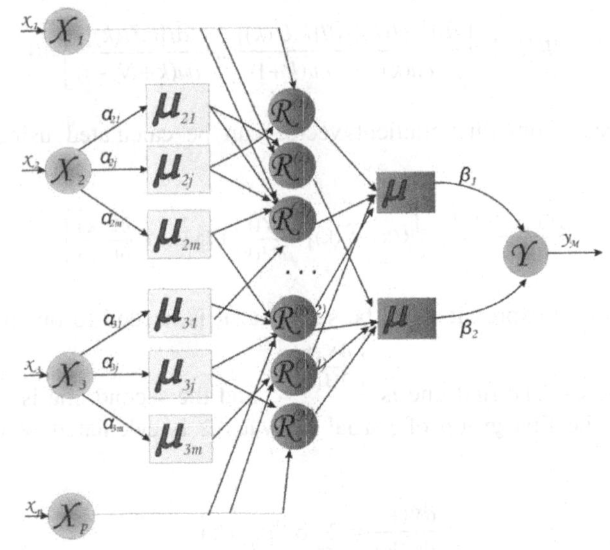

Fig. 2. Structure of the proposed Semi Fuzzy Neural Network

4 Implicit Generalized Predictive Control

Using the proposed SFNN model, the *Optimization Algorithm* computes the future control actions at each sampling period, by minimizing the proposed by Clarke et al [16] cost function which is typical for Generalized Predictive Control (GPC):

$$J(k,u(k)) = \sum_{i=N_1}^{N_2} (r(k+i) - \hat{y}(k+i))^2 + \rho\sum_{i=1}^{N_u} \Delta u(k+i-1)^2 \quad (11)$$

where $\hat{y}$ is the predicted model output, r is the system reference, u is the control action, N_1 is the minimum prediction horizon, N_2 is the maximum prediction horizon, N_u is the control horizon and ρ is the weighting factor penalizing changes in the control actions.

In this paper it is used Levenberg-Marquardt optimization algorithm, which uses the approximated Hessian and the information in the gradient, taking into account some regularization factors. The algorithm iterates using the following general equation:

$$x^{(k+1)} = x^{(k)} - [H^{-1}(x^{(k)}) + \lambda E]\nabla Q(x^{(k)}) \quad (12)$$

where H is the Hessian, E is the identity matrix and λ is the Levenberg-Marquardt parameter, which adjust the direction of movement to extremes, from gradient method at a great value ($\lambda > 10^3$) to the Newton optimization method when $\lambda = 0$.

When the criterion function is a quadratic one and there are no constraints on the control action, the cost function can be minimized analytically. If the criterion **J** is minimized with respect to the future control actions **u**, then their optimal values can be calculated by applying the condition:

$$\nabla J[k,U(k)] = \left[\frac{\partial J[k,U(k)]}{\partial u(k)}, \frac{\partial J[k,U(k)]}{\partial u(k+1)}, \ldots, \frac{\partial J[k,U(k)]}{\partial u(k+N_u-1)}\right] = 0 \quad (13)$$

Each element from this gradient vector can be calculated using the following equation:

$$\frac{\partial J[k,U(k)]}{\partial U(k)} = \left[-2\left[R(k) - \hat{Y}(k)\right]^T \frac{\partial \hat{Y}(k)}{\partial U(k)} + 2\rho\, \hat{U}(k)^T \frac{\partial \hat{U}(k)}{\partial U(k)}\right] \quad (14)$$

From the above expression can be seen that it is needed to obtain two groups of partial derivatives. The first one is $\left[\frac{\partial \hat{Y}(k)}{\partial U(k)}\right]$, and the second one is $\left[\frac{\partial \hat{U}(k)}{\partial U(k)}\right]$. Each element from the first group of partial derivatives is calculated with the following equations:

$$\frac{\partial \hat{y}(k)}{\partial u(k)} = \sum_{i=1}^{N} b_1^{(i)} \overline{\mu}_y^{(i)}(k) \quad (15)$$

$$\frac{\partial \hat{y}(k+1)}{\partial u(k)} = \sum_{i=1}^{N}\left[a_1^{(i)} \frac{\partial \hat{y}(k)}{\partial u(k)} + b_2^{(i)} \right] \overline{\mu}_y^{(i)}(k+1) \tag{16}$$

$$\frac{\partial \hat{y}(k+N_2)}{\partial u(k)} = \sum_{i=1}^{N}\left[a_1^{(i)} \frac{\partial \hat{y}(k+N_2-1)}{\partial u(k)} + a_2^{(i)} \frac{\partial \hat{y}(k+N_2-2)}{\partial u(k)} \right] \overline{\mu}_y^{(i)}(k+N_2) \tag{17}$$

The second group partial derivatives have the following form:

$$\frac{\partial \hat{U}(k)}{\partial U(k)} = \begin{bmatrix} \frac{\partial \Delta u(k)}{\partial u(k)} & \cdots\cdots & \frac{\partial \Delta u(k)}{\partial \Delta u(k+N_u-1)} \\ \cdots\cdots & \cdots\cdots & \cdots\cdots \\ \frac{\partial \Delta u(k+N_u-1)}{\partial u(k)} & \cdots\cdots & \frac{\partial \Delta u(k+N_u-1)}{\partial \Delta u(k+N_u-1)} \end{bmatrix} \tag{18}$$

Since $\Delta u(k)=u(k)-u(k-1)$, this is:

$$\frac{\partial \hat{U}(k)}{\partial U(k)} = \begin{bmatrix} 1 & 0 & 0 & 0 & 0 \\ -1 & 1 & 0 & 0 & 0 \\ 0 & -1 & 1 & 0 & 0 \\ 0 & 0 & -1 & 1 & 0 \\ 0 & 0 & 0 & -1 & 1 \end{bmatrix} \tag{19}$$

So, this is the way to obtain the first derivative that is needed for the first order gradient optimization algorithms. But, the Newton algorithm belongs to the second order gradient optimization algorithms. So, it is needed to calculate the second partial derivatives of the cost function. In this case (13) can be rewritten as follows:

$$H[k,U(k)]=[\frac{\partial^2 J[k,U(k)]}{\partial u^2(k)}, \frac{\partial^2 J[k,U(k)]}{\partial u^2(k+1)}, \ldots, \frac{\partial^2 J[k,U(k)]}{\partial u^2(k+N_u-1)}] \tag{20}$$

Each element from this vector is calculated as follows:

$$H[k,U(k)] = \left[-2\left[R(k)-\hat{Y}(k)\right]^T \frac{\partial \hat{Y}(k)}{\partial U(k)} + 2\rho \hat{U}(k)^T \frac{\partial \hat{U}(k)}{\partial U(k)} \right]' = \tag{21}$$

$$= \left[-2\left(\left[R(k)-\hat{Y}(k)\right]^T\right)' \frac{\partial \hat{Y}(k)}{\partial U(k)} - 2\left[R(k)-\hat{Y}(k)\right]^T \left(\frac{\partial \hat{Y}(k)}{\partial U(k)}\right)' + 2\rho\left(\hat{U}(k)^T\right)' \frac{\partial \hat{U}(k)}{\partial U(k)} + 2\rho \hat{U}(k)^T \left(\frac{\partial \hat{U}(k)}{\partial U(k)}\right)' \right]$$

Since $\left(\frac{\partial \hat{Y}(k)}{\partial U(k)}\right)' = 0$ and $\left(\frac{\partial \hat{U}(k)}{\partial U(k)}\right)' = 0$, finally:

$$H[k, U(k)] = -2\left(\frac{\partial \hat{Y}}{\partial U(k)}\right)^2 + 2\rho\left(\frac{\partial \hat{U}(k)}{\partial U(k)}\right)^2 \tag{22}$$

After calculating the Hessian matrix it is need to obtain its inverse matrix, and then according to (12) it can be written:

$$\Delta u(k) = \Delta u(k+1) + \rho^{-1}[H^{-1}(x^{(k)}) + \lambda E] * \left[e(k+N_1)\frac{\partial \hat{y}(k+N_1)}{\partial u(k)} + ... + e(k+N_2)\frac{\partial \hat{y}(k+N_2)}{\partial u(k)} \right] \tag{23}$$

5 Simulation Results

To investigate the modeling potentials of the proposed SFNN model, Mackey-Glass (MG) and Rossler chaotic system series benchmark models, have been used. The used time series will not converge or diverge, and the trajectory is highly sensitive to initial conditions. The MG time series is described by the time-delay differential equation:

$$x(i+1) = \frac{x(i) + ax(i - s)}{(1 + x^c(i - s)) - bx(i)} \tag{24}$$

where a=0.2; b=0.1; C=10; and x(0)=0.1 and s= 17s. The results are given on Fig.3 and Fig. 4.

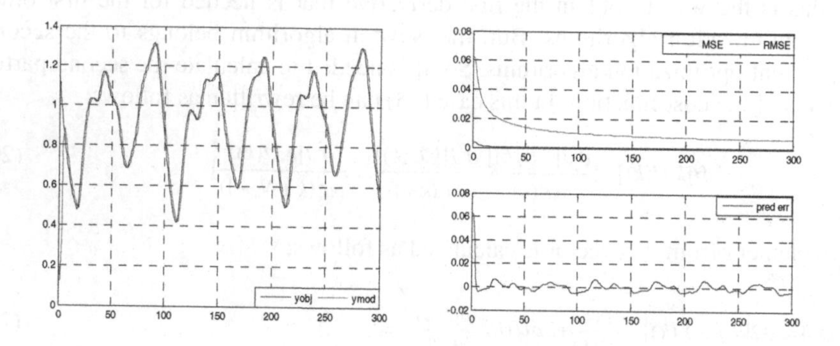

Fig. 3. Validation of the proposed SFNN - type I by using Mackey-Glass chaotic time series

On Figure 3 are shown results on SFNN Type I validation by using Mackey-Glass chaotic time series. As it can be seen, the proposed model structure predicts accurately the generated time series, with minimum prediction error and fast transient response of the RMSE, reaching value closer to zero. The values of the RMSE and MSE in the 50-th time step are respectively 0,014 and 0.00019.

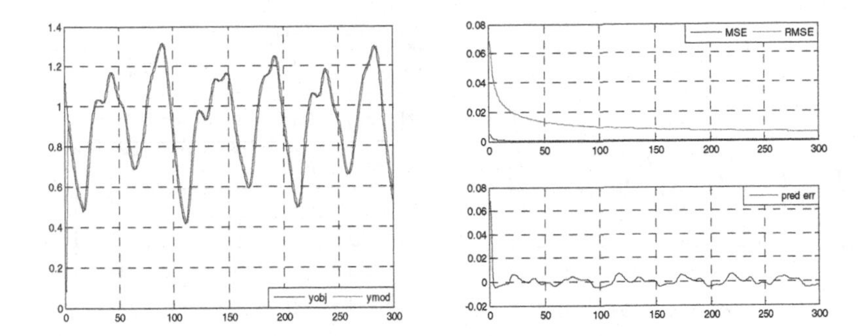

Fig. 4. Validation of the proposed SFNN – type II by using Mackey-Glass chaotic time series

Similarly, on Fig. 4 are shown results on SFNN Type II validation by using Mackey-Glass chaotic time series.

Another test of the proposed SFNN model is made with Rossler chaotic time series. These series is described by three coupled first-order differential equations:

$$\frac{dx}{dt} = -y - z$$
$$\frac{dy}{dt} = x + ay \qquad (25)$$
$$\frac{dz}{dt} = b + z(x - c)$$

where a=0.2; b=0.4; c=5.7; initial conditions x0=0.1; y0=0.1; z0=0.1. The validation of SFNN Type I and SFNN Type II models with Rossler chaotic time series are given on Fig.5 and Fig 6.

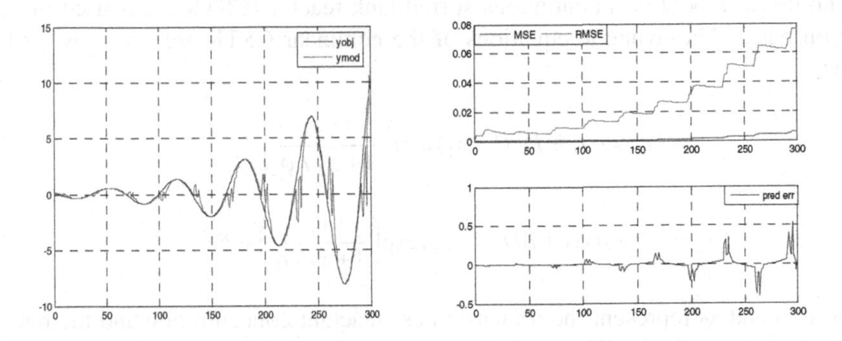

Fig. 5. Validation of the proposed SFNN - type I by using Rossler chaotic time series

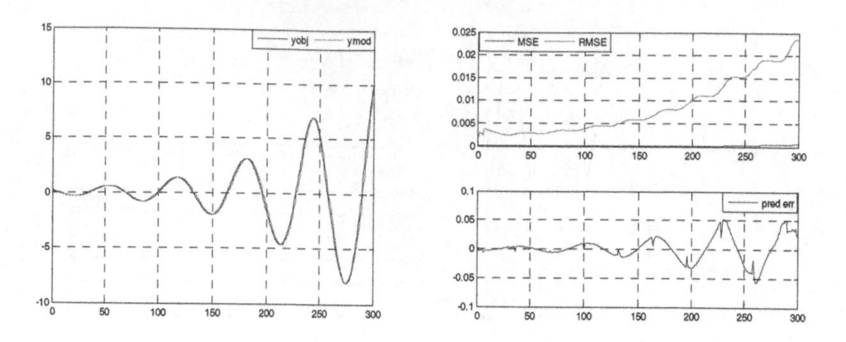

Fig. 6. Validation of the proposed SFNN - type II by using Rossler chaotic time series

Table 1. SFNN Type I and SFNN Type II Comparison

Mackey-Glass chaotic time series					Rossler chaotic time series				
Steps	SFNN Type I		SFNN Type II		Steps	SFNN Type I		SFNN Type II	
	MSE	RMSE	MSE	RMSE		MSE	RMSE	MSE	RMSE
50	0,00019	0,014	0,000168	0,013	50	$8{,}06.10^{-5}$	0,00898	$8{,}3.10^{-6}$	0,0029
100	$9{,}8.10^{-5}$	0,0099	$8{,}85.10^{-5}$	0,0093	100	0,000109	0,0104	$1{,}4.10^{-5}$	0,0037
150	$6{,}81.10^{-5}$	0,0083	$6{,}16.\ 10^{-5}$	0,0079	150	0,00038	0,0195	$3{,}45.10^{-5}$	0,0058
200	$5{,}33.10^{-5}$	0,0073	$4{,}85.\ 10^{-5}$	0,007	200	0,00109	0,0331	0,000104	0,0102
250	$4{,}43.10^{-5}$	0,0067	$4{,}05.\ 10^{-5}$	0,006	250	0,00263	0,0531	0,00023	0,0150
300	$3.85.10^{-5}$	0,0062	$3{,}5.\ 10^{-5}$	0,0051	300	0,00595	0,0772	0,00055	0,0234

The values of MSE and RMSE in interval of 50 steps are summarized in Table 1. From these data it can be concluded that the SFNN Type II model is more accurate than the SFNN Type I model, ie the case that u(k), u(k-1) are fuzzyfied inputs.

Additionally, to investigate the performance of the designed SFNN model, it was incorporated into GPC scheme. To evaluate the obtained model based control strategy, a nonlinear model of a continuous stirred tank reactor (CSTR), was used in a simulation batch. The dynamic equations of the nonlinear CSTR are given by [16] as follow:

$$\dot{x}_1 = -x_1 + D_a(1-x_1)\exp\left(\frac{x_2}{1+x_2/\phi}\right) \tag{26}$$

$$\dot{x}_2 = -(1+\delta)x_2 + BD_a(1-x_1)\exp\left(\frac{x_2}{1+x_2/\phi}\right) + \delta u \tag{27}$$

where x_1 and x_2 represent the dimensionless reactant concentration and the reactor temperature, respectively. The control action u is dimensionless cooling jacket temperature. The physical parameters in the CSTR model equations are Da, φ, B and δ which correspond to the Damkhler number, the activated energy, the heat of reaction and the heat transfer coefficient, respectively.

The number of the Gaussian fuzzy membership functions for each input variable is chosen to be equal to three. Their initial parameters are normalized and uniformly distributed into the universe of discourse ϵ [0, 1]. The SFNN based GPC controller has the following parameters: N1=1; Np=5; Nu=3; ρ=0.08. The results are shown in Fig. 7.

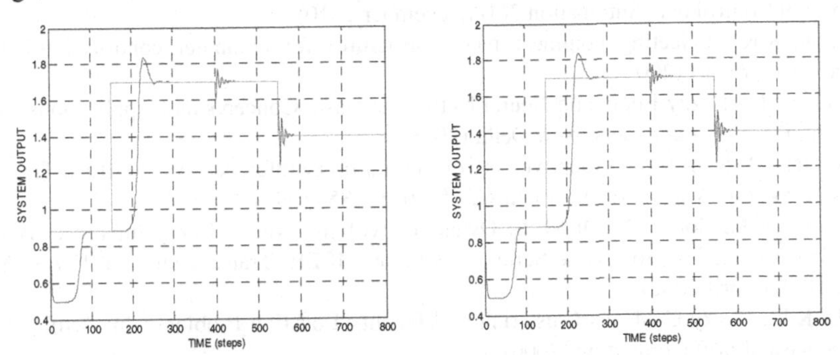

Fig. 7. Transient process response in case of variable system reference and plant parameter changes – SFNN Type I (left) and SFNN Type II (right)

6 Conclusions

In this paper it is presented implicit Generalized Predictive Controller (GPC) based on Semi Fuzzy Neural Network (SFNN) model. To demonstrate the effectiveness of the proposed SFNN model is made two groups of tests. First, the model is used to predict Mackey-Glass and Rossler chaotic system series. The simulation results shows that the SFNN predicts accurately the generated time series, with minimum prediction error and fast transient response of the RMSE, reaching value closer to zero. The main advantage of SFNN is that operated by a small number of rules and respectively has a smaller number of parameters for learning. Thus, it carries out the modeling of non-linear systems with considerably less calculation in comparison with the classical FNN. Furthermore, SFNN has other advantages - it is not require a priori data and is not bound by additional procedures, such as clustering. This makes it suitable for real-time applications such as predictive controllers.

Secondly, the designed SFNN was incorporated into GPC scheme and the obtained model based control strategy is tested to control CSTR. The results show that the predictive controller based on proposed SFNN model works precisely in a cases of a variable system reference and plant parameter changes. It is important to note, that the considered two types of SFNN models have insignificant differences in the accuracy of prediction of chaotic system series, but generally provide identical control of CSTR.

Acknowledgement. The research work reported in the paper is partly supported by the project ***AComIn*** "Advanced Computing for Innovation", grant 316087, funded by the FP7 Capacity Programme (Research Potential of Convergence Regions).

References

1. Qin, S.J., Badgwell, T.A.: A survey of industrial model predictive control technology. Control Eng. Pract. 11, 733–764 (2003)
2. Holkar, K.S., Waghmare, L.M.: An Overview of Model Predictive Control. International Journal of Control and Automation 3(4) (December 2010)
3. Pearson, R.K.: Selecting nonlinear model structures for computer control. Journal of Process Control 13 (2003)
4. Alavala, C.R.: Fuzzy Logic and Neural Networks: Basic Concepts and Applications. New Age Int. Pvt. Ltd. Publishers, New Delhi (2008)
5. Roger Jang, J.-S.: ANFIS: Adaptive-network-based fuzzy inference system. IEEE Transactions on System, Man, and Cybernetics 23(5), 665–685 (1993)
6. Kasabov, N.K., Song, Q.: DENFIS: Dynamic Evolving Neural-Fuzzy Inference System and Its Application for Time-Series Prediction. IEEE Transactions on Fuzzy Systems 10(2) (April 2002)
7. Žalik, K.R.: Fuzzy C-Means Clustering and Facility Location Problems. In: Artificial Intelligence and Soft Computing (2006)
8. Gallucc, L., et al.: Graph based k-means clustering. Signal Processing 92(9), 1970–1984 (2012)
9. Kasabov, N.K., Filev, D.: Evolving Intelligent Systems: Methods, Learning & Applications. In: International Symposium of Evolving Fuzzy Systems (September 2006)
10. Angelov, P.: Autonomous Learning Systems. Wiley (2013)
11. Allende-Cid, H., et al.: Self-Organizing Neuro-Fuzzy Inference System. In: Iberoamerican Congress on Pattern Recognition CIARP, pp. 429–436 (2008)
12. Ferreyra, A., de Jesus Rubio, J.: A new on-line self-constructing neural fuzzy network. In: Proceedings of the 45th IEEE Conference on Decision & Control, San Diego, CA, USA, December 13-15 (2006)
13. Yu, W., et al.: System Identification using hierarchical fuzzy neural networks with stable learning algorithm. Journal of Intelligent & Fuzzy Systems 18, 171–183 (2007)
14. Gegov, A.: Complexity Management in Fuzzy Systems. STUDFUZZ, vol. 211. Springer, Heidelberg (2007)
15. Gegov, A., Gobalakrishnan, N.: Advanced Inference in Fuzzy Systems by Rule Base Compression. Mathware & Soft Computing 14, 201–216 (2007)
16. Ray, W.H.: Advanced Process Control. McGraw-Hill, New York (1981)
17. Clarke, D.W., Mohtai, C., Tuffs, P.S.: Generalized predictive control – part I. The basic algorithm. Automatica 23(2), 137–148 (1987)

Comparison of Neural Networks with Different Membership Functions in the Type-2 Fuzzy Weights

Fernando Gaxiola, Patricia Melin, and Fevrier Valdez

Tijuana Institute of Technology, Tijuana México
fergaor_29@hotmail.com, pmelin@tectijuana.mx,
fevrier@tectijuana.edu.mx

Abstract. In this paper a comparison of the triangular, Gaussian, trapezoidal and generalized bell membership functions used in the type-2 fuzzy inference systems, which are applied to obtain the type-2 fuzzy weights in the connection between the layers of a neural network. We used two type-2 fuzzy systems that work in the backpropagation learning method with the type-2 fuzzy weight adjustment. We change the type of membership functions of the two type-2 fuzzy systems. The mathematical analysis of the proposed learning method architecture and the adaptation of the type-2 fuzzy weights are presented. The proposed method is based on recent methods that handle weight adaptation and especially fuzzy weights. In this work neural networks with type-2 fuzzy weights are presented. The proposed approach is applied to the case of Mackey-Glass time series prediction.

1 Introduction

Neural networks have been applied in several areas of research, like in the time series prediction area, which is the study case for this paper, especially in the Mackey-Glass time series [10].

The approach presented in this paper works with type-2 fuzzy weights in the neurons of the hidden and output layers of the neural network used for prediction of the Mackey- Glass time series. These interval type-2 fuzzy weights are updated using the backpropagation learning algorithm. We used two type-2 inference systems with Triangular, Trapezoidal, Gaussian and Generalized Bell membership functions.

This paper is focused in the managing of weights, because on the practice of neural networks, when performing the training of neural networks for the same problem is initialized with different weights or the adjustment are in a different way each time it is executed, but at the end is possible to reach a similar result [5] [12] [13][14].

The next section explains the proposed method and the problem description. Section 3 describes the neural network with type-2 fuzzy weights proposed in this paper. Section 4 presents the simulation results for the proposed method. Finally, in section 5 some conclusions are presented.

P. Angelov et al. (eds.), *Intelligent Systems'2014*,
Advances in Intelligent Systems and Computing 322, DOI: 10.1007/978-3-319-11313-5_62

2 Proposed Method and Problem Description

The objective of this work is to use interval type-2 fuzzy sets to generalize the backpropagation algorithm to allow the neural network to handle data with uncertainty, and a comparison of the type of membership functions in the type-2 fuzzy systems, which can betriangular, Gaussian, trapezoidal and generalized bell. The Mackey-Glass time series (for τ=17) is utilized for testing the proposed approach.

The updating of the weights will be done differently to the traditional updating of the weights performed with the backpropagation algorithm (Fig. 1).

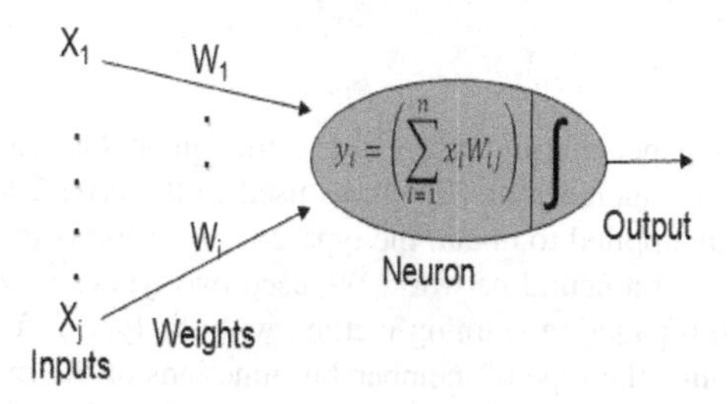

Fig. 1. Scheme of current management of numerical weights (type-0) for the inputs of each neuron

We developed a method for adjusting weights to achieve the desired result, searching for the optimal way to work with type-2 fuzzy weights [8].

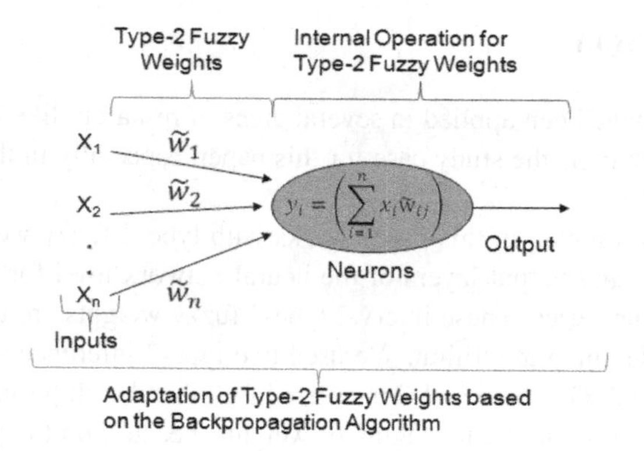

Fig. 2. Schematics of each neuron with the proposed management of weights using interval type-2 fuzzy sets

We used the sigmoid activation function for the hidden neurons and the linear activation function for the output neurons, and we utilized this activation functions because these functions have obtained good results in similar approaches.

3 Neural Network Architecture with type-2 Fuzzy Weights

The proposed neural network architecture with interval type-2 fuzzy weights (see Fig. 3) is described as follows:

Layer 0: Inputs.

$$x = [x_1, x_2, \cdots, x_n] \quad (1).$$

Phase 1: Interval type-2 fuzzy weights for the connection between the input and the hidden layer of the neural network.

$$\widetilde{w}_{ij} = [\overline{w}_{ij}, \underline{w}_{ij}] \quad (2).$$

Where $\widetilde{w}_{ij}$ are the weights of the consequents of each rule of the type-2 fuzzy system with inputs (current type-2 fuzzy weight, change of weight) and output (new fuzzy weight) [3].

Phase 2: Equations of the calculations in the hidden neurons using interval type-2 fuzzy weights.

$$Net = \sum_{i=1}^{n} x_i \widetilde{w}_{ij} \quad (3)$$

Phase 3: Equations of the calculations in the output neurons using interval type-2 fuzzy weights.

$$Out = \sum_{i=1}^{n} y_i \widetilde{w}_{ij} \quad (4)$$

Phase 4: Obtain a single output of the neural network.

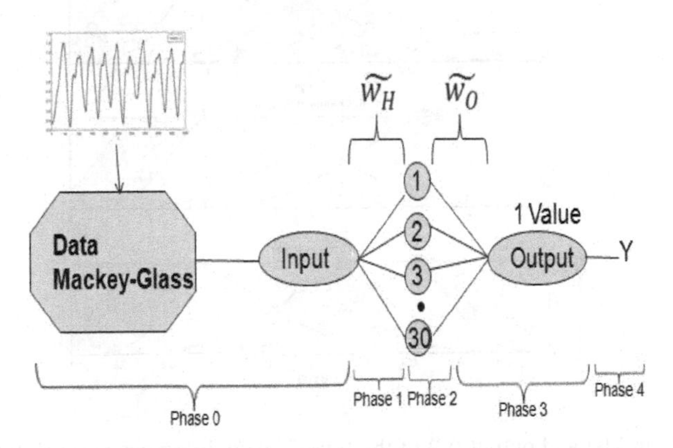

Fig. 3. Proposed neural network architecture with interval type-2 fuzzy weights

We considered a neural network architecture with 1 neuron in the output layer and 30 neurons in the hidden layer.

This neural network uses two type-2 fuzzy inference systems, one in the connections between the input neurons and the hidden neurons, and the other in the connections between the hidden neurons and the output neuron. In the hidden layer and output layer of the network we are updating the weights using the two type-2 fuzzy inference system that obtains the new weights in each epoch of the network on base at the backpropagation algorithm [6] [7].

The two type-2 fuzzy inference systems have the same structure and consist of two inputs (the current weight in the actual epoch and the change of the weight for the next epoch) and one output (the new weight for the next epoch) (see Fig. 4) [2][11][15].

We used two membership functions with their corresponding range for delimiting the inputs and outputs of the two type-2 fuzzy inference systems.

We used triangular, trapezoidal, Gaussian and generalized bell membership functions (for example, see Fig. 5 and Fig. 6).

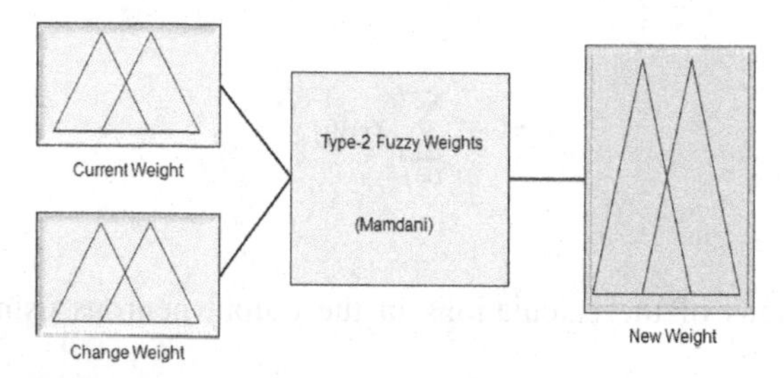

Fig. 4. Structure of the two type-2 fuzzy inference systems that were used to obtain the type-2 fuzzy weights in the hidden and output layer

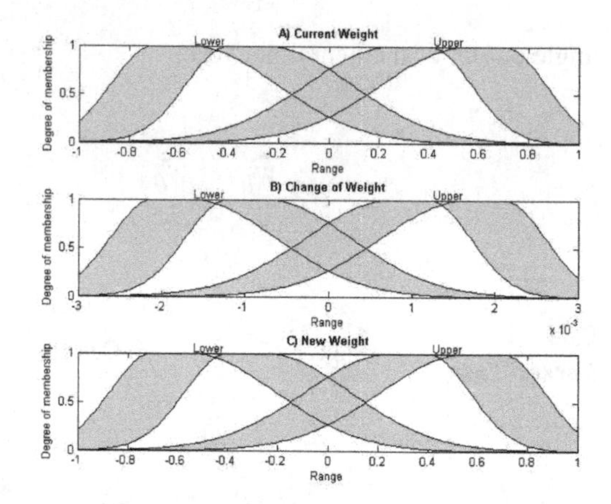

Fig. 5. Inputs (A and B) and output (C) of the type-2 fuzzy inference systems that were used to obtain the type-2 fuzzy weights in the hidden layer

We obtain the two type-2 fuzzy inference systems manually and a footprint of uncertainty of 15 % in the triangular membership functions and the others type-2 fuzzy inference systems are obtained to change the type of membership function in the inputs and output.

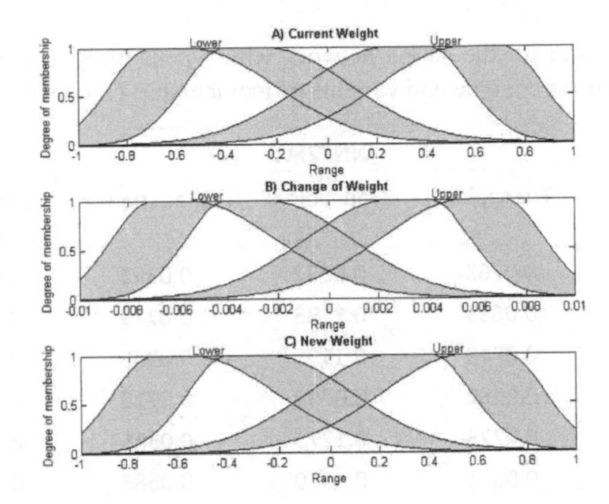

Fig. 6. Inputs (A and B) and output (C) of the type-2 fuzzy inference system that were used to obtain the type-2 fuzzy weights in the output layer

The two type-2 fuzzy inference systems used the same six rules, the four combinations of the two membership functions and two rules added for null change of the weight (see Fig. 7).

1. (Current_Weight is lower) and (Change_Weight is lower) then (New_Weight is lower)
2. (Current_Weight is lower) and (Change_Weight is upper) then (New_Weight is lower)
3. (Current_Weight is upper) and (Change_Weight is lower) then (New_Weight is upper)
4. (Current_Weight is upper) and (Change_Weight is upper) then (New_Weight is upper)
5. (Current_Weight is lower) then (New_Weight is lower)
6. (Current_Weight is upper) then (New_Weight is upper)

Fig. 7. Rules of the type-2 fuzzy inference system used in the hidden and output layer for the neural network with type-2 fuzzy weights

4 Simulation Results

We performed experiments in time-series prediction, specifically for the Mackey-Glass time series (τ=17). We used 100 epochs and $1x10^{-8}$ for network error in the experiments [1] [4] [9].

We present the obtained results of the experiments performed with the neural network with type-2 fuzzy weights (NNT2FW) in the different types: triangular, trapezoidal, two-sided Gaussian, and generalized bell; These results are achieved without optimizing of the neural network and the type-2 fuzzy systems, which means that all parameters of the neural network and the range and values of the membership functions of the type-2 fuzzy systems are established empirically. The average error was obtained of 30 experiments.

In Table 1, we present the prediction error obtained with the results achieved as output. The best prediction error is of 0.0391 and the average prediction error is of 0.0789. The APE represents the average prediction error of the experiments.

Table 1. Prediction error for the neural network with type-2 fuzzy weights for Mackey-Glass time series, with the different type and variants of membership functions

	NNT2FW			
No. Experiment	Triangular	Trapezoidal	Gaussian	Generalized Bell
1	0.0768	0.0892	0.0391	0.0947
2	0.0658	0.1133	0.0779	0.1454
3	0.0719	0.1195	0.0565	0.1772
4	0.0552	0.0799	0.0718	0.1165
5	0.0728	0.1273	**0.0355**	0.1870
6	0.0661	0.1070	0.0688	0.1356
7	0.0808	0.1511	0.0796	0.1586
8	0.0686	0.1320	0.0589	0.1156
9	0.0888	0.0927	0.0454	0.1642
10	0.0866	0.1490	0.0693	0.1275
APE	0.0789	0.1272	**0.0659**	0.1574

5 Conclusions

In this paper, we presented a comparison of learning method that updates weights (type-2 fuzzy weights) in each connection between the neurons of the layers of neural network using a type-2 fuzzy inference system with triangular, trapezoidal, Gaussians and generalized bell membership functions.

Additionally, the neurons work internally with the type-2 fuzzy weights and therefore, obtaining results at the output of each neuron of the neural network. The modifications performed in the neural network that allows working with type-2 fuzzy weights provide at the neural network greater robustness and less susceptibility at the noise in the data of the time series.

The best prediction error of 0.0391 of the neural network with type-2 fuzzy weights for the Mackey-Glass time series obtained with Gaussians membership functions is better than the best prediction error of 0.0552, 0.0799 or 0.0947 obtained with triangular, trapezoidal or generalized bell, respectively (as shown in Table 1).

This result is good considering that the used parameters for the neural networks at the moment are determined in an empirical way.

References

1. Abiyev, R.H.: A Type-2 Fuzzy Wavelet Neural Network for Time Series Prediction. In: García-Pedrajas, N., Herrera, F., Fyfe, C., Benítez, J.M., Ali, M. (eds.) IEA/AIE 2010, Part III. LNCS, vol. 6098, pp. 518–527. Springer, Heidelberg (2010)
2. Castillo, O., Melin, P.: A review on the design and optimization of interval type-2 fuzzy controllers. Applied Soft Computing 12(4), 1267–1278 (2012)
3. Castro, J., Castillo, O., Melin, P., Rodríguez-Díaz, A.: A Hybrid Learning Algorithm for a Class of Interval Type-2 Fuzzy Neural Networks. Information Sciences 179(13), 2175–2193 (2009)
4. Castro, J.R., Castillo, O., Melin, P., Mendoza, O., Rodríguez-Díaz, A.: An Interval Type-2 Fuzzy Neural Network for Chaotic Time Series Prediction with Cross-Validation and Akaike Test. In: Castillo, O., Kacprzyk, J., Pedrycz, W. (eds.) Soft Computing for Intelligent Control and Mobile Robotics. SCI, vol. 318, pp. 269–285. Springer, Heidelberg (2010)
5. Fletcher, R., Reeves, C.M.: Function Minimization by Conjugate Gradients. Computer Journal 7, 149–154 (1964)
6. Gaxiola, F., Melin, P., Valdez, F.: Backpropagation Method with Type-2 Fuzzy Weight Adjustment for Neural Network Learning. In: Annual Meeting of the North American Fuzzy Information Processing Society (NAFIPS), pp. 1–6 (2012)
7. Gaxiola, F., Melin, P., Valdez, F.: Genetic Optimization of Type-2 Fuzzy Weight Adjustment for Backpropagation in Ensemble Neural Network. In: Castillo, O., Melin, P., Kacprzyk, J. (eds.) Recent Advances on Hybrid Intelligent Systems. SCI, vol. 451, pp. 159–172. Springer, Heidelberg (2013)
8. Jang, J.S.R., Sun, C.T., Mizutani, E.: Neuro-Fuzzy and Soft Computing: a Computational Approach to Learning and Machine Intelligence, p. 614. Ed. Prentice Hall (1997)
9. Karnik, N., Mendel, J.: Applications of Type-2 Fuzzy Logic Systems to Forecasting of Time-Series. Information Sciences 120(1-4), 89–111 (1999)
10. Martinez, G., Melin, P., Bravo, D., Gonzalez, F., Gonzalez, M.: Modular Neural Networks and Fuzzy Sugeno Integral for Face and Fingerprint Recognition. Advances in Soft Computing 34, 603–618 (2006)
11. Melin, P.: Modular Neural Networks and Type-2 Fuzzy Systems for Pattern Recognition. SCI, vol. 389. Springer, Heidelberg (2012)
12. Moller, M.F.: A Scaled Conjugate Gradient Algorithm for Fast Supervised Learning. Neural Networks 6, 525–533 (1993)
13. Phansalkar, V.V., Sastry, P.S.: Analysis of the Back-Propagation Algorithm with Momentum. IEEE Transactions on Neural Networks 5(3), 505–506 (1994)
14. Powell, M.J.D.: Restart Procedures for the Conjugate Gradient Method. Mathematical Programming 12, 241–254 (1977)
15. Sepúlveda, R., Castillo, O., Melin, P., Montiel, O.: An Efficient Computational Method to Implement Type-2 Fuzzy Logic in Control Applications. In: Analysis and Design of Intelligent Systems using Soft Computing Techniques, pp. 45-52 (2007)

Multi-dimensional Fuzzy Modeling with Incomplete Fuzzy Rule Base and Radial Basis Activation Functions

Gancho Vachkov[1], Nikolinka Christova[2], and Magdalena Valova[2]

[1] School of Engineering and Physics,
The University of the South Pacific (USP),
Suva, Fiji
gancho.vachkov@gmail.com
[2] Department of Automation of Industry,
University of Chemical Technology and Metallurgy,
Sofia, Bulgaria
nchrist@uctm.edu, m_valova@abv.bg

Abstract. A new type of a fuzzy model is proposed in this paper. It uses a reduced number of fuzzy rules with respective radial basis activation functions. The optimal number of the rules is defined experimentally and their locations are obtained by clustering or by PSO optimization procedure. All other parameters are also optimized in order to produce the best model. The obtained model is able to work with sparse data in the multidimensional experimental space. As a proof a synthetic example, as well as a real example of a 5-dimensional sparse data have been used. The results obtained show that the PSO optimization of the fuzzy rule locations is a better approach than the clustering algorithm, which utilizes the distribution of the available experimental data.

Keywords: Fuzzy Model, Incomplete Fuzzy Rule Base, Radial Basis Functions, Activation Function, Clustering, Multidimensional Sparse Data.

1 Introduction

Fuzzy modeling is widely accepted as a universal approach to creating highly nonlinear models of real industrial and other processes and systems, based on available experimental data [21]. The main strength of the fuzzy models [17] is their ability to incorporate two different types of available information about the process, namely numerical (experimental) data as well as accumulated knowledge about the process in the form of a Fuzzy Rule Base (FRB). The most often used fuzzy model is the classical Takagi-Sugeno (TS) fuzzy model introduced in 1978 [17], which has been proved to be a universal approximator. It uses numerical outputs of the fuzzy rules, called singletons.

The structure of the fuzzy model is built from three blocks, as follows: a Fuzzy Rule Base, a Parameter Base (PB) and a Computational Unit (CU).

The FRB is a collection of all available fuzzy rule bases in if-then format. The PB can be considered as a data base that saves all the tuning parameters of the fuzzy

P. Angelov et al. (eds.), *Intelligent Systems'2014*,
Advances in Intelligent Systems and Computing 322, DOI: 10.1007/978-3-319-11313-5_63

model, namely: the parameters, defining the membership functions for all inputs and the parameters (coefficients) of all the singletons.

As it is well known from the literature [6], [17], [21] the CU uses the collected information from FRB and PB and the available information from the measured Inputs in order to perform the following 3 computation steps: *Fuzzification*, *Fuzzy Inference* and *Defuzzification*.

Fuzzification is a step that converts the real values from the numerical inputs into fuzzy membership values, based on the given shapes and locations of the pre-defined membership functions; *Fuzzy Inference* is the step that defines the activation degree (firing degree) for each of the fuzzy rules in FRB. Here the following *t-norm* operations are the most used: *Min* operation and *Product* operation; Finally the *Defuzzification* step converts the internal fuzzy information to a real numerical output of the fuzzy model. Here the most used method is the *weighted average* of the numerical outputs from all fuzzy rules (the singletons), based on their respective firing degree.

Despite the clear merits of the fuzzy model as universal approximator, it has also some demerits. The biggest of them is that the fuzzy model is not quite a suitable tool for modeling of high-dimensional (multi-input) processes. The reason is that the number of fuzzy rules in FRB and the number of parameters in PB (locations and boundaries of the membership functions) are growing exponentially with the growth of the number of inputs. This means that the total number of all tuning parameters of the fuzzy model is easily becoming too large, which needs a powerful and good method for optimization.

In addition, we have often the case of *sparse* and insufficient number of experimental data [11]. As a result the number of all collected input-output data is less than the total number of the tuning parameters of the model, which makes impossible to find an exact analytical solution to the problem.

To alleviate the problem of the growing tuning parameters of the fuzzy model, we propose in this paper the use of incomplete FRB and Radial Basis Activation (RBA) functions [8], [13,14,15] with one-dimensional width, which are able to approximate to a good extent the activation surface of the original fuzzy model. All the details are given in the rest of the paper, including a synthetic numerical example, as well as an example for creating *5*-dimensional model, based on a small number and sparse input-output data.

2 Background and Calculation Problems of the Fuzzy Models

Let us assume that a set $D=\{d_1,d_2,...,d_M\}$ of M experimental data is available from a real nonlinear process with K inputs and one output. Each single experiment is saved as one input-output pair: $[\mathbf{X},Y]=[x_1,x_2,...,x_K;Y]$. Then the purpose of the fuzzy model is to create a nonlinear function $Y_m=F(\mathbf{X})$ that minimizes the rooted mean square error:

$$RMSE=\sqrt{\sum_{i=1}^{M}(Y_i-Y_{im})^2\Big/M} \tag{1}$$

The classical TS fuzzy model uses a FRB with N fuzzy rules. The number N is calculated as a *product* of the number of the membership functions MF, defined for each input, as follows:

$$N = \prod_{i=1}^{K} MF_i \tag{2}$$

Such FRB is further on called *complete* FRB. An example of a complete FRB with N = 5x5 = 25 fuzzy rules for a 2-dimensional fuzzy model is shown in Fig. 1. Here $MF_1 = MF_2 = 5$ and the fuzzy rules are shown as rhomboid-type curve symbols.

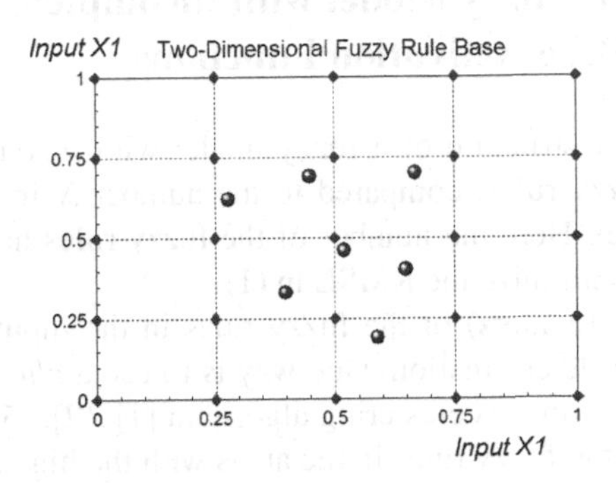

Fig. 1. Example of one Complete FRB with N=25 fuzzy rules and one Incomplete FRB with N=7 fuzzy rules of a 2-dimensional Fuzzy Model

It is obvious that the number N in (2) is growing exponentially with increasing the number of inputs of the fuzzy model. This makes implausible the use of the classical fuzzy models with complete FRM.

Another problem arising in computation of the multidimensional fuzzy model with complete FRB is that quite often we have sparse data in the K-dimensional input space that does not provide sufficient information for determining all the parameters of the fuzzy model. In other words we are facing the problem of insufficient information for parameter estimation. One solution to this problem is to intentionally decrease the number N of the fuzzy rules, by using some heuristics and preliminary information about the available set of experimental data. Here the simplest way to do this is to *cluster* the available experimental data into a small number of clusters [2], [5], [18], [19], [20] and to place the centers of the fuzzy rules into the cluster centers. Such example is shown in the FRB from the above Fig. 1, where the original number of *N=25* fuzzy rules is reduced to another FRB with only *N=7* rules (shown as ball-type curve symbols), after clustering of the experimental data. The newly obtained FRB with the reduced number of rules is further on called *incomplete* FRB.

Here it is worth noting that the incomplete FRB does not follow the orthogonal principle for creating the FRB in the classical fuzzy model, namely the notion of "neighboring rules" on each input is no more valid. In addition, the concept of fixed number of membership functions for each input will be no more needed.

We propose in this paper a replacement of the membership functions and the orthogonal (complete) FRB with incomplete FRB that consists of activation functions in the form of Radial Basis Functions (RBF) [13,14,15,19]. The functional equivalence between radial basis function networks and fuzzy inference systems has been proven in [14]. As shown in the sequel of the paper, the proposed modification of the fuzzy model with RBF activation functions reduces the computational costs and makes it possible to create a fuzzy model by utilizing the available sparse data in the high dimensional input space.

3 The Proposed Fuzzy Model with Incomplete Fuzzy Rule Base and Radial Base Activation Functions

We propose here a new structure of a fuzzy model with incomplete FRB that has a reduced number of fuzzy rules, compared to the number N in (2) from the classical orthogonal fuzzy model. Here the number of the fuzzy rules has to be decided experimentally, in order to minimize the RMSE in (1).

As for the locations (centers) of the fuzzy rules in the input space, there are two different ways for their determination. One way is to use a *clustering approach*, such as the well-known Fuzzy-means clustering algorithm [1], [2], [5], [6], [10], [16], [18]. It places the centers of the fuzzy rules in the areas with the highest density (concentration) of the input data. Here we assume that such area is the most important for revealing the true nonlinearity of the model, even if this is not always the case.

Another way to determine "the best" locations of the fuzzy rules is to use a suitable multidimensional optimization method, such as the popular Particle Swarm Optimization (PSO) [3], [7] that is used in this paper.

The structure of the fuzzy rules is also changed in this proposed fuzzy model, namely we are no more dealing with predefined membership functions for each input, but rather use *one* RBF for each fuzzy rule that represents the activation function for this rule. We call this RBF a *Radial Basis Activation* (RBA) function, as follows:

$$Act_j = \exp\left[-\sum_{i=1}^{K} (x_i - c^j_{oi})^2 \Big/ 2\sigma^2 \right], \; j=1,2,...,N \tag{3}$$

Here $Act_j \in [0,1]$ is the activation (firing) degree of the j-th fuzzy rule, $x^j_{oi}, i=1,2,...,K$ is the center of the j-th fuzzy rule ($j=1,2,\ldots,N$) and σ is one-dimensional parameter that defines the *width* (spread) of the RBA function.

We have investigated the shape of the activation functions in the case of the classical fuzzy rules that consist of K membership functions, in order to analyze the approximation ability of the newly proposed RBA functions in (3). The analysis is made for the two cases of fuzzy inference, namely: *Min*-operation and *Product* operation. All the results are graphically shown in Fig. 2, 3, 4 and 5 for a 2-dimensional example of a fuzzy model with Triangular and with Gaussian membership functions.

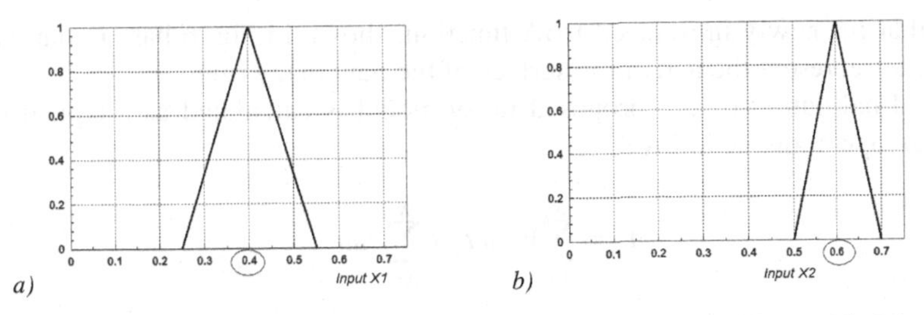

Fig. 2. Two *Triangular* Membership Functions of a 2-dimensional Fuzzy Model

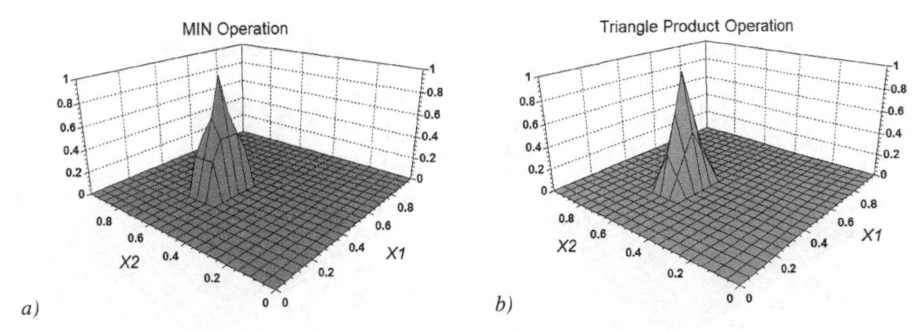

Fig. 3. Activation Functions for *Min* and *Product* operation

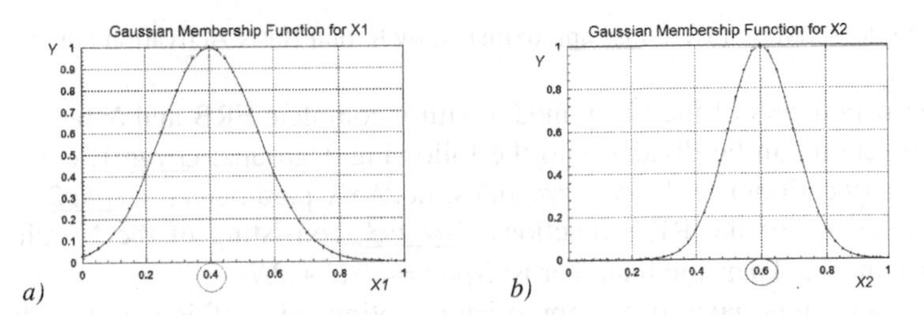

Fig. 4. Two *Gaussian* Membership Functions of a 2-dimensional Fuzzy Model

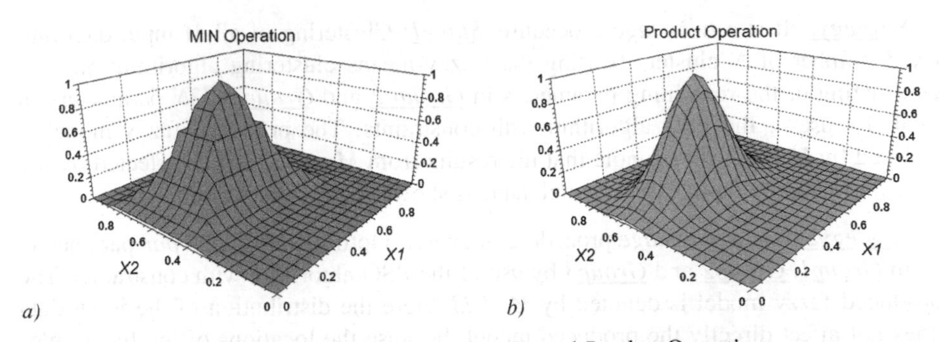

Fig. 5. Activation Functions for *Min* and *Product* Operation

Fig. 6 depicts the RBA function for one fuzzy rule and its difference with the original activation function, produced in Fig. 5 for the case of Product operation. It is clear that the newly introduced RBA function, shown in Fig. 6 has similar shape and is able to represent the activation surface of the new fuzzy rule.

The final output of our proposed fuzzy model is calculated as a typical *weighted average* operation, as follows:

$$Y_m = \sum_{j=1}^{N} V_j Act_j \Big/ \sum_{j=1}^{N} Act_j , \qquad (4)$$

where V_j , $j = 1,2,...,N$ is are the values of the singletons for all rules.

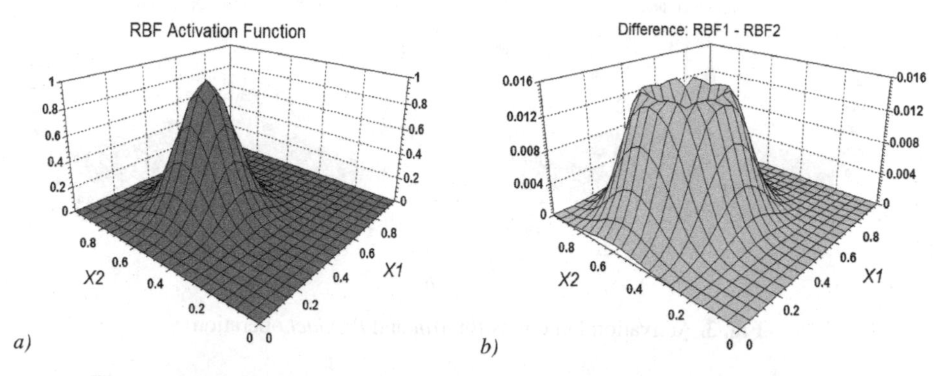

Fig. 6. Activation Response approximated by Radial Basis Activation Function

All the parameters of the fuzzy model with incomplete FRB and N radial basis activation function can be divided into the following 3 groups: <u>*Group*1</u>, consisting the coordinates (locations) of the N fuzzy rules, i.e. NxK parameters; <u>*Group2*</u>, consisting of the widths σ of the RBA functions; <u>*Group3*</u>, consisting of the N values of all singletons in (4). Their total number is: $Npar = N\text{x}K + 2N$.

These parameters have to be appropriately optimized. In this paper we have used two strategies for optimization, as follows:

<u>*Strategy1*</u>. It is a *two-stage* procedure. <u>*Stage1*</u>: Clustering of all M input data into fixed number of N clusters b using the Fuzzy-means clustering algorithm; <u>*Stage2*:</u> Optimizing of the remaining parameters in <u>*Group 2*</u> and <u>*Group 3*</u> (*2N* parameters in total) by use of the PSO algorithm with constraints; The produced fuzzy model is denoted by *Model2*. Please note that the results from *Model 2* will be affected by the distribution of the input experimental data, as shown in the next section.

<u>*Strategy2*</u>. It is a *one-stage* procedure, namely: Optimizing of all *Npar* parameters from <u>*Group1*</u>, <u>*Group2*</u> and <u>*Group3*</u> by use of the PSO algorithm with constraints. The produced fuzzy model is denoted by *Model3*. Here the distribution of the input data does not affect directly the produced model, because the locations of the fuzzy rules (RBA functions) will be found automatically by the optimization algorithm.

4 Simulation Results of a Test Example

In order to investigate the performance of the proposed fuzzy model, we have constructed a *2*-dimensional highly nonlinear test example, as shown in the next Fig. 7.

The experimental data have been generated from this example and separated as two data sets, as shown in Fig. 8. As seen from the figure, the *Training* data set consists of $M=427$ semi-random data, while the set of $M=240$ random data has been used as *Test* data set.

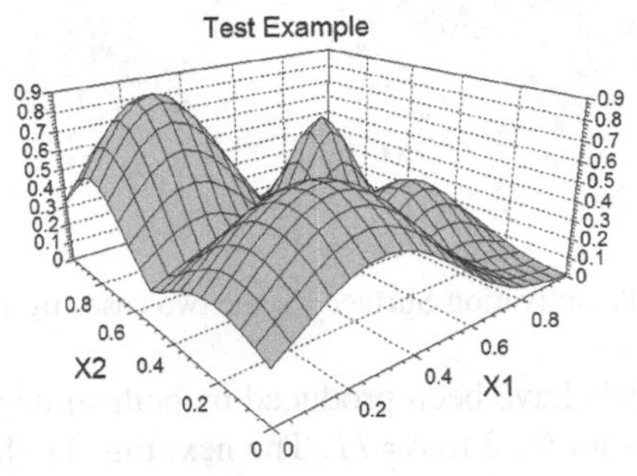

Fig. 7. Test Example and the Training Data set

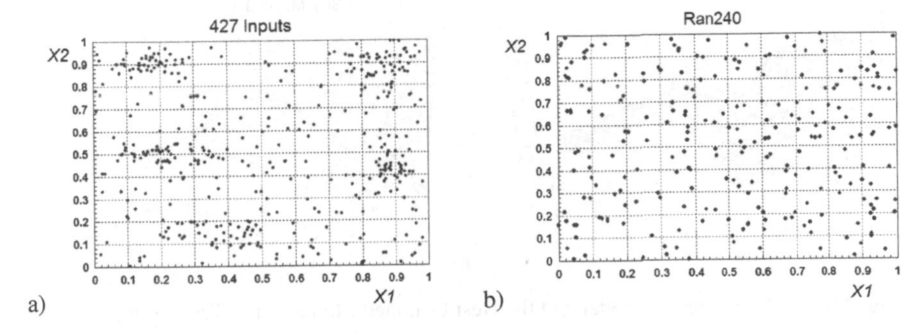

Fig. 8. Training Data Set and a Test Data Set

The *Strategy1* (Clustering plus PSO Optimization) was used to produce the *Model2* with $N=5$ fuzzy rules and the details are shown in Fig. 9*a* and Fig. 10*a*.

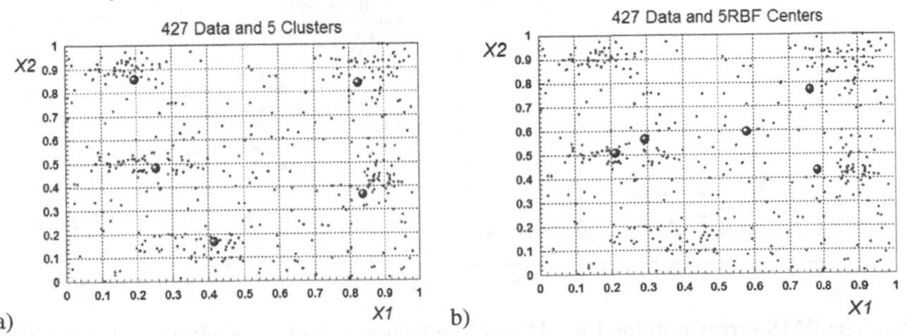

Fig. 9. Centers of the 5 RBFs obtained by Clustering (*Strategy1*) and Optimization (*Strategy2*)

Fig. 9*b* and Fig. 10*b* represent the details from creating the *Model3* with the same number of fuzzy rules, by using the *Strategy2* (PSO optimization of all *3* Groups parameters).

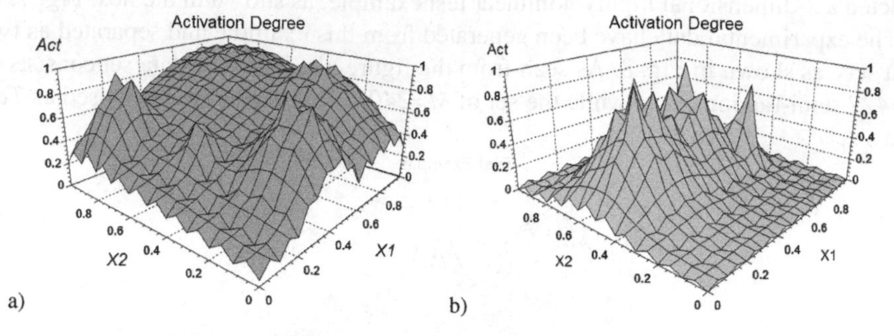

Fig. 10. Activation Surface for the two cases from Fig. 8.

Different fuzzy models have been produced by both strategies by varying the number of the fuzzy rules from $N=2$ to $N=11$. The next Fig. 11 shows the result from one intermediate model, as well as the best prediction model.

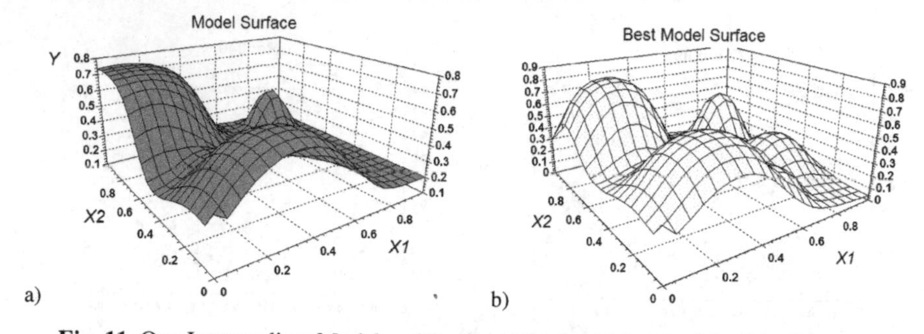

Fig. 11. One Intermediate Model and the Best Obtained Model of the Test Example

The RMS error for the training data set for Model2 and Model3 are shown in Fig. 12*a*. It is seen that *Strategy2* (*Model3*) shows better approximation results. Fig. 12*b* shows the RMS error of *Model3* for the *Training* Data set and the *Test* Data set.

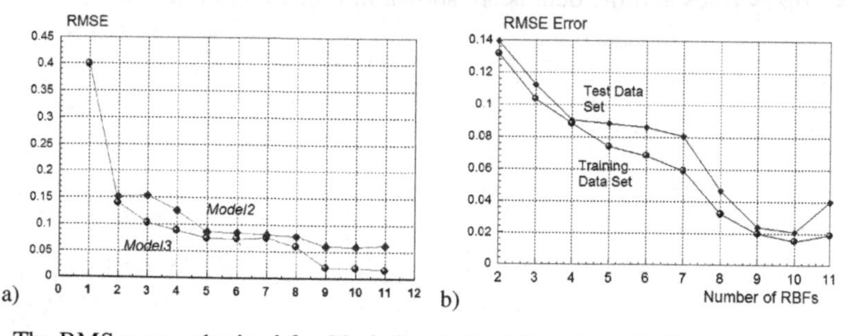

Fig. 12. The RMS errors obtained for *Model2* and *Model3* and the RMS error for the *Training* Data Set and the *Test* Data Set

5 Example for Creating a Multi-Dimensional Fuzzy Model by Use of Sparse Experimental Data

The proposed methodology has been applied for development of multi-dimensional fuzzy models based on sparse experimental data taken from a biotechnological process with anaerobic digestion. It is an effective process for treatment of different agricultural, municipal and industrial wastes. It combines environmental depollution (ecological aspect) with production of renewable energy – biogas, the main component of which is methane [12].

All the available experimental data (*M=79*) are divided into two sets – training and test data sets with 70 and 9 experiments respectively. The developed models have 5 input variables and one output variable. The obtained RMS errors during the training of the two kind models (*Model2* and *Model3*) for different number of RBFs are depicted in Fig. 13.

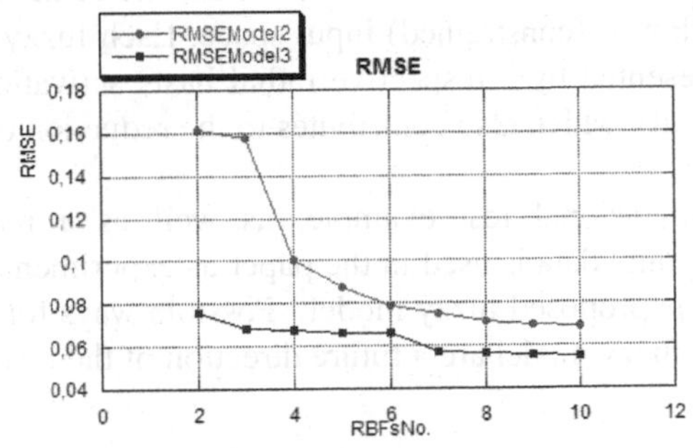

Fig. 13. RMSE errors during training

The original (experimental) outputs, the modeled outputs and the absolute errors for the created best model are illustrated on Fig. 14.

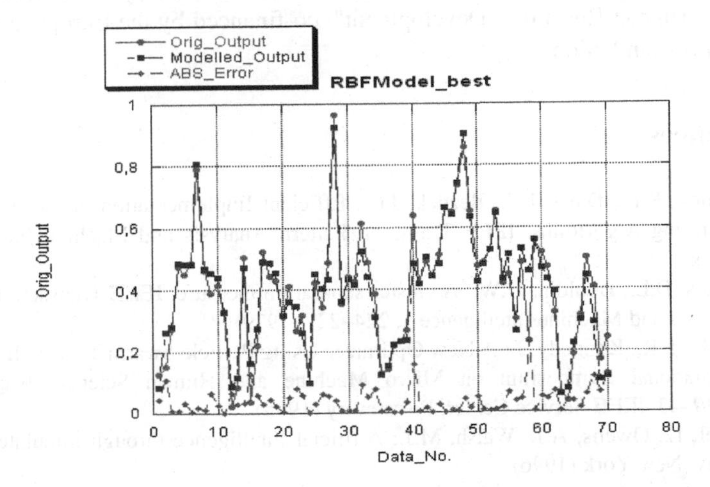

Fig. 14. Original Output, Modeled Output and Absolute Error for the training data set

The validation of the best created model has been also carried out. The RMSE of the test data set was 0.0636, while the RMSE for the training set was 0.0570.

6 Conclusions

The new proposed fuzzy model in this paper has a different structure, compared with the classical TS fuzzy model. It allows reducing the total number of parameters that have to be tuned by utilizing the information from the available sparse data in the multidimensional input space. The use of incomplete fuzzy rule base gives significant reduction of the number of parameters (locations of the fuzzy rules). Then the optimal locations can be found by using one of the two proposed methods, namely clustering or PSO optimization. It is experimentally proven in the paper that the PSO optimization produces a better model, because it does not take into account the structure and the distribution of the experimental data. Here the centers of the fuzzy rules are freely moved within the allowed (constrained) input space. Each fuzzy rule in the proposed fuzzy model is represented by a respective radial basis activation function with one common width (spread), which also contributes to the reduction of the total number of parameters.

A synthetic 2-dimensional test example, as well as a real example with 5-dimensional experimental data is used in the paper as experimental proof for the practical application of the proposed fuzzy model. Possible ways for improving the accuracy of the proposed fuzzy model are a future direction of the research.

Acknowledgment. This work has been financially supported by the project BG051PO001-3.3.06-0014 "Center of mathematical modeling and computer simulation for training and development of young researchers", UCTM – Sofia, under the scheme BG051PO001-3.3.06 "Support for the development of PhD students, post graduate students and young scientists" with the financial support of Operational Programme "Human Resources Development", co-financed by the European Social Fund of the European Union.

References

1. Cannon, R.L., Dave, J.V., Bezdek, J.C.: Efficient Implementation of the Fuzzy c-Means Clustering Algorithms. IEEE Trans. on Pattern Analysis and Machine Intelligence 8(2) (1986)
2. Davies, D.L., Bouldin, D.W.: A cluster separation measure. IEEE Transactions on Pattern Analysis and Machine Intelligence 1, 224–227 (1979)
3. Eberhart, R., Kennedy, J.: A New Optimizer Using Particle Swarm Theory. In: Proc. of 6th International Symposium on Micro Machine and Human Science, Nagoya, Japan, pp. 39–43. IEEE Service Center, Piscataway (1995)
4. Fogel, L., Owens, A.J., Walsh, M.J.: Artificial Intelligence through Simulated Evolution. Wiley, New York (1996)

5. Guo, P., Chen, C.L., Lyu, M.R.: Cluster Number Selection for a Small Set of Samples Using the Bayesian Ying-Yang Model. IEEE Trans. Neural Networks. 13(3), 757–763 (2002)
6. Ismail, M.A., Selim, S.Z.: Fuzzy *c*-means: Optimality of solutions and effective termination of the algorithm. Pattern Recognition 19, 481–485 (1986)
7. Kennedy, J., Eberhart, R.: Particle swarm optimization. In: Proceedings of IEEE International Conference on Neural Networks, pp. 1942–1948. IEEE Press, Piscataway (1995)
8. Orr, M.J.L.: Regularisation in the Selection of Radial Basis Function Centres. Neural Computation 7(3), 606–623 (1995)
9. Pal, N.R., Bezdek, J.C.: On Cluster Validity for the Fuzzy C-Means Model. IEEE Trans. Fuzzy Systems 3(3), 370–379 (1995)
10. Pedrycz, W., Waletzky, J.: Fuzzy clustering with partial supervision. IEEE Trans. Syst. Man Cybern. Part B Cybern. 27, 787–795 (1997)
11. Kacprzyk, J., Fedrizzi, M.: Developing a fuzzy logic controller in case of sparse testimonies. Int. Journal of Approximate Reasoning 12(3/4), 221–236 (1995)
12. Chorukova, E., Simeonov, I.: Monitoring System of Pilot Scale Bioreactor for Anaerobic Digestion of Organic Wastes. In: Proc. of the Int. Symposium on Control of Industrial and Energy Systems, Bankja, Bulgaria, November 7-8 (2013) (in Bulgarian)
13. Powell, M.J.D.: Radial basis functions for multivariable interpolation: a review. In: Algorithms for Approximation, pp. 143–167. Clarendon Press, Oxford (1987)
14. Roger Jang, J.S., Sun, C.T.: Functional equivalence between radial basis function networks and fuzzy inference systems. IEEE Transactions on Neural Networks 4(1), 156–159 (1993)
15. Schilling, R.J., Carroll, J.J.: Approximation of Nonlinear Systems with Radial Basis Function Neural Networks. IEEE Trans. on Neural Networks 12(1) (2001)
16. Das, S., Abraham, A.: Automatic Clustering Using An Improved Differential Evolution Algorithm. IEEE Transactions on Systems, Man, And Cybernetics—Part A: Systems And Humans 38(1), 218–237 (2008)
17. Takagi, T., Sugeno, M.: Fuzzy identification of systems and its applications to modeling and control. IEEE Trans. Syst., Man, Cybern. 15, 116–132 (1985)
18. Wang, X., Wang, Y., Wang, L.: Improving fuzzy c-means clustering based on feature-weight learning. Pattern Recognit. Lett. 25, 1123–1132 (2004)
19. Webb, A., Shannon, S.: Shape-adaptive radial basis functions. IEEE Trans. Neural Networks 9 (1998)
20. Xie, X.L., Beni, G.A.: Validity measure for fuzzy clustering. IEEE Trans. Pattern Anal. Mach. Intell. 3, 841–846 (1991)
21. Xu, C., Shin, Y.C.: Intelligent Systems Modeling, Optimization, and Control (Automation and Control Engineering Series) Summary. CRC Press (2008)

Application of Artificial Neural Networks for Modelling of Nicolsky-Eisenman Equation and Determination of Ion Activities in Mixtures

Józef Wiora[1], Dariusz Grabowski[2], Alicja Wiora[1], and Andrzej Kozyra[1]

[1] Institute of Automatic Control, Silesian University of Technology, ul. Akademicka 16, 44-100 Gliwice, Poland

[2] Institute of Electrical Engineering and Computer Science, Silesian University of Technology, ul. Akademicka 10, 44-100 Gliwice, Poland

{jozef.wiora,dariusz.grabowski,alicja.wiora,andrzej.kozyra}@polsl.pl

Abstract. The paper deals with the problem of ion activity determination for a mixture by means of ion-selective electrodes. Mathematical model of the analysed phenomenon is described by the Nicolsky-Eisenman equation, which relates activities of ions and ion-selective electrode potentials. The equation is strongly nonlinear and, especially in the case of multi-compound assays, the calculation of ion activities becomes a complex task. Application of multilayer perceptron artificial neural networks, which are known as universal approximators, can help to solve this problem. A new proposition of such network has been presented in the paper. The main difference in comparison with the previously proposed networks consists in the input set, which includes not only electrode potentials but also electrode parameters. The good network performance obtained during training has been confirmed by additional tests using measurement results and finally compared with the original as well as the simplified analytical model.

Keywords: Artificial neural network, Potentiometry, Ion-selective electrode.

1 Introduction

Determination of ion activities in a mixture is frequently performed using ion-selective electrodes (ISEs). The measurement is conducted by immersing the electrodes in a sample and measuring the potential difference between at least one ISE and one reference electrode, which together create an electrochemical cell. Despite their name, selectivity of ISEs is limited and their potential is dependent not only on an ion of interest, known as the primary ion, but also on other ions present in the sample, known as the interfering ions. Such situation is often modelled by the Nicolsky-Eisenman (N-E) equation which binds activities of ions with ISE potentials. The equation is nonlinear and especially in the case of multi-compound assays, the calculation of ion activity on the base of electrode potentials becomes a complex task.

P. Angelov et al. (eds.), *Intelligent Systems'2014*,
Advances in Intelligent Systems and Computing 322, DOI: 10.1007/978-3-319-11313-5_64

Analytical calculations of the activities may be impossible when more electrodes are used. In such cases, the values are often determined numerically using some iterative methods, which require rather more computing power. It may be a drawback if such algorithm is implemented in an embedded system. A method which requires less computing power during recalculations of measurement results consists in approximation by an artificial neural network (ANN). Applications of ANNs for working with ISEs are known in the literature [1,2,3], but here the concept is different. In the cited works, ANNs are trained individually for each ISE. The ANNs include not only a model, but also parameters describing the model. Because ISEs are not repetitive and their properties vary with time, the training process should be periodically repeated and done for every electrode. The training process is usually time-consuming and expensive. In this paper, we separate the mathematical model from parameters describing an individual ISE. The ANN is then trained only one time and used for all similar electrodes. The ANN describes only model behaviour, whereas parameters are introduced to its inputs. In the subsequent sections, we examine the concept using a simple case in order to easily compare results obtained by our ANN with the analytical solutions.

2 Background

Potential of an ISE measuring activity of ions in a mixture is usually described by the Nicolsky-Eisenman equation [4,5,6,7,8,9]:

$$E_i = E_i^\circ + S_i \cdot \lg \left[a_i + \sum_{j \neq i} K_{ij} (a_j)^{z_i/z_j} + L_i \right] \tag{1}$$

where E° is the cell constant which is constant at a fixed temperature and pressure; $S_i = 2.303 \frac{RT}{z_i \mathrm{F}}$ is the Nernstian slope; R – gas constant; T – thermodynamic temperature; z – charge number of a given ion; F – Faraday constant; lg – decimal logarithm; j – interfering ion; a – activity of a given ion; K_{ij} – potentiometric selectivity coefficient for the primary ion i against the interfering ion j; and L is electrode practical lower limit of detection. The Nernstian slope at temperature $T = 25°\mathrm{C}$ has its theoretical value of $S_i = 59.16\,\mathrm{mV}/z_i$, but in real measurement circuits it is slightly different. Therefore, the slope, together with other parameters (E°, S, K, L), is determined experimentally by calibration. There are a couple of calibration methods commonly accepted and recommended by IUPAC [10]. Other methods, optimised for special cases, have also been elaborated, *e.g.* for an array of ISEs [11] or best-fitting their response [12].

Ion activity is a quantity which describes ability of ions to react with other ions. The ion activity is related to its concentration by the activity coefficient γ:

$$a = \gamma \frac{c}{c^\circ} \tag{2}$$

where c° is standard concentration having the same unit as c and the value equal to one. To predict the activity coefficient value of an ion in a mixture, the Pitzer model is adequate [13,14]. As the concentration one can use molarity (molar concentration) expressed in mol/L or molality expressed in mol/kg. It is possible to recalculate molality to molarity and also molal activity coefficient to molar one knowing molar mass of all components and density of the mixture.

Sometimes, instead of activities, it is more flexible to operate on their negative logarithm. By the analogy to pH, the pX value is used. It is defined as:

$$\mathrm{pX} = -\lg a_X \tag{3}$$

For a given ion, letter X is replaced by its chemical symbol, *e.g.* pCa, pLi.

A multicomponent measurement preformed using ISEs is conducted by immersing a couple of ISEs in a sample. In such a way, a set of potentials is obtained. To calculate the activity (or concentration) of a given ion, a set of N-E equations (1), each equation describes one electrode, has to be solved. In order to do it, parameters of the electrodes have to be prior known. Determination of activities can be done analytically or using a numerical method. The calibration is a time-consuming task, therefore some authors have proposed application of ANNs to bind potentials with activities and omit the calibration step and solving. Unfortunately, the ANN has to be trained, so the calibration task has been moved to the training process, which can be even more time-consuming. Electrode parameters are not stable over time, therefore both calibration and training have to be periodically re-executed. Because number of standard mixtures needed for training is (or should be) much higher than for calibration, we state that it is possible to introduce the calibration parameters together with potentials into the ANN inputs and next calculate activities. The concept is schematically presented in

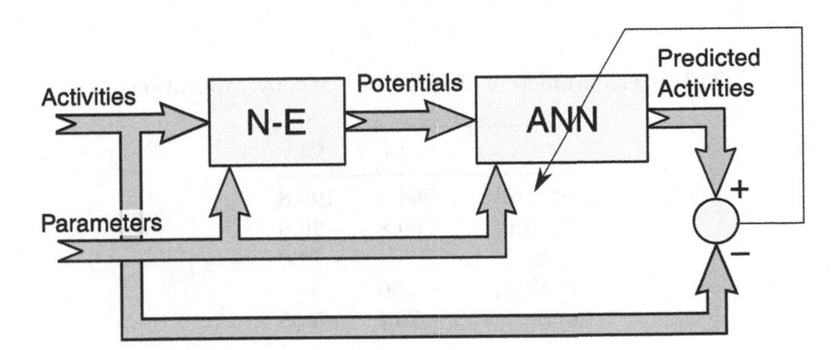

Fig. 1. Training of ANN. Activities and parameters are generated within a given range. Potentials are calculated according to N-E equation. Predicted activities are compared with setted activities and the obtained differences are used for optimization of ANN.

Fig. 1. During the training process, the same parameters of ISEs are sent to the N-E model and to the ANN. Then the potentials of ISEs are calculated using the model and according to the given activities chosen from the training set. Next, the obtained potentials are sent to the ANN as the second part of the training set. Differences between the predicted and real activities are the basis for the ANN weight optimization.

3 Experimental and Analytical Verification

The presented concept has been verified using a case study. The goal of the study was comparison of results obtained by the ANN organised in such unconventional way with results obtained in typical way, *i.e.* by solving the set of N-E equations. Two ISEs have been chosen: the first one is selective to an univalent cation – lithium (Li), and the second one to a divalent cation – calcium (Ca). For two electrodes, the task of finding formulas describing activities is moderate difficult; for three electrodes – very difficult and their mathematical representation unreadable. Cations with various valency have been chosen as more interesting.

3.1 Measurements

All solutions were prepared using pure p. a. chlorides produced by POCH, Gliwice, Poland, in Class A measurement glassware. Measurements were performed using the data acquisition system and the ISEs described in details elsewhere [11,12,15]. In order to make the measurement conditions similar to an industrial or environmental environment, the system was not put into a Faraday cup and, therefore, electromagnetic interferences disturbed measured potentials.

In order to calibrate the ISEs, nine pure standard solutions of Li and nine of Ca having molar concentration of $10^{-1}, 10^{-2}, \ldots, 10^{-9}$ mol/L were prepared. The concentrations were next recalculated to activities using the Pitzer model of activity coefficient. Obtained calibration parameters have been presented in Tab. 1. The slopes and standard potentials of electrodes were calculated using

Table 1. Parameters of ISEs obtained by calibration

i	Li	Ca
E_i° , mV	264.0	199.8
S_i , mV	59.8	29.9
$\lg K_{i,\text{Li}}$	—	−1.30
$\lg K_{i,\text{Ca}}$	−1.80	—
$\lg L_i$	−5.12	−5.60

pairs of the most concentrated pure solutions, respectively for each electrode. Selectivity coefficients were determined using the Separate Solution Method I [10] at $a = 0.01$. Because the prepared standards have activities different than 0.01, potentials were interpolated using two closest solutions. Practical limits of detections were obtained by best-fitting of results (minimizing the sum of the squares of the errors for potentials) in a spreadsheet.

Additional 13 mixtures of Li and Ca have been prepared (Tab. 3) and ISE potentials have been read in the same way as during calibration. The measurements have been used to compare results obtained by ANN with analytical solutions.

3.2 Analytical Solution of the Activities

The analysed case simplifies the N-E Eq. (1) to:

$$\begin{cases} E_{\mathrm{Li}} = E^{\circ}_{\mathrm{Li}} + S_{\mathrm{Li}} \lg\left(a_{\mathrm{Li}} + K_{\mathrm{Li,Ca}}\sqrt{a_{\mathrm{Ca}}} + L_{\mathrm{Li}}\right) \\ E_{\mathrm{Ca}} = E^{\circ}_{\mathrm{Ca}} + S_{\mathrm{Ca}} \lg\left(a_{\mathrm{Ca}} + K_{\mathrm{Ca,Li}} a^2_{\mathrm{Li}} + L_{\mathrm{Ca}}\right) \end{cases} \tag{4}$$

In order to solve the set of equations, it is convenient to introduce the following variables:

$$\xi_{\mathrm{Li}} = 10^{\frac{E_{\mathrm{Li}} - E^{\circ}_{\mathrm{Li}}}{S_{\mathrm{Li}}}} \quad \text{and} \quad \xi_{\mathrm{Ca}} = 10^{\frac{E_{\mathrm{Ca}} - E^{\circ}_{\mathrm{Ca}}}{S_{\mathrm{Ca}}}} \tag{5}$$

then Eq. (4) has the form without logarithm:

$$\begin{cases} \xi_{\mathrm{Li}} = a_{\mathrm{Li}} + K_{\mathrm{Li,Ca}}\sqrt{a_{\mathrm{Ca}}} + L_{\mathrm{Li}} \\ \xi_{\mathrm{Ca}} = a_{\mathrm{Ca}} + K_{\mathrm{Ca,Li}} a^2_{\mathrm{Li}} + L_{\mathrm{Ca}} \end{cases} \tag{6}$$

From that, one can solve the formula describing the activities:

$$\begin{cases} a_{\mathrm{Li}} = \dfrac{(\xi_{\mathrm{Li}} - L_{\mathrm{Li}}) \pm K_{\mathrm{Li,Ca}}\sqrt{(\xi_{\mathrm{Ca}} - L_{\mathrm{Ca}})(1 + K^2_{\mathrm{Li,Ca}} K_{\mathrm{Ca,Li}}) - K_{\mathrm{Ca,Li}}(\xi_{\mathrm{Li}} - L_{\mathrm{Li}})^2}}{1 + K^2_{\mathrm{Li,Ca}} K_{\mathrm{Ca,Li}}} \\ a_{\mathrm{Ca}} = \dfrac{\begin{array}{c}(\xi_{\mathrm{Ca}} - L_{\mathrm{Ca}})(1 + K^2_{\mathrm{Li,Ca}} K_{\mathrm{Ca,Li}}) - K_{\mathrm{Ca,Li}}(1 - K^2_{\mathrm{Li,Ca}} K_{\mathrm{Ca,Li}})(\xi_{\mathrm{Li}} - L_{\mathrm{Li}})^2 + \\ \mp K_{\mathrm{Ca,Li}} 2K_{\mathrm{Li,Ca}}(\xi_{\mathrm{Li}} - L_{\mathrm{Li}})\sqrt{(\xi_{\mathrm{Ca}} - L_{\mathrm{Ca}})(1 + K^2_{\mathrm{Li,Ca}} K_{\mathrm{Ca,Li}}) - K_{\mathrm{Ca,Li}}(\xi_{\mathrm{Li}} - L_{\mathrm{Li}})^2}\end{array}}{(1 + K^2_{\mathrm{Li,Ca}} K_{\mathrm{Ca,Li}})^2} \end{cases} \tag{7}$$

Selectivity coefficients are usually less than one, therefore we can assume that $0 < K^2_{\mathrm{Li,Ca}} K_{\mathrm{Ca,Li}} \ll 1$ and then:

$$\begin{cases} a_{\mathrm{Li}} = (\xi_{\mathrm{Li}} - L_{\mathrm{Li}}) \pm \underbrace{K_{\mathrm{Li,Ca}}\sqrt{(\xi_{\mathrm{Ca}} - L_{\mathrm{Ca}}) - \underbrace{K_{\mathrm{Ca,Li}}(\xi_{\mathrm{Li}} - L_{\mathrm{Li}})^2}_{\text{negligible influence}}}}_{\text{influence of the interfering ion}} \\ a_{\mathrm{Ca}} = (\xi_{\mathrm{Ca}} - L_{\mathrm{Ca}}) - \underbrace{K_{\mathrm{Ca,Li}}(\xi_{\mathrm{Li}} - L_{\mathrm{Li}})^2}_{\text{influence of the interfering ion}} + \\ \qquad \underbrace{\mp 2K_{\mathrm{Ca,Li}} K_{\mathrm{Li,Ca}}(\xi_{\mathrm{Li}} - L_{\mathrm{Li}})\sqrt{(\xi_{\mathrm{Ca}} - L_{\mathrm{Ca}}) - K_{\mathrm{Ca,Li}}(\xi_{\mathrm{Li}} - L_{\mathrm{Li}})^2}}_{\text{negligible influence}} \end{cases} \tag{8}$$

Interfering ions always make that the observed activity, before taking the interference into account, seems to be higher than it is, therefore the mathematical solution with the plus sign should be rejected. After removing not important constituents, we obtain:

$$\begin{cases} a_{\mathrm{Li}} = (\xi_{\mathrm{Li}} - L_{\mathrm{Li}}) - K_{\mathrm{Li,Ca}}\sqrt{(\xi_{\mathrm{Ca}} - L_{\mathrm{Ca}})} \\ a_{\mathrm{Ca}} = (\xi_{\mathrm{Ca}} - L_{\mathrm{Ca}}) - K_{\mathrm{Ca,Li}}(\xi_{\mathrm{Li}} - L_{\mathrm{Li}})^2 \end{cases} \tag{9}$$

3.3 Description of the Applied ANN

On the base of the Stone-Weierstrass theorem, it was proved that multilayer perceptron (MLP) neural networks with one hidden layer having sigmoid activation functions are universal approximators. It means that they can approximate arbitrary well any continuous nonlinear mapping if the training data set is consistent and there are enough neurons in the hidden layer. Since the first proof given in [16], many practical applications of MLP networks to nonlinear system modelling have been reported [17,18,19]. Some other approaches to solve this problem with the help of neural networks are also known [20,21,22,23].

In the paper, a two-layer feedforward MLP neural network with nonlinear neurons (hyperbolic tangent activation functions) in the hidden layer and linear neurons in the output layer has been applied to fit the mapping between activities of ions and potential differences as well as some other parameters (see Sect. 2). In general, the number of the ANN inputs and outputs is determined by the problem under consideration. In our case, the network includes 6 inputs and 2 outputs (Fig. 2). The number of neurons in the hidden layer must be set experimentally. During simulations, a few cases have been checked, namely 5, 10, 20, 30 and 50 hidden neurons.

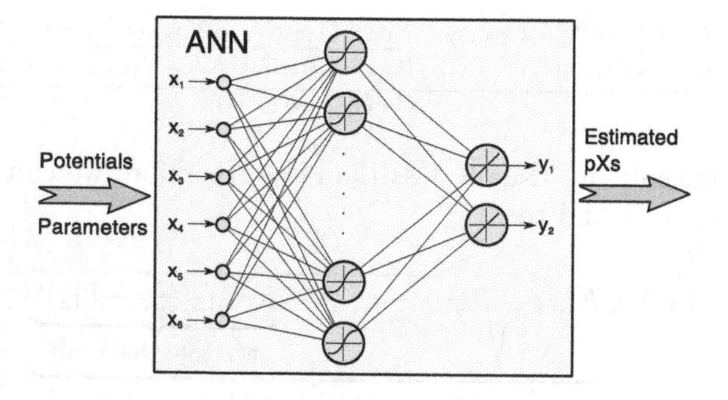

Fig. 2. Structure of the neural network used for solution of the problem

The original data set that has been obtained using the simulation model consists of 8100 samples defining ANN inputs, *i.e.* $x_1 = \lg(\xi_{\text{Li}})$, $x_2 = \lg(\xi_{\text{Ca}})$, $x_3 = -\lg(K_{\text{Li,Ca}})$, $x_4 = -\lg(K_{\text{Ca,Li}})$, $x_5 = -\lg(L_{\text{Li}})$, $x_6 = -\lg(L_{\text{Ca}})$, and desired outputs, *i.e.* $d_1 = \text{pLi}$ and $d_2 = \text{pCa}$. The data set has been divided into training (5670 samples), validation (1215 samples) and testing (1215 samples) data subsets. The data set has been generated using the N-E equation for assumed range of all input variables: $x_1 \in [-6.2, -0.5]$, $x_2 \in [-6.2, -0.5]$, $x_3 \in [2.9, 5.9]$, $x_4 \in [1.6, 4.6]$, $x_5 \in [1.5, 6.5]$, $x_6 \in [1.5, 6.5]$. As a result, the desired output signals d_1 and d_2 fall between 0.5 and 6.5.

The network has been trained using the above data set and the classical Levenberg-Marquardt backpropagation algorithm. The mean-squared error of the network output signal during the learning process for the hidden layer with

10 neurons has been presented in Fig. 3. The more precise insight in the accuracy of the mapping realized by the ANN gives the error histogram presented in Fig. 4. It can be noticed that most errors fall between −1.0 and 1.0. Results obtained for other numbers of hidden neurons have been summarized in Tab. 2. A good compromise between the network complexity and performance is reached for 10 hidden neurons.

Table 2. Mean squared error and absolute error for selected numbers of hidden neurons

Number of hidden neurons	MSE	Absolute error Δ_{min}	Δ_{max}
5	0.122	−0.92	2.65
10	0.060	−1.02	2.09
20	0.058	−1.58	1.91
30	0.059	−1.16	1.93
50	0.048	−1.32	1.64

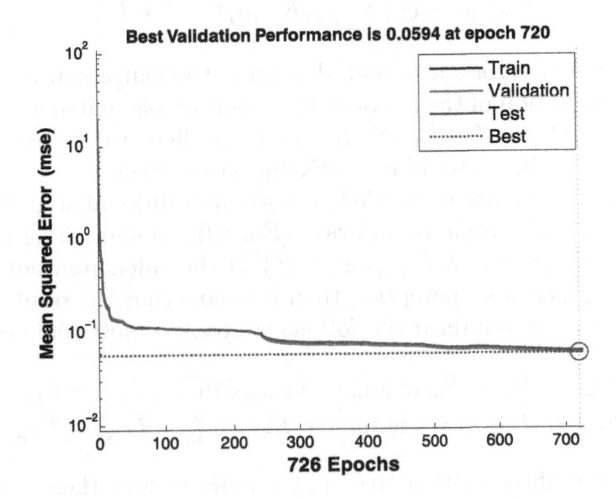

Fig. 3. Mean squared error during the learning process of the neural network

3.4 Checking of Result Correctness

Properties of the N-E equation causes that variability of ISE potential is limited. For a given ISE, its potential changes according to:

$$\frac{E_i}{S_i} > \frac{E_i^\circ}{S_i} + \lg\left(L_i\right) \tag{10}$$

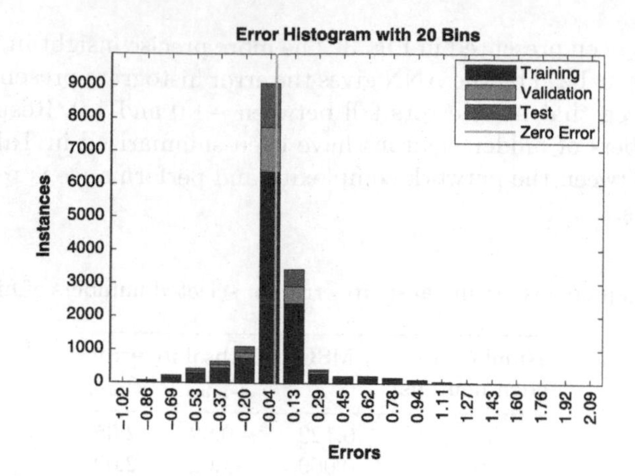

Fig. 4. Error histogram for all data samples obtained from simulation after training ANN

and in presence of interfering ions, the potential is even more limited:

$$\frac{E_i}{S_i} > \frac{E_i^{\circ}}{S_i} + \lg\left[\sum{}_{j\neq i} K_{ij}(a_j)^{z_i/z_j} + L_i\right] \tag{11}$$

Due to measurement errors, a potential outside the range can be recorded. In such case, determination of the ion activity is impossible and analytical solution gives complex numbers. The ANN, however, predicts values and there is no information whether the determined activities are correct.

For the reason, the authors decided to add an additional step. Using results calculated by ANN or simplified solution (Eq. (9)), potentials of each ISE are calculated according to the N-E equation (1). If the calculated potential differs significantly from measured potential, then it means that the result is incorrect. For our study, we have obtained the following absolute potential errors:

$$\Delta E_{\mathrm{Li}} = \left|E_{\mathrm{Li}}^{\circ} + S_{\mathrm{Li}} \lg\left(a_{\mathrm{Li}} + K_{\mathrm{Li,Ca}}\sqrt{a_{\mathrm{Ca}}} + L_{\mathrm{Li}}\right) - E_{\mathrm{Li}}\right| \tag{12}$$

$$\Delta E_{\mathrm{Ca}} = \left|E_{\mathrm{Ca}}^{\circ} + S_{\mathrm{Ca}} \lg\left(a_{\mathrm{Ca}} + K_{\mathrm{Ca,Li}}a_{\mathrm{Li}}^{2} + L_{\mathrm{Ca}}\right) - E_{\mathrm{Ca}}\right| \tag{13}$$

The ANN and simplified solution give approximate results, therefore some value of ΔE_{Max} is allowed. If the error is grater than ΔE_{Max} value, it is supposed that the obtained activity is determined incorrectly. The cut-off value has been determined experimentally: 25 mV for univalent ISEs and 12.5 mV for divalent ISEs as a half of the slope value.

4 Results and Discussions

The results of measurements described in Sect. 3.1 have been used to check the real performance of the ANN. The error histogram for the measurement data set has been presented in Fig. 5. It contains information only for measurement samples for which the range assumptions taken for the ANN training input data set (see Sect. 3.3) have been fulfilled. It can be noticed that errors fall between −0.7 and 0.2 what is consistent with the errors obtained for simulation data (Fig. 4). The regression plot shown in Fig. 6 displays the ANN outputs with respect to values obtained during measurements (targets). In the ideal case, the data should fall along a 45 degree line, where the network outputs are equal to the targets. The mapping realized by the ANN is reasonably good.

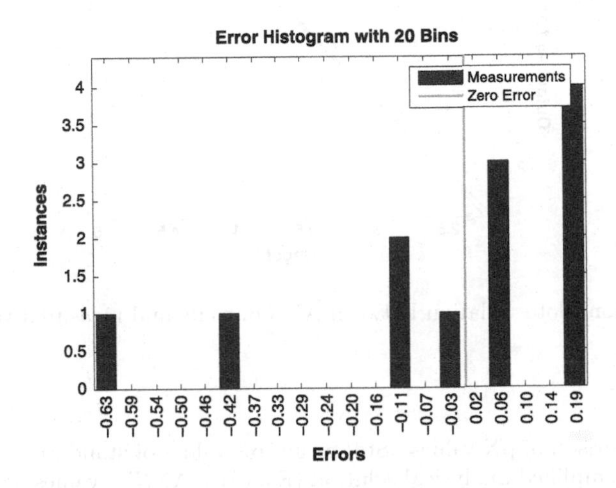

Fig. 5. Error histogram for data samples obtained from measurements and processed by the trained ANN

Results for mixtures have been put together in Tab. 3. Mixtures #1 and #2 have their activities below limits of detections, therefore both simplified analytical solution and ANN are incorrect, whereas accurate analytical solution gives complex numbers what suggests that the measured potentials exceed allowed range. Solutions #3, #7, #8, #12, and #13 are pure and predictions of pXs of the ions, which are not present, are also incorrect. Results for mixtures #4—#9 have very good conformity between both methods even though the results predict the pXs of standards moderately, but acceptably for typical industrial and environmental applications. The differences may be due to existence of measurement errors of potentials. The predicted values for mixtures #10 and #11 are reasonable even though the accurate analytical result for Li in #10 is undetermined. Results obtained by simplified solutions (Eq. (9)) comparing with unsimplified ones (Eq. (7)) give the same values within the precision of two digits after point except values marked by the star – columns 3 and 4 in the table.

Additional correctness checking has allowed us to point out the results which are determined with significant errors. The errors have been caused by limited accuracy of potential measurements and activities which are difficult to determine due to limits of detections or presence of interfering ions. The results flagged by □ in Tab. 3 occur for the cases easy to predict taking into account above

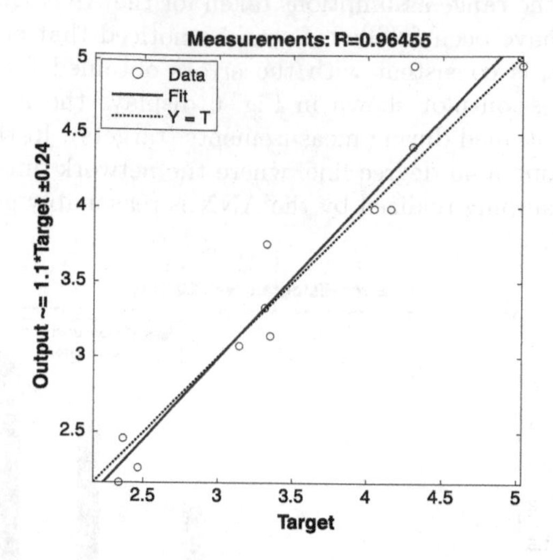

Fig. 6. Regression plot – relation between ANN outputs and measured values (target)

Table 3. Comparison of pX values. 'Std' means pX values of standards; 'Ana' – values obtained using simplified analytical solution (Eq. (9)); 'ANN' – values obtained by the ANN. Values of 10.00 mean that the ions are not present. Unsatisfied results marked as *italic*; good result as **bold**. The star (*) before number means, that the analytical result, calculated by the unsimplified equation (Eq. (7)), is undetermined. The square (□) before number means, that the test for the ANN result fails.

	Std		Ana		ANN		|ANN−Ana|		|Ana−Std|	
#	pLi	pCa	pLi	pCa	pLi	pCa	pLi	pCa	pLi	pCa
col	1	2	3	4	5	6	7	8	9	10
1	*8.12*	*8.51*	10.00	*6.52	□-5.81	□232.59	*15.81*	*226.07*	*1.88*	*1.99*
2	*7.12*	*7.51*	10.00	*6.60	□-5.60	□227.07	*15.60*	*220.47*	*2.88*	0.91
3	4.01	*10.00*	4.21	6.03	4.12	6.18	0.09	0.15	0.20	*3.97*
4	4.31	5.03	*4.42	4.95	4.41	4.96	**0.01**	**0.01**	0.11	0.08
5	3.31	4.16	3.29	4.00	**3.33**	3.99	**0.04**	**0.01**	**0.02**	0.16
6	2.34	3.35	2.12	3.17	2.17	3.14	0.05	**0.03**	0.22	0.18
7	*10.00*	3.94	□*3.47	4.01	□4.89	4.01	*1.42*	**0.00**	*6.53*	0.07
8	*10.00*	4.93	□*3.90	4.98	□5.31	5.02	*1.41*	**0.04**	*6.10*	0.05
9	4.31	4.05	□*3.43	3.99	□4.96	3.99	*1.53*	**0.00**	0.88	0.06
10	3.33	3.14	*3.66	3.13	3.76	3.07	0.10	0.06	0.33	**0.01**
11	2.36	2.46	2.30	2.34	2.46	2.26	0.16	0.08	0.06	0.12
12	*10.00*	1.58	□*2.32	1.56	□3.69	1.43	*1.37*	0.13	*7.68*	**0.02**
13	1.10	*10.00*	1.16	*4.47	1.00	□6.92	0.16	*2.45*	0.06	*5.53*

discussion – nobody could expect better results for these cases. Only pCa result for #13 needs to be flagged – the test would fail if the cut-off value was smaller.

5 Conclusions

The presented results have proved that it is possible to use ANNs to model the N-E equation to calculate activities of ions in a mixture. The application of the ANN has been organised in a way different from typical approach – here, the parameters of the model are given to ANN inputs instead of including them into the ANN. The obtained quality of results can be considered to be consistent with the experience gained in the past during research experiments involving ion activity determination.

Acknowledgements. The authors would like to thank the Polish Ministry of Science and Higher Education for financial support.

References

1. Baret, M., Massart, D.L., Fabry, P., Conesa, F., Eichner, C., Menardo, C.: Application of neural network calibrations to an halide ISE array. Talanta 51, 863–877 (2000)
2. Mimendia, A., Legin, A., Merkoçi, A., del Valle, M.: Use of sequential injection analysis to construct a potentiometric electronic tongue: Application to the multidetermination of heavy metals. Sensors and Actuators B: Chemical 146, 420–426 (2010)
3. Gutiérrez, M., Alegret, S., del Valle, M.: Bioelectronic tongue for the simultaneous determination of urea, creatinine and alkaline ions in clinical samples. Biosensors and Bioelectronics 23, 795–802 (2008)
4. Midgley, D., Torrance, K.: Potentiometric Water Analysis, 2nd edn. John Wiley & Sons, Inc., Chichester (1991)
5. Kane, P., Diamond, D.: Determination of ion-selective electrode characteristics by non-linear curve fitting. Talanta 44(4), 1847–1858 (1997)
6. Amman, D.: Ion-Selective Microelectrodes. Springer, Heidelberg (1986)
7. IUPAC: Potentiometric selectivity coefficients of ion-selective electrodes. Part I. Inorganic cations (technical report). Pure Appl. Chem. 72(10), 1851–2082 (2000)
8. Simon, W., Ammann, D., Oehme, M., Morf, W.E.: Calcium-selective electrodes. Annals of the New York Academy of Sciences 307(1), 52–70 (1978)
9. Martnez-Barrachina, S., Alonso, J., Matia, L., Prats, R., del Valle, M.: Determination of trace levels of anionic surfactants in river water and wastewater by a flow injection analysis system with on-line preconcentration and potentiometric detection. Analytical Chemistry 71(17), 3684–3691 (1999)
10. Inczédy, J., Lengyel, T., Ure, A.M.: The Orange Book: Compendium of Analytical Nomenclature. Definitive rules 1997, 3rd edn. IUPAC. Blackwell Science Ltd., Oxford (1998)
11. Kozyra, A., Wiora, J., Wiora, A.: Calibration of potentiometric sensor arrays with a reduced number of standards. Talanta 98, 28–33 (2012)

12. Wiora, J., Wiora, A.: A system allowing for the automatic determination of the characteristic shapes of ion-selective electrodes. In: Pisarkiewicz, T. (ed.) Optoelectronic and Electronic Sensors VI. Proceedings of SPIE, vol. 6348 (2006)
13. Pitzer, K.S.: Activity Coefficients in Electrolyte Solution, 2nd edn. CRC Press, Boca Raton (1991)
14. Bastkowski, F., et al.: Pitzer ion activities in mixed electrolytes for calibration of ion-selective electrodes used in clinical chemistry. Accred. Qual. Assur. 18, 469–479 (2013)
15. Wiora, A., Wiora, J., Kozyra, A.: Dynamic models of ion-selective electrodes and their interface electronics. Metrology and Measurement Systems 13(4), 421–432 (2006)
16. Hornik, K., Stinchcombe, M., White, H.: Multilayer feedforward networks are universal approximators. Neural Networks 2(5), 359–366 (1989)
17. Malinowski, A., Cholewo, T.J., Zurada, J.M., Aronhime, P.B.: Inverse mapping with neural network for control of nonlinear systems, vol. 3, pp. 453–456 (1996)
18. Kim, T., Adali, T.: Approximation by fully complex multilayer perceptrons. Neural Computation 15(7), 1641–1666 (2003)
19. Pei, J.S., Wright, J., Smyth, A.: Mapping polynomial fitting into feedforward neural networks for modeling nonlinear dynamic systems and beyond. Computer Methods in Applied Mechanics and Engineering 194(42-44), 4481–4505 (2005)
20. Grabowski, D., Walczak, J.: Generalized spectrum analysis by means of neural networks. In: Rutkowski, L., Kacprzyk, J. (eds.) Neural Networks and Soft Computing. Advances in Soft Computing, pp. 704–709 (2003)
21. Oh, S.K., Pedrycz, W.: Self-organizing polynomial neural networks based on polynomial and fuzzy polynomial neurons: Analysis and design. Fuzzy Sets and Systems 142(2), 163–198 (2004)
22. Ilin, R., Kozma, R., Werbos, P.: Beyond feedforward models trained by backpropagation: A practical training tool for a more efficient universal approximator. IEEE Transactions on Neural Networks 19(6), 929–937 (2008)
23. Cimino, M., Pedrycz, W., Lazzerini, B., Marcelloni, F.: Using multilayer perceptrons as receptive fields in the design of neural networks. Neurocomputing 72(10-12), 2536–2548 (2009)

Part XIV
Classification and Clustering

An Interval-Valued Fuzzy Classifier Based on an Uncertainty-Aware Similarity Measure

Anna Stachowiak, Patryk Żywica,
Krzysztof Dyczkowski, and Andrzej Wójtowicz

Faculty of Mathematics and Computer Science,
Adam Mickiewicz University,
Umultowska 87, 61-614 Poznań, Poland
{aniap,bikol,chris,andre}@amu.edu.pl

Abstract. In this paper we propose a new method for classifying uncertain data, modeled as interval-valued fuzzy sets. We develop the notion of an interval-valued prototype-based fuzzy classifier, with the idea of preserving full information including the uncertainty factor about data during the classification process. To this end, the classifier was based on the uncertainty-aware similarity measure, a new concept which we introduce and give an axiomatic definition for. Moreover, an algorithm for determining such a similarity value is proposed, and an application to supporting medical diagnosis is described.

1 Introduction

In this paper we present the novel concept of an interval-valued fuzzy classifier for supporting decision-making processes based on imprecise and incomplete (uncertain) data. The main aim was to develop a comprehensive, consistent and effective approach that enables the modeling and processing of input data, and then the presentation of results, to be done in a way that preserves the valuable information concerning the amount of uncertainty at each stage of the process.

Our theoretical framework is the well-developed interval-valued fuzzy set (IVFS) theory, introduced by Zadeh in [1] as a natural extension of fuzzy set theory. While a fuzzy set models a gradual, but precise, degree of truth of a statement, IVFSs make it possible to add uncertainty about that degree. IVFS $\widetilde{A}$ is defined by a pair of fuzzy sets $\underline{A}, \overline{A} : X \to [0,1]$ such that for each $x \in X$ an interval $[\underline{A}(x), \overline{A}(x)]$ is understood to contain the true, incompletely known membership degree. The length of this interval reflects the amount of uncertainty about an element x, taking values from 0 when $\underline{A}(x) = \overline{A}(x)$ to 1 when $[\underline{A}(x), \overline{A}(x)] = [0,1]$. Framework of IVFS was designed to be able to deal with incomplete data. In cases where data is missing, we assign the unit interval as the membership degree, meaning that all membership degrees are equally possible.

An IVFS is also a special case of type-2 fuzzy set (also introduced by Zadeh [1]), known by the name of interval type-2 fuzzy set. Another equivalent notion is the Atanassov intuitionistic fuzzy set theory, which specifies a membership

P. Angelov et al. (eds.), *Intelligent Systems'2014*,
Advances in Intelligent Systems and Computing 322, DOI: 10.1007/978-3-319-11313-5_65

function and a non-membership function separately (for a comparison, see [2]). All of these theories have proved their usefulness in many areas of soft computing such as fuzzy control, image processing, financial prediction, decision-making, computer vision, medical diagnosis, etc. [3, 4, 5, 6, 7]. In what follows we will contribute to medical decision support systems by constructing an interval-based fuzzy classifier.

Hereafter by $\mathcal{FS}$ we denote the set of all fuzzy sets, by $\mathcal{IVFS}$ we denote the set of all interval-valued fuzzy sets, and by $\mathcal{I}$ we denote the set of all intervals in $[0,1]$. Formally a classifier can be defined as a function $D : \mathcal{O}^n \to \mathcal{C}$ where $\mathcal{O}$ is a set of possible values of classification features (it may be for example a subset of R or $\mathcal{I}$), $\mathcal{C}$ is the set of all possible m-element vectors which coordiantes correspond to particular classes and values reflect the membership to these classes, n is the number of classification features and m is the number of classes. As stated in [8], designing a classifier means "finding good D". D can be specified functionally or as a computer algorithm.

We can define different sets of labeled vectors in $\mathcal{C}$, resulting in different types of classifiers, e.g.:

1. crisp classifier:
$$\mathcal{C}_{cr} = \{y \in \{0,1\}^m\}$$
2. probabilistic classifier:
$$\mathcal{C}_{pr} = \{y \in [0,1]^m : \sum_{i=1}^{m} y_i = 1\}$$
3. fuzzy classifier:
$$\mathcal{C}_{fu} = \{y \in [0,1]^m\}$$

There are many books on the theory and applications of fuzzy classification and pattern recognition. A very good introduction to designing fuzzy classifiers, with a very large subject bibliography, can be found in [8]; also worth examining are [9] and [10]. More information on the construction and use of crisp and probabilistic classifiers can be found in [11, 12].

In this article we introduce a new type of classifier:

4. interval-valued fuzzy classifier:
$$C_{iv} = \{y \in \mathcal{I}^m\}$$

A popular approach is to design, for each class, one prototype vector which represents the entire class. A very important element of such classification process is the construction of suitable vectors to describe the classes. Classifiers of this type are called prototype-based classifiers. Prototypes can be formed using clustering algorithms, such as k-means, or can result from the application of expert knowledge. We have taken the second approach, and through the use of IVFSs we were able to model experts knowledge along with its subjectivity, uncertainty and information deficiency.

In the next section we introduce in detail the idea of an uncertainty-aware similarity measure, which forms a central concept of a prototype-based classifier and enables the proper comparison of uncertain data. We propose an efficient algorithm for determining its value using the notion of relative cardinality of IVFSs. The third section is devoted to an interval-based classifier. In Section 4 we apply the ideas presented to the problem of supporting ovarian tumor diagnosis. Finally we state some conclusions.

2 Similarity Measure for Uncertain Data

The similarity measure is a central concept of prototype-based classifiers that estimate the class label of a test sample based on the similarities between the test sample and a set of given prototypes.

In the following we briefly review some of the similarity measures for IVFSs known from the literature, and next we introduce the concept of an uncertainty-aware similarity measure, together with an effective algorithm for computing its value.

2.1 An Overview of Similarity Measures for IVFSs

A common approach for measuring similarity involves the use of a distance metric. For example, in [13] the following similarity measure based on a normalized Hamming distance was proposed:

$$sim_H(\widetilde{A}, \widetilde{B}) = 1 - \frac{1}{2n} \sum_{i=1}^{n} \left(|\underline{A}(x_i) - \underline{B}(x_i)| + |\overline{A}(x_i) - \overline{B}(x_i)| \right) . \tag{1}$$

Another well-known approach is the Jaccard coefficient extended to IVFSs in the following way [14, 15]:

$$sim_J(\widetilde{A}, \widetilde{B}) = \frac{\sum_{i=1}^{n} \min(\overline{A}(x_i), \overline{B}(x_i)) + \sum_{i=1}^{n} \min(\underline{A}(x_i), \underline{B}(x_i))}{\sum_{i=1}^{n} \max(\overline{A}(x_i), \overline{B}(x_i)) + \sum_{i=1}^{n} \max(\underline{A}(x_i), \underline{B}(x_i))} \tag{2}$$

Both sim_H and sim_J measure similarity with a single, scalar value. In the method proposed by Bustince [4] the similarity value is defined as an interval:

$$\begin{aligned} sim_B(\widetilde{A}, \widetilde{B}) &= \left[S_L(\widetilde{A}, \widetilde{B}), S_U(\widetilde{A}, \widetilde{B}) \right] \\ S_L(\widetilde{A}, \widetilde{B}) &= t\left(Inc_L(\widetilde{A}, \widetilde{B}), Inc_L(\widetilde{B}, \widetilde{A}) \right) \\ S_U(\widetilde{A}, \widetilde{B}) &= t\left(Inc_U(\widetilde{A}, \widetilde{B}), Inc_U(\widetilde{B}, \widetilde{A}) \right) \\ Inc_L(\widetilde{A}, \widetilde{B}) &= \inf_{x \in X} \left\{ 1, \min\left(1 - \underline{A}(x) + \underline{B}(x), 1 - \overline{A}(x) + \overline{B}(x)\right) \right\} \\ Inc_U(\widetilde{A}, \widetilde{B}) &= \inf_{x \in X} \left\{ 1, \max\left(1 - \underline{A}(x) + \underline{B}(x), 1 - \overline{A}(x) + \overline{B}(x)\right) \right\} \end{aligned} \tag{3}$$

where t is a t-norm i.e. an increasing, commutative and associative mapping $t : [0,1]^2 \to [0,1]$ satisfying $t(1, x) = x$ for all $x \in [0,1]$. More examples of IVFS similarity measures can be found in [14].

2.2 An Uncertainty-Aware Similarity Measure

The property of reflexivity of a similarity measure, natural and unquestionable in classical set theory, is not so obvious when comparing IVFSs with positive uncertainty value. Consider an extreme case of two different concepts represented by two one-element totally uncertain IVFS $\widetilde{A} = \widetilde{B} = {}^{[0,1]}/_{x}$. Mentioned IVFS, according to all classical similarity measures, including sim_H, sim_J and sim_B, are definitely identical ($sim_H(\widetilde{A},\widetilde{B}) = sim_J(\widetilde{A},\widetilde{B}) = 1$, $sim_B(\widetilde{A},\widetilde{B}) = [1,1]$). But is that also true for concepts they represent? In view of the total lack of knowledge, such a claim has no basis. In fact, the "real" membership degrees may be totally different for $\widetilde{A}$ and $\widetilde{B}$, and consequently a possible value of $sim(\widetilde{A},\widetilde{B})$ may be anywhere in the range $[0,1]$. Ignoring the uncertainty of $\widetilde{A}$ and $\widetilde{B}$ may have particularly adverse consequences when such a similarity measure is applied to practical problems. In this paper we emphasize the role and importance of the uncertainty factor for making well-informed decisions. For this reason we propose the notion of an uncertainty-aware similarity measure, described by the following definition.

Definition 1. *A mapping $sim_u : \mathcal{IVFS} \times \mathcal{IVFS} \to \mathcal{I}$ is said to be an uncertainty-aware similarity measure if it satisfies the following conditions for all $\widetilde{A},\widetilde{B},\widetilde{C},\widetilde{D} \in \mathcal{IVFS}$:*

1. $sim_u(\widetilde{A},\widetilde{B}) = [1,1] \Leftrightarrow \widetilde{A} = \widetilde{B}\ \&\ \widetilde{A},\widetilde{B} \in \mathcal{FS}$
2. $sim_u(\widetilde{A},\widetilde{B}) = sim_u(\widetilde{B},\widetilde{A})$
3. *if* $\widetilde{A} \subseteq \widetilde{B} \subseteq \widetilde{C}$ *then* $sim_u(\widetilde{A},\widetilde{C}) \leq sim_u(\widetilde{A},\widetilde{B})$ *and* $sim_u(\widetilde{A},\widetilde{C}) \leq sim_u(\widetilde{B},\widetilde{C})$
4. *if* $\widetilde{A} \sqsubseteq \widetilde{B}$ *and* $\widetilde{C} \sqsubseteq \widetilde{D}$ *then* $sim_u(\widetilde{A},\widetilde{C}) \preceq sim_u(\widetilde{B},\widetilde{D})$

where

- $[a,b] \leq [c,d] \Rightarrow a \leq c\ \&\ b \leq d$,
- $\widetilde{A} \subseteq \widetilde{B} \Leftrightarrow \underline{A}(x) \leq \underline{B}(x)\ \&\ \overline{A}(x) \leq \overline{B}(x), \forall x \in X$,
- $[a,b] \preceq [c,d] \Rightarrow a \geq c\ \&\ b \leq d$,
- $\widetilde{A} \sqsubseteq \widetilde{B} \Leftrightarrow \underline{A}(x) \geq \underline{B}(x)\ \&\ \overline{A}(x) \leq \overline{B}(x), \forall x \in X$.

Conditions 2 and 3 are standard symmetry and monotonicity properties. Conditions 1 and 4 reflect the goal of preserving the uncertainty value – when comparing uncertain objects, the final result should also be uncertain. Moreover, in 4 we require monotonicity with respect to uncertainty – the more certain are the compared IVFSs, the more certain is the resulting similarity measure sim_u.

In [15], and then in [16], [17] similarity for IVFSs was considered in the spirit of Definition 1. In the following we present an uncertainty-aware similarity measure of IVFSs based on the notion of relative cardinality of fuzzy sets, i.e. on the fact that for any $A, B \in \mathcal{FS}$:

$$sim(A,B) = \sigma(A \cap_t B | A \cup_t B) \tag{4}$$

where

$$\sigma(A|B) = \frac{\sigma(A \cap_t B)}{\sigma(B)}$$

is a relative cardinality of fuzzy sets. The scalar cardinality of fuzzy set $\sigma(A)$ is typically defined by the so-called sigma-count ([18, 19]):

$$\sigma(A) = \sum_{x \in X} A(x).$$

Moreover, sum $A \cup_t B$ and intersection $A \cap_t B$ are defined as:

$$\begin{aligned} (A \cup_t B)(x) &= t^* (A(x), B(x)) \\ (A \cap_t B)(x) &= t (A(x), B(x)) \end{aligned}$$

where as a t-norm t we can take for example a minimum t-norm $t_\wedge(a,b) = a \wedge b$, a product t-norm $t_{\text{prod}}(a,b) = a \cdot b$ or a Łukasiewicz t-norm $t_Ł(a,b) = 0 \vee (a+b-1)$. A t-conorm t^* is a dual operation such that $t^*(a,b) = 1 - t(1-a, 1-b)$.

To introduce a formula for an uncertainty-aware similarity measure, we first construct fuzzy representation sets of $\widetilde{A}$ and $\widetilde{B}$, where a representation set is defined as:

$$Rep(\widetilde{A}) = \{A \in \mathcal{FS} \mid \forall_{x \in X} \underline{A}(x) \leq A(x) \leq \overline{A}(x)\}.$$

Then the uncertainty-aware similarity measure sim_σ based on the notion of the relative cardinality, is defined in the following way.

Definition 2. *The cardinality-based uncertainty-aware similarity measure of two IVFSs $\widetilde{A} = (\underline{A}, \overline{A})$ and $\widetilde{B} = (\underline{B}, \overline{B})$, with a t-norm t, is an interval defined as:*

$$sim_\sigma(\widetilde{A}|\widetilde{B}) = \left[\underline{sim}_\sigma(\widetilde{A}|\widetilde{B}), \overline{sim}_\sigma(\widetilde{A}|\widetilde{B})\right] \tag{5}$$

where

$$\begin{aligned} \underline{sim}_\sigma(\widetilde{A}|\widetilde{B}) &= \min_{\substack{A \in Rep(\widetilde{A}) \\ B \in Rep(\widetilde{B})}} \sigma(A \cap_t B | A \cup_t B) \\ \overline{sim}_\sigma(\widetilde{A}|\widetilde{B}) &= \max_{\substack{A \in Rep(\widetilde{A}) \\ B \in Rep(\widetilde{B})}} \sigma(A \cap_t B | A \cup_t B) \end{aligned}$$

The formula given by Definition 2 is t-norm dependent. In the present paper we consider two of the most widely used t-norms: minimum and product. An effective algorithm for calculating a value of (5) for a minimum t-norm was given in [15]. In the following we present the algorithm for the case of the product t-norm that we introduced in [20].

The presented approach to measuring similarity is a key concept in the interval-valued classifier described in the next section. It allows the proper and effective comparison of imprecise and incomplete data, without losing information about the amount of uncertainty contained in that data.

Algorithm 1. Algorithms for computing lower (left) and upper (right) bounds of relative cardinality $\sigma_{IF}(\widetilde{A}|\widetilde{B})$ introduced in [20].

Lower bound:

1: $n \leftarrow \sum_{x\in X} \underline{A}(x) \cdot \underline{B}(x)$
2: $d \leftarrow \sum_{x\in X} \underline{B}(x)$
3: **for all** $x \in X$ in descending order of $\underline{A}(x)$ **do**
4: $r \leftarrow \frac{n}{d}$
5: $n \leftarrow n + \underline{A}(x) \cdot (\overline{B}(x) - \underline{B}(x))$
6: $d \leftarrow d + \overline{B}(x) - \underline{B}(x)$
7: **if** $r \geq \frac{n}{d}$ **then**
8: **return** r
9: **end if**
10: **end for**

Upper bound:

1: $n \leftarrow \sum_{x\in X} \overline{A}(x) \cdot \underline{B}(x)$
2: $d \leftarrow \sum_{x\in X} \underline{B}(x)$
3: **for all** $x \in X$ in ascending order of $\overline{A}(x)$ **do**
4: $r \leftarrow \frac{n}{d}$
5: $n \leftarrow n + \overline{A}(x) \cdot (\overline{B}(x) - \underline{B}(x))$
6: $d \leftarrow d + \overline{B}(x) - \underline{B}(x)$
7: **if** $r \leq \frac{n}{d}$ **then**
8: **return** r
9: **end if**
10: **end for**

3 Interval-Valued Fuzzy Classifier

Our classifier is designed to deal with situations in which both the classified objects and the classes themselves are imprecise, subjective and/or incomplete. In such cases, the resulting classification would also be imprecise or incomplete. In our approach we take full account of these features of the data. The proposed classifier will be able to classify objects coded as an IVFS into classes which are also described in that way. Moreover, classification will also be described in interval-valued fashion. The final classification may be obtained with the use of the *score function* proposed in [21].

More formally, the problem can be formulated in the following way. Given a set $\mathcal{O}$ of objects to classify, described as an IVFS, compute for each of them the IVFS in the domain of set $\mathcal{C}$ of all possible classes and its interval-valued membership which describe the degree to which the object belongs to given class.

In this paper we assume that class prototypes as well as objects are coded as IVFSs. For instance class $c \in \mathcal{C}$ is coded as IVFS $\widetilde{iv}(c)$, and object $o_i \in \mathcal{O}$ as $\widetilde{iv}(o_i)$. The intuition behind the proposed classifier is to use an uncertainty-aware similarity measure to compute the similarities between objects and class prototypes and then use them as membership degrees in the resulting classification. Formally, the assignment of object $o_i \in \mathcal{O}$ to classes using the singleton notation can be stated as follows:

$$\widetilde{A}_{o_i} = \sum_{c\in\mathcal{C}} {}^{sim_{IF}(\widetilde{iv}(c),\widetilde{iv}(o_i))}/_{c}$$

It should be noted that uncertainty-aware similarity plays a fundamental role in our method. Thus it is crucial to use a similarity measure applicable to the problem being solved. $sim_{IF}(\widetilde{iv}(c), \widetilde{iv}(o_i)$ is an interval, hence the resulting classification $\widetilde{A}_{o_i}$ is an IVFS.

4 Application to Supporting Ovarian Tumor Diagnosis

The proposed classifier is demonstrated on a real data set from the field of medicine. Ovarian tumors are currently one of the most deadly diseases among women. According to recent statistics, annual numbers of new cases and deaths in the USA amount to 22,000 and 14,000 [22].

Two main groups of tumors are discriminated: malignant and benign. Such a division turns the diagnostic process into a classic binary classification. Each of these groups subdivides into histopathological types. This means that the problem might be expanded to a multi-class classification.

The correct discrimination between ovarian tumors is a key issue, because it determines the method of treatment. For this reason, a number of preoperative models for malignancy discrimination have been developed over the past two decades. They range from basic scoring systems [23, 24] to formal mathematical models, in particular rule-based schemes [25] and machine learning techniques [26, 27]. Unfortunately, in external evaluation the efficacy of predictions of such models rarely exceeds 90%, in terms of both sensitivity and specificity [28, 29]. Therefore, there is still a need to develop an effective preoperative model for inexperienced gynecologists.

In our previous research we have indicated the imprecision of data obtained by a gynecologist during examinations [30]. Many features are undoubtedly objective, such as levels of blood markers. Some examinations, however, require assessment on the part of the gynecologist, who can be a source of subjectivity – the more experienced he/she is, the more confident the result. In particular, this is the case when an ultrasonographer examines a tumor. Furthermore, in some cases we also have to deal with lack of data. Some examinations might be omitted by a physician, either for medical reasons or due to the their unavailability. These attributes may be conveniently modeled using IVFS theory.

The interval-valued classifier described above was applied to the problem of supporting ovarian tumor diagnosis. In this problem we try to assign the best matching histopathological profile of a tumor using the data available before an operation.

Both patient and histopathological profiles were coded as IVFSs. For the sake of simplicity, only four histopathological types were modeled for the present example. Two of them were benign – *Endometrioid cyst* and *Mucinous cystadenoma* – and two malignant – *Serous adenocarcinoma* and *Undifferentiated carcinoma* – referred to further as HP 1, HP 6, HP 21 and HP 25 respectively. Characteristics of class prototypes were obtained from an experts knowledge and partially from analysis of historical medical data. Among more than fifty attributes describing patient, five were arbitrarily selected: age, size of papillary projections (PAP), blood serum levels of CA-125 and HE4 tumor markers, and resistive index (RI). These attributes may be more or less subjective or imprecise. Moreover, some data may be not available at all. Based on a survey among gynecologists at Poznań University of Medical Sciences, characteristics of those parameters were obtained. A patients age is known precisely, while blood serum levels of tumor markers are subject to some uncertainties. Resistive index and

Table 1. Profiles of ovarian tumor histopathological type coded as IVFS

HP type	AGE	PAP	CA125	HE4	RI
1	[0.27, 0.64]	[0, 0.27]	[0, 0.04]	[0, 0.03]	[0.49, 0.78]
6	[0.29, 0.72]	[0, 0.14]	[0, 0.18]	[0.01, 0.06]	[0.22, 0.83]
21	[0.47, 0.76]	[0, 0.52]	[0.3, 1]	[0.12, 0.9]	[0.23, 0.56]
25	[0.39, 0.77]	[0, 0.58]	[0.15, 0.98]	[0.04, 0.62]	[0.27, 0.45]

Table 2. Profiles of patients coded as IVFS

Patient #	Postoperative diagnosis	AGE	PAP	CA125	HE4	RI
1	HP 1	[0.78, 0.78]	[0.00, 0.38]	[0.00, 0.06]	[0.00, 1.00]	[0.00, 1.00]
2	HP 2	[0.14, 0.14]	[0.35, 0.85]	[0.05, 0.15]	[0.00, 1.00]	[0.60, 1.00]
3	HP 21	[0.62, 0.62]	[0.00, 0.25]	[0.95, 1.00]	[0.95, 1.00]	[0.00, 1.00]

size of papillary projections were indicated as subjective attributes in the survey, thus their value is uncertain. Moreover, values of the last three attributes were not always available.

Real data based on histopathological profiles and example patients are presented in Tables 1 and 2. Note that patients missing data were replaced with the unit interval [0, 1].

Now a classification using the cardinality-based uncertainty-aware similarity measure formulated by (5) can be computed. By definition, the third patients classification is the following:

$$\widetilde{A}_{o_3} = {}^{sim_u(\widetilde{iv}(o_3),\widetilde{iv}(hp_3))}/_{hp_3} + {}^{sim_u(\widetilde{iv}(o_3),\widetilde{iv}(hp_6))}/_{hp_6} + {}^{sim_u(\widetilde{iv}(o_3),\widetilde{iv}(hp_{21}))}/_{hp_{21}} + {}^{sim_u(\widetilde{iv}(o_3),\widetilde{iv}(hp_{25}))}/_{hp_{25}}$$

If we use the minimum t-norm for calculation of similarity, the final classification is as follows:

$$\widetilde{A}_{o_3} = {}^{[0.07,0.48]}/_{hp_1} + {}^{[0.08,0.52]}/_{hp_6} + {}^{[0.21,0.99]}/_{hp_{21}} + {}^{[0.13,0.90]}/_{hp_{25}} \tag{6}$$

and in the case of the product t-norm:

$$\widetilde{A}_{o_3} = {}^{[0.50,0.98]}/_{hp_1} + {}^{[0.31,0.98]}/_{hp_6} + {}^{[0.54,0.99]}/_{hp_{21}} + {}^{[0.47,0.99]}/_{hp_{25}} . \tag{7}$$

The interpretation of classification from (6) is as follows. The possibility that this patient should be diagnosed as HP 1 is [0.07, 0.48], which is a low score. By contrast, the membership of class HP 21 is [0.21, 0.99]. To choose the best matching class we can use the score function defined in [21] as

$$\mathrm{score}([a,b]) = a + b - 1 .$$

Using this score, it is easy to see that class HP 21 is the best match for this patient (with a score of 0.20, compared with -0.45, -0.4 and 0.03). It is worth noting that this patient was postoperatively diagnosed with a type HP 21 tumor.

Tables 3 and 4 contain both classifications for the patients represented by interval and score result.

Table 3. Classification obtained using minimum t-norm

Patient #	HP1		HP6		HP21		HP25	
	interval	score	interval	score	interval	score	interval	score
1	[0.08, 0.93]	0.01	[0.09, 0.97]	0.06	[0.11, 0.91]	0.02	[0.09, 0.96]	0.05
2	[0.17, 0.85]	0.02	[0.10, 0.79]	-0.11	[0.09, 0.81]	-0.10	[0.10, 0.83]	-0.07
3	[0.07, 0.48]	-0.45	[0.08, 0.52]	-0.4	[0.21, 0.99]	0.20	[0.13, 0.90]	0.03

Table 4. Classification obtained using product t-norm

Patient #	HP1		HP6		HP21		HP25	
	interval	score	interval	score	interval	score	interval	score
1	[0.50, 0.98]	0.48	[0.32, 0.99]	0.31	[0.28, 0.99]	0.27	[0.25, 0.99]	0.24
2	[0.54, 0.99]	0.53	[0.40, 0.99]	0.39	[0.22, 0.99]	0.21	[0.20, 0.99]	0.19
3	[0.50, 0.98]	0.48	[0.31, 0.98]	0.29	[0.54, 0.99]	0.53	[0.47, 0.99]	0.48

5 Conclusions

The main contribution of this paper is the concept of an uncertainty-aware similarity measure, which was axiomatically defined and for which an efficient algorithm was given. Based on this measure an interval-valued fuzzy classifier was constructed and applied to real-life data concerning patients with ovarian tumors. The uncertainty-aware similarity measure proved to be particularly useful for supporting medical diagnosis, where uncertainty and incompleteness of information are common and inevitable features. We obtained promising results showing that IVFS theory is convenient for modeling and processing such data, and our experience suggests that practitioners prefer informative, even if uncertain, feedback rather than excessively precise data. In our further research we plan to improve the classification process and to thoroughly investigate the properties of the uncertainty-aware similarity measure, including the influence of the t-norm used in the similarity formula.

References

[1] Zadeh, L.A.: The concept of a linguistic variable and its application to approximate reasoning–I. Information Sciences 8(3), 199–249 (1975)
[2] Atanassov, K.T.: Intuitionistic fuzzy sets. Fuzzy Sets and Systems 20(1), 87–96 (1986)
[3] Deschrijver, G., Kerre, E.: Advances and challenges in interval-valued fuzzy logic. Fuzzy Sets and Systems 157, 622–627 (2006)
[4] Bustince, H.: Indicator of inclusion grade for interval valued fuzzy sets. Application to approximate reasoning based on interval-valued fuzzy sets. International Journal of Approximate Reasoning 23, 137–209 (2000)
[5] Wu, D.: On the fundamental differences between interval type-2 and type-1 fuzzy logic controllers. IEEE Transactions on Fuzzy Systems 20, 832–848 (2012)
[6] Yager, R.R.: Fuzzy subsets of type II in decisions. Journal of Cybernetics 10, 137–159 (1980)
[7] Karnik, N.N., Mendel, J.M.: Applications of type-2 fuzzy logic systems to forecasting of time-series. Information Sciences 120, 89–111 (1999)
[8] Bezdek, J., Keller, J., Krisnapuram, R., Pal, N.: Fuzzy Models and algorithms for pattern recognition and image processing. Springer (2005)
[9] Kuncheva, L.: Fuzzy Classifier Design. STUDFUZZ, vol. 49. Springer, Heidelberg (2000)
[10] Höppner, F., Klawonn, F., Kruse, R., Runkler, T.: Fuzzy Cluster Analysis: Methods for Classification, Data Analysis and Image Recognition. Wiley (1999)
[11] Duda, R., Hart, P., Stork, D.: Pattern Classification, 2nd edn. Wiley (2000)
[12] Bishop, C.: Pattern Recognition and Machine Learning. Information Science and Statistics. Springer (2006)
[13] Zeng, W., Li, H.: Relationship between similarity measure and entropy of interval valued fuzzy sets. Fuzzy Sets and Systems 157, 1477–1484 (2006)
[14] Wu, D., Mendel, J.M.: A comparative study of ranking methods, similarity measures and uncertainty measures for interval type-2 fuzzy sets. Information Sciences 179, 1169–1192 (2009)
[15] Nguyen, H.T., Kreinovich, V.: Computing degrees of subsethood and similarity for interval-valued fuzzy sets: Fast algorithms. In: Proc. of the 9th International Conference on Intelligent Technologies, InTec 2008, pp. 47–55 (2008)
[16] Wu, D., Mendel, J.: Efficient algorithms for computing a class of subsethood and similarity measures for interval type-2 fuzzy sets. In: FUZZ-IEEE, pp. 1–7 (2010)
[17] Stachowiak, A., Dyczkowski, K.: A similarity measure with uncertainty for incompletely known fuzzy sets. In: Proceedings of the 2013 Joint IFSA World Congress NAFIPS Annual Meeting, pp. 390–394 (2013)
[18] Luca, A.D., Termini, S.: A definition of nonprobabilistic entropy in the setting of fuzzy sets theory. Information and Computation 20, 301–312 (1972)
[19] Szmidt, E., Kacprzyk, J.: Entropy for intuitionistic fuzzy sets. Fuzzy Sets and Systems 118, 467–477 (2001)
[20] Żywica, P., Stachowiak, A.: A new method for computing relative cardinality of intuitionistic fuzzy sets. In: Atanassov, K., et al. (eds.) New Developments in Fuzzy Sets, Intuitionistic Fuzzy Sets, Generalized Nets and Related Topics. IBS PAN – SRI PAS, Warsaw (to appear, 2014)
[21] Chen, S., Tan, J.M.: Handling multicriteria fuzzy decision-making problems based on vague set theory. Fuzzy Sets and Systems 67, 163–172 (1994)

[22] Siegel, R., Ma, J., Zou, Z., Jemal, A.: Cancer statistics, 2014. CA: A Cancer Journal for Clinicians 64(1), 9–29 (2014)
[23] Alcázar, J.L., Mercé, L.T., Laparte, C., Jurado, M., López-Garcia, G.: A new scoring system to differentiate benign from malignant adnexal masses. Obstetrical & Gynecological Survey 58(7), 462–463 (2003)
[24] Szpurek, D., Moszyński, R., Ziętkowiak, W., Spaczyński, M., Sajdak, S.: An ultrasonographic morphological index for prediction of ovarian tumor malignancy. European Journal of Gynaecological Oncology 26(1), 51–54 (2005)
[25] Timmerman, D., Testa, A.C., Bourne, T., Ameye, L., Jurkovic, D., Van Holsbeke, C., Paladini, D., Van Calster, B., Vergote, I., Van Huffel, S., et al.: Simple ultrasound-based rules for the diagnosis of ovarian cancer. Ultrasound in Obstetrics & Gynecology 31(6), 681–690 (2008)
[26] Timmerman, D., Testa, A.C., Bourne, T., Ferrazzi, E., Ameye, L., Konstantinovic, M.L., Van Calster, B., Collins, W.P., Vergote, I., Van Huffel, S., et al.: Logistic regression model to distinguish between the benign and malignant adnexal mass before surgery: a multicenter study by the International Ovarian Tumor Analysis Group. Journal of Clinical Oncology 23(34), 8794–8801 (2005)
[27] Van Calster, B., Timmerman, D., Nabney, I.T., Valentin, L., Testa, A.C., Van Holsbeke, C., Vergote, I., Van Huffel, S.: Using Bayesian neural networks with ARD input selection to detect malignant ovarian masses prior to surgery. Neural Computing and Applications 17(5-6), 489–500 (2008)
[28] Valentin, L., Hagen, B., Tingulstad, S., Eik-Nes, S.: Comparison of 'pattern recognition' and logistic regression models for discrimination between benign and malignant pelvic masses: a prospective cross validation. Ultrasound in Obstetrics & Gynecology 18(4), 357–365 (2001)
[29] Van Holsbeke, C., Van Calster, B., Valentin, L., Testa, A.C., Ferrazzi, E., Dimou, I., Lu, C., Moerman, P., Van Huffel, S., Vergote, I., et al.: External validation of mathematical models to distinguish between benign and malignant adnexal tumors: a multicenter study by the International Ovarian Tumor Analysis Group. Clinical Cancer Research 13(15), 4440–4447 (2007)
[30] Wójtowicz, A., Żywica, P., Szarzyński, K., Moszyński, R., Szubert, S., Dyczkowski, K., Stachowiak, A., Szpurek, D., Wygralak, M.: Dealing with Uncertainty in Ovarian Tumor Diagnosis. In: Atanassov, K., et al. (eds.) New Developments in Fuzzy Sets, Intuitionistic Fuzzy Sets, Generalized Nets and Related Topics. IBS PAN – SRI PAS, Warsaw (to appear, 2014)

Differential Evolution Based Nearest Prototype Classifier with Optimized Distance Measures and GOWA

David Koloseni and Pasi Luukka

Laboratory of Applied Mathematics,
Lappeenranta University of Technology,
P.O. Box 20, FI-53851,
Lappeenranta, Finland
{david.koloseni,pasi.luukka}@lut.fi

Abstract. Nearest prototype classifier based on differential evolution algorithm, pool of distances and generalized ordered weighted averaging is introduced. Classifier is based on forming optimal ideal solutions for each class. Besides this also distance measures are optimized for each feature in the data sets to improve recognition process of which class the sample belongs. This leads to a distance vectors, which are now aggregated to a single distance by using generalized weighted averaging (GOWA). In earlier work simple sum was applied in the aggregation process. The classifier is empirically tested with seven data sets. The proposed classifier provided at least comparable accuracy or outperformed the compared classifiers, including the earlier versions of DE classifier and DE classifier with pool of distances.

Keywords: Classification, Differential evolution, Pool of distances, generalized ordered weighted averaging operator.

1 Introduction

The nearest prototype classifiers have shown increasing interest among researchers. The nearest prototype classifier (NPC) is intuitive, simple and mostly pretty accurate as stated in [1] by Bezdek and Kuncheva. In [2], an evolutionary nearest prototype classifier (ENPC) was introduced where no parameters are involved, therefore eliminates the problems that classical methods have in tuning and searching for the appropriate values. The ENPC algorithm is based on the evolution of a set of prototypes that can execute several operators in order to increase their quality, and with a high classification accuracy emerging for the whole classifier. In [3], a large-scale classification by a prototype based nearest neighbor approach was presented. The approach has several benefits such parameter free, implicitly able to handle large numbers of classes, avoids overfitting to noise and generalizes well to previously unseen samples and has very low computational costs during testing.

P. Angelov et al. (eds.), *Intelligent Systems'2014*,
Advances in Intelligent Systems and Computing 322, DOI: 10.1007/978-3-319-11313-5_66

Differential Evolution (DE) [4], has gained popularity for solving classification problems due to its robustness, simplicity, easy implementation and fast convergence. DE algorithm has been used in classification tasks as automatic classification in database [5],[6],[7], satellite image registration [8] and, optimizing positioning of prototypes [9], [10], [11] to mention few. In DE nearest prototype classifier [13], the Differential evolution (DE) algorithm [12], [4] was used to optimize the selection of a suitable distance measure from a created pool of distances. The distances [13] were vector based, and the optimal distance found was applied to the whole data set. This was further generalized in [18] where distance optimization was taken in the feature level instead of data set level. This work was based on the fact that different features in the data set can have different optimal distances.

The ordered weighted averaging (OWA) operator [21] was originally introduced by Yager in 1988. It has been a subject of rigorous research ever since. It has been applied in various areas in decision making. In classification problems OWA is still relatively new acquaintance and seems to be discovered just recently. To mention some work, Usunier et al [14], applied OWA in ranking with pairwise classification. In similarity classifier OWA was applied by Luukka and Kurama in [16] to aggregate similarities. To clustering methods OWA was applied by Cheng et al [15] where OWA was applied to K-means clustering. To combine information from three classifiers OWA was applied by Danesh et al. [15] in text classification problem.

This article extends the work done in [18]. After determining the optimal distance measures for each feature, a total distance measure is needed obtained by combining the normalized distance values from each individual feature. Earlier a simple sum was used to aggregate these into a single distance value. Obviously this is not the ideal way of doing it and in current work this is done by computing the generalized ordered weighted averaging (GOWA) operator for aggregation of all the normalized distances. This GOWA operator is applied together with regular monotonic increasing (RIM) quantifiers where suitable quantifier function is examined and optimal parameter for quantifier is found.

The data sets applied in this work are taken from the UCI-Repository of Machine Learning data set [19]. These include three artificial data sets and four real world data sets.

2 Preliminaries

In this section we briefly introduce the main methods which are needed to build the differential evolution classifier with generalized ordered weighted averaging operators. We first start by introducing the differential evolution algorithm, then we proceed with the Ordered Weighted Averaging Operators (OWA) and introduce Generalized Ordered Weighted Averaging (GOWA) operator. Besides this we also briefly go through quantifier guided aggregation and introduce the weighting scheme applied in GOWA by using quantifier guided aggregation.

2.1 Differential Evolution

The DE algorithm [12], [4] was introduced by Storn and Price in 1995 and it belongs to the family of Evolutionary Algorithms (EAs). The design principles of DE are simplicity, efficiency, and the use of floating-point encoding instead of binary numbers. As a typical EA, DE has a random initial population that is then improved using selection, mutation, and crossover operations. Several ways exist to determine a stopping criterion for EAs but usually a predefined upper limit G_{max} for the number of generations to be computed provides an appropriate stopping condition. Other control parameters for DE are the crossover control parameter CR, the mutation factor F, and the population size NP.

In each generation G, DE goes through each D dimensional decision vector $\boldsymbol{v}_{i,G}$ of the population and creates the corresponding trial vector $\boldsymbol{u}_{i,G}$ as follows in the most common DE version, DE/rand/1/bin [4]:

$$
\begin{array}{l}
r_1, r_2, r_3 \in \{1, 2, \dots, NP\}, \text{(randomly selected,} \\
\quad \text{except mutually different and different from } i) \\
j_{rand} = \text{floor}\,(rand_i[0,1) \cdot D) + 1 \\
\text{for}(j = 1; j \leq D; j = j + 1) \\
\{ \\
\quad \text{if}(rand_j[0,1) < CR \vee j = j_{rand}) \\
\qquad u_{j,i,G} = v_{j,r_3,G} + F \cdot (v_{j,r_1,G} - v_{j,r_2,G}) \\
\quad \text{else} \\
\qquad u_{j,i,G} = v_{j,i,G} \\
\}
\end{array}
$$

In this DE version, NP must be at least four and it remains fixed along with CR and F during the whole execution of the algorithm. Parameter $CR \in [0, 1]$, which controls the crossover operation, represents the probability that an element for the trial vector is chosen from a linear combination of three randomly chosen vectors and not from the old vector $\boldsymbol{v}_{i,G}$. The condition "$j = j_{rand}$" is to make sure that at least one element is different compared to the elements of the old vector. The parameter F is a scaling factor for mutation and its value is typically $(0, 1+]$[1]. In practice, CR controls the rotational invariance of the search, and its small value (*e.g.*, 0.1) is practicable with separable problems while larger values (*e.g.*, 0.9) are for non-separable problems. The control parameter F controls the speed and robustness of the search, *i.e.*, a lower value for F increases the convergence rate but it also adds the risk of getting stuck into a local optimum. Parameters CR and NP have the same kind of effect on the convergence rate as F has.

After the mutation and crossover operations, the trial vector $\boldsymbol{u}_{i,G}$ is compared to the old vector $\boldsymbol{v}_{i,G}$. If the trial vector has an equal or better objective value, then it replaces the old vector in the next generation. This can be presented as follows (in this paper maximization of objectives is assumed) [4]:

$$
\boldsymbol{v}_{i,G+1} = \begin{cases} \boldsymbol{u}_{i,G} \text{ if } & f(\boldsymbol{u}_{i,G}) \geq f(\boldsymbol{v}_{i,G}) \\ \boldsymbol{v}_{i,G} \text{ otherwise} \end{cases} .
$$

[1] Notation means that the upper limit is about 1 but not strictly defined.

DE is an elitist method since the best population member is always preserved and the average objective value of the population will never get worse.

2.2 Generalized Weighted Averaging Operator

In [20] introduced a class of aggregation operator called generalized ordered weighted averaging (GOWA) operators. The GOWA operator is an extension of the OWA operator [21] with addition of a parameter controlling the power to which the argument values are raised. The GOWA operator can be defined as follows:

Definition 1. *A generalized ordered weighted averaging (GOWA) operator of dimension m is a mapping $R^m \to R$ that has associated weighting vector $W = [w_1, w_2, ..., w_m]$ of dimension m with*
$\sum_{i=1}^{m} w_i = 1$ *,* $w_i \in [0,1]$ *and* $1 \leq i \leq m$ *if*

$$GOWA(a_1, a_2, ..., a_m) = \Big(\sum_{i=1}^{m} w_i b_i^{\lambda}\Big)^{\frac{1}{\lambda}} \tag{1}$$

where b_i is the largest ith element of the collection of objects $a_1, a_2, ..., a_m$. The function value $GOWA(a_1, a_2, ..., a_m)$ determines the aggregated values of arguments $a_1, a_2, ..., a_m$ and, λ is a parameter such that $\lambda \in (-\infty, \infty)$.

If $\lambda = 1$ then the GOWA operator is reduced to OWA operator [21] as

$$OWA(a_1, a_2, ..., a_m) = \Big(\sum_{i=1}^{m} w_i b_i\Big) \tag{2}$$

In this article we create a vector of distances $d_1, d_2, ..., d_m$ to which we apply the GOWA operator

$$GOWA(d_1, d_2, ..., d_m) = \Big(\sum_{i=1}^{T} w_i D_i^{\lambda}\Big)^{\frac{1}{\lambda}} \tag{3}$$

where D_i is the largest ith element of the collection of objects $d_1, d_2, ..., d_m$

2.3 Quantifier Guided Aggregation

The weights can be generated in different ways e.g., by using linguistic quantifiers introduced in [21]. In this article we applied the linguistic quantifier called regular increasing monotone (RIM) quantifier, which is introduced below following the work of [21].

Definition 2. *A fuzzy subset Q of the real line is called a Regular Increasing Monotone (RIM) quantifier if it satisfies following conditions*

1. *$Q(0) = 0$,*
2. *$Q(1) = 1$,*
3. *If $x \geq y$ then $Q(x) \geq Q(y)$.*

The weights for the GOWA operator are computed using linguistic quantifiers introduced by Yager [21], $w_i = Q(\frac{i}{m}) - Q(\frac{i-1}{m})$ where $i = 1, 2, ..., m$ and Q is assumed to be Regular Increasing Monotone (RIM) quantifier. The use of GOWA with RIM quantifiers is that they can be adopted for our purpose of studying the effect of aggregation operators on distance measures. These weights must satisfy the conditions of GOWA operator. There are several ways of choosing the quantifier function. After preliminary screening we ended up using following exponential linguistic quantifier, which was found to work reasonably well. For exponential linguistic quantifier, the quantifier function Q can be written as:

$$Q(r) = e^{-\alpha r} \tag{4}$$

The weights associated to the exponential quantifier are computed by using the RIM type of procedure now as

$$w_i = e^{-\alpha(\frac{i}{m})} - e^{-\alpha(\frac{i-1}{m})}, i = 1, 2, ..., m \tag{5}$$

3 Differential Evolution Classifier with Optimized Distances for Features and Aggregation by GOWA

In this article, we present a classification method using a DE algorithm [12], which is based on minimizing erroneous classification costs (objective function). In the process, we split the data into a training set and testing set. The splitting of the data is done in such a way that half of the data are used in the training set and the other half in the testing set. We use the training set to maximize the classification accuracy by minimizing the classification error in the objective function.

$$\text{minimize} : 1 - \frac{\sum_{j=1}^{m} B(x_j)}{m} \tag{6}$$

where

$$B(x_j) = \begin{cases} 1 \text{ if } g(x_j) = T(x_j) \\ 0 \text{ if } g(x_j) \neq T(x_j) \end{cases}$$

and we denote $T(x_j)$ as the true class from sample x_j and

$$g(x_j) = \{i \mid \min_{i=1,\ldots,N} d(\mathbf{x_j}, \mathbf{y_i})\} \tag{7}$$

meaning that we decide that the vector x_j belongs to class i if $d(\mathbf{x_j}, \mathbf{y_i})$ is minimum, with the sample vector $\mathbf{x_j}$ and the ideal candidate $\mathbf{y_i}$ for class i. If the sample vector class $g(x_j)$ is equal to the true class $T(x_j)$, then the sample is correctly classified and $B(x_j)$ hence gets the value 1 otherwise 0. The sum of $B(x_j)$ gives the total number of samples correctly classified where as m is the total number of samples in the training set. Total number of classes is denoted by N.

Here we propose to use GOWA operator in the computation of $d(\mathbf{x_j}, \mathbf{y_i})$, which is the aggregated value of the distance vector where distances are calculated from vectors $\mathbf{x_j}$ and $\mathbf{y_i}$ respectively. Let us represent each object from the training dataset by a vector of type $(\mathbf{x_j}, \mathbf{y_i}) = (x_{j1}, x_{j2}, ..., x_{jT}; y_i)$ and $\mathbf{y_i} = (y_{i1}, y_{i2}, ..., y_{iT})$ where there are N different classes of objects $\mathbf{y_i}, i = 1, 2, ..., N$ and T is the total number of features in the data set. The aggregated distance $\mathbf{d}(\mathbf{x_j}, \mathbf{y_i})$ in equation (7) is now gained by using the GOWA operator which can be given as:

$$\mathbf{d}(\mathbf{x_j}, \mathbf{y_i}) = GOWA(d_{k,1}, d_{k,2}, ..., d_{k,T}) \tag{8}$$

$$= \Big(\sum_{i=1}^{T} w_i D_i^{\lambda} \Big)^{\frac{1}{\lambda}} \tag{9}$$

where

$$\begin{aligned} d_{k,1} &= d_{k,1}(x_{j1}, y_{i1}) \\ d_{k,2} &= d_{k,2}(x_{j2}, y_{i2}) \\ &\vdots \\ d_{k,T} &= d_{k,T}(x_{jT}, y_{iT}) \end{aligned}$$

The distance vector which is aggregated with GOWA $(d_{k,1}, d_{k,2}, ..., d_{k,T})$ is computed so that $d_{k,i}$ can be any of the distance from the pool of distances, which is given in Table 1. Subindex k denotes the k :th distance and subindex i denotes particular feature for which distance is to be calculated.

Table 1. Pool of distances

k	Distance
1	$d_1(x, y) = (\lvert x - y\rvert^{r_1})$; , $r_1 \in [1, \infty)$
2	$d_2(x, y) = \lvert x - y\rvert^{r_2}/\max\{\lvert x\rvert, \lvert y\rvert\}; r_2 \in [1, \infty)$
3	$d_3(x, y) = \lvert x - y\rvert^{r_3}/1 + \min\{\lvert x\rvert, \lvert y\rvert\}$;, $r_3 \in [1, \infty)$
4	$d_4(x, y) = \lvert x - y\rvert/1 + \max\{\lvert x\rvert, \lvert y\rvert\}$
5	$d_5(x, y) = \lvert x - y\rvert/[1 + \lvert x\rvert + \lvert y\rvert]$
6	$d_6(x, y) = \lvert x/[1 + \lvert x\rvert] - y/[1 + \lvert y\rvert]\rvert$
7	$d_7(x, y) = r_4(x - y)^2/(x + y)$; , $r_4 \in (0, \infty)$
8	$d_8(x, y) = \lvert x - y\rvert/(1 + \lvert x\rvert)(1 + \lvert y\rvert)$

In the pool of distances we used eight possible distance measures to choose from. This means that for each feature $1, 2, \cdots, T$ we also optimized the choice of the particular distance which we applied when computing the distance vector $(d_{k,1}, d_{k,2}, ..., d_{k,T})$. Here the choice for distance in is optimized by optimizing the parameter $k = \{1, 2, ..., 8\}$ for each feature. Since in GOWA we order the elements (now distances in our case) the ordered distance are now represented in D_i

where i denotes the ith largest element of the distance vector $(d_{k,1}, d_{k,2}, ..., d_{k,T})$. Besides this we have coefficient λ, which coming from the GOWA and can be selected from $\lambda \in (-\infty, \infty)$ and $\lambda \neq 0$. The decision to apply GOWA operator in the aggregation is the fact that, it is a generalization of OWA, which makes its use justifiable in a sense that we should get at least as good results as with OWA and in principle we have a possibility to find better results through generalization.

For differential evolution algorithm which is incorporated here. The aim is to optimize the following subcomponents; 1) ideal candidates for each class, 2) selection of proper distance for each feature, 3) free parameters related to distance measures 4) Possible other free parameters. First part consist of N dimensional class vectors $\{y_1, y_2, \cdots, y_N\}$. Second part, selection of proper distance for each feature is $T-$dimensional vector consisting of $\{k_1, k_2 \ldots, k_T\}$ optimal selection of distance measure. Third part is parameters related to distance measures, which is now $\{p_{1,1}, p_{1,2}, p_{1,3}, p_{1,4}, \cdots, p_{1,T}, p_{2,T}, p_{3,T}, p_{4,T}\}$ which is due to the fact that in our current pool of distance we have four distance which are parametric in nature and the parameter needs to be optimized for each of them. Besides this any of the distance can be selected to any of the feature giving this way $4T$ possible parameters to optimize in this part. By allowing this kind of flexibility, we make it easier for the DE to change the parameter. More information about the parameter in this context is given in to [18]. Besides this in our currently presented classifier we have other free parameters $\{\alpha\}, \{\lambda\}$ which are now coming from the usage of GOWA and quantifier guided aggregation. Here the α is parameter for the exponential quantifier (5)and λ is the generalization parameter in GOWA. In order to do this with DE our vector to be optimized $(\mathbf{v}_{i,G})$ now gets the following form:

$$\begin{aligned} \mathbf{v}_{i,G} = & \{\{y_1, y_2, \cdots, y_N\}, \{k_1, k_2 \ldots, k_T\}, \\ & \{p_{1,1}, p_{1,2}, p_{1,3}, p_{1,4}, \cdots, p_{1,T}, p_{2,T}, p_{3,T}, p_{4,T}\}, \{\alpha\}, \{\lambda\}\} \end{aligned} \quad (10)$$

The core procedure for this classification method is described in Algorithm 1.

4 Data Sets and Classification Results

4.1 Data Sets

Data sets which were applied to test the classifier were taken from the UCI machine learning data repository [19]. Three artificial data sets, which are designed for specific type of test problems was chosen and four real world data sets. Artificial data sets where taken from the Monks problem and simply named as Monk1, Monk2 and Monk3. The MONK's problem were the basis of a first international comparison of learning algorithms. One significant characteristic of this comparison is that it was performed by a collection of researchers, each of whom was an advocate of the technique they tested (often they were the creators of the various methods). In this sense, the results are less biased than in comparisons

Require: $Data[1,\ldots,T]$, $[S_1, S_2, ..., S_T]$ $y_1[1,\ldots,T]$,$y_2[1,\ldots,T]$,...,$y_N[1,\ldots,T]$, $p1[1,\ldots,T]$, $p2[1,\ldots,T]$,$p3[1,\ldots,T]$,$p4[1,\ldots,T]$ $center$=$[y_1; y_2; ...; y_N]$
for $j = 1$ to T **do**
  for $i = 1$ to N **do**
    if $switch(j) == 1$ **then**
      $d(:,i,j) = dist1(data, repmat(center(i,j), T, 1), p1(j))$
    else if $switch(j) == 2$ **then**
      $d(:,i,j) = dist2(data, repmat(center(i,j), T, 1), p2(j))$
    else if $switch(j) == 3$ **then**
      $d(:,i,j) = dist3(data, repmat(center(i,j), T, 1), p3(j))$
    else if $switch(j) == 4$ **then**
      $d(:,i,j) = dist4(data, repmat(center(i,j), T, 1))$
    else if $switch(j) == 5$ **then**
      $d(:,i,j) = dist5(data, repmat(center(i,j), T, 1))$
    else if $switch(j) == 6$ **then**
      $d(:,i,j) = dist6(data, repmat(center(i,j), T, 1), p4(j))$
    else if $switch(j) == 7$ **then**
      $d(:,i,j) = dist7(data, repmat(center(i,j), T, 1))$
    else
      $d(:,i,j) = dist8(data, repmat(center(i,j), T, 1))$
    end if
  end for
end for
$D = scale(d)$
for $i = 1$ to T **do**
  $d_{total}(:,i) = GOWA(D(:,:,i), w, \lambda);$
end for
for $i = 1$ to $length(d_{total})$ **do**
  $class(:,i) = find(d_{total}(i,j) == min(d_{total}(i,:)));$
end for

Fig. 1. *The core procedure of the classification method*

performed by a single person advocating a specific learning method, and more accurately reflect the generalization behavior of the learning techniques as applied by knowledgeable users. Short description of the real world data set with the main properties is given next.

Australian Credit Approval Data
This data set concerns credit card applications. It has 2 classes with 690 instances and 14 number of attributes. Eight (8) of the features are numerical and six (6) are categorical features. Classes represent the information whether the application for credit should be approved or not.

Balance-Scale Data
This data set was generated to model psychological experimental results. Each example is classified as having the balance scale tip to the right, tip to the left, or be balanced. The attributes are the left weight, the left distance, the right weight and the right distance. Total of 5 different attributes were measured for this purpose and number of instances were 625.

Bank Note Authentication
Data were extracted from images that were taken from genuine and forged banknote-like specimens. For digitization, an industrial camera usually used for print inspection was used. The final images have 400x 400 pixels. Due to the

object lens and distance to the investigated object gray-scale pictures with a resolution of about 660 dpi were gained. Wavelet Transform tool were used to extract features from images.

Echocardiogram Data
In this data set one sample is results from echocardiogram measurements of one patient who has recently suffered an acute heart attack. Measurements are taken from echocardiograms, which are ultrasound measurements of the heart itself. The goal of physicians using these measurements is to predict a patient's chances of survival. Experimental work is being preformed to determine if an echocardiogram, in conjunction with other measures, could be used to predict whether or not a patient would survive for longer than a certain time period. This data can give means for predicting future heart attacks in former heart patients. Fundamental properties of the data sets are summarized in Table 2.

Table 2. Properties of the data sets

Name	Nb of classes	Nb of attributes	Nb of instances
Australian	2	14	690
Balance scale	3	4	625
Bank note authentication	2	5	1372
Echocardiogram	2	12	132
Monk1	2	7	432
Monk2	2	7	432
Monk3	2	7	432

4.2 Classification Results

The data sets were subjected to 1000 generations ($G_{max} = 1000$) and divided into 30 times random splits of testing sets and learning sets ($N = 30$) based on which mean accuracies and variances were then computed. Classification results were compared to previous versions of the differential evolution classifiers. The original DE classifier was first presented in [10] and DE classifier with pool of distances (DEPD classifier) in [18]. This way we can easily compare how GOWA operator is now changing the results with the classifier. These results are represented under DEPD with GOWA classifier label in Table 3. For each data set, the mean classification accuracy with 99% confidence interval using student t distribution $\mu \pm t_{1-\alpha}S_{\mu}/\sqrt{n}$ was also calculated and these are presented in Table 3.

Classification results with the highest mean accuracy is highlighted in boldface in Table 3. As can be seen from the table in five data sets from seven DEPD with GOWA managed to get the highest mean classification accuracy. With one data set (Echocardiogram) original DE classifier managed to get highest mean accuracy and with one (Bank note authentication) DEPD classifier. On

Table 3. Comparison of the results of the data sets from the proposed method with other classifiers

Data	DE classifier	DEPD classifier	DEPD with GOWA classifier
Australian	66.78 ± 1.27	82.94 ± 1.65	$\mathbf{83.29 \pm 2.21}$
Balance scale	88.66 ± 1.09	91.30 ± 0.14	$\mathbf{91.40 \pm 0.87}$
Bank note authentication	99.02 ± 0.50	$\mathbf{99.61 \pm 0.18}$	99.57 ± 0.24
Echocardiogram	$\mathbf{88.81 \pm 2.92}$	86.55 ± 3.04	88.08 ± 2.08
Monk1	76.71 ± 1.38	81.31 ± 0.77	$\mathbf{82.64 \pm 1.06}$
Monk2	88.13 ± 0.77	95.40 ± 0.55	$\mathbf{96.13 \pm 0.72}$
Monk3	87.73 ± 1.06	95.52 ± 0.59	$\mathbf{96.90 \pm 0.52}$

average variances were a little higher using DEPD with GOWA classifier, but not significantly. These results show that GOWA operator can be a useful addition to the classification process with some data sets.

5 Conclusions

In this paper we have introduced a DEPD with GOWA classifier. In DEPD with GOWA we create a pool of distances which is used to optimized the distance measure for each feature. In doing this we create a distance vector between the ideal solution and sample we want to classify. This distance vector is then aggregated into a single measure by using generalized ordered weighting averaging (GOWA). Classification results from this aggregation was compared to normalized sum (DEPD classifier). Results were computed for three artificial data sets and four real world data sets. Mean accuracies were highest in five cases out of seven. This shows that there exists classification problems where this type of aggregation can be more useful than using a sum. The proposed classifier provided at least comparable accuracy to compared classifiers. This clearly shows that this type of aggregation can be useful if properly performed.

References

1. Bezdek, J.C., Ludmila, I.K.: Nearest Prototype Classifier Designs: An experimental study. International Journal of Intelligent Systems 16(12), 1445–1473 (2001)
2. Fernando, F., Pedro, I.: Evolutionary Design of Nearest Prototype Classifers. Journal of Heuristics, 431–545 (2004)
3. Wohlhart, P., Kostinger, M., Donoser, M., Roth, P.M., Bischof, H.: Optimizing 1-Nearest Prototype Classifiers. In: 2013 IEEE Conference on Computer Vision and Pattern Recognition (CVPR), pp. 460–467. IEEE (2013)
4. Storn, R., Price, K.V., Lampinen, J.: Differential Evolution- A practical Approach to Global Optimization. Springer (2005)
5. De Falco, I.: Differential Evolution for Automatic Rule Extraction from Medical Databases. Applied Soft Computing 13(2), 1265–1283 (2013)
6. De Falco, I.: A Differential Evolution-based System Supporting Medical Diagnosis through Automatic Knowledge Extraction from Databases. In: 2011 IEEE International Conference on Bioinformatics and Biomedicine (BIBM), pp. 321–326. IEEE (2011)

7. De Falco, I., Della Cioppa, A., Tarantino, E.: Automatic Classification of Hand-segmented Image Parts with Differential Evolution. In: Rothlauf, F., et al. (eds.) EvoWorkshops 2006. LNCS, vol. 3907, pp. 403–414. Springer, Heidelberg (2006)
8. De Falco, I., Della Cioppa, A., Maisto, D., Tarantino, E.: Differential Evolution as a Viable Tool for Satellite Image Registration. Applied Soft Computing 8(4), 1453–1462 (2008)
9. Luukka, P., Lampinen, J.: A Classification Method Based on Principal Component Analysis and Differential Evolution Algorithm Applied for Predition Diagnosis from Clinical EMR Heart Data sets. In: Computational Intelligence in Optimization: Applications and Implementations. Springer (2010)
10. Luukka, P., Lampinen, J.: Differential Evolution Classifier in Noisy Settings and with Interacting Variables. Applied Soft Computing 11, 891–899 (2011)
11. Triguero, I., Garcia, S., Herrera, F.: Differential Evolution for Optimizing the Positioning of Prototypes in Nearest Neighbor Classification. Pattern Recognition 44, 901–916 (2011)
12. Storn, R., Price, K.V.: Differential Evolution: A Simple and Efficient Heuristic for Global Optimization over Continuous Space. Journal of Global Optimization 11(4), 341–359 (1997)
13. Koloseni, D., Lampinen, J., Luukka, P.: Optimized Distance Metrics for Differential Evolution based Nearest Prototype Classifier. Expert Systems With Applications 39(12), 10564–10570 (2012)
14. Usunier, N., Buffoni, D.: Gallinari,Patrick: Ranking with Ordered Weighted Pairwise Classification. In: 26th Annual International Conference on Machine Learning, pp. 1057–1064. ACM (2009)
15. Cheng, C.-H., Jia-Wen, W., Ming-Chang, W.: OWA-Weighted Based Clustering Method for Classification Problem. Expert Systems with Applications 36(3), 4988–4995 (2009)
16. Luukka, P., Kurama, O.: Similarity Classifier with Ordered Weighted Averaging Operators. Expert Systems with Applications 40, 995–1002 (2013)
17. Danesh, A., Behzad, M.: Omid, Fi.: Improve Text Classification Accuracy Based on Classifier Fusion Methods. In: 10th International Conference on Information Fusion, pp. 1–6. IEEE Press (2007)
18. Koloseni, D., Lampinen, J., Luukka, P.: Optimized Distance Metrics for Differential Evolution based Nearest Prototype Classifier. Expert Systems With Applications 40(10), 4075–4082 (2013)
19. Newman, D.J., Hettich, S., Blake, C.L., Merz, C.J.: UCI Repository of machine learning databases. University of California, Department of Information and Computer Science, Irvine (1998), `http://www.ics.uci.edu/~mlearn/MLRepository.html`
20. Yager, R.R.: Generalized OWA Aggregation Operators. Fuzzy Optimization and Decision Making 3(1), 93–107 (2004)
21. Yager, R.R.: On Ordered Weighted Averaging Operators in Multicriteria Decision Making. IEEE Transactions on Systems, Man,Cybernetcs 18, 183–190 (1988)

Differential Evolution Classifier with Optimized OWA-Based Multi-distance Measures for the Features in the Data Sets

David Koloseni[1,2], Mario Fedrizzi[3,4], Pasi Luukka[1,4], Jouni Lampinen[5,6], and Mikael Collan[4]

[1] Laboratory of Applied Mathematics, Lappeenranta University of Technology, P.O. Box 20, FI-53851, Lappeenranta, Finland
[2] University of Dar es salaam, Department of Mathematics, P.O. Box 35062, Dar es salaam,Tanzania
[3] Department of Industrial Engineering, University of Trento, Via Inama 5, I-38122, Trento, Italy
[4] School of business, Lappeenranta University of Technology, P.O. Box 20, FI-53851, Lappeenranta, Finland
[5] Department of Computer Science, University of Vaasa, P.O. Box 700, FI-65101, Vaasa, Finland
[6] Department of Computer Science, VSB-Technical University of Ostrava, 17. listopadu 15,70833, Ostrava-Poruba, Czech Republic
{david.koloseni,pasi.luukka,mikael.collan}@lut.fi
{jouni.lampinen@uwasa.fi,mario.fedrizzi@unitn.it}

Abstract. This paper introduces a new classification method that uses the differential evolution algorithm to feature-wise select, from a pool of distance measures, an optimal distance measure to be used for classification of elements. The distances yielded for each feature by the optimized distance measures are aggregated into an overall distance vector for each element by using OWA based multi-distance aggregation.

Keywords: Classification, Differential evolution, Pool of distances, Multi-distances.

1 Introduction

Evolutionary Algorithms (EAs) have significant advantages over many classification methods [1]. One evolutionary algorithm, Differential Evolution (DE) [2], has gained popularity due to its robustness, simplicity, easy implementation and fast convergence. DE has been applied i.e. to bankruptcy prediction [3], online feature analysis [4], scheduling algorithms based on fuzzy constraints [5], fuzzy clustering techniques [6].

When optimizing the performance of a system, the goal is to find a set of values of the system parameters for which the overall performance of the system will be the best under some given conditions. Several research suggests that better classification accuracy has been obtained by applying other distance

P. Angelov et al. (eds.), *Intelligent Systems'2014*,
Advances in Intelligent Systems and Computing 322, DOI: 10.1007/978-3-319-11313-5_67

measures than the most commonly applied Euclidean distance e.g., Manhattan distances and Minkowski distance were used in [7], and Mahalanobis distance and Gower distances were used in [8]. A relaxational metric adaptation algorithm that adjusts the locations of patterns in the input space in such a way that similar patterns are close together while dissimilar patterns are far away was applied in [9]. Weighted Kullback-Leibler divergence and distance measure based on Bayesian criterion was used in [10]. Energy distance, outline distance, tree distance and depth annotation distance were used for exploratory search of RNA motifs in [11]. Enclosed area distance, pattern distance and dynamic time warping distance in evolutionary time series segmentation was applied [12]. However, these studies have mostly been tested with a few different distance measure and none of them have focused on distances at feature level but on data set level.

In [13] and [14] DE based classifier use Minkowski metric for computing the distance between the class vectors and the sample to be classified. In [15] a pool of distance measures were created and particular optimize the parameters related to the selection of the suitable measure. In 1988, Professor Yager R.R introduced an aggregation operator known as ordered weighted averaging (OWA) operator [16]. In OWA operator one crucial issue is on how to determine the associated weights [17]. In our current work the weights are generated by using linguistic quantifiers [16], the classification of these linguistic quantifiers are Regular Increasing monotone (RIM), regular decreasing monotone (RDM) and Regular unimodal (RUM) [18]. A Multi-distance function has been recently presented [19],[20] [21]. OWA based multi-distance functions can be extended to combine the distance values of all pairs of elements in the collection into ordered weighted averaging based multi-distances. These distances solve the problem of aggregating pairwise values for the construction of multi-distance function.

In [15], the differential evolution algorithm [1], [2] was used to optimize the selection of distance-measure used in classification, by considering several distance-measures (from a pool of distance-measures). The distances considered in [15] are vector based distances, and the optimal distance found was applied for the whole data set. Data sets are usually not single-feature sets, but consist of multiple features (i.e. several measurements for each object), this means that it is not necessary that a selected optimal distance-measure is actually optimal for all features of a given data set. Feature-level optimization of the used distance-measures was presented in [22]. After obtaining an optimized distance-measure for each data-set feature the distances for each feature of an element were aggregated by using the sum of the distances.

This article extends the work of earlier studies by introducing an enhanced method to obtain a total distance (distance vector) for each object. The total distance is obtained by combining the normalized distance values from each individual feature by computing an ordered weighted average (OWA) based multi-distance aggregation of all normalized distances.

The data sets for demonstrating the proposed approach and for its empirical evaluation are taken from the UCI-Repository of Machine Learning data set [23].

2 OWA Operator and Multi-distances

In this section we briefly introduce the OWA-operator and how to generate the weights for this operator. We also briefly review OWA based multi-distances [19].

2.1 Weights Generations Methods in OWA

In 1988, Ronald R. Yager introduced a new aggregation operator, called the ordered weighted averaging operator (OWA) [16], and gave the following definition.

Definition 1. *An ordered weighted averaging (OWA) operator of dimension m is a mapping* $R^m \to R$ *that has associated weighting vector* $W = [w_1, w_2, ..., w_m]$ *of dimension m with*

$\sum_{i=1}^{m} w_i = 1$, $w_i \in [0, 1]$ *and* $1 \leq i \leq m$

such that:

$$OWA(a_1, a_2, ..., a_m) = \sum_{i=1}^{m} w_i b_i \tag{1}$$

where b_i *is the largest ith element of the collection of objects* $a_1, a_2, ..., a_m$. *The function value* $OWA(a_1, a_2, ..., a_m)$ *determines the aggregated values of arguments* $a_1, a_2, ..., a_m$.

Yager [16] introduced a measure called orness which characterize the OWA operator. In below definition to a weighting vector $W = [w_1, w_2, ..., w_m]^T$, the orness is maximal (i.e., $orness(w) = 1$) when the OWA operator behaves like the maximum operator. Orness is minimal (i.e., $orness(w) = 0$) when the OWA operator behaves like the minimum operator. Then for any W the orness, $\beta(W)$ is always in the interval $[0, 1]$.

Definition 2. *Given a weighting vector* $W = [w_1, w_2, ..., w_m]^T$, *the measure of orness of the OWA aggregation operator for W is defined as*

$$\beta(W) = \frac{1}{m-1} \sum_{i=1}^{m} (m - i) w_i. \tag{2}$$

The weights can be generated in different ways e.g., by using linguistic quantifiers introduced in [16]. In this article we concentrate on linguistic quantifiers called regular increasing monotone (RIM) quantifiers, which are introduced below following the work of [16].

Definition 3. *A fuzzy subset Q of the real line is called a regular increasing monotone (RIM) quantifier if it satisfies following conditions*

1. $Q(0) = 0$,
2. $Q(1) = 1$,
3. *If* $x \geq y$ *then* $Q(x) \geq Q(y)$.

The weights for the OWA operator are computed using linguistic quantifiers introduced by Yager [16], $w_i = Q(\frac{i}{m}) - Q(\frac{i-1}{m})$ where $i = 1, 2, ..., m$ and Q is assumed to be regular increasing monotone (RIM) quantifier. The use of OWA with RIM quantifiers is that they can be adopted for our purpose of studying the effect of aggregation operators on distance measures. These weights must satisfy the conditions of OWA operator. In this article, the following five (5) weighting schemes are used. All the generated functions for the aggregation procedure used in this case the first four are RIM quantifiers. Besides these four we also tested O'Hagan method for weight generation, which is based on optimizing the entropy for specified orness value.

1. **Basic linguistic quantifier:**

$$Q(r) = r^{\alpha}, \alpha \geq 0 \tag{3}$$

and its corresponding weight is computed by using the relation

$$w_i = \left(\frac{i}{m}\right)^{\alpha} - \left(\frac{i-1}{m}\right)^{\alpha}, i = 1, 2, ..., m \tag{4}$$

2. **Quadratic linguistic quantifier:**

$$Q(r) = \left(\frac{1}{1 - \alpha r^{\frac{1}{2}}}\right), \alpha \geq 0 \tag{5}$$

and its corresponding weight is computed by using the relation

$$w_i = \left(\frac{1}{1 - \alpha \cdot (\frac{i}{m})^{\frac{1}{2}}}\right) - \left(\frac{1}{1 - \alpha \cdot (\frac{i-1}{m})^{\frac{1}{2}}}\right), i = 1, 2, ..., m \tag{6}$$

The quadratic linguistic quantifier has the form $Q(r) = (\frac{1}{1-\alpha \cdot r^{\gamma}})$ where α controls the maximum value of the weight generating function and γ is the ratio between the maximum and minimum value of the generating function, the mean is obtained if we set the value of $\gamma = \frac{1}{2}$.

3. **Exponential linguistic quantifier:**

$$Q(r) = e^{-\alpha r} \tag{7}$$

The weights associated to the exponential quantifier are computed by

$$w_i = e^{-\alpha(\frac{i}{m})} - e^{-\alpha(\frac{i-1}{m})}, i = 1, 2, ..., m \tag{8}$$

4. **Trigonometric linguistic quantifier:**

$$Q(r) = \arcsin(\alpha r) \tag{9}$$

In this case, the weights are computed by using the relation,

$$w_i = arcsin\left(\alpha \cdot \left(\frac{i}{m}\right)\right) - arcsin\left(\alpha \cdot \left(\frac{i-1}{m}\right)\right) \tag{10}$$

where $i = 1, 2, ..., m$

O'Hagan Method in Weighting Generation

In 1988 O'Hagan [24] introduced another technique for computing the weights. The procedure for obtaining the aggregation assumes a predefined degree of orness and then the weights are obtained in such a way that maximizes the entropy. The solution is based on the constrained optimization problem

$$\text{maximize}: -\sum_{i=1}^{m} w_i ln(w_i)$$

subject to

$$\beta = \frac{1}{n-1}\sum_{i=1}^{m}(n-1)w_i$$

$$\sum_{i=1}^{m} w_i = 1 \text{ and } w_i \geq 0.$$

The above constrained optimization problem can be solved in different ways. In this article we use the analytical solution introduced by [25] for stepwise generation of the weights.

a. if $m = 2$ implies that $w_1 = \alpha$ and $w_2 = 1 - \alpha$
b. if $\alpha = 0$ or $\alpha = 1$ implies that the corresponding weighting vectors are $w = (0, ...0, 1)$ or $w = (1, 0, ..., 0)$ respectively.
c. if $m \geq 3$ and $0 \leq \alpha \leq 1$ then, we have,

$$w_i = \left(w_1^{m-i} \cdot w_m^{i-1}\right)^{\frac{1}{m-1}}$$

$$w_m = \frac{((m-1)\cdot\alpha - m).w_1 + 1}{(m-1)\cdot\alpha + 1 - m\cdot w_1}$$

$$w_1[(m-1)\cdot\alpha + 1 - m\cdot w_1]^m = ((m-1)\cdot\alpha)^{m-1}\cdot[((m-1)\cdot\alpha - m)\cdot w_1 + 1]$$

For $m \geq 3$, the weights are computed by first obtaining the first weight followed by the last weight of the weighting vector before other weights are computed.

2.2 Multi-distances

Next we introduce the definition of multi-distance given in [20] and how the OWA operator can be applied with this concept.

Definition 4. *Multi-distance is a representation of the notion of multi argument distances. The set X is a union of all m-dimensional lists of elements of X, multi distance is defined as a function $D : X \to [0, \infty)$ on a non empty set X provided that the following properties are satisfied for all m and $x_1, x_2, ..., x_m, y \in X$*

c1. $D(x_1, x_2, ..., x_m) = 0$ *if and only if* $x_i = x_j$ *for all* $i, j = 1, 2, ..., m$
c2. $D(x_1, x_2, ..., x_m) = D(x_{\sigma(1)}, x_{\sigma(2)}, ..., x_{\sigma(m)})$ *for any permutation* σ *of* $i, j = 1, 2, ..., m$
c3. $D(x_1, x_2, ..., x_m) \leq D(x_1, y) + D(x_2, y) + ... + D(x_m, y)$.

We say that D is a strong multi-distance if it satisfies $c1, c2$, and

$c3^\star$ $D(\boldsymbol{x}_1, \boldsymbol{x}_2, ..., \boldsymbol{x}_m) \leq D(\boldsymbol{x}_1, \boldsymbol{y}) + D(\boldsymbol{x}_2, \boldsymbol{y}) + ... + D(\boldsymbol{x}_m, \boldsymbol{y})$. for all $\boldsymbol{x}_1, \boldsymbol{x}_2, ..., \boldsymbol{x}_m, \boldsymbol{y} \in X$

In application context, the estimation of distances between more than two elements of the set X can be constructed using multi-distances by means of OWA functions [16]. Martin and Mayor proposed the use of the following characterization

$$D_w(x_1, x_2, ..., x_m) = OWA_w(d(x_1, x_2), d(x_2, x_3), ..., d(x_{m-1}, x_m)) \tag{11}$$

where the elements $x_1, x_2, ..., x_m$ belong to set X.

In this article we create a vector $\mathbf{d}$ consists of distances, $\mathbf{d} = (d_1, d_2, ..., d_m)$ to which we apply multi-distance

$$D_w(d_1, d_2, ..., d_m) = OWA_w(d(d_1, d_2), d(d_2, d_3), ..., d(d_{m-1}, d_m)) \tag{12}$$

where we use Minkowsky measure for a particular distance in D_w.

3 Differential Evolution Classifier with Optimized OWA Based Multi-distances for Features in the Data Sets

3.1 Differential Evolution

The DE algorithm [1], [2] is an evolutionary optimization algorithm. The DE algorithm belongs to the class of stochastic population based global optimization algorithms. The design principles of DE are simplicity, robustness, efficiency, and the use of floating-point encoding instead of binary numbers for representing internally the solution candidates for the optimization problem to be solved. As a typical EA, DE starts with a randomly generated initial population of candidate solutions for the optimization problem to be solved, which is then improved using selection, mutation and crossover operations. Several ways exist to determine a stopping criterion for EAs but usually a predefined upper limit G_{max} for the number of generations to be computed provides an appropriate stopping condition. Other control parameters for DE are the crossover control parameter CR, the mutation factor F, and the population size NP.

In each generation G, DE goes through each D dimensional decision vector $\boldsymbol{v}_{i,G}$ of the population and creates the corresponding trial vector $\boldsymbol{u}_{i,G}$ as follows in the most common DE version, DE/rand/1/bin [2]:

$$u_{j,i,G} = \begin{cases} v_{j,i,G} = v_{j,r3,G} + F \cdot (v_{j,r1,G} - v_{j,r2,G}) \\ \text{if} \quad rand_j[0,1) \leq CR \vee j = j_{rand} \\ v_{j,i,G}, \qquad \text{otherwise} \end{cases}$$

where $i = 1, 2, ..., NP$, $j = 1, 2, ..., D$ $j_{rand} \in 1, 2, ..., D$, random index, chosen for once each i. $r1, r2, r3 \in 1, 2, ..., NP$, randomly selected, excepted for $r1 \neq r2 \neq r3 \neq i$ $CR \in [0, 1]$ and $F \in (0, 1+]$

In this DE version, *NP* must be at least four and it remains fixed along *CR* and F during the whole execution of the algorithm. Parameter $CR \in [0, 1]$, which controls the crossover operation, represents the probability that an element for the trial vector is chosen from a linear combination of three randomly chosen vectors and not from the old vector $\boldsymbol{v}_{i,G}$. The condition "$j = j_{rand}$" is to make sure that at least one element is different compared to the elements of the old vector. The parameter F is a scaling factor for mutation and its value is typically $(0, 1+]$[1]. In practice, *CR* controls the rotational invariance of the search, and its small value (*e.g.*, 0.1) is practicable with separable problems while larger values (*e.g.*, 0.9) are for non-separable problems. The control parameter F controls the speed and robustness of the search, *i.e.*, a lower value for F increases the convergence rate but it also adds the risk of getting stuck into a local optimum. Parameters *CR* and *NP* have the same kind of effect on the convergence rate as F has.

After the mutation and crossover operations, the trial vector $\boldsymbol{u}_{i,G}$ is compared to the old vector $\boldsymbol{v}_{i,G}$. If the trial vector has an equal or better objective value, then it replaces the old vector in the next generation. This can be presented as follows (in this paper minimization of objectives is assumed) [2]:

$$\boldsymbol{v}_{i,G+1} = \begin{cases} \boldsymbol{u}_{i,G} \text{ if } \quad f(\boldsymbol{u}_{i,G}) \leq f(\boldsymbol{v}_{i,G}) \\ \boldsymbol{v}_{i,G}, \text{otherwise} \end{cases} . \tag{13}$$

DE is an elitist method since the best population member is always preserved and the average objective value of the population will never get worse. The objective function, f, is minimized by applying the number of incorrectly classified learning set samples. After the optimization process the final solution, defining the optimized classifier, is the best member of the last generation's, G_{max}, population, the individual $\boldsymbol{v}_{i,G_{max}}$. The best individual is the one providing the lowest objective function value and therefore the best classification performance for the learning set.

The control parameters of the DE algorithm were set as follows: *CR*=0.9 and F=0.5 which were applied for all classification problems. *NP* was chosen so that it was six times the size of the optimized parameters, also the number of generations used is $G_{max} = 1000$. These values are the same as those used by [13],

[1] Notation means that the upper limit is about 1 but not strictly defined.

[14]. However, these selections were mainly based on general recommendations and practical experiences with the usage of DE, (e.g. [13] and [14]) and no systematic investigations were performed to find the optimal control parameter values. Therefore further classification performance improvements by may be possible by finding better control parameter settings.

3.2 Nearest Prototype DE Classifier with Pool of Distances and Multi-distance

In this article, we propose a classification method using a DE algorithm [1] to minimize erroneous classification costs (objective function). In the process, we split the data into a training set and testing set. The splitting of the data is done in such a way that half of the data are used in the training set and the other half in the testing set. We use the training set to maximize the classification accuracy so that the objective function is minimal.

$$\text{minimize} : 1 - \frac{\sum_{j=1}^{m} B(x_j)}{m} \tag{14}$$

where

$$B(x_j) = \begin{cases} 1 \text{ if } g(x_j) = T(x_j) \\ 0 \text{ if } g(x_j) \neq T(x_j) \end{cases}$$

and we denote $T(x_j)$ as the true class from sample x_j and

$$g(x_j) = \{i \mid \min_{i=1,\dots,N} d(\mathbf{x_j}, \mathbf{y_i})\} \tag{15}$$

meaning that we decide that the vector x_j belongs to class i if $d(\mathbf{x_j}, \mathbf{y_i})$ is minimum, with the sample vector $\mathbf{x_j}$ and the ideal candidate $\mathbf{y_i}$ for class i. If the ideal vector class $g(x_j)$ is equal to the true class $T(x_j)$, then the sample is correctly classified and $B(x_j)$ hence gets the value 1 otherwise 0. The sum of $B(x_j)$ gives the total number of samples correctly classified where as m is the total number of samples in the training set. Total number of classes is denoted by N.

Here we propose to use multi-distance in the computation of $d(\mathbf{x_j}, \mathbf{y_i})$. Let us represent each object from the training dataset by a vector of type $(\mathbf{x_j}, \mathbf{y_i}) = (x_{j1}, x_{j2}, ..., x_{jT}; y_i)$ and $\mathbf{y_i} = (y_{i1}, y_{i2}, ..., y_{iT})$ where there are N different classes of objects $\mathbf{y_i}, i = 1, 2, ..., N$ and T is the total number of features in the data set. The distance $\mathbf{d}(\mathbf{x_j}, \mathbf{y_i})$ in equation (15) is from the multi-distance that is

$$\mathbf{d}(\mathbf{x_j}, \mathbf{y_i}) = D_w(d_{k,1}, d_{k,2}, ..., d_{k,T}) \tag{16}$$

where

$$\begin{aligned} d_{k,1} &= d_{k,1}(x_{j1}, y_{i1}) \\ d_{k,2} &= d_{k,2}(x_{j2}, y_{i2}) \\ &\vdots \\ d_{k,T} &= d_{k,T}(x_{jT}, y_{iT}) \end{aligned}$$

$d_{k,i}$ can be any of the distance from a pool of distances, D_w is a multi-distance equation (12), and k is a parameter to be optimized, $k = \{1, 2, ..., 8\}$ and comes from the pool of distances [26]. In the current work, this pool of distances is given in Table 1.

Table 1. Pool of distances

k	Distance
1	$d_1(x, y) = (\lvert x - y\rvert^{r_1});\ , r_1 \in [1, \infty)$
2	$d_2(x, y) = \lvert x - y\rvert^{r_2}/\max\{\lvert x\rvert, \lvert y\rvert\}; r_2 \in [1, \infty)$
3	$d_3(x, y) = \lvert x - y\rvert^{r_3}/1 + \min\{\lvert x\rvert, \lvert y\rvert\}\ ;, r_3 \in [1, \infty)$
4	$d_4(x, y) = \lvert x - y\rvert/1 + \max\{\lvert x\rvert, \lvert y\rvert\}$
5	$d_5(x, y) = \lvert x - y\rvert/[1 + \lvert x\rvert + \lvert y\rvert]$
6	$d_6(x, y) = \lvert x/[1 + \lvert x\rvert] - y/[1 + \lvert y\rvert]\rvert$
7	$d_7(x, y) = r_4(x - y)^2/(x + y);\ , r_4 \in (0, \infty)$
8	$d_8(x, y) = \lvert x - y\rvert/\ (1 + \lvert x\rvert)(1 + \lvert y\rvert)$

As can be seen from Table 1, besides optimizing the selection of distance measure from eight possible choices, we also have parameters in distances which need to be optimized. In our earlier work [22] we optimized and examined the parameters with three different approaches. In this work we chose the the parameter vector where parameter values are not restricted but each parameter to be of length equal to the number of features in the data set at hand [22]. The underlying idea is that we apply elementwise distance measures and aim to optimize the selection for proper elementwise distance measures and their possible free parameters. The distances are then normalized and summed into a single value. These distance vectors are then aggregated to produce overall distance value by OWA based multi-distance aggregation. The optimization problem can now be summarized so that the DE is searching an optimal vector for $v_{i,G}$ which is optimize by minimizing equation (14) and

$$\mathbf{v}_{i,G} = \{\{y_1, y_2, \cdots, y_N\}, \{\alpha\}, \{p\}, \{S_1, S_2 \ldots, S_T\}, \\ \{p_{1,1}, p_{1,2}, p_{1,3}, p_{1,4}, \cdots, p_{1,T}, p_{2,T}, p_{3,T}, p_{4,T}\}\} \tag{17}$$

where $\{y_1, y_2, \cdots, y_N\}$ are T - dimensional class vectors that are to be optimized for the current data set. The parameters $\{p\}$ come from the multi-distance, the distance measure we used for multi-distance has the form $d = |x - y|^p$ in equation (15). Parameter $\{\alpha\}$, which we are optimizing is for the particular weight generation scheme applied. In the current method either for the RIM quantifier or for the O'Hagan method. For optimal selection of distance measure for each feature an optimal solution has to be found from eight possible choices for T possible features. Hence for optimal selection of distance measure, DE is now seeking a switch vector $\{S_1, S_2, ..., S_T\}$. Here vector $\{S_1, S_2, ..., S_T\}$ contains the selection of a particular measure to a particular feature and is the integer value within $[1, 8]$. Additionally, this we also create a parameter vector which contains the optimal parameters for a particular distance given in a pool of distances. The parameter vector is of the form $\{p_{1,1}, p_{1,2}, p_{1,3}, p_{1,4}, \cdots, p_{1,T}, p_{2,T}, p_{3,T}, p_{4,T}\}$. Thus, for example, for distance d_1 the parameter vector would be $r_1 = \{p_{1,1}, \cdots, p_{1,T}\}$ and that if, for the iteration i optimal distance for feature three would be d_2 with parameter $p_{2,3}$ and if in iteration $i+1$ we find that d_1 is the best choice for feature three then its parameter would be $p_{1,3}$, so that we do not restrict the parameter values vectorwise but allows each of the four parameters $\{r_1, r_2, r_3, r_4\}$ to be vectors of length T. By allowing this kind of flexibility, we make it easier for the DE to change the parameter. More information about the parameter in this context is given in to [22]. The core procedure for this classification method is described in Algorithm 1.

4 Classification Results

The data sets for experimentation with the proposed approach were taken from the UCI machine learning data repository [23]. The data sets were subjected to 1000 generations ($G_{max} = 1000$) and divided into 30 times random splits of testing sets and learning sets ($N = 30$) based on which mean accuracies and variances were then computed. The data sets are briefly introduced below.

Australian Credit Approval Data. This data set concerns credit card applications. It has 2 classes with 690 instances and 14 number of attributes. Eight (8) of the features are numerical and six (6) are categorical features. Classes represent the information whether the application for credit should be approved or not.

Credit Approval Data. This data set concerns credit card applications. The main decision to be made is whether the credit card application should be approved or not. The data set consists of 690 instances and has 15 attributes. There are six numerical attributes and nine categorical attributes.The data set is extended version of the Australian data set and it has one more feature.

German Credit Data. In the German credit data set, the task is to classify customers as good (1) or bad (2) in other words they are creditworthy or not, depending on 20 features about them and their bank accounts. Information form 1000 customers is included in the data set.

Next we analyze the results obtained from experimental studies in which we compare results from the proposed method with earlier versions of the DE classifier [13] and the DE classifier with pool of distances (DEPD) [22]. Additionally, all five weight generation schemes for OWA are compared.

Table 2 presents mean classification accuracies with 99% confidence interval calculated for DE classifier with OWA based multi-distance aggregation and pool of distances. In OWA we also examined five different weight generation schemes. The OWA based multi-distance with different weights generation schemes are:

basic linguistic quantifier, quadratic linguistic quantifier, exponential linguistic quantifier, trigonometric linguistic quantifier and the O'Hagan method.

Improvement in accuracy compared to our previous method in [22] was over 2% for the O'Hagan method in Australian data set. In the credit approval data set, there an improvement of over 2% with the O'Hagan method, while for the German data set there is an improvement of over 1% with quadratic linguistic quantifier scheme. The comparison with the DE classifier shows that there is an increase of almost 20% with the O'Hagan method for the Australian data set and more than 2% for credit approval with the O'Hagan method and in case of the German data set quadratic linguistic scheme outperformed DE classifier by more than 2%.

Table 2. Comparison of the results of the data sets from the proposed method with other classifiers

Method	Australian	Credit approval	German
Basic	82.75 ± 3.54	84.46 ± 1.66	72.20 ± 0.85
Quadratic	82.50 ± 4.44	83.96 ± 2.45	74.07 ± 0.79
Exponential	84.67 ± 1.62	84.06 ± 2.23	73.94 ± 0.85
Trigonometric	84.24 ± 1.60	84.47 ± 1.14	73.46 ± 1.04
O'Hagan	85.25 ± 1.37	84.89 ± 1.07	73.10 ± 1.40
DE classifier	66.78 ± 1.37	81.99 ± 1.61	71.18 ± 0.90
DEPD	82.94 ± 1.65	82.50 ± 1.69	73.37 ± 0.86

5 Conclusions

In this paper we have proposed a DE classifier with multi-distances, in which the idea of multi-distance measure is used in the aggregation of the distances. We also determine the optimal distance measures for each feature together with their optimal parameters and combine all featurewisely determined distance measures to form a single total distance measure by using Ordered weighting averaging (OWA) multi-distances, which was applied for the final classification decisions. The experimental results were computed for three different benchmarking data sets. The proposed method provided at least comparable accuracy or outperformed the compared classifiers, including the original version of DE classifier. This clearly shows that this type of aggregation can be useful if properly performed.

In our future research directions it is possible to extend this work for optimal selection of the distance measure from a predefined pool of distance measures with Fermat multi-distance or Sum-based multi-distance and to apply a suitable aggregation method based on ordered weighting averaging (OWA) multi-distances functions and can be extended to combine the distance values of all pairs of elements in the collection into ordered weighted averaging based multi-distances to mention few possibilities.

Acknowledgment. The work of Jouni Lampinen was supported by the IT4Innovations Centre of Excellence project. CZ.1.05/1.1.00/02.0070 funded by the Structural Funds of EU and Czech Republic Ministry of Education.

References

1. Storn, R., Price, K.V.: Differential Evolution - A simple and Efficient Heuristic for Global Optimization over Continuous Space. Journal of Global Optimization. 11, no 4,(1997) 341–359.
2. Storn, R., Price, K.V.: Lampinen, J.: Differential Evolution- A practical Approach to Global Optimization. Springer.(2005)
3. Chauhan, N., Ravi, V., Chandra, D.K.: Differential Evolution Trained Wavelet Neural Networks: Application to bBankruptcy Prediction in Banks, Expert Systems with Applications, 36, 4, 7659–7665 (2009)
4. Velayutham C.S., Kumar, S.: Differential Evolution Based On-Line Feature Analysis in an Asymmetric Subsethood Product Fuzzy Neural Network. 3316, 959–964.(2004)
5. Qiao, F., Zhang, G.: A Fuzzy Differential Evolution Scheduling Algorithm Based on Grid. Bulletin of Electrical Engineering and Informatics. 1 (4), 279–284 (2012)
6. Gomez-Skarmeta A.F., Delgado, M., Vila M.A.: About the Use of Fuzzy Clustering Techniques for Fuzzy Model Identification. Fuzzy Sets and Systems. 106 (2), 179–188 (1999)
7. Shahid, R., Bertazzon, S., Knudtson, M.L. and Ghali, W.A.: Comparison of Distance Measures in Spatial Analytical Modeling for Health Service Planning. BMC Health Services Research 9:200 (2009)
8. Dettmann, E., Becker, C., Schmeiser, C.: Distance Functions for Matching in Small Samples. Computational Statistics and Data Analysis. 55, 1942-1960 (2011)
9. Chang, H., Yeung, D.Y., Cheung, W. K.: Relaxational Metric Adaptation and its Application to Semi-Supervised Clustering and Content-Based Image Retrieval. Pattern Recognition. 39, 10 1905–1917 (2006)
10. Ogawa, A., Takahashi, S.: Weighted Distance Measures for Efficient Reduction of Gaussian Mixture Components in HMM-Based Acoustic Model. In: 2008 IEEE International Conference on Acoustics, Speech and Signal Processing, pp 4173–4176 (2008)
11. Schonfeld, J., Ashlock, D.: Evaluating Distance Measures for RNA Motif Search. In: 2006 IEEE Congress on Evolutionary Computation. pp 2331–2338 (2006)
12. Yu, J., Yin, J, Zhang, J.: Comparison of Distance Measures in Evolutionary Time Series Segmentation, In: 3rd International Conference on Natural Computation, pp 456–460 (2007)
13. Luukka, P., Lampinen, J.: Differential Evolution Classifier in Noisy Settings and with Interacting Variables. Applied Soft Computing. 11 , 891–899 (2011)
14. Luukka, P., Lampinen, J.: A Classification Method Based on Principal Component Analysis and Differential Evolution Algorithm Applied for Prediction Diagnosis from Clinical EMR Heart Data sets. Computational Intelligence in Optimization: Applications and Implementations. Springer.(2010)
15. Koloseni, D., Lampinen, J., Luukka, P.: Optimized Distance Metrics for Differential Evolution based Nearest Prototype Classifier, Expert Systems With Applications, vol. 39 (12), 10564–10570 (2012)

16. Yager, R.R.: On Ordered Weighted Averaging Operators in Multicriteria Decision Making, IEEE Transactions on Systems, Man,Cybernetcs, 18, 183–190 (1988)
17. Luukka, P, Kurama, O. : Similarity Classifier with Ordered Weighted Averaging Operators. Expert Systems with Applications, Vol. 40, 995–1002 (2013)
18. Yager, R.R.: Connectives and Quantifiers in Fuzzy Sets, Fuzzy Sets and Systems, 40, 39–75 (1991)
19. Martin, J., Mayor, G.: Some Properties of Multi-argument Distances and Fermat Multi-distance, IPMU2010,(2010) pp 703–711.
20. Martin, J., Mayor, G.: Multi-argument Distances. Fuzzy Sets Systems. 167, 92-100 (2011)
21. Martin, J., Mayor, G.: Regular Multidistances, XV Congreso Espagnol sobre Tecnologiasy Logica Fuzzy, ESTYLF 2010, Huelva, 297-301 (2010)
22. Koloseni, D., Lampinen, J., Luukka, P.: Optimized Distance Metrics for Differential Evolution based Nearest Prototype Classifier, Expert Systems With Applications. 40 (10), (2013) 4075–4082.
23. Newman, D.J., Hettich, S., Blake, C.L., Merz, C.J. , UCI Repository of machine learning databases [http://www.ics.uci.edu/~mlearn/MLRepository.html]. Irvine, CA:University of California, Department of Information and Computer Science.(1998)
24. O'Hagan, M.: Aggregating Template or Rule Antecedents in Real Time Expert Systems with Fuzzy Set Logic. In: 22nd Annual IEEE Asilomar Conference on Signals, Systems and Computers, pp 681-689. Pacific Grove, California (1988)
25. Füller, R., Majlender, P.: An Analytical Approach for Obtaining Maximal Entropy OWA Operator Weights. Fuzzy sets and Systems, 124, 53–57 (2001)
26. Bandemer, H. and Näther, W.: Fuzzy Data Analysis. Kluwer Academic Publishers.(1992)

Intuitionistic Fuzzy Decision Tree: A New Classifier

Paweł Bujnowski, Eulalia Szmidt, and Janusz Kacprzyk

Systems Research Institute, Polish Academy of Sciences,
ul. Newelska 6, 01–447 Warsaw, Poland
and
Warsaw School of Information Technology, ul. Newelska 6, 01-447 Warsaw, Poland
pbujno@gmail.com, {szmidt,kacprzyk}@ibspan.waw.pl

Abstract. We present here a new classifier called an intuitionistic fuzzy decision tree. Performance of the new classifier is verified by analyzing well known benchmark data. The results are compared to some other well known classification algorithms.

1 Introduction

Decision trees, recursively partitioning a space of instances (observations), are very popular classifiers with well known advantages. Quinlan the ID3 algorithm [22] is a source of many other approaches which have been developed along that line (cf. [27]).

Looking for more stable, and effective methods to extract knowledge in uncertain classification problems, classical (crisp) decision trees were extended to fuzzy decision trees (Janikow [17], Olaru et al. [21], Yuan and Shaw [41], Marsala [19], [20]).

Having in mind advantages of the intuitionistic fuzzy sets introduced by Atanassov [1], [2], [3] (A-IFSs for short), the next natural step was to apply these sets while building the trees.

In this paper we propose a new intuitionistic fuzzy decision tree classifier. The data is expressed by means of intuitionistic fuzzy sets. Also the measures constructed for the intuitionistic fuzzy sets are applied while making decisions how to split a node while expanding the tree. The intuitionistic fuzzy tree proposed here is an extension of the fuzzy ID3 algorithm [7].

To verify the potential of the new algorithm, an analysis of well known benchmark data is providing. The results are compared to other commonly used algorithms.

2 A Brief Introduction to A-IFSs

One of the possible generalizations of a fuzzy set in X (Zadeh [42]) given by

$$A^{'} = \{< x, \mu_{A'}(x) > | x \in X\} \tag{1}$$

where $\mu_{A'}(x) \in [0,1]$ is the membership function of the fuzzy set $A^{'}$, is an A-IFS (Atanassov [1], [2], [3]) A is given by

$$A = \{< x, \mu_A(x), \nu_A(x) > | x \in X\} \tag{2}$$

P. Angelov et al. (eds.), *Intelligent Systems'2014*,
Advances in Intelligent Systems and Computing 322, DOI: 10.1007/978-3-319-11313-5_68

where: $\mu_A : X \to [0,1]$ and $\nu_A : X \to [0,1]$ such that

$$0 \leq \mu_A(x) + \nu_A(x) \leq 1 \tag{3}$$

and $\mu_A(x)$, $\nu_A(x) \in [0,1]$ denote a degree of membership and a degree of non-membership of $x \in A$, respectively. (An approach to the assigning memberships and non-memberships for A-IFSs from data is proposed by Szmidt and Baldwin [29]).

Obviously, each fuzzy set may be represented by the following A-IFS: $A = \{< x, \mu_{A'}(x), 1 - \mu_{A'}(x) > | x \in X\}$.

An additional concept for each A-IFS in X, that is not only an obvious result of (2) and (3) but which is also relevant for applications, we will call (Atanasov [2])

$$\pi_A(x) = 1 - \mu_A(x) - \nu_A(x) \tag{4}$$

a *hesitation margin* of $x \in A$ which expresses a lack of knowledge of whether x belongs to A or not (cf. Atanassov [2]). It is obvious that $0 \leq \pi_A(x) \leq 1$, for each $x \in X$.

The hesitation margin turns out to be important while considering the distances (Szmidt and Kacprzyk [30], [31], [33], entropy (Szmidt and Kacprzyk [32], [34]), similarity (Szmidt and Kacprzyk [35]) for the A-IFSs, etc. i.e., the measures that play a crucial role in virtually all information processing tasks (Szmidt [28]).

The hesitation margin turns out to be relevant for applications – in image processing (cf. Bustince et al. [15], [14]) and the classification of imbalanced and overlapping classes (cf. Szmidt and Kukier [36], [37], [38]), group decision making (e.g., [4]), genetic algorithms [24], negotiations, voting and other situations (cf. Szmidt and Kacprzyk papers).

3 Intuitionistic Fuzzy Decision Tree – New Algorithm Description

The source *ID3* tree introduced by Quinlan [22] was extended in many ways, among others, by the soft decision tree introduced by Baldwin et al. [7] which gave inspiration to the intuitionistic fuzzy decision tree proposed here.

The methods presented here make use of numeric attributes but they can also be applied to the nominal attributes (the algorithm is even simpler then).

Intuitionistic fuzzy sets are used for data representation. Next, the new idea of deriving intuitionistic fuzzy sets in each node was applied as potentially giving the most accurate results.

The most important step in generating a decision tree is splitting the nodes which demands to point out the best attributes for splitting. Picking up the attributes influences accuracy of a decision tree, and its interpretation properties. In the tree presented here intuitionistic fuzzy entropy was used (Szmidt and Kacprzyk[32]) as a counterpart of "information gain" [22].

Below the most important components of the algorithm are described.

Fuzzy Partitions of the Attribute Values (Granulation)
Replacing a continuous domain with a discrete one, i.e., the idea of a universe partition (granulation), has been extended to fuzzy sets by Ruspini [25]. Here the idea was used

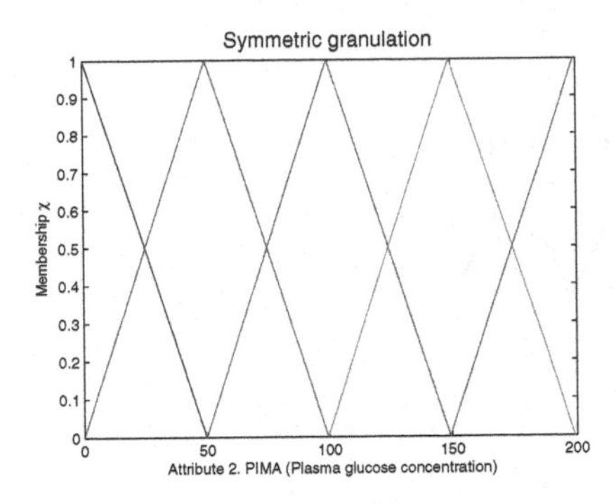

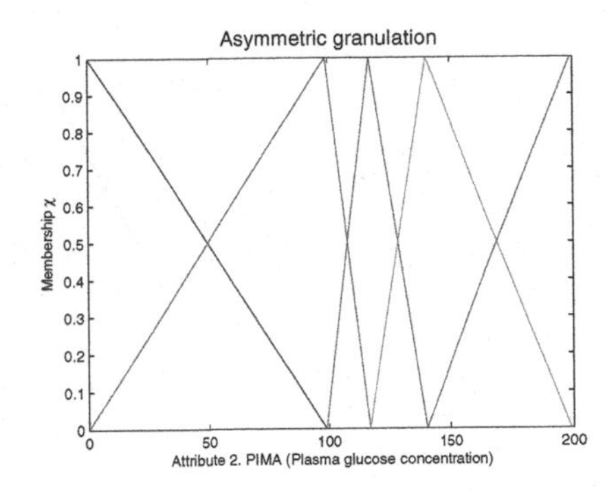

Fig. 1. Example of symmetric fuzzy partitioning, and asymmetric fuzzy partitioning (on attribute 2 "Plasma glucose concentration" of benchmark "Pima Diabetes" with 5 fuzzy sets)

to partition a universe of each attribute by introducing a set of triangular fuzzy sets such that for any attribute value the sum of memberships of the partitioning fuzzy sets is 1.

To be more precise, the membership $\chi_{j,k}(o_{ij})$ of the i-th observation (instance) o_{ij} in respect to the j-th attribute to the triangular fuzzy sets k and $k+1$ (where $k = 1, \ldots, p$) is:

$$\chi_{j,k}(o_{ij}) + \chi_{j,k+1}(o_{ij}) = 1, \quad k = 1, \ldots, p-1, \tag{5}$$

and for the j-th attribute A_j we have $o_{ij} \in A_j,\ i = 1, \ldots, n,\ j = 1, \ldots, m$.
From (5) it follows that the sum of the membership values for an observation o_{ij} is one (the sum results from only two neighboring fuzzy sets).

Remark. We use symbol χ for the membership values for the purpose of granulation so to make a difference between membership values resulting from the attribute granulation (χ) and the membership values of the intuitionistic fuzzy sets μ.

Two types of granulation are used:
– symmetric granulation with evenly spaced triangular fuzzy sets (symmetric fuzzy partitions), and
– asymmetric granulation with unevenly spaced triangular fuzzy sets (asymmetric fuzzy partitions such that each partition contains equal number of data points) [5,25].
An example of symmetric fuzzy partitioning (symmetric granulation), and asymmetric fuzzy partitioning (asymmetric granulation) is shown in Fig. 1. The two kinds of partitioning are illustrated on attribute 2 of the "PIMA Diabetes" problem with 5 fuzzy sets. Fuzzy partitioning (triangular fuzzy sets) is a starting point to assign nodes in a soft ID3 decision tree - cf. Fig. 2.

Now we will present a fuzzy generalization of *ID3* algorithm [7].

Fuzzy *ID3* Algorithm
The following database is considered

$$T = \{o_i =< o_{i,1}, \ldots, o_{i,m} > \mid i = 1, \ldots, n\}, \tag{6}$$

where $o_{i,j}$ is a value of the j-th attribute $A_j, j = 1, \ldots, m$, for the i-th instance. We assume that $o_{i,j}$ are crisp.

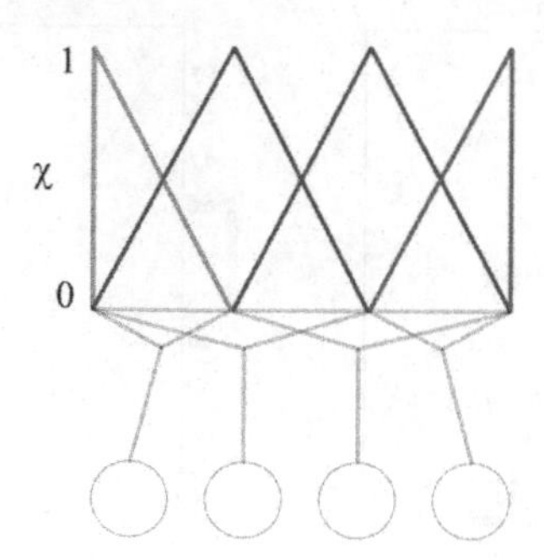

Fig. 2. Constructing nodes in a soft ID3 tree resulting from a fuzzy partitioning

We consider top down approach of generating a fuzzy *ID3* decision tree from data, i.e., the root contains all the instances at the beginning. Each node is split by partitioning its instances. A node becomes a leaf if all the attributes are used in the path considered or if all its instances are from a unique class.

The rules can represent splitting the nodes in a decision tree. Assume that P_j is a partition set of the attribute space Ω_j $(j = 1, \ldots, m)$, and that partition of each attribute is via triangular fuzzy sets. Let $P_{\chi_{j,k}} \in P_j$ be the k-th partitioning fuzzy set expressed by a triangular membership function $\chi_{j,k}$ being a component of the partition of the j-th attribute. The following rule expresses conjunction of the fuzzy conditions along the path from the root to a tree node

$$B \equiv P_{\chi_{j1}} \wedge \cdots \wedge P_{\chi_{jN}} \tag{7}$$

where $P_{\chi_{jr}}$ are triangular fuzzy sets, and its set of indices represented by the subsequence (j_r) is in a considered rule a result of pointing up a pair: (1) a unique attribute numbers j, and (2) one from the k triangular fuzzy sets for each attribute partitioning. Formula (7) expresses a conjunction of the conditions which are to be fulfilled for an instance o_i so that it were present in a considered node. Database $T = \{o_i,\ i = 1, \ldots, n\}$ generates a *support* for B (7) given as:

$$w(B) = \sum_{i=1}^{n} \prod_{j_r} Prob(P_{\chi_{j_r}} | o_i) \tag{8}$$

where $Prob(P_{\chi_{j_r}} | o_i)$ is a probability defined on the fuzzy set $P_{\chi_{j_r}}$ provided the observation o_i. It is easily calculated using the membership function $\chi_{j_r(o_i)}$.

Consider $\{C_l,\ l = 1, \ldots, h\}$ a set of decision classes. Formula (8) is also used for generating support for a given decision class, e.g., C_x in a given node, namely

$$Prob(C_x | B) = \frac{w(C_x \wedge B)}{\sum_{l=1}^{h} w(C_l \wedge B)} = \frac{w(C_x \wedge B)}{w(B)}. \tag{9}$$

To split a node (starting from a root) it is necessary to evaluate the attributes' abilities to generate a next level with the child nodes. A potential possibility of an attribute A for producing child nodes $A_s, s = 1, \ldots p$ is tested by calculating its classical entropy:

$$I(A_s) = -\sum_{l=1}^{h} Prob(C_l|A_s) \, log(Prob(C_l|A_s)), \; s = 1, \ldots p., \tag{10}$$

The common entropy for an attribute A is the following weighted mean value:

$$I(A) = \frac{\sum_{s=1}^{p} w(A_s) \cdot I(A_s)}{\sum_{s=1}^{p} w(A_s)} \tag{11}$$

It is assumed in (10) and (11) that A_s represents a rule from the root to the s-th child node.

The above formulas make it possible to generate the nodes in a fuzzy *ID3* tree [7].

Deriving Intuitionistic Fuzzy Sets from Data

Using intuitionistic fuzzy sets we will present now a generalization of the previously described soft *ID3* approach.

Assume that an attribute A, splitting a node into the child nodes $A_s, s = 1, \ldots p$, is tested. For simplicity we assume only two decision classes C^+ and C^-. Support for these classes in each node is

$$\begin{array}{l} for\ class\ C^+ : w(C^+ \wedge A_1), w(C^+ \wedge A_2), \cdots, w(C^+ \wedge A_p) \\ for\ class\ C^- : w(C^- \wedge A_1), w(C^- \wedge A_2), \cdots, w(C^- \wedge A_p). \end{array} \tag{12}$$

Independently for each class their frequencies for the verified splitting are calculated (proportions between support of a class in the child nodes and its cardinality in the parent node)

$$\begin{array}{l} p(C^+|A_s) : \frac{w(C^+ \wedge A_1)}{w(C^+ \wedge A)}, \frac{w(C^+ \wedge A_2)}{w(C^+ \wedge A)}, \cdots, \frac{w(C^+ \wedge A_p)}{w(C^+ \wedge A)} \\ p(C^-|A_s) : \frac{w(C^- \wedge A_1)}{w(C^- \wedge A)}, \frac{w(C^- \wedge A_2)}{w(C^- \wedge A)}, \cdots, \frac{w(C^- \wedge A_p)}{w(C^- \wedge A)}. \end{array} \tag{13}$$

The relative frequencies $p(C^+|A_i)$ and $p(C^-|A_i)$ (13) make it possible to use the algorithm given in [6,7] to construct independently fuzzy sets representing the classes C^+, and C^-. The fuzzy sets obtained for C^+, and C^- are abbreviated Pos^+ and Pos^-, respectively. In the fuzzy *ID3* tree [7] the fuzzy sets $Pos^+(A_s)$ and $Pos^-(A_s)$, $s = 1, \ldots, p$ are tested by a classical entropy (10) - (11) to assess the attributes.

For the purpose of the algorithm proposed here we use the fuzzy model (expressed by Pos^+ and Pos^-) to construct intuitionistic fuzzy model (details are presented in Szmidt and Baldwin [29]). Intuitionistic fuzzy model of the data in the child nodes A_s, $s = 1, \ldots p$ (due to [29]) is expressed by the following intuitionistic fuzzy terms

$$\begin{array}{l} \pi(A_s) = Pos^+(A_s) + Pos^-(A_s) - 1 \\ \mu(A_s) = Pos^+(A_s) - \pi(A_s) \\ \nu(A_s) = Pos^-(A_s) - \pi(A_s). \end{array} \tag{14}$$

In effect each child node s is described by the following intuitionistic fuzzy set

$$< A_s, \mu(A_s), \nu(A_s), \pi(A_s) >, \; s = 1, \ldots, p \tag{15}$$

where μ describes support for the class C^+; ν describes support for the class C^-; π expresses lack of knowledge concerning μ and ν.

Characteristic of an instance o_i at node A_s can be expressed as well in terms of intuitionistic fuzzy sets

$$\chi_{A_s}(o_i) \cdot < \mu(A_s), \nu(A_s), \pi(A_s) >, \; i = 1, \ldots, n,$$

where χ_{A_s} is a membership function at node A_s expressed by the product in (8). Having in mind the property (5) we can obtain full information value of an instance o_i while partitioning A and obtaining in result the child nodes $\{A_s, \; s = 1, \ldots, p\}$:

$$\chi_{A_s}(o_i) \cdot < \mu(A_s), \nu(A_s), \pi(A_s) > + \chi_{A_{s+1}}(o_i) \cdot < \mu(A_{s+1}), \nu(A_{s+1}), \pi(A_{s+1}) > . \tag{16}$$

In the algorithm proposed for assessing and choosing the attributes while splitting the nodes in the intuitionistic fuzzy decision tree, either (15) or (16) may be used.

Selection of an Attribute to Split a Node

While expanding a tree – a crisp, fuzzy or intuitionistic fuzzy tree, the crucial step is splitting a node into children nodes. To split a node an attribute is selected on the basis of its "information gain". Different measures may be used to assess "information gain". We use here an intuitionistic fuzzy entropy [32].

Intuitionistic fuzzy entropy $E(x)$ of an intuitionistic fuzzy element $x \in A$ is [32]:

$$E(x) = \frac{\min\{l_{IFS}(x, M), l_{IFS}(x, N)\}}{\max\{l_{IFS}(x, M), l_{IFS}(x, N)\}}, \tag{17}$$

where M, N are the intuitionistic fuzzy elements ($< \mu, \nu, \pi >$) fully belonging (M) or fully not belonging (N) to a set considered

$$M = < 1, 0, 0 >$$
$$N = < 0, 1, 0 >,$$

$l_{IFS}(\cdot, \cdot)$ is the normalized Hamming distance [31,33]:

$$l_{IFS}(x, M) = \tfrac{1}{2}(|\mu_x - 1| + |\nu_x - 0| + |\pi_x - 0|)$$

$$l_{IFS}(x, N) = \tfrac{1}{2}(|\mu_x - 0| + |\nu_x - 1| + |\pi_x - 0|).$$

Other intuitionistic fuzzy measures may be used to evaluate the attributes (cf. [39], [40]) but due to the space limitation here we discuss entropy only.

Intuitionistic fuzzy entropy of an intuitionistic fuzzy set with n elements: $X = \{x_1, \ldots, x_n\}$ is [32]:

$$E(X) = \frac{1}{n} \sum_{i=1}^{n} E(x_i). \tag{18}$$

We make use of the intuitionistic fuzzy representations (12)–(15) of the possible child nodes derived while testing attribute A to compute intuitionistic fuzzy entropy $E(A_s)$ (17) in a child node A_s, $s = 1, \ldots, p$.

Total intuitionistic fuzzy entropy of an attribute A is abbreviated $E(W_A)$ whereas entropy of a child node – $E(A_s)$. Total intuitionistic fuzzy entropy of A is a sum of the weighted intuitionistic fuzzy entropy measures of all the child nodes A_s, $s = 1, \ldots, p$, with the weights reflecting supports (cardinalities) of the nodes:

$$E(W_A) = \frac{\sum_{s=1}^{p} w(A_s)E(A_s)}{\sum_{s=1}^{p} w(A_s)}. \tag{19}$$

An alternative way to (19) of calculating $E(W_A)$ is by applying a weighted intuitionistic fuzzy representation of each instance o_i (16) while partitioning an attribute A. Next, using (18), a total intuitionistic fuzzy entropy is calculated for a chosen attribute. This method was applied in the numerical experiments (cf. Section 4).

An attribute for which total intuitionistic fuzzy entropy is minimal is selected for splitting a node.

A process of generating intuitionistic fuzzy decision tree is in Fig. 3.

Classification of the Instances

A description of a leaf in a soft tree is via a proportion of the classes considered. As a single instance usually belongs to several leaves, we need aggregated information about total degree of membership of a single observation to each class.

Here, to classify the instances we use measure *SUM* being a sum of the products of the instance membership values at leafs and support for a class considered in these leafs [7]. Total support of the observation o_i, $i = 1, \ldots, n$, for a class C is:

$$supp(C|o_i)_{SUM} = \sum_{j=1}^{L} supp(C|T_j) \cdot \chi(T_j|o_i), \tag{20}$$

where: $\{T_j : j = 1, \ldots, L\}$ – a set of the leafs; L – the number of the leafs; $supp(C|T_j)$ – a support of the classes considered in the j-th leaf; $\chi(T_j|o_i)$ – a membership value of the observation o_i (it is a result of the partitioning of the universe attributes), different for each leaf, fulfilling: $\sum_{j=1}^{L} \chi(T_j|o_i) = 1$.

4 Numerical Experiments

The classification abilities of the new intuitionistic fuzzy decision tree have been compared with other well known algorithms.

The following measures were used in the process of the comparison:
- total proper identification of the instances belonging to the classes considered,
- the area under ROC curve [16].

Behavior of the intuitionistic fuzzy decision tree proposed here was compared with other classifiers, namely:

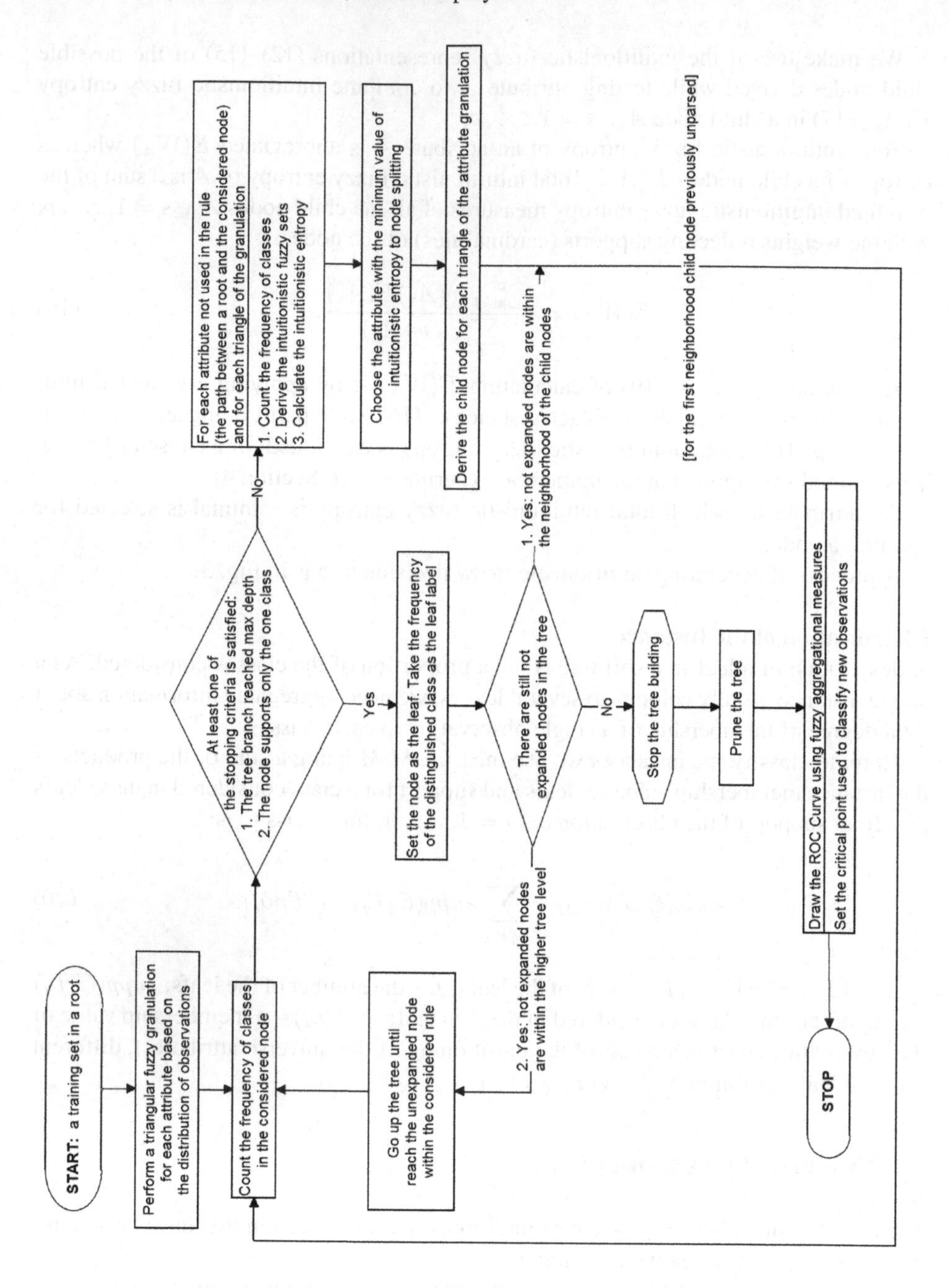

Fig. 3. A flowchart representing a process of generating intuitionistic fuzzy tree

Table 1. "Glass" benchmark data – comparison of the intuitionistic fuzzy decision tree and other classifiers

Algorithm	**Classification accuracy ($\bar{x} \pm \sigma$) w % for all classes**
$RandomForest$	77.05 ± 8.22
IFS tree (K, asym)	75.16 ± 6.21 $(*)$
$pruned\,IFS\ tree\ (K, asym)$	71.92 ± 6.30 $(-)$
$SDT\ (refitting)$	$71.09 \pm b.d.$ $(b.d.)$
$NBTree$	70.95 ± 9.95 $(-)$
$SDT\ (backfitting)$	$70.91 \pm b.d.$ $(b.d.)$
LMT	68.17 ± 9.91 $(-)$
$J48\ (unpruned\,C4.5)$	68.07 ± 9.54 $(-)$
$J48\ (pruned\,C4.5)$	67.61 ± 9.26 $(-)$
$MultilayerPerceptron$	65.96 ± 9.11 $(--)$
$LogisticModelTree$	63.92 ± 8.81 $(--)$

- ***J48*** – implementation of the crisp tree proposed by Quinlan *C4.5* ([23]),
- ***LMT*** (*Logistic Model Tree*) – a hybrid tree building the logistic models at the leaves ([18]),
- ***NBTree*** – hybrid decision tree building the Bayes classifiers at the leaves,
- ***RandomForest*** – here consisting of 10 decision trees which nodes are generated on the basis of a random set of attributes ([11]),
- ***MultilayerPerceptron*** – neural network,
- ***Logistic*** – logistic regression,
- ***Soft Decision Trees (SDT)*** – proposed by Olaru and Wehenkel [21].

The evaluation of the above algorithms (excluding Soft Decision Trees (SDT) which results are presented in [21]) was performed using WEKA (http://www.cs.waikato.ac.-nz/ml/weka/).

We present here the results obtained by intuitionistic fuzzy decision tree for "Glass" benchmark data (http://archive.ics.uci.edu/ml/datasets.html). The dataset contains 214 instances, 10 numerical attributes, 6 classes (4th class empty). Simple cross validation method is used with 10 experiments of 10-fold cross validation (giving 100 trees). An average value of the accuracy measures, and of their standard deviations is calculated for each experiment. So to compare an average accuracy of the new intuitionistic fuzzy decision tree with other classifiers, *t-Student* test was used (Table 1). One minus in Table 1 means that the (worse) result was obtained by a classifier while using classical *t-Student* test, two minuses mean using corrected *t-Student* test (for cross validation).

Analyzing the results for accuracy (Table 1) we may notice that the intuitionistic fuzzy decision tree turned out a better classifier for "Glass" benchmark data than other crisp and soft decision trees, even better than *Multilayer Perceptron* and *Logistic Model Tree*, and placed itself on the second position being only a little worse than *Random Forest*.

Final ranking of the tested algorithms (Table 2) in respect to all the examined data sets ("PIMA", "Ionosphere", "Sonar", "Wine", "Glass", "Iris" – http://archive.ics.-uci.edu/ml/datasets.html, and two real data sets describing incomes, and a children

Table 2. Ranking of the verified algorithms

Algorithm	**Ranking of the results**			
	Accuracy in respect to all classes		**AUC ROC**	
	$\bar{x} \pm \sigma$	$median$	$\bar{x} \pm \sigma$	$median$
LMT	3.0 ± 1.9	2.5	2.2 ± 1.1	2.0
IFS tree	2.8 ± 1.0	3.0	2.4 ± 1.3	3.0
$RandomForest$	4.4 ± 2.6	4.0	3.2 ± 1.9	3.0
$MultilayerPerceptron$	3.8 ± 2.6	3.5	4.0 ± 1.0	4.0
$Logistic$	4.6 ± 3.7	3	4.2 ± 2.6	5.0
$NBTree$	5.5 ± 1.6	5.0	5.2 ± 0.8	5.0
$SDT\ (backfitting)$	6.7 ± 1.2	6.5	$b.d.$	$b.d.$
$SDT\ (refitting)$	6.5 ± 2.2	7.0	$b.d.$	$b.d.$
$J48\ (C4.5)$	6.9 ± 1.0	7.0	6.8 ± 0.4	7.0

illness) was done taking into account the average values with standard deviations ($\bar{x} \pm \sigma$), and the medians.

Results in Table 2 are presented in increasing order due to the median of the measure *AUC ROC*, and next, due to the median of the percentage of the proper identification of the classes. We can see that the second position of the intuitionistic fuzzy decision tree is only worse from a very effective hybrid tree *LMT*. It is worth stressing that a little worse than the intuitionistic fuzzy decision tree turned out *RandomForest* and *MutilayerPerceptron*.

Assessing the results of the verified algorithms (Table 2), besides the median, it is worth noticing as well the mean values and standard deviations of the ranking. Standard deviation for intuitionistic fuzzy tree is low in respect of both measures considered (because for each data set considered the tree did not obtained poor results). Again, it is worth emphasizing that the standard deviation of the accuracy is lower for intuitionistic fuzzy tree than for logistic regression (*Logistic*), random forest (*RandomForest*), and neural network (*MultilayerPerceptron*).

Finally, in many applications when comprehensibility and transparency to the human being is relevant, the proposed classifier, as a tree type classifier, can be a properer, if not the best choice.

5 Conclusions

We have proposed an extension of the fuzzy ID3 decision tree algorithm, namely, a new intuitionistic fuzzy decision tree. The new classifier was tested on well known benchmark examples giving very encouraging results.

Acknowledgment. Partially supported by the Ministry of Science and Higher Education Grant UMO-2012/05/B/ST6/03068.

References

1. Atanassov, K.: Intuitionistic Fuzzy Sets. VII ITKR Session. Sofia (June 1983), (Deposed in Central Sci.- Techn. Library of Bulgarian Academy of Sciences., 1697/84)
2. Atanassov, K.: Intuitionistic Fuzzy Sets: Theory and Applications. Springer (1999)
3. Atanassov, K.: On Intuitionistic Fuzzy Sets Theory. Springer (2012)
4. Atanassova, V.: Strategies for Decision Making in the Conditions of Intuitionistic Fuzziness. In: Int. Conf. 8th Fuzzy Days, Dortmund, Germany, pp. 263–269 (2004)
5. Baldwin, J.F., Karale, S.B.: A symmetric Triangular Fuzzy Sets for Classification Models. In: Palade, V., Howlett, R.J., Jain, L. (eds.) KES 2003. LNCS, vol. 2773, pp. 364–370. Springer, Heidelberg (2003)
6. Baldwin, J.F., Lawry, J., Martin, T.P.: A mass assignment theory of the probability of fuzzy events. Fuzzy Sets and Systems 83, 353–367 (1996)
7. Baldwin, J.F., Lawry, J., Martin, T.P.: Mass Assignment Fuzzy ID3 with Applications. In: Unicom Workshop on Fuzzy Logic Applications and Future Directions, London (1997)
8. Bartczuk, Ł., Rutkowska, D.: A New Version of the Fuzzy-ID3 Algorithm. In: Rutkowski, L., Tadeusiewicz, R., Zadeh, L.A., Żurada, J.M. (eds.) ICAISC 2006. LNCS (LNAI), vol. 4029, pp. 1060–1070. Springer, Heidelberg (2006)
9. Benbrahim, H., Bensaid, A.: A comparative study of pruned decision trees and fuzzy decision trees. NAFIPS 2000, 227–231 (2000)
10. Bezdek, J.C.: Pattern Recognition with Fuzzy Objective Function Algorithms. Kluwer Academic Publishers, Dordrecht (1981)
11. Breiman, L.: Random Forests. Machine Learning 45(1), 5–32 (2001)
12. Breiman, L., Friedman, J.H., Olsen, R.A., Stone, C.J.: Classification and Regression Trees. Wadsworth, Belmont (1984)
13. Bujnowski, P.: Using intuitionistic fuzzy sets for constructing decision trees in classification tasks. PhD dissertation, IBS PAN, Warsaw (2013) (in Polish)
14. Bustince, H., Mohedano, V., Barrenechea, E., Pagola, M.: Image thresholding using intuitionistic fuzzy sets. In: Atanassov, K., Kacprzyk, J., Krawczak, M., Szmidt, E. (eds.) Issues in the Representation and Processing of Uncertain and Imprecise Information. Fuzzy Sets, Intuitionistic Fuzzy Sets, Generalized Nets, and Related Topics, EXIT, Warsaw (2005)
15. Bustince, H., Mohedano, V., Barrenechea, E., Pagola, M.: An algorithm for calculating the threshold of an image representing uncertainty through A-IFSs. In: IPMU 2006, pp. 2383–2390 (2006)
16. Hand, D.J., Till, R.J.: A simple generalization of the area under the ROC curve for multiple class classification problems. Machine Learning 45, 171–186 (2001)
17. Janikow, C.Z.: Fuzzy Decision Trees: Issues and Methods. IEEE Transactions on Systems, Man, and Cybernetics 28(1), 1–14 (1998)
18. Landwehr, N., Hall, M., Frank, E.: Logistic Model Trees. Machine Learning 95(1-2), 161–205 (2005)
19. Marsala, C.: Fuzzy decision trees to help flexible querying. Kybernetika 36(6), 689–705 (2000)
20. Marsala, C., Bouchon-Meunier, B.: An adaptable system to construct fuzzy decision tree. In: NAFIPS 1999, pp. 223–227 (1999)
21. Olaru, C., Wehenkel, L.: A complete fuzzy decision tree technique. Fuzzy Sets and Systems, pp. 221–254. Elsevier (2003)
22. Quinlan, J.R.: Induction of decision trees. Machine Learning 1, 81–106 (1986)
23. Quinlan, J.R.: C4.5: Programs for Machine Learning. Morgan Kaufman Publishers, Inc., San Mateo (1993)

24. : Generalized net model of intuitionistic fuzzy logic control of genetic algorithm parameters. Notes on Intuitionistic Fuzzy Sets 19(2), 1310–4926 (2013) ISSN 1310-4926
25. Ruspini, E.H.: A New Approach to Clustering. Information and Control 15, 22–32 (1969)
26. Rutkowski, L.: Artificial intelligence methods and techniques. PWN, Warszawa, pp. 237–307 (2009) (in Polish)
27. Safavian, S.R., Landgrebe, D.: A survey of decision tree classifier methodology. IEEE Trans. Systems Man Cybernet. 21, 660–674 (1991)
28. Szmidt, E.: Distances and Similarities in Intuitionistic Fuzzy Sets. Springer (2014)
29. Szmidt, E., Baldwin, J.:: Intuitionistic Fuzzy Set Functions, Mass Assignment Theory, Possibility Theory and Histograms, 2006 IEEE WCCI, 237–243 (2006)
30. Szmidt, E., Kacprzyk, J.: On measuring distances between intuitionistic fuzzy sets. Notes on IFS 3(4), 1–13 (1997)
31. Szmidt, E., Kacprzyk, J.: Distances between intuitionistic fuzzy sets. Fuzzy Sets and Systems 114(3), 505–518 (2000)
32. Szmidt, E., Kacprzyk, J.: Entropy for intuitionistic fuzzy sets. Fuzzy Sets and Systems 118, 467–477 (2001)
33. Szmidt, E., Kacprzyk, J.: Distances Between Intuitionistic Fuzzy Sets: Straightforward Approaches not work. In: 3rd International IEEE Conference Intelligent Systems, London, pp. 716–721 (May 2006)
34. Szmidt, E., Kacprzyk, J.: Some Problems with Entropy Measures for the Atanassov Intuitionistic Fuzzy Sets. In: Masulli, F., Mitra, S., Pasi, G. (eds.) WILF 2007. LNCS (LNAI), vol. 4578, pp. 291–297. Springer, Heidelberg (2007)
35. Szmidt, E., Kacprzyk, J.: A New Similarity Measure for Intuitionistic Fuzzy Sets: Straightforward Approaches not work. In: 2007 IEEE Conf. on Fuzzy Systems, pp. 481–486 (May 2007a)
36. Szmidt, E., Kukier, M.: Classification of Imbalanced and Overlapping Classes using Intuitionistic Fuzzy Sets. In: IEEE IS 2006, London, pp. 722–727 (2006)
37. Szmidt, E., Kukier, M.: A New Approach to Classification of Imbalanced Classes via Atanassov's Intuitionistic Fuzzy Sets. In: Wang, H.-F. (ed.) Intelligent Data Analysis: Developing New Methodologies Through Pattern Discovery and Recovery, pp. 85–101. Idea Group (2008)
38. Szmidt, E., Kukier, M.: Atanassov's intuitionistic fuzzy sets in classification of imbalanced and overlapping classes. In: Chountas, P., Petrounias, I., Kacprzyk, J. (eds.) Intelligent Techniques and Tools for Novel System Architectures. SCI, vol. 109, pp. 455–471. Springer, Heidelberg (2008)
39. Szmidt, E., Kacprzyk, J., Bujnowski, P.: Measuring the Amount of Knowledge for Atanassov's Intuitionistic Fuzzy Sets. In: Fanelli, A.M., Pedrycz, W., Petrosino, A. (eds.) WILF 2011. LNCS (LNAI), vol. 6857, pp. 17–24. Springer, Heidelberg (2011)
40. Szmidt, E., Kacprzyk, J., Bujnowski, P.: How to measure the amount of knowledge conveyed by Atanassov's intuitionistic fuzzy sets. Information Sciences 257, 276–285 (2014)
41. Yuan, Y., Shaw, M.J.: Induction of fuzzy decision trees. Fuzzy Sets and Systems 69, 125–139 (1996)
42. Zadeh, L.A.: Fuzzy sets. Information and Control 8, 338–353 (1965)

Characterization of Large Target Sets with Probabilistic Classifiers

Ray-Ming Chen

Department of Computer Science,
Friedrich-Alexander-Universität Erlangen-Nürnberg,
Martenstraße 3, 91058 Erlangen, Germany
ray.chen@fau.de

Abstract. The usual characterization of a target set based on an equivalence relation via an information system only involves one instance of data of the information system and one universe. In this paper, I will give an approach to characterize a large target set that would involve several instances and several universes via probabilistic classifiers. First of all, I reduce the large target set to smaller pieces that could be characterized by the universes in hand, and then combine all the characterizations to form a rough set for the large target set. Furthermore, I also devise some ways to choose the optimal reduction(s) and characterization(s) for such large target set.

1 Introduction

In the usual characterization of a target set, the target set is always assumed to be a subset of a given universe. However, if there is a new object that is larger than the universe, then one has to retrieve all other information contained in other universes to see whether there is a way to characterize this new large target set. It is not always the case that one could be able to collect all the information regarding any target sets. If there is no direct information to characterize a large target set, alternatively one could divide a large target set into pieces that could be characterized by utilizing some information in hand. Furthermore, the data in a information system might have more than just one instance. For example, if one is doing a survey, he would ask many people for the same questions-which creates more than one instance of data.

In this paper, we will consider a large target set that will involve more than one universe and more than one instance. We will show how to form a set of classifiers to characterize such large target set and how to decide the optimal characterization among all the possible choices.

The idea is to reduce the large target set to several smaller pieces that could then be characterized by some universes and instances in hand and then combine all the characterizations to form the rough set of the large target set. Since there are many ways to reduce the large target set, in Section 5, we come up with some criteria to decide which reduction and combined rough set is the optimal for the large target set.

P. Angelov et al. (eds.), *Intelligent Systems'2014*,
Advances in Intelligent Systems and Computing 322, DOI: 10.1007/978-3-319-11313-5_69

In Section 3, we study one universe and a set of probabilistic instances. The target set studied in this section is still assumed to be smaller than the universe. Then we show how to characterize such target via a probabilistic rough-set space. In Section 4, we will show how to characterize a large target set with several universes and instances via a probabilistic product rough-set space.

In the real world, new objects or new species are discovered everyday. To know them, one needs to utilize the information in hand. The approach we offer here gives a way to extend our capability to accommodate some large objects and characterize them by the information one has.

2 Notation

For any set S, we use $\mathcal{P}(S)$ to denote its power set and $|S|$ to denote the its size. In order to specify the element in a finite set, for any finite set V, we associate a bijective function $\sigma(V) : \{1, 2, ..., |V|\} \to V$ to fix its representation and use the abbreviation $V_{\sigma(V)} \equiv (\sigma(V)(1), \sigma(V)(2), ..., \sigma(V)(|V|))$ for its ordered set. We also use the abbreviation $|V|^{\downarrow}$ to denote the set $\{1, 2, ..., |V|\}$. Furthermore, for any ordered set $Y = (Y_1, Y_2, ..., Y_n)$, we use the projection notation $\rho^{(m)}(Y)$ to denote Y_m, where $n \geq m$. We use the notation $\prod$ for either the product of sets or numbers.

3 Probabilistic Classifier Space

Let us consider only one universe and several instances (with a probabilistic distribution) of an information system. Let U be a universe. Let a specification $\mathcal{A}(U) = \{a_1, a_2, ..., a_k\}$ be a set of attributes related to U. Let $[a_i]$ be the domain of the attribute values of a_i. Firstly, we represent an information system in the form of a function f.

Definition 1. For each $f : U \to [a_1] \times [a_2] ... \times [a_k]$, we call it an instance of $\mathcal{A}(U)$. Furthermore, define $\rho^{(n)}(f) : U \to [a_n]$ with $(\rho^{(n)}(f))(u) := \rho^{(n)}(f(u))$.

Definition 2. We call any set $\mathbb{F}(U) = \{f_1, f_2, ..., f_k\}$, which consists of instances of $\mathcal{A}(U)$, an outcome of $\mathcal{A}(U)$.

Example 1. Let us show some information systems in the form of instances. Suppose $U = \{u_1, u_2, u_3, u_4, u_5, u_6, u_7, u_8\}$, $\mathcal{A}(U) = \{temperature, weight\}$ with $[temperature] = \{high, medium, low\}$ and $[weight] = \{heavy, light\}$. Let the instances be $\mathbb{F}(U) = \{f_1, f_2, f_3\}$, where f_1, f_2 and f_3 are defined as follows:

Table 1. Information Systems under Instances f_1, f_2 and f_3

Universe	$\rho^{(1)}(f_1)$	$\rho^{(2)}(f_1)$	$\rho^{(1)}(f_2)$	$\rho^{(2)}(f_2)$	$\rho^{(1)}(f_3)$	$\rho^{(2)}(f_3)$
u_1	high	heavy	high	light	high	heavy
u_2	high	heavy	low	heavy	low	light
u_3	medium	light	high	light	high	light
u_4	low	heavy	medium	light	medium	heavy
u_5	medium	light	low	heavy	high	light
u_6	low	heavy	medium	heavy	low	light
u_7	low	light	medium	heavy	high	light
u_8	low	light	medium	light	medium	heavy

Secondly, one assigns a probability (or weight) distribution to the set of instances. This assignment could be objective or subjective. It represents the existence of the probability or weight among the information systems.

Definition 3. For any probability distribution $p^{\mathcal{A}(U)} : \mathbb{F}(U) \to [0,1]$, we call it an outcome distribution of $\mathcal{A}(U)$.

We will call each ordered set $(U, \mathcal{A}(U), \mathbb{F}(U), p^{\mathcal{A}(U)} : \mathbb{F}(U) \to [0,1])$ a U-state. Each instance f indeed induces an equivalent relation. This equivalent relation will then decide a partition of the universe and such partition will serve as a classifier for the target sets. Let $u, v \in U$ be arbitrary. Define a f-induced equivalent relation $u \sim_f v$ iff $f(u) = f(v)$ and define an equivalent class $[u]_f = \{v \in U : v \sim_f u\}$. Define a f-induced partition of U to be $\langle f \rangle \equiv U/\sim_f = \{[u]_f : u \in U\}$ We call $PS_{\mathbb{F}(U)} = \{\langle f \rangle : f \in \mathbb{F}(U)\}$ a $\mathbb{F}(U)$-induced partition space. Next, we define its probability distribution.

Definition 4. Define a function $Par : \mathbb{F}(U) \to PS_{\mathbb{F}(U)}$ with $Par(g) := \langle g \rangle$ and define $Par^{-1} : PS_{\mathbb{F}(U)} \to \mathcal{P}(\mathbb{F}(U))$ with $Par^{-1}(Y) := \{g \in \mathbb{F}(U) : \langle g \rangle = Y\}$.

Definition 5. Define a $\mathbb{F}(U)$-induced partition probability distribution $p^{\mathbb{F}(U)}$: $PS_{\mathbb{F}(U)} \to [0,1]$ to be $p^{\mathbb{F}(U)}(D) := \sum\limits_{f \in Par^{-1}(D)} p^{\mathcal{A}}(f)$.

Definition 6. Each $(PS_{\mathbb{F}(U)}, p^{\mathbb{F}(U)} : PS_{\mathbb{F}(U)} \to [0,1])$ is called a $\mathbb{F}(U) - induced$ probabilistic classifier space.

Example 2. Let us continue Example 1. Suppose $\rho^{\mathcal{A}} : \mathbb{F}(U) \to [0,1]$ is assigned as follows. Then based on our definitions, one computes $PS_{\mathbb{F}(U)}$ and $\rho^{\mathbb{F}(U)}$.

Table 2. Probabilistic Classifier Space: $(PS_{\mathbb{F}(U)}, p^{\mathbb{F}(U)})$

$\mathbb{F}(U)$	$p^{\mathcal{A}}$	$\mathbb{F}(U)$-induced partition space, $PS_{\mathbb{F}(U)}$	Partition Probability
f_1	$\frac{1}{5}$	$P_1 = \{\{u_1, u_2\}, \{u_3, u_5\}, \{u_4, u_6\}, \{u_7, u_8\}\}$	$p^{\mathbb{F}(U)}(P_1) = \frac{1}{5}$
f_2	$\frac{3}{7}$	$P_2 = \{\{u_1, u_3\}, \{u_2, u_5\}, \{u_6, u_7\}, \{u_4, u_8\}\}$	$p^{\mathbb{F}(U)}(P_2) = \frac{3}{7}$
f_3	$\frac{13}{35}$	$P_3 = \{\{u_1\}, \{u_2, u_6\}, \{u_3, u_5, u_7\}, \{u_4, u_8\}\}$	$p^{\mathbb{F}(U)}(P_3) = \frac{13}{35}$

Definition 7. Let X be a target set. Let its Pawlak rough sets, which is characterized by the classifier $\mathcal{K} \in PS_{\mathbb{F}(U)}$, be represented by an ordered set $X_{\mathcal{K}} = (lb_{\mathcal{K}}(X), ub_{\mathcal{K}}(X))$, where $lb_{\mathcal{K}}(X)$ is the $\mathcal{K}$-characterized lower bound of X and $ub_{\mathcal{K}}(X)$ is the $\mathcal{K}$-characterized upper bound of X.

We have formed the probabilistic classifier space $(PS_{\mathbb{F}}(U), p^{\mathbb{F}(U)} : PS_{\mathbb{F}}(U) \rightarrow [0,1])$. The next step is to look at the characterizations of a target set X via these probabilistic classifiers.

Definition 8 (Rough Space). Given a target set X and an outcome $\mathbb{F}(U)$ of $\mathcal{A}(U)$, define the rough space $Ch^U_X(\mathbb{F}(U)) = \{X_{\mathcal{K}} : \mathcal{K} \in PS_{\mathbb{F}(U)}\}$.

A rough space contains all the possible characterized rough sets of a target set. Unlike the deterministic information system which characterize a target set with exactly one ordered pair, a rough space generalizes Pawlak rough sets when the classifiers are induced by probabilistic information systems.

Definition 9. Define a function $RS_X : PS_{\mathbb{F}(U)} \rightarrow Ch^U_X(\mathbb{F}(U))$ with $RS_X(\mathcal{K}) := X_{\mathcal{K}}$ and define $RS^{-1}_X : Ch^U_X(\mathbb{F}(U)) \rightarrow \mathcal{P}(PS_{\mathbb{F}(U)})$ with $RS^{-1}_X(Y) := \{\mathcal{K} \in PS_{\mathbb{F}(U)} : X_{\mathcal{K}} = Y\}$.

Now we want to define a rough distribution $p^X : Ch^U_X(\mathbb{F}(U)) \rightarrow [0,1]$ by

Definition 10 (Rough Distribution). $p^X(Y) := \sum_{\mathcal{K} \in RS^{-1}_X(Y)} p^{\mathbb{F}(U)}(\mathcal{K})$.

Example 3. Suppose the target set $X = \{u_1, u_2, u_3, u_5, u_6, u_8\}$ and the classifiers in Example 2. Then one characterizes X and form the rough space.

Table 3. Pawlak Characterizations

characterizations	Pawlak rough sets
$X_{\langle f_1 \rangle}$	$(\{u_1, u_2, u_3, u_5\}, \{u_1, u_2, u_3, u_4, u_5, u_6, u_7, u_8\})$
$X_{\langle f_2 \rangle}$	$(\{u_1, u_2, u_3, u_5\}, \{u_1, u_2, u_3, u_4, u_5, u_6, u_7, u_8\})$
$X_{\langle f_3 \rangle}$	$(\{u_1, u_2, u_6\}, \{u_1, u_2, u_3, u_4, u_5, u_6, u_7, u_8\})$

Let R_1 denote $(\{u_1, u_2, u_3, u_5\}, \{u_1, u_2, u_3, u_4, u_5, u_6, u_7, u_8\})$ and R_2 denote $(\{u_1, u_2, u_6\}, \{u_1, u_2, u_3, u_4, u_5, u_6, u_7, u_8\})$. Then the rough space $Ch^U_X(\mathbb{F}(U))$ is $\{R_1, R_2\}$ and its rough distribution $p^X : Ch^U_X(\mathbb{F}(U)) \rightarrow [0,1]$ is $p^X(R_1) = \frac{22}{35}$ and $p^X(R_2) = \frac{13}{35}$.

So far we have shown how to characterize a target set X by a probabilistic rough space $(Ch^U_X(\mathbb{F}(U)), p^X : Ch^U_X(\mathbb{F}(U)) \rightarrow [0,1])$ with a probabilistic classifier space $(PS_{\mathbb{F}(U)}, p^{\mathbb{F}(U)} : PS_{\mathbb{F}(U)} \rightarrow [0,1])$ induced by a U-state.

4 Multiple Universes and Instances for Large Target Sets

Sometimes the target set X is too large, to characterize it, one would need to reduce X to pieces to accommodate the universes and information in hand. Let $\mathbb{U} = \{U_1, U_2, ..., U_n\}$ be a set of universes. Let $\mathbb{A}(\mathbb{U}) = \{\mathcal{A}(U_1), \mathcal{A}(U_2), ..., \mathcal{A}(U_n)\}$ be a set of specifications. Let $\mathfrak{F}^{\mathbb{U}} = \{\mathbb{F}(U_1), \mathbb{F}(U_2), ..., \mathbb{F}(U_n)\}$ be a set of outcomes of $\mathbb{A}(\mathbb{U})$. Let $PS^{\mathbb{U}} = \{PS_{\mathbb{F}(U)} : \mathbb{F}(U) \in \mathfrak{F}^{\mathbb{U}}\}$. Let $p^{\mathbb{A}(\mathbb{U})} = \{p^{\mathcal{A}(U_1)}, p^{\mathcal{A}(U_2)}, ..., p^{\mathcal{A}(U_n)}\}$ be a set of outcome distributions.

Example 4. Let $\mathbb{U} = \{U_1 = \{a, b, c, d, e\}, U_2 = \{b, d, e, f, g, k\}\}$, $\mathbb{A}(\mathbb{U}) = \{\mathcal{A}(U_1), \mathcal{A}(U_2)\}, \mathfrak{F}^{\mathbb{U}} = \{\mathbb{F}(U_1) = \{f_{11}, f_{12}, f_{13}\}, \mathbb{F}(U_2) = \{f_{21}, f_{22}, f_{23}, f_{24}\}\}$ and $PS^{\mathbb{U}} = \{PS_{\mathbb{F}(U_1)}, PS_{\mathbb{F}(U_2)}\}$. Let $p^{\mathbb{A}(\mathbb{U})} = \{p^{\mathcal{A}(U_1)}, p^{\mathcal{A}(U_2)}\}$, in which one has $p^{\mathcal{A}(U_1)}(f_{11}) = \frac{1}{3}, p^{\mathcal{A}(U_1)}(f_{12}) = \frac{1}{2}, p^{\mathcal{A}(U_1)}(f_{13}) = \frac{1}{6}; p^{\mathcal{A}(U_2)}(f_{21}) = \frac{2}{5}, p^{\mathcal{A}(U_2)}(f_{22}) = \frac{1}{3}, p^{\mathcal{A}(U_2)}(f_{23}) = \frac{2}{9}$ and $p^{\mathcal{A}(U_2)}(f_{24}) = \frac{2}{45}$. It goes the following characterizations:

Table 4. Multiple Probabilistic Characterizations

Partition Space,$PS_{\mathbb{F}(U_1)}$	$p^{\mathbb{F}(U_1)}$	Partition Space $PS_{\mathbb{F}(U_2)}$	$p^{\mathbb{F}(U_2)}$
$\langle f_{11}\rangle = \{\{a, d\}, \{b\}, \{c, e\}\}$	$\frac{1}{3}$	$\langle f_{21}\rangle = \{\{b, d\}, \{e, k\}, \{f, g\}\}$	$\frac{2}{5}$
$\langle f_{12}\rangle = \{\{a, b, e\}, \{c, d\}\}$	$\frac{1}{2}$	$\langle f_{22}\rangle = \{\{b, e\}, \{d, f, g, k\}\}$	$\frac{1}{3}$
$\langle f_{13}\rangle = \{\{b, c\}, \{a, d, e\}\}$	$\frac{1}{6}$	$\langle f_{23}\rangle = \{\{b, f, g\}, \{d, e, k\}\}$	$\frac{2}{9}$
		$\langle f_{24}\rangle = \{\{b\}, \{e, k\}, \{d, g\}, \{f\}\}$	$\frac{2}{45}$

We say a target set X is large in $\mathbb{U}$ iff for all $U \in \mathbb{U}, X \not\subseteq U$ and $X \subseteq \cup\mathbb{U}$. Every ordered set $S = (S_1, S_2)$ with $S_1 \subseteq S_2$ is called a rough set. In order to characterize large target set, we need to define some operations over rough sets. For any rough set $S = (S_1, S_2)$, define the lower bound $lb(S) = S_1$ and the upper bound $ub(S) = S_2$.

Definition 11. For any rought sets S and R, define a rough union operation $\cup^r$ by $S \cup^r R := (lb(S) \cup lb(R), ub(S) \cup ub(R))$.

Let $\mathbb{R} = \{R_1, R_2, ..., R_n\}$ be a set of rough sets. Define $\cup^r\mathbb{R} := (\cup\{lb(R) : R \in \mathbb{R}\}, \cup\{ub(R) : R \in \mathbb{R}\})$. Since we are going to reduce the large target set X to smaller pieces, the relation between the pieces and the universes and classifiers must be specified. An ordered set $Y = (Y_1, Y_2, ..., Y_n)$ is an ordered cover of $Z = (Z_1, Z_2, ..., Z_n)$(or $Z \propto Y$) iff $Z_i \subseteq Y_i$ for all $i \in \{1, 2, ..., n\}$. If Y is not an ordered cover of Z, we use the notation $Z \not\propto Y$. Let us take Example 4 for example. Then $(\{a, b\}, \{d, g\}) \propto (U_1, U_2)$, but $(\{d, g\}, \{a, b\}) \not\propto (U_1, U_2)$.

Definition 12. For any set S, we define the set of the covers in S as $Cov(S) := \{Y \in \mathcal{P}(S) : \cup Y = S.\}$

Every cover in X indeed associates a way to reduce the large target set X and $Cov(X)$ is the set of all the possible reductions. However, not all the reductions are feasible for the characterizations. It will depend on the universes and information available.

Definition 13. For every $J \in Cov(X)$, let $J_{\sigma(J)}$ be the ordered set of J ordered by $\sigma(J)$. We call $J_{\sigma(J)}$ an ordered covered cover of X(with respect to $\mathbb{U}$) if and only if there exists a $\mathbb{C} \in \prod_{1}^{|J|}\mathbb{U}$, where $\rho^{(m)}(\mathbb{C}) \neq \rho^{(n)}(\mathbb{C})$ for all $m, n \in |\mathbb{C}|^{\downarrow}$, such that $(\rho^{(1)}(J_{\sigma(J)}), \rho^{(2)}(J_{\sigma(J)}), ..., \rho^{(|J|)}(J_{\sigma(J)})) \propto (\rho^{(1)}(\mathbb{C}), \rho^{(2)}(\mathbb{C}), ..., \rho^{(|J|)}(\mathbb{C}))$. If $J_{\sigma(J)}$ is an ordered covered cover of X, we say J is a covered cover of X. We call such $\mathbb{C}$ an ordered covering cover of $J_{\sigma(J)}$ and define the set of all the ordered covering covers of $J_{\sigma(J)}$ to be $J^{\propto}_{\sigma(J)}$ Furthermore, define the set of all the covered covers of X to be $CCov(X)$ and the set of all the ordered covered covers of X to be $CCov^{\sigma}(X)$.

Example 5. Take Example 4 for example. Let the target set be $X = \{a, b, d, g\}$. We have every covered covers of X and the set of its covering covers:

Table 5. Ordered Covering Covers of Covered Covers of X

$CCov(X)$	$CCov^{\sigma}(X)$	ordered covering covers
$J1 = \{\{a\}, \{b, d, g\}\}$	$J1_{\sigma(J1)} = (\{a\}, \{b, d, g\})$	$J1^{\propto}_{\sigma(J1)} = \{\mathbb{C} \equiv (U_1, U_2)\}$
$J2 = \{\{a, b\}, \{d, g\}\}$	$J2_{\sigma(J2)} = (\{a, b\}, \{d, g\})$	$J2^{\propto}_{\sigma(J2)} = \{\mathbb{C} \equiv (U_1, U_2)\}$
$J3 = \{\{a, d\}, \{b, g\}\}$	$J3_{\sigma(J3)} = (\{a, d\}, \{b, g\})$	$J3^{\propto}_{\sigma(J3)} = \{\mathbb{C} \equiv (U_1, U_2)\}$
$J4 = \{\{a, b, d\}, \{g\}\}$	$J4_{\sigma(J4)} = (\{a, b, d\}, \{g\})$	$J4^{\propto}_{\sigma(J4)} = \{\mathbb{C} \equiv (U_1, U_2)\}$
$J5 = \{\{a, b\}, \{b, d, g\}\}$	$J5_{\sigma(J5)} = (\{a, b\}, \{b, d, g\})$	$J5^{\propto}_{\sigma(J5)} = \{\mathbb{C} \equiv (U_1, U_2)\}$
$J6 = \{\{a, b, d\}, \{b, g\}\}$	$J6_{\sigma(J6)} = (\{a, b, d\}, \{b, g\})$	$J6^{\propto}_{\sigma(J6)} = \{\mathbb{C} \equiv (U_1, U_2)\}$
$J7 = \{\{a, d\}, \{b, d, g\}\}$	$J7_{\sigma(J7)} = (\{a, d\}, \{b, d, g\})$	$J7^{\propto}_{\sigma(J7)} = \{\mathbb{C} \equiv (U_1, U_2)\}$
$J8 = \{\{a, b, d\}, \{d, g\}\}$	$J8_{\sigma(J8)} = (\{a, b, d\}, \{d, g\})$	$J8^{\propto}_{\sigma(J8)} = \{\mathbb{C} \equiv (U_1, U_2)\}$
$J9 = \{\{a, b, d\}, \{b, d, g\}\}$	$J9_{\sigma(J9)} = (\{a, b, d\}, \{b, d, g\})$	$J9^{\propto}_{\sigma(J9)} = \{\mathbb{C} \equiv (U_1, U_2)\}$

Since each ordered covering cover is an ordered set of universes, it induces an ordered set of partition spaces. These partition spaces will serve as the classifiers to characterize the large target set X. Let $J \in CCov(X)$ be arbitrary. Let $\mathbb{C} \in J^{\propto}_{\sigma(J)}$ be arbitrary. Up to now, we have characterized the reduced parts of the large target set X. The next step is to merge them to form a rough set for X.

Definition 14. Define $CH^{\mathbb{C}}_{J_{\sigma(J)}} := \prod^{n \in |\mathbb{C}|^{\downarrow}} Ch^{\rho^{(n)}(\mathbb{C})}_{\rho^{(n)}(J_{\sigma(J)})}(\mathbb{F}(\rho^{(n)}(\mathbb{C})))$.

Let $S = (S_1, S_2, ..., S_n)$ be an ordered set with each element being an ordered set. We define an ordered union operation $\overset{\circ}{U}S := \cup^{r}\{S_1, S_2, ..., S_n\}$ and this will be applied to combine a set of rough sets to form a combined rough space $CRS_{J_{\sigma(J)}}(\mathbb{C}) = \{\mathring{\cup} Y : Y \in CH^{\mathbb{C}}_{J_{\sigma(J)}}\}$.

Definition 15. Define $OU^{\mathbb{C}}_{J_{\sigma(J)}} : CH^{\mathbb{C}}_{J_{\sigma(J)}} \rightarrow CRS_{J_{\sigma(J)}}(\mathbb{C})$ by $OU^{\mathbb{C}}_{J_{\sigma(J)}}(\boldsymbol{R}) := \mathring{\cup}\boldsymbol{R}$.

Definition 16. Define $OU^{\mathbb{C}^{-1}}_{J_{\sigma(J)}} : CRS_{J_{\sigma(J)}}(\mathbb{C}) \rightarrow \mathcal{P}(CH^{\mathbb{C}}_{J_{\sigma(J)}})$ by $OU^{\mathbb{C}^{-1}}_{J_{\sigma(J)}}(Y) := \{\boldsymbol{R} \in CH^{\mathbb{C}}_{J_{\sigma(J)}} : OU^{\mathbb{C}}_{J_{\sigma(J)}}(\boldsymbol{R}) = Y\}$.

Definition 17. We then define a product probability $pp^{\mathbb{C}}_{J_{\sigma(J)}} : CH^{\mathbb{C}}_{J_{\sigma(J)}} \rightarrow [0,1]$ by $pp^{\mathbb{C}}_{J_{\sigma(J)}}(\boldsymbol{R}) := \prod^{n \in |\mathbb{C}|^{\downarrow}} p^{\rho^{(n)}(J_{\sigma(J)})}(\rho^{(n)}(\boldsymbol{R}))$.

Then the probability distribution $\tau\tau^{\mathbb{C}}_{J} : CRS_{J_{\sigma(J)}}(\mathbb{C}) \rightarrow [0,1]$ is defined as follows:

Definition 18. $\tau\tau^{\mathbb{C}}_{J}(Y) := \sum^{\boldsymbol{R} \in OU^{\mathbb{C}^{-1}}_{J_{\sigma(J)}}(Y)} pp^{\mathbb{C}}_{J_{\sigma(J)}}(\boldsymbol{R})$.

Let us take $\mathbb{C} = (U_1, U_2) \in J5^{\propto}_{\sigma(J5)}$ in Example 5 for example. In this case, the large target set X is reduced to P_1 and P_2, where P_1 is $\rho^{(1)}(J5_{\sigma(J5)})$ and P_2 is $\rho^{(2)}(J5_{\sigma(J5)})$. Then one has $Ch^{U_1}_{P_1}(\mathbb{F}(U_1)) = \{\boldsymbol{R}_1 = (\{b\}, \{a,b,d\}), \boldsymbol{R}_2 = (\emptyset, \{a,b,e\}), \boldsymbol{R}_3 = (\emptyset, \{a,b,c,d,e\})\}$ and $p^{P_1}(\boldsymbol{R}_1) = \frac{1}{3}, p^{P_1}(\boldsymbol{R}_2) = \frac{1}{2}, p^{P_1}(\boldsymbol{R}_3) = \frac{1}{6}$; $Ch^{U_2}_{P_2}(\mathbb{F}(U_2)) = \{\boldsymbol{H}_1 = (\{b,d\}, \{b,d,f,g\}), \boldsymbol{H}_2 = (\emptyset, \{b,d,e,f,g,k\}), \boldsymbol{H}_3 = (\{b,d,g\}, \{b,d,g\})\}$, furthermore, $p^{P_2}(\boldsymbol{H}_1) = \frac{2}{5}, p^{P_2}(\boldsymbol{H}_2) = \frac{5}{9}, p^{P_2}(\boldsymbol{H}_3) = \frac{2}{45}$. Based on the this result, we have the characterization space of $J5$ as follows:

Table 6. Multiple Probabilistic Characterization Space:$J5$

$Ch^{U_1}_{P_1}(\mathbb{F}(U_1)) \times Ch^{U_2}_{P_2}(\mathbb{F}(U_2))$	$CRS_{J5}(\mathbb{C})$	$\tau\tau^{\mathbb{C}}_{J5}$
$((\{b\},\{a,b,d\}),(\{b,d\},\{b,d,f,g\}))$	$(\{b,d\},\{a,b,d,f,g\})$	$\frac{1}{3} \times \frac{2}{5} = \frac{2}{15}$
$((\{b\},\{a,b,d\}),(\emptyset,\{b,d,e,f,g,k\}))$	$(\{b\},\{a,b,d,e,f,g,k\})$	$\frac{1}{3} \times \frac{5}{9} = \frac{5}{27}$
$((\{b\},\{a,b,d\}),(\{b,d,g\},\{b,d,g\}))$	$(\{b,d,g\},\{a,b,d,g\})$	$\frac{1}{3} \times \frac{2}{45} = \frac{2}{135}$
$((\emptyset,\{a,b,e\}),(\{b,d\},\{b,d,f,g\}))$	$(\{b,d\},\{a,b,d,e,f,g\})$	$\frac{1}{2} \times \frac{2}{5} = \frac{1}{5}$
$((\emptyset,\{a,b,e\}),(\emptyset,\{b,d,e,f,g,k\}))$	$(\emptyset,\{a,b,d,e,f,g,k\})$	$\frac{1}{2} \times \frac{5}{9} = \frac{5}{18}$
$((\emptyset,\{a,b,e\}),(\{b,d,g\},\{b,d,g\}))$	$(\{b,d,g\},\{a,b,d,e,g\})$	$\frac{1}{2} \times \frac{2}{45} = \frac{1}{45}$
$((\emptyset,\{a,b,c,d,e\}),(\{b,d\},\{b,d,f,g\}))$	$(\{b,d\},\{a,b,c,d,e,f,g\})$	$\frac{1}{6} \times \frac{2}{5} = \frac{1}{15}$
$((\emptyset,\{a,b,c,d,e\}),(\emptyset,\{b,d,e,f,g,k\}))$	$(\emptyset,\{a,b,c,d,e,f,g,k\})$	$\frac{1}{6} \times \frac{5}{9} = \frac{5}{54}$
$((\emptyset,\{a,b,c,d,e\}),(\{b,d,g\},\{b,d,g\}))$	$(\{b,d,g\},\{a,b,c,d,e,g\})$	$\frac{1}{6} \times \frac{2}{45} = \frac{1}{135}$

Example 6. Repeating the above process for other covered covers of X, we have the following table ($J5$ is shown above):

$CRS_{J1_{\sigma(J1)}}(\mathbb{C})$	$\tau\tau_{J1}^{\mathbb{C}}$	$CRS_{J2_{\sigma(J2)}}(\mathbb{C})$	$\tau\tau_{J2}^{\mathbb{C}}$
$(\{b,d\},\{a,b,d,f,g\})$	$\frac{2}{15}$	$(\{b\},\{a,b,d,f,g\})$	$\frac{2}{15}$
$(\emptyset,\{a,b,d,e,f,g,k\})$	$\frac{5}{9}$	$(\{b\},\{a,b,d,f,g,k\})$	$\frac{1}{9}$
$(\{b,d,g\},\{a,b,d,g\})$	$\frac{2}{135}$	$(\{b\},\{a,b,d,e,f,g,k\})$	$\frac{2}{27}$
$(\{b,d\},\{a,b,d,e,f,g\})$	$\frac{4}{15}$	$(\{b,d,g\},\{a,b,d,g\})$	$\frac{2}{135}$
$(\{b,d,g\},\{a,b,d,e,g\})$	$\frac{4}{135}$	$(\emptyset,\{a,b,d,e,f,g\})$	$\frac{1}{5}$
		$(\emptyset,\{a,b,d,e,f,g,k\})$	$\frac{5}{18}$
		$(\{d,g\},\{a,b,d,e,g\})$	$\frac{1}{45}$
		$(\emptyset,\{a,b,c,d,e,f,g\})$	$\frac{1}{15}$
		$(\emptyset,\{a,b,c,d,e,f,g,k\})$	$\frac{5}{54}$
		$(\{d,g\},\{a,b,c,d,e,g\})$	$\frac{1}{135}$

$CRS_{J3_{\sigma(J3)}}(\mathbb{C})$	$\tau\tau_{J3}^{\mathbb{C}}$	$CRS_{J4_{\sigma(J4)}}(\mathbb{C})$	$\tau\tau_{J4}^{\mathbb{C}}$
$(\{a,d\},\{a,b,d,f,g\})$	$\frac{28}{135}$	$(\{a,b,d\},\{a,b,d,f,g\})$	$\frac{28}{135}$
$(\{a,d\},\{a,b,d,e,f,g,k\})$	$\frac{1}{9}$	$(\{a,b,d\},\{a,b,d,f,g,k\})$	$\frac{1}{9}$
$(\{a,b,d\},\{a,b,d,g\})$	$\frac{2}{135}$	$(\{a,b,d\},\{a,b,d,g\})$	$\frac{2}{135}$
$(\emptyset,\{a,b,c,d,e,f,g\})$	$\frac{14}{45}$	$(\emptyset,\{a,b,c,d,e,f,g\})$	$\frac{56}{135}$
$(\emptyset,\{a,b,c,d,e,f,g,k\})$	$\frac{1}{6}$	$(\emptyset,\{a,b,c,d,e,f,g,k\})$	$\frac{2}{9}$
$(\{b\},\{a,b,c,d,e,g\})$	$\frac{1}{45}$	$(\emptyset,\{a,b,c,d,e,g\})$	$\frac{4}{135}$
$(\emptyset,\{a,b,d,e,f,g\})$	$\frac{14}{135}$		
$(\emptyset,\{a,b,d,e,f,g,k\})$	$\frac{1}{18}$		
$(\{b\},\{a,b,d,e,g\})$	$\frac{1}{135}$		

$CRS_{J6_{\sigma(J6)}}(\mathbb{C})$	$\tau\tau_{J6}^{\mathbb{C}}$	$CRS_{J7_{\sigma(J7)}}(\mathbb{C})$	$\tau\tau_{J7}^{\mathbb{C}}$
$(\{a,b,d\},\{a,b,d,f,g\})$	$\frac{28}{135}$	$(\{a,b,d\},\{a,b,d,f,g\})$	$\frac{2}{15}$
$(\{a,b,d\},\{a,b,d,e,f,g,k\})$	$\frac{1}{9}$	$(\{a,d\},\{a,b,d,e,f,g,k\})$	$\frac{5}{27}$
$(\{a,b,d\},\{a,b,d,g\})$	$\frac{2}{135}$	$(\{a,b,d,g\},\{a,b,d,g\})$	$\frac{2}{135}$
$(\emptyset,\{a,b,c,d,e,f,g\})$	$\frac{56}{135}$	$(\{b,d\},\{a,b,c,d,e,f,g\})$	$\frac{1}{5}$
$(\emptyset,\{a,b,c,d,e,f,g,k\})$	$\frac{2}{9}$	$(\emptyset,\{a,b,c,d,e,f,g,k\})$	$\frac{5}{18}$
$(\{b\},\{a,b,c,d,e,g\})$	$\frac{4}{135}$	$(\{b,d,g\},\{a,b,c,d,e,g\})$	$\frac{1}{45}$
		$(\{b,d\},\{a,b,d,e,f,g\})$	$\frac{1}{15}$
		$(\emptyset,\{a,b,d,e,f,g,k\})$	$\frac{5}{54}$
		$(\{b,d,g\},\{a,b,d,e,g\})$	$\frac{1}{135}$

$CRS_{J8_{\sigma(J8)}}(\mathbb{C})$	$\tau\tau_{J8}^{\mathbb{C}}$	$CRS_{J9_{\sigma(J9)}}(\mathbb{C})$	$\tau\tau_{J9}^{\mathbb{C}}$
$(\{a,b,d\},\{a,b,d,f,g\})$	$\frac{2}{15}$	$(\{a,b,d\},\{a,b,d,f,g\})$	$\frac{2}{15}$
$(\{a,b,d\},\{a,b,d,f,g,k\})$	$\frac{1}{9}$	$(\{a,b,d\},\{a,b,d,e,f,g,k\})$	$\frac{5}{27}$
$(\{a,b,d\},\{a,b,d,e,f,g,k\})$	$\frac{2}{27}$	$(\{a,b,d,g\},\{a,b,d,g\})$	$\frac{2}{135}$
$(\{a,b,d,g\},\{a,b,d,g\})$	$\frac{2}{135}$	$(\{b,d\},\{a,b,c,d,e,f,g\})$	$\frac{4}{15}$
$(\emptyset,\{a,b,c,d,e,f,g\})$	$\frac{4}{15}$	$(\emptyset,\{a,b,c,d,e,f,g,k\})$	$\frac{10}{27}$
$(\emptyset,\{a,b,c,d,e,f,g,k\})$	$\frac{10}{27}$	$(\{b,d,g\},\{a,b,c,d,e,g\})$	$\frac{4}{135}$
$(\{d,g\},\{a,b,c,d,e,g\})$	$\frac{4}{135}$		

5 Optimal Problem

Normally there are more than one covered cover of the target set X. One faces the problem which covered cover does the characterization better. In this section, we come up with two approaches for such optimal problem: Shannon entropy approach and average precision approach. One can even assign some weights to these approaches or modify the criteria of the choice to accommodate his needs.

5.1 Shannon Entropy Approach

In this approach, one only cares about the amount of information contained in the characterizations. The values of the characterizations play no role. For any probability distribution p, let $Entropy(p)$ be its Shannon entropy. Then one finds the best covered cover J of a target set X based on $\underset{J \in CCov(X)}{min} \underset{\mathbb{C} \in J^{\propto}_{\sigma(J)}}{min} Entropy(\tau\tau^{\mathbb{C}}_{J})$.

Example 7. Take Example 6 for example.

$CCov(X)$	$Entropy(\tau\tau^{\mathbb{C}}_{J})$
$J1$	$\frac{74}{45}log_2 3 - \frac{1}{9}log_2 5 - \frac{20}{27} \doteq 1.6077$
$J2$	$\frac{143}{90}log_2 3 + \frac{2}{27}log_2 5 + \frac{4}{27} \doteq 2.8385$
$J3$	$\frac{13}{6}log_2 3 + \frac{2}{3}log_2 5 - \frac{28}{45}log_2 7 - \frac{28}{45} \doteq 2.6130$
$J4$	$\frac{8}{3}log_2 3 + \frac{2}{3}log_2 5 - \frac{28}{45}log_2 7 - \frac{88}{45} \doteq 2.0722$
$J5$	$\frac{17}{10}log_2 3 - \frac{1}{9}log_2 5 + \frac{2}{9} \doteq 2.6587$
$J6$	$\frac{8}{3}log_2 3 + \frac{2}{3}log_2 5 - \frac{28}{45}log_2 7 - \frac{88}{45} \doteq 2.0722$
$J7$	$\frac{17}{10}log_2 3 - \frac{1}{9}log_2 5 + \frac{2}{9} \doteq 2.6587$
$J8$	$\frac{94}{45}log_2 3 + \frac{2}{27}log_2 5 - \frac{32}{27} \doteq 2.2976$
$J9$	$\frac{11}{5}log_2 3 - \frac{1}{9}log_2 5 - \frac{10}{9} \doteq 2.1178$

One finds the best covered cover of X is $J1$ and the optimal probabilistic rough space $(CRS_{J1_{\sigma(J1)}}(\mathbb{C}), \tau\tau^{\mathbb{C}}_{J1})$.

5.2 Average Precision Approach

This approach will take the precision of each characterization into consideration. Based on the concept of an expected value, we define the average precision of a covered cover J of X with a covering cover $\mathbb{C}$

Definition 19. $AP_J(\mathbb{C}) = \sum^{R \in CRS_{J_{\sigma(J)}}} \tau\tau^{\mathbb{C}}_{J_{\sigma(J)}}(R) \times \frac{|\rho^{(2)}(R) - \rho^{(1)}(R)|}{|\rho^{(2)}(R)|}$.

Then one finds the best covered cover J of X based on $\underset{J \in CCov(X)}{max} \underset{\mathbb{C} \in J^{\propto}_{\sigma(J)}}{max} AP_J(\mathbb{C})$.

Example 8. Take Example 6 for example.

Table 7. Average Precision

$CCov(X)$	Average Precision, $AP(\mathbb{C})$
$J1$	$\frac{2}{15}\times\frac{2}{5}+\frac{5}{9}\times 0+\frac{2}{135}\times\frac{3}{4}+\frac{4}{15}\times\frac{2}{6}+\frac{4}{135}\times\frac{3}{5}=\frac{77}{450}\doteq 0.1711$
$J2$	$\frac{2}{15}\times\frac{1}{5}+\frac{1}{9}\times\frac{1}{6}+\frac{2}{27}\times\frac{1}{7}+\frac{2}{135}\times\frac{3}{4}+\frac{1}{5}\times 0+\frac{5}{18}\times 0+$ $\frac{1}{45}\times\frac{2}{5}+\frac{1}{15}\times 0+\frac{5}{54}\times 0+\frac{1}{135}\times\frac{2}{6}=\frac{1109}{14175}\doteq 0.0782$
$J3$	$\frac{28}{135}\times\frac{2}{5}+\frac{1}{9}\times\frac{2}{7}+\frac{2}{135}\times\frac{3}{4}+\frac{14}{45}\times 0+\frac{1}{6}\times 0+\frac{1}{45}\times\frac{1}{6}+$ $\frac{14}{135}\times 0+\frac{1}{18}\times 0+\frac{1}{135}\times\frac{1}{5}=\frac{619}{4725}\doteq 0.1310$
$J4$	$\frac{28}{135}\times\frac{3}{5}+\frac{1}{9}\times\frac{3}{6}+\frac{2}{135}\times\frac{3}{4}+\frac{56}{135}\times 0+\frac{2}{9}\times 0+\frac{4}{135}\times 0=\frac{43}{225}\doteq 0.1911$
$J5$	$\frac{2}{15}\times\frac{2}{5}+\frac{5}{27}\times\frac{1}{7}+\frac{2}{135}\times\frac{3}{4}+\frac{1}{5}\times\frac{2}{6}+\frac{5}{18}\times 0+\frac{1}{45}\times\frac{3}{5}+$ $\frac{1}{15}\times\frac{2}{7}+\frac{5}{54}\times 0+\frac{1}{135}\times\frac{3}{6}=\frac{61}{315}\doteq 0.1937$
$J6$	$\frac{28}{135}\times\frac{3}{5}+\frac{1}{9}\times\frac{3}{7}+\frac{2}{135}\times\frac{3}{4}+\frac{56}{135}\times 0+\frac{2}{9}\times 0+\frac{4}{135}\times\frac{1}{6}=\frac{5333}{28350}\doteq 0.1881$
$J7$	$\frac{2}{15}\times\frac{3}{5}+\frac{5}{27}\times\frac{2}{7}+\frac{2}{135}\times 1+\frac{1}{5}\times\frac{2}{7}+\frac{5}{18}\times 0+\frac{1}{45}\times\frac{3}{6}+$ $\frac{1}{15}\times\frac{2}{6}+\frac{5}{54}\times 0+\frac{1}{135}\times\frac{3}{5}=\frac{2293}{9450}\doteq 0.2426$
$J8$	$\frac{2}{15}\times\frac{3}{5}+\frac{1}{9}\times\frac{3}{6}+\frac{2}{27}\times\frac{3}{7}+\frac{2}{135}\times 1+\frac{4}{15}\times 0+\frac{10}{27}\times 0+\frac{4}{135}\times\frac{2}{6}=\frac{5443}{28350}\doteq 0.1920$
$J9$	$\frac{2}{15}\times\frac{3}{5}+\frac{5}{27}\times\frac{3}{7}+\frac{2}{135}\times 1+\frac{4}{15}\times\frac{2}{7}+\frac{10}{27}\times 0+\frac{4}{135}\times\frac{3}{6}=\frac{179}{675}\doteq 0.2652$

One finds the best covered cover of X is $J9$ and the optimal probabilistic rough space $(CRS_{J9_{\sigma(J9)}}(\mathbb{C}), \tau\tau^{\mathbb{C}}_{J9})$.

6 Conclusions and Future Work

I have extended the characterization problem to a large target set with several universes and instances. Such extension enables one to know a new large object by the information one has in hand. There is still some room for further research; for example, the way to combine the rough sets of smaller covered covers, the criteria used to decide the optimal covered cover(s) of the large target set and the assignment of probability distribution or weight to an outcome of a specification.

References

1. Pawlak, Z.: Rough Sets. International Journal of Computer and Information Sciences 11(5) (1982)
2. Pawlak, Z.: Rough Sets: theoretical aspects of reasoning about data. Springer Science+Business Media Dordrecht (1991)
3. Nguyen, H.S., Skowron, A.: Rough Sets: From Rudiments to Challenges. In: Rough Sets and Intelligent Systems - Professor Zdzisław Pawlak in Memoriam, vol. 1, ch. 3. Springer, Heidelberg (2013)
4. Lotfi, A.: Zaden, Fuzzy Sets. Information and Control 8, 338–353 (1965)
5. Chen, R.-M.: Data-based approximation of intuitionistic fuzzy target sets. Notes on Intuitionistic Fuzzy Sets 20(2) (2014)

Data-Based Approximation of Fuzzy Target Sets

Ray-Ming Chen

Department of Computer Science,
Friedrich-Alexander-Universität Erlangen-Nürnberg,
Martenstraße 3, 91058 Erlangen, Germany
ray.chen@fau.de

Abstract. A target set in a usual characterization is always assumed to be a crisp set and so is the the classifier. In this paper, I give an approach to characterize a target set which is a fuzzy set (or a fuzzy target set) based on fuzzy classifiers induced by data in an information system. I also give a way to compare the efficiency of different information systems in characterizing fuzzy target sets.

1 Introduction

In the usual characterization problem, the target set is a crisp set. However, when the target set itself is a fuzzy set, how can we characterize it? In this paper, I come up with an approach to approximate fuzzy target sets and fuzzy target subsets via reasoning on given data in an information system. Furthermore, I assign some orderings to decide which information systems characterize a given fuzzy target sets or subsets better. I also give a way to classify information systems. This paper will generalize my presentation in 18th ICIFS 2014 [3]. In that paper I consider intuitionistic fuzzy target sets, while in this paper, fuzzy target sets. The theories regarding fuzzy sets and intuitionistic fuzzy sets are constructed by Zaden [1] and Atanassov [2], respectively.

In Section 2, I will show how to form fuzzy sets from the data in an information system and how to form a classifier based on these fuzzy sets. Membership depicts the relation between an element and its universe. In the classical sense, it is always assumed that the membership is decidable, i.e., either a set belongs to a given universe or not. But in a computational sense, such membership is not always observable or decidable as an algorithm is not provided. One has to estimate the membership indirectly via some devised algorithms. In this paper, I utilize the information of an information system to estimate such membership. Such estimation implicitly assumes that the elements in an universe behave similarly to the universe, i.e., if one observes or expects the universe to have property A, then its elements are also expected to inherit property A.

Then, in Section 3, I will define the characterization of a fuzzy target set via a fuzzy classifier. I will also provide other ways of characterizations in Section 5. All the properties regarding these characterizations would be studied in Section 4 and 6. In the real world, we have to apply fuzziness in many applications and fields. Fuzzy sets are prevailing everywhere and an approach to characterize or

P. Angelov et al. (eds.), *Intelligent Systems'2014*,
Advances in Intelligent Systems and Computing 322, DOI: 10.1007/978-3-319-11313-5_70

approximate such objects is necessary. The main purpose of this paper is to offer and explore some approaches to facilitate such characterization.

2 Fuzzy Classifier: Data-Driven Fuzzy Sets

In the usual characterization problem, the target set is supposed to be a crisp set. In this paper, we extend the target set to a fuzzy set, a fuzzy subset or even a set of fuzzy subsets. In a real world, one normally has to do his reasoning based on available data. The question is: given an information system, how can one classify fuzzy target sets based on this? First of all, we have to construct a classifier based on the available data. Then one can use this classifier to classify the fuzzy target sets. For any ordered set $Y = (Y_1, Y_2, ..., Y_k)$, we use the projection notation $\rho^{(m)}(Y)$ to denote Y_m, if $m \leq k$.
Let U be a universe. Let $\mathfrak{A} = \{a_1, a_2, ..., a_n\}$ be a set of attributes and $\mathbb{A} = \{\mathcal{B}_1, \mathcal{B}_2, ..., \mathcal{B}_n\}$ be a set of domains of attribute values (numerical or non-numerical) of $\mathfrak{A}$. Let $\mathcal{N} = \{1, 2, ..., n\}$. Let $\mathcal{I}^{\mathfrak{A}} : U \rightarrow \prod_{i \in \mathcal{N}} \mathcal{B}_i$ be an information system which assigns the attribute values to each element $u \in U$. Then one forms the data-driven fuzzy sets (i.e., a membership function) $\mathcal{M}^{\mathcal{I}^{\mathfrak{A}}} : \prod_{i \in \mathcal{N}} \mathcal{B}_i \rightarrow [0, 1]^U$, where $[0, 1]^U$ denotes all the fuzzy sets with the domain U, as follows:

$$(\mathcal{M}^{\mathcal{I}^{\mathfrak{A}}}(\mathcal{A}))(u) := \frac{|\{m \in \mathcal{N} : \rho^{(m)}(\mathcal{I}^{\mathfrak{A}}(u)) = \mathcal{A}(m)\}|}{|\mathfrak{A}|},$$

where $\mathcal{A}(m)$ denotes the m'th component of $\mathcal{A}$. $\mathcal{M}^{\mathcal{I}^{\mathfrak{A}}}(\mathcal{A})$ represents U in terms of $\mathcal{A}$. If $\mathcal{I}^{\mathfrak{A}}$ is known from the context, then I simply use $\mathcal{M}(\mathcal{A})$ to denote $\mathcal{M}^{\mathcal{I}^{\mathfrak{A}}}(\mathcal{A})$.

Now I want to classify any arbitrary fuzzy target set in terms of all the $\mathcal{M}^{\mathcal{I}^{\mathfrak{A}}}(\mathcal{A})$. With this specification-based classifiers, one can classify all the fuzzy target set via these classifiers. Let the $\mathcal{I}^{\mathfrak{A}}$-induced classifier be $\mathcal{K}_{\mathcal{I}^{\mathfrak{A}}} = \{\mathcal{M}^{\mathcal{I}^{\mathfrak{A}}}(\mathcal{A}) : \mathcal{A} \in \prod_{i \in \mathcal{N}} \mathcal{B}_i\}$. $\mathcal{K}_{\mathcal{I}^{\mathfrak{A}}}$ will serve as a classifier to classify the fuzzy target sets. Each fuzzy set $f \in [0, 1]^U$ can be interpreted as U is represented in some attributes in some information systems. The characterization of f by $\mathcal{K}_{\mathcal{I}^{\mathfrak{A}}}$ will show the relation between its attributes related to f and the ones in the current information system.

Example 1. *Suppose the universe* $U = \{u_1, u_2, u_3, u_4, u_5, u_6\}$, $\mathfrak{A} = \{a_1, a_2, a_3\}$, $\mathbb{A} = \{\mathcal{B}_1, \mathcal{B}_2, \mathcal{B}_3\}$, *where* $\mathcal{B}_1 = \{1, 2\}, \mathcal{B}_2 = \{\mathit{low}, \mathit{medium}, \mathit{high}\}, \mathcal{B}_3 = \{\mathit{yes}, \mathit{no}\}$. $\prod_{i \in \{1,2,3\}} \mathcal{B}_i = \{\mathcal{A}_1, \mathcal{A}_2, ..., \mathcal{A}_{12}\}$ *and the information system* $\mathcal{I}^{\mathfrak{A}}$ *are shown as follows:*

Table 1. Information System $\mathcal{I}^{\mathfrak{A}}$

Universe	Attribute a_1	Attribute a_2	Attribute a_3
u_1	1	high	yes
u_2	2	low	no
u_3	1	low	no
u_4	2	medium	yes
u_5	1	high	yes
u_6	1	low	yes

Based on the information system $\mathcal{I}^{\mathfrak{A}}$, one forms the fuzzy knowledge of each element in U. Take $\mathcal{A}_5 = \{1, low, yes\}$ for example. The universe U would be represented by $\mathcal{A}_5$ as follows:

$$\mathcal{M}^{\mathcal{I}^{\mathfrak{A}}}(\mathcal{A}_5) = (\frac{|\{1, yes\}|}{3}, \frac{|\{low\}|}{3}, \frac{\{1, low\}}{3}, \frac{|\{yes\}|}{3}, \frac{|\{1, yes\}|}{3}, \frac{|\{l, low, yes\}|}{3})$$

If one continues the same process, he has the $\mathcal{I}^{\mathfrak{A}}$-induced classifier $\mathcal{K}_{\mathcal{I}^{\mathfrak{A}}}$ as follows:

Table 2. $\mathcal{I}$-induced classifier $\mathcal{K}_{\mathcal{I}^{\mathfrak{A}}}$

Attributes	(a_1, a_2, a_3)	Classifier $\mathcal{K}_{\mathcal{I}^{\mathfrak{A}}}$
$\mathcal{A}_1$	(1, high, yes)	$\mathcal{M}(\mathcal{A}_1) = (1, 0, \frac{1}{3}, \frac{1}{3}, 1, \frac{2}{3})$
$\mathcal{A}_2$	(1, high, no)	$\mathcal{M}(\mathcal{A}_2) = (\frac{2}{3}, \frac{1}{3}, \frac{2}{3}, 0, \frac{2}{3}, \frac{1}{3})$
$\mathcal{A}_3$	(1, medium, yes)	$\mathcal{M}(\mathcal{A}_3) = (\frac{2}{3}, 0, \frac{1}{3}, \frac{2}{3}, \frac{2}{3}, \frac{2}{3})$
$\mathcal{A}_4$	(1, medium, no)	$\mathcal{M}(\mathcal{A}_4) = (\frac{1}{3}, \frac{1}{3}, \frac{2}{3}, \frac{1}{3}, \frac{1}{3}, \frac{1}{3})$
$\mathcal{A}_5$	(1, low, yes)	$\mathcal{M}(\mathcal{A}_5) = (\frac{2}{3}, \frac{1}{3}, \frac{2}{3}, \frac{1}{3}, \frac{2}{3}, 1)$
$\mathcal{A}_6$	(1, low, no)	$\mathcal{M}(\mathcal{A}_6) = (\frac{1}{3}, \frac{2}{3}, 1, 0, \frac{1}{3}, \frac{2}{3})$
$\mathcal{A}_7$	(2, high, yes)	$\mathcal{M}(\mathcal{A}_7) = (\frac{2}{3}, \frac{1}{3}, 0, \frac{2}{3}, \frac{2}{3}, \frac{1}{3})$
$\mathcal{A}_8$	(2, high, no)	$\mathcal{M}(\mathcal{A}_8) = (\frac{1}{3}, \frac{2}{3}, \frac{1}{3}, \frac{1}{3}, \frac{1}{3}, 0)$
$\mathcal{A}_9$	(2, medium, yes)	$\mathcal{M}(\mathcal{A}_9) = (\frac{1}{3}, \frac{1}{3}, 0, 1, \frac{1}{3}, \frac{1}{3})$
$\mathcal{A}_{10}$	(2, medium, no)	$\mathcal{M}(\mathcal{A}_{10}) = (0, \frac{2}{3}, \frac{1}{3}, \frac{2}{3}, 0, 0)$
$\mathcal{A}_{11}$	(2, low, yes)	$\mathcal{M}(\mathcal{A}_{11}) = (\frac{1}{3}, \frac{2}{3}, \frac{1}{3}, \frac{2}{3}, \frac{1}{3}, \frac{2}{3})$
$\mathcal{A}_{12}$	(2, low, no)	$\mathcal{M}(\mathcal{A}_{12}) = (0, 1, \frac{2}{3}, \frac{1}{3}, 0, \frac{1}{3})$

This following example gives an application of data-based approximation of a fuzzy target set.

Example 2. *Suppose there is a panel consisting of five judges $U = \{u_1, u_2, u_3, u_4, u_5\}$ in a gymnastic competition $\mathfrak{A} = \{$unven bars, balance beam, floor exercises$\}$, and the scoring techniques are $\mathbb{A} = \{\mathcal{B}_1 = \{b_{11} \equiv$ high bar, $b_{12} \equiv$ low bar$\}, \mathcal{B}_2 = \{b_{21} \equiv$ one arm, $b_{22} \equiv$ two arms, $b_{23} \equiv$ one leg$\}, \mathcal{B}_3 = \{b_{31} \equiv$ bounce, $b_{32} \equiv$ spring$\}\}$. Then there are $\prod_{i \in \{1,2,3\}} \mathcal{B}_i = \{\mathcal{A}_1, \mathcal{A}_2, ..., \mathcal{A}_{12}\}$ possible performances* *for the judges. Each judge assigns a value $k \in [0,1]$ to a performance and each $\mathcal{M}(\mathcal{A}_i)$ records the judges' scores with respect to the performance $\mathcal{A}_i$. The results are detailed as follows:*

Table 3. Performance-induced Classifier $\mathcal{K}_{\mathcal{I}^{\mathfrak{A}}}$

Performances	Classifier (scores), $\mathcal{K}_{\mathcal{I}^{\mathfrak{A}}}$
$\mathcal{A}_1 = (b_{11}, b_{21}, b_{31})$	$\mathcal{M}(\mathcal{A}_1) = (0.98, 0.92, 0.97, 0.88, 1.00)$
$\mathcal{A}_2 = (b_{11}, b_{21}, b_{32})$	$\mathcal{M}(\mathcal{A}_2) = (0.78, 0.87, 0.67, 0.88, 0.59)$
$\mathcal{A}_3 = (b_{11}, b_{22}, b_{31})$	$\mathcal{M}(\mathcal{A}_3) = (0.45, 0.76, 0.89, 0.77, 0.59)$
$\mathcal{A}_4 = (b_{11}, b_{22}, b_{32})$	$\mathcal{M}(\mathcal{A}_4) = (0.98, 0.89, 0.88, 0.98, 0.81)$
$\mathcal{A}_5 = (b_{11}, b_{23}, b_{31})$	$\mathcal{M}(\mathcal{A}_5) = (0.29, 0.65, 0.32, 0.70, 0.19)$
$\mathcal{A}_6 = (b_{11}, b_{23}, b_{32})$	$\mathcal{M}(\mathcal{A}_6) = (0.44, 0.35, 0.87, 0.40, 0.92)$
$\mathcal{A}_7 = (b_{12}, b_{21}, b_{31})$	$\mathcal{M}(\mathcal{A}_7) = (0.44, 0.45, 0.45, 0.42, 0.46)$
$\mathcal{A}_8 = (b_{12}, b_{21}, b_{32})$	$\mathcal{M}(\mathcal{A}_8) = (0.44, 0.25, 0.37, 0.41, 0.40)$
$\mathcal{A}_9 = (b_{12}, b_{22}, b_{31})$	$\mathcal{M}(\mathcal{A}_9) = (0.94, 0.35, 0.87, 0.80, 0.42)$
$\mathcal{A}_{10} = (b_{12}, b_{22}, b_{32})$	$\mathcal{M}(\mathcal{A}_{10}) = (0.84, 0.75, 0.67, 0.80, 0.92)$
$\mathcal{A}_{11} = (b_{12}, b_{23}, b_{31})$	$\mathcal{M}(\mathcal{A}_{11}) = (0.94, 0.95, 0.97, 0.90, 0.92)$
$\mathcal{A}_{12} = (b_{12}, b_{23}, b_{32})$	$\mathcal{M}(\mathcal{A}_{12}) = (0.41, 0.35, 0.29, 0.40, 0.46)$

Now a female gymnast is expected to get $(0.78, 0.88, 0.68, 0.77, 0.81)$ out of the five judges. Then based on her imperfect information, she could further estimate the closest approximation to his expectation via our approach.

3 Characterizations

Let $[0,1]^U$ denote all the membership functions from the domain U to the range $[0,1]$ and let $\mathbf{0}, \mathbf{1}$ be the membership functions $\mathbf{0}(u) = 0$ and $\mathbf{1}(u) = 1$ for all $u \in U$, respectively.

Definition 1. *For all $\alpha, \beta \in [0,1]^U$, define $\alpha \cup^* \beta$, $\alpha \cap^* \beta$ and $\mathbf{1} -^* \alpha$ as follows:*

1. $(\alpha \cup^* \beta)(n) := max\{\alpha(n), \beta(n)\}$.
2. $(\alpha \cap^* \beta)(n) := min\{\alpha(n), \beta(n)\}$.
3. $(\mathbf{1} -^* \alpha)(n) := 1 - \alpha(n)$.

For any $\mathbb{H} \subseteq [0,1]^U$, we use $\cup^*\mathbb{H}$ to denote the fuzzy set $(\cup^*\mathbb{H})(n) := max\{\alpha(n) : \alpha \in \mathbb{H}\}$; similarly, we use $\cap^*\mathbb{H}$ to denote the fuzzy set $(\cap^*\mathbb{H})(n) := mix\{\alpha(n) : \alpha \in \mathbb{H}\}$;

Definition 2. *For all $\alpha, \beta \in [0,1]^U, \alpha \leq \beta$ iff for all $u \in U, \alpha(u) \leq \beta(u)$.*

Now we want to show how to approximate a fuzzy target set based on the $\mathcal{I}^{\mathfrak{A}}$-induced classifier $\mathcal{K}_{\mathcal{I}^{\mathfrak{A}}}$. We call this approach a uniform approximation. First of all, one identifies the relation $(\leq)$ between the fuzzy target set and the fuzzy sets of the classifier. Then one collects all the fuzzy sets that are larger or equal to the target and then intersects all the fuzzy sets in this set to yield an upper bound for the target; similarly, one collects all the fuzzy sets that are smaller or equal to the target and then unions all the fuzzy sets in this set to yield a lower bound for the target.

Definition 3 (Smallest Upper Bound). *Define the upper bound function* $\underline{ub} : [0,1]^U \to [0,1]^U$ *to be*

$$\underline{ub}(\alpha) := \begin{cases} \cap^* \{\beta \in \mathcal{K}_{\mathcal{I}^{\mathfrak{A}}} : \alpha \leq \beta\}, & \text{if there exists } \beta \in \mathcal{K}_{\mathcal{I}^{\mathfrak{A}}} \text{ such that } \alpha \leq \beta; \\ \mathbf{1}, & \text{otherwise}. \end{cases}$$

Definition 4 (Greatest Lower Bound). *Define the lower bound function* $\overline{lb} : [0,1]^U \to [0,1]^U$ *to be*

$$\overline{lb}(\alpha) := \begin{cases} \cup^* \{\beta \in \mathcal{K}_{\mathcal{I}^{\mathfrak{A}}} : \beta \leq \alpha\}, & \text{if there exists } \beta \in \mathcal{K}_{\mathcal{I}^{\mathfrak{A}}} \text{ such that } \beta \leq \alpha; \\ \mathbf{0}, & \text{otherwise}. \end{cases}$$

Now we can use an ordered pair $(\overline{lb}(\alpha), \underline{ub}(\alpha))$ as bounds for the fuzzy target set α.

Example 3. *The following graph shows the interaction between* α, *the fuzzy target set, and a classifier which consists of a set of data-based-induced fuzzy sets* $\{f_1, f_2, ..., f_8\}$. *The upper bound of* α *is* $\underline{ub}(\alpha) = \alpha^+ = f_1 \cap^* f_2 \cap^* f_4$ *and the lower bound of* α *is* $\overline{lb}(\alpha) = \alpha^- = f_6 \cup^* f_7 \cup^* f_8$.

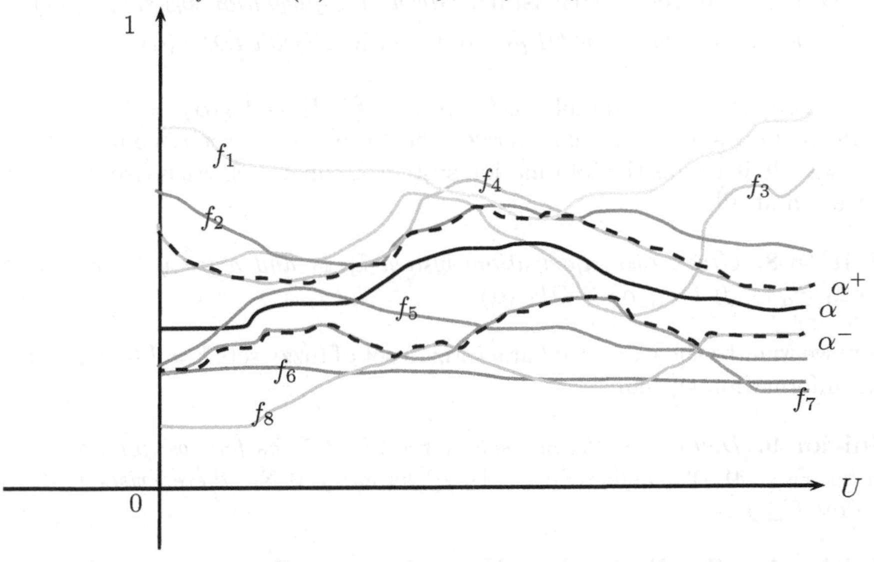

Example 4. *Let us compute the bounds for Example 2. Let the fuzzy target set be* $\alpha = (0.78, 0.88, 0.68, 0.77, 0.81)$. *One then has* $\overline{lb}(\alpha) = \mathcal{M}^{\mathcal{I}^{\mathfrak{A}}}(\mathcal{A}_5) \cap^* \mathcal{M}^{\mathcal{I}^{\mathfrak{A}}}(\mathcal{A}_7) \cap^* \mathcal{M}^{\mathcal{I}^{\mathfrak{A}}}(\mathcal{A}_8) \cap^* \mathcal{M}^{\mathcal{I}^{\mathfrak{A}}}(\mathcal{A}_{12}) = (0.44, 0.65, 0.45, 0.70, 0.46)$ *and* $\underline{ub}(\alpha) = \mathcal{M}^{\mathcal{I}^{\mathfrak{A}}}(\mathcal{A}_1) \cap^* \mathcal{M}^{\mathcal{I}^{\mathfrak{A}}}(\mathcal{A}_4) \cap^* \mathcal{M}^{\mathcal{I}^{\mathfrak{A}}}(\mathcal{A}_{11}) = (0.94, 0.89, 0.88, 0.88, 0.81)$.

4 Properties I

Definition 5. *$\alpha \in [0,1]^U$ is a crisp fuzzy set iff $\overline{lb}(\alpha) = \alpha = \underline{ub}(\alpha)$.*

Claim 1. *$\overline{lb}(\mathbf{0}) = \mathbf{0}$ and $\underline{ub}(\mathbf{1}) = \mathbf{1}$.*

Claim 2. *For any $\alpha \in [0,1]^U, \overline{lb}(\alpha) \leq \alpha \leq \underline{ub}(\alpha)$.*

Claim 3. *For all $\alpha, \beta \in [0,1]^U$,*

1. *If $\alpha \leq \beta$, then $\overline{lb}(\alpha) \leq \overline{lb}(\beta)$ and $\underline{ub}(\alpha) \leq \underline{ub}(\beta)$;*
2. *$\underline{ub}(\underline{ub}(\alpha)) \geq \underline{ub}(\alpha)$ and $\overline{lb}(\overline{lb}(\alpha)) \leq \overline{lb}(\alpha)$;*
3. *$\overline{lb}(\underline{ub}(\alpha)) \geq \overline{lb}(\alpha)$ and $\underline{ub}(\overline{lb}(\alpha)) \leq \underline{ub}(\alpha)$.*
4. *$\underline{ub}(\alpha \cup^* \beta) \geq \underline{ub}(\alpha) \cup^* \underline{ub}(\beta)$.*

Proof. It follows immediately from the fact that $\{\gamma \in \mathcal{K}_{\mathcal{I}^{\mathfrak{A}}} : \gamma \leq \alpha\} \subseteq \{\gamma \in \mathcal{K}_{\mathcal{I}^{\mathfrak{A}}} : \gamma \leq \beta\}$ and $\{\gamma \in \mathcal{K}_{\mathcal{I}^{\mathfrak{A}}} : \beta \leq \gamma\} \subseteq \{\gamma \in \mathcal{K}_{\mathcal{I}^{\mathfrak{A}}} : \alpha \leq \gamma\}$. □

Claim 4. *$\mathbf{1} -^* \underline{ub}(\alpha) \leq \underline{ub}(\mathbf{1} -^* \alpha)$ and $\mathbf{1} -^* \overline{lb}(\alpha) \geq \overline{lb}(\mathbf{1} -^* \alpha)$.*

Proof. $\underline{ub}(\alpha) \geq \alpha$ iff $\mathbf{1} -^* \underline{ub}(\alpha) \leq \mathbf{1} -^* \alpha \leq \underline{ub}(\mathbf{1} -^* \alpha)$. □

Definition 6. *Define a norm $||.|| : [0,1]^U \to [0, |U|]$ as $||\alpha|| := \sum^{i \in \{1,2,\ldots,|U|\}} \alpha(i)$.*

Definition 7 (Degree of Precison). *Given an information system $\mathcal{I}^{\mathfrak{A}}$, define a degree-of-precision function $DP_{\mathcal{I}^{\mathfrak{A}}} : [0,1]^U \to [0,1]$ with $DP_{\mathcal{I}^{\mathfrak{A}}}(\alpha) := \frac{||\overline{lb}(\alpha)||}{||\underline{ub}(\alpha)||}$.*

Take Example 4 for example. $DP_{\mathcal{I}^{\mathfrak{A}}}(\alpha) = \frac{27}{44}$. If $DP_{\mathcal{I}^{\mathfrak{A}}}(\alpha) = 1$, it means the information system $\mathbb{A}$ characterizes the fuzzy target set α completely. If $DP_{\mathcal{I}^{\mathfrak{A}}}(\alpha) = 0$, it means the information system $\mathbb{A}$ can not characterize the fuzzy target set α at all.

Definition 8. *Given two information system $\mathbb{I}_1, \mathbb{I}_2$ and a fuzzy target set α, define $\mathbb{I}_1 \succeq_\alpha \mathbb{I}_2$ iff $DP_{\mathbb{I}_1}(\alpha) \geq DP_{\mathbb{I}_2}(\alpha)$.*

Now we want to show how to characterize a set of fuzzy sets $\mathbb{F} = \{f_1, f_2, ..., f_k\}$ via an information system $\mathcal{I}^{\mathfrak{A}}$.

Definition 9. *Define a $\leq$-defined subset relation $\subseteq^{\leq}$ as follows: for any arbitrary $\mathbb{F}_1, \mathbb{F}_2 \subseteq [0,1]^U$, define $\mathbb{F}_1 \subseteq^{\leq} \mathbb{F}_2$ iff for all $f \in \mathbb{F}_1$, there exists $f' \in \mathbb{F}_2$ such that $f \leq f'$.*

Definition 10 (Smallest Subset Upper Bound). *Define the upper bound of $\mathbb{F}$ to be $\underline{Ub}(\mathbb{F}) := \{\underline{ub}(f) : f \in \mathbb{F}\}$.*

Definition 11 (Greatest Subset Lower Bound). *Define the lower bound of $\mathbb{F}$ to be $\overline{Lb}(\mathbb{F}) := \{\overline{lb}(f) : f \in \mathbb{F}\}$.*

Example 5. *Let us choose the classifier from Example 1. Then given the target fuzzy subset* $\mathbb{F}$*, one computes its lower and upper bounds as follows:*

Table 4. Bounds for a target fuzzy Subset

target fuzzy subset $\mathbb{F}$	$\overline{Lb}(\mathbb{F})$	$\underline{Ub}(\mathbb{F})$
$(\frac{1}{\sqrt{7}}, \frac{1}{3}, \frac{1}{2}, 0, \frac{1}{7}, \frac{1}{5})$	$(0, 0, 0, 0, 0, 0)$	$(\frac{1}{3}, \frac{1}{3}, \frac{2}{3}, 0, \frac{1}{3}, \frac{1}{3})$
$(\frac{1}{2}, \frac{1}{3}, \frac{2}{3}, \frac{1}{3}, \frac{1}{\sqrt{3}}, \frac{4}{9})$	$(\frac{1}{3}, \frac{1}{3}, \frac{2}{3}, \frac{1}{3}, \frac{1}{3}, \frac{1}{3})$	$(\frac{2}{3}, \frac{1}{3}, \frac{2}{3}, \frac{1}{3}, \frac{2}{3}, 1)$
$(\frac{1}{4}, 1, \frac{3}{4}, \frac{4}{5}, \frac{1}{\sqrt{11}}, \frac{1}{3})$	$(0, 1, \frac{2}{3}, \frac{2}{3}, 0, \frac{1}{3})$	$(1, 1, 1, 1, 1, 1)$

Claim 5. $\overline{Lb}(\mathbb{F}) \subseteq^{\leq} \mathbb{F} \subseteq^{\leq} \underline{Ub}(\mathbb{F})$.

Claim 6. $\{\overline{Lb}(\mathbb{F}) : \mathbb{F} \in \mathcal{P}([0,1]^U)\}$ *and* $\{\underline{Ub}(\mathbb{F}) : \mathbb{F} \in \mathcal{P}([0,1]^U)\}$ *are closed under set operations: union,intersection and complement.*

Claim 7. *If* $\mathbb{F} \subseteq \mathbb{F}'$*, then* $\overline{Lb}(\mathbb{F}) \subseteq \overline{Lb}(\mathbb{F}\prime)$ *and* $\underline{Ub}(\mathbb{F}) \subseteq \underline{Ub}(\mathbb{F}\prime)$.

Given any two subsets $\mathbb{F}_1, \mathbb{F}_2 \subseteq [0,1]^U$ and an information system $\mathcal{I}^{\mathfrak{A}}$, the following indicator tells which subset is characterized better by $\mathcal{I}^{\mathfrak{A}}$:

Definition 12. *Define the precision of* $\mathbb{F} \subseteq [0,1]^U$ *by*

$$\mathbb{D}P_{\mathcal{I}^{\mathfrak{A}}}(\mathbb{F}) := \frac{\sum_{\alpha \in \mathbb{F}} DP_{\mathcal{I}^{\mathfrak{A}}}(\alpha)}{|\mathbb{F}|}.$$

Example 6. *Take Example 4.* $\mathbb{D}P_{\mathcal{I}^{\mathfrak{A}}}(\mathbb{F}) = \frac{1+\frac{4}{11}+\frac{7}{18}}{3} = \frac{347}{594}$.

Definition 13. $\mathbb{F}_1 \geq_{\mathfrak{A}} \mathbb{F}_2$ *iff* $\mathbb{D}P_{\mathcal{I}^{\mathfrak{A}}}(\mathbb{F}_1) \geq \mathbb{D}P_{\mathcal{I}^{\mathfrak{A}}}(\mathbb{F}_2)$.

Example 7. *Take Example 5 for example. Suppose*
$\mathbb{F}' = \{(\frac{1}{2}, \frac{3}{4}, \frac{1}{3}, \frac{1}{3}, \frac{1}{7}, \frac{1}{5}), (\frac{2}{\sqrt{11}}, \frac{2}{3}, \frac{1}{2}, 1, \frac{3}{4}, \frac{2}{5}), (\frac{6}{\sqrt{7}}, \frac{1}{4}, \frac{1}{2}, 1, \frac{2}{5}, \frac{7}{11}), (\frac{1}{6}, \frac{2}{3}, \frac{1}{2}, 0, \frac{2}{7}, \frac{4}{9})\}$.

Given any two information systems $\mathcal{I}^{\mathfrak{A}}$ and $\mathcal{J}^{\mathfrak{B}}$ and a target fuzzy subset $\mathbb{F} \subseteq [0,1]^U$, one can decide which information system characterizes $\mathbb{F}$ better.

Definition 14. $\mathcal{I}^{\mathfrak{A}} \geq_{\mathbb{F}} \mathcal{J}^{\mathfrak{B}}$ *iff* $\mathbb{D}P_{\mathcal{I}^{\mathfrak{A}}}(\mathbb{F}) \leq \mathbb{D}P_{\mathcal{J}^{\mathfrak{B}}}(\mathbb{F})$.

Example 8. *Let us take* $\mathcal{I}^{\mathfrak{A}}$ *in Example 1 and* $\mathbb{F}$ *in Example 5 for example. Let* $J^{\mathfrak{B}}$ *be the following information system:*

Table 5. Information System $\mathcal{J}^{\mathfrak{B}}$

Universe	Attribute b_1	Attribute b_2	Attribute b_3	Attribute b_4
u_1	*hot*	heavy	yes	up
u_2	*hot*	heavy	no	down
u_3	*warm*	heavy	no	up
u_4	*warm*	light	no	down
u_5	*hot*	light	yes	down
u_6	*warm*	heavy	no	up

Table 6. Bounds for a target fuzzy Subset

target fuzzy subset $\mathbb{F}$	$\overline{Lb}(\mathbb{F})$	$\underline{Ub}(\mathbb{F})$
$(\frac{1}{\sqrt{7}}, \frac{1}{3}, \frac{1}{2}, 0, \frac{1}{7}, \frac{1}{5})$	$(0,0,0,0,0,0)$	$(\frac{1}{2}, \frac{1}{2}, \frac{1}{2}, 0, \frac{1}{4}, \frac{1}{2})$
$(\frac{1}{2}, \frac{1}{3}, \frac{2}{3}, \frac{1}{3}, \frac{1}{\sqrt{3}}, \frac{4}{9})$	$(0,0,0,0,0,0)$	$(1,1,1,1,1,1)$
$(\frac{1}{4}, 1, \frac{3}{4}, \frac{4}{5}, \frac{1}{\sqrt{11}}, \frac{1}{3})$	$(0,0,0,0,0,0)$	$(1,1,1,1,1,1)$

Now one computes $\mathbb{DP}_{\mathcal{J}^{\mathfrak{B}}}(\mathbb{F}) = 1$ *and thus* $\mathcal{J}^{\mathfrak{B}} \leq_{\mathbb{F}} \mathcal{I}^{\mathfrak{A}}$.

One can even further extend the target sets to a set of fuzzy subsets $\mathfrak{T} = \{\mathbb{F}_1, \mathbb{F}_2, ..., \mathbb{F}_k\}$. Now one can even classify any set of information systems

5 Alternative Characterizations

Though in Section 3 we provide a uniform approximation for fuzzy target set, there is still room to manipulate. Here we provide another three characterizations. The users or designers could choose the suitable one according to his preferences. For example, if one is not sure about the accuracy of the data or he wants to play safe, then he could adopt the characterization to increase the tolerance of errors by our first alternative. The first alternative uses greatest upper bound and smallest lower bound for the approximation.

Definition 15 (Greatest Upper Bound). *Define the upper bound function* $\overline{ub} : [0,1]^U \to [0,1]^U$ *by*

$$\overline{ub}(\alpha) := \begin{cases} \cup^* \{\beta \in \mathcal{K}_{\mathcal{I}^{\mathfrak{A}}} : \alpha \leq \beta\} & \text{if there exists } \beta \in \mathcal{K}_{\mathcal{I}^{\mathfrak{A}}} \text{ such that } \alpha \leq \beta; \\ \mathbf{1}, \text{ otherwise} . \end{cases}$$

Definition 16 (Smallest Lower Bound). *Define the lower bound function* $\underline{lb} : [0,1]^U \to [0,1]^U$ *by*

$$\underline{lb}(\alpha) := \begin{cases} \cap^* \{\beta \in \mathcal{K}_{\mathcal{I}^{\mathfrak{A}}} : \beta \leq \alpha\} & \text{if there exists } \beta \in \mathcal{K}_{\mathcal{I}^{\mathfrak{A}}} \text{ such that } \beta \leq \alpha; \\ \mathbf{0}, \text{ otherwise} . \end{cases}$$

The second alternative uses greatest upper bound and greatest lower bound for the approximation.

Definition 17 (Greatest Upper Bound). *Define the upper bound function* $\overline{ub} : [0,1]^U \to [0,1]^U$ *by*

$$\overline{ub}(\alpha) := \begin{cases} \cup^* \{\beta \in \mathcal{K}_{\mathcal{I}^{\mathfrak{A}}} : \alpha \leq \beta\} & \text{if there exists } \beta \in \mathcal{K}_{\mathcal{I}^{\mathfrak{A}}} \text{ such that } \alpha \leq \beta; \\ \mathbf{1}, \text{ otherwise} . \end{cases}$$

Definition 18 (Greatest Lower Bound). *Define the lower bound function* $\overline{lb} : [0,1]^U \to [0,1]^U$ *to be*

$$\overline{lb}(\alpha) := \begin{cases} \cup^*\{\beta \in \mathcal{K}_{\mathcal{I}^{\mathfrak{A}}} : \beta \leq \alpha\} & \text{if there exists } \beta \in \mathcal{K}_{\mathcal{I}^{\mathfrak{A}}} \text{ such that } \beta \leq \alpha; \\ \mathbf{0}, \text{ otherwise} . \end{cases}$$

The third alternative is to use smallest upper bound and smallest lower bound for the approximation.

Definition 19 (Smallest Upper Bound). *Define the upper bound function* $\underline{ub} : [0,1]^U \to [0,1]^U$ *by*

$$\underline{ub}(\alpha) := \begin{cases} \cap^*\{\beta \in \mathcal{K}_{\mathcal{I}^{\mathfrak{A}}} : \alpha \leq \beta\} & \text{if there exists } \beta \in \mathcal{K}_{\mathcal{I}^{\mathfrak{A}}} \text{ such that } \alpha \leq \beta; \\ \mathbf{1}, \text{ otherwise} . \end{cases}$$

Definition 20 (Smallest Lower Bound). *Define the lower bound function* $\underline{lb} : [0,1]^U \to [0,1]^U$ *to be*

$$\underline{lb}(\alpha) := \begin{cases} \cap^*\{\beta \in \mathcal{K}_{\mathcal{I}^{\mathfrak{A}}} : \beta \leq \alpha\} & \text{if there exists } \beta \in \mathcal{K}_{\mathcal{I}^{\mathfrak{A}}} \text{ such that } \beta \leq \alpha; \\ \mathbf{0}, \text{ otherwise} . \end{cases}$$

6 Properties II

All the following properties are trivial. I leave the readers to check.

Claim 8. *For all* $\alpha \in [0,1]^U, \underline{lb}(\alpha) \leq \overline{lb}(\alpha) \leq \alpha \leq \underline{ub}(\alpha) \leq \overline{ub}(\alpha)$.

Claim 9. *For all* $\alpha \in [0,1]^U, \underline{lb}(\alpha) = \overline{lb}(\alpha) = \underline{ub}(\alpha) = \overline{ub}(\alpha)$ *iff* $\{\gamma \in \mathcal{K}_{\mathcal{I}^{\mathfrak{A}}} : \gamma \leq \alpha\} = \{\gamma \in \mathcal{K}_{\mathcal{I}^{\mathfrak{A}}} : \alpha \leq \gamma\} = \{\alpha\}$.

Claim 10. *If* $\alpha \leq \beta$, *then* $\underline{lb}(\alpha) \leq \underline{lb}(\beta)$ *and* $\overline{ub}(\alpha) \leq \overline{ub}(\beta)$.

Claim 11. $\mathbf{1} -^* \overline{ub}(\alpha) \leq \mathbf{1} -^* \underline{ub}(\alpha) \leq \mathbf{1} -^* \alpha \leq \underline{ub}(\mathbf{1} -^* \alpha) \leq \overline{ub}(\mathbf{1} -^* \alpha)$.

Claim 12. $\underline{lb}(\mathbf{1} -^* \alpha) \leq \overline{lb}(\mathbf{1} -^* \alpha) \leq \mathbf{1} -^* \alpha \leq \mathbf{1} -^* \overline{lb}(\alpha) \leq \mathbf{1} -^* \underline{lb}(\alpha)$.

7 Conclusions and Future Work

I have extended the characterization problem to a fuzzy target set and come up with a new approach to characterize fuzzy target set with data-driven fuzzy classifiers. Furthermore, the way to compare the efficiency of different information is also constructed. For further research, one can study other alternative ways to form the fuzzy classifier based on the data or the mixture of different characterizations based on some weight or probability distribution. Apart from that, if one wants to lift the assumption that the elements behave uniformly as the universe, then he could devise some approaches to construct the membership element by element. One could also add some learning mechanisms to facilitate such estimation.

References

1. Zaden, L.A.: Fuzzy Sets. Information and Control 8, 338–353 (1965)
2. Atanassov, K.T.: Intuitionistic fuzzy sets. Fuzzy Sets and Systems 20, 87–96 (1986)
3. Chen, R.-M.: Data-based approximation of intuitionistic fuzzy target sets. Notes on Intuitionistic Fuzzy Sets 20(2) (2014)
4. Pawlak, Z.: Rough Sets. International Journal of Computer and Information Sciences 11(5) (1982)
5. Pawlak, Z.: Rough Sets: theoretical aspects of reasoning about data. Springer Science+Business Media Dordrecht (1991)
6. Nguyen, H.S., Skowron, A.: Rough Sets: From Rudiments to Challenges. In: Rough Sets and Intelligent Systems - Professor Zdzisław Pawlak in Memoriam, vol. 1, ch. 3. Springer, Heidelberg (2013)

Part XV
Perception, Judgment, Affect, and Sentiment Analyses

Towards Perception-Oriented Situation Awareness Systems

Gianpio Benincasa, Giuseppe D'Aniello, Carmen De Maio, Vincenzo Loia*, and Francesco Orciuoli

University of Salerno,
Via Giovanni Paolo II, 132. Fisciano (SA) Italy
{gbenincasa,gidaniello,cdemaio,loia,forciuoli}@unisa.it
http://www.unisa.it

Abstract. This paper proposes a new approach for identifying situations from sensor data by using a perception-based mechanism that has been borrowed from humans: sensation, perception and cognition. The proposed approach is based on two phases: low-level perception and high-level perception. The first one is realized by means of semantic technologies and allows to generate more abstract information from raw sensor data by also considering knowledge about the environment. The second one is realized by means of Fuzzy Formal Concept Analysis and allows to organize and classify abstract information, coming from the first phase, by generating a knowledge representation structure, namely lattice, that can be traversed to obtain information about occurring situation and augment human perception. The work proposes also a sample scenario executed in the context of an early experimentation.

Keywords: Computer Perception, Situation Awareness, Ontologies, Fuzzy Formal Concept Analysis, Intelligent Systems.

1 Introduction and Motivation

The emerging paradigm of Internet of Things (IoT) provides an infrastructure that connects things anytime and anywhere, with anything and anyone, by using any network and service. Many environmental sensors have been continuously improved by becoming smaller, cheaper and more intelligent [7]. This trend has led to a widespread adoption of sensors deployed to monitor the environments in which we live. Moreover, personal devices, like smartphones and tablets, are equipped with several sensing capabilities which could be used to monitor users' activities or environmental phenomena.

The exploitation of heterogeneous data collected by different sensors allows applications to provide users with context-aware behaviours and support decision-making processes in several and heterogeneous application domains like, for instance, Emergency Management, Energy Savings, e-Healthcare, Physical and

* Corresponding author.

P. Angelov et al. (eds.), *Intelligent Systems'2014*,
Advances in Intelligent Systems and Computing 322, DOI: 10.1007/978-3-319-11313-5_71

Cyber Security. In this scenario, Situation Awareness represents a powerful paradigm that enable the aforementioned capabilities. Situation Awareness has been defined by Endsley as "the perception of the elements in an environment within a volume of time and space, the comprehension of their meaning, and a projection of their status in the near future" [4]. Traditional models of Situation Awareness foresee the employment of a challenging task, namely Situation Identification.

A widely accepted conceptual model of Situation Identification proposes to: i) start from data collected by a sensor network, ii) derive (from sensor data) more abstracted elements, namely context attributes and iii) identify the occurring situations by considering the presence or absence of specific context attributes.

The complexity of Situation Identification is due to several factors like, for instance, the variety of admissible situations, the uncertainty and imprecision of data, the dynamic nature of the observed environments and so on. In order to face Situation Identification task and provide better support for decision-making processes, many researches have contributed to improve the data acquisition step and the low-level data processing task (e.g., data validation, data fusion, data cleaning, etc.). Other approaches focus on the techniques for identifying situations starting from context attributes and sensor data. It is possible to distinguish among two main categories of approaches [10]: *Specification-based approaches* which represent expert knowledge by means of logic rules and then use reasoning engines in order to infer proper situations from context attributes; *Learning-based approaches* which foresee the use of machine learning and data mining techniques in order to explore relations between sensors data and situations. The first ones are not suitable in presence of complex scenarios (in which it is too difficult modelling all the relations among context attributes and situation). The main drawback of the learning-based approaches, instead, is that the learned models are black boxes, which are not easily understandable by humans.

In this paper, a perception-based approach for Situation Identification is provided. The proposed approach takes care of studies on human perception and mainly leverages on Fuzzy Formal Concept Analysis that offers a solid and flexible framework to analyse and organize abstract data (or context attributes) coming from a previous processing step on sensor data. The proposed approach mostly responds to the need for information awareness, sustaining human operators to deal with Situation Identification.

The main idea underlying the proposed approach is that human perception is based on sensation but also on cognition. Human cognitive processes support the perception by considering and integrating additional knowledge about context and previous experiences together with data coming from sensation. These aspects allow humans to react also when they have incomplete information. In more details, human perception is defined as "the interpretations of sensation, giving them meaning and organization" [5]. As a result, the activity of perception is not rigorous and well-defined: despite the fact that it depends on the elaboration of information acquired by sensing, the results of the perception are not the same for different individuals or even for the same individual at different times. This is because the human perception is affected by external factors that are not related with sensing activities.

Focusing on the role of cognitive processes and on the information level, the proposed approach, as well as human perception, provides several mechanisms, acting at different levels of abstraction, to enable computational cognition for supporting perception. Ontologies and, in general, Semantic Technologies have been exploited to assign meaning to sensor data and connect them with contexts, providing more abstract and manageable pieces of information. Then, Fuzzy Formal Concept Analysis, as we have already introduced above, enables the construction of knowledge structures, namely Fuzzy Concept Lattices, which allow to assign meaning to the abstract pieces of information and support more complex forms of reasoning on them. Another work [1] propose a perception-based approach focused on sensor data without considering the perception activity at higher levels.

The structure of the paper is as follows: Section 2 provides an overview of the proposed approach. Section 3 describes the process of context attributes identification. Section 4 describes the application of the Fuzzy Formal Concept Analysis for Situation Identification. Section 5 provides a sample scenario in the domain of Road Traffic Management, where the proposed approach is applied. Section 6 concludes the work.

2 Overall Picture of the Approach

The proposed approach, based on Situation Awareness, aims at supporting decision-making processes and it is inspired by human-like concepts of *perception* and *sensing*. The main idea is to simulate methods for analysis and measurements similar to those of humans [8]. More specifically, the approach is based on a general human-oriented perception model, as defined by Matlin and Foley in [5], consisting of three processes (as shown in Fig. 1):

- *Sensation* is the basic experience generated as stimuli fall on our sensory systems;
- *Perception* is the interpretation of the sensations, giving them meaning and organization;
- *Cognition* includes the acquisition, retrieval and use of the information.

These three processes are not clearly separated. For instance, *perception* strongly depends on the *cognition* process, because it relies on beliefs, goals, interests, preferences, etc. The workflow depicted in Fig. 1 instantiates these human-oriented processes in an artificial perception system. In detail, the *sensation* process is implemented as a data acquisition process by means of sensors which gather raw data from the ecosystem. The main capability of the *sensation* process is to elaborate sensor data, by performing operations like data cleaning, data validation, low-level data fusion and data storing. In the *perception* process, raw data are processed in order to give them meaning and organization. We divide this process in two steps: *low-level perception* and *high-level perception*. The

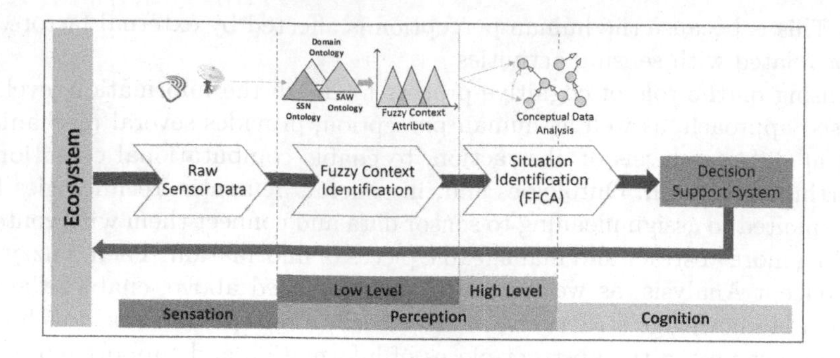

Fig. 1. Human oriented approach for environmental perception

former involves the early processing and representation of information from the various sensory capabilities. The *high-level perception*, instead, involves taking a more global view of this information, extracting meaning from the *low-level* representation, and making sense of situations at a conceptual level [2].

The *low-level perception* phase has the aim of identifying all relevant context features of the environment which will be used for Situation Identification. In this step, raw sensor data are semantically described with respect to a formal model defined by means of ontologies. For every kind of sensor deployed in the environment, it is necessary to define a mapping function that transforms sensor data in concepts of the semantic model. This mapping function allows the abstraction from specific and technical details of sensors to overcome potential interoperability issues. This mapping function can be defined both by domain experts or it can be learned by means of machine learning algorithms. Moreover, the proposed semantic model allows the description of the environment and of the surrounding context in which sensors operate. This allows to represent important information for enabling cognitive processes at perception level: in fact, when a human perceives some information from the environment, he/she uses other environmental information and background knowledge in order to contextualise the perceived information. These human-like cognitive processes are enabled by means of the description of the context in which the sensors are deployed and the description of the shared domain knowledge. The output of this step consists of the semantic representation of sensor data and surrounding context, according to the ontological model described in Section 3.1. Next, the semantically annotated sensor data have to be processed in order to represent them in a human-like way, by considering that humans do not think in term of measurements but in term of perceptions. An essential difference between measurements and perceptions is that in general, measurements are crisp, whereas perception are fuzzy. Crisp information are not suitable to reflect the reasoning activity and the process of concepts formation in humans, because these two tasks are based on information which are fuzzy [11]. For this reason, in the proposed approach, the main features measured by sensors and represented with ontological concepts, are transformed into fuzzy attributes and into the corresponding fuzzy values.

The aim of the *high-level perception* phase is to identify the current situation. A situation can be considered as the highest level of generalization of a context: a situation, in fact, allows eliciting the most important information from a context. For this reason, the situation is identified by means of fuzzy context attribute values which have been identified in the previous process. The identification of the current situation may involve the use of other information, like domain knowledge, historical data, or information provided by users. Let us remark, in fact, that Situation Identification process is both a perception and cognitive process. The identification of situations in the high-level perception involves the exploitation of domain knowledge, surrounding context of gathered data, background knowledge and it involves human-like cognitive processes (e.g. reasoning on domain knowledge in order to contextualise the perceived information). Furthermore, the identification of a situation depends on beliefs, interests and goals of people. For these reason, in Fig. 1, the Situation Identification process is situated among perception and cognition phases. For identifying current situations, the approach relies on Fuzzy Formal Concept Analysis (FFCA) [3] described in Section 4.1.

Being able to perceive the current situation in a way similar to people, it is possible to support decision making processes even in case where people can not directly determine the state of the environment. This could happen, for instance, when it is too dangerous for people to be into the environment (e.g., hazardous or contaminated area, disasters, fire, flooding, etc.), or there are too much information to process (e.g., if too sensors are deployed, it is difficult for people to be able to observe them all for taking decisions). In such cases, it is possible to create rules which automatically operate some actions on the ecosystem in order to act accordingly with the current *situation* (e.g., activating alarm in case of emergency; communicating information to security staff, etc.). The following sections provide details on each phase of the approach.

3 Low-level Perception: From Sensor Data to Fuzzy Context

In the *low-level perception* phase, the data gathered by sensors are transformed in features of a *context* that will be used for identify the current *situation.* To achieve this, sensor data are represented by means of a semantic model (Section 3.1); then, they are transformed in fuzzy value by means of fuzzy controllers. These fuzzy values, along with the relevant environmental features, represent the current state of the *context.*

3.1 Semantic Model for Representing Ecosystem

This section describes the model for the semantic description of a dynamic ecosystem. The main objective of the model is to support the transformation process of raw sensor data in concepts close to the human perception or close to their abstraction process. In order to achieve this, it is necessary to model the overall ecosystem. First of all, it is needed to model the sensing devices, their capabilities, their characteristics and the observations which they produce, in order to provide additional information related to sensor data which are useful for improving Situation Identification.

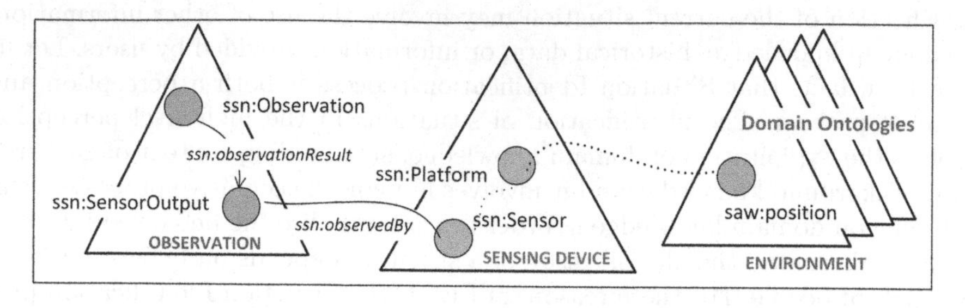

Fig. 2. Ontologies for modelling a dynamic ecosystem

The model foresees three main elements for describing the ecosystem: an ontology for describing sensing devices, an ontology for describing observations made by those devices and a set of ontologies for modelling the surrounding environment. Fig. 2 shows these three elements and the respective main classes and main properties of the ontologies used to describe them. In a more detailed way, the model is logically divided in three parts:

- *Sensors*: represents the sensing devices that are deployed in the environment. These sensors could be physically attached to the objects of the environment, or they could belong to people moving in it; in some cases, these sensing devices may refer to logical devices (e.g., humans can be considered sensing devices). The *Semantic Sensor Network Ontology*[1] (SSN) is used for describing sensor information.
- *Observation*: represents the measures made up by the sensing devices. They are characterized by a value representing the result, timing information representing the instant of measurement and the characteristic that has been observed (e.g., a position, a temperature, etc.). Observations are represented by populating SSN ontology.
- *Environment*: represents the surrounding environment. It describes all the relevant information that can influence the perception, like position of the sensor, time, weather, pressure, physical phenomena and so on. Due to the heterogeneity and complexity of information which have to be represented, these concepts are modelled by means of several domain-specific ontologies. For instance, *Time Ontology*[2] can be used for describing temporal information; *Geonames*[3] ontology for describing geospatial information.

[1] Semantic Sensor Network Ontology
http://www.w3.org/2005/Incubator/ssn/ssnx/ssn
[2] http://www.w3.org/TR/owl-time/
[3] http://www.geonames.org/ontology/documentation.html

3.2 Identification of Context Attributes

In the *low-level perception* phase, the sensor data, represented by means of the semantic model, have to be processed in order to identify the dominant features of the ecosystem which allow, in the high-level perception phase, the identification of the current situation. The set of all relevant features of the environment represent the *context*, while each feature is a *context attribute*. The set of all *context attributes* in a context, varies with respect to the set of available sensors and depends on the specific application domain of the system.

There exist two type of context attributes:

- *fuzzy context attribute* represents the fuzzy value related to an observation made by a sensor (e.g., if an anemometer measures a wind speed of 15m/s, the corresponding fuzzy attribute will be `wind_speed = strong_wind`).
- *environmental context attribute* represents a feature that is related to the surrounding environment in which the sensors are deployed. They are represented by concepts of the *Environment* ontologies in the semantic model (see section 3.1).

A context is represented by a vector $C = [fc_1, fc_2, ...fc_n, ec_1, ec_2, ...ec_m]$ of context attributes, in which $fc_i, i = 1..n$ are the n fuzzy context attributes and $ec_j, j = 1..m$ are the m environmental context attributes. Each fuzzy context attribute fc_i is calculated by means of an agent called $F_{fc_i}agent$; this agent, at a specific time interval, calculates the new fuzzy context attribute value. It reads the values gathered by sensors and then calculates the corresponding fuzzy value. Another agent, namely $Cagent$, is responsible for updating vector C: it interacts periodically with the existing $F_{fc_i}agents$ in order to request the up-to-date values of each fc_i. Subsequently, each $F_{fc_i}agent$ replies with a message stating the up-to-date fuzzy value for context attribute fc_i. Moreover, the $Cagent$ is responsible for gathering values for environmental context attributes ec_j, by retrieving them from the semantic model. These concepts are retrieved by means of SPARQL queries. Fig. 3 shows an example of a SPARQL query for retrieving the position of a sensor that has made an observation with URI `<http://www.example.com/obs_Id_XXX>`. Moreover, this sensor is deployed on a platform (`ssn:Platform`) and this platform is attached (property `unisaw:attachedWith`) to a physical object (class `saw:PhysicalObject`). Lastly, this physical object is situated in a specific position (property `saw:hasPosition`) that constitutes the result of the query.

4 High-level Perception: From Fuzzy Context to Situation

Formal Concept Analysis (FCA) is a technique of data analysis. Recent studies [9], [3], [6], integrate FCA and fuzzy techniques in order to deal with uncertain and vague information. In particular, this approach exploits a fuzzy extension of FCA, (FFCA) [3]. The use of FFCA in the sensor network domain is considered appropriate for characterizing environmental situation of ecosystems. In

```
SELECT ?sensor ?location
WHERE {
        ?sensor a ssn:SensingDevice.
        ?observation ssn:observedBy ?sensor.
        ?sensor ssn:hasDeployment ?deployment.
        ?deployment ssn:deployedOnPlatform ?platform.
        ?platform unisaw:attachedWith ?physicalObject.
        ?physicalObject a saw:PhysicalObject.
        ?physicalObject saw:hasPosition ?location.
        FILTER( ?observation = <http://www.example.com/obs_Id_XXX>)
}
```

Fig. 3. SPARQL query for retrieving the position of a sensor

particular, FFCA reduces the loss of information by emphasizing approximate relationships between situations and external conditions relative to the real world (represented by *context attributes*) that is unsharp. Definitely, FFCA provides mechanisms sustaining people in perceiving the environment in which they are acting with respect to the occurring situations and making better decisions.

4.1 Fuzzy Formal Concept Analysis-FFCA

Fuzzy Formal Concept Analysis (FFCA) combines fuzzy logic and FCA representing the uncertainty through membership values in the range [0, 1]. FFCA enables the representation of the relationships between objects (extent) and attributes (intent) in a given domain by means of a *formal context.* In particular, a straight mapping creates correspondence between the set of attributes M and fuzzy context attributes defined from sensor data (described in Section 3.2), as well as between the set of objects G and the situation collection. Thus, in a decision-making scenario, $(g, m) \in I$ means that the specific situation g is described by the fuzzy context attributes m (i.e., perception of sensor data evaluated with respect to the situation and the environmental data). The following definitions describe the fuzzy extension of the Formal Concept Analysis.

Definition 1. *A* **Fuzzy Formal Context** *is a triple* $K = (G, M, I = \varphi(G \times M))$*, where G is a set of objects, M is a set of attributes, and I is a fuzzy set on domain* $G \times M$*. Each relation* $(g, m) \in I$ *has a membership value* $\mu(g, m)$ *in [0, 1].*

Definition 2. Fuzzy Representation of Object. *Each object O in a fuzzy formal context K can be represented by a fuzzy set* $\Phi(O)$ *as* $\Phi(O) = A_1(\mu_1)$, $A_2(\mu_2), \ldots, A_m(\mu_m)$*, where* $A_1, A_2, \ldots, A_m$ *is the set of attributes in K and* μ_i *is the membership of O with attribute* A_i *in K.* $\Phi(O)$ *is called the fuzzy representation of O.*

A fuzzy formal context is often represented as a cross-table as shown in Fig. 4. Let us note each cell of the table contains a membership value in [0, 1].

According to fuzzy theory, the definition of Fuzzy Formal Concept is given as follows [12].

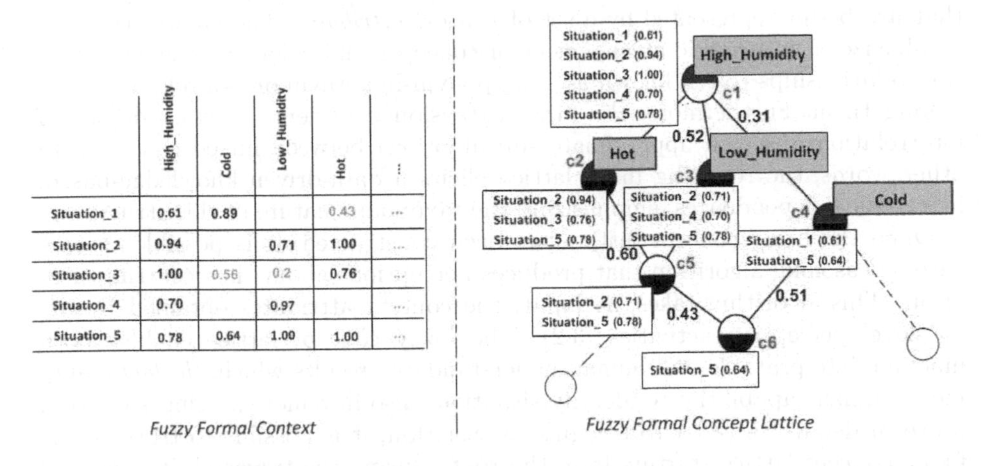

	High_Humidity	Cold	Low_Humidity	Hot	...
Situation_1	0.61	0.89		0.43	...
Situation_2	0.94		0.71	1.00	...
Situation_3	1.00	0.56	0.2	0.76	...
Situation_4	0.70		0.97		...
Situation_5	0.78	0.64	1.00	1.00	...

Fig. 4. Fuzzy formal context and the relative concept lattice with threshold T = 0.6

Definition 3. Fuzzy Formal Concept. *Given a fuzzy formal context K=(G, M, I) and a confidence threshold T, we define* $A^* = \{m \in M \mid \forall\, g \in A: \mu(g,m) \geq T\}$ *for* $A \subseteq G$ *and* $B^* = \{g \in G \mid \forall\, m \in B: \mu(g,m) \geq T\}$ *for* $B \subseteq M$. *A fuzzy formal concept (or fuzzy concept) of a fuzzy formal context K with a confidence threshold T is a pair* $(A_f = \varphi(A), B)$, *where* $A \subseteq G$, $B \subseteq M$, $A^*=B$ *and* $B^*=A$. *Each object* $g \in \varphi(A)$ *has a membership* μ_g *defined as*

$$\mu_g = \min{}_{m \in B}\, \mu(g,m)$$

where $\mu(g,m)$ *is the membership value between object g and attribute m, which is defined in I. Note that if* $B=\{\,\}$ *then* $\mu_g = 1$ *for every g. A and B are the extent and intent of the formal concept* $(\varphi(A), B)$ *respectively.*

In Fig. 4(a) the fuzzy formal context has a confidence threshold T=0.6, so all the relationship with membership values less than 0.6 are not shown.

Definition 4. *Let* (A_1, B_1) *and* (A_2, B_2) *be two fuzzy concepts of a fuzzy formal context (G, M, I).* $(\varphi(A_1), B_1)$ *is the* ***Subconcept*** *of* $(\varphi(A_2), B_2)$, *denoted as* $(\varphi(A_1), B_1) \leq (\varphi(A_2), B_2)$, *if and only if* $\varphi(A_1)\ \varphi(A_2)$ $(B_2 \subseteq B_1)$. *Equivalently,* (A_2, B_2) *is the* ***Superconcept*** *of* (A_1, B_1).

For instance, let us observe in Fig. 4, the concept c5 is subconcept of concepts c2 e c3. Equivalently the concepts c2 e c3 are superconcept of concept c5.

Definition 5. ***A Fuzzy Concept Lattice*** *of a fuzzy formal context K with a confidence threshold T is a set F(K) of all fuzzy concepts of K with the partial order* $\leq$ *with the confidence threshold T .*

Fig. 4 shows an example of lattice coming from the related table, with threshold $T = 0.6$. Let us note each node (i.e., a formal concept) includes the situation that are better represented by a set of *context attributes*. The lattice structure emphasizes a taxonomic arrangement of concepts and evidences the subsumption relationships (often known as a "hyponym-hypernym or is-a relationship") among them. Furthermore, the fuzzy extension considers a certain degree of interrelations (i.e., an approximate subsumption) between linked concepts. In other words, the resulting fuzzy lattice elicits a data-driven knowledge-based, hierarchical dependences, emphasizing the taxonomic nature of this structure.

Once the fuzzy concept lattice has been constructed, it is possible to execute a reasoning algorithm that produces perceptions about the occurring situation. This algorithm takes, as input, the context attributes obtained by the low-level perception activities (3.2). The fuzzy concept lattice enables both machine-interpratable and human-understandable results which, *de facto*, augment human capabilities to identify situations also in remote sensing scenarios. More in details, in order to recognise a situation, it is possible to traverse the fuzzy concept lattice starting from the root. During the traversal, it is needed to compare the input context attributes with the information at each node in the lattice (fuzzy concept), moving downwards until the attributes associated to the currently visited concept are equal to the input context attributes. When this concept is found, the objects associated to it are recognised as occurring situations. The weight associated to each situation (object) in the fuzzy concept lattice, provides information on how a situation is more likely than another.

5 Sample Scenario: Road Traffic Management

In order to evaluate the applicability of the proposed approach, let us consider the following road traffic management (RTM) scenario (see Fig. 5). The aim is to support road traffic operators in the identification of the cause related to a traffic jam. The result of the above task enhances decision making processes, helping human operators to be aware of an accident in shorter time and preventing them from being overloaded with too much information. Let us suppose that every vehicle is monitored by a GPS sensor that allows retrieving the current position and speed. This allows estimating the presence and the severity of the traffic jam. A traditional Decision Support System (DSS) may have difficulty in the identification of the cause of this traffic jam (accident, road works, etc.). This is due to the lack of support, in the Situation Identification process, for background knowledge and contextual information, which can be typically exploited by human operator. For instance, the human operator, by using her background domain knowledge (e.g., she knows that at 8.00 am is a rush hour, so traffic is predictable), or by using contextual information (e.g., she knows that a football game begins shortly), is able to *perceive* and *deduce* the right cause of the traffic jam and to apply the most suitable intervention strategy (e.g., redirect traffic on different roads, close a road, etc.).

In the sample scenario, the FFCA classifier is trained with historical training data about traffic and environmental information related to the considered road. The classifier is able to distinguish these situations: `Accident`, `Predictable_`

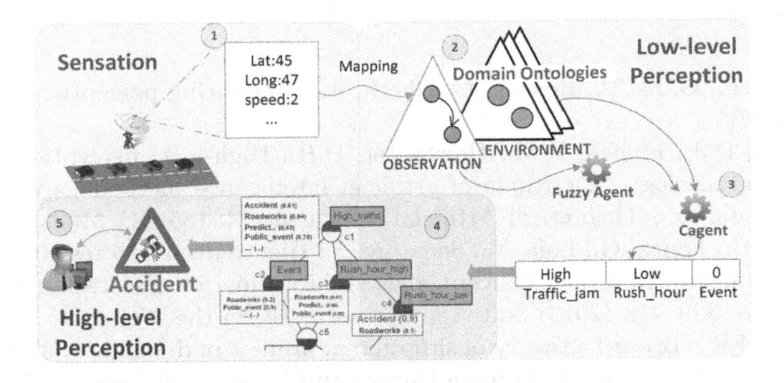

Fig. 5. Sample scenario

`Traffic_in_Rush_Hour`, `Public_event`, `Roadworks`. Fig. 5 shows the overall process of the sample scenario. In the step 1, the GPS sensors collect position and speed of vehicles. These data are represented by means of Observation ontologies (step 2), while Domain ontologies represent environmental information. In particular, they represents that, at the time in which data are gathered, no public event occur in short time and it is not a rush hour. In the step 3, the *Cagent* produces the following vector C: $C(traffic\_jam = high; rush\_hour = low; event = 0)$. The $traffic\_jam$ attribute is a fuzzy context attribute calculated by a $fuzzy\_agent$ by considering speed and position of several vehicles; $rush\_hour$ is a fuzzy context attribute that considers current time and domain knowledge. In step 4, the C vector is used for querying the FFCA classifier. In particular, let us consider the following query: $Q = \{Traffic\_Jam\_High, Rush\_Hour\_Low\}$ Among the recognisable situations, the output of the classifier for the query is: $Accident = 0.9$ $RoadWorks = 0.1$. This allows to alert human operators (step 5) that, with a high probability, an accident occurred; so the operators can rapidly adopt the adequate intervention strategies.

6 Final Remarks

This paper provides a novel approach, based on the concept of perception, for Situation Awareness supporting Decision Support Systems in heterogeneous domains. The approach consists of two main phases: Low-level Perception and High-level Perception. In Low-level Perception phase, the raw data are transformed in contexts attributes that will be used by the High-level Perception phase. The High-level perception phase allows to recognize the situations using a fuzzy concept lattice. A sample scenario, reported in this work, provided promising results. The approach is composed by two main phases In future works the authors will enhance the proposed approach for handling highly dynamic environments in which the sensors may provide uncertain information.

References

1. Barnaghi, P., Ganz, F., Henson, C., Sheth, A.: Computing perception from sensor data (2012)
2. Chalmers, D.J., French, R.M., Hofstadter, D.R.: High-level perception, representation, and analogy: A critique of artificial intelligence methodology. Journal of Experimental and Theoretical Artificial Intelligence 4, 185–211 (1992)
3. De Maio, C., Fenza, G., Loia, V., Senatore, S.: Hierarchical web resources retrieval by exploiting fuzzy formal concept analysis. Information Processing & Management 48(3), 399–418 (2012) Soft Approaches to IA on the Web
4. Endsley, M.R.: Toward a theory of situation awareness in dynamic systems. Human Factors: The Journal of the Human Factors and Ergonomics Society 37(33), 32–64 (1995)
5. Fole, H.J., Matlin, M.W. (eds.): Sensation and Perception. Allyn and Bacon, Newton (1997)
6. Jelsteen, J., Evangelin, D., Alice Pushparani, J., Nelson Samuel Jebastin, J.: Ontology learning process using fuzzy formal concept analysis. International Journal of Engineering Trends and Technology 4(2), 148–152 (2013)
7. Perera, C., Zaslavsky, A.B., Compton, M., Christen, P., Georgakopoulos, D.: Semantic-driven configuration of internet of things middleware. CoRR abs/1309.1515 (2013)
8. Robertsson, L., Iliev, B., Palm, R., Wide, P.: Perception modeling for human-like artificial sensor systems. Int. J. Hum.-Comput. Stud. 65(5), 446–459 (2007)
9. Tho, Q., Hui, S., Fong, A.C.M., Cao, T.H.: Automatic fuzzy ontology generation for semantic web. IEEE Transactions on Knowledge and Data Engineering 18(6), 842–856 (2006)
10. Ye, J., Dobson, S., McKeever, S.: Situation identification techniques in pervasive computing: A review. Pervasive and Mobile Computing 8(1), 36–66 (2012)
11. Zadeh, L.A.: A new direction in AI - toward a computational theory of perceptions. In: Reusch, B. (ed.) Fuzzy Days 2001. LNCS, vol. 2206, p. 628. Springer, Heidelberg (2001)
12. Zhou, B., Hui, S., Chang, K.: A formal concept analysis approach for web usage mining. In: Intelligent Information Processing II, IFIP International Federation for Information Processing, vol. 163, pp. 437–441. Springer (2005)

Textual Event Detection Using Fuzzy Fingerprints

Luís Marujo[1,2], Joao Paulo Carvalho[1], Anatole Gershman[2],
Jaime Carbonell[2], João P. Neto[1], and David Martins de Matos[1]

[1] INESC-ID, IST – Universidade de Lisboa, Portugal
[2] Languages Tecnologies Institute, Carnegie Mellon University, Pittsburgh, USA
{Lmarujo,anatoleg,jgc}@cs.cmu.edu,
{joao.carvalho,joao.neto,david.matos}@inesc-id.pt

Abstract. In this paper we present a method to improve the automatic detection of events in short sentences when in the presence of a large number of event classes. Contrary to standard classification techniques such as Support Vector Machines or Random Forest, the proposed Fuzzy Fingerprints method is able to detect all the event classes present in the ACE 2005 Multilingual Corpus, and largely improves the obtained G-Mean value.

Keywords: Event Detection, Support Vector Machines, Fuzzy Fingerprints, ACE2005.

1 Introduction

Automatic event detection is an important Information Extraction task that can be used to help finding specific content of interest to a user. By event detection we refer to the ability to properly classify text excerpts according to specific categories such as "Meet", Transport", "Attack", etc. For example, the following news excerpt "The construction of the high speed train line from Madrid to Lisbon, scheduled to start operation in 2017 has been cancelled" should be automatically detected as a "Transport" event.

Event detection has been largely explored in the context of Question Answering, Topic Detection and Tracking (TDT), and Summarization (see Section 2). Here we specifically address the problem of single event detection when in the presence of a large number of classes' scenario. So we have specific case of a single-label/multi-class classification problem, where each sentence in a document is classified to one target event or to "no event".

In our experiments we use the ACE 2005 Multilingual Corpus [2], which was specifically developed with this task in mind. Even if this corpus was manually annotated for 27 different single label event types, usually only a few event types are used due to the (arguably) insufficient number of instances necessary to train more traditional classifiers. For example, in [13], only 6 events types are used due to the difficulties in obtaining results with more classes when using Support Vector Machines (SVMs) [7] or Random Forest [9]; i.e., less than 20% of the possible event types are used.

P. Angelov et al. (eds.), *Intelligent Systems'2014*,
Advances in Intelligent Systems and Computing 322, DOI: 10.1007/978-3-319-11313-5_72

In this paper we propose the use of Fuzzy Fingerprints [15] as a mechanism to improve automatic event detection for a large number of classes. When applying Fuzzy Fingerprints to the ACE 2005 Multilingual Corpus, it was possible to detect up to 100% of the event types, obtaining a much higher G – mean [12,23], an assessment measure especially adequate to imbalanced multiclass classification problems, when comparing to the best prior event detection method, the SVMs [13].

In order to obtain the results, we started by replicating and confirming the work described by Naughton, and then improved their results by adding several new features to the used machine learning algorithms. We achieved a 4.6% improvement in F1 scores over the results reported by [13]. Then we created a Fuzzy Fingerprints Event library, and adapted the similarity score proposed in [5] to retrieve events in order to improve the results when using all the event types in the ACE 2005 corpus.

This paper is organized as follows: Section 2 introduces the related work. The corpus used to train and evaluate the methods is presented in Section 3. Section 4 details the Machine learning methods used to detect events, and the Fingerprint event detection method is presented in Section 5. Evaluation and results are shown in Section 6. Finally, Section 7 presents the overall conclusions and discusses future work.

2 Related Work

In the late 1990s, the event detection problem was addressed under the Topic Detection and Tracking (TDT) trend [6,8,24,25]. The TDT project was divided into 2 primary tasks: First Story Detection or New Event Detection (NED), and Event Tracking. In the NED task, the goal was to discover documents that discuss breaking news articles from a news stream. The Event Tracking task was focused on the tracking of articles describing the same event or topic over time. More recent work using the TDT datasets [1,22,18] on Event Threading tried to organize news articles about armed clashes into a sequence of events, but still assumed that each article described a single event. Passage Threading [1] extends the event threading by relaxing the one event per news article assumption and uses a binary classifier to identify "violent" paragraphs.

Even though the TDT program ended in 2004, new event detection research followed. Automatic Content Extraction (ACE) is the most pertinent example for this work. The goal of ACE Event is detection and recognition of events in text. In addition to the identification events, the ACE 2005 task identifies participants, relations, and attributes of each event. This extraction is an important step towards the overarching goal of building a knowledge base of events [4]. The most recent research [20,21] explores bootstrapping techniques and cross-document techniques augmenting the ACE 2005 with other corpora, including MUC-6 (Message Understanding Conference). In this work, the identification task is treated as a word classification task using 34 labels (33 event types + off-event). It is also common to use various lexical features, WordNet[1] synonyms, dependency parsing (only to identify relations), and named-entities extraction [20,21].

[1] http://wordnet.princeton.edu

3 Corpus

The ACE2005 Corpus was created for the ACE evaluations, where an event is defined as a specific occurrence involving participants described in a single sentence. The corpus has a total of 12,298 sentences. Each sentence is identified with event types or off event. There are 33 event types: Be-Born, Marry, Divorce, Injure, Die, Transport, Transfer-Ownership, Transfer-Money, Start-Org, Merge-Org, Declare-Bankruptcy, End-Org, Attack, Demonstrate, Meet, Phone-Write, Start-Position, End-Position, Nominate, Elect, Arrest-Jail, Release-Parole, Trial-Hearing, Charge-Indict, Sue, Convict, Sentence, Fine, Execute, Extradite, Acquit, Appeal, Pardon.

From these 33 events, only the following 6 have a high number of instances or sentences in the corpus: Die, Attack, Transport, Meet, Injure, and Charge-Indict. These are the only ones used in the previous referred works.

About 16% of the sentences contain at least 1 event, and 15% of those sentences are classified as multi-event (or multi-label); for example, the sentence "three people died and two were injured when their vehicle was attacked" involves 4 event types (or one event with 4 event type labels). This means that multi-event sentences correspond to only 2.50% of the corpus, and were removed since we are addressing single label classification. Also, six event types only occur in multi-event sentences, which means they are not present in the used test and training datasets. Finally, the event Extradite only occurs once in the corpus, and was removed since it is not possible to separate 1 instance into test and training sets. As a result, the dataset used in this work contains 26 different event types.

4 Machine Learning Event Detection

A state-of-art way to solve a text multi-class problem, like single event detection, is to use Support Vector Machines (SVM) techniques [26]. Random Forests (RF) [9] are also seen as an alternative to SVM because they are considered one of the most accurate classifier [17]. RF has also advantages on datasets where the number of features is larger than the number of observations [17]. In our dataset, the number of features extracted is between two and three times more than the total number of examples of events.

We followed a fairly traditional approach of training an SVM and RF classifier to classify each sentence into an event label. We used Weka [11] implementations of SVM and RF. These implementation enabled us to test several features, which to the best of our knowledge, have not been used for this purpose. These features include the use of signal words, sentiment analysis, etc. Some of these features lead to significant improvements in the previous G-mean results.

4.1 Feature Extraction

The spoken transcripts documents found in the ACE2005 corpus contain raw Automatic Speech Recognized (ASR) single-case words with punctuation. This means that the transcriptions were either manually produced or were generated by a standard ASR with minimal manual post-processing. Absence of capitalization is known to negatively influence the performance of parsing, sentence boundaries identification, and NLP tasks in general. Recovery of capitalization entails determining the proper capitalization of words from the context. This task was performed using a discriminative approach described in [3]. We capitalized every first letter of a word after a full stop, exclamation, and question mark. After true-casing, we automatically populate three lists for each article: list K of key phrases, list V of verbs, and list E of named entities. The key phrase extraction is performed using a supervised automatic key phrase extraction method [10]. Verbs are identified using Stanford Parser, and named-entities using Stanford NER. This extraction is performed over all English documents of the corpus. The K, V, and E lists are used in the extraction of lexical features and dependency parsing-based features. The lists K and V were also augmented using WordNet synsets to include less frequent synonyms. Furthermore, we manually created list M of modal verbs, and list N of negation terms.

The feature space for the classification of sentences consists of all entries in the lists V, E, K, M, and N which are corpus specific. The value of each feature is the number of its occurrence in the sentence. These numbers indicate the description of events by numbering the number of participants, actions, locations, and temporal information. We have also explored other uncommon types of features: Rhetorical Signals [10] and Sentiment Scores [14]. Finally, we removed all features with constant values across classes. This process reduced by half the number of features and improved the classification results.

5 Fingerprint Event Detection

In this work, we propose the use of an adaptation of the Fuzzy Fingerprints classification method described in [5,15] to tackle the problem of Event detection. In [15] the authors approach the problem of text authorship detection by using the crime scene fingerprint analogy to claim that a given text has its authors writing style embedded in it. If the fingerprint is known, then it is possible to identify whether a text whose author is unknown, has a known author's fingerprint on it.

The algorithm itself works as follows:

1) Gather the top-*k* [17] word frequencies in all known texts of each known author;

2) Build the fingerprint by applying a fuzzifying function to the top-*k* list. The fuzzified fingerprint is based on the word order and not on the frequency value;

3) Perform the same calculations for the text being identified and then compare the obtained text fuzzy fingerprint with all available author fuzzy fingerprints. The most similar fingerprint is chosen and the text is assigned to the fingerprint author;

The proposed fuzzy fingerprint method for Event Detection, while similar in intention and form, differs in a few crucial steps.

Firstly it is important to establish the parallel between the context of author identification and Event Detection. Instead of author fingerprints, in this work we are looking to obtain the fingerprints of Events. Once we have an event fingerprint library, each unclassified sentence can be processed and compared to the fingerprints existing in the event library.

Secondly, a different criterion was used in ordering the top-*k* words for the fingerprint. While in [15] it is used the word frequency as the main feature to create and order the top-k list, here we use an adaptation of an Inverse Document Frequency (*idf*) technique, aiming at reducing the importance of frequent terms that are common across several events.

Finally, the similarity score differs from the original, based on the fact that the source sentence are, by design, very short texts, while the original Fuzzy Fingerprint method was devised to classify much longer texts (newspaper articles, books, etc. ranging from thousands to millions of characters)[9]. Here we propose the use of a score with values between 0 and 1, where the lowest score indicates that the sentence in question is not related to the topic fingerprint, and the highest value indicates that the sentence and the event fingerprints are the same.

5.1 Building the Event Fingerprint Library

In order to build the Event fingerprint library, the proposed method goes over the event training set, which is composed of 80% of the sentences annotated as describing the event in question, and counts the word frequency. Only the top-*k* most frequent words are considered.

The main difference between the original method and the one used here is due to the small size of each sentence: in order to make the different event fingerprints as unique as possible, its words should also be as unique as possible. Therefore, in addition to counting each word occurrence, we also account for of its Inverse Topic Frequency (itf), an adaptation of the traditional inverse document frequency – idf:

$$itf_v = \frac{N}{n_v}, \tag{1}$$

where N is the cardinality of the event fingerprint library (i.e., the total number of events), and n_v becomes the number of fingerprint events where the specific word v is present. The ITF allows moving the position of common words down on top-*k* list, therefore decreasing their relevance.

After obtaining the top-*k* list for a given event, we follow the original method and apply a fuzzy membership function to build the fingerprint. The selected membership function is a Pareto-based linear function, where 20% of the top k elements assume 80% of the membership degree:

$$\mu(i) = \begin{cases} 1-(1-b)\frac{i}{k}, & i \leq a \\ a\left(1-\frac{i-a}{k-a}\right), & i > a \end{cases} \quad , a, b = 0.2 \tag{2}$$

The fingerprint is a k sized bi-dimensional array, where the first column contains the list of the top-*k* words ordered by their tf-ITF, and the second column contains the word i membership value $\mu(i)$ obtained by the application of (2).

5.2 Retrieving the Sentence-to-Event Score

In the original method, checking the authorship of a given document would proceed as follows: build the document fingerprint (using the exact procedure described above); compare the document fingerprint with each fingerprint present in the library and choose the highest score. Within the Event detection context, such approach would not work due to the very small number of words contained in one sentence since it simply does not make sense to count the number of individual word occurrences. Therefore we developed a Sentence-to-Event score (S2E) that tests how much a sentence fits to a given event fingerprint. The S2E function (3), provides a normalized value ranging between 0 and 1, that takes into account the size of the (preprocessed) sentence (i.e., its number of features). In the present work, the features are simply the set of words present in the sentence. Not even stop-word removal is performed (empirical results shown that the best results were obtained without stop-words removal, or imposing a minimum word size).

$$S2E(\Phi, S) = \frac{\sum_v \mu_\Phi(v) : v \in (\Phi \cap S)}{\sum_{i=0}^{j} \mu_\Phi(w_i)} \tag{3}$$

In (3), Φ is the Event fingerprint, S is the set of words of the sentence, $\mu_\Phi(v)$ is the membership degree of word v in the event fingerprint, and j is the is the number of features of the sentence. Essentially, S2E divides the sum of the membership values $\mu_\Phi(v)$ of every word v that is common between the sentence and the event fingerprint, by the sum of the top j membership values in $\mu_\Phi(w_i)$ where $w \in (\Phi)$. Eq. (3) will tend to 1 when most to all words of the sentence belong to the top words of the fingerprint, and tend to 0 when none or very few words of the sentence belong to the bottom words of the fingerprint.

6 Evaluation and Results

6.1 Evaluation Metrics

The standard evaluation metrics used in text classification tasks are Precision (P_i), Recall (R_i), and F1-measure ($F1_i$), where i is the class index. Precision is the fraction of sentences correctly classified (a.k.a. true positives, tp) over all sentences classified with the same event label, i.e., the sum of tp with false positives (fp):

$$P_i = \frac{\#tp}{\#tp + \#fp}$$

Recall is the fraction of sentences belonging to an event label that were successfully identified:

$$R_i = \frac{\#tp}{\#tp + \#fn}$$

The F1-measure combines the precision and recall in the following way:

$$F1_i = \frac{2\, P_i R_i}{P_i + R_i}$$

The disadvantage of these metrics is in the sensibility to imbalanced distribution of the data. The average F1-measure is not the best evaluation metric for datasets with several classes, because it is possible to obtain high F1-measure values while still failing to detect a relevant number of classes. To overcome this limitation, Kubat et al. proposed the G-mean metric (4)[12] to evaluate imbalanced binary classification problems. The extension of G-mean to imbalanced multiclass classification problems was proposed by Sun et al. in [23]. G-mean is defined as the geometric mean of the recall values R_i, and therefore has the disadvantage of assuming the value zero when at least one recall value R_i is zero. To overcome this limitation, we introduce a smoothing G-Mean version, the SG-Mean (5). A smoothing constant (e.g., $\delta = 0{,}001$) added to each R_i solves the problem of multiplication by zero if a class is not detected. With this metric it is possible to evaluate the performance of a method while still considering the loss of classes.

$$G - Mean = \left(\prod_{i=1}^{n} R_i\right)^{\frac{1}{n}} \tag{4}$$

$$SG - Mean = \left(\prod_{i=1}^{n} R_i + \delta\right)^{\frac{1}{n}}, \delta > 0 \tag{5}$$

To complement these metrics, we also report the number of classes that the methods fail to detect ($\#R_i=0$).

6.2 Results

The SVM performed better than Random Forest to detect events in low to medium number of classes. For these reason we chose SVM to investigate the inclusion of the additional features over the baseline set proposed by Naughton [3]. We have also investigated the influence of the new features introduced in this work by using all features except for the ones under test. These novel features raised the G-Mean scores by 16.6% (Table 1) when detecting six events. The average F1 value was also improved to 0.496.

The inclusion of the dependency parse based features raised the G-Mean score by 16,3 %, which is the highest contribution among the new features. The second best result, using domain-Id features, is 2,93%. The relevance based features, such as the sentiment analysis, and rhetorical features had the lowest contribution with respectively 2,7% and 1,7%. As expected, the introduction of new features reduced the recall of the majority class (no-event or off event) between -1,9% and -1,3%, but improved the recall of the remaining labels. The exception to this fact is the detection of "Die" events that was also penalized. This can be explained in part by the imbalanced distribution of the event types, which bias the classifier towards more frequent event types. In this case, the classifier is biased towards "Attack" events, which is three times more frequent than "Die".

When increasing the number of event types to cover all the 26 events present in the ACE 2005 database, the SVM performed very poorly, failing to detect 11 of the 26 events. This implied a G-Mean = 0, and the F1 decreased to 0.241. The SG-Mean was also a rather poor 0.0029.

Table 1. Feature Extraction analysis using R_i results in ACE 2005 using SVM with improved features

Labels	All Features	All - Rhetorical Signals	All - Dependency Parsing	All - Sentiment Analysis	All - Domain Id	Baseline Features
Injure	0,444	0,444	0,333	0,444	0,444	0,444
Transport	0,233	0,219	0,123	0,233	0,247	0,164
Attack	0,435	0,413	0,449	0,406	0,449	0,406
N	0,932	0,930	0,935	0,936	0,930	0,948
Meet	0,583	0,583	0,500	0,583	0,583	0,542
Charge-Indict	0,444	0,444	0,444	0,444	0,333	0,222
Die	0,375	0,375	0,375	0,333	0,375	0,417
G-Mean	**0,456**	0,448	0,392	0,444	0,443	0,391

Several tests were done in order to find the best Fuzzy Fingerprint parameters. The best empirical results led to the inclusion of all words in the fingerprints (i.e., include stop words and small sized words). The fingerprint size *K* was optimized for the best SG-mean. Figs 1-3 show the obtained SG-Mean and number of undetected event classes for 6 and 26 events for several values of *K*.

With 6 events, the best G-Mean=0.564 and SG-Mean=0.565, were obtained for *K*=200, and represent an improvement of around 25% when compared to the best SVM result. However, the F1=0.367, was lower.

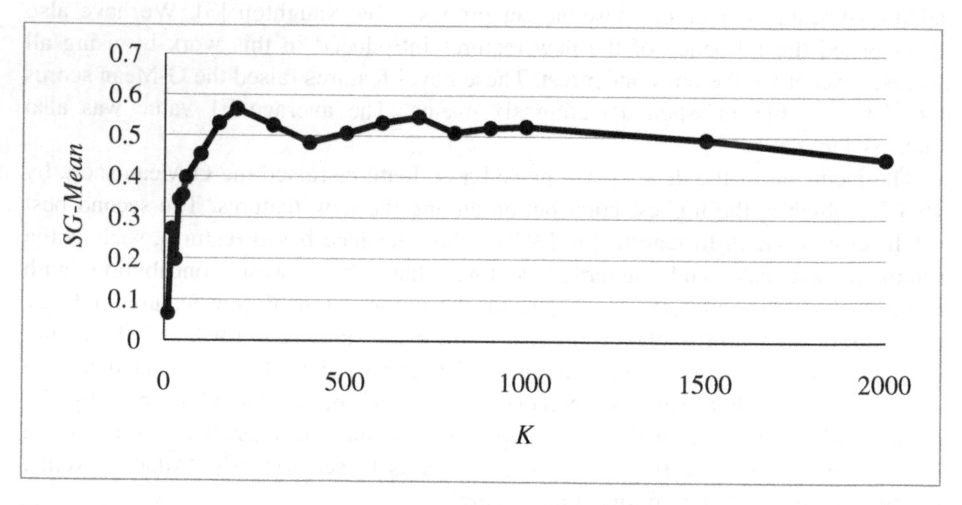

Fig. 1. SG-Mean results in the ACE 2005 with 6 events using Fuzzy Fingerprints with several *K* values

When tested for the 26 events database, the Fuzzy Fingerprints method is able to detect all the classes for the values of K>=400, while the SVM only detects 15 out of 26 classes. The best SG-Mean is 0.441, a 15x improvement over the best SVM result. The best F1=0.244 is similar for both methods.

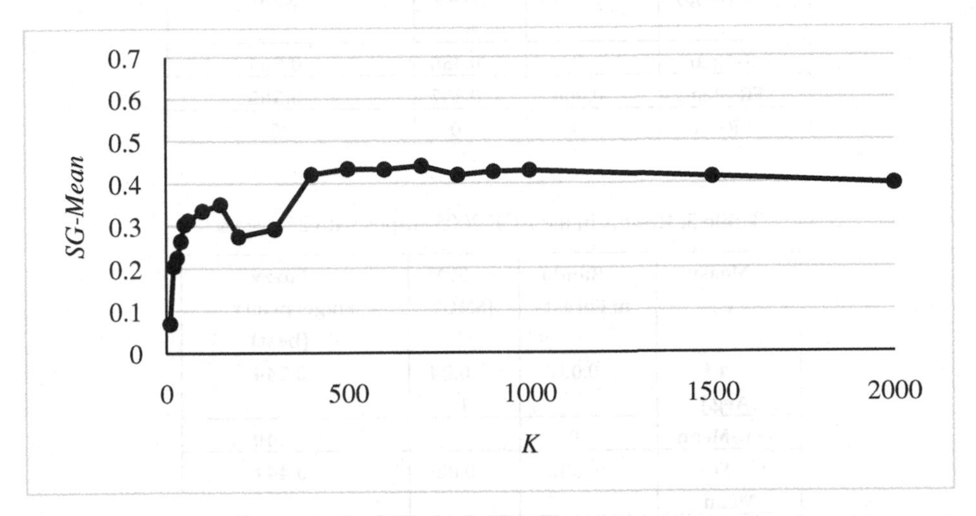

Fig. 2. SG-Mean results in the ACE 2005 with 26 events using Fuzzy Fingerprints with several K values

Finally, Tables 4 and 5 show the comparative results (including RF) for the 6 and 26 events ACE2005 database. Best results are shown in bold.

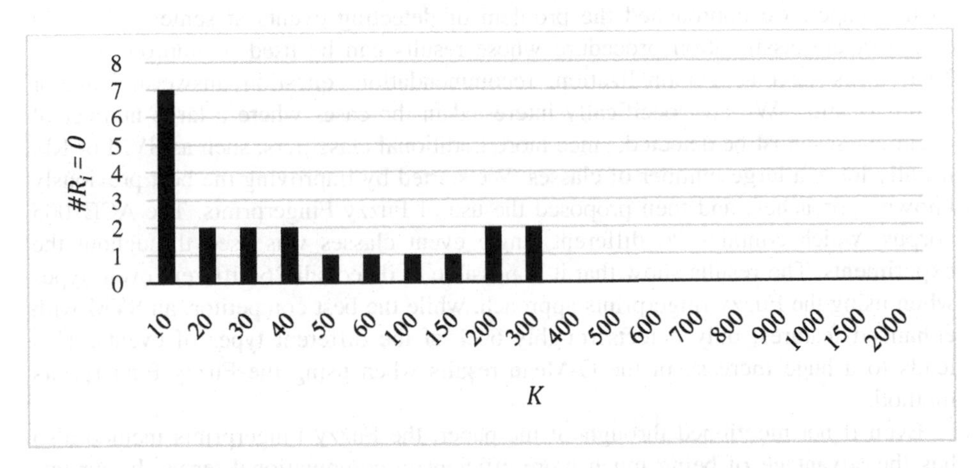

Fig. 3. #Ri=0 (number of event classes missed) results in the ACE 2005 with 26 events using Fuzzy Fingerprints with several K values

Table 2. Results in the ACE 2005 corpus with 6 events

Measure	Random Forest	SVM (SMO)	Fuzzy Fingerprints (best)
F1 (Avg.)	0.371	**0.496**	0.367
G-Mean	0	0.456	**0.564**
SG-Mean	0.008	0.457	**0.565**
#R$_i$ = 0	4	**0**	**0**

Table 3. Results in the ACE 2005 corpus with 26 events

Measure	Random Forest	SVM (SMO)	Fuzzy Fingerprints (best)
F1 (Avg.)	0.037	**0.241**	**0.244**
G-Mean	0	0	**0.440**
SG-Mean	0.002	0.029	**0.441**
#R$_i$ = 0	23	11	**0**

7 Conclusions and Future Work

In this paper, we approached the problem of detecting events at sentence level, a single label classification procedure whose results can be used to improve several NLP tasks such as personalization, recommendation, question answering, and/or summarization. We are specifically interested in the cases where a large number of event classes must be detected, since more traditional classifiers, such as SVM or RF, usually loose a large number of classes. We started by improving the best previously known approaches, and then proposed the use of Fuzzy Fingerprints. The ACE2005 corpus, which contains 26 different single event classes was used throughout the experiments. The results show that it is possible to detect all 26 different event types when using the Fuzzy fingerprints approach, while the best competitor, an SVM with enhanced features, only detects roughly 60% of the different types of events. This leads to a huge increase in the G-Mean results when using the Fuzzy Fingerprints method.

Even if not mentioned throughout the paper, the Fuzzy Fingerprints method also has the advantage of being much more efficient in computational terms. In our test conditions, it is more than 20x faster than SVM when classifying the 26 event types.

The application of the Fuzzy Fingerprints is still in an early development phase. Future work includes using advanced features such as keyphrases to build the fingerprints, and also the fuzzification of keyphrases. It is also in our plans to apply the method to the detection of multiple events in single sentences.

Acknowledgements. This work was supported by national funds through FCT Fundação para a Ciencia e a Tecnologia, project PTDC/IVC-ESCT/4919/2012, project PEstOE/EEI/LA0021/2013, and through the FCT-Carnegie Mellon Portugal Program under grant SFRH/BD/33769/2009.

References

1. Feng, A., Allan, J.: Finding and linking incidents in news. In: Proceedings of the Sixteenth ACM Conference on Conference on Information and Knowledge Management, pp. 821–830. ACM (2007)
2. Walker, C., Strassel, S., Medero, J.: ACE2005 Multilingual training Corpus. LDC 2006 (2006)
3. Batista, F., Moniz, H., Trancoso, I., Mamede, N.: Bilingual Experiments on Automatic Recovery of Capitalization and Punctuation of Automatic Speech Transcripts. IEEE Transactions on Audio, Speech, and Language Processing 20(2), 474–485 (2012)
4. Ji, H., Grishman, R.: Knowledge base population: Successful approaches and challenges. In: Proceedings of the 49th Annual Meeting of the Association for Computational Linguistics (ACL 2011), Portland, Oregon, USA, pp. 1148–1158 (2011)
5. Rosa, H., Batista, F., Carvalho, J.P.: Twitter Topic Fuzzy Fingerprints. In: Proc. of the WCCI2014 – World Congress of Computational Intelligence. IEEE Xplorer, Beijing (2014)
6. Allan, J., Carbonell, J., Doddington, G., Yamron, J., Yang, Y., Archibald, B., Scudder, M.: Topic Detection and Tracking Pilot Study Final Report. In: Proceedings of the Broadcast News Transcription and Understanding Workshop (1998)
7. Platt, J.C.: Fast training of support vector machines using sequential minimal optimization. In: Advances in Kernel Methods, pp. 185–208 (1999)
8. Carbonell, J., Yang, Y., Laerty, J., Brown, R., Pierce, T., Liu, X.: CMU Approach to TDT: Segmentation, Detection, and Tracking. In: Proceedings of the 1999 Darpa Broadcast News Conference (1998)
9. Carvalho, J.P.: On the Semantics and the Use of Fuzzy Cognitive Maps and Dynamic Cognitive Maps in Social Sciences. Fuzzy Sets and Systems 214, 6–19 (2013)
10. Breiman, L.: Random forests. Machine learning 45(1), 5–32 (2001)
11. Marujo, L., Gershman, A., Carbonell, J., Frederking, R., Neto, J.P.: Supervised Topical Key Phrase Extraction of News Stories using Crowdsourcing, Light Filtering and Co-reference Normalization Pre-Processing. In: Proceedings of 8th International Conference on Language Resources and Evaluation (LREC) (2012)
12. Hall, M., Frank, E., Holmes, G., Pfahringer, B., Reutemann, P., Witten, I.H.: The WEKA Data Mining Software: An Update. SIGKDD Explorations 11(1) (2009)
13. Kubat, M., Matwin, S.: Addressing the curse of imbalanced training sets: one-sided selection. In: Proceedings of ICML, pp. 179–186 (1997)
14. Naughton, M., Strokes, N., Carthy, J.: Investigating Statistical Techniques for Sentence-Level Event Classification. In: Proceedings of the 22nd International Conference on Computational Linguistics, vol. 1. Association for Computational Linguistics (2008)
15. Thelwall, M., Buckley, K., Paltoglou, G., Cai, D.: Sentiment strength detection in short informal text. Journal of the American Society for Information Science and Technology 61(12), 2544–2558 (2010)

16. Homem, N., Carvalho, J.P.: Authorship Identification and Author Fuzzy Fingerprints. In: Proc. of the NAFIPS2011 - 30th Annual Conference of the North American Fuzzy Information Processing Society. IEEE Xplorer (2011)
17. Homem, N., Carvalho, J.P.: Finding top-k elements in data streams. Information Sciences 180(24), 4958–4974 (2010)
18. Diaz-Uriarte, R., De Andres, S.A.: Gene selection and classification of microarray data using random forest. BMC Bioinformatics 7(1), 3 (2006)
19. Nallapati, R., Feng, A., Peng, F., Allan, J.: Event threading within news topics. In: Proceedings of the Thirteenth ACM conference on Information and knowledge management - CIKM 2004, pp. 446–453. ACM Press, New York (2004)
20. Saurí, R., Knippen, R., Verhagen, M., Pustejovsky, J.: Evita: a robust event recognizer for QA systems. HLT/EMNLP2005
21. Liao, S., Grishman, R.: Filtered ranking for bootstrapping in event extraction. In: Proceedings of the 23rd International Conference on Computational Linguistics (Coling 2010), Beijing, pp. 680–688 (August 2010)
22. Liao, S., Grishman, R.: Using document level cross-event inference to improve event extraction. In: Proceedings of the 48th Annual Meeting of the Association for Computational Linguistics (ACL 2010), Uppsala, Sweden, pp. 789–797 (July 2010)
23. Hong, Y., Zhang, J., Ma, B., Yao, J.: Using cross-entity inference to improve event extraction. In: Proceedings of the 49th Annual Meeting of the Association for Computational Linguistcs (ACL 2011), Portland, Oregon, USA, vol. 4, pp. 1127–1136 (2011)
24. Sun, Y., Kamel, M.S., Wang, Y.: Boosting for learning multiple classes with imbalanced class distribution. In: Proc. of ICDM 2006, pp. 592-602. IEEE (2006)
25. Yang, Y., Carbonell, J., Brown, R., Pierce, T., Archibald, B., Liu, X.: Learning approaches for detecting and tracking news events. IEEE Intelligent Systems and their Applications 14(4), 32–43 (1999)
26. Yang, Y., Pierce, T., Carbonell: A Study of Retrospective and On-line Event Detection. In: Proceedings of the 21st Annual International ACM SIGIR (1998)
27. Yang, Y., Liu, X.: A re-examination of text categorization methods. In: Proceedings of the 22nd Annual International ACM SIGIR Conference on Research and Development in Information Retrieval, pp. 42–49. ACM (1999)

Inferring Drivers Behavior through Trajectory Analysis

Eduardo M. Carboni and Vania Bogorny

Universidade Federal de Santa Catarina (INE/UFSC), Brazil
{eduardo,vania}@inf.ufsc.br

Abstract. Several works have been proposed for both collective and individual trajectory behavior discovery, as flocks, outliers, avoidance, chasing, etc. In this paper we are especially interested in abnormal behaviors of individual trajectories of drivers, and present an algorithm for finding anomalous movements and categorizing levels of driving behavior. Experiments with real trajectory data show that the method correctly finds driving anomalies.

Keywords: trajectory data mining, abnormal trajectories, trajectory behavior, driver classification, abrupt movements.

1 Introduction

Everyday thousands of people are victims of traffic accidents that are directly related to the behavior of drivers. According to the *World Health Organization*, the total number of road traffic deaths worldwide is around 1.24 million per year [14]. The top causes of accidents are related to excess of speed, drunk drivers, unsafe lane changes, improper turns, street racing and others [15]. Driver behaviors can affect not only traffic, but pedestrians crossing a street, passengers in a bus or taxi, and the transportation of delicate products like fruits and vegetables. According to the Brazilian food supply company (ANVISA), it is stated that around 30% of fruits and vegetables are damaged during transportation, because of driving behavior.

Several works have been proposed for driver behavior analysis in simulation systems [2, 3, 4, 5, 6, 7, 8], [13]. In [2], [4], [8], for instance, sensors and simulators are used to recognize the movements of drivers as pass over, change lane, acceleration. In [3], [5], [7] drivers are classified in levels of danger considering characteristics of a vehicle in relation to the distance and speed of vehicles in the neighborhood. A more recent work classifies drivers based on simulation data using excess of speed, car out of lane, abrupt swings on the wheel and abrupt changes in throttle and brake pedals [13].

With the popularity of mobiles devices such as GPS and cell phones, large amounts of real traces are available to analyze the behavior of drivers. Only a few works in the literature consider trajectories of moving objects, i.e., real trajectories of drivers. Existing works that consider trajectories basically look for general patterns or trajectory outliers, and do neither analyze driving behavior nor classify drivers in levels of danger [9, 10]. In [9] the aim is to find reckless taxi drivers based on the speed of the taxi

P. Angelov et al. (eds.), *Intelligent Systems'2014*,
Advances in Intelligent Systems and Computing 322, DOI: 10.1007/978-3-319-11313-5_73

and the region where the taxi is passing. In [10] the focus is on abnormal trajectories of taxis that deviate the standard route from origin and destination, where the standard route represents the path followed by the majority of taxis.

The discovery of anomalous driving in advance can help companies to advise drivers about their behavior or to keep them out of such jobs. Indeed, it may help to prevent accidents and to reduce waste in food supply. In this paper we focus on real trajectories of drivers, and propose an algorithm (that is an evolution of a previous short work [11]) to identify anomalous behavior based on abrupt movements of individual trajectories, and classify drivers in levels of danger. In summary, we make the following contributions in relation to existing works: (i) find abrupt movements based on abrupt accelerations, decelerations and direction changes in the driver real trajectory; (ii) discover events close to subtrajectories with anomalous movements; (iii) analyze repetitive (frequent) anomalous movements in individual trajectories of the same object; (iv) analyze common abrupt movements between different trajectories, i.e., if the anomalies happened at the same spatial location; (v) compare anomalous movement subtrajectories with road network characteristics, such as maximal speed limit; and (vi) classify drivers in levels of danger.

The rest of the paper is organized as follows: Section 2 presents the related work, Section 3 presents the main definitions and the proposed algorithm, Section 4 presents experiments with real data and Section 5 concludes the paper.

2 Related Work

Several works have focused on driver behavior analysis in simulation systems, considering sensors and simulators. Pentland et al. [2], for instance, proposed a model for human driving behavior analysis in order to predict sequences of behaviors in a few seconds. In a computer graphic simulation, using a car with sensors, the driving control of steering angle and steering velocity, speed, and acceleration are analyzed. The main objective is to recognize, in advance, the future actions of a driver, but not to detect anomalies in movements. Gindele et al. [8] developed a model that estimates the behavior of vehicles using sensors on wheels. In relation to other vehicles in the neighborhood, the following behaviors are estimated: free_ride (when there is no vehicle in front), following (when there is a vehicle ahead on the same lane), acceleration_phase (the vehicle accelerates to become fast enough to pass the vehicle in front), sheer_out (the vehicle keeps accelerating and changes lane for overtaking), overtake (the vehicle changes lane until it is far enough from the other vehicle to sheer back in) and sheer_in (the vehicle moves back to the original lane and changes to back to free_ride or following). The focus of this work is on analyzing driving movements in relation to other vehicles, and not in discovering abrupt movements or classifying drivers in levels of danger. Sathyanarayana et al. [4] analyses speed, steering wheel angle and brake/acceleration pedal counts to find different maneuvers. With the detected sequence of movements the proposed method discovers three different maneuvers (left turn, right turn and lane change). In this work the only objective is to find different maneuvers.

In the set of works that analyze driving behavior using simulators and sensors instead of real trajectories, some of them classify the behavior of drivers. Imamura et al. [5] classifies drivers in *normal* and *abnormal,* using a correlation between steering wheel operation and vehicle velocity in a driver simulator system. Inata et al. [6] proposed a method to find anomalous behavior of drivers based on speed, distance from neighbor vehicles, and acceleration and deceleration measured by sensors on the pedals. Rigolli et al. [3] classifies drivers as *aggressive*, *safe* and *cautious*. The analysis is performed for each vehicle in relation to the vehicles in the neighborhood, considering the speed of a vehicle in relation to the speed and the distance of the objects in the neighborhood. The normal speed of a vehicle should be similar to the speed of the vehicles in the neighborhood. So if the vehicles in the neighborhood have speed around 100km/h and one vehicle is at speed 150km/h, the faster is classified as *aggressive*. In summary, Rigolli defines the behavior of drivers in relation to other drivers considering distance and speed, while we look for abrupt movements in individual trajectories of each driver.

Among the works developed in simulation systems, the work of Quintero et al. [7] is the closest to our approach. The objective is to discover driving faults as excess of speed, movement out of lane, abrupt swings on the wheel, and abrupt changes in throttle and brake pedals, generating a percentage of errors. This work is extended in Quintero et al. [13], where the percentage of faults is used to classify the drivers in levels of danger: *moderate* and *aggressive*.

The previous works have been developed for driving behavior analysis in simulation systems using different types of sensors. So far, only a few works were developed for driving behavior analysis in GPS trajectories. Verroios et al. [12], for instance, analyzes cars with dangerous behavior in order to send alerting messages to vehicles in the neighborhood. A dangerous behavior can be, for instance, a car entering a main road with high speed. The focus is not in discovering types of dangerous behaviors but in the communication protocols, the format of the messages and their content, and which cars send and receive messages. The message is automatically send by the car with anomalous behavior to all cars that may collide.

Liao et al. [9] and Zhang et al. [10] look for anomalies in taxi trajectories. Liao et al. detects reckless behaviors of taxi drivers considering speed, time, position and passenger loading information. If the speed of a taxi is either higher or lower than the normal speed of the region (extracted from other taxi trajectories that pass at the same region) at the same period (morning, morning_rush_hour, noon, afternoon, afternoon_rush_hour, night, late_night) the taxi driver is considered *abnormal*. In [14], the space is split into a grid. The trajectories that have the same origin and destination should move through the same cells. The majority of the trajectories that move along the same cells are considered a normal behavior, while the outliers are considered anomalous.

Although the previously detailed works analyze several characteristics of driving, most of them have not been developed for real trajectories. Apart from these existing approaches, there are commercial tools as [16] which evaluate the behavior of drivers. These tools, in general, evaluate the driver based on individual movements, and do not compare a behavior to other trajectories or external events, as proposed in this paper.

In this work we propose to find abrupt movements without considering pedal sensors and without considering the behavior of objects in the neighborhood, but simply analyzing the trace of the moving object. In summary, we analyze the following properties of individual trajectories to classify the driver in levels of danger: abrupt movements including acceleration, deceleration and curves, the reason of the abrupt movements (e.g. external events that can affect the movement as a traffic jam or a radar), repetitive abrupt movements, and the speed at the abrupt movement in relation to the speed of the road network.

3 Finding Anomalous Driving Behavior

In this section we first present some basic definitions (Section 3.1) and a two-step algorithm for discovering anomalous driving (Section 3.2).

3.1 Main Definitions

We start with the basic definitions for trajectories that are well known: point, trajectory and subtrajectory.

Definition 1. Point. A point p is a triple (x,y,t), where x and y are the latitude and longitude that represent space and t is the timestamp in which the point has been collected.

Definition 2. Trajectory. A trajectory T is an ordered list of points $\langle p_1, p_2, p_3, ..., p_n \rangle$, where $p_j = (x_j,y_j,t_j)$ and $t_1 < t_2 < t_3 < ... < t_n$.

In general, a trajectory does not present the same behavior during the complete trajectory. Therefore, we analyze trajectory parts, i.e., the subtrajectories.

Definition 3. Subtrajectory. A subtrajectory s of T is a list of points $<p_k, p_{k+1}, \ldots, p_l>$, where $p_k \subset T$ and $k \geq 1$ and $l \leq n$.

The first analysis for characterizing driving behavior is to look for subtrajectories with abrupt movement. Here we consider as abrupt movement any subtrajectory with abrupt acceleration, abrupt deceleration or abrupt direction change. *Acceleration* in Physics is defined as the variation of speed divided by the variation of time. We define as *abrupt* the acceleration where the variation of speed divided by the variation of time is higher than a given threshold called minimal acceleration *minA*.

Definition 4. Abrupt Acceleration. The acceleration from a point p_i to a point p_j of a trajectory, where $t_j > t_i$, is considered *abrupt* if $\frac{v_{pj} - v_{pi}}{t_j - t_i} > minA$, and $minA > 0$.

Similarly, we define as *abrupt* a negative acceleration which is higher than a minimal deceleration threshold, called *minD*.

Definition 5. Abrupt Deceleration. A deceleration from a point p_i to a point p_j of a trajectory, where $t_j > t_i$, is abrupt if $\frac{v_{pj} - v_{pi}}{t_j - t_i} *(-1) > minD$, and $minD > 0$.

The third analysis is related to abrupt direction change. We consider a direction change as abrupt when it makes the object feel uncomfortable. While in Carboni and Bogorny et al. [11] we considered as abrupt direction change a turn in high speed, here we make use of the centripetal force, which is well defined in physics. Centripetal force is a force that keeps a body moving with a uniform speed along a circular path and is directed along the radius towards the center [1]. In this work we define abrupt direction change when the centripetal acceleration is higher than a given threshold, called *minC*.

Definition 6. Abrupt direction change. Given v_p as the speed of the moving object at point p and r as the radius of the curve, a direction change is abrupt if and only if the centripetal acceleration $\frac{{v_p}^2}{r} > minC$.

With the previous definitions we are able to find subtrajectories with *abrupt movement*.

Definition 7. Abrupt movement. A trajectory has abrupt movements when it has at least one subtrajectory with abrupt acceleration, abrupt deceleration or abrupt direction change.

Having defined *abrupt movement* we start a deeper analysis on these movements, looking for some characteristics that may justify such behavior. In this analysis we consider three main features:

F1: the existence of previously known *episodes/events* in the same area of the *abrupt movement,* which could be the reason for the anomaly.

In this paper we consider as event or episode a place that is previously known as the possible cause of an anomaly, like a traffic light, a police office, a blitz of even a pedestrian cross, etc.

F2: the *speed* of the moving object when the abrupt movement starts in relation to the speed of the road network, i.e., if the speed is similar to the road speed or if it is above the maximum limit.

F3: if the *abrupt movement* in different trajectories occurs at the same spatial area, i.e., different trajectories share an abrupt movement.

Based on the previous features we define four categories of drivers:

Level 1 (Careful driver): a trajectory is of a careful driver when it does not present abrupt behavior. Although someone may complain that it makes no sense to discover careful drivers, we claim that it is very interesting for applications where the company may want to give a reward or compliment to the good drivers.

Level 2 (Distracted driver): A distracted driver has subtrajectories with anomalous behavior, but only at places with events (F1) OR at places where other trajectories present similar behavior (F3).

Level 3 (Dangerous driver): a driver is considered dangerous when he/she has subtrajectories with anomalous movement in places *without events* OR when he/she has more than one subtrajectory with abrupt movements which do not overlap anomalous movements of other trajectories.

Level 4 (Very dangerous driver): a driver is considered *very dangerous* when it has subtrajectories with speed above the street maximum speed limit (F2) and when he/she has subtrajectories with one of the following behaviors: (i) several subtrajectories with anomalous behavior, (ii) anomalous subtrajectories which do not intersect abrupt subtrajectories of other objects, (iii) anomalous subtrajectories in places *without* events.

3.2 The Proposed Algorithm

In this paper we propose a two-step algorithm for discovering anomalous driving behaviors: first it identifies *abrupt movements* (abnormal subtrajectories) based on abrupt acceleration, deceleration and direction change; second, it analyzes the area where abrupt movements happened, the speed of the trajectory and the maximal speed of the road in order to classify the drivers.

As the analyzed movements are very *short*, the subtrajectories with anomalous behavior are normally only a few points. By considering abrupt movements between every two points only, noise can be introduced. By considering too many points (as four or more) the abrupt movement may not be captured. So after some analysis and experiments on real trajectory data, we consider in our algorithm that at least three consecutive points should have abrupt change of behavior for a subtrajectory to be characterized with abnormal movement. Another important issue is that abrupt movements can be captured well for trajectories with frequently sampled points, like 1 or 2 seconds. A dataset with sampling rate as 30 seconds, for instance, would not reveal anomalous movements, unless the data were previously interpolated.

The pseudo code of the algorithm is split in two main steps: *findAbrupt,* which is shown in Part 1 and *driverClassifier* in Part 2.

```
Part 1: findAbrupt
Input:
(01) T      // set of trajectories
(02) minA   //minimal acceleration
(03) minD   //minimal deceleration
(04) minC   //minimal direction change (centripetal acceleration)
Method:
(05) for(i=0;i<= count(T.tid);i++){              // for each trajectory
(06)   for(p=0,p< trajectory.size - 2,p++){    // for each point
(07)     if(((v_p+2-v_p+1)/(t_p+2-t_p+1))>minA
            AND((v_p+1-v_p)/(t_p+1-t_p))>minA)
(08)       abruptList.add((p),(p+1));
(09)     if(((v_p+2-v_p+1)/(t_p+2-t_p+1)*-1)>minD
            AND ((v_p+1-v_p)/(t_p+1-t_p)*(-1))>minD)
(10)       abruptList.add((p),(p+1));
(11)     r= getRadius(p, p+1, p+2);
(12)     if((v_p²/r)>minC AND (v_p+1²/r)>minC AND(v_p+2²/r)>minC)
(13)       abruptList.add((p),(p+1));
(14) }}return abruptList();
```

Part 1 of the algorithm has as input the set of trajectories T (line 1) and the thresholds for acceleration, deceleration and direction change (lines 2, 3 and 4). For each trajectory (line 5) the algorithm analyses the points (line 6) in order to find anomalies. If there is a subtrajectory of at least three points with abrupt acceleration (line 7), it is stored in an abrupt movement behavior list (line 8). The same test is performed to find subtrajectories with abrupt deceleration (lines 9 and 10). The next step (line 11) is to find the radius of the trajectory turns to analyze abrupt curves (line 12).

Figure 1 shows how the radius is computed. We consider 3 sequential points p_1, p_2 and p_3. From these points the line segments $\overrightarrow{p_1p_2}$ and $\overrightarrow{p_2p_3}$ are created. Two perpendicular lines $\overrightarrow{l_1}$ and $\overrightarrow{l_2}$ are created crossing the centroid of $\overrightarrow{p_1p_2}$ and $\overrightarrow{p_2p_3}$. The point where $\overrightarrow{l_1}$ and $\overrightarrow{l_2}$ intersect each other is the center of the curve. The distance from the intersection point to p_2 is the radius of the curve. Having computed the radius the algorithm computes the centripetal acceleration of the movement to discover the subtrajectories with abrupt direction change and adds these subtrajectories to the list of anomalous movements (line 13). It finishes returning the list of abrupt movements.

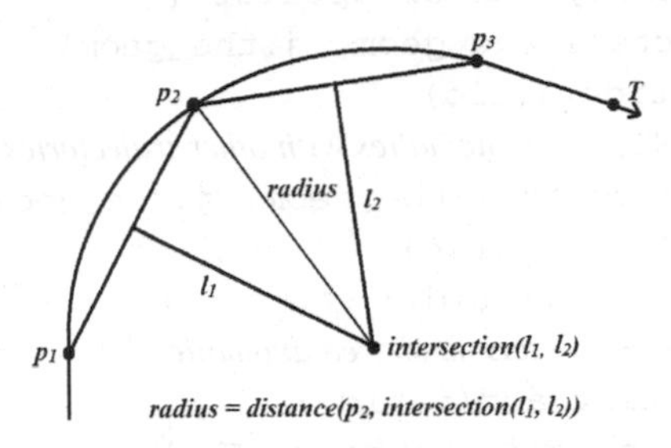

Fig. 1. Radius of the curve

The second part of the algorithm (part 2) receives as input the set of abrupt subtrajectories, the set of streets R and the set of events E. We make a buffer (line 04) around the abrupt movements in order to increase the area of the anomalies. As the abrupt movements are very few trajectory points, a buffer is needed to capture the intersections with other anomalies at the same place and with the roads and events. After some experiments we found 10 meters as a good measure to overlap anomalies.

The algorithm starts comparing each anomalous subtrajectory (line 5) with all other anomalous subtrajectories (line 7). If a trajectory has several subtrajectories with anomalous behavior (line 10) it is added to a repetitive anomalies list (line 11). If a trajectory has subtrajectories with abrupt movements where no other trajectory presents similar anomaly (line 12), this trajectory is added to the list of trajectories with individual anomalies (line 13). The next step is to verify if the anomalous subtrajectories intersect events (line 14). If this is not the case, the trajectory is added to a nonevent anomalous list (line 15). If the anomalous subtrajectory has speed higher than the maximum speed limit of the street where the object is moving (line 16), then this subtrajectory is added to a speed abrupt list (line 17). It is important to notice that so far we compare the speed of the trajectory of the moving object in relation to the speed of the road only when an abrupt movement happens. The intent is to discover if an abrupt movement of deceleration or direction change was made to suddenly move according to the maximum speed of the road network.

Having analyzed the anomalous subtrajectories, the algorithm starts classifying the trajectories. A trajectory is classified as *Level 4* (Very Dangerous Driver) at lines 18 and 19, as trajectory of *Level 3* (Dangerous Driver) at lines 20 and 21, as trajectory *Level 2* (Distracted driver) at lines 22 and 23 or as a trajectory *Level 1* (Careful driver) at lines 24 and 25.

```
Part 2: driverClassifier
Input:
(01)  abruptList;  // subtrajectories with abrupt movements
(02)  R;                    // set of streets
(03)  E;                  // set of events
Method:
(04)buffer (abruptList.the_geom, 10);
(05)for each anomaly i ∈ abruptList{//for each abrupt subtrajectory
(06)   IND = TRUE; // individual anomalies
(07) for each anomaly j ∈ abruptList {
(08)   if(intersects(i.the_geom, j.the_geom)
          AND i.tid<>j.tid)
(09)    IND = FALSE;    // anomalies with other trajectories
(10)   if(not intersects(i.the_geom, j.the_geom)
          AND i.tid = j.tid)
(11)    repeatList.add(i.tid);         }
(12) if (IND = TRUE) // has no shared anomalies
(13)   trajectoryList.add(i.tid);
(14) if(not intersects(i.the_geom, E.the_geom)//check events
(15)   nonEventList.add(i.tid);
(16) if(intersects(i.the_geom,R.the_geom)
       AND i.speed > R.speed) // check speed
(17)   speedList.add(i.tid);
(18) if(i.tid in(speedList.tid) AND i.tid in (
   nonEventList.tid,repeatList.tid,trajectoryList.tid))
(19)    level.add(i.tid,'LEVEL4'); // Very Dangerous Driver
(20) elseif(i.tid in(nonEventList.tid, repeatList.tid))
(21)    level.add(i.tid,'LEVEL3');     // Dangerous Driver
(22) elseif(i.tid not in (nonEventList.tid)
          AND i.tid not in (trajectoryList.tid))
(23)    level.add(i.tid,'LEVEL2');  // Distracted Driver
(24) else
(25)    level.add(i.tid,'LEVEL1');  // Careful Driver
(26)}return level();
```

4 Experimental Results

In this section we present experimental results with real trajectories of cars collected in the city of Florianopolis, Brazil. The dataset consists of 33 trajectories with points collected at intervals of 1 second. For this experiment we have the set of streets of the city and a set of events as traffic lights, schools, crosswalks, speed bumps. We considered five different values for the thresholds *minA, minD,* and *minC*, as shown in Table 1. These thresholds were defined starting with small values (Exp1) and increasing until almost no anomalies were found (Exp5). Acceleration starts with 3 m/s^2 up to 7 m/s^2 and deceleration and direction changes increase twice as much (double of acceleration). Deceleration is a movement that is more abrupt than acceleration, since suddenly braking a car when it is at high speed is more abrupt than to accelerate. After some tests we came to the conclusion that both direction change (centripetal acceleration) and deceleration can be two times greater than acceleration to find abrupt movements. This can help to automatically define these two parameters.

Table 1. Experimental results for 5 sets of parameter values

Thresholds	Exp1	Exp2	Exp3	Exp4	Exp5
minA	$3_{m/s^2} \Leftrightarrow 10.8_{Km/h}$	$4_{m/s^2} \Leftrightarrow 14.4_{Km/h}$	$5_{m/s^2} \Leftrightarrow 18.0_{Km/h}$	$6_{m/s^2} \Leftrightarrow 21.6_{Km/h}$	$7_{m/s^2} \Leftrightarrow 25.2_{Km/h}$
minD	$6_{m/s^2} \Leftrightarrow 21.6_{Km/h}$	$8_{m/s^2} \Leftrightarrow 28.8_{Km/h}$	$10_{m/s^2} \Leftrightarrow 36.0_{Km/h}$	$12_{m/s^2} \Leftrightarrow 42.2_{Km/h}$	$14_{m/s^2} \Leftrightarrow 50.4_{Km/h}$
minC	$6_{m/s^2} \Leftrightarrow 21.6_{Km/h}$	$8_{m/s^2} \Leftrightarrow 28.8_{Km/h}$	$10_{m/s^2} \Leftrightarrow 36.0_{Km/h^2}$	$12_{m/s^2} \Leftrightarrow 42.2_{Km/h}$	$14_{m/s^2} \Leftrightarrow 50.4_{Km/h}$
Results					
Anomalous Trajectories	21	16	8	7	2
Number of Anomalies	103	42	19	11	3
CAREFUL	12	17	25	26	31
DISTRACTED	2	2	0	0	0
DANGEROUS	3	6	5	6	2
VERY DANGEROUS	16	8	3	1	0

For each set of parameters (each column in Table 1), we show the number of *trajectories with anomalies*, the total *number of anomalies* in all trajectories, and the driver *classification levels*. One can notice that for the lower parameter values (Exp1), 21 anomalous trajectories were found with a total of 103 anomalous movements. For this experiment, 12 drivers were classified as careful, 2 as distracted, 3 as dangerous and 16 as very dangerous. The number of anomalies decreases as the thresholds for abrupt movements increase. In Exp5, for instance, only 2 trajectories presented anomalous behavior with a total of 3 anomalies. Most drivers (31) were classified as careful (without anomalies) and only two as dangerous.

Notice that as the values of the parameters to measure abrupt movements increase (from Exp1 to Exp5), the number of *anomalous trajectories* reduces (from 21 to 2), as well as the number of *very dangerous* drivers (from 16 to 0). As a consequence, the number of *careful drivers* increases (from 12 to 31). However, it is worth mentioning that the higher the parameter values, the lower is the number of anomalous movements; but the movements that are still discovered with higher thresholds, are much more abrupt. For instance, the 2 dangerous drivers in Exp 5 make abrupt movements two times greater than the 3 dangerous drivers in Exp1, because in Exp5 the parameter values are minA=7m/s, minD=14m/s and minC=14m/s, while in Exp1, minA=3m/s and minD=6m/s and minC=6m/s.

Figure 2 shows part of the trajectory dataset where most anomalies happened. Figure 2 (left) shows a satellite image of the area where the trajectories were collected and *A*, *B*, and *C* are places where the anomalous movements were known a priori. *A* is a place with an event (traffic light), *B* is a strong curve followed by an event (traffic light) and *C* is a strong curve. The algorithm correctly found the previously known abrupt movements for the three first experiments in Table 1 (Exp1, Exp2 and Exp3), with acceleration varying from 3m/s to 5m/s and deceleration and direction change varying from 6m/s to 10m/s.

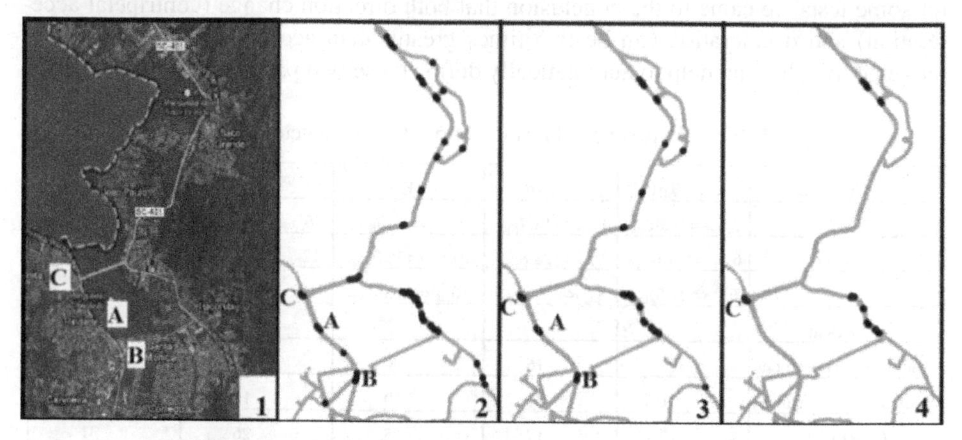

Fig. 2. Results for three different sets of parameter values

Figure 2 (2) shows the result for Exp2, where the lighter color (yellow) represents the trajectories with anomalous subtrajectories and the black colors represent the abrupt movements (anomalous subtrajectories). Figure 2 (3) shows the result for parameter values of Exp3, where one can notice that the anomalies *A*, *B* and *C* were still found. The fourth set of parameter values in Table 1 (Exp4) was too high to detect the anomalies *A* and *B*, so only the abrupt curve *C* was detected.

In order to illustrate some anomalous trajectories in detail, we show a dangerous and very dangerous trajectory. Figure 3 (left) shows the area of the trajectory on a map, Figure 3 (center) shows a very dangerous driver (tid 14) and Figure 3(right) a dangerous driver (tid 18) from the set of Exp3. In yellow are all trajectories and in light gray the two anomalous trajectories. In Figure 3(center) e_1 is a subtrajectory with abrupt deceleration intersecting a traffic light (event); s_1 is an abrupt deceleration starting with excess of speed. The speed at that point was 78 km/h on a road where the limit is 60 km/h. The other anomaly at u_1 is an anomaly without event. So the two last anomalies characterize a *very dangerous* driver.

In Figure 3(right) the anomalous subtrajectories are highlighted in black (u_2 and u_3) and were at places without events and not above speed limit, therefore characterizing a *dangerous* driver.

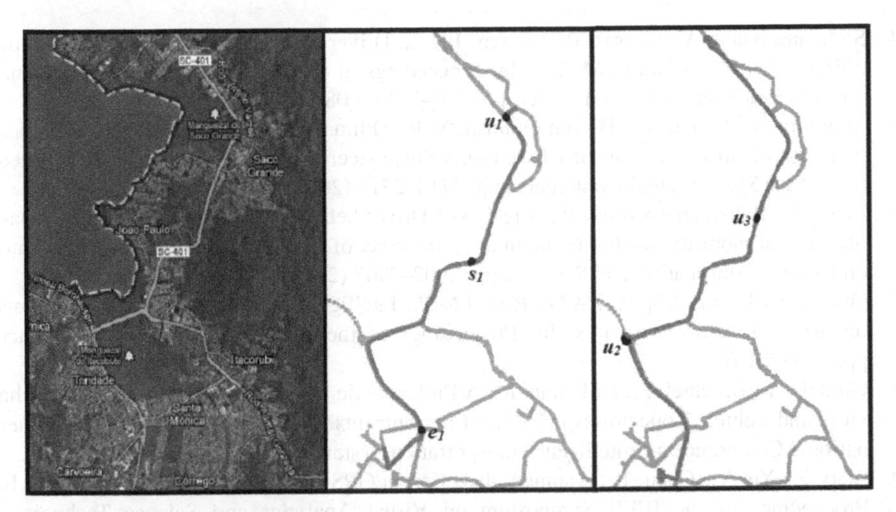

Fig. 3. Very Dangerous and Dangerous trajectories

5 Conclusions and Future Works

Trajectory behavior analysis is becoming very useful in several application domains. In this paper we presented a two-step algorithm to measure the behavior of drivers. First, the algorithm finds abrupt movements considering abrupt accelerations, decelerations, and abrupt changes of direction. Second, according to different characteristics related to the abrupt movements as repetitive anomalies, events/episodes, and speed above street speed limit the driver is classified in levels of danger. Initial experiments were performed with real data of car trajectories, where part of the anomalous movements was previously known. The algorithm correctly found the anomalous movements.

As future works we intent do perform more experiments with real data, to evaluate other characteristics of movement, evaluate the picks with higher speed in the complete trajectory, and define other features for detecting driving anomalous behavior.

Acknowledgments. Authors of this paper would like to thank CNPQ and EU project FP7-PEOPLE SEEK (N. 295179 http://www.seek-project.eu) for partially supporting this research.

References

1. Lommel, E.: Experimental physics. In: Paul, K. (ed.) Trench, Trübner & Company (1899)
2. Pentland, A., Liu, A.: Modeling and prediction of human behavior. Neural Computation 11, 229–242 (1999)
3. Rigolli, M., Williams, Q., Gooding, M.J., Brady, M.: Driver behaviouralclassification from trajectory data. Proceedings of the IEEE Intelligent Transportation Systems 6, 889–894 (2005)

4. Sathyanarayana, A., Boyraz, P., Hansen, J.H.L.: Driver behavior analysis and route recognition by Hidden Markov Models. In: Proceedings of the IEEE International Conference on Vehicular Electronics and Safety, pp. 276–281 (2008)
5. Imamura, T., Yamashita, H., bin Othman, M.R., Othman, Z.Z., Miyake, T.: Driving behavior classification river sensing based on vehicle steering wheel operations. In: Proceedings of the SICE Annual Conference, pp. 2714–2718 (2005)
6. Inata, K., Raksincharoensak, P., Nagai, M.: Driver behavior modeling based on database of personal mobility driving in urban area. In: Proc. of the IEEE International Conference on Control, Automation and Systems, pp. 2902–2907 (2008)
7. Quintero, M.C.G., López, J.A.O., Rúa, J.M.P.: Intelligent erratic driving diagnosis based on artificial neural networks. In: Proceedings of the IEEE ANDESCON Conference, pp. 1–6 (2010)
8. Gindele, T., Brechtel, S., Dillmann, R.: A Probabilistic Model for Estimating Driver Behaviors and Vehicle Trajectories in Traffic Environments. In: Proceedings of the IEEE International Conference on Intelligent Transportation Systems, pp. 1625–1631 (2010)
9. Liao, Z., Yu, Y., Chen, B.: Anomaly detection in GPS data based on visual analytics. In: Proceedings of the IEEE Symposium on Visual Analytics and Science Technology, pp. 51–58 (2010)
10. Zhang, D., Li, N., Zhou, Z.-H., Chen, C., Sun, L., Li, S.: iBAT: detecting anomalous taxi trajectories from GPS traces. In: Proceedings of ACM UbiComp (2011)
11. Carboni, E.M., Bogorny, V.: Identificando comportamentos anômalos em trajetórias de objetos móveis. In: Proceedings of the XII Simposio Brasileiro de Geoinformatica (GEOINFO), pp. 141–146 (2011)
12. Verroios, V., Vicente, C.M., Delis, A.: Alerting for vehicles demonstrating hazardous driving behavior. In: Proceedings of the 11th ACM International Workshop on Data Engineering for Wireless and Mobile Access (MobiDE 2012), pp. 15–22 (2012)
13. Quintero, M., Lopez, C.G., Pinilla, J.O., Driver, A.C.C.: behavior classification model based on an intelligent driving diagnosis system. In: Proceedings of the 15th International IEEE Conference Intelligent Transportation Systems (ITSC), pp. 894–899 (2012)
14. WHO - World Health Organization, Global status report on road safety 2013 (2013), `http://www.who.int/entity/violence_injury_prevention/road_safety_status/2013/en/index.html`
15. Accident and Traffic Rules & Regulation: Top 25 Causes of Car Accidents (2012), `http://accidentviews.wordpress.com/2012/02/14/top-25-causes-of-car-accidents/`
16. AA Car Insurance, `http://www.theaa.com/insurance/car-insurance.jsp`.

Framework for Customers' Sentiment Analysis

Catarina Marques-Lucena[1,2], João Sarraipa[1,2],
Joaquim Fonseca[3], António Grilo[3], and Ricardo Jardim-Gonçalves[1,2]

[1] DEE, FCT, Universidade Nova de Lisboa, 2829 Caparica, Portugal
[2] Center of Technology and Systems, CTS, UNINOVA, 2829-516 Caparica, Portugal
[3] UNIDEMI, FCT, Universidade Nova de Lisboa, 2829 Caparica, Portugal

Abstract. Web software usage used as a vehicle of communication, where sentiments related to different subjects are shared through the web, has had a steep increase in recent years. For example, customers write about characteristics that they like the most, and they dislike, influencing others to use a specific commented service or product. Thus, the content of these comments could be used to address customers' sentiments in such way that would increase companies' services quality. However, since the output of such kind of web software is essentially unstructured data it is difficult for company managers to have access to that knowledge, and consequently reason through it accordingly. In this paper, the authors present a framework for customers' sentiment analysis, which has the ability to automatically acquire knowledge from web software. Thus, it addresses the target of customer's sentiments with the purpose of supporting companies' managers in their decision-making to improve their products and services.

1 Introduction

Social software tools typically handle the capturing, storing and presentation of communication and focus on establishing and maintaining a connection among users [1]. Due to its current increase of usage (e.g. facebook, twiter), it has become a big data source. Big data is a term applied to data sets whose size is beyond the ability of commonly used software tools to capture, manage, and process the data within a tolerable elapsed time [2]. Social software focuses on human communication, thus the vast majority of the information is in human readable format only. As a result, researchers are actively contributing to the appearance and enhancement of technological solutions to handle such data sets with the goal to supply new mechanisms for enterprise's management [3]. This results in the increase of companies' use of web software to promote services and products, which is also often used by their customers to post their comments expressing sentiments as response [4]. Consequently, web software is commonly accepted as a communication bridge between customers and companies, where users' opinion become a major criterion for the improvement of the quality of services. Blogs, review sites and micro blogs provide a good understanding of the reception level of the products and services [5]. A topic that is currently getting much attention is how to use electronic customers comments to increase the quality of

P. Angelov et al. (eds.), *Intelligent Systems'2014*,
Advances in Intelligent Systems and Computing 322, DOI: 10.1007/978-3-319-11313-5_74

companies' services and products in order to increase the competitiveness level [6] [4]. Additionally, a relevant challenge is how to handle real-time data analysis. Thus, it is desirable to automatically provide to a company manager the formal knowledge about the following questions: '*What is the comment about?*'; '*What is the polarity of the comment? Is it positive or negative?*'

The response to the previous questions would allow a company to improve the quality of the offered services and/or products by reacting to customers' sentiments, and avoiding to spend, for instance, a large amount of money on customers satisfaction surveys. In addition, the use of surveys, if they are not in the form of open question, may narrow companies' range of understanding and actions. An example is the open question: '*Which are the things that you dislike the most*', and the more narrow question '*Did you like the staff?*'. Based on the second question, the manager can only make fact-based decisions relative to the staff. The free opinion sharing (first question) provided by customers sentiments, besides helping to identify key improvement variables, can also be used as a rumour detection. As an instance, if several customers write something like: '*The food is very bad and I advise everyone to never go to that place*'.

A quick detection of this rumour will allow the company manager to react accordingly and avoid major losses. Another advantage resulting of such sentiment analysis is the potential for 'spying' the market competition. Most of the web softwares allow access to the html source. Thus, by accessing to the comments posted in the competition page it is possible for a company to track customers' preferences and act in accordance to that background. To address this challenge, the authors propose a framework for automatic customers' sentiment analysis.

To introduce this framework, the authors start with a brief survey about the necessity of transforming web software content into useful and formal knowledge. Then, the framework to improve customers' satisfaction is presented. An example of this framework application is also provided in a form of a methodology. Finally, some conclusions and future work are presented.

2 From Unstructured Data to Improve Customers Experience

In contrast to earlier times, when finding sources of information was the key problem to companies and individuals, most of today's information society challenges are related to the creation and employment of mechanisms to search and retrieve relevant data from the huge quantity of available information. This is aligned with the McAfee and Brynjolfsson statement, which says that digital data are so easily shared and replicated that the amount of data and information available for organizations analysis is exploding [7]. Thus, this data or information must be analysed to be retrieved as useful and formal knowledge, which then can be used to take the most suitable decisions [8]. As a response, the big data initiative seeks to glean intelligence from data and translate that into business advantage.

According with the result of a survey conducted in 2012 by NewVantage Partners [9] there are seven groups of tangible benefits to achievable by big data initiatives.

On the top of Dig Data initiatives benefits rate are: (1) Better, fact-based decision-making (22%); and (2) Improve customers' experience (22%). These are the main characteristics that the proposed framework intends to address. However, it also intends, even (mostly) indirectly, to address the other advantages of the study: Increased sales, new product innovations, reduced risk, more efficient operations and higher quality products and services.

The key value of big data is to speed up the time-to-answer period, allowing an increase in the pace of decision making at both the operational and tactical levels [10][11]. To achieve fact-based decision-making, it is necessary to acquire useful knowledge, thus knowledge management is required. Knowledge can be defined as the human expertise stored in a person's mind, gained through experience, and interaction with the person's environment [12]. Authors' contributions enable the extraction and analysis of customer's comments. This analysis is focused on the comment's sentiments. Sentiment analysis is associated to the automatic analysis of evaluative data and tracking of the predictive judgments [13]. This big data usage would allow organizations managers to take decisions based on evidences rather than intuition [7], being the evidence related to the customer satisfaction level of a specific service or product. Thus, algorithms that perform better decision-making support, every time the customer respond to or ignore a provided recommendation, should be implemented. This is aligned to the companies' necessity of focusing in enhancing their knowledge asset in order to innovate [14] and survive in the current competitive markets.

2.1 Contextual Decision-Making Approach

To acquire useful knowledge for companies usage, at first, is necessary to establish the data to be collect. A formal data collection process is necessary since it ensures that the collected data is both defined and accurate and that subsequent decisions based on arguments embodied in the findings are valid [15]. Considering the web software, the data collection is usually done by html processing of the source.

Following, it is necessary the data analysis to be returned as useful knowledge. Amongst others outputs, data analysis can be applied to find hidden patterns in data, evaluating significant data and interpretation of results [16]. The output of data analysis is formalized useful knowledge to be used in the decision-making process. Examples of data analysis outputs are reports, relations, trends, customer segmentation and patterns. Thus, decision-making based on the gathered knowledge can be made.

Decision-making can be regarded as the cognitive process resulting in the selection of a specific resource or action among several alternative scenarios. Every decision making process produces a final choice [17]. Simon described the decision-making process as consisting of three phases: intelligence, design and choice [18]. Context is any information that can be used to characterize the situation of an entity. An entity is a person, place, or object that is considered relevant to the interaction between a user and an application, including the user and application themselves [19].

Making context based decisions will reduce the effort to coordinate tasks and resources by providing a context in which to interpret utterances and to anticipate actions [20]. Thus, knowledge provided by customers is used to achieve fact-based reasoning and inference, agility, reactiveness, innovation, context awareness, decision intelligence and competitive intelligence. Thus, making context-based decisions would contribute to an improved customer experience.

3 Framework for Customers' Sentiment Analysis

A large amount of information available on web software is only accessible through presentation-oriented HTML pages. However, such pages often follow a global layout template which facilitates the retrieval of information [21]. Companies' web software layouts usually integrate a customers' review/comments area, where customers can express their sentiments related to the used products or services. These comments are the input of the proposed framework for customers sentiment analysis presented at Fig. 1. This framework encompasses **data collection**, **pre-processing** and sentiment **analysis** to be returned as **useful knowledge** and finally to support companies **decision-making**. Thus, after the collection of the data, the next step encompasses the pre-processing of HTML content to return customers comments in the textual form.

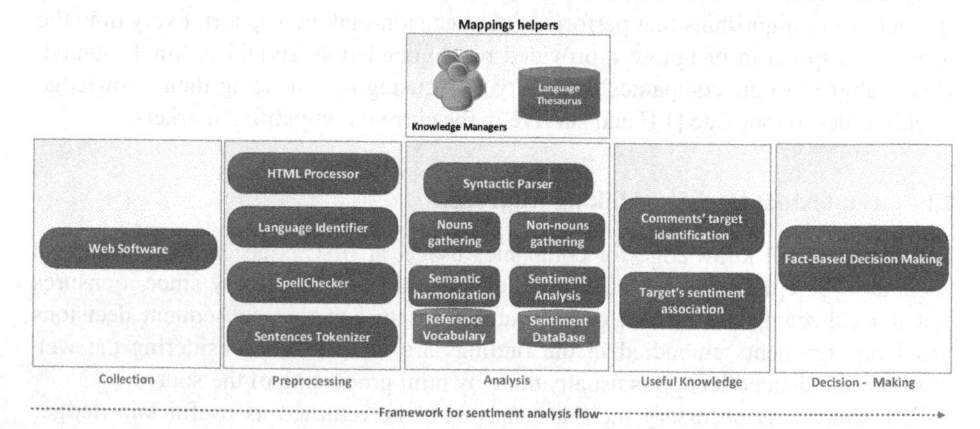

Fig. 1. Web software knowledge management framework

The essential issue in sentiment analysis is the identification of how sentiments are expressed in texts and whether the expressions indicate favourable or unfavourable opinions toward the subject [4]. A typical approach to sentiment analysis is to start with a lexicon of positive and negative words and phrases [22]. However, in order to obtain an accurate sentiment analysis is required a sentence-level or even phrase-level sentiment analysis. Without phrase-level sentiment analysis, the association of the extracted sentiment to a specific topic is difficult and most of the sentiment extraction algorithms perform poorly in this respect [23]. For that reason, in authors' proposed framework, prior to comments' sentiment analysis is made a pre-processing that encompasses: language detection, spell checker and sentences tokenization, and which output allows sentence level sentiment analysis.

Since, in this framework we are interested in customers' reviews, which can be written both in the services (or products) providers' native language or in a foreign language, after HTML processing, is made a language identification. Automatic language identification is the process of using a computer system to identify the language of a spoken utterance [24]. An accurate language identification can facilitate the use of background information about the language and use of more specialized approaches in many natural language processing tasks dealing with a collection or a stream of texts, each of which can be written in a different language [25].

Since many web sites contain no spell checker, comments written by customers can have spelling errors. The existence of these errors makes more difficult the task of natural language processing. For that reason, in authors' proposed framework was incorporated a spell check prior to the comments analysis.

Most processing tasks in Natural Language Processing (NLP), presupposes a preliminary phase of tokenization, after which the input sequence is provided with its individual tokens explicitly identified and isolated from each other [26]. Tokenization is the process of breaking up the given text into units called tokens. The tokens may be words, numbers or punctuation marks. Tokenization does this task by locating word boundaries. The ending point of a word and beginning of the next word is called word boundaries [27].

There is a great interest of using syntactic information as part of an information retrieval strategy. Humans gather information not only based on the meaning of the words, but also based on the structure in which words are put together [28]. The knowledge of customers' comments syntax can be used to increase the performance of sentiment analysis task. Accordingly with [29], parsing intends the recognition and explanation of concatenation patterns that conform to a system of rules (grammar) and are presented to the analyser as a string (sentence) of consecutive units (words) which correspond to individuals of a pre-established set (vocabulary). One of the possible outputs of parsing process are parsing trees. These consists in an ordered, rooted tree that represents the syntactic structure of a string (sentence) according to some formal grammar [30].

At this point, it is possible to extract the nouns, adverbs, verbs and adjectives of customers' comments. The nouns will be assumed as the target of customers' sentiments and the rest of grammatical terms will be used to extract the polarity of the comment. To the sentiment polarity extraction it is used a dictionary of the comments' language with polarities associated to each word.

After the semantic harmonization process, the output is a common syntax and semantics to describe the customers' comments targets. The semantic harmonization process is implemented by merging or mapping the knowledge of the various sets of terms with the existing version of a semantic reference vocabulary [31]. If there is an existing semantic reference vocabulary knowledge base of companies' market, this can be adopted or improved. Otherwise, a reference knowledge base must be achieved, using MENTOR [32], as an example.

To increase knowledge alignment efficiency with the domain of interest it is used a thesaurus to establish mappings, which normally are made manually by domain experts [33]. The use of such established mappings increase the probability of finding more synonyms, which will result in a wider lexicon to which the framework is able to handle.

The knowledge output of the framework is a common vocabulary to identify the targets of the comments, the adjectives adverbs and verbs, and the polarity associated. Thus, companies' managers will know: where to act (comments' target with negative polarity associated) and how to act (comments' adjectives, adverbs and verbs). This will result in a fact-based decision making that enables companies services and products quality improvement.

4 Portuguese Hospitality Industry Customers' Comments Analysis

Semantik project aims to develop a new platform of services, in a Software as a Service (SaaS) logic, to allow the usage of structured, semi-structure and unstructured data from the web to serve enterprises management necessities. It is proposed the use of semantic web (Web 3.0) methodologies and technologies to help enterprises to identify, analyse and classify information in a more correct and effective way. Semantik will allow enterprises to take advantage of the enormous amount of available information on web using automated collection and presentation of data. Information from various sources is organized in a proposed decision support framework including intelligence classification according to the business model, signal strength and management level. We choose hospitality industry because web technologies had an enormous impact on marketing and sales strategies. The sole objective of Semantik is to help enterprises to become more competitive in the global market by making an efficient use of all generated information from external agents.

In the scope of Semantik, the authors developed a methodology to analyse customers' comments. The output of the methodology are both the targets and their associated polarity. Based on that information, the manager can improve the decision-making by directly address customers' opinions.

4.1 Methodology for Portuguese Customers' Comments Analysis

In this section it is presented a methodology for customers' sentiment analysis of the Portuguese hospitality industry (see Fig. 2). This methodology accomplishes complete useful knowledge acquisition from web software data. Booking (http://www.booking.com/) is the web software used to analyse hospitality industry customers' satisfaction. The input of this methodology is Booking's web software whose layout integrates a customers' review/comments area (see step 1 of Fig. 2), where customers post their opinion about the services provided.

The first step of the proposed method encompasses the collection of the data. To do that, some HTML processing is required. The step two begins the analysis of the comments with the language processing and spell checker application. The comments used in the next steps are the ones written in Portuguese. The use of a language identifier was no need, since the comments' language are in the HTML metadata. In this methodology were only tested the Portuguese comments, thus a spell checker based on Open Office Portuguese dictionary was used.

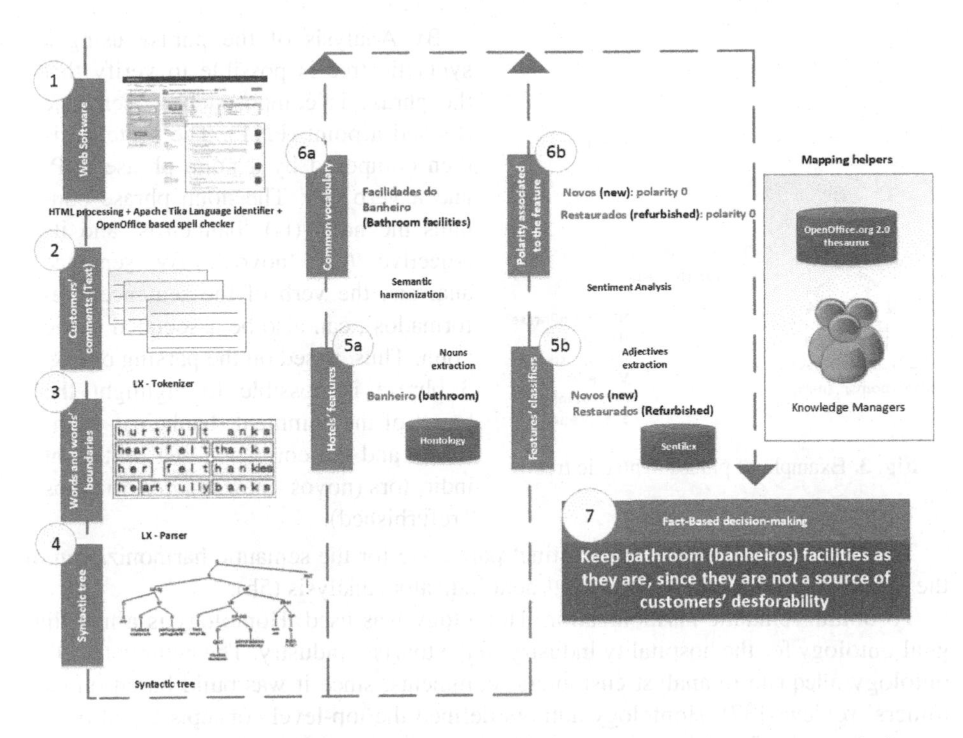

Fig. 2. Methodology for Portuguese hospitality industry customers' comments analysis

In the third step is made the tokenization of the extracted sentences to allow the application of the syntactic parser. This is made with LX-Tokenizer [34], which is a tokenizer that takes into account Portuguese non-trivial cases that involve ambiguous strings. The outputs of LX-Tokenizer application to the positive comment are the correspondent words and words' boundaries.

Then, in the fourth step, it is made the syntactic parsing of the comments through the LX-Parser. LX-parser is a freely available on-line service for constituency parsing of Portuguese sentences. The parser was trained and evaluated over CINTIL-Treebank, a treebank produced from the output of a deep processing grammar by manually selecting the correct parse for a sentence from among all the possible parses produced by the grammar [35] [36]. The parser produces several outputs that consists in the several phrases that compose the full customer comment. This functionality allows the isolation of the several targets of interest, since they usually are in separated phrases or tree branches, and individually analysis.

Taking as an example the comment 'Banheiros novos. Reformados', which translation is 'New bathrooms. Refurbished', the application of the syntactic parser is presented in Fig. 3.

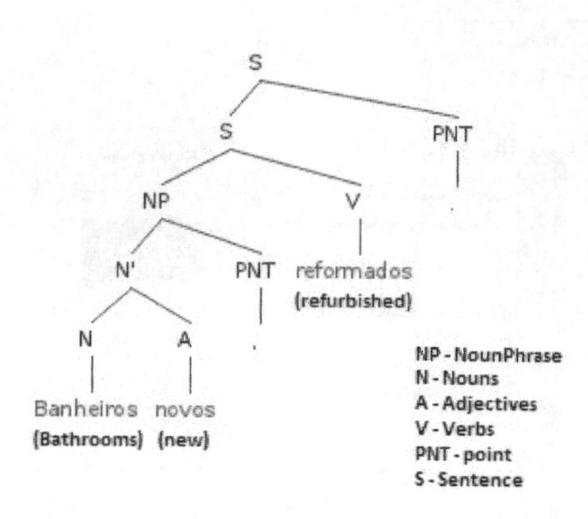

Fig. 3. Example of phrase syntactic tree

By Analysis of the phrase using a syntactic tree is possible to verify that the phrase is composed by a sentence (S) and a point (PNT). The sentence is then composed by a noun phrase (NP) and a verb (V). The noun phrase contains the noun (N) 'banheiros' and its adjective (A) 'novos'. By sentence analysis, the verb of the sentence, 'reformados', can also be associated to the noun. Thus, based on the parsing of Fig. 3 phrase is possible to highlight the target of the comment (banheiro - bathroom) and the corresponding sentiment indicators (novos - new, and reformados - refurbished).

The fifth step is dived in two distinct parts, one for the semantic harmonization of the targets (5a) and other for the sentiment indicators analysis (5b).

To obtain semantic harmonization, Hontology was used. Hontology is a multilingual ontology for the hospitality industry in the tourism industry. The authors find this ontology adequate to analyst customers' comments, since it was build based on customers' reviews [37]. Hontology authors defined the top-level concepts based on the customer needs. From these concepts, the customers have a broad and depth vision of the domain, which they are looking for information. The core concepts of Hontology are: Accommodation, Facility, Room, Service/Staff and Guest Type.

Other advantage of Hontology usage consists in the hierarchical representation of the hospitality industry lexicon. Considering the concept 'Banheiro', if a customer express dissatisfaction with the bathroom condition, he could be showing dissatisfaction by 'Balança' - Bathroom scale, 'Banheira' - Bathtub, 'Duche' - Shower ..., because they are part of 'Banheiro' (lower level in the taxonomy). Thus, for the hotel manager, it is possible to conclude that one or more constituents of the bathroom facilities must be changed, since the customer negative comment encompasses all of them.

As can be observed in Fig. 4, for each noun detected in the sentence the methodology try to do the mapping with the Hontology concepts. If the mapping does not exist, the algorithm increases the probability of the mapping by using a list of synonyms of the noun. The list of synonyms is retrieved from OpenThesaurusPT, which is an open source project to the development of a synonyms dictionary for the Portuguese language1. In the last case, if the mapping still not existing, it will be a knowledge engineer to do the mapping manually. Fig. 5 shows the mapping between the concepts 'Banheiros' (Bathroom) and 'Facilidades do Banheiro' (Bathroom Facilities).

[1] http://openthesaurus.caixamagica.pt/

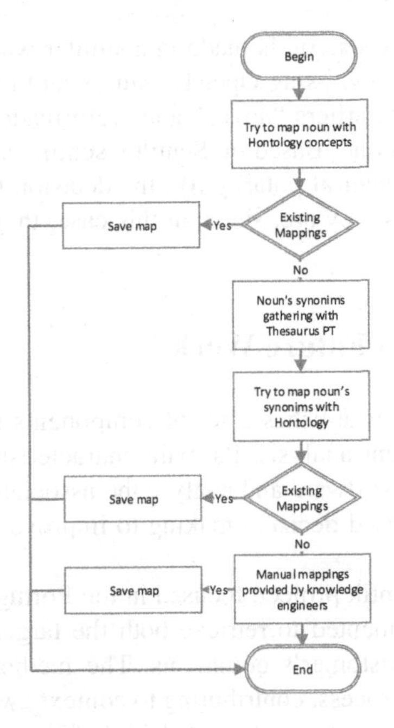

Fig. 4. Semantic Harmonization flow

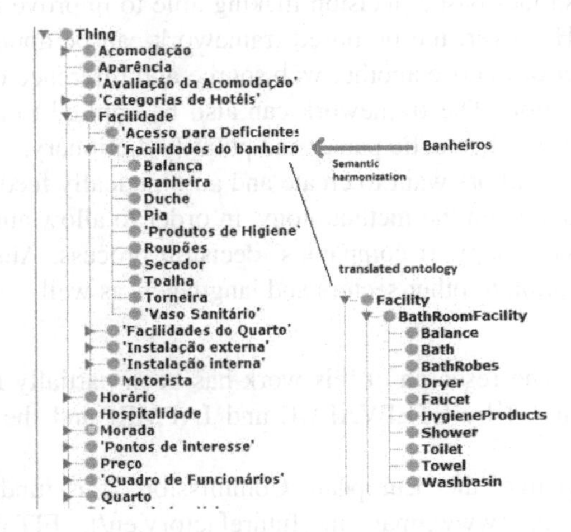

Fig. 5. Semantic harmonization between 'banheiros' and 'facilidades do banheiro'

For comments polarity identification, authors opt to use a list of adjectives, verbs and adverbs and their polarities, namely Sentilex. Sentilex was built with a methodology for automatically enlarging a Portuguese sentiment for mining social judgments. Sentilex is a Portuguese lexicon with 7.014 lemmas and 82.347 inflected forms [38]. The polarity attributed by sentilex is: positive (1), negative (-1), or neutral (0). Being the Sentilex inputs the previous comment sentiment indicators. The mapping of the classifiers with the sentilex lexicon is made in a similar way of the harmonization of the nouns with the Hontology, using OpenThesaurus and manual mappings (see Fig. 4). Giving as inputs the classifiers "novos" and "reformados", the attributed polarity from sentilex is 0 for each one. Based on Sentilex sentiment retrieval, and since both sentiment indicators have neutral polarity (0), the decision to make (step seven) is to keep bathroom facilities as they are, since, in this case, they are not a source of customers' unfavourability.

5 Conclusions and Future Work

The proposed Framework establishes a set of components that aim to guide companies to customer's sentiment analysis. Its main characteristic is the ability of gather customer's data from web software and analyse the associated sentiments. This allow companies to make fact-based decision making to improve the services and products provided.

In the example of Semantik project, focused in the Portuguese hospitality industry, a methodology was implemented to retrieve both the targets and target's associated polarities from booking customer's comments. The methodology output can, then, feed the decision-making process, contributing to context awareness decisions. It aims to help hospitality companies to analyse and identify useful knowledge in a more effective and establish fact-based decision-making able to improve the quality of services and products. However, the proposed framework can be applied to several domains is only a matter of choose another web source and reference ontology to do the vocabulary harmonization. The framework can also be applied to another languages by changing the tokenizer, syntactic parser and polarity dictionary.

As future work, the authors want to create and automatically feed an ontology with the gathered knowledge from the methodology, in order to allow automatic reasoning and recommendations to support companies' decision process. Authors what to test the framework application to other sectors and languages, as well.

Acknowledgements. The research of this work has been partially funded by project SEMANTIK, co-financed by STEPVALUE and IAPMEI and the European Funds QREN COMPETE.

We also acknowledge the European Commission, for funding the projects IMAGINE nr 285132, (www.imag ine-futurefactory.eu/), FITMAN nr 604674 (http://www.fitman-fi.eu), and OSMOSE nr 610905 (http://www.osmose-project.eu), and Fundação da Ciência e Tecnologia for funding our research units.

References

1. Allen, C.: Tracing the Evolution of Social Software (2004), http://www.lifewithalacrity.com/2004/10/tracing_the_evo.html (accessed April 2014)

2. Gold, M.K.: Debates in the Digital Humanities. University of Minnesota Press (2012)
3. Jardim-Goncalves, R., Grilo, A.: Building information modeling and interoperability. Autom. Constr. 19, 387 (2010)
4. Nasukawa, T., Yi, J.: Sentiment Analysis: Capturing Favorability Using Natural Language Processing. In: Proc. 2nd Int. Conf. Knowl. Capture, pp. 70–77. ACM, New York (2003)
5. Jadeja, A., Jeet Rajput, I.: Feature Based Sentiment Analysis on Customer Feedback: A Survey. Int. J. Eng. (2013)
6. Gamon, M.: Sentiment classification on customer feedback data: noisy data: large feature vectors, and the role of linguistic analysis (2003)
7. McAfee, A., Brynjolfsson, E.: Big Data: The Management Revolution (cover story). Harv. Bus. Rev. 90, 60–68 (2012)
8. Montoyo, A., Martínez-Barco, P., Balahur, A.: Subjectivity and sentiment analysis: An overview of the current state of the area and envisaged developments. Decis. Support. Syst. 53, 675–679 (2012), doi:10.1016/j.dss.2012.05.022
9. NewVantage Partners, Big Data Executive Survey: Creating a Big Data Environment to Accelerate Business Value (2012)
10. Capgemini (2012) The Deciding Factor: Big Data & Decision Making (accessed April 2014), `http://www.capgemini.com/resources/the-deciding-factor-big-data-decision-making`
11. NewVantage Partners, Big Data Executive Survey: Themes & Trends (2012)
12. Ackoff, R.L.: From Data to Wisdom. J. Appl. Syst. Anal. 16, 3–9 (1989)
13. Pang, B., Lee, L.: Opinion Mining and Sentiment Analysis. Found. Trends Inf. Retr. 2, 1–135 (2008), doi:10.1561/1500000011
14. Matthews, K., Harris, H.: Maintaining Knowledge Assets. In: Mathew, J., Kennedy, J., Ma, L., Tan, A., Anderson, D. (eds.) Eng. Asset. Manag., pp. 618–626. Springer, London (2006)
15. Sapsford, R., Jupp, V.: Data Collection and Analysis. SAGE Publications (1996)
16. Pawlak, Z.: Rough sets and intelligent data analysis. Inf. Sci (Ny) 147, 1–12 (2002)
17. Reason, J.: Human Error, xv, p. 302. Cambridge University Press, Cambridge [England], New York (1990)
18. Simon, H.A.: The New Science of Management Decisions. Shape Autom. Men Manag. (1960)
19. Abowd, G.D., Dey, A.K., Brown, P.J., Davies, N., Smith, M., Steggles, P.: Towards a Better Understanding of Context and Context-Awareness. In: Gellersen, H.-W. (ed.) HUC 1999. LNCS, vol. 1707, pp. 304–307. Springer, Heidelberg (1999)
20. Gutwin, C., Greenberg, S., Roseman, M.: Workspace Awareness in Real-Time Distributed Groupware: Framework, Widgets, and Evaluation. In: Proc. HCI People Comput. XI, pp. 281–298. Springer, London (1996)
21. Simon, K., Hornung, T., Lausen, G.: Learning Rules to Pre-process Web Data for Automatic Integration. In: 2006 Second Int. Conf. Rules Rule Markup Lang Semant Web, pp. 107–116 (2006), doi:10.1109/RULEML.2006.16
22. Wilson, T., Wiebe, J., Hoffmann, P.: Recognizing Contextual Polarity in Phrase-level Sentiment Analysis. In: Proc. Conf. Hum. Lang. Technol. Empir. Methods Nat. Lang. Process., pp. 347–354. Association for Computational Linguistics, Stroudsburg (2005)
23. Dave, K., Lawrence, S., Pennock, D.M.: Mining the Peanut Gallery: Opinion Extraction and Semantic Classification of Product Reviews. In: Proc. 12th Int. Conf. World Wide Web, pp. 519–528. ACM, New York (2003)

24. Torres-carrasquillo, P.A., Singer, E., Kohler, M.A., Deller, J.R.: Approaches to language identification using Gaussian mixture models and shifted delta cepstral features. In: Proc. ICSLP 2002, pp. 89–92 (2002)
25. Tromp, E., Pechenizkiy, M.: Graph-Based N-gram Language Identification on Short Texts. In: Proc. Twent. Belgian Dutch Conf. Mach. Learn (Benelearn 2011), pp. 27–34 (2011)
26. Branco, A.H., Silva, J.: Tokenization of Portuguese: resolving the hard cases. docs.di.fc.ul.pt (2003)
27. Processing Nlpaonnl, Tokenization: Overview (2014), `http://language.worldofcomputing.net/category/tokenization` (accessed April 2014)
28. Metzler, D.P., Haas, S.W.: The Constituent Object Parser: Syntactic Structure Matching for Information Retrieval. ACM Trans. Inf. Syst. 7, 292–316 (1989), doi:10.1145/65943.65949
29. Von Glasersfeld, E., Pisani, P.P.: The Multistore Parser for Hierarchical Syntactic Structures. Commun. ACM 13, 74–82 (1970), doi:10.1145/362007.362026
30. Chiswell, I., Hodges, W.: Mathematical Logic. Oxford University Press, New York (2007)
31. Agostinho, C., Sarraipa, J., Goncalves, D., Jardim-Goncalves, R.: Tuple-Based Semantic and Structural Mapping for a Sustainable Interoperability. In: Camarinha-Matos, L. (ed.) Technol. Innov. Sustain, pp. 45–56. Springer, Heidelberg (2011)
32. Sarraipa, J., Jardim-Gonçalves, R., Steiger-Garção, A.: MENTOR: an enabler for interoperable intelligent systems. Int. J. Gen. Syst. 39, 557–573 (2010)
33. Jardim-Goncalves, R., Coutinho, C., Cretan, A., Bratu, B.: A framework for sustainable interoperability of negotiation processes. Control Probl. Manuf. 14, 1258–1263 (2012)
34. Branco, A., Silva, J.: Evaluating Solutions for the Rapid Development of State-of-the-Art POS Taggers for Portuguese. LREC (2004)
35. Silva, J., Reis, R., Gonçalves, P., Branco, A.: LX-Parser and LX-DepParser: Online Services for Constituency and Dependency Parsing. inf.pucrs.br
36. Silva, J., Branco, A., Castro, S., Reis, R.: Out-of-the-box robust parsing of Portuguese. Process. Port. (2010)
37. Chaves, M.S., de Freitas, L.A., Vieira, R.: Hontology: A Multilingual Ontology for the Accommodation Sector in the Tourism Industry. In: Filipe, J., Dietz, J.L.G. (eds.) KEOD, pp. 149–154. SciTePress (2012)
38. Silva, M.J., Carvalho, P., Sarmento, L.: Building a Sentiment Lexicon for Social Judgement Mining. In: Caseli, H., Villavicencio, A., Teixeira, A., Perdigão, F. (eds.) PROPOR 2012. LNCS, vol. 7243, pp. 218–228. Springer, Heidelberg (2012)

Author Index

Printed in the United States
By Bookmasters